The Early Proterozoic Trans-Hudson Orogen of North America

Edited by

John F. Lewry
Department of Geology, University of Regina, Regina, Saskatchewan S4S 0A2, Canada

and

Mel R. Stauffer
University of Saskatchewan, Saskatoon, Saskatchewan S7N 0W0, Canada

Geological Association of Canada Special Paper 37
1990

Canadian Cataloguing in Publication Data

Main entry under title:

The Early proterozoic Trans-Hudson Orogen of North America

(Geological Association of Canada special paper ; 37)
Papers presented at a symposium held in Saskatoon, Sask., in May 1987.
Includes bibliographical references and index.
Bibliography: p. 1–13.
ISBN 0-919216-44-7

1. Orogeny. 2. Geology, Structural — Prairie Provinces. 3. Geology, Structural — Quebec (Province). 4. Geology, Structural — South Dakota. 5. Geology, Stratigraphic — Proterozoic. I. Lewry, J. F. II. Stauffer, M. R. III. Geological Association of Canada. IV. Series.

QE621.5.N6E37 1990 551.8'2'0971 C91–090054–X

Geological Association of Canada
Department of Earth Sciences
Memorial University of Newfoundland
St. John's, Newfoundland
Canada A1B 3X5

The GEOLOGICAL ASSOCIATION OF CANADA is Canada's national society for the geosciences. It was established in 1947 to advance geology and its understanding among both professionals and the general public. The GAC membership of 3000 includes representatives of all geological disciplines from across Canada and many parts of the world employed in government, industry and academia. There are specialist divisions for environmental earth sciences, geophysics, mineral deposits, paleontology, Precambrian, sedimentology, tectonics, and volcanology. Regional sections of GAC have been set up in Victoria, Vancouver, Edmonton, Winnipeg, and St. John's, and there are affiliated groups in Toronto and the Maritimes.

GAC activities include the organization and sponsorship of conferences, seminars, short courses, field trips, lecture tours, and student and professional awards and grants. The Association publishes the quarterly journal *Geoscience Canada* and the quarterly newsletter *Geolog*, a *Special Paper* series, Short Course Notes, and several other continuing series. GAC also maintains liaison with other earth science societies and provides advice to government and the public on geological issues. The Association was incorporated under the Canada Corporation Act in January, 1984.

For information contact: Geological Association of Canada, Department of Earth Sciences, Memorial University of Newfoundland, St. John's, Newfoundland, Canada A1B 3X5

L'ASSOCIATION GÉOLOGIQUE DU CANADA est la société nationale canadienne pour les sciences de la Terre. Créée en 1947, elle a comme double objectif de faire progresser la géologie et de sensibiliser les spécialistes et les membres du grand public aux sciences de la Terre. Ses trois mille membres représentent toutes les disciplines géologiques; ils viennent de toutes les régions du Canada et de nombreux autres pays; ils oeuvrent dans le secteur public, dans le secteur industriel et dans le monde universitaire L'Association comprend des divisions de spécialistes en géophysique, en gisements minéraux, en paléontologie, en Précambrien, en sédimentologie, en tectonique, en volcanologie et en sciences de la Terre touchant à l'environnement. Des sections régionales existent à Victoria, Vancouver, Edmonton, Winnipeg et St. John's, et des groupes affiliés se sont constitués à Toronto et dans les provinces maritimes.

Les activités de l'Association géologique du Canada comprennent l'organisation et le parrainage de conférences, de colloques, de cours de formation de courte durée, de visites sur le terrain et de tournées de conférences. Elle décerne des octrois et des bourses aux étudiants et aux personnes qui travaillent dans le domaine des sciences de la Terre. L'Association publie un journal trimestriel, *Geoscience Canada*, et un bulletin trimestriel d'information, *Geolog*, une série de mémoires, des notes de cours et diverses autres séries de publications. Elle assure en outre la liaison avec d'autres sociétés en sciences de la Terre et fournit des conseils au gouvernement et au grand public sur des questions géologiques. L'Association a été constituée en corporation en janvier 1984, en vertu de la Loi sur les corporations canadiennes.

Pour renseignements: Association géologique du Canada, Department of Earth Sciences, Memorial University of Newfoundland, St. John's, Terre-Neuve, Canada A1B 3X5

Typesetting: C.M. Whittaker & Associates, Ltd., Orleans, Ontario
Printing: The Runge Press Ltd., Ottawa, Ontario

Contents

WESTERN CRATON

EASTERN SEGMENT

COMPARATIVE STUDIES IN OTHER OROGENS

Preface

The Early Proterozoic Trans-Hudson Orogen is Canada's most recently recognized and defined major orogenic belt. Unconstrained by man-made borders, it extends in subsurface into north-central United States (as indicated in several papers in this volume): contiguous elements may possibly continue through central Greenland and even into northern Scandinavia. Its conceptual beginnings occurred in the mid-1970s as a result of the accumulation and compilation of several decades of geological work, mainly by the Department of Energy and Mines in Saskatchewan and Manitoba, with considerable input from the University of Regina and the University of Saskatchewan. Although presented to the world via numerous reports and papers in the late 1970s and early 1980s by geologists from these institutions, the name "Trans-Hudson Orogen" was suggested en passant by Paul Hoffman of the Geological Survey of Canada in a 1981 paper contained in a Geological Survey of Canada symposium volume entitled *Proterozoic Basins of Canada*.

The present volume is the outgrowth of a very successful Geological Association of Canada symposium on the Trans-Hudson Orogen held in Saskatoon, Saskatchewan, in May 1987. This symposium marked the "coming of age" of the orogen and its firm establishment as a major geological entity in the evolution of North America. The present volume includes only about half of the papers presented at the symposium and the reader is referred to the Geological Association of Canada – Mineralogical Association of Canada – Canadian Society of Petroleum Geologists Joint Annual Meeting 1987 *Program with Abstracts* volume for abstracts of other papers.

As in major orogenic belts of younger age, the Trans-Hudson Orogen encompasses variably reworked continental basement gneisses and a wide variety of metasedimentary, metavolcanic, and intrusive rocks whose physical and chemical characteristics and spatiotemporal relationships can readily be accommodated within plate tectonic models of geodynamic evolution: indeed, such an approach is followed by most papers in this volume. However, any new, rapidly developing topic of study is almost inevitably prone to proliferation of a complex and often contradictory terminology as various workers approach the subject from a different geographic base and with differing viewpoints. Terms are proposed and are either retained or dropped as the subject matures and a conceptual consensus is approached. Study of the Trans-Hudson Orogen, now barely in the "adolescent" stage, is still plagued by a plethora of competing terms, even for many first-order descriptive aspects. No attempt is made in this volume to enforce a particular set of terms; thus, in many cases, the reader will encounter different names used by different authors for the same rock unit, geological subdivision, or boundary. We hope that continued advances in knowledge will soon lead to general acceptance of a more uniform and simple set of names: perhaps the present volume will help to stimulate that process.

Stockwells' decades-old subdivision of the Canadian Shield into "structural provinces" represented a major advance in understanding the Precambrian development of North America. However, the past 20 years has seen an even more powerful development — the increasing recognition that the North American Craton comprises a succession of orogens, with older ones acting as cratons to newer ones. Subdivision of the entire shield, including its subsurface extension, into sequential mobile belts is not yet quite complete, but with the recognition of the Trans-Hudson Orogen the day is probably fast approaching when geologists will no longer make much use of the structural province concept, instead referring to locations by using the name of the appropriate orogen or its adjacent bounding cratons.

As editors and primary organizers of the original symposium, we are grateful for the help provided by numerous individuals and organizations. Publication costs were in part borne by the Natural Sciences and Engineering Research Council of Canada, the Geological Survey of Canada, and the Howard Street Robinson Fund. The International Geological Correlation Project (IGCP) 215 and the Geological Association of Canada cosponsored the symposium and the former provided travel funds for several European contributors. Translation of abstracts to French was done by André Ciesielski to whom we owe a debt of gratitude for undertaking this task at a time when it was needed on short notice. We would like to thank the following for reviewing manuscripts: K. Arndsdell, W.R.A. Baragar, M.E. Bickford, W. Brisbin, C. Brooks, H. Charlesworth, W. Church, G. Clark, L.C. Coleman, K.D. Collerson, D. Dunlop, G. Edwards, E. Froese, W. Fyson, D. Gendzwill, C. Gower, A.G. Green, E. Irving, R. Kerrich, K. Kyser, F. Langord, A. LeCheminant, R. Macdonald, D. McRitchie, J.C. Moore, E. Nisbet, G. Parslow, J. Patchett, Z. Petermann, D. Sangster, F. Schwerdtner, R.J. Scoates, P. Simony, P. Sims, W.R. van Schmus, B.R. Watters, and W. Weber.

J.F. Lewry *M.R. Stauffer*

The Trans-Hudson Orogen: Extent, Subdivision, and Problems

John F. Lewry
Department of Geology, University of Regina,
Regina, Saskatchewan S4S 0A2, Canada

Kenneth D. Collerson
Earth Sciences Board of Studies,
University of California, Santa Cruz, California 95064, U.S.A.

Lewry, J.F. and Collerson, K.D., The Trans-Hudson Orogen: extent, subdivision, and problems, in Lewry, J.F. and Stauffer, M.R., eds., The Early Proterozoic Trans-Hudson Orogen of North America: Geological Association of Canada, Special Paper 37, p. 1–14.

Abstract

The Trans-Hudson Orogen, a major component of the Early Proterozoic Pan-American orogenic system, is here considered to extend from South Dakota, across Hudson Bay, and into Greenland and Labrador. Nomenclature, the extent and character of major internal zones of this complex orogen, their relation to bounding cratons, and problems of definition and interpretation are reviewed. It is suggested that traditional "provincial" divisions of the Canadian Shield are outmoded and should be abandoned.

As is true of many Phanerozoic orogens, much of the Trans-Hudson Orogen is "ensialic," comprising reworked margins of three or more bounding Archean continental paleoplatforms that were amalgamated ca. 1.9–1.8 Ga ago. Intervening juvenile Early Proterozoic crust is extensively preserved only in the Reindeer Zone, a collage of accreted magmatic arcs, volcanogenic clastics, and later molasse deposits in the western part of the orogen. The extent of formerly intervening oceanic lithosphere and details of relative plate motions leading to orogeny are largely matters of speculation. Nonetheless, clear distinction between asymmetrically reworked, ancient, continental margins and juvenile arc tectonostratigraphic assemblages; recognition of ophiolitic remnants; features of igneous rock distribution and petrochemistry; and thermotectonic evolution (e.g., evidence of extreme shortening, overthrusting, and nappe - transcurrent fault dismemberment of accreted terranes) strongly suggest evolution within a collisional plate tectonic regime. Some characteristic features of such geodynamic evolution (e.g., subduction-zone accretionary complexes and mélanges) are undocumented. This does not necessarily imply significant differences between Phanerozoic and Early Proterozoic lithospheric evolution.

Résumé

L'orogène Trans-Hudson, constitue une des composantes principales du système orogénique panaméricain d'âge Protérozoïque Inférieur; il s'étend du Dakota du Sud vers la baie d'Hudson, le Groënland et le Labrador. On révise la nomenclature, létendue et le caractère des zones internes principales de cette orogène complexe, ses relations avec les cratons voisins et les problèmes de définition et d'interprétation. On propose que les divisions traditionnelles "provinciales" du bouclier canadien soient abandonnées étant donné leur désuétude.

De même que pour les orogènes phanérozoïques, une grande partie de l'orogène Trans-Hudson est 'ensialique', comprenant des marges reprises d'au moins trois plateformes continentales archéennes adjaçentes amalgamées vers 1.8–1.9 Ga. Une croûte juvénile d'âge Protérozoïque Inférieur est largement préservée dans la zone Reindeer, formée par accrétion d'arcs magmatiques, de sédiments volcanogéniques et de molasses tardives situées dans la partie ouest de l'orogène. La participation d'une lithosphere océanique ainsi que les mouvements de plaques donnant lieu à l'orogénèse sont pures spéculations. Néanmoins, une distinction nette entre les marges continentales anciennes reprises asymétriquement et les assemblages tectonostratigraphiques d'arcs juvéniles, l'identification locale de lambeaux d'ophiolites, la pétrochimie et la distribution des roches ignées et l'évolution thermotectonique (i.e., évidences de raccourcissements extrêmes, charriages et démembrement par failles coulissantes et nappes de terrains accrétés) suggère une évolution dans un régime de collision de plaques tectoniques. Les caractéristiques d'une telle évolution géodynamique (i.e., zone de subduction, complexes d'accrétion et mélanges) n'ont pas été identifiées empêchant toute comparaison entre l'évolution lithosphérique Phanérozoïque et Protérozoïque Inférieur.

INTRODUCTION

The Trans-Hudson Orogen (Hoffman, 1981) is one of the best-preserved and best-exposed Early Proterozoic orogenic belts in the world. It is the dominant component of a more extensive orogenic system (Hoffman, 1988, this volume) that in North America includes not only the Wopmay (Hoffman, 1980), Penokean (Van Schmus, 1980), and Ketilidian (Allaart, 1976) orogens but also several younger Early Proterozoic belts, lying south of the Superior and Wyoming cratons (Fig. 1), which are obscured by ca. 1.0 Ga Grenvillian orogenic overprinting and/or by Phanerozoic cover. The evolution of this system coincides with a major episode of North American continental accretion, and, indeed, with a more general Laurentian supercontinent assembly in the latter part of the Early Proterozoic (Hoffman, 1988). Available data strongly suggest that this "Pan-American" orogenic system evolved via essentially "uniformitarian" plate tectonic processes: phases of initial continental rifting and passive margin development, ensuing subduction, arc generation and closure, juvenile terrane accretion, and terminal continental suturing at collisional plate boundaries can all be documented, albeit with varying degrees of confidence. Ophiolite remnants are now well documented (Schulz, 1987; St-Onge and Lucas, this volume), and preorogenic seafloor spreading and subsequent closure may have been comparable in magnitude to that attending the evolution of Phanerozoic orogens (Dunsmore and Symons, this volume).

Our purposes here are to develop a general context for other papers in this volume, and to provoke debate on more contentious aspects of subdivision and interpretation of the Trans-Hudson Orogen. Thus, we briefly review the continental framework and major elements of the orogen, comment on internal lithotectonic subdivision and nomenclature, and, in particular, emphasise current problems (including the definition and actual extent of the Trans-Hudson Orogen), many of which are addressed more fully by other authors. Our approach is somewhat different from that followed in the comprehensively referenced overview of the "Churchill Province" presented by Hoffman (this volume) and in places we intentionally adopt the role of "devil's advocate" with respect to some aspects of his interpretation.

GENERAL SUBDIVISION AND NOMENCLATURE OF THE PAN-AMERICAN SYSTEM

Modern advances in understanding the Precambrian craton of North America have rendered much of the traditional "provincial" subdivision and related nomenclature of Stockwell (1961, 1970), based on broad structural trends and isotopic closure ages, redundant save for historical reference. Such division was of great value and is to some degree still valid with respect to the little reworked Archean Slave, Superior, Wyoming, and North Atlantic (Nain) cratons. Elsewhere, however, its utility is now minimal, most notably in regions significantly affected by Early Proterozoic orogeny, such as the "Churchill Province." Therefore, we favour abandonment of this old provincial classification and adoption of a scheme of subdivision and nomenclature more in tune with crustal evolutionary principles applied in the modern analysis of younger orogens and their bounding cratons.

We suggest that the term "Pan-American System" (Lewry *et al.*, 1987) might now be informally adopted to encompass all of the Early Proterozoic orogenic belts of North America, including Wop-

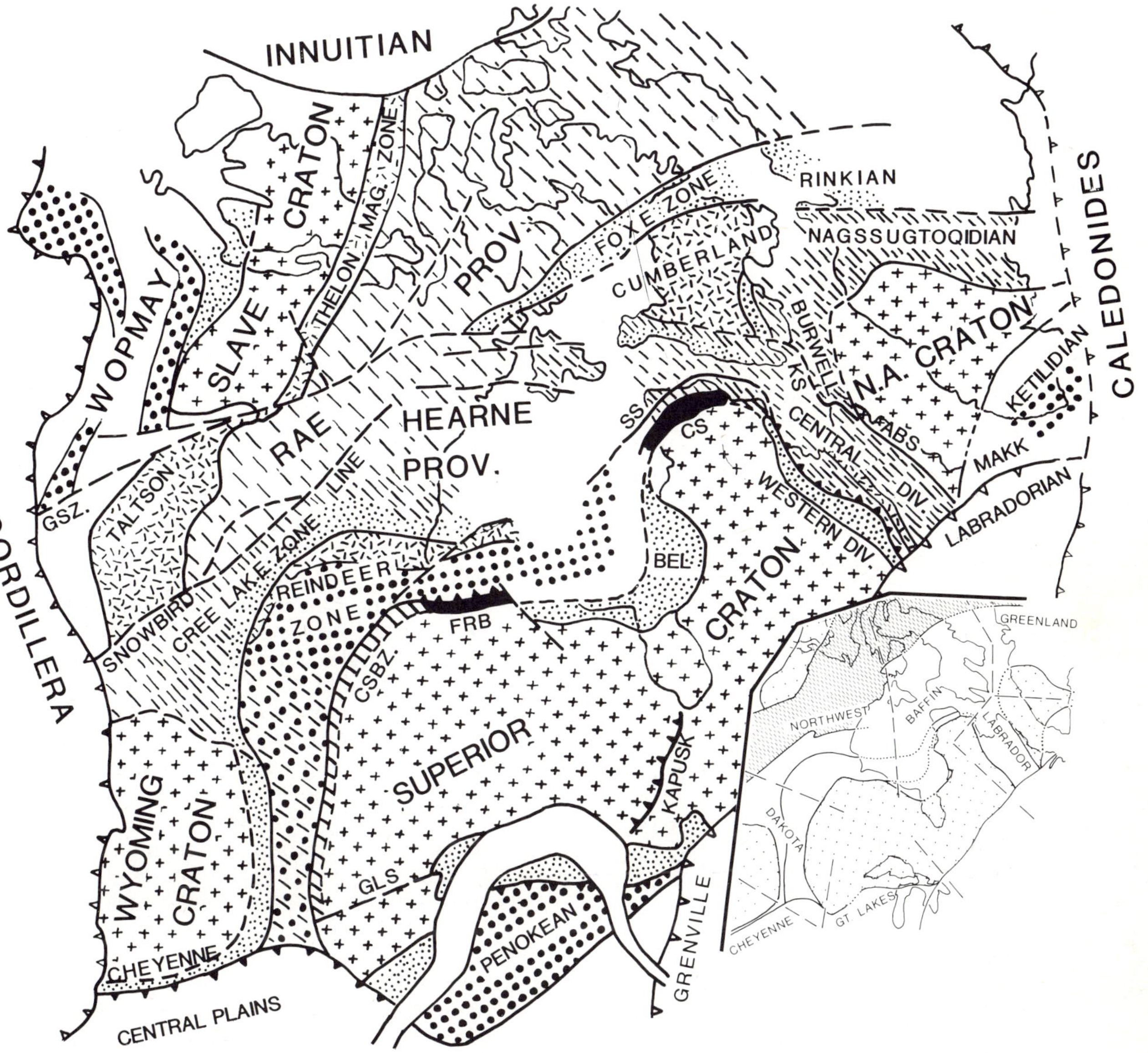

Figure 1. *Major elements of the Trans-Hudson Orogen, other Early Proterozoic orogenic belts, and Archean cratons in North America. Inset shows the proposed nomenclature of orogenic segments used in this paper. Abbreviations: ABS, Abloviak shear zone; KS, Komaktorvik shear zone; SS, Sugluk suture; CS, Cape Smith Belt; BEL, Belcher Islands area; FRB, Fox River belt; GLS, Great Lakes shear zone; GSZ, Great Slave shear zone; CSBZ, Churchill–Superior Boundary Zone; and MAKK, Makkovik belt. Legend: crosses, little reworked Archean cratons; oblique dash, Archean continental crust variably reworked during Early Proterozoic orogeny and with scattered remnants of Early Proterozoic inboard supracrustals (closer dashes indicate generally comprehensive high-grade reworking); stipple, Early Proterozoic foreland fold/thrust belts and other areas preserving abundant Early Proterozoic inboard supracrustal rocks; random ticks, major continental magmatic arc batholiths; heavy dots, mostly accreted juvenile Early Proterozoic arc terranes (mixture of heavy dot and oblique dash in southern part of Reindeer zone indicates likely occurrence of both allochthonous juvenile elements and highly reworked (Archean) continental crust); black, predominantly mafic–ultramafic allochthons thrust over the margins of the Superior craton (mainly ophiolitic in the Cape Smith Belt, possibly representing foredeep volcanics and intrusives in the Fox River belt).*

may, Thelon, Trans-Hudson, Ketilidian, Penokean, Labradorian, and Central Plains orogens and their extensions. Following common usage in other orogens, this system can conveniently be divided into a number of geographic "segments" (Fig. 1, inset) corresponding to broad transects across the orogenic belts and their bounding Archean cratons. Such segments, and the names we have assigned to them, have little geological significance; they are adopted only for convenient geographic reference, in place of old "provincial" subdivisions. Alternative nomenclature might well be preferred.

In the following review, the term "craton" refers specifically to coherent areas of Archean continental lithosphere that underwent little or no significant thermotectonic or isotopic reworking during Early to Middle Proterozoic orogeny. The term "paleoplatform," on the other hand, is applied more widely to apparently formerly coherent Archean continental lithospheric plate elements, regardless of the degree of subsequent Proterozoic reworking. Thus, for example, the Superior paleoplatform is defined as encompassing not only the Superior craton but also contiguous, more reworked extensions of the latter into adjacent ensialic zones of the encircling Early Proterozoic orogens. We acknowledge that this violates, to some degree, the commonly accepted definition of a platform but, lacking a suitable alternative, prefer to eschew the designation "province" to avoid confusion with the old Stockwellian terms. For the most part, the term "zone" is used for lithotectonically distinct major internal subdivisions of the orogen, representing either reactivated portions of bounding continental platforms or accreted juvenile terrane components. Despite some resultant confusion with respect to terminology, the terms "domain," "belt," and "block" are retained for local internal lithostructural subdivisions of such zones, where these are entrenched in the literature. The term "terrane" is used in the sense defined by Coney *et al.* (1980) and Williams and Hatcher (1982). We concede some remaining terminological inconsistency, considering this to be presently unavoidable and inconsequential.

EXTENT AND SUBDIVISION OF THE TRANS-HUDSON OROGEN

As popularly defined, the Trans-Hudson Orogen extends in subsurface from South Dakota, where it is apparently transected by juvenile terranes of the slightly younger (ca. 1.8–1.7 Ga) Central Plains Orogen (Sims and Peterman, 1986), northward to the edge of the Canadian Shield (Green *et al.*, 1985a,b; Klasner and King, 1986, this volume) and thence through the exposed shield and across Hudson Bay at least as far as the Cape Smith Belt, in northern Ungava.

It is currently debatable whether the Trans-Hudson Orogen should be regarded as including orogenic elements north and east of Cape Smith, i.e., the Foxe and Cumberland zones (as defined below) in south and central Baffin Island, the Rinkian and Nagssugtoqidian zones of central Greenland, and orogenic zones of the Labrador segment (Fig. 1). Hoffman (1987, 1988, this volume) presents cogent arguments for the exclusion of these northern and eastern orogenic elements, placing primary emphasis on the location of possible sutures (which in some cases are rather speculative) and the degree of continuity of probable juvenile terranes in thus limiting the extent of the orogen.

We are more inclined to emphasise the undoubted continuity of intense Hudsonian thermotectonism and attendant magmatism, rather than the continuity of sutures and juvenile terranes, or the lack thereof. There are many uncertainties as to the age, character, and continuity of lithotectonic elements in the Baffin–Labrador–Greenland segments; however, the same could be said, until quite recently, of complex Phanerozoic mobile belts, such as the Appalachian and Cordilleran orogens. Although there may well be significant lateral variation in the preservation and age of juvenile terrane elements, peak orogeny, and terminal collisional suturing (cf. Hoffman, this volume and references therein), available age data nonetheless suggest that initial active-margin arc magmatism in much of the Trans-Hudson Orogen occurred ca. 1.9 Ga, and that terminal collision and suturing of bounding continental platforms and intervening juvenile terranes occurred within a relatively short interval, ca. 1.85–1.80 Ga (Gordon *et al.*, this volume; Machado, this volume). Such time spans are comparable to, and perhaps even shorter than, those involved in the development of well-documented Phanerozoic belts, such as the Appalachian-Caledonian orogen, in which at least three distinct major orogenic episodes, related to successive phases of accretionary suturing, are documented. At present, therefore, we are not persuaded that there is much advantage in regarding northern and eastern orogenic components as being other than parts of a single, essentially continuous, complex orogen. The coverage of papers in this volume reflects this viewpoint.

As defined above, the Trans-Hudson Orogen is thus considered to incorporate variably reactivated margins of at least three, and possibly more, formerly independent(?) Archean continental paleoplatforms (together with their inboard epicontinental cover and fringing rifted- to passive-margin miogeoclinal wedges), and a variety of intervening accreted Early Proterozoic juvenile terranes. The distribution and character of these lithotectonic elements are reviewed below. The lithotectonic subdivision and nomenclature in western parts of the orogen (Dakota and Northwest segments, Fig. 1, inset) are relatively well established and subject to only minor dissension, whereas in eastern parts of the orogen (Baffin, Greenland, and Labrador segments) they are less certain and subject to considerable debate, as intimated above. Therefore, these two regions are discussed separately.

WESTERN OROGENIC SEGMENTS

Northwest segment

Superior paleoplatform margin. In the Northwest segment, the documented extent of the Superior paleoplatform is abruptly delimited by the northern–western margin of the Churchill–Superior Boundary Zone (although presently retaining this name, we think that it might be replaced by a shorter name, such as the Nelson zone?). This zone, generally regarded as an integral part of the Trans-Hudson Orogen (cf. Weber, this volume), is relatively narrow and mainly comprises reworked Archean basement gneisses derived from deep crustal Pikwitonei granulite facies protoliths of the contiguous Superior craton. The Fox River belt, forming much of the northern part of the boundary zone, is a 15–20 km wide, steeply north-dipping homoclinal sequence of pillowed to massive, tholeiitic to komatiitic basalts, apparently coeval differentiated mafic–ultramafic sills, and subordinate intercalated metasediments, all at low to very low metamorphic grade. Sediments in the lowest parts of the succession may be miogeoclinal but it has been suggested that most of the sedimentary sequence comprises foredeep clastics (Hoffman, 1988, this volume; Weber, this volume). Although little internal deformation is documented, it seems likely that the Fox River belt is internally imbricated (Weber, this volume). It is overthrust by highly deformed, high-grade "Kisseynew gneisses" to the north and may be in tectonic contact with underlying reworked Archean gneisses of the boundary zone to the south (Bleeker, this volume; Weber, this volume). However, the age, geotectonic setting, and degree of allochthoneity of the Fox River belt are not yet fully resolved.

The "Thompson belt," forming the north-northeast-trending western part of the boundary zone, includes varied Lower Proterozoic metasediments and mafic–ultramafic lavas generally correlated with those of the Fox River belt. Associated serpentinitic ultramafic bodies are currently regarded as dismembered but essentially autochthonous intrusions. However, uncertain age and structural relationships do not preclude early emplacement of both supracrustals and ultramafic rocks in allochthonous slices, which may also incorporate reworked basement (Green *et al.*, 1985a,b; Bleeker, this volume).

The western margin of the Thompson belt is markedly discordant to gross lithotectonic trends in juvenile terranes of the Reindeer Zone, to the west, and evidently represents a major crustal discontinuity. However, the kinematic history and overall sense of displacement along this junction remain somewhat contentious. Recent kinematic studies document dominantly vertical, Superior-side-up displacement along ductile to brittle shear zones subparallel to the junction, within the boundary zone (Bleeker, this volume; Fueten and Robin, 1989). Such data might be viewed as evidence of similar major displacement along the junction itself. However, Bleeker (this volume) interprets late, oblique en echelon folds within the boundary zone as indicating significant transpressive strike-slip displacement.

The western margin of the zone has clearly had a long, complex kinematic evolution: we retain the view (cf. Lewry, 1981) that it is difficult to explain apparent truncation of lithotectonic domains in the Reindeer Zone without major transform/transcurrent displacement along the western margin of the Churchill–Superior Boundary Zone at some stage in its evolution. Sinistral offset of aeromagnetic trends across the unexposed Owl River shear zone (which continues on trend with the western margin of the Churchill–Superior Boundary Zone, within the Reindeer Zone to the northeast) is consistent with this thesis. Current evidence favours interpretation of the boundary discontinuity as evolving via early shortening and overthrusting followed by sinistral displacement in a transpressive regime (Green *et al.*, 1985a; Bleeker, this volume).

The northern and western margins of the boundary zone are thus everywhere defined by faults and shear zones across which no continuity of Archean basement has been established. Adjacent internal parts of the orogen are therefore regarded as being paleogeographically "suspect" with respect to the ancient Superior continental margin.

Reindeer Zone: an accreted juvenile terrane collage?. The Reindeer Zone (name modified after Stauffer (1984)) is defined as including all lithotectonic elements lying between the Churchill–Superior Boundary Zone and the Wathaman–(Chipewyan) Batholith. The zone comprises a lithologically and structurally complex collage of arc-related volcanic and plutonic rocks, coeval or derived volcanogenic clastics, subordinate later arkosic molasse assemblages (Stauffer, this volume), and crustal melt fractions, and can be divided areally into a number of more-or-less distinct lithostructural domains. U–Pb zircon geochronology (Baldwin *et al.*, 1985; Lewry *et al.*, 1987; Van Schmus *et al.*, 1987; Gordon *et al.*, this volume) has yielded ages of between 1910 and 1830 Ma for felsic arc volcanics and plutons: ages of volcanic rocks are mostly within the older part of this range (ca. 1910–1875 Ma). Major and trace element discrimination data from volcanics and plutons (Watters and Pearce, 1987; Syme, this volume; Thom *et al.*, this volume) and preliminary Nd, Pb, and Sr isotopic data (e.g., Chauvel *et al.*, 1987) from varied rock types suggest that most of the Reindeer Zone collage evolved within an essentially oceanic, subduction-related arc setting, with subordinate (sedimentary?) input from Archean sources. Underlying Archean continental basement is documented only in three localities: the Sahli "granite" in the Hanson Lake block (Bell and Macdonald, 1982; Van Schmus *et al.*, 1987), the Iskwatikan Complex in western Glennie domain, and the core of the Hunter Bay dome just to the south (cf. Lewry *et al.*, this volume, and figures and references therein). In all of these cases, the Archean rocks occur at the deepest locally exposed structural levels and it is possible that tectonically underthrust Archean basement might underlie much of the juvenile collage (Lewry *et al.*, this volume).

Most of the Reindeer Zone can be considered to have an exotic, or at least "suspect," Early Proterozoic paleogeographic provenance relative to adjacent ancient continental margins to the northwest: no preorogenic continuity across the straddling Wathaman Batholith (Fig. 1) has been established and, as indicated above, the character of original, and present, basement underlying the Early Proterozoic juvenile assemblages is generally uncertain.

The Wathaman–(Chipewyan) Batholith (Lewry *et al.*, 1981; Fumerton *et al.*, 1984; Meyer, 1987; Halden *et al.*, this volume) is a remarkably homogeneous, dominantly megacrystic, monzogranite–granodiorite body, dated at ca. 1.85 Ga (Lewry *et al.*, 1987), extending for at least 900 km along the regional strike. It is everywhere emplaced between reworked Archean continental crust and arc-related domains of the Reindeer Zone, to the northwest and southeast, respectively, and is generally interpreted as a continental margin magmatic arc.

The southeastern margin of the batholith is characterised by an extensive metasomatic and injection aureole. In Manitoba, the northern contact is also largely intrusive; whereas in Saskatchewan, its contacts with the ensialic Peter Lake domain and Cree Lake Zone are almost entirely delimited by mylonitic gneisses of the Needle Falls – Parker Lake shear system. The latter comprises two well-defined northeast-trending "en echelon" ductile shears, the Needle Falls and Parker Lake shear zones, which are linked along the southern margin of the Peter Lake domain by a much broader, east–west trending, splay zcne of inhomogeneous shear. Restricted but unequivocal kinematic data from the southern part of the Needle Falls shear zone (Stauffer and Lewry, 1988) indicate slightly oblique, dextral strike-slip displacement with a subordinate east-side-up vertical component.

Rae-Hearne paleoplatform. Continuous, variably reworked Archean continental crust to the north and west of the Wathaman Batholith encompasses much of the former western and northern Churchill Province, the Slave craton, and extensions of the latter beneath the fold–thrust belt of the Wopmay Orogen. This Archean basement is locally overlain by and infolded with remnants of inboard Lower Proterozoic epicontinental, rifted margin(?), miogeoclinal, and foreland basin cover.

The "Slave–Churchill" boundary has recently been the focus of renewed investigation and debate. Previous interpretation of the "Thelon Front" as being an Early Proterozoic cryptic suture (Gibb and Thomas, 1977; Gibb, 1978) is now supported by recent data and reinterpretation (see discussion and references in Hoffman (1988, this volume)), which largely answer objections to this thesis previously expressed by Lewry *et al.* (1985) and others. Thus, we now accept the view that the Slave paleoplatform represents a former, essentially independent, Archean "minicontinent" flanked by Early Proterozoic collisional orogens both to the west (Wopmay Orogen) and east (Thelon Orogen). Major orogenic activity in the Thelon Orogen (ca. 1.98–1.89 Ga) was a little earlier than, and not directly related to, the Trans-Hudson Orogen.

Interpretation of the Archean terrain lying between the Wathaman Batholith and the eastern boundary of Slave craton has been in a state of flux over the past decade, with concomitant confusing changes in nomenclature. The region is transected by a major tectonic break, segments of which have been recognised locally as the Virgin River (Lewry and Sibbald, 1977), Black Lake (Gilboy, 1980), and Tulemalu (Tella and Eade, 1986) shear zones: this discontinuity is now collectively termed the Snowbird "line" (Hoffman, 1987, this volume; Lewry, 1987) or "tectonic zone" (Hoffman, 1988).

The variably reworked paleoplatform lying between the Snowbird line and the eastern margin of Slave craton is now termed the Rae province: it has previously been called the Amer Lake zone (Lewry *et al.*, 1985) or North Keewatin zone (Hoffman, 1987) and includes the previously defined Western Craton (Lewry and Sibbald, 1980) and Taltson, Queen Maud, Armit Lake, and Committee Bay blocks (Heywood and Schau, 1978; Lewry *et al.*, 1985). The region to the southeast of the Snowbird line, now termed the Hearne province (Hoffman,

1988), includes the Cree Lake Zone, and Tulemalu and Ennadai blocks (Heywood and Schau, 1978; Lewry *et al.*, 1985), and was formerly the South Keewatin zone of Hoffman (1987).

We accept the terms Rae province, Hearne province, and Snowbird line as defined by Hoffman (1988, this volume), albeit with some interpretative reservations. "The interpretation of the Snowbird zone as a suture remains hypothetical" (Hoffman, 1988): indeed, there is currently little firm evidence of a related magmatic arc or other "geological" indication of plate subduction–collision along the zone. Although Ross *et al.* (1988) have documented extensive Early Proterozoic plutonism (Rimbey zone) adjacent to subsurface extensions of the Snowbird line in Alberta, this has not yet been convincingly proven to constitute a continental magmatic arc: in the exposed shield, plutons such as the Junction Granite (Lewry and Sibbald, 1977) represent crustal melts intruded into the shear zone. Moreover, postulated northeasterly continuity of the Snowbird line into the Baker Lake area and across Hudson Bay (Hoffman, this volume) is debatable.

Preliminary kinematic data from the Virgin River segment (Carolan and Collerson, 1988) suggest dominant dextral transcurrent movement along this portion of the Snowbird line: it may represent an entirely intracontinental Early Proterozoic transform, possibly involving reactivation of an older, Archean crustal break. We note that although the Snowbird line appears to be represented by granulite facies mylonites in the north (see Hoffman, this volume), it is largely defined by lower amphibolite facies mylonites (involving retrogression of granulite facies gneisses of Rae province) in its Saskatchewan segments (Lewry and Sibbald, 1977; Carolan and Collerson, 1988).

Thus, whether Rae and Hearne provinces behaved as essentially independent minicontinents during all or part of the Early Proterozoic, or comprised a single coherent continental platform throughout passive- to active-margin evolution preceding development of the Trans-Hudson Orogen, is largely a matter of speculation. We therefore use the term "Rae–Hearne paleoplatform" to denote the entire area included in the Rae and Hearne provinces of Hoffman (this volume), leaving open the question of whether this comprised one or two separate continental entities during the evolution of the Trans-Hudson Orogen.

The extent and character of Early Proterozoic reactivation differ markedly in different parts of the Rae–Hearne paleoplatform. In the Northwest segment, only the highly reactivated southern part of Hearne province, long defined as the Cree Lake Zone, and the adjoining, enigmatic Peter Lake domain are here regarded as integral ensialic components of the Trans-Hudson Orogen. The remainder of the paleoplatform represents generally less reactivated "hinterland" or lies within the recently defined Thelon Orogen.

The Cree Lake Zone (Lewry and Sibbald, 1977, 1980; Lewry *et al.*, 1985) mainly comprises upper amphibolite to granulite facies granitoid orthogneisses and migmatitic paragneisses, with subordinate mafic gneisses deriving from both metavolcanic and plutonic protoliths. Most of this complex probably represents reworked Archean basement, though some paragneisses, migmatite neosomal components, and homogeneous late-kinematic granites are probably Early Proterozoic in age (Bickford *et al.*, 1988). The Archean basement is generally thought to have been unconformably overlain by, and is now complexly infolded with, Lower Proterozoic supracrustal cover rocks of the Wollaston Group. The latter are most completely preserved in the Wollaston domain (eastern part of Cree Lake Zone) and are broadly interpreted as being a passive margin sequence. However, lowermost arkosic rocks and intercalated mafic volcanics exposed in the eastern part of the domain may represent an earlier rifted margin assemblage, and upper feldspathic and/or calcareous quartzites and "meta-arkoses" may possibly represent synorogenic foreland basin, rather than passive margin, sediments. The Wollaston Group is considered by almost all local workers to be of Early Proterozoic age but this is not conclusively documented. Such uncertainty as to age does not detract from fundamental differences between the Cree Lake and Reindeer Zones or from the geotectonic importance of their mutual boundary.

Early work in Saskatchewan and Manitoba suggested that most of the Cree Lake Zone underwent complex multiphase deformation and medium- to high-grade thermal reworking during the Early Proterozoic. If so, the whole zone should be considered part of the Trans-Hudson Orogen. Such interpretation depends heavily on the argument (cf. Lewry and Sibbald, 1980) that (a) the junction between the Wollaston domain and the adjacent Mudjatik domain, to the west, is lithologically and thermotectonically gradational; (b) some of the complexly deformed and metamorphosed supracrustal belts in the west are correlative with the Wollaston Group; and (c) the main thermotectonic overprint must, thus, be of Early Proterozoic age, albeit mainly superimposed on a prior high-grade Archean terrain. However, Annesley and Madore (1989) have recently suggested that the Wollaston–Mudjatik boundary might be a major tectonic discontinuity. Moreover, lithological correlation between the two domains is not unequivocally established: some, and possibly most, of the supracrustal remnants within the western Cree Lake Zone may be Archean, a possibility conceded by Lewry and Sibbald (1980). Furthermore, the extension of low-grade Archean metavolcanics of the Kaminak Group into otherwise high-grade terrain of the northern Cree Lake Zone (Macdonald and Chiarenzelli, 1986; Reilly, 1989), and the occurrence, in northeast Saskatchewan and northwest Manitoba, of low-grade, Lower Proterozoic metasediments correlative with the Hurwitz Group to the north, cast doubt on the supposed Early Proterozoic age of high-grade thermotectonic events, at least in the northern part of the zone. Preliminary U–Pb zircon and Sm–Nd isotopic data from southern Mudjatik domain (Bickford *et al.*, 1987, 1988) confirm the extensive occurrence of Archean rocks but place few unequivocal constraints on the age of the subsequent thermotectonic overprint, apart from indicating that it must predate ca. 1840 Ma.

An associated problem concerns the relative age of the Wollaston Group and the undoubtedly Lower Proterozoic, generally lower metamorphic grade, Hurwitz Group to the north. The latter includes lower continental clastics, a well-developed shelf–miogeoclinal sequence, and upper foredeep to molasse basin deposits (Aspler *et al.*, 1989). Structurally, the Hurwitz Group forms part of a late Early Proterozoic foreland thrust–fold belt with dominant northwesterly vergence (Aspler *et al.*, 1989), suggesting a close relationship to the Trans-Hudson Orogen. Remnants of the Hurwitz Group in northeastern Saskatchewan are also documented as allochthonous slices with northwesterly thrust vergence (Reilly, 1989). It remains uncertain, however, whether this group is the same age as, or somewhat younger than, the Wollaston Group.

Apparently on the basis of such uncertainties, Hoffman (1988, this volume) includes only the Wollaston and Seal River domains (the latter defined only in Manitoba) within the orogen proper and considers the remainder of the Cree Lake Zone (including Mudjatik and Virgin River domains in Saskatchewan) to be an essentially cratonic part of Hearne province. We are not yet ready to embrace this view; however, debate on this problem might be resolved by ongoing coordinated isotopic, metamorphic, and structural investigations in the Cree Lake Zone (Carolan and Collerson, 1988; Orrell *et al.*, 1988).

Dakota segment

Subdivision and interpretation of the character and geotectonic evolution of this largely unexposed orogenic segment depend almost entirely on subsurface geophysical and drill core evidence. Available data suggest little significant difference between the exposed part of the Churchill–Superior Boundary Zone and its unexposed southern extension (Green *et al.*, 1985a,b; Klasner and King, 1986, this volume). The nature of the unexposed southern continuation of the Reindeer Zone is, however, subject to much uncertainty and debate.

Reindeer Zone. Sm-Nd isotopic data obtained for basement core samples from supposed extensions of the Reindeer Zone in south-central Saskatchewan (Collerson *et al.*, 1988, 1989) yield estimated crustal residence ages ranging from 3.4 to 2.4 Ga, though some of the same cores give Early Proterozoic U–Pb zircon ages. This suggests that much of the unexposed southern continuation of the zone comprises Archean crust (albeit variably reworked by Hudsonian thermotectonism) or Early Proterozoic melts derived largely from Archean crustal sources. There is little indication of widespread preservation of juvenile Early Proterozoic crust. Such evidence of extensive reworked Archean crust in the basement of south-central Saskatchewan calls into question the validity of both subsurface extrapolation of lithotectonic elements based on geophysical data (Green *et al.*, 1985a,b) and estimates of juvenile crust formation during the Early Proterozoic (Patchett and Arndt, 1986). It also imposes significant constraints on tectonic models of evolution of this part of the Trans-Hudson Orogen. Such data can, however, be reconciled with the essentially thin-skinned tectonic model for the exposed Reindeer Zone proposed by Lewry *et al.* (this volume), which suggests the possible continuous or semicontinuous extension of Archean crust (of the Superior paleoplatform?) beneath juvenile Early Proterozoic allochthons.

Wyoming craton boundary zones. The Trans-Hudson Orogen in the Dakota segment is bounded to the west by the Wyoming craton. Although isotopic systems are progressively reset towards the southern and eastern margins of the latter, exposed portions of this Archean continental platform display only limited Early Proterozoic tectonometamorphic overprinting, in marked contrast to the Cree Lake Zone. Only a narrow marginal zone (corresponding approximately to Zone V of Klasner and King (1986, this volume)), exposed only in the Black Hills of South Dakota (Redden *et al.*, this volume), can be considered to be an ensialic component of the Trans-Hudson Orogen. The ancient continental margin of the Wyoming paleoplatform may coincide approximately with the Central Great Plains electrical conductivity anomaly (Camfield and Gough, 1977; Green *et al.*, 1985a; Klasner and King, 1986, this volume); however, the precise geological significance of this prominent geophysical feature (and for the most part its exact location) remains obscure.

The nature of the unexposed junction between the Wyoming and Rae–Hearne paleoplatforms is largely a matter of conjecture. Recent speculation has focused both on the "Vulcan low," a buried crustal discontinuity, defined by gravity and magnetic anomalies, that extends through southern Alberta and southwestern Saskatchewan, and on the "Great Falls tectonic zone" farther south. Initially interpreted as a Middle Proterozoic continental rift, the Vulcan low has recently been viewed as a possible locus of (either Early Proterozoic or Archean?) collisional suture between the Hearne and Wyoming paleoplatforms (Hoffman, 1988; Ross *et al.*, 1988). Differing published views of these discontinuities are reviewed by Hoffman (this volume).

EASTERN OROGENIC SEGMENTS

Circum-Superior ensialic zones

The Superior paleoplatform was probably once fringed by a continuous Lower Proterozoic continental terrace wedge, remnants of which are now preserved as little-deformed autochthonous cover and well-developed foreland fold/thrust belts of the Labrador Trough and Richmond Gulf – Belcher Islands – Sutton Inlier regions (BEL in Fig. 1). A thin autochthonous sequence also underlies allochthons of the Cape Smith belt. These inboard supracrustal zones, together with thermotectonically and isotopically reworked marginal parts of the Superior paleoplatform (together essentially comprising the "Circum-Superior Belt" of Baragar and Scoates (1981)), are all external components of the Trans-Hudson Orogen.

Circum-Superior supracrustal rocks include early continental to rifted margin assemblages, well-developed passive-margin miogeoclinal successions, and locally preserved younger foredeep and foreland basin flysch–molasse sequences. Several later papers discuss relations in the Labrador Trough. However, perhaps the most complete sequence, directly comparable to Phanerozoic continental margin sequences, is that of the Richmond Gulf – Belcher Islands area (e.g., Ricketts and Donaldson, 1981), where a well-developed passive-margin quartzite–carbonate sequence is overlain by active-margin volcanogenic flysch deposits derived from an emergent arc to the northwest; these, in turn, are overlain by arkosic molasse.

Progressively reworked Archean basement of the Superior paleoplatform extends beneath these inboard assemblages and, in places, possibly for some distance into the higher grade crystalline "internides" of the orogen. Such zones of significantly reworked basement, as in the Churchill–Superior Boundary Zone, are generally rather narrow and are terminated by major tectonic sutures and/or the margins of juvenile Early Proterozoic terrane elements. This contrasts markedly with the broad extent of brittle to ductile tectonometamorphic reworking of continental paleoplatforms on the northern and northwestern sides of the orogen. This overall gross orogenic asymmetry is also reflected internally in the character and distribution of Early Proterozoic plutonism (cf. Lewry *et al.*, 1981, 1985) and might reasonably be interpreted as reflecting dominant subduction polarity at consuming plate boundaries. There is abundant evidence of major shortening, telescoping, and the cratonward transport of semiautochthonous to thoroughly allochthonous elements across the margins of the Superior paleoplatform, most notably in the largely allochthonous Cape Smith Belt, which includes obducted ophiolitic sheets (Hoffman, 1985; St-Onge and Lucas, this volume).

Other orogenic zones in the Baffin and Greenland segments

Preferences as to the subdivision and nomenclature of northern, largely ensialic components of the Trans-Hudson Orogen in the Baffin and Greenland segments are to some extent dependent on differing views of the likely eastward extension of the Rae and Hearne provinces, and the geotectonic significance and eastward continuity of the Snowbird line. It is now generally agreed that northern parts of Melville Peninsula and Baffin Island ("Prince Albert Fold Belt" of Jackson and Taylor (1972)) are underlain by mildly reworked Archean hinterland to the Trans-Hudson Orogen, representing an eastward extension of Rae province. To the south of this, the Foxe Zone ("Foxe Fold Belt" of Jackson and Taylor (1972)) of southern Melville Peninsula and central Baffin Island includes highly remobilised Archean basement and inboard miogeoclinal to foredeep Lower Proterozoic supracrustal rocks of the Piling–Penrhyn groups. The metamorphic grade rises from greenschist facies in the north to upper amphibolite – granulite facies in the south, with concomitant progressive involvement of reactivated basement in refolded nappe cores and sheath folds (see references in Hoffman (this volume)). This zone has many, albeit perhaps superficial and misleading, similarities to the Cree Lake Zone; however, previously postulated continuity between the two, through western Hudson Bay (Lewry *et al.*, 1985), is currently argued to be unlikely (Hoffman, 1988, this volume). Eastward continuity of the Foxe Zone with the Rinkian belt of the Greenland segment, on the other hand, seems more probable (Grocott and Pulvertaft, this volume).

The little-known, high-grade, predominantly granitoid terrane of south-central Baffin Island, lying south of the Foxe Zone, is here provisionally termed the Cumberland Zone. Sparse data suggest that it largely comprises reworked deep crustal Archean continental basement extensively intruded by Early Proterozoic charnockite–granite plutons of batholithic dimensions. Components of the latter are dated at 1.9–1.87 Ga (Pidgeon and Howie, 1979). In southern Baffin Island, the Dorset fold belt comprises amphibolite to granulite facies, miogeoclinal to flyschoid supracrustal rocks, of probable Early Proterozoic age, infolded with granitoid gneisses, which probably represent Archean basement. To the east, supracrustal rocks of the Hoare Bay Group are of uncertain age.

Available data impose few firm constraints on the interpretation of the Cumberland Zone and its relation to other parts of the Trans-Hudson system. Hoffman (1988, this volume) suggests that the Snowbird line extends eastward across Hudson Bay to the south of Baffin Island and that the Cumberland Zone, therefore, represents an extension of Rae province. This implies essential continuity of Archean continental basement throughout the Foxe and Cumberland zones and the absence of any collisional plate boundary north of the proposed eastward extension of the Snowbird line (Hoffman, 1988). A further corollary is that the miogeoclinal Piling–Penrhyn groups must then have been rather distally related to a continental margin now represented by the Sugluk suture and/or extension of the Snowbird line or, alternatively, represent deposition within an "oroclinal" basin (Hoffman, this volume). However, Grocott and Pulvertaft (this volume) argue that Early Proterozoic thermotectonism in the Rinkian Zone of Greenland was related to collisional suturing of the North Atlantic craton to a northern continental platform (presumably an extension of Rae province) and that the Nagssugtoqidian Zone, to the south, incorporates a cryptic suture. Westward extrapolation into the Baffin segment suggests that the likely extension of this postulated suture might lie not far south of the Foxe Zone, within the Cumberland Zone. Alternatively, it must swing abruptly southwards towards the Komaktorvik shear zone in eastern Labrador segment (Hoffman, this volume; Wardle *et al.*, this volume).

As a possible working alternative to the interpretation given by Hoffman (this volume), we suggest that the Cumberland Zone of Baffin Island (and, by extrapolation, the Central Division of the Labrador segment, discussed below) might represent highly reactivated and partially dismembered extensions of the North Atlantic craton rather than of Rae province. Such a hypothesis implies only two major collisional sutures in the eastern part of the Trans-Hudson Orogen: one between the former Superior and North Atlantic paleoplatforms (including Sugluk suture and the junction between the Western and Central divisions in the Labrador segment), the other extending westward from the Nagssugtoqidian somewhere between the Foxe and Cumberland zones (or within the latter). Work by Ryan *et al.* (1988) in the Central Division of the Labrador segment provides some support for this thesis.

Labrador segment

Subdivision and interpretation of the Labrador – New Quebec portion of the Pan-American orogenic system have recently undergone major reevaluation. The nomenclature of internal elements is somewhat controversial, as indeed is the question of whether the elements should, in fact, be considered as parts of the Trans-Hudson Orogen. Our review mainly follows Wardle *et al.* (this volume), both with respect to the general internal subdivision and nomenclature, and with respect to regarding this segment as being an integral part of the Trans-Hudson Orogen.

The "Western Division" of the orogen (essentially coincident with the previously recognised Labrador Trough and the "New Quebec Orogen" of Hoffman (this volume)) is a well-preserved, west-verging foreland fold–thrust belt. Most workers distinguish a western autochthonous zone (Schefferville Zone of Wardle *et al.* (this volume)) succeeded to the east by an imbricated, semiautochthonous to allochthonous fold–thrust zone (Howse Zone of Wardle *et al.* (this volume)). This, in turn, is overthrust by allochthonous sheets of the Doublet and Laporte terranes. The supracrustal succession in the two western zones includes early continental rift and passive margin marine shelf sediments, later black shales, turbidites and volcanics now generally interpreted as foredeep facies, and upper molasse sediments. The entirely allochthonous and thrust-bounded Doublet terrane consists mainly of mafic pillow basalts and pyroclastics intruded by gabbroic and ultramafic sills: this sequence has primitive oceanic tholeiite characteristics and is interpreted as being an early rifted margin assemblage, possibly representing a transitional continental margin or true oceanic environment. The structurally overlying Laporte terrane comprises a monotonous arkosic/greywacke turbidite assemblage, which may represent either an inboard passive margin continental rise prism (Wardle and Bailey, 1981) or an outboard fore-arc basin fill (Machado *et al.*, 1989; Van der Leeden *et al.*, this volume).

The Central Division of the Labrador segment consists mainly of variably reworked Archean continental crust, locally with intercalated remnants of probable Lower Proterozoic miogeoclinal metasediments. However, the western part also includes the ca. 1811 Ma, late-orogenic De Pas Batholith, which is similar in character to the somewhat older

Wathaman–Chipewyan Batholith of the western Trans-Hudson Orogen. It is now generally accepted that Central Division Archean gneisses have been thrust over rocks of the Western Division and that their junction is related to cryptic collisional suturing.

The boundary between the Central and Eastern divisions is defined by granulite-facies mylonitic gneisses of the huge Abloviak shear zone. Kinematic indicators within this major deep crustal discontinuity indicate predominantly transcurrent sinistral displacement with significant west-side-up vertical displacement (Wardle *et al.*, this volume). Throughout most of its extent, the Eastern Division consists only of a relatively narrow foreland zone, which includes variably reworked basement of the North Atlantic craton and overlying shelf to basinal metasediments of the Ramah Group. In the north, however, this foreland is overthrust by highly strained, upper amphibolite to granulite facies rocks of the Komaktorvik Zone, interpreted as being Archean basement affected by intense Hudsonian reworking. The Burwell Zone, lying between the Komaktorvik Zone and the northern part of the Abloviak shear zone, is tentatively interpreted as being "Nain-type" Archean crust affected by a moderate Hudsonian overprint.

It is now widely accepted that the Labrador segment evolved via oblique convergence and collision between the Superior continental margin and eastern ensialic elements of the orogen (Korstgaard *et al.*, 1987): the extent of prior oceanic lithosphere subducted at associated plate boundaries is a matter of speculation. There is also considerable uncertainty regarding the affinity of the Archean continental terrain of the Central Division and, therefore, some controversy as to the number of collisional sutures incorporated within the Labrador segment.

Most modern workers agree that the boundary between the Western and Central divisions is related to collision involving eastward subduction of the Superior plate, overthrusting of upper plate Archean crust, and possible development of a collisional magmatic welt represented by the De Pas Batholith. The suture is commonly placed at the western margin of the Laporte terrane, such a location being favoured by occurrence of remnants of rocks of oceanic affinities in this zone and evidence that the Central Division represents uplifted deep crust.

The significance of the Abloviak and Komaktorvik shear zones remains uncertain. Hoffman (1988, this volume) interprets both as being collisional sutures between formerly independent North Atlantic, Rae, and Burwell minicontinents and proposes the name Torngat Orogen for the zone incorporating them. A corollary of this interpretation is that continuation of the collisional suture postulated by Grocott and Pulvertaft (this volume) to occur south of the Rinkian belt may be represented by the Komaktorvik shear zone. At present, however, firm evidence of collisional suturing (e.g., magmatic arc) of the North Atlantic craton, Central Division, and Burwell zone is undocumented: the Abloviak and Komaktorvik shear zones may simply represent major transcurrent or transform intraplate dislocations of a greater North Atlantic paleoplatform. For example, Ryan *et al.* (1988) document early leucogabbro–anorthosite complexes and migmatitic gneisses cut by metamorphosed diabase dykes in part of the Central Division just east of the Labrador–Quebec border: they suggest that the leucogabbro–anorthosites are time equivalents of similar "Fisknaesset-type" Archean layered complexes in the North Atlantic craton of eastern Labrador and Greenland and that the diabase dykes are time equivalents of the 2300–2400 Ma old swarm that intrudes the North Atlantic craton. Such correlation implies extension of reworked crust of the North Atlantic craton to the west of the Abloviak shear zone and suggests that the latter is an intracontinental feature, rather than a suture.

DISCUSSION

The Trans-Hudson Orogen was somewhat casually defined less than a decade ago. Since then, great advances in the documentation and understanding of this crustal entity have been made, as demonstrated by the contents of this volume. A growing body of lithostratigraphic, structural, metamorphic, geochemical, and isotopic data permits definition of most of the critical geotectonic elements and spatiotemporal relationships that typify younger orogenic belts developed by subduction/collision at convergent plate boundaries. There can now be little serious dispute that the geodynamic processes that shaped the orogen involved relative motion of major lithospheric plates and amalgamation of relatively small, previously dispersed Archean continents in a relatively short time span.

Most of the presently preserved orogen comprises variably reworked margins of pre-existing Archean continental plates. Intervening juvenile arc terranes and oceanic crustal remnants are presently documented only in the exposed Reindeer Zone of northern Saskatchewan–Manitoba, in the Cape Smith "klippe," and possibly as minor remnants in western parts of the Labrador segment. However, this predominantly "ensialic" character does not necessarily imply significant differences in juvenile crust formation or orogenic geodynamics compared with Phanerozoic/Recent orogens. Preservation of juvenile crust at the present erosional level of the

Appalachian–Caledonian orogen is little greater than in the Trans-Hudson, and would probably be less with but a few more kilometres of erosion. Continuing postcollisional convergence of the Indian and Central Asian plates beneath the Himalaya clearly demonstrates the process of ejection of formerly intervening juvenile arc terranes as high-level allochthons, which will be largely eliminated by subsequent erosion. The effects of Early Proterozoic plate convergence are unlikely to have been greatly different. Indeed, given similar strong convergence and generally smaller continental plates, such overthrusting and stripping of juvenile terranes might have been even more pronounced. The present lack of documentation of fore-arc accretionary wedge and mélange assemblages may simply be a function of such processes rather than being indicative of significant differences in crustal geodynamics.

Resolution of relative plate motions involved in various stages of evolution of the Trans-Hudson Orogen is at a preliminary stage. As indicated here, we are not even sure how many independent Archean minicontinents were actually involved in the Early Proterozoic assembly of the North American craton. Hoffman (this volume) postulates at least seven: we are persuaded of no more than four (Superior, Slave, North Atlantic, and Rae–Hearne–Wyoming). In eastern parts of the orogen (Labrador segment), relatively clear movement indicators and (limited) age data permit tentative interpretation of some phases of relative plate motion (Korstgaard *et al.*, 1987; Wardle *et al.*, this volume). In the west and north, present constraints indicate northerly or northwesterly convergence of the Superior paleoplatform but otherwise permit only speculative modeling. Such speculation is both entertaining and valuable; at this stage, however, we caution against overenthusiastic, uncritical endorsement of such models, including our own!

ACKNOWLEDGEMENTS

This review draws on the efforts of many workers: we hope that we represent their views fairly. We particularly thank Pat Bickford, Charley Gower, Paul Hoffman, Tony Lecheminant, Bob Macdonald, Dave MacRitchie, Bruce Ryan, Tom Sibbald, Mel Stauffer, Dick Wardle, and Werner Weber for critical contributions, stimulating discussion on varied aspects of the orogen, and/or useful criticism of early versions of this synthesis. Our work has been supported for many years by financial assistance from Saskatchewan Energy and Mines, Natural Sciences and Engineering Research Council of Canada operating grants to both authors, and, most recently, National Science Foundation support to K.D. Collerson.

REFERENCES

Allaart, J.H., 1976, Ketilidian mobile belt in South Greenland, *in* Escher, A. and Watt, W.S., eds., Geology of Greenland: Geological Survey of Greenland, Copenhagen, p. 120–151.

Annesley, J.R. and Madore, C., 1989, The Wollaston Group and its underlying Archean basement in Saskatchewan: Saskatchewan Geological Survey, Miscellaneous Report 89–4, p. 87–91.

Aspler, L.B., Bursey, T.L. and Miller, A.R., 1989, Sedimentology, structure and economic geology of the Poorfish–Windy thrust-fold belt, Ennadai Lake area, District of Keewatin, and the shelf to foredeep transition in the foreland of Trans-Hudson Orogen: Geological Survey of Canada, Paper 89-1C, p. 143–155.

Baldwin, D.A., Syme, E.C., Zwanzig, H.V., Gordon, T.M., Hunt, P.A. and Stevens, R.D., 1985, U–Pb zircon ages from the Lynn Lake and Rusty Lake metavolcanic belts, Manitoba: two ages of Proterozoic magmatism: Canadian Journal of Earth Sciences, v. 24, p. 1053–1063.

Baragar, W.R.A. and Scoates, R.F.J., 1981, The Circum-Superior belt: a Proterozoic plate margin?, *in* Kroner, A., ed., Precambrian Plate Tectonics: Elsevier, Amsterdam, p. 297–330.

Bell, K. and Macdonald, R., 1982, Geochronological calibration of the Precambrian Shield in Saskatchewan: Saskatchewan Geological Survey, Miscellaneous Report 82-4, p. 17–22.

Bickford, M.E., Van Schmus, W.R., Collerson, K.D. and Macdonald, R., 1987, U–Pb zircon geochronology project: new results and interpretation: Saskatchewan Geological Survey, Miscellaneous Report 87-4, p. 76-79.

Bickford, M.E., Collerson, K.D. and Lewry, J.F., 1988, U–Pb geochronology and isotopic studies of the Trans-Hudson Orogen in Saskatchewan: Saskatchewan Geological Survey, Miscellaneous Report 88-4, p. 112–116.

Bleeker, W., 1990, New structural–metamorphic constraints on Early Proterozoic oblique collision along the Thompson Nickel Belt, Manitoba, Canada, *in* Lewry, J.F. and Stauffer, M.R., eds., The Early Proterozoic Trans-Hudson Orogen of North America: Geological Association of Canada, Special Paper 37, p. 57–73.

Camfield, P.A. and Gough, D.I., 1977, A possible Proterozoic plate boundary in North America: Canadian Journal of Earth Sciences, v. 14, p. 1229–1239.

Carolan, J. and Collerson, K.D., 1988, The Virgin River Shear Zone in the Careen Lake area: field relationships and kinematic indicators: Saskatchewan Geological Survey, Miscellaneous Report 88-4, p. 92–96.

Chauvel, C., Arndt, N.T., Kielinzcuk, S. and Thom, A., 1987, Formation of 1.9 Ga old continental crust. I. Nd isotopic data: Canadian Journal of Earth Sciences, v. 24, p. 396–406.

Collerson, K.D., Van Schmus, W.R., Lewry, J.F. and Bickford, M.E., 1988, Buried Precambrian basement in south-central Saskatchewan: provisional results from Sm–Nd model ages and U–Pb zircon geochronology: Saskatchewan Geological Survey, Miscellaneous Report 88-4, p. 142–151.

Collerson, K.D., Lewry, J.F., Van Schmus, W.R. and Bickford, M.E., 1989, Sm–Nd isotopic restraints on the age of buried Precambrian basement in central and southern Saskatchewan: implications for diamond exploration: Saskatchewan Geological Survey, Miscellaneous Report 89-4, p. 168–171.

Coney, P.J., Jones, D.L. and Monger, J.W.H., 1980, Cordilleran suspect terranes: Nature, v. 288, p. 329–333.

Dunsmore, D.J. and Symons, D.T.A., 1990, Paleomagnetism of the Lynn Lake gabbros in the Trans-Hudson Orogen and closure of the Superior and Slave cratons, *in* Lewry, J.F. and Stauffer, M.R., eds., The Early Proterozoic Trans-Hudson Orogen of North America: Geological Association of Canada, Special Paper 37, p. 215–228.

Fueten, F. and Robin, P.F., 1989, Structural petrology along a transect across the Thompson Belt, Manitoba: dip-slip at the western Churchill–Superior Boundary: Canadian Journal of Earth Sciences, v. 26, p. 1976–1989.

Fumerton, S.L., Stauffer, M.R. and Lewry, J.F., 1984, The Wathaman Batholith: largest known Precambrian pluton: Canadian Journal of Earth Sciences, v. 21, p. 1082–1097.

Gibb, R.A., 1978, Slave–Churchill collisional tectonics: Nature, v. 27, p. 50–52.

Gibb, R.A. and Thomas, M.D., 1977, The Thelon Front: a cryptic suture in the Canadian Shield?: Tectonophysics, v. 38, p. 211–222.

Gilboy, C.F., 1980, Bedrock compilation geology: Stony Rapids area (NTS 74P); 1:250,000 scale preliminary map: Saskatchewan Geological Survey, Miscellaneous Report 80-4.

Gordon, T.M., Hunt, P.A., Bailes, A.H. and Syme, E.C., 1990, U–Pb ages from the Flin Flon and Kisseynew belts, Manitoba: chronology of crust formation at an Early Proterozoic accretionary margin, in Lewry, J.F. and Stauffer, M.R., eds., The Early Proterozoic Trans-Hudson Orogen of North America: Geological Association of Canada, Special Paper 37, p. 177–199.

Green, A.H., Hajnal, Z. and Weber, W., 1985a, An evolutionary model of the western Churchill Province and western margin of the Superior Province in Canada and the north-central United States: Tectonophysics, v. 116, p. 281–322.

Green, A.H., Weber, W. and Hajnal, Z., 1985b, Evolution of Proterozoic terrains beneath the Williston Basin: Geology, v. 13, p. 624–628.

Grocott, J. and Pulvertaft, T.C.R., 1990, The Early Proterozoic Rinkian Belt of central West Greenland, in Lewry, J.F. and Stauffer, M.R., eds., The Early Proterozoic Trans-Hudson Orogen of North America: Geological Association of Canada, Special Paper 37, p. 443–463.

Halden, N.M., Clark, G.S., Corkery, M.T., Lenton, P.G. and Schledewitz, D.C.P., 1990, Trace element and Rb–Sr whole rock isotopic constraints on the origin of the Chipewyan, Thorsteinson and Baldock batholiths, in Lewry, J.F. and Stauffer, M.R., eds., The Early Proterozoic Trans-Hudson Orogen of North America: Geological Association of Canada, Special Paper 37, p. 201–214.

Heywood, H.H. and Schau, M., 1978, A subdivision of the northern Churchill structural province: Geological Survey of Canada, Paper 78-1A, p. 139–143.

Hoffman, P.F., 1980, Wopmay Orogen: a Wilson Cycle of Early Proterozoic age in the northwest of the Canadian Shield, in Strangway, D.W., ed., The Continental Crust and its Mineral Deposits: Geological Association of Canada, Special Paper 20, p. 523–549.

Hoffman, P.F., 1981, Autopsy of Athapuscow Aulacogen: a failed arm affected by three collisions, in Campbell, F.H.A., ed., Proterozoic Basins of Canada: Geological Survey of Canada, Paper 81-10, p. 97–102.

Hoffman, P.F., 1985, Is the Cape Smith Belt (northern Québec) a klippe?: Canadian Journal of Earth Sciences, v. 22, p. 1361–1369.

Hoffman, P.F., 1987, Continental transform tectonics: Great Slave shear zone (1.9 Ga), northwest Canada: Geology, v. 15, p. 85–88.

Hoffman, P.F., 1988, United plates of America, the birth of a craton: Early Proterozoic assembly and growth of Laurentia: Annual Review of Earth and Planetary Sciences, v. 16, p. 543–603.

Hoffman, P.F., 1990, Subdivision of the Churchill province and extent of the Trans-Hudson Orogen, in Lewry, J.F. and Stauffer, M.R., eds., The Early Proterozoic Trans-Hudson Orogen of North America: Geological Association of Canada, Special Paper 37, p. 15–39.

Jackson, G.D. and Taylor, F.C., 1972, Correlation of major Aphebian rock units in the northeastern Canadian Shield: Canadian Journal of Earth Sciences, v. 9, p. 1650–1659.

Klasner, J.S. and King, E.R., 1986, Precambrian basement geology of North and South Dakota: Canadian Journal of Earth Sciences, v. 23, p. 1083–1102.

Klasner, J.S. and King, E.R., 1990, A model for tectonic evolution of the Trans-Hudson Orogen in the Dakotas, in Lewry, J.F. and Stauffer, M.R., eds., The Early Proterozoic Trans-Hudson Orogen of North America: Geological Association of Canada, Special Paper 37, p. 271–285.

Korstgaard, J., Ryan, B. and Wardle, R., 1987, The boundary between Proterozoic and Archean crustal blocks in central West Greenland and northern Labrador, in Park, R.G. and Tarney, J., eds., Evolution of the Lewisian and Comparable Precambrian High Grade Terrains: Geological Society of London, Special Publication 27, p. 247–259.

Lewry, J.F., 1981, Lower Proterozoic arc–microcontinent collisional tectonics in the western Churchill Province: Nature, v. 294, p. 69–72.

Lewry, J.F., 1987, The Trans-Hudson Orogen: extent, subdivisions and problems: Geological Association of Canada – Mineralogical Association of Canada, Program with Abstracts, v. 12, p. 67.

Lewry, J.F. and Sibbald, T.I.I., 1977, Variation in lithology and tectonometamorphic relationships in the Precambrian basement of northern Saskatchewan: Canadian Journal of Earth Sciences, v. 14, p. 1453–1467.

Lewry, J.F. and Sibbald, T.I.I., 1980, Thermotectonic evolution of the Churchill Province in northern Saskatchewan: Tectonophysics, v. 68, p. 45–82.

Lewry, J.F., Stauffer, M.R. and Fumerton, S.L., 1981, A Cordilleran-type batholithic belt in the Churchill Province in northern Saskatchewan: Precambrian Research, v. 14, p. 277–313.

Lewry, J.F., Sibbald, T.I.I. and Schledewitz, D.C.P., 1985, Variation in character of Archean rocks in the western Churchill Province and its significance, in Ayres, L.D., Thurston, P.C., Card, K.D. and Weber, W., eds., Evolution of Archean Supracrustal Sequences: Geological Association of Canada, Special Paper 28, p. 239–261.

Lewry, J.F., Macdonald, R., Livesey, C., Meyer, M.T., Van Schmus, W.R. and Bickford, M.E., 1987, U–Pb geochronology of accreted terranes in the Trans-Hudson Orogen in northern Saskatchewan, in Pharaoh, T.C., Beckinsale, R.D. and Rickards, R., eds., Geochemistry and Mineralization of Proterozoic Volcanic Suites: Geological Society of London, Special Publication 33, p. 147–166.

Lewry, J.F., Thomas, D.J., Macdonald, R. and Chiarenzelli, J., 1990, Structural relations in accreted terranes of the Trans-Hudson Orogen, Saskatchewan: telescoping in a collisional regime?, in Lewry, J.F. and Stauffer, M.R., eds., The Early Proterozoic Trans-Hudson Orogen of North America: Geological Association of Canada, Special Paper 37, p. 75–94.

Macdonald, R. and Chiarenzelli, J.R., 1986, A U–Pb zircon date for the Ennadai Group: Saskatchewan Geological Survey, Miscellaneous Report 86-4, p. 112–113.

Machado, N., 1990, A comparison of U–Pb ages for the Labrador Trough in Quebec and the Trans-Hudson Orogen in Manitoba and Saskatchewan, in Lewry, J.F. and Stauffer, M.R., eds., The Early Proterozoic Trans-Hudson Orogen of North America: Geological Association of Canada, Special Paper 37, p. 433–441.

Machado, N., Goulet, N. and Gariépy, C., 1989, U–Pb geochronology of reactivated Archean basement and of Hudsonian metamorphism in the northern Labrador Trough: Canadian Journal of Earth Sciences, v. 26, p. 1–15.

Meyer, M.T., 1987, Geochronology and geochemistry of the Wathaman Batholith, the remnant of an Early Proterozoic continental arc in the Trans-Hudson Orogen, Saskatchewan, Canada, unpublished M.Sc. thesis, University of Kansas, 107 p.

Orrell, S.E., Tangeman, J.A. and Bickford, M.E., 1988, Geology of Early Proterozoic gneissic rocks, Black Birch Lake area: Saskatchewan Geological Survey, Miscellaneous Report 88-4, p. 84–91.

Patchett, P.J. and Arndt, N.T., 1986, Nd isotopes and tectonics of 1.9–1.7 Ga crustal genesis: Earth and Planetary Science Letters, v. 78, p. 329–338.

Pidgeon, R.T. and Howie, R.A., 1979, U–Pb age of zircon from a charnockitic granulite from Pangnirtung on the east coast of Baffin Island: Canadian Journal of Earth Sciences, v. 12, p. 1046–1047.

Redden, J.A., Peterman, Z.E., Zartman, R.E. and DeWitt, S., 1990, U–Th–Pb geochronology and preliminary interpretation of Precambrian tectonic events in the Black Hills, South Dakota, in Lewry, J.F. and Stauffer, M.R., eds., The Early Proterozoic Trans-Hudson Orogen of North America: Geological Association of Canada, Special Paper 37, p. 229–251.

Reilly, B.A., 1989, Reconnaissance bedrock mapping, Misaw Lake area: Saskatchewan Geological Survey, Miscellaneous Report 89-4, p. 5–10.

Ricketts, B.D. and Donaldson, J.A., 1981, Sedimentary history of the Belcher Group of Hudson Bay, in Campbell, F.H.A., ed., Proterozoic Basins of Canada: Geological Survey of Canada, Paper 81-10, p. 235–254.

Ross, G.M., Parrish, R.R., Bowring, S.A. and Tankard, A.J., 1988, Tectonics of the Canadian Shield in the Alberta subsurface: Geological Association of Canada – Mineralogical Association of Canada – Canadian Society of Petroleum Geologists, Program with Abstracts, p. A106.

Ryan, B., Lee, D. and Dunphy, D., 1988, The discovery of probable Archean rocks within the Labrador arm of the Trans-Hudson Orogen near the Labrador–Quebec border (NTS 14D/3,4,5 and 24A/1.8): Newfoundland Department of Mines, Mineral Development Division, Report 88-1, p. 1–14.

Schulz, K.J., 1987, An Early Proterozoic ophiolite in the Penokean Orogen: Geological Association of Canada – Mineralogical Association of Canada, Program with Abstracts, p. 87.

Sims, P.K. and Peterman, Z.E., 1986, Early Proterozoic Central Plains Orogen: a major buried structure in the north-central United States: Geology, v. 14, p. 488–491.

Stauffer, M.R., 1984, Manikewan: an Early Proterozoic ocean in central Canada, its igneous history and orogenic closure: Precambrian Research, v. 25, p. 257–281.

Stauffer, M.R., 1990, The Missi Formation: an Aphebian molasse deposit in the Reindeer Zone of the Trans-Hudson Orogen, Canada, in Lewry, J.F. and Stauffer, M.R., eds., The Early Proterozoic Trans-Hudson Orogen of North America: Geological Association of Canada, Special Paper 37, p. 121–141.

Stauffer, M.R. and Lewry, J.F., 1988, Kinematic investigation of part of the Needle Falls shear zone: Saskatchewan Geological Survey, Miscellaneous Report 88-4, p. 156–160.

Stockwell, C.H., 1961, Structural provinces, orogenies and time classification of rocks of the Canadian Shield, in Lowden, J.A., ed., Age Determinations by the Geological Survey of Canada: Geological Survey of Canada, Paper 61-17, p. 108–118.

Stockwell, C.H., 1970, Geology of the Canadian Shield, in Douglas, R.J.W., ed., Geology and Economic Minerals of Canada: Geological Survey of Canada, Economic Geology Report 1, p. 44–45.

St-Onge, M.R. and Lucas, S.B., 1990, Evolution of the Cape Smith Belt: Early Proterozoic continental underthrusting, ophiolite obduction and thick-skinned folding, in Lewry, J.F. and Stauffer, M.R., eds., The Early Proterozoic Trans-Hudson Orogen of North America: Geological Association of Canada, Special Paper 37, p. 313–351.

Syme, E.C., 1990, Stratigraphy and geochemistry of the Lynn Lake and Flin Flon metavolcanic belts, Manitoba, in Lewry, J.F. and Stauffer, M.R., eds., The Early Proterozoic Trans-Hudson Orogen of North America: Geological Association of Canada, Special Paper 37, p. 143–161.

Tella, S. and Eade, K.E., 1986, Occurrence and possible tectonic significance of high-pressure granulite fragments in the Tulemalu fault zone, District of Keewatin, N.W.T., Canada: Canadian Journal of Earth Sciences, v. 23, p. 1950–1962.

Thom, A., Arndt, N. and Stauffer, M.R., 1990, Flin Flon and west La Ronge belts, Saskatchewan: products of Proterozoic subduction-related volcanism, in Lewry, J.F. and Stauffer, M.R., eds., The Early Proterozoic Trans-Hudson Orogen of North America: Geological Association of Canada, Special Paper 37, p. 163–175.

Van der Leeden, J., Belanger, M., Danis, D., Girard, R. and Martelain, J., 1990, Lithotectonic domains in the high-grade terrain east of the Labrador Trough (Quebec), in Lewry, J.F. and Stauffer, M.R., eds., The Early Proterozoic Trans-Hudson Orogen of North America: Geological Association of Canada, Special Paper 37, p. 371–386.

Van Schmus, W.R., 1980, Chronology of igneous rocks associated with the Penokean Orogeny in Wisconsin, in Morey, G.B. and Hansen, G.N., eds., Selected Studies of Archean Gneisses and Lower Proterozoic Rocks, Southern Canadian Shield: Geological Society of America, Special Paper 182, p. 159–168.

Van Schmus, W.R., Bickford, M.E., Lewry, J.F. and Macdonald, R., 1987, U–Pb geochronology in the Trans-Hudson Orogen, northern Saskatchewan, Canada: Canadian Journal of Earth Sciences, v. 24, p. 407–424.

Wardle, R.J. and Bailey, D.G., 1981, Early Proterozoic sequences in Labrador, in Campbell, F.H.A., ed., Proterozoic Basins of Canada: Geological Survey of Canada, Paper 81-10, p. 331–359.

Wardle, R.J., Ryan, B., Nunn, G.A.G. and Mengel, F.C., 1990, Labrador segment of the Trans-Hudson Orogen: crustal development through oblique convergence and collision, in Lewry, J.F. and Stauffer, M.R., eds., The Early Proterozoic Trans-Hudson Orogen of North America: Geological Association of Canada, Special Paper 37, p. 353–369.

Watters, B.R. and Pearce, J., 1987, Metavolcanic rocks of the La Ronge Domain in the Churchill Province, Saskatchewan: geochemical evidence for a volcanic arc origin, in Pharaoh, T.C., Beckinsale, R.D. and Rickards, R., eds., Geochemistry and Mineralization of Proterozoic Volcanic Suites: Geological Society of London, Special Publication 33, p. 167–182.

Weber, W., 1990, The Churchill–Superior Boundary Zone, southeast margin of the Trans-Hudson Orogen: a review, in Lewry, J.F. and Stauffer, M.R., eds., The Early Proterozoic Trans-Hudson Orogen of North America: Geological Association of Canada, Special Paper 37, p. 41–55.

Williams, H. and Hatcher, R.D., Jr., 1982, Suspect terranes and accretionary history of the Appalachian Orogen: Geology, v. 10, p. 530–536.

Subdivision of the Churchill Province and Extent of the Trans-Hudson Orogen[1]

Paul F. Hoffman
Continental Geoscience Division, Geological Survey of Canada,
Ottawa, Ontario K1A 0E8, Canada

Hoffman, P.F., Subdivision of the Churchill Province and extent of the Trans-Hudson Orogen, in Lewry, J.F. and Stauffer, M.R., eds., The Early Proterozoic Trans-Hudson Orogen of North America: Geological Association of Canada, Special Paper 37, p. 15–39.

Abstract

The Churchill Province is a tectonic hinterland welded by Early Proterozoic collision zones to the Archean forelands of the Superior, Nain, Slave, and Wyoming provinces. The Churchill hinterland is itself composed of three Archean provinces separated by crustal breaks that may represent Early Proterozoic sutures. The three Archean provinces of the Churchill hinterland are named the Hearne, Rae, and Burwell provinces. The amalgamation of the seven Archean provinces occurred between 1.97 and 1.82 Ga, and collectively they form the Archean protocraton of Laurentia.

The various Early Proterozoic orogenic belts that weld the Archean provinces evolved as discrete entities, each with its own history and character. The Trans-Hudson Orogen welds the northwest margin of the Superior Province to the Hearne Province. Its fundamental subduction polarity was northwest-dipping and it is distinguished from the other orogens by the preservation of intraoceanic island arcs in northern Manitoba and Saskatchewan, and by the obduction of ophiolites in northern Quebec. The suture zone represented by the Trans-Hudson Orogen may be linked with that of the New Quebec orogen, which welds the northeast margin of the Superior Province to the Rae Province. The New Quebec orogen records a dextral component of transcurrent shear combined with a northeast-dipping subduction polarity, and may have resulted from clockwise rotation of the southeast arm of the Rae Province in response to collisional indentation of the Churchill hinterland by the Superior Province. The Baffin orogen evolved from a wedge-shaped basin that opened as the southeast arm of the Rae Province rotated away from the northeast arm. The Cumberland batholith intrudes the ancestral Baffin basin and may be a magmatic arc developed above a northwest-dipping subduction zone that accommodated convergence with the Burwell Province. Collision between the Burwell and Rae provinces destroyed the basin. The Nain Province is welded to the back of the Burwell Province by the Nagssugtoqidian orogen of West Greenland. South of the Burwell Province, the Nain Province is welded directly to the Rae Province by the Torngat, orogen of Labrador. The Baffin, Torngat, and Nagssugtoqidian orogens all have northwest

[1]Geological Survey of Canada Contribution No. 48087.

dipping subduction polarities and may be approximately coeval, although the exact sequence of suturing is uncertain. The tectonic contact between the southwest arm of the Rae Province and the Hearne Province to the south is the Snowbird line, an anastomosing system of granulite-grade mylonites containing bodies of layered anorthosite–gabbro–pyroxenite. The tectonic polarity and significance of the Snowbird line are controversial. It may be a suture zone that was reactivated during collision in the Trans-Hudson Orogen to the southeast. The boundary between the Hearne and Wyoming provinces is the subsurface Great Falls tectonic zone. The Thelon orogen forms the northwest boundary of the Churchill hinterland and evolved from an east-dipping subduction zone, which led to collision at 1.97 Ga between the Slave and Rae provinces.

Previously, the Churchill Province was regarded as a reactivated hinterland related to a single collision zone between the Churchill and Superior provinces. Here, as many as seven former microcontinents and a multiplicity of collision zones are envisaged.

Résumé

La province de Churchill est un arrière-pays tectonique soudé aux avants-pays Archéen des provinces Supérieur, Nain, Slave et Wyoming, par des zones de collision Protérozoïque Inférieur. L'arrière-pays Churchill est composé de trois provinces archéennes nommément Hearne, Rae et Burwell, séparées par des ruptures de la croûte représentant des sutures Protérozoïque Inférieur. L'accrétion des sept provinces archéennes s'est produite entre 1.97 et 1.82 Ga pour former le proto-craton Archéen de Laurentia.

Les différentes ceintures orogéniques qui relient les provinces archéennes ont eu des évolutions distinctes, avec une histoire et des caractères propres. L'orogène Trans-Hudson relie la marge nord-ouest de la province Supérieur à la province Hearne. Elle se distingue des autres orogènes par le plongement nord de la polarité de subduction, la préservation d'arcs insulaires intra-océaniques au nord du Manitoba et de la Saskatchewan et l'obduction d'ophiolites au nord du Québec. La zone de suture que représente l'orogène Trans-Hudson pourrait être reliée à l'orogène du Nouveau Québec qui relie la marge nord-est de la province Supérieur à la province Rae. L'orogene du Nouveau-Québec montre une composante de cisaillement décrochant dextre combinée à une polarité de subduction nord-est et pourrait être le résultat d'une rotation dextre du bras sud-est de la province Rae en réponse à une indentation de l'arrière-pays Churchill par le Supérieur pendant la collision. L'orogène Baffin a évolué à partir d'un bassin fermé qui s'est ouvert au fur à mesure que le bras sud-est de la province Rae s'éloignait par rotation du bras nord-est. Le batholite Cumberland recoupe le bassin ancien de Baffin et pourrait être relié à un arc magmatique qui s'est développé au-dessus d'une zone de subduction à vergence nord provoquant la convergence vers la province Burwell. La collision entre les provinces Rae et Burwell a détruit le bassin. La province du Nain est reliée à l'arrière de la province Burwell par l'orogène Nagssugtoqidian à l'ouest du Groenland. Au sud de la province Burwell, la province du Nain est reliée directement à la province Rae par l'orogène Torngat située au Labrador. Les orogènes de Baffin, Torngat et Nagssugtoqidian ont des polarités de subduction plongeant vers le nord-ouest et pourraient être synchrones bien que les séquences exactes de sutures sont incertaines. Le contact tectonique entre le bras sud-est de la province Rae et la province Hearne au sud est constitué par la ligne Snowbird, un système anastomosé de mylonites au faciès granulite contenant des masses litées d'anorthosites–gabbros–pyroxénites. La polarité tectonique et la signification de la ligne Snowbird sont controversées; il pourrait s'agir d'une zone de suture réactivée pendant une collision dans l'orogène Trans-Hudson au sud-est. La frontière entre les provinces Wyoming et Hearne est formée par la zone tectonique de Great Falls, actuellement enfouie sous la surface. L'orogène de Thelon constitue la limite nord-ouest de l'arrière-pays Churchill et s'est formée à partir d'une zone de subduction plongeant vers l'est et donnant lieu vers 1.97 Ga à la collison entre les provinces Rae et Slave.

Antérieurement, la province Churchill était vu comme un arrière-pays réactivé et relié à une zone de collision simple entre les provinces Supérieur et Churchill. Ici on envisage un modèle impliquant sept micro-continents anciens et de multiples zones de collision.

INTRODUCTION

The Churchill Structural Province (Fig. 1), as defined by Stockwell (1961), has been described as "that part of the Canadian Shield remaining after the adjacent, better defined, younger and older structural provinces have been delineated" (Davidson, 1972, p. 382). Since 1972, a growing body of isotopic data has demonstrated the fundamental duality of the province. In northern Manitoba and Saskatchewan (Fig. 2), a zone up to 400 km wide contains juvenile Early Proterozoic crust that lithologically resembles oceanic island arcs and accretionary prisms (Bailes, 1971; Sangster, 1972, 1978; Watters and Armstrong, 1985; Van Schmus and Schledewitz, 1986; Baldwin *et al.*, 1987; Chauvel *et al.*, 1987; Gaskarth and Parslow, 1987; Hegner and Hulbert, 1987; Van Schmus *et al.*, 1987; Watters and Pearce, 1987; Syme, this volume; Thom *et al.*, this volume). However, the rest of the province is underlain by Archean crust that preserves infolded erosional remnants of Early Proterozoic cover of epicontinental facies (Lewry *et al.*, 1985).

A unifying feature of the Churchill Province is that both the Archean and Early Proterozoic domains were more severely deformed and metamorphosed at 1.9–1.8 Ga than the adjacent Superior, Wyoming, Slave, and Nain provinces (Fig. 1). The deformation and metamorphism that distinguished the Churchill Province was referred to as the Hudsonian orogeny by Stockwell (1961). A leader in Canadian Shield geology for half a century, Stockwell (1961, 1982) was more concerned with using orogenies as a basis for dividing Precambrian time than with understanding the specific causes of orogenies. In the tradition of Élie de Beaumont, Leopold Kober, and Hans Stille (Greene, 1982; Sengör, 1984), he imagined orogenies as manifestations of tectonic convulsions of continental, even global, scale.

Since the ascendancy of mobilism in the 1960s, concern has shifted away from correlation of orogenies toward understanding their causal dynamics, principally arc–continent and continent–continent collisions. Dewey and Burke (1973) made the first comprehensive interpretation of the Churchill Province in mobilistic terms. Adopting the hypothesis of Gibb and Walcott (1971) that the Superior–Churchill province boundary zone is a Himalayan-type collision zone of Hudsonian age, they proposed that an Archean hinterland underwent Hudsonian reworking in a Tibetan-type plateau environment related to the collision. In their model (Dewey and Burke, 1973, Fig. 3), the reworked hinterland extends to the Slave Province in the northwest, to at least Baffin Island in the north, and to the Nain Province (as defined by Taylor (1971)) in the northeast. Thus, the Slave–Churchill and Nain–Churchill boundaries were viewed as limits of intraplate crustal thickening, not as Hudsonian sutures (cf. Thomas and Kearey, 1980; Lewry *et al.*, 1985; Wardle *et al.*, this volume).

The Tibetan plateau, however, contains a number of pre-Cenozoic sutures (Allegre *et al.*, 1984; Dewey

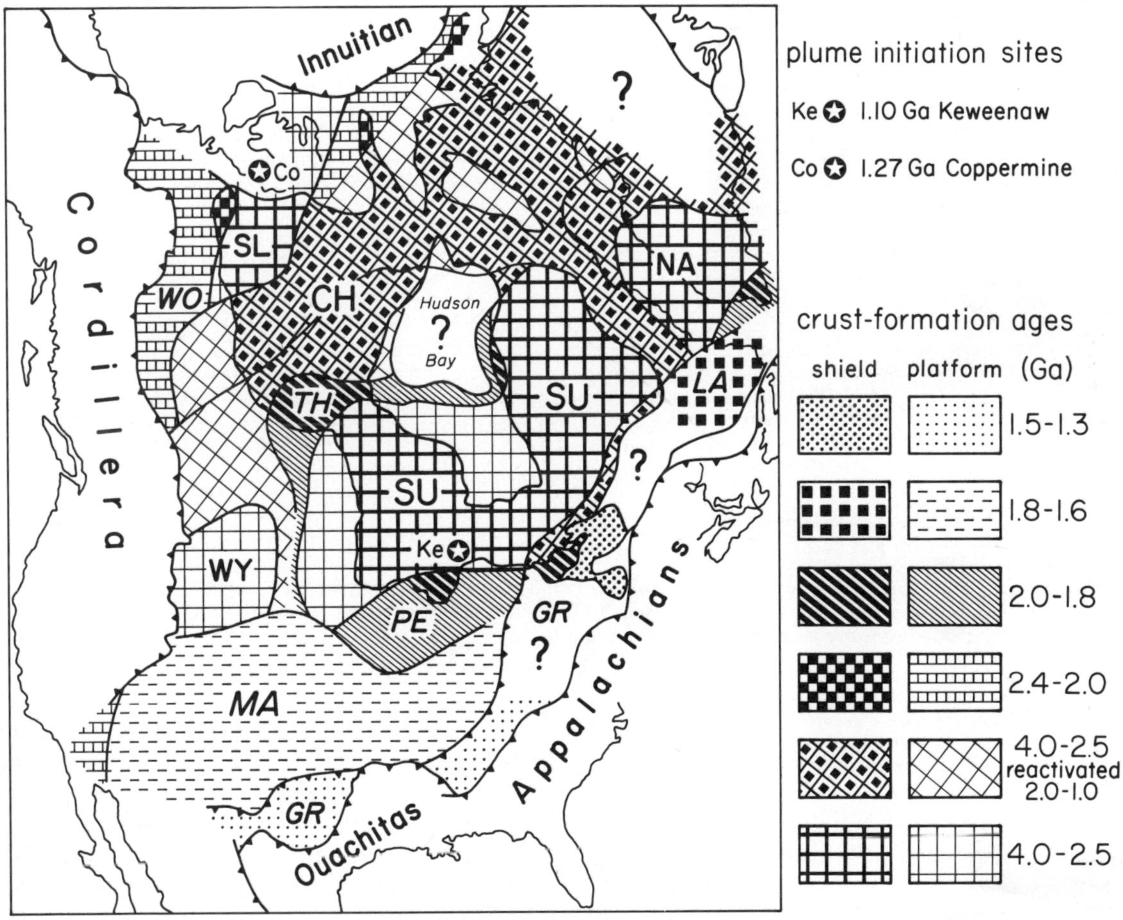

Figure 1. *The North American craton, showing ages of crust–mantle separation inferred from combined Sm–Nd, Pb–Pb, and U–Pb isotopic data from the shield and platform. Abbreviations: CH, Churchill hinterland; GR, Grenville orogen; LA, Labradorian orogen; MA, Mazatzal (including Central Plains) orogen; NA, Nain Province; PE, Penokean orogen; SL, Slave Province; SU, Superior Province; TH, Trans-Hudson Orogen; WO, Wopmay orogen; and WY, Wyoming Province. Breaks in the pattern indicate Archean province boundaries within the Churchill hinterland. Greenland is positioned in a predrift reconstruction after Rowley and Lottes (1988).*

et al., 1988). By analogy, pre-Hudsonian sutures (some subsequently reactivated) within and bounding the northern Churchill Province are not implicitly ruled out. In the last decade, geophysical, geological, and geochronological evidence has increasingly suggested the existence of such sutures, meaning that the northern Churchill Province, indeed most of the Canadian Shield, represents a tectonic collage of microcontinents amalgamated in the Early Proterozoic (Gibb, 1978; Gibb *et al.*, 1978, 1983; Sharpton *et al.*, 1987; Hoffman, 1988a, 1989). This evidence is reviewed here and a new tectonic subdivision of the Churchill Province is presented accordingly.

Many hold that tectonic subdivisions should eschew genetic implications and seek stability in nomenclature. I disagree on both counts, believing that this approach encourages classification for its own sake, at the expense of what geoscience should really be about – understanding what happened

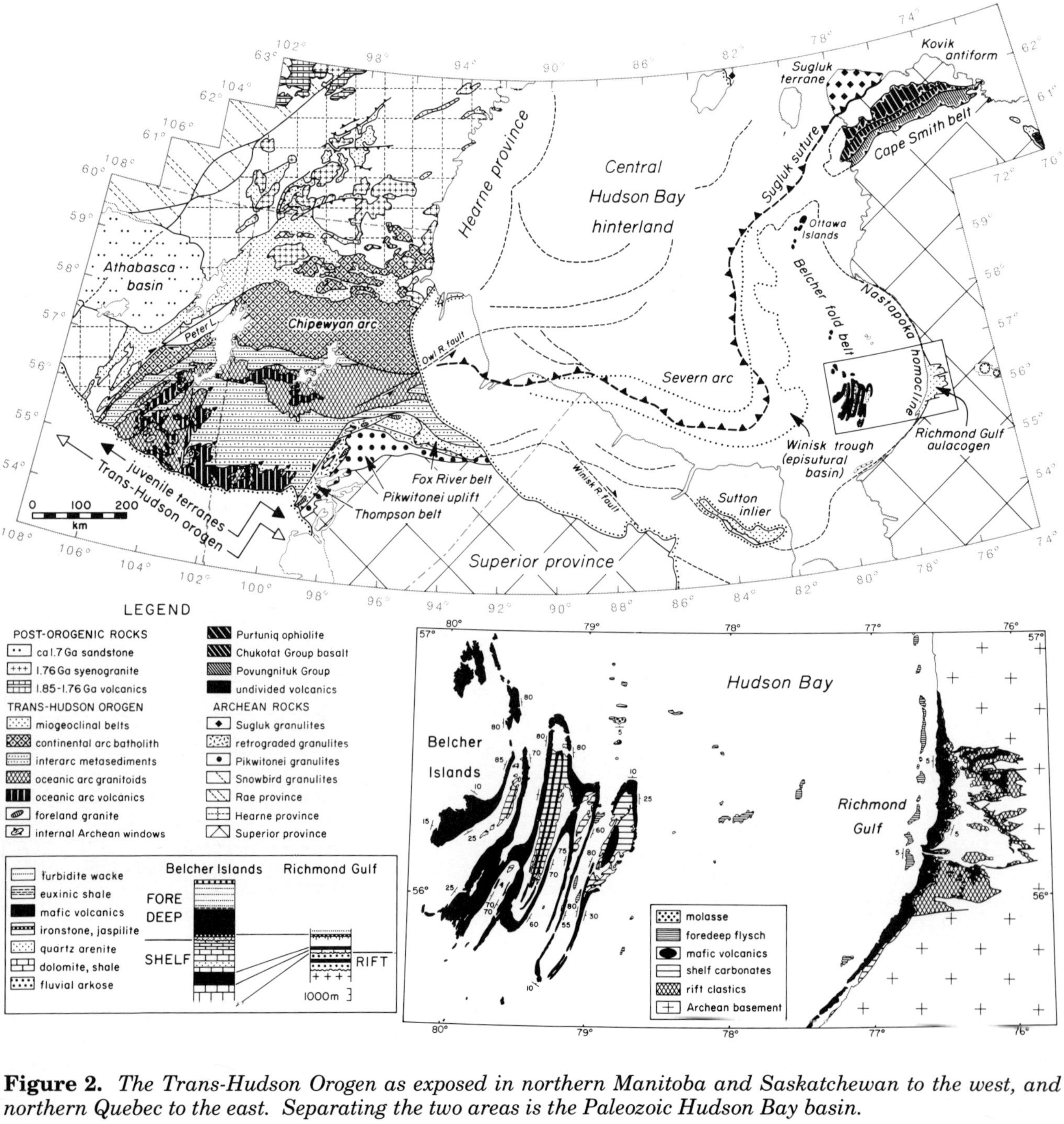

Figure 2. *The Trans-Hudson Orogen as exposed in northern Manitoba and Saskatchewan to the west, and northern Quebec to the east. Separating the two areas is the Paleozoic Hudson Bay basin.*

and why. Accordingly, the proposed subdivision seeks not to mask the important problems but to expose them. If successful, it should hasten its own demise.

TRANS-HUDSON OROGEN

The popularity of the name Trans-Hudson Orogen accounts for most of the references to the paper (Hoffman, 1981) in which the name was proposed, as an afterthought, without precise definition. The intent was to identify the Early Proterozoic collision zone between the Superior Province and the Archean domains of the northwestern Churchill Province (Lewry, 1981; Stauffer, 1984; Green *et al.*, 1985b). The name alludes to the fact that the orogen crosses beneath the Paleozoic Hudson Bay basin (Gibb *et al.*, 1983). In its type area in northern Manitoba and Saskatchewan (Fig. 2), the orogen is divisible into external zones representing the deformed Archean continental margins and an internal zone that exposes juvenile Proterozoic crust. The external zones include miogeoclinal (continental terrace and foredeep) strata and granitoid intrusions of Early Proterozoic age, and retrograded Archean basement. The internal zone contains the relics of Stauffer's (1984) "Manikewan ocean" but differs from his "Reindeer zone," which includes the Chipewyan (Wathaman) batholith. The latter is interpreted as an Andean-type magmatic arc developed on Archean crust (Ray and Wanless, 1980; Lewry *et al.*, 1981; Fumerton *et al.*, 1984; Meyer, 1987; Thom *et al.*, this volume). The presence of such an arc exclusively on the northwestern continental margin implies an overall subduction polarity whereby the Superior Province served as the foreland of the orogen and the northwestern Churchill Province as the hinterland. Tectonic shortening of the miogeocline in the Thompson foreland belt began before the Molson dykes were intruded at 1.88 Ga (Bleeker, this volume). The Manikewan ocean closed between 1.85 and 1.83 Ga, the cessation of arc magmatism (Ray and Wanless, 1980; Meyer, 1987) and the age of molassic sedimentation in the internal zone (Gordon *et al.*, 1987) respectively.

Northeast of the type area (Fig. 2), gravity and magnetic anomalies suggest that the orogen projects diagonally across Hudson Bay (Gibb *et al.*, 1978; Mukhopadyay and Gibb, 1981; Coles and Haines, 1982; Gibb, 1983; Geological Survey of Canada, 1987; Roksandic, 1987; Sharpton *et al.*, 1987). The southeastern external zone (Baragar and Scoates, 1981, 1987; Arndt *et al.*, 1987) is discontinuously exposed in the Thompson belt (Bleeker, this volume), the Fox River belt (Scoates, 1981), the Sutton inlier (Bostock, 1971), the Richmond Gulf aulacogen and overlying Nastapoka homocline (Chandler, 1984, 1988; Chandler and Parrish, 1989), the Belcher fold belt (Jackson, 1960; Ricketts and Donaldson, 1981; Ricketts, 1981; Ricketts *et al.*, 1982), the Ottawa Islands (Baragar and Scoates, 1987; Baragar, 1984; Roksandic, 1987), and the Cape Smith thrust–fold belt (St-Onge and Lucas, this volume). The configuration of the northwestern external zone beneath the Hudson Bay basin is conjectural, but a prominent positive magnetic anomaly (Severn arc, Fig. 2) may track the eastward extension of the Chipewyan arc (Coles and Haines, 1982). Thus, the "Central Hudson Bay hinterland" (Roksandic, 1987) may be underlain by Archean crust linking the Hearne Province with the Sugluk terrane, or it may be composed of Early Proterozoic crust that was amalgamated before the subduction of Manikewan oceanic lithosphere, which gave rise to the magmatic arc. Northeast of the Hudson Bay basin, the internal zone is represented by the Sugkluk suture (Fig. 2), which separates Kovik antiform, a continuation of the Superior Province, from the Sugluk terrane (Doig, 1987; St-Onge and Lucas, 1990). The northwest-dipping Sugluk suture may be the root zone of the ophiolite-bearing Cape Smith thrust–fold belt (Hoffman, 1985; Lucas, 1989; St-Onge and Lucas, this volume). However, thrusting began in the Cape Smith belt before 1.88 Ga, but the Sugluk terrane experienced felsic magmatism and granulite-grade metamorphism at 1.83–1.82 Ga (Parrish, 1989). The links between the Trans-Hudson Orogen and the Early Proterozoic orogens northeast of the Superior Province are discussed later.

South of its type area, the Trans-Hudson Orogen is believed to extend in the subsurface between the Superior and Wyoming provinces (Fig. 1). This is inferred from electrical conductivity anomalies (Camfield and Gough, 1977; Handa and Camfield, 1984; Jones and Savage, 1986), magnetic and gravity anomalies (Green *et al.*, 1985a,b; Klasner and King, 1986; Thomas *et al.*, 1987), drillhole data (Klasner and King, 1986), and contrasting Nd model ages for the two Archean provinces (Peterman and Futa, 1987). The internal zone of juvenile Proterozoic crust must pinch to the south as Archean Nd model ages characterize the basement across central and southern Saskatchewan (Collerson *et al.*, 1989). At its southern limit, the Trans-Hudson Orogen is truncated by the Central Plains orogen (Sims and Peterman, 1986), equivalent to the 1.79–1.63 Ga Mazatzal orogen (Fig. 1).

SNOWBIRD LINE

The Snowbird line (Hoffman, 1988a) divides the Archean hinterland of the Trans-Hudson Orogen into the Hearne and Rae provinces (Fig. 3). It was first recognized by Taylor (1963) at Snowbird Lake (latitude 61°N, longitude 103°W) as a zone of mylonite many kilometres in width containing bodies of layered gabbro–anorthosite–pyroxenite. Subsequent work has shown that an anastomosing system of granulite-facies mylonites and associated anorthosite–gabbro–pyroxenite bodies extends for 1500 km from the edge of the shield at longitude 109°W to Hudson Bay at longitude 88°W (Lewry and Sibbald, 1977; Gilboy, 1980; Schau *et al.*, 1982; Tella and Eade, 1986; Hanmer, 1987; Gordon, 1988; Tella and Annesley, 1988). Steep horizontal gradients in the gravity and magnetic anomaly fields, coincident with the mylonite zone, extend southwestward from the shield for at least 750 km and intersect the Rocky Mountains in Alberta at about latitude 52°20′N (Geological Survey of Canada, 1987; Sharpton *et al.*, 1987; Ross *et al.*, *in prep.*). The steep gravity gradient also extends eastward across the northern Hudson Bay basin and converges with the Sugluk suture in Hudson Strait. An inlier of mylonitic gabbroic anorthosite exposed on Walrus Island (latitude 63°30′N, longitude 83°30′W) is the only exposure of the Snowbird line between Daly Bay and the Sugluk terrane (Heywood and Sanford, 1976). Mylonitic layering dips gently to the southeast around Daly Bay (Gordon, 1988) and to the northeast on Walrus Island, but kinematic information is lacking.

The southwestern part of the Snowbird line was first interpreted as a geosuture on the basis of gravity anomalies (Walcott and Boyd, 1971; Gibb and Halliday, 1974). In the subsurface, this interpretation is suppported by the truncation of the 1.99–1.91 Ga Taltson magmatic zone and by granites of the 1.82–1.78 Ga Rimbey magmatic zone (Fig. 3), which imply southeast-dipping subduction beneath the Hearne Province (Ross *et al.*, *in prep.*). In the shield, folding and thrusting of Early Proterozoic sedimentary cover of the Hearne and Rae provinces (Hurwitz and Amer groups respectively) may be related to shortening orthogonal to the Snowbird line (cf. Patterson, 1986), but no magmatic arc related to the Snowbird line has been found (Fig. 4). The Virgin River – Black Lake segment has been described as an intracontinental reactivation front related to the parallel Trans-Hudson Orogen (Lewry and Sibbald, 1980), but elsewhere the Snowbird line does not parallel the Trans-Hudson Orogen (Fig. 3), suggesting an independent origin.

Age constraints on the Snowbird line also suggest a difference between the shield and the subsurface. Nowhere has the age of granulite-facies mylonitization been determined directly. In the area of the Baker basin (Fig. 4), the Snowbird line is stitched by a posttectonic suite of lamprophyre dykes, pyroxene–syenite stocks, and related alkaline volcanic and associated terrestrial sedimentary rocks inferred to be about 1.85 Ga (Blake, 1980; LeCheminant *et al.*, 1981, 1987a, 1988). Accordingly, the high-grade mylonitization and juxtaposition of the Hearne and Rae provinces in the shield would be older than 1.85 Ga. However, the plate convergence implied by the Rimbey magmatic zone (cf. Ross *et al.*, *in prep.*) would be younger than 1.82 Ga. Near the edge of the shield, the main high-grade (dextral) shearing in the Virgin River shear zone predates the 1.82-Ga Junction granite, which may be related to the Rimbey arc and is itself transected by relatively narrow low-grade shear zones (Carolan *et al.*, 1989). The age relations suggest a complex history for the Snowbird line, possibly involving an initial juxtaposition of the Hearne and Rae provinces prior to 1.85 Ga, followed by reactivation resulting from collision between the Hearne and Superior provinces. It would be expected that the reactivation would be compressional opposite the northwest promontory of the Superior Province. Consequent southwest-directed extrusion of the Hearne Province might have caused dextral–oblique opening of a wedge-shaped ocean basin along the subsurface Snowbird line, analogous to the Neogene opening of the South China Sea as proposed by Tapponnier *et al.* (1982). The subsequent closure of the new basin, possibly linked to the collision of the Hearne and Wyoming provinces, may be recorded by the Rimbey magmatic arc.

Figure 3. *Tectonic elements of the western Canadian Shield and adjacent platform (modified and extended from Ross* et al. *(in prep.)). Archean provinces (1–6): 1, Superior; 2, Wyoming; 3, Medicine Hat; 4, Hearne; 5, Rae; and 6, Slave, including Simpson Islands terrane (SN). Zones of Archean crustal reactivation (7–10): 7, Thompson belt; 8, Queen Maud uplift; 9, Peter Lake terrane; and 10, thrust–fold belts involving Early Proterozoic cover. 11, mylonitic mafic–felsic granulites of uncertain age occurring along the Snowbird line. 12, metasedimentary–volcanic rocks of Early Proterozoic age in the Lacombe belt of Hearne Province. 13, volcanic (1.93 Ga) and sedimentary rocks of Wilson Island Group (WN) within and adjacent to Great Slave Lake shear zone (GSLsz). Early Proterozoic terranes (14–16): 14, Buffalo Head, Chinchaga, and Thorsby (2.3–1.9 Ga crystallization ages); 15, Hottah (1.95–1.90 Ga arc built on 2.3–2.1 Ga crust); and 16, Nahanni (pre-1.6 Ga). Early Proterozoic continental magmatic arcs (17–20): 17, Thelon–Taltson and Ksituan (both 2.00–1.90 Ga); 18, Great Bear (1.88–1.84 Ga) and Fort Simpson (ca. 1.85 Ga); 19, Wathaman–Chipewyan (1.85 Ga); and 20, Rimbey (1.82–1.78 Ga). Internal zones of the Trans-Hudson Orogen (21–22): 21, subsurface extension (Archean subcrop); and 22, juvenile crust exposing 1.91–1.85 Ga island arcs and interarc metasediments. 23, edge of platform (including Middle–Late Proterozoic) cover. 24, edge of Cordilleran deformation.*

The age of initial juxtaposition of the Hearne and Rae provinces (if they were ever separated) presumably postdates the Taltson magmatic arc (1.99–1.91 Ga) and predates the Baker basin (1.85 Ga). In the Slave Province (which collided with the Rae Province at 1.97 Ga), regional northwest–southeast crustal shortening, responsible for northwest-directed thrusting of the Great Slave Supergroup, oblique cross-folding of the Coronation Supergroup of Wopmay orogen, and folding of the southwest arm of the Goulburn Supergroup (Fig. 5), occurred at 1.88 Ga (Hoffman, 1981, 1988b; Hoffman *et al.*, 1988). This shortening event predates the Hearne–Superior collision and may be a distant manifestation of the Rae–Hearne collision at the Snowbird line.

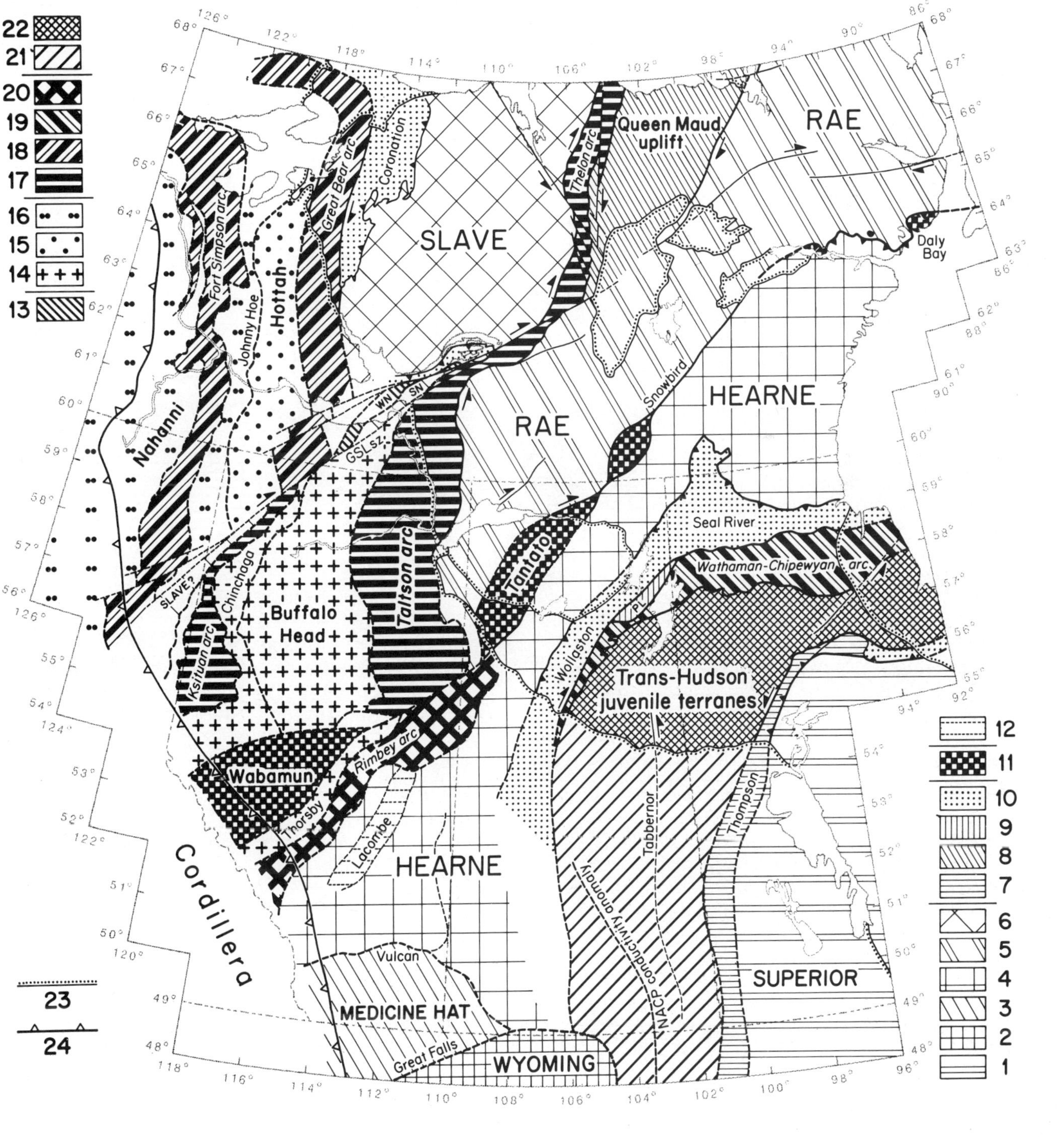

HEARNE PROVINCE

The name Hearne Province (Fig. 3) is proposed for the area of Archean crust bounded on the southeast by the Trans-Hudson Orogen, on the southwest by the Wyoming Province, and on the north and northwest by the Rae Province. The name is derived from Hearne Bay on Nueltin Lake (latitude 60°N, longitude 99°W). Half a century before Franklin's first Arctic expedition, Samuel Hearne (1745–1792) travelled overland from Fort Churchill on Hudson Bay to Coronation Gulf, returning via Great Slave Lake, to investigate aboriginal reports of copper near the mouth of the Coppermine River. An employee of the Hudson Bay Company, Hearne was the first European to traverse, in geological terms, the Hearne, Rae, and Slave provinces, and the Snowbird, Thelon, Wopmay, and Taltson orogenic belts.

The Hearne Province appears to be entirely underlain by Archean crust. The exposed Archean rocks include large areas of upper-amphibolite- and granulite-grade gneisses. A greenschist- to lower-amphibolite-grade terrain in the north-central Hearne Province comprises belts of mafic-through-

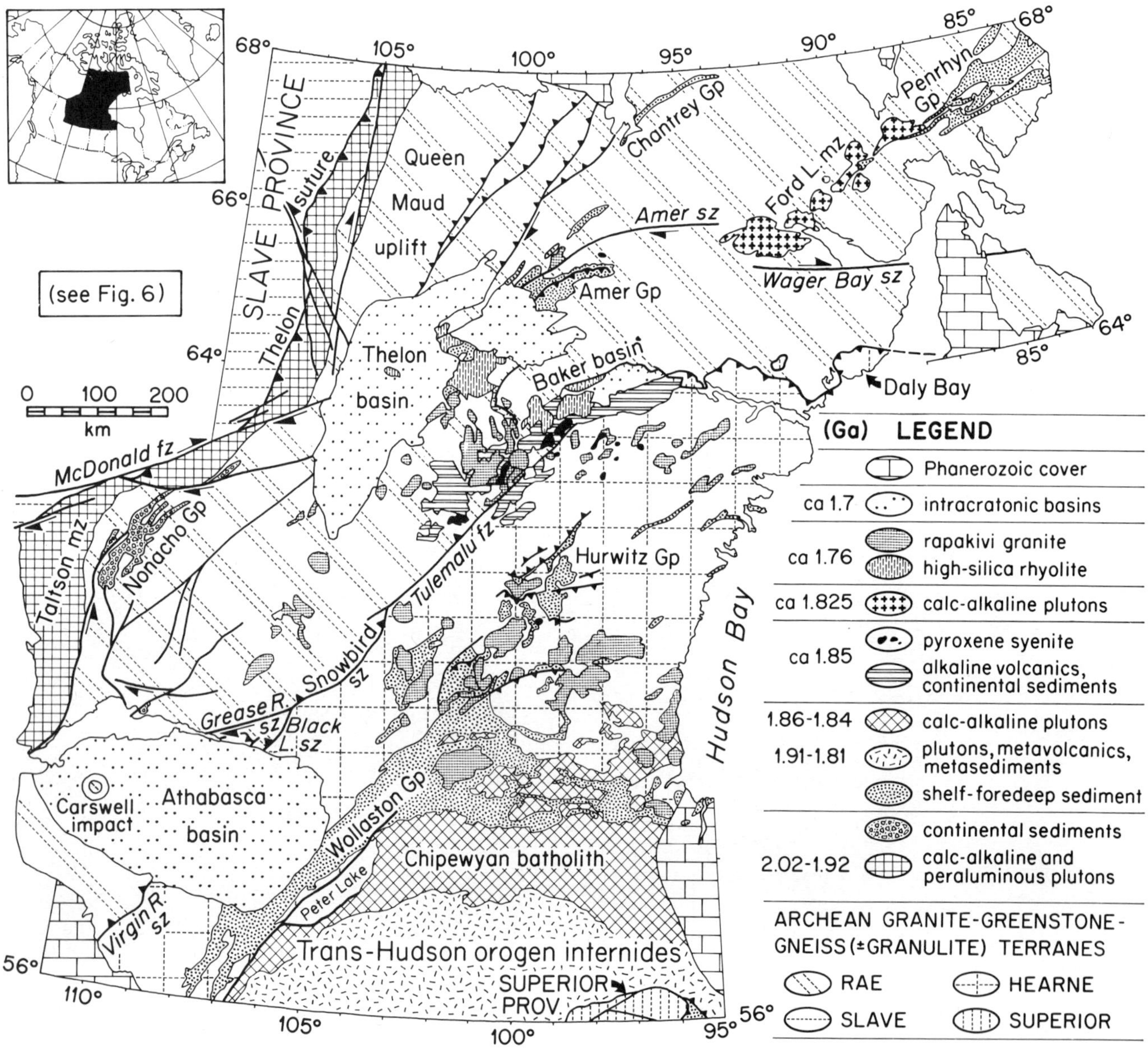

Figure 4. *Parts of the Rae and Hearne provinces, separated by the Snowbird line. Note overlap of the 1.85(?) Ga Baker basin across the Snowbird line, providing a minimum age for juxtaposition of the two Archean provinces.*

felsic volcanic and associated sedimentary rocks, intruded by mafic-through-felsic plutons (Davidson, 1970a, 1972; Eade, 1978; Lewry *et al.*, 1978; Schledewitz, 1978; Park and Ralser, 1990). The volcanic and related plutonic rocks resemble island arcs compositionally, and occur in northeast-trending belts. U–Pb zircon ages of 2.68–2.70 Ga date felsic volcanism in two of the belts (Chiarenzelli and Macdonald, 1986; Mortensen and Thorpe, 1988). One belt southeast of Baker basin is intruded by an alkaline igneous complex, composed of nepheline syenite, ijolite–pyroxenite, and carbonatite, and has a Sm/Nd age of 2.54 ± 0.8 Ga (Davidson, 1970b; Cavell *et al.*, 1987; P.A. Cavell, personal communication, 1987). Sparse, widely scattered U–Pb zircon ages for the high-grade gneisses range from 3.48 to 2.67 Ga (Wanless, 1979; Eade, 1986). Zircons from drill core of granitic basement to the Athabasca basin near the boundary between the Hearne Province and the Wollaston belt are interpreted as having igneous crystallization ages of about 2.60–2.65 Ga (Krogh and Clark, 1987). Sm–Nd model ages for drill cores of basement gneiss in southwestern Saskatchewan and adjacent Alberta are 3.2–2.6 Ga (Frost and Burwash, 1986). A composite Sm–Nd model age of 2.9 Ga was obtained from samples collected in Saskatchewan along the edge of the shield across the Hearne Province and adjacent parts of the Rae Province (McCulloch and Wasserburg, 1978).

More than 40 erosional remnants of Early Proterozoic sedimentary cover, up to 135 km in length, occur in the northern Hearne Province (Fig. 4). Their distribution coincides broadly with the area of lower grade basement rocks. The regional stratigraphy and lithofacies of the cover rocks are described by Bell (1970) and Eade and Chandler (1975). The main stratigraphic unit, the Hurwitz Group, contains a prominent supermature quartzarenite (Kinga Formation) that is underlain

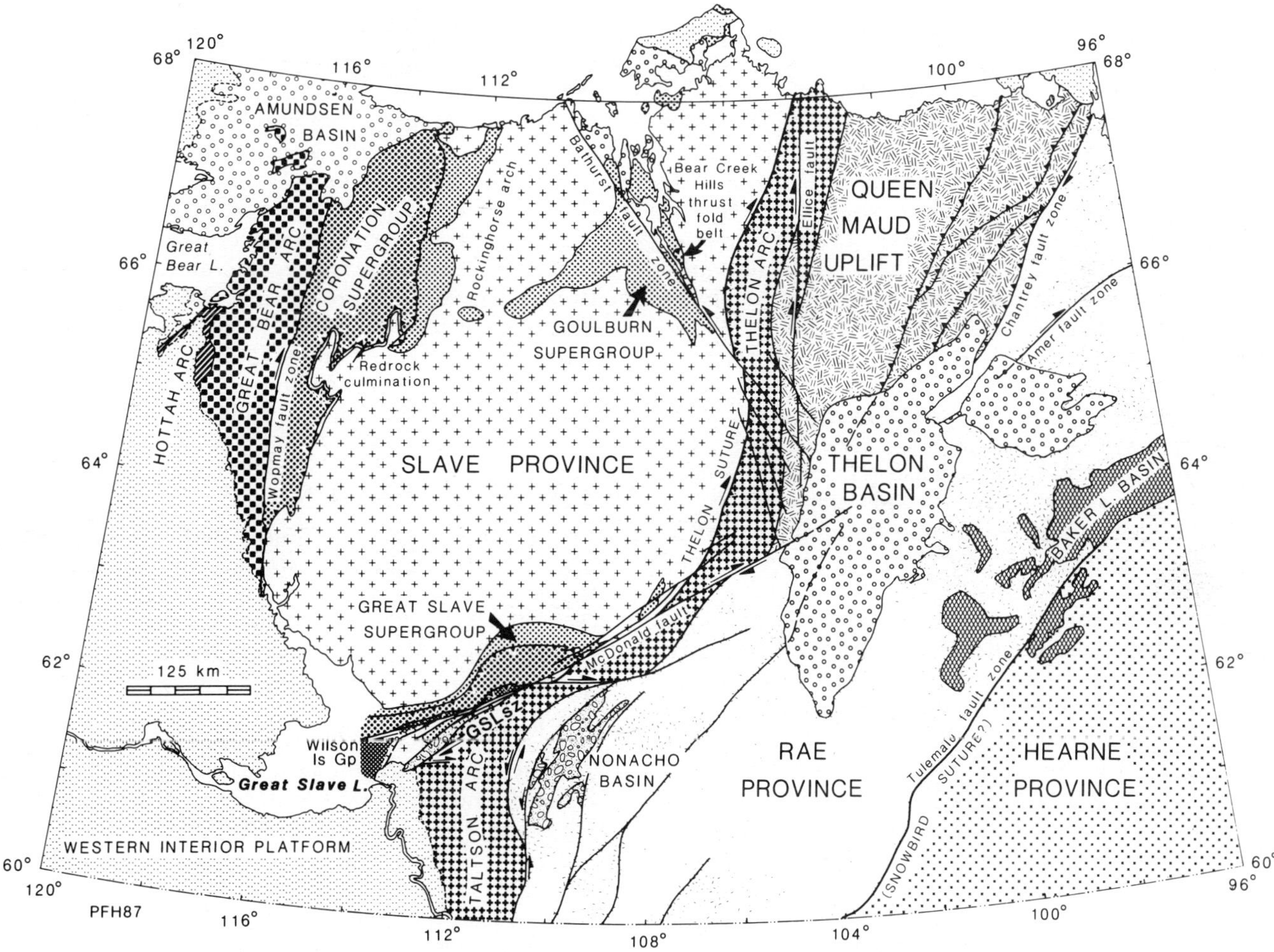

Figure 5. *Northwestern Canadian Shield showing elements of the Thelon orogen separating the Rae and Slave provinces. The Wopmay orogen flanks the Slave Province to the west. GSLsz, Great Slave Lake shear zone.*

by conglomerate (Padlei Formation) and overlain successively by ferruginous pelite (Ameto Formation), locally hosting mafic flows and sills; stromatolitic argillaceous dolomite (Watterson Formation); greywacke–siltstone (Ducker Formation); and feldspathic arenite (Tavani Formation). The Hurwitz Group erosionally overlies a north-northeasterly trending swarm of porphyritic mafic dykes (Kaminak dykes) that cut the late Archean alkaline igneous complex. The dykes have a minimum age of about 2.4 Ga, based on K–Ar whole-rock dates of relatively unaltered, chilled margin samples (Wanless *et al.*, 1973; Christie *et al.*, 1975). Coarse immature clastic sequences of limited distribution occur stratigraphically below (Montgomery Lake Group) and above (Kiyuk Group, Aspler *et al.* (1989)) the Hurwitz Group. An additional sequence of feldspathic and dolomitic sediments (Bate Lake sediments) is of uncertain stratigraphic position relative to the Hurwitz Group (Eade and Chandler (1975) and references therein). The ages of the Hurwitz Group and associated sequences relative to the Early Proterozoic strata of the northwestern externides of the Trans-Hudson Orogen (Wollaston and Seal River groups) are unknown.

In the northern part of the Hearne Province (Fig. 4), the Hurwitz Group occurs in cuspate basement-cover synclines of northeasterly to easterly trend, in part bounded by northwest- to north-dipping basement thrusts (Davidson, 1970a; Heywood, 1973; Eade, 1974; Tella *et al.*, 1986). To the south, the Hurwitz Group folds trend easterly and basement thrusts dip south (Eade, 1973, 1974). In between, the Hurwitz Group folds have an irregular basin-and-dome geometry (Eade, 1971, 1974; Eade and Chandler, 1975), believed to be a product of interference between sets of older northeasterly, and younger easterly to southeasterly trending folds and thrusts (Eade, 1974). The older set is orthogonal to the Snowbird line; the younger to the adjacent segment of the Trans-Hudson Orogen (Fig. 4).

Folding of the Hurwitz Group and associated strata predated the intrusion of epizonal rapakivi granite bodies in the southern District of Keewatin. One such body yields a Rb–Sr isochron age of 1.76 ± 0.2 Ga (recalculated from Wanless and Eade (1975)), coeval with similar granites and related rhyolites southwest of the Baker basin (Fig. 4), which have U–Pb zircon ages of 1.78–1.74 Ga (LeCheminant *et al.*, 1987a; Loveridge *et al.*, 1988).

BOUNDARY BETWEEN THE HEARNE AND WYOMING PROVINCES

Nd crust-formation ages for the buried southern part of the Hearne Province are mostly about 2.8 Ga (Frost and Burwash, 1986), whereas those of the Wyoming Province exceed 3.1 Ga (Peterman and Futa, 1987). The boundary between them may correspond to the Great Fall tectonic zone (O'Neill and Lopez, 1985; O'Neill *et al.*, 1988), a 1.8 Ga shear zone that defines the southern limit of the Medicine Hat block (Fig. 3). The Medicine Hat block is distinguished by northwesterly trending geopotential anomalies bounded on the north by a discontinuity known as the Vulcan low (Ross *et al.*, *in prep.*). Green *et al.* (1985a) identified the Vulcan low as the boundary between the Hearne and Wyoming provinces, and Hoffman (1988a) interpreted it as a south-dipping suture on the basis of gravimetric and seismic reflection profiles. However, Nd isotopic data show that the Medicine Hat block has 2.9–2.8 Ga crust-formation ages more like those of the Hearne Province. Moreover, if the parallel magnetic high north of the Vulcan low signals a related magmatic arc (cf. Ross *et al.*, in prep.), then a north-dipping suture is implied. A 2.7–2.6 Ga age for the proposed suture is indicated by U–Pb zircon dating of basement cores from the magnetic high, the Vulcan low, and the northern margin of the Medicine Hat block (Ross *et al.*, *in prep.*). Accordingly, the preferred interpretation is that the Great Falls tectonic zone represents a ca. 1.8 Ga boundary between the Hearne and Wyoming provinces, and that the Vulcan low is a late Archean suture within the southern Hearne Province. There is evidence that the Vulcan low was reactivated during deposition of the Middle Proterozoic Belt–Purcell Supergroup (Kanasewich, 1968), the Late Proterozoic Windermere Supergroup (Lis and Price, 1976), and the early Paleozoic Cordilleran passive margin sequence (Bond and Kominz, 1984).

THELON OROGEN

The Thelon orogen (Hoffman, 1988a) is interpreted as a dextral–oblique collision zone between the Slave (foreland) and Rae (hinterland) provinces (Fig. 3). Postcollisional indentation by the relatively rigid Slave Province was accommodated by oblique shortening across the Queen Maud uplift concurrent with dextral shear on the Great Slave Lake shear zone (Hoffman, 1987a; Hanmer, 1988), and by younger conjugate strike slip on the McDonald and Bathurst fault systems (Gibb, 1978; Gibb and Thomas, 1977). The Thelon orogen comprises the following exposed tectonic elements (Fig. 5): (1) remnant foreland thrust–fold belts involving pre- and synorogenic sediments correlative with the autochthonous Great Slave and Goulburn supergroups (Hoffman, 1981, 1988b; Grotzinger and McCormick, 1988; Johnson, *in press*; Tirrul and Grotzinger, *in press*); (2) a zone of dextrally sheared Archean rocks along the eastern margin of the

Slave Province, in which ca. 1.9 Ga medium-pressure metamorphism overprints late Archean low-pressure metamorphism (Henderson *et al.*, 1982; Thompson *et al.*, 1986); (3) the composite Taltson–Thelon magmatic zone, which underwent 1.99–1.96 Ga subduction(?)-related plutonism coeval with granulite-grade metamorphism, followed by 1.96–1.91 Ga collision-induced anatectic plutonism concurrent with unroofing (Bostock *et al.*, 1987; van Breemen *et al.*, 1987; James *et al.*, 1988; Van Breemen and Henderson, 1988; Thériault and Bostock, 1989); and (4) the Queen Maud uplift, a presumed Tibetan-type hinterland plateau underlain by Archean gneisses of granulite and upper amphibolite grades (Fraser, 1978; Heywood and Schau, 1978) and bounded on the east by dextral–oblique (back) thrusts (Tella *et al.*, 1984). U–Pb zircon ages for volcanic ash beds of 1.97 Ga near the base of the Bear Creek foredeep (Goulburn Supergroup) and 1.92 Ga for the foredeep flysch–molasse transition, respectively, establish the contemporaneity between foredeep sedimentation on the Slave Province margin and Taltson–Thelon magmatism on the leading edge of the Rae Province (Bowring and Grotzinger, 1989). The Thelon magmatic zone extends northeastward across the Arctic platform at least as far as central Ellesmere Island, according to U–Pb and Sm–Nd isotopic data (Frisch and Hunt, 1988; Reichenbach and Parrish, 1988; E. Hegner, personal communication, 1989) and distinctive magnetic anomalies (Geological Survey of Canada, 1987).

The coeval magmatic evolution of the Taltson and Thelon arcs implies that the Slave Province and the Buffalo Head terrane converged and collided with the Rae Province contemporaneously (Fig. 3). However, the Buffalo Head terrane is characterized by abundant 2.3–1.9 Ga granitic rocks (Ross *et al.*, *in prep.*), whereas the few plutonic rocks in the Slave Province in that age range belong to distinctive 2.2–2.0 Ga alkaline–peralkaline suites (Davidson, 1978; Bowring *et al.*, 1984; Cavell and Baadsgaard, 1986; Roscoe *et al.*, 1987). If, as seems probable, the Slave Province and Buffalo Head terrane were joined prior to their collision with the Rae Province, the nature of their mutual boundary is conjectural. In any event, the Buffalo Head terrane does not appear to be directly related to the Hottah terrane, which collided with the Slave Province at 1.89 Ga, about 80 Ma after the Slave–Rae collision (Bowring and Podosek, 1989; Housh *et al.*, 1989). The presence of 2.4–2.0 Ga granites intrusive into Archean rocks of the Rae Province east of the Taltson magmatic zone (Van Schmus *et al.*, 1986; Bostock and Loveridge, 1988) suggests that east-dipping subduction beneath the western margin of the Rae Province began long before the arrival of the Slave Province.

RAE PROVINCE

The name Rae Province (Hoffman, 1988a) was proposed for the area of Archean crust extending between the Hearne and Slave provinces in the northwest, and between the Superior and Nain provinces in the northeast. The name is derived from Rae Strait (latitude 69°N, longitude 95°W) in the north-central part of the province. The strait, which separates King William Island from Boothia Peninsula, is the key to the "Northwest Passage" and ignorance of its existence doomed the fabled Royal Navy expedition led by Franklin. John Rae (1813–1893), a Hudson Bay Company employee, was an early exponent of Inuit travelling techniques and his extensive traverses along the Arctic coast led to the first evidence bearing on the fate of the Franklin expedition.

The Rae Province has three arms, each of which consists of Archean basement and subordinate infolded keels of Early Proterozoic platformal sedimentary cover. The southwest arm is flanked by the Snowbird line and the Taltson magmatic zone (Fig. 3), the southeast arm by the New Quebec and Torngat orogens (Fig. 6), and the northeast arm by the Thelon magmatic zone and the Foxe fold belt (Fig. 6).

The best exposed transect of the southwest arm lies north of the Athabasca basin (Fig. 4) along the north shore of Lake Athabasca (Gilboy, 1980; Macdonald and Broughton, 1980; Nielsen *et al.*, 1981; Macdonald and Slimmon, 1984; Slimmon, 1987). Midway along the transect, 3.0 Ga basement is overlain by mixed quartzite, marble, and metavolcanic rocks (Murmack Bay Group), which are intruded by 2.4–2.0 Ga granite plutons (Van Schmus *et al.*, 1986). Close to the Taltson magmatic zone, an extensive 2.6 Ga granitic terrain (Van Schmus *et al.*, 1986; Stevenson *et al.*, 1987) is intruded by 2.4–2.3 Ga granite plutons (Bostock and Loveridge, 1988) and overlain unconformably by sinistrally wrenched fanglomeratic and lacustrine sediments of the Nonacho Group (Aspler and Donaldson, 1985). All ductile deformation in the transect area, including extensive mylonitization, predates the basal 1.7 Ga sandstones of the Athabasca basin (Cumming *et al.*, 1987).

Archean rocks in the central part and northeast arm of the Rae Province consist of extensive felsic gneisses and relatively narrow belts of greenschist- to lower amphibolite-grade volcanic and sedimentary strata. Northwest of the Baker basin, the Archean supracrustal belts (Ketyet and Woodburn groups) are derived from basalt, rhyolite, dacite, and komatiite, and sediments including conglomerate, quartzite, iron formation, and greywacke (Annesley, 1981; Ashton, 1982; Schau, 1983; LeCheminant *et al.*, 1984; Tella, 1984; Taylor, 1985). Preliminary U–Pb zircon dating indicates that the felsic gneisses range from 3.3 to 2.8 Ga, and that there were two episodes of felsic magmatism at 2.8 and 2.6 Ga (Ashton, 1985; LeCheminant *et al.*, 1987a). Lithologically similar 2.9–2.8 Ga supracrustal belts (Prince Albert and Mary River groups) occur within, and trend parallel to, the northeast arm of the Rae Province (Jackson and Taylor, 1972; Schau, 1982; Frisch, 1982). Reconnaissance isotopic studies indicate major plutonism at 2.7 Ga and crust of unknown extent that is older than the supracrustal belts.

Infolded keels of Early Proterozoic platformal sedimentary cover (Amer and Chantrey groups) trend northeastward from the Thelon basin (Tippett

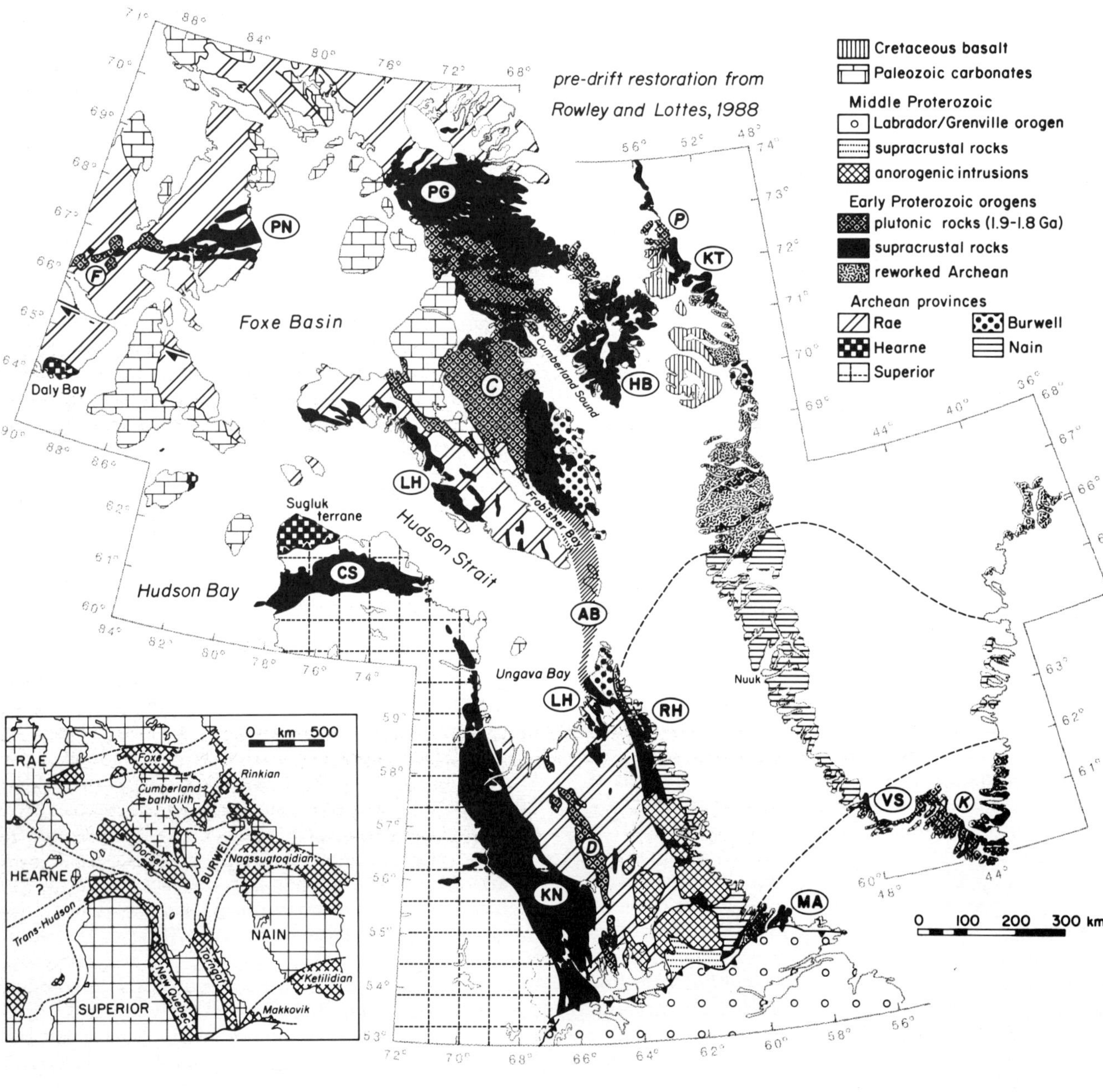

Figure 6. *Geology of parts of eastern Canada and southern Greenland showing inferred relations (inset) between various Archean provinces and Early Proterozoic orogens in a predrift reconstruction. Supracrustal sequences: AB, Tasiuyak gneiss (partly cozonal with Abloviak shear zone); CS, Cape Smith klippe; HB, Hoare Bay Group; KN, Kaniapiskau Supergroup; KT, Karrat Group; LH, Lake Harbour Group; MA, Moran and Aillik groups; PG, Piling Group; PN, Penrhyn Group; RH, Ramah Group; and VS, strata marginal to Ketilidian orogen. Early Proterozoic plutonic suites: C, Cumberland batholith; D, De Pas batholith; F, Ford Lake suite; K, Ketilidian batholith; and P, Proven granite.*

and Heywood, 1978; Frisch and Patterson, 1983; Tella *et al.*, 1983, 1984; LeCheminant *et al.*, 1984; Frisch *et al.*, 1985; Patterson, 1986). The internal stratigraphy of the Amer Group, in particular, resembles that of the Hurwitz Group of the Hearne Province. Quartzarenite near the base is overlain successively by ferruginous terrigenous sediments, siliceous dolomite, and euxinic pelite that grades upward with increasing proportions of feldspathic siltstone and sandstone. The largest Amer Group belt, 125 km long and up to 50 km wide, is a refolded imbricate thrust stack of uncertain structural vergence in which the cover is, in part, tectonically interleaved with thin sheets of basement (Tippett and Heywood, 1978; Patterson, 1986). Metamorphic grade decreases southward across the belt from sillimanite–kyanite–muscovite to chlorite–muscovite bearing assemblages. Deformation and metamorphism of the Amer Group likely predate the ca. 1.85 Ga alkaline magmatic suite of the Baker basin region, and definitely predate the ca. 1.76 Ga suite of rapakivi granites and related rhyolite ash flows (LeCheminant *et al.*, 1981).

The inferred integrity of the southeast arm of the Rae Province (Fig. 6) is obscured by Hudson Strait and its connecting waterways, the reconnaissance level of geologic mapping, the paucity of isotopic geochronology, and the lack of high-resolution aeromagnetic coverage north of the mainland. Nevertheless, the entire southeast arm of the province is believed to be underlain by Archean crust. Felsic gneisses, partly of granulite grade, structurally underlie deformed and metamorphosed (upper amphibolite to granulite grade) rocks of the Lake Harbour Group (Fig. 6), which comprises quartzite, marble, sulphidic pelite, metagreywacke, and minor mafic layers (Davison, 1959; Emslie, 1963, 1964; Heywood, 1967; Blackadar, 1967a,b; Jackson and Taylor, 1972; Heywood and Sanford, 1976; Taylor, 1979). The Lake Harbour Group is presumed to be Early Proterozoic and the associated felsic orthogneisses of southern Baffin Island (Dorset belt of Fig. 6) have 3.5–2.3 Ga Nd model ages, indicating an extensive Archean basement (Jackson *et al.*, 1990). South of Ungava Bay, migmatitic gneisses on strike with the De Pas batholith (1.83–1.81 Ga) contain 2.9–2.7 Ga zircons (Machado *et al.*, 1989) and Middle Proterozoic anorogenic intrusions have Sr, Nd, and Pb isotopic evidence for an Archean crustal component (Ashwal *et al.*, 1986; Wooden *et al.*, 1987). On the east side of the De Pas batholith, the Archean gneisses also host a 2.32 Ga bimodal plutonic–volcanic suite (van der Leeden *et al.*, this volume).

The continuity of the southeast arm of the province is supported by gross structural trends, which curve from easterly in southern Foxe Basin, southeasterly along Hudson Strait, to southerly approaching the Grenville orogen (Fig. 6). The trends are compatible with deformation due to indentation by the Superior Province (cf. Gibb, 1983) and consequent clockwise rotation of the southeast arm of the Rae Province. Such a rotation (analogous to the Tertiary counterclockwise rotation of Sardinia–Corsica and its subsequent collision with Italy as a result of the Alpine indentation of northwest Europe by the Apulian promontory) would have been accompanied by the opening of a wedge-shaped basin, open to the east, between the southeast and northeast arms of the Rae Province. The Early Proterozoic sediments of the Foxe fold belt (Penrhyn, Piling, and Hoare Bay groups, Fig. 6) may have been deposited in such a basin. Development of the Cumberland batholith and the flanking Dorset and Foxe fold belts may chronicle the destruction of the basin upon the arrival of the Burwell and Nain provinces.

NEW QUEBEC OROGEN

The New Quebec orogen (Hoffman, 1988a) is interpreted as a dextral–oblique collision zone between the Superior Province and the overriding southeast arm of the Rae Province (Kearey, 1976; Thomas and Kearey, 1980; Wardle and Bailey, 1981; LeGallais and Lavoie, 1982; Wardle, 1982; Wardle *et al.*, 1986; Perreault *et al.*, 1988; Boone and Hynes, this volume; Poirier *et al.*, this volume; van der Leeden *et al.*, this volume). Autochthonous strata (Knob Lake Group) in the Superior Province foreland are flanked by a southwest-vergent thrust–fold belt, the Archean hinterland of which experienced synorogenic metamorphism and plutonism (Fig. 7). The venerable term “Labrador trough” refers particularly to the foreland thrust–fold belt of the orogen and the adjacent autochthonous strata.

The autochthonous strata consist of three unconformity-bounded sequences (Wardle and Bailey, 1981; LeGallais and Lavoie, 1982; Clark, 1984, 1988). The middle sequence onlaps the Archean foreland everywhere except for the Cambrien (named from Cambrien Lake) aulacogen (Fig. 7), where the lower sequence extends into the foreland. In the aulacogen and the thrust–fold belt, the lower sequence comprises fluvial clastics (Seward Subgroup), marine-shelf quartzite and dolomite (Pistolet Subgroup), and euxinic shale with turbidites deposited in an intrashelf basin or ancestral foredeep (Swampy Bay Subgroup). In the thrust–fold belt, the basinal strata of the Swampy Bay Subgroup pass upward into dolomite (Denault or Abner Formation), representing a shoaling of the basin culminating in a peritidal carbonate platform (Hoffman and Grotzinger, 1988). Mafic alkaline and tholeiitic volcanics, with corresponding sills, occur in allochthonous correlatives of the Seward and Swampy Bay subgroups respectively. Rhyolites in equivalents of the latter(?) are 2.14 Ga (Clark, 1984). The middle sequence begins with sediments deposited on a subsiding submarine ramp (Wishart quartzite, Ruth shale, Sokoman ironstone). They are overlain by euxinic shale and turbidites (Menihek Formation), interpreted as a foredeep flysch (Hoffman, 1987b). In the thrust–fold belt, alkaline volcanics found locally in the Sokoman ironstone are 1.88 Ga, and mafic sills related to tholeiitic volcanics in equivalents of the Menihek flysch are 1.88–1.87 Ga (Chevé and Machado, 1988; Machado *et al.*, 1989; R. Parrish, personal communication, 1988). The upper sequence, a fluvial molasse of westerly derivation (Tamarack River or Chioak Formation), occurs on the autochthon but is truncated by the frontal thrust system.

Thrust sheets of Knob Lake Group strata are overridden by an allochthon composed of mafic–ultramafic sills (Montagnais Group) with screens of mafic volcanics and fine-clastic sediments (Doublet Group), and a structurally higher allochthon predominantly composed of stratified arkosic semipelite (Laporte Group). The Montagnais and Doublet groups have traditionally been grouped together within or above the Menihek flysch, but

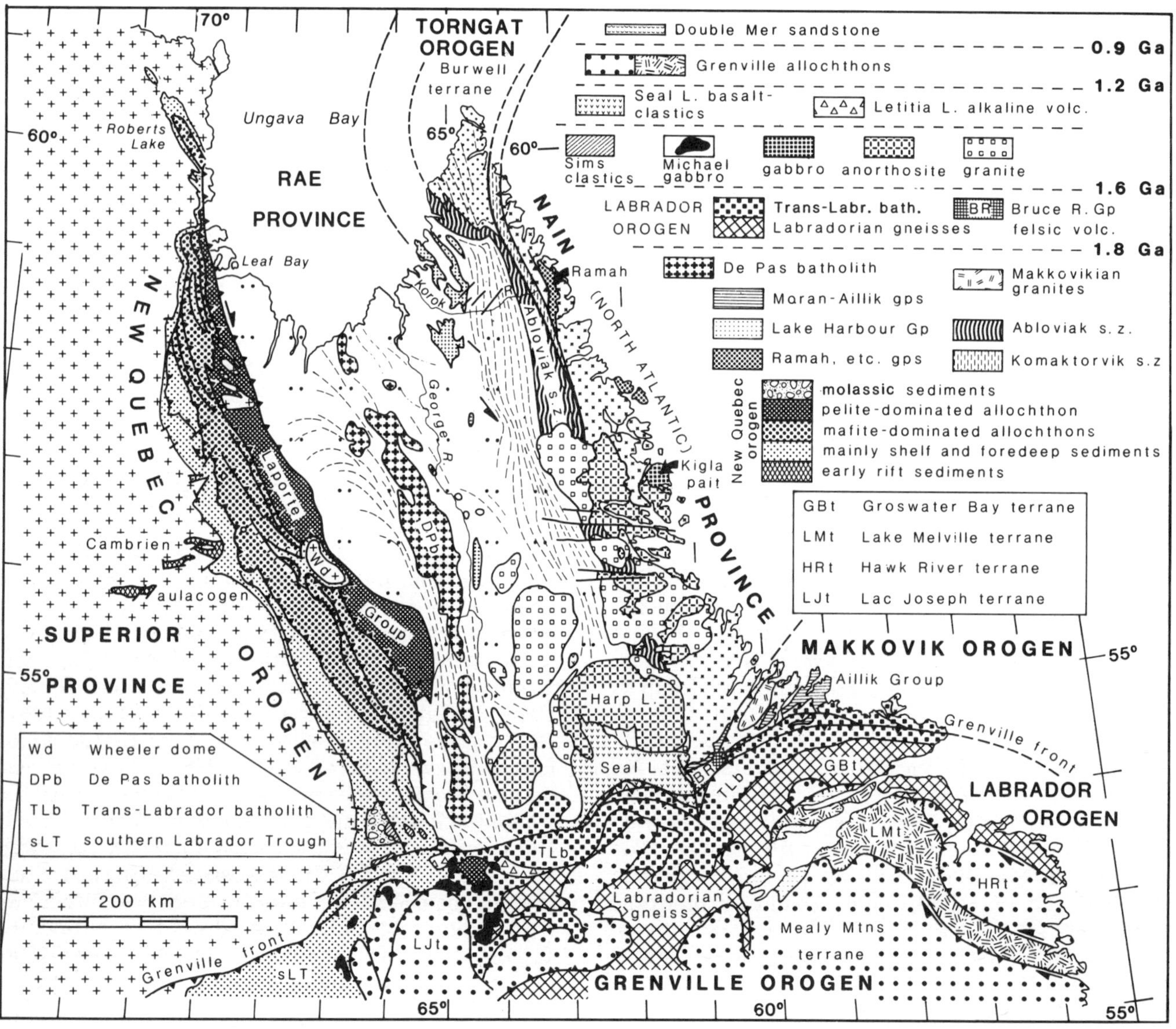

Figure 7. *Geology of northeastern Quebec and Labrador showing collision zones between Superior and Rae provinces (New Quebec orogen) and between Rae and Nain provinces (Torngat orogen). Makkovik and Labrador orogens involved accretion of Early Proterozoic crust from the south. The Grenville orogen is defined by northwest-directed thrusting of Middle Proterozoic age.*

this lacks geochronological support and has no structural rationale. The Laporte Group has been interpreted either as a continental-rise prism deposited outboard of the Knob Lake Group (Wardle and Bailey, 1981) or as a fore-arc basin on the leading edge of the overriding plate (van der Leeden *et al.*, this volume). Each of the two allochthons is equivalent to a Cordilleran "suspect terrane" (Coney *et al.*, 1980). Recumbent antiforms of Archean basement veneered by quartzite and marble occur fault bounded within the Laporte Group allochthon, but there is disagreement as to whether they lie structurally above or below the Laporte Group. All this means that there is much uncertainty as to where in the Labrador trough to draw a suture or sutures.

In the hinterland east of the Labrador trough, the 1.83–1.81 Ga De Pas batholith (Fig. 7) is interpreted as a magmatic arc cozonal with, but partly younger than, a granulite-grade dextral shear zone involving Archean rocks and probable equivalents of the Laporte Group (Taylor, 1979; van der Leeden *et al.*, this volume). In the north, a dextral–oblique thrust separates the Laporte Group allochthon, or Laporte terrane, from an overriding complex in which granulite-grade metamorphism is related to 1.85–1.83 Ga tonalite and granite plutons that intrude Early Proterozoic(?) metasediments and metavolcanics, and Archean gneisses (Perrault *et al.*, 1988; Machado *et al.*, 1989; Poirier *et al.*, this volume). Contemporaneous cooling of the metamorphic hinterland and the foreland thrust–fold belt (1.79 Ga monazites, 1.77 Ga titanites, 1.74 Ga rutiles) implies that collision and thrusting onto the Superior Province occurred between 1.87 and 1.79 Ga (Machado *et al.*, 1989) — 1.87 Ga being the age of mafic sills that are cut by thrusts and subsequently metamorphosed, and 1.79 Ga being the age of metamorphism that postdates thrusting. The observation that thrusting began before 1.88 Ga in the Cape Smith belt (Parrish, 1989; St-Onge and Lucas, this volume) but after 1.87 Ga in the New Quebec orogen accords with the idea that the New Quebec orogen resulted from clockwise rotation of the southeast arm of the Rae Province in response to collisional indentation by the Ungava promontory of the Superior Province.

Near the south end of the New Quebec orogen (Fig. 7), the Archean hinterland has been thrust westward, truncating the thin-skinned foreland thrust–fold belt (Wardle *et al.*, this volume). The late west-vergent structures trend parallel and may be related to the Torngat orogen, 300 km to the east, implying that the New Quebec orogen antedates the Torngat orogen. This is consistent with the proposition that sedimentation and deposition in the Foxe fold belt are related to collisions of the Superior and Burwell–Nain provinces, respectively, with the Rae Province.

TORNGAT OROGEN

The Torngat orogen (Hoffman, 1988a) is interpreted as a sinistral–oblique collision zone between the Nain Province and the overriding southeast arm of the Rae Province (Taylor, 1979; Korstgard *et al.*, 1988; Ermanovics *et al.*, 1989). The Torngat and New Quebec orogens have opposite senses of transcurrent shear, making it unlikely that they are coeval. The Torngat orogen consists of three zones (Fig. 7): (1) a relatively narrow (less than 20 km) zone of east-vergent thrusts of Archean basement and Early Proterozoic cover (Ramah Group) marginal to the Nain Province, (2) a continuous medial high-grade (10 kbar (1 GPa), 800°C) zone, 5–40 km wide, in which sinistral transcurrent mylonites (Abloviak shear zone) are approximately cozonal with a distinctive light-grey garnetiferous diatexite (Tasiuyak gneiss) derived from euxinic pelite and semipelite, and (3) a broad zone (over 60 km) of protomylonitic orthogneisses (Archean and Early Proterozoic?) and amphibolite- to granulite-grade paragneiss, quartzite, and marble correlated with the Lake Harbour Group (Taylor, 1979).

The Ramah Group consists of shelf and overlying foredeep sediments, cut by mafic sills, that were metamorphosed to greenschist and higher metamorphic grades, and cut by east-vergent pre- and postmetamorphic thrusts (Morgan, 1979, 1981; Knight and Morgan, 1981; Ermanovics *et al.*, 1989). The younger set of thrusts ramp down into the Archean basement within 20 km of the Abloviak shear zone, and postdate the sinistral transcurrent mylonitic fabric of the shear zone. Ermanovics *et al.* (1989) infer that the Abloviak transcurrent shear zone postdates the Ramah Group, a relationship previously in question. The protolith of the Tasiuyak gneiss may have been a deeper water equivalent of the Ramah and/or Lake Harbour groups. Its continuity for over 1200 km along strike suggests that it represents continental rise or oceanic sediments marking the inferred suture zone between the Rae Province to the west and the Nain and Burwell provinces to the east (Fig. 6). If the inferred ocean was of significant width, subduction-related orthogneisses of Early Proterozoic age should occur in the Rae hinterland west of the Tasiuyak gneiss.

BURWELL PROVINCE

Near the northern tip of Labrador, the Tasiuyak gneiss belt hooks to the west, diverging from the Komaktorvik shear zone, which continues northward (Fig. 7). The marked negative magnetic anomaly associated with the Tasiuyak gneiss continues northward across Hudson Strait to southeast Baffin Island (Fig. 6). The Komaktorvik shear zone is interpreted as an extension of the Nagssugtoqidian thrust zone (Korstgard *et al.*, 1988) that bounds the north margin of the Nain (North Atlantic) Province in West Greenland. Both are composed of Archean gneisses that experienced Early Proterozoic metamorphism and ductile thrusting toward the Nain foreland. The Komaktorvik shear zone is marked by large bodies of Archean(?) anorthosite (Taylor, 1979). Between the Tasiuyak gneiss belt and the Komaktorvik shear zone in Labrador is a block of variably trending gneisses named the "Burwell domain" by Korstgard *et al.* (1988). Hoffman (1988a) referred to the same block as the "Burwell terrane," acknowledging that the Tasiuyak gneiss belt and the Komaktorvik shear zone might have evolved from distinct sutures.

Kalsbeek *et al.* (1987, 1988) interpret the Nagssugtoqidian orogen as a ca. 1.85 Ga suture zone between the Nain Province and a small, relatively rigid, Archean block to the north exposed in Disko Bay (latitude 70°N, longitude 51°W). The Rinkian orogen (Grocott and Pulvertaft, this volume) lies north of the Archean Disko block. Here, it is proposed that the Burwell terrane and the Disko block were formerly intregal parts of a single Archean microcontinent, the Burwell Province (Fig. 6).

BAFFIN OROGEN

The Cumberland charnockite–monzogranite batholith (Jackson and Morgan, 1978; Jackson *et al., in press*) underlies a vast area of southern Baffin Island and adjacent west Greenland (Fig. 6). Zircon ages of 1.90–1.85 Ga have been obtained for charnockites and enderbites of the Cumberland batholith (Pidgeon and Howie, 1975; Henderson, 1985a; Jackson *et al., in press*). Concordant monazite ages of 1.85–1.83 Ga, about 20 Ma younger than zircon ages for the same rock (Jackson *et al., in press*), suggest slow cooling from peak metamorphic (emplacement?) temperatures estimated to be 750–800°C at less than 6 kbar (600 MPa) (Cortemiglia *et al.*, 1985). The batholith intrudes and is flanked by three fold belts composed mainly of Early Proterozoic(?) sedimentary rocks that have undergone amphibolite- to granulite-grade metamorphism (Jackson and Taylor, 1972; Jackson and Morgan, 1978). On the southwest, the Lake Harbour Group defines the Dorset fold belt (Davison, 1959; Blackadar, 1967a,b). On the north, the Piling Group defines the Foxe fold belt (Morgan, 1983; Henderson, 1985a,b; Henderson *et al.*, 1989). On the east is an unnamed fold belt that includes the Hoare Bay Group (Jackson and Taylor, 1972) and rocks north of Frobisher Bay that represent a continuation of the Tasiuyak gneiss of the Torngat orogen (Fig. 6). "Baffin orogen" would be a useful and appropriate name incorporating the Cumberland batholith, the three flanking fold belts, and their extensions (as described below) across Hudson Strait, Davis Strait, and Foxe Basin.

A prominent linear magnetic low (Geological Survey of Canada, 1987), corresponding to the Tasuiyak gneiss, trends north across the mouth of Hudson Strait (Fig. 6). It is flanked by positive anomalies corresponding to the granulite orthogneisses that flank the Tasiuyak gneiss in northern Labrador. The negative anomaly strikes toward Resolution Island, south of the mouth of Frobisher Bay. The island is geologically unmapped but a drill core 13 km east of the island bottomed in a "light grey garnetiferous gneiss" (MacLean and Falconer, 1977), probably the Tasiuyak gneiss. Frobisher Bay is the locus of a major northwest-trending sinistral shear zone, indicated by the deflections in trend of the adjacent gneisses (cf. Blackadar, 1967a). The shear zone was also the locus of post-Ordovician normal(?) faulting. The continuation of the Tasiuyak gneiss north of the shear zone may correspond to the paragneiss belt east of the Cumberland batholith between Frobisher Bay and Cumberland Sound (Fig. 6). The paragneiss (Unit 7 of Blackadar (1967a)) is a "light grey weathering, garnetiferous paragneiss, commonly spangled with graphite" and has a "well laminated, granular appearance, accentuated by innumerable rusty streaks." The description is a perfect match for the Tasiuyak gneiss (Unit Afg of Taylor (1979, p. 37–38)), which characterizes the Abloviak shear zone for a strike length of 500 km in Labrador. The tonalitic ortho(?)gneiss (Unit 4 of Blackadar (1967a)) that lies east of the paragneiss belt in southern Baffin Island (Hall Peninsula) may belong to the Burwell Province (Fig. 6).

The affinity of the Hoare Bay Group (Fig. 6) remains uncertain. Most of it is composed of amphibolite- to granulite-grade metagreywacke and pelite, with subordinate mafic metavolcanics (Jackson and Taylor, 1972). Its protolith was like that of the Tasiuyak(?) gneiss on Hall Peninsula and it occupies the same position relative to the charnockite–monzogranite batholith. However, the Hoare Bay Group is less strongly sheared and metamorphosed overall than the Tasiuyak gneiss. These attributes of the Hoare Bay Group, combined with its relatively wide area of preservation attained at the expense of the Archean(?) tonalite orthogneiss to the south, suggest that it may have been deposited in a remnant basin resulting from incomplete suturing of the Burwell and Rae provinces.

The Dorset fold belt (Fig. 6) is a system of gently plunging, northwest-trending folds, developed in amphibolite- to granulite-grade paragneisses (Lake Harbour Group), derived, in decreasing order of abundance, from greywacke, euxinic pelite, marble, quartzite, and mafic–ultramafic sheets (Davison, 1959; Blackadar, 1967a,b; Jackson and Taylor, 1972). The internal stratigraphy of the Lake Harbour Group has not been worked out, but marble and pelite overlie granulite gneisses unconformably(?) on Hudson Strait, south of Lake Harbour (Davison, 1959). The Dorset belt may have originated as an orocline related to clockwise rotation of the southeast arm of the Rae Province in response to collisional indentation by the Superior Province, and was subsequently subjected to sinistral transpression during collisional indentation by the Burwell Province.

The Foxe–Rinkian fold belt (Fig. 6) defines the northwest margin of the Baffin orogen and extends from Melville Peninsula (Penrhyn Group), across Baffin Island (Piling Group), to West Greenland (Karrat Group). The belt widens northeastward and appears to pinch out west of Foxe Basin, although a structural continuation in Archean rocks between the Penrhyn and Amer groups (Fig. 4) cannot be ruled out. The Penrhyn, Piling, and Karrat groups all contain a lower platformal assemblage consisting of quartzite, pelite, and marble in variable proportions, and an upper sequence in which ferruginous pelite, locally enclosing mafic–ultramafic sheets, passes upward into a thick metagreywacke flysch (Jackson and Taylor, 1972; Henderson, 1983, 1985a,b, 1988; Morgan, 1983; Grocott and Pulvertaft, this volume). The passage from platformal to flyschoid sequences signals a major increase in subsidence attributable to extension and/or flexure of the Archean basement. The structural histories of the three exposed segments of the belt differ significantly in detail, but all experienced early stretching, followed by northwest-directed thin-skinned thrusting and northwest–southeast thick-skinned shortening (Henderson, 1984; Henderson *et al.*, 1988, 1989; Grocott and Pulvertaft, this volume). A subhorizontal stretching fabric, strongest west of Foxe Basin, is oriented parallel to the strike of the belt (Henderson, 1981). On Baffin Island, an early shear zone juxtaposed sillimanite-grade rocks of the lower assemblage on its footwall against biotite-grade schists of the flyschoid sequence on its hanging wall (Henderson *et al.*, 1989), suggesting that the early shear zone and associated metamorphism are extensional in origin. In Greenland, the early extension predates the oldest recognized metamorphism (Grocott and Pulvertaft, this volume).

Metasediments similar to the Piling and/or Lake Harbour groups occur as enclaves within the Cumberland batholith (Jackson and Taylor, 1972). The Karrat Group occurs both north and south of the Proven charnockite–leucogranite complex (Grocott and Pulvertaft, this volume), which has a Rb–Sr isochron age of 1.86 ± 0.025 Ga and appears to be an extension of the Cumberland batholith in the Rinkian belt (Fig. 6). On Melville Peninsula, the Penrhyn Group and its basement are intruded by numerous relatively small bodies of aplitic and pegmatitic granodiorite (Henderson, 1983, 1988). To the southwest, the Ford Lake magmatic zone (Figs. 4 and 6) comprises a dozen or more zoned plutons ranging in composition from gabbro to granite (LeCheminant *et al.*, 1987b). The plutons have U–Pb zircon ages of 1.83–1.82 Ga (significantly younger than the Cumberland batholith, but approximately coeval with the De Pas batholith of the New Quebec orogen) and were emplaced during the waning stages of metamorphism and deformation of the Penrhyn Group (Henderson, 1983; LeCheminant *et al.*, 1987b). The Ford Lake plutonic suite has the character and setting of a continental magmatic arc (LeCheminant *et al.*, 1987b), but to which continental margin it might be related is not obvious.

The Baffin orogen is distinguished by an extremely high thermal regime, manifested by extensive charnockitic plutonism and relatively high grades of metamorphism in the marginal fold belts. Temperatures peaked above 750°C at depths of less than 20 km. The thermal regime may be a cumulative effect of a rapid sequence of events involving (1) lithospheric stretching between the southeast and northeast arms of the Rae Province during collisional indentation by the Superior Province, (2) northwest-dipping subduction at the Tasiuyak suture and consequent arc magmatism (Cumberland batholith) accommodating convergence of the Burwell and/or Nain provinces, and (3) lithospheric shortening and possible delamination (cf. Dewey, 1988; England and Houseman, 1988; Sandiford, 1989) during collision of the Burwell and/or Nain provinces with the Rae Province.

DISCUSSION

A common view of the Trans-Hudson Orogen is that it extends northeastward from its type area around the Hudson Bay basin and continues across southern Baffin Island to central Greenland (Lewry *et al.*, 1985; Van Schmus *et al.*, 1987; Lewry and Collerson, this volume). In this view, an arm of the orogen extends southward between the Superior and Nain provinces. Although this view is valid as demarcating a zone of intense "Hudsonian" metamorphism and deformation, it is questionable with respect to implying a continuous geosuture. It is as the term "Alpine" (i.e., referring to a system of Tertiary mountain belts extending from Gibraltar to New Zealand) is to the Alps (i.e., a particular collision zone in south-central Europe).

The approach adopted in this paper is to attempt to identify and delineate particular orogens, spatially related to specific sutures and temporally related to particular collision events. Accordingly, the Trans-Hudson Orogen can be inferred to extend only as far to the northeast as Hudson Strait, where it possibly merges with the Snowbird suture. If the Dorset fold belt is continuous with the fold belt involving the Lake Harbour Group west of the Abloviak shear zone in Labrador (Jackson and Taylor, 1972), then the suture associated with the Trans-Hudson Orogen cannot extend across Baffin Island to Greenland. This conclusion is supported by the continuation of the Tasiuyak gneiss (a suture zone linking the Torngat and Baffin orogens) across the mouth of Hudson Strait (Geological Survey of Canada, 1987).

The suture zone represented by the Trans-Hudson Orogen may be linked with that of the New Quebec orogen, conforming with the venerable concept of a Circum-Ungava orogen (Wilson, 1968; Dimroth *et al.*, 1970). However, the orogens should be named separately (like the Appalachians, the Ouachitas, and the Cordillera) because there is structural and geochronological evidence that orogenic events in the Cape Smith belt and Labrador trough were not strictly contemporaneous. Thrusting began before 1.88 Ga in the Trans-Hudson Orogen (Cape Smith belt, Parrish (1989); Thompson belt, Bleeker (this volume)), but after 1.88 Ga in the New Quebec orogen (Machado *et al.*, 1989). The late (D3) cross folds in the Cape Smith belt are subparallel with the (D1) thrusts and (D2) basement folds of the Labrador Trough, consistent with the inference that collision began on the northwest margin of the Superior Province and followed on the northeast margin.

CONCLUSIONS

A new subdivision of the Churchill Province is advanced. It follows from the proposition that the province is a composite structure comprised of the Trans-Hudson Orogen — a 1.9–1.8 Ga collision zone characterized by accreted and/or obducted oceanic crustal relics — and a composite Archean hinterland. The Churchill hinterland is distinguished from other Archean provinces by 2.0–1.8 Ga modifications attributed to inward-dipping subduction zones and collisional indentation by first the Slave, then the Superior, and finally(?) the Nain province. The respective collision zones bounding the hinterland are the Thelon orogen (1.97–1.91 Ga), the Trans-Hudson and New Quebec orogens (1.88–1.79 Ga), and the Torngat and Nagssugtoqidian orogens.

The Churchill hinterland is provisionally divided into the Burwell, Rae, and Hearne provinces. The former two are separated by the Baffin orogen; the latter two by the Snowbird line. Evolutionary models are suggested in which the Snowbird and Baffin orogens originated (or were reactivated) as wedge-shaped basins formed as a result of lateral "extrusion" during collisional indentation of the Churchill hinterland by the Superior Province (cf. Tapponnier *et al.*, 1982, 1986). Indentation by the Ungava promontory of the Superior Province caused the southeast arm of the Rae Province to rotate clockwise, opening the ancestral Baffin basin in its wake, until it collided obliquely with the northeast margin of the Superior Province, forming the New Quebec orogen. Subsequent compression of the Baffin orogen may have been caused by the arrival of the Burwell and Nain provinces from the southeast. Thus, the geosuture corresponding to the Trans-Hudson Orogen is linked to the New Quebec orogen and does not extend across southern Baffin Island to West Greenland.

ACKNOWLEDGEMENTS

This paper is an outgrowth of research for the Decade of North American Geology volume on the Precambrian craton of Canada and Greenland. It stems from the need for a suitable nomenclature for Early Proterozoic orogenic belts to be described in that volume. The following are thanked for discussions and/or preprints: Larry Aspler, Normand Bégin, Pat Bickford, Wouter Bleeker, Hewitt Bostock, Sam Bowring, Fred Chandler, Jeff Chiarenzelli, Tom Clark, Ken Collerson, Ron Doig, Ken Eade, Al Fraser, Tom Frisch, Terry Gordon, Normand Goulet, Simon Hanmer, Ernst Hegner, Jack Henderson, Garth Jackson, Tom Krogh, Tony LeCheminant, John Lewry, Dale Loveridge, Steve Lucas, Bob Macdonald, Nuno Machado, Jim Mortensen, Randy Parrish, John Peirce, Serge Perreault, Chris Roddick, Gerry Ross, Dave Scott, Marc St-Onge, Subhas Tella, Reg Thériault, Otto van Breemen, John van der Leeden, Randy Van Schmus, and Dick Wardle. I am particularly grateful to Tom Clark for preprints of his compilation maps of the Labrador trough. Ken Card and Hans Hofmann dissuaded me from proposing the name Keewatin province on the grounds that it was preempted by Lawson's "Keewatin series." The original manuscript was critically read by Fred Taylor and Charlie Gower.

REFERENCES

Allegre, C.J. *et al.*, 1984, Structure and evolution of the Himalaya–Tibet orogenic belt: Nature, v. 307, p. 17–22.

Annesley, I.R., 1981, Field characteristics, petrology, and geochemistry of the Amer Lake ultramafic metavolcanics, District of Keewatin: Geological Survey of Canada, Paper 81-1A, p. 275–279.

Arndt, N.T., Brügmann, G.E., Lehnert, K., Chauvel, C. and Chappell, B.W., 1987, Geochemistry, petrogenesis and tectonic environment of Circum-Superior belt basalts, Canada, in Pharoah, T.C., Beckinsale, R.D. and Rickard, D., eds., Geochemistry and Mineralization of Proterozoic Volcanic Suites: Blackwell, Oxford, p. 133–145.

Ashton, K.E., 1982, Further geological studies of the "Woodburn Lake Group" northwest of Tehek Lake, District of Keewatin: Geological Survey of Canada, Paper 82-1A, p. 151–157.

Ashton, K.E., 1985, Archean orthoquartzites from the Churchill Structural Province near Baker Lake, N.W.T.: Geological Association of Canada – Mineralogical Association of Canada, Program with Abstracts, v. 10, p. A2.

Ashwal, L.D., Wooden, J.L. and Emslie, R.F., 1986, Sr, Nd, and Pb isotopes in Proterozoic intrusives astride the Grenville Front in Labrador: implications for crustal contamination and basement mapping: Geochimica et Cosmochimica Acta, v. 50, p. 2571–2585.

Aspler, L.B. and Donaldson, J.A., 1985, The Nonacho basin (Early Proterozoic), Northwest Territories, Canada: sedimentation and deformation in a strike-slip setting, in Biddle, K.T. and Christie-Blick, N., eds., Strike-slip Deformation, Basin Formation, and Sedimentation: Society of Economic Paleontologists and Mineralogists, Special Publication 37, p. 193–209.

Aspler, L.B., Bursey, T.L. and Miller, A.R., 1989, Sedimentology, structure, and economic geology of the Poorfish–Windy thrust–fold belt, Ennadai Lake area, District of Keewatin, and the shelf to foredeep transition in the foreland of Trans-Hudson Orogen: Geological Survey of Canada, Paper 89-1C, p. 143–155.

Bailes, A.H., 1971, Preliminary compilation of the geology of the Snow Lake – Flin Flon – Sherridon area: Manitoba Mines Branch, Geological Paper 1/71, 27 p.

Baldwin, D.A., Syme, E.C., Zwanzig, H.V., Gordon, T.M., Hunt, P.A. and Stevens, R.D., 1987, U–Pb zircon ages from the Lynn Lake and Rusty Lake metavolcanic belts, Manitoba: two ages of Proterozoic magmatism: Canadian Journal of Earth Sciences, v. 24, p. 1053–1063.

Baragar, W.R.A., 1984, Pillow formation and layered flows in the Circum-Superior belt of eastern Hudson Bay: Canadian Journal of Earth Sciences, v. 21, p. 781–792.

Baragar, W.R.A. and Scoates, R.F.J., 1981, The Circum-Superior belt: a Proterozoic plate margin?, in Kröner, A., ed., Precambrian Plate Tectonics: Elsevier, Amsterdam, p. 297–330.

Baragar, W.R.A. and Scoates, R.F.J., 1987, Volcanic geochemistry of the northern segments of the Circum-Superior belt of the Canadian Shield, in Pharoah, T.C., Beckinsale, R.D. and Rickard, D., eds., Geochemistry and Mineralization of Proterozoic Volcanic Suites: Blackwell, Oxford, p. 113–131.

Bell, R.T., 1970, The Hurwitz Group, a prototype for deposition on metastable cratons, in Baer, A.J., ed., Symposium on Basins and Geosynclines of the Canadian Shield: Geological Survey of Canada, Paper 70-40, p. 159–169.

Blackadar, R.G., 1967a, Geological reconnaissance, southern Baffin Island, District of Franklin: Geological Survey of Canada, Paper 66-47, 32 p. (with three maps, scale 1:506,880).

Blackadar, R.G., 1967b, Geology of Mingo Lake – Macdonald Lake map-area, Baffin Island, District of Franklin: Geological Survey of Canada, Memoir 345, 54 p. (with Map 1185A, scale 1:253,440).

Blake, D.H., 1980, Volcanic rocks of the Paleohelikian Dubawnt Group in the Baker Lake – Angikuni Lake area, District of Keewatin, N.W.T.: Geological Survey of Canada, Bulletin 309, 39 p.

Bleeker, W., 1990, New structural–metamorphic constraints on Early Proterozoic oblique collision along the Thompson Nickel Belt, Manitoba, Canada, in Lewry, J.F. and Stauffer, M.R., eds., The Early Proterozoic Trans-Hudson Orogen of North America: Geological Association of Canada, Special Paper 37, p. 57–73.

Bond, G.C. and Kominz, M.A., 1984, Construction of tectonic subsidence curves for the early Paleozoic miogeocline, southern Canadian Rocky Mountains: implications for subsidence mechanisms, age of breakup, and crustal thinning: Geological Society of America Bulletin, v. 95, p. 155–173.

Boone, E. and Hynes, A., 1990, A structural cross-section of the Northern Labrador Trough, New Quebec, in Lewry, J.F. and Stauffer, M.R., eds., The Early Proterozoic Trans-Hudson Orogen of North America: Geological Association of Canada, Special Paper 37, p. 387–396.

Bostock, H.H., 1971, Geological notes on Aquatuk River map-area, Ontario: Geological Survey of Canada, Paper 70-42, 57 p. (with map, scale 1:500,000).

Bostock, H.H. and Loveridge, W.D., 1988, Geochronology of the Taltson magmatic zone and its eastern cratonic margin, District of Mackenzie, in Radiogenic Age and Isotopic Studies. Report 2: Geological Survey of Canada, Paper 88-2, p. 59–65.

Bostock, H.H., van Breemen, O. and Loveridge, W.D., 1987, Proterozoic geochronology in the Taltson magmatic zone, N.W.T., in Radiogenic Age and Isotopic Studies. Report 1: Geological Survey of Canada, Paper 87-2, p. 73–80.

Bowring, S.A. and Grotzinger, J.P., 1989, Implications of new U–Pb dating and stratigraphic correlations for current tectonic models for Wopmay orogen and Thelon tectonic zone: Geological Association of Canada – Mineralogical Association of Canada, Program with Abstracts, v. 14, p. A74.

Bowring, S.A. and Podosek, F.A., 1989, Nd isotopic evidence from Wopmay orogen for 2.0–2.4 Ga crust in western North America: Earth and Planetary Science Letters, v. 94, p. 217–230.

Bowring, S.A., Van Schmus, W.R. and Hoffman, P.F., 1984, U–Pb zircon ages from Athapuscow aulacogen, east arm of Great Slave Lake, N.W.T., Canada: Canadian Journal of Earth Sciences, v. 21, p. 1315–1324.

Camfield, P.A. and Gough, D.I., 1977, A possible Proterozoic plate boundary in North America: Canadian Journal of Earth Sciences, v. 14, p. 1229–1238.

Carolan, J., Collerson, K.D., Grimisch, G. and Lewry, J.F., 1989, Kinematic indicators in the Virgin River shear zone and implications for tectonic juxtaposition of the Rae and Hearne cratons in Saskatchewan: Geological Association of Canada – Mineralogical Association of Canada, Program with Abstracts, v. 14, p. A9.

Cavell, P.A. and Baadsgaard, H., 1986, Geochronology of the Big Spruce Lake alkaline intrusion: Canadian Journal of Earth Sciences, v. 23, p. 1–10.

Cavell, P.A., Wijbrans, J.R. and Baadsgaard, H., 1987, Preliminary results of an isotopic study of the Kaminak Lake Archean terrane (N.W.T.): Geological Association of Canada – Mineralogical Association of Canada, Program with Abstracts, v. 12, p. 29.

Chandler, F.W., 1984, Metallogenesis of an early Proterozoic foreland sequence, eastern Hudson Bay, Canada: Journal of the Geological Society of London, v. 141, p. 299–313.

Chandler, F.W., 1988, Geology of the Early Proterozoic Richmond Gulf graben, east coast of Hudson Bay, Canada: Geological Survey of Canada, Bulletin 362, 76 p.

Chandler, F.W. and Parrish, R.R., 1989, Age of the Richmond Gulf Group and implications for rifting in the Trans-Hudson Orogen, Canada: Precambrian Research, v. 44, p. 277–288.

Chauvel, C., Arndt, N.T., Kielinczuk, S. and Thom, A., 1987, Formation of Canadian 1.9 Ga old continental crust. I. Nd isotopic data: Canadian Journal of Earth Sciences, v. 24, p. 396–406.

Chevé, S.R. and Machado, N., 1988, Reinvestigation of the Castignon Lake carbonatite complex, Labrador trough, New Quebec: Geological Association of Canada – Mineralogical Association of Canada, Program with Abstracts, v. 13, p. 20.

Chiarenzelli, J.R. and Macdonald, R., 1986, A U–Pb zircon date for the Ennadai Group: Saskatchewan Geological Survey, Miscellaneous Report 86-4, p. 112–113.

Christie, K.W., Davidson, A. and Fahrig, W.F., 1975, The paleomagnetism of Kaminak Dikes: no evidence of significant Hudsonian plate motion: Canadian Journal of Earth Sciences, v. 12, p. 2048–2064.

Clark, T., 1984, Géologie de la région du Lac Cambrien, Territoire du Nouveau-Québec: Ministère de l'Énergie et des Ressources, ET 83-02, 71 p. (with map, scale 1:50,000).

Clark, T., 1988, Stratigraphie, pétrographie et pétrochimie de la formation de fer de Baby dans la région du Lac Hérodier (Fosse du Labrador): Ministère de l'Énergie et des Ressources, ET 87-13, 36 p.

Coles, R.L. and Haines, G.V., 1982, Regional patterns in magnetic anomalies over Hudson Bay: Canadian Journal of Earth Sciences, v. 19, p. 1116–1121.

Collerson, K.D., Van Schmus, W.R., Bickford, M.E. and Lewry, J.F., 1989, The buried basement in central and southern Saskatchewan: evidence from Sm–Nd model ages and U–Pb zircon geochronology: Geological Association of Canada – Mineralogical Association of Canada, Program with Abstracts, v. 14, p. A74.

Coney, P.J., Jones, D.L. and Monger, J.W.H., 1980, Cordilleran suspect terranes: Nature, v. 288, p. 329–333.

Cortemiglia, G.C., Messiga, B. and Terranova, R., 1985, Caratteri geologici e petrografici dell'area del Summit Lake nell'-Isola di Baffin (Arcipelago Artico Canadese): Bolletina Societa Geologica Italiana, v. 104, p. 123–142.

Cumming, G.L., Krstic, D. and Wilson, J.A., 1987, Age of the Athabasca Group, northern Alberta: Geological Association of Canada – Mineralogical Association of Canada, Program with Abstracts, v. 12, p. 35.

Davidson, A., 1970a, Precambrian geology, Kaminak Lake maparea, District of Keewatin: Geological Survey of Canada, Paper 69-51, 27 p. (with Map 1285A, scale 1:250,000).

Davidson, A., 1970b, Kaminak Lake alkalic complex, District of Keewatin (55L): Geological Survey of Canada, Paper 70-1A, p. 135–137.

Davidson, A., 1972, The Churchill Province, _in_ Price, R.A. and Douglas, R.J.W., eds., Variations in Tectonic Styles in Canada: Geological Association of Canada, Special Paper 11, p. 381–433.

Davidson, A., 1978, The Blachford Lake intrusive suite: an Aphebian alkaline plutonic complex in the Slave Province, Northwest Territories: Geological Survey of Canada, Paper 78-1A, p. 119–127.

Davison, W.L., 1959, Geology, Lake Harbour, Baffin Island, District of Franklin, Northwest Territories: Geological Survey of Canada, Map 29-1958, scale 1:63,360.

Dewey, J.F., 1988, Extensional collapse of orogens: Tectonics, v. 7, p. 1123–1139.

Dewey, J.F. and Burke, K.C.A., 1973, Tibetan, Variscan, and Precambrian basement reactivation: products of continental collision: Journal of Geology, v. 81, p. 683–692.

Dewey, J.F., Shackleton, R.M., Chang Chengfa and Sun Yiyin, 1988, The tectonic evolution of the Tibetan Plateau: Philosophical Transactions of the Royal Society of London, Series A, v. 327, p. 379–413.

Dimroth, E., Baragar, W.R.A., Bergeron, R. and Jackson, G.D., 1970, The filling of the Circum-Ungava geocyncline, _in_ Baer, A.J., ed., Symposium on Basins and Geosynclines of the Canadian Shield: Geological Survey of Canada, Paper 70-40, p. 45–142.

Doig, R., 1987, Rb–Sr geochronology and metamorphic history of Proterozoic to early Archean rocks north of the Cape Smith Belt, Quebec: Canadian Journal of Earth Sciences, v. 24, p. 813–825.

Eade, K.E., 1971, Geology of Ennadai Lake map-area, District of Keewatin: Geological Survey of Canada, Paper 70-45, 19 p. (with Map 24-1970, scale 1:250,000).

Eade, K.E., 1973, Geology of Nuelton Lake and Edehon Lake (west half) map-areas, District of Keewatin: Geological Survey of Canada, Paper 72-21, 29 p. (with Maps 3/4-1972, scale 1:250,000).

Eade, K.E., 1974, Geology of Kognak River area, District of Keewatin, Northwest Territories: Geological Survey of Canada, Memoir 377, 66 p. (with Map 1364A, scale 1:250,000).

Eade, K.E., 1978, Notes on metamorphism in southern District of Keewatin, _in_ Fraser, J.A. and Heywood, W.W., eds., Metamorphism in the Canadian Shield: Geological Survey of Canada, Paper 78-10, p. 191–194.

Eade, K.E., 1986, Precambrian geology of the Tulemalu Lake – Yathkyed Lake area, District of Keewatin: Geological Survey of Canada, Paper 84-11, 31 p. (with Map 1604A, scale 1:250,000).

Eade, K.E. and Chandler, F.W., 1975, Geology of Watterson Lake (west half) map-area, District of Keewatin: Geological Survey of Canada, Paper 74-64, 10 p. (with Map 7-1974, scale 1:250,000).

Emslie, R.F., 1963, Michikamau Lake, east half, Quebec–Newfoundland: Geological Survey of Canada, Paper 63-20, 4 p. (with Map 31-1963, scale 1:253,440).

Emslie, R.F., 1964, Geology, Kasheshibaw Lake, west half, Newfoundland–Quebec: Geological Survey of Canada, Map 3-1964, scale 1:253,440 (with marginal notes).

England, P.C. and Houseman, G.A., 1988, The mechanics of the Tibetan Plateau: Philosophical Transactions of the Royal Society of London, Series A, v. 326, p. 301–320.

Ermanovics, I.F., Van Kranendonk, M., Corriveau, L., Mengel, F., Bridgwater, D. and Sherlock, R., 1989, The boundary zone of the Nain–Churchill provinces in the North River – Nutak map areas, Labrador: Geological Survey of Canada, Paper 89-1C, p. 385–394.

Fraser, J.A., 1978, Metamorphism in the Churchill Province, District of Mackenzie, _in_ Fraser, J.A. and Heywood, W.W., eds., Metamorphism in the Canadian Shield: Geological Survey of Canada, Paper 78-10, p. 195–202.

Frisch, T., 1982, Precambrian geology of the Prince Albert Hills, western Melville Peninsula, Northwest Territories: Geological Survey of Canada, Bulletin 346, 70 p. (with map, scale 1:250,000).

Frisch, T. and Hunt, P.A., 1988, U–Pb zircon and monazite ages from the Precambrian shield of Ellesmere and Devon islands, Arctic Archipelago, _in_ Radiogenic Age and Isotopic Studies. Report 2: Geological Survey of Canada, Paper 88-2, p. 117–125.

Frisch, T. and Patterson, J.G., 1983, Preliminary account of the geology of the Montresor River area, District of Keewatin: Geological Survey of Canada, Paper 83-1A, p. 103–108.

Frisch, T., Annesley, I.R. and Gittins, C.A., 1985, Geology of the Chantrey belt and its environs, lower Hayes River and Darby Lake map-areas, northern District of Keewatin: Geological Survey of Canada, Paper 85-1A, p. 259–266.

Frost, C.D. and Burwash, R.A., 1986, Nd evidence for extensive Archean basement in the western Churchill Province, Canada: Canadian Journal of Earth Sciences, v. 23, p. 1433–1437.

Fumerton, S.L., Stauffer, M.L. and Lewry, J.F., 1984, The Wathaman batholith: largest known Precambrian pluton: Canadian Journal of Earth Sciences, v. 21, p. 1082–1097.

Gaskarth, J.W. and Parslow, G.R., 1987, Proterozoic volcanism in the Flin-Flon greenstone belt, east-central Saskatchewan, Canada, _in_ Pharoah, T.C., Beckinsale, R.D. and Rickard, D., eds., Geochemistry and Mineralization of Proterozoic Volcanic Suites: Blackwell, Oxford, p. 183–200.

Geological Survey of Canada, 1987, Magnetic anomaly map of Canada, fifth edition: Geological Survey of Canada, Map 1255A, scale 1:5,000,000.

Gibb, R.A., 1978, Slave–Churchill collision tectonics: Nature, v. 271, p. 50–52.

Gibb, R.A., 1983, Model for suturing of Superior and Churchill plates: an example of double indentation tectonics: Geology, v. 11, p. 413–417.

Gibb, R.A. and Halliday, D.W., 1974, Gravity measurements in southern District of Keewatin and southeastern District of Mackenzie, with maps: Earth Physics Branch, Gravity Map Series 124-131, Ottawa, 36 p.

Gibb, R.A. and Thomas, M.D., 1977, The Thelon Front: a cryptic suture in the Canadian Shield?: Tectonophysics, v. 38, p. 211–222.

Gibb, R.A. and Walcott, R.J., 1971, A Precambrian suture in the Canadian Shield: Earth and Planetary Science Letters, v. 10, p. 417–422.

Gibb, R.A., Thomas, M.D., Kearey, P. and Tanner, J.G., 1978, Gravity anomalies at structural province boundaries of the Canadian Shield: products of Proterozoic plate interaction: GeoSkrifter, v. 10, p. 21–57.

Gibb, R.A., Thomas, M.D., Lapointe, P.L. and Mukhopadyay, M., 1983, Geophysics of proposed Proterozoic sutures in Canada: Precambrian Research, v. 19, p. 349–384.

Gilboy, C.F., 1980, Bedrock compilation geology: Stony Rapids area: Saskatchewan Geological Survey, Preliminary Geological Map, scale 1:250,000.

Gordon, T.M., 1988, Precambrian geology of the Daly Bay area, District of Keewatin: Geological Survey of Canada, Memoir 422, 21 p.

Gordon, T.M., Hunt, P.A., Loveridge, W.D., Bailes, A.H. and Syme, E.C., 1987, U–Pb zircon ages from the Flin Flon and Kisseynew belts, Manitoba: chronology of Early Proterozoic crust formation: Geological Association of Canada – Mineralogical Association of Canada, Program with Abstracts, v. 12, p. 48.

Green, A.G., Weber, W. and Hajnal, Z., 1985a, Evolution of Proterozoic terrains beneath the Williston Basin: Geology, v. 13, p. 624–628.

Green, A.G., Hajnal, Z. and Weber, W., 1985b, An evolutionary model of the western Churchill Province and western margin of the Superior Province in Canada and the north-central United States: Tectonophysics, v. 116, p. 281–322.

Greene, M.T., 1982, Geology in the nineteenth century: changing views of a changing world: Cornell University Press, Ithaca, NY, 324 p.

Grocott, J. and Pulvertaft, T.C.R., 1990, The Early Proterozoic Rinkian belt of central West Greenland, in Lewry J.F. and Stauffer, M.R., eds., The Early Proterozoic Trans-Hudson Orogen of North America: Geological Association of Canada, Special Paper 37, p. 443–463.

Grotzinger, J.P. and McCormick, D.S., 1988, Flexure of the early Proterozoic lithosphere and the evolution of Kilohigok basin (1.9 Ga), northwest Canadian Shield, in Kleinspehn, K. and Paola, C., eds., New Perspectives in Basin Analysis: Springer-Verlag, Heidelberg, p. 405–430.

Handa, S. and Camfield, P.A., 1984, Crustal electrical conductivity in north-central Saskatchewan: the North American Central Plains anomaly and its relation to a Proterozoic plate margin: Canadian Journal of Earth Sciences, v. 21, p. 533–543.

Hanmer, S.K., 1987, Granulite facies mylonites: a brief structural reconnaissance north of Stony Rapids, northern Saskatchewan: Geological Survey of Canada, Paper 87-1A, p. 563–572.

Hanmer, S.K., 1988, Great Slave Lake shear zone, Canadian Shield: reconstructed vertical profile of a crustal-scale fault zone: Tectonophysics, v. 149, p. 245–264.

Hegner, E. and Hulbert, L., 1987, Sm–Nd ages and origin of mafic–ultramafic intrusions from the Proterozoic Lynn Lake domain, Manitoba, Canada: Geological Association of Canada – Mineralogical Association of Canada, Programs with Abstracts, v. 12, p. 54.

Henderson, J.B., Thompson, P.H. and James, D.T., 1982, The Healey Lake map area and the Thelon front problem, District of Mackenzie: Geological Survey of Canada, Paper 82-1A, p. 191–195.

Henderson, J.R., 1981, Structural analysis of sheath folds with horizontal X-axes, northeast Canada: Journal of Structural Geology, v. 3, p. 203–210.

Henderson, J.R., 1983, Structure and metamorphism of the Aphebian Penrhyn Group and its Archean basement complex in the Lyon Inlet area, Melville Peninsula, District of Franklin: Geological Survey of Canada, Bulletin 324, 50 p. (with map, scale 1:100,000).

Henderson, J.R., 1984, Description of a virgation in the Foxe Fold Belt, Melville Peninsula, Canada, in Kröner, A. and Greiling, R., eds., Precambrian Tectonics Illustrated: Schweizerbart'sche, Stuttgart, p. 251–261.

Henderson, J.R., 1985a, Geology, Ekalugad Fiord – Home Bay, District of Franklin: Geological Survey of Canada, Map 1606A, scale 1:250,000.

Henderson, J.R., 1985b, Geology, McBeth Fiord – Cape Henry Kater, District of Franklin: Geological Survey of Canada, Map 1605A, scale 1:250,000.

Henderson, J.R., 1988, Geology, southeastern Melville Peninsula, District of Franklin, Northwest Territories: Geological Survey of Canada, Map 1655A, scale 1:100,000.

Henderson, J.R., Grocott, J., Henderson, M.N., Falardeau, F. and Heijke, P., 1988, Results of fieldwork in Foxe Fold Belt near Dewar Lakes, Baffin Island, N.W.T.: Geological Survey of Canada, Paper 88-1C, p. 101–108.

Henderson, J.R., Grocott, J., Henderson, M.N. and Perreault, S., 1989, Tectonic history of the Lower Proterozoic Foxe–Rinkian belt in central Baffin Island, N.W.T.: Geological Survey of Canada, Paper 89-1C, p. 185–197.

Heywood, W.W., 1967, Geological notes, northeastern District of Keewatin and southern Melville Peninsula, District of Franklin, Northwest Territories: Geological Survey of Canada, Paper 66-40, 20 p. (with Map 14-1966, scale 1:506,880).

Heywood, W.W., 1973, Geology of Tavani map-area, District of Mackenzie: Geological Survey of Canada, Paper 72-47, 14 p. (with Map 6-1972, scale 1:250,000).

Heywood, W.W. and Sanford, B.V., 1976, Geology of Southampton, Coats, and Mansel islands, District of Keewatin, Northwest Territories: Geological Survey of Canada, Memoir 382, 35 p. (with Map 1404A, scale 1:250,000).

Heywood, W.W. and Schau, M., 1978, A subdivision of the northern Churchill Structural Province: Geological Survey of Canada, Paper 78-1A, p. 139–143.

Hoffman, P.F., 1981, Autopsy of Athapuscow Aulacogen: a failed arm affected by three collisions, in Campbell, F.H.A., ed., Proterozoic Basins of Canada: Geological Survey of Canada, Paper 81-10, p. 97–101.

Hoffman, P.F., 1985, Is the Cape Smith belt (northern Quebec) a klippe?: Canadian Journal of Earth Sciences, v. 22, p. 1361–1369.

Hoffman, P.F., 1987a, Continental transform tectonics: Great Slave Lake shear zone (1.9 Ga), northwest Canada: Geology, v. 15, p. 785–788.

Hoffman, P.F., 1987b, Early Proterozoic foredeeps, foredeep magmatism, and Superior-type iron-formations of the Canadian Shield, in Kröner, A., ed., Proterozoic Lithospheric Evolution: American Geophysical Union, Geodynamics Series, v. 17, p. 85–98.

Hoffman, P.F., 1988a, United plates of America, the birth of a craton: Early Proterozoic assembly and growth of Laurentia: Annual Review of Earth and Planetary Sciences, v. 16, p. 543–603.

Hoffman, P.F., 1988b, Geology and tectonics, east arm of Great Slave Lake, Northwest Territories: Geological Survey of Canada, Map 1628A, scale 1:250,000.

Hoffman, P.F., 1989, Precambrian geology and tectonic history of North America, in Bally, A.W. and Palmer, A.R., eds., The Geology of North America – An Overview: Geological Society of America, The Geology of North America, v. A, p. 447–512.

Hoffman, P.F. and Grotzinger, J.P., 1988, Abner/Denault reef complex (2.1 Ga), Labrador trough, N.E. Quebec, in Geldsetzer, H.H.J., James, N.P. and Tebbutt, G.E., eds., Reefs: Canada and Adjacent Areas: Canadian Society of Petroleum Geologists, Memoir 13, p. 49–54.

Hoffman, P.F., Tirrul, R., King, J.E., St-Onge, M.R. and Lucas, S.B., 1988, Axial projections and modes of crustal thickening, eastern Wopmay orogen, northwest Canadian Shield, in Clark, S.P., Jr., Burchfiel, B.C. and Suppe, J., eds., Processes in Continental Lithospheric Deformation: Geological Society of America, Special Paper 218, p. 1–19.

Housh, T., Bowring, S.A. and Villeneuve, M., 1989, Lead isotopic study of arc magmatism within Early Proterozoic Wopmay orogen, NW Canada: importance of continental crust: Journal of Geology, v. 97, p. 735–747.

Jackson, G.D., 1960, Belcher Islands, Northwest Territories: Geological Survey of Canada, Paper 60-20, 13 p. (with map, scale 1:126,720).

Jackson, G.D. and Morgan, W.C., 1978, Precambrian metamorphism on Baffin and Bylot islands, in Fraser, J.A. and Heywood, W.W., eds., Metamorphism in the Canadian Shield: Geological Survey of Canada, Paper 78-10, p. 249–267.

Jackson, G.D. and Taylor, F.C., 1972, Correlation of major Aphebian rock units in the northeastern Canadian Shield: Canadian Journal of Earth Sciences, v. 9, p. 1650–1669.

Jackson, G.D., Frisch, T., Hegner, E. and Hunt, P.A., 1990, U–Pb geochronology, Nd model ages and tectonics of the eastern Arctic Canadian Shield: Geological Survey of Canada, Current Activities Forum 1990, Program with Abstracts, p. 12.

Jackson, G.D., Hunt, P.A., Loveridge, W.D. and Parrish, R.R., *in press*, Geochronology of Baffin Island, in Radiogenic Age and Isotopic Studies. Report 3: Geological Survey of Canada, Paper 89-2.

James, D.T., van Breemen, O. and Loveridge, W.D., 1988, Early Proterozoic U–Pb zircon ages for granitoid rocks from the Moraine Lake transect, Thelon tectonic zone, Artillery Lake area, NWT, in Radiogenic Age and Isotopic Studies. Report 2: Geological Survey of Canada, Paper 88-2, p. 67–72.

Johnson, B.J., *in press*, Stratigraphy and structure of the Early Proterozoic Wilson Island Group, East Arm thrust–fold belt, N.W.T.: Canadian Journal of Earth Sciences.

Jones, A.G. and Savage, P.J., 1986, North American Central Plains conductivity anomaly goes east: Geophysical Research Letters, v. 13, p. 685–688.

Kalsbeek, F., Pidgeon, R.T. and Taylor, P.N., 1987, Nagssugtoqidian mobile belt of West Greenland: a cryptic 1850 Ma suture between two Archaean continents – chemical and isotopic evidence: Earth and Planetary Science Letters, v. 85, p. 365–385.

Kalsbeek, F., Taylor, P.N. and Pidgeon, R.T., 1988, Unreworked Archean basement and Proterozoic supracrustal rocks from northeastern Disko Bugt, West Greenland: implications for the nature of Proterozoic mobile belts in Greenland: Canadian Journal of Earth Sciences, v. 25, p. 773–782.

Kanasewich, E.R., 1968, Precambrian rift: genesis of strata-bound ore deposits: Science, v. 161, p. 1002–1005.

Kearey, P., 1976, A regional structural model of the Labrador Trough, northern Quebec, from gravity studies, and its relevance to continental collision in the Precambrian: Earth and Planetary Science Letters, v. 28, p. 371–378.

Klasner, J.S. and King, E.R., 1986, Precambrian basement geology of North and South Dakota: Canadian Journal of Earth Sciences, v. 23, p. 1083–1102.

Knight, I. and Morgan, W.C., 1981, The Aphebian Ramah Group, northern Labrador, in Campbell, F.H.A., ed., Proterozoic Basins of Canada: Geological Survey of Canada, Paper 81-10, p. 313–330.

Korstgard, J., Ryan, B. and Wardle, R.J., 1988, The boundary between Proterozoic and Archaean crustal blocks in central West Greenland and northern Labrador, in Park, R.G. and Tarney, J., eds., Evolution of the Lewisian and Comparable High-grade Terrains: Blackwell, Oxford, p. 247–259.

Krogh, T.E. and Clark, L.A., 1987, Zircon dating of sub-Athabasca granitoid rocks, Saskatchewan: Geological Association of Canada – Mineralogical Association of Canada, Program with Abstracts, v. 12, p. 64.

LeCheminant, A.N., Ianelli, T.R., Zaitlin, B. and Miller, A.R., 1981, Geology of Tebesjuak Lake map area, District of Keewatin: a progress report: Geological Survey of Canada, Paper 81-1B, p. 113–128.

LeCheminant, A.N., Jackson, M.J., Galley, A.G., Smith, S.L. and Donaldson, J.A., 1984, Early Proterozoic Amer Group, Beverly map-area, District of Keewatin: Geological Survey of Canada, Paper 84-1B, p. 159–172.

LeCheminant, A.N., Roddick, J.C. and Henderson, J.R., 1987a, Geochronology of Archean and Early Proterozoic magmatism in the Baker Lake – Wager Bay region, N.W.T.: Geological Association of Canada – Mineralogical Association of Canada, Program with Abstracts, v. 12, p. 66.

LeCheminant, A.N., Roddick, J.C., Tessier, A.C. and Bethune, K.M., 1987b, Geology and U–Pb ages of early Proterozoic calc-alkaline plutons northwest of Wager Bay, District of Keewatin: Geological Survey of Canada, Paper 87-1A, p. 773–782.

LeCheminant, A.N., Miller, A.R. and LeCheminant, G.M., 1988, Early Proterozoic alkaline igneous rocks, District of Keewatin, Canada: petrogenesis and mineralization, in Pharoah, T.C., Beckinsale, R.D. and Rickard, D.T., eds., Geochemistry and Mineralization of Proterozoic Volcanic Suites: Blackwell, Oxford, p. 219–240.

LeGallais, C.J. and Lavoie, S., 1982, Basin evolution of the Lower Proterozoic Kaniapiskau Supergroup, central Labrador miogeocline (Trough), Quebec: Bulletin of Canadian Petroleum Geology, v. 30, p. 150–166.

Lewry, J.F., 1981, Lower Proterozoic arc–microcontinent collisional tectonics in the western Churchill Province: Nature, v. 294, p. 69–72.

Lewry, J.F. and Collerson, K.D., 1990, The Trans-Hudson Orogen: extent, subdivision, and problems, in Lewry, J.F. and Stauffer, M.R., eds., The Early Proterozoic Trans-Hudson Orogen of North America: Geological Association of Canada, Special Paper 37, p. 1–14.

Lewry, J.F. and Sibbald, T.I.I., 1977, Variation in lithology and tectonometamorphic relationships in the Precambrian basement of northern Saskatchewan: Canadian Journal of Earth Sciences, v. 14, p. 1453–1467.

Lewry, J.F. and Sibbald, T.I.I., 1980, Thermotectonic evolution of the Churchill Province in northern Saskatchewan: Tectonophysics, v. 68, p. 45–82.

Lewry, J.F., Sibbald, T.I.I. and Rees, C.J., 1978, Metamorphic patterns and their relation to tectonism and plutonism in the Churchill Province in northern Saskatchewan, in Fraser, J.A. and Heywood, W.W., eds., Metamorphism in the Canadian Shield: Geological Survey of Canada, Paper 78-10, p. 139–154.

Lewry, J.F., Stauffer, M.R. and Fumerton, S., 1981, A Cordilleran-type batholithic belt in the Churchill Province in northern Saskatchewan: Precambrian Research, v. 14, p. 277–313.

Lewry, J.F., Sibbald, T.I.I. and Schledewitz, D.C.P., 1985, Variation in character of Archean rocks in the western Churchill Province and its significance, in Ayres, L.D., Thurston, P.C., Card, K.D. and Weber, W., eds., Evolution of Archean Supracrustal Sequences: Geological Association of Canada, Special Paper 28, p. 239–261.

Lis, M.G. and Price, R.A., 1976, Large-scale block faulting during deposition of the Windermere Supergroup in southeastern British Columbia: Geological Survey of Canada, Paper 76-1A, p. 135–136.

Loveridge, W.D., Eade, K.E. and Roddick, J.C., 1988, A U–Pb age on zircon from a granite pluton, Kamilukuak Lake area, District of Keewatin, establishes a lower limit for the age of the Christopher Island Formation, Dubawnt Group, in Radiogenic Age and Isotopic Studies. Report 1: Geological Survey of Canada, Paper 87-2, p. 67–72.

Lucas, S.B., 1989, Structural evolution of the Cape Smith thrust belt and the role of out-of-sequence faulting in the thickening of mountain belts: Tectonics, v. 8, p. 655–676.

Macdonald, R. and Broughton, P., 1980, Geological map of Saskatchewan, provisional edition 1980: Saskatchewan Geological Survey, two sheets, scale 1:1,000,000.

Macdonald, R. and Slimmon, W.L., 1984, Bedrock compilation, greater Beaverlodge area (NTS 74N-6 to -11): Saskatchewan Geological Survey, Miscellaneous Report 84-4 (with map, scale 1:100,000).

Machado, N., Goulet, N. and Gariépy, C., 1989, U–Pb ages of reactivated Archean basement and of Hudsonian metamorphism in the northern Labrador Trough: Canadian Journal of Earth Sciences, v. 26, p. 1–15.

'MacLean, B. and Falconer, R.K.H., 1977, Baffin Island shelf – shallow corehole drilling, 1976: Geological Survey of Canada, Paper 77-1B, p. 125–127.

McCulloch, M.T. and Wasserburg, G.J., 1978, Sm–Nd and Rb–Sr chronology of continental crust formation: Science, v. 200, p. 1003–1011.

Meyer, M.T., 1987, Geochronology and geochemistry of the Wathaman Batholith, the remnant of an Early Proterozoic continental-arc in the Trans-Hudson Orogen, Saskatchewan, Canada, unpublished M.Sc. thesis, University of Kansas, Lawrence, 107 p.

Morgan, W.C., 1979, Geology, Nachvak Fiord – Ramah Bay, Newfoundland–Quebec: Geological Survey of Canada, Map 1469A, scale 1:50,000.

Morgan, W.C., 1981, Geology, Bears Gut – Saglek Fiord, Newfoundland: Geological Survey of Canada, Map 1478A, scale 1:50,000.

Morgan, W.C., 1983, Geology, Lake Gillian, District of Franklin: Geological Survey of Canada, Map 1560A, scale 1:250,000.

Mortensen, J.K. and Thorpe, R.I., 1988, U–Pb zircon ages of felsic volcanic rocks in the Kaminak Lake area, District of Keewatin, in Radiogenic Age and Isotopic Studies. Report 1: Geological Survey of Canada, Paper 87-2, p. 123–128.

Mukhopadyay, M. and Gibb, R.A., 1981, Gravity anomalies and deep structure of eastern Hudson Bay: Tectonophysics, v. 72, p. 43–60.

Nielsen, P.A., Langenberg, C.W., Baadsgaard, H. and Godfrey, J.D., 1981, Precambrian metamorphic conditions and crustal evolution, northeastern Alberta, Canada: Precambrian Research, v. 16, p. 171–193.

O'Neill, J.M. and Lopez, D.A., 1985, Character and regional significance of Great Falls tectonic zone, east-central Idaho and west-central Montana: American Association of Petroleum Geologists Bulletin, v. 69, p. 437–447.

O'Neill, J.M., Duncan, M.S. and Zartman, R.E., 1988, An Early Proterozoic gneiss dome, Highland Mountains, Montana: Geological Society of America, Abstracts with Programs, v. 20, p. 460–461.

Park, A.F. and Ralser, S., 1990, Further contributions to the stratigraphy and structure of the southwest part of the Tavani map area, District of Keewatin: Geological Survey of Canada, Paper 90-1C.

Parrish, R.R., 1989, U–Pb geochronology of the Cape Smith belt and Sugluk block, northern Quebec: Geoscience Canada, v. 16, p. 126–130.

Patterson, J.G., 1986, The Amer Belt: remnant of an Aphebian foreland fold and thrust belt: Canadian Journal of Earth Sciences, v. 23, p. 2012–2023.

Perreault, S., Hynes, A. and Machado, N., 1988, Timing in the tectonic evolution of the NE segment of the Labrador Trough, Kuujjuaq, northern Quebec: Geological Association of Canada, Program with Abstracts, v. 13, p. A97.

Peterman, Z.E. and Futa, K., 1987, Is the Archean Wyoming Province exotic to the Superior craton? Evidence from Sm–Nd model ages of basement cores: Geological Society of America, Abstracts with Programs, v. 19, p. 803.

Pidgeon, R.T. and Howie, R.A., 1975, U–Pb age of zircon from a charnockite granulite from Pangnirtung on the east coast of Baffin Island: Canadian Journal of Earth Sciences, v. 12, p. 1046–1047.

Poirier, G., Perreault, S. and Hynes, A., 1990, Nature of the eastern boundary of the Labrador Trough near Kuujjuaq, Quebec, in Lewry, J.F. and Stauffer, M.R., eds., The Early Proterozoic Trans-Hudson Orogen of North America: Geological Association of Canada, Special Paper 37, p. 397–411.

Ray, G.E. and Wanless, R.K., 1980, The age and geological history of the Wollaston, Peter Lake, and Rottenstone domains in northern Saskatchewan: Canadian Journal of Earth Sciences, v. 17, p. 333–347.

Reichenbach, I. and Parrish, R.R., 1988, Age of crystalline basement in BC and NWT, Canada, and CO and AK, USA, as inferred from U–Pb zircon geochronology of diatremes: Geological Society of America, Abstracts with Programs, v. 20, p. 110.

Ricketts, B.D., 1981, A submarine fan – distal molasse sequence of middle Precambrian age, Belcher Islands, Hudson Bay: Bulletin of Canadian Petroleum Geology, v. 29, p. 561–582.

Ricketts, B.D. and Donaldson, J.A., 1981, Sedimentary history of the Belcher Group of Hudson Bay, in Campbell, F.H.A., ed., Proterozoic Basins of Canada: Geological Survey of Canada, Paper 81-10, p. 235–254.

Ricketts, B.D., Ware, M.J. and Donaldson, J.A., 1982, Volcaniclastic rocks and volcaniclastic facies in the Middle Precambrian (Aphebian) Belcher Group, Northwest Territories, Canada: Canadian Journal of Earth Sciences, v. 19, p. 1275–1294.

Roksandic, M.M., 1987, The tectonics and evolution of the Hudson Bay region, in Beaumont, C. and Tankard, A.J., eds., Sedimentary Basins and Basin-forming Mechanisms: Canadian Society of Petroleum Geologists, Memoir 12, p. 507–518.

Roscoe, S.M., Henderson, M.N., Hunt, P.A. and van Breemen, O., 1987, U–Pb zircon age of an alkaline granite body in the Booth River intrusive suite, in Radiogenic Age and Isotopic Studies. Report 1: Geological Survey of Canada, Paper 87-2, p. 95–100.

Ross, G.M., Parrish, R.R., Villeneuve, M.E. and Bowring, S.A., *in prep.*, Tectonic assembly of Precambrian basement beneath the Alberta basin, western Canada.

Rowley, D.B. and Lottes, A.L., 1988, Plate-kinematic reconstructions of the North Atlantic and Arctic: Late Jurassic to present: Tectonophysics, v. 155, p. 73–120.

Sandiford, M., 1989, Horizontal structures in granulite terrains: a record of mountain building or mountain collapse?: Geology, v. 17, p. 449–452.

Sangster, D.F., 1972, Isotopic studies of ore-leads in the Hanson Lake – Flin Flon – Snow Lake mineral belt, Saskatchewan and Manitoba: Canadian Journal of Earth Sciences, v. 9, p. 500–513.

Sangster, D.F., 1978, Isotopic studies of ore leads of the circum-Kisseynew volcanic belt of Manitoba and Saskatchewan: Canadian Journal of Earth Sciences, v. 15, p. 1112–1121.

Schau, M., 1982, Geology of the Prince Albert Group in parts of Walker Lake and Laughland Lake map-areas, District of Keewatin: Geological Survey of Canada, Bulletin 337, 62 p. (with map, scale 250,000).

Schau, M., 1983, Baker Lake map-area (56D): Geological Survey of Canada, Open File 883 (with map, scale 1:125,000, with marginal notes).

Schau, M., Tremblay, F. and Christopher, A., 1982, Geology of Baker Lake map-area, District of Keewatin: a progress report: Geological Survey of Canada, Paper 82-1A, p. 143–150.

Schledewitz, D.C.P., 1978, Patterns of regional metamorphism in the Churchill Province of Manitoba (north of 58°), in Fraser, J.A. and Heywood, W.W., eds., Metamorphism in the Canadian Shield: Geological Survey of Canada, Paper 78-10, p. 179–190.

Scoates, R.F.J., 1981, Volcanic rocks of the Fox River belt, northeastern Manitoba: Manitoba Mineral Resources Division, Geological Report GR 81-1, 109 p. (with map, scale 1:50,000).

Sengör, A.M.C., 1984, The Cimmeride orogenic system and the tectonics of Eurasia: Geological Society of America, Special Paper 195, 82 p.

Sharpton, V.L., Grieve, R.A.F., Thomas, M.D. and Halpenny, J.F., 1987, Horizontal gravity gradient: an aid to the definition of crustal structure in North America: Geophysical Research Letters, v. 14, p. 808–811.

Sims, P.K. and Peterman, Z.E., 1986, Early Proterozoic Central Plains orogen: a major buried structure in the north-central United States: Geology, v. 14, p. 488–491.

Slimmon, W.L., 1987, Bedrock compilation geology, Fond-du-Lac sheet (NTS 74O): Saskatchewan Geological Survey, Preliminary Geological Map, scale 1:250,000.

Stauffer, M.R., 1984, Manikewan: an Early Proterozoic ocean in central Canada, its igneous history and orogenic closure: Precambrian Research, v. 25, p. 257–281.

Stevenson, R., Cumming, G.L. and Krstic, D., 1987, Lead isotope geochronology of the Portman Lake area, Northwest Territories: Canadian Journal of Earth Sciences, v. 24, p. 2188–2196.

Stockwell, C.H., 1961, Structural provinces, orogenies, and time classification of rocks of the Canadian Precambrian Shield, in Lowden, J.A., ed., Age Determinations by the Geological Survey of Canada: Geological Survey of Canada, Paper 61-17, p. 108–118.

Stockwell, C.H., 1982, Proposals for time classification and correlation of Precambrian rocks and events in Canada and adjacent areas of the Canadian Shield. Part 1. A time classification of Precambrian rocks and events: Geological Survey of Canada, Paper 80-19, 135 p.

St-Onge, M.R. and Lucas, S.B., 1990, Early Proterozoic collisional tectonics in the internal zone of the Ungava (Trans-Hudson) orogen: Lacs Nuvilik and Sugluk map areas, Quebec: Geological Survey of Canada, Paper 90-1C.

St-Onge, M.R. and Lucas, S.B., 1990, Evolution of the Cape Smith Belt: Early Proterozoic continental underthrusting, ophiolite obduction, and thick-skinned folding, in Lewry, J.F. and Stauffer, M.R., eds., The Early Proterozoic Trans-Hudson Orogen of North America: Geological Association of Canada, Special Paper 37, p. 313–351.

Syme, E.C., 1990, Stratigraphy and geochemistry of the Lynn Lake and Flin Flon metavolcanic belts, Manitoba, in Lewry, J.F. and Stauffer, M.R., eds., The Early Proterozoic Trans-Hudson Orogen of North America: Geological Association of Canada, Special Paper 37, p. 143–161.

Tapponnier, P., Peltzer, G., Le Dain, A.Y., Armijo, R. and Cobbold, P., 1982, Propagating extrusion tectonics in Asia: new insights from simple experiments with plasticine: Geology, v. 10, p. 611–616.

Tapponnier, P., Peltzer, G. and Armijo, R., 1986, On the mechanics of the collision between India and Asia, in Coward, M.P. and Ries, A.C., eds., Collision Tectonics: Blackwell, Oxford, p. 115–157.

Taylor, F.C., 1963, Snowbird Lake map-area, District of Mackenzie: Geological Survey of Canada, Memoir 333, 23 p. (with Map 1138A, scale 1:253,440).

Taylor, F.C., 1971, A revision of Precambrian structural provinces in northeastern Quebec and northern Labrador: Canadian Journal of Earth Sciences, v. 8, p. 579–584.

Taylor, F.C., 1979, Reconnaissance geology of a part of the Precambrian Shield, northeastern Quebec, northern Labrador and Northwest Territories: Geological Survey of Canada, Memoir 393, 99 p. (with Maps 1428A–1444A (17 sheets), scale 1:250,000).

Taylor, F.C., 1985, Precambrian geology of the Half Way Hills area, District of Keewatin: Geological Survey of Canada, Memoir 415, 19 p. (with map, scale 1:50,000).

Tella, S., 1984, Geology of the Amer Lake (NTS 66H), Deep Rose Lake (NTS 66G), and parts of the Pelly Lake (NTS 66F) map-areas, District of Keewatin: Geological Survey of Canada, Open File 1043 (with map, scale 1:250,000, with marginal notes).

Tella, S. and Annesley, I.R., 1988, Hanbury Island shear zone: a deformed remnant of a ductile thrust, District of Keewatin: Geological Survey of Canada, Paper 88-1C, p. 283–290.

Tella, S. and Eade, K.E., 1986, Occurrence and possible significance of high-pressure granulite fragments in the Tulemalu fault zone, District of Keewatin, N.W.T.: Canadian Journal of Earth Sciences, v. 23, p. 1950–1962.

Tella, S., Ashton, K.E., Thompson, D.L. and Miller, A.R., 1983, Geology of the Deep Rose Lake map-area, District of Keewatin: Geological Survey of Canada, Paper 83-1A, p. 403–409.

Tella, S., Thompson, D.L. and James, D.T., 1984, Geology of parts of the Deep Rose Lake and Pelly Lake map-areas, District of Keewatin: Geological Survey of Canada, Paper 84-1A, p. 313–322.

Tella, S., Annesley, I.R., Borradaile, G.J. and Henderson, J.R., 1986, Precambrian geology of Tavani, Marble Island and Chesterfield Inlet map-areas, District of Keewatin: a progress report: Geological Survey of Canada, Paper 86-13, 20 p.

Thériault, R.J. and Bostock, H.H., 1989, Nd isotopic studies in the ca. 1.9 Ga Taltson magmatic zone, N.W.T.: Geological Association of Canada – Mineralogical Association of Canada, Program with Abstracts, v. 14, p. A10.

Thom, A., Arndt, N.T., Chauvel, C. and Stauffer, M., 1990, Flin Flon and Western La Ronge belts, Saskatchewan: products of Proterozoic subduction-related volcanism, in Lewry, J.F. and Stauffer, M.R., eds., The Early Proterozoic Trans-Hudson Orogen of North America: Geological Association of Canada, Special Paper 37, p. 163–175.

Thomas, M.D. and Kearey, D., 1980, Gravity anomalies, block faulting, and Andean-type tectonism in the eastern Churchill Province: Nature, Physcial Science, v. 283, p. 61–63.

Thomas, M.D., Sharpton, V.L. and Grieve, R.A.F., 1987, Gravity patterns and Precambrian structure in the North American Central Plains: Geology, v. 15, p. 489–492.

Thompson, P.H., Culshaw, N., Buchanan, J.R. and Manojlovic, P., 1986, Geology of the Slave Province and Thelon tectonic zone in the Tinney Hills – Overby Lake (west half) map area, District of Mackenzie: Geological Survey of Canada, Paper 86-1A, p. 275–289.

Tippett, C.R. and Heywood, W.W., 1978, Stratigraphy and structure of the northern Amer Group (Aphebian), Churchill Structural Province, District of Keewatin: Geological Survey of Canada, Paper 78-1B, p. 7–11.

Tirrul, R. and Grotzinger, J.P., *in press*, Early Proterozoic collisional orogeny along the northern Thelon tectonic zone, N.W.T., Canada: evidence from foreland: Tectonics.

van Breemen, O. and Henderson, J.B., 1988, U–Pb zircon and monazite from the eastern Slave Province and Thelon tectonic zone, Artillery Lake, N.W.T., in Radiogenic Age and Isotopic Studies. Report 2: Geological Survey of Canada, Paper 88-2, p. 73–83.

van Breemen, O., Henderson, J.B., Loveridge, W.D. and Thompson, P.H., 1987, U–Pb zircon and monazite geochronology and zircon morphology of granulites and granite from the Thelon Tectonic Zone, Healey Lake and Artillery Lake map-areas, N.W.T.: Geological Survey of Canada, Paper 87-1A, p. 783–801.

van Breemen, O., Thompson, P.H., Hunt, P.A. and Culshaw, N., 1988, U–Pb zircon and monazite geochronology from the northern Thelon Tectonic Zone, District of Mackenzie, in Radiogenic Age and Isotopic Studies. Report 1: Geological Survey of Canada, Paper 87-2, p. 81–94.

van der Leeden, J., Bélanger, M., Danis, D., Girard, R. and Martelain, J., 1990, Lithotectonic domains in the high-grade terrain east of the Labrador Trough (Quebec), *in* Lewry, J.F. and Stauffer, M.R., eds., The Early Proterozoic Trans-Hudson Orogen of North America: Geological Association of Canada, Special Paper 37, p. 371–386.

Van Schmus, W.R. and Schledewitz, D.C.P., 1986, U–Pb zircon geochronology of the Big Sand Lake area, northeastern Manitoba: Manitoba Mineral Resources Division, Report of Field Activities 1986: p. 207–210.

Van Schmus, W.R., Person, S.S., Macdonald, R. and Sibbald, T.I.I., 1986, Preliminary results from U–Pb zircon geochronology of the Uranium City region: Saskatchewan Geological Survey, Miscellaneous Report 86-4, p. 108–111.

Van Schmus, W.R., Bickford, M.E., Lewry, J.F. and Macdonald, R., 1987, U–Pb geochronology in the Trans-Hudson Orogen, northern Saskatchewan, Canada: Canadian Journal of Earth Sciences, v. 24, p. 407–424.

Walcott, R.I. and Boyd, J.B., 1971, The gravity field of northern Alberta, and part of Northwest Territories and Saskatchewan, with maps: Earth Physics Branch, Gravity Map Series 103-111, Ottawa, 13 p.

Wanless, R.K., 1979, Geochronology of Archean rocks of the Churchill Province: Geological Association of Canada – Mineralogical Association of Canada, Program with Abstracts, v. 4, p. 85.

Wanless, R.K. and Eade, K.E., 1975, Geochronology of Archean and Proterozoic rocks in the southern District of Keewatin: Canadian Journal of Earth Sciences, v. 12, p. 95–114.

Wanless, R.K., Stevens, R.C., Lachance, G.R. and Delabio, R.N., 1973, Age determinations and geological studies: Geological Survey of Canada, Paper 73-2, p. 42–48.

Wardle, R.J., 1982, Geology of the south-central Labrador Trough: Newfoundland Mineral Development Division, Maps 82-5 and 82-6, scale 1:100,000 (and cross-sections).

Wardle, R.J. and Bailey, D.G., 1981, Early Proterozoic sequences in Labrador, *in* Campbell, F.H.A., ed., Proterozoic Basins of Canada: Geological Survey of Canada, Paper 81-10, p. 331–358.

Wardle, R.J., Rivers, T., Gower, C.F., Nunn, G.A.G. and Thomas, A., 1986, The northeastern Grenville Province: new insights, *in* Moore, J.M., Davidson, A. and Baer, A.J., eds., The Grenville Province: Geological Association of Canada, Special Paper 31, p. 13–29.

Wardle, R.J., Ryan, B., Nunn, G.A.G. and Mengel, F.C., 1990, Labrador segment of the Trans-Hudson Orogen: crustal development through oblique convergence and collision, *in* Lewry, J.F. and Stauffer, M.R., eds., The Early Proterozoic Trans-Hudson Orogen of North America: Geological Association of Canada, Special Paper 37, p. 353–369.

Watters, B.R. and Armstrong, R.L., 1985, Rb–Sr study of metavolcanic rocks from the La Ronge and Flin Flon domains, northern Saskatchewan: Canadian Journal of Earth Sciences, v. 22, p. 452-463.

Watters, B.R. and Pearce, G.R., 1987, Metavolcanic rocks of the La Ronge domain in the Churchill Province, Saskatchewan: geochemical evidence for a volcanic arc origin, *in* Pharoah, T.C., Beckinsale, R.D. and Rickard, D., eds., Geochemistry and Mineralization of Proterozoic Volcanic Suites: Blackwell, Oxford, p. 167–182.

Wilson, J.T., 1968, Comparison of the Hudson Bay arc with some other features, *in* Beals, C.S., ed., Science, History, and Hudson Bay: Department of Energy, Mines and Resources, Ottawa, p. 1015–1033.

Wooden, J.L., Ashwal, L.D., Wiebe, R.A. and Emslie, R.F., 1987, Regional Pb isotope systematics in Proterozoic intrusives, Nain Province, Labrador: Eos, v. 68, p. 1519.

The Churchill–Superior Boundary Zone, Southeast Margin of the Trans-Hudson Orogen: A Review

Werner Weber
Manitoba Geological Services, 535-330 Graham Avenue,
Winnipeg, Manitoba R3C 4E3, Canada

Weber, W., The Churchill–Superior Boundary Zone, southeast margin of the Trans-Hudson Orogen: a review, in Lewry, J.F. and Stauffer, M.R., eds., The Early Proterozoic Trans-Hudson Orogen of North America: Geological Association of Canada, Special Paper 37, p. 41–55.

Abstract

Geological and geochronological data indicate that Early Proterozoic thermotectonic overprinting of rocks in the Churchill–Superior Boundary Zone was related to collisional tectonics during a late compressional phase (ca. 1820–1720 Ma) of development of the Trans-Hudson Orogen. Recently obtained U–Pb zircon ages indicate that most supracrustals and mafic–ultramafic igneous rocks in the Churchill–Superior Boundary Zone (with the possible exception of some Thompson belt supracrustals) are not related to initial rifting prior to development of the Trans-Hudson Orogen, as suggested earlier. Instead, lithological, geochemical, and geochronological data suggest that they developed during a late phase of synorogenic rifting of the Superior plate. This rifting may have occurred in a marginal basin or back-arc basin tectonic setting, contemporaneous with island-arc volcanism in the Reindeer Zone.

Geological and geophysical data suggest that, during collision, the Reindeer Zone was underthrust by Superior Province and that the edge of the Superior plate created during initial breakup of the Archean protocraton lies beneath Proterozoic rocks, a significant distance northwest of the present surface contact between Churchill and Superior provinces.

Résumé

Les données géologiques et géochronologiques indiquent qu'une surimpression thermotectonique d'âge Protérozoïque Inférieur sur des roches situées au voisinage de la frontière Churchill–Supérieur est reliée à une tectonique de collision durant la phase compressive (vers 1720–1820 Ma) de l'orogène Trans-Hudson. Des âges U–Pb récents indiquent que la plupart des roches supracrustales et basiques/ultrabasiques ignées de la frontière Churchill–Supérieur (exception possible des roches supracrustales de la ceinture Thompson) ne font pas partie d'un rift précédant le développement de l'orogène Trans-Hudson, tel que proposé précédemment. Les lithologies, la géochimie et la géochronologie suggèrent un développement durant une phase tardive de rift syn-orogénique de la plaque Supérieur formé dans un bassin marginal ou dans une tectonique de bassin d'arrière-arc, contemporain avec le volcanisme d'arc insulaire de la zone Reindeer.

Les données géologiques et géophysiques suggèrent que durant la collision, la zone de Reindeer a été charriée sur la province Supérieur et que la marge de la plaque Supérieur créée durant la dislocation initiale du protocraton Archéen se trouve actuellement sous les roches protérozoïques à bonne distance au nord-ouest du contact Churchill–Supérieur.

INTRODUCTION

The southeastern marginal zone of the Trans-Hudson Orogen is the Churchill–Superior Boundary Zone (Fig. 1). It is a complex transition between the Early Proterozoic Reindeer Zone and the Archean Superior craton. The Churchill–Superior Boundary Zone is divided into three main segments: the Thompson belt, the Split Lake block, and the Fox River belt (Weber and Scoates, 1978; Fig. 2). It was affected by strong Hudsonian deformation and metamorphism, which overprinted east-trending Archean structures and Archean mineral assemblages. This overprint is particularly strongly developed in the Thompson belt. The Churchill–Superior Boundary Zone contains Early Proterozoic supracrustal rocks that are part of the "Circum-Superior belt" (Baragar and Scoates, 1981).

Distinctive aeromagnetic and regional gravity signatures permit southward extrapolation of the Thompson belt beneath Paleozoic cover through southwest Manitoba and for another 400 km into the United States (Green *et al.*, 1985).

GEOLOGY OF NORTHWEST SUPERIOR PROVINCE

The Churchill–Superior Boundary Zone forms the margin of the northwestern Superior Province in Manitoba (Weber and Scoates, 1978). Contact with the Reindeer Zone of the Churchill Province is everywhere defined by major faults, some of which are probably thrusts (Fig. 2). At its eastern and southern margins, the Churchill–Superior Boundary Zone grades into the northwestern part of the Superior craton, which, in the exposed Shield in Manitoba, consists of the Pikwitonei granulite domain and the Gods Lake greenstone–granitoid domain (Fig. 2). Contact between these two domains is defined by an orthopyroxene isograd, which reflects prograde, late Archean metamorphism (Weber and Scoates, 1978; Hubregtse, 1980; Heaman *et al.*, 1986a). This part of the Superior Province represents an oblique cross-section of Archean continental crust with a relief of about 20 km over a distance of about 100 km (Weber and Mezger, *in press*).

Superior Province craton

Pikwitonei domain. The Pikwitonei granulite domain comprises enderbite, opdalite, and derived orthogneiss with trails of high-grade equivalents of Gods Lake domain greenstone belt rocks, e.g., mafic to ultramafic metavolcanics, banded oxide facies iron formation, and minor siliceous, aluminous, and calcareous metasediments (Weber and Scoates, 1978; Hubregtse, 1980).

The age of high-grade metamorphism at Cauchon Lake (Fig. 2, location C), near the southeastern margin of the Pikwitonei granulite domain, has been determined at 2695 ± 2 Ma and 2637 ± 2 Ma (U–Pb zircon ages, Heaman *et al.* (1986a)). These two ages can be correlated in the field with two deformational events (D_1, D_2 of Hubregtse (1980)). The orthopyroxene isograd was established during the younger D_2 event.

Geothermobarometry and garnet U–Pb geochronology (Mezger *et al.*, 1989a, 1990) confirm a high-grade event(s) in the Cauchon Lake area between 2700 and 2640 Ma with temperatures of about 750°C and pressures of about 7 kbar (700 MPa). At Natawahunan Lake, *P–T* conditions were slightly higher, 820°C and 7.5–8 kbar (750–800 MPa), and farther north, at the contact with the Churchill–Superior Boundary Zone, pressures were 9 kbar (900 MPa). U–Pb zircon and garnet ages of pegmatites in the Cauchon Lake area (Mezger *et al.*, 1989a) register a retrogressive event under amphibolite facies conditions (D_{2B}, Hubregtse, 1980) at ca. 2600 Ma. U–Pb rutile ages of 2.3–2.4 Ga from Cauchon and Natawahunan lakes indicate conditions of 425°C and 4 kbar (400 MPa) (Mezger *et al.*, 1989b), implying that the granulites were closer to the surface. Based on the anticlockwise pressure–temperature–time path, near isobaric cooling path, slow cooling rate, and multiple thermal events within about 150 Ma, Mezger *et al.* (1990) suggest that the granulites formed in a long-lived Andean-type continental margin, with heating before and during tectonic loading, rather than in a fold and thrust belt typified by loading before heating.

The granulites must have been at the surface before 1883 Ma, the minimum age of the Ospwagan group in the Thompson belt, because this group was deposited on (retrogressed) Pikwitonei-type granulites (Bleeker and Macek, 1988; Weber and Mezger, *in press*). Also, the observation that the 1883 Ma Molson dykes have the same glassy (unrecrystallized) contacts in both the Pikwitonei domain and the lower metamorphic grade Gods Lake domain suggests that the granulites were at the same high crustal level as the Gods Lake domain prior to intrusion of the Molson swarm.

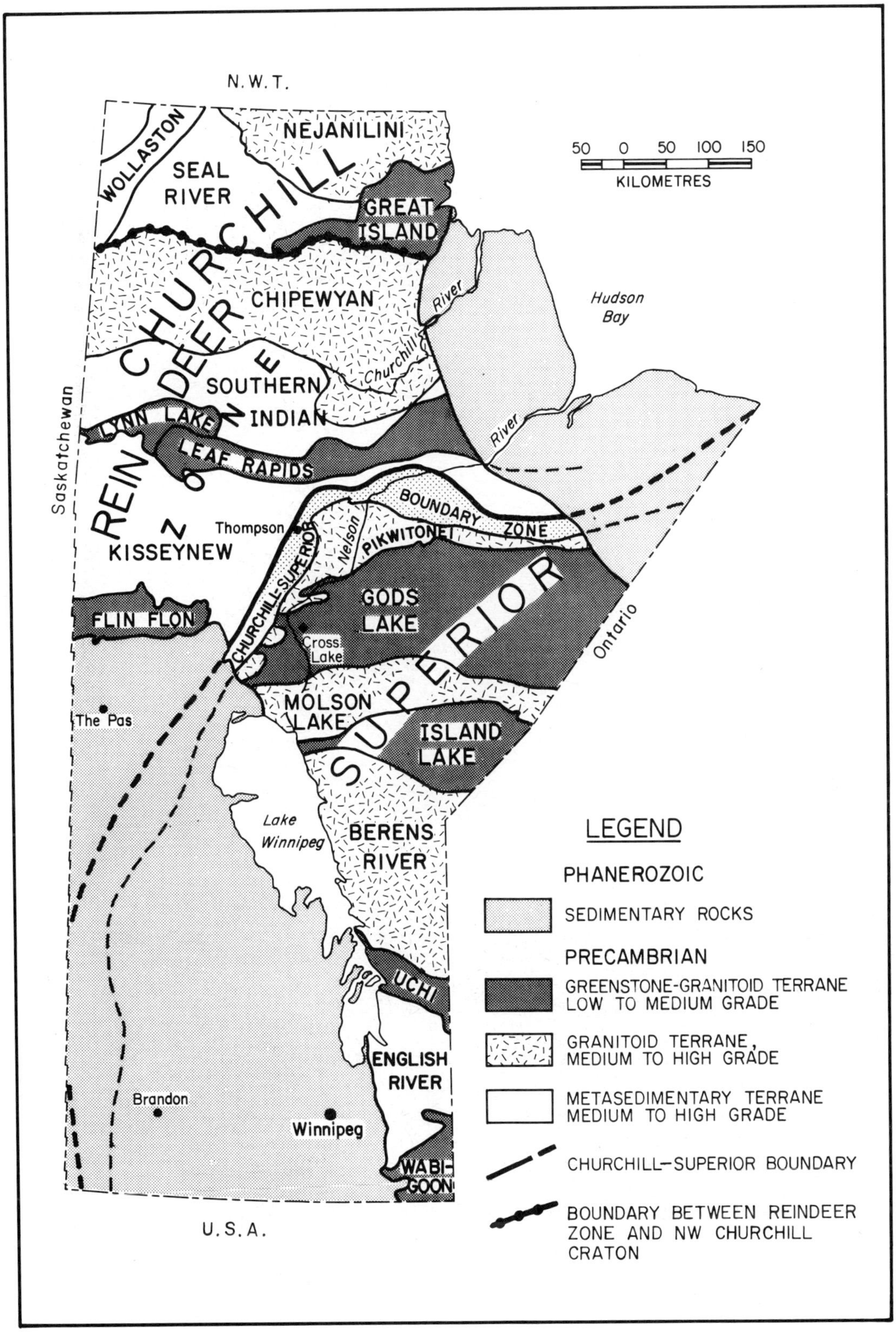

Figure 1. *Principal geological domains in Manitoba.*

The granulites may have been brought to the surface as a result of extension, which led to tectonic thinning, and eventual rifting and breakup of the Archean protocontinent, thus starting the Trans-Hudson orogenic cycle. Alternatively, an unknown thrusting event, similar to that which produced the Kapuskasing structure, could have brought the granulites to shallow crustal levels prior to rifting.

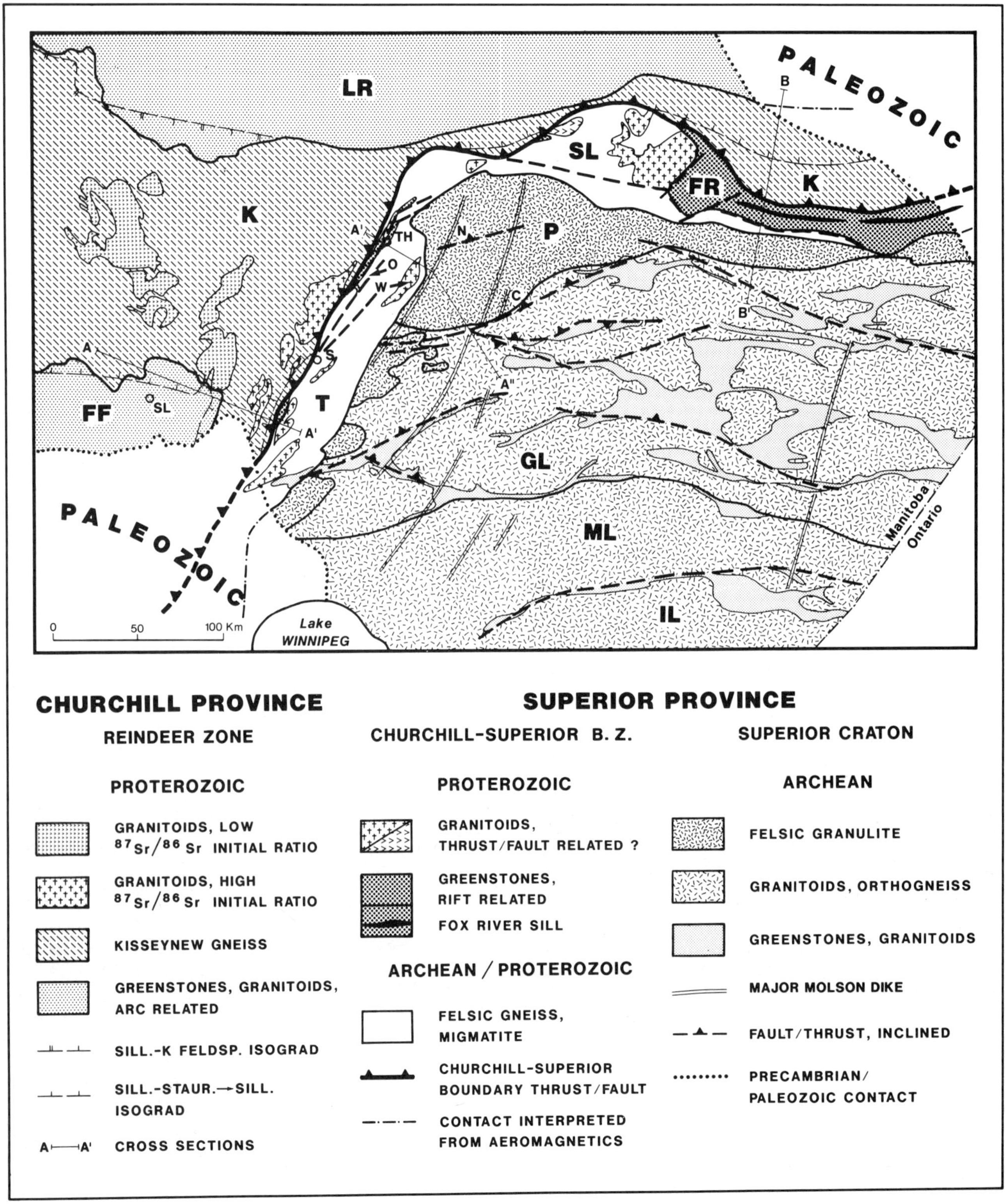

Gods Lake domain. The Gods Lake domain, southeast of the Pikwitonei domain (Figs. 1 and 2), comprises about 80% granitoid rocks and about 20% greenstones within greenschist to upper amphibolite facies. The granitoids are mainly gneissic tonalites and granodiorites, and related orthogneisses. The greenstones form east-trending, relatively thin (maximum 50 km wide), continuous and discontinuous belts and also occur as inclusions in granitoid rocks. The supracrustal rocks are grouped into (1) an older, 2.83–2.7 Ga tholeiitic to calc-alkaline, subaqueous volcanic succession and associated sediments, and (2) a younger, 2.71–2.69 Ga, unconformably overlying subaerial, sedimentary, and high-potassium volcanic succession (Hubregtse, 1985; Corkery *et al.*, 1988).

Granitoid rocks formed during three main periods: at ca. 3.0 Ga, between 2760 and 2712 Ma, and between 2702 and 2650 Ma (Corkery *et al.*, 1988).

A northwesterly increase in metamorphic grade, from greenschist – lower amphibolite facies (3.5 kbar (350 MPa)) to upper amphibolite facies (5 kbar (500 MPa)) in the Gods Lake domain, and to granulite facies (7–9 kbar (700–900 MPa)) in the Pikwitonei domain (Weber and Scoates, 1978; Mezger *et al.*, 1989a), and concomitant increase in the ratio of granitoids to greenstones suggest that this area represents an oblique cross-section of Archean continental crust exposing a relief of about 20 km over a distance of 100 km (Weber and Mezger, *in press*).

Churchill–Superior Boundary Zone

Lithologies and metamorphism. The exposed portion of the Churchill–Superior Boundary Zone in Manitoba includes the Thompson belt, Split Lake block, and Fox River belt (Fig. 2).

The Thompson belt consists of several blocks bounded by mylonitic zones that strike northeast, slightly oblique both to the regional strike of the belt and to the fault contact with the Reindeer Zone. These fault blocks comprise rocks of different Hudsonian metamorphic grade. For instance, the Burntwood River fault separates the Thompson block, characterized by metamorphic conditions of 700–750°C and 6–7 kbar (600–700 MPa) (Bleeker, this volume), from the lower metamorphic grade Pipe Lake block, characterized by conditions of 600–660°C and 5–6 kbar (500–600 MPa) (Bleeker, this volume). Kinematic indicators in the block bounding mylonites suggest a Superior-side-up sense of movement (Bleeker, 1988; Fueten and Robin, 1989). This subvertical block movement was possibly synchronous with the generation of domes and basins, which may have formed during regional transpression with a sinistral horizontal component (Bleeker, this volume).

Most of the Thompson belt, Split Lake block, and southern margin of Fox River belt comprise polymetamorphic migmatitic gneisses, which are probably largely derived from Pikwitonei-type granulites through selective retrogression and recrystallization, migmatization, and structural reconstitution under amphibolite-grade conditions during Hudsonian orogeny. This interpretation is based on the observed tectonometamorphic gradation between the Churchill–Superior Boundary Zone and Pikwitonei domain (Weber, 1976; Hubregtse, 1977, 1978) and preservation of relicts of only partially overprinted (retrogressed) granulites in the Churchill–Superior Boundary Zone, e.g., at Paint Lake, 15 km south of Thompson (Russell, 1981), and at Split Lake (Corkery, 1985).

However, some amphibolite-grade gneisses lack such granulite-grade structural or mineralogical relicts and may never have reached granulite facies conditions. Such gneisses occur at Paint Lake (unit 3, Charbonneau *et al.*, 1979), Sasagiu Rapids (Fig. 2, location S), and Manassan River, 20 km southwest of Thompson. U–Pb zircon data indicate that they represent higher crustal equivalents of Pikwitonei granulites, as exposed in the northern part of the Gods Lake domain.

Reworked gneisses in the Churchill–Superior Boundary Zone were affected by both a metamorphic–metasomatic and a structural overprint. In the transition zone, the metamorphic overprint extended farther into the Pikwitonei domain than the structural overprint. It involved hydration of granulite facies assemblages (in part associated with Hudsonian granitoid melts), resulting in the breakdown of orthopyroxene, calcic plagioclase, and magnetite, and in their recrystallization under greenschist facies to blue–green amphibole + quartz, albitic plagioclase, and biotite. Farther west in the transition zone, the S_2 layering of the Archean gneisses was overprinted by folds with north-northeast-trending, steeply dipping axial

Figure 2. *Geology of the Thompson region (modified after Manitoba Mineral Resources Division (1979)). Metamorphic isograds after Zwanzig (this volume). Only major Molson dykes are shown. Domain abbreviations in bold lettering: FF, Flin Flon belt; FR, Fox River belt; K, Kisseynew domain; GL, Gods Lake domain; IL, Island Lake domain; LR, Leaf Rapids domain; ML, Molson Lake domain; P, Pikwitonei domain; SL, Split Lake block; and T, Thompson belt. Locations: C, Cauchon Lake; N, Natawahunan Lake; O, Ospwagan Lake; S, Sasagiu Rapids (north end of Setting Lake); SL, Snow Lake; TH, Thompson; and W, Wintering Lake. A–A′ and A′–A″, and B–B′ are section lines of cross-sections shown in Figures 3 and 4 respectively.*

planes, resulting in transposed north-northeast-striking S_3 layering (Hubregtse, 1980). This structural overprint was associated with hydration and metamorphic recrystallizaton under amphibolite facies conditions, resulting, for example, in the formation of garnet + quartz porphyroblasts from orthopyroxene. It was also accompanied by potassium metasomatism, leading to widespread biotite formation, and involved migration of alkalis into shear zones along the eastern margin of the Thompson belt. This produced gneisses, with augen up to 20 cm long, at Apussigamasi Lake (Weber and Scoates, 1978) and between Apussigamasi Lake and the Grass River (Manitoba Mineral Resources Division, 1979). Hudsonian granitoid rocks are associated with these augen gneiss zones, suggesting that they, and larger plutons such as the Wintering granodiorite (Fig. 2, location W), represent shear- or thrust-generated melts. Hudsonian granitoid melts intruded reworked gneisses throughout the belt to a variable degree, mainly in the form of thin sheets but locally as larger bodies, such as the Mystery Lake granite, just north of Thompson, and the Fox Lake granite, at the western margin of the Split Lake block.

The aeromagnetic map shows the zone of magnetite breakdown due to hydration along the eastern margin of the Thompson belt by an abrupt change from the Archean east to east-northeast-trending magnetic units to a north-northeast-trending, 10–20 km wide belt of very low magnetic response (cf. Kornik and MacLaren, 1966; Kornik, 1969). West of the transition zone, the aeromagnetic signatures indicate variably magnetic north-northeast-trending units. These include strongly magnetic Proterozoic supracrustals and granulite relicts.

Proterozoic supracrustal rocks. Proterozoic igneous and sedimentary supracrustal rocks occur as infolded and/or fault-bounded belts and slivers in migmatitic gneisses of the Churchill–Superior Boundary Zone along the western margin of the Thompson belt, in the Thompson area, at Ospwagan Lake – Pipe Lake and Setting Lake, and in the Split Lake block at Assean Lake (Fig. 2). These supracrustals include mafic and minor ultramafic flows, serpentinites, and ultramafic amphibolites (Stephenson, 1974; Scoates *et al.*, 1977; Peredery *et al.*, 1982). The latter are considered to represent deformed and segmented sills. Nickel mineralization is spatially and genetically related to serpentinite (Peredery *et al.*, 1982). A thin felsic volcanic unit was recently found at Ospwagan Lake (Bleeker and Macek, 1988). Associated metasedimentary rocks comprise silicate-, carbonate-, oxide-, and sulphide-facies banded iron formations, metapelites, meta-arenites, and dolomitic marble with a basal conglomeratic unit (Bleeker and Macek, 1988).

Low-grade supracrustal rocks in the Fox River belt overlie, or were thrust over, migmatitic gneisses of the Churchill–Superior Boundary Zone (Weber and Scoates, 1978). They consist of a north-facing, homoclinal succession, which is lithologically and compositionally similar to Proterozoic supracrustal assemblages in the Thompson belt. The succession is composed of ultramafic to mafic, largely komatiitic volcanics, and siltstone, shale, iron formation, dolomite, and greywacke (Scoates, 1981). The most prominent geological unit in the Fox River belt is a 250 km long stratiform, mafic/ultramafic intrusion, the Fox River Sill (Fig. 2; Scoates, 1990).

Proterozoic supracrustal rocks in the Fox River belt were considerably less metamorphosed and deformed than those in the Thompson belt. Metamorphic grade ranges from prehnite–pumpellyite to lower greenschist facies (Weber and Scoates, 1978; Scoates, 1981).

Geochronology. Recent U–Pb geochronology (Machado *et al.*, 1987; Machado, this volume) has established three Hudsonian magmato-metamorphic events in the Thompson belt.

The Wintering Lake granite (Fig. 2, location W) yielded a concordant monazite–zircon age of 1822 Ma, which is interpreted as being the age of intrusion (Machado *et al.*, 1987). This age is similar to U–Pb zircon ages of 1845^{+52}_{-19}, 1809^{+13}_{-10}, and 1819^{+24}_{-13} Ma for anatectic granitoids in the Kisseynew domain of the Reindeer Zone (Gordon *et al.*, this volume), and the age of 1815 Ma for peak metamorphism in this domain (Gordon *et al.*, 1987, this volume). Peak metamorphism in the Setting Lake area of the Thompson belt has a similar age of 1809 ± 14 Ma (Machado, this volume). These ages are interpreted as dating the collision of Superior Province craton with the Reindeer Zone.

Pegmatites at Paint Lake, Pipe Lake openpit mine, and Thompson openpit mine yielded zircon ages of 1786^{+2}_{-5}, 1768, and 1771 ± 2 Ma (Krogh *et al.*, 1985; Machado *et al.*, 1987), placing a minimum age on high-grade metamorphism or possibly dating the start of metamorphic relaxation after peak metamorphism. This second event is also recorded by concordant monazite ages of 1780 and 1770 Ma in migmatitic gneiss at Manassan River, 20 km southwest of Thompson, and in the Thompson openpit mine (Machado *et al.*, 1987, 1990).

A third event is suggested by a concordia intercept zircon age of 1726 ± 12 Ma for a granitic sill at Sasagiu Rapids near Setting Lake (Fig. 2, location S) (Machado, this volume) and a corresponding sphene age of 1737 ± 5 Ma in the host migmatitic gneiss (Machado *et al.*, 1987). This event may have been of regional extent because similar sphene ages of 1720 and 1743 Ma (Machado *et al.*, 1987) were obtained from migmatitic gneiss and amphibolite inclusions, respectively, in the Thompson openpit mine. These data suggest that granitoid melts were generated over a period of ca. 100 Ma. Further work is required to explain the generation of melt over such a long period after initial collision.

U–Pb zircon data indicate that migmatitic gneisses and amphibolites from the Thompson openpit mine, and source rocks of the Mystery Lake granite (10 km north of Thompson) and Wintering Lake granite went through a 2.64–2.7 Ma metamorphic event that is approximately time equivalent to the high-grade metamorphism in the Pikwitonei granulite domain (cf. Machado *et al.*, 1987). In contrast, U–Pb zircon data from migmatitic gneiss at Manassan River, 20 km southwest of Thompson, and at Sasagiu Rapids do not record this event but an older event at ca. 3.0 Ga (Machado *et al.*, 1987, 1990). This suggests that these gneisses did not experience the high-grade Pikwitonei event, but may be correlative with higher crustal level granitoids of the Gods Lake domain from which 3.0 Ga granitoids are known (Corkery *et al.*, 1988). Machado *et al.* (1988) report that the 3.0 Ga granitoids were intruded by granites that contain zircon relicts, indicating that the melt was derived from 2.6 Ga Pikwitonei-type granulites: this suggests that the 3.0 Ga granitoids are structurally underlain by granulites, either due to a primary lower–upper crust relationship or due to thrust tectonics. The latter interpretation would support the tectonic model proposed for the Thompson belt (Fig. 3).

The Fox River Sill has yielded a U–Pb zircon age of 1883 ± 1.5 Ma (Heaman *et al.*, 1986b). The same

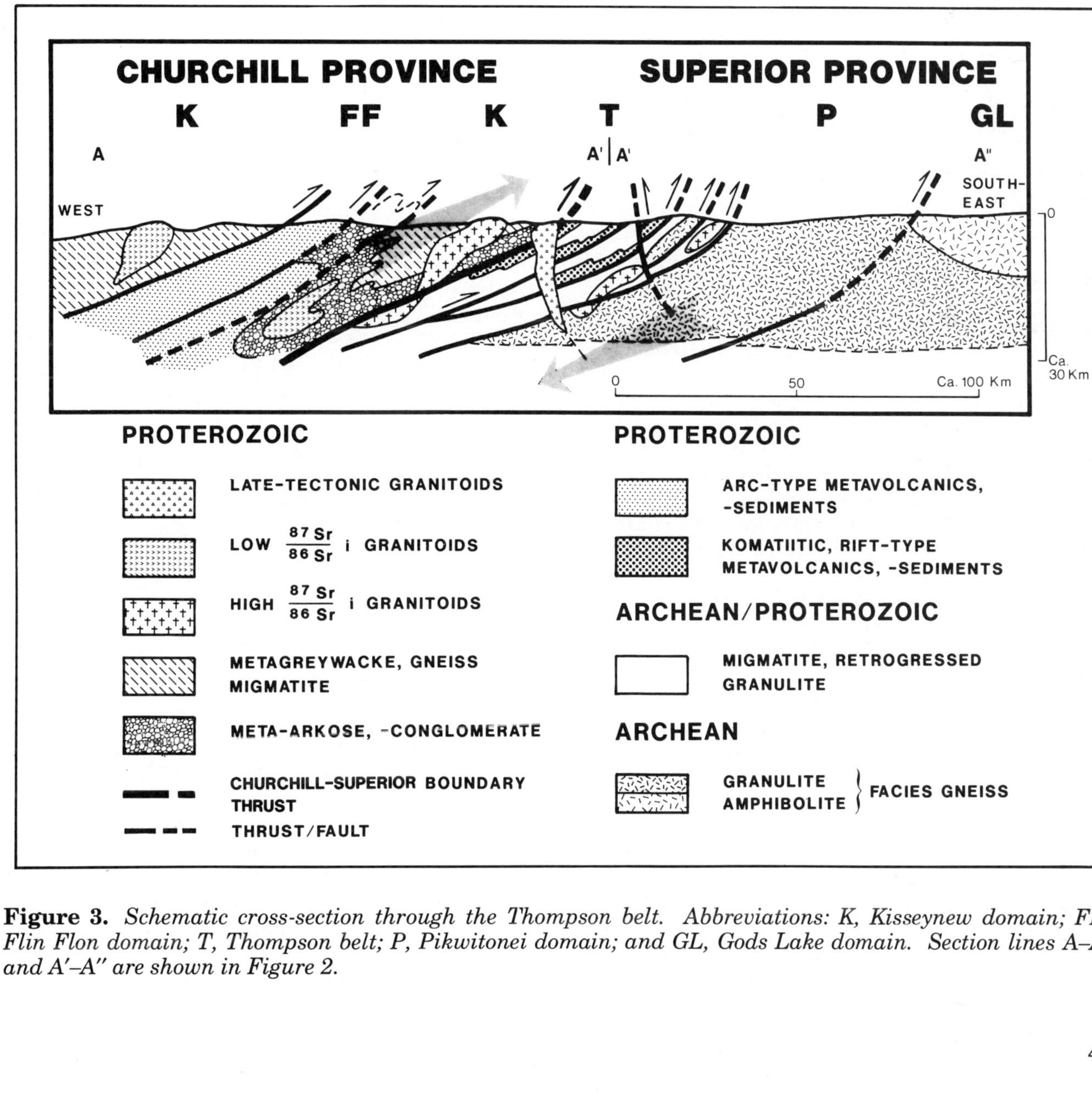

Figure 3. *Schematic cross-section through the Thompson belt. Abbreviations: K, Kisseynew domain; FF, Flin Flon domain; T, Thompson belt; P, Pikwitonei domain; and GL, Gods Lake domain. Section lines A–A′ and A′–A″ are shown in Figure 2.*

age was obtained from the Molson dyke swarm (Scoates and Macek, 1978), considered to be comagmatic with the Fox River Sill on the basis of chemical similarities (Scoates, 1981). Two dykes were dated: the Cuthbert Lake dyke, of mafic to ultramafic composition, yielded an age of 1883 ± 2 Ma, and the gabbroic Cross Lake dyke gave an age of 1884 ± 1.5 Ma (Heaman *et al.*, 1986b).

Thompson belt supracrustal rocks have been correlated with those of the Fox River belt (Baragar and Scoates, 1981; Scoates, 1981) on the basis of lithological similarities in the supracrustal successions, the similar chemical composition of the igneous rocks, and the parallelism in orientation of the north-northeast-trending Molson dyke swarm and the Thompson belt. However, recent fieldwork (Macek, 1987; Bleeker and Macek, 1988) suggests that metasediments of Pipe Lake openpit mine are older than supracrustal rocks of the Fox River belt because they have gone through a period of metamorphism and deformation prior to the intrusion of mafic dykes considered to be part of the Molson swarm. Available radiometric data, although not precise enough to establish definite age relationships, are consistent with this conclusion: Thompson belt nickeliferous sulphides have yielded Pb–Pb ages of 2320 ± 30 Ma (Cumming *et al.*, 1982) and metasedimentary rocks have given Rb–Sr whole-rock ages of 1935 ± 155 Ma (Cranstone and Turek, 1976) and ca. 1.8-2.1 Ga (Brooks and Theyer, 1981).

Tectonic significance of Proterozoic supracrustal rocks in the Churchill–Superior Boundary Zone. Proterozoic supracrustal rocks in the Churchill-Superior Boundary Zone have been described as being part of the "Circum-Superior belt" (Baragar and Scoates, 1981) and interpreted as being rift sequences because included mafic–ultramafic magmatic rocks are chemically similar to those emplaced in rift environments (Baragar and Scoates, 1981; Scoates, 1981). However, these rocks cannot all be related to "initial" rifting, as was suggested by Green *et al.* (1985), because the magmatic rocks that have yielded precise ages, such as the 1883 ± 1.5 Ma Fox River Sill and comagmatic Molson dykes, are younger or equal in age to felsic arc–volcanic rocks in the Reindeer Zone, dated at 1915$^{+7}_{-6}$ and 1910$^{+15}_{-10}$ Ma in the Lynn Lake belt, 1878 ± 3 Ma in the Rusty Lake belt (Baldwin *et al.*, 1987), and 1886 ± 1 Ma in the Flin Flon belt (Syme *et al.*, 1987). U–Pb zircon data from Saskatchewan indicate that all felsic volcanism occurred between 1885 and 1875 Ma (Van Schmus *et al.*, 1987). Because these magmatic rocks are considered to have formed as a result of subduction of Proterozoic oceanic crust (Syme, this volume), initial rifting must be older. Consequently, the dated rift sequences, the Fox River Sill and Molson dykes, with the probable exception of some older Thompson belt supracrustal rocks, must be related to a younger period of extension near the margin of the Superior craton. This extensional regime was contemporaneous with subduction in the Reindeer Zone and was followed by compression 40 Ma later, when the 1822 Ma Wintering Lake granite intruded, during collision between the Reindeer Zone and Superior craton.

Hoffman (1988) has suggested that much of the magmatism along the northwest margin of the Superior craton is foredeep magmatism related to buckling of Archean crust by Hudsonian thrust sheets advancing onto the Superior craton. However, there is no stratigraphic evidence, in metasediments of the Churchill–Superior Boundary Zone, of advancing thrust sheets from the Reindeer Zone. Only the facies of metaturbidites and fluvial deposits in the adjacent Kisseynew domain are similar to foredeep sediments. However, those metaturbidites were derived from adjacent active arc volcanoes (Bailes, 1980), dated at 1915–1878 Ma: the fluvial sediments are younger and are derived from continental arc magmatism at 1865–1850 and 1830 Ma (Zwanzig, this volume). Thus, the Kisseynew gneisses seem to be too young to have been the (indirect) cause of 1883 Ma magmatism in the Churchill–Superior Boundary Zone.

New trace-element data from mafic magmatic rocks of the Churchill–Superior Boundary Zone, from Ospwagan Lake, Assean Lake, and Fox River belt (Halden, 1988), indicate that these rocks were derived from a depleted mantle source, consistent with a model of a marginal basin or back-arc spreading tectonic setting. The back-arc spreading model would imply that this magmatism is related to subduction, presumably in the Reindeer Zone. Such a model is strongly supported by the identical U–Pb zircon ages of the Flin Flon arc volcanics and Churchill–Superior Boundary Zone magmatism, and implies that a possible subduction zone was dipping towards the Superior craton at that time.

A possible scenario might be as follows: the northwest margin of the Superior protocontinent evolved initially as a passive continental margin. During this process extensional tectonics, associated with low-angle normal faults in the upper crust, produced crustal thinning and isostatic uplift of lower crust (cf. Wernicke, 1985). Subsequently, the thinned crust rifted. Thinned crust near the new rifted continental margin subsided due to thermal subsidence, leading to the formation of marginal basins (cf. Bond and Kominz, 1988). The thinned continental margin was underplated by mafic igneous material of oceanic origin that intruded and extruded along the margin.

GEOMETRY OF THE CONTACT BETWEEN THE SUPERIOR CRATON AND REINDEER ZONE

Early development of plate tectonic models, and proposals that the contact between the Superior and Churchill provinces represents a collision suture (cf. Gibb and Walcott, 1971; Dewey and Burke, 1973; Gibb, 1975; Gibb *et al.*, 1983), promoted discussion of the three-dimensional geometry of this junction. Such discussion was heavily dependent on geological criteria derived from better documented potential analogs, such as the Ivrea zone (e.g., Berckhemer, 1968; Giese, 1968), and the assumption that available data from the Churchill–Superior Boundary Zone warranted such a comparison. This led to the proposal (Weber, 1980; Fountain and Salisbury, 1981) that the Superior Province had overthrust the Churchill Province during the Hudsonian orogeny, resulting in the exposure of deep crustal Pikwitonei granulites through erosion and isostatic rebound of a crust thickened by the plate collision. Pactunc and Baer (1986), on the basis of geothermobarometry studies, suggested the formation of fault blocks during early Hudsonian crustal thinning and rifting, followed by tilting and stacking of the blocks onto the Churchill Province. However, careful evaluation of geological data and new geophysical data give increasing evidence that the Superior Province was overthrust by the Churchill Province.

Geological data

Thompson belt. Proterozoic granitoid rocks in the Kisseynew domain within 20 km of the Thompson belt (Fig. 2) yield high $^{87}Sr/^{86}Sr$ initial ratios of 0.7056–0.7071 (Cranstone and Turek, 1976), whereas those elsewhere in the Kisseynew domain have initial ratios of 0.702–0.7038 (Clark *et al.*, 1974; Josse *et al.*, 1974). This suggests that the former granitoids must have been derived from melting of Archean crust lying beneath Proterozoic rocks and, thus, implies that at least part of the Reindeer Zone was underthrust by rocks of the Archean Superior craton (Fig. 3). Alternatively, the granitoids could have been derived through melting of Proterozoic metasedimentary rocks with a large component of Archean detritus; however, isotopically analysed Kisseynew domain metasediments yield low $^{87}Sr/^{86}Sr$ initial ratios, indicating that the metasediments comprise detritus with a short crustal history, i.e., they are of Proterozoic age (Clark *et al.*, 1974; Clark, 1980, 1981). Underthrusting of the Reindeer Zone by the Superior craton, and the associated tectonic intercalation of various thrust slices involving low-grade metamorphic supracrustal rocks, could have provided enough water to produce retrogression of granulites and the formation of Proterozoic granitoids, dykes, and plutons in the Thompson belt and Split Lake block. As described earlier, there is field geological evidence that strong shearing, possibly related to detachment at the base of a main thrust complex, occurred along the northern part of the eastern margin of the Thompson belt and that associated granitoid rocks, such as the Wintering Lake granite, may have been thrust generated (Fig. 3).

Since the oral presentation of this paper, detailed structural and metamorphic studies in the Thompson openpit mine have yielded strong evidence (Bleeker, this volume) indicating that the Churchill–Superior Boundary Zone was formed through thrusting of the Reindeer Zone onto the Superior craton. The best evidence is the clockwise sense of the *P–T–t* path (Bleeker, this volume, Fig. 5) during Hudsonian metamorphism, which indicates that rocks in the Churchill–Superior Boundary Zone have been metamorphosed as a result of tectonic loading.

Subvertical movements along faults (Fig. 3) in the Thompson belt (Bleeker, 1988, this volume; Fueten and Robin, 1989) probably reflect block tectonics related to younger isostatic readjustments. Structural data from the Phanerozoic succession of southwestern Manitoba (cf. McCabe, 1971) suggest that such movement along the Churchill–Superior Boundary Zone continued through the Phanerozoic and may have been a significant contributing factor in the development of the Williston and Hudson Bay sedimentary basins.

Fox River belt. Geological data indicate the presence of a major, north dipping reverse fault or thrust zone forming the contact between the supracrustals of the Fox River belt and the Kisseynew gneisses (Scoates, 1981). The tectonic nature of the contact is indicated by the difference in grade of metamorphism between the two terrains (cf. Fig. 4): the Fox River supracrustals contain prehnite and pumpellyite, typical of the subgreenschist facies, whereas the overlying Kisseynew gneisses are in amphibolite facies (Weber and Scoates, 1978). The tectonic nature of the contact is further documented by the fact that the metasedimentary rocks in these two Proterozoic supracrustal domains have completely different source areas, as indicated by the high $^{87}Sr/^{86}Sr$ initial ratio in the Fox River belt (G.S. Clark, personal communication) and low $^{87}Sr/^{86}Sr$ initial ratio in the Kisseynew gneiss (Clark, 1980). Metasedimentary rocks of the Thompson belt are also characterized by high initial ratios (Brooks and Theyer, 1981).

A northerly dip of the fault separating Fox River belt rocks from Kisseynew gneisses is indicated by northeasterly dips of metamorphic layering of 50–80° just north of the fault at Stephens Lake, and northerly dips of 50° in layering of the most northerly exposed Fox River volcanic rocks (Scoates *et al.*, 1981; Scoates, 1981). Further evidence for this fault contact being a thrust is the presence of a zone of lower grade metamorphic, staurolite-bearing Kisseynew metasediments, which are structurally overlain by higher metamorphic grade, sillimanite-bearing gneisses (Figs. 2 and 4) along the fault between Split Lake and Fox River. Such zones with inverted isograds are typical of the base of major thrust sheets (cf. Windley, 1983). This inverted isograd extends along most of the Churchill–Superior boundary fault (Fig. 2).

The contact of the Fox River supracrustal succession with the underlying gneisses is unexposed and has not been intersected by drillholes; thus, the exact nature of the contact is unknown. However, cataclastic augen gneiss occurs just south of the inferred contact (Scoates and Weber, 1977), suggesting a tectonic contact. In addition, thick platform carbonates occurring at the base of the Early Proterozoic succession in the Sutton inlier (Bostock, 1971), which represents the eastern extension of the Fox River belt, are absent in the Fox River area, suggesting that the basal part of the stratigraphic succession has been cut out. This implies that the lower contact of the belt is also a thrust. Southeastwards thrusting is also indicated by the occurrence of southeast and east- to southeast-

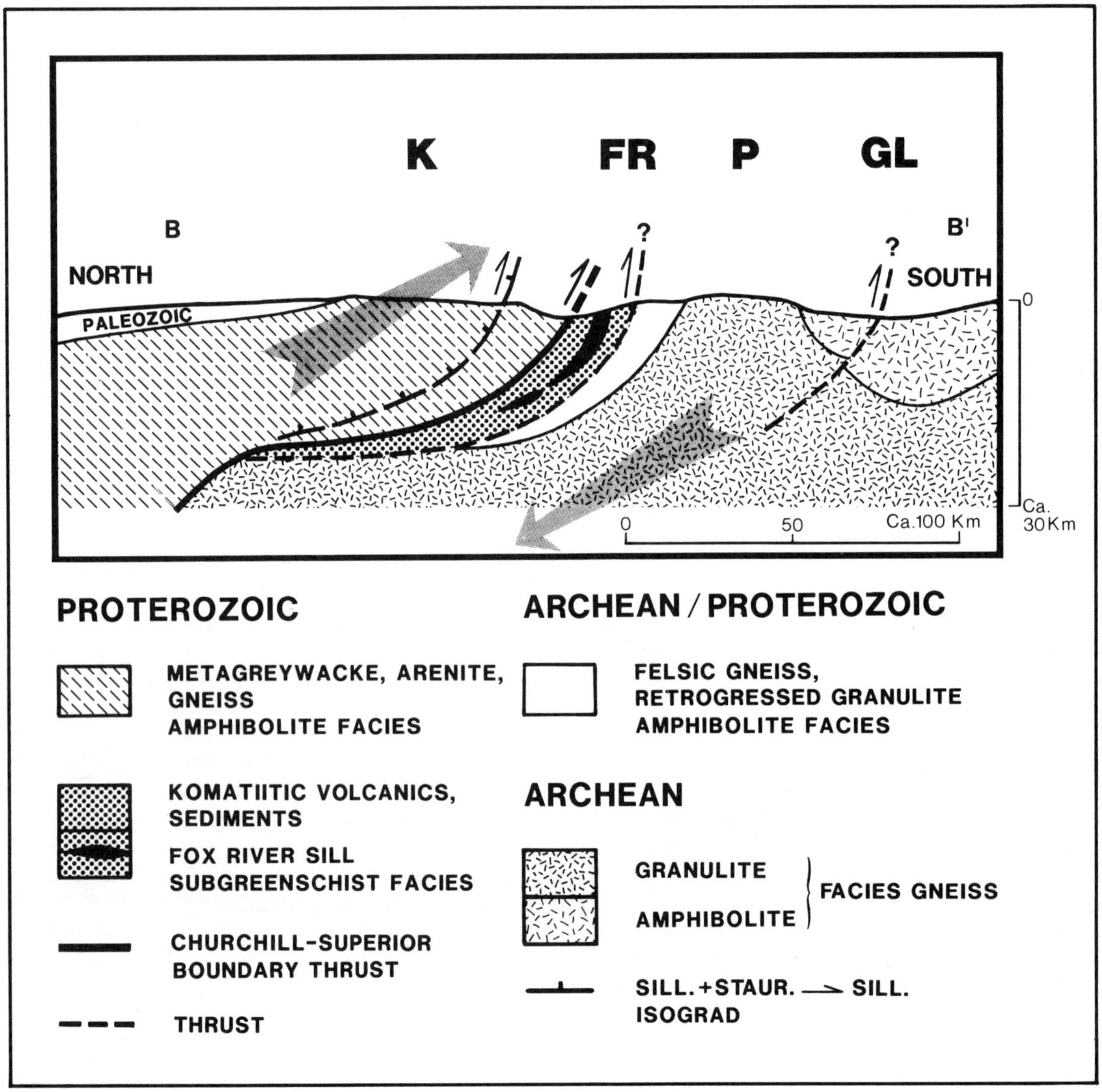

Figure 4. *Schematic cross-section through the Fox River belt. Abbreviations: K, Kisseynew domain; FR, Fox River belt; P, Pikwitonei domain; and GL, Gods Lake domain. Section line B–B′ is shown in Figure 2.*

trending, northerly dipping mylonite zones in the northwest Superior Province (Fig. 2). Some of these appear to displace 1883 Ma Molson dykes; for example, the Nelson River dyke appears to be offset by a mylonite zone northeast of Sipiwesk Lake (Manitoba Energy and Mines, 1979), suggesting that mylonite zones represent foreland or back-arc (?) shear zones and that the crust of northwestern Superior Province may have been shortened during this thrusting event.

Geophysical data

Gravity. A comparison of the geological map (Manitoba Energy and Mines, 1979) and the gravity map (Fig. 5) indicates that regional positive gravity anomalies in the northwestern Superior Province are related to granulites of the Pikwitonei domain, as noted by Gibb (1968). The gravity signature coinciding with exposed granulites extends into the Fox River area up to about 80 km north of the surficial contact between Archean gneisses and Fox River supracrustals. This may be interpreted to mean that granulites underlie Fox River belt supracrustal rocks and Kisseynew gneisses to at least 80 km north of this contact.

A strong linear positive gravity anomaly is superimposed on the regional positive gravity anomaly: this corresponds with the Fox River Sill. However, the maximum of the linear anomaly is offset by 5–7 km to the north of the sill subcrops (Manitoba Energy and Mines, 1979). This, and the fact that a steep horizontal gravity gradient is associated with the southern margin of the Fox River Sill gravity anomaly, suggests that the sill has a northerly dip. This observation is in agreement with northerly dips of supracrustal strata indicated by geological data.

Geophysical and geological data thus indicate that the Kisseynew gneiss is allochthonous and may have been thrust 80 km or more over the Superior Province margin in the Fox River region (Fig. 4). This interpretation could be tested by seismic surveys.

Gravity (Fig. 5) and aeromagnetic data in the Lake Winnipegosis – Pelican Lake area, in the southern extension of the Thompson belt beneath Paleozoic cover, suggest a tectonic relationship between the Superior Province and Reindeer Zone similar to that in the Fox River area. A narrow, linear, discontinuous, strongly magnetic anomaly at least 100 km long indicates an ultramafic body similar to the Fox River Sill, beneath Paleozoic cover. Drilling has confirmed that ultramafic rocks are the cause of this anomaly (drillhole DENBY #1, Cominco, Manitoba Energy and Mines, canceled assessment files). A positive gravity anomaly is also associated with the body (Fig. 5), but the peak value is displaced 5–10 km to the west, suggesting a westerly dip. A very steep horizontal gradient is associated with the eastern margin of the positive gravity anomaly. This supports the interpretation of a westerly dip for the ultramafic body. The aeromagnetic expression of this body suggests that it has not been folded. This implies that the southern extension of the Thompson belt is less deformed than the exposed portion and may be more like the Fox River belt.

Seismic. Reflection seismic data from the subsurface extension of the Thompson belt in southwestern Manitoba at latitude 49°30′ (Green *et al.*, 1985) indicate westerly dipping (up to 36°) reflectors in the Churchill–Superior Boundary Zone that have been interpreted by Green *et al.* (1985) as being caused by thrusts within Archean gneisses and/or thrusts separating different lithologies and/or layering within supracrustal sequences. A similar pattern of reflections has been observed on a short seismic reflection section recorded across the Thompson belt in the Thompson area (Green, 1981). The data are consistent with a model in which the Reindeer Zone is underthrust by the Superior Province. Recent refraction seismic data from latitude 49°30′ in southwestern Manitoba and southeastern Saskatchewan (Morel-a-l'Huissier *et al.*, 1987) support this model. Earlier refraction seismic surveys over the exposed Shield (Mereu and Hunter, 1969; Hall and Hajnal, 1973) indicate a thickened crust of 35 km in the Reindeer Zone, adjacent to the Thompson belt, compared with a thickness of approximately 30 km in adjacent areas, again consistent with the proposed model.

SUMMARY AND CONCLUSIONS

The Churchill–Superior Boundary Zone lies at the northwestern margin of the Archean Superior craton and is in fault or thrust contact with Proterozoic rocks of the Reindeer Zone of the Trans-Hudson Orogen. The exposed Churchill–Superior Boundary Zone includes the Thompson belt, Split Lake block, and Fox River belt. Characteristic aeromagnetic and gravity anomaly signatures indicate that the Churchill–Superior Boundary Zone extends beneath Phanerozoic cover into southwestern Manitoba and the United States. It consists dominantly of polymetamorphic migmatitic gneisses, in part derived from 2.6–2.7 Ga Pikwitonei-type granulites and in part from 3.0 Ga upper crustal gneisses through metamorphic and structural overprinting during Hudsonian orogeny resulting from collision between the Archean Superior craton and Proterozoic island-arc terrain of the Reindeer Zone. Precise U–Pb zircon, monazite, and sphene ages identify three Hudsonian metamorphic episodes: (1) 1820 Ma: peak conditions of middle to

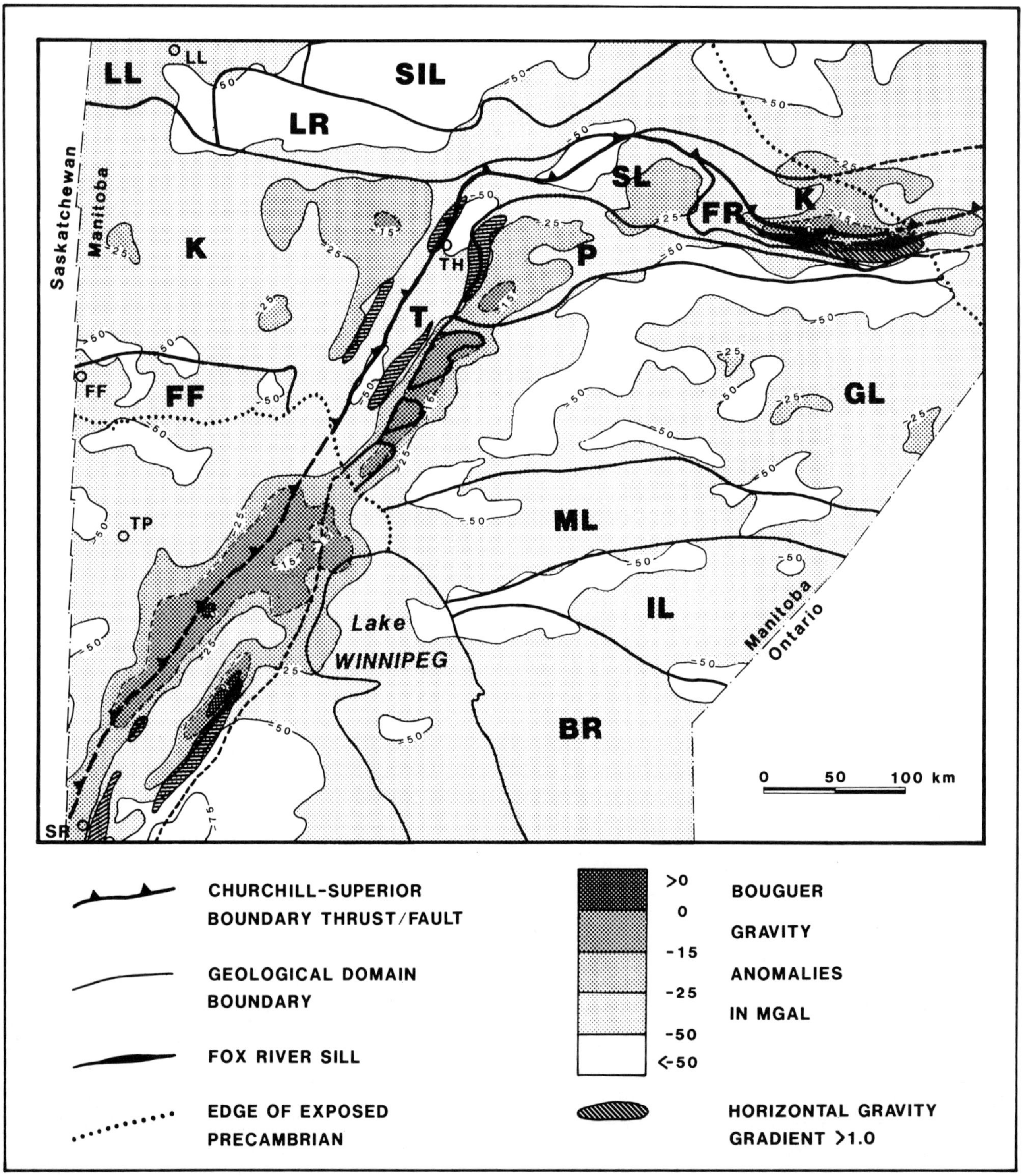

Figure 5. *Bouguer anomaly map of the Thompson region (after Energy, Mines and Resources Canada, 1981) and derived horizontal gradient map. Domain abbreviations (in bold lettering) as in Figure 2. Additional domain abbreviations: BR, Berens River domain; LL, Lynn Lake domain; and SIL, Southern Indian Lake domain. Locations: FF, Flin Flon; LL, Lynn Lake; SR, Swan River; TH, Thompson; and TP, The Pas. (1 Gal = 1 cm/s²).*

upper amphibolite facies conditions (700–750°C, 6–7 kbar (600–700 MPa)), with the generation of thrust-related (?) granitoids; (2) 1770-1790 Ma: generation of pegmatoids, possibly due to metamorphic relaxation; and (3) 1720–1740 Ma: last metamorphic event, with local formation of granitoid melts.

Proterozoic metavolcanic and metasedimentary rocks in the Churchill–Superior Boundary Zone are part of the Circum-Superior belt. They occur as highly deformed, infolded and/or fault-bounded belts and slivers in the western part of the Thompson belt, and in the Split Lake block, and make up the much less deformed east-trending Fox River belt. Although the magmatic rocks have rift-type chemical affinities, those of the 1883 ± 2 Ma Fox River Sill and the comagmatic Molson dyke swarm cannot be part of the initial rifting as previously proposed because they are coeval with, or younger than, arc volcanic rocks in the Trans-Hudson Orogen. Instead, lithological, geochronological, and geochemical data are consistent with a model of synorogenic rifting, in which igneous material of oceanic origin underplated a tectonically thinned Superior Province continental margin, and derived magmas were intruded and extruded into a marginal basin or back-arc basin tectonic setting. The latter implies a contemporaneous subduction zone in the Reindeer Zone island-arc terrain, dipping towards the Superior craton.

Geophysical and geological data suggest that the Archean Superior craton and adjacent Churchill–Superior Boundary Zone underthrust the Proterozoic Reindeer Zone, thereby producing the thermotectonically reworked rocks of the Churchill–Superior Boundary Zone.

The main geological evidence for underthrusting along the Thompson belt is the occurrence of Proterozoic granitoid rocks with high $^{87}Sr/^{86}Sr$ initial ratios of greater than 0.705 in the Reindeer Zone within 20 km of the Thompson belt. This suggests that Archean rocks underlay the eastern margin of the Reindeer Zone during granite generation. Geological evidence for underthrusting along the Fox River belt includes northerly dipping planar structures in the stratigraphically uppermost portions of the Fox River belt and in the adjacent overlying Kisseynew gneisses, and the fact that lower greenschist to prehnite–pumpellyite facies Fox River belt rocks are structurally overlain by upper amphibolite facies Kisseynew gneisses characterized by inverted metamorphic isograds at the (structural) base.

The main geophysical evidence for underthrusting derives from gravity data. Gravity anomaly patterns typical of Pikwitonei granulites extend north of the Fox River belt, suggesting that Pikwitonei granulites dip north beneath the Fox River belt. A positive gravity anomaly related to the Fox River Sill is characterized by a steep horizontal gradient along its southern margin and its peak is displaced 5–7 km north with respect to the sill subcrops; this suggests that the sill dips to the north.

Recognition that the northwestern Superior Province margin dips underneath the Reindeer Zone, and the fact that the rift related magmatic rocks in the Churchill–Superior Boundary Zone are not related to initial rifting, suggests that the edge of the Archean plate (created through the breakup of an Archean protocontinent prior to initiation of the Trans-Hudson Orogen) lies beneath Proterozoic rocks an unknown distance northwest of the present surface contact between the Churchill and Superior provinces.

ACKNOWLEDGEMENTS

The author wishes to thank the following colleagues who have contributed over the last 12 years to the better understanding of the complex Churchill–Superior Boundary Zone and to this paper: R.F.J. Scoates, J.J.M.W. Hubregtse, J.J. Macek, H.V. Zwanzig, T.M. Corkery, and W.D. McRitchie, presently or formerly with the Manitoba Geological Survey; N. Halden, University of Manitoba; W. Bleeker, University of New Brunswick; A. Green, O.R. Eckstrand, and C. Van Staal, Geological Survey of Canada; N. Machado, L. Heaman, and T. Krogh, Royal Ontario Museum; W. Peredery, R. Somerville, and M. Toderian, INCO Ltd.; and F. Fueten, University of Toronto. R.F.J. Scoates, J.F. Lewry, H.V. Zwanzig, W.D. McRitchie, and W.R.A. Baragar reviewed and improved the manuscript. Thanks are also extended to Shirley Weselak and Leah Chudy for their efforts to read my writing. Responsibility for the interpretations presented in this paper lies entirely with the author.

REFERENCES

Bailes, A.H., 1980, Origin of Early Proterozoic volcaniclastic turbidites, south margin of the Kisseynew sedimentary gneiss belt, File Lake, Manitoba: Precambrian Research, v. 12, p. 197–225.

Baldwin, D.A., Syme, E.C., Zwanzig, H.V., Gordon, T.M., Hunt, P.A. and Stevens, R.P., 1987, U–Pb zircon ages from the Lynn Lake and Rusty Lake metavolcanic belts, Manitoba: two ages of Proterozoic magmatism: Canadian Journal of Earth Sciences, v. 24, p. 1053–1063.

Baragar, W.R.A. and Scoates, R.F.J., 1981, The Circum-Superior belt: a Proterozoic plate margin?, *in* Kröner, A., ed., Precambrian Plate Tectonics: Elsevier, Amsterdam, p. 297–330.

Berckhemer, H., 1968, Topographie des "Ivrea-Korpers" abgeleited aus seismischen und gravimetrischen Daten: Symposium "Zone Ivrea-Verbano": Schweizerische Mineralogische und Petrographische Mitteilungen, v. 48, p. 235–246.

Bleeker, W., 1988, Structural investigations in the Thompson nickel belt, Manitoba: Geological Survey of Canada, Current Activity Forum 1988, Program with Abstracts.

Bleeker, W., 1990, New structure – metamorphic constraints on Early Proterozoic oblique collision along the Thompson Nickel Belt, Manitoba, Canada, in Lewry, J.F. and Stauffer, M.R., eds., The Early Proterozoic Trans-Hudson Orogen of North America: Geological Association of Canada, Special Paper 37, p. 57–73.

Bleeker, W. and Macek, J.J., 1988, Thompson nickel belt project – Pipe pit mine: Manitoba Energy and Mines, Report of Field Activities 1988, p. 111–115.

Bond, G.C. and Kominz, M.A., 1988, Evolution of thought on passive continental margins from the origin of geosynclinal theory (~1860) to the present: Geological Society of America Bulletin, v. 100, p. 1909–1933.

Bostock, H.H., 1971, Geological notes on Aquatuk River map area, Ontario, with emphasis on the Precambrian rocks: Geological Survey of Canada, Paper 79-42.

Brooks, C. and Theyer, P., 1981, Rb/Sr geochronology in the Thompson belt, Manitoba: implication for Aphebian crustal development and metallogenesis: Canadian Journal of Earth Sciences, v. 18, p. 932–943.

Charbonneau, R., Scoates, R.F.J. and Macek, J.J., 1979, Thompson Nickel belt project: Manitoba Mineral Resources Division, Report of Field Activities 1979, p. 20–24.

Clark, G.S., 1980, Rubidium–strontium geochronology in the Lynn Lake greenstone belt, northwestern Manitoba: Manitoba Department of Energy and Mines, Geological Paper GP80-2, 9 p.

Clark, G.S., 1981, Rubidium–strontium geochronology in the Churchill structural province, northern Manitoba: Manitoba Mineral Resources Division, Report of Field Activities 1981, p. 97–99.

Clark, G.S., Anderson, R. and McRitchie, W.D., 1974, Total rock Rb–Sr ages from the Burntwood area: Centre for Precambrian Studies, University of Manitoba, Annual Report 1974, Part 2, Research, p. 51–52.

Corkery, M.T., 1985, Geology of the Lower Nelson River Project area: Manitoba Energy and Mines, Geological Report GR82-1.

Corkery, M.T. and Cameron, H.D.M., 1987, Cross Lake supracrustal investigations: Manitoba Energy and Mines, Report of Field Activities 1987, p. 135.

Corkery, M.T., Lenton, P.G., Breedveld, M. and Davis, D.W., 1988, Cross Lake geological investigations: Manitoba Energy and Mines, Report of Field Activities 1988, p. 106–110.

Cranstone, D.A. and Turek, A., 1976, Geological and geochronological relationships of the Thompson nickel belt, Manitoba: Canadian Journal of Earth Sciences, v. 13, p. 1058–1069.

Cumming, G.L., Eckstrand, O.R. and Peredery, W.V., 1982, Geochronological interpretations of lead isotope ratios in nickel sulphides of the Thompson belt, Manitoba: Canadian Journal of Earth Sciences, v. 19, p. 2306–2324.

Dewey, J.F. and Burke, K.C.A., 1973, Tibetan, Variscan, and Precambrian basement reactivation: products of continental collision: Journal of Geology, v. 81, p. 683–692.

Energy, Mines and Resources Canada, 1981, Gravity map of Canada 1980, manuscript maps nos. 52090, 52096, 56090, 56096. Gravity and Geodynamics Division, Earth Physics Branch.

Fountain, D.M. and Salisbury, M.H., 1981, Exposed cross-sections through the continental crust: implications for crustal structure, petrology and evolution: Earth and Planetary Science Letters, v. 56, p. 263–277.

Fueten, F. and Robin, P.-Y.F., 1989, Structural petrology along a transect across the Thompson belt, Manitoba: dip-slip at the western Churchill–Superior boundary: Canadian Journal of Earth Sciences, v. 26, p. 1976–1989.

Gibb, R.A., 1968, A geological interpretation of the Bouguer anomalies adjacent to the Churchill–Superior boundary in northern Manitoba: Canadian Journal of Earth Sciences, v. 5, p. 439–453.

Gibb, R.A., 1975, Collision tectonics in the Canadian Shield?: Earth and Planetary Science Letters, v. 27, p. 378–382.

Gibb, R.A. and Walcott, R.I., 1971, A Precambrian suture in the Canadian Shield: Earth and Planetary Science Letters, v. 10, p. 417–422.

Gibb, R.A., Thomas, M.P., Lapointee, P.C. and Mukhopadhyay, M., 1983, Geophysics of proposed Proterozoic sutures in Canada: Precambrian Research, v. 19, p. 349–394.

Giese, P., 1968, Die Struktur der Erdkruste im Bereich der Ivrea-Zone: Schweizerische Mineralogische und Petrographische Mitteilungen, v. 48, p. 261–284.

Gordon, T.M., Lemkow, D.R. and Roddick, J.C., 1987, Metamorphism of the Kisseynew sedimentary gneiss belt, Manitoba: thermal evolution of an Early Proterozoic interarc basin: Geological Association of Canada – Mineralogical Association of Canada, Program with Abstracts, v. 12, p. 48.

Gordon, T.M., Hunt, P.A., Bailes, A.H. and Syme, E.C., 1990, U–Pb ages from the Flin Flon and Kisseynew belts, Manitoba: chronology of crust formation at an Early Proterozoic accretionary margin, in Lewry, J.F. and Stauffer, M.R., eds., The Early Proterozoic Trans-Hudson Orogen of North America: Geological Association of Canada, Special Paper 37, p. 177–199.

Green, A.G., 1981, Results of a seismic reflection survey across the fault zone between the Thompson nickel belt and the Churchill tectonic province, northern Manitoba: Canadian Journal of Earth Sciences, v. 18, p. 13–25.

Green, A.G., Hajnal, Z. and Weber, W., 1985, An evolutionary model of the western Churchill Province and western margin of the Superior Province of Canada and the north-central United States: Tectonophysics, v. 116, p. 281–322.

Halden, N.M., 1988, Tectonic setting of Circum-Superior mafic and ultramafic magmatism in Manitoba: Manitoba Energy and Mines, Report of Field Activities 1988, p. 116–123.

Hall, D.H. and Hajnal, Z., 1973, Deep crustal seismic studies in Manitoba: Bulletin of the Seismological Society of America, v. 63, p. 885–910.

Heaman, L.M., Machado, N., Krogh, T. and Weber, W., 1986a, Preliminary U–Pb zircon results from the Pikwitonei granulite domain, Manitoba: Geological Association of Canada – Mineralogical Association of Canada, Program with Abstracts, p. 79.

Heaman, L.M., Machado, N., Krogh, T.E. and Weber, W., 1986b, Precise U–Pb zircon ages for the Molson dyke swarm and the Fox River Sill: implications for Early Proterozoic crustal evolution in NE Manitoba, Canada: Contributions to Mineralogy and Petrology, v. 94, p. 82N-89.

Hoffman, P.F., 1988, Proterozoic foredeeps, foredeep magmatism and Superior-type iron formations in the Canadian Shield, in Kröner, A., ed., Proterozoic Lithospheric Evolution: American Geophysical Union, Geodynamics Series. (In press.)

Hubregtse, J.M.W., 1977, Sipiwesk Lake – Wintering Lake area: Manitoba Mineral Resources Division, Report of Field Activities 1977, p. 73-79.

Hubregtse, J.M.W., 1978, Sipiwesk Lake – Landing Lake – Wintering Lake area: Manitoba Mineral Resources Division, Report of Field Activities 1978, p. 54–62.

Hubregtse, J.M.W., 1980, The Archean Pikwitonei granulite domain and its position at the margin of the northwestern Superior Province, central Manitoba; Manitoba Department of Energy and Mines, Geological Paper GP80-3, 16 p.

Hubregtse, J.M.W., 1985, Geology of the Oxford Lake – Carrot River area: Manitoba Energy and Mines, Geological Services, Geological Report GR83-1A, 73 p.

Josse, G., Clark, G. and Bailes, A.H., 1974, Rubidium–strontium age determinations from the File–Morton–Woosey lakes area of the Flin Flon volcanic belt: Centre for Precambrian Studies, University of Manitoba, Annual Report 1974, p. 40–50.

Kornik, L.J., 1969, An aeromagnetic study of the Moak Lake – Setting Lake structure in northern Manitoba: Canadian Journal of Earth Sciences, v. 6, p. 373-381.

Kornik, L.J. and MacLaren, A.S., 1966, Aeromagnetic study of the Churchill–Superior boundary in northern Manitoba: Canadian Journal of Earth Sciences, v. 3, p. 547–557.

Krogh, T., Heaman, L., Machado Fernandez, N. and Weber, W., 1985, U–Pb geochronology program: Thompson belt and Pikwitonei domain: Manitoba Energy and Mines, Report of Field Activities 1985, p. 183–184.

Macek, J.J., 1987, Geological mapping at Pipe mine: Manitoba Energy and Mines, Report of Field Activities 1987, p. 136–137.

Machado, N., Krogh, T.E. and Weber, W., 1990, U–Pb geochronology of basement gneisses in the Thompson belt (Manitoba): evidence for pre-Kenoran and Pikwitonei-type crust and Early Proterozoic basement reactivation in the western margin of the Archean Superior Province: Canadian Journal of Earth Sciences, v. 27, p. 794–802.

Machado, N., 1990, Evaluation of two Early Proterozoic orogens in the Canadian Shield: a comparison of U–Pb ages for the Labrador Trough (Quebec), the Thompson Belt and the Trans-Hudson Orogen (Manitoba and Saskatchewan), in Lewry, J.F. and Stauffer, M.R., eds., The Early Proterozoic Trans-Hudson Orogen of North America: Geological Association of Canada, Special Paper 37, p. 433–441.

Machado, N., Heaman, L., Krogh, T.E. and Weber, W., 1987, U-Pb geochronology program: Thompson belt – northern Superior Province: Manitoba Energy and Mines, Report of Field Activities 1987, p. 138–140.

Machado, N., Heaman, L. and Weber, W., 1988, U–Pb geochronology program: Thompson belt and northwest Superior Province: Manitoba Energy and Mines, Report of Field Activities 1988, p. 124.

Manitoba Mineral Resources Division, 1979, Geological map of Manitoba: Scale 1:1,000,000, Map 79-2.

McCabe, H.R., 1971, Stratigraphy of Manitoba, an introduction and review, in Turnock, A.C., ed., Geoscience Studies in Manitoba: Geological Association of Canada, Special Paper 9, p. 167–188.

Mereu, R.T. and Hunter, J.A., 1969, Crustal and upper mantle structure under the Canadian Shield from Project Early Rise data: Bulletin of the Seismological Society of America, v. 59, p. 147–165.

Mezger, K., Hanson, G.N. and Bohlen, S.R., 1989a, U–Pb systematics of garnet: dating the growth of garnet in the late Archean Pikwitonei granulite domains at Cauchon and Natawahunan lakes, Manitoba, Canada: Contributions to Mineralogy and Petrology, v. 101, p. 136–148.

Mezger, K., Hanson, G.N. and Bohlen, S.R., 1989b, U–Pb ages of metamorphic rutiles: application to the cooling history of high grade terranes: Earth and Planetary Science Letters, v. 96, p. 106–118.

Mezger, K., Bohlen, S.R. and Hanson, G.N., 1990, Metamorphic history of the Archean Pikwitonei granulite domain and the Cross Lake subprovince, Superior Province, Manitoba, Canada: Journal of Petrology, v. 31, p. 483–517.

Morel-a-l'Huissier, P., Green, A.G. and Pike, C.J., 1987, Crustal refraction surveys across the Trans-Hudson Orogen/Williston basin of south-central Canada: Journal of Geophysical Research, v. 92(137), p. 6403–6420.

Pactunc, A.D. and Baer, A.J., 1986, Geothermobarometry of the northwestern margin of the Superior Province: implications for its tectonic evolution: Journal of Geology, v. 194, p. 381–394.

Peredery, W.V. and the geological staff of Inco Metals Company, Thompson, Manitoba, 1982, Geology and nickel sulphide deposits of the Thompson belt, Manitoba, in Hutchinson, R.W., Spence, C.D. and Franklin, J.M., eds., Precambrian Sulphide Deposits (H.S. Robinson Memorial Volume): Geological Association of Canada, Special Paper 25, p. 165–209.

Russell, J.K., 1981, Metamorphism of the Thompson nickel belt gneisses: Canadian Journal of Earth Sciences, v. 18, p. 191–209.

Scoates, R.F.J., 1981, The volcanic rocks of the Fox River belt, northeastern Manitoba: Manitoba Mineral Resources Division, Geological Report GR81-1, 109 p.

Scoates, R.F.J., 1990, The Fox River Sill, northeastern Manitoba – a major stratiform intrusion: Manitoba Geological Services, Geological Report GR82-3.

Scoates, R.F.J. and Macek, J.J., 1978, Molson dyke swarm: Manitoba Mineral Resources Division, Geological Paper 78-1, 53 p.

Scoates, R.F.J. and Weber, W., 1977, Fox River regional correlation, part A:5 Dafoe – Fox River: Manitoba Mineral Resources Division, Report of Field Activities 1977, p. 62, 63.

Scoates, R.F.J., Macek, J.J. and Russell, J.K., 1977, Thompson nickel belt project: Manitoba Resources Division, Report of Field Activities 1977, p. 47–54.

Scoates, R.F.J., Corkery, M.T., Macek, J.J., Trueman, D.L. and Weber, W., 1981, Geology of the western part of the Fox River belt: Manitoba Mineral Resources Division, Map GR81-1. (Also enclosed in Manitoba Mineral Resources Division, Geological Report GR81-1.)

Stephenson, J.F., 1974, Geology of the Ospwagan Lake (east half) area: Manitoba Mines Branch, Publication 74-1, 67 p.

Syme, E.C., 1990, Stratigraphy and geochemistry of two Proterozoic arcs: Lynn Lake and Flin Flon metavolcanic belts, Manitoba, in Lewry, J.F. and Stauffer, M.R., eds., The Early Proterozoic Trans-Hudson Orogen of North America: Geological Association of Canada, Special Paper 37, p. 143–161.

Syme, E.C., Bailes, A.H., Gordon, T.M. and Hunt, P.A., 1987, U–Pb zircon geochronology in the Flin Flon belt: age of Amisk volcanism: Manitoba Energy and Mines, Report of Field Activities 1987, p. 105–107.

Van Schmus, W.R., Bickford, M.E., Lewry, J.F. and MacDonald, R., 1987, U–Pb geochronology in the Trans-Hudson Orogen, northern Saskatchewan, Canada: Canadian Journal of Earth Sciences, v. 24, p. 407–424.

Weber, W., 1976, Cauchon, Partridge Crop and Apussigamasi lakes area: Manitoba Mineral Resources Division, Report of Field Activities 1976, p. 54–57.

Weber, W., 1980, The Pikwitonei granulite domain: edge of the Superior Province craton (abstract): Transactions of the American Geophysical Union, v. 61(17), p. 386–397.

Weber, W. and Mezger, K., *in press*, An oblique cross section of Archean continental crust at the northwestern margin of the Superior Province, Manitoba, Canada, in Salisbury, M.H. and Fountain, D.M., eds., Exposed Cross-sections of the Continental Crust: Kluwer Academic Publishers, Dordrecht, Netherlands.

Weber, W. and Scoates, R.F.J., 1976, Mystery Lake (63P-13): Manitoba Mineral Resources Division, Preliminary Map 1976 T-1, scale 1:50,000.

Weber, W. and Scoates, R.F.J., 1978, Archean and Proterozoic metamorphism in the northwestern Superior Province and along the Churchill–Superior boundary, Manitoba, in Fraser, J.A. and Heywood, W.W., eds., Metamorphism in the Canadian Shield: Geological Survey of Canada, Paper 78-10, p. 5–16.

Wernicke, B., 1985, Uniform-sense normal simple shear of the continental lithosphere: Canadian Journal of Earth Sciences, v. 22, p. 108–135.

Windley, B.F., 1983, Metamorphism and tectonics of the Himalaya: Journal of the Geological Society of London, v. 140, p. 849–865.

Zwanzig, H.V., 1990, Stratigraphy and tectonics of the Kisseynew complex, Manitoba, in Lewry, J.F. and Stauffer, M.R., eds., The Early Proterozoic Trans-Hudson Orogen of North America: Geological Association of Canada, Special Paper 37, p. 95–120.

New Structural–Metamorphic Constraints on Early Proterozoic Oblique Collision Along the Thompson Nickel Belt, Manitoba, Canada

Wouter Bleeker
Department of Geology, University of New Brunswick,
Box 4400, Fredericton, New Brunswick E3B 5A3, Canada

Bleeker, W., New structural–metamorphic constraints on Early Proterozoic oblique collision along the Thompson Nickel Belt, Manitoba, Canada, in Lewry, J.F. and Stauffer, M.R., eds., The Early Proterozoic Trans-Hudson Orogen of North America: Geological Association of Canada, Special Paper 37, p. 57–73.

Abstract

The Thompson Nickel Belt of northern Manitoba represents an Early Proterozoic collision zone between juvenile Churchill Province terranes and the margin of the Archean Superior plate. Recently proposed plate tectonic models for the assembly of this part of the Canadian Shield have primarily been based on geophysical and geochronological data, and large-scale geological relationships. This study presents new, detailed structural and metamorphic constraints and reviews existing data. An evolutionary model is proposed that satisfies both regional-scale and outcrop-scale observations, and clarifies the complex structure of the belt.

Résumé

La ceinture Thompson Nickel au nord du Manitoba représente une zone de collision Protérozoïque Inférieur située entre des terranes juvéniles de la province de Churchill et la marge de la plaque archéenne du Supérieur. Le modèle de tectonique de plaques récemment proposé pour l'accrétion de cette partie du bouclier canadien est basé sur des données géophysiques et géochronologiques et sur les relations géologiques à grande échelle. Cette étude présente des données structurales et métamorphiques détaillées auxquelles le modèle est soumis et révise les données existantes. On propose un modèle évolutif qui satisfait aux observations de terrain à petites et grandes échelles et clarifie la structure complexe de la ceinture.

INTRODUCTION

The Thompson Nickel Belt (hereafter referred to as TNB) of northern Manitoba (Fig. 1) has long been considered one of the major sutures in the Canadian Shield (Gibb and Walcott, 1971; Gibb, 1975). It represents part of the Churchill–Superior Boundary Zone (Weber and Scoates, 1978; Green *et al.*, 1985), which separates unreworked >2.5 Ga Archean crust of the Superior craton from largely juvenile 1.9–1.7 Ga Early Proterozoic crust (Chauvel *et al.*, 1987) of the internal zone of the Trans-Hudson Orogen (Lewry *et al.*, 1985; Hoffman, 1988b).

Early attempts to subdivide the shield into structural provinces focused attention on this junction as the probable locus of a major province boundary (for a review see Bell (1971)). Interest was enhanced by the discovery of nickel deposits along the Setting Lake – Moak Lake belt (Zurbrigg, 1963; Coats *et al.*, 1972), more recently called the TNB. Subsequent geological mapping, and geochronological and geophysical studies firmly established the existence of a major structural boundary, although different workers, relying on different criteria, generally proposed different locations for it. Recent work strongly favours a plate tectonic interpretation of the boundary (Gibb and Walcott, 1971; Gibb, 1975, 1983; Lewry, 1981; Green *et al.*, 1985; Hoffman, 1988b), and it is interesting to note that former confusion as to the exact location of the Churchill–Superior boundary is a natural consequence of the gently dipping attitude of the plate contact in a collisional orogen (Fig. 2). Different methods will identify different attributes and depth

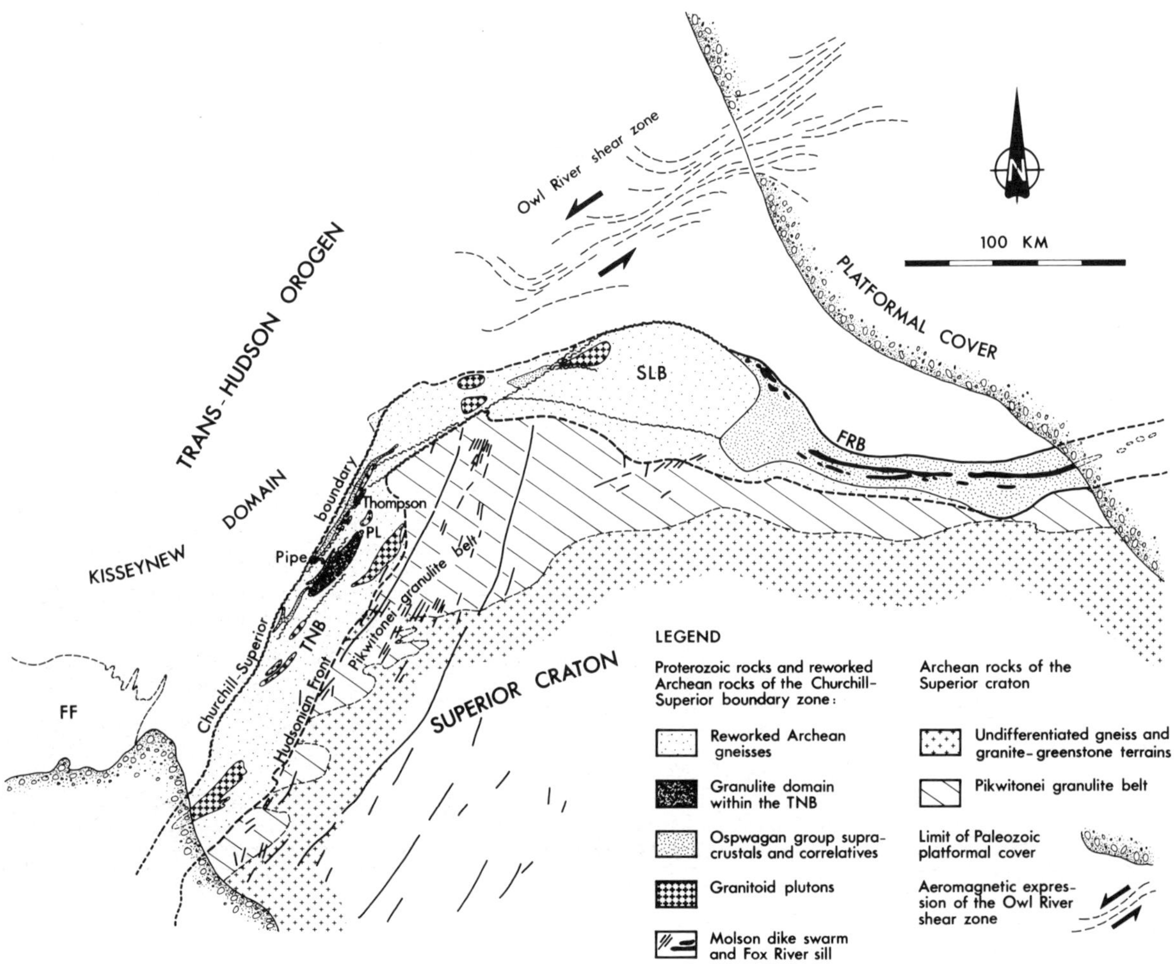

Figure 1. *Simplified geological map of the northwest part of the Superior craton, Churchill–Superior Boundary Zone, and adjacent internal zone of the Trans-Hudson Orogen (Churchill Province). The Churchill–Superior Boundary Zone comprises the Thompson Nickel Belt (TNB), Split Lake Block (SLB), and Fox River Belt (FRB). Abbreviations: FF, Flin Flon – Snow Lake volcanic belt; PL, Paint Lake granulite domain.*

levels of the gently dipping plate contact and will, therefore, yield divergent results.

Plate tectonic modeling of the assembly of this part of the shield (Gibb, 1983; Green *et al.*, 1985; Hoffman, 1988b) has led to the view that the TNB represents an early collision zone, due to north–south or northwest–southeast oblique convergence between the Superior plate and juvenile Trans-Hudson terranes, which subsequently evolved into a sinistral transform boundary (Lewry, 1981).

The present study was initiated when development of an open pit at Thompson Mine created exceptional new outcrops in the otherwise poorly exposed TNB. The new structural and metamorphic data generated from these exposures, and from other selected areas, allow testing and refining of tectonic models for the belt.

REGIONAL GEOLOGICAL SETTING

The TNB is a 25–35 km wide belt, along the western margin of the Superior craton (Fig. 1), that is dominated by polymetamorphic gneisses. It has strong gravity and magnetic expressions (Kornik and MacLaren, 1966; Gibb, 1968; Kornik, 1969), attributes that permit delineation of the belt and its southern extension below platformal cover, as far as South Dakota (Green *et al.*, 1979). To the north-northeast, the belt appears structurally continuous with the Owl River shear zone (Bell, 1966), whose aeromagnetic expression records minimum sinistral displacement on the order of 100 km (Fig. 1), twice Bell's original estimate. Lithologically, however, the Churchill–Superior Boundary Zone swings to the east and has its extension in the Split Lake Block and the Fox River Belt (Fig. 1; Weber and Scoates, 1978). Like the TNB, the Split Lake Block is dominated by polymetamorphic gneisses, whereas the Fox River Belt consists of a homoclinal, steeply north-dipping sequence of Aphebian supracrustals and related intrusives (Baragar and Scoates, 1981), which have been correlated with similar supracrustals in the TNB, i.e., the Ospwagan group (Scoates *et al.*, 1977).

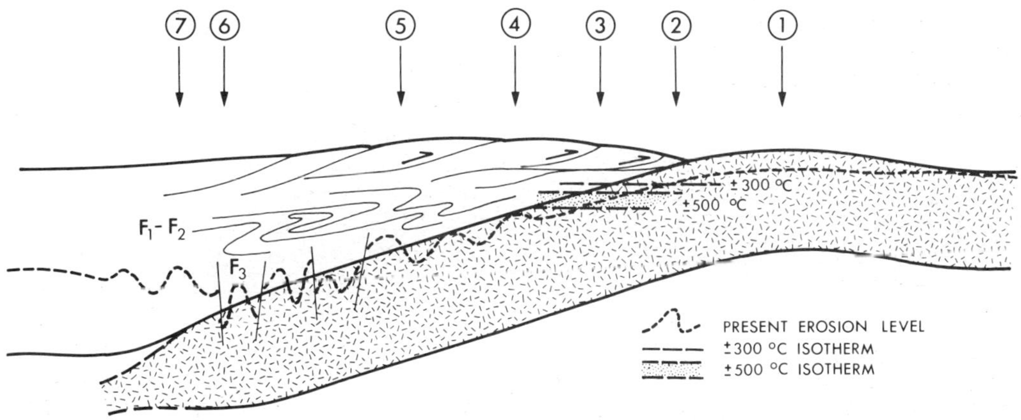

Figure 2. *Generalized section through a collisional orogen, as applied to the TNB (section viewed toward the north-northeast). An overthrust/nappe pile, comprising detached basement slices and Aphebian supracrustals, is emplaced across the Superior plate margin. For simplicity, late, high-amplitude, upright F_3 folds and subsequent ductile reverse faults are restored. The present erosion level exposes a deep, cryptic section through the originally northwest-dipping plate contact. Seven "boundaries" can be defined, many of which have played a role in the "Churchill–Superior boundary debate" (cf. Bell, 1971): (1) Outer gravity high over much of the Pikwitonei granulite belt — the Nelson River high (Gibb, 1968). (2) Intersection between the present erosion level and the ±300°C (transient) Hudsonian isotherm; west of this boundary, K–Ar biotite systems show complete resetting. (3) Where the ±500°C isotherm is crossed, K–Ar hornblende systems show resetting and magnetite-consuming reactions in the granulites cause an extensive aeromagnetic low. This is the present (arbitrary) definition of the eastern boundary (Kornik and MacLaren, 1966). (4) Upright F_3 folds attain sufficient intensity to dominate the aeromagnetic fabric of the belt. (5) First Aphebian supracrustals — a fundamental boundary, but obscure in the TNB because of poor exposure, basement involvement in thrusting, and tight upright F_3 folding. (6) Most westerly occurrence of (reworked) basement gneisses at the surface. (7) Leading, thinned edge of the Superior plate in the subsurface, probably corresponding with the strong gravity gradient between the TNB low and the inner Burntwood River high (Gibb, 1968). The western boundary of the TNB has been defined by boundary (6), but data available at present suggest that the leading edge of the Superior plate may project farther west in the subsurface.*

Unreworked Archean crust to the southeast of the TNB includes low- to medium-grade granite–greenstone and gneiss terrains, and the high-grade Pikwitonei granulite belt (Fig. 1). The latter shows sufficient similarities to the lower grade terrains to suggest that it represents deeper level exhumation of an overall gneiss and granite–greenstone crust (Roussel, 1965; Weber and Scoates, 1978; Hubregtse, 1980). The Pikwitonei granulite belt, which has an associated linear gravity high for much of its length (Gibb, 1968), parallels nearly the entire southeastern margin of the Churchill–Superior Boundary Zone.

Archean crust of the northwestern Superior craton is further characterized by mafic/ultramafic dykes of the 1883 Ma (Heaman *et al.*, 1986) Molson swarm (Ermanovics and Fahrig, 1975; Scoates and Macek, 1978), which can be followed into the TNB (Cranstone and Turek, 1976; this study), but are absent in Aphebian crust to the northwest.

On the northwest side, the Churchill–Superior Boundary Zone is bordered by a collage of Early Proterozoic terranes (Lewry, 1981; Green *et al.*, 1985; Lewry *et al.*, 1985, 1987; Hoffman, 1988b) of which the Kisseynew domain is in direct contact with the TNB. The Kisseynew domain probably represents the remnants of a back-arc or interarc basin (Hoffman, 1988b).

RELATIONSHIPS ALONG THE SOUTHEASTERN BOUNDARY OF THE THOMPSON NICKEL BELT: THE HUDSONIAN FRONT

K–Ar mineral dating on samples from the Pikwitonei granulites has generally yielded (late) Kenoran ages (e.g., Lowdon, 1962; Wanless *et al.*, 1966; Ermanovics and Wanless, 1983), indicating that the granulite belt experienced major uplift to shallow crustal levels well before the Hudsonian event. Closer to the TNB boundary, K–Ar ages start to show consistent resetting to Hudsonian ages.

Molson dykes provide excellent markers for recording Hudsonian metamorphic and deformational gradients across the eastern TNB boundary. Just within the Pikwitonei granulite belt, the dykes have well-developed chilled margins (Fig. 3a) and many appear to have intruded along cataclastic faults. The dykes contain small xenoliths of fault breccia and major dykes show bimodal north-northwest and north-northeast trends. Contact zones of north-northwest trending Molson dykes are consistently sheared and associated pseudotachylites (Fig. 3b,c) crosscut chilled margins. Movement along these brittle and brittle–ductile shear zones was sinistral strike slip and predated the thermal peak of the Hudsonian event, since just within the TNB these shear zones and associated pseudotachylites show static recrystallization and overgrowth of upper greenschist to lower amphibolite facies assemblages (Fig. 3b,c). Molson dykes apparently intruded just prior to, or possibly during, active strike-slip faulting, which seems incompatible with the hypothesis that they are related to rifting of the Archean plate (Hubregtse, 1980; Green *et al.*, 1985), marking initiation of the Wilson cycle that ended with the Hudsonian event. Moreover, recognition of 1.91 Ga island-arc volcanics in the Lynn Lake Belt (Baldwin *et al.*, 1987) shows that active convergence and lithosphere consumption within the internal zone of the Trans-Hudson Orogen occurred well before intrusion of the 1883 Ma Molson dyke swarm. Finally, mafic dykes within the TNB, which can be correlated with the Molson swarm on the basis of their chemistry and field relationships, cut through all Aphebian supracrustals and appear even to postdate earliest Hudsonian fold structures (Bleeker, 1990).

The pattern of K–Ar ages and the dyke – host rock relations indicate that the Molson dykes intruded relatively cold granulites. Thus, the suggestion that the Pikwitonei granulites were thrust to shallow crustal levels during Hudsonian collision (Fountain and Salisbury, 1981; Paktunc and Baer, 1986) is probably incorrect. Furthermore, the vergence suggested by that model is the opposite of that observed in the TNB (this study) and the

Figure 3. *Photomicrographs of diagnostic microstructures. Samples (a) through (d) are from Sipiwesk Lake, along the eastern boundary of the TNB. (a) Chilled margin of Molson dyke against two pyroxene–magnetite–plagioclase granulite; plane polarized light, scale bar is 2.0 mm. (b) Recrystallized pseudotachylite from brittle–ductile shear zone. Foam-textured quartz (left side) is part of a recrystallized porphyroclast; crossed polars, scale bar is 0.5 mm. (c) Recrystallized pseudotachylite similar to (b), but crosscutting Molson dyke. Recrystallized matrix, mainly hornblende–plagioclase, is overgrown by garnet poikiloblasts (black patches). Sharp contact with subophitic dyke rock at right side of photo; crossed polars, scale bar is 0.5 mm. (d) Euhedral garnet, with vermicular quartz inclusions, overgrowing Molson dyke igneous assemblage. Large horizontal crystal is a relict plagioclase phenocryst; unpolarized light, scale bar is 0.5 mm. (e) Muscovite deformed by F_3 microfold; note nearly complete lack of recovery in the strongly bent hinge, qualitatively indicating relatively low retrograde temperatures; crossed polars, scale bar is 0.1 mm, Ospwagan group semipelitic schist, Manasan Quarry. (f) Retrograde muscovite nest after peak-metamorphic sillimanite–quartz aggregate, partially overgrown by syn- to late-F_3 andalusite (grey porphyroblast at tip of arrow); crossed polars, scale bar is 1.0 mm, Ospwagan group metapelite, Pipe Mine.*

Trans-Hudson Orogen as a whole (Lewry *et al.*, 1985; Hoffman, 1988b).

The eastern boundary of the TNB has been defined on the basis of a continuous total field aeromagnetic low (Kornik and MacLaren, 1966; Kornik, 1969), separating high-intensity, east–west trending aeromagnetic fabric of the Pikwitonei granulite belt from lower intensity, northeast–southwest trending aeromagnetic fabric of the TNB. This is an arbitrary but convenient definition in that the low can be imaged accurately and coincides approximately with eastern limits of Hudsonian deformation and recognizable metamorphic overprinting. Thin-section observations show that incipient Hudsonian metamorphism eliminates the high magnetite content of the granulites via biotite-, hornblende-, or garnet-forming reactions. This thermal front, indicated by the lower am-

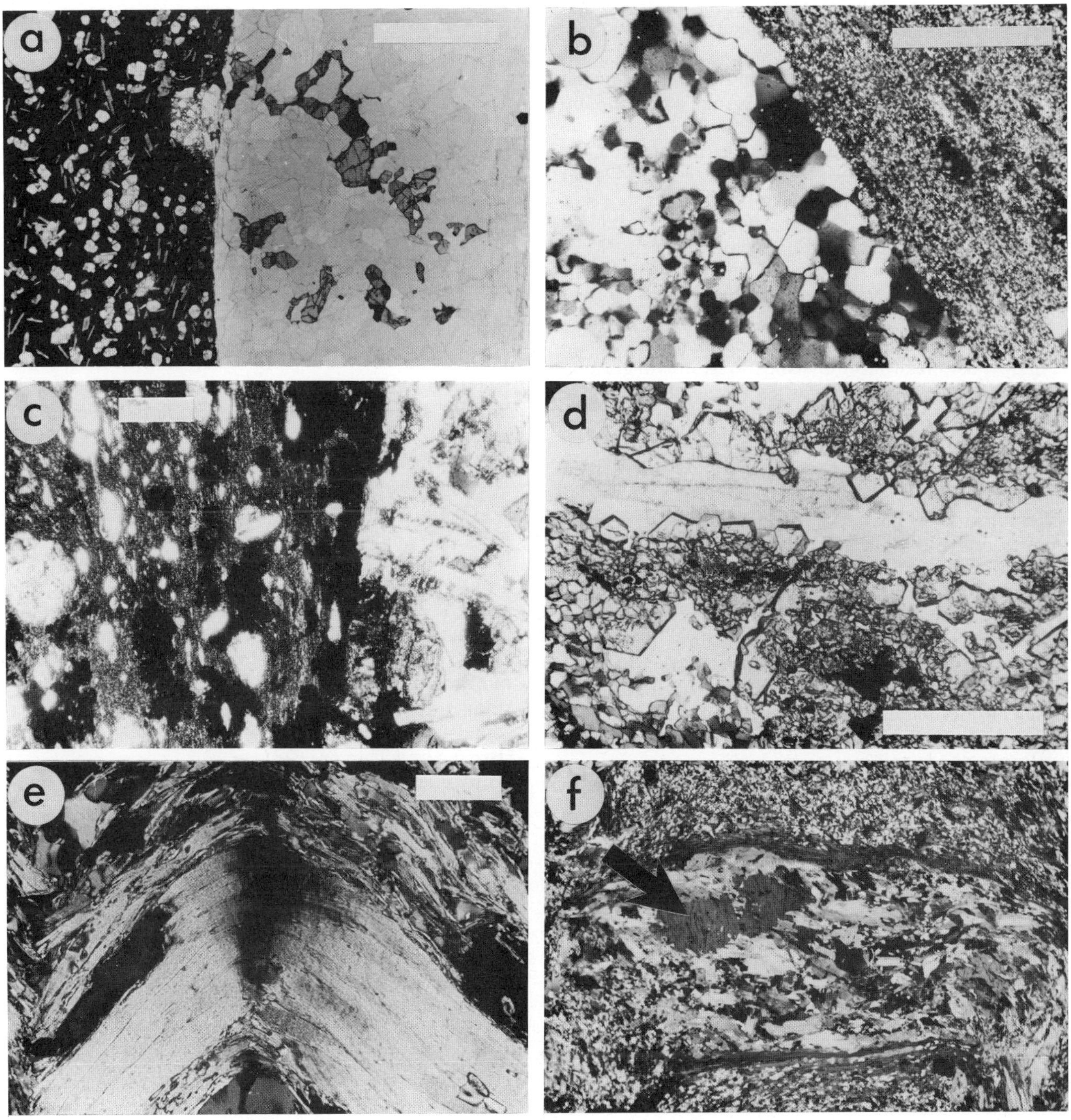

phibolite facies magnetite-consuming reactions, is situated east of the penetrative Hudsonian deformation front. This explains the sharp truncation of Kenoran aeromagnetic fabrics, with no indication of dragging Kenoran anomalies into the transposing grain of the TNB.

Where aqueous fluids were introduced, incipient Hudsonian metamorphism led to static retrogression of pyroxenes to fine-grained hornblende–quartz aggregates, with minor biotite and destruction of magnetite. Minor epidote is present close to the front. Brown–green hornblende, a late phase in the granulite assemblage, is relatively unaffected, although it generally shows strong exsolution of ilmenite platelets, a marginal colour change toward blue–green hornblende and marginal corrosion by biotite.

Where aqueous fluids did not penetrate the granulites, high-grade assemblages are well preserved and are separated on an outcrop scale from hydrated and retrogressed equivalents by sharp fronts. These hydration fronts overprint Molson dykes and can commonly be related to subhorizontal pegmatite dykes or veinlets. However, in mafic granulites, abundant fine-grained garnets occur locally in irregular patches that cut across the Kenoran compositional layering. Within the garnet-rich patches, orthopyroxene has reacted out or is rimmed by clinopyroxene, and textures indicate that garnet is the product of the pressure-sensitive reaction

plagioclase + orthopyroxene $\rightarrow$ garnet $\pm$ clinopyroxene + quartz

Resulting garnet–clinopyroxene–plagioclase–quartz±hornblende assemblages, which also overprint Molson dykes (Fig. 3d), are interpreted to be due to loading of Archean basement by a Hudsonian overthrust slab.

Well-preserved Pikwitonei granulite basement containing Molson dykes can be traced far into the TNB (Rance, 1966; Cranstone and Turek, 1976; this study) and has now been identified in basement windows as far west as the western boundary of the TNB.

GEOLOGY OF THE THOMPSON NICKEL BELT

The TNB is largely underlain by highly strained amphibolite to granulite facies polymetamorphic ortho- and paragneisses of Archean age (Cranstone and Turek, 1976; Machado *et al.*, 1987). These are generally considered to represent variably reworked and retrogressed Archean basement rocks, comparable to the Pikwitonei granulites (Cranstone and Turek, 1976; Weber and Scoates, 1978). However, some of the migmatitic amphibolite facies gneisses may represent reworking of Archean amphibolite facies protoliths. Prior to the Hudsonian collision, granulite basement was at shallow depths, but not necessarily exposed everywhere at the surface. At this stage, actual exposed basement may, in part, have comprised amphibolite facies migmatites, passing into granulites at a depth of several kilometres. Such an interpretation would greatly reduce the amount of aqueous fluids necessary to completely rehydrate the basement.

Aphebian supracrustals, ultramafic bodies with associated Ni deposits, mafic and ultramafic dykes and sills, and granitoid plutons are all volumetrically minor components of the TNB (Coats *et al.*, 1972; Weber and Scoates, 1978; Peredery and Geological Staff, 1982).

Aphebian metasediments and metavolcanics, referred to as the Ospwagan group (Scoates *et al.*, 1977), occur along the western margin of the TNB in narrow, structurally controlled belts (Fig. 1). Metasediments consist of arenites and wackes, impure carbonates and marbles, ferruginous or carbonaceous pelites, chert, thin but widespread silicate, oxide and sulphide facies iron formations with associated apatite-rich horizons, and greywackes. Ospwagan group volcanics range in composition from ultramafic to mafic (Stephenson, 1974; Peredery, 1979; Paktunc, 1984; Baragar and Scoates, 1987) and appear to stratigraphically overlie the sediments. The absence of felsic or intermediate volcanics is striking and has hindered dating of the Ospwagan group.

Unlike the polymetamorphic gneisses, Ospwagan group supracrustals record a single Hudsonian *P–T* loop, with peak-metamorphic conditions ranging from lower to upper amphibolite facies, i.e., staurolite grade to garnet–sillimanite–K-feldspar grade. Earlier documentation of (prograde?) greenschist facies rocks (e.g., Weber and Scoates, 1978) could not be substantiated. The metasediments are cut by probable Molson dykes, bracketing their age between 2.4 and 1.88 Ga. Sr-isotope systematics suggest a narrower 2.1 to 1.88 Ga range (Brooks and Theyer, 1981).

Little is known about the tectonic environment in which the Ospwagan group accumulated. The complexly folded supracrustals, affected by at least one set of large recumbent structures, could be partly allochthonous or para-autochthonous. The apparent absence of coarse clastics at the base, the typical association of clastic and chemical sediments in a succession of rather thin units, and the nature and stratigraphic level of the volcanic suite are not characteristic of a typical passive margin succession. Similar but better preserved Early Proterozoic sequences elsewhere along the Circum-Superior belt have been interpreted to be partly the product of foredeep development (Hoffman, 1988a). Extending this interpretation to parts of the Ospwagan group implies that loading of the Superior margin had begun prior to intrusion of the Molson dykes.

Several minor granitoid plutons occur along the TNB. Some have migmatitic envelopes and contain garnet and muscovite in addition to biotite; they appear anatectic in origin and are coeval with the Hudsonian thermal peak. Their intrusive relationships and single-phase upright fabrics (S_3) indicate that they postdate early collisional tectonism (F_1) but are overprinted by late upright structures (F_3). U–Pb zircon dating has been hindered by Archean inheritance, but one pluton yielded a concordant monazite age of 1822 ± 3 Ma (Machado *et al.*, 1987).

RELATIONS ALONG THE WESTERN BOUNDARY OF THE THOMPSON NICKEL BELT: THE CHURCHILL–SUPERIOR BOUNDARY FAULT

Reworked Archean basement and Ospwagan group supracrustals are generally in abrupt structural contact with Kisseynew domain paragneisses along the western boundary of the TNB (Fig. 1; cf. Rance, 1966). The original structural position of these gneisses relative to TNB rocks is not clear. In one place at least, an association of TNB-like gneisses with Ospwagan group supracrustals is known to occur in drill core approximately 10 km west of the present idealized boundary (INCO staff, personal communication). If the poorly exposed Churchill–Superior boundary fault originated as one of the many dip-slip faults that dominate the western TNB, it is possible that the leading edge of the Superior plate occurs at least 10–20 km (and possibly much farther) west in the subsurface (cf. Fig. 2, boundary 6 and 7). Although the origin and early history of the boundary fault are unclear at present, abundant observations indicate that the fault was subsequently overprinted by late, brittle–ductile and brittle transcurrent displacements.

DEFORMATION HISTORY

Structural mapping and microstructural analysis of Ospwagan group supracrustals and associated gneisses in the Thompson Open Pit have revealed a complex tectono-metamorphic evolution (Fig. 4), which was generally confirmed by investigations

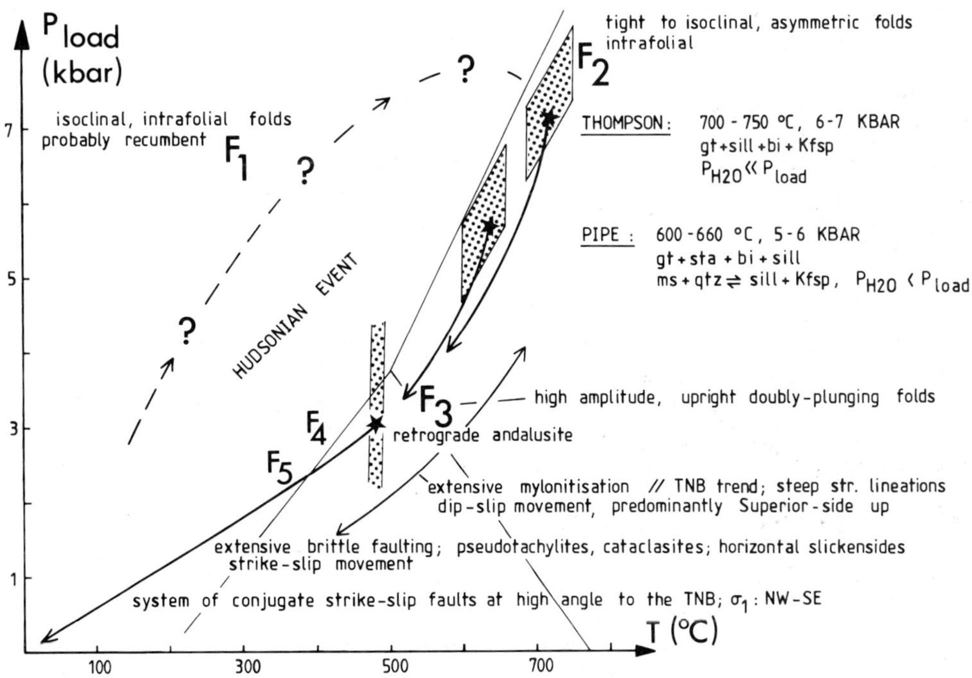

Figure 4. *Preliminary* P–T-deformation loop for the Hudsonian event as derived for Thompson and Pipe mines. Quantitative P–T data are based on garnet–biotite thermometry (calibrations used: Ferry and Spear (1978) and Thompson (1976)) and plagioclase–garnet–sillimanite–quartz barometry (calibration used: Newton and Haselton (1981), with correction noted by E. Froese (personal communication)). P_{H_2O} *calculations are based on the equilibria biotite–sillimanite–quartz–garnet–K-feldspar–vapour (Schmid and Wood, 1976; Russel, 1981) for Thompson Mine metapelites, and muscovite–quartz–sillimanite–K-feldspar–vapour (Skippen, 1977) for Pipe Mine. Aluminum silicate phase diagram after Holdaway (1971). (Note: 1 bar = 100 kPa.)*

elsewhere along the TNB. This structural history is summarized below.

F_1 isoclinal folds, often intrafolial and observed on a scale of 10 cm to 50 m, may represent one or several fold generations that all developed somewhere along the prograde path (Fig. 4). A layer-parallel differentiated foliation is associated with F_1.

F_2 tight to isoclinal folds were developed under medium- to high-grade conditions. F_1-F_2 interference patterns are generally of the mushroom type (Fig. 5). F_1 and F_2 may represent different events but could be products of a single progressive deformation in which the older folds (F_1) have undergone substantial hinge rotation before being overprinted by F_2. Both F_1 and F_2 were probably recumbent and, although F_2 developed under medium- to high-grade conditions, the complete F_1–F_2 deformational phase has been overprinted by the ultimate thermal peak.

Basement gneisses associated with the supracrustals show a similar F_1–F_2 history and overlie the supracrustals in Thompson Mine (Fig. 6). Archean structures are completely obliterated and their gneissosity, the gneiss–metasediment contact, and highly deformed mafic dykes are all parallel to the layer-parallel transposition foliation (S_0/S_2) in the

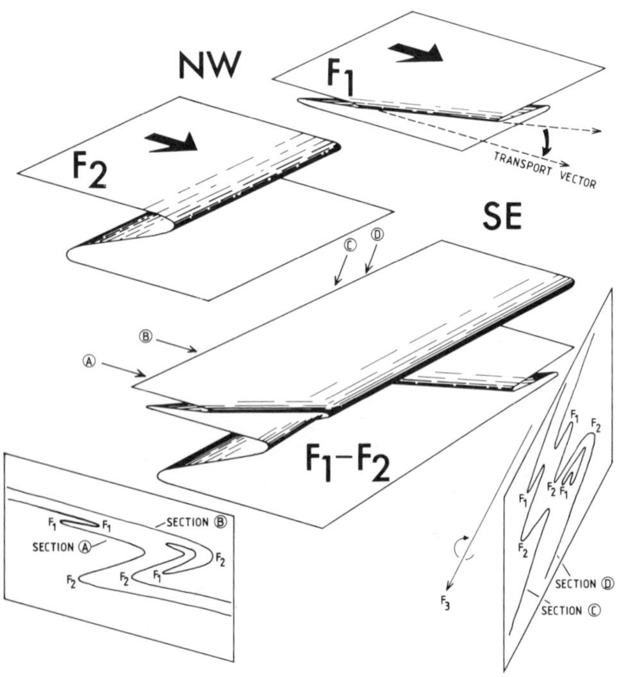

Figure 5. *F_1–F_2 interference structures at Thompson Mine after unfolding of upright F_3 folds. Note that only section (D) is fully diagnostic. Interference would often go unnoticed in general sections such as (A) and (C). Similar F_1–F_2 structures were observed at Pipe Mine.*

metasediments. Locally, however, there are domains of well-preserved basement gneisses displaying palimpsest Kenoran structures and textures, and little deformed Molson dykes that have retained their angular relationships with the gneissic foliation (Fig. 7a). These domains escaped F_1–F_2 deformation and the most attractive explanation is that they represent windows of autochthonous basement, structurally underlying an F_1–F_2 overthrust or nappe pile. Exposure of both structural levels is explained by subsequent high-amplitude upright folding (F_3) and ductile reverse faulting.

Nearly upright, high-amplitude, doubly plunging F_3 folds vary from open to tight and are penetrative throughout the TNB (Figs. 6 and 8). F_3 axial planes dip at 65–90° to the southeast and strike northeast, making a consistent 10–20° angle with the trend of the belt (Fig. 8). Most of the macrostructures associated with the mine sites, such as the doubly plunging F_3 antiform at Thompson Mine (Fig. 6), are F_3 folds. Much of the complexity of the TNB is a function of the tight, penetrative F_3 folding, which transposed earlier recumbent geometries into present near-vertical attitudes. F_3 folds developed somewhere along the retrograde uplift path (Fig. 4). Intense F_3 folding generally correlates with intense retrogression and F_3-folded micas show little or no recovery or kink band boundary migration (Fig. 3e). Andalusite porphyroblasts developed locally at the expense of sillimanite (Fig. 3f) during late stages of F_3 deformation.

F_3 is locally overprinted by two or more fold generations. However, these appear to have little influence on the overall geometry.

Widespread mylonitization was coeval with and outlasted F_3 folding. Fabrics formed during this phase were subsequently overprinted by brittle–ductile and brittle faulting (cf. Fig. 4). Both mylonite and subsequent cataclastic zones are subparallel to the trend of the belt (Fig. 8, plot 8). Contrary to expectation, the mylonite zones show evidence of dip-slip rather than strike-slip movements (Fig. 7b–e). Downdip stretching lineations are ubiquitous throughout much of the western TNB and kinematic indicators record that the dominant movement was ductile reverse faulting (see also Fueten *et al.* (1986)). Cumulative displacements across the mylonite zones, together with F_3 folding, cause the local and often abrupt metamorphic gradients across the western TNB. For instance, the difference in peak metamorphic grade between Thompson and Pipe mines (Fig. 4) is probably due to their location on opposite sides of one of the main dip-slip mylonite zones, the Burntwood River lineament. Such local gradients are thus a late feature unrelated to the overall structural and metamorphic asymmetry of the belt.

Mylonite zones were overprinted by pseudotachylite-generating brittle–ductile and brittle faults and in this stage movement was dominantly strike slip (Fig. 7f,g). Both sinistral and dextral displacements have been recorded and where overprinting relations could be observed, dextral displacements were youngest.

The latest phase of faulting is recorded in a conjugate system of transcurrent faults oriented at a high angle to the belt, the dextral and sinistral sets striking N80°E–N110°E and N150°E–N180°E respectively. They are characterized by high-level, calcite-cemented fault breccias, horizontal slickenside striae, subvertical intersections ($\sigma 2$), and a northwest-trending acute bisectrix ($\sigma 1$).

COMPARISON WITH THE TWO-STAGE COLLISION–TRANSFORM FAULTING MODEL

Structural and metamorphic data are consistent with an early collisional event (F_1–F_2). Vergence is not easily resolved from individual F_1 and F_2 folds because of poor exposure and subsequent deformation. However, overall asymmetry and facing of the belt (Fig. 2) are indicated by (1) the geometrical relationships along the eastern boundary, (2) a general increase in metamorphism and deformation in a northwesterly direction across the belt, (3) the steep southeasterly dips of F_3 axial planes (explained below), (4) the absence of any magmatic expression of southeast-facing subduction below the Superior plate (cf. Dewey and Burke, 1973), (5) the asymmetry of the gravity profile (Gibb, 1968), and (6) the presence of shallow westerly dipping reflectors (Green, 1981). The overall asymmetry fits the larger scale asymmetry of the Trans-Hudson Orogen (Lewry *et al.*, 1985; Hoffman, 1988b) and indicates that the Superior plate collided with, and was overridden by, juvenile terranes of the Trans-Hudson Orogen.

Evidence of postulated later sinistral transform faulting (Lewry, 1981; Green *et al.*, 1985) is less obvious. The suggestion that the TNB was the locus of

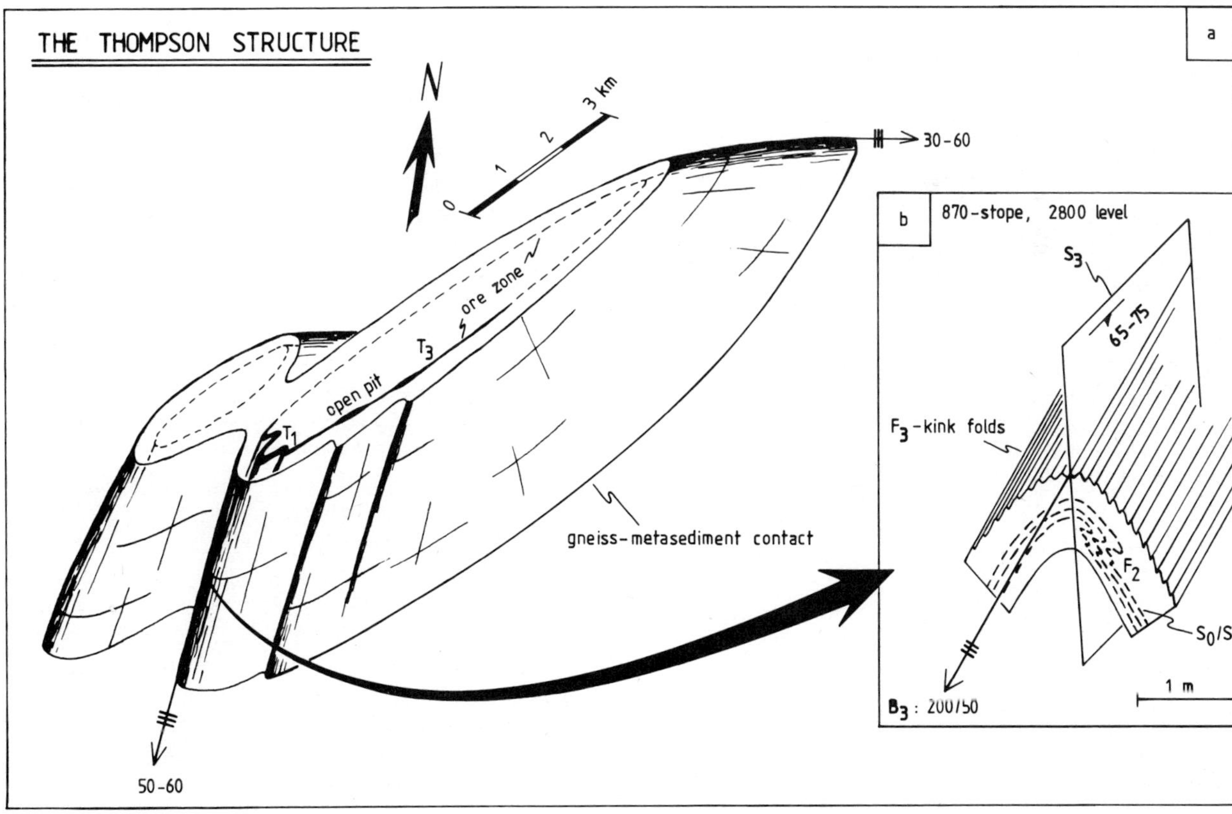

Figure 6. *Preliminary sketch of the Thompson macrostructure. Location of open pit and shafts of underground mine (T_1 and T_3) are indicated. The macrostructure is an F_3 doubly plunging antiform with a subsidiary closure to the southwest of the main hinge at T_1. The core is occupied by upper amphibolite facies Ospwagan group metasediments, whereas completely transposed Archean gneisses constitute the outer limbs. This downward-facing gneiss–metasediment package represents either the overturned limb of a large-scale recumbent fold (F_1) or a detached slab of gneisses emplaced on top of Ospwagan group metasediments.*

a
b
c
d
e
C
S
f
g

important sinistral transcurrent movements was first offered by Bell (1966) on the basis of the aeromagnetic signature of the Owl River shear zone (Fig. 1). No supporting field data have ever been presented, but the near-colinearity of this major shear zone with the TNB, the strong linear character of the TNB itself, and the overall north–south to northwest–southeast convergence indicated in the Trans-Hudson Orogen as a whole (Hoffman, 1988b) relative to the orientation of the TNB are compelling general arguments for the late transform nature of this boundary.

The most obvious structural features likely to have recorded this displacement, i.e., the mylonite zones subparallel to the trend of the belt, indicate dip-slip rather than strike-slip movement.

However, evidence for the occurrence of the sinistral transform faulting is found in the nearly upright F_3 folds. Their doubly plunging axes define girdles that outline the average macroscopic F_3 axial plane for each domain (Fig. 8, plots 2 through 7). This feature is extremely constant throughout the TNB and defines an en echelon pattern in which individual axial planes make a 10–20° angle with the trend of the belt (Fig. 8). The pattern is further accentuated by the elliptical shape of granitoid plutons and of the granulite domain on Paint Lake (cf. Figs. 1 and 8). The latter could represent a window of deeper Pikwitonei-like basement in the axial culmination of a regional F_3 antiform. Support for the penetrative en echelon pattern is provided by the aeromagnetic fabric of the TNB, which shows a similar obliquity to the trend of the belt (cf. Kornik, 1969).

The en echelon pattern of F_3 folds is consistent with sinistral ductile wrenching along a collision zone consisting of relatively rigid basement overlain by a structural pile with a strong subhorizontal anisotropy (e.g. Wilcox *et al.*, 1973), i.e., the large-scale geometry that is envisaged to have resulted from the F_1–F_2 collisional phase. In such a situation, wrenching would be accommodated differently in the basement than in the overlying structural pile. The rigid basement, already characterized by generally steep anisotropies, is likely to have deformed along discrete shear zones probably with a main shear zone along the leading edge of the underthrust plate (continuous with the Owl River shear zone?). The overlying F_1–F_2 nappe pile, with its subhorizontal anisotropy, accommodated the wrenching by forming en echelon F_3 folds, which subsequently rotated to their present trend. As a consequence, a zone of detachment must have developed between the basement and the overlying F_3 folded nappe pile to accommodate strain incompatibilities. Two independent lines of evidence support the existence of such a detachment zone. Tight, high-amplitude F_3 folds indicate considerable post-F_2 shortening across the TNB; much of the necessary extension must have taken place vertically and deformation of a thick crustal section by such folds would thus have produced an unrealistic relief. Consequently, F_3 folds have to be confined to a superficial layer, i.e., the F_1–F_2 nappe pile and possibly the directly underlying basement. Secondly, seismic reflection profiling (Green, 1981) suggests that the steep surface geology does not extend to great depth and is underlain by basement characterized by shallow northwest-dipping reflectors.

Wrench folds generated in such a setting will have axial planes perpendicular to the enveloping surface of the rigid basement, e.g., horizontal strata overlying a horizontal rigid basement will develop doubly plunging wrench folds with vertical axial planes. Axial planes of F_3 folds in the TNB dip consistently to the southeast at 65–90° (Figs. 6 and 8), suggesting that the enveloping surface of the rigid basement, i.e., the Superior plate, dips gently in a westerly direction.

AN EVOLUTIONARY MODEL

Figure 9 illustrates a proposed evolutionary model that attempts to accommodate data available at present. The initial collision geometry is shown in Figure 9a, with an overthrust prism advancing onto the Archean foreland. The prism is characterized by brittle thrusts at high structural levels and ductile

Figure 7. *Critical mesoscopic structures illustrating aspects of the structural evolution. (a) Outcrop of well-preserved basement on Ospwagan Lake, between Thompson and Pipe mines, showing weakly deformed Molson dykes with a preserved angular relationship to the gneissic foliation. (b) Completely reworked basement gneisses along the Burntwood lineament, one of the major mylonite zones in the western TNB. Strongly doubly plunging folds are associated with downdip stretching lineations (cf. Fig. 8, plot 8). (c) Mylonitic gneisses along the Burntwood lineament. Subhorizontal section through a vertical sheath fold. (d) Vertical (proto) sheath folds with associated vertical stretching lineation (parallel to pencil), Pipe Mine. (e) Vertical section through protomylonitic granitoid (Burntwood lineament north of Thompson; view toward the north-northeast), displaying a strong C–S fabric indicating ductile reverse movement with the southeast side up relative to the northwest side. (f) Late brittle–ductile array of dextral R-shears, with associated pseudotachylite veinlets (at tip of solid arrow); displacement is dextral. Churchill–Superior boundary fault, Ospwagan Lake. (g) Pseudotachylite veinlets (at tips of solid arrows) associated with brittle, sinistral overprinting along the Burntwood mylonite zone.*

thrusts/recumbent F_1–F_2 folds at deeper structural levels. F_1 folds probably developed at intermediate levels and were subsequently overprinted by F_2 under near-peak metamorphic conditions. Involvement of superficial basement in the thrusting/recumbent folding resulted in slices of completely transposed gneisses that locally overlie Ospwagan group supracrustals, as in the Thompson structure (Fig. 6). Brittle and brittle–ductile strike-slip faulting occurred in the Archean foreland ahead of the advancing overthrust prism following, and possibly during, Molson dyke intrusion. These shear zones were subsequently overprinted by the thermal peak following the final emplacement of the overthrust slab over the eastern boundary zone. Figure 9a further shows the Fox River Belt in its precollisional geometry for comparison.

After initial crustal thickening, the collision zone was overprinted by the thermal peak of metamorphism with accompanying anatexis in the lower levels of the thickened crust (Fig. 9b). Granitoids such as the Mystery Lake and Wintering Lake

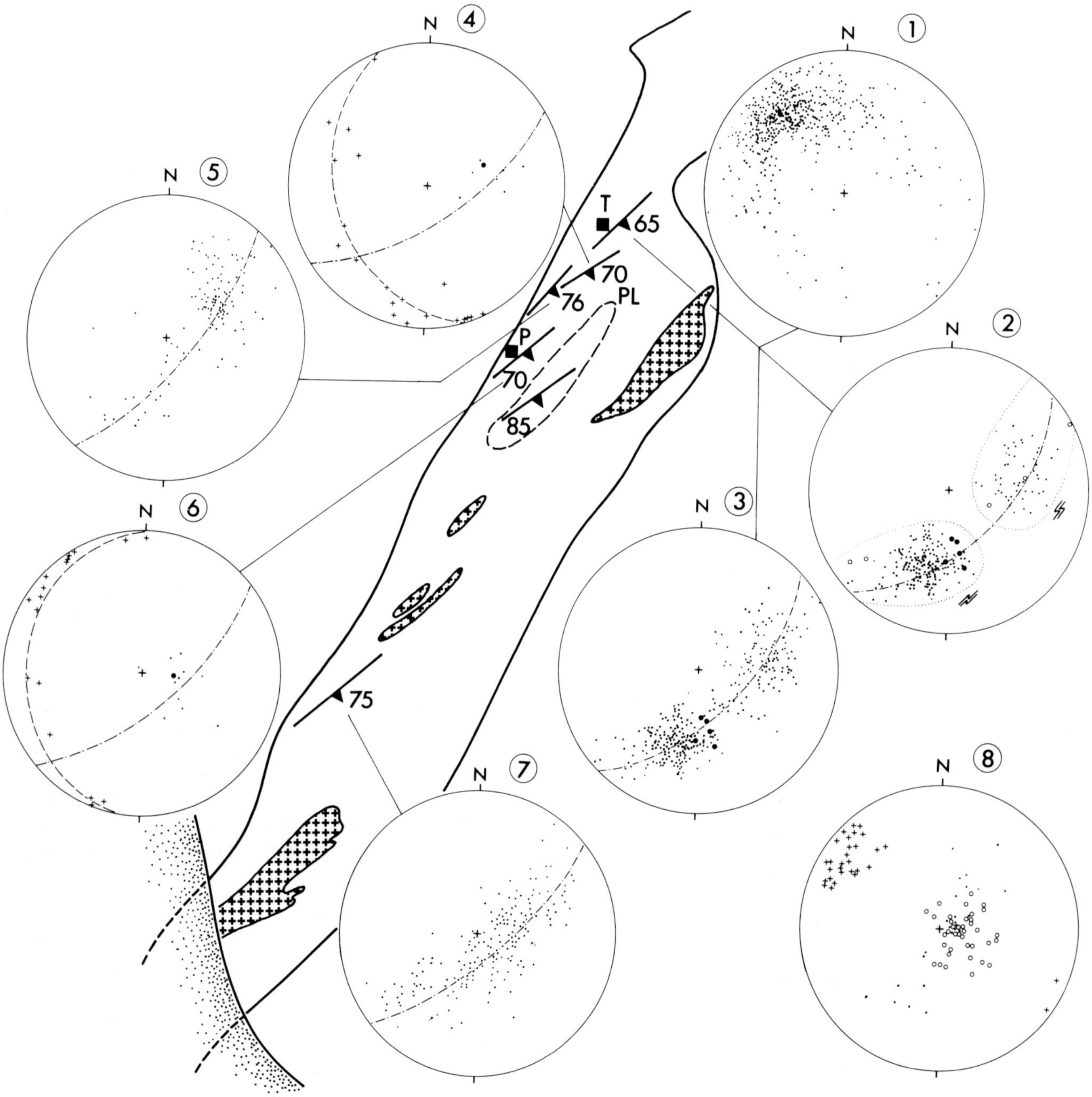

plutons, due north and 25 km southeast of Thompson (Fig. 1), respectively, intruded around 1.82 Ga (Machado *et al.*, 1987), indicating that thermal relaxation had taken place by that time. The approximate present erosion level, which exposes a deep and rather cryptic section through the belt, is indicated in Figure 9b. In its present position, the Superior plate shows nearly complete rebound, with strongly deformed Ospwagan supracrustals preserved only in narrow, structurally controlled belts along the western boundary of the TNB.

Oblique collision gradually evolved into sinistral transform motion (cf. Green *et al.*, 1985) due to increasing resistance to further underthrusting by the buoyant Superior plate below the thickened collision zone (Fig. 9c). En echelon F_3 folds are attributed to sinistral wrenching of the overthrust/nappe pile with its strong subhorizontal anisotropy. The overall kinematic framework probably still involved a considerable component of orthogonal convergence, resulting in sinistral transpression. The model predicts that F_3 folding should be contemporaneous with large-scale displacements along the Owl River shear zone.

Figure 9d shows the phase of extensive ductile reverse faulting contemporaneous with and following on F_3. Where vertical differential movements were penetrative, doubly plunging F_3 hinge lines rotated to steeper plunges approaching locally subvertical sheath-like geometries. There is a general correlation between the degree to which penetrative, downdip stretching lineations developed and the steepness of plunges of meso- and macroscopic F_3 folds. Deeper structural levels show up in the culminations of regional doubly plunging F_3 antiforms, locally exposing deeper and better preserved granulite facies basement. Both F_3 folding and ductile reverse faulting contributed to large vertical differential movements, causing locally steep and abrupt metamorphic gradients across the western TNB.

The dominance of ductile reverse faulting and absence of extensive ductile strike-slip shear zones appear to contradict the hypothesis of large-scale sinistral transform motion. However, recent thermal, seismic, and structural data from the San Andreas Fault (Brune *et al.*, 1969; Lachenbruch and Sass, 1980; Mount and Suppe, 1987) indicate that very low strength of the main fault zone results in a reorientation of the regional $\sigma 1$ into a near-perpendicular orientation with respect to the fault, such that subhorizontal overlying strata are deformed by thrusts and upright folds that trend nearly parallel to the main fault (Mount and Suppe, 1987). The western TNB may provide a deeply dissected, Early Proterozoic analog for such neotectonics. Another factor that has probably contributed to the dominantly dip-slip movements along the mylonite zones of the TNB is isostatic rebound of the edge of the Superior plate in an overall transpressive regime.

Figure 9d also shows the Fox River Belt in its postcollisional geometry. Relative timing of overthrusting in the Fox River Belt is unknown, but it may have been contemporaneous with sinistral transpression along the TNB. The steeply north-dipping homoclinal sequence (Baragar and Scoates, 1981) is best explained as an imbricate stack of listric thrusts that flatten out into a basal thrust at depth. The highly sinusoidal boundary of the Fox

Figure 8. *Synopsis of TNB structural data. Lower hemisphere, equal-area projections. Plot 1: Poles to composite layer-parallel transposition foliation* ($\pi S_{0/2}$) *in metasediments and gneisses, Thompson Open Pit* (N = 420). Distribution comprises a strong unimodal maximum and two weak great circles, reflecting the macrostructure of the Thompson Mine environment (cf. Fig. 6). πS plots elsewhere in the TNB are essentially similar. Plot 2: F_3 *fold axes, Thompson Open Pit (small dots,* N = 205), five axes of a single mesoscopic doubly plunging F_3 *fold (open circles), and five macroscopic axes calculated for the main antiformal* F_3 *hinge at* T_1 *(solid circles). Note overall great-circle distribution, strong asymmetry grouping, and the central gap, all consistent with the macrostructure as shown in Figure 6. The great circle outlines the average macroscopic* F_3 *axial plane* (S_3, *dash–dot great circle). Plot 3: All linear elements for the Thompson Open Pit* (N = 276), i.e., $F_1...F_5$ *fold axes and associated lineations. Because the* F_3 *macrostructure is tight, all pre-*F_3 *fold axes are rotated toward the* S_3 *girdle, whereas all post-*F_3 *linear elements will be controlled by attitudes of the limbs of* F_3 *and thus plot close to the* S_3 *girdle. Therefore, this plot defines the average macroscopic* S_3 *plane just as well, without differentiating fold generations. Plot 4: Structural data for* F_3 *macrostructure at Manasan Quarry, defining local* S_3 *plane.* $\pi S_{0/2}$ *(crosses,* N = 20) and associated π girdle (dashed great circle), mesoscopic F_3 *axes (small dots,* N = 7), and macroscopic F_3 *axis (solid circle, defined by* π*-girdle axis). Macroscopic* S_3 *axial plane (dash–dot great circle) is estimated from known strike and* F_3*-axis data. Plot 5: Fold-axis data for Ospwagan Lake* (N = 116). The distribution again defines the macroscopic S_3 *plane. Plot 6: Structural data for* F_3 *macrostructure at Pipe Mine (symbols as in plot 4; crosses,* N = 17; small dots, N = 11). Plot 7: Fold-axis data for the Wabowden area (N = 185; Rance, 1966). Plot 8: Synoptic data on mylonite zones in the western TNB. Poles to mylonitic foliation (crosses, N = 29), associated downdip stretching lineations (open circles, N = 38), and fold axes of associated (proto) sheath folds (small dots, N = 25). The macroscopic F_3 *axial planes* (S_3), *as defined by plots 2 through 7, outline a penetrative en echelon pattern of nearly upright* F_3 *doubly plunging folds throughout the TNB.*

River Belt with Archean foreland, which appears to conceal the Split Lake fault (Fig. 1), strongly suggests a shallow-dipping lower enveloping surface (i.e., basal thrust) to the Fox River Belt supracrustals.

Figure 9e shows the Superior plate and marginal TNB in a stage of nearly complete rebound. Brittle–ductile and brittle faults overprinted the mylonite zones and abundant pseudotachylites were generated. Kinematic indicators point at a transition to strike-slip movements, with sinistral displacements being overprinted by dextral displacements. The last phase of faulting occurred along a conjugate system of strike-slip faults at a high angle to the belt (Fig. 9e) and is probably related to residual southeast–northwest convergence between the Superior plate and the amalgamated Churchill Province. Finally, northwest-trending dykes of the Mackenzie swarm, approximately 1270 Ma in age (Heaman and LeCheminant, 1988), crosscut the boundary without any offset (Manitoba Mineral Resources Division, 1979).

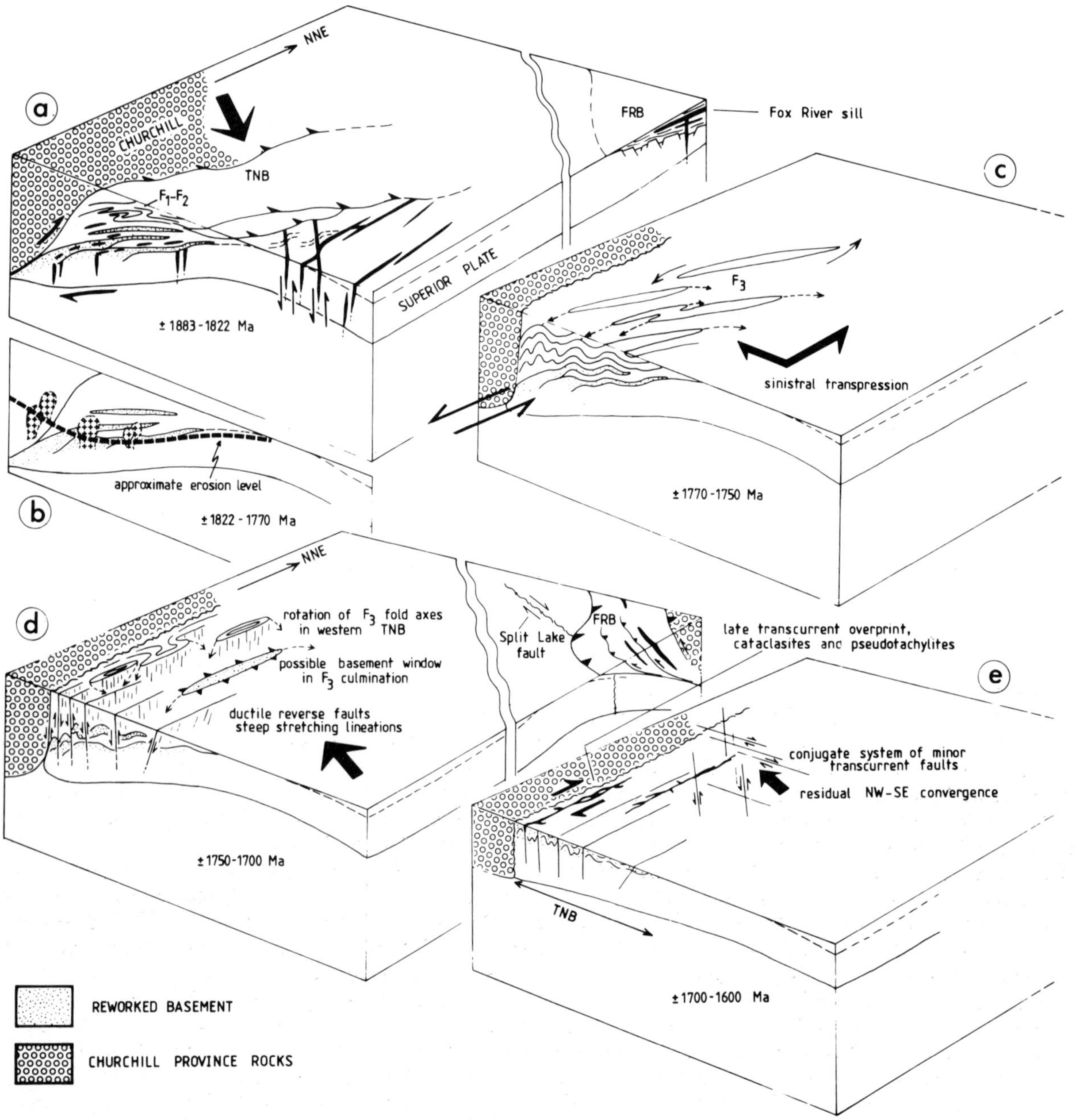

Figure 9. *Block diagrams illustrating the tectonic evolution of the northwestern part of the Superior craton, its margin, and adjacent internal zone of the Trans-Hudson Orogen (Churchill Province). Consistent scaling has not been attempted. Refer to text for explanation.*

CONCLUSIONS AND DISCUSSION

The proposed evolutionary model (Fig. 9) retains many features of the model put forward by Green *et al.* (1985), but changes several critical relationships and fills in much needed detail. The model is, at least qualitatively, compatible with all data available at present.

A critical feature of the model is the early collisional geometry (Figs. 2 and 9a) deduced from varied geological, geochronological, and geophysical data. Key evidence supporting such geometry is provided by (1) the overall structural and geophysical asymmetry of the belt, (2) the structural–metamorphic scenario (Fig. 4), and (3) recognition of basement windows as far west as along the western boundary of the TNB. Furthermore, interpretation of the en echelon pattern of nearly upright F_3 folds as sinistral wrench folds (Fig. 9c) is critically dependent on the former existence of an overthrust/nappe pile with a subhorizontal anisotropy, underlain by rigid basement.

Ospwagan group supracrustals do not resemble a typical passive margin sequence. They possibly represent, in part, the products of foredeep development. As the metasediments are cut by Molson-like dykes, the latter interpretation implies that tectonic loading and subsidence of the Superior-plate margin was initiated prior to 1883 Ma, and that generation of the Molson dyke swarm was contemporaneous with, and possibly related to, the collisional process. A causal relationship between the collision and the Molson dyke swarm is suggested by (1) parallelism of the majority of the dykes with the TNB and the Pikwitonei granulite belt, (2) the observation that the dykes appear to be most numerous close to (Scoates and Macek, 1978) and within the TNB, and (3) structural observations indicating that the dykes intruded after the onset of Hudsonian deformation. The early collisional F_1–F_2 phase is overprinted by the thermal peak of Hudsonian metamorphism. Nearly upright, en echelon F_3 folds formed when the tectonic regime evolved from an oblique collision zone into a transpressive sinistral transform boundary (cf. Lewry, 1981; Green *et al.*, 1985). These folds, attributed to wrenching of an overthrust/nappe pile with a strong subhorizontal anisotropy that was overlying a rigid basement, dominate the structure of the TNB, transpose earlier recumbent geometries into nearly vertical attitudes, and probably developed contemporaneously with large-scale sinistral displacements along the Owl River shear zone.

F_3 folding was followed by extensive ductile reverse faulting. Recent work on the San Andreas Fault (Mount and Suppe (1987) and references therein) may suggest that such a deformation regime is not necessarily incompatible with strike-slip faulting. It requires, however, strong decoupling of transcurrent and compressive deformation due to the presence of a nearly frictionless main fault zone. No such main fault zone, which must have accommodated most of the transcurrent movement, has yet been identified. It may be situated just west of the present Churchill–Superior boundary fault along a linear zone of elongated granitoid bodies (Manitoba Mineral Resources Division, 1979). The en echelon aeromagnetic fabric of the TNB appears to bend into the north-northeast trending fabric of this lineament, which, if extended to the north-northeast, joins up with the Owl River shear zone.

ACKNOWLEDGEMENTS

This research is part of a Ph.D. study supported by the Geological Survey of Canada, and a contribution to Canada–Manitoba Mineral Development Agreement 1984–1989. Development of the model owes much to discussions with Cees van Staal, Roger Eckstrand, Terry Gordon, Edgar Froese, and Paul Hoffman (Geological Survey of Canada); Frank Fueten (University of Toronto); and especially Paul F. Williams (University of New Brunswick). Comments by Cees van Staal, Paul Williams, and Roger Eckstrand on the first draft improved this paper, as did final review by Werner Weber, John Lewry, and an anonymous referee. I acknowledge logistic support from the Manitoba Geological Survey and thank Jozef Macek and Werner Weber for showing me critical localities in the Thompson and Pikwitonei belts. INCO staff geologists are thanked for their support and access to INCO property. Ansel Murphy and Scott Sale provided thin sections and Angel Gomez made final drafts of some figures.

REFERENCES

Baldwin, D.A., Syme, E.C., Zwanzig, H.V., Gordon, T.M., Hunt, P.A. and Stevens, R.D., 1987, U–Pb zircon ages from the Lynn Lake and Rusty Lake metavolcanic belts, Manitoba: two ages of Proterozoic magmatism: Canadian Journal of Earth Sciences, v. 24, p. 1053–1063.

Baragar, W.R.A. and Scoates, R.F.J., 1981, The Circum-Superior belt: a Proterozoic plate margin?, *in* Kröner, A., ed., Precambrian Plate Tectonics: Elsevier, Amsterdam, p. 297–330.

Baragar, W.R.A. and Scoates, R.F.J., 1987, Volcanic geochemistry of the northern segments of the Circum-Superior belt of the Canadian Shield, *in* Pharaoh, T.C., Beckinsale, R.D. and Rickard, D., eds., Geochemistry and Mineralization of Proterozoic Volcanic Suites: Geological Society Special Publication No. 33, p. 113–131.

Bell, C.K., 1966, Churchill–Superior boundary in northeastern Manitoba: Geological Survey of Canada, Paper 66-1, p. 133–136.

Bell, C.K., 1971, History of the Churchill–Superior boundary in Manitoba, *in* Turnock, A.C., ed., Geoscience Studies in Manitoba: Geological Association of Canada, Special Paper 9, p. 5–10.

Bleeker, W., 1990, Evolution of the Thompson Nickel Belt and its nickel deposits, Manitoba, Canada, unpublished Ph.D. thesis, University of New Brunswick.

Brooks, C. and Theyer, P., 1981, Rb/Sr geochronology in the Thompson Nickel Belt, Manitoba: implication for Aphebian crustal development and metallogenesis: Canadian Journal of Earth Sciences, v. 18, p. 932–943.

Brune, J.N., Henyey, T.L. and Roy, R.F., 1969, Heat flow, stress and rate of slip along the San Andreas fault, California: Journal of Geophysical Research, v. 74, p. 3821–3827.

Chauvel, C., Arndt, N.T., Kielinzcuk, S. and Thom, A., 1987, Formation of 1.9 Ga old continental crust. I. Nd isotopic data: Canadian Journal of Earth Sciences, v. 24, p. 396–406.

Coats, C.J.A., Quirke, T.T., Bell, C.K., Cranstone, D.A. and Campbell, F.H.A., 1972, Geology and mineral deposits of the Flin Flon, Lynn Lake and Thompson areas, Manitoba, and the Churchill–Superior front of the western Precambrian shield: International Geological Congress, XXIV Session, Guidebook, Field Excursion A31–C31.

Cranstone, D.A. and Turek, A., 1976, Geological and geochronological relationships of the Thompson Nickel Belt, Manitoba: Canadian Journal of Earth Sciences, v. 13, p. 1058–1069.

Dewey, J.F. and Burke, K.C.A., 1973, Tibetan, Variscan and Precambrian basement reactivation: Journal of Geology, v. 81, p. 683–692.

Ermanovics, I.F. and Fahrig, W.F., 1975, The petrochemistry and paleomagnetism of the Molson dykes, Manitoba: Canadian Journal of Earth Sciences, v. 12, p. 1564–1575.

Ermanovics, I.F. and Wanless, R.K., 1983, Isotopic age studies and interpretation of Superior Province in Manitoba: Geological Survey of Canada, Paper 82-12, 22 p.

Ferry, J.M. and Spear, F.S., 1978, Experimental calibration of the partitioning of Fe and Mg between biotite and garnet: Contributions to Mineralogy and Petrology, v. 66, p. 113–117.

Fountain, D.M. and Salisbury, M.H., 1981, Exposed cross-section through the continental crust: implications for crustal structure, petrology and evolution: Earth and Planetary Science Letters, v. 56, p. 263–277.

Fueten, F., Robin, P.Y.F. and Pickering, M.E., 1986, Deformation in the Thompson belt: a progress report: Geological Survey of Canada, Paper 86-1B, p. 797–809.

Gibb, R.A., 1968, A geological interpretation of the Bouguer anomalies adjacent to the Churchill–Superior boundary in northern Manitoba: Canadian Journal of Earth Sciences, v. 5, p. 439–453.

Gibb, R.A., 1975, Collision tectonics in the Canadian Shield?: Earth and Planetary Science Letters, v. 27, p. 378–382.

Gibb, R.A., 1983, Model for suturing of Superior and Churchill plates: an example of double indentation tectonics: Geology, v. 11, p. 413–417.

Gibb, R.A. and Walcott, R.I., 1971, A Precambrian suture in the Canadian Shield: Earth and Planetary Science Letters, v. 10, p. 417–422.

Green, A.G., 1981, Results of a seismic reflection survey across the fault zone between the Thompson Nickel Belt and the Churchill tectonic province, northern Manitoba: Canadian Journal of Earth Sciences, v. 18, p. 13–25.

Green, A.G., Cumming, G.L. and Cedarwell, D., 1979, Extension of the Superior–Churchill boundary zone into southern Canada: Canadian Journal of Earth Sciences, v. 16, p. 1691–1701.

Green, A.G., Hajnal, Z. and Weber, W., 1985, An evolutionary model of the western Churchill Province and western margin of the Superior Province of Canada and the north-central United States: Tectonophysics, v. 116, p. 281–322.

Heaman, L.M. and LeCheminant, A.N., 1988, U–Pb baddeleyite ages of the Muskox Intrusion and Mackenzie Dyke Swarm, N.W.T., Canada: Geological Association of Canada – Mineralogical Association of Canada, Program with Abstracts, v. 12, p. 53.

Heaman, L.M., Machado, N., Krogh, T. and Weber, W., 1986, Precise U–Pb zircon ages for the Molson dyke swarm and the Fox River sill: implications for Early Proterozoic crustal evolution in NE Manitoba, Canada: Contributions to Mineralogy and Petrology, v. 94, p. 82–89.

Hoffman, P.F., 1988a, Proterozoic foredeeps, foredeep magmatism and Superior-type iron formations in the Canadian Shield, _in_ Kröner, A., ed., Proterozoic Lithospheric Evolution: American Geophysical Union, Geodynamics Series, v. 17, p. 85–98.

Hoffman, P.F., 1988b, United plates of America, the birth of a craton: Early Proterozoic assembly and growth of Laurentia: Annual Review of Earth and Planetary Sciences, v. 16, p. 543–603.

Holdaway, M.J., 1971, Stability of andalusite and the aluminum silicate phase diagram: American Journal of Science, v. 271, p. 97–131.

Hubregtse, J.J.M.W., 1980, The Archean Pikwitonei granulite domain and its position at the margin of the northwestern Superior Province, central Manitoba: Manitoba Mines Branch, Geological Paper 80-3, 16 p.

Kornik, L.J., 1969, An aeromagnetic study of the Moak Lake – Setting Lake structure in northern Manitoba: Canadian Journal of Earth Sciences, v. 6, p. 373–381.

Kornik, L.J. and MacLaren, A.S., 1966, Aeromagnetic study of the Churchill–Superior boundary in northern Manitoba: Canadian Journal of Earth Sciences, v. 3, p. 547–557.

Lachenbruch, A.H. and Sass, J.H., 1980, Heat flow and energetics of the San Andreas fault zone: Journal of Geophysical Research, v. 85, p. 6185–6222.

Lewry, J.F., 1981, Lower Proterozoic arc–microcontinent collisional tectonics in the western Churchill Province: Nature, v. 294, p. 69–72.

Lewry, J.F., Sibbald, T.I.I. and Schledewitz, D.C.P., 1985, Variation in character of Archean rocks in the western Churchill Province and its significance, _in_ Ayres, L.D., Thurston, P.C., Card, K.D. and Weber, W., eds., Evolution of Archean Supracrustal Sequences: Geological Association of Canada, Special Paper 28, p. 239–261.

Lewry, J.F., Macdonald, R., Livesey, C., Meyer, M., Van Schmus, R. and Bickford, M.E., 1987, U–Pb geochronology of accreted terranes in the Trans-Hudson Orogen, northern Saskatchewan, Canada, _in_ Pharaoh, T.C., Beckinsale, R.D. and Rickard, D., eds., Geochemistry and Mineralization of Proterozoic Volcanic Suites: Geological Society Special Publication No. 33, p. 147–166.

Lowdon, J.A., 1962, Age determinations by the Geological Survey of Canada: Geological Survey of Canada, Paper 61-17, p. 50.

Machado, N., Heaman, L.M., Krogh, T.E. and Weber, W., 1987, U–Pb geochronology program: Thompson Belt – Northern Superior Province: Manitoba Energy and Mines, Report of Field Activities 1987.

Manitoba Mineral Resources Division, 1979, Geological map of Manitoba, scale 1:1,000,000: Map 79-2.

Mount, V.S. and Suppe, J., 1987, State of stress near the San Andreas fault: implications for wrench tectonics: Geology, v. 15, p. 1143–1146.

Newton, R.C. and Haselton, H.T., 1981, Thermodynamics of the garnet–plagioclase–Al_2SiO_5–quartz geobarometer, _in_ Newton, R.C., Navrotsky, A. and Wood, B.J., eds., Thermodynamics of Minerals and Melts: Springer, New York, p. 129–145.

Paktunc, A.D., 1984, Petrogenesis of ultramafic and mafic rocks of the Thompson Nickel Belt, Manitoba: Contributions to Mineralogy and Petrology, v. 88, p. 348–353.

Paktunc, A.D. and Baer, A.J., 1986, Geothermobarometry of the northwestern margin of the Superior Province: implications for its tectonic evolution: Journal of Geology, v. 194, p. 381–394.

Peredery, W.V., 1979, Relationships of ultramafic amphibolites to metavolcanic rocks and serpentinites in the Thompson belt, Manitoba: Canadian Mineralogist, v. 17, p. 187–200.

Peredery, W.V. and Geological Staff, 1982, Geology and nickel sulphide deposits of the Thompson belt, Manitoba, *in* Hutchinson, R.W., Spence, C.D. and Franklin, J.M., eds., Precambrian Sulphide Deposits (H.S. Robinson Memorial Volume): Geological Association of Canada, Special Paper 25, p. 165–209.

Rance, H., 1966, Superior–Churchill structural boundary, Wabowden, Manitoba, unpublished Ph.D. thesis, University of Western Ontario.

Roussel, D.H., 1965, Geology of the Cross Lake area: Province of Manitoba, Department of Mines and Natural Resources, Publication 62-4, 69 p.

Russel, J.K., 1981, Metamorphism of the Thompson Nickel Belt gneisses: Canadian Journal of Earth Sciences, v. 18, p. 191–209.

Schmid, R. and Wood, B.J., 1976, Phase relations in granulitic metapelites from the Ivrea–Verbano Zone: Contributions to Mineralogy and Petrology, v. 54, p. 255–279.

Scoates, R.F.J. and Macek, J.J., 1978, Molson Dyke Swarm: Manitoba Mines Branch, Geological Paper 78-1, 53 p.

Scoates, R.F.J., Macek, J.J. and Russel, J.K., 1977, Thompson Nickel Belt project: Manitoba Mineral Resources Division, Report of Field Activities, p. 47–53.

Skippen, G.B., 1977, Dehydration and decarbonation equilibria, *in* Greenwood, H.J., ed., Short Course in Application of Thermodynamics to Petrology and Ore Deposits: Mineralogical Association of Canada, p. 66–82.

Stephenson, J.F., 1974, Geology of the Ospwagan Lake (east half) area: Manitoba Department of Mines, Publication 74-1, 67 p.

Thompson, A.B., 1976, Mineral reactions in pelitic rocks. II. Calculation of some P–T–X (Fe–Mg) phase relations: American Journal of Science, v. 276, p. 425–454.

Wanless, R.K., Stevens, R.D., Lachance, G.R. and Rimsaite, J.Y.H., 1966, Age determinations and geological studies: Geological Survey of Canada, Paper 65-17, p. 55–57.

Weber, W. and Scoates, R.F.J., 1978, Archean and Proterozoic metamorphism in the northwestern Superior Province and along the Churchill–Superior boundary, Manitoba, *in* Fraser, J.A. and Heywood, W.W., eds., Metamorphism in the Canadian Shield: Geological Survey of Canada, Paper 78-10, p. 5–16.

Wilcox, R.E., Harding, T.P. and Seely, D.R., 1973, Basic wrench tectonics: American Association of Petroleum Geologists, v. 57 (1), p. 74–96.

Zurbrigg, H.F., 1963, Thompson Mine geology: Canadian Institute of Mining and Metallurgy Bulletin, v. 56, p. 451–460.

Structural Relations in Accreted Terranes of the Trans-Hudson Orogen, Saskatchewan: Telescoping in a Collisional Regime?

John F. Lewry
Department of Geology, University of Regina,
Regina, Saskatchewan S4S 0A2, Canada

David J. Thomas and Robert Macdonald
Saskatchewan Geological Survey, Toronto-Dominium Building,
1914 Hamilton Street, Regina, Saskatchewan S4P 4V4, Canada

J. Chiarenzelli
Department of Geology, University of Kansas,
Lawrence, Kansas 66045, U.S.A.

Lewry, J.F., Thomas, D.J., Macdonald, R. and Chiarenzelli, J., Structural relations in accreted terranes of the Trans-Hudson Orogen, Saskatchewan: telescoping in a collisional regime?, in Lewry, J.F. and Stauffer, M.R., eds., The Early Proterozoic Trans-Hudson Orogen of North America: Geological Association of Canada, Special Paper 37, p. 75–94.

Abstract

The Reindeer Zone, western Trans-Hudson Orogen, is mainly a collage of juvenile, accreted(?) Early Proterozoic oceanic arcs and arc-related terranes. Structural relations and strain variation in the Saskatchewan part of the zone suggest that the main tectonism involved extreme crustal shortening, fold/thrust stacking, and the transportation of major allochthonous sheets, some evidently soled by initially flat-lying, regionally extensive high-strain zones. These early structures are refolded about more upright north-trending folds and later, regionally more important, northeast-trending folds to produce the present gross lithostructural patterns. Early structures were modified and, in many places, oversteepened during this refolding and also disrupted by late faulting. Modification of the prior domainal subdivision of the Reindeer Zone, involving revised definition and nomenclature of lithostructural elements, is proposed to more accurately reflect gross "three-dimensional" structural geometry.

La Ronge Domain is a stack of variably strained northwest/north-dipping tectonostratigraphic packages, most of which are bounded by high-strain zones. Glennie Domain largely comprises refolded nappe sheets soled by early mylonitic porphyroclastic gneisses/schists (Guncoat and Nistowiak gneisses). Early folds in much of Kisseynew Domain were probably initially highly overturned to recumbent. Flin Flon Domain is viewed as an imbricate thrust wedge soled by the Sturgeon Weir thrust and overlain by high-grade nappes of the Kisseynew Domain.

Windows of Archean basement are exposed at the lowest structural levels in the Glennie Domain and Hanson Lake Block, below allochthonous Early Proterozoic nappe sheets and their high-strain thrust soles. Thus, it seems possible that the Reindeer Zone is partly or wholly underthrust by lower plate continental lithosphere emplaced during terminal continental collision subsequent to arc formation. Seismic investigation and dating/isotopic characterisation of late-/post-kinematic granites could test this possibility.

Résumé

La zone Reindeer, dans l'orogène Trans-Hudson ouest, est un collage juvénile ou accrété? de terranes de type arc océanique Protérozoïque Inférieur. Les relations structurales et les variations de la déformation dans la partie Saskatchewan de la zone suggèrent que la tectonique principale implique un raccourcissement crustal important, des empilements de pli-charriages et un transport des écailles allochtones principales, sous lesquelles se trouvent des zones régionales sub-horizontales à haute déformation. Ces structures anciennes sont replissées par des plis nord et tardivement par des plis nord-est, régionalement plus important produisant le patron lithostructural actuel. Ces structures ont été redressées durant la deuxième phase de pli et disloquées par des failles tardives. On propose une modification des subdivisions en domaines de la zone Reindeer, impliquant une redéfinition et une nomenclature des éléments lithostructuraux illustrant plus précisément la géométrie structurale tri-dimentionnelle.

Le domaine La Ronge, est un empilement d'entités tectonostratigraphiques variablement déformé, penté nord/nord-ouest et bordé par des zones de haute déformation. Le domaine Glennie est principalement composé d'écailles de nappes replissées sous lesquelles se trouvent des schistes/gneiss porphyroclastiques et mylonitiques anciens (gneiss Guncoat et Nistowiak). Les plis anciens du domaine de Kisseynew ont probablement été déversés et couchés initialement. Le domaine Flin-Flon est vu comme un coin charrié et imbriqué situé au-dessus du charriage Sturgeon Weir et recouvert par les nappes à haut métamorphisme du domaine Kisseynew.

Des fenêtres de socle archéen affleurent dans les domaines structuraux inférieurs dans le domaine Glennie et le block Hanson Lake, sous les écailles de nappes allochtones Protérozoïque Inférieur et leurs semelles à haute déformation. Il est donc possible que la zone Reindeer soit en tout ou en partie sous-charriée par une plaque de lithosphère continentale mise en place pendant la collision continentale finale postérieure à la formation d'arcs. Des investigations séismiques et des datations des granites tardi et post-cinématiques pourraient permettre de tester cette hypothèse.

INTRODUCTION

Most of the Trans-Hudson Orogen comprises thermotectonically reworked Archean continental basement and inboard, predominantly miogeoclinal, Lower Proterozoic supracrustal rocks, intruded by Early Proterozoic plutons of various ages (cf. Lewry *et al.*, 1985; Hoffman, 1988, this volume; Lewry and Collerson, this volume). However, available petrochemical, radiometric age, and other isotopic data indicate that the Reindeer Zone (Stauffer, 1984), exposed in Saskatchewan and Manitoba, consists almost entirely of juvenile, accreted and possibly exotic Early Proterozoic crustal elements (Chauvel *et al.*, 1987; Lewry *et al.*, 1987; Van Schmus *et al.*, 1987; Watters and Pearce, 1987; Syme, this volume; Thom *et al.*, this volume).

The Reindeer Zone (Fig. 1) is a collage of seemingly related lithotectonic domains mainly comprising subduction-generated(?) arc volcanic and plutonic rocks; associated, coeval and/or derived volcanogenic clastic assemblages; and subordinate, younger molasse sediments. To the west and north, this collage is separated from the ensialic Peter Lake Domain and eastern Cree Lake Zone (Wollaston Domain) by the straddling, ca. 1850 Ma, synkinematic Wathaman Batholith (Lewry *et al.*, 1981; Fumerton *et al.*, 1985; Meyer, 1987): in Saskatchewan, the northwest margin of this huge, remarkably homogeneous plutonic complex is almost entirely defined by post-Wathaman mylonitic gneisses of the Needle Falls – Parker Lake shear zone. The southeast margin of the Reindeer Zone is also delimited by major tectonic discontinuities defining the western and northern margins of the adjacent ensialic Churchill–Superior Boundary Zone. In the Reindeer Zone, Archean continental basement is presently documented only in the Hanson Lake Block (Bell and Macdonald, 1982; Van Schmus *et al.*, 1987) and in the Iskwatikan Lake/Hunter Bay area of western Glennie Domain (Chiarenzelli *et al.*, 1987; Chiarenzelli and Lewry, 1988). Nd, Pb, and Sr isotopic data (Bell and Macdonald, 1982; Chauvel *et al.*, 1987; Todt *et al.*, *in press*) and petrochemical data (e.g., Watters and Pearce, 1987; Syme, this volume) indicate that Early Proterozoic volcanic and plutonic rocks were, with few exceptions, predominantly mantle derived, with only minor, probably sedimentary, input from Archean crustal sources. Some plutons may represent penecontemporaneous remelting of basal parts of juvenile arc complexes. However, some late- to posttectonic plutons west of the Churchill–Superior Boundary Zone have isotopic features suggesting derivation from the melting of much older basement (Weber, this volume, and references therein).

The Reindeer Zone can be divided conveniently into several lithostructural domains (Macdonald and Broughton, 1980; Lewry, 1981a; Lewry *et al.*, 1985) (Fig. 1) and subdomains (Lewry and Slimmon, 1985; Coombe *et al.*, 1986). Such domainal differentiation markedly influenced the development of evolutionary models that invoked plate subduction, arc–microcontinent collision, and transform movements prior to terminal continent–continent collision in this part of the orogen (Lewry, 1981a; Lewry *et al.*, 1981; Stauffer, 1984; Green *et al.*, 1985a,b) and that emphasised the geotectonic significance of major tectonic domainal boundaries, such as the Stanley and Tabbernor shear zones. However, U–Pb zircon ages and other isotopic data (Chauvel *et al.*, 1987; Van Schmus *et al.*, 1987) indicate that such models require significant amendment in at least some respects. In particular, such data do not support the thesis that the Glennie Domain is, wholly or in part, an accreted Archean microcontinent (Lewry, 1981a).

Nonetheless, previously recognised domainal subdivision of the Reindeer Zone remains essentially valid and is largely retained in this paper, albeit primarily for purposes of convenient descriptive reference. It is clear, however, that such "two-dimensional" subdivision does not adequately reflect, or sufficiently emphasise, the large-scale "three-dimensional" structural geometry of the collage, many aspects of which are of critical kinematic and geotectonic significance. This paper discusses such structural geometry, and its age and geotectonic implications, as seen within the Reindeer Zone in Sas-

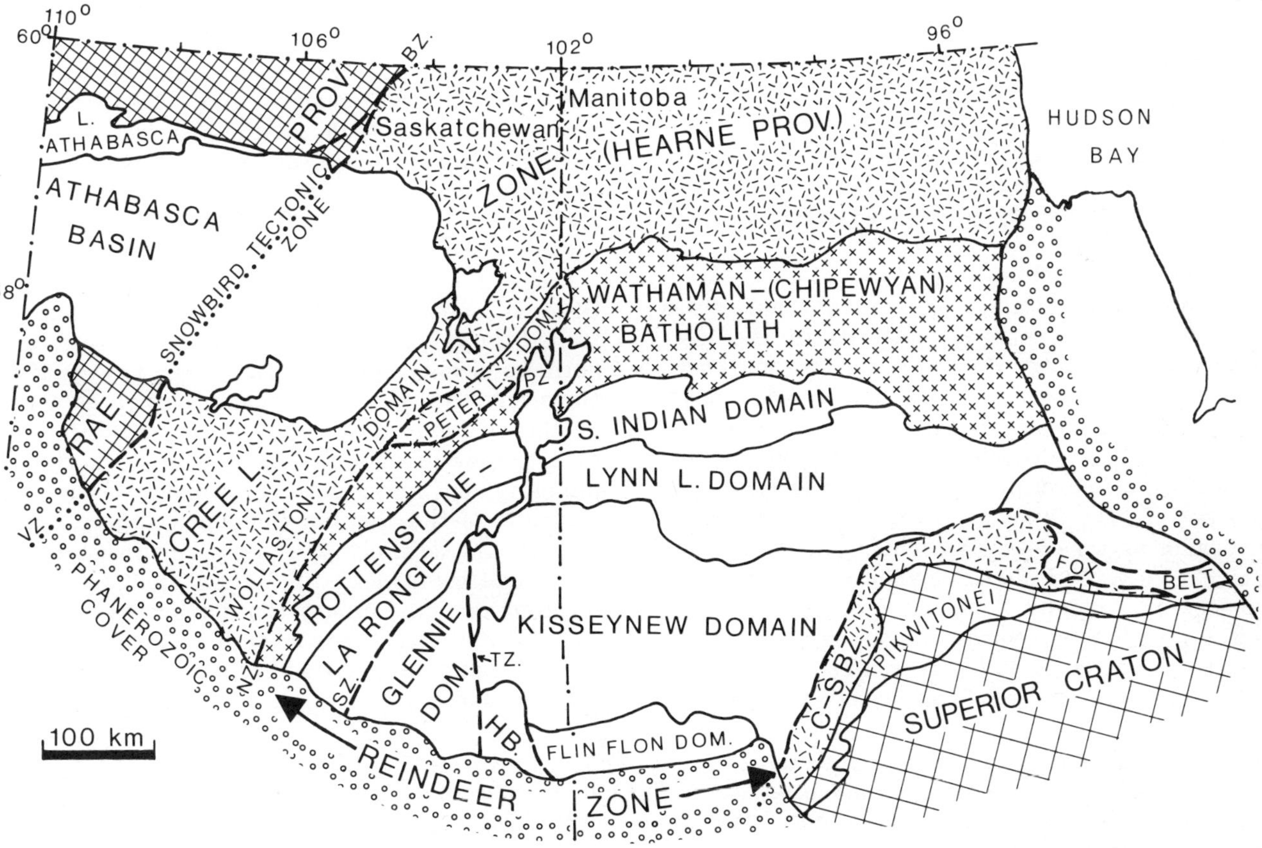

Figure 1. *Previously published lithotectonic domains of the Reindeer Zone, and other elements of the Trans-Hudson Orogen in Saskatchewan and Manitoba. Ornamented Precambrian components are mainly "ensialic" elements of the orogen and bounding cratons. Unornamented domains are mostly juvenile Early Proterozoic crustal elements. Heavy dashed lines indicate major ductile shear zones or brittle–ductile faults. Abbreviations: C–S BZ, Churchill–Superior Boundary Zone; HB, Hanson Lake Block; TZ, Tabbernor fold/fault zone; SZ, Stanley fault/shear zone; NZ and PZ, Needle Falls and Parker Lake shear zones; and VZ and BZ, Virgin River and Black Lake shear zones (parts of Snowbird tectonic zone).*

katchewan. A provisional, revised subdivision of the collage, which emphasises gross three-dimensional structural features, is outlined. Evidence is presented to support the view that much of the juvenile collage was tectonically dismembered subsequent to accretion and was emplaced as a series of allochthonous, regional-scale, early ductile nappes and imbricated, flat-lying thrust sheets developed in an essentially thin-skinned tectonic regime. This regime is consistent with massive crustal shortening in a convergent plate-margin setting, presumably during terminal continent–continent collision.

The following review represents only a preliminary interpretation of the structural geometry and tectonic evolution of the Reindeer Zone, based on a general regional analysis of available (and in many respects inadequate) published data and amplified by more detailed local studies by the authors and others. However, the proposed model incorporates predictions that are amenable to testing by further detailed field, isotopic, and geophysical investigations.

GENERAL STRUCTURAL GEOMETRY AND TECTONIC SEQUENCE IN THE REINDEER ZONE

Late fold phases

From available data, it is evident that gross lithological and structural patterns in the Saskatchewan part of the Reindeer Zone are markedly controlled by two sets of relatively late, large amplitude regional folds (only the last of these is shown in Fig. 2). Although the various domains differ somewhat in overall structural geometry, a common set of late, large amplitude, open to closed, flexural or flattened–flexural folds, generated about upright or steeply inclined northeast-trending axial surfaces, is evident throughout much of the Reindeer Zone. For descriptive convenience, these are hereafter referred to as D_4 structures, although, in fact, a variable number of earlier sets of kinematic structures are documented in different parts of the zone. In Glennie, Kisseynew, and Flin Flon domains and the Hanson Lake Block, D_4 structures refold upright to overturned, close to tight, generally north- to northwest-trending D_3 folds to produce regional type 1 or 2 interference patterns. The Glennie Domain is typified by type 1 "dome and basin" interference patterns, with local complication and variation probably deriving both from a significant D_4 flexural strain component and from the local involvement of pre-D_3 major fold closures. In the Kisseynew Domain, D_3/D_4 interference is more commonly of type 2 or 3; differentiation of D_3 structures and earlier tight to isoclinal major folds is locally uncertain. In the La Ronge Domain, D_4 regional folding is obvious only in the southwest, where a major D_4 closure occupying much of the Nemeiben Zone (and defining the so-called "La Ronge Horseshoe") refolds "at least" two earlier generations of tight to isoclinal folds to produce complex interference patterns of variable geometry (cf. Lewry and Slimmon, 1985; Coombe *et al.*, 1986). Elsewhere in the La Ronge and Rottenstone domains, D_4 deformation may be manifested primarily in large-scale rotational oversteepening of earlier planar structures and perhaps by late transpressive movement along subvertical high-strain zones such as the Birch Rapids straight belt (Lewry, 1984; Lewry and Slimmon, 1985).

There is considerable local variation in the attitude of major and minor D_4 fold hinge lines, controlled largely by plunge reversal across axial surfaces of major D_3 structures (e.g., Lewry, 1977). However, the general regional plunge is to the northeast and in Saskatchewan regional lithostructural trends are dominated by a broad northeast-plunging D_4 anticlinorium, which is cored by elements of the Glennie Domain (compare Figs. 1 and 2). Moderately to steeply northwest-dipping structural components of the La Ronge and Rottenstone domains occupy the northwestern limb of this regional structure and thus, in part, must represent higher structural levels than much of the Glennie Domain.

Figure 2. *Revised subdivision and nomenclature of major three-dimensional structural components of the Reindeer Zone in Saskatchewan, as proposed in this paper. Ornamentation mostly differentiates gross lithostructural sheets (levels) rather than lithology, but molasse-type arkosic assemblages at all structural levels are specifically indicated by fine stipple. Random tick ornament indicates Archean basement windows of the Sahli Granite (in Hanson Lake Block) and those underlying the Cartier Sheet, northeast of Lac La Ronge. Thick barbed lines are major refolded thrust soles. Thick unbarbed continuous lines indicate late (D_4) regional fold axial traces – synforms, antiforms, and plunge direction are indicated by conventional symbols. Abbreviations: N.B.B., previously defined "Nut Bay Belt" (now part of the Wapassini Sheet); N.E., previously defined "Nemeiben Zone" (also part of the Wapassini Sheet); NG, Nistowiak (mylonitic) Gneisses; HG, Hunter Bay (mylonitic) Gneisses – probable lateral equivalents of the Nistowiak Gneisses; B.R.S.B., Birch Rapids "straight belt"; and S.Z., Stanley shear/fault zone. Lines a–a and b–b mark the location of sections shown in Figure 9.*

Early structures and definition of major lithostructural sheets

Although major advances have recently been made, details of early tectonic events and related structural geometry remain poorly elucidated in much of the Reindeer Zone. Available data nonetheless allow a general interpretation of early kinematic history and gross structural geometry.

Complex polyphase folding, evident throughout most of the Reindeer Zone, commonly involves at least two sets of "pre-D_3" structures. Aside from the Flin Flon Domain and part of the La Ronge Magmatic Belt (i.e., the previously defined Central Metavolcanic Belt of Coombe *et al.* (1986)), all parts of the zone display one or more early sets of tight to isoclinal folds, with coeval foliation and linear

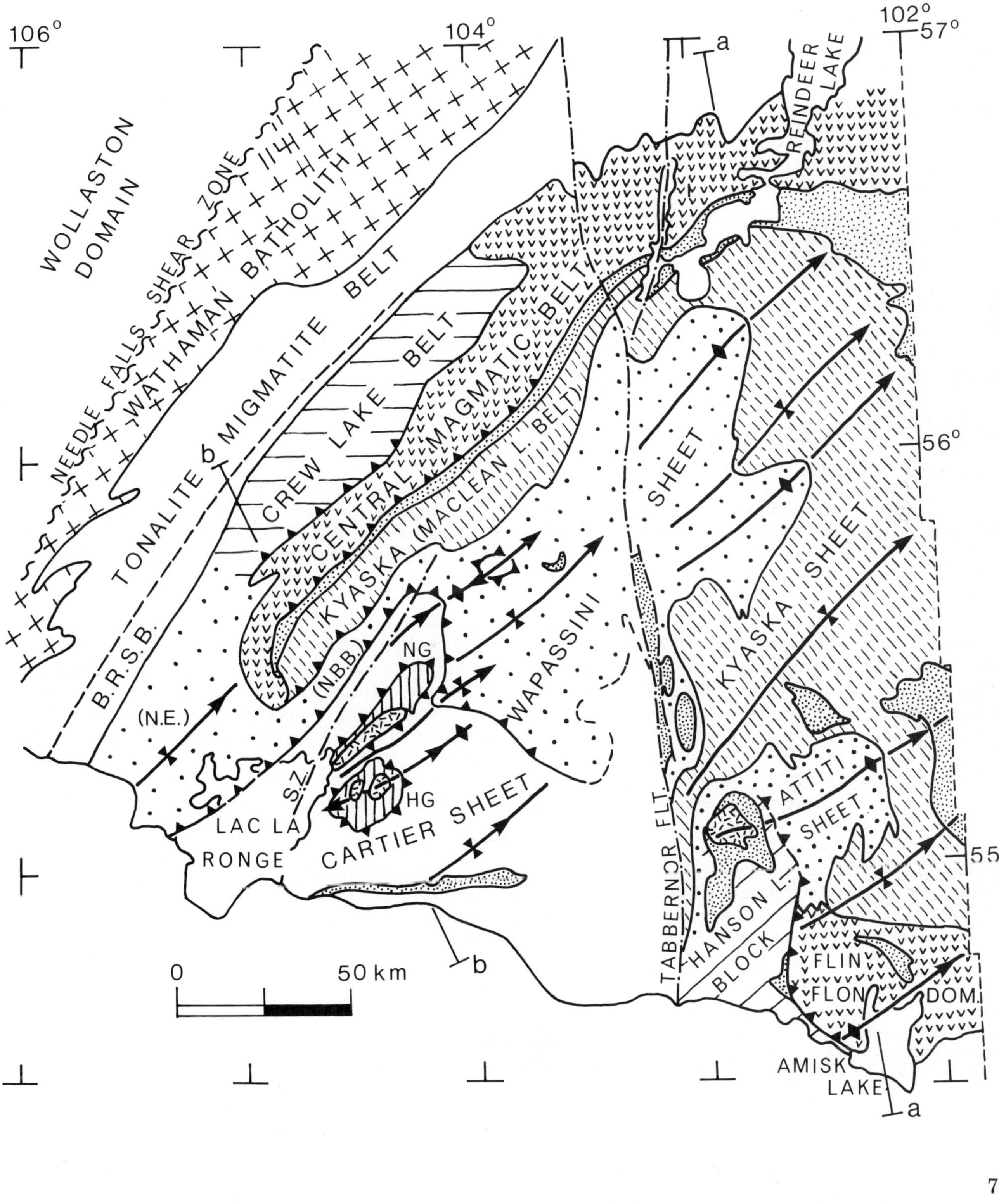

fabric elements developed on all scales. Typically, early gneissic foliation and coeval intrafolial isoclines (D_1 structures) are refolded by tight to isoclinal D_2 major and minor folds, with variable transposition of primary layering and D_1 structures into the S_2 foliation direction. For example, the western part (MacLean Lake Belt) of the newly designated Kyaska Sheet (Figs. 2 and 3) is dominated by doubly plunging D_2 isoclines with moderate to steeply northwest-dipping axial surfaces: locally, earlier fabrics are transposed into a strongly developed S_2 foliation defining ductile high-strain attenuation zones commonly localised along overturned D_2 fold limbs (Abraham, 1985; Thomas, 1986). A similar sequence is recorded in much of the Glennie Domain (e.g., Lewry, 1977) and Kisseynew Domain (e.g., Gilboy, 1980; Ashton *et al.*, 1986, 1987). Elsewhere, as in the Rottenstone Domain and higher grade northeastern parts of the La Ronge Domain, several essentially coplanar, and commonly coaxial isoclinal fold and foliation sets, probably representing a progressive pre-D_3 strain continuum, are recognised (e.g., Lewry *et al.*, 1981). However, in the Flin Flon Domain and the lower grade parts of the La Ronge Magmatic Belt, only one set of "pre-D_3" structures is commonly documented and there is little evidence of early large-scale isoclinal fold development (e.g., Stauffer and Mukherjee, 1971; Thomas, 1985, 1986).

As described below, a number of ductile high-strain zones, some of regional extent and varying in present attitude from subhorizontal to subvertical, are documented as having formed during these early deformational events. Some define major tectonic boundaries between domains of contrasting lithologic character, structural style, and apparent bulk strain; others occur partly or entirely within previously defined domains.

When viewed in relation to the geometry of later D_3 and D_4 folds, the character, distribution, and attitude of the regionally developed early (D_1 and D_2) planar structures, folds, and coeval high-strain zones suggest that they reflect a complex stack of initially highly overturned to flat-lying, more or less discrete, early fold nappes or thrust sheets. Many presently identified major allochthonous sheets are soled by discrete ductile-strain discontinuities, major mylonitic gneiss, and tectonic schist zones, or otherwise show evidence of high-strain gradients towards their soles. This early flat-lying complex was refolded by mesoscopic to regional D_3 and D_4 folds and oversteepened in regional fold limbs: in places it is also disrupted by late, subvertical, ductile to brittle fault zones. Although most of the early structural sheets show prominent internal multiphase isoclinal–intrafolial folding and coeval foliation development on all scales, their boundaries, even where not presently documented as high-strain zones, tend to be structurally simple, with limited tectonic interdigitation attributable to pre-D_3 folding.

The distribution of presently defined major structural units, with attendant revised nomenclature for the Reindeer Zone, is shown in Figure 2. Many sheets differentiated here have more or less distinct gross lithological character and in whole or in part correspond to previously defined domains or subdomains. Others do not, or incorporate parts of two or more previously recognised lithotectonic entities; in such cases, they are assigned new names.

Figure 3. *More detailed geology of the southwestern part of the area shown in Figure 2, covering parts of previously defined Rottenstone, La Ronge, and Glennie domains (Fig. 1). A–A′, B–B′, and C–C′ are lines of the sections shown in Figure 4. Heavy solid lines show regional D_4 fold axial traces with arrows indicating the plunge. Ornamentation (also applies to subsequent text figures, mostly): random tick, Archean basement of the Iskwatikan Lake dome (IC) and core of the Hunter Bay dome to the south, underlying Nistowiak (NG) and Hunter Bay (HG) mylonitic gneisses; discontinuous squiggle, mylonitic gneisses/schists of the Nistowiak – Hunter Bay and Guncoat Gneiss (GG) nappe soles; vertical lines, metavolcanics and volcanogenic metasediments within the Cartier Sheet and other parts of southern Glennie Domain; oblique lines, orthogneisses and plutonic rocks of the Wapassini Sheet, structurally overlying Guncoat Gneisses, including continuations in the Nut Bay Belt and Nemeiben Zone (see Fig. 2); coarse stipple, undifferentiated supracrustals in the Wapassini Sheet; fine dashes following trend lines, "Kisseynew-type" volcanogenic metasediments and other supracrustals/migmatites of the Kyaska Sheet (previously "MacLean Lake Belt" in this area) — overlying the Wapassini Sheet; heavy crosses, plutons of the Central Magmatic and Crew Lake belts; "vee" pattern, mainly volcanic rocks of the Central (La Ronge) Magmatic Belt, structurally overlying the Kyaska Sheet; fine stipple, "molasse-type" meta-arkosic assemblages throughout the area, including the McLennan Group of the upper part of the Kyaska Sheet and similar rocks at other structural levels, all stratigraphically younger than other supracrustal and metavolcanic sequences; and unornamented (save for structural trend lines), undifferentiated orthogneisses and plutons of the Cartier Sheet, overlying the Nistowiak nappe sole and underlying the Guncoat Gneiss sole zone. Also unornamented, apart from trend lines, are metasediments of the Crew Lake Belt, structurally overlying the Central Magmatic Belt but with an equivocal relation to the southwestern part of the Wapassini Sheet (Nemeiben Zone), and rocks of the Tonalite–Migmatite Belt northwest of the Birch Rapids straight belt.*

For the most part, Figure 2 emphasises only gross structural "levels," rather than lithological subdivisions. However, the general occurrence of meta-arkosic molasse assemblages (cf. Stauffer, this volume) is specifically shown because their distribution is thought to be of particular structural significance. We do not thereby imply that such rocks (including the Sickle, Missi, McLennan, and Sherridon groups and similar assemblages) necessarily all represent remnants of a coeval, formerly continuous stratigraphic unit, and, indeed, consider this to be unlikely. However, in all cases where

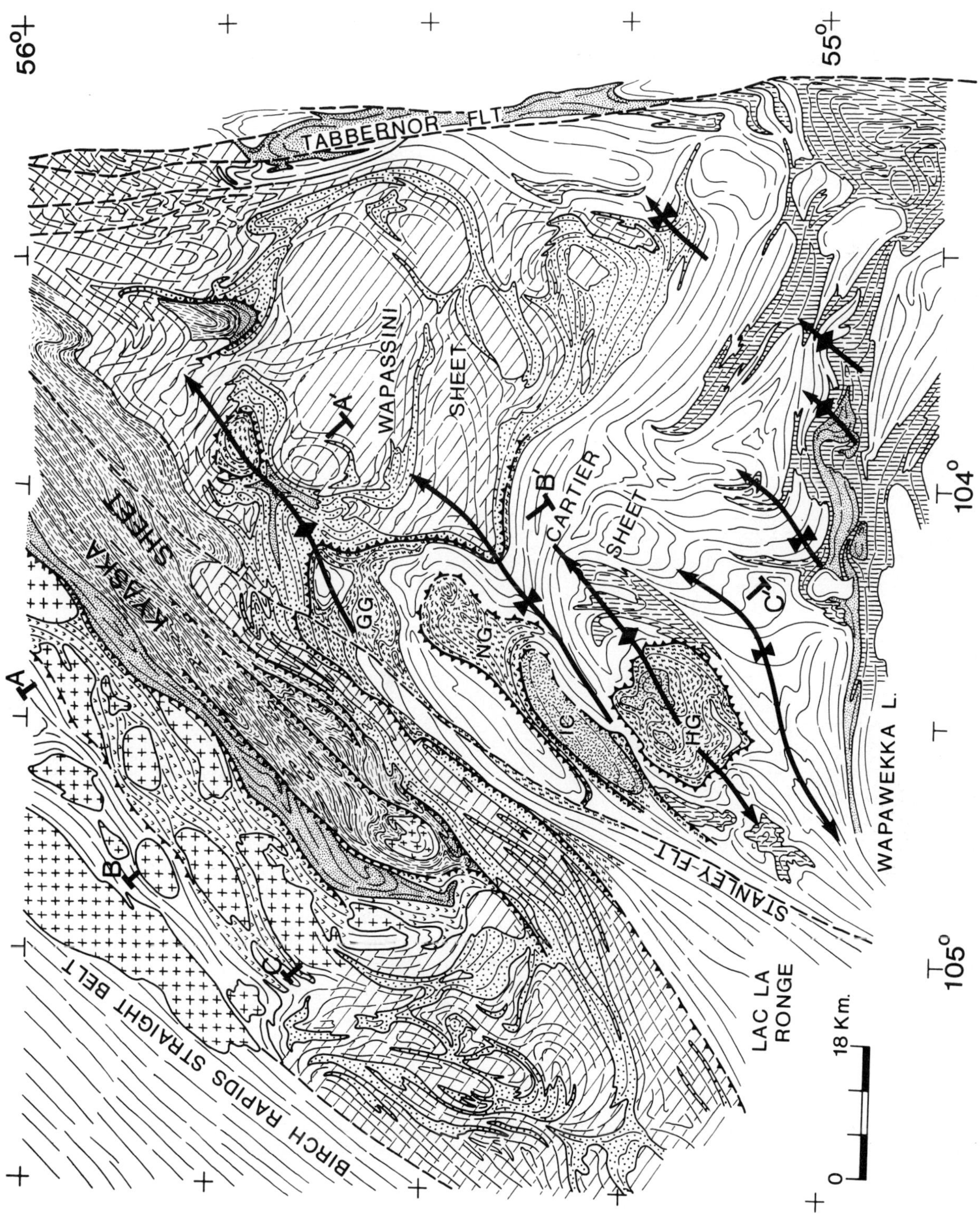

their relative age is "unequivocally" documented, such rocks postdate arc volcanic and related volcanogenic metagreywacke sequences: none are clearly documented as predating arc volcanism. Such evidence, plus distinctive common characteristics indicating that the meta-arkose assemblages were deposited in a similar early orogenic setting, suggests that they may be at least "broadly" coeval. Thus, they can reasonably be taken to represent a crude lithostratigraphic marker at the regional scale. Nonetheless, such molasse assemblages occur at various structural levels within the complex, implying large-scale tectonic interleaving and imbrication.

The foregoing broad synopsis of structural relations in the Reindeer Zone is amplified below in a detailed discussion of three principal areas in Saskatchewan.

SOUTHWEST LA RONGE DOMAIN AND GLENNIE DOMAIN

Probable tectonic stacking of more or less discrete, early kinematic lithostructural sheets is displayed in a transect across central parts of the La Ronge Domain and the adjoining Glennie Domain (Figs. 3 and 4). From northwest to southeast, the sequence from highest to lowest structural levels is discussed in the following sections.

Tonalite–Migmatite Belt

The presently defined extent of this belt (Fig. 2), which forms the southeastern part of the Rottenstone Domain, follows previous usage (e.g., Lewry *et al.*, 1981; Lewry and Slimmon, 1985). It consists mainly of varied trondhjemitic, tonalitic, granodioritic, and subordinate granitic plutonic rocks, and related injection migmatites incorporating a variety of subordinate supracrustal paleosomal components: metamorphic grade is generally within the upper amphibolite facies. Several phases of tight to isoclinal–intrafolial fold and coeval foliation development, spanning an extended period of plutonic neosome injection, are documented throughout the belt (Lewry *et al.*, 1981, and references therein). Tectonic foliation and transposed layering are generally subvertical or dip moderately to steeply to the northwest throughout much of the Rottenstone Domain. However, in the

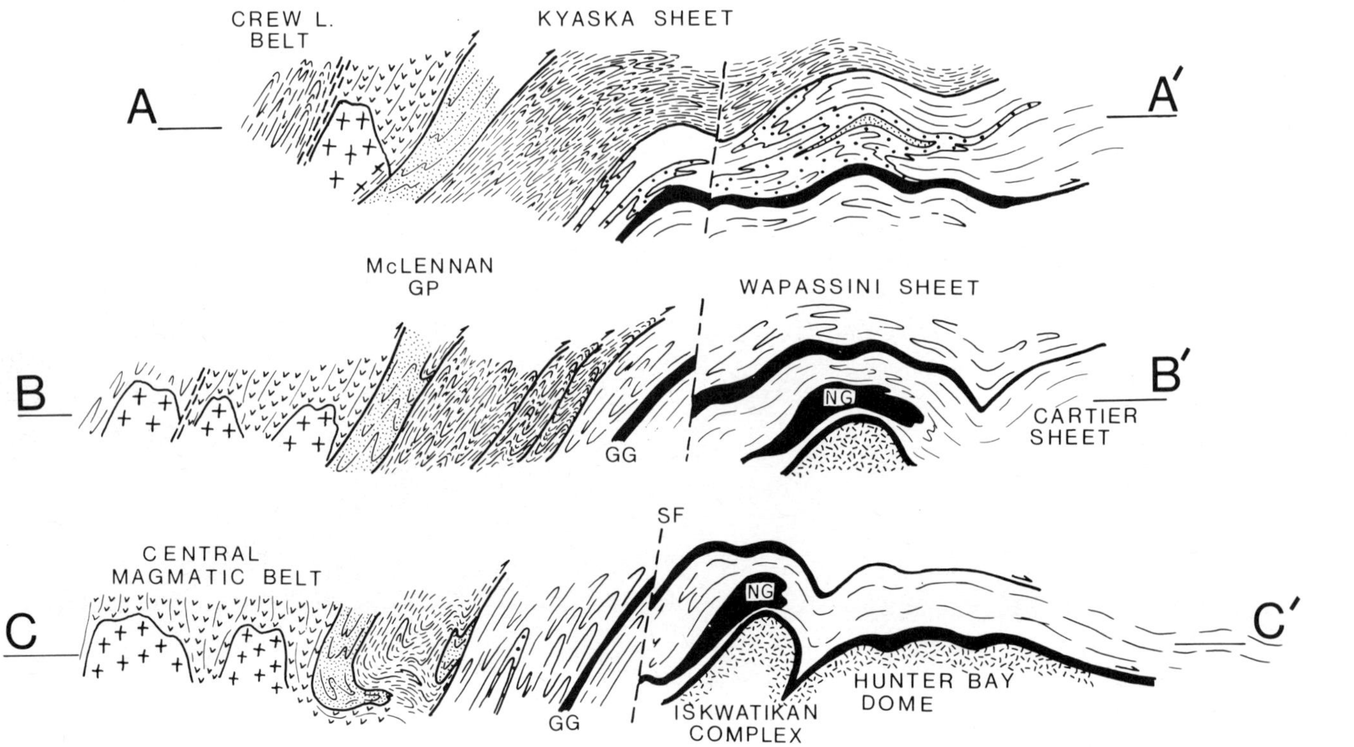

Figure 4. *Generalised serial cross-sections across La Ronge and Glennie domains, showing the relation of major lithostructural sheets. Ornamentation as in Figure 3, save that mylonites of the Nistowiak – Hunter Bay – Guncoat gneisses are shown here in solid black. Orthogneisses and paragneisses of the Wapassini Sheet are indicated separately, to portray early structures within the nappe sheet, only in section A–A′.*

Reindeer Lake area (Fig. 5), early planar structures generally dip more gently to the north or northwest and are refolded on both a large and small scale by a late fold set (D_5?) developed about subvertical northerly trending axial surfaces spatially related to the Tabbernor fold–fault zone (e.g., Gilboy, 1980; Lewry *et al.*, 1980).

The Tonalite–Migmatite Belt is lithologically transitional into less migmatised, lower grade rocks of the La Ronge Domain (e.g., Lewry *et al.*, 1981; Coombe *et al.*, 1986). In the southwest, however, the junction between the two domains is defined by the eastern boundary of the Birch Rapids Straight Belt, a broad subvertical zone of inhomogeneous, probably transpressive, ductile high strain and local later brittle cataclasis (Lewry and Slimmon, 1985). Farther to the northeast, high-strain zones occur locally in the vicinity of the Tonalite–Migmatite/La Ronge Domain boundary but are not generally considered to represent major crustal discontinuities (Lewry *et al.*, 1980; Ray, 1981; Gilboy, 1982).

This belt lies to the northwest of the region on which our analysis is principally focused and is not discussed further.

Crew Lake Belt

This part of the La Ronge Domain (Lewry, 1984; Coombe *et al.*, 1986) is a lithologically and, in some respects, structurally distinct package, overlying the La Ronge Magmatic Belt to the southeast. The junction with the latter is documented locally as a diffuse to sharply defined high-strain zone but elsewhere is defined only on lithological criteria (e.g., Macdonald, 1984; Thomas, 1984, 1985; Harper, 1985). The Crew Lake Belt mainly comprises isoclinally folded mixed metasediments intruded by granitic plutons. The metamorphic grade is generally in the middle to upper amphibolite facies.

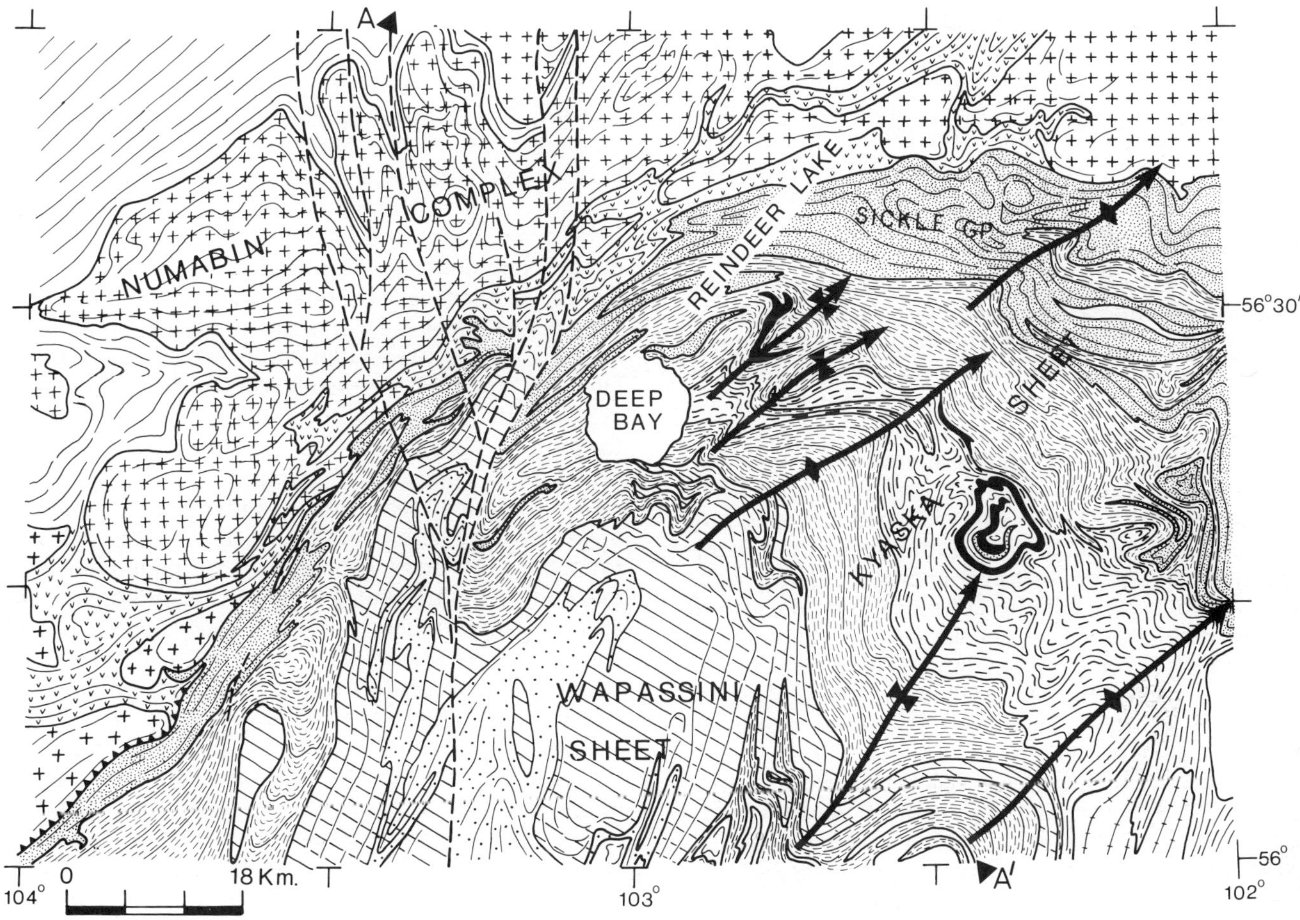

Figure 5. *General geology of the South Reindeer Lake area (northern part of Fig. 2). Apart from Deep Bay, at the south end of Reindeer Lake, lake outlines are omitted to avoid obscuring the interpreted geology. Ornamentation mostly as in Figure 3, save that different lithostratigraphic units of the "Burntwood Supergroup" are differentiated with fine and heavy dashes to emphasise early recumbent major fold closures in the Kyaska Sheet. A–A′ indicates the approximate line of the cross-section shown in Figure 6b.*

La Ronge Magmatic Belt

As defined here, this belt includes the previously designated Central Metavolcanic Belt and Numabin Complex (Coombe *et al.*, 1986) of the La Ronge Domain, which, though differing both in relative areal extent of volcanic and plutonic rocks and in style and intensity of deformation, form an essentially coherent, continuous lithotectonic entity. The Central Metavolcanic Belt is a predominantly volcanoplutonic slab generally displaying relatively low, but heterogeneous, internal strain: the metamorphic grade is in the upper greenschist to low amphibolite facies. To the northeast, it grades into the predominantly plutonic, higher grade, more highly deformed Numabin Complex (Fig. 5).

Kyaska Sheet

This newly defined sheet includes both the MacLean Lake Belt (Macdonald and Broughton, 1980; Lewry and Slimmon, 1985) of the southeastern La Ronge Domain, and most of the Kisseynew Domain. Lithologically, it comprises most of the volcanogenic Burntwood Supergroup and complexly infolded meta-arkosic Sickle–Sherridon groups of the Kisseynew Domain, as well as essentially contiguous volcaniclastics, volcanogenic metagreywackes, and polymictic metaconglomerates of the "MacLean Lake gneisses" (probable Burntwood equivalents) and meta-arkosic rocks of the McLennan Group (Sickle equivalent?) in the MacLean Lake Belt.

It is now evident that the McLennan Group is younger than the La Ronge metavolcanics. In the Mackay Lake area, near the southwest end of its outcrop area north of Lac La Ronge (Figs. 2 and 3), it has been shown to "stratigraphically" overlie the latter with pronounced unconformity (Lewry, 1986; Thomas, 1987). Generally, however, the group "structurally underlies" the metavolcanics, commonly with sharp tectonic contact. Thus, over much of its extent, the junction between McLennan Group meta-arkoses of the Kyaska Sheet and the structurally overlying La Ronge Magmatic Belt is abruptly defined by the McLennan Lake Tectonic Zone (Lewry, 1983; Thomas, 1984, 1985), a subvertical to steeply northwest-dipping ductile high-strain zone, which can now be traced from Milner Lake, in the southwest, to north of South Waddy Lake (Redhill Lake Tectonic Zone of Harper (1985)). Farther northeast, the junction is only lithologically defined at present; however, high-strain zones are documented locally at or close to the contact (e.g., Johnston, 1969, 1970; Fox and Johnston, 1981): we speculate that the junction in this region might be a complexly imbricated thrust zone.

The initial attitude of the McLennan Lake Tectonic Zone is unknown, and the sense of early displacement is not clearly established. Local kinematic indicators suggest late "normal" downthrow to the northwest (Thomas, 1986), but the structure may have been initiated as a thrust.

Wapassini Sheet

The Wapassini Sheet structurally underlies the Kyaska Sheet. It is defined here as incorporating not only the previously named "Wapassini Allochthon" (Lewry, 1984; Lewry and Slimmon, 1985) of the Glennie Domain but also the structurally and lithologically contiguous Nut Bay Belt and Nemeiben Zone of the La Ronge Domain. From Lac La Ronge northward and eastward at least as far as Drinking Lake (central part of the Glennie Domain), the Wapassini Sheet is soled by the Guncoat Gneisses, a package of spectacular early kinematic, biotitic, plagioclase–porphyroclastic tectonic schists and mylonitic gneisses largely derived from already highly migmatised pelitic gneisses and pelite diatexite protoliths (Lewry, 1981b). The upper boundary of this ductile high-strain unit is defined as the Guncoat thrust (Lewry and Slimmon, 1985). The Wapassini Sheet is structurally continuous through the Nut Bay Belt, hitherto considered to be an eastern component of the La Ronge Domain, and to the southwest can be traced with little or no discontinuity into the southwestern part of the Nemeiben Zone around the hinge of the "La Ronge Horseshoe." Thus, we suggest that the Wapassini Sheet is an enormous allochthonous slice overlain by central elements of the La Ronge Domain and, in the east, tectonically emplaced across still lower crustal elements of the Glennie Domain. Throughout its extent, the Wapassini Sheet is a complex of strongly foliated plutonic rocks, migmatites, and subordinate, mainly metavolcanic, supracrustal rocks. The relative proportion of these lithological components varies markedly in different parts of the sheet, but individual major rock units and early isoclinal fold trends can be traced on a regional scale from the Glennie Domain into the Nut Bay Belt and thence into the Nemeiben Zone.

Although regional relations indicate that the Wapassini Sheet largely underlies both the Central Magmatic Belt and MacLean Lake Belt (Kyaska Sheet), the precise relationship between its southwestern extension, in the Nemeiben Zone, and these more northerly elements of the La Ronge Domain is unclear. Poorly constrained lithostructural trends suggest possible continuity between the Nemeiben Zone and the southwestern part of the Crew Lake Belt, implying that the latter might represent the overturned limb of a late (D_4?) regional fold cored by the La Ronge Magmatic Belt. However, such a closure is not presently recognised within the La

Ronge Magmatic Belt and resolution of this problem must await further analysis.

Cartier Sheet

The Cartier Sheet may incorporate much of the Glennie Domain structurally underlying the Guncoat mylonitic gneisses. In the southwest, it includes most of the "Iskwatikan Complex" (Lewry and Slimmon, 1985) and the "Meraste Zone" of the Lac La Ronge area (Coombe *et al.*, 1986), which was previously defined as a component of the La Ronge Domain.

In the southwestern Glennie Domain, the Cartier Sheet is soled by an early, complexly refolded, ductile high-strain zone defined by the dominantly mylonitic, porphyroclastic Nistowiak – Hunter Bay gneisses. The Nistowiak Gneisses (Lewry, 1984; Lewry and Slimmon, 1985; Chiarenzelli and Chiarenzelli, 1986; Chiarenzelli *et al.*, 1987; Chiarenzelli and Lewry, 1988) are a zone of early (pre-D_3) ductile protomylonitic to ultramylonitic porphyroclastic and "straight" gneisses derived mainly from quartz dioritic to granitic protoliths. The zone also includes subordinate feldspar–porphyroclastic tectonic schists (similar to the Guncoat Gneisses), pelitic gneisses, calc-silicates, and amphibolitic intercalations. It is structurally sandwiched by less-strained plutonic rocks, subordinate paragneisses, and migmatites. The Hunter Bay Gneisses (Lewry, 1984; Lewry and Slimmon, 1985) are very similar in character, differing, if at all, only in the proportion of different protoliths involved in the mylonitic assemblage. They are exposed mainly in an irregular, more-or-less closed ring around the flanks of an open domal structure in the Hunter Bay area (east of Lac La Ronge) and are again apparently underlain and overlain by less deformed, predominantly plutonic gneisses. The geometry of imposed D_3/D_4 fold interference in the Nistowiak – Hunter Bay area is imperfectly resolved; however, the general similarity of the Nistowiak and Hunter Bay mylonitic packages, apparent continuity between the two along Pothole Bay in northeastern Lac La Ronge, and current interpretation of the character of late refolding strongly suggest that they represent parts of a single, formerly continuous, flat-lying, high-strain zone, occurring at a deeper structural level than the Guncoat Gneiss zone: this high-strain zone defines the base of the Cartier nappe sheet.

Rocks underlying the Cartier Sheet in the Iskwatikan–Nistowiak lakes and Hunter Bay areas represent the deepest structural levels presently recognised in the La Ronge and Glennie domains. Recent mapping and preliminary radiometric age data (Chiarenzelli *et al.*, 1987; Chiarenzelli and Lewry, 1988) indicate that the core of the Iskwatikan Lake dome, immediately underlying the Nistowiak Gneisses (Fig. 3), mostly comprises Archean basement gneisses. This "Iskwatikan Lake Complex" comprises tonalitic gneisses and migmatites intruded by a variably strained megacrystic augen granite gneiss. The megacrystic granite yields a U–Pb zircon age of ca. 2500 Ma and preliminary data suggest that the tonalitic gneisses are at least 2800 Ma old. Parts of the flanking mylonitic Nistowiak Gneisses are also documented as being derived from Archean protoliths, though present data suggest that most are derived from Early Proterozoic rocks. It should be noted that part of the Nistowiak Gneiss outcrop area comprises a "bulbous" closed structure separated from structurally underlying mylonites defining the main Archean–Proterozoic contact by a narrow belt of less highly strained rocks (Fig. 4). Chiarenzelli *et al.* (1987) tentatively interpret this structure as being a major sheath fold, which ultimately roots in the main detachment zone.

The Cartier Sheet again comprises varied, foliated to undeformed plutonic rocks; subordinate mixed supracrustals; and migmatites. A distinction between the Wapassini and Cartier sheets, and crustal components underlying them, cannot be reliably established on the basis of lithological character but is dependent on the recognition of the mylonitic packages forming their soles. At present, the latter are defined only in the western Glennie Domain. In the southern and eastern Glennie Domain, and areas east of the Tabbernor zone, such structural subdivision must await further field investigation. For the present, therefore, components structurally underlying the Kyaska Sheet in these areas are only speculatively included within either the Wapassini or the Cartier sheets, or are assigned names of only local significance.

Internal features and relationship of sheets

Aside from parts of the Central Magmatic Belt, all major tectonic sheets defined above display high internal strain, including tight to isoclinal folding and coeval/coaxial foliation and lineation of at least two generations. In the La Ronge Domain, gross lithological layering, tectonic foliation, and axial surfaces of the early fold sets dip moderately to steeply to the northwest. A prominent, generally downdip stretching lineation is widely developed throughout all sheets, including much of the Central Magmatic Belt. To the southeast, passing into the Glennie Domain, early gneissose foliation and axial surfaces of isoclinal folds within the Kyaska, Wapassini, and Cartier sheets flatten out and are thereafter complexly refolded by upright D_3 and D_4 mesoscopic to macroscopic folds. Locally, the early stretching lineation is also evidently refolded by the later fold sets. In most areas that have been mapped in some detail, junctions between major lithostructural sheets are marked by abrupt ductile to brittle high-strain zones, which are also refolded. Elsewhere, abrupt tectonic discontinuities are undocumented, but junctions are either sharp lithological breaks or abruptly transitional in character. Present relations between the lithotectonic sheets defined above are shown in generalised, but representative, structural sections across the southwest La Ronge Domain and Glennie Domain (Fig. 4). The attitudes of these stacked allochthons prior to late regional refolding are not well constrained. Present attitudes and internal structural geometry, however, strongly suggest that they were emplaced as generally flat-lying to gently dipping sheets and that their present steeply homoclinal attitude in the La Ronge Domain is the result of late refolding and rotational oversteepening.

SOUTH REINDEER LAKE AREA

In the South Reindeer Lake area (Fig. 5), gross layering and early planar fabrics in much of the Tonalite–Migmatite Belt and underlying Central Magmatic Belt dip moderately or gently to the north or northwest, typically at 40° or less. Dips generally steepen southwards, toward the contact between the Central Magmatic Belt and the McLennan–Sickle Group.

The Kyaska Sheet – incorporating psammopelitic to pelitic metagreywackes of the Burntwood Supergroup ("Kisseynew gneisses") and younger McLennan–Sickle Group meta-arkoses structurally underlying volcanics and plutons of the Central Magmatic Belt – is folded about a series of open, gently northeast-plunging D_4 major folds in the nose of a regional D_4 anticlinorium, which dominates the northern part of the Kisseynew Domain (Gilboy, 1980). To the southwest, the core of this structure is occupied by granitoids and migmatites apparently representing an extension of the Wapassini Sheet. The D_4 structures refold overturned to recumbent, tight to isoclinal folds and coeval foliation of at least two, and possibly more, generations (Gilboy, 1980), producing complex fold interference patterns defined both by lithological units and by early foliation.

Junction relations between major lithostructural sheets in the South Reindeer area are poorly known. The boundary between the Kyaska Sheet and overlying Central Magmatic Belt is complex and poorly understood. Major ductile high-strain zones occur along both the upper (northwestern) and the lower (southeastern) margins of the Numabin Complex in the Numabin Bay region (Lewry *et al.*, 1980), but otherwise no obvious high-strain contacts are documented in this region. The junction of the Kyaska and Wapassini sheets is locally affected by large-scale isoclinal fold digitation, but otherwise the nature of this boundary is poorly documented. Johnston (1983) recognised lithological and structural discontinuity between "Upper Kisseynew Gneisses" and "Lower Kisseynew Gneisses" in the area southwest of Reindeer Lake, which he interpreted as a tectonically modified unconformity. This supposed break coincides with the sole of what is now defined as the Kyaska Sheet: there seems to be little persuasive evidence of stratigraphic unconformity and the junction may, in fact, represent an entirely tectonic break.

Figure 6a shows a generalised down-plunge view of the Kyaska and Wapassini sheets south of Reindeer Lake, approximately normal to gently plunging D_4 fold axes. Parts of the section are poorly constrained, insofar as extension of some early isoclinal closures into the plane of projection is uncertain. However, it provides a fair representation of early structural relations in these essentially recumbent sheets. Figure 6b shows a more extensive schematic structural section across the whole area of Figure 5, utilising data drawn from the down-plunge construction: it involves some simplification and adjustment of closure directions of early isoclines to accommodate variable plunge, and is again not rigorously constrained. Nonetheless, it is considered to be a reasonable portrayal of gross structural relations and indicates the essentially flat-lying attitude of major lithostructural sheets and early internal isoclinal folds across the area.

FLIN FLON - HANSON LAKE AREA

Although the Flin Flon – Hanson Lake – Wildnest Lake region (Fig. 7) has been subject to more intensive geological study over the past four decades than perhaps any other comparable area in the Saskatchewan Shield, many critical aspects of its geology remain uncertain and controversial. In particular, the precise relation between low-grade supracrustals of the Flin Flon – (Snow Lake) Domain and more highly deformed, high-grade gneisses of the Kisseynew Domain (across the so-called "Kisseynew Front") and Hanson Lake Block is not yet completely resolved. We consider that the following interpretation of major lithotectonic relationships is consistent both with available published data from the area and with the more general structural model presented in this paper.

In the Kisseynew Domain and the Hanson Lake Block (Fig. 7), early foliation and isoclinal folds of one or more generations (D_1/D_2?) are refolded by close to isoclinal north-trending D_3 folds, which, in turn, are refolded by dominant open to closed, generally upright northeast-trending D_4 structures to produce complex large- and small-scale interference patterns (e.g., Pearson, 1973; Ashton *et al.*, 1986, 1987). The Sahli – Hanson Lake "dome," which dominates the structure of the Hanson Lake Block, appears to be primarily a D_3/D_4 interference feature. Most of the deformation sequence in the Kisseynew Domain and Hanson Lake Block occurred under medium- to high-grade metamorphic conditions, which outlasted D_4 folding (e.g., Sibbald, 1978; Ashton *et al.*, 1986, 1987).

In the Flin Flon Domain, late northeasterly regional folds again refold north-trending structures, which are correlated with the D_3 fold set elsewhere in the region (Pearson, 1973). However, "pre-D_3" deformation is relatively weak, with no indication of early isoclinal folding (Stauffer and Mukherjee, 1971), and predates peak low- to medium-grade metamorphism.

East of the Sahli – Hanson Lake regional culmination, D_4 major folds generally plunge gently to moderately to the northeast, with local plunge reversal and offsetting of axial planes across D_3 axial surfaces; the plunge increases to the south and is locally subvertical in parts of the Flin Flon Domain. Successively deeper structural levels are thus exposed towards the core of the Sahli – Hanson Lake dome and the area east of this culmination provides an oblique section through the early structural pile.

Within that pile, the structurally highest lithotectonic element now preserved is the Kyaska Sheet, here again incorporating most of the Kisseynew Domain and comprising migmatitic supracrustal rocks of the Burntwood Supergroup complexly interfolded with major granite–granodiorite sheets and Sherridon Group meta-arkoses. It structurally overlies both low-grade metavolcanics of the Flin Flon Domain, along the north-dipping southern limb of the Wildnest Lake synform, and higher grade, predominantly hornblendic gneisses of the Attiti Lake area, here termed the "Attiti Sheet" for local descriptive convenience. Low-grade rocks of the Flin Flon Domain are thrust over higher grade gneisses of the Hanson Lake Block along the Sturgeon Weir thrust (MacQuarrie, 1979; Ashton *et al.*, 1987). The deepest structural levels, exposed in the core of the Sahli – Hanson Lake dome, include not only reworked Archean monzocharnockites of the Sahli and MacMillan Point bodies but also quartzofeldspathic gneisses (meta-arkoses?) and pelitic gneisses underlying hornblende gneisses of

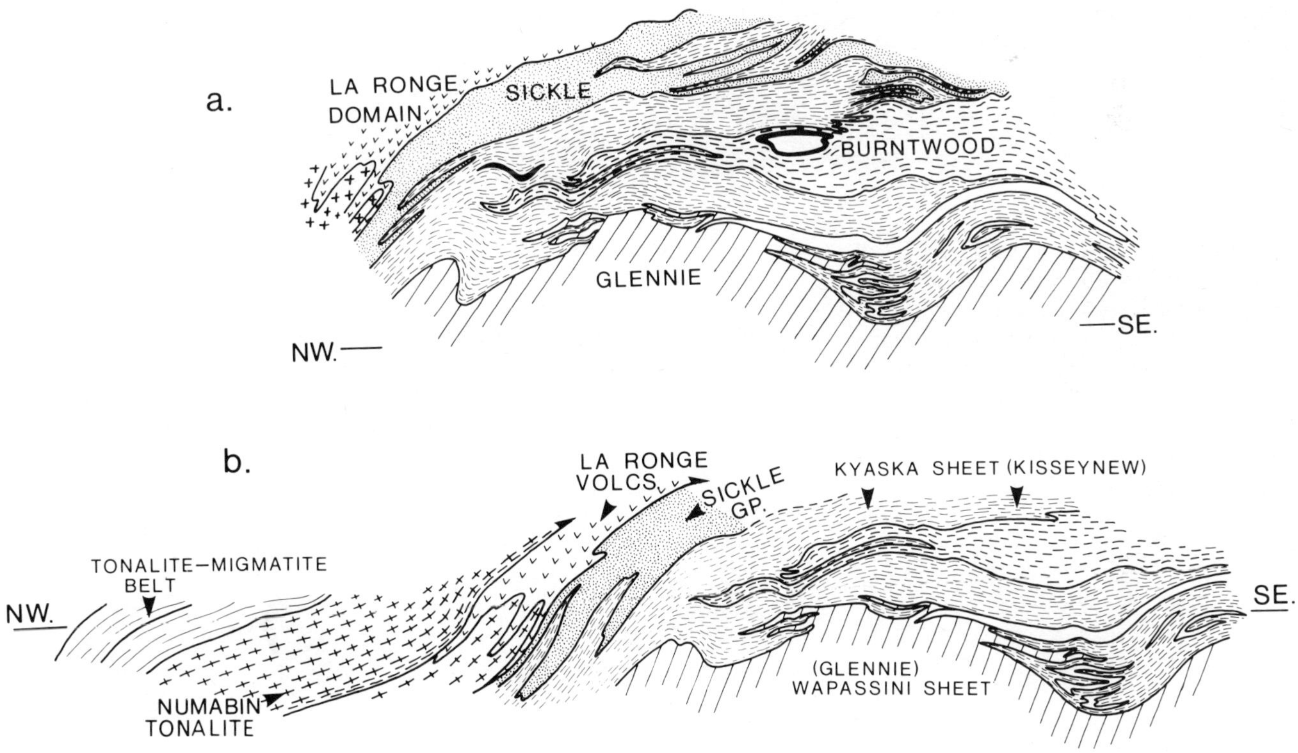

Figure 6. *(a) Generalised down-plunge view of major structures in the Kyaska and Wapassini sheets, south of Reindeer Lake (Fig. 5), constructed mainly from the mapping of Gilboy (1980). (b) Generalised structural section across the South Reindeer Lake area (line A–A′ in Fig. 5), utilising data drawn from various sources, including the down-plunge sections used for Figure 6a.*

the Attiti Sheet. Meta-arkoses and possible meta-arkoses thus occur throughout the pile from the lowest to highest structural levels.

Figure 8 is a composite schematic structural section generated from detailed down-plunge sections through the area east of the Sahli – Hanson Lake domal culmination and other available structural data. The section includes minor speculative interpretation, but is nonetheless considered to be a valid portrayal of fundamental lithotectonic relations in the area. It suggests that the Flin Flon Domain might be an internally imbricated thrust wedge driven out ahead of, and partly overridden by, the isoclinally folded Kyaska nappe sheet. Both

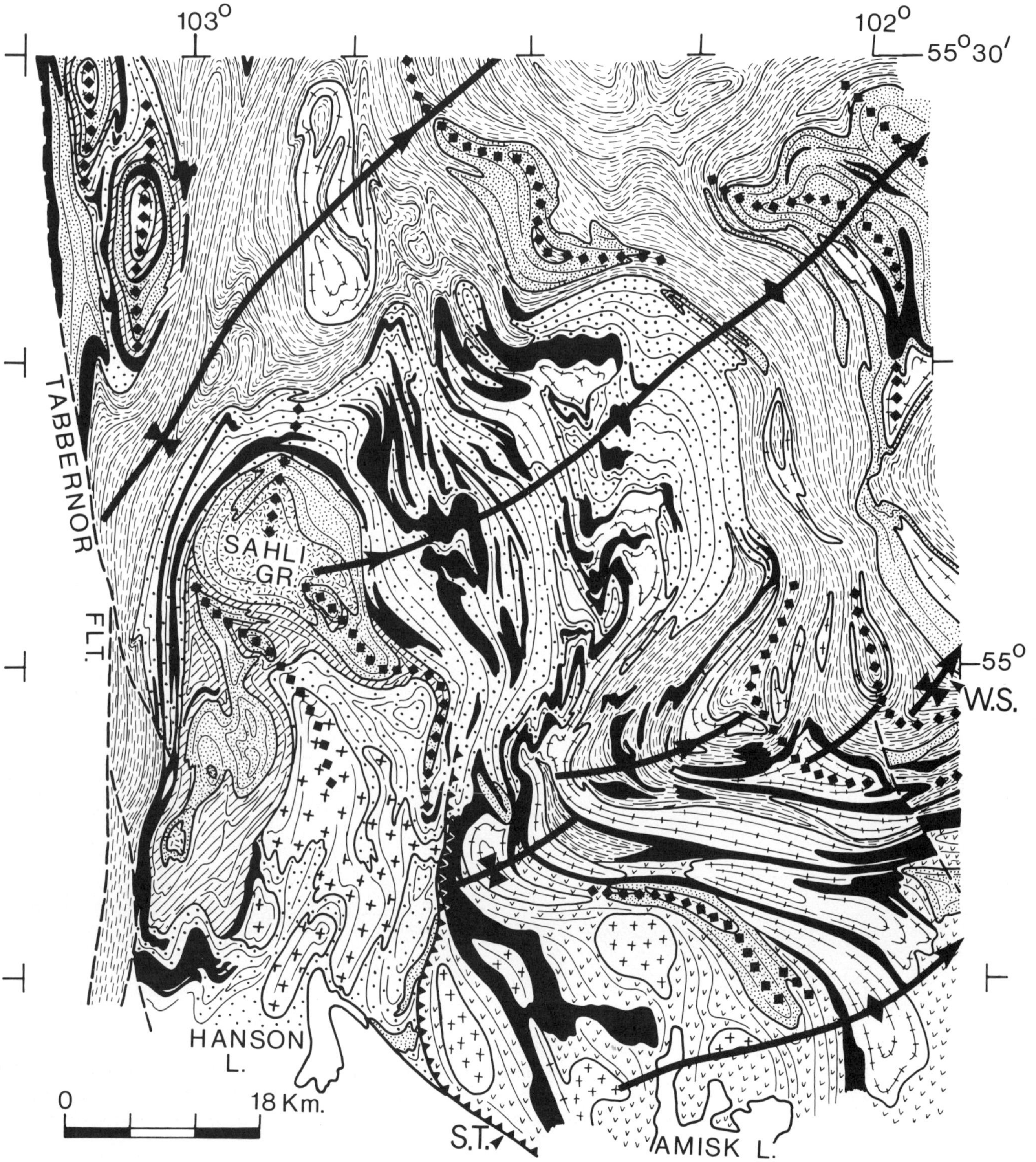

of these are tectonically emplaced across lower structural levels occupying the core of the Sahli – Hanson Lake dome. It seems likely that the Sturgeon Weir thrust extends northwards as a presently undefined, ductile high-strain zone and that the Flin Flon metavolcanics are laterally continuous with part of the hornblende gneisses forming the Attiti Sheet, an interpretation supported by the mapping of Ashton *et al.* (1987).

Preliminary reexamination (Lewry and Macdonald, 1988) of "porphyroblastic gneisses" (Macdonald, 1974) in the Pelican–Mirond lakes area, to the north of and structurally somewhat above the Sahli monzocharnockite (Fig. 7), indicates that these rocks are mostly protomylonitic to locally ultramylonitic porphyroclastic gneisses derived from tonalitic–granodioritic plutonic rocks, xenolithic hornblende gneiss screens, and later granite/pegmatite neosome. They are very similar in character to the Nistowiak Gneisses and are thus provisionally interpreted as incorporating a major high-strain zone separating Early Proterozoic allochthons from underlying Archean basement (and other rocks) in the core of the Hanson Lake Block. The full extent and nature of strain variation within these gneisses are currently undetermined. However, one very high-strain mylonitic–ultramylonitic zone appears to immediately underlie pelitic gneisses of the Kyaska Sheet northwest of the Sahli inlier: the southern continuation of this zone is thought to lie within the "Hornblende Gneiss Series" of Sibbald (1978).

The overall regional distribution and geometry of early porphyroclastic–mylonitic gneisses in the

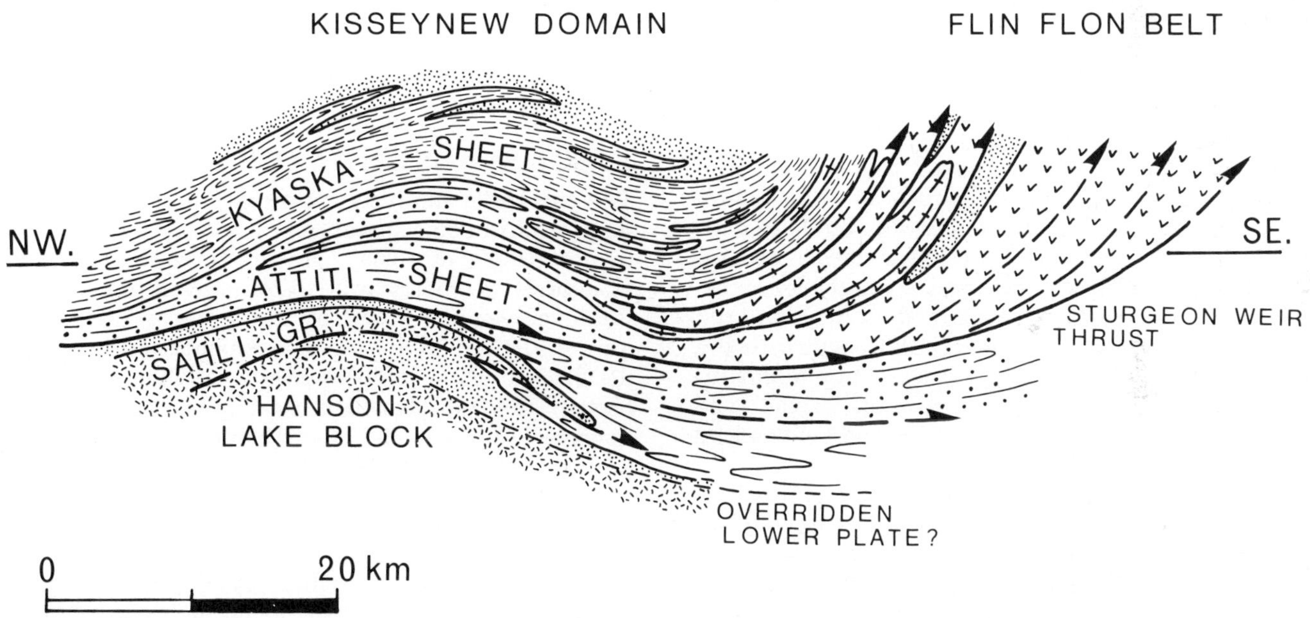

Figure 8. *Schematic and highly interpretive structural cross-section, approximately from northwest to southeast, across the region shown in Figure 7. The section draws on available structural data and more detailed down-plunge sections that were constructed for the area east of the Sahli domal culmination. Details are discussed in the text.*

Figure 7. *General geology of the Flin Flon – Hanson Lake – Kisseynew region (southeastern part of the area shown in Fig. 2). Refer to Figure 2 for the extent and nomenclature of broad geographic/structural subdivisions. Ornamentation: "vee" pattern, undivided low-grade metavolcanics of the Flin Flon Domain; crosses, late granite plutons; trend lines with cross tick, early deformed plutonic sheets; black, mainly mafic volcanics of the Flin Flon Domain and possibly equivalent higher grade amphibolites in other domains; fine dash following trend lines, metasedimentary gneisses of the Kyaska Sheet (old Kisseynew Domain); coarse stipple, undifferentiated, generally hornblendic supracrustal and plutonic gneisses of the Attiti Sheet and parts of the Hanson Lake Block (including protomylonitic–mylonitic gneisses in Hanson Lake Block); fine stipple, meta-arkoses and possibly meta-arkosic quartzofeldspathic gneisses; oblique lines, pelitic gneisses of the Hanson Lake Block; fine random tick, Archean basement in Hanson Lake Block; heavy solid lines, late ("D_4") fold axial surface traces; and discontinuous heavy squares, traces of probable D_3 fold closures. Abbreviations: S.T., Sturgeon Weir thrust and W.S., Wildnest Lake synform.*

Sahli – Hanson Lake area is not well established. Tentative extrapolation based on reinterpretation of lithological descriptions in previous Saskatchewan Geological Survey reports suggests that they might form a closed belt that wraps entirely around the Sahli – Hanson Lake dome. If so, the Sturgeon Weir thrust might represent a major splay thrust rooting in the deeper level regional nappe sole defined by such rocks.

We propose that the Kisseynew "Front" is broadly coincident with a zone of listric thrust imbrication between the Flin Flon thrust wedge and the overriding Kyaska nappe sheet (Fig. 8). General structural considerations suggest that such thrusts are likely to occur at the margins of major plutonic sheets, and/or within the intervening, less competent, intensely foliated supracrustal screens, and that they may shallow downwards toward the underlying Sturgeon Weir thrust and its postulated northern extension. These proposals clearly need to be tested by further field studies; however, recent documentation of internal thrust imbrication of the Snow Lake volcanics in Manitoba (Zwanzig, 1983, 1984, this volume; Gordon *et al.*, this volume, and references therein) and the results of recent mapping in Saskatchewan (Ashton *et al.*, 1987) are both consistent with the model presented here. Stauffer and Mukherjee (1971) interpret a number of curving brittle–ductile fault zones in the Flin Flon area as being late structures, essentially coeval with the D_4 regional folding: we suggest that they might instead represent earlier listric thrusts of the type predicted above, albeit reactivated and refolded by the D_4 structures.

The relatively steep metamorphic gradient across the "Kisseynew Front" may, in part, be due to telescoping of early synkinematic metamorphic zones across the predicted imbrication zone between the Flin Flon thrust wedge and the hotter overriding Kyaska nappe sheet. Kyanite-bearing assemblages in the southern Kisseynew Domain documented by Ashton *et al.* (1986, 1987) may, as those authors imply, reflect early crustal thickening related to nappe emplacement. It is clear, however, that peak low- to medium-pressure parageneses, ranging from upper greenschist facies in the Flin Flon Domain to upper amphibolite – lower granulite facies in areas to the north and west, must postdate early nappe emplacement and listric thrusting because metamorphic isograds transect major structural boundaries and fold axial surfaces. Thus, the Kisseynew "Front" is viewed here as a complex multiphase tectonometamorphic transition zone involving both early- and late-kinematic structures and metamorphic overprint.

As in the case of the Iskwatikan Lake Complex, Archean basement "windows" of the Sahli and MacMillan Point bodies occur at the deepest exposed structural levels. These inliers may represent simple domal interference culminations of continuous continental basement underlying the entire thrust wedge/nappe pile or, alternatively, they may represent a refolded early thrust slice, as speculatively suggested in Figure 8. Possibly extensive underlying Archean basement, suggested by the structural interpretation (Figs. 8 and 9), is not apparently reflected in the discriminant geochemistry of the Amisk Group arc volcanics or in sparse Nd, Pb, and Sr isotopic data from the area, both of which suggest only minor involvement of pre-Early Proterozoic basement in magma genesis and sediment derivation. Such a discrepancy can be reconciled by postulating either that basement inliers represent an areally restricted microcontinental sliver or that the arc assemblages in the overlying nappe pile were tectonically emplaced across a lower-plate continental margin subsequent to their generation and eruption in an oceanic or transitional lithospheric setting.

DISCUSSION AND CONCLUSIONS

The foregoing analysis represents only a preliminary, and in some instances rather speculative, interpretation of the gross three-dimensional structural geometry and tectonic evolution of the Reindeer Zone. Much of that interpretation requires further testing and documentation by more detailed analysis of presently available data and further field studies in critical areas. We argue, however, that a model of early nappe emplacement and possible major crustal imbrication explains observed large-scale lithostructural relationships and available detailed structural data from the Reindeer Zone better than any other interpretation presented to date. Many hitherto ill-explained features of the zone, including some problems of long standing, such as that of the "Kisseynew Front," can be most satisfactorily accommodated within such a model. Furthermore, the thesis incorporates several specific or implicit predictions that are amenable to testing.

The interpretation of relations in the Flin Flon – Hanson Lake area predicts the occurrence of early imbricate listric thrusts, both in the general vicinity of the Kisseynew "Front" and also possibly within the low-grade Flin Flon – Snow Lake "thrust wedge." Such thrusts should be manifested as ductile to brittle high-strain zones (depending on local metamorphic grade) refolded both by the regional northeast-trending D_4 folds and possibly also by the earlier northerly trending (D_2 or D_3 ?) fold set in the area. Thrusts in the vicinity of the Kisseynew "Front" should involve the emplacement of early high-grade metamorphic zones across lower grade rocks to the south. However, reaction isograds related to the late low pressure – high temperature metamorphism affecting the "Kisseynew Gneisses" should transect the early thrusts. Some high-strain zones interpreted by Stauffer and Mukherjee (1971) as being essentially coeval with northeasterly D_4 folding might, on reexamination, prove to be much older features or may represent reactivation of earlier folded thrusts.

The model also suggests that the Flin Flon volcanics might, in part, be equivalent to, and pass laterally into, higher grade hornblendic gneisses of the Attiti Sheet and that the underlying Sturgeon Weir thrust should be traceable northwards as an early ductile high-strain zone and/or as a splay from a lower thrust detachment that is refolded around the Sahli – Hanson Lake dome. These predictions can be tested by careful field investigation north and northwest of the Granite Lake area and detailed remapping of porphyroclastic gneiss belts throughout the Hanson Lake Block and Attiti Lake area.

Recognition of major ductile high-strain zones developed during early phases of the structural evolution of the Reindeer Zone, and their potential significance, is a recent development. Such zones have thus far been clearly defined only in the western Glennie Domain and adjacent La Ronge Domain and even here require further detailed study and analysis. Refinement of the regional model presented here requires that the full extent of the Guncoat and Nistowiak/Hunter Bay "sole" zones be traced out and that they be subjected to detailed kinematic analysis. More generally, future field studies must include documentation of strain gradients associated with early deformational events in an effort to locate similar high-strain zones elsewhere in the Glennie and Kisseynew domains.

The current lack of such strain data from most of the Reindeer Zone renders any direct comparison with nappe complexes described in other orogenic belts somewhat speculative. However, the situation in the Reindeer Zone might have some structural parallels to well-documented relationships in the Helvetic zone of the Alps (e.g., Siddans, 1983). The latter comprises a pile of ductile fold nappes, the soles of which are marked by very high-strain gradients. In places, this has produced mylonitic gneisses forming well-defined soles to individual nappe sheets. Elsewhere, strain gradients are lower and overturned limbs are merely attenuated without the development of discrete shear-zone discontinuities. The Reindeer Zone differs most notably from the Helvetic nappe pile in that it mainly incorporates arc volcanics and plutons, rather than inboard passive margin supracrustals.

The structural interpretation presented here implies that allochthons of the Reindeer Zone have been driven southwards or southeastwards across underthrust Archean crust of the convergent lower-plate Superior passive margin, and might, therefore, now be largely or entirely underlain by Archean basement, of which the Iskwatikan, Hunter Bay, and Sahli complexes are the only presently exposed windows. Alternatively, the Sahli and Iskwatikan – Hunter Bay inliers may represent only small detached microcontinental slivers caught up in the arc complex. Aside from a

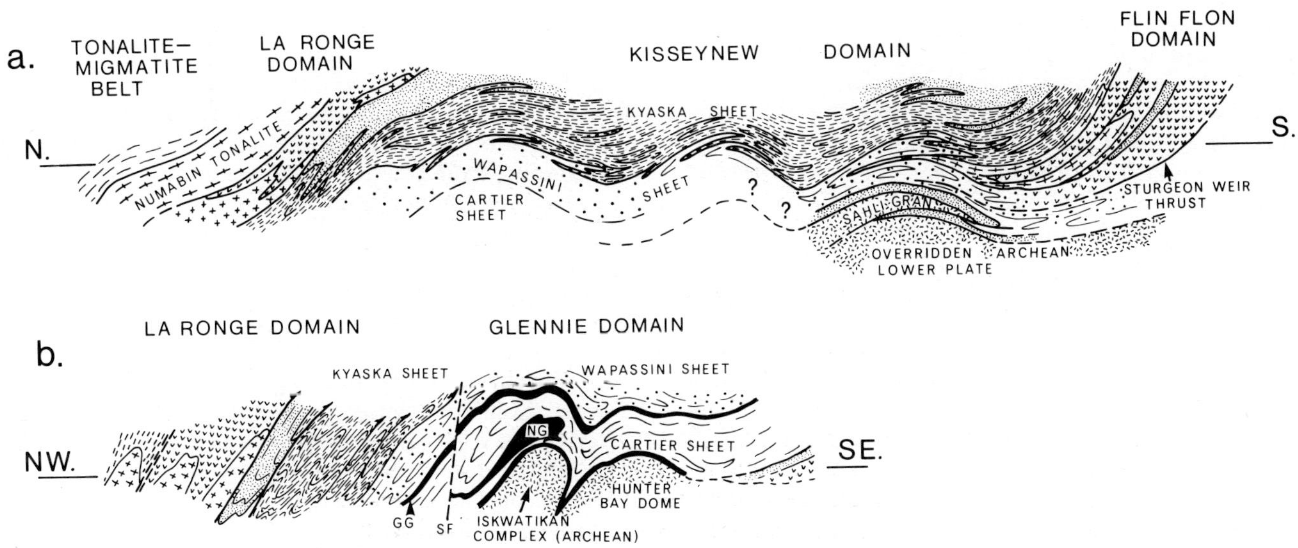

Figure 9. *Schematic interpretation of regional structural relations across the Reindeer Zone east (a) and west (b) of the Tabbernor zone, along section lines indicated in Figure 2. These sections represent an interpretive synthesis of sections in previous figures.*

continued search for other exposed Archean remnants, two lines of investigation might provide data concerning the extent of underthrust basement:

(1) Isotopic characterisation of late- to postkinematic granites and pegmatites: Isotopic data derived from early arc volcanics and plutons indicate merely that such rocks were generated in an essentially oceanic setting prior to major plate convergence and collisional orogeny. Only the later granitic melt fractions are likely to reveal possible derivation from subsequently underthrust Archean continental crust.

(2) Detailed geophysical investigation of deep crustal structure in the Reindeer Zone: Green *et al.* (1985a,b) document seismic evidence of gently dipping structures and thin-skinned tectonics in the vicinity of the Churchill–Superior Boundary Zone. Reflection and refraction seismic profiling farther west might permit testing of the essentially thin-skinned model presented here and provide evidence regarding the extent of underlying Archean continental crust.

Details of the relation of regional metamorphic patterns (Lewry *et al.*, 1978) to the postulated kinematic evolution of the Reindeer Zone are not addressed in this paper, mainly because the metamorphic history and $P/T/t$ regimes are imperfectly understood over much of the region (but see Gordon *et al.* (this volume)). In general, documented metamorphic zonation and thermal history can be reasonably well explained within the structural context presented here. However, several major problems are apparent. For example, there is a general decline in metamorphic grade in the southern Glennie Domain, the reverse of what might be expected in terms of the broad structural pattern and predicted structural level of this region. Available data suggest that this anomaly cannot be explained by late retrograde overprinting of an earlier, deep crustal level, high-grade terrain. Two possible explanations appear to be consistent with our present structural model: (1) that the southern Glennie Domain comprises a downturned stack of refolded nappe sheets originally overlying the Kyaska and Wapassini sheets, and (2) that this region comprises a huge upturned fold/thrust imbricate stack that was never overridden by major nappe sheets defined to the north and west. Preliminary results of current detailed remapping of parts of the southern Glennie Domain (Delaney, 1988; Slimmon, 1988; Thomas, 1988) suggest that the second of these alternatives is more likely. However, neither the structure nor the metamorphic history of the southern Glennie Domain is yet well documented and only further work will determine whether the anomalous patterns in this area are fatal to, or simply impose important constraints on, the model presented here.

In previous interpretations of structural relations in, and the evolution of, the Reindeer Zone (e.g., Lewry, 1981a) southern parts of the Stanley and Tabbernor shear zones were interpreted as being fundamental crustal discontinuities; the former reflecting a possible arc–microcontinent collisional suture, the latter an early transform microplate boundary. The present analysis downgrades the importance of these features. It seems clear that early nappe sheets of the Reindeer Zone extend right across the Tabbernor zone, at least in the north where the Kyaska and Wapassini sheets appear to be only superficially disrupted by splaying, subparallel late faults of the Tabbernor system. In the south, the Tabbernor zone includes a rather more significant tectonic break, in part marked by early ductile mylonites. However, it now seems possible that these mylonites, previously assigned specifically to the Tabbernor zone, might instead represent part of an earlier, refolded and oversteepened, high-strain nappe sole, which is only coincidently coextensive with the Tabbernor zone in this area: there are still indications that elements of the Wapassini, Cartier, and lower structural sheets originally extended on either side of the latter. Thus, a major discontinuity predating nappe transport cannot be convincingly documented at the present level of exposure.

Nonetheless, the Tabbernor zone is a feature of evident regional importance, as previously emphasised. It is traceable as a well-defined geophysical discontinuity well into North Dakota to the south, and as a late brittle fault system hundreds of kilometres to the north (cf. Lewry, 1981a; Green *et al.*, 1985a). Much of the movement on this zone clearly postdates nappe emplacement, possibly reflecting largely transcurrent dismemberment of the Reindeer collage at late stages of oblique terminal collision. It remains possible that such movement reflects reactivation of a much earlier (prenappe) fundamental transform fault, which was overridden during nappe development.

The poorly exposed southern extension of the Stanley zone, through Lac La Ronge, is no longer viewed as a fundamental early crustal break. Its previous interpretation as such (Lewry, 1981a,b) was predicated on the widespread occurrence of early protomylonitic and mylonitic gneisses in this area. Some of these, in the eastern part of La Ronge, are now reinterpreted as being extensions of the Nistowiak – Hunter Bay mylonitic nappe sole. Others, on the western side of the lake, appear to represent a southwestern extension of the Guncoat Gneisses.

Radiometric age data indicate that the La Ronge – Lynn Lake and Flin Flin – Snow Lake volcanic arcs are of similar age. It has been variously suggested that (1) they represent dismembered segments of the same initially continuous arc system (e.g., Green *et al.*, 1985a,b), (2) they developed as essentially coeval but independent arc systems, and (3) they differ somewhat in age of initial generation, within the error limits imposed by dating (e.g., Lewry, 1981a).

The tectonic model offered here imposes few constraints with respect to these alternatives. Indeed, we suggest that the magnitude of probable crustal shortening, stacking, and dismemberment of the juvenile terrane collage is such as to make it unlikely that this problem will ever be fully resolved.

ACKNOWLEDGEMENTS

The work and ideas of many past and present members of the Saskatchewan Geological Survey, on which much of this paper is based, are gratefully acknowledged. We thank K. Arnsdell and an unknown reviewer for constructive comments and Paul Hoffman, as always, for stimulating discussion and support. This work was financially supported both by the Saskatchewan Department of Energy and Mines and by a Natural Sciences and Engineering Research Council of Canada operating grant to J.F. Lewry.

REFERENCES

Abraham, A.P.G., 1985, Geology of the Otter Lake area and its relation to the southeastern complex of the western Churchill Province, Saskatchewan, unpublished M.Sc. thesis, University of Regina, 237 p.

Ashton, K.E., Wheatley, K.J., Moser, D., Paul, D. and Wilcox, K.J., 1986, The Kisseynew Gneisses of Saskatchewan: update: Saskatchewan Geological Survey, Miscellaneous Report 86-4, p. 145–148.

Ashton, K.E., Wilcox, K.J., Wheatley, K.J., Paul, D. and de Tombe, J., 1987, The boundary zone between the Flin Flon Domain, Kisseynew Gneisses and Hanson Lake Block, northern Saskatchewan: Saskatchewan Geological Survey, Miscellaneous Report 87-4, p. 131–134.

Bell, K. and Macdonald, R., 1982, Geochronological calibration of the Precambrian Shield in Saskatchewan: Saskatchewan Geological Survey, Miscellaneous Report 82-4, p. 17–22.

Chauvel, C., Arndt, N.T., Kielinzcuk, S. and Thom, A., 1987, Formation of Canadian 1.9 Ga old continental crust. I. Nd isotopic data: Canadian Journal of Earth Sciences, v. 24, p. 396-406.

Chiarenzelli, J.R. and Chiarenzelli, V.K., 1986, Nistowiak Lake tectonic studies (part of NTS 73P-7): Saskatchewan Geological Survey, Miscellaneous Report 86-4, p. 65–71.

Chiarenzelli, J.R. and Lewry, J.F., 1988, The Nistowiak Gneisses: evidence of a collisional suture in the Trans-Hudson Orogen, northern Saskatchewan: Geological Association of Canada – Mineralogical Association of Canada – Canadian Society of Petroleum Geologists, Program with Abstracts, v. 13, p. A20.

Chiarenzelli, J.R., Lewry, J.F. and Landon, M., 1987, Bedrock geology, Iskwatikan Lake area: evidence for Hudsonian juxtaposition of Proterozoic and Archean rocks along a ductile detachment surface: Saskatchewan Geological Survey, Miscellaneous Report 87-4, p. 46–51.

Coombe, W., Lewry, J.F. and Macdonald, R., 1986, Regional geological setting of gold in the La Ronge Domain, Saskatchewan, *in* Clark, L.A., ed., Gold in the Western Shield: Canadian Institute of Mining and Metallurgy, Special Volume 38, p. 26–56.

Delaney, G.D., 1988, Bedrock geological mapping, Brownell Lake area: Saskatchewan Geological Survey, Miscellaneous Report 88-4, p. 8–19.

Fox, J.S. and Johnston, W.G.Q., 1981, Komatiites, boninites and tholeiitic picrites in the Central La Ronge Metavolcanic Belt, Saskatchewan and Manitoba, and their possible economic significance: Canadian Institute of Mining and Metallurgy Bulletin, July 1981, p. 73–82.

Fumerton, S.L., Stauffer, M.R. and Lewry, J.F., 1985, The Wathaman Batholith: largest known Precambrian pluton: Canadian Journal of Earth Sciences, v. 21, p. 1082–1097.

Gilboy, C.F., 1980, Geology of the Reindeer Lake South (southeast) area: Saskatchewan Department of Mineral Resources, Report 198, 62 p.

Gilboy, C.F., 1982, Geology of an area around Rottenstone and Dobbin lakes (part of NTS area 74A): Saskatchewan Department of Mineral Resources, Report 193, 68 p.

Gordon, T.M., Hunt, P.A., Bailes, A.H. and Syme, E.C., 1990, U–Pb zircon ages from the Flin Flon and Kisseynew belts, Manitoba: chronology of Early Proterozoic crust formation at an Early Proterozoic accretionary margin, *in* Lewry, J.F. and Stauffer, M.R., eds., The Early Proterozoic Trans-Hudson Orogen of North America: Geological Association of Canada, Special Paper 37, p. 177–199.

Green, A.G., Hajnal, Z. and Weber, W., 1985a, An evolutionary model of the western Churchill Province and western margin of the Superior Province in Canada and the north-central United States: Tectonophysics, v. 116, p. 281–322.

Green, A.G., Weber, W. and Hajnal, Z., 1985b, Evolution of Proterozoic terranes beneath the Williston Basin: Geology, v. 13, p. 624–628.

Harper, C., 1985, Bedrock geological mapping, Waddy–Tower lakes area: Saskatchewan Geological Survey, Miscellaneous Report 85-4, p. 6–17.

Hoffman, P.F., 1988, United plates of America, the birth of a craton: Annual Review of Earth and Planetary Sciences, v. 16, p. 543–603.

Hoffman, P.F., 1990, Subdivision of the Churchill province and extent of the Trans-Hudson Orogen, *in* Lewry, J.F. and Stauffer, M.R., eds., The Early Proterozoic Trans-Hudson Orogen of North America: Geological Association of Canada, Special Paper 37, p. 15–39.

Johnston, W.G.Q., 1969, Geology of the eastern portion of the Waddy Lake area, Saskatchewan: Saskatchewan Department of Mineral Resources, Report 127, 44 p.

Johnston, W.G.Q., 1970, Geology of the May Lake area (east half), Saskatchewan: Saskatchewan Department of Mineral Resources, Report 130, 31 p.

Johnston, W.G.Q., 1983, Geology of the Southend area: Saskatchewan Energy and Mines, Open File Report OF 82-4, 151 p.

Lewry, J.F., 1977, Geology of the Glennie Lake area, Saskatchewan: Saskatchewan Department of Mineral Resources, Report 143, 59 p.

Lewry, J.F., 1981a, Lower Proterozoic arc–microcontinent collisional tectonics in the western Churchill Province: Nature, v. 294, p. 69–72.

Lewry, J.F., 1981b, La Ronge Project. II. Geology of the Stanley shear zone: Saskatchewan Geological Survey, Miscellaneous Report 81-4, p. 28–33.

Lewry, J.F., 1983, Character and structural relations of the "McLennan Group" meta-arkoses, McLennan–Jaysmith lakes area: Saskatchewan Geological Survey, Miscellaneous Report 83-4, p. 49–55.

Lewry, J.F., 1984, Bedrock compilation, Lac La Ronge and Wapawekka areas: Saskatchewan Geological Survey, Miscellaneous Report 84-4, p. 34-41.

Lewry, J.F., 1986, Bedrock geological mapping, Mackay Lake North: Saskatchewan Geological Survey, Miscellaneous Report 86-4, p. 48–51.

Lewry, J.F. and Collerson, K.D., 1990, The Trans-Hudson Orogen: extent, subdivision, and problems, *in* Lewry, J.F. and Stauffer, M.R., eds., The Early Proterozoic Trans-Hudson Orogen of North America: Geological Association of Canada, Special Paper 37, p. 1–14.

Lewry, J.F. and Macdonald, R., 1988, Observations on deformation in the Glennie Domain and Hanson Lake Block: Saskatchewan Geological Survey, Miscellaneous Report 88-4, p. 35–41.

Lewry, J.F. and Slimmon, W.L., 1985, Compilation bedrock geology, Lac La Ronge, NTS area 73P/73I: Saskatchewan Department of Energy and Mines, Report 225 (1:250,000 scale map with marginal notes).

Lewry, J.F., Sibbald, T.I.I. and Rees, C.J., 1978, Metamorphic patterns and their relation to tectonism and plutonism in the Churchill Province in northern Saskatchewan, *in* Fraser, J.A. and Heywood, W.W., eds., Metamorphism in the Canadian Shield: Geological Survey of Canada, Paper 78-10, p. 139–154.

Lewry, J.F., Curry, A. and Roberts, K., 1980, Geology of an area around Ghana Lake, Saskatchewan: Saskatchewan Department of Mineral Resources, Report 192, 29 p.

Lewry, J.F., Stauffer, M.R. and Fumerton, S.L., 1981, A Cordilleran-type batholithic belt in the Churchill Province in northern Saskatchewan: Precambrian Research, v. 14, p. 277–313.

Lewry, J.F., Sibbald, T.I.I. and Schledewitz, D.C.P., 1985, Reworking of Archean basement in the western Churchill Province and its significance, *in* Ayres, L.D., Thurston, P.C., Card, K.D. and Weber, W., eds., Evolution of Archean Supracrustal Sequences: Geological Association of Canada, Special Paper 28, p. 239–261.

Lewry, J.F., Macdonald, R., Livesey, C., Meyer, M., Van Schmus, W.R. and Bickford, M.E., 1987, U–Pb geochronology of accreted terranes in the Trans-Hudson Orogen in northern Saskatchewan, Canada, *in* Pharaoh, T.C., Beckinsale, R.D. and Rickard, D., eds., Geochemistry and Mineralization of Proterozoic Volcanic Suites: Geological Society of London, Special Publication 33, p. 147–166.

Macdonald, R., 1974, Pelican Narrows (west) area: Saskatchewan Department of Mineral Resources, p. 30–37.

Macdonald, R., 1984, Review geology, Forbes Lake area: Saskatchewan Geological Survey, Miscellaneous Report 84-4, p. 50–52.

Macdonald, R. and Broughton, P., 1980, Geological map of Saskatchewan, provisional edition: Saskatchewan Geological Survey, scale 1:1,000,000.

MacQuarrie, R.R., 1979, Geological re-investigation mapping, Birch Portage South, NTS area 63L-I55: Saskatchewan Geological Survey, Miscellaneous Report 79-10, p. 29–38.

Meyer, M., 1987, Geochemistry and geochronology of the Wathaman Batholith, the remnant of an Early Proterozoic continental arc, Saskatchewan, Canada, unpublished M.Sc. thesis, University of Kansas, 50 p.

Pearson, D.E., 1973, The location and structure of the Precambrian Kisseynew Gneiss Domain of northern Saskatchewan: Canadian Journal of Earth Sciences, v. 9, p. 1235–1249.

Ray,G.E., 1981, Geology of part of the Foster Lake (south) – La Ronge (northwest) area: Saskatchewan Department of Mineral Resources, Report 185, 31 p.

Sibbald, T.I.I., 1978, Geology of the Sandy Narrows (east) area, Saskatchewan: Saskatchewan Department of Mineral Resources, Report 170.

Siddans, A.W.B., 1983, Finite strain patterns in some alpine nappes: Journal of Structural Geology, v. 5, p. 441–448.

Slimmon, W.L., 1988, Bedrock geological mapping, Gee Lake area: Saskatchewan Geological Survey, Miscellaneous Report 88-4, p. 26–31.

Stauffer, M.R., 1984, Manikewan: an Early Proterozoic ocean in central Canada, its igneous history and orogenic closure: Precambrian Research, v. 25, p. 257–281.

Stauffer, M.R., 1990, The Missi Formation: an Aphebian molasse deposit in the Reindeer Lake Zone of the Trans-Hudson Orogen, *in* Lewry, J.F. and Stauffer, M.R., eds., The Early Proterozoic Trans-Hudson Orogen of North America: Geological Association of Canada, Special Paper 37, p. 121–141.

Stauffer, M.R. and Mukherjee, A.C., 1971, Superimposed deformations in the Missi metasedimentary rocks near Flin Flon, Manitoba: Canadian Journal of Earth Sciences, v. 8, p. 217-242.

Syme, E.C., 1990, Stratigraphy and geochemistry of the Lynn Lake and Flin Flon metavolcanic belts, Manitoba, *in* Lewry, J.F. and Stauffer, M.R., eds., The Early Proterozoic Trans-Hudson Orogen of North America: Geological Association of Canada, Special Paper 37, p. 143–161.

Thom, A., Arndt, N.T. and Stauffer, M.R., 1990, Flin Flon and west La Ronge belts, Saskatchewan: products of Proterozoic subduction-related volcanism, *in* Lewry, J.F. and Stauffer, M.R., eds., The Early Proterozoic Trans-Hudson Orogen of North America: Geological Association of Canada, Special Paper 37, p. 163–175.

Thomas, D.J., 1984, Geological mapping, Star Lake area: Saskatchewan Geological Survey, Miscellaneous Report 84-4, p. 21–33.

Thomas, D.J., 1985, Geological mapping, Roundish-Bervin lakes area: Saskatchewan Geological Survey, Miscellaneous Report 85-4, p. 18–27.

Thomas, D.J., 1986, Bedrock geological mapping, Esmay Lake area: Saskatchewan Geological Survey, Miscellaneous Report 86-4, p. 19–31.

Thomas, D.J., 1987, Bedrock geological mapping, Bing Lake area: Saskatchewan Geological Survey, Miscellaneous Report 87-4, p. 18–27.

Thomas, D.J., 1988, Bedrock geological mapping, Palf Lake area: Saskatchewan Geological Survey, Miscellaneous Report 88-4, p. 20–25.

Todt, W.A., Arndt, N.T. and Hofmann, A.W., *in press*, Formation of 1.9 Ga old continental crust. 2. Pb isotopic data: Canadian Journal of Earth Sciences.

Van Schmus, W.R., Bickford, M.E., Lewry, J.F. and Macdonald, R., 1987, U–Pb geochronology in the Trans-Hudson Orogen, northern Saskatchewan, Canada: Canadian Journal of Earth Sciences, v. 24, p. 407–424.

Watters, B.R. and Pearce, J.A., 1987, Metavolcanic rocks of the La Ronge Domain in the Churchill Province, Saskatchewan: geochemical evidence for a volcanic arc origin, *in* Pharaoh, T.C., Beckinsale, R.D. and Rickard, D., eds., Geochemistry and Mineralization of Proterozoic Volcanic Suites: Geological Society of London, Special Publication 33, p. 167–182.

Weber, W., 1990, The Churchill–Superior Boundary Zone, southeast margin of the Trans-Hudson Orogen: a review, *in* Lewry, J.F. and Stauffer, M.R., eds., The Early Proterozoic Trans-Hudson Orogen of North America: Geological Association of Canada, Special Paper 37, p. 41–55.

Zwanzig, H.V., 1983, Kisseynew Project: Lobstick Narrows: Manitoba Department of Energy and Mines, Mineral Resources Division, Report of Field Activities 1983, p. 15–22.

Zwanzig, H.V., 1984, Kisseynew Project: Lobstick Narrows – Cleunion Lake, Puffy Lake and Nokomis areas: Manitoba Department of Energy and Mines, Mineral Resources Division, Report of Field Activities 1984, p. 38–45.

Zwanzig, H.V., 1990, Kisseynew gneiss belt in Manitoba: stratigraphy, structure, and tectonic evolution, *in* Lewry, J.F. and Stauffer, M.R., eds., The Early Proterozoic Trans-Hudson Orogen of North America: Geological Association of Canada, Special Paper 37, p. 95–120.

Kisseynew Gneiss Belt in Manitoba: Stratigraphy, Structure, and Tectonic Evolution

H.V. Zwanzig
Manitoba Energy and Mines, 535-330 Graham Avenue,
Winnipeg, Manitoba R3C 4E3, Canada

Zwanzig, H.V., Kisseynew gneiss belt in Manitoba: stratigraphy, structure, and tectonic evolution, in Lewry, J.F. and Stauffer, M.R., eds., The Early Proterozoic Trans-Hudson Orogen of North America: Geological Association of Canada, Special Paper 37, p. 95–120.

Abstract

The Kisseynew belt is an Early Proterozoic metamorphic complex in the juvenile internal zone of the Trans-Hudson Orogen. It is dominated by metasedimentary and intrusive rocks closely related to surrounding volcano-plutonic belts, which are interpreted as remnants of magmatic arcs. Orthogneisses (1830–1890 Ma) are infrastructural extensions of arc massifs. Widespread sedimentary gneisses (Burntwood Suite), interpreted as volcaniclastic turbidite fan deposits, extend south onto the margin of the low-grade Flin Flon volcanic belt. High-energy turbidites, breccias, conglomerates, and mafic–ultramafic volcanic rocks on the north flank of the Kisseynew belt were probably deposited at an extensional basin margin, or possibly in an accretionary complex. Alluvial deposits overlie turbidites on the flanks of the Kisseynew belt, and overlap the volcanic belts with angular unconformity. The older (Sickle) alluvial suite has a felsic magmatic provenance, probably a continental arc at the present site of the 1855 Ma Chipewyan batholith. The younger (Missi) suite contains 1830 Ma subaerial volcanic units, also suggesting deposition in a continental arc.

The core of the belt was migmatized and injected by syntectonic 1815 Ma, S-type plutons under middle crustal conditions, probably during terminal collision in which the internal zone was compressed between Archean basement rocks of the Superior and Hearne provinces. Early isoclinal folds, thrusts, and large sheetlike structures were refolded by upright and recumbent synmetamorphic structures. Fabric orientation data and structure sections suggest that nappes were driven north and south over the Kisseynew flanks due to tectonic wedging of competent orthogneisses in the infrastructure, beneath a melt-softened sediment complex.

Late-metamorphic upright folds and east-side-up faults are similar to late collisional structures on the impinging margin of the Superior craton.

Résumé

La ceinture Kisseynew est un complexe métamorphique Protérozoïque Inférieur situé dans la zone juvénile interne de l'orogène Trans-Hudson. La ceinture est dominée par des roches méta-sédimentaires et intrusives reliées à des ceintures volcano-plutoniques adjacentes interprétées comme des vestiges d'arcs magmatiques. Des orthogneiss âgés de 1830–1890 Ma constituent l'extension sous-jacente de l'édifice d'arc. Des gneiss très répandus d'origine sédimentaire sont interprétés comme des dépôts deltaïques de turbidites volcano-clastiques s'étendant vers le sud sur la marge de la ceinture volcanique peu métamorphisée de Flin Flon. Des turbidites à haute énergie, des brèches, des conglomérats et des roches volcaniques basiques/ultrabasiques provenant du flanc nord de la ceinture Kisseynew se sont probablement déposées près des marges d'un bassin en extension ou dans un complexe accrétionnaire. Sur les flancs de la ceinture Kisseynew, des dépôts alluvionnaires surmontent des turbidites et chevauchent les ceintures volcaniques avec une discordance angulaire. La séquence alluvionnaire ancienne de Sickle a une source magmatique acide, probablement un arc continental situé là où est présentement le batholite Chipewyan âgé de 1855 Ma. La suite de Missi, plus jeune, comporte des unités volcaniques sub-aériennes âgées de 1830 Ma suggérant aussi les dépôts dans un arc continental.

Le centre de la ceinture a été migmatisé et injecté par des plutons syn-tectoniques de type S (1815 Ma) au niveau de la mésozone probablement pendant la dernière collision entre les roches de socle archéennes des provinces Supérieur et Hearnes. Des plis isoclinaux, des charriages et des grandes structures en écailles ont été replissés par des structures redressées et couchées syn-métamorphiques. L'orientation des textures et les coupes structurales suggèrent que les nappes ont été poussées vers le nord et le sud sur les flancs de la ceinture Kisseynew à cause d'un écaillage des orthogneiss compétents dans le sous-bassement situé sous un complexe sédimentaire anatectique.

Des plis tardi-métamorphiques redressés et des failles effondrées à l'ouest sont similaires aux structures tardi-collision situées sur la marge du craton Supérieur.

INTRODUCTION

The Kisseynew gneiss belt is a complex of high-grade metasedimentary and meta-igneous rocks in Manitoba and Saskatchewan. It lies within the Reindeer Zone (Stauffer, 1984) of the Trans-Hudson Orogen (Hoffman, 1981), a zone that has been interpreted as a composite terrane (Lewry *et al.*, 1987) assembled during the Hudsonian orogeny from Early Proterozoic juvenile magmatic arcs, now granitic and volcanic belts (Stauffer *et al.*, 1975; Watters, 1984, 1985; Syme, 1985, this volume), and related sedimentary basins, now gneissic domains (Fig. 1). This terrane apparently collided with Archean rocks of the Superior craton and the Hearne Province, but related suture zones are ill-defined. The Kisseynew belt is the telescoped remnant of the largest of the proposed arc-related basins.

In this paper, the lithostratigraphy, selected chemistry, detrital modes, and structure of the Kisseynew belt and critical adjoining parts of the volcanic belts are discussed, using data mainly from Manitoba. These data suggest that domains of sedimentary, volcanic, and plutonic rocks in the Reindeer Zone developed adjacent to each other or as superimposed magmatic belts during evolution of Early Proterozoic crust, and are not unrelated exotic terranes. However, composite structure sections reveal a structural style consistent with crustal telescoping and collisional tectonics.

The main part of the belt is 300 km long and 150 km wide, but, with its narrow eastern and western arms, the whole gneissic domain extends 800 km across the central Reindeer Zone and along the Churchill–Superior Boundary Zone. The main part is bounded to the north by the Lynn Lake and Leaf Rapids volcano-plutonic domains, to the south by the Flin Flon volcanic belt, and to the southeast by mylonite zones along the Churchill–Superior boundary. The predominantly granitoid Glennie Lake Domain lies to the west, in Saskatchewan (Lewry, 1981). Boundaries between the Kisseynew belt and the volcanic domains vary in character from one area to another; faults are common, but elsewhere lithologic units extend across the boundary (Bailes, 1980), which is then marked only by an increase in metamorphic grade toward the interior of the belt (Bailes and McRitchie, 1978; Froese and Moore, 1980). Overall, the boundary is best defined by a sillimanite-forming reaction isograd (Fig. 2).

The western arm of Kisseynew-type gneisses, the McLean Lake Belt (Macdonald, 1980), extends from

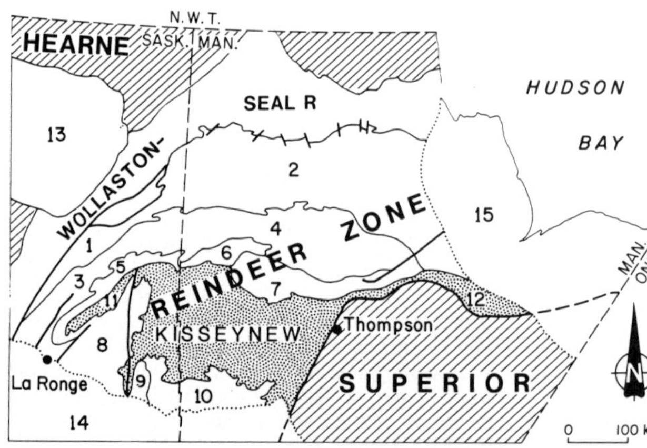

Figure 1. *Geological setting of the Kisseynew belt (stippled) and other lithostructural domains in the Trans-Hudson Orogen between Archean provinces (lined). The Wollaston – Seal River hinterland (undivided) and Reindeer internal zone are shown. The following subdivisions of the Reindeer internal zone are shown: 1–2, Wathaman–Chipewyan batholith; 3–4, Rottenstone – Southern Indian gneissic domain; 5, La Ronge volcanic belt; 6, Lynn Lake volcanic belt; 7, Leaf Rapids – Rusty Lake domain; 8, Glennie Lake Domain; 9, Hanson Lake domain; 10, Flin Flon volcanic belt; 11, McLean Lake Belt; and 12, eastern Kisseynew belt. Younger cover includes the following: 13, Athabasca sandstone; and 14–15, Phanerozoic rocks.*

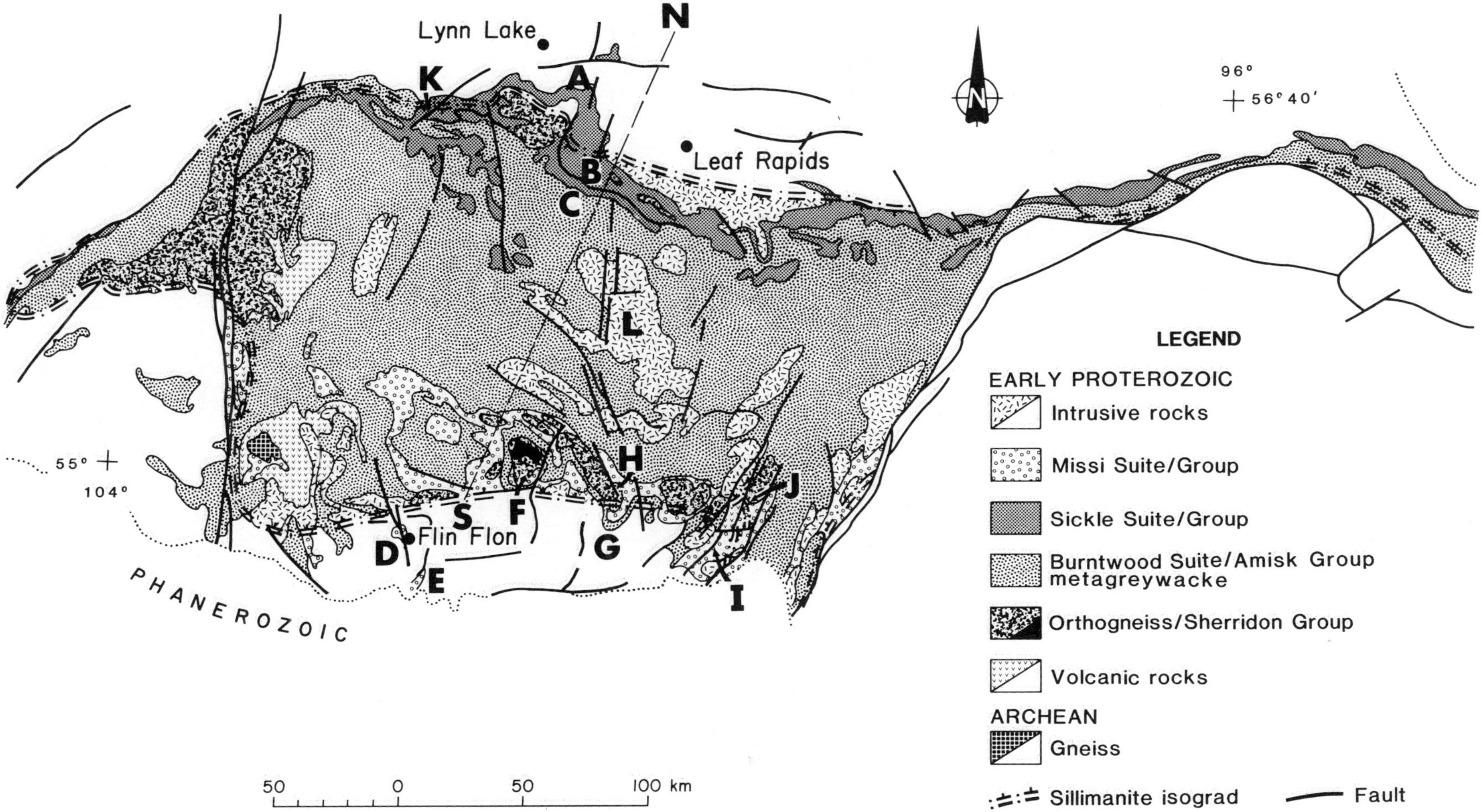

Figure 2. *Simplified geology of the Kisseynew belt, northern Glennie Lake and Hanson Lake domains. Positions of the sillimanite isograd and major faults are approximate. Also shown are locations of stratigraphic columns A–J from Figures 3 and 5, and the structure section N–S from Figure 19. Legend: K, Laurie Lake; B, Granville Lake; L, Highrock Lake; and I, Wekusko Lake.*

Reindeer Lake nearly to Lac La Ronge, Saskatchewan, and is bounded by the La Ronge volcanic belt and the Glennie Lake Domain. The eastern arm extends beneath Ordovician cover of the Hudson Bay basin. It is bounded to the north by the Wathaman–Chipewyan batholith (Lewry *et al.*, 1981; Halden *et al.*, this volume) and to the south by the Superior craton. At the Churchill–Superior boundary, Kisseynew gneisses with low initial $^{87}Sr/^{86}Sr$, corresponding to sediments derived from a juvenile Proterozoic source, are faulted against Archean basement and Proterozoic high-grade (Ospwagan Group) and weakly metamorphosed (Fox River) sedimentary cover with an Archean detrital component and high initial $^{87}Sr/^{86}Sr$ (Brooks and Theyer, 1981; Weber, this volume). This sedimentary cover, abundant associated volcanic rocks, and coeval intrusions formed in rift-related (Scoates, 1981) and probably arc-related (Halden, 1988) basins fringing the Superior craton. The Churchill–Superior boundary is interpreted as a suture zone in plate tectonic models of the Trans-Hudson Orogen (Gibb, 1968; Green *et al.*, 1985).

LITHOSTRATIGRAPHY

The Kisseynew belt contains varied plutonic rocks, and five supracrustal suites derived by high-grade metamorphism from stratigraphic groups recognized in adjacent lower grade domains. The supracrustal components are as follows: (1) fine-grained amphibolites and felsic gneisses equivalent to metavolcanic and volcaniclastic rocks of the Amisk Group in the Flin Flon belt; (2) Sherridon gneisses: mostly quartz-rich rocks of uncertain origin; (3) Burntwood Suite: graphite-bearing garnet–biotite gneisses, migmatites, and amphibolites, equivalent to greywacke–mudstone turbidites in the upper part of the Amisk Group; (4) Sickle Suite: quartzofeldspathic gneisses interpreted as non-marine deposits forming the distal facies of the Sickle Group alluvial sandstones and conglomerates; and (5) Missi Suite: quartz-rich sedimentary gneiss and metavolcanic rocks, equivalent to the Missi Group and also of continental origin.

Stratigraphic relationships of these suites are established in marginal parts of the Kisseynew belt and adjacent parts of the volcanic belts, where primary structures are preserved. Amisk Group-equivalent gneisses, the oldest strata recognized in the south, extend across the sillimanite isograd from the Kisseynew belt into the lower grade Flin Flon belt. The Burntwood Suite is the oldest unit exposed in most parts of the belt (Gilbert *et al.*, 1980; Lenton, 1981): its base is not generally exposed but it is locally underlain (stratigraphically or structurally) by metavolcanic rocks that in the south, are equivalent to the Amisk Group. The suite is unconformably or disconformably overlain by the Sickle Suite in the north (Gilbert *et al.*, 1980) and the Missi Suite in the south (Zwanzig and Lenton, 1987). These two younger metasedimentary gneiss suites have not been identified in direct mutual contact and their relative age is uncertain. Similar rocks also occur adjacent to the Churchill–Superior boundary (Albino and Macek, 1981), and, in Saskatchewan, in the Glennie Lake Domain (e.g., Delaney, 1987) and southeast of the La Ronge volcanic belt, where they are known as the McLennan Group (Lewry, 1983).

The plutonic component of the Kisseynew belt comprises large intrusive bodies ranging from diorite to leucogranite, and abundant smaller masses of leucotonalite to leucogranite, probably representing anatectic melt fractions. The granitic rocks have various ages relative to the supracrustal rocks and to phases of deformation. They display a range of deformational textures from pretectonic (granoblastic or gneissic) to late tectonic (massive or foliated). Pretectonic intrusions display a systematic transition from moderately foliated bodies in the volcanic belts to orthogneisses on the flanks of the Kisseynew belt and coarsely recrystallized meta-igneous rocks in the centre.

Amisk Group-equivalent gneisses

On the south flank of the Kisseynew belt, mafic to felsic metavolcanic rocks, and gneisses of similar composition, occur in saddle-shaped structures between granitoid orthogneiss domes, as narrow mantles around some domes, and as long rafts within them. Pre-Missi zircon ages of orthogneisses indicate an Amisk age for the inclusions. These Amisk Group-equivalent rocks are fine grained and locally retain pillow structures near the Flin Flon belt. Coarser, commonly layered amphibolites occur as a minor component of the Kisseynew belt up to 40 km north of the volcanic belt. Some are metagabbros; others are interpreted as highly recrystallized Amisk basalts. The amphibolites are locally interlayered with felsic and biotite-rich gneisses interpreted as metavolcanic and metasedimentary rocks.

Recognition of rocks equivalent to the Amisk Group in the Kisseynew belt is important for mineral exploration: the Puffy Lake gold deposit is interpreted to be hosted by these rocks. Some epidote–diopside–calcite-bearing amphibolites are interpreted as weakly altered basalts, and anthophyllite-bearing rocks as products of intense alteration.

Sherridon gneisses

The Sherridon gneisses, or Sherridon Group (Robertson, 1953), consist of "a succession of siliceous, pelitic and calc-silicate gneisses interlayered with amphibolite" (Froese and Goetz, 1981), which outcrop in a hook-shaped structure about 15 km long and 5 km wide on the south flank of the Kisseynew belt. These gneisses are coarsely recrystallized and generally retain no primary features. They are interpreted to derive from both supracrustal and plutonic protoliths (Zwanzig and Lenton, 1987). Coarsely garnetiferous, quartz-rich gneisses and local zones of garnet–anthophyllite–cordierite gneiss probably represent premetamorphic alteration products. Surrounding the Sherridon gneisses are orthogneisses that were considered to be older rocks (part of the Nokomis Group) by Robertson (1953). Ashton and Froese (1988) proposed an alternate interpretation in which the Sherridon gneisses are high-grade equivalents of the Amisk Group. This paper suggests that the Sherridon structure is a sheath fold that forms a window to a level below some of the granitic orthogneisses, an interpretation consistent with an older age for the Sherridon gneisses.

Burntwood Suite

The Burntwood River Metamorphic Suite (Gilbert *et al.*, 1980), or Burntwood Suite of this paper, is the most extensive supracrustal component in the Kisseynew belt. It contains migmatites and biotite gneisses, generally with garnet porphyroblasts and minor graphite, and locally with sillimanite and cordierite porphyroblasts. The gneisses were derived from greywacke–mudstone turbidites and locally retain graded bedding and other diagnostic primary structures. In many locations, a thin amphibolite unit derived from sediments, lavas, and intrusive rocks occurs at or near the top of the Burntwood Suite. In the central Kisseynew belt, only the Burntwood Suite and granitic rocks are exposed; on the flanks, these are interfolded with other supracrustal units.

Much of the Burntwood Suite was originally mapped as Nokomis Group (Robertson, 1953) and, more recently, as Amisk Group (Bailes, 1980; Ashton *et al.*, 1987) on the south flank, and as Wasekwan Group on the north flank (Campbell, 1972); in each case, it was given the name of the volcanic succession in the adjacent low-grade domain. Upon completion of mapping in the Kisseynew domain (Baldwin *et al.*, 1979; Gilboy, 1980), such nomenclature implied that all volcanic rocks and the Burntwood Suite in the Reindeer Zone in Manitoba were of equivalent age. This paper retains original terminology in low-grade domains, but only a tentative age correlation is made between the Burntwood Suite and adjacent volcanic rocks, which represent deposition in a different environment and may have been structurally juxtaposed. The Burntwood Suite has undergone such high-grade metamorphism and strong deformation that stratigraphic relations can no longer be clearly recognized. Thus, it is not given group status in the Kisseynew belt (North America Commission on Stratigraphic Nomenclature, 1983).

On parts of the north flank, the Burntwood Suite, although folded, is interpreted as a fining- and thinning-upward succession constituting a turbidite basin facies (Gilbert *et al.*, 1980). At Granville Lake, the upper, more pelitic part of the section contains two narrow amphibolite units derived from volcanic and intrusive rocks, and grades rapidly up into a lens of coarsening-upward conglomerate 35 km long and a maximum of 250 m thick. The conglomerate may have been deposited in a prograding marine channel or upper fan environment on the north side of the Kisseynew basin. It is unconformably or disconformably overlain by the Sickle Suite.

North of this main turbidite succession, the Burntwood Suite comprises fine- and coarse-grained metaturbidites, conglomerates, metabasalts, ultramafic rocks, carbonates, cherts, and sedimentary breccias (Zwanzig, 1981). This assemblage is characterized by rapid facies changes and is disrupted by large folds and regional faults that also involve the Sickle Suite (Fig. 3). It contains thin units of angular mudstone clasts in a greywacke matrix: other breccias have metasedimentary and metavolcanic blocks largely altered to calc-silicate rocks. Thick beds of high-energy turbidite contain large floating blocks of sedimentary rocks, basalt, and rhyolite and may have been deposited at the foot of a marine scarp. The metabasalts comprise 300 m of pillowed flows underlain by gabbro and metagreywacke, and overlain by thin units of ultramafic tuff–breccia, chert, carbonate, and solid sulphide iron formation. This succession is interpreted as part of a subaqueous, mafic volcanic platform developed at the northern boundary of the Kisseynew belt. It is also overlain by the Sickle Suite with apparent disconformity. The basalts were variously considered as "upfaulted...pre-Sickle rocks" (Norman, 1934) or "Sickle series" (Cranstone, 1968), but their distinctive geochemistry is identical to correlative basalts within the thin amphibolite units in the main Burntwood Suite turbidite succession.

Migmatites derived from the turbidites predominate in the Kisseynew belt. The best preserved paleosome has repetitive layering defined by reverse metamorphic grading; protoliths can be recognized as graded greywacke–mudstone beds (Fig. 4A). The most mobilized rocks form agmatite

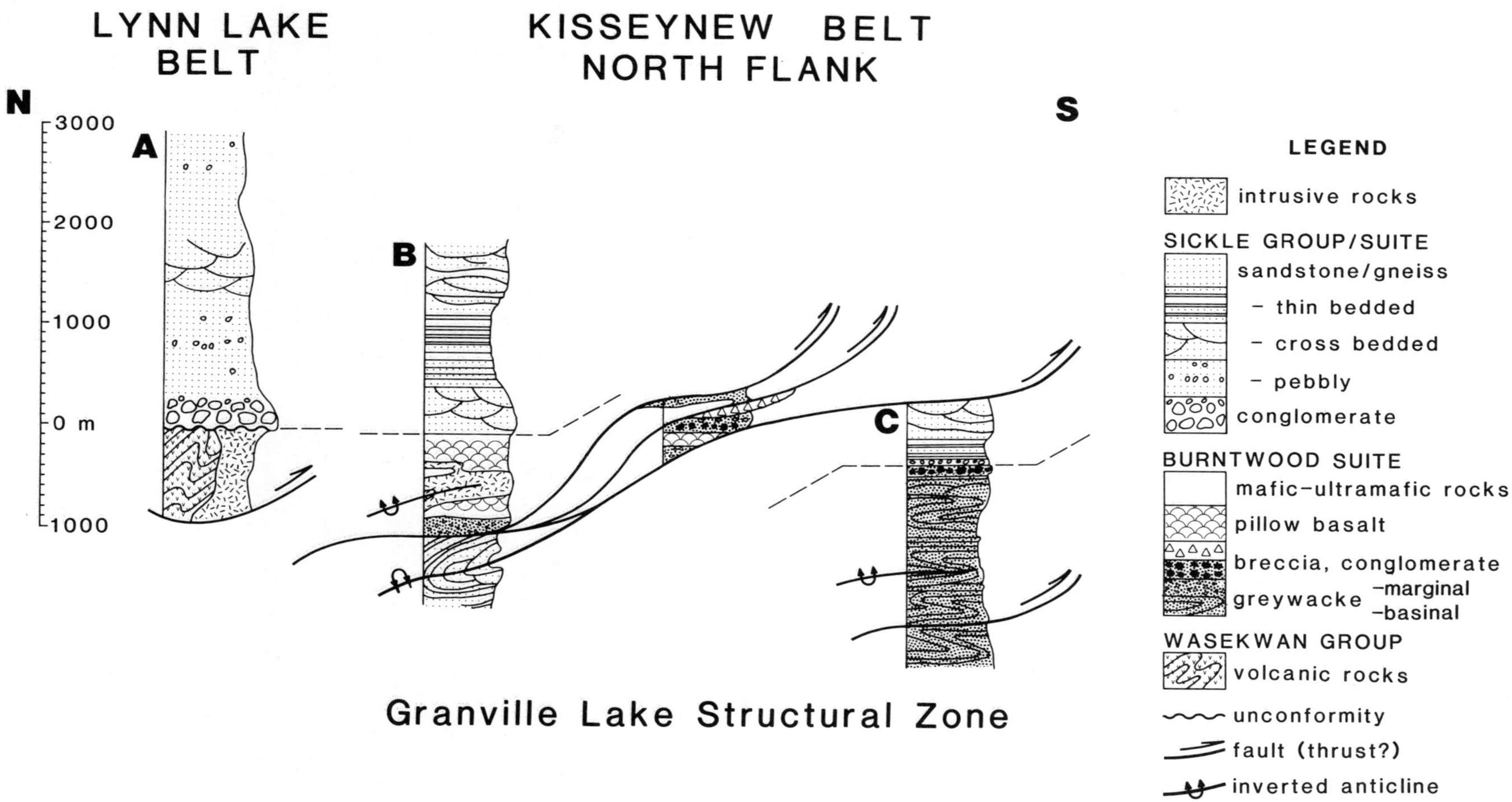

Figure 3. *Stratigraphic and structural interpretation of the Granville Lake structural zone from the Kisseynew belt north flank. Columns represent adjacent fault slices. Breccia zones may represent early growth faults at the basin margin overprinted by post-Sickle thrusts and recumbent folds. Locations of columns are shown in Figure 2.*

(diatexite, Fig. 4B) or granitoid sheets with schlieren of restite. Veins and patches of neosome are seriate pegmatitic and commonly contain garnet or sillimanite, or garnet–cordierite, and minor amounts of graphite. Larger bodies with the same leucogranitic composition as the veins are abundant in the central part of the belt.

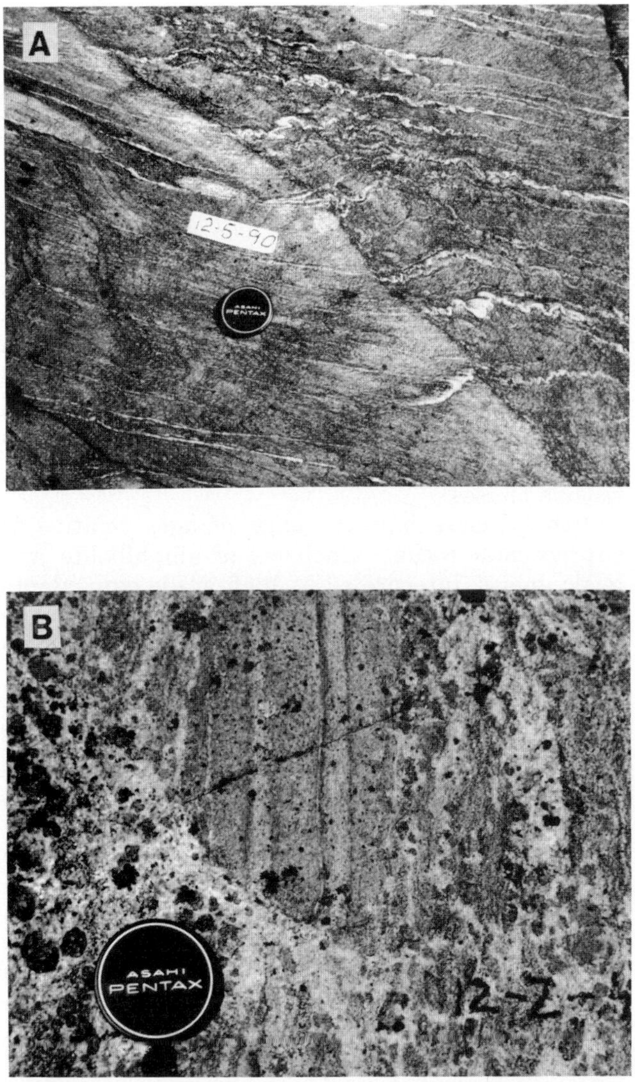

Figure 4. *Burntwood Suite: (A) Graded metagreywacke–mudstone beds (diagonal) showing open F_2 folds, axial planar S_2 foliation (parallel with tape), and veins of leucosome (white) with small, late folds (F_3–F_4). The tape is 10 cm long. (B) High-grade migmatite (diatexite) showing relict bedding (layering) in paleosome and coarse garnet porphyroblasts (black) with little evidence of strain in leucosome.*

The amphibolite at the top of the Burntwood Suite in the main part of the Kisseynew belt is a distinctive unit in which black, hornblende-rich layers alternate with green, diopside-rich layers (Schledewitz, 1972; Baldwin, 1974; Zwanzig and Wielezynski, 1975; Lenton, 1981; Zwanzig, 1984a). In some areas, calc-silicate rock, iron-rich rock, impure marble, protoquartzite, or chert are intercalated with the amphibolite. Felsic units within Fe-rich amphibolite host small stratiform gold deposits on the Kisseynew south flank (Ostry, 1986). The various rock types at the top of the Burntwood Suite are interpreted as sediment-starved basin deposits and igneous rocks.

Sickle Suite

The Sickle Metamorphic Suite (Gilbert *et al.*, 1980; Lenton, 1981; Zwanzig *et al., in prep.*), or Sickle Suite in this paper, comprises quartz-rich gneisses containing magnetite and, generally, potassium feldspar: these derived from lithic sandstone, calcareous sandstone, arkose, conglomerate, and argillaceous siltstone. Lower grade equivalents in the Lynn Lake volcanic belt belong to the Sickle Group (Norman, 1934). The Sickle Suite extends from the eastern arm of the Kisseynew belt, along the north flank, 600 km west to Reindeer Lake. The correlative McLennan Group extends an additional 200 km southwest in Saskatchewan.

Sickle-age rocks form an overlap succession that lies unconformably on the Wasekwan Group and Pool Lake intrusive suite in the Lynn Lake belt (Manitoba Energy and Mines, 1986a); they lie with structural conformity (stratigraphic disconformity or unconformity?) on Burntwood Suite amphibolite and turbidite-derived gneisses in the Kisseynew belt (Fig. 3). The low-grade Sickle Group is interpreted as an alluvial fan deposit (Zwanzig *et al., in prep.*). It is over 3900 m thick and comprises a basal conglomerate overlain by arkosic sandstone. Southwards, towards the Kisseynew belt, the sandstone succession becomes generally fine grained, argillaceous, or calc-siliceous. Bedding is thinner and the basal conglomerate tapers out as the lower contact becomes structurally conformable. Tabular crossbeds at Granville Lake suggest southerly sediment transport. Consistent regional differences in overall character of Sickle-age rocks suggest that sediments were deposited in several coalescing subbasins, averaging about 100 km in diameter.

The high-grade Sickle Suite is layered, uniform, or migmatitic, but crossbedding and magnetite placers are locally preserved. Neosome is typically pink granite or pegmatite containing magnetite and locally sillimanite or muscovite. Varying occurrence of either amphibole or sillimanite in the paleosome reflects compositional differences: such variations locally define mappable units that are remnants of a stratigraphic succession.

Missi Suite

The Missi Metamorphic Suite (Schledewitz, 1987; Zwanzig and Lenton, 1987), or Missi Suite in this paper, consists of magnetite-bearing quartz-rich gneisses derived from sandstone, and subordinate amphibolites and felsic gneisses derived from volcanic rocks. This suite extends from an area east of Wekusko Lake along the south flank of the Kisseynew belt 200 km west into Saskatchewan. Similar rocks extend over 100 km farther west across the Glennie Lake Domain and occur to the east along the Churchill–Superior Boundary Zone (Albino and Macek, 1981). Much of the Missi Suite was previously mapped as the Sherridon Group (Robertson, 1953; Froese and Goetz, 1981), but is here considered to be equivalent to the lower grade Missi Group (Bailes, 1980; Zwanzig and Lenton, 1987) and to recently mapped Missi Group gneiss in Saskatchewan (Ashton and Wheatly, 1986).

Missi-age rocks form an overlap succession that lies unconformably on subaerially weathered volcanic rocks of the Amisk Group and intrusive rocks in the Flin Flon belt (Bailes and Syme, 1983, 1989); it lies unconformably on granitoid orthogneisses and Amisk-age gneisses (Zwanzig, 1984a; Hunt and Zwanzig, 1990); and, generally with structural conformity (transposed unconformity?), on deep-water turbidites in the southern Kisseynew belt. The most complete sections of the Missi Group are over 2 km thick (Fig. 5). Coarse, pebbly sandstone and conglomerate occur in the south at Athapapuskow Lake (Syme, 1987); predominantly coarse sandstone occurs near Flin Flon. Such rocks are interpreted as alluvial fan deposits (Mukherjee, 1974; Syme *et al.*, 1982).

In the Kisseynew belt, medium-grained gneisses (Missi Suite) interpreted as metasandstones locally contain deformed, large-scale crossbedding or scattered pebbles, but in general are highly recrystallized. Magnetiferous biotite gneiss was presumably derived from a fine-grained distal sandstone facies, possibly deposited in an alluvial plain. Fine-grained, banded, pink-, green-, and grey-weathering rocks were derived from thin-bedded sandstone, calcareous sandstone, and argillaceous siltstones. Protoquartzite, at the east end of the south flank, may have been sorted on a beach or in another shallow-water environment (Bailes, 1985).

Missi Group metavolcanic rocks are most abundant in the Wekusko Lake region, at the east end of the Flin Flon belt (Stockwell, 1937; Gordon and Gall, 1982; Bailes, 1985). They comprise mafic, intermediate, and felsic flows and welded ash-flow tuffs deposited subaerially and interlayered with crossbedded metasandstones. Fine-grained amphibolites and pink-weathering felsic gneisses in the Kisseynew belt, extending from Wekusko Lake northwest to Saskatchewan and forming an integral part of the Missi Suite, are interpreted to correlate with these lower grade volcanic rocks (Zwanzig, 1984a). High-grade metabasalts locally contain layers with large amygdales; some felsic units retain quartz phenocrysts or rare fragmental textures and are chemically identical to ash-flow tuffs at Wekusko Lake.

Orthogneisses

Highly recrystallized intrusive rocks are abundant on the flanks of the Kisseynew belt (Fig. 2). Individual bodies average 15 km long, and occupy closed domal structures and recumbent sheath folds. These orthogneisses are characterized by medium- to fine-grained granoblastic texture, strong foliation, and linear fabric, indicating intense ductile deformation. Igneous textures are best preserved in areas adjacent to low-grade domains. The orthogneisses are predominantly tonalitic, but range from granitic to dioritic. Each body comprises a limited number of homogeneous quartzofeldspathic phases derived from different intrusive phases.

Orthogneisses have a range of ages relative to supracrustal rocks. Enclaves of amphibolite and felsic gneiss, interpreted as high-grade equivalents of the Amisk and Wasekwan groups, locally provide maximum ages. Some orthogneiss bodies are unconformably overlain by the Sickle Suite in the north or the Missi Suite in the south, but their relationship to the Burntwood Suite is unknown. At least one gneissic body has been interpreted to comprise felsic volcanic gneiss of the Missi Suite (Bailes, 1980), and some bodies (magnetite–biotite monzogranite of Schledewitz (1987)) appear to intrude the Missi Suite. The full age range can be established only by more U–Pb geochronology than the three existing pre-Missi dates.

Some orthogneisses were originally mapped as metasedimentary rocks belonging to the Sherridon Group or Nokomis Group (Robertson, 1953; Kornik, 1968), but in recent remapping most of these have been reinterpreted as highly metamorphosed or migmatized intrusive rocks equivalent to plutons in the adjoining volcanic belt. A systematic transition from massive granodiorite and tonalite to remobilized orthogneisses is locally well displayed between the Flin Flon belt and the Kisseynew south flank. Widespread patches of neosome and numerous syntectonic granitic veins and sheets cut the remobilized bodies. Garnet porphyroblasts are common, especially near inclusions or patches of neosome.

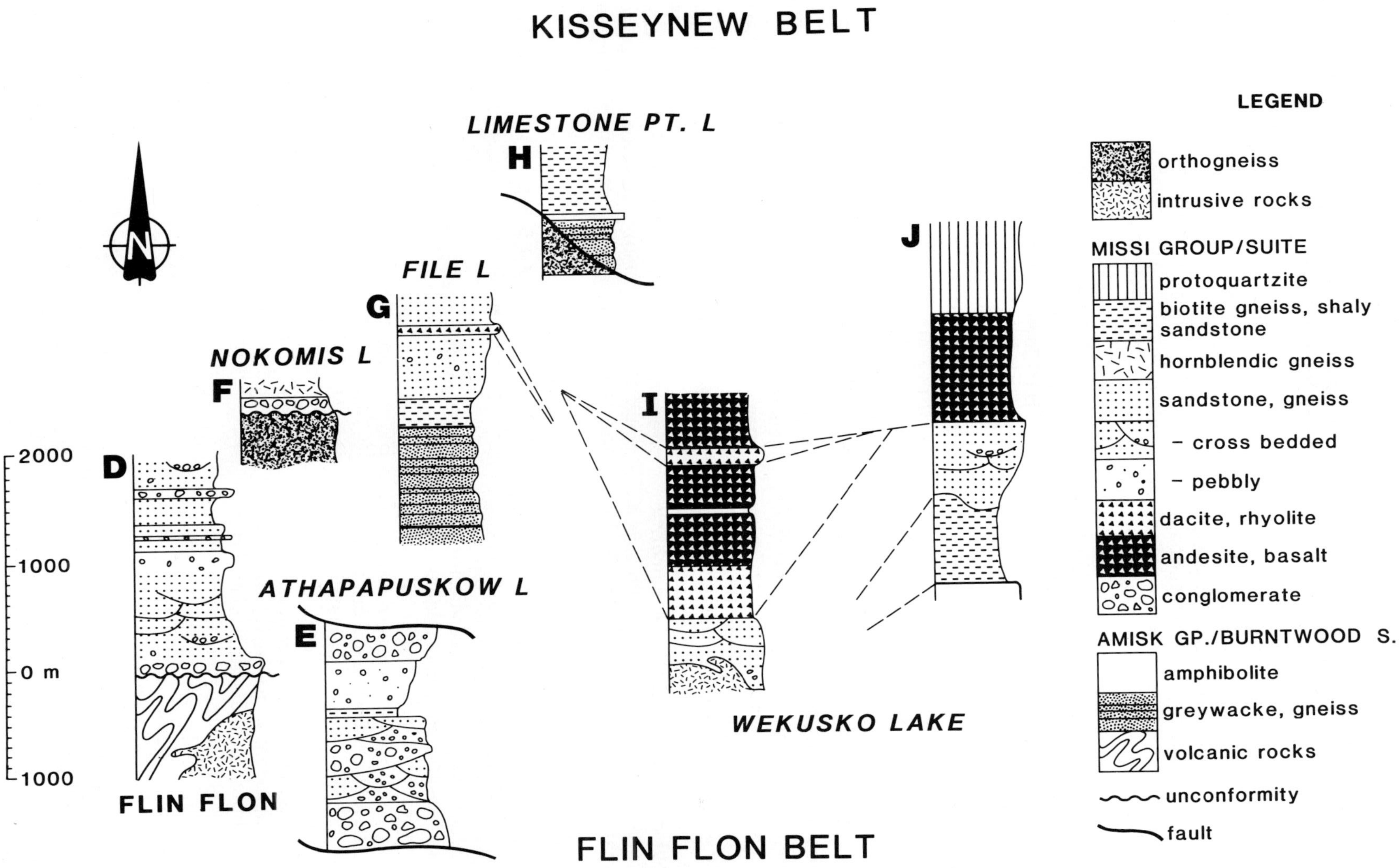

Figure 5. *Schematic stratigraphic sections from the Kisseynew belt south flank and the Flin Flon belt. Locations of columns are shown in Figure 2.*

Highrock Lake tonalite

A second suite of pretectonic intrusions comprises foliated garnetiferous tonalite and granodiorite preferentially emplaced into the core of the Kisseynew belt, primarily as the Highrock Lake batholith (L in Fig. 2). This suite is foliated to gneissic, generally homogeneous, and equigranular. The rocks are medium- to coarse-grained equivalents of the orthogneisses that recrystallized under more static conditions, with retention of original igneous textures. The suite was originally mapped as younger than the Touchbourne enderbites (Baldwin *et al.*, 1979; Manitoba Energy and Mines, 1988), but this relationship is uncertain: gravity anomaly maps (Geological Society of America, 1987) suggest that the Highrock Lake tonalite extends north in subsurface and is equivalent to the Livingston granodiorite (Lenton and Corkery, 1981), which yields a U–Pb age older than the Touchbourne enderbites. The highly discordant zircon age of the tonalite is consistent with either interpretation.

Touchbourne intrusive suite

A third pretectonic intrusive suite comprises (hypersthene-bearing) enderbites and garnetiferous quartz diorites, tonalites, and granodiorites locally associated with metagabbro and ultramafic rocks. These occur in the central part of the Kisseynew belt as large folded sheets and locally as agmatitic blocks in a leucogranite matrix. One U–Pb zircon age indicates that these rocks are coeval with the Missi Suite.

Syntectonic intrusions

Granitic rocks considered to be syntectonic and synmetamorphic include porphyritic granodiorite, occurring as stocks, sheets, and axial planar dyke complexes, and widespread large and small bodies of leucotonalite to leucogranite. These rocks have variable grain size and texture; many are seriate pegmatitic. They represent neosomal mobilizate of the migmatite and locally constitute over 75% of outcrop areas. Thicker sheets commonly have sharply defined, apparently intrusive contacts; other bodies have diffuse boundaries or biotite-rich marginal selvages. Field relations and the presence of garnet, cordierite, and sillimanite in such leucogranitoid rocks suggest that they were generated essentially *in situ* by anatexis.

GEOCHRONOLOGY

Recently published U–Pb concordia upper-intercept ages on zircon and monazite help provide a time frame for the stratigraphy, magmatism, and metamorphism in the Kisseynew gneiss belt. Three magmatic peaks in the Reindeer Zone, dated at approximately 1880, 1855, and 1830 Ma, can be related to depositional ages of the main supracrustal suites (Fig. 6). The approximate age of peak metamorphism, main deformation, and syntectonic granite intrusion is 1815 Ma (Gordon *et al.*, 1987, this volume).

The ages of early volcanic rocks in domains adjoining the Kisseynew belt range from 1910^{+15}_{-10} to 1878 ± 3 Ma (Baldwin *et al.*, 1987; Van Schmus *et al.*, 1987). This includes an 1886 ± 1.3 Ma Amisk Group metarhyolite (Syme *et al.*, 1987), which provides the best age estimate for correlative mafic to felsic gneisses on the south flank of the Kisseynew belt.

Early intrusive rocks in the Lynn Lake, Flin Flon, Kisseynew, and Glennie domains range from 1890^{+8}_{-6} to 1874^{+12}_{-8} Ma (Baldwin *et al.*, 1987; Van Schmus *et al.*, 1987; Hunt and Zwanzig, 1990; Gordon *et al.*, this volume). The dated rocks include 1890 Ma tonalitic and granodioritic orthogneisses and 1874 Ma granitic orthogneiss from the Kisseynew south flank. However, the orthogneisses probably have a wider range of ages: younger phases of granitic gneiss contain inclusions interpreted as Missi Suite.

The Burntwood Suite is equivalent to metaturbidites on the north margin of the Flin Flon belt, which contain abundant felsic volcanic clasts and are interpreted to be deposited during Amisk volcanism (Bailes, 1980). However, the age range of the Amisk Group is not established and the

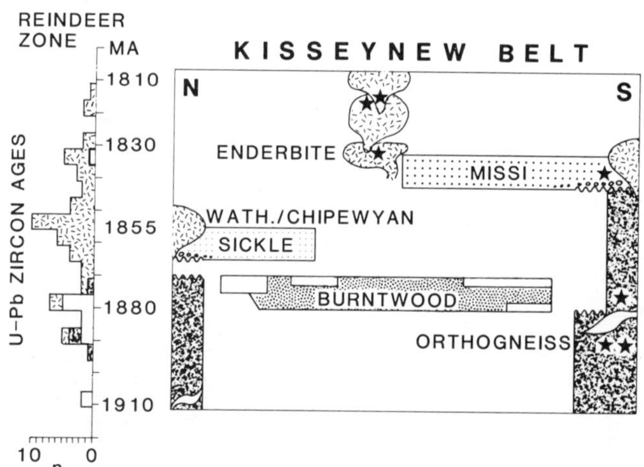

Figure 6. *Schematic, time-stratigraphic cross-section of the Kisseynew belt showing intrusive relationships and peaks in magmatic activity as indicated in a histogram of U–Pb zircon ages of the Reindeer Zone. Patterns and symbols are as in Figure 5; stars represent individual age determinations.*

Burntwood Suite, which overlies the main volcanic succession, may be younger than the dated rhyolite. Other, possibly correlative epiclastic rocks that contain abundant felsic volcanic detritus occur in the Leaf Rapids domain, unconformably above 1878 Ma rhyolite (Baldwin, 1987).

The great magmatic pulse at ca. 1855 Ma provides the minimum age for the Sickle Suite: Livingston granodiorite yielding this age (W.R. Van Schmus, personal communication) intrudes the Sickle Suite on the northern edge of the Kisseynew belt (Lenton and Corkery, 1981). Slightly older magmatic rocks in the Wathaman–Chipewyan domain may have predated the Sickle Suite and provided some of its detritus. A maximum age of 1876^{+8}_{-6} Ma (Baldwin *et al.*, 1987) is provided by the Pool Lake intrusive suite, which is unconformably overlain by the Sickle Group in the Lynn Lake belt.

The next magmatic pulse yields ages between 1830 and 1836 Ma from enderbite in the Kisseynew belt; granitic rocks from the eastern half of the Flin Flon belt, Glennie Lake Domain and La Ronge Domain; and rhyolitic ash-flow tuff from Wekusko Lake (Gordon *et al.*, 1987; Van Schmus *et al.*, 1987). The tuff is part of the Missi Group and equivalent to felsic gneiss in the Missi Suite.

Strong deformation in the Kisseynew belt postdates the 1830 Ma Missi Suite. The main deformation, intrusion of syntectonic granodiorite, leucotonalite to leucogranite and widespread anatexis occurred at ca. 1815 Ma (see Gordon *et al.* (this volume) for a discussion). Older intrusive rocks, including the orthogneisses, are classified as pretectonic in the Kisseynew belt. However, evidence for earlier deformation exists in the Lynn Lake belt (Gilbert *et al.*, 1980) and Flin Flon belt (Bailes and Syme, 1989). Moreover, a deformed, 1832 Ma quartz-diorite dyke in the Southern Indian domain appears to postdate a period of deformation (Van Schmus and Schledewitz, 1986), and an undeformed, 1830 Ma granitic sheet crosscuts the deformed main phase of the Wathaman batholith (Bickford *et al.*, 1986).

TECTONIC SETTINGS

Tectonic settings of sedimentation and magmatism in the Kisseynew belt are inferred from clast provenance, detrital modes, and geochemistry. Clast studies give information about source terrane, but are restricted to metaconglomerates for the high-grade rocks. Detrital modes of metasandstones are available only from low-grade rocks correlative with the gneisses. These modes are plotted on ternary tectonic diagrams after Dickenson *et al.* (1983). Ratios of alkalies against silica for high-grade metasedimentary rocks are plotted on diagrams showing tectonic fields of modern sandstones after Roser and Korsch (1986). Limited data for basalt geochemistry from the north flank are compared with data from the adjacent Lynn Lake volcanic belt. The data indicate fundamental changes in tectonic setting involving (1) early marine sedimentation and volcanism for the Burntwood Suite, apparently in an island-arc environment; (2) later continental-arc conditions, probably in two stages, first Sickle and then Missi; and (3) middle crustal conditions during terminal collision.

Island-arc setting

Metaturbidites equivalent to the Burntwood Suite, at the margin of the Flin Flon belt, contain abundant volcanic clasts and scarce plutonic clasts from an active volcanic source, probably the island arc-volcanoes represented by the Amisk Group (Bailes, 1980; Syme, this volume). Detrital modes of these greywackes and mudstones lie in the fields of undissected and transitional island arcs (Fig. 7).

Primary microclasts are not preserved in the Kisseynew belt, but alkalies versus silica ratios of metagreywackes and mudstones from the Flin Flon belt margin are the same as those of corresponding quartzofeldspathic gneisses and garnet–biotite gneisses from the main Kisseynew belt (Fig. 8), supporting Bailes' (1980) conclusion that the sediments from the two areas were derived from the same source. Compared with the tectonic setting of modern sands, the sediments lie in the field of active continental margins rather than island arcs.

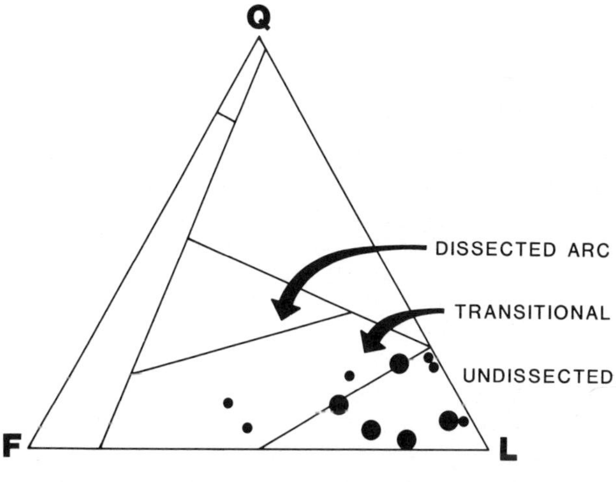

Figure 7. *QFL (quartz – feldspar – lithic clast) plot of framework modes for low-grade metagreywacke, equivalent to Burntwood gneiss. Large dots are samples with less than 50% matrix; small dots are samples with more matrix or with partly obliterated lithic clasts. The data are from Bailes (1980); tectonic fields are after Dickenson* et al. *(1983).*

Moreover, greywacke contains more SiO_2 and less K_2O than mudstone, as expected from an epiclastic or weathered pyroclastic source (Roser and Korsch, 1986). Low initial $^{87}Sr/^{86}Sr$ throughout the Burntwood Suite (Clark, 1981) and high ε_{Nd} of Kisseynew sedimentary gneiss in Saskatchewan (Chauvel *et al.*, 1986) indicate Early Proterozoic source rocks for the sediments. The chemical, petrographic, and isotopic data, together, suggest that the source was an active mature island arc.

Ratios of alkalies versus silica of Burntwood turbidites from the northern part of the Kisseynew belt overlap data from the central part, but some samples are slightly higher in potassium and silica contents (Fig. 8). These data can be interpreted to indicate local input of mature detritus from the north. Cobbles and pebbles at the top of the Burntwood Suite in the north consist of quartzose metasandstone with internal structures, such as shaly partings, magnetite placers, crossbedding, and quartz veins, indicating erosion of a metamorphosed shallow-water deposit. The presence of sedimentary breccias and large blocks suggests that the northern basin margin was unstable during deposition of the Burntwood Suite.

Preliminary geochemical data from Burntwood Suite mafic and ultramafic volcanic rocks at Granville Lake suggest a different tectonic environment than for the volcanic rocks of the Lynn Lake belt, which are similar to lavas from Cenozoic island arcs (Syme, 1985, this volume). The Granville Lake volcanics range from primitive, Mg-rich tholeiites to fractionated ferrobasalt with consistently more TiO_2 than basalts from the Lynn Lake belt (Fig. 9). Basalts from Tod Lake and Laurie Lake, on the Lynn Lake belt boundary, have intermediate TiO_2 values. The Burntwood lavas were probably erupted in a tensional tectonic regime. Negative rare earth element (REE) slopes, however, are consistent with volcanism near an island arc rather than an oceanic ridge (Fig. 10).

Geochemistry of the orthogneisses has not been studied, but their common range from diorite and magnetiferous, hornblende–biotite tonalite to granite suggests a subaluminous calc-alkalic composition.

Continental-arc setting

The Sickle Suite had a different provenance, depositional environment, and tectonic setting than the Burntwood Suite. It comprises alluvial and shallow-water deposits formed in a continental environment above an unconformity. A single detrital mode from the low-grade Sickle Group plots on the edge of the dissected-arc field (Fig. 11). Conglomerate clasts at the base of the succession reflect the composition of underlying rocks. Higher up, the clasts were derived from a felsic sedimentary, volcanic, and hypabyssal intrusive terrane, compris-

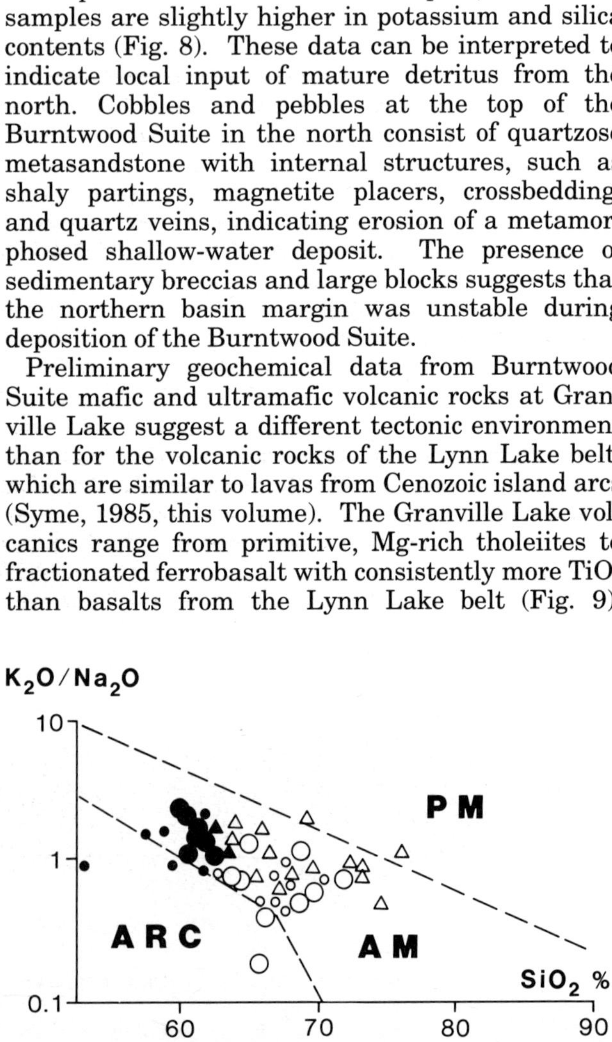

Figure 8. *Plots of alkali ratio versus silica of low-grade metagreywacke (large open dots) – mudstone (large filled dots), high-grade equivalents from the central Kisseynew belt (small open and filled dots) and the north flank (triangles). The tectonic fields determined from modern sands are island arc (ARC), active continental margin (AM), and passive margin (PM). Some of the data are unpublished, other data are from Bailes (1980).*

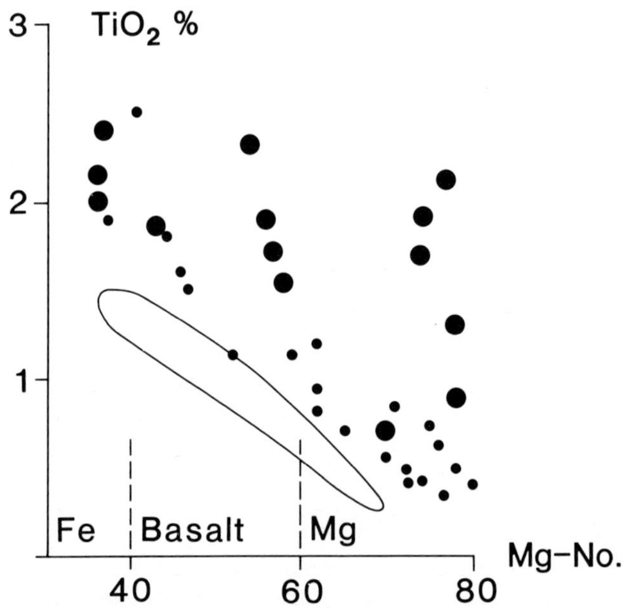

Figure 9. *TiO_2 versus Mg-number of mafic–ultramafic volcanics from various fault slices on the Kisseynew belt north flank: Granville Lake (large dots), Tod Lake and Laurie Lake (small dots). The main field of arc tholeiites from the Lynn Lake belt is outlined for comparison. Mg-number = Mb / (Mg + Fe^{+2}), where 0.85 total iron is the apportioned FeO. Some of the data are unpublished, other data are from Gilbert* et al. *(1980).*

ing, on average, one-third arkosic sandstone; one-third fine-grained granite, porphyry, and rhyolite; and the rest a mixture of quartz, mafic rocks, and other clast types (Zwanzig *et al.*, *in prep.*).

Ratios of alkalies versus silica of the Sickle Suite overlap fields of sandstones from active and passive continental margins (Fig. 12). Their average is the same as ratios for the Chipewyan granite, in the northern Reindeer Zone. Provenance of the Sickle sediments may have been an older felsic volcanic and sedimentary cover, and early hypabyssal intrusions similar in composition to the Chipewyan

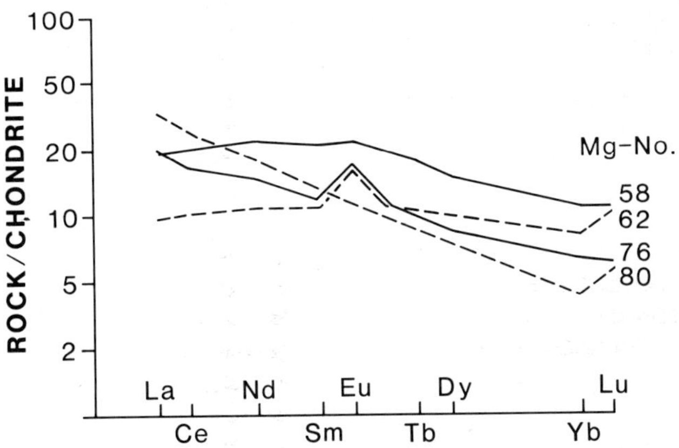

Figure 10. *Chondrite-normalized rare earth element plots of mafic-ultramafic volcanic rocks (amphibolites) from the Kisseynew belt north flank: Granville Lake (solid lines), Tod Lake and Laurie Lake (dashed lines).*

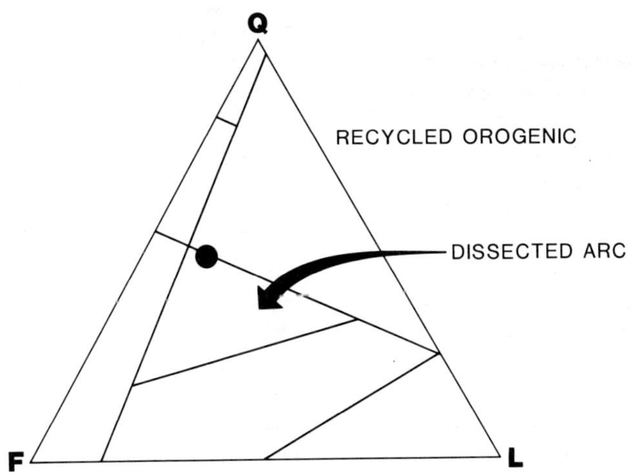

Figure 11. *QFL (quartz – feldspar – lithic clasts) plot of framework mode of Sickle Group low-grade metasandstone with tectonic fields as in Figure 7.*

batholith. Southerly transport in a system of alluvial fans shed over the Kisseynew basin margin is consistent with a source in such a tectonic hinterland. Southerly thrusting, perpendicular to the Sickle outcrop belt, suggests telescoping of a continental arc-related basin during intrusion of the main batholith to the northwest.

The continental environment persisted during deposition of the Missi sandstones and volcanic rocks. The alkali–silica index overlaps the fields of active and passive margins (Fig. 13), but abundant volcanic clasts in the conglomerates and the absence of widespread orthoquartzite and pure carbonate rule out passive margin deposition. The low-grade Missi Group conglomerate in the southwestern Flin Flon belt reflects erosion of a volcanic terrane. Average clast population includes 50% basalt, 17% rhyolite and dacite, 18% sandstone and siltstone, and 9% intrusive rocks (Syme, 1987).

The Missi Group volcanic section at Wekusko Lake incorporates intermediate-K_2O meta-andesite and some trachyandesite (Gordon and Lemkow, 1987). High-grade metavolcanic rocks in the Kisseynew belt include low-TiO_2 basalt and andesite. Major element chemistry of the coeval Touchbourne enderbites indicates a calc-alkalic affinity. These compositions and subaerial flow characteristics are consistent with the suggestion that the Missi Suite was deposited in a continental-arc environment.

Main tectonic setting

The syntectonic leucogranites have a peraluminous mineral content and are interpreted to be derived by large-scale crustal anatexis. This large volume of S-type granite is consistent with a tectonic en-

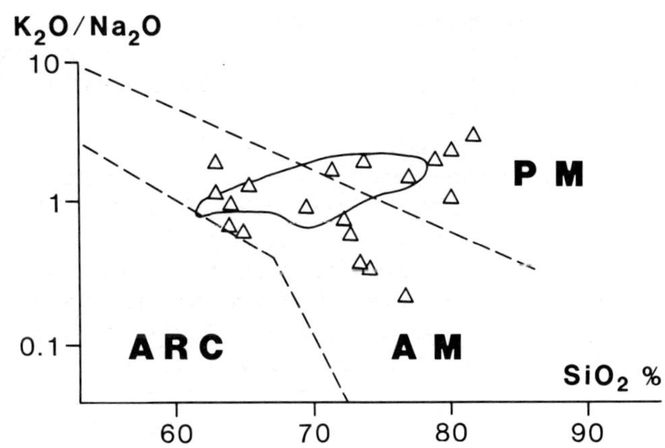

Figure 12. *Plots of alkali ratio versus silica of Sickle Suite sandstone-derived gneiss. The outlined field (solid line) is from the Chipewyan granite; tectonic fields as in Figure 6.*

vironment of continental collision. More definite conclusions are based on structural analysis.

STRUCTURE

The main body of the Kisseynew belt in Manitoba has been divided into three west-northwest trending structural zones: the north flank, the central migmatite belt, and the south flank (Fig. 14). The flanking zones have a recognizable lithostratigraphy that permits delineation of several ages of folds and faults. In the central belt, the stratigraphy is obscure, and only synmetamorphic structures recorded in the foliation and migmatitic layering are obvious. Continuity of structures across the zones, nevertheless, suggests that the same phases of deformation affected the whole belt.

The oldest structures are observed on the north flank, where sedimentary tops are preserved. These structures include postulated synsedimentary normal faults and premetamorphic isoclinal folds. Early nappe-sized structures, formed before peak metamorphism, control the regional west-northwest trend of stratigraphic units. These large folds are considered to have been originally recumbent, but have been generally reoriented by several generations of synmetamorphic, upright and recumbent folds.

The geometry of major and minor synmetamorphic structures has been studied in stereoplots of 900 subareas of about 30 km^2 each. Dominant orientations of foliations, axial planes, and linear structures were plotted on maps, as summarized in Figure 14. Systematic changes in preferred orientation of planar and linear structures from one area to another are used to study the large-scale geometry and movement pattern in the belt (Fig. 15). Geometry, style, and fold sequence is further defined in composite down-plunge structure sections constructed from mesoscopic fabric data and the published maps of Baldwin *et al.* (1979).

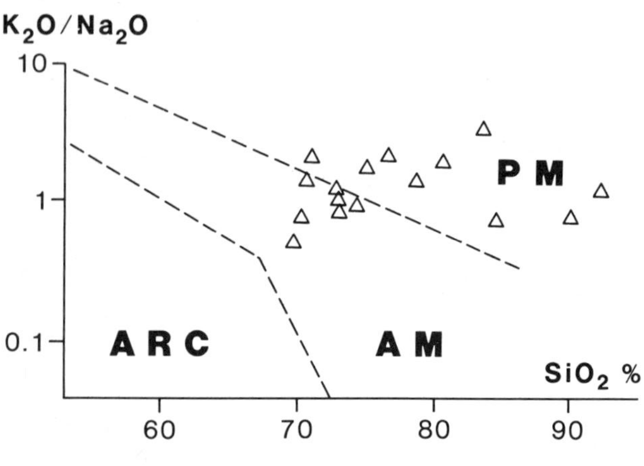

Figure 13. *Plots of alkali ratio versus silica of Missi Suite / Group gneiss / metasandstone; tectonic fields as in Figure 6. Data are from Bailes (1975, 1980, 1985) and Zwanzig (1985).*

Structural history

Variations in structural trends and evidence of cross folding indicate that deformation in the Kisseynew belt was polyphase. Large, sheetlike F_1 structures are deformed by several generations of later folds, but a unique age cannot always be assigned to a specific trend of cross folds. Relative ages of minor structures are generally indeterminate because interference of small folds is rare. Small early folds are lacking or go unnoticed, and large early folds and faults were reactivated or rotated during later deformation. The best criteria for development of a structural chronology are the relationships of folds and shear zones to (1) other structures where relative ages are evident, (2) stratigraphic units and dated intrusive suites, and (3) prograde and subsequent retrograde metamorphic mineral assemblages. The history of deformation compiled from these data involved five stages (Table I), some of which overlap in time or are coeval: relative ages vary from one part of the belt to another.

Premetamorphic structures. The earliest structures are interpreted to be synsedimentary faults in the Burntwood Suite, and post-Sickle folds and thrusts developed along the south shore of Granville Lake, in a belt informally called the Granville Lake structural zone (GLSZ) in Figure 14. This zone is 7 km wide and extends from the Metcalf Bay fault (MFB) for nearly 75 km to the east. Beyond these points, the early structures were largely obliterated by synmetamorphic structures, but 50 km northwest of Granville Lake a similar early fault zone may have been deformed by later structures at Laurie Lake (L, Fig. 14). The early structural history of the GLSZ is uncertain, but the oldest faults are interpreted to be extensional because units are cut out, leaving isolated bodies of metaturbidite and mafic–ultramafic volcanic rocks separated by faults and lenses of breccia. The breccias are sedimentary and largely altered to calc-silicate rocks; they were probably directly related to the faulting. The mixture of blocks in the associated high-energy turbidites includes fragments of the breccia itself, a feature that suggests cannibalistic sedimentation at a marine fault scarp. Some discontinuous units may be olistostromes or broken formations deposited at the base of the scarp. The early structures were reactivated as thrusts, which were later overturned.

The main structures in the GLSZ constitute a post-Sickle thrust and fold complex with three north-facing fault slices (Fig. 3, columns B and C). The southern slice contains the fining-upward basin-facies turbidite succession and minor volcanic rocks overlain by the Sickle Suite. The central structural slice contains the high-energy turbidites, breccias, and discontinuous volcanic units from the basin margin. The most northerly slice is a coherent succession from the fringing volcanic platform, also overlain by the Sickle Suite. The whole F_1 thrust stack was overturned to the north by F_3 back folding and northerly nappe emplacement (Fig. 16).

Thus, the early folds and thrusts apparently had a southerly vergence and strain was concentrated in south-facing fold limbs (Fig. 16b). Some thrust slices have a wide, well-preserved, north-facing panel and a narrower, faulted, south-facing panel. Because folding and thrusting predated overturning of the section, the systematic arrangement of depositional facies in adjacent structural slices indicates that the volcanic belt was thrust over the basin margin, which, in turn, was thrust over the basinal turbidites (cf. Fig. 16a,c). The map pattern of the Lynn Lake – Granville Lake region (Manitoba Energy and Mines, 1986a,b) suggests that the GLSZ is part of a regional system of thrust

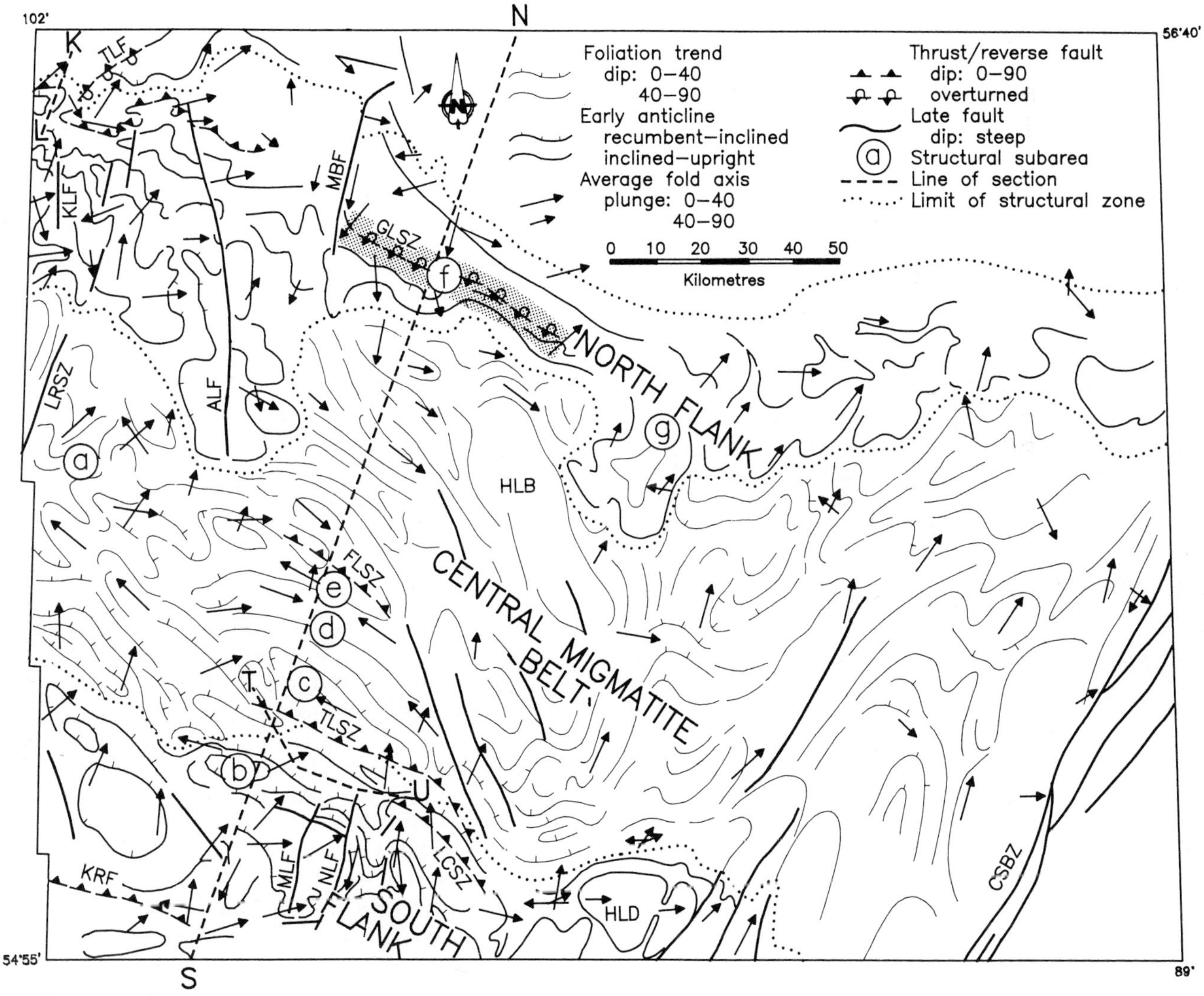

Figure 14. *Synoptic structural map of the main part of the Kisseynew belt in Manitoba showing foliation trend lines in the central migmatite belt and early fold trends in the flanking zones. Shallow-dipping structures are ornamented with ticks. Orientations of linear structures (arrows) were taken from subarea stereograms (not shown). Faults and shear zones are named in the text. The Granville Lake structural zone (GLSZ) is shaded. Also shown are the Highrock Lake batholith (HLB), Herblet Lake dome (HLD), and Churchill–Superior Boundary Zone (CSBZ).*

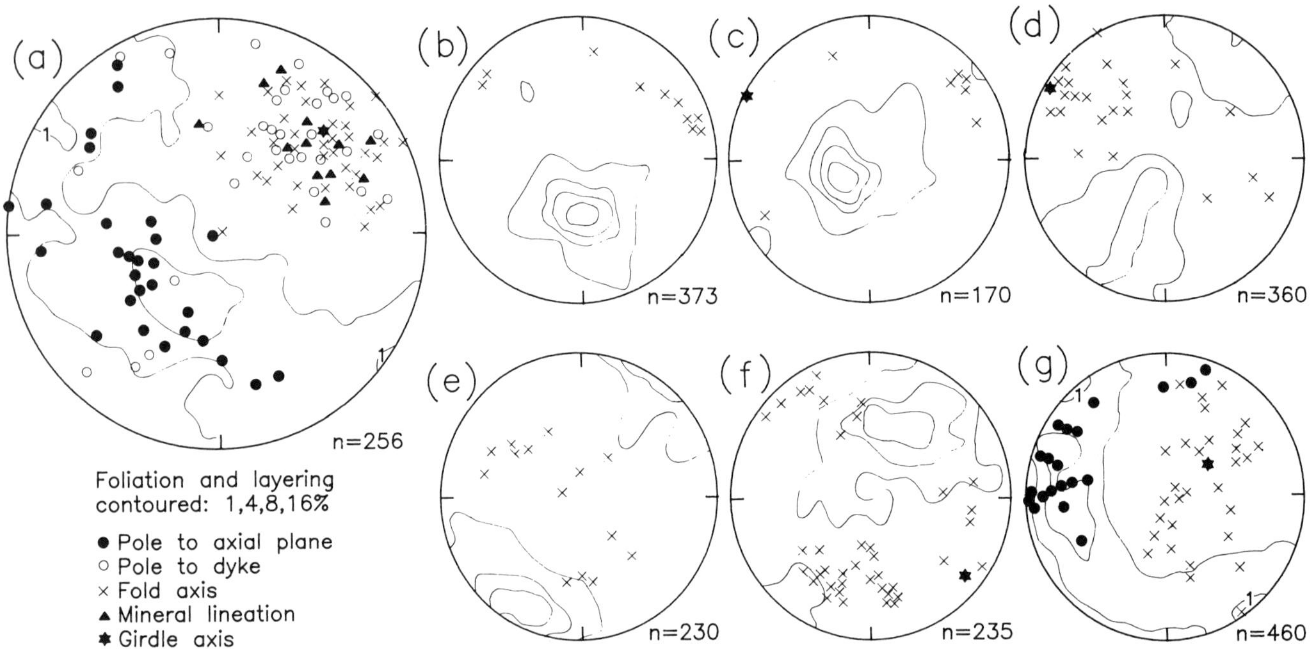

Figure 15. *Synoptic equal-area stereograms corresponding to subareas (a) to (g) in Figure 14;* n = *number of contoured data points.*

faults reversing the earlier normal displacement between the Lynn Lake belt and the Kisseynew basin.

On the south side of the GLSZ, reversals in graded bedding tops indicate F_1 isoclinal folding of the basin-facies turbidites. One of the rarely observed fold closures has an obliquely crosscutting foliation superimposed during later metamorphism. Preservation of the gradually fining and abruptly coarsening successions followed by the Sickle Suite suggests that the exposed stratigraphic level rises towards the north across the structural pile. This feature can be explained if the isoclines represent asymmetric, first-order folds that have longer north-facing limbs than south-facing limbs. The explanation is consistent with the observation that two thirds of the recorded tops face north and that the early vergence was south. Single folds are laterally continuous over several kilometres. Axial traces are 25–200 m apart and there are generally no minor folds. The same type of sharp-hinged, asymmetric, first-order folds have been described from a Mesozoic back-arc basin (Tanner and Macdonald, 1982) and may characterize the structural style produced during the closure of such basins.

Locally, south of Granville Lake, the basin-facies turbidites contain soft-sediment deformation structures, producing repetitions beneath bedding-slip faults. Calc-silicate alteration in the faults suggests deformation under high fluid flux during dewatering. These structures are probably a result of local slumping; they are not sufficiently widespread to have formed in an accretionary com-

Table I
History of deformation

(?)	Possible normal faulting on the north flank, syndepositional with Burntwood sediments and mafic–ultramafic volcanic rocks
F_1	Isoclinal recumbent to upright folding after deposition of Missi and Sickle sediments; thrusting; deep tectonic burial and possible development of early nappes
F_2	Northerly trending upright folding; peak metamorphism; granite intrusion and migmatization dated at ca. 1815 Ma
F_3	Northwest-trending upright and recumbent folding, with rotation of early structures into recumbent and reclined orientations; development of sheath folds and southwest-directed shear zones; back folding on the north flank; emplacement of syntectonic granitoid sheets also dated at ca. 1815 Ma
F_4	Upright, north- to northeast-trending folding, nearly coaxial with F_2-F_3 structures; rotation of early planes into alignment with north-northeast trending shear zones and steepening of early linear structures; uplift and late granitic dyke intrusion
F_5	North- and east-trending block faulting, with development of mylonite and metamorphic retrogression zones during uplift; intrusion of planar pegmatite dykes

plex. However, if they are part of the main F_1 isoclinal folding of the turbidites, this represents an important pre-Sickle compressional deformation, possibly related to accretion and closing of the Kisseynew basin (Zwanzig, 1984b). In Figure 16, the F_1 folds are assumed to be post-Sickle.

In most parts of the Kisseynew belt, graded bedding is not preserved, early thrust faults cannot be identified, and F_1 structures can be traced only where Burntwood or Amisk gneisses are interfolded with one of the younger suites of gneisses. Early anticlines in the flank zones (Fig. 14) are delineated by narrow outcrop belts of older gneisses. These sheetlike isoclines are draped around younger cross folds and oval structures produced by later fold interference (F_2–F_3) or sheath folding. The F_1 structures are considered to have formed as recumbent folds (Bailes, 1980; Gilbert *et al.*, 1980), but an alternate origin cannot be ruled out: some of the isoclines may have formed as upright or inclined immature folds that were overturned and flattened into recumbent, mature folds during progressive deformation (Talbot, 1979). Tectonic burial and high-grade metamorphism probably influenced the strain pattern.

Synmetamorphic structures. The intense deformation that occurred during high-grade metamorphism generated three sets of folds: F_2, F_3, and F_4. These have axial planes that trend north, northwest, and north-northeast respectively. The

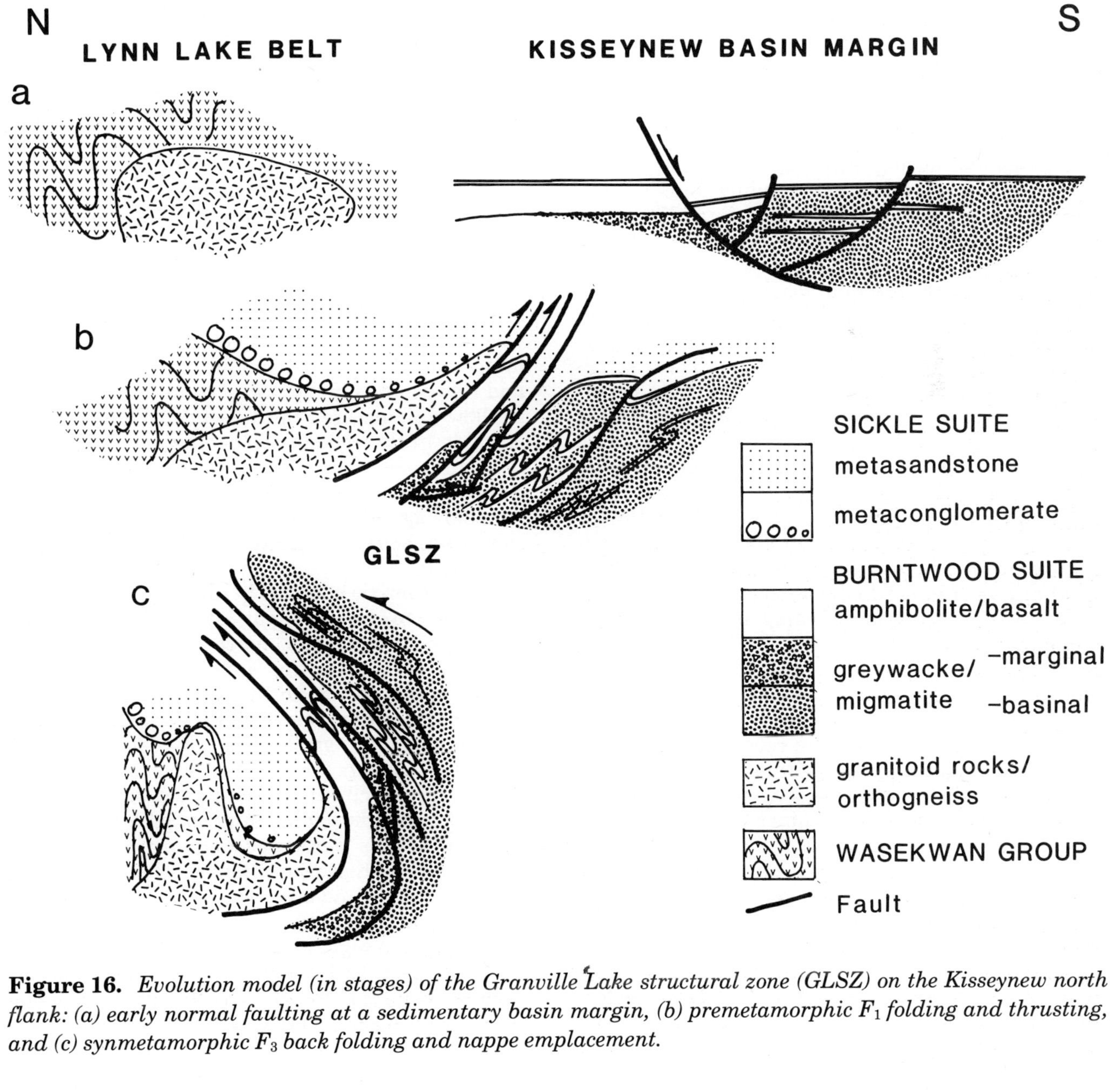

Figure 16. *Evolution model (in stages) of the Granville Lake structural zone (GLSZ) on the Kisseynew north flank: (a) early normal faulting at a sedimentary basin margin, (b) premetamorphic F_1 folding and thrusting, and (c) synmetamorphic F_3 back folding and nappe emplacement.*

structures are associated with mineral alignment of amphibole, sillimanite, and elongate overgrowths on garnet porphyroblasts. Deformation was coeval with development of migmatitic layering (Fig. 4a) and intrusion of granitic sheets. Steeply dipping F_3 and F_4 shear zones contain retrograde muscovite, sillimanite, and hematite developed during the early waning stages of metamorphism.

North-trending F_2 folds and west-northwest trending F_3 folds have been defined on the flanks of the Kisseynew belt and in adjacent parts of the volcanic belts. Large, S-shaped, northeast-plunging, recumbent and reclined folds within the Kisseynew belt may have formed as upright, northerly trending structures that were overturned and flattened during F_3 folding. Large areas on the flanks of the Kisseynew belt have an inverted stratigraphic succession and are interpreted to represent underlimbs of fold nappes. Maximum structural overlap of a single fold is about 15 km. The nappes were probably initiated during F_1 folding, but were transported during F_3 folding at peak metamorphic conditions and during initial uplift. In large parts of the belt, north-northeast trending folds are superimposed on the northwest-trending (F_3) structures, but reverse relationships also exist. There was probably a considerable time overlap in the development of the different fold sets.

In the central migmatite belt, the combined F_2–F_3 geometry has produced northeast-plunging, reclined, recumbent and sheath folds. These structures are commonly refolded about north- to northeast-trending, upright axial planes (F_4). This geometry has been analyzed in a synoptic stereogram (Fig. 15a) of subarea (a) southeast of the Loon River shear zone (LRSZ, Fig. 14). The dominant F_2–F_3 structural trend is west-northwest and the dip is northeast, defining the main concentration of poles to foliation (4% contour) and axial planes of minor folds (filled circles). The concentration of northeast-plunging minor fold axes, crenulations, mineral lineations, and rodding is interpreted as the direction of stretching and tectonic transport (slip line) developed under conditions of high strain.

Nearly coaxial F_4 structures reoriented the F_2–F_3 folds, producing the present girdle distribution with a northeast-plunging girdle axis (Fig. 14a). The secondary concentration of minor-fold axial planes has the steep southeast dip of the Loon River shear zone. The steep and shallow minor folds are not distinguished by age, but were probably formed early in the polyphase fold history and were partly realigned with the LRSZ during F_4. The nearly cylindrical geometry and northeast-stretching direction was maintained during the progressive deformation. Relatively late planar and slightly curved granitic dykes cutting the migmatitic layering occupy extension fractures oriented normal to the earlier stretching lineation. The dykes indicate that northeastward extension continued into the late stages of F_4 folding. Other areas had a similar history of progressive deformation, but the resulting fold patterns are quite variable because of local dominance of one set of folds over another.

The central part of the Kisseynew belt (subareas b–f, Fig. 14) features northwest (F_3) structural trends, shallow northwest–southeast plunging lineations, and variably northeast–southwest plunging lineations (Fig. 15b–f). A systematic change in dip of the foliations and axial planes across the Kisseynew belt defines a large, downward-converging fan structure. Dips are northeast at shallow angles on much of the south flank, as in subarea (b), and in the southwestern part of the central belt (having trend lines with ticks), as in subarea (c). North of this, in subarea (d), dips steepen about a subhorizontal F_3 axis; dips are nearly vertical at the Flatrock Lake shear zone (FLSZ), as in subarea (e). In the central part of the north flank, near the Granville Lake structural zone, subarea (f), the most common dip is moderately southwest. Near the volcanic belts, dips steepen again across narrow downward-diverging fan structures. Steep dips continue across the volcanic belts.

Linear structures were developed in various orientations and were rotated according to their location within the fan structures. The plunge of minor fold axes is predominantly northeast (F_2–F_3) and less commonly northwest–southeast (F_3) in the shallow-dipping area in the south. Plunge is most variable in the zone of steepening and is moderate to steep in subarea (e) upright folding. At Granville Lake, in subarea (f), the plunge concentration is bimodal again and F_3 axes plunge northwest–southeast, but the reclined F_2–F_3 axes plunge south-southwest.

The systematic changes in dip and plunge are interpreted to be a result of southwest-directed transport in the south, northeast–southwest compression in the centre, and northeast transport (back folding) in the north. The axis of fanning is subhorizontal, northwest–southeast, and represents the common F_3 girdle axis (Fig. 15c–f). This geometry suggests large-scale north–south displacement and recumbent folding in the infrastructure, and more upright folding in the suprastructure. As a result, nappes were apparently driven north and south over the Kisseynew flanks. The upright folds (F_3) on the boundaries with the volcanic belts formed under slightly lower grade metamorphic conditions at a higher level.

The western and northeastern parts of the Kisseynew belt feature northeast-trending F_4 fold trains and shear zones alternating with areas in which F_2–F_3 structures are preserved. Large-scale interference structures, such as those at Hall Lake, in subarea (g), are common. Foliation and early minor folds are refolded about a northeast-plunging axis (girdle axis in Fig. 15g). Minor F_2–F_4 fold axes are dispersed in a vertical plane interpreted as the plane of flattening during F_4 folding. Late minor folds (F_4), with characteristics of flexure folds produced by horizontal shortening, are locally abundant: migmatitic layers act as dominant members; high-order folds are fanned about low-order folds. Flattened conglomerate clasts are locally realigned with F_4 axial planes and "faserkiesel," which grew as spheres at the sillimanite isograd (Jackson, 1988), are flattened parallel to F_4 axial planes. Folds commonly die out against earlier fold trains as in experimentally produced superimposed buckle folds (Skjernaa, 1975). All of these data indicate west-northwest–east-southeast compression and vertical thickening during F_4 folding. Slight overturning of axial planes indicates northwest transport.

Late-metamorphic structures. Late, north- and east-trending faults and mylonite zones (Fig. 14) are related to the uplift of highly migmatized blocks or offsets in metamorphic isograds. These faults caused intermediate- to low-grade retrogression, extensive epidotization, hematization, and kink folding. They are local detachment zones for large box-shaped F_5 folds. North-trending faults, such as the Abrey Lake fault (ALF) and Kamuchawie Lake Fault (KLF) in the north, and the Molly Lake fault (MLF) in the south, have sinistral separation of north-dipping units. The Nokomis Lake fault (NLF) has steeply plunging slickenlines and produces a repetition of shallow-dipping units. These faults can be explained with east-side-up block movement and possibly sinistral slip, similar to faults in the Churchill–Superior Boundary Zone (Bleeker, this volume). Other northerly trending faults, such as the Metcalf Bay fault (MBF) and possibly the Tod Lake fault (TLF), suggest west-side-up displacement. The north-dipping Kississing River fault (KRF) has rotated porphyroblasts, suggesting sinistral and reverse (oblique) movement consistent with the F_3 displacement pattern, but mylonite and pseudotachylite in this fault zone indicate that deformation occurred in the ductile to brittle F_5 regime.

Structure sections

Structural style and vergence of the large synmetamorphic folds have been analyzed using down-plunge and diagrammatic structure sections. Southerly vergence was recognized on the south flank, north of the Kississing River fault (KRF, Fig. 14), by Pollock (1965), and on the north flank by Thomas (1980) and Jackson (1988).

A simplified structural cross-section at Laurie Lake, viewed towards the southeast (Fig. 17), shows a large recumbent anticline of uncertain vergence (F_1) developed in the Burntwood Suite and draped

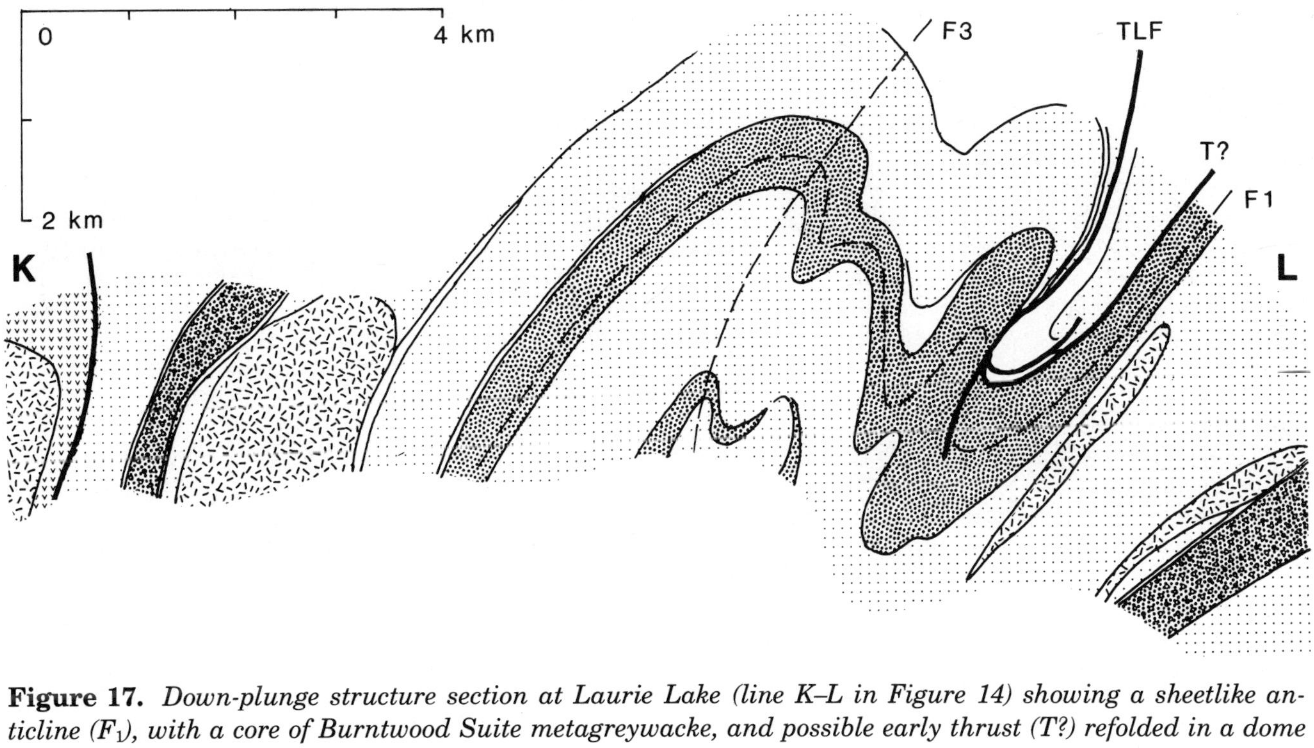

Figure 17. *Down-plunge structure section at Laurie Lake (line K–L in Figure 14) showing a sheetlike anticline (F_1), with a core of Burntwood Suite metagreywacke, and possible early thrust (T?) refolded in a dome (F_3) and saddle structure, and reactivated in the Tod Lake fault (TLF). Legend as in Figure 16.*

over the Sickle Suite in the Laurie Lake dome (F_3) and saddle structure. The southerly vergence was established during F_3 folding; the doming is a result of sheath folding (Jackson, 1988), or was accomplished by complex interference of the F_3 folds with F_2 and F_4 structures. Rocks of the basin-margin facies, feldspathic greywacke, tuff of Gilbert *et al.* (1980), occur north and east of the Laurie Lake dome at a higher structural level. Wasekwan Group rocks from the Lynn Lake belt are faulted against Sickle Group rocks overlying the basin-margin facies. Lenses of amphibolite and inferred faults correlated with the Granville Lake structural zone are folded around the south side of the dome and into the saddle structure. East of the dome, these early faults may have been reactivated in the late-metamorphic Tod Lake fault. A second anticlinal nappe containing Burntwood Suite gneisses in the core lies south of and below the saddle structure. These gneisses resemble the basin-margin facies: possibly the upper nappe overrode them from the south; consequently, it may be an early back fold. East of the Granville Lake structural zone, the back-folded sequence is also involved in cross folding (Schledewitz, 1972).

A structure section from the Takipi Lake area on the south flank is viewed down the shallow northeasterly F_2–F_3 plunge (Fig. 18). This section shows typical closed structures that can be interpreted as tongue-shaped, cross folds or sheath folds. These structures contain tonalitic orthogneisses mantled with amphibolite and Missi Suite gneisses. The orthogneisses are apparently rooted at depth and to the northeast: the plunge diverges downward about some of the larger bodies. Highly attenuated slivers of ductile Burntwood migmatite maintain a relatively uniform thickness for tens of kilometres and, like straight gneisses, are interpreted to have formed in a regime of high, simple shear strain (Ghosh, 1974). Several high-strain zones, including the Takipi Lake shear zone (TLSZ) and the Molly Lake fault (MLF) occur in the section. Large, nearly horizontal, syntectonic (F_3) sheets of leucogranite at Takipi Lake are associated with shear zones. Major oval structures interpreted as sheath folds in the hanging wall of the TLSZ plunge down the northeasterly slip and stretching direction. The hanging wall is interpreted to have been displaced to the southwest (towards the viewer in Fig. 18). The sheath folds have been tightly stacked into a broad arch, across which the ductile shear zones are draped. Open F_4(?) folds near the MLF are S-shaped and indicate northwesterly transport. The open Z-shaped fold above the fault may be a ramp. High strain is indicated elsewhere on the south flank by ribbon gneisses developed from conglomerates (Zwanzig, 1984a) and by textures indicating grain-size reduction in the orthogneisses.

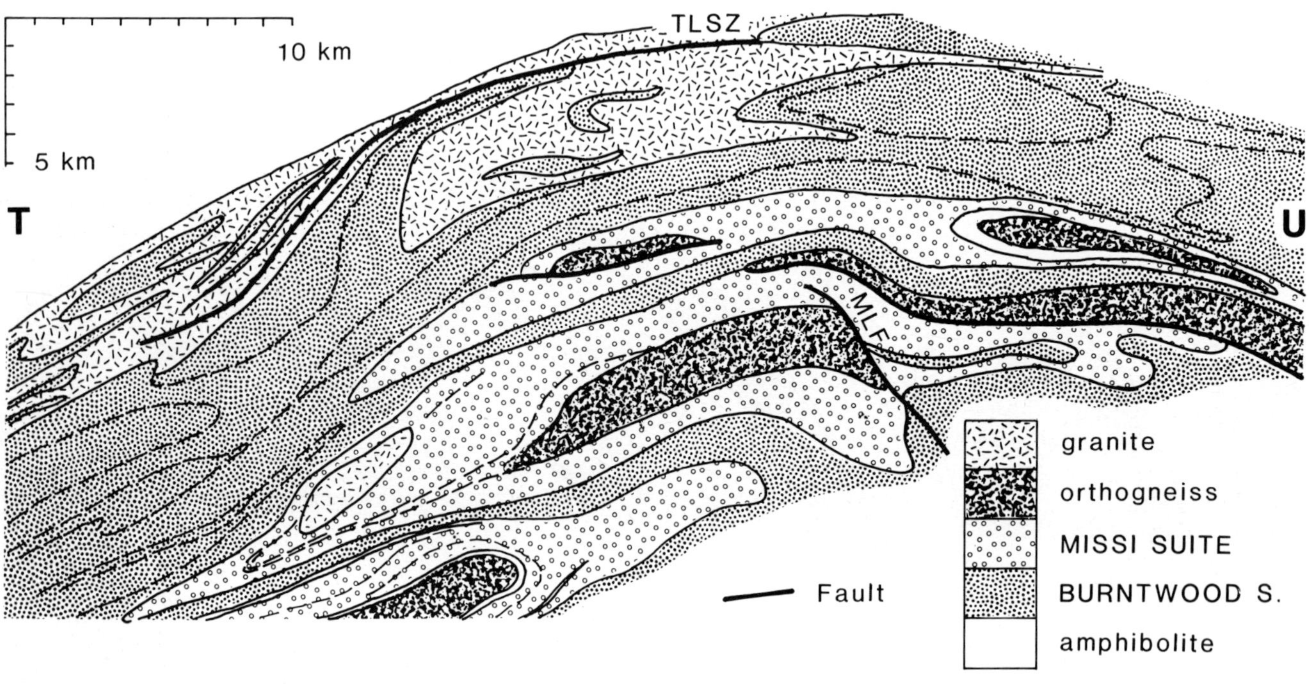

Figure 18. *Down-plunge structure section near Takipi Lake (line T–U in Figure 14) showing sheath-like recumbent F_2–F_3 folds with cores of orthogneiss and mantles of Missi Suite gneisses. Abbreviations: TLSZ, Takipi Lake shear zone; MLF, Molly Lake fault.*

Blastomylonites with downdip lineations are common on the margins of orthogneiss bodies.

A simplified, unbalanced, composite structure section drawn across the centre of the Kisseynew belt illustrates the geometry on a crustal scale (Fig. 19). Where possible, the section was drawn by projecting the map view along the northwest–southeast F_3 plunge. This technique utilizes the 15 km structural relief of the curved F_3 hinge lines. The figure incorporates data up to 30 km from the line of section and the projection was along curved lines; consequently, it is diagrammatic.

The infrastructure is interpreted to consist largely of granitoid orthogneisses. In the southern half of the section, the orthogneisses have infolded slivers of amphibolite derived from Amisk Group volcanic rocks and are mantled with recumbently folded Burntwood and Missi gneisses. The largest structures are interpreted as F_1–F_3 nappes. In the northern half of the section, the orthogneisses extend from the Lynn Lake and Leaf Rapids domains towards the south and, with their discontinuous mantles of amphibolite, are exposed in domal culminations along the large anticline north of the GLSZ. These infrastructural gneisses are overlain by the Sickle Suite in a north-verging, synclinal F_3 nappe that has trapped rocks with well-preserved primary structures in its core (Fig. 16c). Locally, the early foliation (S_1–S_2) that had become south dipping was overprinted by nearly horizontal S_3 foliation. The inverted south limb of the nappe contains the GLSZ, with its F_1 thrust slices of the Burntwood Suite basin-margin facies and well-preserved turbidites. The F_3 anticlinal nappe that overrode the syncline contains the Burntwood Suite migmatites. Much of the upper nappe was eroded, but there may have been 15 km of structural overlap. High-grade, south-dipping migmatites, to the south of this, suggest that metamorphic isograds were inverted.

The composite section indicates further that the migmatites in the central belt were structurally thickened and driven over the rocks on both flanks of the Kisseynew gneiss belt. Rocks from the deepest structural level are on the south margin of the central migmatite belt, in the hanging wall of shallow-dipping, high-strain zones. They comprise the highest grade migmatites (diatexites), abundant sheets of leucogranite, and bodies of enderbite. The transition to upright structures north, across the main fan structure, is interpreted to be a result of deeper levels of exposure in the south than in the north. The infrastructure in the core zone probably rode on the back of lower nappes during its displacement to the southwest. A zone of slightly lower grade rocks comprising Burntwood gneisses with staurolite, and Missi rocks with preserved crossbedding was trapped along the south flank of the Kisseynew belt. This zone forms the downward-diverging fan structure that straddles the KRF area. It may be a large F_3 synform between the recumbent folds in the Kisseynew belt and the upright folds on the margin of the Flin Flon belt.

DISCUSSION AND CONCLUSION

Lithologic suites in the Kisseynew belt were derived by high-grade metamorphism of sedimentary, intrusive, and local volcanic units that are closely related to rocks in adjacent granitic and volcanic belts. Each suite is interpreted to represent a step in the 100 Ma evolution of the Early Proterozoic crust in the Trans-Hudson Orogen from an oceanic island arc to a postcollision continental environment.

Orthogneisses, which occupy large domes and sheath folds on the flanks of the Kisseynew belt, are equivalent to lower grade (1830–1890 Ma) intrusions in the adjacent volcanic belts. Screens of supracrustal gneisses are equivalent to the (1880–1910 Ma) volcanic and sedimentary rocks in those belts. The Sherridon gneisses have an uncertain origin and are interpreted to occupy a sheath fold that provides a window through the orthogneisses. Together, these rocks represent a massif, constructed in several superimposed magmatic arcs. Long-range structural projections suggest that the orthogneisses form the infrastructure to the predominantly sedimentary gneisses throughout much of the Kisseynew belt.

Paragneisses of the Burntwood Suite are widespread throughout the Kisseynew belt in Manitoba and Saskatchewan; they represent turbidites deposited in a deep-water marine environment that extended onto the north margin of the Flin Flon volcanic belt. Mafic–ultramafic volcanism and possible normal faulting on the north flank of the Kisseynew belt are interpreted to postdate the main volcanism in the Lynn Lake belt, but may be coeval with late-stage island-arc volcanism in the Flin Flon belt. This scenario is consistent with an interpretation of the Kisseynew belt as an interarc basin formed by rifting of the Lynn Lake arc massif. However, similar gneissic belts of Archean age have been interpreted by some workers as accretionary complexes (e.g., Percival, 1989). The various models may be tested using single zircon dating.

The Sickle Suite alluvial deposits indicate a crustal buildup to continental thickness during pre-Sickle deformation, but the onset time of closure of the Kisseynew basin is unknown; it was probably no earlier than the 1876–1878 Ma volcanism in the Leaf Rapids domain and intrusion of the Pool Lake intrusive suite. Sickle-age sediments were transported south and deposited, probably at a

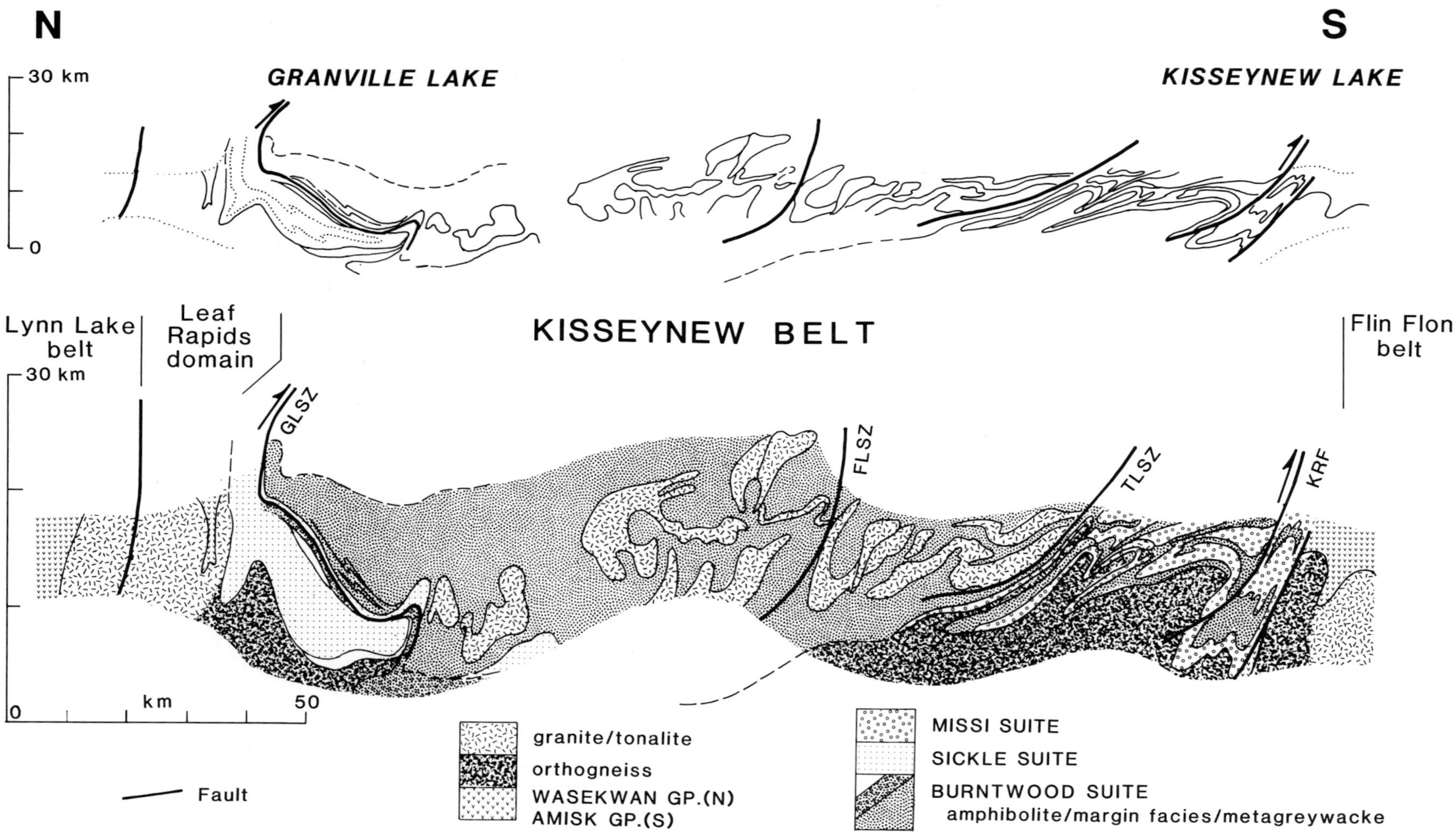

Figure 19. *Schematic structure cross-section of the Kisseynew belt (line N–S in Figures 2 and 14) determined from long-range projections parallel to southeast-trending folds (F_3), using structural relief provided by cross folds (F_4). The section is oblique, perpendicular to low-angle plunge at the base, and moderate plunge higher up. The line drawing shows the true thickness; the patterned version has a ×2 vertical exaggeration.*

mountain front, in a continental arc-related basin. The source of granitic, felsic volcanic, and sedimentary detritus may have been an older felsic cover of the (1855 Ma) Wathaman–Chipewyan batholithic belt, which is the core zone of the continental arc.

The Missi Suite is interpreted as the product of younger, superimposed continental-arc sedimentation and volcanism. It formed a northward-fining, predominantly clastic deposit, eroded from, and shed over, the well-dissected, deformed Flin Flon arc and the Kisseynew belt south flank. The (1830 Ma) Touchbourne enderbites and numerous intrusions in the Flin Flon belt are part of the Missi age magmatism.

Main deformation and peak metamorphism occurred 15–25 Ma after Missi volcanism. The dominant structure is interpreted to reflect late-collisional north–south convergence and uplift between the Hearne and Superior provinces. The divergently overriding nappes on the north and south flanks probably resulted from melt-enhanced deformation with inversion of metamorphic isograds (Hollister and Crawford, 1986). Syntectonic (1810–1820 Ma) S-type plutons and migmatitic neosome occupy gently dipping sheets at deep structural levels and form steep-sided bodies at higher levels. At the outer margin of the Kisseynew belt, the slightly lower grade mantle of the migmatite complex, with preserved sedimentary structures and high-level fault zones (GLSZ, KRF), appears to be trapped in large synforms. These structures are marked by downward-diverging foliation fans.

Crustal shortening produced a structural style related to wedge tectonics in Phanerozoic collision orogens (Price, 1986). The infrastructure of orthogneisses and their attenuated supracrustal cover appears to have been inserted under the central migmatite belt (Fig. 20). The orthogneisses are continuous with lower structural levels in the adjacent volcanic belts, and are probably equivalent to the plutons and narrow volcanic belts in the Glennie Lake Domain, Saskatchewan. Allochthonous sheets, such as those described in the Glennie Lake Domain (Chiarenzelli *et al.*, 1987), may be present at lower crustal levels, but underthrust Archean rocks, as in Saskatchewan, are not indicated by Sr isotopic data in Manitoba, except near the Churchill–Superior boundary.

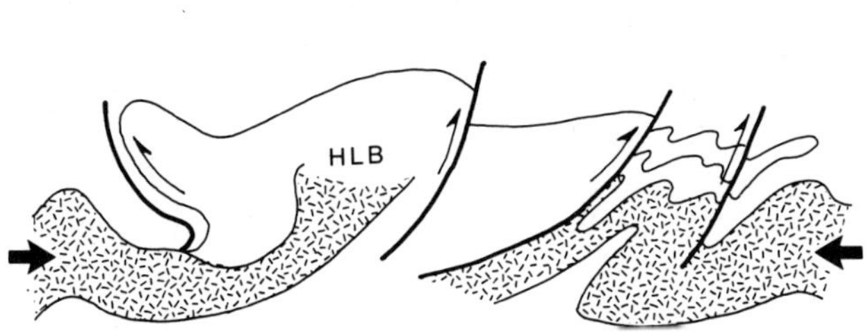

Figure 20. *Model for structural wedging of the Kisseynew belt in which competent orthogneisses and screens of early metavolcanic rocks in the infrastructure (patterned) were inserted beneath the melt-softened sedimentary suprastructure to the north. The northern infrastructure was pried up and remobilized in the Highrock Lake batholith (HLB), and the north flank and GLSZ were back folded.*

Late-metamorphic deformation involved northeast–southwest compression, producing late upright folds and northerly trending faults with east-side-up and probably sinistral displacement. Identical structures in the Churchill–Superior Boundary Zone are interpreted to represent the last stages of oblique collision between the Kisseynew belt and the Superior craton (Bleeker, this volume).

ACKNOWLEDGEMENTS

The Kisseynew belt in Manitoba has been mapped during the last 20 years by Manitoba Energy and Mines to which I am indebted for most of the published and unpublished information used to compile this paper. In particular, I wish to thank D.A. Baldwin, M.T. Corkery, P.G. Lenton, W.D. McRitchie, D.C.P. Schledewitz, W. Weber, E. Froese, and T.M. Gordon. Any conclusions drawn from their work are my responsibility. R. Macdonald and an anonymous reader are thanked for their suggested revisions of the manuscript.

REFERENCES

Albino, K.C. and Macek, J.J., 1981, Thompson Nickel Belt Project – Setting Lake: Manitoba Energy and Mines, Mineral Resources Division, Report of Field Activities 1981, p. 30–36.

Ashton, K.E. and Froese, E., 1988, Could the Sherridon Group at Sherridon be a high-grade equivalent of the Amisk Group?: Manitoba Energy and Mines, Minerals Division, Report of Field Activities 1988, p. 183–184.

Ashton, K.E. and Wheatley, K.J., 1986, Preliminary report on the Kisseynew gneisses in the Kisseynew–Wildnest lakes area, Saskatchewan: Geological Survey of Canada, Paper 86-1B, p. 305–317.

Ashton, K.E., Wilcox, K.H., Wheatley, K.J., Paul, D. and deTombe, J., 1987, The boundary zone between the Flin Flon domain, Kisseynew gneisses and Hanson Lake block, northern Saskatchewan: Saskatchewan Geological Survey, Miscellaneous Report 87-4, p. 131–135.

Bailes, A.H., 1975, Geology of the Guay–Wimapedi lakes area: Manitoba Department of Mines, Resources and Environmental Management, Mineral Resources Division, Geological Services Branch Publication 75-2.

Bailes, A.H., 1980, Geology of the File Lake area: Manitoba Department of Energy and Mines, Mineral Resources Division, Geological Report GR78-1.

Bailes, A.H., 1985, Geology of the Saw Lake area: Manitoba Energy and Mines, Geological Report GR83-2.

Bailes, A.H. and McRitchie, W.D., 1978, The transition from low to high grade metamorphism in the Kisseynew sedimentary gneiss belt, Manitoba, in Fraser, J.A. and Heywood, W.W., Metamorphism of the Canadian Shield: Geological Survey of Canada, Paper 78-10, p. 155–178.

Bailes, A.H. and Syme, E.C., 1983, Flin Flon – Schist Lake Project: Manitoba Department of Energy and Mines, Mineral Resources Division, Report of Field Activities 1983, p. 23–29.

Bailes, A.H. and Syme, E.C., 1989, Geology of the Flin Flon – White Lake area: Manitoba Energy and Mines, Geological Services, Geological Report GR87-1.

Baldwin, D.A., 1974, The geology of the Kadeniuk Lake area: Manitoba Department of Mines, Resources and Environmental Management, Mineral Resources Division, Geological Paper 2/74, p. 10–12.

Baldwin, D.A., 1987, Physical volcanology of the northwest segment of the Karsakuwigamak block, Proterozoic Rusty Lake metavolcanic belt, northern Manitoba, unpublished Ph.D. thesis, University of Manitoba.

Baldwin, D.A., Frohlinger, T.G., Kendrick, G., McRitchie, W.D. and Zwanzig, H.V., 1979, Geology of the Nelson House Pukatawagan region (Burntwood Project): Manitoba Department of Mines, Natural Resources and Environment, Mineral Resources Division, Geological Maps 78-3-1 to 78-3-22.

Baldwin, D.A., Syme, E.C., Zwanzig, H.V., Gordon, T.M., Hunt, P.A. and Stevens, R.P., 1987, U–Pb zircon ages from the Lynn Lake and Rusty Lake metavolcanic belts, Manitoba: two ages of Proterozoic magmatism: Canadian Journal of Earth Sciences, v. 24, p. 1053–1063.

Bickford, M.E., Van Schmus, W.R., Macdonald, R., Lewry, J.F. and Pearson, J.G., 1986, U–Pb zircon geochronology project for the Trans-Hudson Orogen: current sampling and recent results: Saskatchewan Geological Survey, Miscellaneous Report 86-4, p. 101–107.

Bleeker, W., 1990, New structural–metamorphic constraints on Early Proterozoic oblique collision along the Thompson Nickel Belt, Manitoba, Canada, in Lewry, J.F. and Stauffer, M.R., eds., The Early Proterozoic Trans-Hudson Orogen of North America: Geological Association of Canada, Special Paper 37, p. 57–73.

Brooks, C. and Theyer, P., 1981, Rb/Sr geochronology in the Thompson belt, Manitoba: implications for Aphebian crustal development and metallogenesis: Canadian Journal of Earth Sciences, v. 18, p. 932–943.

Campbell, F.H.A., 1972, Stratigraphic and structural studies in the Granville Lake – Lynn Lake region: Manitoba Department of Mines, Resources and Environmental Management, Publication 71-2A.

Chauvel, C., Arnt, N.T., Kielinzcuk, S. and Thom, A., 1986, Formation of Canadian 1.9 Ga old continental crust. I. Nd isotopic data: Canadian Journal of Earth Sciences, v. 24, p. 396–406.

Chiarenzelli, J.R., Lewry, J.F. and Landon, M., 1987, Bedrock geology, Iskwatikan Lake area: evidence for Hudsonian juxtaposition of Proterozoic and Archean rocks along a ductile detachment surface: Saskatchewan Geological Survey, Summary of Investigations 1987, p. 46–51.

Clark, G.S., 1981, Rubidium–strontium geochronology of major tectonic domains in the Churchill structural province, northern Manitoba, a preliminary report: Centre for Precambrian Studies, University of Manitoba, Annual Report for 1980, Paper No. 12, p. 22–28.

Cranstone, D.A., 1968, Geology of the Watt Lake area (east half): Manitoba Department of Mines and Natural Resources, Mines Branch, Publication 61-5.

Corkery, M.T., 1985, Geology of the Lower Nelson River Project area: Manitoba Energy and Mines, Geological Services, Geological Report GR82-1.

Delaney, G.D., 1987, Bedrock geological mapping, Carruthers–Uskik lakes area: Saskatchewan Geological Survey, Miscellaneous Report 87-4, p. 8–17.

Dickenson, W.R., Beard, L.S., Brakenridge, G.R., Erjavec, J.L., Ferguson, R.C., Inman, K.F., Knepp, R.A., Lindberg, F.A. and Ryberg, P.T., 1983, Provenance of North American Phanerozoic sandstones in relation to tectonic setting: Geological Society of American Bulletin, v. 94, p. 222–235.

Froese, E. and Goetz, P.A., 1981, Geology of the Sherridon Group in the vicinity of Sherridon, Manitoba: Geological Survey of Canada, Paper 80-21.

Froese, E. and Moore, J.M., 1980, Metamorphism in the Snow Lake area, Manitoba: Geological Survey of Canada, Paper 78-27.

Geological Society of America, 1987, Gravity anomaly map of North America, northwest sheet 1: Decade of North American Geology publication.

Ghosh, S.K., 1974, Strain distribution in superposed buckling folds and the problem of reorientation of early lineations: Tectonophysics, v. 21, p. 249–272.

Gibb, R.A., 1968, A geological interpretation of the Bouguer anomalies adjacent to the Churchill–Superior boundary in northern Manitoba: Canadian Journal of Earth Sciences, v. 5, p. 439–453.

Gilbert, H.P., Syme, E.C. and Zwanzig, H.V., 1980, Geology of the metavolcanic and volcaniclastic metasedimentary rocks in the Lynn Lake area: Manitoba Department of Energy and Mines, Mineral Resources Division, Geological Paper GP80-1.

Gilboy, C.F., 1980, Geology of the Reindeer Lake South (southeast) area: Saskatchewan Geological Survey, Saskatchewan Mineral Resources, Report 198.

Gordon, T.M. and Gall, Q., 1982, Metamorphism in the Crowduck Bay area, Manitoba: Geological Survey of Canada, Paper 82-1A, p. 197–201.

Gordon, T.M. and Lemkow, D.R., 1987, Geochemistry of the Missi Group volcanic rocks, Wekusko Lake, Manitoba: Geological Survey of Canada, Open File Report 1442.

Gordon, T.M., Hunt, P.A., Loveridge, W.D., Bailes, A.H. and Syme, E.C., 1987, U–Pb zircon ages from the Flin Flon and Kisseynew belts, Manitoba: chronology of early Proterozoic crust formation: Geological Association of Canada – Mineralogical Association of Canada, Program with Abstracts, v. 12, p. 48.

Gordon, T.M., Hunt, P.A., Bailes, A.H. and Syme, E.C., 1990, U–Pb ages from the Flin Flon and Kisseynew belts, Manitoba: chronology of crust formation at an Early Proterozoic accretionary margin, in Lewry, J.F. and Stauffer, M.R., eds., The Early Proterozoic Trans-Hudson Orogen of North America: Geological Association of Canada, Special Paper 37, p. 177–199.

Green, A.G., Hajnal, Z. and Weber, W., 1985, An evolutionary model of the western Churchill Province and western margin of the Superior Province of Canada and the north-central United States: Tectonophysics, v. 116, p. 281–322.

Halden, N.M., 1988, Tectonic setting of Circum-Superior ultramafic and mafic volcanism in Manitoba: Manitoba Energy and Mines, Minerals Division, Report of Field Activities 1988, p. 116–123.

Halden, N.M., Clark, G.S., Corkery, M.T., Lenton, P.G. and Schledewitz, D.C.P., 1990, Trace-element and Rb–Sr whole-rock isotopic constraints on the origin of the Chipewyan, Thorsteinson, and Baldock batholiths, Churchill Province, Manitoba, in Lewry, J.F. and Stauffer, M.R., eds., The Early Proterozoic Trans-Hudson Orogen of North America: Geological Association of Canada, Special Paper 37, p. 201–214.

Hoffman, P.F., 1981, Autopsy of Athapapuskow aulacogen: a failed arm affected by three collisions, in Campbell, F.H.A., ed., Proterozoic Basins of Canada: Geological Survey of Canada, Paper 81-10, p. 97–102.

Hollister, L.S. and Crawford, M.L., 1986, Melt-enhanced deformation: a major tectonic process: Geology, v. 14, p. 558–561.

Hunt, P.A. and Zwanzig, H.V., 1990, Pre-Missi granitoid domes in the Puffy Lake area, Kisseynew gneiss belt, Manitoba, in Radiogenic Age and Isotopic Studies. Report 3: Geological Survey of Canada, Paper 89-2, p. 71–76.

Jackson, S.L., 1988, Metamorphism and structure of the Laurie Lake area, northern Manitoba, unpublished Ph.D. thesis, Queen's University.

Kornik, L.J., 1968, Guay Lake area (west half): Manitoba Department of Mines and Natural Resources, Mines Branch, Publication 64–6.

Lenton, P.G., 1981, Geology of the McKnight–McCallum lakes area: Manitoba Department of Energy and Mines, Mineral Resources Division, Geological Report GR79-1.

Lenton, P.G. and Corkery, M.T., 1981, The lower Churchill River Project (interim report): Manitoba Department of Energy and Mines, Mineral Resources Division, Open File Report OF81-3.

Lewry, J.F., 1981, Lower Proterozoic arc–microcontinent collisional tectonics in the western Churchill Province: Natura, v. 294, p. 69–72.

Lewry, J.F., 1983, Character and structural relations of the 'McLennan Group' meta-arkoses, McLennan–Jaysmith lakes area: Saskatchewan Geological Survey, Miscellaneous Report 83-4, p. 49–55.

Lewry, J.F., Stauffer, M.R. and Fumerton, S., 1981, A Cordilleran-type batholithic belt in the Churchill Province in northern Saskatchewan: Precambrian Research, v. 14, p. 277–313.

Lewry, J.F., Macdonald, R., Livesey, C., Meyer, M., Van Schmus, R. and Bickford, M.E., 1987, U–Pb geochronology of accreted terranes in the Trans-Hudson Orogen, northern Saskatchewan, Canada, in Pharaoh, T.C., Beckinsale, R.D. and Rickard, D., eds., Geochemistry and Mineralization of Proterozoic Volcanic Suites: Geological Society Special Publication No. 33, p. 147–166.

Macdonald, R., 1980, New edition of the geological map of Saskatchewan, Precambrian Shield area: Saskatchewan Geological Survey, Miscellaneous Report 80-4, p. 19–21.

Manitoba Energy and Mines, 1986a, Geological compilation map series, Granville Lake, NTS 64C, 1:250,000 scale, with descriptive notes.

Manitoba Energy and Mines, 1986b, Geological compilation map series. Uhlman Lake, NTS 64B, 1:250,000 scale, with descriptive notes.

Manitoba Energy and Mines, 1988, Geological compilation map, preliminary edition, Kississing, NTS 63N, 1:250,000 scale.

Mukherjee, A.C., 1974, Some aspects of sedimentology of the Missi Group and its environment of deposition: Canadian Journal of Earth Sciences, v. 11, p. 1018–1019.

Norman, G.W.H., 1934, Granville Lake district, northern Manitoba: Geological Survey of Canada, Summary Report 1933, p. 23–41.

North American Commission on Stratigraphic Nomenclature, 1983, North American Stratigraphic Code: American Association of Petroleum Geologists Bulletin, v. 67, p. 841–875.

Ostry, G., 1986, Mineral investigations in the Kisseynew gneiss terrain: Manitoba Energy and Mines, Mineral Division, Report of Field Activities 1986, p. 100–106.

Percival, J.A., 1989, A regional perspective of the Quetico metasedimentary belt, Superior Province, Canada: Canadian Journal of Earth Sciences, v. 26, p. 677–693.

Pollock, G.D., 1965, Geology of the Russick Lake area: Manitoba Department of Mines and Natural Resources, Mines Branch, Publication 63-2.

Price, R.A., 1986, The southeastern Canadian Cordillera: thrust faulting, tectonic wedging, and delamination of the lithosphere, in Platt, J.P., Coward, M.P., Deramond, J. and Hossack, J., eds., Thrusting and Deformation: Journal of Structural Geology, v. 8, p. 239–254.

Robertson, D.S., 1953, Batty Lake map-area, Manitoba: Geological Survey of Canada, Memoir 271.

Roser, B.P. and Korsch, R.J., 1986, Determination of tectonic setting of sandstone–mudstone suites using SiO_2 contents and K_2O/N_2O ratio: Journal of Geology, v. 94, p. 635–650.

Schledewitz, D.C.P., 1972, Geology of the Rat Lake area: Manitoba Department of Mines, Resources and Environmental Management, Mines Branch, Publication 71-2B.

Schledewitz, D.C.P., 1987, Kisseynew Project: Kississing Lake: Manitoba Energy and Mines, Minerals Division, Report of Field Activities 1987, p. 51–54.

Scoates, R.F.J., 1981, Volcanic rocks of the Fox River belt, northeastern Manitoba: Manitoba Department of Energy and Mines, Mineral Resources Division, Geological Report GR81-1.

Skjernaa, L., 1975, Experiments in superimposed buckle folding: Tectonophysics, v. 27, p. 255–270.

Stauffer, M.R., 1984, Manikewan: an Early Proterozoic ocean in central Canada, its igneous history and orogenic closure: Precambrian Research, v. 25, p. 257–281.

Stauffer, M.R., Mukherjee, A.C. and Koo, J., 1975, The Amisk Group: an Aphebian island arc deposit: Canadian Journal of Earth Sciences, v. 12, p. 2021–2035.

Stockwell, C.H., 1937, Gold deposits of the Herb Lake area, northern Manitoba: Geological Survey of Canada, Memoir 208.

Syme, E.C., 1985, Geochemistry of metavolcanic rocks in the Lynn Lake belt: Manitoba Energy and Mines, Geological Report GR84-1.

Syme, E.C., 1987, Athapapuskow Lake project: Manitoba Energy and Mines, Report of Field Activities 1987, p. 207–210.

Syme, E.C., 1990, Stratigraphy and geochemistry of the Lynn Lake and Flin Flon metavolcanic belts, Manitoba, in Lewry, J.F. and Stauffer, M.R., eds., The Early Proterozoic Trans-Hudson Orogen of North America: Geological Association of Canada, Special Paper 37, p. 143–161.

Syme, E.C., Bailes, A.H., Price, D.P. and Ziehlke, D.V., 1982, Flin Flon volcanic belt: geology and ore deposits at Flin Flon and Snow Lake, Manitoba: Geological Association of Canada – Mineralogical Association of Canada, Field Trip Guidebook No. 6.

Syme, E.C., Bailes, A.H., Gordon, T.M. and Hunt, P.A., 1987, U–Pb zircon geochronology in the Flin Flon belt: age of Amisk volcanism: Manitoba Energy and Mines, Minerals Division, Report of Field Activities 1987, p. 105–107.

Talbot, J.C., 1979, Fold trains in a glacier of salt in southern Iran: Journal of Structural Geology, v. 1, p. 5–18.

Tanner, P.W.G. and Macdonald, D.I.M., 1982, Models for the deposition and simple shear deformation of a turbidite sequence in the South Georgia portion of the southern Andes back-arc basin: Journal of the Geological Society of London, v. 139, p. 739–754.

Thomas, M.W., 1980, Structural geology of a gneissic terrain, Laurie Lake, Manitoba, unpublished M.Sc. thesis, University of Manitoba.

Van Schmus, W.R. and Schledewitz, D.C.P., 1986, U–Pb zircon geochronology of the Big Sand Lake area, northern Manitoba: Manitoba Energy and Mines, Report of Field Activities 1986, p. 207–210.

Van Schmus, W.R., Bickford, M.E., Lewry, J.F. and Macdonald, R., 1987, U–Pb geochronology in the Trans-Hudson Orogen, northern Saskatchewan, Canada: Canadian Journal of Earth Sciences, v. 24, p. 407–424.

Watters, B.R., 1984, Geochemical patterns for metavolcanic rocks of the La Ronge domain: Saskatchewan Geological Survey, Miscellaneous Report 84-4, p. 88–91.

Watters, B.R., 1985, Geochemistry of metavolcanic and plutonic rocks, Star Lake and Waddy Lake area: Saskatchewan Geological Survey, Miscellaneous Report 85-4, p. 28–34.

Weber, W., 1990, The Churchill–Superior Boundary Zone, southeast margin of the Trans-Hudson Orogen: a review, _in_ Lewry, J.F. and Stauffer, M.R., eds., The Early Proterozoic Trans-Hudson Orogen of North America: Geological Association of Canada, Special Paper 37, p. 41–55.

Zwanzig, H.V., 1981, Granville Lake Project: Manitoba Department of Energy and Mines, Mineral Resources Division, Report of Field Activities 1981, p. 17–21.

Zwanzig, H.V., 1984a, Kisseynew Project: Lobstick Narrows – Cleunion Lake, Puffy Lake and Nokomis Lake areas: Manitoba Energy and Mines, Mineral Resources Division, Report of Field Activities 1984, p. 38–45.

Zwanzig, H.V., 1984b, Early Proterozoic tectonics of the Kisseynew – Lynn Lake margin, Manitoba: Geological Association of Canada – Mineralogical Association of Canada, Program with Abstracts, v. 9, p. 118.

Zwanzig, H.V., 1985, Kisseynew Project: Lobstick Narrows – Cleunion Lake, Puffy Lake and Nokomis Lake areas: Manitoba Energy and Mines, Geological Services, Mines Branch, Report of Field Activities 1985, p. 51–53.

Zwanzig, H.V. and Lenton, P.G., 1987, Kisseynew Project: Limestone Point Lake – Star Lake areas: Manitoba Energy and Mines, Minerals Division, Report of Field Activities 1987, p. 55–64.

Zwanzig, H.V. and Wielezynski, P., 1975, Geology of the Kamuchawie Lake area: Manitoba Department of Mines, Resources and Environmental Management, Mineral Resources Division, Summary of Geological Field Work 1975, p. 12–15.

Zwanzig, H.V., Syme, E.C. and Gilbert, H.P., _in_ prep., Geology of the Sickle Group, upper Wasekwan Group metasedimentary rocks and Black Trout diorite in the Lynn Lake area: Manitoba Energy and Mines, Geological Report.

The Missi Formation: an Aphebian Molasse Deposit in the Reindeer Lake Zone of the Trans-Hudson Orogen, Canada

Mel R. Stauffer
University of Saskatchewan,
Saskatoon, Saskatchewan S7N 0W0, Canada

Stauffer, M.R., The Missi Formation: an Aphebian molasse deposit in the Reindeer Lake Zone of the Trans-Hudson Orogen, Canada, in Lewry, J.F. and Stauffer, M.R., eds., The Early Proterozoic Trans-Hudson Orogen of North America: Geological Association of Canada, Special Paper 37, p. 121–141.

Abstract

The Missi Formation consists of what have previously been called the Missi, Sherridon, Sickle, and McLennan groups, in the Flin Flon, Kisseynew, and La Ronge domains of the Reindeer Lake Zone of the Trans-Hudson Orogen in northern Saskatchewan and Manitoba. It is a thick, synorogenic molasse blanket that lies on a variety of Aphebian volcanic arc and sedimentary interarc materials over a present area of >60,000 km^2. On the north and south flanks of the outcrop area (Sickle and Missi groups respectively), it consists of coarse clastic fluvial rocks entirely. In the south-central portion (old Sherridon group), it includes not only coarse clastic rocks but also carbonates and mudstones (amphibolites) thought to be of marine or lacustrine origin, and both mafic and felsic volcanic rocks have been reported in some places.

In the western part of the Flin Flon Domain, the Missi Formation is divided into two tectonically controlled fining-upwards sequences, the Beaverdam and Cliff Lake members. On the basis of grain sizes and bed forms, each member is thought to represent the transition from proximal fan to braided stream conditions, with abrupt uplift between deposition of the members. The considerable amount of hematite (specularite to earthy) occurring in thin laminae in many of the sandstone beds suggests that the Missi Formation is a redbed.

Missi rocks, like those beneath them, have undergone polyphase deformation, resulting in a complex array of faults and, in the Flin Flon Domain, a sequence of at least three phases of superposed folds. Schistosity and/or lineation are locally well developed and finite strains range from so low as to have no visual affect on the rock to so high that deformed clasts give the rock nearly the appearance of banded gneiss.

The Missi Formation is thought to have been deposited in a compressional tectonic environment, mainly as a series of coalescing alluvial fans to braided stream wash, during the transition from island arcs to tectonic mountains brought about by the Hudsonian orogenic event.

Résumé

La formation de Missi est composée de ce qui a précédemment été appelé groupe de Missi, Sherridon, Sickle et McLennan dans les domaines Flin-Flon, Kisseynew et La Ronge de la zone Reindeer Lake de l'orogène Trans-Hudson au nord du Manitoba et de la Saskatchewan. Il s'agit d'une séquence épaisse de molasses syn-orogéniques recouvrant sur 60 000 km^2 une variété de matériaux d'arcs volcaniques et de sédiments interarcs aphébiens. Du côté nord et sud de la région (groupe de Sickle et Missi respectivement), il s'agit de roches clastiques fluviatiles grossières. Dans la partie sud centrale (groupe de Sherridon), il s'agit de roches clastiques grossières, de carbonates et de mudstones (amphibolites) considérées d'origine lacustre et marine et de volcanites acides et basiques signalées localement.

Dans la partie ouest du domaine Flin-Flon, la formation de Missi est divisée en deux séquences granoclassées controlées tectoniquement, les membres Beaverdam et Cliff Lake. En se basant sur la granulométrie et la forme des lits, chaque membre est considéré représentant la transition de delta proximal variant à des cours d'eau anastomosés avec une remontée abrupte entre la déposition de chaque membre. La quantité important d'hématite (de spécularite à hématite terreuse) qui apparaît en fine laminae dans beaucoup de lits grèseux suggère que la formation de Missi est un redbed.

Les roches de Missi et celles sous-jaçentes, ont subi une déformation polyphasée résultant en un système complexe de failles et dans le domaine Flin-Flon, en une séquence de trois phases de plis superposés. La schistosité et/ou les linéations sont localement bien développées; la déformation finie varie de faible (aucun effet visuel sur les roches) à fort (les cailloux déformés donnent à la roche l'aspect de gneiss rubannes).

On considère que la formation de Missi a été déposée dans un environnement tectonique compressif, principalement en une série de deltas alluvionnaires coalescents passant à des rivières anastomosées, pendant la transition arcs insulaires/formation de montagnes de l'orogénèse Hudsonienne.

INTRODUCTION

The word "missi" means "big" in the Cree language and was used by Bruce (1918) as the name of a series of metaconglomerates and metasandstones that he found in the western part of a greenstone belt now known as the Flin Flon Domain in the Reindeer Lake Zone of the Trans-Hudson Orogen (Fig. 1). The name was taken from Missi Island in Amisk Lake (Fig. 1d), which, however, contains no Missi rocks.

Most early workers used the name "Missi Series" (e.g., Alcock, 1923; Ambrose, 1936a,b; Tanton, 1941; Stockwell, 1946, 1960; Byers, 1953; Byers and Dahlstrom, 1954; Byers *et al.*, 1965). However, later workers (e.g., Bailes, 1971; Stauffer and Mukherjee, 1971; Chandler, 1980; Syme, 1987) used the name "Missi group." The term "group" was used informally because there has been no formal subdivision into formations. It is proposed here to formalize the name "Missi Formation" for what was previously designated "group" or "series," and toward this end, Missi rocks in the western part of the Flin Flon Domain are divided here into two distinct members ("Beaverdam Member" and the overlying "Little Cliff Member").

Missi rocks occur in three major, partly fault-bounded, structurally complex synclinal basins at the western end of the Flin Flon Domain (Fig. 1d), in a large complex structural basin at the eastern end of the Flin Flon Domain (Fig. 1c) (also see Shanks and Bailes (1977)), and in several small fault slices and/or synclinal patches not shown on Figure 1. Rocks probably correlative with the Missi occur elsewhere in the Reindeer Lake Zone (e.g., the Sherridon and Sickle groups in the Kisseynew Domain, Fig. 1c). Together, these seem to form a single (although complex) syntectonic molasse deposit, and it is proposed here to extend the name "Missi Formation" to all such rocks.

The Missi Formation in the Flin Flon Domain unconformably overlies the Amisk group of mainly metavolcanics and associated pretectonic intrusions. These and the Missi are intruded by a variety of syntectonic to nearly posttectonic, mainly granitoid intrusive bodies (Fig. 2). The Amisk group contains a high proportion of pillow lavas and is considered to be of island-arc origin (Bailes, 1971; Stauffer *et al.*, 1975; Walker and Watters, 1982; Stauffer, 1984). Missi rocks are considered to be fluvial deposits (Byers and Dahlstrom, 1954; Bailes, 1971; Stauffer and Mukherjee, 1971), and details are discussed later in this paper. Classification of Missi sandstones is made difficult by the pervasive presence of chlorite (locally biotite), some of which probably has been derived from the metamorphic breakdown of feldspar (Mukherjee, 1971), thus increasing the apparent original clay fraction. Nonetheless, most of the Missi material was probably deposited as sublithic arenite to subarkose, with minor quartz arenite and lithic to arkosic arenite beds (using the classification of Pettijohn *et al.* (1972)).

All rocks in the Flin Flon Domain are deformed; those of Missi age and earlier have undergone polyphase deformation (Stauffer and Mukherjee, 1971; Bailes and Syme, 1987). Also, metamorphism, primarily coeval with the P_2 deformation of Stauffer and Mukherjee (1971), has resulted in metamorphic grades ranging from subgreenschist at the southern end of the Athapap Basin (Fig. 1d) to amphibolite facies at the southern boundary of the Kisseynew Domain (e.g., Ambrose, 1936a; Stauffer and Mukherjee, 1971; Stauffer *et al.*, 1975). The grade also increases to amphibolite facies around the larger syntectonic granitoid plutons (e.g., Heywood, 1966; Froese, 1969; Longiaru, 1980).

A synthesis of Rb/Sr, U/Pb, and Pb/Pb dates indicates that Amisk volcanism and early intrusion occurred at about 1850–1890 Ma and that post-Missi, synorogenic intrusion and metamorphism culminated prior to 1800 Ma (Mukherjee *et al.*, 1971; Bell *et al.*, 1975; Sangster, 1978; Bell and Macdonald, 1982; Stauffer, 1984; Syme *et al.*, 1987). The time of deposition of the Missi Formation is

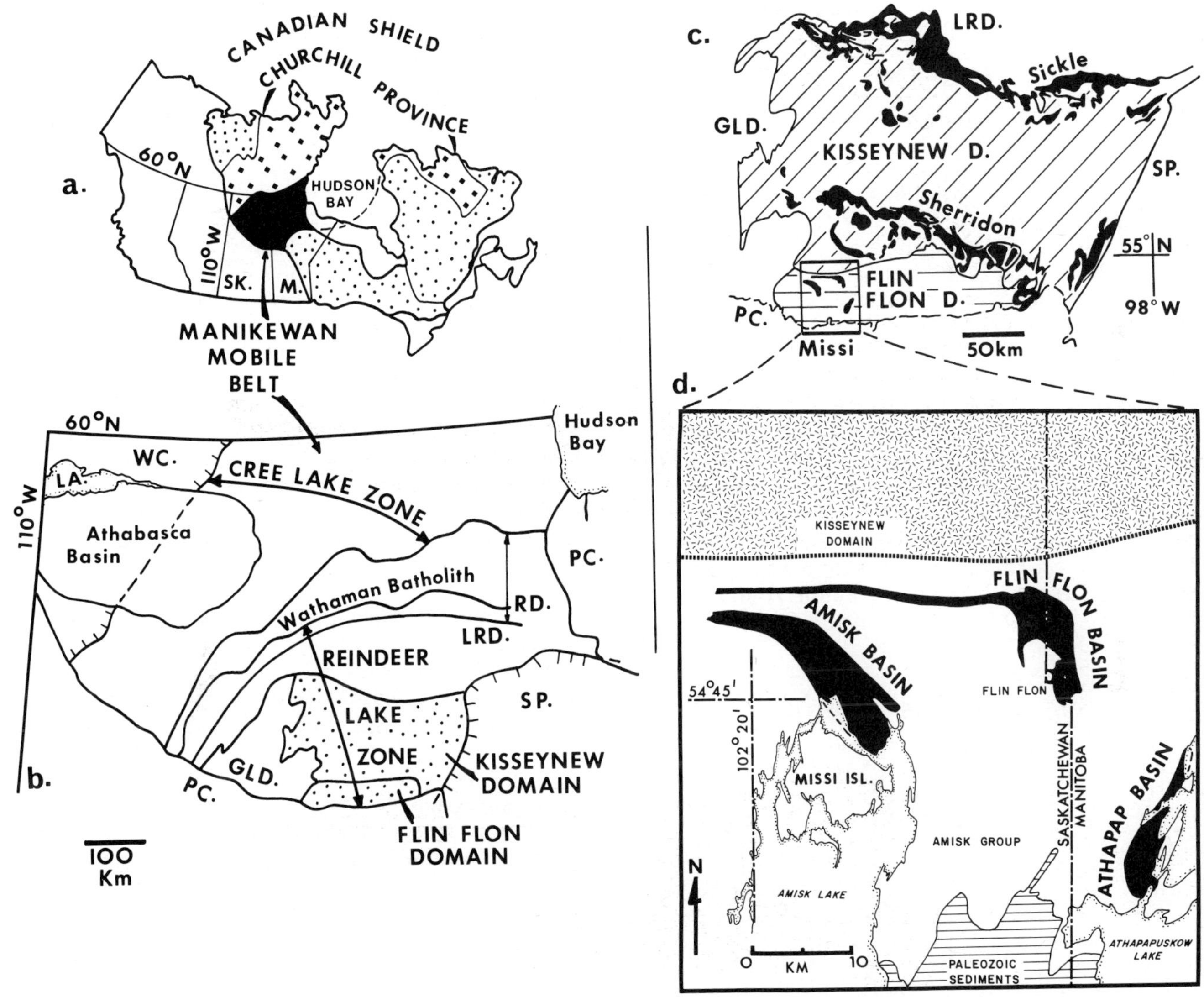

Figure 1. *(a) Location of Manikewan Mobile Belt in the southern part of the Churchill Province of the Canadian Shield. Abbreviations: SK, Saskatchewan and M, Manitoba. Modified from Stauffer (1984). (b) Subdivision of the Manikewan Mobile Belt into the miogeoclinal Cree Lake Zone and eugeoclinal Reindeer Lake Zone, which is further divided into the Rottenstone Domain (RD) (including the Wathaman Batholith), the La Ronge Domain (LRD), the Glennie Lake Domain (GLD), and the Kisseynew and Flin Flon domains (indicated by the stipple pattern), which are shown in more detail in (c). The Western Craton (WC) (recently renamed the Amer Lake Zone by Lewry* et al. *(1985)) and the Superior Province (SP) all contain Archean rocks and bound the Manikewan Mobile Belt. The Athabasca Basin contains essentially flat-lying clastics of Middle Proterozoic age, and flat-lying Phanerozoic rocks (PC) cover the mobile belt to the south and east. Abbreviation: LA, Lake Athabasca. Modified from many sources, especially Lewry and Sibbald (1977) and Stauffer (1984). (c) Distribution of Missi rocks in the Flin Flon Domain, correlative Sherridon rocks in the southern Kisseynew Domain, and similar Sickle rocks in the northern Kisseynew Domain. Compiled and modified from Macdonald and Broughton (1980), McRitchie* et al. *(1979), and Bailes (1971). (d) Locations of the Amisk, Flin Flon, and Athapap Missi basins in the western Flin Flon Domain.*

thus between 1800 and 1850 Ma, but given that a number of other events (such as pre-Missi erosion and post-Missi deformation) also occurred during this time span, the actual time available for Missi deposition in the Flin Flon Domain may be considerably less.

SUB-MISSI UNCONFORMITY (FLIN FLON DOMAIN)

Pre-Missi rocks commonly show evidence of weathering for as much as 10 m below the Missi contact. This evidence includes paleospheroidal weathering with well-developed "onion-skin" structure (Fig. 3a–c) and darkening of the rock between the spheroids or, in their absence, of the rock as a whole (this darkening is best seen in lower greenschist facies rocks). Such darkened rock represents a paleoregolith that has been metamorphosed along with the rest of the rocks so that it now appears essentially as a more highly altered material (mainly chlorite).

These products of weathering have been reworked so that the lower metre or so of the Missi Formation locally contains numerous clasts identical to the spheroids, some still displaying onion-skin texture. Also, in some places, the lower metre or less of Missi basal conglomerates has a matrix rich in dark, chlorite-rich, quartz-free material similar to and derived from the paleoregolith.

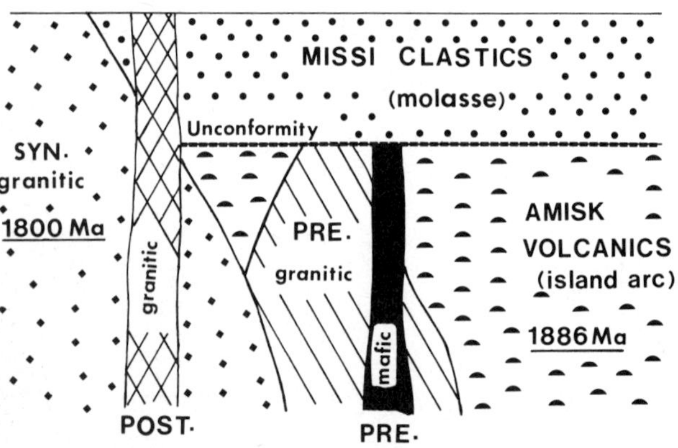

Figure 2. *Schematic section showing the sequence of major rock units in the western part of the Flin Flon Domain. Pre., syn., and post. refer to the time of intrusion relative to polyphase deformation of Amisk and Missi rocks.*

This evidence of pre-Missi *in situ* weathering is important because it shows that uplift of the Amisk submarine volcanic rocks and subaerial weathering began some significant period of time prior to the development of Missi-depositing streams. Just how long a period of time this was is indeterminable but, by comparison with modern weathering, it must have been on the order of at least several million years. Also, weathering would have occurred in areas of high elevation during deposition by Missi streams, the weathered rock to be later covered by laterally expanding and thickening Missi deposits.

There has been some debate as to whether the sub-Missi unconformity is an angular unconformity or merely a disconformity. Stockwell (1946, 1960), for example, suggested that the Missi overlies the Amisk with marked angular unconformity. Byers and Dahlstrom (1954) concede local discordance but disclaim the presence of angular unconformity. Stauffer and Mukherjee (1971) agreed with Byers and Dahlstrom because all phases of the polyphase tectonic deformation occur in both Amisk and Missi rocks. Part of the problem is that local discordance between Amisk and Missi rocks is explainable without major angular unconformity by several highly probable and/or observed phenomena: (1) local high initial dips in the Amisk rocks, (2) local buttress unconformities with initial dip to the Missi rocks, (3) disharmonic folding between the two rock units, and (4) local shearing and displacement along the contact.

Nonetheless, considerable discordance is recognizable, especially at the southwestern corner of the Flin Flon Basin (Figs. 4 and 5a) (also see Bailes and Syme (1983)) where over 2 km of Amisk volcanic rocks are truncated by basal Missi rocks in the core of an F_1 syncline. This truncation is unlikely to be a décollement due to disharmonic folding because (1) the Missi/Amisk contact in this region shows few signs of shearing; in fact, it displays some of the best-preserved spheroidally weathered zones; and (2) the Missi should be the more tightly folded as it is of weaker lithology and is better bedded. The discordance is much more likely to be a result of pre-Missi erosional removal of Amisk material that was undergoing early F_1 folding (Fig. 4c). Missi rocks were then deposited during continued F_1 folding, completing the formation of the Burley Lake F_1 syncline (Fig. 4d) (later refolded by F_2 folds).

It seems clear that considerable uplift, weathering, erosion, and some folding occurred prior to and

Figure 3. *(a–c) Field photographs of paleospheroidal weathering in a pre-Missi diabase dyke immediately beneath the Missi Formation. The intensity of weathering increases from (a) to (c). Note the onion-skin texture, especially in the upper left-hand corner of (b) and in the central part of (c). Also note the darkening of the rock between spheroids, especially in (c). The photograph was taken at the southwest end of the Flin Flon Basin (Fig. 1d). (d–e) Deformed Missi conglomerate. (d) S_2 stretched pebbles oblique to bedding in unit 4, Flin Flon Basin. (e) F_3 folded pebbles originally stretched in S_2, unit 1, Amisk Basin (same scale as (d)).*

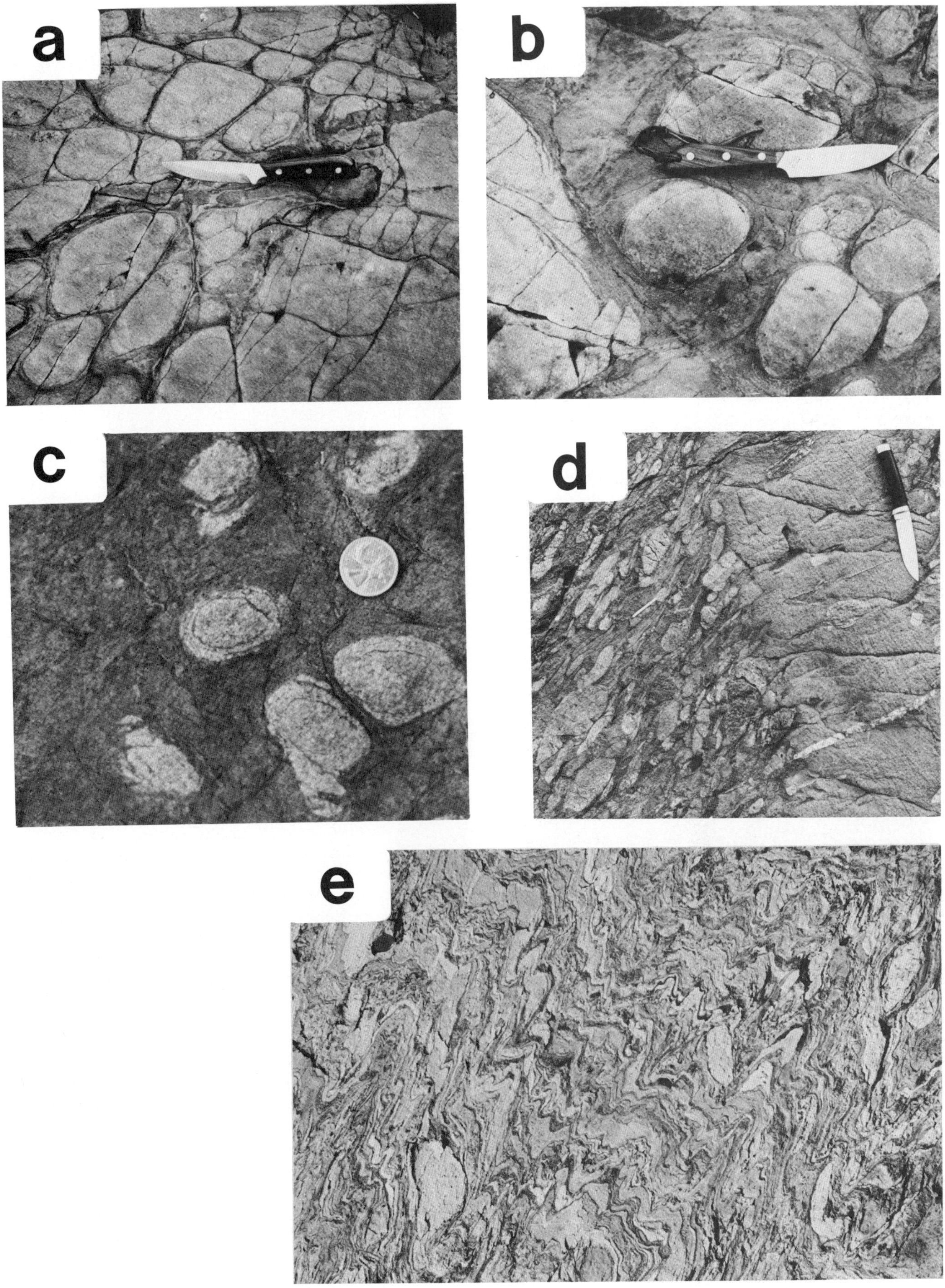
a
b
c
d
e

during Missi deposition; thus, the Missi is synorogenic.

STRUCTURE AND STRATIGRAPHY OF THE FLIN FLON BASIN

General

Missi rocks in the Flin Flon Basin (Fig. 1d) are the best exposed, the most easily accessible, stratigraphically the most complete, and the most studied in the orogen. For these reasons, the Flin Flon Basin is designated as the type area for the Missi Formation (despite Bruce's (1918) study on Amisk Lake) and is described here in detail.

The structural geology and preliminary stratigraphy were described by Stauffer and Mukherjee (1971), Stauffer (1974), and Stauffer and Burnett (1979). Three phases of large-scale folding, along

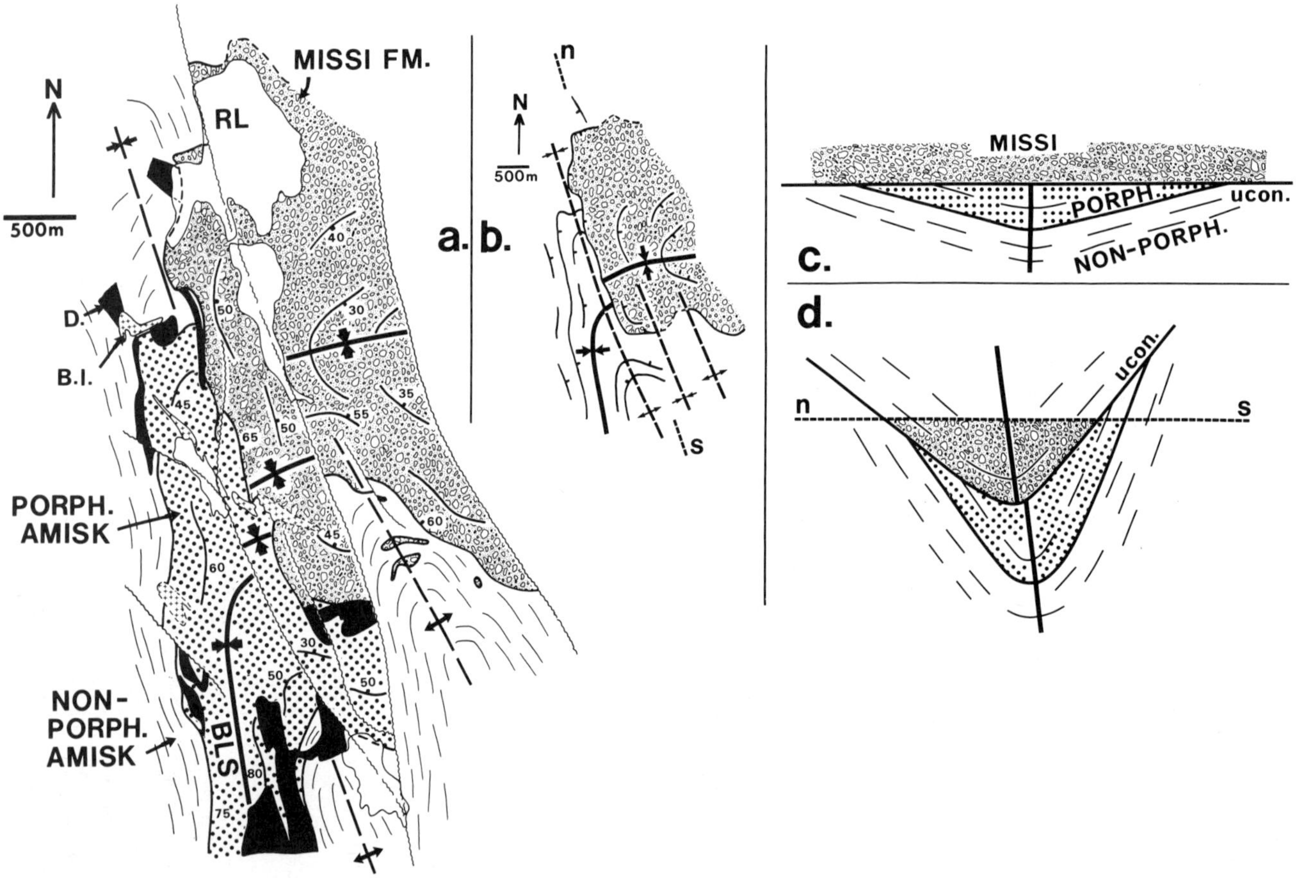

Figure 4. *(a) Geology at the southwestern corner of the Flin Flon Basin (location is shown on Fig. 5a), showing the continuation of the F_1 Burley Lake syncline (BLS) through the Missi/Amisk contact. The offset in the syncline's axial trace at the contact is probably due, in part, to its approximate positioning, but could be due mainly to the Amisk being folded prior to Missi deposition. Two major stratigraphic units are shown in the Amisk volcanics: a unit dominated by porphyritic mafic flows (plagioclase and hornblende phenocrysts) (dot pattern), and an underlying unit dominated by nonporphyritic mafic flows (short thin lines are parallel to the local strike). The Missi Formation is shown by the "gravelly" texture. Bed trend lines with local average dips are shown for both Amisk and Missi rocks (all beds shown are "right-way" up). Abbreviations: B.I., post-Missi boundary intrusions (gabbro–diorite mainly); D., pre-Missi diorite (solid black). Wiggly lines are faults; that running along Ross Lake (RL) is the Ross Lake fault. The long-dashed lines are axial traces of F_2 folds. The map is based on data from Stockwell (1960) and Stauffer and Mukherjee (1971), plus new mapping. (b) Simplified version of (a) as it might have appeared before faulting. Symbols as in (a). Line n–s shows the position of the cross-sections in (c) and (d). (c–d) Cross-sections along n–s (shown in (b)) showing the hypothetical evolution of the F_1 Burley Lake syncline, which may have begun before Missi deposition, but continued its development during erosion and Missi deposition.*

with complex faulting, have affected the rocks (Fig. 5a), with pronounced but locally variable amounts of flattening during F_2 folding, resulting in well-developed cleavage and lineation (Fig. 5b). Primary sedimentary features useful for determining the stratigraphic facing are well preserved; otherwise, neither the structure nor the stratigraphy could be worked out properly. Deformation has modified crossbedding angles enough to preclude their use as flow-direction indicators. Likewise, such deformation precludes the possibility of studying clast fabrics from a sedimentological point of view, for clast orientations now fluctuate about the tectonic cleavage, regardless of bed orientation (Stauffer and Burnett, 1979).

Stauffer and Mukherjee (1971) recognized two thick, fining-upwards, and maturing-upwards sequences in the Flin Flon Basin, each with a basal polymictic conglomerate unit. Many rocks contain considerable black hematite, a point of some interest that is discussed later, and most of the clasts in conglomerates and pebbly sandstones are of easily recognizable Amisk rock types or of pre-Missi plutonic material. Type sections have been measured through each of these sequences, although the sections had to be segmented because of the complex structure and locally poor outcrop (Fig. 6). By making frequent small "along-strike" jogs within each segment, better than 75% outcrop coverage was achieved for the sections, a situation far from obtainable in most parts of the Churchill Province. The two sequences are designated here as members, the lower one being named the "Beaverdam Member" and the upper one the "Little Cliff Member." Except for their lack of mudstones, these are similar in many respects to several members of the Tertiary Chuckanut Formation in

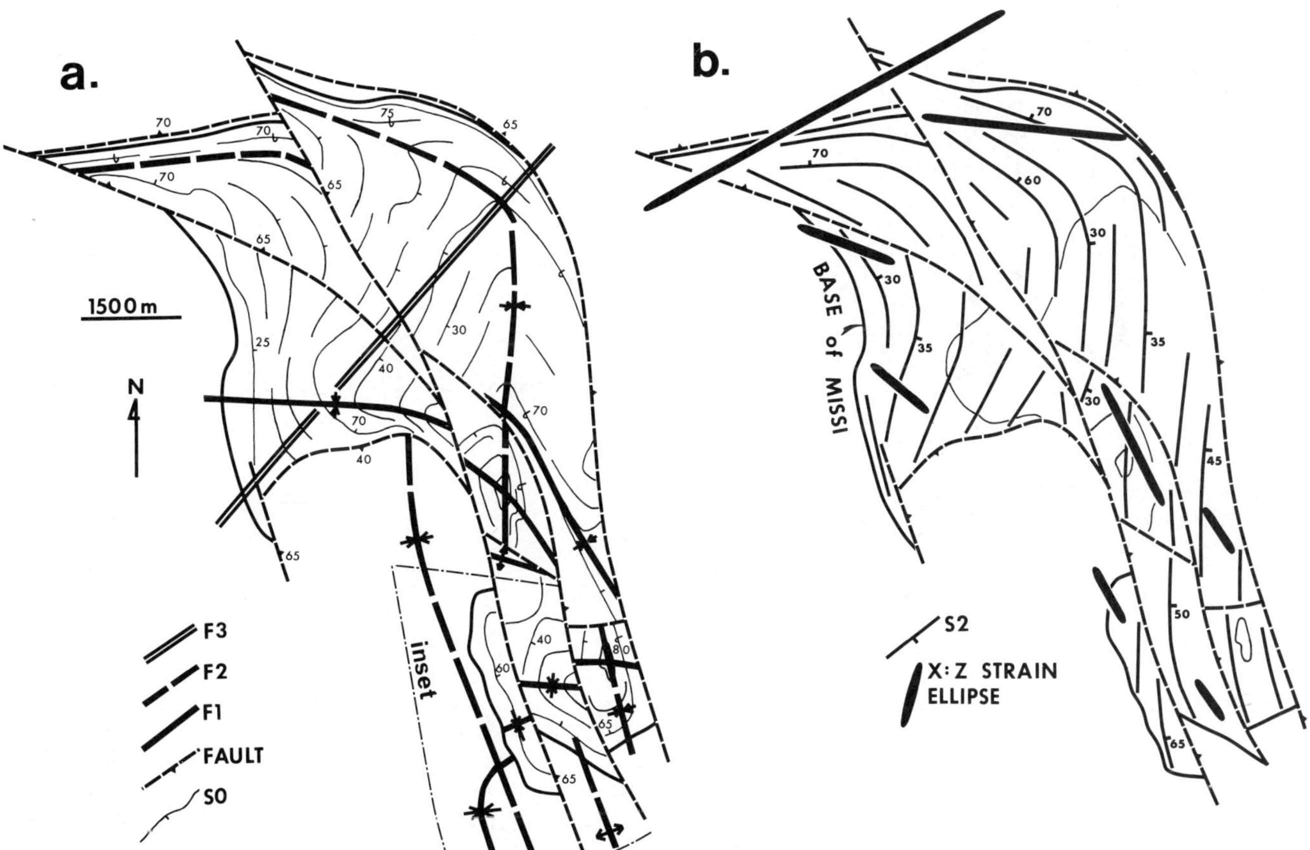

Figure 5. *Structure of the Flin Flon Basin. (a) Polyphase folds (F_1, F_2, F_3), faults, and bedding (S_0) trends. All folds are overturned synforms except for those obviously labelled as antiforms (F_2 folds are overturned to the west). All axial traces are approximate, especially that for the F_3 fold. Modified from Stauffer and Mukherjee (1971). The inset is the region shown in Figure 4. (b) P_2 foliation (S_2) trends and the X–Z strain ellipse ($X > Y > Z$). The long axes of the ellipses give the plunge direction of X, and plunges range from about 30 to 40° southeast, east, or north-northeast. Modified from Stauffer and Burnett (1979).*

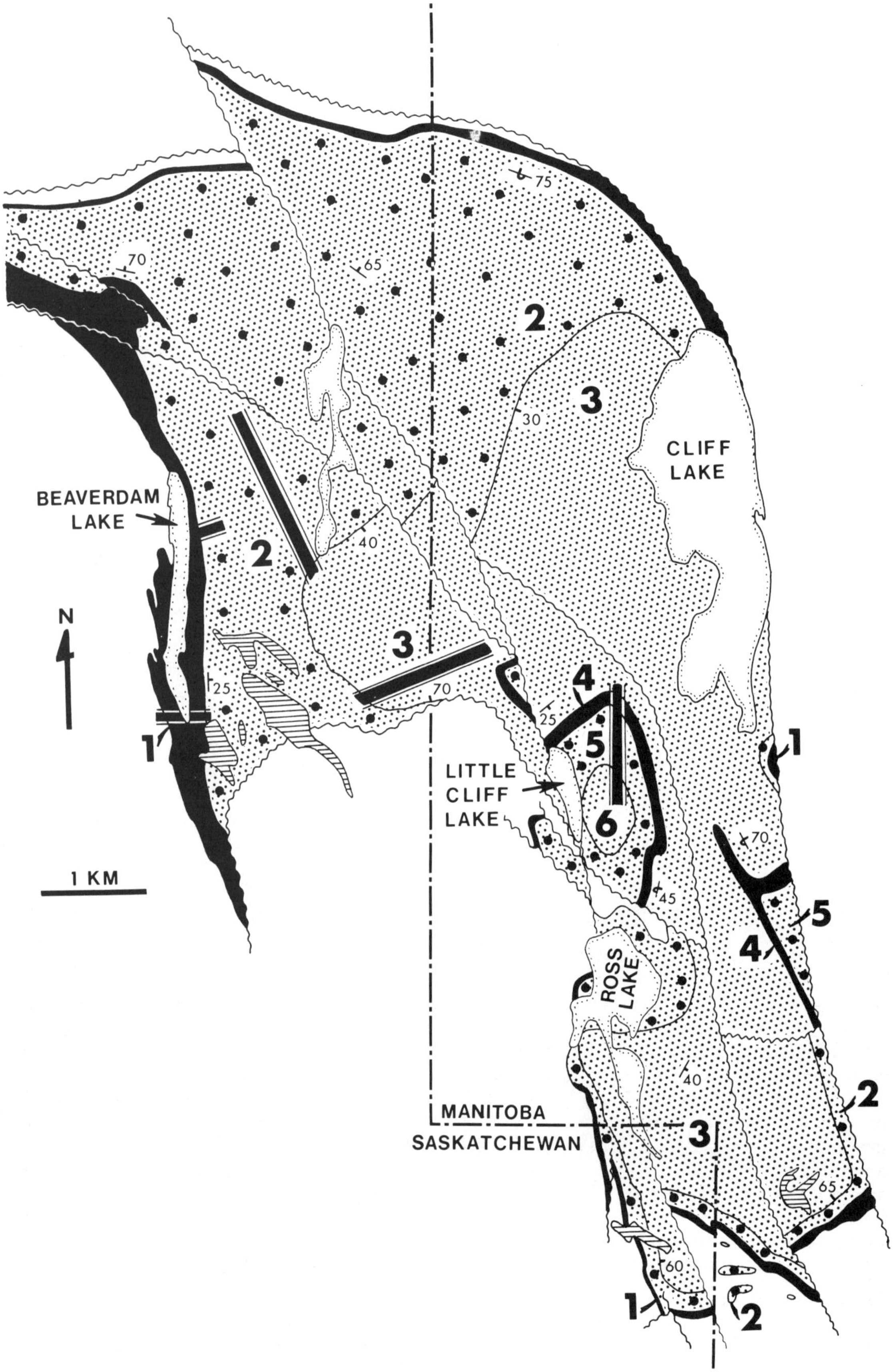
2
3
CLIFF
LAKE
BEAVERDAM
LAKE
2
N
3
4
1
5
1
LITTLE
CLIFF
LAKE
6
1 KM
5
4
ROSS
LAKE
2
MANITOBA
SASKATCHEWAN
3
1
2
75
70
65
30
40
70
25
25
70
45
40
65
60

Washington State, considered by Johnson (1984) to be alluvial fan to braided stream deposits.

Beaverdam Member

Despite its gradational character, the Beaverdam Member can be divided into three distinctive units: unit 1 – a basal, 0–200 m thick, polymictic conglomeratic unit considered here to be an alluvial fan or proximal braided-stream deposit; unit 2 – a gradational overlying unit of pebbly sandstones, several metres to over 600 m thick, considered to be of midfan to braided-stream origin; and unit 3 – an upper unit of clast-poor sandstones, up to 1300 m thick, that probably is a distal braided-stream deposit.

The generalized geology of the type section through the Beaverdam Member is given in Figure 7. This section is typical of the entire basin, although many details vary locally, especially the thickness, metamorphic grade, and amount of strain. The most important lithological features remain the same, i.e., poorly sorted, mainly cobble-rich, polymictic conglomerate at the base; increasing sandstone component upwards; a marked upward decrease in original clast diameter from an average of around 8 cm to an average of about 2 cm; a marked preponderance of mafic clasts near the base with an increasing proportion of quartz-bearing clasts upwards; and, although most sandstones are immature, a general increase in quartz content upwards. Also, crossbedding is rare in the lower parts of the Beaverdam Member, but increases upwards so that many beds in unit 3 display crossbedding. Nearly all the conglomerates are clast supported, with 60–75% clasts. Such conglomerate is not present everywhere at the base of the Missi, however, and in a few places the basal Missi is pebbly sandstone (unit 2).

Unit 1 is similar in its major aspects to Miall's (1977, 1978) Scott-type proximal river model. Internal features consistent with unit 1 being either a proximal alluvial fan or a proximal braided-stream deposit are best seen at the western side of the basin and include (1) the several metre thicknesses of very poorly bedded to unbedded, ungraded, clast supported, poorly sorted pebble–cobble–boulder conglomerates (bed type α, Figs. 8 and 9); (2) the lenticular nature of sandstone beds within the lower parts of the unit (bed type α, Figs. 8 and 9); (3) the lenticular nature of both conglomerate and sandstone beds in the interbedded upper parts of the unit (bed type β, locally α, Fig. 8); and (4) fairly common crossbedding in the upper sandstone lenses. Also, unit 1 is most likely subaerial because there are no associated turbidites or other marine deposits, there are thick, typically fluvial, units above; and unit 1 directly overlies a major subaerial weathering profile. The massive conglomerates with their associated poorly bedded sandstone lenses may have been deposited as debris flows, longitudinal bars, or channel-lag deposits. Although the proportion of fine matrix is considerably lower than in most modern debris flows (e.g., Hooke, 1967; Walker, 1975), similar ancient clast-supported conglomerates have been considered to be debris-flow deposits by some workers (e.g., Miall, 1970). Nonetheless, deposition by proximal braided streams is perhaps more likely, similar to those suggested for some Cretaceous conglomerates in Montana by Ryder and Scholten (1973).

The deposits of unit 2 are much like those of Miall's (1977, 1978) Donjek-type braided river sequence, although the upper parts contain fewer conglomerate beds. Features consistent with this unit having formed in the middle reaches of a braided stream (possibly midfan) include (1) the lenticular nature of generally thin, originally horizontal beds of conglomerate, some of which grade upwards to sandstone with crudely developed horizontal bedding (bed types γ and δ, Fig. 8), probably representing channel-lag or longitudinal bar deposits; (2) the slightly pebbly character of most of the sandstone beds (all are medium to coarse grained); (3) common evidence of channel scouring and moderately common solitary to less common grouped trough crossbeds (bed type ε, and locally θ, Figs. 3 and 9), probably a result of deposition in migrating channels, some of which developed crossbedded dunes near the close of the depositional episode; and (4) a lack of fine sandstone to mudstone and no observed ripple bedding or planar crossbedding.

Unit 3 is virtually identical to Miall's (1977, 1978) South Saskatchewan-type braided river model. Features consistent with this unit having formed as a sandy facies, distal braided-stream deposit include (1) common pebbly scour- to channel-fill sandstone grading upwards to crossbedded dunes (bed types ε and θ); (2) fairly common, thin, gravel channel fill (bed type δ); (3) fairly common well-developed crossbedding characteristic of river dunes (bed types λ, π,and γ), especially near the top of the unit; (4) fairly continuous single-pebble conglomerate to pebble beds, which occur on some

Figure 6. *Stratigraphic map of the Flin Flon Basin showing the positions of the measured type sections (heavy line flanked by thin lines). Rock units numbered 1, 2, and 3 belong to the Beaverdam Member; units 4, 5, and 6 belong to the Little Cliff Member. Structural symbols have been omitted for simplicity, but the rock unit contacts can be compared with bedding trends on Figure 4a. The type sections are marked in the field by rock cairns.*

more-or-less originally nearly horizontal erosion surfaces (as in bed type κ, Fig. 8), perhaps as thin longitudinal bar deposits; (5) lateral gradation within some beds from massive, coarse- to medium-grained and locally pebbly sandstone to finer grained and ultimately crossbedded sandstone (trough to planar, see lower part of bed type λ), probably representing deposition on the side or end of a sandbar (these deposits also display normal vertical grading); and (6) the occasional occurrence of penecontemporaneously folded trough crossbedding (bed type ψ), especially near the top of the unit (a review of the literature indicates that this feature is most common in braided-stream deposits, e.g., Hendry and Stauffer (1975)). Virtually all of the rocks seen in unit 3 appear to be channel-fill (including dunes) and bar deposits; no features were observed that are clearly overbank, backswamp, or eolian in origin.

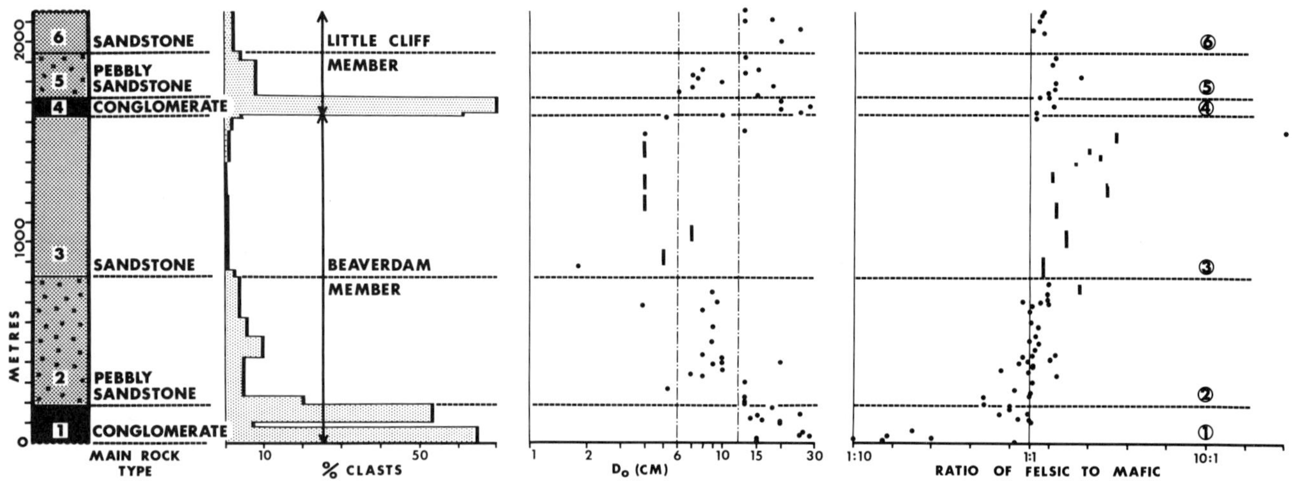

Figure 7. *Stratigraphic compilation of all segments of the type section, indicating textures and mineral compositions of clasts. Percentages of clasts were determined by point counting several square metres of each of hundreds of outcrop surfaces; percentages are purely two dimensional and no attempt was made to make the numbers volumetrically significant. Data were then combined into groups representing tens of metres of section, depending on the consistency of the data. This graph clearly displays the basis for subdividing the Missi Formation. Overall, unit 1 contains more than 50% clasts; unit 2, 5–7%; unit 3, <1%; unit 4, >50%; unit 5, about 10%; and unit 6, about 3%. Every unit, however, contains both conglomerates (or at least clast-rich zones) and sandstones with various proportions of scattered clasts.*

The average diameter (D_o) of the 10 largest clasts was also determined at each of hundreds of localities and is shown on a logarithmic scale. Again, data have been grouped according to consistency. In some cases, due to scarcity, clasts were obtained from several outcrops over tens of metres of thickness. In the conglomerates, at least, all 10 measurements come from within several square metres. These data were very difficult to obtain because of the "stretched" nature of the deformed clasts. In most places, the size of clasts was calculated from measurements on a single planar surface by taking into account the present deformed shapes and the geometry of the strain ellipsoid as determined by Stauffer and Burnett (1979). The original diameter was then calculated as the diameter of a sphere with the same volume as the now ellipsoidal deformed clast. Again, maximum clast size varies markedly between units: unit 1 is consistently over 12 cm; unit 2 is mainly from 6 to 12 cm; unit 3 is mainly less than 6 cm; unit 4 is mainly more than 12 cm; and unit 5 and 6 clasts commonly also have maximum diameters of greater than 12 cm, but the average is much less than that for unit 4 clasts. In general, the size of the average clast in an outcrop is about one third that of the largest clasts.

The ratio of felsic to mafic clasts was obtained by determining the compositions of at least 50 and mostly about 150 clasts, either within a few square metres or, where necessary, over up to several tens of metres thickness. Almost all clasts are placed easily in one of these two categories. The mafics include mainly fine-grained mafic volcanic rocks, but also some medium-grained rocks of possible intrusive origin (there is almost no observable quartz in hand specimens of any rocks of this group). Felsic clasts include locally variable proportions of vein quartz, rhyolite to dacite, and granite to granodiorite, with some other types present. Some units can be distinguished fairly easily on the basis of clast maturity. Unit 1 nearly everywhere has a felsic/mafic ratio of <1, especially in the lower parts where it is generally about 1:8; unit 2 fluctuates about 1:1 (1:4 to 4:1); unit 3 is everywhere greater than 2:1 (mainly about 3:1 and ranges up to 6:1); and units 4, 5, and 6 all vary from about 1:1 to 4:1.

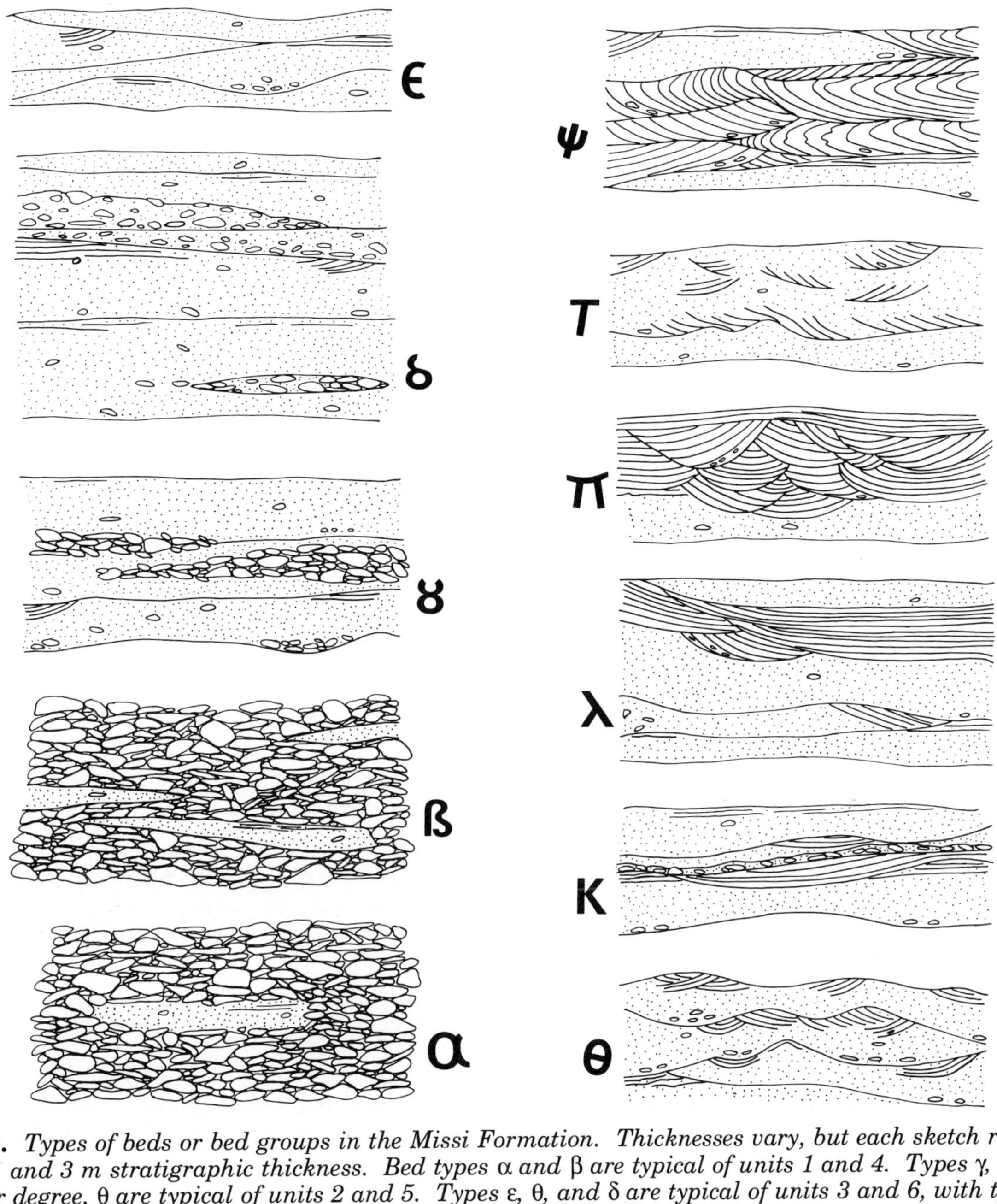

Figure 8. *Types of beds or bed groups in the Missi Formation. Thicknesses vary, but each sketch represents between 1 and 3 m stratigraphic thickness. Bed types* α *and* β *are typical of units 1 and 4. Types* γ, δ, ε, *and, to a lesser degree,* θ *are typical of units 2 and 5. Types* ε, θ, *and* δ *are typical of units 3 and 6, with types* λ, κ, π, γ, *and* ψ *also being fairly common. Whether crossbedding is obvious in an outcrop or not depends partly on the presence of laminations containing hematite and biotite or chlorite, e.g., compare type* ε *with* θ *and type* π *with* τ. *Some beds without such laminations are indistinctly crossbedded by laminations of slightly different grain size (this style of crossbedding tends to be masked by the development of a penetrative metamorphic foliation). These bed types are purely descriptive and refer to what one sees in outcrops of Missi rocks; nonetheless, they correspond approximately to a number of general lithofacies discussed by Miall (1978) and Rust (1978). Missi bed types* α *and* β *are both varieties of their gravel Gm facies; types* ε *and* θ *include Gm and the sandy Sh and St facies; types* ε *and* θ *are transitional between their Gt and Ss facies, with some St tops; types* π, τ, *and* ψ *are all varieties of their St facies; type* κ *is complex and includes at least lithofacies Gm, Sh, Se, St, and Ss; and type* λ *includes St, Sh, and possibly Sp facies. There are a few bed types different than any of these, for example, several thin beds within unit 6 contain small-scale sigmoidal crossbedding, but this feature is rare. Also, the alignment of clasts shown in several of the bed types is tectonic not primary; approximate parallelism with bedding is common due to complex relationships between bedding, cleavage, lineation, and outcrop surfaces.*

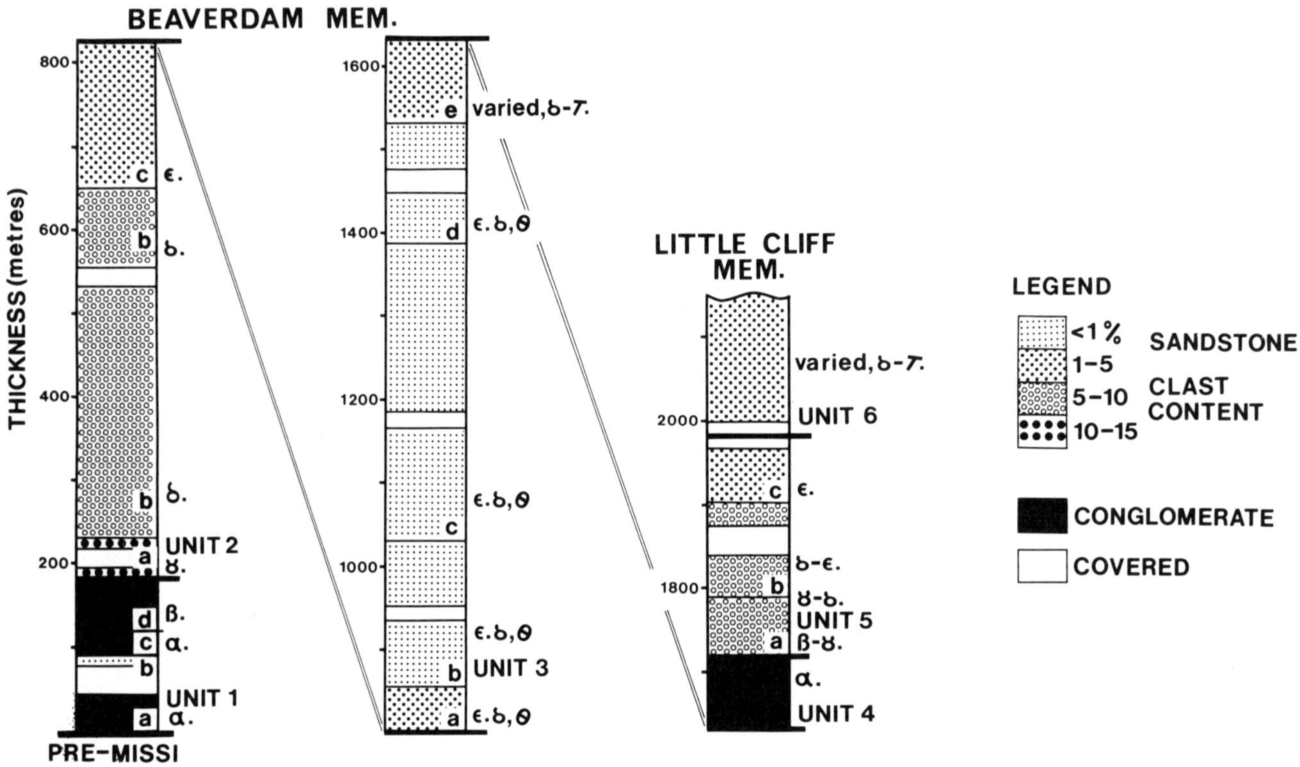

Figure 9. *Subdivisions within the Beaverdam and Little Cliff members at type sections in the Flin Flon Basin, showing vertical gradations in clast content and bed types (see Fig. 8).*

Little Cliff Member

The Little Cliff Member can also be divided into three distinctive fining-upwards and maturing-upwards units (Fig. 7 and 9): unit 4 – a basal, 30–90 m thick, polymictic conglomerate unit similar in most respects to unit 1 of the Beaverdam Member and, like it, considered to be a Scott-type proximal braided-stream deposit; unit 5 – a gradational, overlying unit of pebbly sandstone greater than 200 m thick that is similar to unit 2 of the Beaverdam Member and, like it, considered to be a Donjek-type braided-stream deposit; and unit 6 – an upper unit of sandstone greater than 100 m thick that, like the similar unit 3, is probably a South Saskatchewan-type distal braided-stream deposit.

The Little Cliff Member is very similar to the Beaverdam Member in its clast content, clast sizes, bed types, and fining-upward nature. The major difference is in maturity, as indicated by the following: (1) the Beaverdam Member grades from having a high mafic clast content near its base (felsic/mafic clast ratio as low as 1:10, Fig. 6) to having a high felsic clast content near its top (felsic/mafic clast ratio as high as 14:1); (2) the Little Cliff Member, on the other hand, has much less variability and the felsic/mafic clast ratio is everywhere between 1:1 and 4:1, with no obvious vertical trend (Fig. 6); and (3) the upper part of the Beaverdam Member contains quartzite beds (unit 3 and, more rarely, unit 2), but none were seen in the rather limited exposures of the upper Little Cliff Member (unit 6).

Original thickness

Phase 2 deformation, which produced schistosity, grain elongation, and large folds (Stauffer and Mukherjee, 1971), also modified bed thicknesses throughout the Flin Flon Basin. This was studied for the type section by applying Schwerdtner's (1978) method to the strain data obtained by Stauffer and Burnett (1979). Table I shows that most of the units have been thinned and that the Missi at the type section in the Flin Flon Basin was originally about 3780 m thick, compared with a present value of 2150 m.

AMISK BASIN

Lithology and stratigraphy

This basin (Fig. 1d) has been described by Byers and Dahlstrom (1954), Ahuja (1970), and Longiaru (1980). It contains mainly units 1 and 3 of the

Beaverdam Member (Fig. 10a). Unit 2 is only locally developed and is too thin to map. Units 1 and 3 are considerably thicker here than in the Flin Flon Basin but have also been tectonically thinned. Strain analysis has not been done in the Amisk Basin; however, an "eyeballed" estimate suggests that the original stratigraphic thickness was at least double the present 2000 m for unit 1 and a minimum of 3500 m for unit 3.

In most respects, these units are lithologically identical to their equivalents at the type section in the Flin Flon Basin, except that unit 1 conglomerates at Amisk Lake locally contain considerably larger clasts (up to an original diameter of 1 m for granitoid clasts). Also, unit 1 thins dramatically to the south between the mainland and the small islands to the southeast (Fig. 10a), where pebbly sandstone locally overlies paleoweathered Amisk rocks.

Structure of Amisk Basin

The Amisk Basin is bounded on the west by the MacDonald Creek fault and on the east by the Mosher Lake fault (although possibly in part by intrusive contact with the Reynard Lake pluton) (Fig. 10). Rocks within the basin have been subject to polyphase deformation, which is partly correlative with that in the Flin Flon Basin. Unconformable contacts with paleoweathered Amisk rocks occur at the extreme northwestern and southeastern ends of the basin.

The earliest recognizable fold is the Magdalen Lake syncline, which appears to be equivalent to the F_2 folds at Flin Flon. Like them, it has an axial-plane schistosity formed, in part, by stretched clasts. This schistosity (S_2) is generally parallel or subparallel to bedding except where it crosses the

Table I
Missi Formation, Flin Flon Basin

Unit	Present thickness (m) at type section		Estimated original thickness (m)
1	186		439
2	637		1860
3	808		759
4	91		64
5	228		269
6	199	(minimum)	390
Total	2149		3781

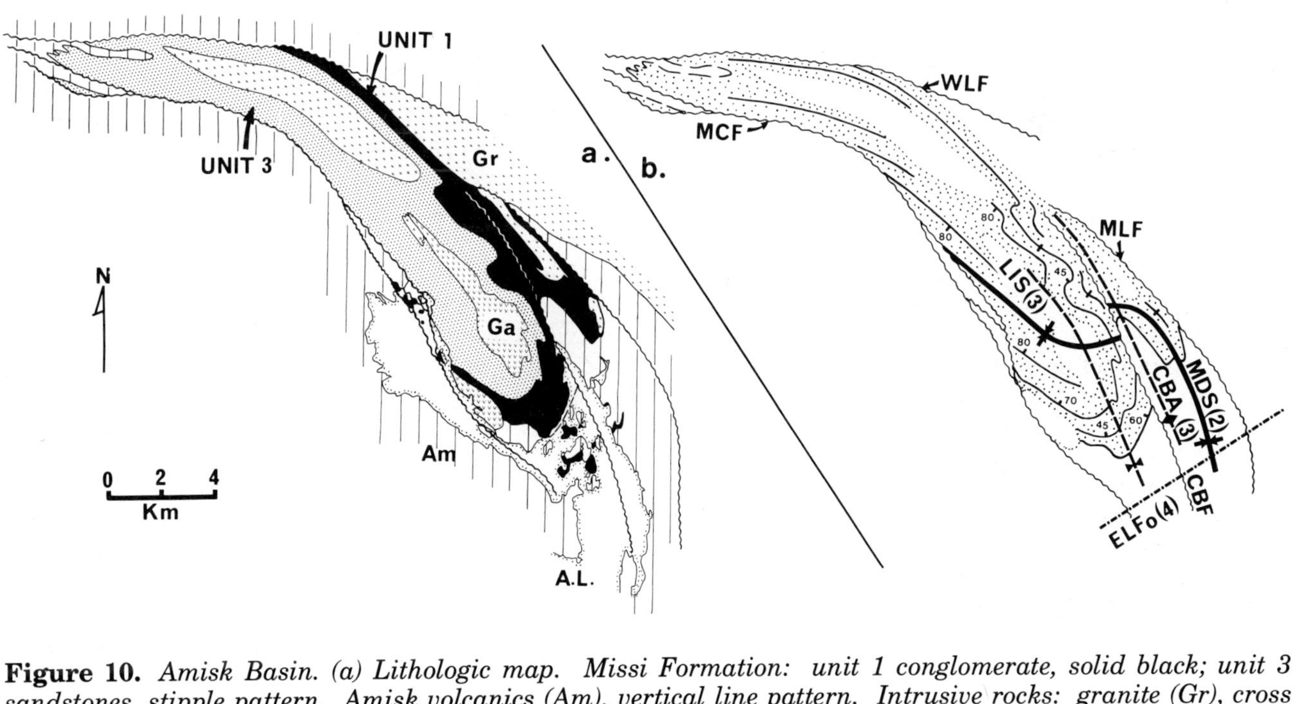

Figure 10. *Amisk Basin. (a) Lithologic map. Missi Formation: unit 1 conglomerate, solid black; unit 3 sandstones, stipple pattern. Amisk volcanics (Am), vertical line pattern. Intrusive rocks: granite (Gr), cross pattern (the pluton containing "Gr" is the Reynard Lake pluton); gabbro (Ga), "V" pattern. Wiggly lines are faults. Abbreviation: AL, Amisk Lake. (b) Structural geology. Faults (wiggly lines): MCF, MacDonald Creek fault; CBF, Comeback Bay fault; MLF, Mosher Lake fault; and WLF, Wilson Lake fault. F_2 fold: MDS (2), Magdalen Lake syncline, heavy black axial trace. F_3 folds: CBA (3), Comeback Bay anticline; LIS (3), Lookout Island syncline (3), heavy dashed axial traces. F_4 fold: ELF_o (4), Embury Lake synform, dash–dot axial trace. Stipple pattern shows the Missi Formation. Solid curving lines are bedding trend lines with local average dips indicated.*

nose of the Magdalen Lake syncline (Fig. 11). Schistosity is so well developed in the northwestern part of the basin that only rare outcrops clearly display bedding.

This fold is truncated by the bounding faults mentioned above, as well as by the Comeback Bay fault (Fig. 10b). These faults all dip subvertically or steeply to the north or northeast, and numerous minor to map-scale folds (F_3) are associated with them (e.g., Comeback Bay anticline and Lookout Island syncline (Figs. 10b and 11)). F_3 folds distort S_2, especially in the southern part of the basin, locally producing spectacular outcrop patterns in unit 1 conglomerate (Fig. 3e). An S_3 crenulation cleavage is developed locally, again mainly in the southern part of the basin (Fig. 11), where it is subvertical and axial planar to minor F_3 folds.

Analysis of minor folds, C–S fabrics, and streaky lineations in the fault zones indicates an oblique-slip displacement. The slip vector plunges about 30°SSE at the southern part of the basin, and the east side has moved up and to the northwest, similar to the movement deduced for faults in the Flin Flon Basin by Stauffer and Mukherjee (1971). Displacements of stratigraphic contacts indicate net slips of about 2.3 km for the Comeback Bay fault (Longiaru, 1980), at least 9 km for the MacDonald Creek fault (Fig. 12), and at least 4 km for the Mosher Lake fault (Fig. 12).

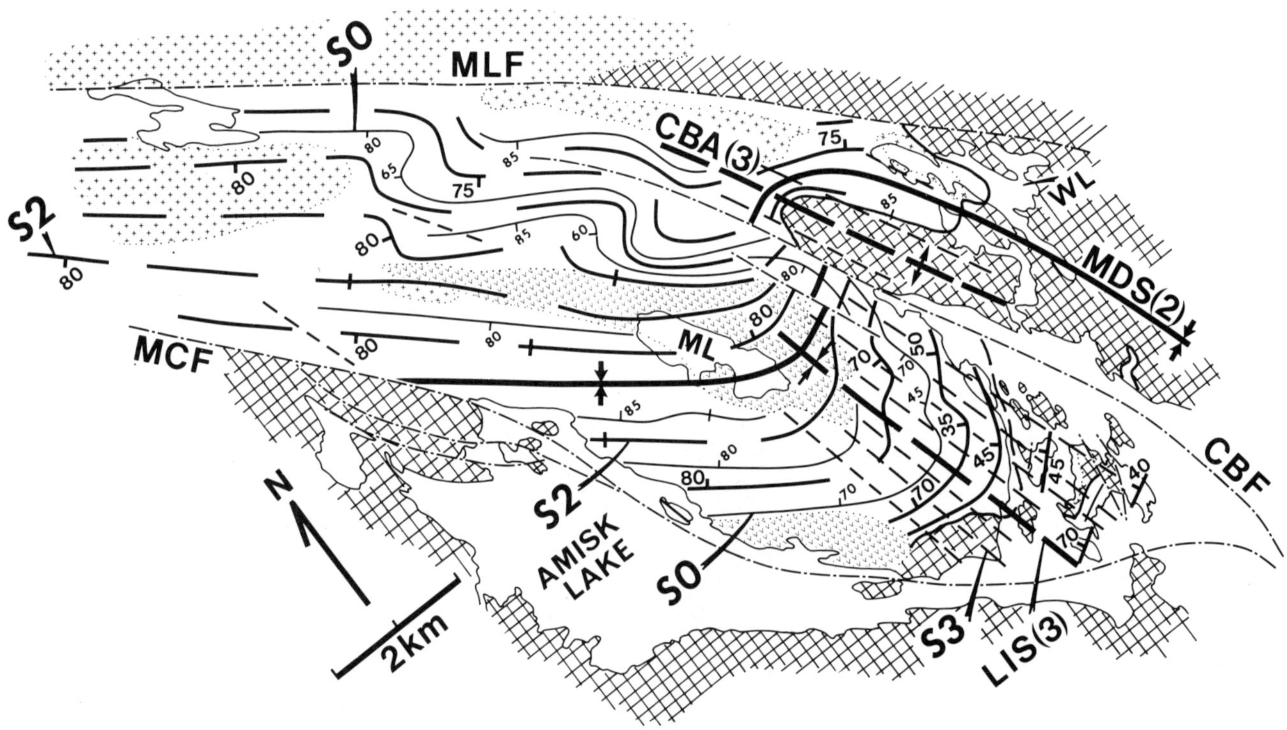

Figure 11. *Structure – surface trend lines in the Amisk Basin. S_0 (light continuous lines), bedding (all dips shown are right-way up); S_2 (medium, long dash), schistosity; S_3 (light, short-dashed line), crenulation cleavage to local schistosity. Dips shown are local averages; S_3 is subvertical. Amisk volcanics (and related intrusions) are cross-hatched. Post-Missi granitoids are indicated by a small cross pattern. Post-Missi gabbro is indicated by a small "V" pattern. Faults: MCF, MacDonald Creek fault; CBF, Comeback Bay fault; and MLF, Mosher Lake fault. F_2 fold: MDS (2), Magdalen Lake syncline. F_3 folds: LIS (3), Lookout Island syncline; CBA (3), Comeback Bay anticline. Abbreviations: WL, Wolverine Lake; ML, Magdalen Lake.*

ATHAPAP BASIN

Lithology and stratigraphy

The Athapap Basin (Fig. 1d) is fault bounded on the east and west sides (Fig. 13a), and contains several smaller faults within it. Buckham (1940) produced a crude lithologic map, Syme (1987) measured a stratigraphic section across the basin, and Bailes and Syme (1987) mapped its northern extension, where the Missi's unconformable contact with Amisk rocks is locally displayed. The Athapap Basin, or more precisely Athapap fault slice, is poorly exposed, and there remain several structural and stratigraphic uncertainties; also, the nature of its southern boundary with older rocks is unknown.

Missi rocks within the basin can be correlated with units 1 to 4 of the Flin Flon Basin, but there are three lithological differences: (1) clast size, (2) the proportion of ironstone clasts, and (3) the presence of earthy hematite.

In unit 1, clasts are somewhat larger at Athapap (up to 65 cm, averaging about 20 cm, compared with a maximum of 40 cm at Flin Flon and 1 m at Amisk Lake). In unit 4, clasts at Athapap are also slightly larger (up to 45 cm, averaging about 15 cm, compared with an average of <10 cm at Flin Flon).

Athapap conglomerate units contain significantly more iron-bearing clasts (jasper and taconite). At Flin Flon and Amisk Lake, jasper clasts are unknown and taconite clasts (generally small) seldom make up more than 2% of an outcrop. The jasper clasts are particularly noticeable at Athapap, and are amongst the largest present in some outcrops, especially in unit 4. Together, the iron-bearing clasts make up about 10% of unit 4 and about 5% of unit 1 rocks at Athapap. All other clast types can be correlated with Amisk volcanic or pre-Missi in-

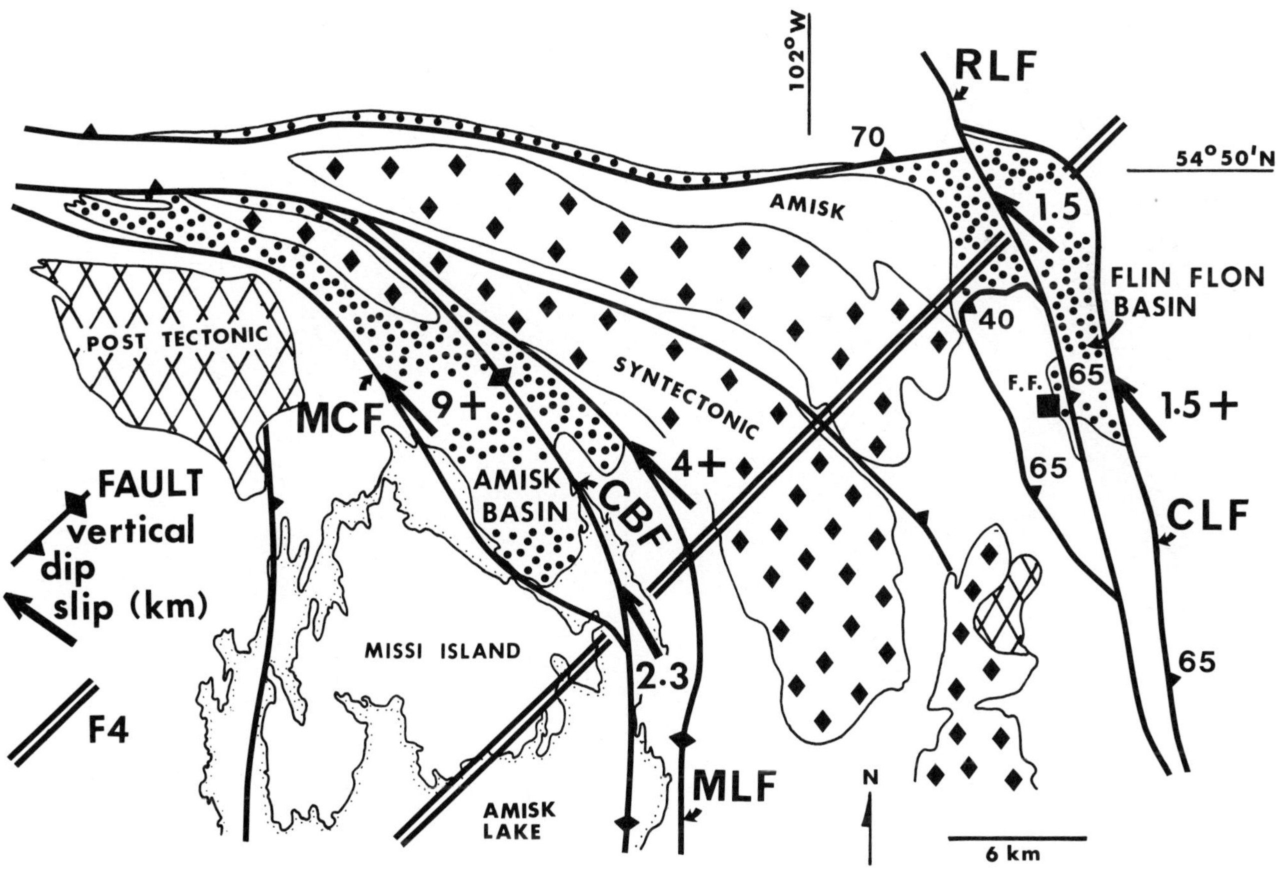

Figure 12. *Major faults (heavy lines) between the Flin Flon and Amisk basins. Faults: RLF, Ross Lake fault; CLF, Cliff Lake fault; MLF, Mosher Lake fault; CBF, Comeback Bay fault; and MCF, MacDonald Creek fault. Intrusive bodies: posttectonic granitoids are cross-hatched; syntectonic granitoids are indicated by large diamond texturing. Missi rocks are indicated by a large dot pattern. Amisk rocks are untextured. The eastern side of this figure is based on data from Stauffer and Mukherjee (1971). The heavy double line is the axial trace of the F_3 Embury Lake fold (a neutral fold).*

trusive rock types; however, no jasper- or taconite-bearing rocks have yet been found in the surrounding terrain, and they, at least, may have an exotic source.

At Athapap, many of the rocks contain a reddish to purplish matrix, most of which is earthy hematite, although some fine-grained jasper grains are present. Elsewhere, hematite occurs as specularite.

The unit 1/Amisk contact is not exposed within the Athapap Basin; thus, the measured thickness of 610 m for unit 1 is a minimum; unit 2 is the only unit fully exposed, with a thickness of 570 m; unit 3 is partly fault bounded at 910 m minimum thick-

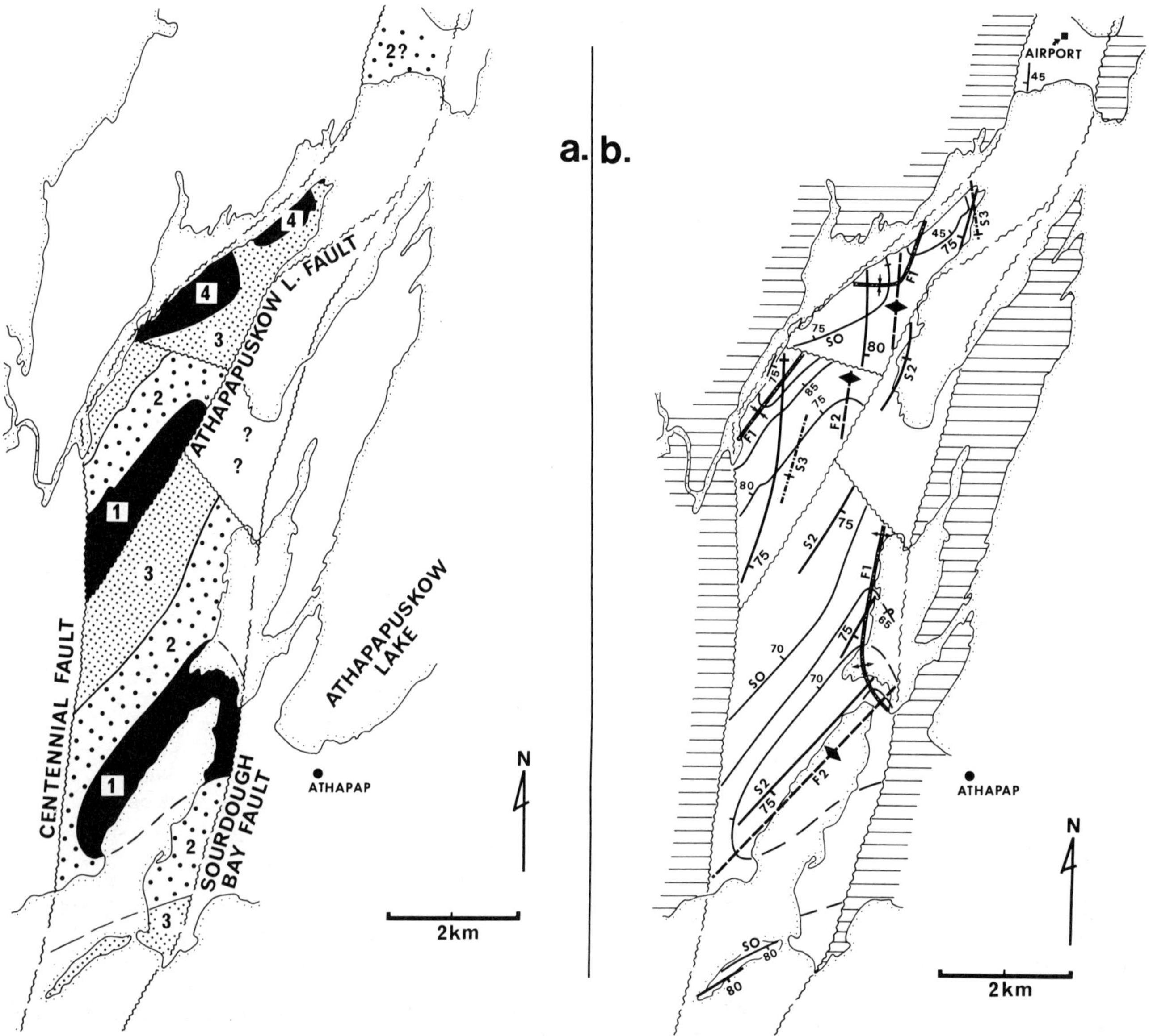

Figure 13. *Geology of the Athapap Basin. (a) Lithological units 1–4 of the Missi Formation. The region labelled 2? at the north end is poorly exposed pebbly sandstone that is probably part of unit 2. The untextured part marked by ?? in the central east contains very poorly exposed conglomerates that could be part of either unit 1 or unit 2, or perhaps even conglomerate beds within other units. (b) Structural map.* S_0, *bedding trend lines with local average dips (right-way up unless otherwise indicated).* S_2, *schistosity.* F_1 *axial traces are indicated by a heavy line with white dashes.* F_2 *axial traces are indicated by a heavy dashed line. The adjacent Amisk volcanic and plutonic rocks are shown by horizontal texturing. The locations of the Flin Flon airport and the Athapap railroad siding are shown.*

ness; and the top of unit 4 is not exposed, resulting in a minimum thickness of 430 m. The total exposed thickness in the Athapap Basin is 2520 m. Deformation is not nearly as strong at Athapap as at Flin Flon or Amisk Lake, and is sufficiently weak that at some outcrops the rocks do not appear to be deformed. Although no strain analyses were conducted, an "eyeballed" estimate suggests less than 50% average flattening and perhaps about 30% tectonic thinning of units, so that the original minimum total thickness would have been somewhat more than 3500 m.

Figure 13a shows two regions with question marks, the northern one (2?) contains pebbly sandstones, probably of unit 2, and much less likely of unit 5. The untextured patch in the east-central area contains very poorly exposed, small outcrops of mainly conglomerate, but it is unknown whether these are part of unit 1, unit 4, or conglomerate beds within other units.

Structure

The tectonic geometry and history of the Athapap Basin remains uncertain, partly due to poor exposure. The bounding faults (Centennial and Sourdough, Fig. 13a) have stratigraphic throws in excess of 3.5 km each, and the interconnecting Athapapuskow fault has a stratigraphic throw of nearly 2 km. Slip directions, however, are unknown. Two periods of folding appear to have affected the rocks (Fig. 13b). The earliest has no related axial-plane foliation and is correlated with F_1 in the Flin Flon Basin. The superimposed folds are associated with a weakly to strongly developed axial-plane schistosity and are correlated with F_2 structures in the Flin Flon Basin. A subvertical S_3 crenulation is only locally developed (Fig. 13b).

CORRELATION

Correlation among Missi basins

Although members and units are correlated among the different Missi structural basins, this should not be taken to mean, for example, that unit 1 necessarily represents the same alluvial fan in the separate basins or even that unit 3 was deposited throughout from the same braided stream. The correlations made here are homotaxial and refer to the fact that similar lithological units occur in similar stratigraphic positions and thus represent similar events occurring at about the same time. It seems likely that two major uplift–erosion events are recorded by the two fining-upwards members mapped in the Flin Flon and Athapap basins, and that these events were coeval between the basins, whether or not the basins were even connected. Also, it is possible that the present structural basins occur where the Missi was originally thickest, such as might be expected in regions of large alluvial fan buildup. Braided-stream deposits, such as units 3 and 6, would have been more widespread; this probably explains why small outcrop patches of Missi in the Flin Flon Domain are commonly sandstone or pebbly sandstone rather than massive conglomerate.

Sherridon–Missi correlation

The Missi Formation in the Flin Flon Domain has been correlated by most workers (e.g., Froese and Goetz, 1981; Zwanzig, 1984) with the later-named Sherridon group (Bateman and Harrison, 1945) in the Kisseynew Domain to the north (Fig. 1c) ever since Bailes (1971) analysed a series of possible correlations (outlined by Harrison (1951)) between rocks in the two domains.

Such correlation is followed here (Fig. 14), with a further suggestion that the name "Sherridon group" now be dropped. In keeping with standard stratigraphic practice, the earlier name, "Missi," should be used exclusively, regardless of whether or not its designation as a "Formation" is accepted (as suggested here).

There are several lithological differences between Missi rocks in the Flin Flon and southern Kisseynew domains. However, quartz-rich, coarse clastic rocks (or their metamorphic equivalents) are prevalent in most regions. Generally, there is a decrease in conglomerates to the north. In several Kisseynew Domain Missi outcrop areas, there are a number of amphibolite beds, some of which probably had volcanic protoliths, whereas others may have been mudstones (e.g., Zwanzig, 1984). Felsic bodies, probably representing rhyolite flows and tuffs, occur in some places (e.g., Zwanzig, 1984), and carbonates and calc-silicates have been mapped by Froese and Goetz (1981), Zwanzig (1984), and Schledewitz (1985). If the designation of the Missi as a "Formation" is accepted, then these other rock units become "members." Primary sedimentary structures (other than parallel bedding) are rare in the Kisseynew Domain, probably because of high states of strain.

The presence of intercalated mudstones and carbonates along with coarse clastics in the southern Kisseynew Domain probably represents the transition to alternating fluvial and shallow marine (or perhaps lacustrine) conditions, whereas the Flin Flon Domain appears to have remained fluvial (at least for the preserved Missi section).

In addition to these lithological differences, Missi rocks in the Flin Flon and Kisseynew domains rest on different materials. In the Flin Flon Domain, they rest unconformably on mainly metavolcanic

Amisk group rocks, although metasedimentary rocks (originally greywacke?) are present locally. In the Kisseynew Domain, they rest mainly on metasedimentary rocks (greywacke) of the Nokomis group, considered by Bailes (1971) to be the marine sedimentary equivalent of Amisk volcanic rocks (an interpretation accepted by all later workers).

Sickle–Missi correlation

The northern side of the Kisseynew Domain is nearly a mirror image of the southern side (though more complex (Baldwin *et al.*, 1987)), with the Missi-like Sickle group (metaconglomerates and metasandstones) unconformably overlying volcanic and sedimentary rocks of the Wasekwan group in the La Ronge (Lynn Lake) Domain north of the Kisseynew Domain (McRitchie, 1974; Gilbert *et al.*, 1980; Baldwin *et al.*, 1987), and conformably overlying the predominantly metasedimentary rocks of the Burntwood River metamorphic suite in the northern Kisseynew Domain (Gilbert *et al.*, 1980; Baldwin *et al.*, 1987). Wasekwan, Burntwood River (including Nokomis), and Amisk rocks are all thought to be broadly correlative as no major unconformable breaks have been found as yet (McRitchie, 1974) and because the few U–Pb ages obtained so far suggest that pre-Missi sedimentation, volcanism, and related plutonism occurred throughout the area during the broad time period of about 1910–1875 Ma (e.g., Zwanzig *et al.*, 1985; Baldwin *et al.*, 1987; Syme *et al.*, 1987; Van Schmus *et al.*, 1987). However, details vary locally and volcanism might not have occurred in all areas at the same time (Baldwin *et al.*, 1987). Also, the U–Pb ages of synorogenic plutons in the region span the time period from about 1875 to 1850 Ma, with some postorogenic intrusions yielding ages of <1840 Ma (e.g., Van Schmus and Bickford, 1984; Zwanzig *et al.*, 1985).

Sickle rocks are extremely similar to the Missi rocks in the Flin Flon Domain (cf. Gilbert *et al.* (1980) with descriptions presented here). Undoubtedly, Missi rocks in the Flin Flon Domain and Sickle rocks were deposited in similar fan to fluvial environments, in a similar geotectonic setting, and, at least crudely, over about the same time span (about 1900–1800 Ma as maximum and minimum limits, although they probably are diachronous in detail). Also, they probably were interconnected by the partly marine Missi (Sherridon) rocks, now restricted mainly to the southern Kisseynew Domain as a result of later arching and erosion of the central Kisseynew Domain.

On the basis of the above discussion, I suggest that Sickle rocks also correlate with the Missi, and that because of prior use of the term "Missi" (Bruce, 1918), the term "Sickle" (Norman, 1934) should be dropped. The name "Missi" should be used for the entire blanket of postgeosynclinal fan – fluvial – shallow marine deposits in the Flin Flon, Kisseynew, and La Ronge domains (Fig. 14), including the various subdivisions of the La Ronge Domain (Fig. 14) shown by Baldwin *et al.* (1987), and the

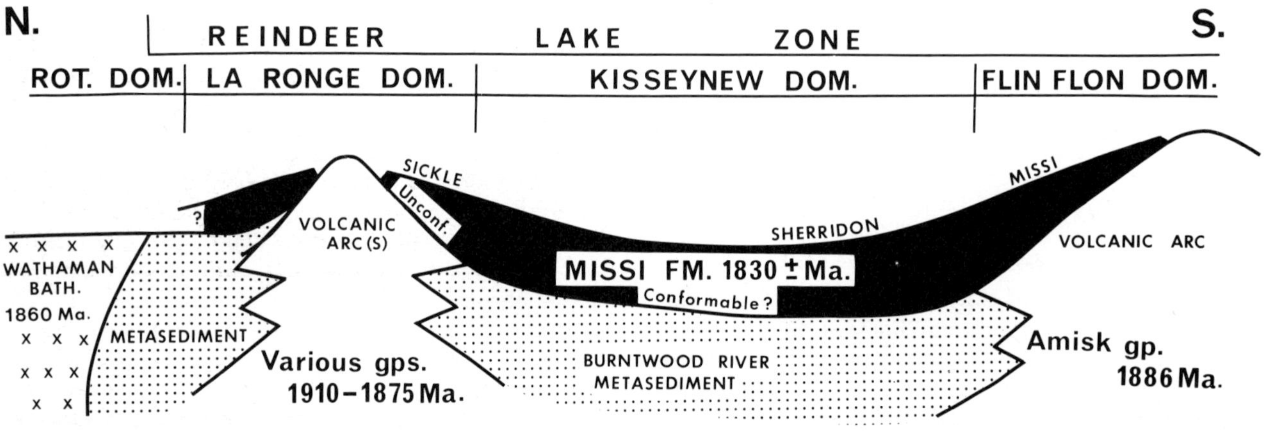

Figure 14. *Generalized cross-section north–south through the Reindeer Lake Zone showing the Missi Formation at the time of deformation, including its regional components and underlying rock materials. Data synthesized from many sources, but mainly Bailes (1971), Lewry and Sibbald (1977), Gilbert* et al. *(1980), Stauffer (1984), Van Schmus* et al. *(1987), and Baldwin* et al. *(1987). Abbreviation: Rot. Dom., Rottenstone Domain. The term "La Ronge Domain" is used in its most general sense to include all volcanic–sedimentary belts between the Kisseynew and Rottenstone domains, and between the Rottenstone and Glennie Lake domains. For simplicity, deformation and intrusive bodies, other than the Wathaman Batholith, have been ignored. The estimated age given for the Missi Formation is probably a late age, and deposition likely began locally several tens of millions of years earlier.*

McLennan group of meta-arkose and conglomerate discussed by Lewry (1983) in the La Ronge Domain, between the Rottenstone and Glennie Lake domains (Fig. 1).

DISCUSSION OF THE GEOTECTONIC SETTING AND ORIGIN OF THE MISSI FORMATION

The Missi Formation defined here rests on a base of volcanic arcs (with intrusions) and interarc materials (Fig. 14) (Bailes, 1971; Stauffer, 1984). Two types of evidence suggest that it was derived almost entirely by erosion of the volcanic arc materials: (1) most clast materials can be matched by rock types in the adjacent volcanic and intrusive terrains, except for the ironstone clasts in the Flin Flon Domain, and the rare clast of recycled Missi; and (2) there is ample evidence of sub-Missi erosion in the volcanic arc regions (e.g., Fig. 4) but not as yet in the interarc region (Kisseynew Domain). Broadly considered, volcanism began by at least 1910 Ma (Baldwin *et al.*, 1987) and continued in different parts of the Reindeer Lake Zone until at least 1875 Ma. Likewise, the tectonic uplifts necessary to initiate subaerial erosion, and subsequent, mainly subaerial, Missi deposition, must have occurred over a protracted time period; nonetheless, the Missi forms a mappable lithologic entity above the arc and interarc materials, even though it is probably diachronic.

The volcanic arcs are interpreted to have been similar to those related to modern-day subduction zones (e.g., Bailes, 1971; Stauffer, 1984; Thom *et al.*, this volume). Continued and ongoing plate collision may have been responsible for uplift and initial (pre-Missi) deformation of these volcanic arcs, whereas the interarc region remained somewhat more stable. The Missi was deposited in a continuously dynamically active setting, with collision, deformation, and deposition occurring simultaneously until final closure of the Reindeer Lake Zone.

Finally, the considerable amount of hematite present (~3% overall, locally >30%) qualifies the Missi as a redbed deposit (Stauffer, 1974; Chandler, 1980). Metamorphism has altered the original earthy hematite to specularite everywhere except in parts of the Athapap Basin. Redbeds commonly are thought to have formed at low latitude, and the Missi Formation may be no exception as the paleomagnetic pole position determined by Park (1975) for a phase of post-Missi, but locally deformed, igneous rocks places the paleoequator close to the centre of the Flin Flon Domain.

ACKNOWLEDGEMENTS

The author wishes to thank the numerous graduate students who have worked on the Missi Formation while at the University of Saskatchewan, especially Sam Longiaru, Amar Mukherjee, Suraj Ahuja, Jim Reynolds, and Andy Burnett. The various projects that led to this paper were financed by grants from the Natural Sciences and Engineering Research Council of Canada. Also, many thanks to Don Winston, University of Montana, and Bob Macdonald, Saskatchewan Geological Survey, for their valuable and constructive comments.

REFERENCES

Ahuja, S.P., 1970, Structural analysis of the Amisk and Missi rocks north of Amisk Lake, Saskatchewan, unpublished M.Sc. thesis, University of Saskatchewan.

Alcock, F.J., 1923, Flin Flon map-area, Manitoba and Saskatchewan: Geological Survey of Canada, Summary Report 1922, p. 1–36.

Ambrose, J.W., 1936a, Progressive kinetic metamorphism in the Missi Series, near Flin Flon, Manitoba: American Journal of Science, v. 32, p. 257–286.

Ambrose, J.W., 1936b, Structures in the Missi Series, near Flin Flon, Manitoba: Transactions of the Royal Society of Canada, v. 30, sec. 4, p. 81–98.

Bailes, A.H., 1971, Preliminary compilation of the geology of the Snow Lake – Flin Flon – Sherridon area: Manitoba Department of Mines and Natural Resources, Mines Branch, Geology Paper 1/71.

Bailes, A.H. and Syme, E.C., 1983, GS-5 Flin Flon – Schist Lake project, *in* Manitoba Energy and Mines, Report of Field Activities 1983: Manitoba Energy and Mines, p. 23–29.

Bailes, A.H. and Syme, E.C., 1987, Geology of the Flin Flon – White Lake area: Manitoba Energy and Mines, Geological Map GR87-1-1.

Baldwin, D.A., Syme, E.C., Zwanzig, H.V., Gordon, T.M., Hunt, P.A. and Stevens, R.D., 1987, U–Pb ages from the Lynn Lake and Rusty Lake metavolcanic belts, Manitoba: two ages of Proterozoic magmatism: Canadian Journal of Earth Sciences, v. 24, p. 1053–1063.

Bateman, J.D. and Harrison, J.M., 1945, Sherridon map-area: Geological Survey of Canada, Paper 45–15.

Bell, K. and Macdonald, R., 1982, Geochronological calibration of the Precambrian Shield in Saskatchewan, *in* Summary of Investigations 1982, Saskatchewan Geological Survey: Saskatchewan Energy and Mines, Report 82-4, p. 17–22.

Bell, K., Blenkinsop, J. and Moore, J.M., 1975, Evidence for a Proterozoic greenstone belt from Snow Lake, Manitoba: Nature, v. 258, p. 698–701.

Bruce, E.L., 1918, Amisk–Athapapuskow lake district (Saskatchewan–Manitoba): Geological Survey of Canada, Memoir 105.

Buckham, A.F., 1940, Athapapuskow Lake: Geological Survey of Canada, Map 807A.

Byers, A.R., 1953, Missi series, Amisk Lake area, northern Saskatchewan: Transactions of the Royal Society of Canada, v. 40, sec. 4, p. 1–10.

Byers, A.R. and Dahlstrom, C.D.A., 1954, Geology and mineral deposits of the Amisk–Wildnest lakes area, Saskatchewan: Saskatchewan Department of Mineral Resources, Report 14.

Byers, A.R., Kirkland, S.J.T. and Pearson, W.J., 1965, Geology and mineral deposits of the Flin Flon area, Saskatchewan: Saskatchewan Department of Mineral Resources, Report 62.

Chandler, F.W., 1980, Proterozoic redbed sequences of Canada: Geological Survey of Canada, Bulletin 311.

Froese, E., 1969, Metamorphic rocks of the Coronation mine and surrounding area, in Byers, A.R., ed., Symposium on the Geology of Coronation Mine, Saskatchewan: Geological Survey of Canada, Paper 68-5, p. 55–77.

Froese, E. and Goetz, P., 1981, Geology of the Sherridon group in the vicinity of Sherridon, Manitoba: Geological Survey of Canada, Paper 78-27, 20 p.

Gilbert, H.P., Syme, E.C. and Zwanzig, H.V., 1980, Geology of the metavolcanic and metavolcaniclastic metasedimentary rocks in the Lynn Lake area: Manitoba Department of Energy and Mines, GP80-1.

Harrison, S.M., 1951, Precambrian correlation and nomenclature, and problems of the Kisseynew gneisses, in Manitoba: Geological Survey of Canada, Bulletin 20, 53 p.

Hendry, H.E. and Stauffer, M.R., 1975, Penecontemporaneous recumbent folds in trough cross-bedding of Pleistocene sands in Saskatchewan, Canada: Journal of Sedimentary Petrology, v. 45, p. 932–943.

Heywood, W.W., 1966, Ledge Lake area, Manitoba and Saskatchewan: Geological Survey of Canada, Memoir 337.

Hooke, R. LeB., 1967, Processes on arid-region alluvial fans: Journal of Geology, v. 75, p. 438–460.

Johnson, S.Y., 1984, Stratigraphy, age, and paleogeography of the Eocene Chuckanut Formation, northwest Washington: Canadian Journal of Earth Sciences, v. 21, p. 92–106.

Lewry, J.F., 1983, Character and structural relations of the 'McLennan Group' meta-arkoses, McLennan–Jaysmith lakes area, in Summary of Investigations 1983: Saskatchewan Geological Survey, p. 49–55.

Lewry, J.L. and Sibbald, T.I.L., 1977, Variations in lithology and tectonometamorphic relationships in the Precambrian basement of northern Saskatchewan: Canadian Journal of Earth Sciences, v. 14, p. 1453–1467.

Lewry, J.L., Sibbald, T.I.I. and Schledewitz, D.C.P., 1985, Variation in character of Archean rocks in the western Churchill Province and its significance, in Ayres, L.D., Thurston, P.C., Card, K.D. and Weber, W., eds., Evolution of Archean Supracrustal Sequences: Geological Association of Canada, Special Paper 28, p. 239–261.

Longiaru, S.J., 1980, Structure and metamorphism of the northeast Amisk Lake area, Saskatchewan, unpublished M.Sc. thesis, University of Saskatchewan.

Macdonald, R. and Broughton, P., compilers, 1980, Geological map of Saskatchewan (scale 1:1,000,000), provisional edition: Saskatchewan Department of Mineral Resources.

McRitchie, W.D., 1974, The Sickle–Wasekwan debate – a review: Manitoba Department of Mines, Resources and Environmental Management, Mines Branch, Geological Paper 1/74, 23 p.

McRitchie, W.D., Schledewitz, D.C.P., Weber, W., Scoates, R.F.J., McCabe, H.R. and Bannatyne, B.B., compilers, 1979, Geological map of Manitoba (scale 1:1,000,000): Manitoba Mineral Resources Division, Map 79-2.

Miall, A.D., 1970, Devonian alluvial fans, Prince of Wales Island, arctic Canada: Journal of Sedimentary Petrology, v. 40, p. 556–571.

Miall, A.D., 1977, A review of the braided-river depositional environment: Earth Science Reviews, v. 13, p. 1–62.

Miall, A.D., 1978, Lithofacies types and vertical profile models in braided river deposits: a summary, in Miall, A.D., ed., Fluvial Sedimentology: Canadian Society of Petroleum Geologists, p. 597–604.

Mukherjee, A.C., 1971, The Precambrian geology of the Flin Flon area, northern Saskatchewan and Manitoba, Canada, unpublished Ph.D. thesis, University of Saskatchewan.

Mukherjee, A.C., Stauffer, M.R. and Baadsgaard, H., 1971, The Hudsonian orogeny near Flin Flon, Manitoba: a tentative interpretation of Rb/Sr and K/Ar ages: Canadian Journal of Earth Sciences, v. 8, p. 939–946.

Norman, G.W.H., 1934, Granville Lake district, northern Manitoba: Geological Survey of Canada, Summary Report 1933, pt. C, p. 23–41, with Map 301A (Granville Lake).

Park, J.K., 1975, Paleomagnetism of the Flin Flon – Snow Lake greenstone belt, Manitoba and Saskatchewan: Canadian Journal of Earth Sciences, v. 8, p. 1272–1290.

Pettijohn, F.J., Potter, P.E. and Siever, R., 1972, Sand and Sandstone: Springer-Verlag, New York, 618 p.

Rust, B.R., 1978, Depositional models for braided alluvium, in Miall, A.D., ed., Fluvial Sedimentology: Canadian Society of Petroleum Geologists, p. 605–625.

Ryder, R.T. and Scholten, R., 1973, Syntectonic conglomerates in southwestern Montana: their nature, origin, and tectonic significance: Geological Society of America Bulletin, v. 84, p. 773–796.

Sangster, D.F., 1978, Isotopic studies of ore-leads of the circum-Kisseynew volcanic belt of Manitoba and Saskatchewan: Canadian Journal of Earth Sciences, v. 15, p. 1112–1121.

Schledewitz, D.C.P., 1985, GS-13 Kisseynew project: Kississing Lake, Big Island – Yakushavich Island region, in Manitoba Energy and Mines, Report of Field Activities 1985: Manitoba Energy and Mines, p. 54–56.

Schwerdtner, W.M., 1978, Determination of stratigraphic thickness of strained units: Canadian Journal of Earth Sciences, v. 15, p. 1379–1380.

Shanks, R.J. and Bailes, A.H., 1977, GS-16 "Missi Group" rocks, Wekusko Lake area, in Manitoba Mineral Resources Division, Report of Field Activities 1977: Manitoba Energy and Mines, p. 83–87.

Stauffer, M.R., 1974, Geology of the Flin Flon area: a new look at the sunless city: Geoscience Canada, v. 1(3), p. 30–35.

Stauffer, M.R., 1984, Manikewan: an early Proterozoic ocean in central Canada, its igneous history and orogenic closure: Precambrian Research, v. 25, p. 257–281.

Stauffer, M.R. and Burnett, A.I., 1979, Down-plunge viewing: a rapid method for estimating the strain ellipsoid for large clasts in deformed rocks: Canadian Journal of Earth Sciences, v. 16, p. 290–304.

Stauffer, M.R. and Mukherjee, A., 1971, Superimposed deformations in the Missi meta-sedimentary rocks near Flin Flon, Manitoba: Canadian Journal of Earth Sciences, v. 8, p. 217–242.

Stauffer, M.R., Mukherjee, A.C. and Koo, J., 1975, The Amisk Group: an Aphebian(?) island-arc deposit: Canadian Journal of Earth Sciences, v. 12, p. 2021–2035.

Stockwell, G.H., 1946, Flin Flon – Mandy area, Manitoba and Saskatchewan: Geological Survey of Canada, Paper 46-14.

Stockwell, G.H., 1960, Flin Flon – Mandy area, Manitoba and Saskatchewan: Geological Survey of Canada, Map 1078A.

Syme, E.C., 1987, GS-7 Athapapuskow project, in Manitoba Energy and Mines, Report of Field Activities 1987: Manitoba Energy and Mines, p. 30–39.

Syme, E.C., Bailes, A.H., Gordon, T.M. and Hunt, P.A., 1987, U–Pb zircon geochronology in the Flin Flon belt: age of Amisk volcanism, in Manitoba Energy and Mines, Report of Field Activities 1987: Manitoba Energy and Mines, p. 105–107.

Tanton, T.L., 1941, Flin Flon, Saskatchewan and Manitoba: Geological Survey of Canada, Map 632A.

Thom, A., Arndt, N.T., Chauvel, C. and Stauffer, M., 1990, Flin Flon and Western La Ronge belts, Saskatchewan: products of Proterozoic subduction-related volcanism, in Lewry, J.F. and Stauffer, M.R., eds., The Early Proterozoic Trans-Hudson Orogen of North America: Geological Association of Canada, Special Paper 37, p. 163–175.

Van Schmus, W.R. and Bickford, M.E., 1984, Preliminary U–Pb age data for the Trans-Hudson Orogen in northern Saskatchewan, in Summary of Investigations 1984, Saskatchewan Geological Survey: Saskatchewan Energy and Mines, Report 84-4, p. 81–85.

Van Schmus, W.R., Bickford, M.E., Lewry, J.F. and McDonald, R., 1987, U–Pb geochronology in the Trans-Hudson Orogen, northern Saskatchewan, Canada: Canadian Journal of Earth Sciences, v. 24, p. 407–424.

Walker, D. and Watters, B.R., 1982, Geochemistry of metavolcanic rocks, Amisk Lake west area, in Summary of Investigations 1982, Saskatchewan Geological Survey: Saskatchewan Energy and Mines, Report 82-4, p. 24–30.

Walker, R.G., 1975, Conglomerate: sedimentary structures and facies models, in Depositional Environments as Interpreted from Primary Sedimentary Structures and Stratification Sequences: Society of Economic Paleontologists and Mineralogists, Notes from Short Course No. 2, p. 133–161.

Zwanzig, H.V., 1984, GS-12 Kisseynew project: Lobstick Narrows – Cleunion Lake, Puffy Lake and Nokomis Lake areas, in Manitoba Energy and Mines, Report of Field Activities 1984: Manitoba Energy and Mines, p. 38–45.

Zwanzig, H.V., Parker, J.S.D., Schledewitz, D.C.P. and Van Schmus, W.R., 1985, GS-1 Lynn Lake regional compilation and geochronology, in Manitoba Energy and Mines, Report of Field Activities 1985: Manitoba Energy and Mines, p. 6–8.

Stratigraphy and Geochemistry of the Lynn Lake and Flin Flon Metavolcanic Belts, Manitoba

Eric C. Syme
Manitoba Energy and Mines, 535-330 Graham Avenue,
Winnipeg, Manitoba R3C 4E3, Canada

Syme, E.C., Stratigraphy and geochemistry of the Lynn Lake and Flin Flon metavolcanic belts, Manitoba, in Lewry, J.F. and Stauffer, M.R., eds., The Early Proterozoic Trans-Hudson Orogen of North America: Geological Association of Canada, Special Paper 37, p. 143–161.

Abstract

A significant component of juvenile Proterozoic crust in the Trans-Hudson Orogen consists of linear belts of metamorphosed volcanic and volcaniclastic rocks; in Manitoba, these include the Lynn Lake Belt (part of the larger La Ronge – Lynn Lake Domain) and the Flin Flon Belt. These metavolcanic belts are similar to Cenozoic intraoceanic arc assemblages in terms of lithologic constitution, environment of deposition, and major-, trace-, and rare-earth element contents.

Wasekwan Group metavolcanic rocks in the southern portion of the Lynn Lake Belt comprise a mafic platform of tholeiitic and "primitive" (magnesian) calc-alkaline basalts, locally overlain by felsic complexes and a dominantly andesitic calc-alkaline suite. Various tholeiitic basalt units overlie the felsic complexes. The northern part of the belt, characterized by abundant aluminous basalts, consists of a sequence of (oldest to youngest) rhyolite, evolved basalt and andesite, turbidites, less fractionated basalt and high-magnesia basalt, and mafic volcanogenic sediments. Tentative stratigraphic correlation suggests that the high-alumina suite overlies the southern succession.

The western part of the Flin Flon Belt is marked by a collage of fault-bounded blocks, precluding a regional stratigraphic synthesis. Intact stratigraphic sequences within fault blocks differ widely in thickness (up to 7.5 km), lithology, and environment of deposition. Predominant units are basalt and basaltic andesite with arc tholeiite trace-element characteristics; subordinate ferrobasalt units are the products of considerable fractionation. A lithologically, magnetically, and geochemically distinct group of basalts has trace-element characteristics more akin to back-arc lavas, and is invariably in fault contact with the arc-type basalts.

Island arc-type mafic volcanic rocks in both the Lynn Lake and Flin Flon belts are characterized by relative depletion in high field strength (HFS) elements (P, Hf, Zr, Ti, Y) and enrichment in large ion lithophile elements (Rb, Ba, K, Th, Sr). The Lynn Lake Belt appears to be geochemically more evolved, with calc-alkaline basalts and andesites comprising a significant portion of the sequence. Most basalts in the Flin Flon Belt contain lower HFS-element and rare-earth element abundances than their counterparts in the Lynn Lake Belt.

Résumé

Les ceintures linéaires de roches volcaniques et volcano-clastiques métamorphisées sont une composante importante de la croûte juvénile Protérozoïque de l'orogène Trans-Hudson; au Manitoba, il s'agit des ceintures Lynn Lake (une partie du domaine La Ronge – Lynn Lake) et Flin-Flon. Ces ceintures métavolcaniques sont semblables aux assemblages d'arcs intra-océaniques Cénozoïque par les lithologies, les environnements sédimentaires et les compositions en éléments majeurs, traces et terres rares.

Les roches métavolcaniques du groupe Wasekwan dans la partie sud de la ceinture Lynn Lake comprennent une plateforme de tholeiites et de basaltes calco-alcalins "primitifs" (magnésiens), localement surmontés par des complexes acides et une suite calco-alcaline andésitique dominante. La partie nord de la ceinture, caractérisée par des basaltes alumineux abondants, comprend du plus vieux au plus jeune, des rhyolites, des basaltes évolués et des andésites, des turbidites, des basaltes moins fractionnés et magnésiens et des sédiments volcanogéniques basiques. Un essai de corrélation stratigraphique montre que la suite alumineuse surmonte la séquence sud.

La partie ouest de la ceinture Flin-Flon est un collage de blocs faillés ne permettant pas une synthèse stratigraphique régionale. Les séquences stratigraphiques intactes des blocs faillés diffèrent quant à l'épaisseur (jusqu'à 7,5 km), aux lithologies et aux environnements sédimentaires. Les basaltes et les andésites basaltiques caractérisées par des éléments traces de tholeiites d'arcs dominent; des unités de ferrobasaltes subordonnées sont un produit de fractionnement important. Un groupe de basaltes qui se distinguent par les lithologies, le magnétisme et la géochimie ont des éléments traces caractéristiques des laves d'arrière-arcs invariablement en contact faillé avec les basaltes d'arc.

Les volcanites basiques d'arc insulaire des ceintures Lynn Lake et Flin-Flon montrent un déficit relatif en éléments HFS (P, Hf, Zr, Ti, Y) et un enrichissement en éléments LIL (Rb, Ba, K, Th, Sr). La ceinture Lynn Lake semble plus évoluée géochimiquement avec des basaltes calco-alcalins et des andésites dominantes. La plupart des basaltes de la ceinture Flin-Flon sont plus pauvres en éléments HFS et REE que leurs équivalents de la ceinture Lynn Lake.

INTRODUCTION

The Trans-Hudson Orogen of Manitoba and Saskatchewan contains lithotectonic components resembling those in younger orogens developed by collisional tectonics at convergent plate boundaries (e.g., Lewry and Sibbald, 1980; Lewry, 1981; Lewry *et al.*, 1981; Van Schmus *et al.*, 1987). Metavolcanic belts within the orogen have the potential to reveal much about the tectonic environment in which they were emplaced: the stratigraphic and, more quantitatively, geochemical aspects of volcanic rocks in younger orogens are known and thus provide a basis for comparison. Recent multielement analyses of metavolcanic rocks in the orogen (Watters, 1981, 1984, 1985, 1986; Syme, 1985) combined with modern trace-element tectonic discriminants tend to confirm previous interpretations based primarily on major-element compositions (e.g., Bell *et al.*, 1975; Stauffer *et al.*, 1975) that the belts represent island-arc assemblages. Detailed geological mapping of the belts in Manitoba, however, has shown that the volcanic stratigraphy and structure are considerably more complex than previously recognized (Gilbert *et al.*, 1980; Bailes and Syme, 1987; Syme, 1987). This paper summarizes the stratigraphic and geochemical characteristics of metavolcanic rocks in the Lynn Lake and Flin Flon belts to document the diversity that occurs, so that meaningful comparisons can be made with the volcanic components of younger orogens. Trace-element compositions of the metavolcanic rocks are tested for compatibility with an island-arc model.

Information presented in this paper is based on 1:50,000 mapping by the Manitoba Geological Survey in the Lynn Lake Belt and 1:20,000 mapping in the Flin Flon Belt. The geochemical data base includes, for the Lynn Lake Belt, 172 major-element (plus Ni and Cr) analyses of volcanic rocks, 68 trace-element analyses (Rb, Ba, Sr, Nb, Zr, Y, Ga, V, Cu, Zn, and Pb), and 45 rare-earth element (REE), U, Hf, Th, and Sc analyses. From Flin Flon, there are currently 134 major-element plus Rb, Ba, Sr, Zr, Y, Ni, Cr, Cu, Zn, and Pb analyses of volcanic rocks, and 20 REE, V, Hf, Th, Sc, Ta, and Nb analyses. The least altered samples were chosen for analysis; collection usually took place after mapping was completed and the best outcrops had been identified. Most samples in the Flin Flon area were collected by drilling. Types of alteration and expected element mobility are discussed by Syme (1985). All analysed rocks from the Lynn Lake Belt have recrystallized to form upper greenschist to middle amphibolite facies assemblages; those from Flin Flon are at subgreenschist to lower greenschist facies.

Geological setting

Within the Thelon segment (Lewry, 1984) of the Trans-Hudson Orogen, the region between the Churchill and Superior platforms is termed the Reindeer Lake Zone (Stauffer, 1984). This zone is a complex of distinct terranes comprising metavolcanic, metasedimentary, and granitoid rocks (Fig. 1). Relatively low-grade metavolcanic and metasedimentary rocks, intruded by discrete plutons, constitute the La Ronge – Lynn Lake, Rusty Lake, Flin Flon, and Hanson Lake belts. Upper amphibolite grade metasedimentary gneisses, derived from graphitic greywacke–mudstone turbidites, mafic volcaniclastic rocks, and overlying sandstone–conglomerate, make up the Kisseynew and Southern Indian – Rottenstone belts. The greywackes were probably deposited contemporaneously with the flanking volcanic rocks (Bailes, 1971, 1980; McRitchie, 1974). Wathaman–Chipewyan and Baldock batholiths (Fig. 1) are composite intrusions of predominantly granitic or granodioritic composition.

The Lynn Lake Belt can be considered a discrete entity, distinct geologically, geochemically, and geochronologically (Baldwin *et al.*, 1987) from the adjacent La Ronge and Rusty Lake belts (Fig. 1). The Lynn Lake Belt is structurally simpler than the Flin Flon Belt and contains a definable regional stratigraphy. The Flin Flon Belt comprises a number of structurally juxtaposed stratigraphic packages, none of which corresponds even in a general way to the sequence at Lynn Lake. All of the information in this paper is drawn from the western part of the Flin Flon Belt, adjacent to the Saskatchewan border. Detailed studies in progress (Bailes, 1986, 1987) indicate that the stratigraphic relationships in the Snow Lake area, 100 km east of Flin Flon, are distinct from those at Flin Flon.

LYNN LAKE BELT

The Lynn Lake Belt, comprising metamorphosed volcanic, sedimentary, and plutonic rocks, extends 130 km east from the Manitoba–Saskatchewan border and has a maximum width of 60 km. The belt is part of the La Ronge – Lynn Lake Domain (Van Schmus *et al.*, 1987) extending a further 270 km southwest to La Ronge, Saskatchewan (Fig. 1). Volcanic and volcaniclastic rocks are termed Wasekwan Group (Bateman, 1945; Campbell, 1969) and have a U–Pb zircon age of 1910 Ma (Baldwin *et al.*, 1987). Wasekwan Group supracrustal rocks were intruded by synvolcanic plugs, and were sub-

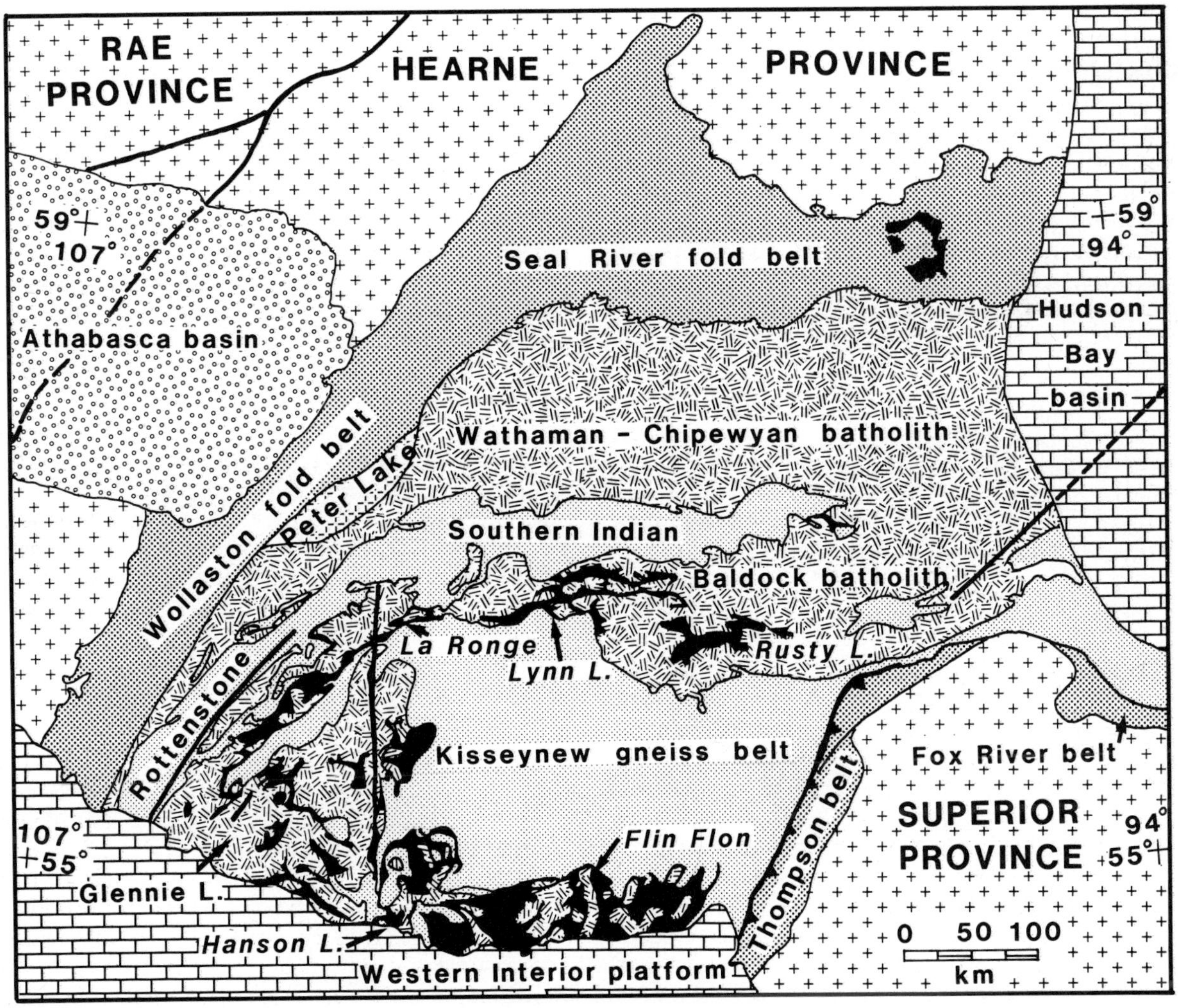

Figure 1. *Tectonic components of the Trans-Hudson Orogen in northern Manitoba and Saskatchewan (after Hoffman, 1988); metavolcanic belts shown in black.*

sequently folded, faulted, and intruded by larger mafic and felsic plutons (1876 Ma; Baldwin *et al.*, 1987). Unconformably overlying these rocks are alluvial sandstone and conglomerate of the Sickle Group (Norman, 1933; Campbell, 1969). Regional metamorphism in the upper greenschist to upper amphibolite facies and further deformation and plutonism took place after deposition of the Sickle Group.

The Wasekwan Group in the Lynn Lake area occupies two northeast-trending structural belts (the Northern and Southern belts), separated by large oval tonalite–granodiorite plutons (Fig. 2). Variations in volcanic stratigraphy occur laterally, between partly overlapping volcanic units within belts, and across strike, between the separate structural belts (Gilbert *et al.*, 1980). Smaller contiguous and isolated areas of greenstone occur in the predominantly granitoid terrane to the south.

Southern Belt

The Southern Belt contains what are interpreted to be the oldest rocks in the Lynn Lake area: aphyric quartz-normative tholeiitic basalts in a unit 26 km long and 2 km thick (Cockeram Lake tholeiite; Syme (1985)), interpreted as a deformed subaqueous shield volcano. These basalts, together with primitive (magnesian) porphyritic calc-alkaline basalts from two eruptive centres, one south of Lynn Lake and one at the base of the Wasekwan Group at Fox Mine (Fig. 2), form a mafic substrate throughout the Southern Belt (Fig. 3). Local intercalation of the tholeiitic and primitive calc-alkaline rocks indicates contemporaneous eruption of these two chemical types.

The basalts are overlain by discontinuous units of sedimentary rocks, including up to 1500 m of turbidite in the western part of the Southern Belt (Fig. 2). Elsewhere, felsic complexes and a calc-alkaline sequence more than 2500 m thick overlie the basaltic platform (Fig. 3). The contact between primitive calc-alkaline basalts and overlying felsic rocks varies from conformable to possibly disconformable. The calc-alkaline sequence (Hughes Lake sequence; Syme (1985)) overlies Cockeram Lake tholeiitic basalt (Fig. 3); the contact is lithologically abrupt. Geographically separate segments of the calc-alkaline suite each display trends to more siliceous compositions with time. For example, in a dominantly andesitic segment, the stratigraphically highest member is dacitic; whereas in another segment, a thick basalt – basaltic andesite sequence is overlain by andesite and capped by rhyolite.

Throughout the entire Southern Belt, felsic complexes and the calc-alkaline suite are overlain by units of basalt, which vary widely in form, extent, petrography, and chemistry. Near the probable stratigraphic top of the Southern Belt succession, mafic rocks are intercalated with felsic volcanic rocks, tuffs, fine-grained sedimentary rocks, and iron formation.

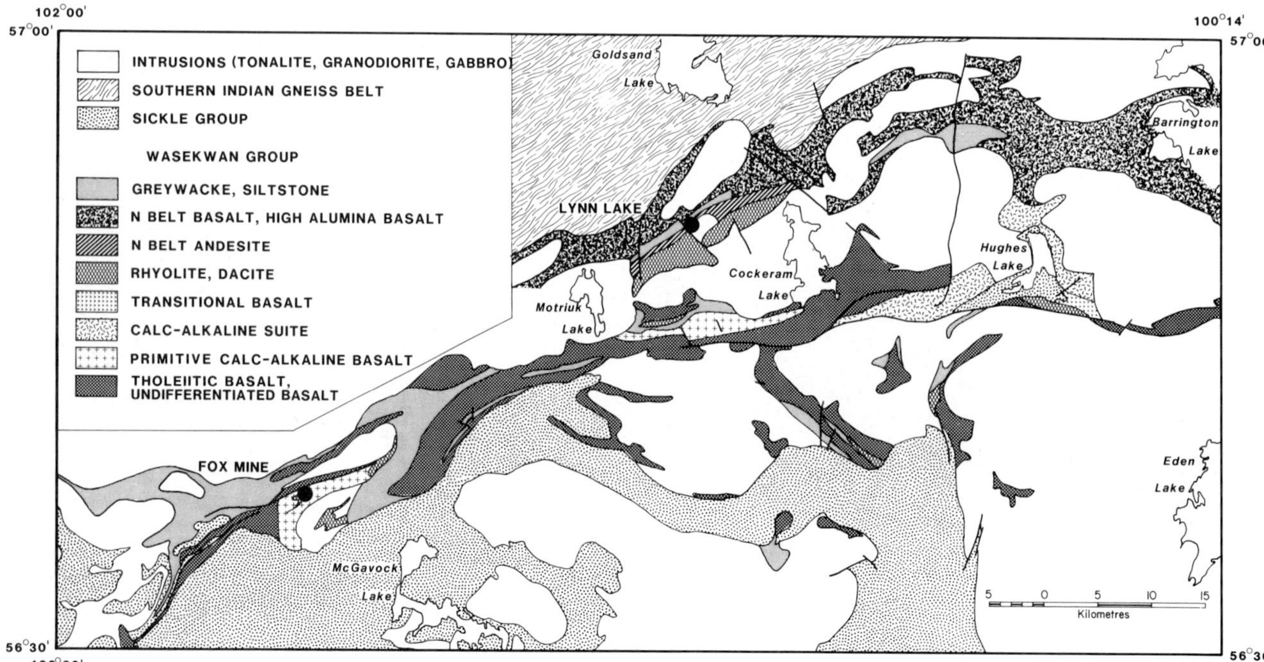

Figure 2. *Generalized geology of the Lynn Lake Belt.*

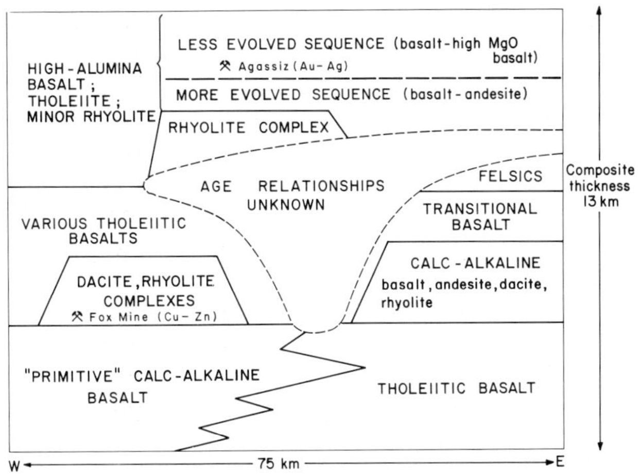

Figure 3. *Simplified chemostratigraphic relationships of metavolcanic rocks in the Lynn Lake Belt.*

Northern Belt

The Northern Belt consists largely of mafic volcanic rocks intercalated with subordinate felsic and intermediate units. The belt is characterized by an abundance of highly aluminous, commonly plagioclase – phyric basalt flows and breccias. These rocks are intercalated with flows with "normal" alumina contents. The 7–8 km thick Northern Belt sequence is divided into five major components (Gilbert *et al.*, 1980) comprising, from oldest to youngest, rhyolite pyroclastic rocks and flows (Division A); high-alumina basalt, basalt, andesite, and rhyolite (Division B); dominantly turbiditic sedimentary rocks (Division C); more primitive high-alumina basalt, basalt, and high-magnesia basalt (Division D); and basaltic tuff and volcanogenic sediments (Division E).

Table I

Average compositions of representative metavolcanic rocks, Lynn Lake Belt

	1	2	3	4	5	6	7	8
SiO_2(wt.%)	52.14	49.23	52.52	49.82	58.15	58.12	76.13	77.20
TiO_2	1.04	0.96	0.97	0.78	0.79	0.76	0.15	0.15
Al_2O_3	14.62	15.00	19.18	19.37	18.48	16.59	12.62	11.54
Fe_2O_3	2.77	3.39	1.90	2.00	1.37	1.54	0.60	0.52
FeO	8.23	6.74	7.81	8.22	5.52	5.06	0.83	1.69
MnO	0.20	0.19	0.20	0.19	0.14	0.12	0.07	0.04
MgO	5.21	7.34	3.03	4.64	2.35	3.89	0.31	0.66
CaO	10.04	10.70	8.93	10.00	8.93	7.30	1.10	0.96
Na_2O	3.16	2.44	2.92	2.74	2.66	4.11	3.93	3.17
K_2O	0.34	1.43	0.81	0.36	0.46	0.59	2.87	3.29
P_2O_5	0.15	0.26	0.22	0.10	0.20	0.24	0.01	0.04
H_2O	1.28	1.59	1.16	1.48	0.75	1.48	0.67	0.41
S	0.08	0.01	0	0	0	0.03	0.01	0
CO_2	0.60	0.43	0.17	0.24	0.10	0.21	0.59	0.23
Rb(ppm)	6	23	9	2	10	9	36	71
Ba	183	583	129	48	169	250	1066	936
Sr	211	678	363	209	314	375	68	140
Th	0.7	3.0	0.8	0.5	1.0	2.4	8.3	9.8
Nb	5	7	6	5	8	8	13	14
Hf	1.7	1.4	2.1	0.9	ND	ND	ND	ND
Zr	75	112	104	54	117	136	105	209
Y	30	21	23	16	25	25	51	38
Ni	46	53	5	13	3	59	4	4
Cr	79	276	15	36	17	64	9	10
No of samples	7	4	6	6	3	8	6	3

Notes: Columns.
(1) Cockeram Lake tholeiite, Southern Belt.
(2) McVeigh Lake "primitive" calc-alkaline basalt, Southern Belt.
(3) Northern Belt Division B high-alumina basalt.
(4) Northern Belt Division D high-alumina basalt.
(5) Northern Belt Division B high-alumina andesite.
(6) Hughes Lake calc-alkaline andesite, Southern Belt.
(7) Calc-alkaline rhyolite, Southern Belt.
(8) Lynn Lake rhyolite (Division A, Northern Belt).

ND, not determined. Individual analyses used to calculate averages are detailed in Syme (1985). Analytical methods: SiO_2, Al_2O_3, TiO_2, and P_2O_5 by colorimetry; Fe_2O_3, FeO, CaO, MgO, Na_2O, K_2O, MnO, Ni, and Cr by atomic absorption (Manitoba Energy and Mines Laboratory); trace elements by X-ray fluorescence spectroscopy (Memorial University); and REEs, Hf, and Th by instrumental neutron activation analysis (Nuclear Activation Services).

Direct stratigraphic correlation of the Northern Belt with any of the rock units in the Southern Belt is precluded by their physical separation. Details of the geochemistry of some rocks in the uppermost part of the Southern Belt, however, strongly indicate that these rocks are equivalent to lithologies in the Northern Belt (Syme, 1985). This relationship would then indicate that the Northern Belt rocks form the upper portion of the Wasekwan Group in the Lynn Lake Belt (Fig. 3).

Geochemistry

The geochemical characteristics of some of the major units in the Wasekwan Group sequence are summarized below; the geochemistry of all stratigraphic units is detailed in Syme (1985).

Aphyric Cockeram Lake tholeiite, forming the base of the exposed Wasekwan Group (Fig. 3), is subalkaline with significant normative hypersthene and quartz; an average composition is listed in Table I. It displays a range of FeO* (total iron expressed as FeO)/MgO ratios and plots in the tholeiitic field on a Jensen cation diagram (Fig. 4a). Low Mg numbers (36–61), relatively high REE abundance (Fig. 5a), and low Ti/Zr (average 83, compared with 100 in midoceanic-ridge basalt (MORB); Sun *et al.* (1979)) suggest that it has experienced some crystal fractionation, probably involving pyroxene. Low contents of Ni and Cr are consistent with olivine and pyroxene fractionation. The absence of an Eu anomaly (Fig. 5a) indicates that plagioclase was not a fractionating phase. The major- and trace-element composition of this basalt suggests that it is similar to a fractionated island-arc tholeiite (discussed in a subsequent section).

Two separate bodies of pyroxene (pseudomorph) – phyric primitive calc-alkaline basalt occur in the Southern Belt, one of which is clearly coeval with the aphyric tholeiite described above (Fig. 3). Some of the more magnesian members of this suite at Fox Mine are essentially metapyroxenites. Crystal settling of pyroxene within single massive flows and even within individual pillows has produced pronounced phase and chemical differentiation. The suite is characterized by the absence of iron enrichment (Fig. 6). These basalts have high Th contents (1.8–3.2 ppm; averages in Table I) relative to tholeiitic basalts (less than 1 ppm). Elevated Th concentrations are also characteristic of the andesitic calc-alkaline sequence (2.3–3.5 ppm). The abundance of Th increases with Zr, and a clear separation of tholeiitic and calc-alkaline rocks on a Th–Zr plot (Syme, 1985) further serves to characterize this suite as calc-alkaline. Considering the relatively magnesian nature of the suite (Table I), these rocks are termed "primitive" calc-alkaline. REE patterns are weakly to relatively strongly light rare-earth element (LREE) enriched (Fig. 5b), despite relatively high MgO contents.

The Hughes Lake calc-alkaline suite is at least 2500 m thick and comprises basalt, andesite, dacite, and rhyolite. The absence of iron enrichment on AFM, MgO versus FeO* (Fig. 6), and Jen-

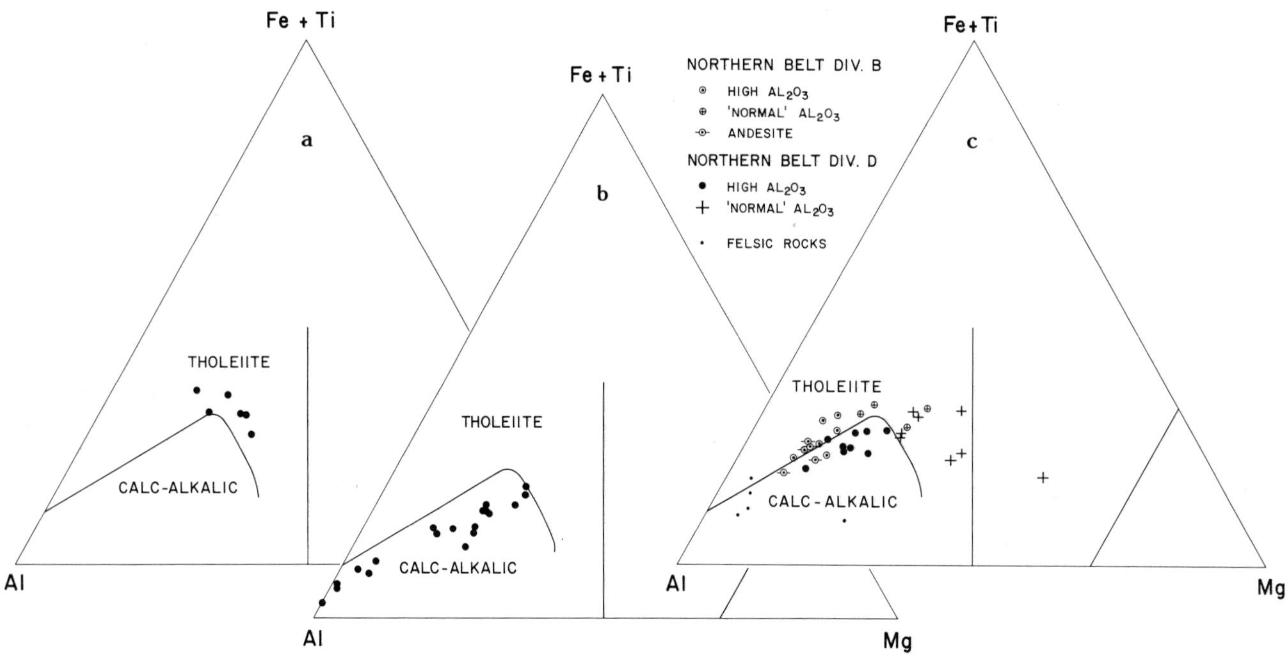

Figure 4. *Chemical variation of Wasekwan Group rocks on Jensen (1976) cation plots: (a) Cockeram tholeiite, (b) Hughes Lake calc-alkaline suite, and (c) Northern Belt.*

sen (Fig. 4b) plots identifies this suite as calc-alkaline. Individual REE patterns of calc-alkaline basalt, andesite, and dacite are virtually identical (Fig. 5c), with marked LREE enrichment and negligible Eu anomalies. The average andesite composition (Table I) is similar to island-arc andesite (Syme, 1985), but with higher Ni, Cr, Zr, and La_N/Yb_N. Rhyolites in the calc-alkaline sequence (Table I) have almost flat REE patterns with pronounced negative Eu anomalies (Fig. 5d).

Mafic rocks in the Northern Belt can be subdivided on the basis of Al_2O_3 content, with those containing more than 17 wt.% Al_2O_3 termed high-alumina basalts and those with less than 17 wt.% termed "normal"-alumina basalt. The high-alumina nature of the majority of flows in the Northern Belt distinguishes it from the Southern Belt.

Northern Belt basalts in both Division B and Division D display mild Fe enrichment (Fig. 7) when "normal" and high-alumina types are considered together; individually, however, the high-alumina rocks show trends of steadily decreasing FeO* with decreasing MgO. On a Jensen plot (Fig. 4c), the entire compositional range from basalt through andesite in Division B falls mainly in the tholeiitic field; Division D high-alumina basalts fall in the calc-alkaline field despite their being part of an overall mild Fe-enrichment trend. The Northern Belt suite is clearly different in rock types and composition from the unambiguously calc-alkaline suite at Hughes Lake (Table I).

All of the stratigraphically higher Division D rocks appear chemically less evolved than the earlier Division B rocks, which contain higher incompatible element abundances. Division D basalts are characterized by higher MgO contents and

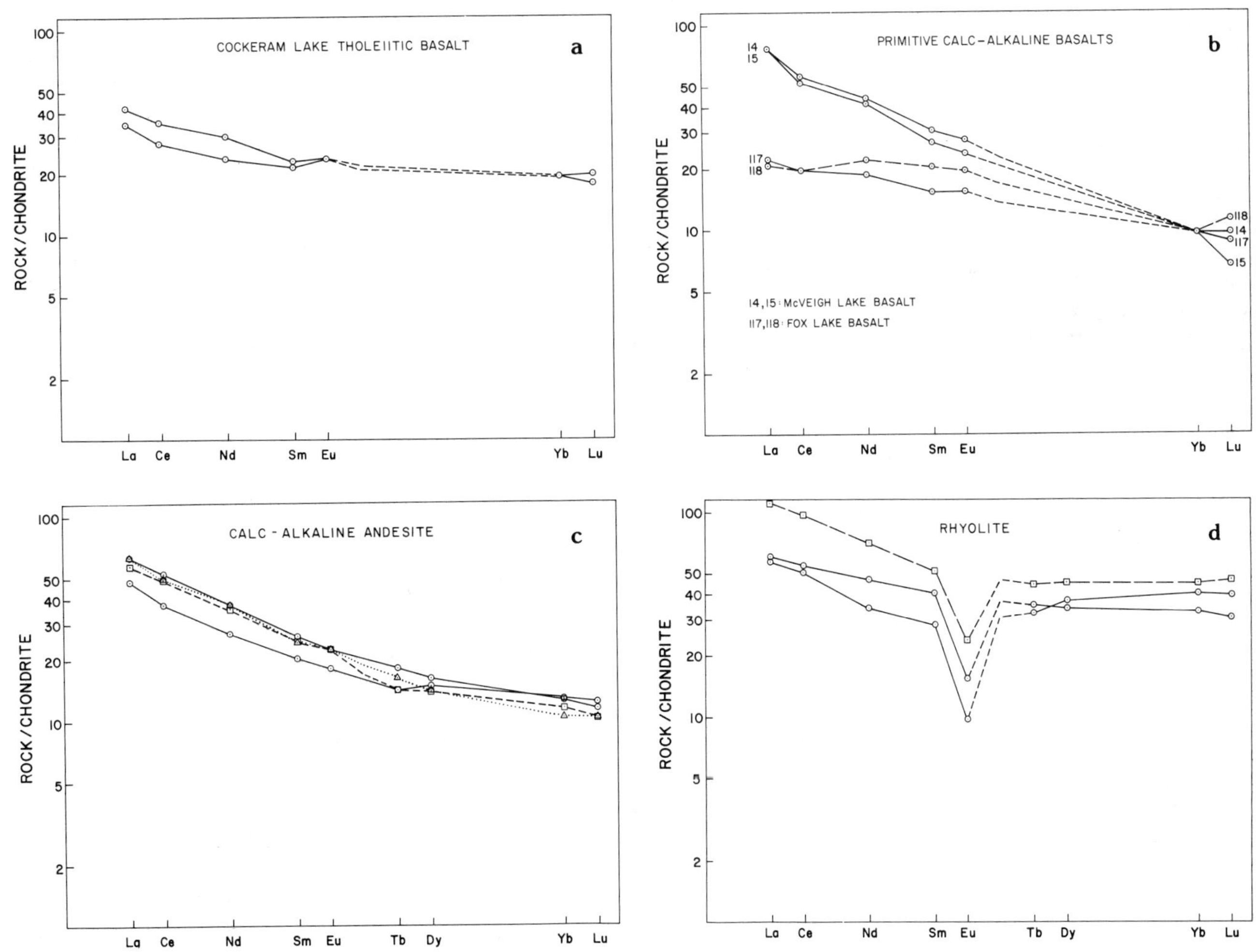

Figure 5. *Chondrite-normalized REE plots of representative Wasekwan Group rocks from the Southern Belt: (a) Cockeram tholeiite, (b) "primitive" calc-alkaline basalts, (c) Hughes Lake calc-alkaline andesite and dacite, and (d) calc-alkaline rhyolite. Normalizing values are those of Sun and Nesbitt (1978); Tb value from Haskin* et al. *(1968). REE analyses are listed in Syme (1985).*

higher Mg numbers than Division B basalts, at equivalent Al_2O_3; this relationship holds for both the high-alumina basalts and "normal"-alumina basalts. Division D high-alumina basalts also contain less SiO_2, LIL elements (K, Sr, Rb, Ba, Th, and U), and high field strength (HFS) elements (P, Ti, Nb, Hf, Zr, and Y) than Division B high-alumina basalts; they also contain more compatible transition metals (Ni, Cr, and V) (Table I; also Syme (1985)). Many of the same characteristics also serve to distinguish between Division B and D rocks with "normal"-alumina contents. Division D contains all of the high-MgO (greater than 10% MgO) basalts in the Northern Belt, including the high-MgO (12–20% MgO) rocks in the "Agassiz metallotect" (Fedikow, 1986).

High-alumina basalts of the Northern Belt are distinguished from all other basalt units in the Lynn Lake region by their very low transition-metal contents: Ni (0–30 ppm), Cr (7–90 ppm), and, to a lesser extent, V (160–254 ppm). In this respect, they closely resemble high-alumina basalts of island arcs (Gunn, 1976; Gorton, 1977; Dixon and Batiza, 1979; Meijer and Reagan, 1981).

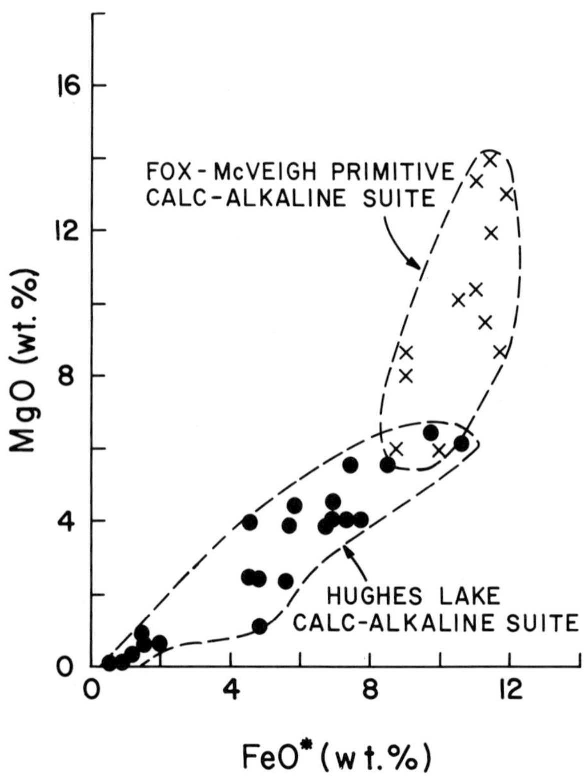

Figure 6. *MgO versus FeO* (i.e., total iron expressed as FeO) for the "primitive" calc-alkaline (Fox–McVeigh) suite and dominantly andesitic Hughes Lake calc-alkaline suite; note the absence of iron enrichment in both.*

High-alumina basalts are characterized by weak to moderate LREE enrichment, relatively low heavy rare-earth element (HREE) abundances, and positive Eu anomalies, which decrease progressively with an increase in absolute REE abundance (Fig. 8a). "Normal"-alumina basalts show generally flat REE patterns and very weak to moderate positive Eu anomalies (Fig. 8b). In both high-alumina and "normal"-alumina suites, Division D has lower total REE concentrations than Division B.

High-alumina andesites in Division B have fractionated REE patterns virtually identical to the most fractionated high-alumina basalts (Fig. 8c). There are significant differences between the Hughes Lake calc-alkaline andesites and the Northern Belt high-alumina andesites – the calc-alkaline rocks have lower Al_2O_3, FeO*, CaO, FeO*/MgO, and Mg/Ni and higher MgO, K_2O, P_2O_5, Ni, Cr, Zr, Ba, and V than Division B andesites (Table I; Syme, 1985).

Lynn Lake rhyolite (Division A of the Northern Belt) displays a fractionated REE pattern characterized by LREE enrichment, a flat HREE pattern, and negative Eu anomalies in two of the three samples analysed (Fig. 8d).

In summary, the stratigraphic and geochemical features of the Northern Belt define an "inverted" sequence of rhyolite, overlain by evolved basalt and andesite (with relatively high incompatible element abundances), overlain in turn by less evolved basalt and high-magnesia basalt (with lower incompatible element abundances). This relationship, coupled with the REE characteristics of the rocks,

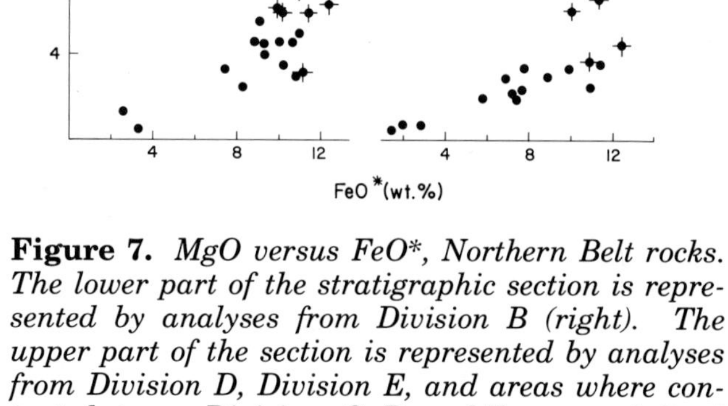

Figure 7. *MgO versus FeO*, Northern Belt rocks. The lower part of the stratigraphic section is represented by analyses from Division B (right). The upper part of the section is represented by analyses from Division D, Division E, and areas where contacts between Divisions C, D, and E are not defined (left); most of the analyses are from Division D rocks. Trends for both Division B and Division D display mild Fe enrichment.*

suggests that the Northern Belt may record the progressive emptying of a very large, compositionally zoned magma chamber (Syme, 1985).

FLIN FLON BELT

The Flin Flon Belt is composed of an assemblage of metamorphosed subaqueous and subordinate subaerial volcanic rocks and associated sediments (Amisk Group), an unconformably overlying fluvial–alluvial sequence of sandstones and conglomerate (Missi Group), and a complex array of intrusive rocks. It is bounded to the north by metasedimentary gneisses of the Kisseynew gneiss belt and is overlain to the south by flat-lying Ordovician dolomitic limestones (Fig. 1). The volcanic belt has a length of 250 km and an exposed width of 32–48 km, but its total width, including the area covered by Phanerozoic rocks, appears to be considerably larger (Green *et al.*, 1985).

The Amisk Group at Flin Flon has a U–Pb zircon age of 1886 Ma (Syme *et al.*, 1987) and the Missi Group at Wekusko Lake has been dated at 1832 Ma (Gordon *et al.*, 1987). Plutons in the Flin Flon Belt have U–Pb zircon ages of 1871 Ma, 1830–1836 Ma (Gordon *et al.*, 1987), and 1847 Ma (Syme *et al.*, 1987).

The Flin Flon area contains a much higher proportion of outcrop than the Lynn Lake Belt; consequently, the Amisk Group can be subdivided in considerably more detail than was possible for the Wasekwan Group (Fig. 9).

Five phases of folding (P1 to P5) have been recognized in the Flin Flon – White Lake area (Bailes and Syme, 1987), one of pre-Missi folding and four that postdate the Missi Group. Faults, however, dominate the structural fabric of the area. Some

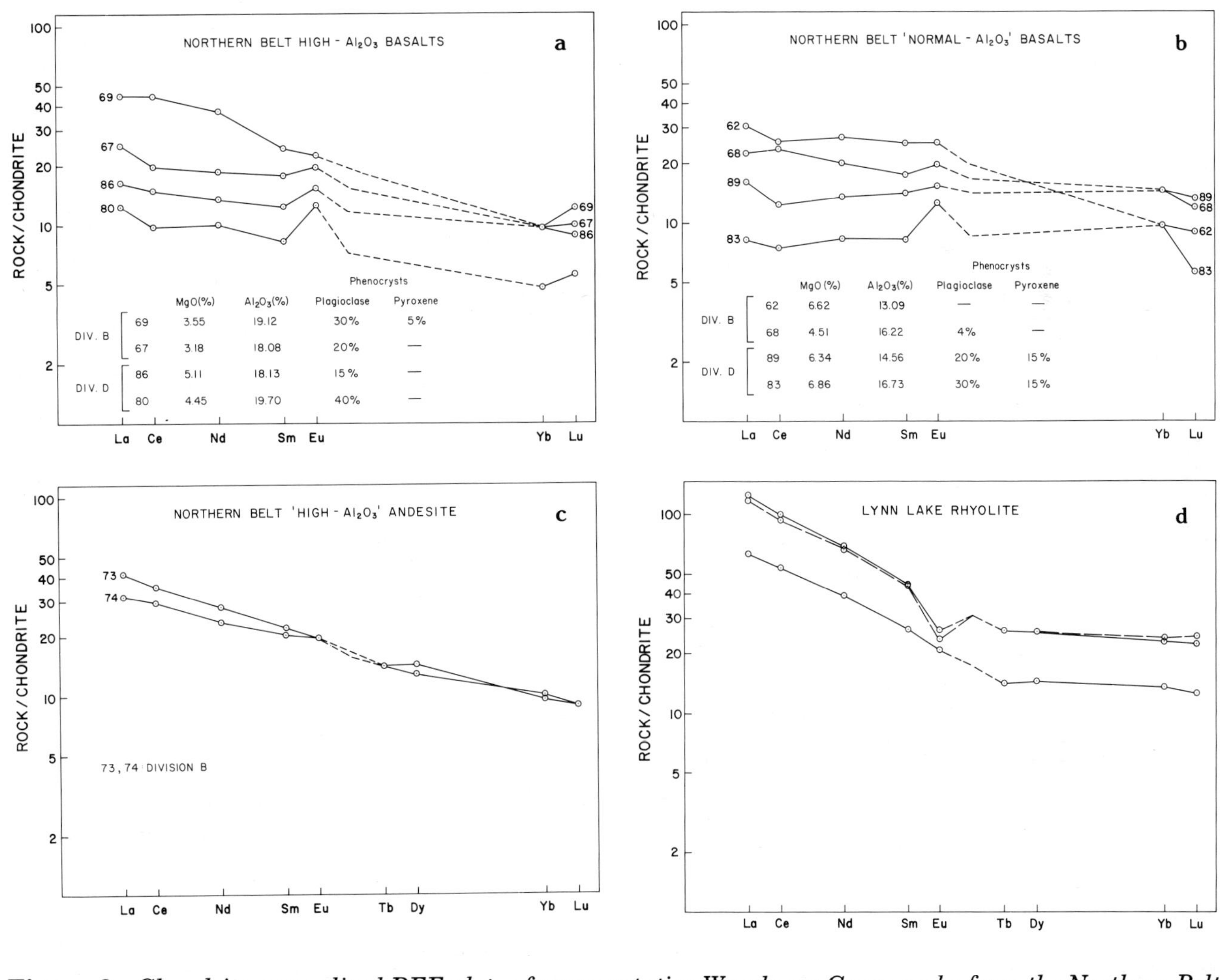

Figure 8. *Chondrite-normalized REE plots of representative Wasekwan Group rocks from the Northern Belt: (a) high-alumina basalts, (b) "normal"-alumina basalts, (c) high-alumina andesites, and (d) rhyolite. REE analyses are listed in Syme (1985).*

faults bound major stratigraphic blocks (Fig. 9). The implications of the fault-block style of deformation are far reaching. Each block has a more or less intact, definable stratigraphy that is unique to that block; correlation of sequences or individual units across block-bounding faults is impossible. Consequently, a comprehensive, composite or regional Amisk Group stratigraphy cannot be established.

Most faults appear to have been initiated at different times during the final, P5, deformation event, and as a result display different degrees of flexure about the P5 Embury Lake antiform (Stauffer and Mukherjee, 1971). The block-bounding faults also juxtapose domains with different styles of deformation, suites of intrusions, and grades of metamorphism. Regional metamorphism in the Flin Flon area increases in grade to the north, from subgreenschist (prehnite–pumpellyite) at Hook Lake to middle greenschist at Ross Lake (Fig. 9).

In the Flin Flon – Athapapuskow Lake area, at least 13 fault-bounded blocks have now been recognized (Bailes and Syme, 1987; Syme, 1987). The stratigraphic sequence in two of these, the Bear Lake block and Hook Lake block, is outlined below (summarized from Bailes and Syme (1989)). These blocks were chosen because they represent some of the thickest intact sequences in the area, and they contain representative lithologies.

Bear Lake block

The Bear Lake block contains a 5 km thick, predominantly east-facing sequence of lava flows and volcaniclastic rocks (Fig. 10). The basal unit comprises 3.3 km of basaltic andesite flows and sub-

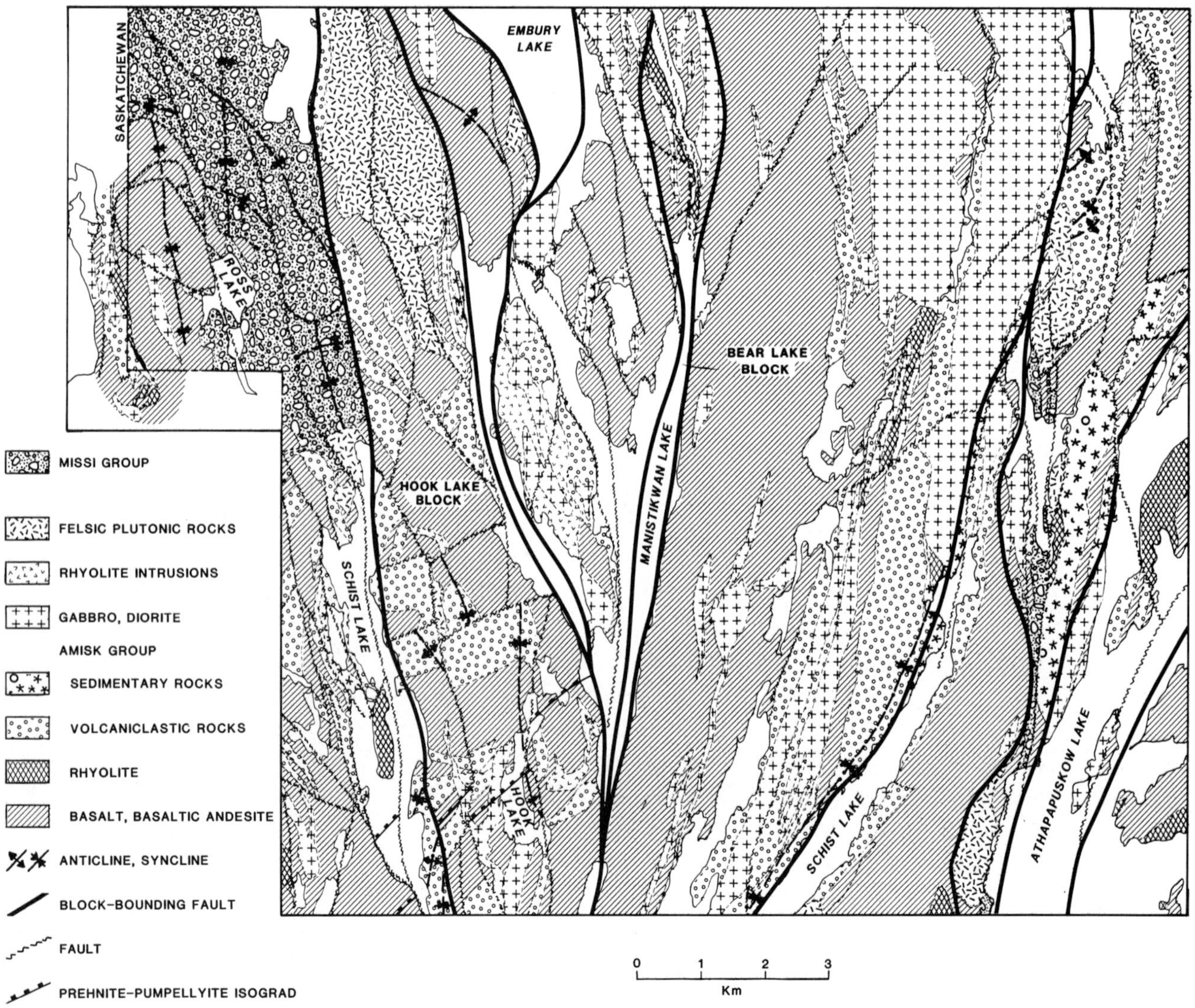

Figure 9. *Generalized geology of the Flin Flon – White Lake area (simplified from Bailes and Syme (1987)). Block-bounding faults are shown as heavy solid lines.*

ordinate pillow fragment breccias. Through the unit there is an upward decrease in average flow thickness and an increase in the proportion of massive to pillowed flows, paralleled by an upward increase in flow vesicularity and the proportion of associated vesicular breccias. The entire unit is interpreted as a shoaling subaqueous shield volcano.

Following the cessation of basaltic andesite volcanism, a large basin (possibly a caldera) formed in what was the summit region of the volcano. A subaqueous rhyolite flow complex up to 250 m thick occurs at the margin of the basin, and is flanked by volcaniclastic basin-fill deposits. These volcaniclastic rocks include (1) heterolithologic breccia derived, in part, from the rhyolite; (2) dacitic tuff and reworked tuff, in large part derived from pyroclastic components of the felsic volcanism; and (3) redeposited andesite lapilli tuff derived from a volcanic centre not exposed in the present sequence. The upper part of the sequence contains fine-grained sedimentary rocks and stratabound massive sulphide deposits interpreted to represent smaller basins within the larger structure.

The succeeding unit comprises massive and pillowed ferrobasalt flows and interflow rhyolite crystal tuffs (Fig. 10). The close association between highly fractionated basalt flows and rhyolite tuff suggests cyclic eruption of rhyolite–basalt pairs from a compositionally zoned, fractionated subvolcanic reservoir.

The topmost major unit in the Bear Lake block comprises 900 m of andesitic tuffs (Fig. 10). These andesitic pyroclastic rocks signal a major compositional and depositional change from the conditions that produced the underlying bimodal basalt–rhyolite unit. The sequence is broadly upward fining and is interlayered near the top with pyritic, graphitic mudstones. Preservation of shards, pumice clasts, and juvenile lithic fragments permits unambiguous interpretation of this unit as pyroclastic; turbidite bed forms indicate deposition from turbulent subaqueous density currents. The features of this unit are consistent with an origin from shallow water or subaerial phreatomagmatic and Plinian eruptions, with contemporaneous gravity flow transport of the pyroclastic materials into an adjacent deep-water basin.

Hook Lake block

The stratigraphic section contained in the Hook Lake block is folded about northwest-trending axial planes and is broken by several faults (Fig. 9). Faults divide the Hook Lake block into five major subblocks, each containing a portion of the complete stratigraphic section (Fig. 11). The composite stratigraphic section is approximately 7.5 km thick, and represents the thickest complete section in the Flin Flon area.

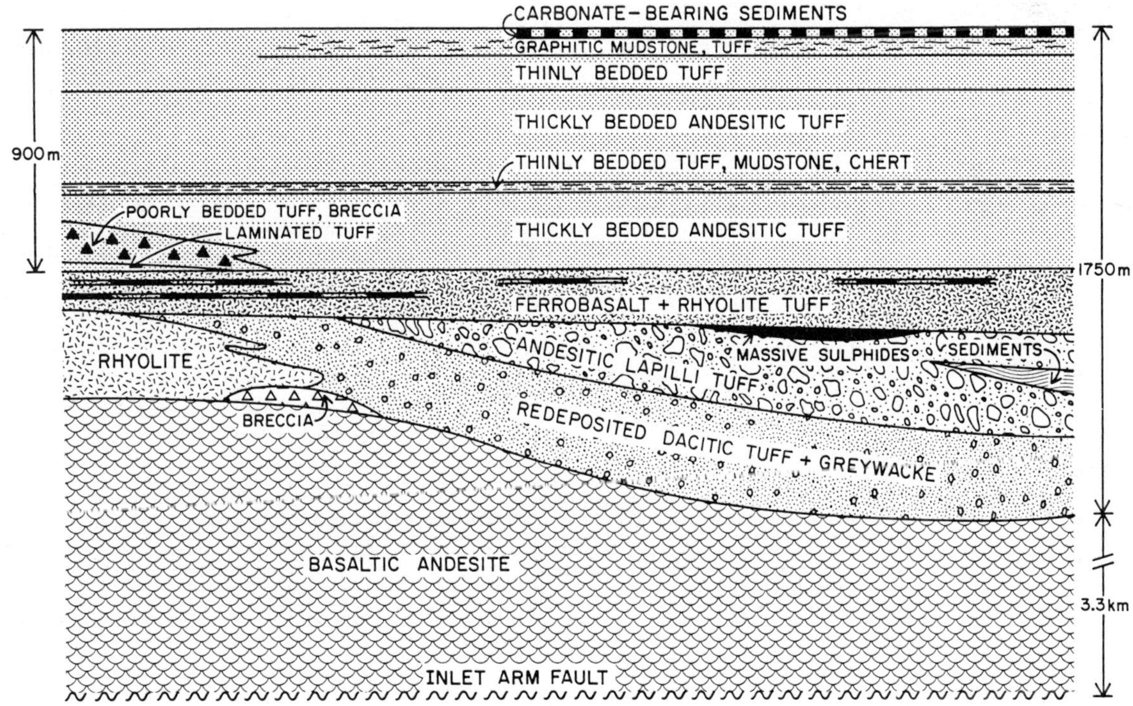

Figure 10. *Stratigraphic relationships of Amisk Group formations in the Bear Lake block, Flin Flon – White Lake area.*

Basalt, basaltic andesite, and rare andesite, in the form of subaqueous lava flows (79% of sequence); pillow fragment breccias (14%); and scoriaceous tuff and breccia (7%) compose the stratigraphic section. The mafic flows occur in mappable units 120–1000 m thick, with each unit defined by phenocryst type and abundance. Contacts between these major units are sharp. Pillow fragment breccia units up to 820 m thick are composed of thick breccia beds; as in the flows, these units can be distinguished on the basis of their phenocryst content.

In contrast to the Bear Lake basaltic andesite, mafic flows in the Hook Lake sequence do not display a well-defined upward increase in vesicularity: strongly amygdaloidal members are present throughout the sequence. The Hook Lake section also does not display any large-scale systematic stratigraphic variations in flow thickness or flow organization. Thick units of pillow fragment breccia (originating from the cold fragmentation of pillowed flows and deposited from subaqueous density currents) are most likely to have accumulated in intravolcanic basins or the lower flanks of subaqueous volcanoes.

Basaltic, scoria-rich tuffs and breccias (Fig. 11) were formed during very shallow water or subaerial

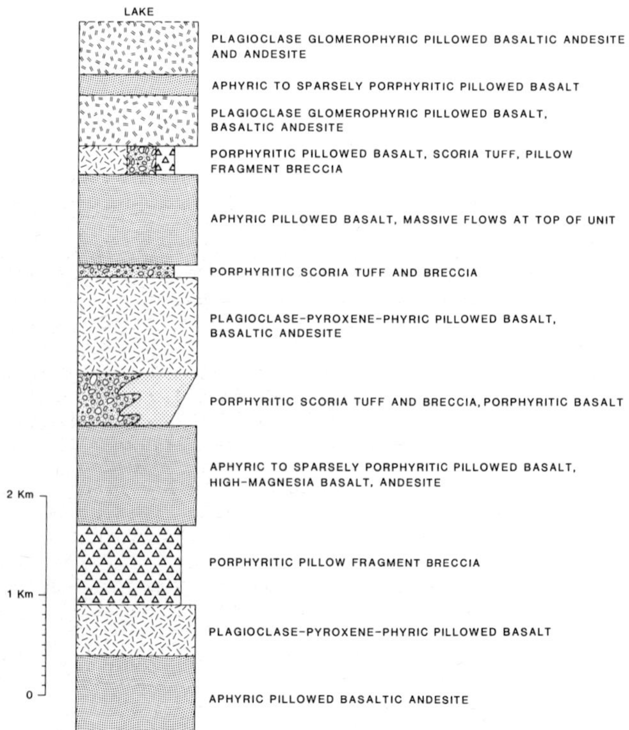

Figure 11. *Composite stratigraphic section of Amisk Group rocks in the Hook Lake block, Flin Flon – White Lake area.*

magmatic eruptions. Turbidite bed forms indicate that the pyroclastic material was transported and deposited by subaqueous density currents. These deposits comprise a small but ubiquitous component of the Hook Lake sequence, indicating that very shallow water or subaerial sources were present during subaqueous accumulation of this 7.5 km thick section.

Contrasts between the Bear Lake and Hook Lake blocks, in terms of stratigraphic components and sequence, inferred styles of eruption, and types of depositional basins, are typical for the 13 fault-bounded blocks recognized in the Flin Flon area. Not only are the blocks lithologically distinct, but the geochemical characteristics of the constituent components are also different.

Geochemistry

Average compositions of selected units in the Flin Flon area are listed in Table II. These units (defined in Bailes and Syme (1987) and Syme (1987)) were chosen to represent the compositional diversity displayed by Amisk Group rocks, particularly in the predominant, mafic portion of the group.

The mafic rocks can be subdivided into three main types: (1) those with high Rb, Ba, and K and low Hf, Zr, Y, Ni, and Cr contents (Bear Lake basaltic andesite, Hidden Lake basalt); (2) basalts characterized by high REEs, P, Hf, Zr, and Y and low Ni and Cr (Scotty Lake and Two Portage Lake basalts); and (3) basalts with low Ba and Th and higher Hf, Zr, Y, Ni, and Cr (Athapapuskow basalt). The implications of the trace-element subdivisions are discussed in the subsequent section.

Each distinct lithostratigraphic component of the Amisk Group can be separated on a variety of major-element variation diagrams (e.g., Fig. 12). The compositional differences correlate with lithologic characteristics of the various units (colour, flow type, flow thickness, vesicularity), so that virtually every major mafic unit in the belt can be mapped as a formation (Bailes and Syme, 1987). Basalts and basaltic andesites in the Amisk Group are dominantly tholeiitic (Fig. 13). Andesite flows (57–62% SiO_2) are extremely rare; some volcaniclastic units have andesitic bulk compositions (Table II), and these occur principally in the Bear Lake block. Rhyolite flows and central vent complexes are 100–900 m thick and compose less than 10% of the Amisk Group at Flin Flon.

Rare-earth element contents of Amisk Group types 1 and 3 mafic flows fall within a relatively narrow range at approximately 10 times chondritic abundances (Fig. 14a,b). The REE patterns are flat, unfractionated, and display neglible Eu anomalies. The REE abundances of type 2, including Scotty Lake and Two Portage Lake basalts, are higher than all other mafic units and chondrite-normalized patterns are slightly LREE enriched (Fig. 14a). High REE abundances, high incompatible trace-element contents, low Ni and Cr, and generally high Fe contents in these rocks suggest that they are the products of significant fractionation.

The REE patterns of rhyolites in the Flin Flon area (Fig. 14c) vary from flat (with slight middle REE depletion) to strongly LREE enriched; most (but not all) have negative Eu anomalies. The variation in REE and trace-element abundances (Table II) probably reflects different source and fractionation paths; similar variation occurs at Lynn Lake.

TECTONIC SETTING

The tectonic environment in which the metavolcanic rocks were emplaced can be interpreted on a broad scale simply by considering the geological setting and nature of the Flin Flon and Lynn Lake belts. Refinement of the interpretation, in terms of an island-arc, transitional, or continental margin model, can be made by considering the trace-ele-

Table II
Average compositions of representative metavolcanic rocks, Flin Flon

	1	2	3	4	5	6	7	8	9	10
SiO_2(wt.%)	53.08	49.98	48.40	51.43	47.43	55.13	56.09	75.65	77.48	78.87
TiO_2	0.47	0.61	1.18	1.48	2.12	0.67	0.42	0.26	0.14	0.13
Al_2O_3	15.43	16.32	16.37	14.21	13.08	17.70	14.64	11.86	11.70	10.90
Fe_2O_3	2.46	2.89	5.52	2.76	2.67	2.70	2.93	0.27	0.08	0.19
FeO	7.74	9.37	6.62	9.84	13.36	5.24	8.31	1.24	0.66	1.03
MnO	0.18	0.25	0.18	0.22	0.22	0.15	0.19	0.02	0.03	0.02
MgO	5.99	5.02	6.49	4.85	5.44	2.78	4.23	0.53	0.01	0.44
CaO	7.53	8.58	8.99	7.80	8.72	6.14	5.82	0.55	1.01	0.43
Na_2O	2.99	3.21	2.80	2.80	2.70	3.66	1.58	2.76	5.80	4.49
K_2O	0.90	0.42	0.26	0.69	0.52	3.09	1.52	5.03	0.39	1.99
P_2O_5	0.14	0.10	0.15	0.20	0.28	0.36	0.29	0.10	0.01	0.03
H_2O	3.15	2.97	3.50	3.39	3.04	2.44	3.54	0.52	0.28	0.46
S	0.03	0.01	0.01	0.06	0.18	0.05	0.03	0.04	0.01	0.01
CO_2	0.23	0.48	0.48	0.26	0.55	0.12	0.69	0.44	0.70	0.47
Other	0.10	ND	0.18	0.08	0.08	0.24	0.12	0.26	ND	0.07
Rb(ppm)	10	3	4	9	3	72	27	57	6	22
Ba	336	219	53	232	192	1054	395	1830	105	403
Sr	184	177	220	205	171	618	395	65	30	90
Th	0.34	0.23	0.2	2.9	0.2	ND	ND	ND	ND	ND
Nb	2	4	8	9	16	ND	ND	ND	ND	ND
Hf	1.4	1.0	2.5	4.0	3.7	ND	ND	ND	ND	ND
Zr	22	19	71	89	82	55	ND	45	75	289
Y	17	15	20	36	33	26	ND	14	39	74
Ni	28	5	105	12	55	15	2	2	4	2
Cr	94	74	274	37	70	53	15	3	26	4
No of samples	9	12	3	6	5	8	4	1	1	3

Notes: Columns:
(1) Bear Lake basaltic andesite, Bear Lake block.
(2) Hidden Lake basalt.
(3) Athapapuskow basalt.
(4) Scotty Lake basalt.
(5) Two Portage Lake ferrobasalt, Bear Lake block.
(6) Vick Lake andesitic tuff, Bear Lake block.
(7) Little Spruce Lake andesitic lapilli tuff, Bear Lake block.
(8) Solodiuk Lake rhyolite, Bear Lake block.
(9) Mine rhyolite (host of the Flin Flon Cu-Zn deposit).
(10) Grassy Narrows rhyolite.

ND, not determined. Individual analyses used to calculate averages are detailed in Bailes and Syme (1989). Analytical methods: SiO_2, Al_2O_3, TiO_2, and P_2O_5 by colorimetry; Fe_2O_3, FeO, CaO, MgO, Na_2O, K_2O, MnO, Ni, and Cr by atomic absorption (Manitoba Energy and Mines Laboratory); trace elements by X-ray fluorescence spectroscopy (University of Ottawa); and REEs, Hf, and Th by instrumental neutron activation analysis (Nuclear Activation Services, Becquerel Laboratories).

ment contents of mafic volcanic rocks in the two belts.

In the very simplest paleogeographic terms, both the Lynn Lake and Flin Flon belts form parts of what were originally major chains of submarine to subaerial volcanoes in an oceanic setting. Both the Amisk and Wasekwan groups consist of volcanic and volcaniclastic rocks deposited in a subaqueous environment. The volcanic rocks are interpreted to have originally formed overlapping volcanic constructs, some of which are flanked by aprons of volcaniclastic rocks and tuff. Locally, there is evidence that portions of the volcanic edifices emerged above sea level to form ephemeral volcanic islands. The Lynn Lake and Flin Flon metavolcanic belts are flanked by deep water, greywacke turbidite basins (Southern Indian and Kisseynew gneiss belts, Fig. 1); the metasedimentary rocks interdigitate with the metavolcanic rocks, suggesting contemporaneous deposition (Bailes, 1980; Gilbert *et al.*, 1980). The basement upon which the metavolcanic rocks and flanking metaturbidites were deposited has not been identified; however, Nd isotope data from the Flin Flon and La Ronge metavolcanic belts indicate that it must have been dominated by juvenile Proterozoic material (Chauvel *et al.*, 1987).

Volcanic rocks from different tectonic settings have distinct MORB-normalized trace-element characteristics (e.g., Pearce, 1982; Ewart and Hawkesworth, 1987); these can be used to interpret the tectonic setting of volcanic rocks from regions where the original tectonic environment has been obscured by subsequent deformation (e.g., Weaver *et al.*, 1982; Pharaoh and Pearce, 1984; Condie, 1986). These MORB-normalized trace-element plots display in a different manner the same information as conventional "tectonic discriminant" diagrams (e.g., Pearce and Cann, 1973); however, many more elements can be represented than is possible on binary or ternary "discriminant" diagrams and relationships between different groups of elements are readily determined.

Amisk Group

MORB-normalized trace-element data for representative Amisk Group basalts (Fig. 15) reflect the threefold subdivision established for these rocks. Bear Lake and Hidden Lake basalts (type 1) are enriched in LIL elements (Rb, Ba, K, Th, Sr) and are moderately to strongly depleted in HFS elements (P, Hf, Zr, Ti, Y), resulting in patterns very similar to an average island-arc tholeiite (Fig. 15a). Decoupling of LIL and HFS elements is a common feature of subduction related magmas (Saunders *et al.*, 1980; Gill, 1981; Tarney *et al.*, 1981). LIL-element enrichment is commonly attributed to transporting these elements to the fore-arc mantle wedge from the subducting slab during dehydration (e.g., Saunders *et al.*, 1980). HFS-element depletion is interpreted to result from either (1) derivation of arc magmas from a depleted lherzolite source (Ewart and Hawkesworth, 1987) or (2) partial melting under high $P(H_2O)$ conditions in subduction zones, stabilizing HFS-element-bearing minor phases (such as sphene, ilmenite, or rutile) in the source residuum and thereby depleting derived melts of these elements (Saunders *et al.*, 1980). Relative abundances of LIL and HFS elements in most Amisk Group basalts, typified by the Bear Lake and Hidden Lake suites, constitute strong

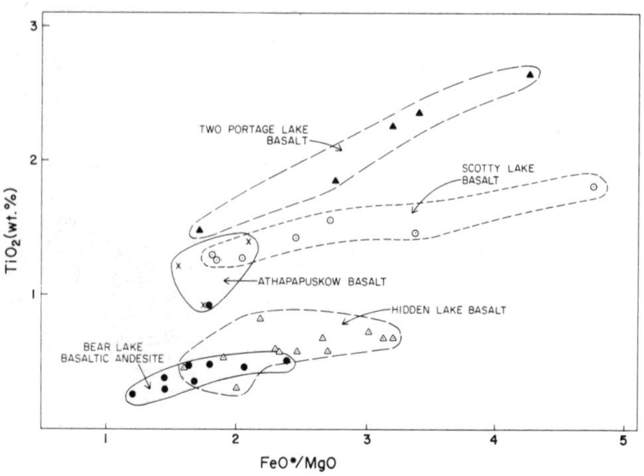

Figure 12. *TiO_2 versus FeO^*/MgO for representative Amisk Group basalts in the Flin Flon area. Most basaltic map units in the Flin Flon Belt have unique major- and trace-element signatures, which aid unit definition and correlation.*

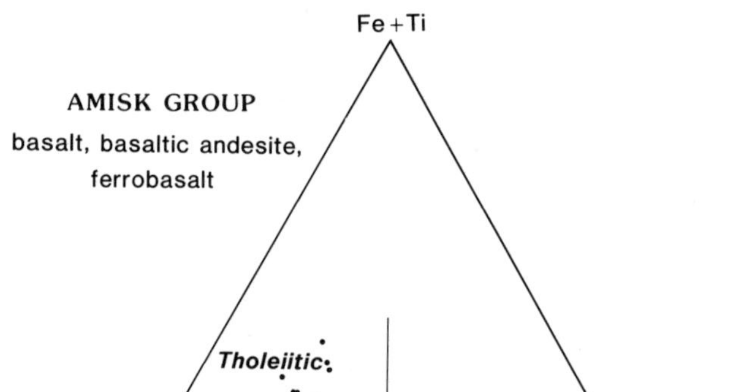

Figure 13. *Chemical variation of Amisk Group basalts and basaltic andesites on a Jensen (1976) cation plot.*

evidence that these Proterozoic volcanic rocks were generated by subduction-related processes similar to those operating in Cenozoic island arcs.

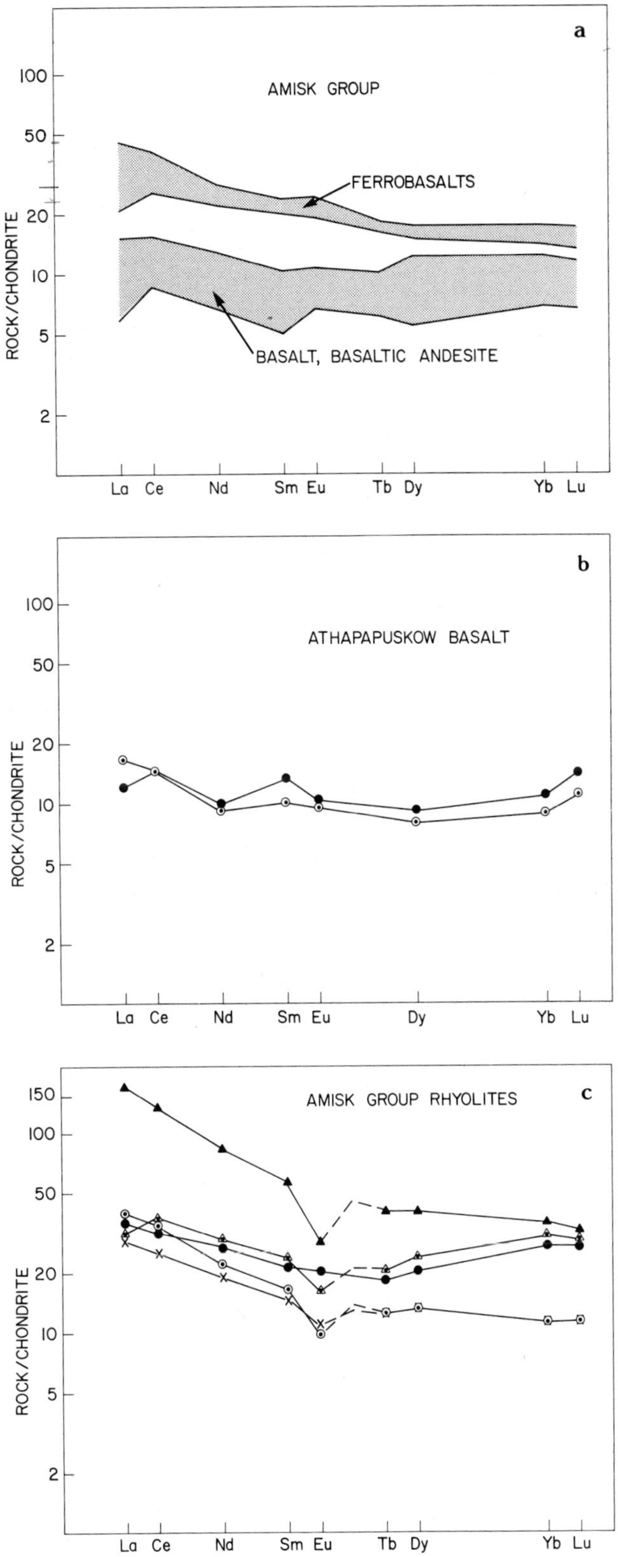

Two Portage Lake and Scotty Lake basalts (type 2) have high incompatible element and REE abundances, and low Ni and Cr (Fig. 15b), identifying these rocks as fractionated basalts. The Two Portage ferrobasalts are intercalated with rhyolite crystal tuff, suggesting that this bimodal suite was derived from a chemically zoned, fractionated magma chamber.

Athapapuskow basalts (type 3) have a trace-element pattern characterized by little enrichment of LIL elements and little depletion of HFS elements, combined with relatively high Ni and Cr (Fig. 15c). This flat pattern is characteristic of basalts erupted at spreading centres; the range of trace-element contents in basalts within a modern back-arc basin displays a similar pattern (Fig. 15c). Chondrite-normalized REE data for Athapapuskow basalts (Fig. 14b) fall within the general range displayed by arc-type Amisk Group basalts (Fig. 14a). Both occur within the range of midoceanic-ridge basalt REE patterns, as in modern island-arc to back-arc suites (Ewart and Hawkesworth, 1987). Basalts with an oceanic, rift-type geochemistry have also been described in the East Amisk Lake area of the Flin Flon Belt in Saskatchewan (Parslow and Gaskarth, 1984). Athapapuskow basalts constitute a distinct and mappable sequence defined by flow morphology and aeromagnetic signature, in addition to geochemical characteristics. The area where these basalts occur is currently the focus of a mapping project (Syme, 1986, 1987), and the reconnaissance analyses summarized here have been augmented in a thorough sampling program. Mapping indicates that the arc-type and "MORB"-type basalts are everywhere in fault contact.

Wasekwan Group

MORB-normalized trace-element data for representative Wasekwan Group rocks from the Lynn Lake Belt are shown in Figure 16. Cockeram Lake tholeiite (Fig. 16a) is enriched in LIL elements and depleted in Hf, Zr, Ti, Y, Ni, and Cr, similar to island-arc tholeiites. Trace-element ratios (previously discussed) and relatively high REE abundances (Figs. 5a, 16a) suggest that, similar to some basalts in the Flin Flon area, this major unit in the Southern Belt represents a somewhat fractionated or evolved basalt. In its ratios of Ti/Zr, Zr/Y, and Hf/Th, Cockeram Lake basalt resembles

Figure 14. *Chondrite-normalized REE plots of representative Amisk Group rocks: (a) ranges for basalt and basaltic andesite (type 1, seven samples), and ferrobasalt (type 2, three samples); (b) Athapapuskow basalt (type 3); and (c) rhyolites. REE analyses are listed in Bailes and Syme (1989) and Syme (unpublished data; Athapapuskow basalt).*

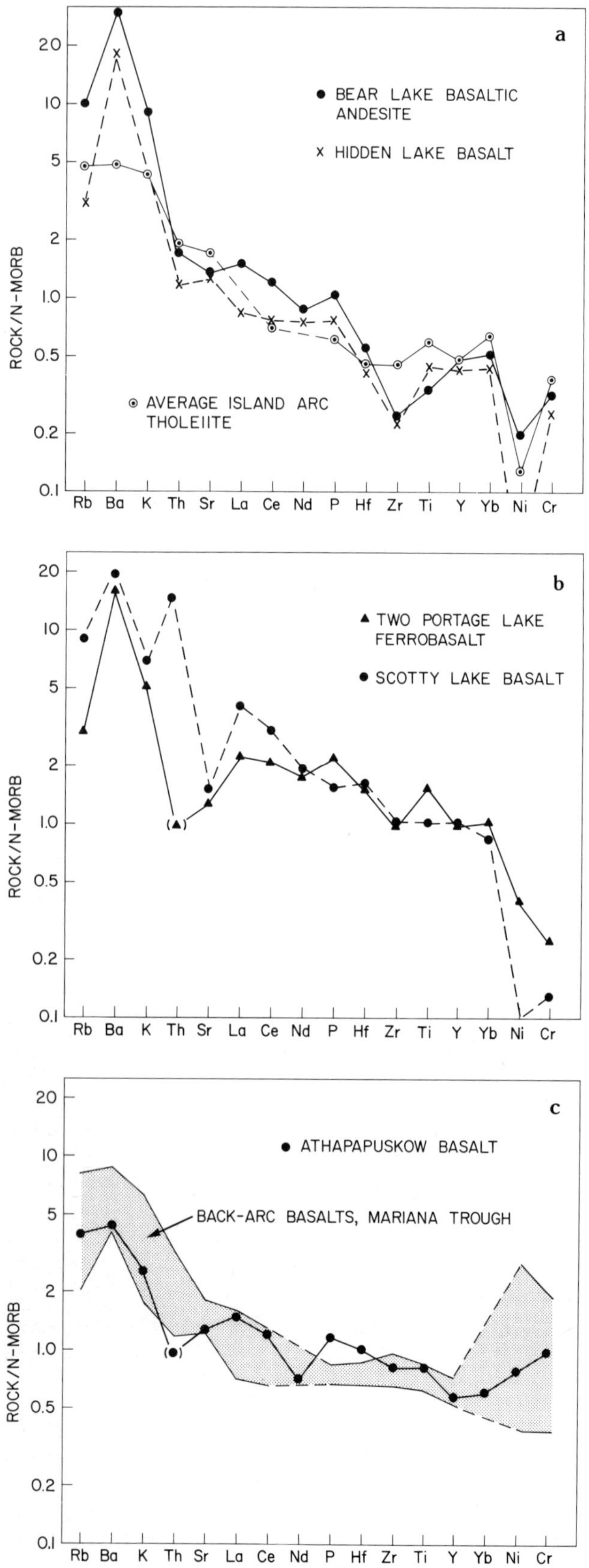

the "medium to high HFS-element" group of Gill (1987), which, together with "low HFS-element" basalts (the classic island-arc tholeiites), characterizes the early stage of development in some modern arcs.

"Primitive" calc-alkaline basalt and Hughes Lake calc-alkaline andesite are moderately to strongly enriched in LIL elements, LREEs, and P relative to N-MORB, are depleted in Hf, Ti, Y, HREEs, and Ni, and have a similar pattern to an "average" island-arc calc-alkaline basalt (Fig. 16b). Calc-alkaline rocks in Cenozoic intraoceanic island arcs commonly display mildly enriched LREE patterns (La_N/Yb_N = 1.5–2.5), with Yb_N 7 to 16 times chondrite (e.g., Arth, 1981; Meijer and Reagan, 1981; Dixon and Stern, 1983), although some arcs display considerably stronger LREE enrichment (La_N/Yb_N to about 6; Gorton (1977)). In contrast, continental margin calc-alkaline rocks are characterized by marked LREE enrichment (e.g., La_N/Yb_N averaging 11.5 in the Central Andes; Dostal *et al.* (1977)). Wasekwan Group calc-alkaline andesites, with La_N/Yb_N = 3.7–6.0 and Yb_N 10.6–12.5 (Syme, 1985), are similar to the more evolved arc lavas. They plot in the "evolved oceanic arc" field in La/Yb versus Sc/Ni and Th plots (fields defined in Condie (1986)).

High-alumina basalts from the Northern Belt display parallel patterns on a MORB-normalized trace-element diagram (Fig. 16c), with the stratigraphically lower Division B basalts clearly more enriched in incompatible elements than the overlying Division D basalts. Both are strongly depleted in Ni and Cr, and show decoupling of LIL and HFS elements, characteristic of island-arc volcanic rocks.

DISCUSSION

The Lynn Lake and Flin Flon metavolcanic belts display considerable internal variation in terms of both lithology and geochemistry. Most of the mafic volcanic components in the Wasekwan and Amisk groups can be interpreted, on the basis of major- and trace-element compositions, to have been emplaced in an island-arc environment. Some of the basaltic piles are, however, clearly fractionated

Figure 15. *MORB-normalized trace-element patterns for representative Amisk Group rocks: (a) average Bear Lake basaltic andesite and Hidden Lake basalt (type 1), with average island-arc tholeiite (Pearce, 1982) for comparison; (b) average Scotty Lake and Two Portage Lake ferrobasalts (type 2); and (c) average Athapapuskow basalt (type 3), with the range for back-arc basin basalts in the Mariana Trough (Saunders and Tarney, 1984) shown for comparison. N-MORB normalizing values are those of Saunders and Tarney (1984).*

in that relatively high total incompatible element and REE abundances are superimposed on a basic pattern of LIL and HFS element decoupling.

The Lynn Lake Belt, at 1910 Ma (Baldwin *et al.*, 1987) the oldest of the Proterozoic metavolcanic belts in Manitoba, is geochemically the most evolved. The lowermost preserved portion of the Wasekwan Group consists of contemporaneous magnesian calc-alkaline basalts and fractionated arc-type tholeiite. Calc-alkaline andesitic rocks, which overlie the mafic platform (Fig. 3), have REE abundances that suggest that they were emplaced in an evolved oceanic arc. The temporal relationship of tholeiites and overlying calc-alkaline rocks displayed in the Southern Belt is typical of some modern oceanic arcs that evolve from arc tholeiites variably depleted in incompatible trace elements to more enriched overlying rocks (Gill, 1987). High-alumina basalts in the Northern Belt, possibly forming the upper part of the Wasekwan Group, display an "inverted" sequence from rhyolite, through fractionated basalt, to younger, less fractionated basalt. The fundamental separation of the Lynn Lake metavolcanic belt into a northern, high-alumina belt and a southern, tholeiite plus calc-alkaline belt may reflect a large-scale primary separation of mainly contemporaneous magma types. Chemical zonation parallel to the axes of volcanic belts, resulting in compositional variation across the belts, is a feature of Cenozoic island arcs in an advanced stage of evolution (Mitchell and Reading, 1971; Colley and Warden, 1974). Stratigraphic and geographic chemical variation in the Lynn Lake Belt is broadly comparable to the chemical variation in evolved and tectonically complex Cenozoic island arcs.

The Flin Flon Belt is also marked by internal variation, but appears to represent a less evolved arc than the Lynn Lake Belt. Both the Amisk and the Wasekwan groups plot within the field of arc lavas on a Ti–Zr plot (Fig. 17), but there is very little overlap between the two groups: Amisk Group rocks generally contain slightly less Ti and much less Zr than their Wasekwan Group counterparts. Most Amisk Group basalts (apart from obviously fractionated units) contain very low HFS-element and REE abundances. These basalts are similar to the "low HFS-element" group of island-arc tholeiites, which erupt at an early stage in modern arc development, occur only in arcs associated with

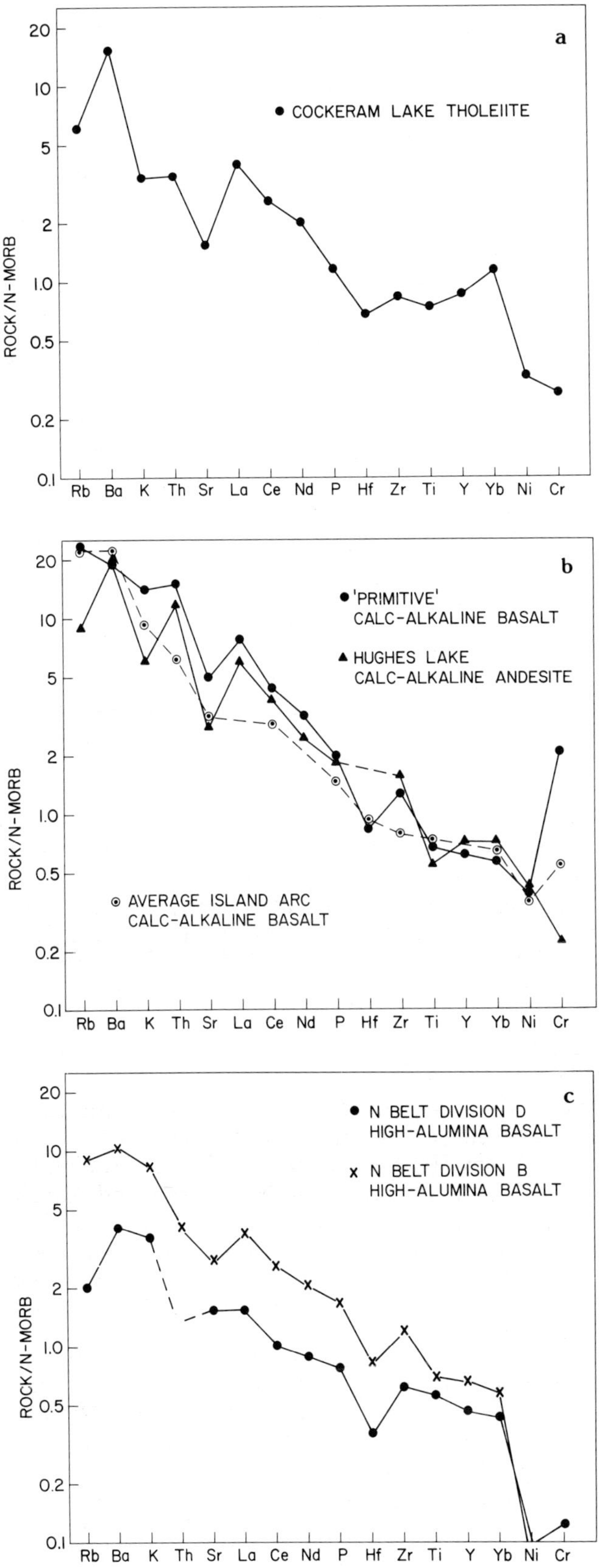

Figure 16. *MORB-normalized trace-element patterns for representative Wasekwan Group rocks: (a) average Cockeram Lake tholeiitic basalt; (b) average "primitive" calc-alkaline basalt and calc-alkaline andesite, with average island-arc calc-alkaline basalt (Pearce, 1982) for comparison; and (c) average high-alumina basalts from Division B and Division D, Northern Belt.*

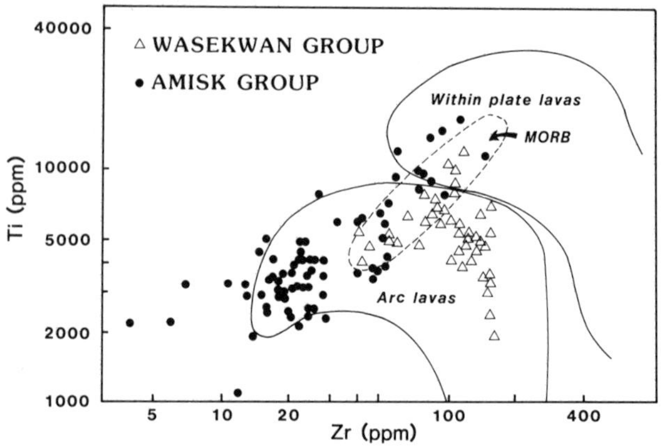

Figure 17. *Ti versus Zr, Amisk and Wasekwan groups. Fields of arc lavas, within-plate lavas, and MORB from Pearce* et al. *(1981). Amisk Group rocks generally contain significantly less Zr than Wasekwan Group rocks; Amisk basalts with higher Ti and Zr are predominantly fractionated type 2 ferrobasalts.*

rapid subduction of old lithosphere, and are often accompanied by the formation of back-arc basins (Gill, 1987).

A significant portion of the Amisk Group in the western part of the belt comprises lithologically, magnetically, and geochemically distinct basalts with trace-element abundances more akin to, for example, back-arc basalts than typical island-arc basalts. Structural juxtaposition of an arc–back-arc pair in the western Flin Flon Belt serves to emphasize the problems of stratigraphic correlation and compilation in this area. As detailed investigations continue in the Flin Flon Belt, there is the strong possibility that other distinct elements of the former arc may be identified.

ACKNOWLEDGEMENTS

The time-consuming process of detailed mapping in the Lynn Lake and Flin Flon belts has been shared over the past decade with my colleagues at the Manitoba Geological Survey: A.H. Bailes, H.V. Zwanzig, and H.P. Gilbert. Many of the ideas presented here have developed through our combined efforts. Drafts of this manuscript have benefited from comments by A.H. Bailes and W. Weber, and careful reviews by J.M. Moore and R.F. Emslie.

REFERENCES

Arth, J.G., 1981, Rare-earth element geochemistry of the island-arc volcanic rocks of Rabaul and Talasea, New Britain: Geological Society of America Bulletin, v. 92, p. 858–863.

Bailes, A.H., 1971, Preliminary compilation of the geology of the Snow Lake – Flin Flon – Sherridon area: Manitoba Mines Branch, Geological Paper 1/71, 27 p.

Bailes, A.H., 1980, Origin of Early Proterozoic volcaniclastic turbidites, south margin of the Kisseynew sedimentary gneiss belt, File Lake, Manitoba: Precambrian Research, v. 12, p. 197–225.

Bailes, A.H., 1986, Chisel–Morgan lakes project: Manitoba Energy and Mines, Report of Field Activities 1986, p. 70–76.

Bailes, A.H., 1987, Chisel–Morgan lakes project: Manitoba Energy and Mines, Report of Field Activities 1987, p. 70–79.

Bailes, A.H. and Syme, E.C., 1987, Geology of the Flin Flon – White Lake area: Manitoba Energy and Mines, Map GR87-1-1.

Bailes, A.H. and Syme, E.C., 1989, Geology of the Flin Flon – White Lake area: Manitoba Energy and Mines, Geological Report GR87-1, 313 p.

Baldwin, D.A., Syme, E.C., Zwanzig, H.V., Gordon, T.M., Hunt, P.A. and Stevens, R.D., 1987, U–Pb zircon ages from the Lynn Lake and Rusty Lake metavolcanic belts, Manitoba: two ages of Proterozoic magmatism: Canadian Journal of Earth Sciences, v. 24, p. 1053–1063.

Bateman, J.D., 1945, McVeigh Lake area, Manitoba: Geological Survey of Canada, Paper 45-14.

Bell, K., Blenkinsop, J. and Moore, J.M., 1975, Evidence for a Proterozoic greenstone belt from Snow Lake, Manitoba: Nature, v. 258, p. 698–701.

Campbell, F.H.A., 1969, Turnbull Lake area: Manitoba Mines Branch, Geological Paper 4/69, p. 15–20.

Chauvel, C., Arndt, N.T., Kielinzcuk, S. and Thom, A., 1987, Formation of Canadian 1.9 Ga old continental crust. I. Nd isotopic data: Canadian Journal of Earth Sciences, v. 24, p. 396–406.

Colley, H. and Warden, A.J., 1974, Petrology of the New Hebrides: Geological Society of America Bulletin, v. 85, p. 1635–1646.

Condie, K.C., 1986, Geochemistry and tectonic setting of Early Proterozoic supracrustal rocks in the southwestern United States: Journal of Geology, v. 94, p. 845–864.

Dixon, T.H. and Batiza, R., 1979, Petrology and chemistry of recent lavas in the northern Marianas: implications for the origin of island arc basalts: Contributions to Mineralogy and Petrology, v. 70, p. 167–181.

Dixon, T.H. and Stern, R.J., 1983, Petrology, chemistry and isotopic composition of submarine volcanoes in the southern Mariana arc: Geological Society of America Bulletin, v. 94, p. 1159–1172.

Dostal, J., Zentilli, M., Caelles, J.C. and Clark, A.H., 1977, Geochemistry and origin of volcanic rocks of the Andes (26°–28°S): Contributions to Mineralogy and Petrology, v. 63, p. 113–128.

Ewart, A. and Hawkesworth, C.J., 1987, The Pleistocene–Recent Tonga-Kermadec arc lavas: interpretation of new isotopic and rare earth data in terms of a depleted mantle source model: Journal of Petrology, v. 28, p. 495–530.

Fedikow, M.A.F., 1986, Geology of the Agassiz stratabound Au–Ag deposit, Lynn Lake, Manitoba: Manitoba Energy and Mines, Open File Report OF85-5, 80 p.

Gilbert, H.P., Syme, E.C. and Zwanzig, H.V., 1980, Geology of the metavolcanic and volcaniclastic metasedimentary rocks in the Lynn Lake area: Manitoba Mineral Resources Division, Geological Paper GP80-1, 118 p.

Gill, J.B., 1981, Orogenic Andesites and Plate Tectonics: Springer-Verlag, New York, 390 p.

Gill, J.B., 1987, Early geochemical evolution of an oceanic island arc and back arc: Fiji and the South Fiji Basin: Journal of Geology, v. 95, p. 589–615.

Gordon, T.M., Hunt, P.A., Loveridge, W.D., Bailes, A.H. and Syme, E.C., 1987, U–Pb zircon ages from the Flin Flon and Kisseynew belts, Manitoba: chronology of Early Proterozoic crust formation: Geological Association of Canada – Mineralogical Association of Canada, Program with Abstracts, v. 12, p. 48.

Gorton, M.P., 1977, The geochemistry and origin of Quaternary volcanism in the New Hebrides: Geochimica et Cosmochimica Acta, v. 41, p. 1257–1270.

Green, A.G., Hajnal, Z. and Weber, W., 1985, An evolutionary model of the western Churchill Province and western margin of the Superior Province in Canada and the north-central United States: Tectonophysics, v. 116, p. 281–322.

Gunn, B.M., 1976, A comparison of modern and Archean oceanic crust and island arc petrochemistry, *in* Windley, B.F., ed., The Early History of the Earth: John Wiley and Sons, Ltd., New York, NY, p. 389–403.

Haskin, L.A., Haskin, M.A., Frey, F.A. and Wildeman, T.R., 1968, Relative and absolute terrestrial abundances of the rare earths, *in* Ahrens, L.H., ed., Origin and Distribution of the Elements: Pergamon Press, New York, NY, p. 889–912.

Hoffman, P.F., 1988, United plates of America, the birth of a craton: Early Proterozoic assembly and growth of Proto-Laurentia: Annual Review of Earth and Planetary Sciences, v. 16, p. 543–603.

Jensen, L.S., 1976, A new cation plot for classifying subalkaline volcanic rocks: Ontario Division of Mines, Miscellaneous Paper 66, 22 p.

Lewry, J.F., 1981, Lower Proterozoic arc–microcontinent collisional tectonics in the western Churchill Province: Nature, v. 294, p. 69–72.

Lewry, J.F., 1984, Continental platforms and suspect terranes in the Lower Proterozoic "Trans-Hudson Orogen": Geological Association of Canada – Mineralogical Association of Canada, Program with Abstracts, v. 9, p. 84.

Lewry, J.F. and Sibbald, T.I.I., 1980, Thermotectonic evolution of the Churchill Province in northern Saskatchewan: Tectonophysics, v. 68, p. 48–82.

Lewry, J.F., Stauffer, M.F. and Fumerton, S.L., 1981, A Cordilleran-type batholithic belt in the Churchill Province in northern Saskatchewan: Precambrian Research, v. 14, p. 277–313.

McRitchie, W.D., 1974, The Sickle–Wasekwan debate: a review: Manitoba Mines Branch, Geological Paper 1/74, 23 p.

Meijer, A. and Reagan, M., 1981, Petrology and geochemistry of the island of Sarigan in the Mariana arc: calc-alkaline volcanism in an oceanic setting: Contributions to Mineralogy and Petrology, v. 77, p. 337–354.

Mitchell, A.H. and Reading, H.G., 1971, Evolution of island arcs: Journal of Geology, v. 79, p. 253–284.

Norman, G.W.H., 1933, Granville Lake District, northern Manitoba: Geological Survey of Canada, Summary Report, pt. C, p. 23–41.

Parslow, G.R. and Gaskarth, J.W., 1984, Geochemistry of the East Amisk area: Saskatchewan Energy and Mines, Open File Report OF84-23, 79 p.

Pearce, J.A., 1982, Trace element characteristics of lavas from destructive plate boundaries, *in* Thorpe, R.S., ed., Andesites: John Wiley and Sons, New York, NY, p. 525–548.

Pearce, J.A. and Cann, J.R., 1973, Tectonic setting of basic volcanic rocks determined using trace element analysis: Earth and Planetary Science Letters, v. 19, p. 290–300.

Pearce, J.A., Alabaster, T., Shelton, A.W. and Searle, M.P., 1981, The Oman ophiolite as a Cretaceous arc-basin complex: evidence and implications: Philosophical Transactions of the Royal Society, v. A300, p. 299–317.

Pharaoh, T.C. and Pearce, J.A., 1984, Geochemical evidence for the geotectonic setting of Early Proterozoic metavolcanic sequences in Lapland: Precambrian Research, v. 25, p. 283-308.

Saunders, A.D. and Tarney, J., 1984, Geochemical characteristics of basaltic volcanism within back-arc basins, *in* Kokelaar, B.P. and Howells, M.F., eds., Marginal Basin Geology: Blackwell Scientific Publications, p. 59–76.

Saunders, A.D., Tarney, J. and Weaver, S.D., 1980, Transverse geochemical variations across the Antarctic Peninsula: implications for the genesis of calc-alkaline magmas: Earth and Planetary Science Letters, v. 46, p. 344–360.

Stauffer, M.R., 1984, Manikewan: an Early Proterozoic ocean in central Canada, its igneous history and orogenic closure: Precambrian Research, v. 25, p. 257–281.

Stauffer, M.R. and Mukherjee, A., 1971, Superimposed deformations in the Missi meta-sedimentary rocks near Flin Flon, Manitoba: Canadian Journal of Earth Sciences, v. 8, p. 217–242.

Stauffer, M.R., Mukherjee, A.C. and Koo, J., 1975, The Amisk Group: an Aphebian (?) island arc: Canadian Journal of Earth Sciences, v. 12, p. 2021–2035.

Sun, S.S. and Nesbitt, R.W., 1978, Petrogenesis of Archean ultrabasic and basic volcanics: evidence from rare earth elements: Contributions to Mineralogy and Petrology, v. 65, p. 301–325

Sun, S.S., Nesbitt, R.W. and Sharaskin, A.Y., 1979, Geochemical characteristics of mid-ocean ridge basalt: Earth and Planetary Science Letters, v. 44, p. 119–138.

Syme, E.C., 1985, Geochemistry of metavolcanic rocks in the Lynn Lake belt: Manitoba Energy and Mines, Geological Report GR84-1, 84 p.

Syme, E.C., 1986, Schist Lake area (Athapapuskow project): Manitoba Energy and Mines, Report of Field Activities 1986, p. 36–42.

Syme, E.C., 1987, Athapapuskow Lake project: Manitoba Energy and Mines, Report of Field Activities 1987, p. 30–39.

Syme, E.C., Bailes, A.H., Gordon, T.M. and Hunt, P.A., 1987, U–Pb zircon geochronology in the Flin Flon belt: age of Amisk volcanism: Manitoba Energy and Mines, Report of Field Activities 1987, p. 105–107.

Tarney, J., Saunders, A.D., Mattey, D.P., Wood, D.A. and Marsh, N.G., 1981, Geochemical aspects of back-arc spreading in the Scotia Sea and Western Pacific: Philosophical Transactions of the Royal Society of London, v. A300, p. 263–285.

Van Schmus, W.R., Bickford, M.E., Lewry, J.F. and MacDonald, R., 1987, U–Pb geochronology in the Trans-Hudson Orogen, northern Saskatchewan, Canada: Canadian Journal of Earth Sciences, v. 24, p. 407–424.

Watters, B.R., 1981, Geochemistry of metavolcanic rocks of the La Ronge Domain: Saskatchewan Geological Survey, Miscellaneous Report 81-4, p. 34–37.

Watters, B.R., 1984, Geochemistry patterns for metavolcanic rocks of the La Ronge Domain: Saskatchewan Geological Survey, Miscellaneous Report 84-4, p. 88–91.

Watters, B.R., 1985, Geochemistry of metavolcanic and plutonic rocks from the Star and Weddy Lake areas: Saskatchewan Geological Survey, Miscellaneous Report 85-4, p. 28–35.

Watters, B.R., 1986, Geochemistry and geochronology of metavolcanic and plutonic rocks in the Central Metavolcanic Belt: Saskatchewan Geological Survey, Miscellaneous Report 86-4, p. 76–83.

Weaver, B.L., Tarney, J., Windley, B.F. and Leake, B.E., 1982, Geochemistry and petrogenesis of Archean metavolcanic amphibolites from Fiskanaesset, S.W. Greenland: Geochimica et Cosmochimica Acta, v. 46, p. 2203–2215.

Flin Flon and Western La Ronge Belts, Saskatchewan: Products of Proterozoic Subduction-related Volcanism

Andrea Thom,[1] Nicholas T. Arndt, and Catherine Chauvel
Max-Planck-Institut für Chemie, Postfach 3060,
D-6500 Mainz, Federal Republic of Germany

Mel Stauffer
Department of Geological Sciences, University
of Saskatchewan, Saskatoon, Saskatchewan S7N 0W0, Canada

Thom, A., Arndt, N.T., Chauvel, C. and Stauffer, M., Flin Flon and Western La Ronge belts, Saskatchewan: products of Proterozoic subduction-related volcanism, in Lewry, J.F. and Stauffer, M.R., eds., The Early Proterozoic Trans-Hudson Orogen of North America: Geological Association of Canada, Special Paper 37, p. 163–175.

Abstract

We report major and trace element data and new Nd isotopic analyses of Early Proterozoic (1.8–1.9 Ga) metavolcanic rocks of the Flin Flon and Western La Ronge belts of the Reindeer Lake Zone in Saskatchewan, Canada.

The petrological features (an association of basalts, andesites, and rhyolites; an abundance of embayed, glomerophyric plagioclase phenocrysts) and the geochemical characteristics of the rocks resemble those of modern subduction-related volcanics. Large-ion lithophile elements (LILE) range from moderately depleted to enriched, whereas heavy rare earth elements (REE) are relatively constant. The light rare earth element enriched samples show negative anomalies of the high-field-strength elements (HFSE; Nb, Hf, Zr, and Ti), whereas samples with flat patterns are characterized by concordant patterns and chondritic Zr/Sm and Ti/Eu ratios. ε_{Nd} values in the volcanic rocks vary widely, from +2 to +5. The volcanic rocks of the Flin Flon belt are interpreted to have formed in an environment like that of a modern intraoceanic island arc, and those in the Western La Ronge belt at an Andean-type continental margin.

---

[1]Present address: Research School of Earth Sciences, Australian National University, P.O. Box 4, Canberra 2601, Australia.

The volcanic rocks formed by a complex process that involved enrichment of mantle with material from the underlying subduction zone. Transportation was in aqueous fluids and the extent of enrichment of each element was proportional to its solubility in the fluid, i.e., LILE > REE > HFSE. The mantle subsequently melted and the magmas fractionated to produce basaltic to rhyolitic magmas. Some felsic lavas formed by remelting of mafic precursors. The low ε_{Nd} values of many lavas indicate the presence of old continental material in their source; presumably subducted sediments derived from adjoining Archean continents.

Résumé

On présente des données importantes sur les éléments majeurs et en trace et sur les analyses de Nd isotopique provenant de roches méta-volcaniques Protérozoïque Inférieur des ceintures Flin-Flon et La Ronge ouest de la zone Reindeer Lake en Saskatchewan, Canada.

Les caractéristiques pétrologiques (association de basaltes, d'andésites et de rhyolites; abondance de phénocristaux de plagioclases gloméro-porphyriques) et géochimiques sont comparables à celles des roches volcaniques modernes associées aux zones de subductions. Les éléments lithophiles (LILE) varient de modérément appauvris à enrichis alors que les terres rares lourdes (HREE) sont relativement constantes. Les échantillons enrichis en LREE montrent des anomalies négatives en éléments HFS (Nb, Hf, Zr et Ti), alors que les échantillons non enrichis montrent des patrons concordants et des rapports Zr/Sm et Ti/Eu chondritiques. Les roches volcaniques, montrent des valeurs d'epsilon Nd variant de +2 à +5. On considère que les roches volcaniques de la ceinture Flin-Flon proviennent d'un environnement d'intra-arc insulaire alors que celles de la ceinture La Ronge ouest proviennent d'une marge continentale andéenne.

Les roches volcaniques se sont formées par des processus complexes impliquant un enrichissement du manteaux par du matériel sous-jaçent aux zones de subductions. Les éléments sont transportés par des fluides aqueux et leurs concentrations sont proportionnelles à leurs solubilités, i.e., LILE > REE > HFSE. Des basaltes aux rhyolites, les roches sont formées par fusion du manteau et fractionnement subséquent des magmas; des laves acides sont formées par refonte des précurseurs basiques. Les valeurs basses d'epsilon Nd de plusieurs laves indiquent la présence de matériel continental ancien dans la source, présumément de sédiments subductés provenant des continents archéens attenants.

INTRODUCTION

Flin Flon and Western La Ronge are two lithotectonic belts containing abundant volcanic rocks that form part of the Reindeer Lake Zone of the Trans-Hudson Orogen, in Saskatchewan and Manitoba, Canada. These belts are two of eight Proterozoic lithotectonic belts arranged broadly parallel to the border of the Archean Hearne Province (Hoffman, 1988) to the northwest (Fig. 1).

Volcanic rocks of the Flin Flon and Western La Ronge belts range in composition from basalt to rhyolite and, as demonstrated later, have major and trace element compositions very similar to those of volcanic rocks at modern oceanic arcs and active continental margins. These features, as well as the location of the belts adjacent to an older cratonic area, provide strong evidence that these rocks erupted in a tectonic environment similar to that at a modern convergent plate boundary. Thus, the volcanic rocks of the Reindeer Lake Zone provide convincing evidence that plate tectonics operated during the Early Proterozoic. The intent of this paper is to document the petrological and chemical aspects of the volcanic rocks that can be used to infer the environment in which they formed, and to establish what information they provide about the tectonic regime and the processes of continent formation that operated in the Precambrian.

GEOLOGICAL BACKGROUND

The Trans-Hudson Orogen is a broad band of Early Proterozoic rocks interpreted as the collision zone between the Superior Province and the Archean provinces to the north and west (Hoffman, 1988). It extends from northern Quebec, across Hudson Bay to Manitoba, Saskatchewan, and the United States midwest, and is made up largely of rocks with crystallization ages between 1.95 and 1.8 Ga (Van Schmus *et al.*, 1987). In the northern part, reworked (metamorphosed, deformed, and remelted) Archean rocks dominate (Patchett and Arndt, 1986; Hoffman, 1988), but in the south, juvenile mantle-derived material is more common.

The Reindeer Lake Zone in northern Saskatchewan and Manitoba is a well-defined tectonic element bounded to the northwest by the Wollaston fold belt (highly deformed and metamorphosed Early Proterozoic sediments overlying Archean basement of the Hearne Province hinterland), to the southeast by the Archean Superior craton, and to the south by Paleozoic sedimentary cover. The geology of the area has been described by Lewry *et al.* (1981) and Stauffer (1984), who recognize eight constituent northeast-trending lithotectonic belts. Most of these consist largely of felsic gneisses, granitoids, and metasedimentary rocks, and only the Flin Flon and Western La Ronge belts contain voluminous metavolcanics.

The ages of rocks in the Reindeer Lake Zone have been shown by U–Pb zircon dating to range from 1.93 to 1.85 Ga (Van Schmus *et al.*, 1987). Only to the northwest, adjacent to the Wollaston Belt, and in some small inliers within the northern Hanson Lake Block and Glennie Domain are Archean ages reported.

PETROGRAPHY AND METAMORPHISM OF THE FLIN FLON AND WESTERN LA RONGE BELTS

In much of the Flin Flon belt, particularly in the area around the Flin Flon mine and town site, the metamorphic grade is lower greenschist facies and primary volcanic structures and textures are well preserved. In mafic units, pillowed and massive flows, in places columnar jointed, can be recognized, together with uncommon volcanogenic sediments. In felsic units, coarse breccias and flow-banding within rhyolite domes and plugs can be mapped in detail. Inspection of thin sections shows most of the rocks to be porphyritic. Plagioclase is the most abundant phenocryst phase in most mafic lavas, often comprising more than 30% of a thin section and forming large, complexly zoned, inclusion-riddled grains; in other samples, relict pyroxene phenocrysts are present. In felsic lavas and hypabyssal rocks, quartz and feldspar are the phenocryst phases. The groundmass is pseudomorphic pyroxene, plagioclase, and opaque minerals in mafic to intermediate rocks, and quartz and feldspar in felsic rocks. With quartz being the sole exception, these minerals are not fresh: felspars are saussuritized, and mafic minerals have been replaced by chlorite, amphibole, epidote, carbonate, and in some cases biotite.

In the Western La Ronge belt, the metamorphic grade is amphibolite facies. The rocks here are completely recrystallized and have been transformed into amphibolites, quartzofeldspathic gneisses, and schists. With the exception of a few pyroxene cores in some coarser grained samples, no relicts of primary minerals or textures are preserved, and, apart from a few poorly preserved pillows and flow breccias, all volcanic structures are obliterated.

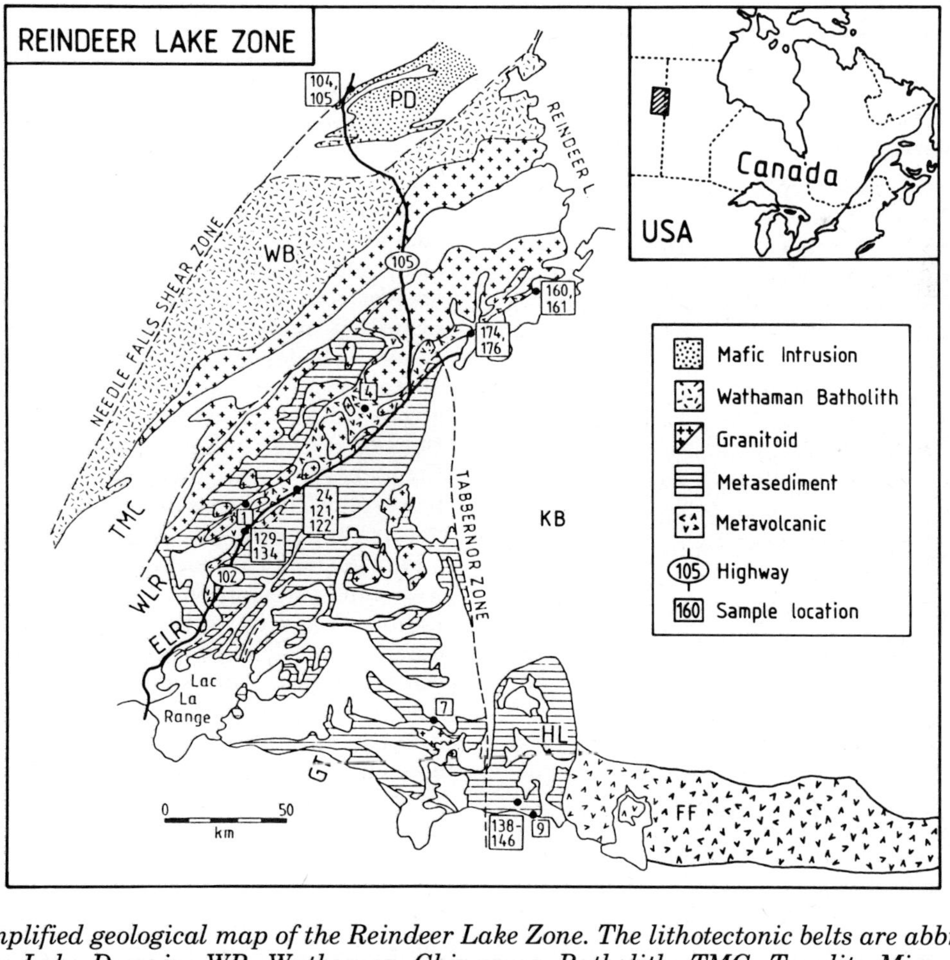

Figure 1. *Simplified geological map of the Reindeer Lake Zone. The lithotectonic belts are abbreviated as follows: PD, Peter Lake Domain; WB, Wathaman–Chipewyan Batholith; TMC, Tonalite–Migmatite Complex; WLR, Western La Ronge belt; ELR, Eastern La Ronge belt; GT, Glennie Terrain; HL, Hanson Lake Block; and FF, Flin Flon belt (from Bickford* et al., *1990).*

CHEMISTRY

Methods

Forty-two volcanic samples from the Flin Flon belt and 15 samples from the Western La Ronge belt were analyzed for major and trace elements. The following methods were used. Major elements were analyzed by X-ray fluorescence spectroscopy (XRF) at the University of Saskatchewan and the University of Freiburg, using the techniques of Nisbet *et al.* (1979) and Norrish and Hutton (1969). Trace elements (other than rare earth elements (REE)) were measured by XRF at Max-Planck-Institut (MPI) using techniques described by Feather and Willis (1976), by spark source mass spectrometry (SSMS) at MPI (Jochum *et al.*, 1988), and by inductive coupled plasma (ICP) at the Centre de Recherches Petrographique et Geochemiques (CRPG) (Govindaraju, 1984). REE and Nd isotopic compositions were analyzed by isotope dilution mass spectrometry at MPI using the methods described by White and Patchett (1984). Representative analyses, recalculated volatile free, and the analytical techniques used are reported in Tables I and II.

Major elements

Within each belt, rock types range from basalt through andesite to rhyolite, and their compositions vary accordingly. SiO_2 ranges from 48 to 79%. Mg numbers (Mg#) vary from a maximum of 0.59 in basalts from the Western La Ronge belt to 0.10 in rhyolites from the Flin Flon belt. More evolved rocks (i.e., andesites, dacites, and rhyolites) tend to be more abundant in the Flin Flon belt.

Two groupings are observed in an AFM diagram (Fig. 2a,b). From both belts, some rocks show an iron-enrichment trend and can be considered tholeiitic; others show no iron enrichment and fall into the calc-alkaline field defined by Irvine and Baragar (1971). However, Figure 2 demonstrates that iron enrichment does not provide a useful parameter to distinguish between different types of volcanic rocks. The various symbols used in the diagrams distinguish between rocks with differing REE patterns, as discussed below.

Compatible trace elements

Relatively low Ni contents (<77 ppm in the Flin Flon belt, <99 ppm in the Western La Ronge belt), consistent with the low Mg numbers, imply that none of the samples represents a primary magma composition. Sc and Cr contents, on the other hand, are relatively high at 55–60 ppm and 378–427 ppm, respectively, and indicate a relatively low degree of fractionation.

Incompatible trace elements

The REE patterns can be used to divide the volcanic rocks into different groups (Fig. 3). In the Western La Ronge belt, three different rock types can be distinguished. Group A has slightly light rare earth element (LREE) depleted, midoceanic-ridge basalt (MORB)-like patterns with abundances of middle and heavy REE about five times those estimated for the primitive mantle (Fig. 3a). Group B basalts and andesites have moderately fractionated patterns with LREE-enrichment and slightly concave-upward middle to heavy REE: in basalts, heavy rare earth elements (HREE) are about three times primitive mantle; in evolved rocks (andesites to rhyolites) four to seven times primitive mantle (Fig. 3b). Group C consists of silicic rocks with patterns characterized by strong enrichment of LREE, but with low concentrations of HREE (lower than primitive mantle values) (Fig. 3c). For normalization in all diagrams, primitive mantle values of Sun and McDonough (1989) have been used.

Three similar magma types are found in the Flin Flon belt. Group A rocks from this area have flat to pronounced U-shaped patterns (Fig. 4a). In basalts, abundances are two to five times primitive mantle. Rhyolites in this group have abundances around 10 times primitive mantle and display negative Eu anomalies. Group B basalts and andesites are variably LREE enriched. The degree of LREE enrichment does not correlate with Mg#, nor with other fractionation indices. HREE abundances in basalts and andesites vary from two to six times primitive mantle (Fig. 4b). As in the Western La Ronge belt, felsic volcanics of Group C (here represented by a single sample) are characterized by strong LREE enrichment, but very low concentrations of HREE (lower than primitive mantle values; Fig. 4c). As is explained below, this grouping on the basis of REE abundance patterns can be used to identify magma types with different petrogeneses.

Petrologically important trace element ratios are well illustrated in the type of diagrams shown in Figures 5 and 6. In these diagrams, the position of a particular element on the x-axis corresponds to its incompatibility, determined using the procedure described by Hofmann (1988). Particularly significant are the relationships between rare earth elements, high-field-strength elements (HFSE) (Nb, Zr, Hf, and Ti), and the large-ion lithophile elements (LILE: Cs, Rb, Ba, K, and Sr). In Figures 5 and 6, we present trace element abundances in samples selected as being characteristic of the six main groups in the two lithotectonic belts. We emphasize two features: (1) enrichment of LILE relative to the REE, and (2) depletion of HFSE relative to the REE. In rocks with LREE enrichment, the HFSE have relatively low normalized concentrations and form negative anomalies. There is an ap-

Table I
Major and trace element concentrations of selected samples from Western La Ronge and Flin Flon

	HUD160 Group A	HUD160[a]	HUD161 A	HUD161[a] A	HUD174 B	HUD130	HUD130[a] B	HUD133 B	HUD4	HUD4[a]	HUD24 C
					Western La Ronge belt						
SiO_2	49.30F		54.40M		50.10F	48.90M		53.90M	73.30N		70.90N
TiO_2	1.15		0.73		0.70	0.37		0.63	0.39		0.32
Al_2O_3	14.50		15.10		15.60	18.80		17.10	12.48		15.60
FeO	12.30		12.60		10.30	10.50		11.60	5.05		2.30
MnO	0.22		0.15		0.10	0.04		0.13	0.06		0.03
MgO	8.50		5.50		8.70	6.80		2.60	1.38		0.99
CaO	10.50		8.20		11.60	11.40		6.40	0.96		2.97
Na_2O	2.90		3.00		2.50	2.02		3.70	4.30		4.40
K_2O	0.51		0.23		0.40	1.30		3.40	1.52		1.96
P_2O_5	0.22		0.20		0.10	0.16		0.51	—		0.11
Total	100.10		100.11		100.10	100.29		99.97	99.44		99.58
LOI	2.25		1.90		2.10	2.10		2.80	1.24		2.03
Mg#	0.55		0.44		0.60	0.54		0.29	0.33		0.43
Zr	62.3		40.3		38.4	22.5		60.5	149.0		138.5
Nb	1.6SS		1.3		1.9	1.5		2.3	4.5		3.8
Sr	164.4		102.0		182.0	516.0		346.0	74.0		246.0
Rb	10.4		2.5		7.7	24.7		65.1	17.5		38.4
Y	29.7		24.5		16.7	11.5		30.4	35.1		5.7
Ba	86.0SS		—		—	343.0SS		—	—		—
Cs	0.23SS		—		—	0.38SS		—	—		—
Sc	57.5		46.7		48.6	41.5		24.2	—		7.6
V	385.3		397.9		343.0	281.4		133.0	7.8		29.1
Cr	142.0		28.7		427.0	119.0		16.8	11.0		31.2
Ni	90.0		17.2		93.1	33.8		4.6	7.0		12.5
La	2.09		2.44		2.23	6.68		11.38	14.16		14.71
Ce	6.98		6.68		5.74	14.16		24.80	29.12		28.13
Nd	6.695	6.687	5.350	5.310	4.850	8.040	8.070	14.720	14.800	15.070	10.940
Sm	2.401	2.403	1.767	1.762	1.650	1.910	1.920	3.620	3.880	3.910	1.890
Eu	0.863		0.661		0.662	0.628		1.120	1.050		0.573
Gd	3.46		2.56		2.35	1.99		4.08	4.55		1.40
Dy	4.22		3.29		2.82	1.97		4.24	5.13		0.91
Er	2.73		2.28		1.75	1.23		2.81	3.30		0.39
Yb	2.60		2.30		1.68	1.23		2.85	3.31		0.33
Lu	0.394		0.377		0.260	0.196		0.444	0.506		0.054

Table I (continued)

	FF32 Group A	FF40 A	FF49 A	FF52 B	FF52[a] B	FF62 C	FF24	BHVO
					Flin Flon belt			
SiO_2	51.58S	71.24S	55.26S	58.88M		52.17	65.94S	
TiO_2	0.50	0.32	0.37	—		0.56	0.52	
Al_2O_3	16.95	12.21	15.88	15.35		15.37	16.31	
FeO	12.22	6.61	9.92	10.23		11.66	3.68	
MnO	0.23	0.12	0.14	0.13		0.21	0.05	
MgO	6.98	0.72	5.88	4.90		5.62	1.23	
CaO	7.00	7.80	10.52	5.27		11.40	5.04	
Na_2O	4.05	0.71	1.34	2.92		1.58	4.19	
K_2O	0.34	0.15	0.56	1.99		0.70	—	
P_2O_5	0.21	0.15	0.18	0.29		0.72	3.05	
Total	100.04	100.03	100.04	99.97		99.99	100.00	
LOI	3.20	2.10	2.40	3.67		4.27	1.84	
Mg#	0.50	0.16	0.51	0.46		0.46	0.37	
Zr	21.2	108.6	18.5	83.5		21.0	135.0	
Nb	0.4	0.7	0.6	4.0		0.5	4.1	
Sr	128.0	230.0	193.0	28.6		14.2	238.0	
Rb	4.5	6.6	0.6	456.0		533.0	28.4	
Y	19.5	69.8	11.2	30.2		11.7	5.9	
Sc	39.6	19.2	42.6	26.7		45.3	9.0	
V	314.4	—	257.2	315.0		324.7	52.7	
Cr	68.7	—	378.1	—		—	31.2	
Ni	46.5	—	76.6	37.6		15.7	8.8	
La		10.73	2.00[b]	20.36		4.95	15.54	15.54
Ce	6.09[b]	23.28	5.20	43.65		10.94	29.80	38.37
Nd	4.17	14.33	3.61	22.39	22.27	6.57	12.12	24.81
Sm	1.28	4.04	1.07	4.73	4.71	1.60	2.17	6.14
Eu	0.511	1.117	0.444	1.334		0.559	0.683	2.13
Gd	1.77	5.37	1.44	4.37		1.73	1.70	6.13
Dy	2.40	7.20	1.78	4.33		1.95	1.07	5.34
Er	1.73	5.36	1.24	2.73		1.30	0.451	2.56
Yb	1.81	5.72	1.30	2.75		1.29	0.374	2.00
Lu	0.280	0.900	0.200	0.420		0.200	0.0542	0.285

Notes: Major and trace elements are in weight per cent, recalculated to 100% water free. Water is determined as loss on ignition (LOI). Trace element concentrations are in ppm.-, not determined. Mg#, Mg-number = molecular proportion MgO/(MgO + FeOT). FeOT is total Fe as FeO. Major element data are from F, Freiburg (XRF); M, Mainz (XRF); N, Nancy (ICP); and S, Saskatoon (XRF). SS, trace elements by SSMS.

[a] Duplicate analyses.

[b] Unpublished data (C. Chauvel, 1986).

parent genetic relationship between depletion of HFSE and enrichment of LREE. Group A basalts from Flin Flon, which have U-shaped REE patterns (Figs. 3 and 4) and, thus, relative enrichment of LREE, have stronger HFSE anomalies than the MORB-like LREE-depleted basalts from the Western La Ronge belt, which have concordant Hf, Zr, and Ti abundances and puny Nb anomalies. Group C lavas from both belts have large negative Nb and Ti anomalies, but normalized Zr concentrations are higher than normalized Sm concentrations (Figs. 5c and 6c).

The enrichment of the LILE relative to the REE is a persistent characteristic ubiquitous in rocks from all magmatic suites and from both volcanic belts. This feature is most distinct in groups A and B. Flin Flon belt volcanics generally have less elevated LILE concentrations than rocks from the Western La Ronge belt.

In terms of this grouping of elements, Th and U behave in contrasting ways (Fig. 5a,b): U is relatively enriched and plots together with the alkali elements, but Th has concentrations like the REE and plots on extentions of REE trends.

Nd isotopic compositions

Ten new Nd isotopic compositions from volcanic rocks were measured (Table III). These results augment data reported earlier by Chauvel *et al.* (1987). ε_{Nd} (1.9 Ga) values range from +3.3 to +4.7 in the Flin Flon belt, and from +2.1 to +4.7 in the Western La Ronge belt. These ε_{Nd}values are among the highest reported for 1.9 Ga old rocks, being exceeded only by values of +5 for Mg basalts from the Fox River Belt in Manitoba (Chauvel *et al.*, 1987).

In both magmatic belts, ε_{Nd}correlates negatively with LREE enrichment (Fig. 7). Group A rocks have high ε_{Nd} and flat to depleted LREE, whereas group B rocks generally have lower ε_{Nd} commensurate with their enriched LREE. Felsic lavas of group C do not show this relationship: their elevated ε_{Nd} and enriched LREE reflect a petrogenesis distinct from those of the other two magma types.

DISCUSSION

Comparison of the Flin Flon and Western La Ronge belts

The principal differences between the compositions of rocks from the two areas are summarized in Table III. Most important are the generally more fractionated nature of the Flin Flon belt rocks (they are more siliceous and have lower Mg# and Ni contents); higher LILE, REE, and Zr abundances in the Western La Ronge belt; and differences in REE patterns in group A basalts from the two areas (U-shaped patterns in the Flin Flon belt and flat patterns in the Western La Ronge belt).

Table II
Sm–Nd isotopic data

Sample	SiO_2 (wt.%)	Sm (ppm)	Nd (ppm)	$^{147}Sm/^{144}Nd^{a}$	$^{143}Nd/^{144}Nd^{a}$	$\varepsilon_{Nd(T)}^{b}$
			Western La Ronge			
HUD4	73.3	3.91	15.07	0.1569	0.512289 ± 4	+2.67
HUD129	55.7	2.71	11.75	0.1393	0.512086 ± 14	+3.02
HUD130	48.9	1.92	8.07	0.1440	0.512096 ± 18	+2.07
HUD131	49.0	2.03	8.4	0.1459	0.512118 ± 14	+2.06
HUD160	49.3	2.4	6.69	0.2173	0.513132 ± 17	+4.72
HUD174	51.0	1.6	4.65	0.2076	0.512994 ± 14	+4.35
			Flin Flon			
FF52	58.5	4.71	22.28	0.1279	0.511953 ± 11	+3.17
FF63	53.2	1.44	6.12	0.1425	0.512219 ± 6	+4.75
FF95	59.8	3.38	14.31	0.1429	0.512154 ± 11	+3.48
FF96	54.8	3.29	15.66	0.1270	0.511954 ± 6	+3.31

[a] Isotopic data are normalized to La Jolla Standard $^{143}Nd/^{144}Nd$: 0.511860. Errors reported for the isotopic ratios are $2\sigma_m$; the error on the $^{147}Sm/^{144}Nd$ ratios is estimated to be 0.2%.

[b] For age correction $\lambda = 6.54 \times 10^{-12}$ year $^{-1}$ and an age of 1.87 Ga for the Saskatchewan area were used. ε_{Nd} was calculated as $((^{143}Nd/^{144}Nd_T)/(^{143}Nd/^{144}Nd_{CHUR(T)})-1) \times 10^4$ with the chondritic values $^{143}Nd/^{144}Nd$: 0.512638 and $^{147}Sm/^{144}Nd$: 0.1967.

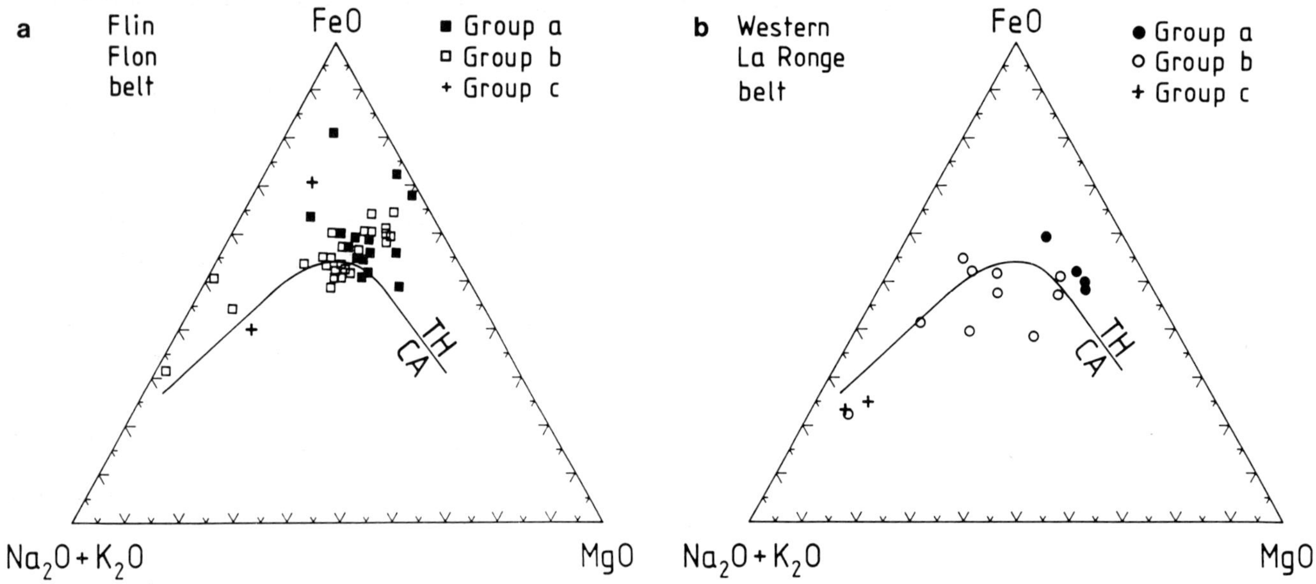

Figure 2. *Compositions of volcanic rocks from (a) the Flin Flon belt and (b) the Western La Ronge belt in alkali–FeO–MgO (AFM) diagrams.*

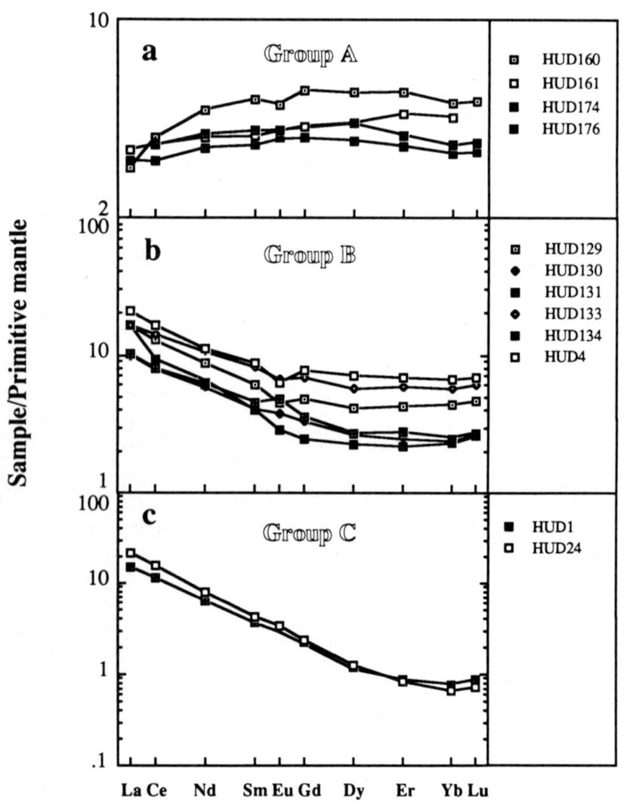

Figure 3. *REE data from the Western La Ronge belt. The data in this diagram, as in Figures 4–6, are normalized using the primitive mantle values of Sun and McDonough (1989).*

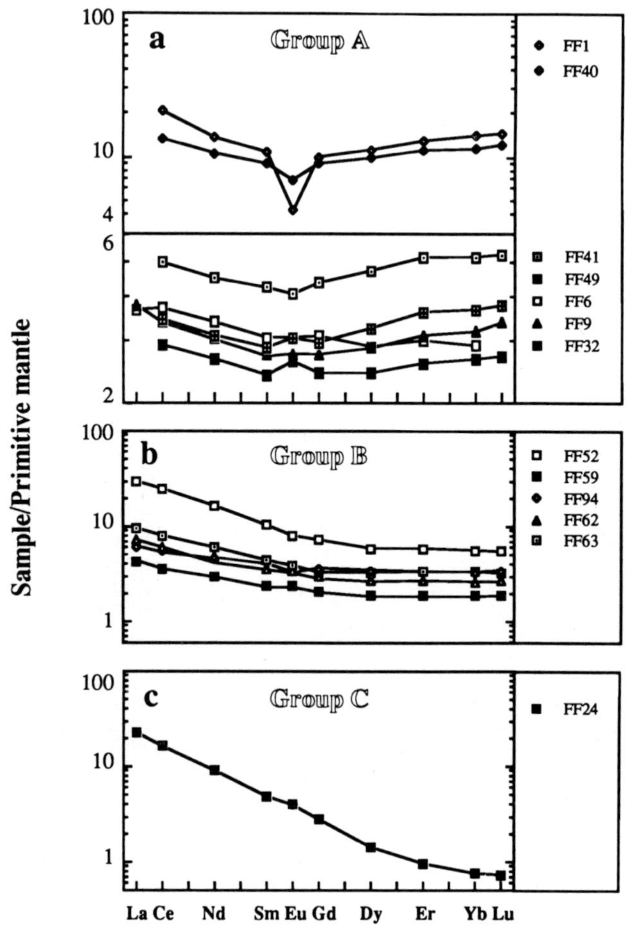

Figure 4. *REE data from volcanic rocks of the Flin Flon belt.*

Comparison with modern volcanics, petrogenesis, and tectonic environment

Volcanic rocks from both areas bear a striking resemblance to rocks from modern convergent plate margins in terms of the range of rock types (Stauffer *et al.*, 1975), their major and trace element chemistry, and their isotopic compositions. Volcanic rocks ranging from basalt through andesite to rhyolite characterize modern island arcs, and these rocks have particular trace element contents, with flat to enriched REE, elevated LILE, and relatively depleted Nb and other HFSE. The U-shaped REE patterns of group A basalts from the Flin Flon belt could be compared with those of boninites, but the Proterozoic rocks do not have the high Mg and Si contents that characterize the modern rocks. ε_{Nd} values in island arcs range from close to zero to about +10, partly overlapping, but generally lower than in MORB. These values are comparable with age-corrected values of +3 to +4 in the Proterozoic Flin Flon belt suite.

Volcanic rocks from modern active continental, Andean-type, margins share many of the features of island-arc volcanics, but include a higher proportion of calc-alkaline rocks (Ewart, 1982) and have more fractionated incompatible elements (e.g., high concentrations of Nb relative to Zr, and of Zr relative to Yb), more elevated alkalis and other incompatible

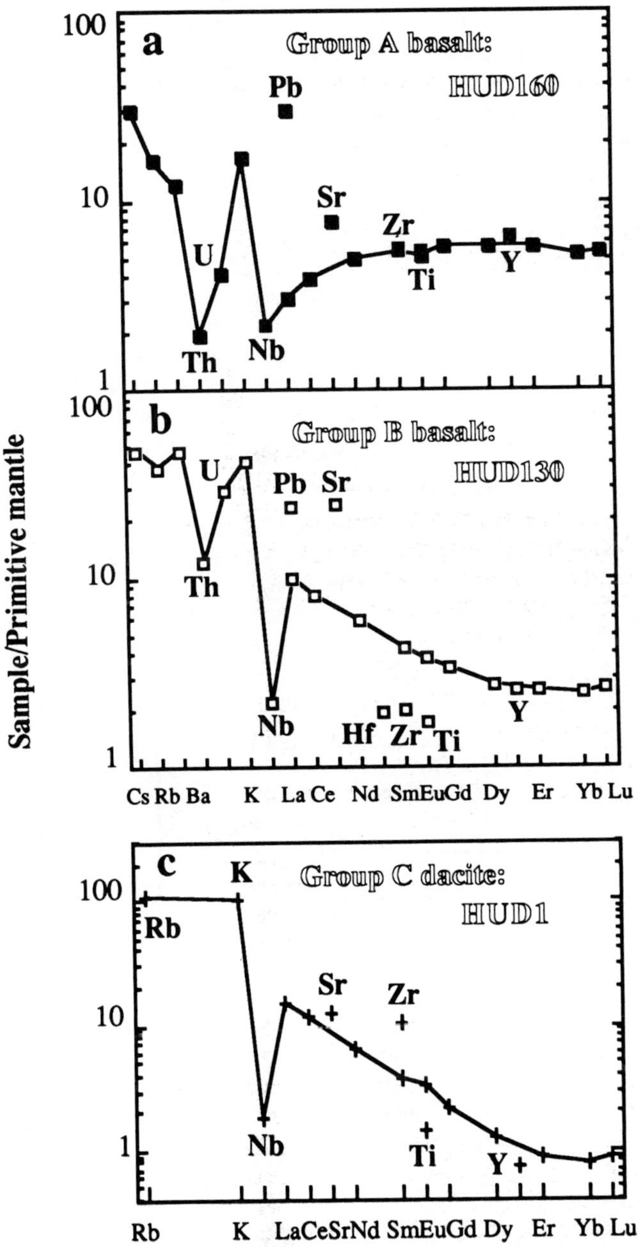

Figure 5. *Trace element data of representative samples from the Western La Ronge belt.*

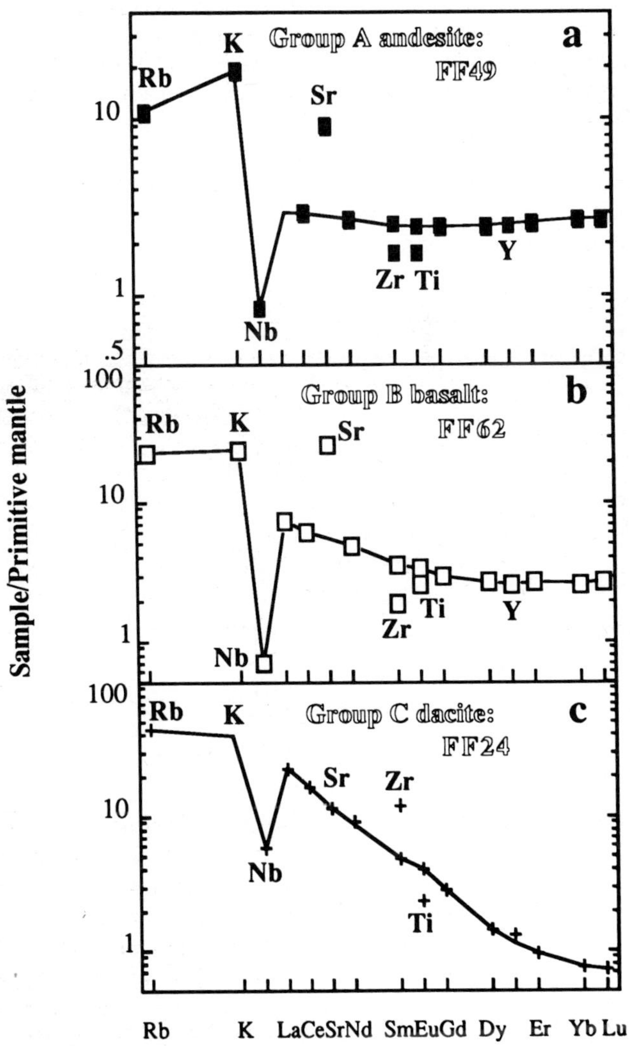

Figure 6. *Trace element data of representative samples from the Flin Flon belt.*

elements (e.g., LREE, Zr), and lower ε_{Nd} values (Baker, 1982; Pearce, 1983). The AFM diagram (Fig. 2) shows that the proportion of rocks with apparent calc-alkaline characteristics is higher in the Western La Ronge belt than in the Flin Flon belt. Selective enrichment of HFSE according to their incompatibility is not a general feature of rocks analyzed for this study, although Zr/Y ratios in calk-alkaline andesites from the Western La Ronge belt are higher than in any samples from the Flin Flon belt (Fig. 8). Watters and Pearce (1987) also reported data from the Waddy Lake and Churchill River areas in the Western La Ronge belt that show high Zr/Y. Lastly, the low ε_{Nd} values (<+3) tend to be more characteristic of lavas from the Western La Ronge belt than the Flin Flon belt.

Thus, in terms of both its position close to the older Archean craton to the northwest, and its chemical and isotopic compositions, the Western La Ronge belt can be compared with a modern active continental margin. This is in agreement with the conclusion of Watters and Pearce (1987), who also argued that magmas in certain sections of the Western La Ronge belt developed in a continental arc. The Flin Flon belt, on the other hand, apparently developed farther from the continent, and has chemical and isotopic features more like those of an intraoceanic arc.

Petrogenesis

The rocks from the volcanic belts of the Reindeer Lake Zone have complicated trace element patterns that reflect a complicated petrogenesis. Particularly important are the differing relative enrichments of HFSE, REE + Th, and LILE, as illustrated in Figures 5 and 6, which highlight the different relative abundances of the three groups of elements. Also important are the ranges of ε_{Nd} values in the two belts.

We explain these features with a model appropriate to the subduction-related tectonic environment inferred above. The model involves partial melting in the mantle wedge overlying a subduction zone. This mantle was initially depleted in incompatible elements, but subsequently was enriched, to varying degrees for different element groups, with material from the subducted plate. The subduction-zone component is introduced into the overlying mantle wedge in aqueous fluids, and the degree of enrichment of each element is related to its solubility in this fluid (e.g., Perfit *et al.*, 1980; Dupuy *et al.*, 1982; White and Patchett, 1984; White and Dupré, 1986). In this model, three element groups are distinguished: highly soluble elements such as LILE and U, which are highly enriched; the less soluble REE and Th, which are less enriched; and the weakly soluble HFSE, which remain in the

Table III
Comparison of chemical features of the Western La Ronge and Flin Flon belts

	Western La Ronge		Flin Flon	
	Group A	Group B	Group A	Group B
Max. Mg-number	0.59	0.56	0.57	0.47
Ni concentration	<99 ppm		<77 ppm	
REE/primitive mantle	5 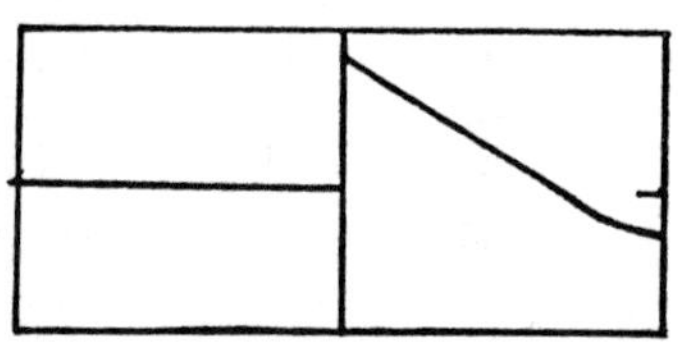		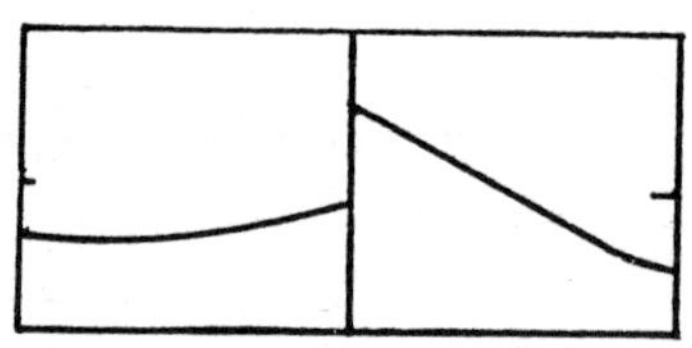	
LILE concentrations relative to REE	Very high		High	
HFSE anomalies relative to REE:				
Nb	Negative	Negative	Negative	Negative
Zr	None	Negative	Negative	Negative
Hf	—	Negative	—	—
Ti	None	Negative	Negative	Negative
ε_{Nd}	+5	+2 to +3	+4	+3 to +5

subducting plate. While the relative proportions of HFSE in the erupted magmas reflect the signature of the original uncontaminated (depleted) mantle, the REE and LILE reflect the amount and composition of the added subduction zone component.

Support for the concept that variation in trace element characteristics arises from admixture of mantle and subduction zone material comes from relationships between specific trace element ratios, particularly the correlation between ε_{Nd} values and Nb/Nb*. Figure 9 shows a negative correlation between the two parameters: with the increasing LREE enrichment that stems from an increasing degree of contamination by the fluid, the negative Nb anomaly becomes larger. Such behaviour is inconsistent with the type of element fractionation that would arise from within-mantle processes because Nb should be more incompatible than Ce during such processes and should correlate positively with the LREE. This decoupling of LREE and HFSE is predicted by the model of contamination of depleted mantle by aqueous fluids, which should be enriched in LREE but have very low concentrations of Nb and other HFSE (Tatsumi *et al.*, 1986). A more detailed, quantitative treatment is presented by Thom (1989).

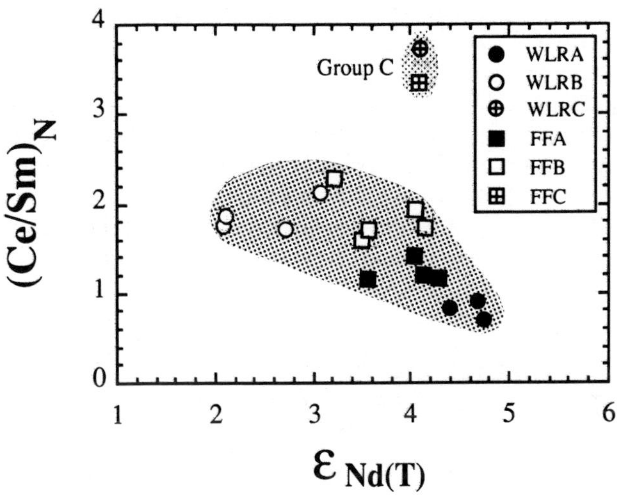

Figure 7. *(Ce/Sm)N versus $\varepsilon_{Nd}(T)$ of volcanic rocks from the Western La Ronge and Flin Flon belts.*

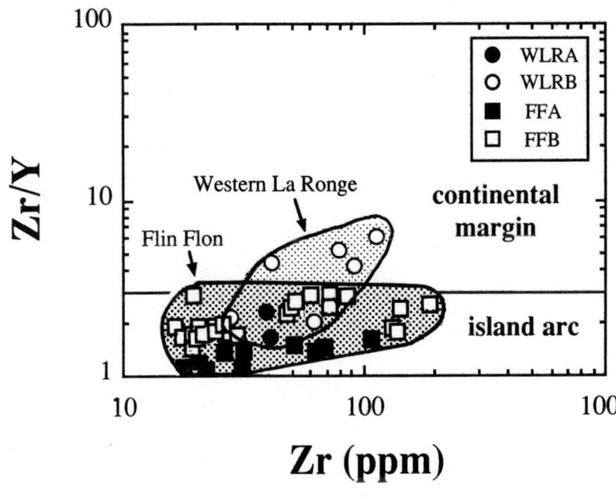

Figure 8. *Zr versus Zr/Y of volcanic rocks from the Western La Ronge and Flin Flon belts.*

Group C dacites from both belts do not exhibit a different relationship between LREE and Nb. Their highly fractionated REE patterns accompanied by very low HREE concentrations (less than primitive mantle values; Figs. 3c and 4c) and their felsic character preclude both their derivation from a peridotitic mantle source and their development as fractionation products of mafic parental magmas. Their chemical compositions, particularly the strong fractionation of REE, and their high SiO_2 contents point instead to derivation by partial melting of a basaltic source that left garnet as a residual phase.

Composition and age of the subducted material

Further information about the nature and relative proportions of mantle and fluid components in the mafic to intermediate volcanic rocks comes from the Nd isotopic data. Following the approach of DePaolo (1981) and Nelson and DePaolo (1984), we assume that the Nd isotopic composition of the mantle

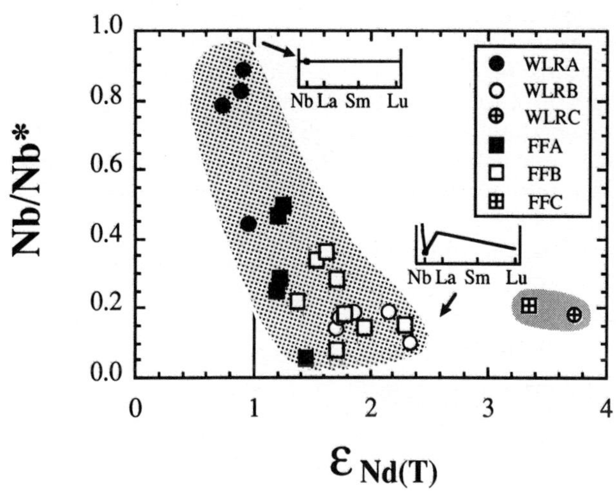

Figure 9. *ε_{Nd} versus Nb/Nb* for volcanic rocks from the Western La Ronge and Flin Flon belts. Nb/Nb* is the analyzed Nb concentration (normalized using primitive mantle values) and Nb* is the value at the Nb position in the diagram determined by extrapolation of the REE spectrum.*

at 1.9 Ga is represented by the most positive ε_{Nd} values measured in basaltic rocks from the area. The highest reported ε_{Nd} values from the Western La Ronge and Flin Flon belts are ~+5, and these are taken to represent the mantle beneath this region 1.9 Ga ago. In contrast, Archean crust at the time of formation of the Trans-Hudson Orogen probably had much lower ε_{Nd} values of between -8 and -12 (Patchett and Arndt, 1986; Chauvel *et al.*, 1987). These values are obtained from composites from Baffin Island (Shaw *et al.*, 1967) and post-Archean average Australian shale (Nance and Taylor, 1976), as analyzed by McCulloch and Wasserburg (1978).

The variation of ε_{Nd} values in the Flin Flon and Western La Ronge rocks is explained by the introduction of a LREE-enriched component with significantly lower ε_{Nd} than the depleted mantle source. This is supported by a negative correlation between $(Ce/Sm)_N$ and ε_{Nd} in the basalts and andesites from both belts (Fig. 7). At Flin Flon, the introduced fluid can be recognized on the basis of high $(Ce/Sm)_N$, negative HFSE anomalies, and relatively low ε_{Nd} values in group A and group B rocks. In contrast, the isotopic compositions and geochemical signatures of group A rocks from Western La Ronge are consistent with their derivation from Proterozoic mantle adulterated with very little contaminant, and only the group B rocks from this belt appear significantly contaminated. According to the quantitative modelling of Thom (1989), the chemical and isotopic compositions of these rocks can be explained by the addition of small amounts (<1.5%) of aqueous fluid released by dehydration of metavolcanic and metasedimentary rocks in the subducted slab. The compositions of these fluids are controlled partly by the proportions and compositions of the altered sedimentary and volcanic rocks, and partly by the nature of the metamorphic phases that formed during the dehydration. The lower ε_{Nd} values are explained by the presence of sediments containing a large Archean component that presumably was derived from cratons positioned to the northwest or south during the initial stages of Trans-Hudson orogenesis (Lewry *et al.*, 1981; Chauvel *et al.*, 1987; Hoffman, 1988).

CONCLUSIONS

(1) Volcanic rocks of the Flin Flon belt have lithological and chemical features like those of a modern intraoceanic island arc; Western La Ronge belt volcanics resemble a modern Andean-type continental margin.

(2) The chemical and isotopic characteristics of these rocks are explained by the introduction of fluids derived by dehydration of the upper part of a subducted slab into variably depleted mantle. This contaminated mantle then partially melted to form the parental magmas of the volcanic rocks. Transportation in aqueous fluids resulted in differing degrees of enrichment of elements with differing solubilities: the degree of enrichment of LILE was greater than that of the REE and Th. Insoluble HFSE were left in the subducting slab.

(3) Nd isotopic data indicate that the subduction component in the Western La Ronge belt volcanics contained some significantly older continental material, presumably sediments from the adjoining Hearne or Superior cratons.

ACKNOWLEDGEMENTS

We would like to thank the following people who assisted in various ways: John Lewry and Pat Bickford helped with sample collection; Klaus-Peter Jochum, Kerstin Lehnert (Max-Planck-Institut) and Euan Nisbet (University of Saskatchewan) assisted with the analyses; Gerhild Feyerherd and Petra Lehmann did part of the drafting; and Randy van Schmus and Jon Patchett reviewed the manuscript. The research was sponsored by the Deutsche Forschungsgemeinschaft, as part of Project No. Kr 590/11-3, and funded in part by a Natural Sciences and Engineering Research Council of Canada grant to Stauffer.

REFERENCES

Baker, P.E., 1982, Evolution and classification of orogenic volcanic rocks, *in* Thorpe, R.S., ed., Andesites: Orogenic Andesites and Related Rocks: John Wiley & Sons, Chichester, p. 11–23.

Bickford, M.E., Collerson, K.D., Lewry, J.F., Van Schmus, W.R. and Chiarenzelli, J.R., 1990, Proterozoic collisional tectonism in the Trans-Hudson Orogen, Saskatchewan: Geology, v. 18, p. 14–18.

Chauvel, C., Arndt, N.T., Kielinczuk, S. and Thom, A., 1987, Formation of Canadian 1.9 Ga old continental crust. I. Nd isotopic data: Canadian Journal of Earth Sciences, v. 24, p. 396–406.

DePaolo, D.J., 1981, Neodymium isotopes in the Colorado Front Range and crust–mantle evolution in the Proterozoic: Nature, v. 291, p. 193–196.

Dupuy, C., Dostal, J., Marcelot, G., Bougault, H., Joron, J.L. and Treuil, M., 1982, Geochemistry of basalts from central and southern New Hebrides arc: implication for their source rock composition: Earth and Planetary Science Letters, v. 60, p. 207–225.

Ewart, A., 1982, The mineralogy and petrology of Tertiary–Recent orogenic volcanic rocks: with special reference to the andesitic-basaltic compositional range, *in* Thorpe, R.S., ed., Andesites: Orogenic Andesites and Related Rocks: John Wiley & Sons, Chichester, p. 25–98.

Feather, C.E. and Willis, J.P., 1976, A simple method for background and matrix correction of spectral peaks in trace element determination by X-ray fluorescence spectrometry: X-ray Spectrometry, v. 5, p. 41–48.

Govindaraju, K., 1984, Compilation of working values and sample description for 170 international reference samples of mainly silicate rocks and minerals: Geostandards Newsletter, v. 8.

Hoffman, P.F., 1988, United plates of America, the birth of a craton: early Proterozoic assembly and growth of Proto-Laurentia: Annual Review of Earth and Planetary Sciences, v. 16, p. 79.

Hofmann, A.W., 1988, Chemical differentiation of the Earth: the relationship between mantle, continental crust, and oceanic crust: Earth and Planetary Science Letters, v. 90, p. 297–314.

Irvine, T.N. and Baragar, W.R.A., 1971, A guide to the chemical classification of the common volcanic rocks: Canadian Journal of Earth Sciences, v. 8, p. 523–548.

Jochum, K.P., Seufert, H.M., Midinet-Best, S., Rettmann, E., Schönberger, K. and Zimmer, M., 1988, Multi-element analysis by isotope dilution – spark source mass spectrometry (ID–SSMS): Fresenius Zeitschrift Analytische Chemie, v. 331, p. 104–110.

Lewry, J.F., Stauffer, M.R. and Fumerton, S.L., 1981, A Cordilleran-type batholithic belt in the Churchill Province in northern Saskatchewan: Precambrian Research, v. 14, p. 277–313.

McCulloch, M.T. and Wasserburg, G.J., 1978, Sm–Nd and Rb–Sr chronology of continental crust formation: Science, v. 200, p. 1103–1111.

Nance, W.B. and Taylor, S.R., 1976, Rare earth patterns and crustal evolution – Australian post-Archean sedimentary rocks: Geochimica et Cosmochimica Acta, v. 40, p. 1539–1551.

Nelson, B.K. and DePaolo, D.J., 1984, Origin of ~1.7 Ga greenstone volcanic successions in southwestern North America and the isotopic evolution of Proterozoic mantle: Nature, v. 312, p. 143–146.

Nisbet, E.G., Dietrich, V.J. and Esenwein, A., 1979, Routine trace element determination in silicate minerals and rocks by X-ray fluorescence: Fortschritte Mineralogie, v. 57, p. 264–279.

Norrish, K. and Hutton, J.T., 1969, An accurate X-ray spectrographic method for the analysis of a wide range of geological samples: Geochimica et Cosmochimica Acta, v. 33, p. 431–453.

Patchett, P.J. and Arndt, N.T., 1986, Nd isotopes and tectonics of 1.9–1.7 Ga crustal genesis: Earth and Planetary Science Letters, v. 78, p. 329–338.

Pearce, J.A., 1983, Role of the sub-continental lithosphere in magma genesis at active continental margins, _in_ Thorpe, R.S., ed., Andesites: Orogenic Andesites and Related Rocks: John Wiley & Sons, Chichester, p. 525–548.

Perfit, M.R., Gust, D.A., Bence, A.E., Arculus, R.J. and Taylor, S.R., 1980, Chemical characteristics of island-arc basalts: implications for mantle sources: Chemical Geology, v. 30, p. 227–256.

Shaw, D.M., Reilly, G.A., Muysson, J.R., Pattenden, G.E. and Campbell, F.E., 1967, An estimate of the chemical composition of the Canadian Precambrian Shield: Canadian Journal of Earth Sciences, v. 4, p. 829–853.

Stauffer, M.R., 1984, Manikewan: an early Proterozoic ocean in central Canada, its igneous history and orogenic closure: Precambrian Research, v. 25, p. 257–281.

Stauffer, M.R., Mukherjee, A.C. and Koo, J., 1975, The Amisk Group: an Aphebian (?) island arc deposit: Canadian Journal of Earth Sciences, v. 12, p. 2021–2035.

Sun, S.S. and McDonough, W.F., 1989, Chemical and isotopic systematics of oceanic basalts: implications for mantle compositions and processes, _in_ Saunders, A.D. and Norry, M.J., eds., Magmatism in the Ocean Basins: Geological Society Special Publication, v. 42, p. 313–345.

Tatsumi, Y., Hamilton, D.L. and Nesbitt, R.W., 1986, Chemical characteristics of fluid phase released from a subducted lithosphere and origin of arc magmas: evidence from high-pressure experiments and natural rocks: Journal of Volcanology and Geothermal Research, v. 29, p. 293–309.

Thom, A., 1989, Petrology and geochemistry of middle Proterozoic volcanics of the Reindeer Lake Terrain in northern Saskatchewan and Manitoba (Canada), unpublished Ph.D. thesis, Jahannes-Gutenberg University, Mainz, 198 p. (In German.)

Van Schmus, W.R., Bickford, M.E., Lewry, J.F. and Macdonald, R., 1987, U–Pb geochronology of the Trans-Hudson Orogen, northern Saskatchewan, Canada: Canadian Journal of Earth Sciences, v. 24, p. 407–424.

Watters, B.R. and Pearce, J.A., 1987, Metavolcanic rocks of the La Ronge Domain in the Churchill Province, Saskatchewan: geochemical evidence for a volcanic arc origin, _in_ Pharaoh, T.C., Beckinsal, R.D. and Rickard, D., eds., Geochemistry and Mineralization of Proterozoic Volcanic Suites: Geological Society Special Publication, No. 33, p. 167–182.

White, W.M. and Dupré, B., 1986, Sediment subduction and magma genesis in the Lesser Antilles: isotopic and trace element constraints: Journal of Geophysical Research, v. 91, p. 5927–5941.

White, W.M. and Patchett, P.J., 1984, Hf–Nd–Sr isotopes and incompatible element abundances in island arcs: implications for magma origins and mantle–crust evolution: Earth and Planetary Science Letters, v. 67, p. 167–185.

U–Pb Ages from the Flin Flon and Kisseynew Belts, Manitoba: Chronology of Crust Formation at an Early Proterozoic Accretionary Margin[1,2]

T.M. Gordon and P.A. Hunt
Geological Survey of Canada, 601 Booth Street,
Ottawa, Ontario K1A 0E8, Canada

A.H. Bailes and E.C. Syme
Manitoba Energy and Mines, 555-330 Graham Avenue,
Winnipeg, Manitoba R3C 4E3, Canada

Gordon, T.M., Hunt, P.A., Bailes, A.H. and Syme, E.C., U–Pb ages from the Flon Flon and Kisseynew belts, Manitoba: chronology of crust formation at an Early Proterozoic accretionary margin, in Lewry, J.F. and Stauffer, M.R., eds., The Early Proterozoic Trans-Hudson Orogen of North America: Geological Association of Canada, Special Paper 37, p. 177–199

Abstract

Formation and assembly of the Proterozoic crust now preserved in the internal zone of the Trans-Hudson Orogen occurred over an 85 Ma period. The earliest magmatism was island-arc volcanic activity in the Lynn Lake belt at 1.91 Ga and island-arc and oceanic-rift volcanism in the Flin Flon belt at 1.89 Ga. At this time, volcanogenic greywacke turbidites were shed into adjacent basins, forming protolith to the Kisseynew sedimentary gneiss belt. Following isoclinal folding, Lynn Lake belt volcanic rocks were intruded by tonalite and granodiorite. This magmatism was synchronous with continental volcanism in the Rusty Lake belt at 1.875 Ga, suggesting that accretion of these arcs to the Hearne craton had occurred by this time. Emplacement of the I-type Wathaman–Chipewyan batholith and smaller bodies in the Flin Flon belt followed at about 1.86–1.85 Ga, postdating early folding in the Flin Flon belt. Uplift and erosion resulted in continental molasse sedimentation beginning at 1.85 Ga. Sills in the Kisseynew belt were emplaced at 1.83 Ga, essentially synchronously with fluvial–alluvial sedimentation, subaerial volcanism,

[1]Geological Survey of Canada Contribution No. 32388.
[2]Manitoba Energy and Mines, Mineral Resources Division contribution to the Canada–Manitoba Mineral Development Agreement (1984–1989).

and plutonism in the eastern part of the Flin Flon belt. Metamorphism in the Kisseynew sedimentary gneiss belt began during the waning stages of this magmatism, and was accompanied by at least two major phases of folding. The thermal culmination resulted in anatexis in the Kisseynew belt at approximately 1.815 Ga. Subsequent uplift carried Kisseynew belt rocks through 650°C at 1.805 Ga and 280°C at 1.71 Ga. The latest movement on large-scale faults in the region postdates regional metamorphism.

Résumé

La formation et l'accrétion de la croûte protérozoïque préservée dans les zone internes de l'orogène Trans-Hudson s'est effectuée sur 85 Ma. Dans la ceinture Lynn Lake, un volcanisme d'arc insulaire vers 1,91 Ga et un volcanisme d'arc insulaire et de rift océanique dans la ceinture Flin-Flon vers 1,89 Ga constitue un magmatisme primaire. Parallèllement, des turbidites (grauwackes volcanogéniques) ont été déposées dans des bassins adjaçents, formant le protolithe de la ceinture de paragneiss de Kisseynew. Après un plissement isoclinal, les roches volcaniques de Lynn Lake sont recoupées par des tonalites et des granodiorites. Ce magmatisme est synchrone du volcanisme continental de la ceinture Rusty Lake vers 1,875 Ga suggérant que l'accrétion de ces arcs au craton Hearne s'est produit en même temps. L'emplacement du batholite de type-I Wathaman–Chipewyan et de petits stocks dans la ceinture Flin-Flon suit vers 1,86–1,85 Ga, postérieurement au plissement ancien de la ceinture. Remontée et érosion donnent lieu à une sédimentation de molasses continentales vers 1,85 Ga. Des sills sont mis en place vers 1,83 Ga dans la ceinture de Kisseynew, synchrone de la sédimentation fluviatile et alluvionnaire, du volcanisme sub-aérien et du plutonisme à l'est de la ceinture Flin-Flon. Le métamorphisme de la ceinture de gneiss sédimentaire de Kisseynew commence durant le déclin de ce plutonisme et est accompagné d'au moins deux phases majeures de plis. Le pic thermal provoque une anatexie dans la ceinture de Kisseynew vers 1,815 Ga. Des remontées subséquentes porte les roches de la ceinture à 650°C vers 1,805 Ga et 280°C à 1,71 Ga. Les derniers mouvements le long des grands accidents de la région sont postérieurs au métamorphisme régional.

INTRODUCTION

The Trans-Hudson Orogen in the southwestern Canadian Shield contains a 500 km wide belt of mainly Early Proterozoic rocks flanked on the northwest and southeast by Archean cratons (Fig. 1). The internal zone can be divided into several lithotectonic domains (Macdonald (1987) and references therein; Hoffman (1988)). On the northwest, the Wollaston and Seal River fold belts consist of Archean basement and overlying Proterozoic clastic sedimentary rocks, which have been interpreted as early rift and miogeoclinal deposits (Coombe, 1979; Ray and Wanless, 1980). The Hanson Lake block, Glennie Lake domain, and Peter Lake domain contain structurally imbricated Archean and Proterozoic gneisses (Chiarenzelli *et al.*, 1987; Lewry *et al.*, 1987; Van Schmus *et al.*, 1987; Chiarenzelli and Lewry, 1988). The central and southeast areas are underlain by east–west trending Early Proterozoic volcanic belts and sedimentary gneiss belts. From north to south, they are the Southern Indian sedimentary gneiss belt, the Lynn Lake and Rusty Lake volcanic belts, the Kisseynew sedimentary gneiss belt, and the Flin Flon volcanic belt. The Wathaman–Chipewyan and Baldock batholiths (McRitchie, 1977; Lenton and Corkery, 1981; Lewry *et al.*, 1981; Schledewitz, 1983; Fumerton *et al.*, 1984) intrude the northern domains.

Tectonic interpretations of the region (Lewry and Sibbald, 1980; Ray and Wanless, 1980; Lewry, 1981; Stauffer, 1984; Green *et al.*, 1985; Hoffman, 1988) suggest that the volcanic and sedimentary gneiss belts originated as island arcs and related basins accreted to the southeast margin of the Hearne–Wyoming craton, whereas the Wathaman–Chipewyan and Baldock batholiths developed as a continental magmatic arc above a north-dipping subduction zone.

Earlier isotopic studies have been reviewed by Sangster (1978), who had previously (Sangster, 1972) been the first to postulate an Early Proterozoic age for rocks in the region.

This paper presents the geological setting and new U–Pb zircon and monazite ages of volcanic, plutonic, and anatectic rocks in the Flin Flon volcanic belt and Kisseynew sedimentary gneiss belt in Manitoba. It revises and adds to the data of Gordon *et al.* (1987). Analytical techniques are described in the Appendix and results are presented in tables and figures within the text. These ages, along with others (Baldwin *et al.*, 1987; Lewry *et al.*, 1987; Van Schmus *et al.*, 1987; Delaney *et al.*, *in press*), establish the temporal relationships of the main magmatic and deformational events in the Trans-Hudson Orogen. The tectonic implications of these data are summarized at the conclusion of this paper.

FLIN FLON BELT

Geological setting

The Flin Flon belt is composed of an assemblage of subaqueous and subaerial volcanic rocks and associated sedimentary rocks, an overlying sequence of terrestrial sedimentary rocks, and a complex suite of intrusive rocks. It is bounded to the north by metasedimentary gneisses of the Kisseynew belt and is overlain to the south by flat-lying Ordovician dolomitic limestones. The volcanic belt has a length of 250 km and an exposed width of 32–48 km, but its total width, including the area covered by Paleozoic rocks, may be considerably larger (Green *et al.*, 1985; Blair *et al.*, 1988).

The oldest rocks in the Flin Flon belt are volcanic rocks assigned to the Amisk Group by Bruce (1918), who named them for a locality on Amisk Lake, Saskatchewan. Most of the rocks were deposited in a shallow to moderately deep marine environment, but locally, as at Amisk Lake in Saskatchewan, subaerial environments have been recognized (Ayres, 1978).

Following intrusion of porphyritic felsic plutons and differentiated gabbroic rocks, the Amisk Group was deformed, eroded, and overlain by sandstones and conglomerates that were deposited in a fluvial–alluvial environment. Bruce (1918) named these clastic rocks the Missi Group, after Missi Island on Amisk Lake. The Missi Group consists of metamorphosed sandstone, pebbly sandstone, and polymictic conglomerate at Flin Flon, and includes volcanic rocks at the eastern end of the belt.

The supracrustal sequences were subsequently metamorphosed and affected by at least two phases of deformation. Both syntectonic and posttectonic plutons are found in the Flin Flon belt.

Structural geology

At Flin Flon, Amisk Group rocks were tightly folded, faulted, and eroded prior to deposition of the Missi Group. Three further phases of folding affecting the Missi Group have been recognized (Stauffer and Mukherjee, 1971; Stauffer, 1974). In the File Lake and Snow Lake areas to the east, two regional folding episodes postdate deposition of the Missi Group (Bailes, 1980a; Froese and Moore, 1980). The earliest of these produced recumbent isoclinal folds, which were subsequently refolded about northeasterly plunging axes.

The Flin Flon area is dominated by relatively straight, steeply dipping, east to north-northwest

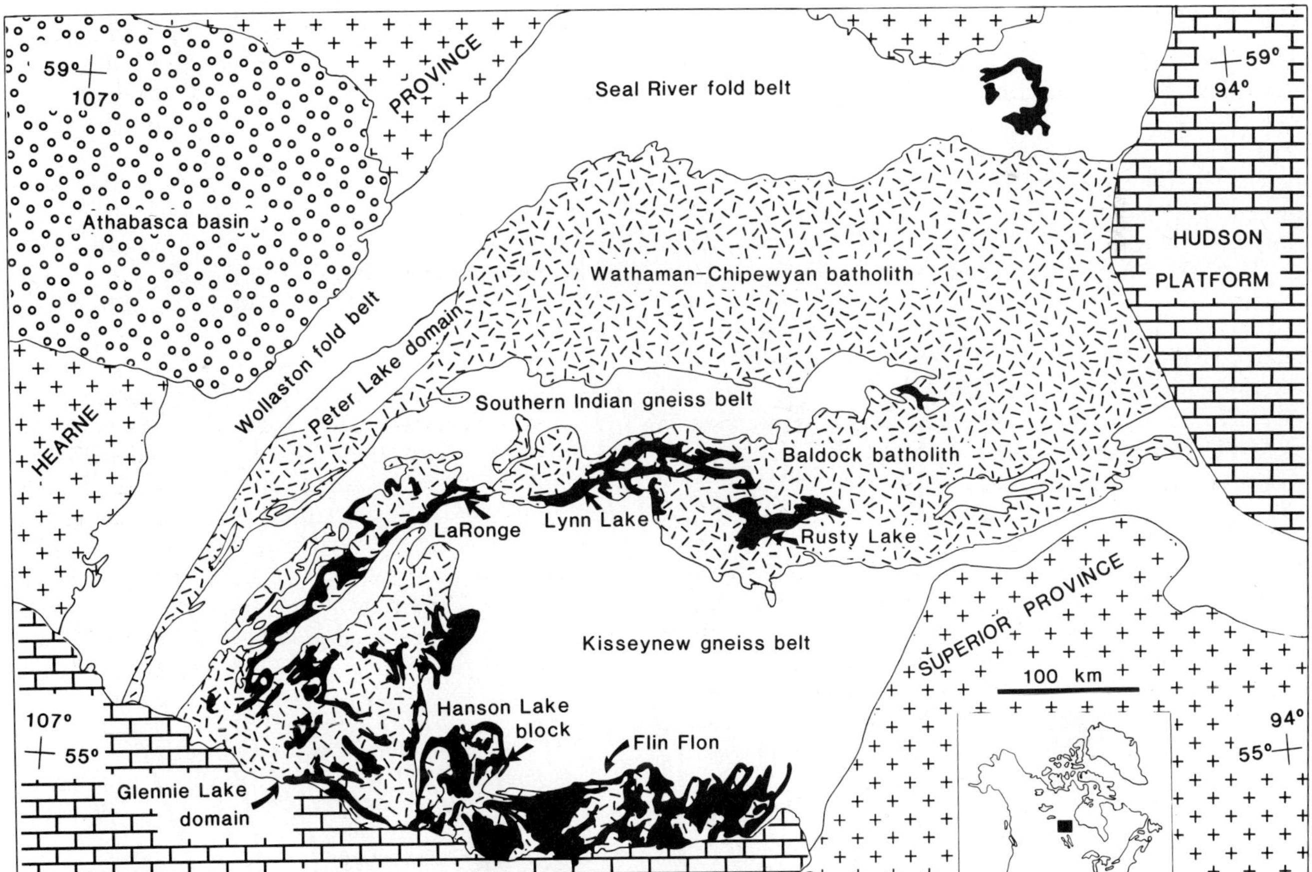

Figure 1. *Lithotectonic domains of the Trans-Hudson Orogen in Manitoba and Saskatchewan as defined in references cited in the text. Volcanic belts are shown in black, granitic rocks in a crosshatched pattern, Archean cratons by vertical crosses, and sedimentary belts by the absence of a pattern.*

trending faults (Bailes and Syme, 1987, *in press*). Two groups of faults are recognized, those that bound major stratigraphic blocks and those that occur within the blocks (Figs. 2 and 3). Stratigraphic correlations cannot be made across block-bounding faults, whereas either correlation or similarity of sequences can be observed across internal faults. Many faults are not simple isolated structures, but tend to splay or occur in clusters, resulting in numerous lensoid fault-bounded domains. With the exception of early, folded faults, most appear to have been initiated during the latest folding event. The younger faults postdate the regional metamorphism and most plutons.

At the eastern end of the Flin Flon belt, the earliest recognized fault is the McLeod Road thrust, which places Amisk Group metavolcanic rocks and Missi Group metasedimentary rocks structurally above Amisk metasedimentary rocks (Froese and Moore, 1980; Galley *et al.*, 1988). Younger north to north-northeast trending faults postdate regional metamorphism (Gordon and Gall, 1982; Bailes, 1987; Bailes and Syme, 1987, *in press*).

Metamorphism

The grade of metamorphism generally increases to the north, towards the Kisseynew belt. In the Flin Flon area, the grade of metamorphism ranges from subgreenschist (prehnite–pumpellyite) to upper greenschist; whereas at Snow Lake, it is lower to upper amphibolite facies (sillimanite–garnet–biotite). The prograde disappearance of staurolite in aluminous sedimentary rocks approximately coincides with a facies change from dominantly volcanic to dominantly sedimentary rocks, and, hence, marks the northern boundary of the volcanic belt.

Regional metamorphism is ascribed to a single thermal event, which reached its highest tempera-

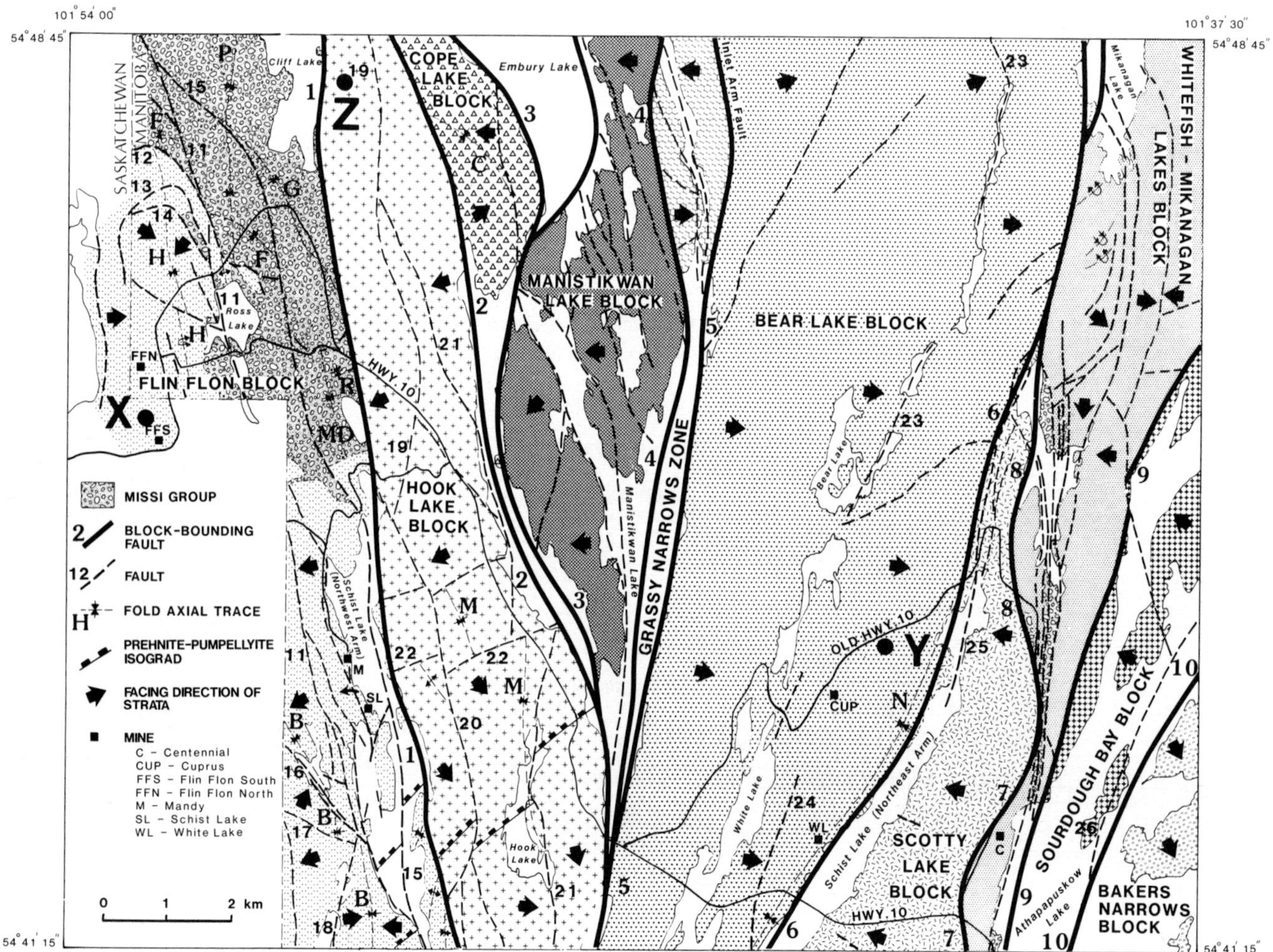

Figure 2. *General geology and fault blocks in the Flin Flon – White Lake area. Sample locations: X, South Main rhyolite domes; Y, Amisk Group rhyolite; and Z, Cliff Lake pluton.*

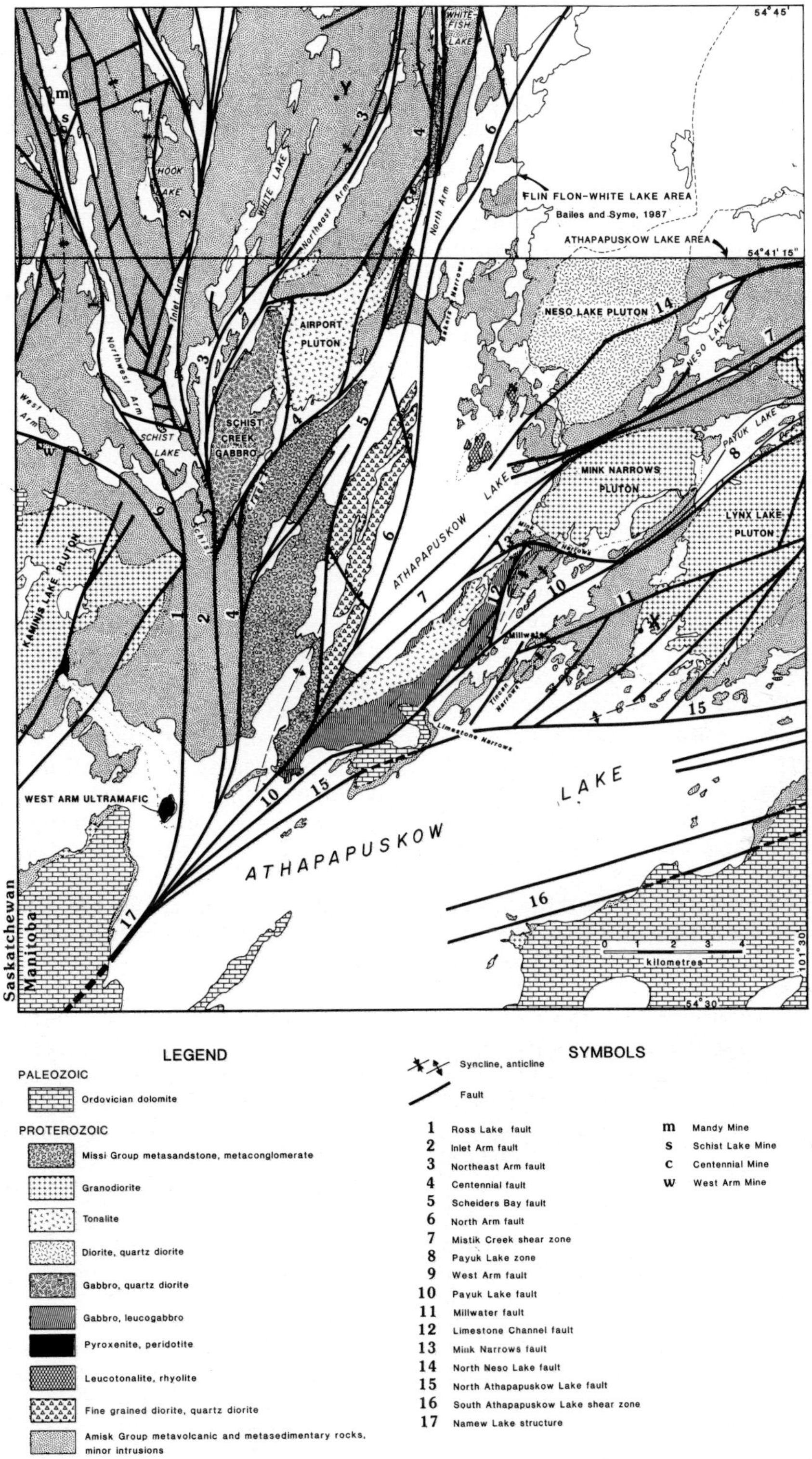

Figure 3. *Simplified geological map of the Schist Lake – Athapapuskow Lake area, with the locations of the analyzed samples: X, Lynx Lake pluton; Y, Amisk Group rhyolite crystal tuff.*

tures in the Kisseynew belt. The final stages of metamorphism postdate the two major folding episodes, but predate development of local northeast-plunging open folds and the youngest faults.

Amisk Group

In the Flin Flon area, Amisk Group rocks occur in a number of distinct stratigraphic sequences within individual fault blocks (Figs. 2 and 3) (Bailes and Syme, 1987, *in press*). The volcanic sequences are dominated by subaqueous flows and related breccias of basalt and basaltic andesite composition. These form units up to 3.3 km thick that are probably portions of major shield volcanoes. Rhyolite flows occur as sporadic, discrete bodies of two morphologic types, massive flows with associated breccia and composite types with massive lobes in a microbreccia matrix. Mafic volcaniclastic rocks vary widely in phenocryst population, texture, grain size, and bedding characteristics; most were deposited from subaqueous density currents. Intermediate volcaniclastic rocks occur only in the Bear Lake fault block, and a particularly thick unit is interpreted to fill a deep intravolcanic basin or a large caldera structure. Felsic volcaniclastic rocks are not common. Minor rock types include volcanic conglomerate, oxide facies iron formation, and carbonate-rich sedimentary rocks. Greywacke, siltstone, and mudstone are an integral, although relatively minor, component of the Amisk Group at Flin Flon. The Hook Lake fault block (Fig. 2) contains the greatest reconstructed section in the Flin Flon area, comprising about 7.5 km of flows, related breccias, and subordinate tuff.

At Snow Lake, at the eastern end of the Flin Flon belt, major block-bounding faults of the type that characterize the Flin Flon area are absent. The Snow Lake section differs from the section at Flin Flon in that it contains substantial thicknesses of felsic volcanic rocks and a thick upper sequence of volcanogenic greywacke turbidites. A 5 km thick, north-facing section of Amisk Group rocks is located between the Ham Lake pluton and Wekusko granite (Fig. 4). It consists of a lower half composed of subaqueous flows and related breccias of basaltic and basaltic andesite composition and an upper half of thickly bedded, heterolithologic and monolithologic mafic breccia, with intercalated units of felsic breccia and tuff and subaqueous mafic porphyritic flows. The volcanogenic greywacke turbidites at the top of the Amisk Group are interpreted to have been derived from contemporaneous upslope Amisk volcanoes (Bailes, 1980a,b). These greywacke turbidites can be traced to the north into more highly metamorphosed and recrystallized paragneisses belonging to the Kisseynew belt. A large tonalite sill intrudes the lower portion of the volcanic sequence and is interpreted to be subvolcanic and to have provided heat for a large-scale hydrothermal system, which altered a significant portion of the overlying volcanic sequence (Walford and Franklin, 1982; Bailes 1986, 1987; Bailes *et al.*, 1987).

Recent petrological and geochemical studies (Syme, 1985, 1987, this volume; Chauvel *et al.*, 1987; Gaskarth and Parslow, 1987; Watters and Pearce, 1987) show that the volcanic rocks are comparable to both Cenozoic island-arc and oceanic, rift-type environments.

Geochronology of the Amisk Group. The South Main rhyolite domes at Flin Flon were sampled in an attempt to determine the age of the Amisk Group at Flin Flon. Two samples yielded only a few zircons, which had unsatisfactory analyses. Unfortunately, no useful age data were obtained.

A sample of Amisk Group rhyolite from the Bear Lake fault block (Bailes and Syme, 1987, *in press*) was obtained from a 1.5 m thick rhyolite crystal tuff bed within a series of ferrobasalt flows (Fig. 3). Details of the lithology, sample location, and local stratigraphy have been published elsewhere (Syme *et al.*, 1987).

The zircons are euhedral with acute terminations, thus possessing the typical morphology of zircons of volcanic origin. They are visibly devoid of inherited cores. The most concordant analysis gives a $^{207}Pb/^{206}Pb$ minimum age of 1886 Ma, identical to the 1886 ± 2 Ma concordia intercept obtained from three zircon fractions (Fig. 5). Analytical data for these and subsequently described samples are given in Tables I and II.

This age is comparable to zircon ages of 1888–1876 Ma obtained from volcanic rocks in the La Ronge domain, Glennie Lake domain, and Hanson Lake block by Van Schmus *et al.* (1987) and Lewry *et al.* (1987). The Amisk Group at Flin Flon is intermediate in age between the 1910 Ma Wasekwan Group at Lynn Lake and 1878 Ma volcanic rocks in the Rusty Lake belt (Baldwin *et al.*, 1987). Precise dating techniques now permit temporal definition and possible subdivision of these belts, which are all broadly comparable in age and geochemical characteristics.

Missi Group

The Missi Group is more than 1900 m thick at Flin Flon and consists of metamorphosed sandstone, pebbly sandstone, and polymictic conglomerate that lies with profound angular unconformity on previously folded Amisk Group rocks and contained intrusions. Missi Group rocks contain primary structures such as trough crossbedding, pebble lags, heavy-mineral placers, conglomerate lenses, and multiple scours that are consistent with deposition

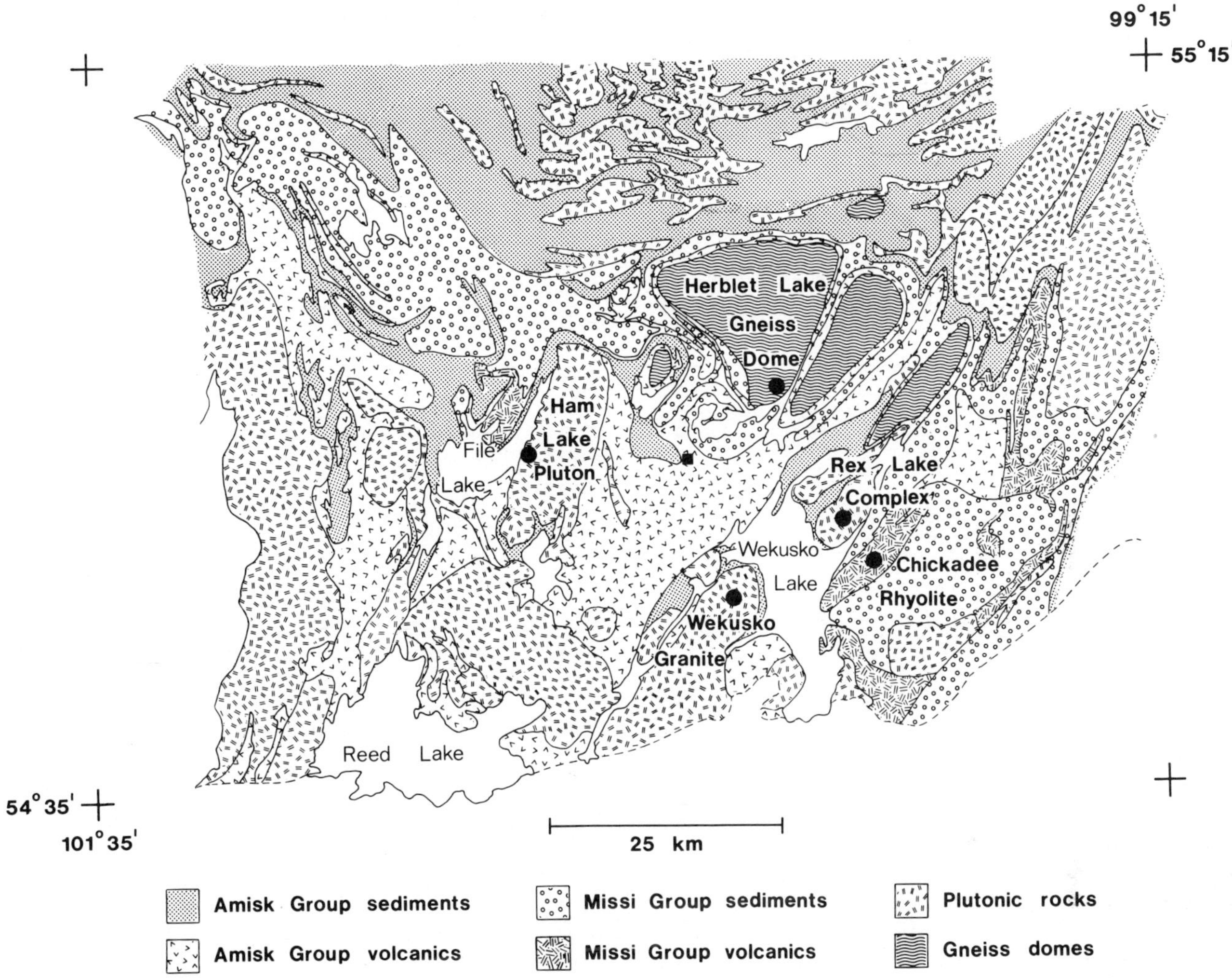

Figure 4. *General geology and sample locations in the Snow Lake region. The town of Snow Lake is shown as a black square.*

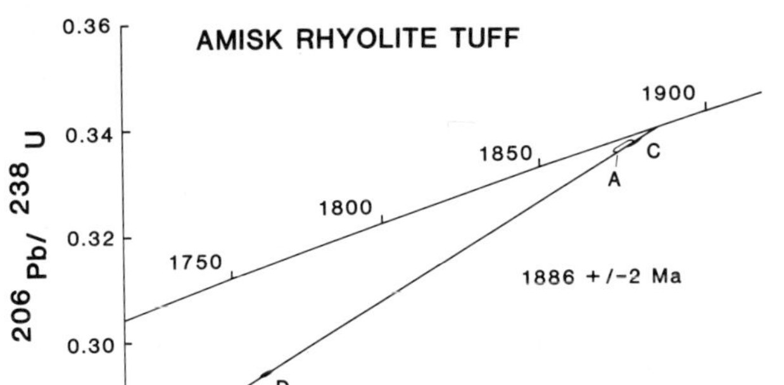

Figure 5. *Concordia diagram showing data from the Amisk Group rhyolite crystal tuff at Flin Flon.*

by fluvial processes, probably within a subaerial alluvial fan. Clasts in the conglomerate include abundant Amisk Group mafic and felsic volcanic rocks, fine- to medium-grained mafic to felsic high-level intrusive rocks, and minor medium- to coarse-grained granitoid rocks, quartz, jasper, iron formation, chert, and epidosite. Intraformational sandstone clasts form a significant part of conglomerates in the upper part of the group. Detrital quartz in conglomerate and sandstone was derived from quartz veins and granitoid plutons; detrital feldspar is altered, Fe stained, and derived mainly from plutonic rocks.

At least 2.5–3 km of Amisk section is truncated at the basal Missi unconformity west and south of Ross Lake. A hematiferous regolith several metres thick is developed in Amisk volcanic and intrusive rocks at all exposed, unfaulted contacts with the Missi.

Table I
Analytical data

Fraction, size[a]	Weight (mg)	U (ppm)	Pb[b] (ppm)	$^{206}Pb^{c}/^{204}Pb$	$Pb_c{}^{d}$ (pg)	$^{208}Pb^{b}$ (%)	$^{206}Pb^{e}/^{238}U$	$^{207}Pb^{e}/^{235}U$	$^{207}Pb^{e}/^{206}Pb$	$^{207}Pb/^{206}Pb$ age(Ma)[f]
					Flin Flon belt					
Amisk Group rhyolite tuff										
A -62+44 NM2 abr	0.0062	104.5	36.6	706	20	7.8	0.33619 (0.18)	5.34142 (0.20)	0.11523 (0.08)	1883.5 (1.5)
C -74+62 NM2 abr	0.0165	161.4	56.5	2261	25	7.7	0.33687 (0.10)	5.36085 (0.11)	0.11542 (0.04)	1886.4 (0.7)
D -62+44 NM2 abr	0.0121	249.1	76.4	2057	27	8.0	0.29400 (0.09)	4.66663 (0.11)	0.11512 (0.04)	1881.8 (0.7)
Chickadee Lake rhyolite										
A -105+74 NM co abr	0.0454	371.1	121.2	3383	98	8.8	0.31145 (0.12)	4.8065 (0.14)	0.11193 (0.05)	1830.9 (0.8)
B -105+74 Mag co abr	0.0256	293.8	96.4	2783	53	8.6	0.31347 (0.31)	4.8400 (0.32)	0.11198 (0.05)	1831.8 (0.9)
C -74+62 NM co abr	0.0624	321.8	104.7	2254	175	8.9	0.30971 (0.19)	4.7825 (0.20)	0.11200 (0.06)	1832.1 (1.2)
D -74+62 Mag	0.0332	327.4	96.9	2876	67	8.7	0.28233 (0.25)	4.3573 (0.26)	0.11193 (0.05)	1831.0 (1.0)
Cliff Lake pluton[g]										
A -105+74 NM co abr	0.184	714.4	215.2	5769	205	8.0	0.28964 (0.25)	4.51717 (0.25)	0.11311 (0.05)	1850.0 (0.9)
B -74+62 NM co	0.285	815.3	219.7	4667	440	7.8	0.25981 (0.25)	4.02906 (0.25)	0.11247 (0.05)	1839.7 (0.9)
C -62+44 NM co	0.249	1069.6	288.0	4184	599	8.4	0.25789 (0.25)	3.96742 (0.25)	0.11158 (0.05)	1825.3 (0.9)
E -74+62 NM co abr	0.0418	746.7	218.7	5386	103	7.9	0.28170 (0.25)	4.38727 (0.25)	0.11296 (0.05)	1847.5 (0.9)
Lynx Lake pluton										
A -74+62 NM0 abr	0.0360	395.7	125.2	5580	50	7.1	0.30715 (0.08)	4.78165 (0.10)	0.11291 (0.03)	1846.8 (0.5)
C +74 Mag0 abr	0.0333	377.1	114.3	3968	59	6.6	0.29576 (0.08)	4.60339 (0.10)	0.11289 (0.04)	1846.4 (0.7)
D -74+62 Mag0 abr	0.0306	390.2	122.5	5073	45	7.2	0.30447 (0.08)	4.73793 (0.10)	0.11286 (0.03)	1846.0 (0.6)
Rex Lake pluton										
A -74+62 Mag co	0.033	552.5	164.7	3408	101	4.2	0.29861 (0.24)	4.5901 (0.25)	0.11149 (0.05)	1823.8 (1.0)
B -105+74 Mag co abr	0.0663	510.4	156.6	6205	106	4.2	0.30755 (0.18)	4.7317 (0.19)	0.11158 (0.05)	1825.3 (0.9)
D -105+74 NM co abr	0.0481	473.7	147.8	7273	62	4.4	0.31223 (0.12)	4.8104 (0.14)	0.11174 (0.04)	1827.9 (0.7)
Wekusko granite										
A -105+74 NM1	0.0412	548.7	171.9	4151	108	4.5	0.31284 (0.19)	4.8275 (0.20)	0.11192 (0.05)	1831.8 (0.8)
D -74+62 NM1 abr	0.0442	482.9	155.4	1858	233	4.8	0.32027 (0.20)	4.9526 (0.22)	0.11215 (0.07)	1834.6 (1.3)
C -74+62 NM1	0.0400	584.8	191.3	3536	136	4.8	0.32586 (0.20)	5.0339 (0.22)	0.11204 (0.05)	1832.7 (0.9)
B -105+74 NM1 abr	0.0386	700.2	222.2	5565	97	4.7	0.31643 (0.12)	4.8762 (0.14)	0.11177 (0.04)	1828.3 (0.7)
Ham Lake pluton[h]										
A +149 abr	1.099	1193.4	341.0	4587	3088	4.0	0.28753 (0.25)	4.34045 (0.25)	0.10948 (0.05)	1790.8 (0.8)
B +149 abr	1.781	1011.0	296.8	6545	3001	4.0	0.29552 (0.25)	4.45678 (0.25)	0.10938 (0.05)	1789.1 (0.8)
C -149+105 NM co abr	1.539	1012.9	291.8	5160	2638	4.0	0.28958 (0.25)	4.40222 (0.25)	0.11026 (0.05)	1803.7 (0.8)
D -149+105 Mag co abr	0.493	607.1	166.0	4730	574	4.0	0.27516 (0.25)	4.14463 (0.25)	0.10924 (0.05)	1786.9 (0.8)
E -105+74 abr	1.224	1092.0	316.5	3881	3686	4.5	0.28985 (0.25)	4.42019 (0.25)	0.11060 (0.05)	1809.3 (0.8)
F -149+105 NM4 abr	0.254	925.2	282.7	5770	787	4.1	0.30659 (0.25)	4.69604 (0.25)	0.11109 (0.05)	1817.3 (0.8)
G -105+74 NM2 abr	0.211	830.4	259.8	7036	491	4.3	0.31330 (0.25)	4.79556 (0.25)	0.11101 (0.05)	1816.0 (0.8)

Kisseynew belt

Herblet Lake gneiss dome										
A -74+62 NM4	0.023	493.4	151.3	2986	74	3.8	0.30856 (0.28)	4.78240 (0.29)	0.11241 (0.05)	1838.7 (0.9)
B -105+74 NM4 abr	0.022	585.5	191.8	2538	103	5.9	0.32199 (0.26)	5.04454 (0.28)	0.11363 (0.06)	1858.2 (1.0)
C -74+62 NM4 abr	0.020	418.9	139.6	3258	53	5.8	0.32759 (0.34)	5.17389 (0.35)	0.11455 (0.05)	1872.8 (0.9)
D -105+74 NM4 abr[i]	0.008	500.0	160.0	1246	54	5.3	0.32494 (0.17)	5.11104 (0.20)	0.11408 (0.07)	1865.4 (1.2)
Kisseynew enderbite										
A -74+62 NM0	0.32	560.8	180.8	2109	172	5.6	0.31845 (0.12)	4.87847 (0.14)	0.11111 (0.06)	1817.6 (1.1)
B +105-74 NM0	0.0695	580.5	191.6	8603	96	6.1	0.32429 (0.11)	4.98351 (0.13)	0.11145 (0.04)	1823.3 (0.7)
C -105+74 NM0 abr	0.0355	485.3	163.3	3079	115	7.6	0.32531 (0.12)	5.00667 (0.14)	0.11162 (0.05)	1826.0 (0.9)
D +105 Mag0 abr	0.0808	448.4	146.0	13050	57	5.1	0.32322 (0.11)	4.97375 (0.13)	0.11606 (0.04)	1825.7 (0.7)
A monazite	0.1124	1405.0	6070.7	15076	212	92.8	0.32238 (0.16)	4.90965 (0.17)	0.11046 (0.04)	1806.9 (0.7)
B monazite	0.0908	1231.7	5588.7	34690	65	93.2	0.32237 (0.16)	4.90681 (0.18)	0.11040 (0.04)	1805.9 (0.7)
Kisseynew tonalite										
A -105+74 NM0 abr	0.0682	797.6	244.0	3266	329	2.2	0.31303 (0.16)	4.80171 (0.17)	0.11125 (0.05)	1820.0 (1.0)
B +105 NM0 abr	0.0446	698.1	218.1	5862	107	2.4	0.31886 (0.12)	4.91634 (0.13)	0.11183 (0.04)	1829.3 (0.8)
C +105 Mag0 abr	0.0703	603.5	188.7	3683	231	2.7	0.31806 (0.16)	4.91699 (0.17)	0.11212 (0.05)	1834.1 (0.9)
D -105+74 Mag0	0.0382	694.1	209.7	3625	143	2.4	0.30872 (0.16)	4.74701 (0.18)	0.11152 (0.05)	1824.3 (0.9)
A monazite	0.0384	6040.3	6182.2	8812	538	69.5	0.32456 (1.6)	4.97152 (1.6)	0.11109 (0.04)	1817.4 (0.7)
Kisseynew granodiorite										
A -105+74 NM co abr	0.273	735.6	232.3	3815	103	6.5	0.30905 (0.11)	4.71158 (0.13)	0.11057 (0.05)	1808.8 (0.8)
B -105+74 Mag co abr	0.0522	883.3	273.2	2372	369	7.3	0.30020 (0.16)	4.55103 (0.18)	0.10995 (0.07)	1798.6 (1.2)
C -74+62 NM co abr	0.0266	696.1	220.8	3387	107	6.6	0.31010 (0.16)	4.72792 (0.18)	0.11058 (0.05)	1808.9 (0.9)
D -74+62 Mag co	0.0685	738.6	216.3	799	1140	8.4	0.28102 (0.16)	4.27392 (0.23)	0.11030 (0.17)	1804.4 (3.0)
A monazite	0.0359	7157.7	7729.2	33525	155	71.3	0.32252 (0.20)	4.90408 (0.21)	0.11028 (0.04)	1804.0 (0.7)
Kisseynew granite										
A +105 NM3 abr	0.0618	1257.7	381.1	10070	154	0.5	0.31574 (0.12)	4.81396 (0.14)	0.11058 (0.04)	1808.9 (0.7)
B -105+74 NM3 abr	0.0641	1535.1	466.6	8401	230	1.9	0.31208 (0.16)	4.77814 (0.17)	0.11104 (0.04)	1816.5 (0.7)
C -105+74 NM3	0.0687	1333.8	391.2	3310	528	2.5	0.30218 (0.16)	4.59761 (0.17)	0.11035 (0.05)	1805.2 (1.0)
D -74+62 NM3	0.0170	1348.5	400.5	530	83	2.1	0.30442 (0.16)	4.63490 (0.17)	0.11043 (0.04)	1806.4 (0.8)
A monazite	0.027	11433.4	8878.5	5925	1060	60.2	0.32180 (0.15)	4.89854 (0.16)	0.11040 (0.04)	1806.1 (0.8)

[a] Sizes (-74+62) refer to the length aspect of zircons in micrometres. Abbreviations: abr, abraded; Mag co, magnetic zircons separated using a cobalt magnet; NM co, nonmagnetic zircons separated using a cobalt magnet; NM1, nonmagnetic cut with Frantz at 1° side slope; and Mag0, magnetic cut with Frantz at 0° side slope.

[b] Radiogenic Pb.

[c] Measured ratio, corrected for spike and fractionation.

[d] Total common Pb in analysis corrected for fractionation and spike.

[e] Corrected for blank Pb and U, and common Pb; errors in parentheses are 1 σ in per cent.

[f] Corrected for blank and common Pb, errors in parentheses are 1 σ in Ma.

[g] Cliff Lake pluton fractions A,B, and C used 208 spike for analysis.

[h] Ham Lake fractions A,B,C,D, and E used 208 spike for analysis.

[i] Zircon sample weight, U, and Pb compositions estimated; decay constants used are those of Steiger and Jager (1977).

Table II
Results of regression analysis using the method of York (1969)

Sample	Points used	Upper intercept (Ma)	Lower intercept (Ma)	MSWD
Flin Flon belt				
Amisk Group rhyolite tuff	A,C,D	1886^{+2}_{-2}	57	2.95
Chickadee Lake rhyolite	A,B,C,D	1832^{+2}_{-2}	11	0.26
Cliff Lake pluton	A,B,C,E	1874^{+32}_{-25}	257	17.9
Lynx Lake pluton	A,C,D	1847^{+4}_{-4}	9	0.45
Rex Lake pluton	A,B,D	1832^{+4}_{-4}	158	1.14
Wekusko granite	A,B,C,D	1834^{+8}_{-6}	159	8.82
Ham Lake pluton	A,B,C,D, E,F,G	1830^{+27}_{-19}	410	29.7
Kisseynew belt				
Herblet Lake gneiss dome	A,B,C,D	1890^{+12}_{-9}	803	2.8
Kisseynew enderbite	A,B,C,D	1830^{+11}_{-5}	622	7.4
Kisseynew tonalite	A,B,C,D	1841^{+35}_{-15}	482	30.9
Kisseynew granodiorite	A,B,C,D	1814^{+17}_{-11}	236	16.3
Kisseynew granite	A,B,C,D	1816^{+23}_{-12}	242	43.5

Note: MSWD, mean square of weighted deviates.

At Snow Lake, the Missi Group is a monotonous sequence of metamorphosed lithic arenites (Froese and Moore, 1980). Conglomerate, although characteristic of the Missi Group at Flin Flon, is notably lacking at Snow Lake. East of Wekusko Lake (Fig. 4), the Missi Group consists of crossbedded sandstones, wackes, conglomerates, and volcanic rocks (Stockwell, 1937). Similar rocks to the east in the Saw Lake area have also been assigned to the Missi Group (Bailes, 1985). Field evidence indicates that the volcanic rocks were deposited subaerially (Shanks and Bailes, 1977; Gordon and Gall, 1982). They range from basalt to rhyolite, and have compositions characteristic of destructive plate margins (Gordon and Lemkow, 1987).

Stauffer (1974, 1988) interprets the Missi Group as a fluvial–alluvial redbed deposit. The presence of granitic clasts and the basal angular unconformity indicates that deposition of the Missi Group postdated a major period of plutonism, deformation, uplift, and erosion, which affected the Amisk Group.

Geochronology of the Missi Group. Samples were collected from two quartz–feldspar porphyry bodies on the east shore of Wekusko Lake. Stockwell (1937) interpreted these as stocks, but several contain evidence of an extrusive origin (Gordon and Gall, 1982; Galley *et al.*, 1988). Several zircon fractions from these hydrothermally altered rocks were analyzed, but are highly discordant and fail to define a well-constrained discordia line. No precise age information can be obtained from these analyses.

The Chickadee Lake rhyolite sample is from a 30 m thick welded ash flow (Shanks and Bailes, 1977; Gordon and Gall, 1982) within a series of rhyolitic flows and volcaniclastic rocks in the Missi Group east of Wekusko Lake (Fig. 4). The flow is on the southeast limb of the Herb Lake syncline (Stockwell, 1937; Bell, 1978) immediately above an amygdaloidal mafic flow within the rhyolitic portion of the section. The welded texture and association with fluviatile sediments indicate that it was deposited subaerially. Flattened fragments up to 5 mm × 50 mm are defined by subtle changes in grain size and mineralogy. The rock contains 4%, 0.5–3 mm feldspar phenocrysts and approximately 2%, 1 mm quartz phenocrysts within a recrystallized, fine-grained mosaic of quartz and feldspar grains. Secondary white mica and magnetite are disseminated throughout the matrix.

Zircons selected for analysis were euhedral and generally clear and colourless, with only a few broken ends. Analyses have similar $^{207}Pb/^{206}Pb$ ages and define a linear array with an upper intercept of 1832 ± 2 Ma (Fig. 6a). This is interpreted as the depositional age of the ash flow.

The volcanic sequence containing the Chickadee Lake rhyolite overlies a series of magnetite-bearing crossbedded sandstones and polymictic conglomerates that are correlative with the Missi Group at Flin Flon (Stockwell, 1937). Subaerially deposited volcanic rocks have also been mapped within the Missi Group in the Saw Lake area to the east (Bailes, 1985). Both volcanic sequences are probably equivalent in age; hence, the youngest volcanic activity in the Flin Flon belt and deposition of the Missi Group at the eastern end of the belt are constrained by the 1832 Ma date. The Missi Group was affected by the major folding episodes recognized at Snow Lake (Froese and Moore, 1980; Bailes, 1985), so this datum also provides the earliest age for the main deformational events in this region.

The Rex Lake pluton, Wekusko granite, and sills within the Kisseynew belt (described below) have similar ages. An igneous event at 1.83 Ga apparently resulted in volcanism and plutonism at the eastern end of the Flin Flon belt as well as plutonic activity within the Kisseynew belt.

Quartzofeldspathic terrigenous sedimentary rocks unconformably overlie volcanic rocks elsewhere in the Trans-Hudson Orogen. These include the Sickle Group in the Lynn Lake and northern Kisseynew belts, the McLennan Group in the La Ronge belt, and the Ourom Lake meta-arkose in the Glennie Lake domain. Because of their lithological similarity, a number of workers have assumed that they were deposited contemporaneously. Delaney *et al.* (*in press*), however, recently obtained an 1850 ± 1 Ma age for the Ourom Lake meta-arkose in the Glennie Lake domain. This age, constrained by U–Pb dating of single detrital zircon grains and a crosscutting pluton, is clearly older than the 1832 ± 2 Ma Missi Group age reported here, and indicates that the various terrigenous sequences may not be time equivalent.

Flin Flon belt intrusive rocks

Porphyritic felsic intrusions characterized by strongly xenolithic margins, abundant dykes, and alteration in adjacent Amisk Group volcanic rocks are amongst the earliest intrusive rocks in the Flin Flon belt. They include the Cliff Lake pluton at Flin Flon and the Sneath Lake pluton at Snow Lake. Both are considered to have been emplaced during Amisk volcanism.

Several major differentiated gabbroic sills intrude Amisk Group rocks throughout the Flin Flon belt. All display strong Fe enrichment and are clearly tholeiitic. If they have similar ages, they represent a major mafic magmatic episode prior to Missi deposition.

Evidence for post-Amisk, pre-Missi felsic intrusive activity includes an exposed unconformity between Missi Group rocks and a granitoid complex 4 km northwest of Flin Flon, the presence of granitic pebbles in Missi conglomerate, and abundant quartz and feldspar grains of igneous origin in Missi Group arenite.

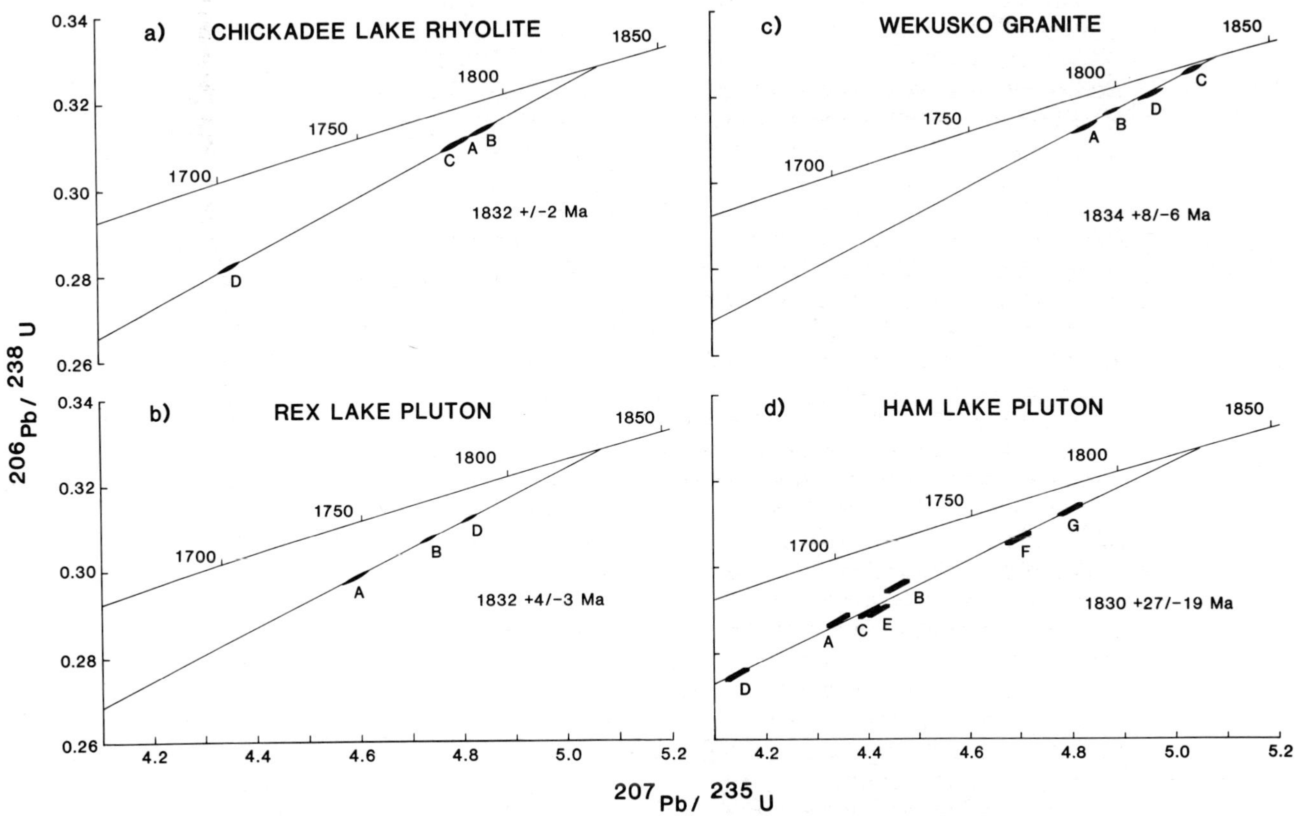

Figure 6. *Concordia diagrams showing data from rocks at the east end of Flin Flon belt.*

Demonstrable post-Missi intrusions include the large granodiorite bodies between Flin Flon and Amisk Lake, a series of mafic dykes near the Manitoba–Saskatchewan boundary (Boundary Intrusions), and several small granitoid plutons east of Wekusko Lake. In addition, a number of plutons are thought to be pre- to syntectonic, and hence emplaced before and during deformation of the Missi Group. These include the Ham Lake pluton, Rex Lake pluton, and Wekusko granite (Cerny *et al.*, 1981; Halden, 1987).

Geochronology of Flin Flon belt intrusive rocks. The Cliff Lake pluton (Fig. 2) is a multiphase, quartz-megacrystic tonalite stock that occurs in the Hook Lake fault block, east of Cliff Lake (Bailes and Syme, 1987, *in press*). A number of features in the intrusion, including magmatic stoping of wall rocks, mineralized breccia pipes (Baldwin, 1980), and abundant Amisk xenoliths, suggest that the Cliff Lake tonalite is a high-level intrusion. It is emplaced entirely within Amisk Group rocks and is cut by a number of small mafic intrusions. No similar intrusive rocks cut the Missi Group; hence, it is interpreted as being pre-Missi and probably syn-Amisk in age.

The sample was obtained from the western side of the pluton in unit 45c of Bailes and Syme (1987, *in press*). This is one of the younger phases of the pluton, characterized by 25–40%, 2–5 cm, rounded, recrystallized mafic xenoliths. It is composed of 50–60%, 0.5–4 mm subhedral plagioclase; 30–40%, 1–8 mm quartz megacrysts and interstitial grains; and 2–35% amphibole, chlorite, epidote, and/or biotite. Accessories include Fe–Ti oxides, apatite, tourmaline, allanite, and zircon. Plagioclase and Fe–Ti oxides show moderate alteration.

Zircons selected for analysis were generally euhedral and lacked visible cores or inclusions. The analyses were performed early in this study (ca. 1985), when larger amounts of sample were required. The zircons are high in U and relatively discordant, resulting in an imprecise age of 1874^{+32}_{-25} Ma (Fig. 7). The 1.87 Ga age is, however, consistent with the 1.88 Ga U–Pb zircon age reported by MacQuarrie (1980) for this pluton. Both results reinforce the field interpretation that the Cliff Lake pluton was related to Amisk volcanism.

The Lynx Lake pluton (Syme, 1987; Syme *et al.*, 1987) is a 3.8 km × 8.6 km, ovoid, concentrically zoned pluton emplaced in Amisk Group rocks on the north shore of Athapapuskow Lake (Fig. 3). The pluton truncates folded Amisk strata and is cut by the Payuk Lake and Millwater faults. The core phase of the pluton is a muscovite–biotite granodiorite with interstitial microcline. The sample was obtained from the margin phase, a medium-grained biotite–hornblende granodiorite with poikilitic perthite.

Analyzed zircons had sharp terminations and an aspect ratio of 2:1. Crystals had typical igneous zoning, without visible evidence of inheritance, although their clarity was only moderately good. Three analyses (Fig. 8) define a chord with an upper intercept age of 1847^{+4}_{-4} Ma.

On the basis of its lack of foliation and absence of crosscutting dykes, the Lynx Lake pluton is considered one of the youngest plutons in the Flin Flon area. The age places constraints on some of the structures in the Millwater fault block: folds must be pre-1847 and faults post-1847 Ma. Plutons of comparable age occur throughout the Glennie Lake and La Ronge domains and form the 1834–1859 Ma Wathaman and Chipewyan batholiths (Zwanzig *et al.*, 1985; Van Schmus and Schledewitz, 1986; Lewry *et al.*, 1987; Van Schmus *et al.*, 1987). South of the Flin Flon belt, a 55 km × 30 km sub-Paleozoic granitoid complex has yielded a similar date of 1845^{+10}_{-8} Ma (Blair *et al.*, 1988). Younger plutons occur in the Snow Lake area (1832–1835 Ma) and at Phantom Lake in Saskatchewan (1820 Ma, R. MacQuarrie, personal communication).

The Rex Lake plutonic complex (Cerny *et al.*, 1981) is a roughly circular composite pluton, 8 km in diameter, intruding Amisk Group rocks on the north shore of Wekusko Lake (Fig. 4). From oldest to youngest, the intrusive phases are gabbro, diorite and quartz diorite, granodiorite, and granite. A northeasterly trending foliation, defined by biotite orientation, indicates that the pluton was emplaced prior to the F2 deformation of Froese and Moore (1980). The sample was collected from the central part of the pluton, a light pink, medium-grained, foliated biotite granodiorite, in map unit R of Cerny *et al.* (1981).

The analyzed zircons were euhedral with 2:1 aspect ratios. They were clear to slightly cloudy,

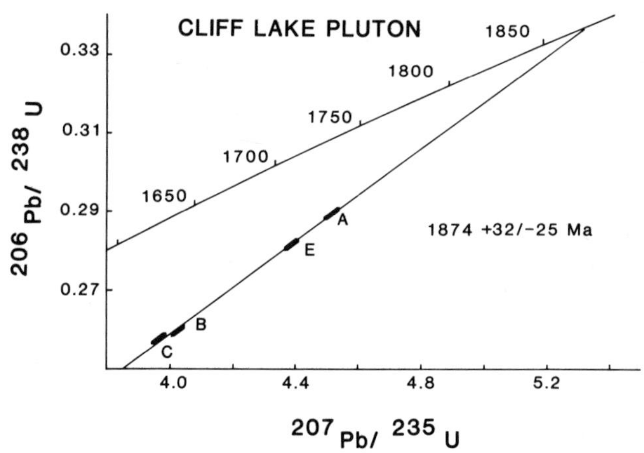

Figure 7. *Concordia diagram showing data from the Cliff Lake pluton at Flin Flon.*

and occasionally yellowish in colour. Igneous zoning was common, but no inherited cores were visible. Three analyses define a chord with an 1832^{+4}_{-3} Ma concordia intercept (Fig. 6b), indistinguishable from that of the Missi Group volcanic rocks described above.

The Rex Lake pluton is thus part of the 1830–1835 Ma igneous suite at the eastern end of the Flin Flon belt. It is tempting to speculate that this multiphase pluton may have formed a feeder system to contemporaneous Missi volcanic rocks.

The Wekusko granite forms an oval-shaped body approximately 8 km in diameter, located on the southwest shore of Wekusko Lake (Fig. 4). It has intruded Amisk Group rocks, although xenoliths are rare. The lack of a tectonic fabric led Cerny *et al.* (1981) to suggest that it was emplaced after regional deformation. The sample was collected from the central part of the intrusion. It is a light pink, medium-grained, biotite–hornblende granite. Zircon, allanite, apatite, and magnetite are accessory minerals.

The zircons analyzed were euhedral with good terminations, had aspect ratios of 3:1, and were clear to slightly cloudy. Although most crystals contained fractures, no cores were observed. Four analyses are slightly to moderately discordant and define a concordia intercept age of 1834^{+8}_{-6} Ma, indistinguishable from the ages of the Missi volcanic rocks and Rex Lake pluton described above (Fig. 6c). If the Wekusko granite is, indeed, posttectonic (Cerny *et al.*, 1981), then deformation of the Missi Group took place within a few million years of its deposition. The interpretation of a posttectonic age is based solely on the absence of a foliation, however, so the possibility remains that the Wekusko granite is coeval with the Missi Group, and, hence, pretectonic in age.

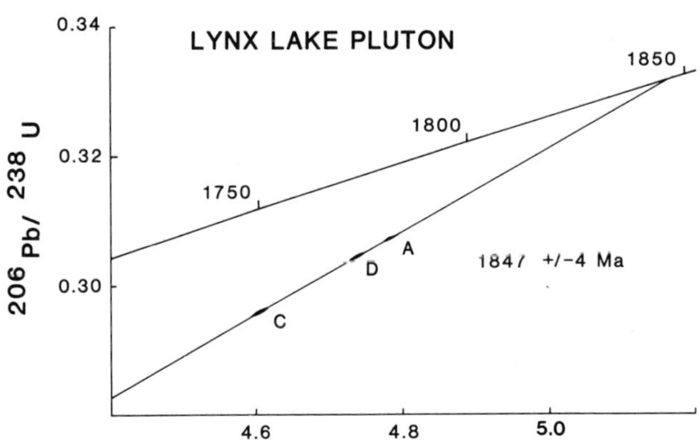

Figure 8. *Concordia diagram showing data from the Lynx Lake pluton on Athapapuskow Lake.*

The Ham Lake pluton (Bailes, 1980a, 1986) is a 6.5 km × 18 km, north-northeast trending batholith on the east shore of File Lake (Fig. 4). It is massive to gneissic, medium- to coarse-grained, and composed of granodiorite, leucotonalite, and tonalite. Amisk Group xenoliths are abundant on its eastern margin. Its emplacement postdates the earliest folding episode (D1 of Bailes (1980a)), but predates the second deformation (D2) and the main episode of regional metamorphism. The sample is a pink-weathering, coarse-grained, biotite–hornblende granodiorite. Accessory minerals include apatite, sphene, and zircon.

The zircons were moderately cloudy, with distinct igneous zoning and some visible inherited cores. The most concordant results, F and G (Fig. 6d), were obtained from strongly abraded zircons. The $^{207}Pb/^{206}Pb$ age of the most concordant analysis is 1817 Ma, within the uncertainty of the 1830^{+27}_{-19} Ma upper intercept of the scattered linear array. The range in age uncertainty spans the 1835–1860 Ma time interval of widespread plutonism in the Trans-Hudson Orogen.

KISSEYNEW GNEISS BELT

Geological setting

The term "Kisseynew gneisses" was originally applied by Bruce (1918) to the complex of gneissic rocks immediately north of the Flin Flon belt at Kisseynew Lake. This definition was expanded by Harrison (1951) and McRitchie (1974) to include high-grade rocks north of the Flin Flon belt, south of the Lynn Lake and Rusty Lake belts, and west of the Thompson lineament. On the west, the Sturgeon–Weir and Tabernor fault zones separate the Kisseynew gneiss belt from the Hanson Lake block and Glennie Lake domain. Most of the belt is underlain by migmatite and anatectic plutons derived from greywacke turbidites, but the flanks are more complex. The relationship between the Kisseynew belt and the adjacent volcanic belts has been controversial and remains the subject of ongoing research.

The southern margin of the Kisseynew belt is a 20–50 km wide complex of interlayered quartzofeldspathic metasedimentary rocks, metavolcanic amphibolite and felsic gneiss, orthogneiss, turbidite-derived paragneiss, and metamorphic rocks of unknown protolith. These units are deformed into complicated fold interference patterns, including gneiss domes and basins. Reviews of structural and stratigraphic interpretations of the region can be found in Bailes (1971), Froese (1984), and Macdonald (1987).

Locally, Amisk Group and Missi Group rocks can be traced from the Flin Flon belt into the southern margin of the Kisseynew belt (Bailes, 1980b; Froese and Moore, 1980; Ashton, 1987). In these areas, the boundary between the belts coincides with a volcanic to sedimentary facies change, an increase in metamorphic grade from a chlorite–muscovite zone to a garnet–biotite–sillimanite zone, and a change in structural style. Elsewhere, faults and narrow mylonite zones separate the two belts.

Within the Kisseynew belt, rocks of the Nokomis Group and Sherridon Group (Robertson, 1953; Froese and Goetz, 1981) have been correlated with Amisk Group sedimentary rocks and the Missi Group respectively. Recent studies (McRitchie, 1985, 1986; Ashton *et al.*, 1987; Schledewitz, 1987; Zwanzig and Lenton, 1987) suggest, however, that the problem requires reexamination.

Regardless of internal stratigraphic complexities, most workers now favour Bailes' (1971) interpretation that the heterogeneous southern flank of the Kisseynew gneiss belt is an extension of the Flin Flon belt, metamorphosed at deeper levels and thrust southward toward the lower grade rocks.

Regional mapping (Baldwin *et al.*, 1979) and geophysical compilations (Committee for the Magnetic Anomaly Map of North America, 1987) show that, in contrast with the southern flank, the 100 km wide central region of the Kisseynew gneiss belt has a relatively homogeneous protolith. This area is underlain by cordierite–garnet paragneiss, migmatite, and anatectic rocks derived from greywacke–shale turbidites (Bailes and McRitchie, 1978; Lenton, 1981). The rocks are characteristically graphitic, poor in sillimanite and K-feldspar, and contain varying proportions of anatectic leucosome material. McRitchie (1974) assigned the name "Burntwood River Supergroup" to (1) these rocks, (2) greywacke-derived Kisseynew gneisses, and (3) the Nokomis Group. Lenton (1981) subsequently revised the name to "Burntwood Metamorphic Suite."

The central part of the Kisseynew gneiss belt is interpreted as a large sedimentary basin fed by turbidity currents from the adjacent volcanic arcs (Bailes, 1971).

Structural geology

On the southern flank of the Kisseynew belt, the earliest recognizable structures are minor isoclinal folds and south- to southwest-verging recumbent folds, involving both Amisk and Missi (or Nokomis and Sherridon) rocks (Zwanzig, 1983, 1984; McRitchie, 1985; Ashton and Wheatley, 1986; Ashton *et al.*, 1986). In places, it has been possible to ascertain that such structures predate peak metamorphism (Bailes, 1975, 1980a, 1985; Schledewitz, 1987). This deformation may have been synchronous with the development of east-trending phase 1 folds at Flin Flon (Stauffer and Mukherjee, 1971); the McLeod Road thrust at Snow Lake (Froese and Moore, 1980); and south-verging recumbent structures east of Wekusko Lake (Frarey, 1950; Gordon and Gall, 1982).

Within the central Kisseynew belt, the oldest folds are isoclinal and were developed synchronously with upper amphibolite facies metamorphism (Bailes and McRitchie, 1978; Lenton, 1981). South of the Baldock batholith, folds associated with the earliest foliation verge north to northeast (Schledewitz, 1972; Elphick, 1972; Manitoba Energy and Mines, 1986a,b). However, along the western portion of the Kisseynew belt – Lynn Lake belt boundary, early structures appear to verge to the south (Jackson, 1988).

Subsequent folding was synchronous with peak metamorphism and is variable in style and orientation. Although most workers report several phases of deformation, at least one generation of northeasterly plunging folds is usually present. During the late stages of metamorphism, retrograde minerals define the foliations associated with local folds and faults. The youngest faults postdate regional metamorphism and include the major fault zones forming the eastern and western boundaries of the Kisseynew belt.

Metamorphism

The metamorphic grade increases from subgreenschist or greenschist facies in the flanking volcanic belts to upper amphibolite to granulite facies in the Kisseynew belt (Bailes and McRitchie, 1978). Within the central portion of the Kisseynew belt, the metamorphic grade is remarkably uniform. Jackson and Gordon (1986), Jackson (1988), and Gordon (*in press*) estimate metamorphic conditions at 5.5 kbar (550 MPa) and 750°C across the belt. Measured radiogenic heat production and estimated crustal thicknesses are inadequate to generate the elevated thermal gradient, suggesting that either an elevated mantle heat flow or heat transport by rising plutons is responsible for the metamorphic event (Gordon, 1988, *in press*).

Kisseynew belt intrusive rocks

Most of the intrusive rocks recognized along the southern flank of the Kisseynew belt are orthogneisses. The oldest of these are metagabbros and metatonalites. Younger units, ranging in composition from granite to tonalite, form the cores of structural domes (Schledewitz, 1987; Zwanzig and Lenton, 1987). In the western part of the area, these plutons postdate an early phase of isoclinal folding, and predate the main metamorphism (Schledewitz, 1987). The eastern domes contain both orthogneiss and paragneiss and were structurally emplaced during regional metamorphism (Bailes, 1975). The youngest intrusive rocks are relatively undeformed, postmetamorphic, leucocratic granites and pegmatites.

Within the core of the belt, Baldwin *et al.* (1979) mapped four major plutonic units with a known sequence of emplacement. The oldest intrusions are hypersthene-bearing, intermediate to mafic, sheets and sills. Coarsely inequigranular garnetiferous tonalite to granodiorite gneiss postdates the hypersthene-bearing sills and occurs with younger granitoids in large irregular bodies, particularly in the Highrock Lake area. Peraluminous granodiorite and granite are the most widespread igneous rocks in areas underlain by the Burntwood Metamorphic Suite. They occur as segregations, sills, dykes, and irregular nebulitic bodies with inclusions of paragneiss.

Other igneous bodies mapped by Baldwin *et al.* (1979) include monzonite, mafic tonalite, syenite, and pegmatite. Except for undeformed pegmatite dykes, there are insufficient data to determine the sequence of intrusion of these bodies.

Geochronology of Kisseynew belt intrusive rocks. The Herblet Lake gneiss dome (Bailes, 1975; Froese and Moore, 1980) is an 18 km wide, roughly triangular structure in the southern flank of the Kisseynew belt, 15 km northeast of the town of Snow Lake (Fig. 4). The gneiss dome comprises layered leucocratic granitoid oligoclase–quartz gneiss, layered amphibolite, and massive pink granitoid oligoclase–quartz–microcline gneiss. It is rimmed by a 200–500 m wide band consisting of amphibolite of uncertain origin and Amisk Group felsic metavolcanic gneiss. Layered rocks within the gneiss dome are interpreted as being of supracrustal origin, and are, in part, lithologically similar to Amisk Group metavolcanic rocks rimming the dome (Fedikow and Malis, 1987). Bailes (1975) suggests that rocks of the gneiss dome were remobilized and emplaced during regional metamorphism.

The sample was collected from the southern part of the core of the dome (Fig. 4), in map unit 8a of Froese and Moore (1980), which corresponds to map unit C3 of Bailes (1975). The rock is a massive pink hornblende–biotite granodiorite with accessory garnet, magnetite, apatite, and zircon.

Zircons selected were euhedral, clear, and colourless, with both sharp and slightly rounded terminations. Igneous zoning was visible in some grains, but no inherited cores were observed. The most concordant analysis gives a $^{207}Pb/^{206}Pb$ minimum age of 1873 Ma. Four analyses are scattered about a chord giving a concordia intercept age of 1890^{+8}_{-6} Ma (Fig. 9).

The igneous zoning observed in the zircons, coupled with the slightly rounded nature of many terminations, is consistent with this unit being a metamorphosed plutonic rock. The 1890 Ma age is thus interpreted as the time of igneous crystallization of the orthogneiss, and is similar to the 1886 Ma age of Amisk volcanism at Flin Flon. These results support the interpretation that the southern flank of the Kisseynew belt is a metamorphosed and highly deformed extension of the Flin Flon belt.

The sample of enderbite is from a 1 km thick sill of unit 9 of Baldwin *et al.* (1979), collected on the File River, 3 km south of Burntwood Lake (Fig. 10). The sill intrudes metagreywacke and migmatite of the Burntwood Metamorphic Suite. A weak foliation and metamorphic segregations indicate a premetamorphic emplacement age for this body.

The rock is a brown-weathering, weakly foliated, medium-grained rock with minor, 2–6 cm, medium- to coarse-grained plagioclase–hypersthene segregations. Its mineralogical composition is approximately 60% sodic andesine, 20% biotite, 10% quartz, and 10% hypersthene, with traces of opaques, zircon, and monazite. Hypersthene is locally altered along parting planes to an unidentified mineral, probably serpentine.

The analyzed zircons were clear and colourless, with inclusions in only 5–10% of the grains. Igneous zoning was present, but no inherited cores were observed. The most concordant fraction has a $^{207}Pb/^{206}Pb$ age of 1826 Ma. A chord fit to four analyses gives a concordia intercept age of 1830^{+11}_{-5} Ma (Fig. 11a). Two monazite analyses are concordant at 1806 ± 2 Ma. A K–Ar biotite age of 1719 ± 21 Ma has also been obtained from this sample (Hunt and Roddick, *in press*).

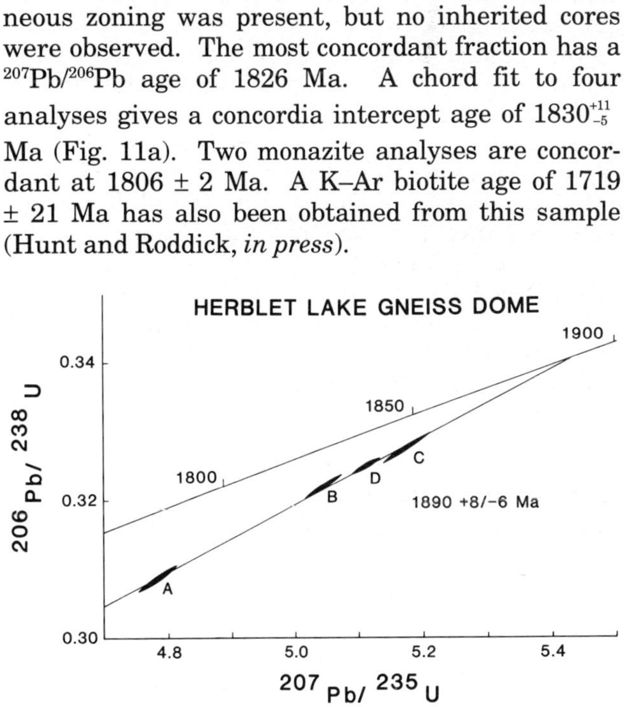

Figure 9. *Concordia diagram showing data from the Herblet Lake gneiss dome on the southern flank of the Kisseynew gneiss belt.*

The rounded terminations on the zircon grains probably resulted from the metamorphism that produced the orthopyroxene in the sill; thus, the 1830 Ma zircon age is interpreted as the time of igneous intrusion. These sills, therefore, are approximately coeval with the igneous event that produced the Missi Group volcanic rocks and plutons near Snow Lake. The monazite and biotite ages are interpreted as the times at which the rock cooled through 650°C (Parrish, 1988) and 280°C (Harrison and McDougall, 1980). These ages, with analogous data from other Kisseynew intrusions, have been used to constrain a numerical model for postmetamorphic uplift and cooling of the Kisseynew sedimentary gneiss belt (Gordon, *in press*).

The tonalite sample was collected on the south arm of Highrock Lake from an apophysis of the 30–50 km wide tonalite–granodiorite complex southeast of the main body of the lake (Fig. 10). These rocks are part of map unit 10 of Baldwin *et al.* (1979). Crystallization of this material predated the widespread development of granitoid pegmatites and mobilizate in the Burntwood Metamorphic Suite.

Figure 10. *Sample locations for intrusive rocks in the Kisseynew belt.*

The rock is a light grey weathering, homogeneous to coarsely inequigranular garnetiferous tonalite. It has approximately 80% antiperthitic oligoclase, 10% quartz, 5% biotite, and 5% garnet, with traces of opaques, zircon, and monazite. A minor amount of pale green biotite rims some garnet grains.

The analyzed zircons were prismatic, with slightly rounded terminations. They were zoned, yellowish, and only moderately clear. Small black inclusions occurred in 3–5% of the grains. The four isotopic analyses from this sample are scattered about a chord intersecting concordia at 1841^{+35}_{-15} Ma (Fig. 11b). Field relationships require that this unit be younger than the 1830^{+11}_{-5} Ma enderbite; hence, the crystallization age is probably close to the oldest $^{207}Pb/^{206}Pb$ age of 1834 Ma. The 1710 ± 25 Ma K–Ar biotite age (Hunt and Roddick, *in press*) from this sample is indistinguishable from those of other igneous rocks within the Kisseynew belt.

The monazite age of 1817 ± 5 Ma is significantly older than the 1804–1806 Ma ages obtained from the other three samples. This disparity may be due to inheritance of radiogenic lead (Copeland *et al.*, 1988), but could also result from different initial rates of postmetamorphic uplift and cooling. This sample is the most northerly of the four central Kisseynew samples and is from the core of the gneiss belt. Initiation of uplift in the core of the migmatite terrain could have resulted from the higher proportion of low-density granitoid magma in this region.

The granodiorite sample is from an irregular 2 km × 10 km thick lens in migmatites of the Burntwood Metamorphic Suite on the northwest arm of Burntwood Lake (Fig. 10). It is part of unit 11 of Baldwin *et al.* (1979). Near the sample locality, this body intrudes sills of their unit 9.

This rock is a pink weathering, medium-grained, garnetiferous, porphyritic granodiorite. It comprises approximately 75% perthitic microcline, 15% quartz, 5% oligoclase, 5% biotite, and minor garnet, myrmekite, opaques, zircon, and monazite. Secondary white mica occurs in local patches.

The zircons were clear and colourless, with few inclusions. They were prismatic and had rounded terminations. Although some igneous zoning was present, inherited cores were not visible. The most concordant analyses (Fig. 11c) give $^{207}Pb/^{206}Pb$ ages of 1809 Ma. Scatter in the four analyses results in a relatively large uncertainty in the concordia intercept of 1814^{+17}_{-11} Ma. The monazite age of 1804 ± 2 Ma is comparable with two of the other monazite ages reported here. The K–Ar biotite age is 1690 ± 29 Ma (Hunt and Roddick, *in press*).

The field relationships and petrology of this unit indicate that it was derived by partial melting of Burntwood Metamorphic Suite metasedimentary rocks. The rounded zircon terminations indicate some postcrystallization resorption. The granodiorite is thus interpreted as having developed near the maximum temperatures attained in the Kisseynew belt, and, hence, with the granite data given below, constrains the age of peak regional metamorphism.

The granite sample is from a 1 km × 5 km concordant sill in migmatite of the Burntwood River Metamorphic Suite on the northwest arm of Burntwood Lake (Fig. 10). It is from unit 12 of Baldwin *et al.* (1979); part of the widespread system of irregular veins, sills, and larger bodies forming the bulk of the anatectic mobilizate in the Burntwood Metamorphic Suite of the Kisseynew sedimentary gneiss belt. Sillimanite, cordierite, and garnet are common accessory minerals in these rocks. They have been interpreted as representing anatectic melts formed during peak metamorphism (Bailes and McRitchie, 1978).

The rock is a light grey to white weathering, heterogeneous, medium- to coarse-grained garnetiferous granite. It has approximately 75% perthitic orthoclase, 15% quartz, 5% biotite, and 5% amoeboid garnet, with minor amounts of zircon, opaques, monazite, and myrmekite. Traces of secondary white mica occur locally.

The analyzed zircons were colourless and clear. Inclusions and fractures were rare, although igneous zoning was present. All fractions were significantly higher in U (1300 ppm) than other igneous rocks from the Kisseynew belt. The oldest $^{207}Pb/^{206}Pb$ age from this sample is 1817 Ma, whereas the discordia intersection obtained from four samples gives an age of 1816^{+23}_{-12} Ma (Fig. 11d). Monazite is nearly concordant at 1806 ± 2 Ma, whereas the K–Ar biotite age is 1705 ± 36 Ma (Hunt and Roddick, *in press*). These ages are comparable to those obtained from the granodiorite; hence, they are similarly interpreted as the age of peak metamorphism and cooling through 650°C and 280°C respectively.

The zircon, monazite, and biotite K–Ar data from the central Kisseynew belt, along with metamorphic pressure–temperature determinations, have

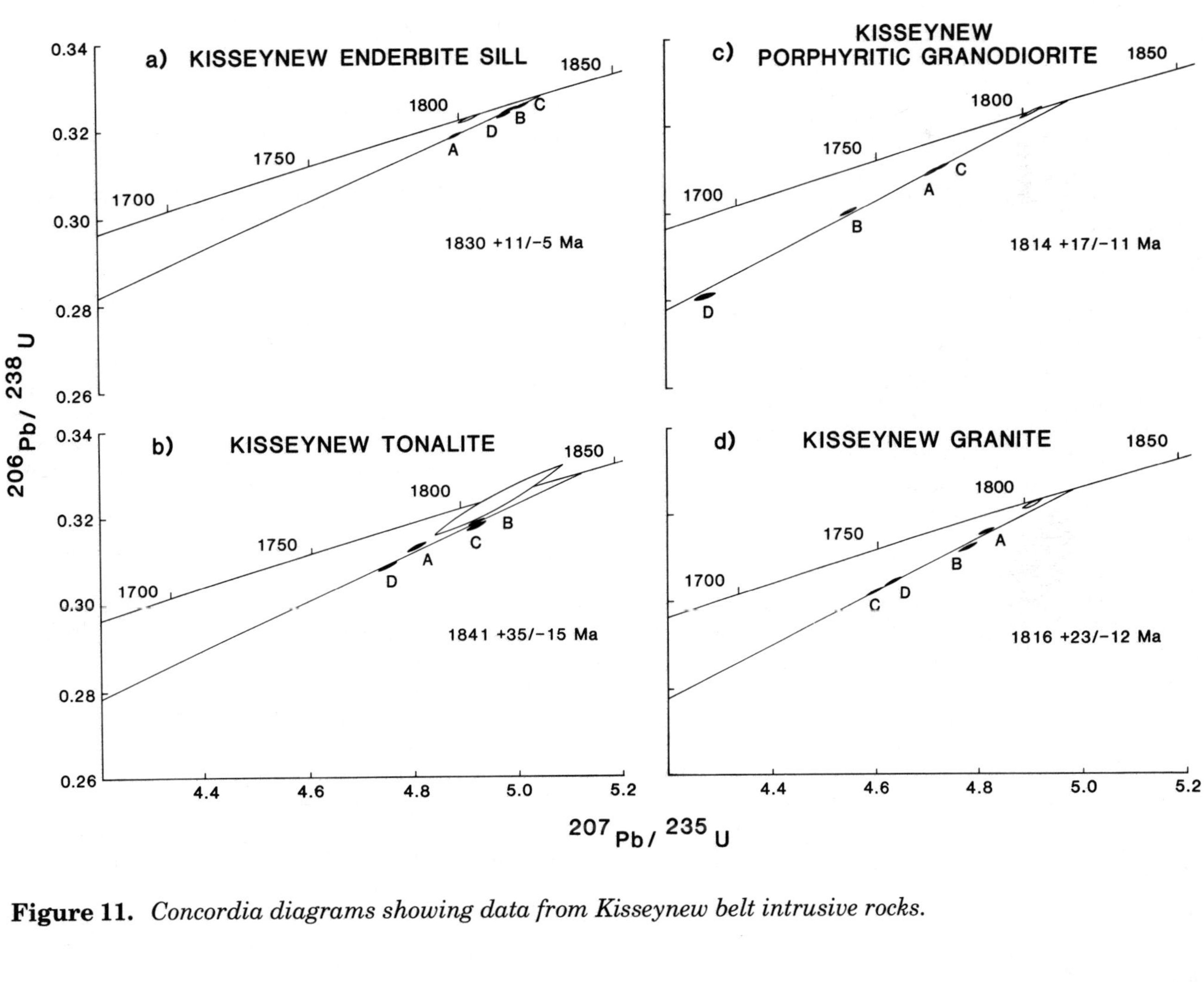

Figure 11. *Concordia diagrams showing data from Kisseynew belt intrusive rocks.*

been used to constrain numerical models for the postmetamorphic uplift of the Kisseynew belt (Gordon, 1988, *in press*). From 1815 to 1705 Ma, the uplift rate is estimated at 60–110 m/Ma. This average is computed over a time period during which the rate probably varied considerably, early uplift being controlled by tectonic processes and later uplift by erosional rebound. This would explain why the average rate is 2 to 10 times slower than rates estimated for various parts of the Coast Mountains of British Columbia during the last 25 Ma (Parrish, 1983) and 10–20 times faster than rates estimated for the southeastern highlands of Australia over the last 250 Ma (Lambeck and Stephenson, 1985).

DISCUSSION

The resistance to resetting of the U–Pb system in zircons, together with the available precision of isotopic analyses, has made it possible to establish the chronology of the assembly of the Manitoba portion of the Trans-Hudson Orogen. Parts of the overall sequence of events have been suggested previously (Lewry and Sibbald, 1980; Ray and Wanless, 1980; Lewry, 1981; Stauffer, 1984; Green *et al.*, 1985). Our data, along with that of Baldwin *et al.* (1987), now provide a quantitative time framework for the main crust-forming events.

At 1.89–1.91 Ga (Fig. 12a), the oldest volcanic rocks formed one or more oceanic arcs while volcanogenic sedimentary rocks were shed into adjacent basins. Locally, crustal spreading produced oceanic rift-type rocks. Sm–Nd analyses of volcanic rocks from the Hanson Lake block, La Ronge belt, and Flin Flon belt are interpreted by Chauvel *et al.* (1987) as indicating only minor Archean crustal contamination, suggesting that the volcanic rocks represent juvenile Proterozoic crust. Stratigraphic continuity between the Flin Flon belt and the southern flank of the Kisseynew belt indicates that these areas were contiguous, but such relationships have not been established for the other lithotectonic domains.

The preaccretionary geometric relationship of the arcs and basins to the Hearne–Wyoming and Superior cratons is not known. The present east–west orientation of tectonic domains provides a first approximation, but, as pointed out by Green *et al.* (1985), this is difficult to justify, particularly in view of the large translations known for Phanerozoic terranes (e.g., Monger *et al.*, 1982). Both the Rusty Lake belt and the Lynn Lake belt are geographically isolated from the Archean craton by ca. 1.86 Ga batholiths, which obscure possible evidence for the kinematics of accretion. Hypotheses of 1.91–1.875 Ga paleogeography and terrane motions are not considered to be testable with currently available data.

The next major magmatic episode took place at approximately 1875 Ma (Fig. 12b) when isoclinally folded volcanic rocks in the Lynn Lake belt were intruded by tonalite and quartz diorite magmas. Several Lynn Lake belt gabbroic bodies with Sm–Nd evidence of an Archean crustal component were also emplaced at about this time (Hegner and Hulbert, 1987). Synchronously, an 8 km thick rhyolitic caldera complex was developed in the Rusty Lake belt, suggesting the presence of a considerable thickness of sialic crust (Baldwin, 1987). The indirect evidence thus implies that the Lynn Lake and Rusty Lake belts were adjacent, and possibly accreted, to the Hearne–Wyoming craton by this time.

There are some constraints on post-1.875 Ga terrane motions. Dunsmore and Symons (1987) interpret their paleomagnetic data as indicating no significant relative rotations or translations of the Churchill and Superior provinces since emplacement of the Lynn Lake belt gabbro bodies. These data suggest that assembly of the central portion of the Trans-Hudson Orogen was substantially complete at, or shortly after, 1.875 Ga. This inference is supported by the fact that the Lynn Lake and Kisseynew belts are stratigraphically overlapped by the Sickle Group, a sequence of polymictic conglomerates and arenites. The minimum age of the Sickle Group is not yet well constrained, but must be greater than the 1.82 Ga age of metamorphism, and is probably close to the 1.85 Ga age of Chipewyan batholith rocks to the north reported by Zwanzig *et al.* (1985).

Cratonization of the region occurred from 1860 to 1845 Ma (Fig. 12c) when widespread plutonic activity produced the Wathaman–Chipewyan batholith and a number of smaller intrusions. Its mineralogy and composition are the same as a Cordilleran I-type batholith, suggesting a subduction-related origin at a craton margin (Fumerton *et al.*, 1984). In Saskatchewan, emplacement of the batholith was preceded by a major episode of isoclinal folding of country rocks on its southeastern edge (Lewry *et al.*, 1981). At this time, Amisk Group rocks in the Flin Flon belt had also undergone isoclinal folding prior to emplacement of the Lynx Lake pluton at 1847 Ma. During the latter stages of this period, uplift and erosion led to the development of continental molasse deposits, at least in the Glennie Lake domain (Delaney *et al.*, *in press*), and possibly elsewhere.

Magmatic activity continued on a reduced scale at ca. 1830 Ma, in the form of sills in the Kisseynew belt and small plutons and subaerial volcanoes within a fluvial–alluvial sedimentary sequence at the eastern end of the Flin Flon belt (Fig. 12d).

This is the youngest episode of magmatism and represents the last major addition of material to the crust within the region.

Between 1830 and 1815 Ma, north–south convergence resulted in thrust faulting in the Flin Flon belt and the development of east–west recumbent isoclinal folds within the Kisseynew belt. High temperature – moderate pressure metamorphism in the core of the gneiss belt began during this period and, at its peak, resulted in widespread anatexis at 1815 Ma (Fig. 12e). The thermal effects extended across the flanks of the Kisseynew belt, leading to the metamorphic zonation now observed in the Lynn Lake and Flin Flon belts. Metamorphism accompanied and outlasted the development of upright northeast-plunging folds with axes parallel to the present Churchill–Superior boundary. Both the early north–south compression and the deformation accompanying metamorphism were essentially intracratonic and represent accommodation of the Proterozoic rocks to late-stage relative motions of the Archean cratons (Hoffman, 1988).

Uplift of the Kisseynew belt began shortly after peak metamorphic conditions were reached, and resulted in the rocks cooling through 280°C at approximately 1705 Ma. The youngest regional-scale deformation took place on northwest to northeast trending faults, which offset isograds and juxtapose rocks of different metamorphic grade. These faults include those forming the east and west boundaries of the Kisseynew belt as well as smaller faults within the Flin Flon belt. The timing and

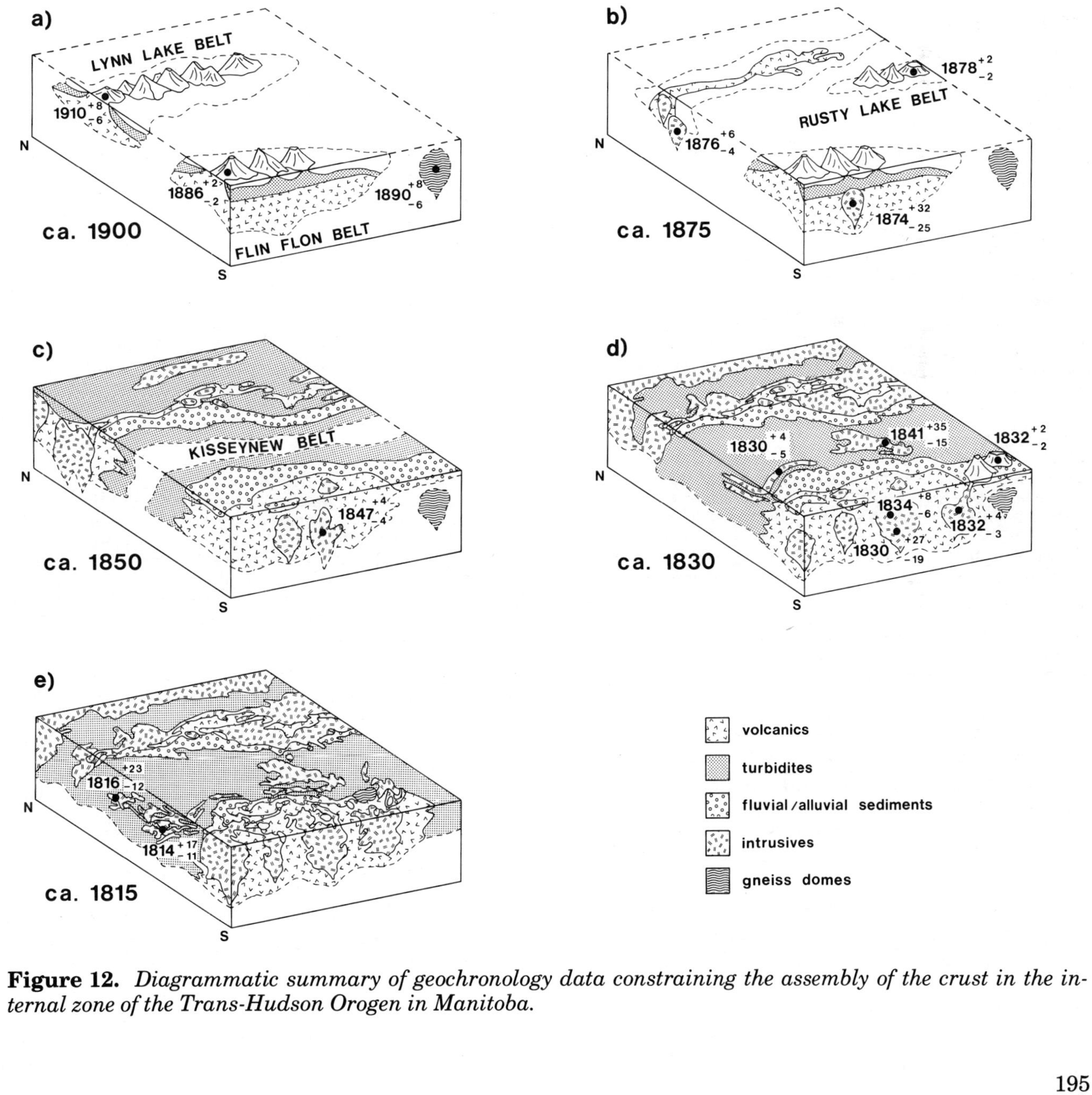

Figure 12. *Diagrammatic summary of geochronology data constraining the assembly of the crust in the internal zone of the Trans-Hudson Orogen in Manitoba.*

kinematics of the major fault zones are being actively investigated (Ashton *et al.*, 1987; Bleeker, 1987).

In summary, assembly of the Proterozoic crust in this part of the Trans-Hudson Orogen took place in three stages over approximately 85 Ma. From ca. 1.9 to 1.87 Ga, juvenile material in the form of volcanic rocks, subvolcanic intrusions, and volcanogenic sedimentary rocks was accreted to the Hearne–Wyoming craton. A continental regime was established at ca. 1.86–1.845 Ga when regional magmatism produced a number of plutons, including a major I-type batholith. This magmatic activity declined until 1.83 Ga when the youngest supracrustal sequence was deposited. During the third and final stage of crustal evolution, anatexis and deformation redistributed pre-existing crust, but added no new material.

ACKNOWLEDGEMENTS

The contributions of Steven Jackson, in terms of field assistance and scientific discussions, are gratefully acknowledged. We thank Katherine Wheatley for her patient help with zircon separation and Deborah Lemkow for cheerfully drafting several versions of some figures. This paper was substantially improved by the thorough and perceptive reviews of Randall Parrish, Edgar Froese, Dave McRitchie, and Pat Bickford. Discussions with many of our colleagues in the Geological Survey of Canada and the Manitoba Minerals Division helped clarify our ideas.

REFERENCES

Ashton, K.E., 1987, Preliminary geological map of the Kisseynew gneisses between the Sturgeon–Weir River and Kisseynew Lake, Saskatchewan: Geological Survey of Canada, Open File 1410.

Ashton, K.E. and Wheatley, K., 1986, Preliminary report on the Kisseynew gneisses in the Kisseynew–Wildnest lakes area, Saskatchewan: Geological Survey of Canada, Paper 86-1B, p. 305–317.

Ashton, K.E., Wheatley, K.J., Moser, D. and Paul, D., 1986, The Kisseynew gneisses of Saskatchewan: update: Saskatchewan Geological Survey, Summary of Investigations 1986, p. 145–148.

Ashton, K.E., Wilcox, K.H., Wheatley, K.J., Paul, D. and de Tombe, J., 1987, The boundary zone between the Flin Flon domain, Kisseynew gneisses and Hanson Lake block, northern Saskatchewan: Saskatchewan Geological Survey, Summary of Investigations 1987, p. 131–134.

Ayres, L.D., 1978, A transition from subaqueous to subaerial eruptive environments in the middle Precambrian Amisk Group at Amisk Lake, Saskatchewan – a progress report: Centre for Precambrian Studies, University of Manitoba 1977 Annual Report, p. 36–51.

Bailes, A.H., 1971, Preliminary compilation of the geology of the Snow Lake – Flin Flon – Sherridon area: Manitoba Mines Branch, Geological Paper 1/71, 27 p.

Bailes, A.H., 1975, Geology of the Guay–Wimapedi lakes area: Manitoba Department of Mines, Resources and Environmental Management, Mineral Resources Division, Geological Services Branch, Publication 75-2, 104 p.

Bailes, A.H., 1980a, Geology of the File Lake area: Manitoba Department of Energy and Mines, Mineral Resources Division, Geological Report 78-1, 134 p.

Bailes, A.H., 1980b, Origin of Early Proterozoic volcaniclastic turbidites, south margin of the Kisseynew sedimentary gneiss belt, File Lake, Manitoba: Precambrian Research, v. 12, p. 197–225.

Bailes, A.H., 1985, Geology of the Saw Lake area: Manitoba Energy and Mines, Geological Services, Geological Report GR83-2, 47 p.

Bailes, A.H., 1986, Chisel–Morgan lakes project: Manitoba Energy and Mines, Minerals Division, Report of Field Activities 1986, p. 70–76.

Bailes, A.H., 1987, Chisel–Morgan lakes project: Manitoba Energy and Mines, Minerals Division, Report of Field Activities 1987, p. 70–79.

Bailes, A.H. and McRitchie, W.D., 1978, The transition from low to high grade metamorphism in the Kisseynew sedimentary gneiss belt, Manitoba, in Fraser, J.A. and Heywood, W.W., eds., Metamorphism in the Canadian Shield: Geological Survey of Canada, Paper 78-10, p. 155–178.

Bailes, A.H. and Syme, E.C., 1987, Geology of the Flin Flon – White Lake area: Manitoba Energy and Mines, Geological Services Branch, Map GR87-1-1.

Bailes, A.H. and Syme, E.C., *in press*, Geology of the Flin Flon – White Lake Area: Manitoba Energy and Mines, Geological Services Branch, Geological Report GR87-1.

Bailes, A.H., Syme, E.C., Galley, A.G. and Skirrow, R.G., 1987, Early Proterozoic volcanism, hydrothermal activity, and associated ore deposits at Flin Flon and Snow Lake, Manitoba: Geological Association of Canada – Mineralogical Association of Canada, Field Trip Guidebook, Trip 1, 95 p.

Baldwin, D.A., 1980, Porphyritic intrusions and related mineralization in the Flin Flon volcanic belt: Manitoba Department of Energy and Mines, Mineral Resources Division, Economic Geology Report ER79-4, 20 p.

Baldwin, D.A., 1987, Proterozoic subaerial volcanism in the Rusty Lake greenstone belt, Ph.D. thesis, University of Manitoba, 438 p.

Baldwin, D.A., Frohlinger, T.G., Kendrick, G., McRitchie, W.D. and Zwanzig, H.V., 1979, Geology of the Nelson House – Pukatawagan region (Burntwood project): Manitoba Department of Mines, Natural Resources and Environment, Mineral Resources Division, Geological Services Branch, Geological Maps 78-3-1 to 78-3-22.

Baldwin, D.A., Syme, E.C., Zwanzig, H.V., Gordon, T.M., Hunt, P.A. and Stevens, R.D., 1987, U–Pb zircon ages from the Lynn Lake and Rusty Lake metavolcanic belts, Manitoba: two ages of Proterozoic magmatism: Canadian Journal of Earth Sciences, v. 24, p. 1053–1063.

Bell, C.K., 1978, Geology, Wekusko Lake map-area, Manitoba: Geological Survey of Canada, Memoir 384, 84 p.

Blair, B., Weber, W., Kornik, L.J. and Gordon, T.M., 1988, Project Cormorant: interpretations of sub-Paleozoic geology of the Cormorant Lake map area from geophysical and drill core data: Geoscience Canada, v. 15, p. 98–100.

Bleeker, W., 1987, Thompson nickel belt: Manitoba Energy and Mines, Minerals Division, Report of Field Activities 1987, p. 171.

Bruce, E.L., 1918, Amisk–Athapapuskow lake district: Geological Survey of Canada, Memoir 105, 91 p.

Cerny, P., Trueman, D.L., Ziehlke, D.V., Goad, B.E. and Paul, B.J., 1981, The Cat Lake – Winnipeg River and the Wekusko Lake pegmatite fields, Manitoba: Manitoba Department of Energy and Mines, Mineral Resources Division, Economic Geology Report ER80-1, 216 p.

Chauvel, C., Arndt, N.T., Kielinzcuk, S. and Thom, A., 1987, Formation of Canadian 1.9 Ga old continental crust. I. Nd isotopic data: Canadian Journal of Earth Sciences, v. 24, p. 396–406.

Chiarenzelli, J.R. and Lewry, J.F., 1988, The Nistowiak gneisses: evidence of a collisional suture in the Trans-Hudson Orogen, northern Saskatchewan?: Geological Association of Canada – Mineralogical Association of Canada – Canadian Society of Petroleum Geologists, Program with Abstracts, v. 13, p. A20.

Chiarenzelli, J.R., Lewry, J.F. and Landon, M., 1987, Bedrock geology, Iskwatikan Lake area: evidence for Hudsonian juxtaposition of Proterozoic and Archean rocks along a ductile detachment surface: Saskatchewan Geological Survey, Summary of Investigations 1987, p. 46–51.

Committee for the Magnetic Anomaly Map of North America, 1987, Magnetic anomaly map of North America: Geological Society of America, 4 sheets, scale 1:5,000,000.

Coombe, W., 1979, Mineral deposits and regional metallogeny, southeastern shield: Saskatchewan Department of Mineral Resources, Summary of Investigations 1979, p. 120–133.

Copeland, P., Parrish, R.R. and Harrison, T.M., 1988, Identification of inherited radiogenic Pb in monazite and its implications for U–Pb systematics: Nature, v. 333, p. 760–763.

Delaney, G., Carr, S. and Parrish, R.R., *in press*, Two U–Pb zircon ages from eastern Glennie Lake domain, Trans-Hudson Orogen, Saskatchewan: Radiogenic Age and Isotopic Studies: Report 2, Geological Survey of Canada, Paper 88-2.

Dunsmore, D.J. and Symons, D.T.A., 1987, Paleomagnetism of the Lynn Lake and Frazer Lake gabbros: a constraint on the geotectonic evolution of the Churchill Province: Geological Association of Canada – Mineralogical Association of Canada, Program with Abstracts, v. 12, p. 39.

Elphick, S.C., 1972, Geology of the Mynarski–Notigi lakes area: Manitoba Department of Mines, Resources and Environmental Management, Mines Branch, Publication 71-2C, 48 p.

Fedikow, M.A.F. and Malis, C., 1987, Mineral occurrence documentation and geochemical surveys in the Snow Lake area: Manitoba Energy and Mines, Minerals Division, Report of Field Activities 1987, p. 91–93.

Frarey, M.J., 1950, Crowduck Bay, Manitoba: Geological Survey of Canada, Map 987A.

Froese, E., 1984, Geology of the Weldon Bay – Fay Lake area, Manitoba: Geological Survey of Canada, Paper 84-1B, p. 355–358.

Froese, E. and Goetz, P., 1981, Geology of the Sherridon Group in the vicinity of Sherridon, Manitoba: Geological Survey of Canada, Paper 80-21, 20 p.

Froese, E. and Moore, J.M., 1980, Metamorphism in the Snow Lake area, Manitoba: Geological Survey of Canada, Paper 78-27, 16 p.

Fumerton, S.L., Stauffer, M.R. and Lewry, J.F., 1984, The Wathaman batholith: largest known Precambrian pluton: Canadian Journal of Earth Sciences, v. 21, p. 1082–1097.

Galley, A.G., Ames, D.E. and Franklin, J.M., 1988, Geological setting of gold mineralization, Snow Lake, Manitoba: Geological Survey of Canada, Open File 1700.

Gaskarth, J.W. and Parslow, G.R., 1987, Proterozoic volcanism in the Flin Flon greenstone belt, east-central Saskatchewan, Canada, in Pharaoh, T.C., Beckinsale, R.D. and Rickard, D., eds., Geochemistry and Mineralization of Proterozoic Volcanic Suites: Geological Society, Special Publication 33, p. 183–200.

Gordon, T.M., 1988, Two-dimensional thermal models of regional metamorphism in the Kisseynew gneiss belt, Manitoba: Geological Association of Canada – Mineralogical Association of Canada – Canadian Society of Petroleum Geologists, Program with Abstracts, v. 13, p. A46.

Gordon, T.M., *in press*, Thermal evolution of the Kisseynew sedimentary gneiss belt, Manitoba: metamorphism at an Early Proterozoic accretionary margin, in Daly, J.S., Cliff, R.A. and Yardley, B.W.D., eds., Evolution of Metamorphic Belts: Geological Society, Special Publication.

Gordon, T.M. and Gall, Q., 1982, Metamorphism in the Crowduck Bay area, Manitoba: Geological Survey of Canada, Paper 82-1A, p. 197–201.

Gordon, T.M. and Lemkow, D.R., 1987, Geochemistry of Missi Group volcanic rocks, Wekusko Lake, Manitoba: Geological Survey of Canada, Open File 1442, 38 p.

Gordon, T.M., Hunt, P.A., Loveridge, W.D., Bailes, A.H. and Syme, E.C., 1987, U–Pb zircon ages from the Flin Flon and Kisseynew belts, Manitoba: chronology of Early Proterozoic crust formation: Geological Association of Canada – Mineralogical Association of Canada, Program with Abstracts, v. 12, p. 48.

Green, A.G., Hajnal, Z. and Weber, W., 1985, An evolutionary model of the western Churchill Province and western margin of the Superior Province in Canada and the north-central United States: Tectonophysics, v. 116, p. 281–322.

Halden, N.M., 1987, Geochemistry of granites in the Snow Lake – Flin Flon area: Manitoba Energy and Mines, Mineral Resources Division, Report of Field Activities 1987, p. 104.

Harrison, J.M., 1951, Precambrian correlation and nomenclature, and problems of the Kisseynew gneisses, in Manitoba: Geological Survey of Canada, Bulletin 20, 53 p.

Harrison, T.M. and McDougall, I., 1980, Investigations of an intrusive contact, northwest Nelson, New Zealand. I. Thermal, chronological, and isotopic constraints: Geochimica et Cosmochimica Acta, v. 44, p. 1985–2003.

Hegner, E. and Hulbert, L., 1987, Sm–Nd ages and origin of mafic–ultramafic intrusions from the Proterozoic Lynn Lake domain, Manitoba, Canada: Geological Association of Canada – Mineralogical Association of Canada, Program with Abstracts, v. 12, p. 54.

Hoffman, P.F., 1988, United plates of America, the birth of a craton: Early Proterozoic assembly and growth of Laurentia: Annual Reviews of Earth and Planetary Sciences, v. 16, p. 543–603.

Hunt, P.A. and Roddick, J.C., *in press*, A compilation of K–Ar ages: Radiogenic Age and Isotopic Studies: Report 2, Geological Survey of Canada, Paper 88-2.

Jackson, S.L., 1988, Metamorphism and structure of the Laurie Lake area, northern Manitoba, Ph.D. thesis, Queen's University.

Jackson, S.L. and Gordon, T.M., 1986, Metamorphic studies in the transition zone between the Lynn Lake greenstone belt and the Kisseynew gneiss belt, Laurie Lake, Manitoba: Geological Survey of Canada, Paper 86-1B, p. 539–546.

Krogh, T.E., 1982, Improved accuracy of U–Pb ages by the creation of more concordant systems using an air abrasion technique: Geochimica et Cosmochimica Acta, v. 46, p. 637–649.

Lambeck, K. and Stephenson, R., 1985, Post-orogenic evolution of a mountain range: southeastern Australian highlands: Geophysical Research Letters, v. 12, p. 801–804.

Lenton, P.G., 1981, Geology of the McNight–McCallum lakes area: Manitoba Department of Energy and Mines, Mineral Resources Division, Geological Survey, Geological Report GR79-1, 39 p.

Lenton, P.G. and Corkery, M.T., 1981, The Lower Churchill River project (interim report): Manitoba Department of Energy and Mines, Mineral Resources Division, Open File Report OF81-3, 23 p.

Lewry, J.F., 1981, Lower Proterozoic arc–microcontinent collisional tectonics in the western Churchill Province: Nature, v. 294, p. 69–72.

Lewry, J.F. and Sibbald, T.I.I., 1980, Thermotectonic evolution of the Churchill Province in northern Saskatchewan: Tectonophysics, v. 68, p. 45–82.

Lewry, J.F., Stauffer, M.R. and Fumerton, S., 1981, A Cordilleran-type batholithic belt in the Churchill Province in northern Saskatchewan: Precambrian Research, v. 14, p. 277–313.

Lewry, J.F., Macdonald, R., Livesey, C., Meyer, M., Van Schmus, R. and Bickford, M.E., 1987, U–Pb geochronology of accreted terranes in the Trans-Hudson Orogen, northern Saskatchewan, Canada, in Pharaoh, T.C., Beckinsale, R.D. and Rickard, D., eds., Geochemistry and Mineralization of Proterozoic Volcanic Suites: Geological Society, Special Publication 33, p. 147–166.

Macdonald, R., 1987, Update on the Precambrian geology and domainal classification of northern Saskatchewan: Saskatchewan Geological Survey, Summary of Investigations 1987, Miscellaneous Report 87-4, p. 87–104.

MacQuarrie, R., 1980, Absolute age of the Flin Flon deposit: Bulletin of the Canadian Institute of Mining and Metallurgy, v. 73, p. 53.

Manitoba Energy and Mines, 1986a, Bedrock geology compilation map series, Granville Lake, NTS 64C, scale 1:250,000.

Manitoba Energy and Mines, 1986b, Bedrock geology compilation map series, Uhlman Lake, NTS 64B, scale 1:250,000.

McRitchie, W.D., 1974, The Sickle–Wasekwan debate – a review: Manitoba Department of Mines, Resources and Environmental Management, Mines Branch, Geological Paper 1/74, 23 p.

McRitchie, W.D., 1977, GS-2 Reindeer Lake – Southern Indian Lake (regional correlation programme): Manitoba Department of Mines, Resources and Environmental Management, Mineral Resources Division, Report of Field Activities 1977, p. 13–18.

McRitchie, W.D., 1985, Kisseynew project: geological reconnaissance of Kisseynew Lake west(63K/13NW): Manitoba Energy and Mines, Geological Services, Mines Branch, Report of Field Activities 1985, p. 57–63.

McRitchie, W.D., 1986, Kisseynew project: geological reconnaissance of Kisseynew Lake west: Manitoba Energy and Mines, Minerals Division, Report of Field Activities 1986, p. 96–99.

Monger, J.W.H., Price, R.A. and Templeman-Kluit, D.J., 1982, Tectonic accretion and the origin of the two major metamorphic and plutonic welts in the Canadian Cordillera: Geology, v. 10, p. 70–75.

Parrish, R.R., 1983, Cenozoic thermal evolution and tectonics of the Coast Mountains of British Columbia. 1. Fission track dating, apparent uplift rates, and patterns of uplift: Tectonics, v. 2, p. 601–631.

Parrish, R.R., 1987, An improved micro-capsule for zircon dissolution in U–Pb geochronology: Chemical Geology (Isotope Geoscience Section), v. 66, p. 99–102.

Parrish, R.R., 1988, U–Pb systematics of monazite and a preliminary estimate of its closure temperature based on natural examples: Geological Association of Canada – Mineralogical Association of Canada – Canadian Society of Petroleum Geologists, Program with Abstracts, v. 13, p. A94.

Parrish, R.R. and Krogh, T.E., 1987, Synthesis and purification of ^{205}Pb for U–b geochronology: Chemical Geology (Isotope Geoscience Section), v. 66, p. 103–110.

Parrish, R.R., Roddick, J.C., Loveridge, W.D. and Sullivan, R.W., 1987, Uranium–lead analytical techniques at the geochronology laboratory, Geological Survey of Canada: Radiogenic Age and Isotopic Studies: Report 1, Geological Survey of Canada, Paper 87-2, p. 3–7.

Ray, G.E. and Wanless, R.K., 1980, The age and geological history of the Wollaston, Peter Lake, and Rottenstone domains in northern Saskatchewan: Canadian Journal of Earth Sciences, v. 17, p. 333–346.

Robertson, D.S., 1953, Batty Lake map-area, Manitoba: Geological Survey of Canada, Memoir 271, 55 p.

Roddick, J.C., 1987, Generalized numerical error analysis with applications to geochronology and thermodynamics: Geochimica et Cosmochimica Acta, v. 51, p. 2129–2135.

Roddick, J.C., Loveridge, W.D. and Parrish, R.R., 1987, Precise U/Pb dating of zircon at the sub-nanogram Pb level: Chemical Geology (Isotope Geoscience Section), v. 66, p. 111–121.

Sangster, D.F., 1972, Isotopic studies of ore-leads in the Hanson Lake – Flin Flon – Snow Lake mineral belt, Saskatchewan and Manitoba: Canadian Journal of Earth Sciences, v. 9, p. 500–513.

Sangster, D.F., 1978, Isotopic studies of ore-leads of the circum-Kisseynew volcanic belt of Manitoba and Saskatchewan: Canadian Journal of Earth Sciences, v. 15, p. 1113–1121.

Schledewitz, D.C.P., 1972, Geology of the Rat Lake area: Manitoba Department of Mines, Resources and Environmental Management, Mines Branch, Publication 71-2B, 57 p.

Schledewitz, D.C.P., 1983, Brochet – Big Sand project: Manitoba Department of Energy and Mines, Mineral Resources Division, Report of Field Activities 1983, p. 5–10.

Schledewitz, D.C.P., 1987, Kisseynew project: Kississing Lake: Manitoba Energy and Mines, Minerals Division, Report of Field Activities 1987, p. 51–54.

Shanks, R.J. and Bailes, A.H., 1977, "Missi Group" rocks, Wekusko Lake area: Manitoba Department of Mines, Resources and Environmental Management, Mineral Resources Division, Report of Field Activities 1977, p. 83–87.

Stauffer, M.R., 1974, Geology of the Flin Flon area: a new look at the Sunless City: Geoscience Canada, v. 1, p. 30–35.

Stauffer, M.R., 1984, Manikewan: an Early Proterozoic ocean in central Canada, its igneous history and orogenic closure: Precambrian Research, v. 25, p. 257–281.

Stauffer, M.R., 1988, The Missi Formation: an Aphebian molasse/red-bed deposit in the Trans-Hudson Orogen: Geological Association of Canada – Mineralogical Association of Canada – Canadian Society of Petroleum Geologists, Program with Abstracts, v. 13, p. A116.

Stauffer, M.R. and Mukherjee, A., 1971, Superimposed deformations in the Missi meta-sedimentary rocks near Flin Flon, Manitoba: Canadian Journal of Earth Sciences, v. 8, p. 217–242.

Steiger, R.H. and Jager, E., 1977, Subcommission on geochronology: convention on the use of decay constants in geo- and cosmochronology: Earth and Planetary Science Letters, v. 36, p. 359–362.

Stockwell, C.H., 1937, Gold deposits of Herb Lake area, northern Manitoba: Geological Survey of Canada, Memoir 208, 46 p.

Sullivan, R.W. and Loveridge, W.D., 1980, Uranium–lead age determinations on zircon at the Geological Survey of Canada: current procedures in concentrate preparation and analysis: Geological Survey of Canada, Paper 80-1C, p. 164.

Syme, E.C., 1985, Geochemistry of metavolcanic rocks in the Lynn Lake belt: Manitoba Energy and Mines, Geological Services Branch, Geological Report GR84-1, 84 p.

Syme, E.C., 1987, Stratigraphy and geochemistry of two Proterozoic arcs: Lynn Lake and Flin Flon metavolcanic belts, Manitoba: Geological Association of Canada – Mineralogical Association of Canada, Program with Abstracts, v. 12, p. 94.

Syme, E.C., 1990, Stratigraphy and geochemistry of the Lynn Lake and Flin Flon metavolcanic belts, Manitoba, in Lewry, J.F. and Stauffer, M.R., eds., The Early Proterozoic Trans-Hudson Orogen of North America: Geological Association of Canada, Special Paper 37, p. 143–161.

Syme, E.C., Bailes, A.H., Gordon, T.M. and Hunt, P.A., 1987, U–Pb zircon geochronology in the Flin Flon belt: age of Amisk volcanism: Manitoba Energy and Mines, Minerals Division, Report of Field Activities 1987, p. 105–107.

Van Schmus, W.R. and Schledewitz, D.C.P., 1986, U–Pb zircon geochronology of the Big Sand Lake area, northern Manitoba: Manitoba Energy and Mines, Minerals Division, Report of Field Activities 1986, p. 207–210.

Van Schmus, W.R., Bickford, M.E., Lewry, J.F. and Macdonald, R., 1987, U–Pb geochronology in the Trans-Hudson Orogen, northern Saskatchewan, Canada: Canadian Journal of Earth Sciences, v. 24, p. 407–424.

Walford, P.C. and Franklin, J.M., 1982, The Anderson Lake Mine, Snow Lake, Manitoba, in Hutchinson, R.W., Spence, C.D. and Franklin, J.M., eds., Precambrian Sulphide Deposits (H.S. Robinson Memorial Volume): Geological Association of Canada, Special Paper 25, p. 481–523.

Watters, B.R. and Pearce, J.A., 1987, Metavolcanic rocks of the La Ronge domain in the Churchill Province, Saskatchewan: geochemical evidence for a volcanic arc origin, in Pharaoh, T.C., Beckinsale, R.D. and Rickard, D., eds., Geochemistry and Mineralization of Proterozoic Volcanic Suites: Geological Society, Special Publication 33, p. 167–182.

York, D., 1969, Least squares fitting of a straight line with correlated errors: Earth and Planetary Science Letters, v. 5, p. 320–324.

Zwanzig, H.V., 1983, Kisseynew project: Lobstick Narrows (parts of 63K/13, 14, and 63N/3,4): Manitoba Department of Energy and Mines, Mineral Resources Division, Report of Field Activities 1983, p. 15–22.

Zwanzig, H.V., 1984, Kisseynew project: Lobstick Narrows – Cleunion Lake, Puffy Lake and Nokomis areas: Manitoba Energy and Mines, Mineral Resources Division, Report of Field Activities 1984, p. 38–45.

Zwanzig, H.V. and Lenton, P. G., 1987, Kisseynew project: Limestone Point – Star Lake areas: Manitoba Energy and Mines, Minerals Division, Report of Field Activities 1987, p. 55–64.

Zwanzig, H.V., Parker, J.S.D., Schledewitz, D.C.P. and Van Schmus, W.R., 1985, Lynn Lake regional compilation and geochronology: Manitoba Energy and Mines, Geological Services, Mines Branch, Report of Field Activities 1985, p. 6–8.

APPENDIX: ANALYTICAL PROCEDURE

Twenty to 30 kg of unweathered rock were collected at each site. Zircon concentrates were prepared using standard procedures involving crushing, sieving, Wilfley table, and heavy-liquid techniques. Magnetic material was removed using a Frantz isodynamic separator. Initial concentrates were subdivided by sieving through nylon screens and passing through a Frantz separator to obtain relatively magnetic and nonmagnetic fractions. All resulting samples were hand picked in ethyl alcohol to obtain the clearest and least-altered zircon for analysis.

The data presented here were collected in the Geological Survey of Canada geochronology laboratory during a period of rapidly evolving analytical techniques. Earlier analyses used methods outlined by Sullivan and Loveridge (1980). Techniques currently employed are reviewed in Parrish *et al.* (1987). Briefly, air abrasion (Krogh, 1982) is used to obtain more concordant material. Blanks are minimized by utilizing microcapsules for sample dissolution (Parrish, 1987). A mixed ^{205}Pb-^{233}U- ^{235}U isotopic tracer (Parrish and Krogh, 1987) provides improved precision and makes it possible to determine the concentration and isotopic composition of Pb in a single measurement. Multicollector mass spectrometry (Roddick *et al.*, 1987) allows the simultaneous collection of precise analyses of all Pb (and U) isotopes. Realistic assignment of analytical uncertainties to computed ages is facilitated using numerical error propagation (Roddick, 1987).

Trace-element and Rb–Sr Whole-rock Isotopic Constraints on the Origin of the Chipewyan, Thorsteinson, and Baldock Batholiths, Churchill Province, Manitoba

N.M. Halden and G.S. Clark
Department of Geological Sciences,
University of Manitoba, Winnipeg, Manitoba R3T 2N2, Canada

M.T. Corkery, P.G. Lenton, and D.C.P. Schledewitz
Manitoba Geological Services, Department of Energy and Mines,
535-330 Graham Avenue, Winnipeg, Manitoba R3C 4E3, Canada

Halden, N.M., Clark, G.S., Corkery, M.T., Lenton, P.G. and Schledewitz, D.C.P., Trace-element and Rb–Sr whole-rock isotopic constraints on the origin of the Chipewyan, Thorsteinson, and Baldock batholiths, Churchill Province, Manitoba, in Lewry, J.F. and Stauffer, M.R., eds., The Early Proterozoic Trans-Hudson Orogen of North America: Geological Association of Canada, Special Paper 37, p. 201-214

Abstract

The Chipewyan Batholith in northern Manitoba represents a major addition of granitic magma to the Proterozoic crust. Chipewyan magmatism shows broad compositional variation in contrast to that of spatially associated bodies, the Thorsteinson and Baldock batholiths, which exhibit narrow ranges in major-element compositions. Discrimination and assessment of the possible tectonic setting on the basis of trace elements has been attempted. On a Rb versus Nb + Y discriminant diagram, the distribution of data points for the Baldock Batholith would be consistent with a volcanic-arc granite setting, and the Thorsteinson Batholith with an intraplate granite setting. The bulk of Chipewyan magmatism cannot be unequivocally assigned to any one tectonic setting.

Rb–Sr isotopic systematics yield whole-rock ages for the Chipewyan seriate granite of 1816 ± 30 Ma and the Katimiwi intrusive complex of 1818 ± 48 Ma, with initial $^{87}Sr/^{86}Sr$ ratios of 0.7021 ± 0.0007 and 0.7034 ± 0.0006 respectively. The Thorsteinson Batholith yields an age of 1713 ± 19 Ma and an initial ratio of 0.7046 ± 0.0012. A composite plot of 28 samples from the Chipewyan Batholith gives an "age" of 1800 ± 32 Ma and an initial ratio of 0.7028 ± 0.0006. The similar Rb–Sr ages for different phases of the Chipewyan Batholith indicate a restricted time span for plutonic activity. The initial $^{87}Sr/^{86}Sr$ ratios are only slightly higher than mantle values for rocks of this age, suggesting that any crustal component had a short residence time.

Résumé

Le batholite Chipewyan au nord du Manitoba, représente un apport majeur de magma granitique à la croûte Protérozoique. Le magmatisme de Chipewyan montre des grandes variations de compositions contrastant avec les massifs voisins, les batholites Thorsteinson et Baldock qui montrent au contraire de faibles écarts des éléments majeurs. Une évaluation des contextes tectoniques possibles basée sur les éléments traces à été réalisée. La répartition des données du batholite Baldock sur un diagramme Rb vs. Nb + Y concorde avec un plutonisme de type arc volcanique et la répartition du batholite Thorsteinson avec un plutonisme intra-plaque. Le magmatisme Chipewyan ne peut être associé à un contexte tectonique particulier.

Les déterminations isotopiques Rb/Sr roche totale donnent 1816 ± 30 Ma pour la série granitique de Chipewyan et 1818 ± 48 Ma pour le complexe intrusif de Katimiwi avec des rapports initiaux $^{87}Sr/^{86}Sr$ de 0,7021 ± 0,0007 et 0,7034 ± 0,0006 respectivement. Le batholite Thorsteinson donne 1713 ± 19 Ma avec un rapport initial de 0,7046 ± 0,0012. Un diagramme composite de 28 échantillons du batholite Chipewyan donne 1800 ± 32 Ma et un rapport initial de 0,7028 ± 0,0006. Les âges Rb/Sr identiques pour différentes phases du batholite Chipewyan indiquent une activité plutonique de courte durée. Les rapports initiaux $^{87}Sr/^{86}Sr$ sont légèrement plus haut que les valeurs mantéliques pour des roches de cet âge indiquant une très courte résidence crustale.

INTRODUCTION

The Chipewyan Batholith is located in the Churchill Province of Manitoba (McRitchie, 1977; Fumerton *et al.*, 1984). The batholith is relatively well exposed at Reindeer Lake, Big Sand Lake, and along the lower Churchill River. Based on field information and correlation with a relatively high airborne magnetic signature, its continuity can be inferred over an area approximately 100 km wide by 400 km long in Manitoba. Including the contiguous Wathaman Batholith in Saskatchewan, the terrane was described by Fumerton *et al.* (1984) as the largest known Precambrian batholith. McRitchie (1977) described the Chipewyan Batholith as coarse grained and porphyritic, consisting of biotite–hornblende, quartz monzonites to granites. Lenton and Corkery (1981) identified additional batholithic units on the southern margin of the Chipewyan Batholith, which they named the Baldock and Thorsteinson batholiths. More recent work by Van Schmus and Schledewitz (1986) described charnockitic and rapakivi phases in the vicinity of Big Sand Lake (Fig. 1).

Both the size and the compositional variability of the Chipewyan Batholith constrain the potential source regions for the batholith and the types of processes responsible for the production of such a large volume of granitic magma. Nd–Sm isotopic work by Patchett and Arndt (1986) suggests that a significant proportion (up to 95%) of granitic rocks in the Churchill Province were derived from magmas extracted from the mantle during the mid-Proterozoic. Work by Chauvel *et al.* (1987) indicates that there may be an Archean sedimentary derived component incorporated in the source region for some magmatic rocks in the Trans-Hudson Orogen. Geochronological data also suggest that magmatic activity spanned a relatively short time. A similar origin, involving derivation from the mantle over a short period of time, has been suggested for the Svecofennian terrane in the Baltic Shield (Patchett and Arndt, 1986; Patchett and Kouvo, 1986). The Central Finnish Batholith is comparable in size to the Chipewyan Batholith and it exhibits a similar complexity in the variety of rock types that constitute the intrusion. Fumerton *et al.* (1984) and Hietanen (1975) propose subduction-type mechanisms for the production of the Chipewyan Batholith and Central Finnish Batholith.

This study, a preliminary investigation of some geochemical and field data, highlights some of the prominent features of the Chipewyan Batholith and some of the adjacent granitic rocks within this region of the Churchill Province. The units discussed in this study are the Chipewyan, Thorsteinson, and Baldock batholiths, described by Lenton and Corkery (1981), and the Katimiwi intrusive complex of Schledewitz (1983). These batholiths and complexes are located along what is essentially the southern margin of the Chipewyan Domain (Zwanzig *et al.*, 1985). Samples used in the discussion of the geochemistry were obtained from the Big Sand Lake – Reindeer Lake and the lower Churchill River areas.

GENERAL GEOLOGICAL SETTING

The major tectonic blocks of the Trans-Hudson Orogen have been described by numerous authors (e.g., McRitchie, 1977; Green *et al.*, 1985; Lewry *et al.*, 1986). For this study, the Chipewyan Batholith (Fig. 1) is considered separately from the western portion of the body in Saskatchewan, which will be referred to as the Wathaman Batholith. The Chipewyan Batholith is bounded in the north by the Wollaston, Seal River, and Great Island domains. The eastern margin is obscured by Phanerozoic rocks of the Hudson Bay Lowlands and exposure is poor (about 2%) along its northern and eastern margins. The Wathaman Batholith is subdivided into the Dobbin and Chipewyan segments (Fumerton *et al.*, 1984). The eastern Chipewyan segment extends

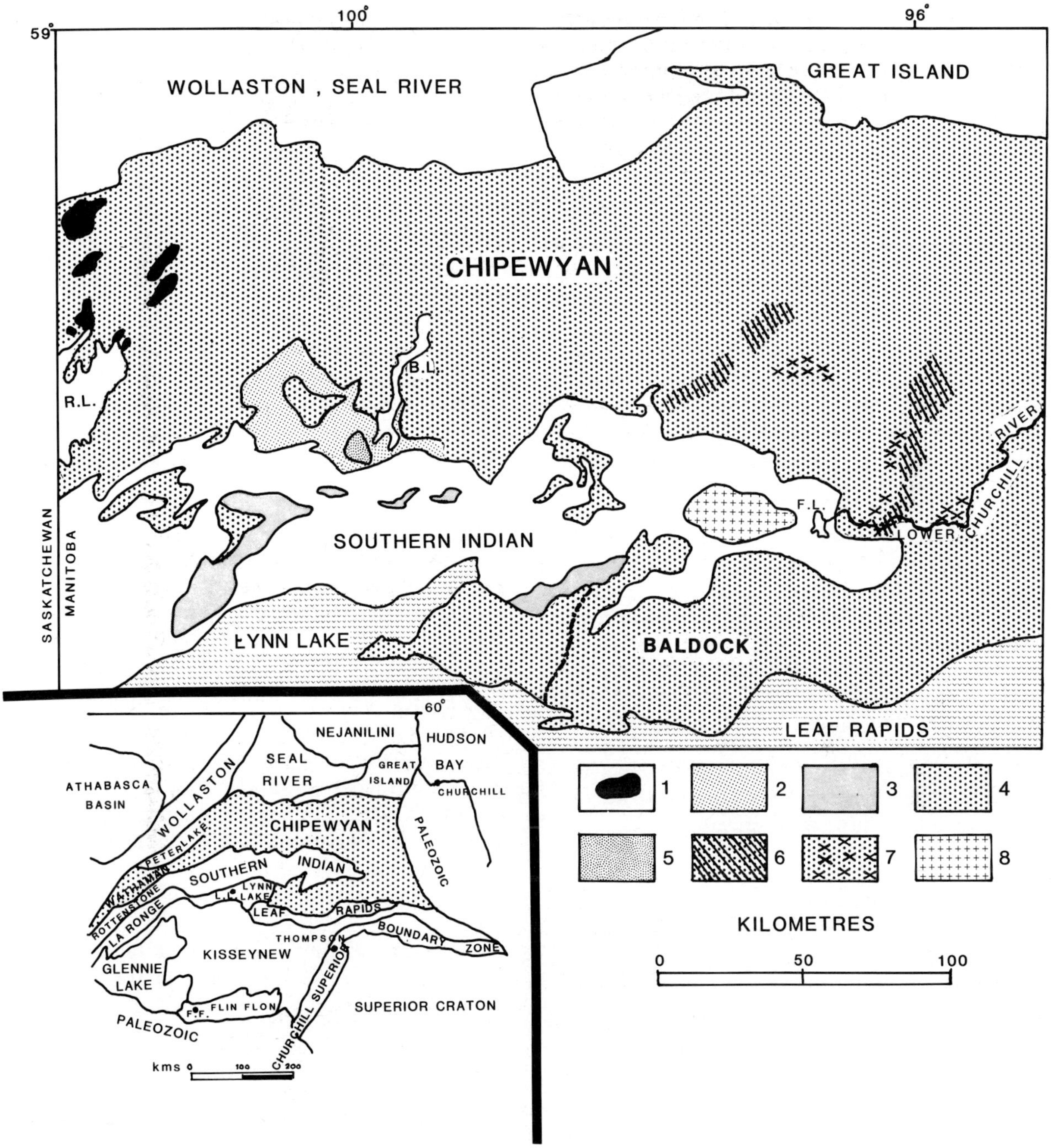

Figure 1. *Geological map of northern Manitoba showing the relationships between the Chipewyan, Baldock, and Thorsteinson batholiths and the location of the Katimiwi intrusive complex. Legend: l, amphibolites and metabasic rocks; 2, Katimiwi intrusive suite; 3, pyroxene-bearing granodiorite to monzogranite, similar to the Katimiwi monzogranite; 4, Chipewyan granites (undifferentiated); 5, Chipewyan granite with rapakivi texture; 6, quartz monzonite; 7, seriate granite; and 8, Thorsteinson granite. Dot–dash line delineates the western extent of the Baldock Batholith. B.L. refers to Big Sand Lake; R.L. refers to Reindeer Lake. The "v" pattern delineates the metavolcanic and intrusive terrane of the Lynn Lake and Leaf Rapids domains. The inset map illustrates the major lithotectonic domains of the Churchill Province in Manitoba and adjacent Saskatchewan. Domain names are those of Zwanzig* et al. *(1985).*

into the main body of the Chipewyan Batholith in the vicinity of Reindeer Lake. The southern contact of the body is with the Southern Indian Domain and the eastern extension of the Lynn Lake Domain. In this latter area, east of the Lynn Lake Domain, the contact between the Chipewyan Batholith and the Baldock Batholith is not well exposed.

Katimiwi intrusive complex

The Katimiwi intrusive complex intrudes hornblende–biotite migmatite and a grey, magnetite–biotite, quartz monzonite of the Southern Indian Domain granitoid gneisses. It is dominantly pyroxene–hornblende monzonite to monzogranite (Schledewitz, 1983). The monzogranite is megacrystic and is the most extensive phase. The megacrysts of microcline and the groundmass feldspars vary from a distinctive olive–brown colour to a variegated pink and brown colour. Hornblende is a common accessory mineral along with rare pyroxene, which is generally altered. A diorite to anorthositic gabbro suite, possibly cognate xenoliths, forms a small component of the Katimiwi intrusive complex.

The contact between the Katimiwi intrusive complex and the Chipewyan Batholith is a hybrid rock of granodioritic composition. Dykes and stocks of the Chipewyan Batholith intrude this hybrid zone. The hybrid zone may have resulted from reactivation of the Katimiwi intrusive complex during emplacement of the Chipewyan granite.

Chipewyan Batholith

Lenton and Corkery (1981) identified three major intrusive and two minor intrusive phases in the lower Churchill River area. The predominant rock type is a magnetite-bearing, K-feldspar megacrystic granite. The second prominent rock is megacrystic quartz monzonite, distinguished from the former granite by its lower quartz content and higher magnetite, sphene, apatite, zircon, and fluorite contents. The contact between these two phases has not been observed, so age relationships are unknown (Lenton and Corkery, 1981). The third major phase is leucocratic, seriate, biotite granite, which intrudes the other two major phases and is more homogeneous and less foliated than the magnetite-bearing granite and quartz monzonite. Its abundance increases to the east, where it is the dominant phase at the contact with the Paleozoic cover. Lenton and Corkery (1981) also described two minor granitic phases; they are both homogeneous and equigranular and vary in composition from granitic to granodioritic. They were emplaced in the form of sheeted complexes of dykes and sills into the three major phases and occasionally contain inclusions of the first two major phases.

Baldock Batholith

This batholith has two major phases and one minor phase as described by Lenton and Corkery (1981). One major phase is a pink, magnetite-bearing, megacrystic granite with minor amounts of hornblende and biotite. There are also variations in the size and distribution of the microcline phenocrysts. The second major phase is pink, magnetite-bearing, leucocratic, seriate, biotite granite similar to the younger phase of the Chipewyan Batholith. The minor phase is quartz monzonite, similar to that of the Chipewyan Batholith, which occurs locally along the northern margin of the Baldock Batholith.

Thorsteinson Batholith and Fidler granodiorite

The Thorsteinson Batholith intrudes supracrustal rocks of the Southern Indian Domain (Fig. 1). It is coarse-grained to pegmatitic, leucocratic, biotite–hornblende granite, and is considered to be the youngest intrusive unit in the region, based on field evidence. This pluton is relatively homogeneous and varies from coarse, quartz-rich, sodic granite with traces of fluorite, to pegmatitic granite on its northern margin. Pegmatite dykes are abundant within the intrusion, but also form an extensive halo of discrete dykes in the surrounding metavolcanic host rocks.

A number of minor intrusions of homogeneous, grey, biotite granite also occur in the Southern Indian Domain, east and north of the Thorsteinson Batholith. These are grouped together under the name Fidler granodiorite. Some bodies of the Fidler granodiorite are cut by the Thorsteinson Batholith, others are found as inclusions within the Thorsteinson Batholith. The age relationship between the Fidler granodiorite and the Chipewyan Batholith is unknown.

GEOCHEMISTRY

Considering the great extent of the magmatism that resulted in the Chipewyan Batholith and spatially related plutons, the present compilation of data represents a preliminary petrogenetic interpretation. Major- and trace-element compositions are presented in Table I, and Rb–Sr whole-rock data are presented in Table II.

Major-element geochemistry

The major phases of the Chipewyan, Baldock, and Thorsteinson batholiths generally show little compositional variability in terms of major-element oxides. The Thorsteinson and Baldock batholiths tend to have more restricted compositional ranges than the Chipewyan Batholith. The QAP plot (Fig. 2) is based on modal analyses (by point counting) of the Chipewyan, Baldock, and Thorsteinson batholiths, and associated granitoid rocks. The majority of the rocks plot within the granite field, although some plot in the granodiorite and monzonite fields.

The Katimiwi intrusive complex can be divided into two groups, one with relatively low SiO_2 (57–59%) and one with relatively high SiO_2 (67–71%). The SiO_2 content of the Chipewyan Batholith ranges from 63 to 72%. On an AFM diagram, most of the analyses fall in the calc-alkaline field (Fig. 3A), whereas a number of the analyses for the Katimiwi suite fall in the tholeiitic field, reflecting a relatively higher Fe content; these are the more magnetite-rich phases. On the Ca–Na–K plot (Fig. 3B), the analyses follow a calc-alkaline trend.

The A/CNK ratios (Fig. 4) for all the units range from 0.80 to 1.06, indicating a spread of compositions from metaluminous to peraluminous. Data for the Wathaman Batholith (Fumerton *et al.*, 1984) show a broader range of A/CNK ratios from 0.60 to 1.17, also indicating, at the higher end, a peraluminous association. The A/CNK–SiO_2 diagram also shows a greater spread of data from the Chipewyan Batholith than from the other plutons.

Trace-element geochemistry

The trace-element geochemistry for the Chipewyan, Baldock, and Thorsteinson batholiths and the Katimiwi intrusive complex is presented in Table I. The K/Ba–SiO_2 and Rb/Sr–SiO_2 plots (Fig. 5A, B) indicate that the Thorsteinson Batholith is the most alkali rich, and probably the most fractionated of the rocks under discussion. The data in Figure 5A,B also demonstrate the restricted compositional range of the Thorsteinson and Baldock batholiths, and to some extent the Katimiwi intrusive complex, compared with the broad compositional range exhibited by the Chipewyan Batholith. Significant spatial separation of the major units precludes any simple genetic correlation. However, the data do tend to suggest fractionation towards the youngest suite of rocks represented by the Thorsteinson Batholith.

Figure 6 illustrates the Rb, Nb, Y trace-element data and SiO_2 data on the tectonic field diagrams of Pearce *et al.* (1984). Figure 6A (Rb versus SiO_2) shows the data from this study in comparison with the data of Fumerton *et al.* (1984). Most of the data from this study plot on and above the boundary separating syncollisional granites (syn-COLG) from volcanic-arc granites (VAG). The majority of the data from the Wathaman Batholith plot well below the field boundary within the VAG field.

Although assigning a tectonic setting on the basis of Figure 6A at this stage is tenuous, the diagram does seem to show that the data of this study and those of Fumerton *et al.* (1984) are different. It would indicate that the Wathaman Batholith was volcanic arc related, whereas Chipewyan magmatism was partly syncollisional in origin.

Figure 6B,C illustrates plots of Nb versus Y and Rb versus Y + Nb, again with the field boundaries of Pearce *et al.* (1984). Data for the Thorsteinson Batholith plot consistently in the within-plate granite (WPG) field, whereas data for the Baldock Batholith and Fidler granodiorite plot consistently within the VAG/syn-COLG and VAG fields. Data for the Chipewyan Batholith and the Katimiwi intrusive complex straddle the field boundaries between these fields and those of the WPG fields.

Pearce *et al.* (1984) illustrate the potential effects of fractional crystallization and crystal accumulation on the spread of data points (see Pearce *et al.*, 1984, Fig. 8). The crystal cumulation trends for feldspars would tend to move points from the WPG field to the VAG field. Depending upon the degree of feldspar fractionation, variable amounts of displacement of data points might be expected. In addition, Pearce *et al.* (1984) also comment upon the accumulation of ferromagnesian minerals; these would have the opposite effect on the data, tending to move the data points from the VAG and syn-COLG fields to the WPG field. The Chipewyan Batholith data could have been affected by fractional crystallization and the accumulation of feldspar and/or ferromagnesian minerals; with the data presently available, it is difficult to quantify the effects of these processes.

Rb–Sr WHOLE-ROCK SYSTEMATICS

A number of recent isotopic studies focused upon the Sm–Nd and U–Pb isotopic systematics of the Trans-Hudson Orogen and other Proterozoic orogenic terranes (Zwanzig *et al.*, 1985; Patchett and Arndt, 1986; Patchett and Kouvo, 1986; Van Schmus *et al.*, 1987). The work of Patchett and Arndt (1986), in particular, indicates that during mid-Proterozoic times a considerable amount of new material was extracted from the mantle and added to the crust over a relatively short period of time. The other studies cited above support at least the chronological basis for this suggestion. This study concentrates upon the Rb–Sr systematics of the southern margin of the Chipewyan batholithic terrane.

The Rb–Sr analytical method has been described by Clark and Weber (1987). The Rb–Sr isotopic data are presented in Table II. Table III is a summary of the isochron ages; isochron plots are illustrated in Figures 7, 8, and 9. The York (1966) least-squares regression method from the program of Brooks *et al.* (1972) was used to calculate the ages and initial ratios. Errors used in the regression were 1.5 and 0.06% (1σ) for $^{87}Rb/^{86}Sr$ and $^{87}Sr/^{86}Sr$, respectively, and 2σ errors are quoted for the ages and initial $^{87}Sr/^{86}Sr$ ratios. Ages are reported using the ^{87}Rb decay constant of 1.42×10^{-11} year^{-1}.

Four areas within, and adjacent to, the Chipewyan Batholith were sampled for the Rb–Sr age study. Two areas represent discrete phases of the batholith: coarse-grained biotite granite from the vicinity of Big Sand Lake and seriate granite from the lower Churchill River area. The Katimiwi pyroxene–hornblende monzonite from the Big Sand Lake area was also dated. For the fourth area, samples were collected on a regional scale in the poorly exposed southeastern part of the batholith. These samples probably do not represent a single discrete phase. The age obtained for these regional samples is 1802 ± 50 Ma, with an initial $^{87}Sr/^{86}Sr$ ratio (*i*) of 0.7028 ± 0.0010 (Fig. 7). The Katimiwi pyroxene–hornblende monzonite and Chipewyan seriate granite give nearly identical ages of 1818 ± 48 Ma (*i* = 0.7034 ± 0.0006) and 1816 ± 30 Ma (*i* = 0.7020 ± 0.0007) respectively (Fig. 8A,B). The coarse-grained biotite granite from Big Sand Lake has a very narrow range of Rb/Sr ratios (about 0.2–0.4), resulting in considerable extrapolation on the isochron plot, which precludes reliable age calculation.

Despite the considerable spatial separation of the Katimiwi intrusive complex and the Chipewyan Batholith and Baldock Batholith of the lower Churchill River area, there appears to be some notable consistency in the data. The similarity of the ages does not permit the resolution of any age differences that might exist for phases within the Chipewyan Batholith. This apparent uniformity in ages, at least for the southern extent of the

Table I

Selected major- and trace-element data from the Chipewyan, Baldock, and Thorsteinson batholiths and the Katimiwi intrusive complex

	CH34	CH36	CH37	CH38	CH39	CH40	CH41	CH43	CH44	CH46	CH48	CH49	CH50	CH53
SiO_2(wt.%)	76.25	73.30	73.70	75.45	75.95	76.45	76.80	74.75	76.45	65.80	61.40	72.60	64.50	65.65
TiO_2	0.08	0.24	0.22	0.28	0.29	0.24	0.16	0.22	0.16	0.62	0.76	0.62	0.97	0.72
Al_2O_3	12.47	13.72	13.36	12.23	11.94	11.77	11.88	13.19	12.33	15.70	16.88	12.61	15.81	15.89
Fe_2O_3	0.48	1.03	1.03	0.82	0.70	0.71	0.57	0.69	0.81	1.82	1.92	1.13	2.42	2.04
FeO	0.48	1.25	1.02	1.10	1.17	1.05	0.92	0.71	0.55	2.34	3.26	1.80	2.07	2.37
MnO	0.02	0.05	0.04	0.08	0.09	0.06	0.10	0.03	0.01	0.06	0.09	0.07	0.09	0.07
MgO	0.11	0.51	0.45	0.20	0.18	0.17	0.09	0.26	0.10	1.61	2.30	0.51	0.96	1.35
CaO	0.46	1.36	1.17	0.61	0.48	0.49	0.62	1.02	0.69	2.82	4.23	1.53	3.01	3.39
Na_2O	2.67	4.09	3.94	3.17	3.10	3.35	3.72	3.13	3.02	3.73	4.05	3.32	3.97	3.67
K_2O	6.08	3.94	4.13	5.37	5.37	5.21	4.78	5.12	5.15	4.00	3.51	4.58	4.70	3.95
P_2O_5	0.00	0.05	0.05	0.03	0.03	0.02	0.01	0.03	0.01	0.25	0.32	0.14	0.29	0.27
H_2O^+	0.22	0.51	0.45	0.54	0.51	0.37	0.27	0.30	0.24	0.55	0.80	0.52	0.46	0.63
Total	99.32	100.05	99.56	99.88	99.81	99.89	99.92	99.45	99.52	99.30	99.52	99.44	99.28	100.00
Rb (ppm)	105	89	106	107	116	91	139	187	162	153	134	194	98	138
Ba	700	1090	1000	580	530	310	80	810	340	1250	1560	930	2800	2230
Sr	83	159	175	47	38	27	15	127	77	556	676	171	478	522
Zr	143	195	153	380	359	319	307	162	175	245	256	363	848	260
Y	35	27	25	63	61	56	96	10	15	26	20	50	73	30
Nb	14	2	7	35	28	31	68	14	4	21	17	45	42	24
FeO* (wt.%)	0.83	2.18	1.95	1.84	0.63	1.70	1.43	1.33	1.28	3.98	4.99	2.82	4.25	4.21
F/F+M	0.886	0.814	0.815	0.906	0.800	0.912	0.945	0.840	0.928	0.715	0.688	0.850	0.819	0.760
Rb/Sr	1.265	0.560	0.606	2.277	3.053	3.370	9.267	1.472	2.104	0.275	0.198	1.135	0.205	0.264
K/Rb	481	367	323	417	384	475	285	227	264	217	217	196	398	238
K/Ba	72.1	30.0	34.3	76.8	84.1	139.5	495.9	52.5	109.6	26.6	18.7	40.9	13.9	14.7

Notes: Sample numbers with the "CH" prefix correspond to the analyses (major elements and Ba) of Lenton and Corkery (1981). Samples CH36 and CH37 are Fidler granodiorite; samples CH38-CH41 are Thorsteinson Batholith; and samples CH59-CH62 are Baldock Batholith. Sample numbers with the "K" and "C" prefixes refer to samples from the Katimiwi intrusive complex and the Chipewyan Batholith from the Big Sand Lake area. Trace-element analyses for this study (Rb, Sr, Y, Nb, and Zr), excluding the Katimiwi intrusive complex, were determined by X-ray fluorescence on an ARL 8420 sequential spectrometer. Analyses were performed on pressed powder pellets produced in a manner similar to that described in Leake *et al.* (1969). Lower limits of detection for Rb, Sr, Y, Nb, and Zr were 2, 2, 2, 5, and 5 ppm respectively. Analytical machine precision (analytical runs, n = 10) for Rb, Sr, Y, Nb, and Zr is ±4, ±4, ±3, ±5, and ±2 ppm respectively. Trace-element analyses for the Katimiwi rocks were done by atomic absorption spectroscopy.

Chipewyan Batholith, is reflected in recent U–Pb zircon ages from the Big Sand Lake area (Zwanzig *et al.*, 1985), although the latter are consistently 2–3% higher than the Rb–Sr ages reported from the present study. Similar discrepancies between Rb–Sr whole-rock ages and U–Pb zircon ages have been noted in other shield terranes (e.g., Bickford and Van Schmus, 1985). The U–Pb zircon age on the Katimiwi pyroxene–hornblende monzonite is 1860 ± 17 Ma (Zwanzig *et al.*, 1985) and there is a U–Pb zircon age available from the Chipewyan porphyritic granite in the Big Sand Lake area of 1857 ± 9 Ma. (Van Schmus and Schledewitz, 1986). The Baldock Batholith has a U–Pb zircon age similar to the Katimiwi intrusive complex and the Chipewyan Batholith (1855 ± 30 Ma) (W.R. Van Schmus and H.V. Zwanzig, personal communication). This would indicate that the southern margin of the Chipewyan Batholith has a consistent chronology, with the Thorsteinson Batholith having the most significantly different age.

To further indicate the uniform nature of the Rb–Sr isotopic systematics of the Chipewyan Batholith, a composite plot of the data for the Katimiwi intrusive complex, Chipewyan seriate granite, Chipewyan biotite granite from Big Sand Lake, and samples taken on a regional basis (excluding the Baldock and Thorsteinson batholiths) is shown in Figure 9. The 28 data points show a very good fit and yield an age of 1800 ± 32 Ma (i = 0.7028 ± 0.0006).

Given the degree of precision in the dating techniques, it might be argued that any events affecting the granites along the southern margin of the Chipewyan Batholith were essentially contemporaneous and that the magmas were derived from what was essentially a homogeneous source with respect to $^{87}Sr/^{86}Sr$.

The Baldock Batholith gives an age of 1802 ± 94 Ma, with i = 0.7024 ± 0.0010 (Fig. 8D), based on the regression of 11 samples. Data for this unit show the most scatter about the regressed line when compared with the other units sampled and they have a

Table I (continued)

CH58	CH59	CH60	CH61	CH62	C139	K036	C136	K321	K235	K200	C92	C89	C141	C95
71.55	70.25	68.60	71.10	69.85	69.10	58.50	71.10	57.85	58.45	67.15	69.85	67.95	63.80	65.90
0.56	0.42	0.41	0.32	0.36	0.63	1.36	0.62	1.53	1.56	0.66	0.44	0.39	0.71	0.73
13.88	14.87	16.28	15.00	15.40	14.70	15.90	13.60	15.16	15.82	14.46	14.52	16.18	16.21	14.97
1.34	1.07	0.96	0.90	1.10	1.19	1.72	1.02	1.87	2.77	1.66	0.54	0.88	2.13	2.08
1.52	1.46	1.62	1.28	1.26	1.82	7.42	1.49	8.69	6.23	3.70	1.96	1.46	2.78	3.06
0.06	0.05	0.03	0.03	0.03	0.04	0.14	0.04	0.17	0.16	0.10	0.05	0.04	0.08	0.06
0.61	0.91	0.95	0.70	0.79	0.86	1.40	0.75	1.68	2.12	0.45	0.56	0.47	1.65	1.76
1.53	2.11	2.81	1.95	1.94	2.02	4.85	1.90	5.29	5.21	2.23	1.85	2.06	3.71	2.86
3.48	3.92	4.57	3.94	3.81	3.49	3.80	3.66	3.64	3.87	3.24	3.46	3.91	3.92	3.70
4.50	4.18	2.91	3.76	4.50	5.24	3.12	4.87	3.00	2.89	5.45	5.20	5.68	3.71	3.52
0.21	0.18	0.15	0.09	0.11	0.26	0.44	0.21	0.52	0.50	0.13	0.15	0.17	0.25	0.37
0.56	0.34	0.30	0.43	0.38	0.33	0.81	0.41	0.55	0.50	0.46	0.32	0.25	0.51	0.65
99.80	99.76	99.59	99.50	99.53	99.68	99.46	99.67	99.95	100.08	99.69	98.90	99.44	99.46	99.66
127	130	82	112	107	189	120	136	85	87	140	116	151	132	141
960	1380	870	1400	1590	1350	2030	1280	126	1700	1860	1870	2230	1990	1000
248	634	561	444	469	324	342	198	21	356	89	252	252	532	620
268	228	168	204	380	238	329	224	379	312	448	417	294	284	320
41	13	8	11	9	32	43	50	41	37	45	36	30	35	25
31	2	2	2	2	23	27	31	19	16	10	10	3	23	23
2.73	2.42	2.48	2.90	2.25	2.89	6.40	2.41	10.37	8.72	5.19	2.45	2.25	4.70	4.93
0.820	0.731	0.726	0.752	0.743	0.773	0.824	0.765	0.863	0.807	0.922	0.817	0.830	0.743	0.739
0.512	0.205	0.146	0.252	0.228	0.583	0.351	0.687	4.048	0.244	1.573	0.460	0.599	0.248	0.227
294	267	295	279	349	230	216	297	293	276	323	372	312	233	207
38.9	25.1	27.8	22.3	23.5	32.2	12.8	31.6	197.6	14.1	24.3	23.1	21.1	25.9	29.2

Table II
Rb – Sr isotopic data

Sample	Rb(ppm)	Sr(ppm)	$^{87}Rb/^{86}Sr$	$^{87}Sr/^{86}Sr^*$
Chipewyan Batholith (southeast)				
139-2	87.6	402.4	0.703	0.7204
78-17	117.0	391.1	0.868	0.7256
256	123.6	250.1	1.433	0.7404
143-1	104.2	164.0	1.844	0.7512
255-1	211.4	191.0	3.225	0.7863
148-1	139.1	105.8	3.835	0.8007
Katimiwi monzocharnockite (Big Sand Lake)				
320-1	8.25	890.9	0.027	0.7044
235-1	64.0	473.1	0.392	0.7131
216-2	77.1	231.9	0.964	0.7290
218-1	102.4	270.2	1.100	0.7322
217	93.3	229.0	1.182	0.7347
200-2	96.6	136.2	2.063	0.7570
Chipewyan seriate granite				
235-1A	109.6	600.4	0.527	0.7157
7500	110.2	563.9	0.565	0.7165
5292-1A	195.4	426.0	1.330	0.7374
5145	177.0	247.8	2.074	0.7575
5297-1A	211.8	211.3	2.914	0.7765
5289	281.8	177.0	4.655	0.8230
5300	273.0	167.9	4.756	0.8265
238-1A	186.4	34.7	16.12	1.1221
Biotite granite (Big Sand Lake)				
5139-7	146.4	545.6	0.776	0.7212
5139-4	146.8	563.6	0.754	0.7222
5139-6	141.8	522.1	0.786	0.7229
5139-2	149.7	556.5	0.778	0.7230
5139-1	145.7	515.9	0.818	0.7238
5139-3	147.8	507.5	0.843	0.7244
5139-8	152.4	522.7	0.844	0.7252
5139-5	163.4	565.4	0.837	0.7252
5138	160.9	418.0	1.114	0.7305
Thorsteinson granite				
210	107.8	190.7	1.642	0.7452
222	92.9	47.4	5.753	0.8460
220	104.6	41.8	7.358	0.8852
219	111.0	30.5	10.82	0.9692
224	120.0	31.1	11.46	0.9844
226	144.6	7.28	66.86	2.3760
Baldock Batholith				
7641	82.1	563.0	0.420	0.7126
7637	106.4	736.2	0.418	0.7130
35-1	86.7	573.5	0.437	0.7134
7647	97.1	624.0	0.452	0.7143
01-1	132.0	632.1	0.602	0.7181
7661	104.9	479.9	0.631	0.7191
7594	112.0	475.1	0.684	0.7211
164-1A	82.3	283.0	0.843	0.7234
7615	111.4	352.0	0.901	0.7248
26-1	134.0	444.9	0.876	0.7268
8224	236.8	172.0	4.022	0.8055

*Normalized to $^{87}Sr/^{86}Sr$ value of 0.1194.

mean square of weighted deviates (MSWD) of 2.4, indicating that some of the points deviate from the regressed line beyond the quoted analytical errors.

The Rb–Sr isochron age for the Thorsteinson Batholith is significantly younger than any of the other units. The age obtained is 1713 ± 19 Ma, with $i = 0.7046 \pm 0.0012$ (Fig. 8C).

Estimating possible crustal residence times of source region materials for the units individually, based upon estimates of their average Rb/Sr ratios and an assumed mantle value at about 1800 Ma of 0.702 for $^{87}Sr/^{86}Sr$, would indicate residence times of no greater than 100 Ma. This is consistent with the suggestion made by Patchett and Arndt (1986) that about 95% of the granitic crust in this part of the Trans-Hudson Orogen was derived over a short time span at around 1860–1800 Ma.

DISCUSSION

The Wathaman–Chipewyan batholithic suite has been referred to as the largest known Precambrian batholith (Fumerton *et al.*, 1984). This is undoubtedly its most spectacular characteristic, but it is also the one likely to generate the most difficulty in genetic interpretation. Characterizing individual magmatic events within the overall context of the batholith would help reveal the nature of the process or processes that might have been responsible for its generation. In many ways, the work

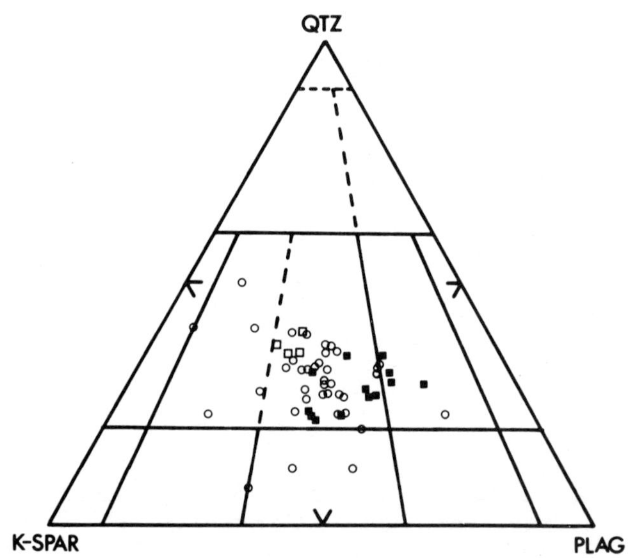

Figure 2. *QAP diagram showing modal analyses of the Baldock Batholith (solid squares), Chipewyan Batholith (circles), and Thorsteinson Batholith (open squares) from the lower Churchill River area. Field boundaries are from Streckeisen (1976); the majority of the rocks are classified as granites, with some granodiorites and monzonites.*

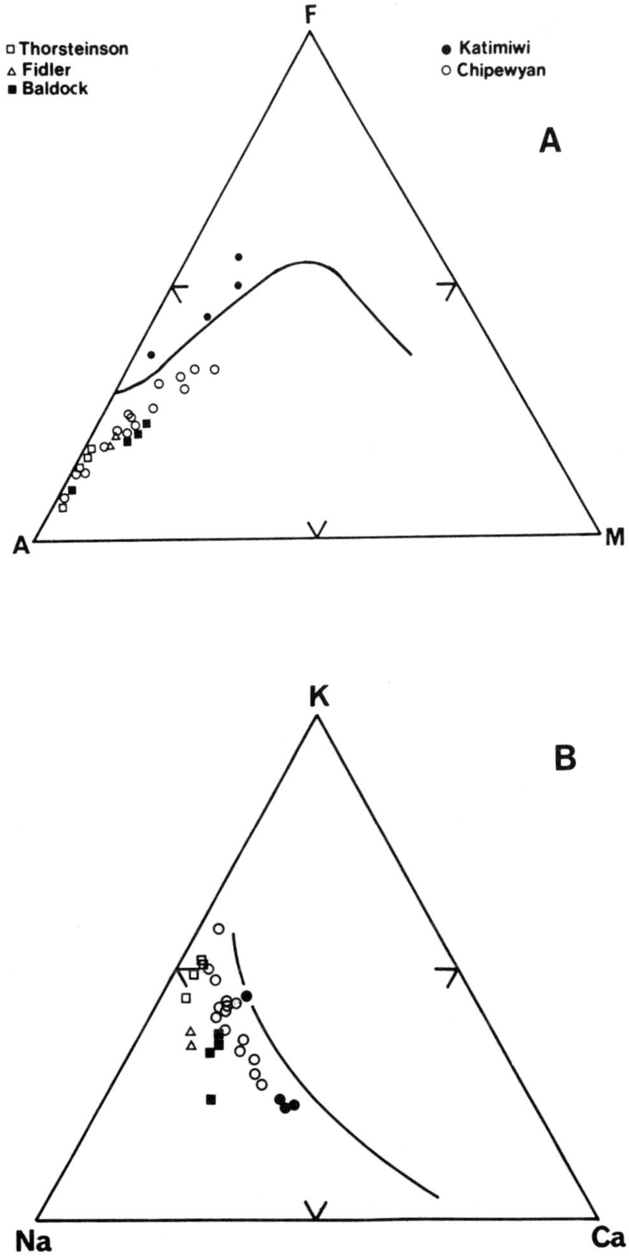

Figure 3. *(A) AFM diagram illustrating that the majority of the rocks fall within the calc-alkaline field and follow a calc-alkaline trend. Four samples from the Katimiwi intrusive complex fall within the tholeiitic field, and contain a higher proportion of magnetite. The field boundary is from Irvine and Baragar (1971); this separates the tholeiitic field (above the line) from the calc-alkaline field (below the line). (B) K–Na–Ca diagram illustrating that the rocks follow a calc-alkaline trend. The trend line is from Nockolds and Allen (1953).*

done by McRitchie (1977), Lenton and Corkery (1981), Schledewitz (1983), Fumerton *et al.* (1984), and Zwanzig *et al.* (1985) in identifying the numerous phases of magmatism in the Chipewyan Batholith serves to highlight the complexity of this task.

Wyllie *et al.* (1976) suggested that "Batholiths are produced from crustal rocks as a normal consequence of regional metamorphism, with the formation of H_2O-undersaturated granite liquid and mobilised migmatites. Some batholiths receive additional contributions of material and heat from the mantle and subducted ocean crust."

The Peruvian Coastal Batholith is a series of intrusive bodies about 1600 km long by about 60 km wide at the present level of erosion. Pitcher (1978) suggests that plate tectonic processes played a significant role in its generation. Hietanen (1975) suggested parallels between the Central Finnish Batholith and K-poor magmas from the northern Sierra Nevada, and suggested that a plate subduction-like process was responsible for the generation of the Central Finnish Batholith. Park *et al.* (1985) discussed relationships between adjacent tectonic units involved in the Svecokarelian orogeny about 1800 Ma ago in eastern Finland and constructed a plate tectonic interpretation for the orogen. This interpretation compares favourably in terms of geochronology and the subdivision of major lithotectonic domains with the evolutionary models of the Churchill Province discussed by Lewry *et al.* (1986) and Green *et al.* (1985).

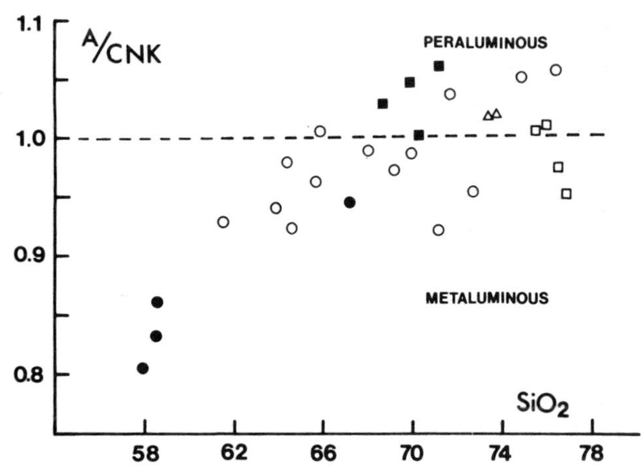

Figure 4. *Plot of A/CNK versus SiO_2 (wt.% oxide). The majority of the rocks from this study are metaluminous. Some of the samples from each suite, except the Katimiwi intrusive complex, are slightly peraluminous. Symbols as in Figure 3. A/CNK refers to whole-rock molar ratios of alumina to lime, soda, and potash ($Al_2O_3/CaO + Na_2O + K_2O$) (Clarke, 1981).*

A potential subduction-related derivation for the Wathaman–Chipewyan batholith was suggested by Fumerton *et al.* (1984). The problems relating to this kind of model are the location of the paleosuture evolving from such a process, the direction of subduction, and the identification of Proterozoic oceanic crust. These problems are further exacerbated by incomplete knowledge of the displacement along faults separating major domains in northern Manitoba.

The isotopic data in this study suggest the derivation of the magmas from material with crustal residence times on the order of 100 Ma. This is consistent with the interpretation that most of the $^{87}Sr/^{86}Sr$ initial ratios would be too high for a single-stage extraction directly from the mantle at about 1800 Ma. The homogeneity of the Sr-isotope initial ratio data shown on Figure 9 may be a feature of the source region. If this is so, then Rb–Sr whole-

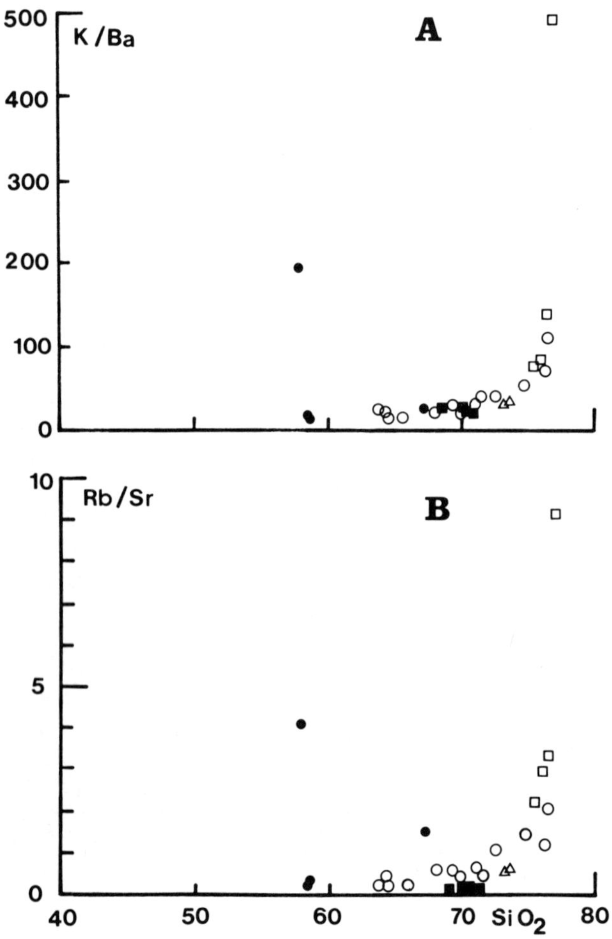

Figure 5. *(A) K/Ba and (B) Rb/Sr versus SiO_2 (wt.% oxide) plots for the rocks in this study. The Thorsteinson Batholith is consistently the most felsic and fractionated of the rocks under discussion. Symbols as in Figure 3.*

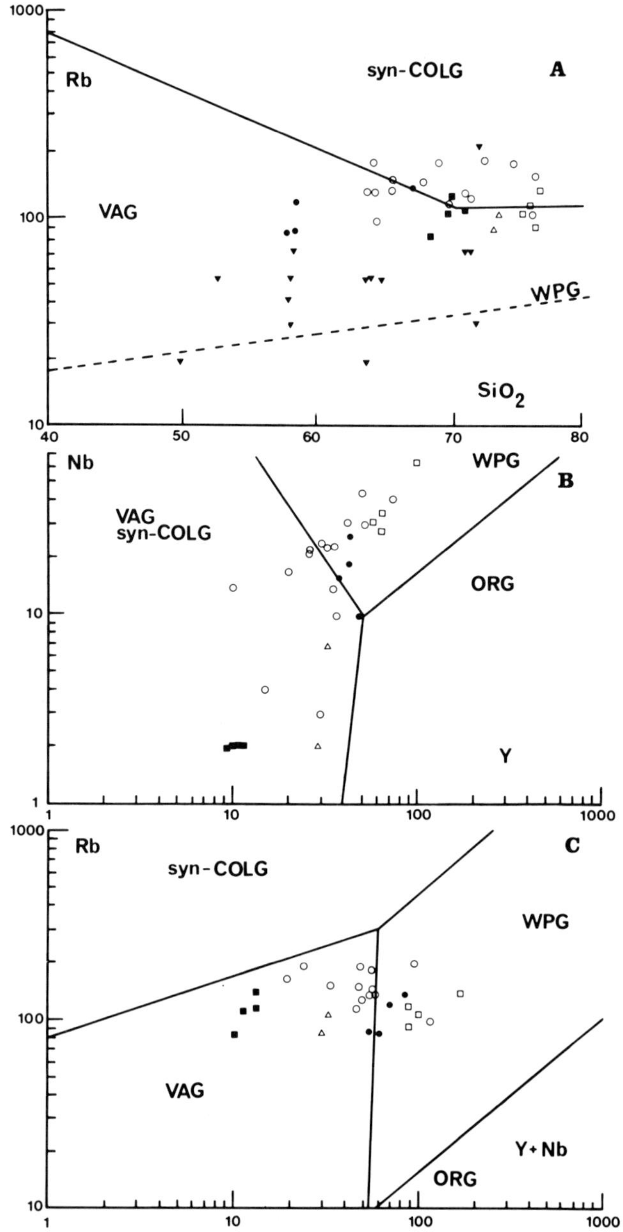

Figure 6. *(A) Log Rb versus SiO_2 (wt.% oxide). Compare the data from this study with the data from Fumerton* et al. *(1984), represented by solid inverted triangles. Field boundaries are from* Pearce et al. *(1984). The majority of the data from this study have greater SiO_2 and Rb values than those from the Wathaman Batholith. (B) Log–log plot of Nb versus Y and (C) log–log plot of Rb versus Y + Nb. Field boundaries are from Pearce* et al. *(1984). The Thorsteinson Batholith is consistently in the WPG (within-plate granite) field; the Baldock Batholith and Fidler granodiorite are consistently in the VAG (volcanic-arc granite) or syn-COLG (syn-collisional granite) fields. The spread of Chipewyan and Katimiwi analytical points probably reflects the effect of the fractional crystallization of feldspar. Symbols as in Figure 3.*

Table III
Rb – Sr isochron age summary for the Chipewyan Batholith and spatially associated plutons

	Age (Ma)	$^{87}Sr/^{86}Sr$ (initial)	MSWD	Number of data points
Chipewyan Batholith (southeast)	1802 ± 50	0.7028 ± 0.0010	0.9	6
Katimiwi intrusive complex	1818 ± 48	0.7034 ± 0.006	0.7	6
Seriate granite (lower Churchill River)	1816 ± 30	0.7020 ± 0.0007	0.9	8
Chipewyan composite	1800 ± 32	0.7028 ± 0.0006	1.8	28
Thorsteinson Batholith	1713 ± 19	0.7046 ± 0.0012	0.4	6
Baldock Batholith	1802 ± 94	0.7024 ± 0.0010	2.4	11

Note: MSWD, mean square of weighted deviates.

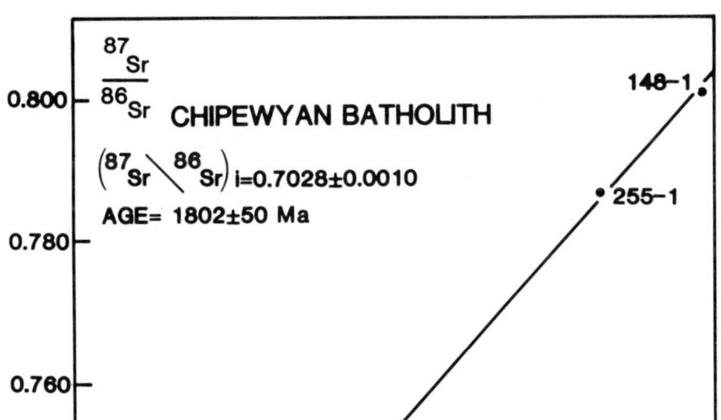

Figure 7. *Rb – Sr whole-rock isochron for the Chipewyan Batholith. Samples were collected on a regional basis. Sample numbers refer to those listed in Table II.*

rock ages of 1820–1800 Ma (notably younger than the U–Pb zircon data) could represent a later closure of the Rb–Sr system just after the crystallization of the bulk of the magma. If extensive partial melting of the lower crust occurred associated with regional metamorphism, then it could have given rise to the Chipewyan Batholith. Whether or not this is related to subduction of oceanic crust is not implicit in the argument, but neither can it be excluded. The southern margin of the Chipewyan Batholith is very extensive and identification of a mechanism whereby isotopic homogenization could be achieved over such a wide area might be problematic.

Pearce *et al.* (1984) do not exclude the application of their diagrams to rocks of Proterozoic age, but it is worth noting that in developing the discriminant diagrams they used data from well constrained and described tectonic settings. Obviously, these constraints and/or descriptions within rocks of Proterozoic age are likely to be contentious. Whether the geological details of Phanerozoic VAG, syn-COLG, and other granitic domains are comparable with those of Proterozoic age remains to be tested. However, the discriminant diagrams offer some hope of distinguishing the tectonic setting of the main magmatic units.

The trace-element discrimination diagrams suggest that it might be possible to classify the Baldock and Thorsteinson batholiths and Fidler granodiorite. Using Pearce *et al.* (1984), the Baldock Batholith and Fidler granodiorite are classified as possible volcanic arc-related granitoids. This is not unreasonable as these units are found as the northern and eastern extensions of the Leaf Rapids

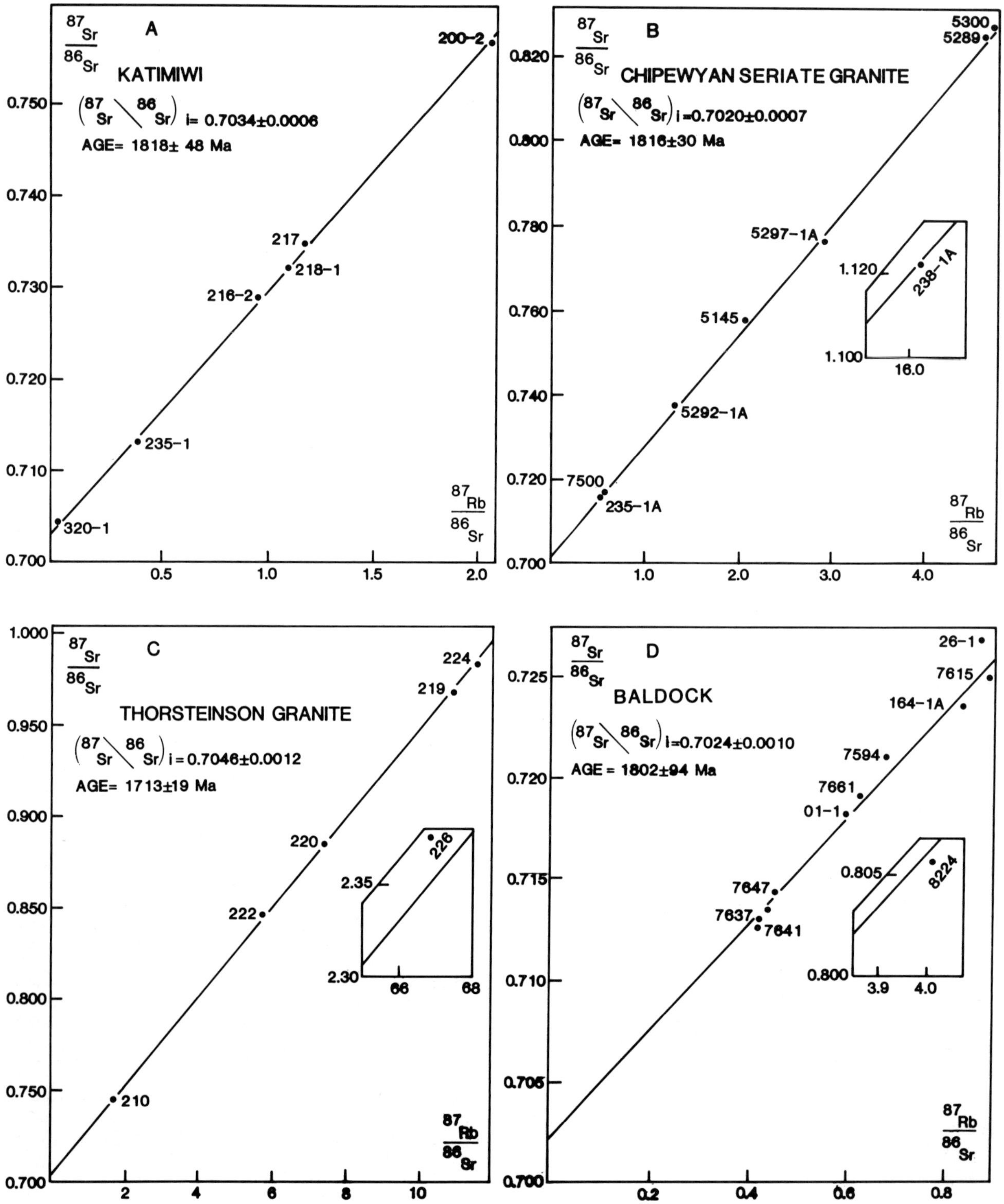

Figure 8. *Rb – Sr whole-rock isochron for (A) the Katimiwi pyroxene–hornblende monzonite, (B) the Chipewyan seriate granite, (C) the Thorsteinson Batholith, and (D) the Baldock Batholith. Sample numbers refer to those listed in Table II.*

Domain (see Fig. 1), which has been interpreted as a marginal basin – volcanic-arc terrane (Syme, 1987).

Data from the Thorsteinson Batholith plot consistently in the within-plate granite field. This batholith is also fluorine enriched (Lenton and Corkery, 1981), a feature characteristic of anorogenic-type granites (cf. Collins *et al.*, 1982; Clemens *et al.*, 1986; Cerny and Meintzer, 1988). Arguably, the emplacement of the Thorsteinson granite may represent the late-stage consolidation of a Proterozoic craton in the area. It has also been suggested that rapakivi phases may be associated with late anorogenic plutonism (cf. Dickinson, 1981; Anderson, 1983), so the presence of some of the rapakivi phases reported by Schledewitz (1983) may take on added significance. The uncertainty of the potential tectonic setting of the bulk of Chipewyan magmatism, as it is presently identified, may be the result of fractional crystallization.

For the generation of the Chipewyan Batholith, a subduction-related plate tectonic process is a reasonable starting hypothesis given the small amount of data presently available. To further build upon this model, consideration must be given to the impact of the location of such processes, not only on the generation of the batholithic terrane but also on the adjacent terrane. Subdivision of the major granitic units into the Wathaman, Chipewyan, Baldock, and Thorsteinson batholiths, and the Katimiwi intrusive complex, represents only a first step in the description of this terrane. Further work will be required to understand the significance of, and connection between, these units. In addition, the possible role of crustal assimilation and subduction of Proterozoic oceanic crust will have to be evaluated and quantified.

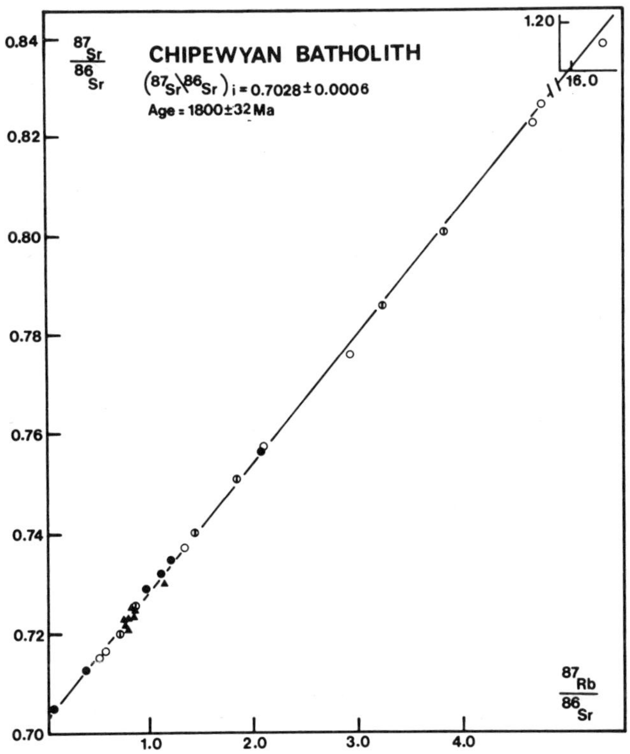

Figure 9. *Composite $^{87}Sr/^{86}Sr$ versus $^{87}Rb/^{86}Sr$ plot. Symbols: Katimiwi intrusive complex, solid circles; Chipewyan seriate granite, lower Churchill River area, barred circles; Big Sand Lake biotite granite, triangles; and Chipewyan Batholith, regional sampling, open circles.*

ACKNOWLEDGEMENTS

This project was funded in part by Natural Sciences and Engineering Research Council of Canada grants A3831 to GSC and A0612 to NMH. P. Beaudoin and W. Blonski assisted with the isotopic and geochemical measurements. We wish to thank P. Cerny, G. Parslow, and two anonymous referees for reviewing an earlier version of the manuscript and offering many helpful suggestions. This paper is published with the permission of the Minister, Manitoba Department of Energy and Mines.

REFERENCES

Anderson, J.1., 1983, Proterozoic anorogenic granite plutonism of North America: Geological Society of America, Memoir 161.

Bickford, M.E. and Van Schmus, W.R., 1985, Resetting of whole rock and mineral Rb–Sr ages by subsequent Proterozoic orogenies: Geological Society of America, Abstracts with Programs, v. 16, p. 523.

Brooks, C., Hart, S.R. and Wendt, I., 1972, Realistic use of two error regression treatments as applied to rubidium–strontium data: Reviews of Geophysics and Space Physics, v. 10, p. 551–577.

Cerny, P. and Meintzer, R.E., 1988, Fertile granites in the Archean and Proterozoic fields of rare-element pegmatites: crustal environment, geochemistry and petrogenetic relationships, in Taylor, R.P. and Strong, D.F., eds., Granite-related Mineral Deposits: Canadian Institute of Mining, Special Volume 39. (In press.)

Chauvel, C., Arndt, N.T., Kielinzcuk, S. and Thom, A., 1987, Formation of Canadian 1.9 Ga old continental crust. I. Nd isotopic data: Canadian Journal of Earth Sciences, v. 24, p. 396–406.

Clark, G.S. and Weber, W., 1987, Rubidium–strontium age and origin of uraniferous granite, Molson Lake – Red Sucker Lake area, northwestern Superior Province, Manitoba: Canadian Journal of Earth Sciences, v. 24, p. 471–478.

Clarke, D.B., 1981, The mineralogy of peraluminous granites: a review: Canadian Mineralogist, v. 19, p. 3–17.

Clemens, J.D., Holloway, J.R. and White, A.J.R., 1986, Origin of an A-type granite: experimental constraints: American Mineralogist, v. 71, p. 317–324.

Collins, W.J., Beams, S.D., White, A.J.R. and Chappell, B.W., 1982, Nature and origin of A-type granites with particular reference to northeastern Australia: Contributions to Mineralogy and Petrology, v. 80, p. 189–200.

Dickinson, W.R., 1981, Plate tectonics through geological time: Philosophical Transactions of the Royal Society of London, v. A301, p. 207–215.

Fumerton, S.1., Stauffer, M.R. and Lewry, J.F., 1984, The Wathaman batholith: largest known Precambrian pluton: Canadian Journal of Earth Sciences, v. 2l, p. 1082–1097.

Green, A.G., Hajnal, Z. and Weber, W., 1985, An evolutionary model of the western Churchill Province and western margin of the Superior Province in Canada and the north-central United States: Tectonophysics, v. 116, p. 28l–322.

Hietanen, A., 1975, Generation of potassium-poor magmas in the northern Sierra Nevada and the Svecofennian of Finland: United States Geological Survey, Journal of Research, v. 3(6), p. 631–645.

Irvine, T.N. and Baragar, W.R.A., 197l, A guide to the chemical classification of common volcanic rocks: Canadian Journal of Earth Sciences, v. 8, p. 523–548.

Leake, B.E., Hendry, G.1., Kemp, A., Plant, A.G., Harvey, P.K., Wilson, J.R., Coats, J.S., Aucott, J.W., Lunel, T. and Howarth, R.J., 1969, The chemical analysis of rock powders by automatic X-ray fluorescence: Chemical Geology, v. 5, p. 7–86.

Lenton, P.G. and Corkery, M.T., 198l, The lower Churchill River project (interim report): Manitoba Department of Energy and Mines, Report OF81-3.

Lewry, J.F., Collerson, K.D., Bickford, M.E. and Van Schmus, W.R., 1986, An evolutionary model of the western Churchill Province and western margin of the Superior Province in Canada and the north-central United States: discussion: Tectonophysics, v. 131, p. 183–197.

McRitchie, W.D., 1977, Reindeer Lake – South Indian Lake, Manitoba: Manitoba Mineral Resources Division, Report GS-2, p.13–18.

Nockolds, S.R. and Allen, R., 1953, The geochemistry of some igneous rock series: Geochimica et Cosmochimica Acta, v. 4, p. 105–142.

Park, A.F., Bowes, D.R., Halden, N.M. and Koistinen, T.J., 1985, Tectonic evolution at an early continental margin: the Svecokarelides, eastern Finland: Geodynamics, v. l(3–5), p. 359–386.

Patchett, P.J. and Arndt, N.T., 1986, Nd isotopes and tectonics of l.9–l.7 Ga crustal genesis: Earth and Planetary Science Letters, v. 78, p. 329–338.

Patchett, J. and Kouvo, O., 1986, Origin of continental crust of l.9–l.7 Ga age: Nd isotopes and U–Pb zircon ages in the Svecokarelian terrain of south Finland: Contributions to Mineralogy and Petrology, v. 92, p. 1–12.

Pearce, J.A., Harris, N.B.W. and Tindle, A.G., 1984, Trace element discrimination diagrams for the interpretation of granitic rocks: Journal of Petrology, v. 4, p. 956–983.

Pitcher, W.S., 1978, The anatomy of a batholith: Journal of the Geological Society of London, v. 135, p. 157–182.

Schledewitz, D.C.P., 1983, Brochet – Big Sand Lake, Manitoba: Manitoba Mineral Resources Division, Report GS-1, p. 5–l0.

Streckeisen, A., 1976, To each plutonic rock its proper name: Earth Science Reviews, v. 12, p. 1–33.

Syme, E.C., 1987, Stratigraphy and geochemistry of two Proterozoic arcs: Lynn Lake and Flin Flon metavolcanic belts, Manitoba: Geological Association of Canada – Mineralogical Association of Canada, Program with Abstracts, v. 12, p. 94.

Van Schmus, W.R. and Schledewitz, D.C.P., 1986, U–Pb zircon geochronology of the Big Sand Lake area, northern Manitoba: Manitoba Mineral Resources Division, Report GS-45, p. 207–210.

Van Schmus, W.R., Bickford, M.E., Lewry, J.F. and Macdonald, R., 1987, U–Pb geochronology in the Trans-Hudson Orogen, northern Saskatchewan, Canada: Canadian Journal of Earth Sciences, v. 24, p. 407–424.

Wyllie, P.J., Huang, W., Stern, C.R. and Maaloe, S., 1976, Granitic magmas: possible and impossible sources, water contents, and crystallisation sequences: Canadian Journal of Earth Sciences, v.13, p. 1007–1019.

York, D., 1966, Least-squares fitting of a straight line: Canadian Journal of Physics, v. 44, p. 1079–1086.

Zwanzig, H.V., Parker, J.S.D., Schledewitz, D.C.P. and Van Schmus, W.R., 1985, Lynn Lake regional compilation and geochronology: Manitoba Mineral Resources Division, Report GS-1, p. 6–8.

Paleomagnetism of the Lynn Lake Gabbros in the Trans-Hudson Orogen and Closure of the Superior and Slave Cratons

D.J. Dunsmore and D.T.A. Symons
Department of Geology, University of Windsor,
Windsor, Ontario N9B 3P4, Canada

Dunsmore, D.J. and Symons, D.T.A., Paleomagnetism of the Lynn Lake gabbros in the Trans-Hudson Orogen and closure of the Superior and Slave cratons, in Lewry, J.F. and Stauffer, M.R., eds., The Early Proterozoic Trans-Hudson Orogen of North America: Geological Association of Canada, Special Paper 37, p. 215–228.

Abstract

The Cu–Ni mineralized Lynn Lake A and EL plugs and the nonmineralized Fraser Lake plug occur in the La Ronge – Lynn Lake greenstone belt of the Trans-Hudson Orogen in the Churchill Province. These massive gabbroic plugs intrude ~1910 Ma Wasekwan volcanics and are cut by ~1876 Ma felsic plutons. Paleomagnetic analysis of 493 specimens from 42 sites in the plugs was done using both alternating-field (AF) and thermal step demagnetization. All three plugs retain the same remanence direction of 157.6°, 59.5° (α_{95} = 4.9°) in domains with AF coercivities of >10 mT and unblocking temperatures of >200°C whether sulphide rich or poor, and their pole position of 84.0°W, 9.0°N (d_p = 5.5°, d_m = 7.4°) falls on the apparent polar wander path for North America at about 1890 Ma. The results indicate that all three plutons are comagmatic and were emplaced at ~1890 Ma with a primary remanence, that the Cu–Ni sulphide mineralization is essentially "syngenetic" and formed at ~1890 Ma also, and that the plugs have not been structurally deformed by relative rotations since emplacement. Analysis of the available pole positions from the Slave, Churchill, and Superior provinces for the 1900–1800 Ma time period strongly suggests that the Slave craton moved ~13°SE and the Superior craton moved ~30°NW into the magmatic arc of the Trans-Hudson Orogen with the concomitant destruction of an ~4000 km wide ocean basin. The separate apparent polar wander paths for the Slave and Superior provinces prior to ~1700 Ma show that closure of the cratons began as early as 2300 Ma.

Résumé

Les amas minéralisés en Cu–Ni de Lynn Lake A et EL et ceux non minéralisées de Fraser Lake se retrouvent dans la ceinture de roche verte La Ronge – Lynn Lake de l'orogénèse Trans-Hudson de la Province Churchill. Ces petits intrusifs gabbroïques massifs recoupent les volcanites de Wasekwan âgés de 1910 Ma et sont recoupés par des plutons acides de 1876 Ma. Une analyse paléomagnétique des amas gabbroïques a été effectuée sur 493 échantillons provenant de 42 sites utilisant la démagnétisation AF et par sauts thermiques. Les trois amas conservent les mêmes directions de rémanence de 157,6°, 59,5° (α_{95} = 4,9°) dans des domaines où la coercitivité AF >10 mT et les T d'ouvertures >200°C, qu'elles soient riches ou pauvres en sulfures; la position des pôles de 84,0°W et 9,0° (d_p = 5,5°, d_m = 7,4°) tombe sur la trajectoire de dérive des pôles de l'Amérique du nord vers 1890 Ma. Les résultats indiquent que les trois *plutons mis en place vers 1890 Ma sont co-magmatiques avec une rémanence primaire, que la minéralisation en sulfures Cu–Ni est essentiellement "syngénétique" aussi vers 1890 Ma et que les amas n'ont pas été déformées par des mouvements relatifs de rotation après leur mise en place. Une analyse des pôles pour les provinces Esclave, Churchill et Supérieur pour la période 1900–1800 Ma suggère que le craton Esclaves a bougé 13° vers le SE et que le craton Supérieur s'est déplacé de 30° vers le NW dans l'arc magmatique de l'orogène Trans-Hudson provoquant la destruction d'un bassin océanique large de 4000 km. Des mouvements des pôles apparents différents pour les provinces Supérieur et Esclaves avant 1700 Ma montrent que la fermeture des cratons commence vers 2300 Ma.*

INTRODUCTION

The Lynn Lake and Fraser Lake gabbros crop out near the town of Lynn Lake in northern Manitoba (Fig. 1). They intrude Aphebian rocks of the La Ronge – Lynn Lake greenstone belt in the Trans-Hudson Orogen of the Churchill Province. This paleomagnetic study was undertaken to examine three aspects of the geology of the gabbros. First, when this study began in 1984, the age of the gabbros was very poorly constrained. Knowing their paleomagnetic pole position could help define their age. Second, the Lynn Lake Gabbro includes the A and EL plugs, which hosted important copper–nickel ore deposits that have been mined by the Sherritt Gordon Company. About 11 other ultrabasic–basic plutons are known in the Lynn Lake belt (Pinsent, 1980), including the Fraser Lake Gabbro, which contains no known economic mineralization despite extensive geophysical surveying and drilling. Paleomagnetism provides a method to see if the mineralized and nonmineralized gabbros are comagmatic and/or relatively rotated and to see if the mineralization is syngenetic or epigenetic. This knowledge could lead to a better exploration rationale. Third, very little paleomagnetic work has been done on rocks from the Churchill Province relative to the bounding Superior and Slave provinces. Additional data might help constrain the various geotectonic models being proposed to explain the evolution of the Trans-Hudson Orogen (e.g., Green *et al.*, 1985; Thomas and Gibb, 1985; Hoffman, 1987).

GEOLOGY

The following summary of the geology of the Lynn Lake region of the Lynn Lake greenstone belt (Fig. 2) is derived primarily from the mapping of Pinsent (1980) and stratigraphic analysis of Gilbert *et al.* (1980).

The basal Wasekwan Group represents at least two cycles of volcanism that may possibly rest on a basement of older granitoid gneisses. The group comprises ~7000 m of dominantly mafic tholeiitic flows and fragmental units intercalated with subordinate felsic volcanic and clastic volcanogenic sedimentary units. Rhyolite and felsic porphyry units in the Wasekwan have yielded a U–Pb zircon age of 1910 ± 13 Ma (Baldwin *et al.*, 1987), Rb–Sr whole-rock ages of 1835 ± 75 and 1790 ± 35 Ma (Clark, 1980), and K–Ar biotite ages of 1730 and 1750 Ma (Moore *et al.*, 1960). Following deposition, the Wasekwan Group was regionally metamorphosed within upper greenschist to middle amphibolite facies levels and severely deformed into isoclinal folds.

A period of mafic plutonism followed in which both differentiated and undifferentiated stocks and sills were emplaced. These plutons range from ultrabasic to basic in composition, with gabbro being the dominant rock type. The Lynn Lake A and EL plugs and the Fraser Lake plug of this study belong to this intrusive event. Two other mafic plutons in the Lynn Lake belt give Pb–Pb galena ages of 1975 and 1794 Ma using the Stacey–Kramer model or 1832 and 1768 Ma using the Cumming–Richards model (Sangster, 1978). The minimum age for the A plug mineralization was determined by Turek (1967), who obtained K–Ar biotite, K–Ar clinopyroxene, and Rb–Sr biotite ages from a crosscutting veinlet of 1669, 1669, and 1668 Ma respectively.

Large discordant intermediate to felsic plutons were emplaced next. Their contacts with the Wasekwan volcanics are generally sharp, whereas those with the mafic plutons may be either sharp, with chilled margins, or gradational hybrid zones (Milligan, 1960). The felsic plutons are massive and range in composition from diorite–tonalite to granite. They have given U–Pb zircon ages of 1876 ± 7 and 1876 ± 8 Ma (Baldwin *et al.*, 1987); Rb–Sr whole-rock ages of 1940 ± 75, 1825 ± 210, and 1768 ± 100 Ma (Clark, 1980); and K–Ar biotite ages of 1640, 1700, and 1730 Ma (Moore *et al.*, 1960; Lowdon *et al.*, 1963).

The Sickle Group unconformably overlies the Wasekwan Group and the mafic and felsic plutons. It comprises about 4000 m of conglomerate and arkosic sandstone. The Sickle rocks, including late pegmatite veinlets, have yielded a Rb–Sr whole-rock age of 1760 ± 85 Ma (Clark, 1980), and K–Ar biotite and muscovite ages of 1720 and 1745 Ma (Lowden *et al.*, 1963). After mild regional deformation, the Sickle Group was intruded by a second series of massive felsic plutons of tonalite to granite composition.

SAMPLING

Sampling of the mafic plutons at Lynn Lake for paleomagnetic study is restricted by the extensive cover of overburden and tailings. Also, therefore, the geology is known mostly from geophysical surveys, drilling, and now-flooded mine workings.

The EL plug (Fig. 3) was sampled at 10 sites (Nos. 1–10) around the present glory hole. The plug is ovoid in plan (300 m × 200 m) and extends vertically to a depth of at least 1.3 km. The perimeter phase is gabbro and diorite that surrounds an interior phase of peridotite and amphibolite. The core is a vertical pipe of massive to disseminated pyrrhotite–pentlandite–chalcopyrite mineralization that has yielded 1.7×10^6 t of 2.1% Ni and 0.8% Cu ore. Although segments have been offset on late faults, the plug does not appear to have been tilted off vertical or bent by postemplacement deformation.

The A plug was sampled at 20 sites (Nos. 11–30) along a 300 m strip running north-northwest from the A shaft to the northwestern contact of the plug and along a 100 m strip running south from the Farley shaft in the middle (Fig. 2). The A plug has a kidney-bean shape in plan (3.4 km × 1.4 km), with sharp vertical contacts. It consists mostly of gabbro and diorite (~80%), with minor quartz–hornblende diorite zones, and it has interior vertical zones of amphibolite, peridotite, and norite. The latter contain 10 discrete mineralized pipes that have been mined to produce 25.8×10^6 t of 0.9% Ni and 0.5% Cu ore. Segments of the plug have been offset on late faults by as much as 570 m (Pinsent, 1980), but there is no evidence for tilting or rotational deformation. Emslie and Moore (1961) suggested that A and EL were once a single layered gabbroic body that had been tilted to the vertical and then faulted into two bodies. Geochemical and structural studies by Vellet (1963) and Pinsent (1977, 1980) indicate that they are two distinct vertical plugs that have not been tilted.

The Fraser Lake plug was sampled at 12 sites (Nos. 31–42), with six in amphibolite along the railway on the eastern side and six in gabbro along the road on the western side (Fig. 2). The plug has a rounded triangular shape in plan (8 km × 5 km),

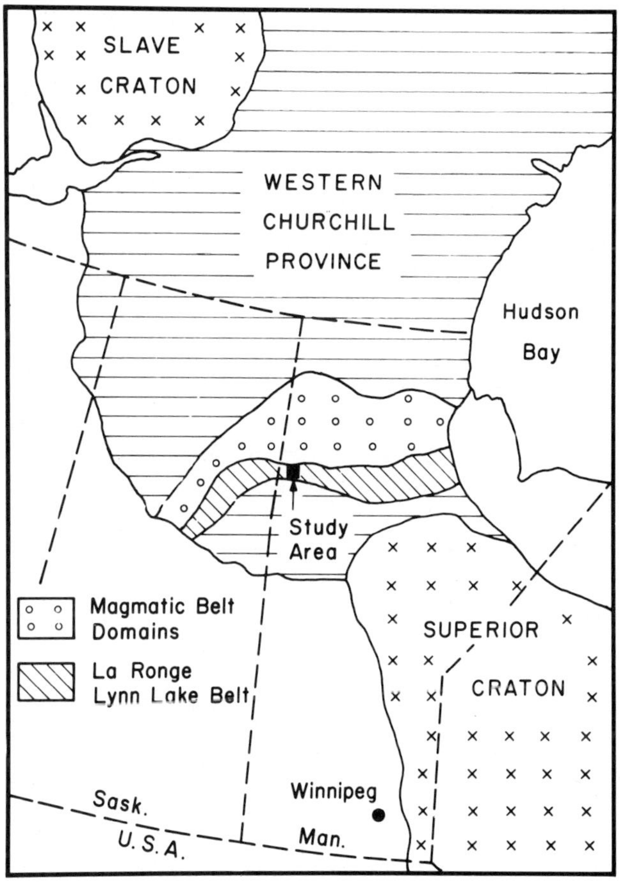

Figure 1. *Location map with major tectonic elements.*

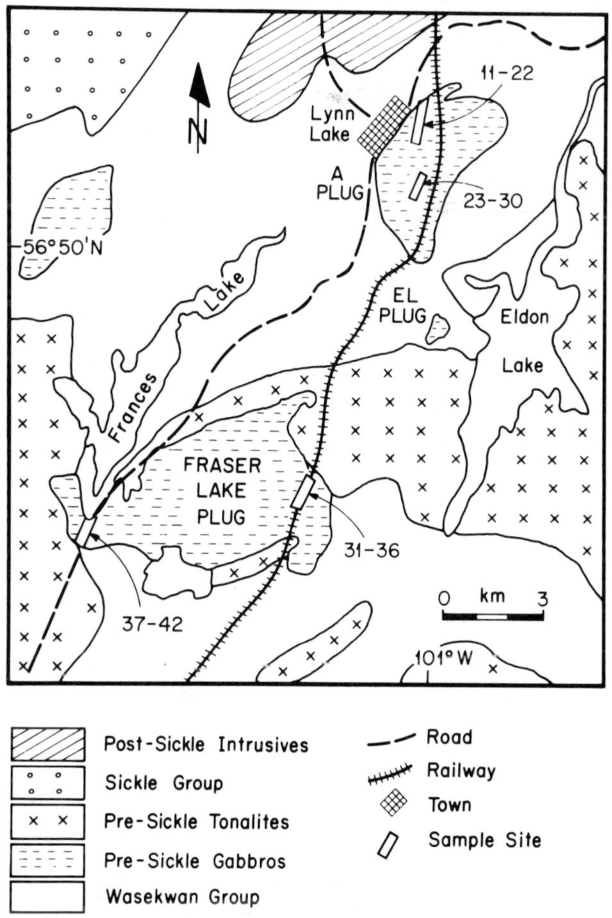

Figure 2. *General geology of the Lynn Lake area (after Pinsent, 1980).*

with vertical contacts along which felsic intrusives were later emplaced in part (Milligan, 1960; Hulbert, 1978). About 90% of the plug is composed of normal gabbro, with some layered gabbro, and the remainder is an interior phase of amphibolite and peridotite. Despite intensive exploration, no economic mineralization has been found. From geochemical evidence, Hulbert (1978) suggested that the rocks now exposed at the surface were formed at a temperature of 1000 ± 45°C and a depth of 3.1 km, whereas those of the A and EL plugs represent shallower and more differentiated intrusive levels.

At each of the 42 sites, five cores were drilled and oriented *in situ* with a sun compass. Two or more specimens were sliced from each core, giving a total of 493 specimens for measurement.

MEASUREMENT

The natural remanent magnetization (NRM) of all specimens was measured on an automated two-axis CTF cryogenic magnetometer inside a shielded room. Two specimens from each site were alternating-field (AF) demagnetized in 10 steps up to 100 mT using a Schonstedt GSD-1 demagnetizer. A further two specimens per site were thermally demagnetized in 12 steps up to 665°C using a Schonstedt TSD-1 demagnetizer. The remaining six or more specimens from each site were AF demagnetized in two steps of 20 mT and 30 or 40 mT. These remaining specimens from three A plug sites (Nos. 21, 22, and 26) and eight Fraser Lake plug sites (Nos. 31–36, 40, and 41) were thermally demagnetized further at temperatures of 500, 550, 600, and 650°C.

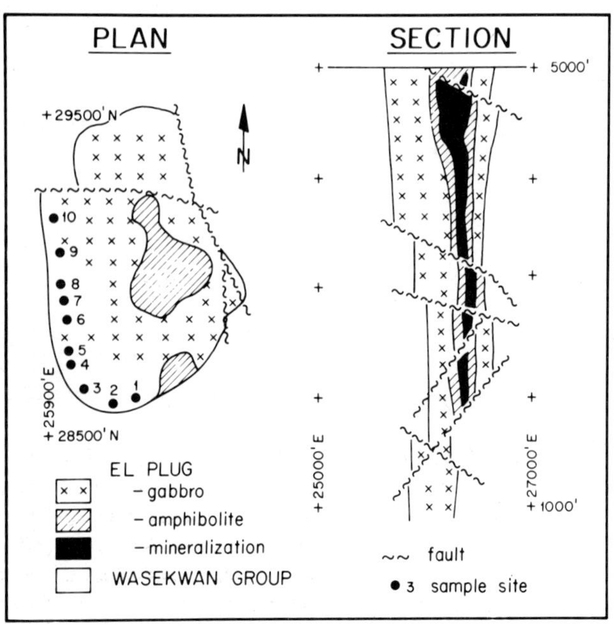

Figure 3. *Geology of the EL plug in plan, with the site locations and in section on a reduced scale.*

STEP DEMAGNETIZATION

The specimens from all three plugs respond to demagnetization in the same fashion by isolating a south-southeast and intermediate-down remanence vector. This direction is isolated throughout the AF coercivity and blocking temperature spectrums. The difference in decay behaviour between the plugs reflects the abundance of sulphides in the specimens. Most Fraser Lake plug specimens have very low sulphide content and quite stable remanence, with an average median destructive field (MDF) of about 25 mT. In contrast, the EL plug specimens were collected adjacent to the ore zone and have obvious disseminated sulphides. These specimens demagnetize easily, with a low MDF of 4.1 mT, and have erratic remanence directions. The specimens from the A plug include some with a very low sulphide content and relatively stable remanence as well as others from the margins of the ore zones with relatively unstable remanence. This mixture results in an intermediate MDF of 7 mT.

Very stable behaviour on AF step demagnetization is shown for an A plug specimen in Figure 4a, where the direction is unchanged from 0 to 100 mT, at which point 30% of the initial NRM remains. Slightly less stable behaviour is shown in Figure 4b for a Fraser Lake plug specimen. Another A plug specimen (Fig. 5a) shows the removal of relatively low stability remanence between 0 and 40 mT, with the same south-southeast and down direction being maintained. Even lower stability is shown by an EL plug specimen (Fig. 5b) that has 80% of its NRM removed by 20 mT. The plot of the step-averaged direction and normalized intensity of all 20 of the AF step demagnetized specimens from the EL plug is shown in Figure 4c. It reveals clearly that the south-southeast and down direction is being sequentially removed both in the low coercivity domains and in the relatively few high coercivity domains.

The same range of remanence behaviours is present on thermal step demagnetization. Figure 6a shows a very stable specimen from the Fraser Lake plug that demagnetizes rapidly at the 585°C Curie temperature of magnetite, indicating the presence of relatively pure pseudosingle domain magnetite. A similar result with a slightly more distributed unblocking temperature range is shown in Figure 6b for an A plug specimen. A second A plug specimen unblocks between 200 and 400°C, but nevertheless holds the same direction (Fig. 7a). Unblocking in this temperature range combined with relatively good AF demagnetization stability

(Fig. 5a) is indicative of a stable pyrrhotite remanence (Schwarz and Vaughan, 1972). An EL plug specimen illustrates very low thermal stability remanence that is common in pyrrhotite where cleaning is mostly complete by the 200°C step (Fig. 7b) without reaching the characteristic south-southeast and down direction.

ANALYSIS

After AF demagnetization of all A and Fraser Lake plug specimens at 20 and 40 mT and EL specimens at 20 and 30 mT, the remanence directions of specimens having a high within-specimen angular standard deviation (ASD ≥ 30°) were rejected as inhomogeneous (Reed, 1959; Harrison, 1980). The remaining vector directions were plotted and density contoured (Kamb, 1959; Stupavsky and Symons, 1982a). For the Fraser Lake plug, the NRM directions give a single component "bull's-eye" anomaly with a very high peak density (Fig. 8a). The 20 and 40 mT treatments retain 68 and 49% of the initial NRM intensity, respectively, and simply further isolate the component to extremely high densities. For the A plug, the NRM plot shows a pronounced lobe into the present Earth's magnetic

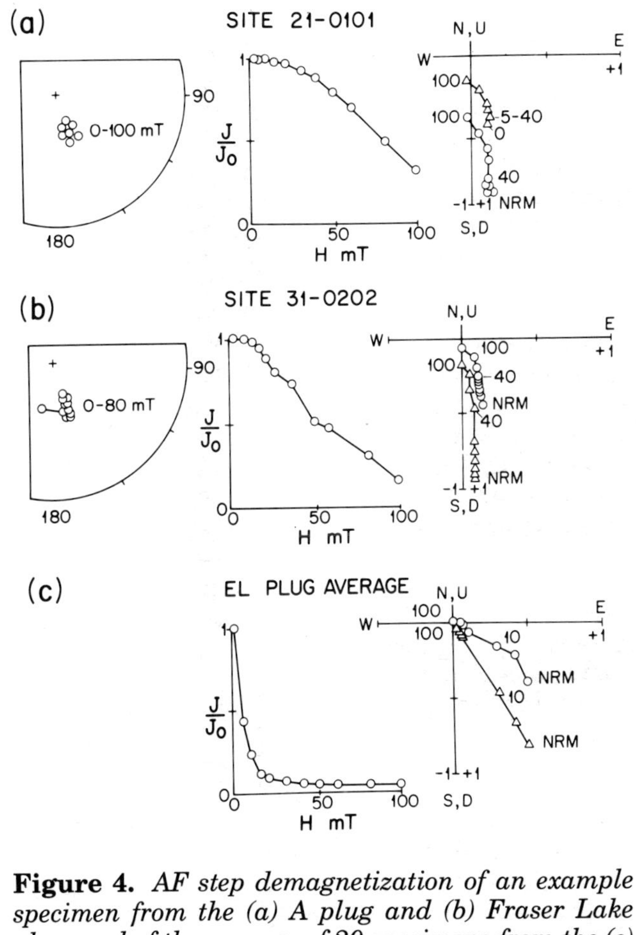

Figure 4. *AF step demagnetization of an example specimen from the (a) A plug and (b) Fraser Lake plug, and of the average of 20 specimens from the (c) EL plug. On the left are segments of equal-area stereograms with the remanence directions going from NRM to 100 mT. All directions are "down" or positive in polarity. In the middle is the intensity decay plot with the measured intensity expressed as a fraction of the original NRM intensity* (J/J_0)*and the demagnetizing field* (H) *in milliteslas (mT). On the right are Zijderveld plots of the directions with the circles plotted in the horizontal (N,E,S,W) plane and the triangles plotted in the vertical (U,E,D,W) plane.*

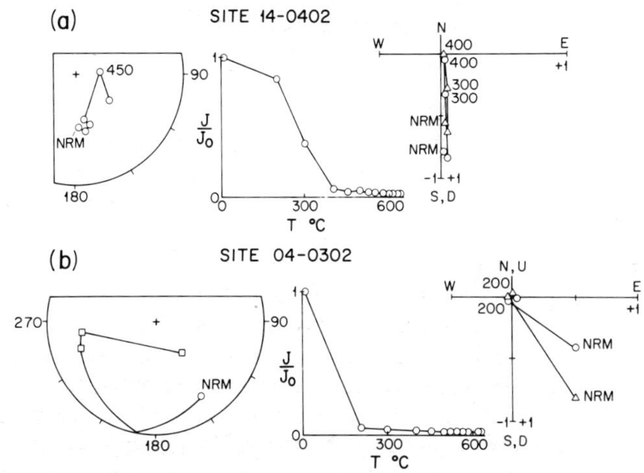

Figure 5. *AF step demagnetization of example specimens from the (a) A plug and (b) Fraser Lake plug. Plotting conventions are as in Figure 4.*

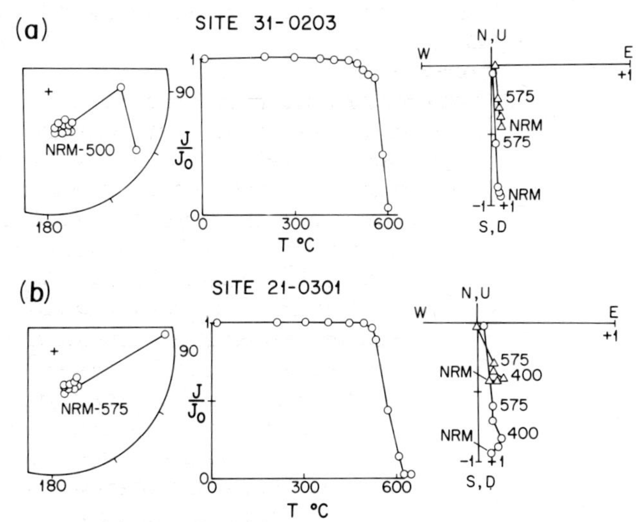

Figure 6. *Thermal step demagnetization of example specimens from the (a) Fraser Lake plug and (b) A plug. Plotting conventions are as in Figure 4.*

field (PEMF) direction, which is removed by the 20 mT cleaning as a one-component bull's-eye anomaly is isolated (Fig. 8b). The A plug peak density at 40 mT of ~13% is lower than the Fraser Lake plug value of ~42%, but it is still highly significant at >>99.99% confidence (Kamb, 1959). For the EL plug, the NRM plot shows two significant lobes, including one in the PEMF direction, as well as considerable scatter (Fig. 8c). Cleaning at 20 and 30 mT gives a clearly significant, but somewhat irregular or less well-defined, anomaly in the south-southeast and down direction also.

Next, the site mean directions were calculated from the specimen directions and, after rejecting sites with Fisher (1953) α_{95} values of ≥ 30°, the site means were combined to get the plug mean directions after different cleaning treatments (Table I). Two significant points emerge. First, the 20 mT, 40 mT, 500°C, and 550°C cleaning processes isolate the same direction in the A plug (Fig. 9a) and in the Fraser Lake plug (Fig. 9b). Second, the same direction is isolated in all three plugs as shown in Figure 9c, where the data for the cleaning treatment giving the best clustering for each site have been used. The pole positions for each of the plugs and for all three plugs combined are given in Table II.

DISCUSSION

It is significant that the same magnetization direction is found in (1) all three plugs, (2) specimens containing minor to high concentrations of magnetite and/or pyrrhotite sulphides, (3) low (<200°C) to high (585°C) blocking temperature domains, and (4) low (<10 mT) to high (>100 mT) coercivity domains. This requires that all specimens acquired their magnetization over a relatively short time period either at the time of emplacement or during subsequent regional metamorphism. The pole position for the gabbros falls directly on an overlapping portion of the apparent polar wander path (APWP) for the North American craton (Irving, 1979; Fig. 10) and indicates a magnetization age of either ~1890 ± 20 or ~1685 ± 15 Ma. The former age fits extremely well with the U–Pb zircon ages that bracket gabbro emplacement to between 1876 and 1910 Ma. The ~1685 Ma age is enigmatic. It fits marginally with the age of the postore biotite veinlet analyzed by Turek (1967), and it is younger than all radiometric ages except for one K–Ar age of 1640 Ma of early experimental vintage (Moore *et al.*, 1960). The remaining K–Ar and Rb–Sr ages are spread between 1700 and 1940 Ma without indicating an obvious time period of metamorphism. Also, if regional metamorphism were responsible, then the temperature would have had to exceed low amphibolite facies levels to reset the most stable domains (Pullaiah *et al.*, 1975). The numerous specimens with "square-shouldered" thermal intensity decay plots (Fig. 6) from the A and Fraser Lake plugs would have had to be heated to close to 585°C to have had their remanence reset. However, the

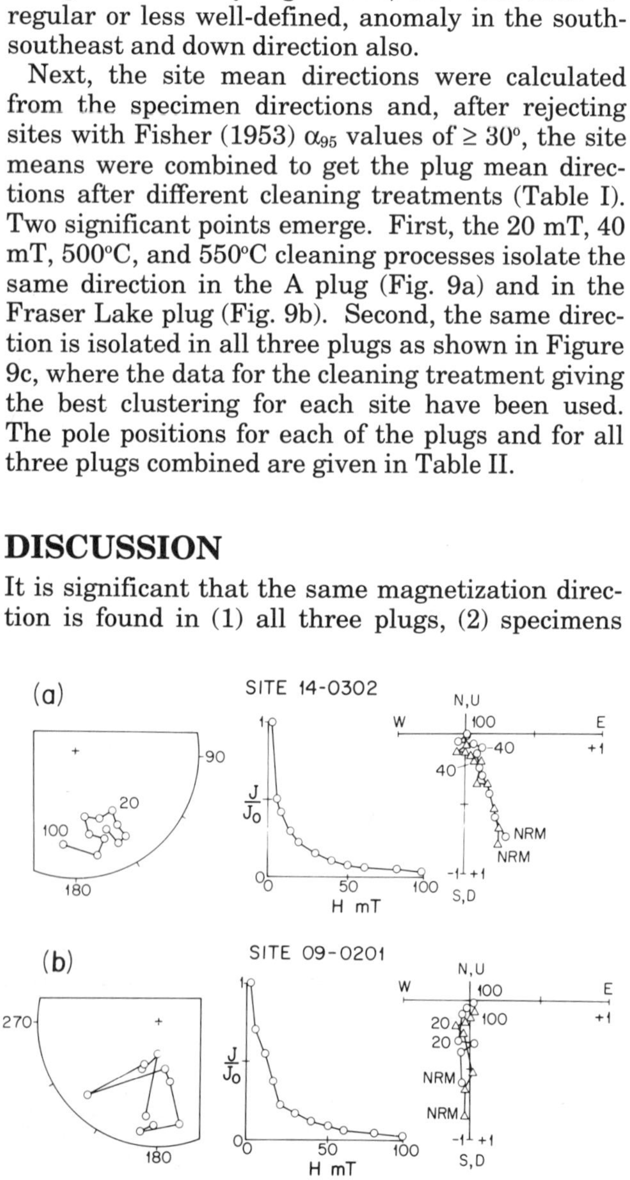

Figure 7. *Thermal step demagnetization of example specimens from the (a) A plug and (b) EL plug. Plotting conventions are as in Figure 4 except for the stereogram in (b), where squares denote upward or negative remanence directions.*

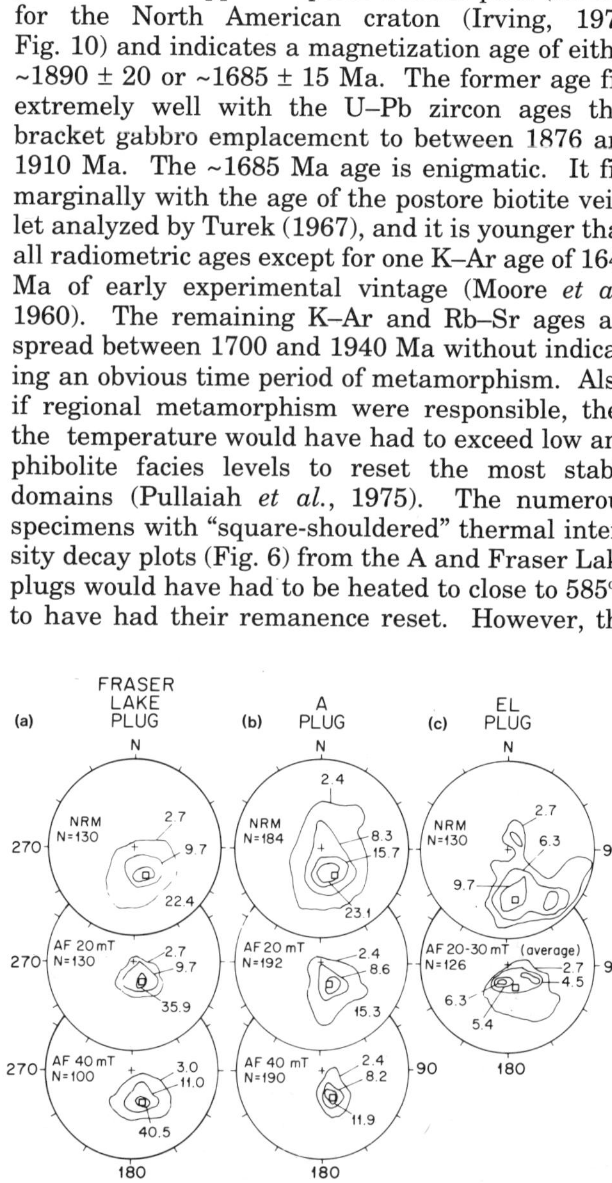

Figure 8. *Contour plots of vector densities on the lower hemisphere of equal-area stereograms for the remanence directions of the NRM and after 20 mT AF and 30 or 40 mT AF cleaning.* N *is the number of specimen vectors. The outer anomaly contour is set at the* P(*F*) = *0.05 density limit, and the values give the contour vector density in per cent.*

Table I
Site and unit mean remanence directions

Plug site	Treat- ment	N	Declination (°)	Inclination (°)	$\boldsymbol{R}$	k	α_{95} (°)
			EL plug				
3	A20	10	123	55	8.979	9	17
4	A20	8	156	30	7.271	10	19
6	A20	7	186	73	6.781	27	12
7	A20	10	177	59	9.323	13	14
8	A20	11	153	50	9.789	8	16
9	A30	11	174	53	9.749	8	17
10	A30	15	138	62	12.428	5	18
	A20	(6)	161	53	5.743	19	13
	OPT	(7)	157	56	6.721	22	13
			A plug				
11	A20	6	156	76	5.747	20	17
12	A20	9	150	65	8.867	60	7
13	A40	6	153	71	5.679	16	18
14	A40	6	182	30	5.760	21	15
16	A40	6	132	72	5.892	46	10
17	A20	6	209	63	5.335	8	26
18	A40	7	167	60	6.681	19	14
21	T550	10	157	57	9.723	33	9
22	T550	10	161	63	9.966	266	3
23	A20	8	162	56	7.453	13	16
24	A20	6	157	68	5.716	18	16
26	A20	6	140	53	5.672	15	18
27	A20	9	154	61	8.022	8	19
28	A40	8	140	49	6.747	6	26
29	A40	7	150	31	6.553	13	17
30	A40	11	140	53	10.672	30	8
	A20	(14)	164	69	13.687	31	7
	A40	(15)	158	60	14.426	24	8
	T500	(3)	159	54	2.965	58	16
	T550	(3)	150	59	2.978	92	13
	OPT	(16)	157	59	15.419	26	7
			Fraser Lake plug				
31	T500	11	148	68	10.938	164	4
32	T440	12	174	66	11.900	110	4
33	A40	11	166	61	10.957	229	3
34	A40	11	161	61	10.983	565	2
35	T550	10	168	61	9.856	63	6
36	T550	10	149	65	9.790	43	8
	A20	(6)	159	70	5.974	194	5
	A40	(6)	172	62	5.964	140	6
	T500	(6)	162	65	5.977	225	5
	T550	(6)	163	65	5.977	219	5
	OPT	(6)	161	64	5.977	222	5
			All sites				
	OPT	(29)	158	60	28.084	31	5

Notes: The "treatment" is designated by A (AF cleaning) followed by the field in mT, T (thermal cleaning) followed by the temperature in °C, and OPT (optimum treatment for each site combined). α_{95} is the radius of the cone of 95% confidence, $\boldsymbol{R}$ is the resultant vector, and k is the precision parameter.

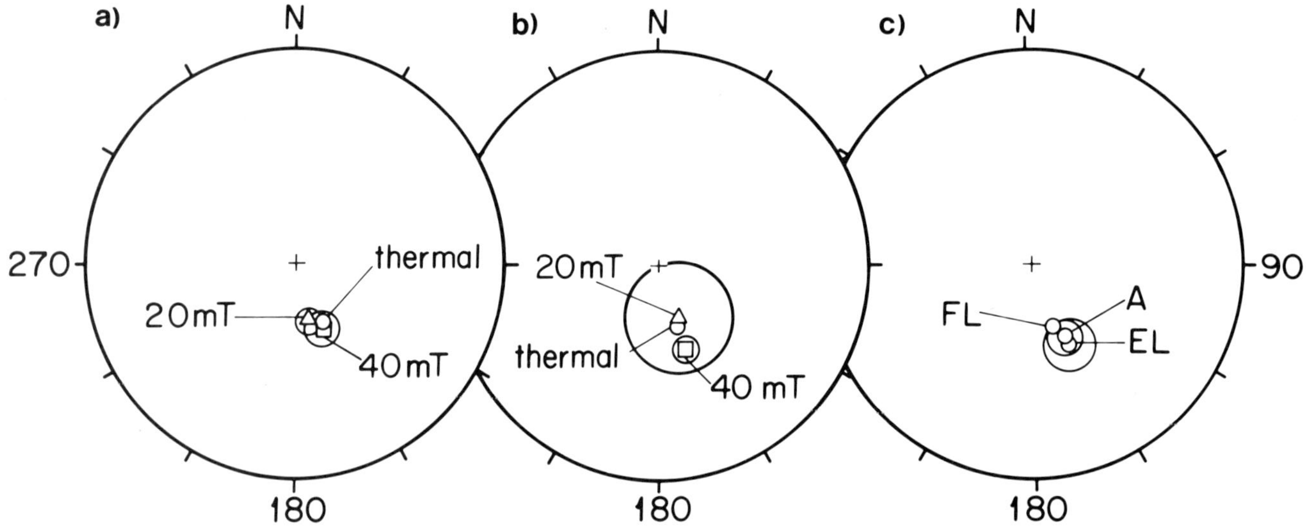

Figure 9. *Mean remanence directions with their cone of 95% confidence as calculated from conventional tiered statistics for the (a) A plug and (b) Fraser Lake plug after various AF (mT) and thermal (°C) cleaning treatments, and for (c) all three plugs based on the treatment giving the best clustering. Refer also to Tables I and II.*

gabbros show no evidence of even low greenschist levels of regional metamorphism (Pinsent, 1980). Also, if remagnetization did occur, the most likely time would have been at 1876 ± 9 Ma, when the large felsic intrusions were emplaced in contact with the Fraser Lake plug and adjacent to the A and EL plugs. Alignment of the low with the high blocking temperature domains implies that any metamorphic event after 1876 Ma was mild. This agrees with Pinsent's (1980) geochemical evidence for shallow emplacement of the gabbros at ~3 km depth, so that even cooling on uplift would be minimal.

Thus, the paleomagnetic results indicate that the A, EL, and Fraser Lake plugs were emplaced contemporaneously at ~1890 ± 20 Ma, that the sulphide mineralization was emplaced at the same time, and that the plutons have not been metamorphosed since then. The results also indicate that the plutons have not been subjected to relative rotation. As Pinsent (1980) argued, they were emplaced into deformed Wasekwan strata as vertical plugs and have been subsequently tilted by no more than 5°. Alignment of the remanence declinations also indicates that the plutons have not been rotated relatively around vertical axes by more than a few degrees. This implies that the plutons were emplaced into a "stable" Wasekwan crust, which was not involved in later localized oroclinal bending.

Table II
Pole positions

Plug	Longitude (°W)	Latitude (°N)	D_p (°)	D_m (°)
EL	83.0	5.6	13.8	19.2
A	83.2	8.8	8.3	11.1
Fraser Lake	88.4	13.7	5.7	7.2
All sites	84.4	9.0	5.5	7.4

Notes: The pole position is calculated from the OPT values presented in Table I and is defined by the radii of its oval of 95% confidence (Fisher, 1953).

GEOTECTONIC COMMENTS

Current geotectonic models for the development of the Trans-Hudson Orogen of the western Churchill Province envisage an Andean-type crustal convergence with the Archean Slave and Superior cratons moving relatively southeast and northwest, respectively, into the orogen (e.g., Lewry, 1981; Stauffer, 1984; Green *et al.*, 1985; Thomas and Gibb, 1985; Hoffman, 1987; Van Schmus *et al.*, 1987). The timing and magnitude of the convergence are poorly constrained at present; however, the paleomagnetic data for the 1825 to 1900 Ma period provide an interesting answer. Many cautionary caveats go with this answer, such as the possibility of unrecognized tectonic rotations or of misassigned magnetization ages.

The poles utilized for the Slave craton (Fig. 11; Table III) include the Seton Formation, for which there is a Rb–Sr isochron of 1870 Ma (Irving and McGlynn, 1979), and the Akaitcho River Formation (Evans *et al.*, 1980) and Kahochella Group (Reid *et al.*, 1981), which are stratigraphically immediately below and above the Seton Formation respectively.

There are obvious concerns with respect to these data. All three poles come from a limited area in the Athapuskow Aulacogen of the Slave Province. Their age assignment is based on only one Rb–Sr isochron. The Kahochella pole includes a vectoral subtraction to correct for nonantiparallel normal and reverse remanence components.

The Superior craton poles have limitations also. The Molson dykes give a U–Pb zircon age of 1884 ± 2 Ma (Machado *et al.*, 1986), but they come from the Thompson mobile belt on the Superior side of the Churchill–Superior boundary so that tectonic rotation is conceivable (Ermanovics and Fahrig, 1975). The Spanish River Complex gives a Rb–Sr isochron age of 1838 ± 95 Ma (Sage, 1983), but its remanence is widely scattered, with a large α_{95} error limit of 25° (Robertson and Watkinson, 1974). The Sudbury norite and granophyre give U–Pb zircon ages of 1850 ± 3 and 1849 ± 3 Ma, respectively (Krogh *et al.*, 1984), and slightly older Rb–Sr isochron ages (Gibbins and McNutt, 1975). Their poles, however, include a structural correction that assumes an initial best-fit dip for the limbs around the basin (Morris, 1984).

The poles for the western Churchill Province include two from northwest of the La Ronge – Lynn Lake belt. These are from the Martin Formation (Evans and Bingham, 1973), whose age assignment at 1835 ± 50 Ma is based on a reset biotite K–Ar age (Irving and McGlynn, 1979), and from the Dubawnt Group (Park *et al.*, 1973), whose assignment is bracketed at ~1825 Ma (Irving and McGlynn, 1979). Another two poles come from southeast of the La Ronge – Lynn Lake belt. Park (1975) reports data from 12 sites in the late-metamorphic gabbroic Boundary Intrusions of the Flin Flon area and from 22 sites in the metamorphosed Amisk Group volcanics, which they intrude. Both units carry a dominant A and minor B magnetization. Park (1975) argued that the B component is older than the A component, primarily on the basis of their pole positions. However, current paths (e.g., Irving, 1979) suggest that the A component, utilized here, is older. These rocks give Sm–Nd, Rb–Sr, and K–Ar ages ranging from 1737 to 1930 Ma (Park, 1975; Chauvel *et al.*, 1987). Other post-Amisk plutons in the area given U–Pb zircon ages between 1800 and 1875 Ma (Watters and Armstrong, 1985), with a

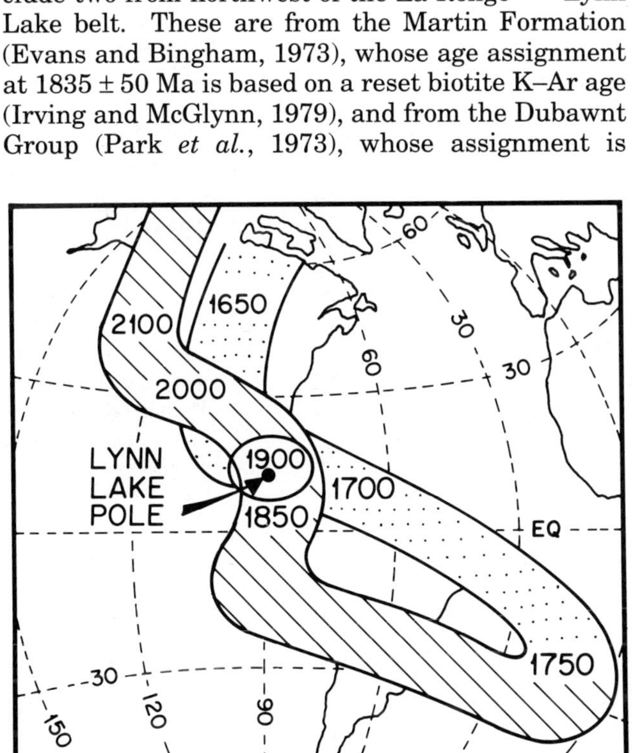

Figure 10. *Pole position for the gabbros in the Lynn Lake area relative to the apparent polar wandering curve of Irving (1979).*

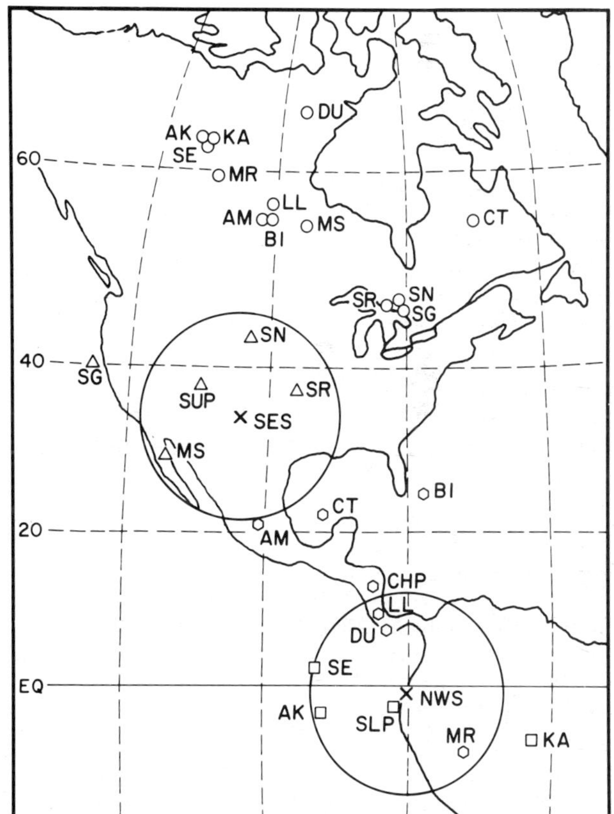

Figure 11. *Pole positions for rock units magnetized between 1900 and 1825 Ma from the Slave (squares), Churchill (hexagons), and Superior (triangles) provinces as listed in Table III. The sampling sites are shown by circles. The mean pole positions for the provinces are labelled SLP, CHP, and SUP respectively. The means with their cone of 95% confidence for poles from northwest and southeast of the Trans-Hudson Orogen suture are crosses labelled NWS and SES respectively.*

Table III
Pole positions for the Trans-Hudson Orogen

Province and units	Age	Pole position Longitude (°W)	Latitude (°N)	α_{95} (°)	Reference
Churchill Province					
(1) Martin Formation (MR)	1835	72	-9	9	Evans and Bingham (1973)
(2) Dubawnt Group (DU)	1825	83	7	8	Park *et al.* (1973)
(3) Lynn Lake gabbros (LL)	1890	84	9	7	This paper
(4) Amisk metavolcanics (AM)	1855	101	21	9	Park (1975)
(5) Boundary Intrusions (BI)	1855	79	26	9	Park (1975)
(6) Castignon Complex (CT)	1875	92	22	14	Park (1977)
Average (CHP)	1875	85	13	14	
Slave Province					
(7) Kahochella Group (KA)	1850	62	-7	9	Reid *et al.* (1981)
(8) Seton Formation (SE)	1870	93	2	6	Irving and McGlynn (1979)
(9) Akaitcho Formation (AK)	1890	92	-4	7	Evans *et al.* (1980)
Average (SLP)	1870	82	-3	28	
Superior Province					
(10) Molson dykes (MS)	1884	114	28	12	Ermanovics and Fahrig (1975)
(11) Spanish River Complex (SR)	1838	96	37	25	Robertson and Watkinson (1974)
(12) Sudbury norite (SN)	1850	103	44	6	Morris (1984)
(13) Sudbury granophyre (SG)	1849	126	40	6	Morris (1984)
Average (SUP)	1855	110	38	14	
Northwest of suture entries 1, 2, 7, 8, and 9 (NWS)	1854	80	-2	14	
Southeast of suture entries 4, 5, and 10–13 (SES)	1855	103	34	14	

Note: The abbreviations in parentheses following the rock units correspond with those in Figure 11.

concentration around 1855 ± 15 Ma (Van Schmus *et al.*, 1987). In addition to the Lynn Lake result described in this paper, the other relevant pole comes from the Castignon Complex in the Labrador Trough of the eastern Churchill Province (Park, 1977). This carbonatite gives a K–Ar age of 1875 ± 50 Ma.

The plot of the pole positions presented in Table III reveals a distinctive pattern (Fig. 11). The Churchill Province poles are in the middle with the Superior poles to the north and the Slave poles to the south. When these pole groups are averaged, the Superior mean pole is significantly different than the Slave mean pole. These mean poles indicate a separation of 47°, implying the subduction and destruction of an oceanic crust that was comparable to the present-day Atlantic. Furthermore, the sense of motion is correct. Using the Churchill mean pole as a reference, the Superior craton moved northwest into the Trans-Hudson Orogen by ~34° and the Slave craton moved southeast into the orogen by ~13°.

Within the western Churchill Province, we find further evidence of the same pattern. The Amisk metavolcanics and Boundary Intrusions from the southeastern side of the La Ronge – Lynn Lake belt give poles that fit best with the Superior Province poles. Conversely, the Martin Formation and Dubawnt Group on the northwest side of the La Ronge – Lynn Lake belt give poles that fit best with the Slave Province poles. This belt, along with the adjacent magmatic arc rocks of the Reindeer – South Indian and Wathaman–Chipewyan domains, is thought to have been generated in a suture zone (e.g., Gibb *et al.*, 1980; Ray and Wanless, 1980; Lewry *et al.*, 1981; Green *et al.*, 1985; Thomas and Gibb, 1985). Comparing poles from northwest of the suture with those from the southeast yields a separation of 40° with the Superior craton moving ~28° northwest and the Slave craton moving ~12° southeast into the suture.

Despite the evident problems with each of the individual pole determinations, all of the poles in concert do favour a convergence model for the creation of the Trans-Hudson Orogen that led to the destruction of a major ocean after ~1860 Ma. Indeed, Stauffer (1984) has suggested on geological grounds that the Trans-Hudson Orogen records the remains of the Manikewan Ocean in which closure was well advanced by 1865 Ma and complete by 1800 Ma. Because the Slave and Superior cratons and their respective mean poles lie on a common great circle, their separation cannot be explained easily by tectonic rotation about a local vertical axis. Rather, separation is best explained by conventional convergent plate tectonics about an Euler pole about 90° away. Green *et al.* (1985, Fig. 10) and Hoffman (1987) suggest a minimum convergence of about 1000 km. They may be unduly conservative because this analysis suggests ~4000 km. If the analysis is valid, and not simply fortuitous, then it is also necessary to construct separate apparent polar wander paths for the Slave and Superior cratons before ~1.7 Ga. Cavanaugh and Seyfert (1977) pointed out that the few older Slave poles appeared discordant on the Superior APWP. They argued for two paths with an east–west offset between them. Irving and McGlynn (1976) suggested the possibility of rotation of the Slave Province about a local vertical axis to fit the poles for the Dogrib dykes and the Big Spruce Complex. Irving (1979) opted to retain a common path for both cratons. Later, Irving *et al.* (1984) compared Early Proterozoic poles for the 2000–2300 Ma interval. They proposed separate Slave and Superior apparent polar wander paths that join at ~1750 Ma, when the two cratons were fixed in their present relative positions.

This analysis argues for separate Slave and Superior apparent polar wander paths that join at ~1700 Ma at the point of overlap in the Irving (1979) path. It implies that the paths are offset along the general northwest–southeast trend from 1800 to 2300 Ma, so that the offset is hard to distinguish. This offset explains why the Coronation Loop from 1825 to 1700 Ma on the Irving (1979) path is defined by younger Slave poles – they have been offset to the southeast, whereas the Superior poles have been offset to the northwest and merged with older poles. It is also consistent with the offset found by Irving *et al.* (1984) in the older 2000–2300 Ma Slave and Superior poles. For the Slave craton, they cite four poles (Sparrow dykes (SD), Indin dykes (ID), Big Spruce Complex (BC 2), and Caribou Lake gabbro (CL)) with a mean of 72°W, 13°N (α_{95} = 9°) for the 2100 ± 75 Ma interval and five poles (Easter Island dyke (ED), Hearn dyke contact (HC), Big Spruce Complex (BC 3), Dogrib dykes (DG), and X dykes (XD)) with a mean of 45°W, 24°S (α_{95} = 24°) for the 2250 ± 75 Ma interval. Thus, the Slave APWP appears to jog northeastward before tracking southward past 2300 Ma (Fig. 12).

The Proterozoic APWP for the Superior craton is anchored by several poles representing more than 150 dykes from the Matachewan–Hearst swarm (Irving and Naldrett, 1977 (MD 1); Schutts and Dunlop, 1981 (MD 2); Ernst and Halls, 1984 (MD 3); Vandall and Symons, 1987 (MD 4); Halls and Palmer, 1988 (MD 5)) that give a mean pole of 62°E, 45°N (α_{95} = 5°) and a U–Pb zircon age of 2452 ± 2 Ma (Heaman, 1988; Smith and Farquhar, 1988). The overlying Huronian Supergroup has yielded Rb–Sr isochron ages of 2215 ± 185 and 2288 ± 87 Ma (Knight and Fairbairn, 1966; Fairbairn *et al.*, 1969). The Huronian poles (Thessalon volcanics

(TV), Symons and O'Leary (1978); Dollyberry volcanics (DB) and "Elliot Lake" volcanics (EV), Stupavsky and Symons (1983); and Gowganda Formation clastics (Symons, 1967 (GF 1); Symons, 1975 (GF 2); Roy and Lapointe, 1976 (GF 3); Morris, 1977 (GF 4, GF 5); Stupavsky and Symons, 1982b (GF 6)) are both pre- and postfolding in origin, but predate Nipissing intrusion. When combined, they give a mean pole of 138°W, 64°N (α_{95} = 21°) with an assigned age of 2275 ± 100 Ma. This pole agrees with a glacial origin for the Gowganda tillite. The Nipissing diabase has given a U–Pb baddeleyite–rutile isochron of 2219 ± 4 Ma (Corfu and Andrews, 1986); Rb–Sr isochrons of 2162 ± 27, 2166 ± 80, and 1700 ± 50 Ma (Van Schmus, 1965; Fairbairn *et al.*, 1969); and K–Ar dates of 2095, 1995, 1660, 1600, 1390, and 1340 Ma (see Symons, 1971). Numerous studies have produced two clusters of poles, each with positive contact tests indicating a primary remanence as discussed by Morris (1979), Stupavsky and Symons (1982b), Roy (1983), and Irving *et al.* (1984). With reservations, the northern cluster in central North America (N1–N4, R + L of Roy (1983)) is considered older, at 2160 ± 60 Ma, and the southern cluster in the Pacific off Peru (N5–N11, M, S of Roy (1983); N13 from Stupavsky and Symons (1982b)) is considered to be a second, younger (2000 Ma (1700 Ma?)), intrusive or

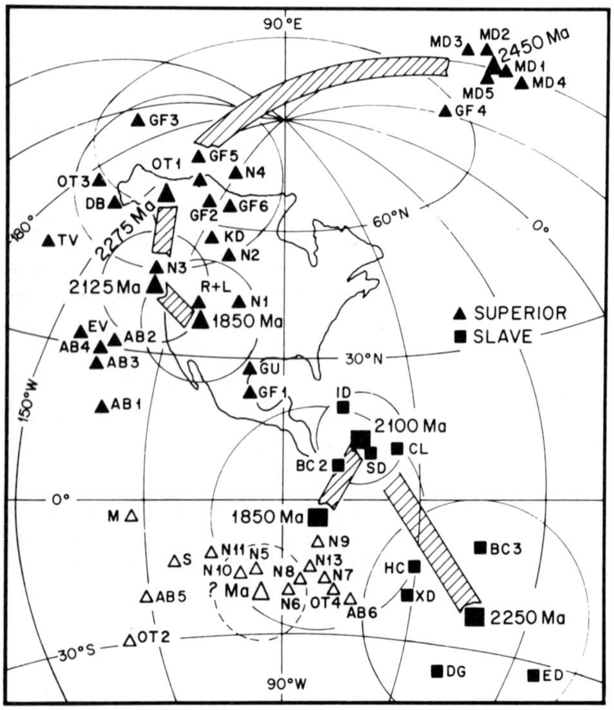

Figure 12. *Polar wander paths for the Slave (squares) and Superior (triangles) cratons during the lower Early Proterozoic. Pole designations are given in the text.*

metamorphic event. The Abitibi diabase dykes are similar to the Nipissing. They give a U/Pb zircon age of 1140 Ma (L. Heaman, personal communication), a Rb–Sr isochron of 2147 ± 68 Ma, a $^{40}Ar/^{39}Ar$ plateau age of 2150 ± 25 Ma, and a wide range of K–Ar whole-rock and mineral ages ranging from 1230 to 2035 Ma (Hanes and York, 1979). They also give a northern cluster of poles (Strangway, 1964 (AB 1); Fahrig *et al.*, 1965 (AB 2); Larochelle, 1966 (AB 3); Irving and Naldrett, 1977 (AB 4)) and southern cluster of poles (Larochelle, 1966 (AB 5, AB 6)). The Otto stock gives a Rb–Sr isochron of 2160 ± 80 Ma (Bell and Blenkinsop, 1976), with primary poles in the northern cluster (Pullaiah and Irving, 1975 (OT 1); Cavanaugh and Seyfert, 1977 (OT 3)) and remagnetization poles in the southern cluster (Pullaiah and Irving, 1975 (OT 2); Cavanaugh and Seyfert, 1977 (OT 4)). The northern cluster also includes two slightly younger poles. These are from the Gunflint iron formation (Symons, 1966 (GU)), which has been dated at 2080 ± 250 Ma by the Sm–Nd method (Stille and Clauer, 1986), and from the Kapuskasing diabase dykes (Halls and Palmer, 1988 (KD)), which have been dated at 2043 ± 14 Ma by several $^{40}Ar/^{39}Ar$ methods (Hanes *et al.*, 1988; Lee *et al.*, 1988). The 13 poles in the northern cluster have a mean of 123°W, 45°N (α_{95} = 12°) with an assigned age of 2125 ± 75 Ma. The 14 poles in the younger southern cluster have a mean of 96°W, 20°S (α_{95} = 10°). Accepting the northern cluster of poles as correct for 2125 ± 75 Ma results in a simple direct northwestward tracking of the APWP for the Superior craton past 2300 Ma (Fig. 12).

Comparing the Slave and Superior apparent polar wander paths indicates that the cratons were separated by a closing ocean basin as far back as 2300 Ma. Because the cratons and the mean pole positions all fall on a common great circle, the separation cannot be easily explained by local or continental rotations. Using the 2250 ± 75 and 2275 ± 100 Ma means, respectively, indicates that the cratons were separated by about 113° along the northwest–southeast trend at this time. From the 2100 ± 75 and 2125 ± 75 Ma means, it appears that the cratons closed by about 60° in the ensuing 150 Ma to give a separation of 53°. Incorporating the southern cluster for the 2125 ± 75 Ma Superior pole into the track would still require the presence of an ocean of about the same magnitude because it is statistically significantly different from the 2100 ± 75 Ma Slave pole. The major problems in sorting out the geotectonics of the closure of the Trans-Hudson Orogen are the dearth of paleomagnetic data from Churchill and Slave provinces, and the uncertain origin of the bimodal age and pole patterns for the Nipissing and Abitibi diabases.

ACKNOWLEDGEMENTS

The authors wish to thank E.A. Timmins for field assistance, Paul Pauliw of Sherritt Gordon Mines Ltd. for maps and other help at Lynn Lake, and the Natural Sciences and Engineering Research Council of Canada, Employment and Immigration Canada, and the University of Windsor for research grants to permit this work to be carried out. The comments of D.J. Dunlop and E. Irving led to substantial improvements in the manuscript.

REFERENCES

Baldwin, D.A., Syme, E.C., Zwanzig, H.V., Gordon, T.M., Hunt, P.A. and Stevens, R.D., 1987, U–Pb zircon ages from the Lynn Lake and Rusty Lake metavolcanic belts, Manitoba: two ages of Proterozoic magmatism: Canadian Journal of Earth Sciences, v. 24, p. 1053–1063.

Bell, K. and Blenkinsop, J., 1976, A Rb–Sr whole-rock isochron from the Otto stock, Ontario: Canadian Journal of Earth Sciences, v. 13, p. 998–1002.

Cavanaugh, M.D. and Seyfert, C.K., 1977, Apparent polar wander paths and the joining of the Superior and Slave provinces during early Proterozoic time: Geology, v. 5, p. 207–211.

Chauvel, C., Arndt, N.T., Kielinczuk, S. and Thom, A., 1987, Formation of Canadian 1.9 Ga old continental crust: Nd isotopic data: Canadian Journal of Earth Sciences, v. 24, p. 396–406.

Clark, G., 1980, Rubidium–strontium geochronology in the Lynn Lake greenstone belt, northwestern Manitoba: Manitoba Department of Energy and Mines, Mineral Resources Division, Geological Paper GP80-2.

Corfu, F. and Andrews, A.J., 1986, A U–Pb age for mineralized Nipissing diabase, Gowganda, Ontario: Canadian Journal of Earth Sciences, v. 23, p. 107–109.

Emslie, R.F. and Moore, J.M., Jr., 1961, Geological studies of the area between Lynn Lake and Fraser Lake: Manitoba Mines Branch Publication 59-4.

Ermanovics, I. and Fahrig, W.F., 1975, The petrochemistry and paleomagnetism of the Molson Dikes, Manitoba: Canadian Journal of Earth Sciences, v. 12, p. 1564–1575.

Ernst, R.E. and Halls, H.C., 1984, Paleomagnetism of the Hearst dike swarm and implications for the tectonic history of the Kapuskasing Structural Zone, northern Ontario: Canadian Journal of Earth Sciences, v. 21, p. 1499–1506.

Evans, M.E. and Bingham, D.K., 1973, Paleomagnetism of the Precambrian Martin Formation, Saskatchewan: Canadian Journal of Earth Sciences, v. 10, p. 1485–1493.

Evans, M.E., Hoye, G.S. and Bingham, D.K., 1980, The paleomagnetism of the Great Slave Supergroup: the Akaitcho River Formation: Canadian Journal of Earth Sciences, v. 17, p. 1389–1395.

Fahrig, W.H., Gaucher, E.H. and Larochelle, A., 1965, Paleomagnetism of diabase dykes of the Canadian Shield: Canadian Journal of Earth Sciences, v. 2, p. 278–298.

Fairbairn, H.W., Hurley, P.M., Card, K.D. and Knight, C.J., 1969, Correlation of radiometric ages of Nipissing diabase and Huronian metasediments with Proterozoic orogenic events in Ontario: Canadian Journal of Earth Sciences, v. 6, p. 489–497.

Fisher, R.A., 1953, Dispersion of a sphere: Proceedings of the Royal Society of London, Series A, p. 295–305.

Gibb, R.A., Thomas, M.D. and Murkhopadhyay, M., 1980, Proterozoic sutures in Canada: Geoscience Canada, v. 7, p. 149–154.

Gibbins, W.A. and McNutt, R.H., 1975, The age of the Sudbury Nickel Irruptive and the Murray Granite: Canadian Journal of Earth Sciences, v. 12, p. 1970–2003.

Gilbert, H.P., Syme, E.C. and Zwanzig, H.V., 1980, Geology of the metavolcanic and volcaniclastic metasedimentary rocks in the Lynn Lake area: Manitoba Department of Energy and Mines, Mineral Resources Division, Geological Paper GP80-1.

Green, A.G., Hajnal, Z. and Weber, W., 1985, An evolutionary model of the western Churchill Province and western margin of the Superior Province in Canada and the north-central United States: Tectonophysics, v. 16, p. 281–322.

Halls, H.C. and Palmer, H.C., 1988, A paleomagnetic study of Precambrian dikes in the vicinity of the Kapuskasing structural zone, Ontario: Lithoprobe-KSZ Workshop '88 Notes, p. 23–36.

Hanes, J.A. and York, D., 1979, A detailed $^{40}Ar/^{39}Ar$ age study of an Abitibi dike from the Canadian Superior Province: Canadian Journal of Earth Sciences, v. 16, p. 1060–1070.

Hanes, J.A., Archibald, D.A., Queen, M. and Lee, J.K.W., 1988, $^{40}Ar/^{39}Ar$ geochronology of diabase dykes: implications for the tectonothermal evolution of the Kapuskasing uplift: Lithoprobe-KSZ Workshop '88 Notes, p. 7–16.

Harrison, C.G.A., 1980, Analysis of the magnetic vector in a single specimen: Geophysical Journal of the Royal Astronomical Society of London, v. 60, p. 489–492.

Heaman, L.M., 1988, A precise U–Pb zircon age for a Hearst dyke: Geological Association of Canada – Mineralogical Association of Canada, Program with Abstracts, v. 13, p. A53.

Hoffman, P.F., 1987, Birth of a craton: Geological Association of Canada – Mineralogical Association of Canada, Program with Abstracts, v. 12, p. 56.

Hulbert, L.J., 1978, Geology of the Fraser Lake gabbro complex, Manitoba, unpublished M.Sc. thesis, University of Regina.

Irving, E., 1979, Paleopoles and paleolatitudes of North America and speculations about dispersed terrains: Canadian Journal of Earth Sciences, v. 16, p. 669–694.

Irving, E. and McGlynn, J.C., 1976, Proterozoic magnetostratigraphy and the tectonic evolution of Laurentia: Philosophical Transactions of the Royal Society, London, v. A280, p. 243–265.

Irving, E. and McGlynn, J.C., 1979, Palaeomagnetism in the Coronation geosyncline and arrangement of continents in the middle Proterozoic: Geophysical Journal of the Royal Astronomical Society, v. 58, p. 309–336.

Irving, E. and Naldrett, A.J., 1977, Paleomagnetism in Abitibi greenstone belt, and Abitibi and Matachewan diabase dikes: evidence of the Archean geomagnetic field: Journal of Geology, v. 85, p. 157–176.

Irving, E., Davidson, A. and McGlynn, J.C., 1984, Paleomagnetism of gabbros of the early Proterozoic Blachford Lake intrusive suite and the Eastern Island dike, Great Slave Lake, NWT: possible evidence for the earliest continental drift: Geophysical Surveys, v. 7, p. 1–25.

Kamb, W.B., 1959, Ice petrofabric observations from Blue Glacier, Washington, in relation to theory and experiment: Journal of Geophysical Research, v. 64, p. 1891–1909.

Knight, C.J. and Fairbairn, H.W., 1966, Rb–Sr isochron ages of volcanic rocks on the north shore of Lake Huron, Ontario, Canada: Geological Society of America, Abstracts with Program, San Francisco, November 1966.

Krogh, T.E., Davis, D.W. and Corfu, F., 1984, Precise U–Pb zircon and baddeleyite ages for the Sudbury area: Ontario Geological Survey, Special Volume 1, p. 431–446.

Larochelle, A., 1966, Paleomagnetism of the Abitibi dyke swarm: Canadian Journal of Earth Sciences, v. 3, p. 671–683.

Lee, J.K.W., Onstott, T.C. and Hanes, J.A., 1988, Comparison of $^{40}Ar/^{39}Ar$ laser microprobe and furnace extraction techniques for biotite and hornblende from the Kapuskasing Structural Zone: a preliminary report: Lithoprobe-KSZ Workshop '88 Notes, p. 245–252.

Lewry, J.F., 1981, Lower Proterozoic arc–microcontinent collisional tectonics in the western Churchill Province: Nature (London), v. 294, p. 69–72.

Lewry, J.F., Stauffer, M.R. and Fumerton, S.L., 1981, A Cordilleran-type batholithic belt in the Churchill Province in northern Saskatchewan: Precambrian Research, v. 14, p. 277–313.

Lowden, J.A., Stockwell, C.H., Tipper, H.W. and Wanless, R.K., 1963, Age determinations and geological studies. Report 3: Geological Survey of Canada, Paper 62-17.

Machado, N., Heaman, L. and Krogh, T.E., 1986, U–Pb zircon ages for the Molson Dike Swarm and the Fox River Sill, NE Manitoba: Geological Association of Canada – Mineralogical Association of Canada, Program with Abstracts, v. 11, p. 97.

Milligan, G.C., 1960, Geology of the Lynn Lake district: Manitoba Mines Branch, Publication 67-1, p. 317.

Moore, J.M., Hart, S.R., Barnett, C.C. and Hurley, P.M., 1960, Potassium–argon ages in northern Manitoba: Geological Society of America Bulletin, v. 71, p. 225–230.

Morris, W.A., 1977, Paleomagnetism of the Gowganda and Chibougamau formations: evidence for 2200 m.y.-old folding and remagnetization event in the Southern Province: Geology, v. 5, p. 137–140.

Morris, W.A., 1979, A positive contact test between Nipissing diabase and Gowganda argillites: Canadian Journal of Earth Sciences, v. 16, p. 607–611.

Morris, W.A., 1984, Paleomagnetic constraints on the magmatic, tectonic, and metamorphic evolution of the Sudbury Basin region: Ontario Geological Survey, Special Volume 1, p. 411–427.

Park, J.K., 1975, Paleomagnetism of the Flin Flon – Snow Lake greenstone belt, Manitoba and Saskatchewan: Canadian Journal of Earth Sciences, v. 12, p. 1272–1290.

Park, J.K., 1977, A reconnaissance paleomagnetic survey of the central Labrador Trough, Quebec: Canadian Journal of Earth Sciences, v. 14, p. 159–174.

Park, J.K., Irving, E. and Donaldson, J.A., 1973, Paleomagnetism of the Precambrian Dubawnt Group: Geological Society of America Bulletin, v. 84, p. 859–870.

Pinsent, R.H., 1977, The Lynn Lake Ni–Cu deposits: Manitoba Mineral Resources Division, Report of Field Activities 1977.

Pinsent, R.H., 1980, Nickel–copper mineralization in the Lynn Lake gabbro: Manitoba Mineral Resources Division, Economic Geology Report ER 79-3.

Pullaiah, G. and Irving, E., 1975, Paleomagnetism of the contact aureole and late dikes of the Otto Stock, Ontario, and its application to Early Proterozoic apparent polar wandering: Canadian Journal of Earth Sciences, v. 12, p. 1609–1618.

Pullaiah, G., Irving, E., Buchan, K.L. and Dunlop, D.J., 1975, Magnetization changes caused by burial and uplift: Earth and Planetary Science Letters, v. 28, p. 133–143.

Ray, G.E. and Wanless, R.D., 1980, The age and geological history of the Wollaston, Peter Lake and Rottenstone domains in northern Saskatchewan: Canadian Journal of Earth Sciences, v. 17, p. 333–347.

Reed, S., 1959, Evaluation of measurement, *in* Marton, L., ed., Methods of Experimental Physics. Volume 6. Part A. Solid State Physics: Academic Press, New York, p. 1–14.

Reid, A.B., McMurry, E.W. and Evans, M.E., 1981, Paleomagnetism of the Great Slave Supergroup, Northwest Territories, Canada: multicomponent magnetization of the Kahochella Group: Canadian Journal of Earth Sciences, v. 18, p. 574–583.

Robertson, W.A. and Watkinson, D.H., 1974, Paleomagnetism of the Spanish River alkalic rock carbonatite complex, Townships 107 and 108, Ontario: Canadian Journal of Earth Sciences, v. 11, p. 795–800.

Roy, J.L., 1983, Paleomagnetism of the North American Precambrian: a look at the data base: Precambrian Research, v. 19, p. 319–348.

Roy, J.L. and Lapointe, P.L., 1976, The paleomagnetism of Huronian red beds and Nipissing diabase; post-Huronian igneous events and apparent polar path for the interval -2300 to -1500 Ma for Laurentia: Canadian Journal of Earth Sciences, v. 13, p. 749–773.

Sage, R.P., 1983, Geology of the Shenago Township alkalic rock complex: Ontario Geological Survey, Open File Report 5415, p. 210.

Sangster, D.F., 1978, Isotopic studies of ore-leads of the Circum-Kisseynew Volcanic Belt of Manitoba and Saskatchewan: Canadian Journal of Earth Sciences, v. 15, p. 1112–1121.

Schutts, L.D. and Dunlop, D.J., 1981, Proterozoic magnetic overprinting of Archean rocks in the Canadian Superior Province: Nature, v. 291, p. 642–645.

Schwarz, E.J. and Vaughan, D.J., 1972, Magnetic phase relations of pyrrhotite: Journal of Geomagnetism and Geoelectricity, v. 22, p. 463–470.

Smith, P.E. and Farquhar, R.M., 1988, Pb isotope systematics in whole rocks and minerals from dykes in the KSZ area: preliminary measurements: Lithoprobe-KSZ Workshop '88 Notes, p. 17–20.

Stauffer, M.R., 1984, Manikewan: an early Proterozoic ocean in central Canada, its igneous history and orogenic closure: Precambrian Research, v. 25, p. 257–281.

Stille, P. and Clauer, N., 1986, Sm–Nd isochron-age and provenance of the argillites of the Gunflint iron formation in Ontario, Canada: Geochimica et Cosmochimica Acta, v. 50, p. 1141–1146.

Strangway, D.W., 1964, Rock magnetism and dike classification: Journal of Geology, v. 72, p. 648–663.

Stupavsky, M. and Symons, D.T.A., 1982a, Isolation of early Paleohelikian remanence in Grenville anorthosites of the French River area, Ontario: Canadian Journal of Earth Sciences, v. 19, p. 819–828.

Stupavsky, M. and Symons, D.T.A., 1982b, Extent of Grenvillian remanence components in rocks of the Southern Province: Canadian Journal of Earth Sciences, v. 19, p. 698–708.

Stupavsky, M. and Symons, D.T.A., 1983, Penokean remagnetization of the basal Huronian Dollyberry Lake basalts near Elliot Lake, Ontario: Canadian Journal of Earth Sciences, v. 20, p. 49–56.

Symons, D.T.A., 1966, A paleomagnetic study on the Gunflint, Mesabi and Cuyuna iron ranges in the Lake Superior region: Economic Geology, v. 61, p. 1336–1361.

Symons, D.T.A., 1967, Paleomagnetism of Precambrian rocks near Cobalt, Ontario: Canadian Journal of Earth Sciences, v. 4, p. 1161–1169.

Symons, D.T.A., 1971, A paleomagnetic study of the Nipissing diabase, Blind River – Elliot Lake area, Ontario: Geological Survey of Canada, Paper 70-63, p. 19–30.

Symons, D.T.A., 1975, Huronian glaciation and polar wander from the Gowganda Formation, Ontario: Geology, v. 3, p. 303–306.

Symons, D.T.A. and O'Leary, R.J., 1978, Huronian polar wander and paleomagnetism of the Thessalon volcanics: Canadian Journal of Earth Sciences, v. 15, p. 1141–1150.

Thomas, M.D. and Gibb, R.A., 1985, Proterozoic plate subduction and collision: processes for reactivation of Archean crust in the Churchill Province, *in* Ayres, L.D., Thurston, P.C., Card, K.D. and Weber, W., eds., Evolution of Archean Supracrustal Sequences: Geological Association of Canada, Special Paper 28, p. 263–279.

Turek, A., 1967, Note on the age of sulphide mineralization at Lynn Lake, Manitoba: Canadian Journal of Earth Sciences, v. 4, p. 572–573.

Vandall, T.A. and Symons, D.T.A., 1987, Paleomagnetism of Archean granites in the Wawa area: further definition of the apparent polar wander path (abstract): Institute on Lake Superior Geology, v. 33, p. 81.

Van Schmus, R., 1965, The geochronology of the Blind River – Bruce Mines area, Ontario, Canada: Journal of Geology, v. 73, p. 755–780.

Van Schmus, W.R., Bickford, M.E., Lewry, J.F. and Macdonald, R., 1987, U–Pb geochronology in the Trans-Hudson Orogen, northern Saskatchewan, Canada: Canadian Journal of Earth Sciences, v. 24, p. 407–424.

Vellet, V., 1963, Geology of the Lynn Lake Ni–Cu deposits: Sherritt Gordon Mines Limited, internal report.

Watters, B.R. and Armstrong, R.L., 1985, Rb–Sr study of metavolcanic rocks from the La Ronge and Flin Flon domains, northern Saskatchewan: Canadian Journal of Earth Sciences, v. 22, p. 452–463.

U–Th–Pb Geochronology and Preliminary Interpretation of Precambrian Tectonic Events in the Black Hills, South Dakota

J.A. Redden, Z.E. Peterman, R.E. Zartman, and E. DeWitt
United States Geological Survey, Denver, Colorado 80225, U.S.A.

Redden, J.A., Peterman, Z.E., Zartman, R.E. and DeWitt, E., U–Th–Pb geochronology and preliminary interpretation of Precambrian tectonic events in the Black Hills, South Dakota, in Lewry, J.F. and Stauffer, M.R., eds., The Early Proterozoic Trans-Hudson Orogen of North America: Geological Association of Canada, Special Paper 37, p. 229–251.

Abstract

Complex sedimentological and tectonic events of Early Proterozoic age (2.5–1.6 Ga) have shaped the geological evolution of the Black Hills, western South Dakota. Rocks of the Wyoming province are represented in the Black Hills uplift by Late Archean (2.5 Ga) granites at Bear Mountain, along the western edge of the uplift, and near Nemo, along the northeastern edge. Initial Early Proterozoic failure of this basement is recognized near Nemo, where coarse clastic rocks derived from the west were deposited in a rift basin. These deposits were intruded by a 1-km-thick, gravity-differentiated metagabbro sill (2.17 Ga), and then folded and eroded prior to development of the main Black Hills basin. Strata of the unconformably overlying main basin are represented by fanglomerate near Nemo, derived from the east and deposited along listric faults, and onlap of similar rocks over Archean basement at Bear Mountain. In the eastern and southeastern Black Hills, shelf quartzite, either equivalent to or younger than the fanglomerate, is overlain by metabasalt, which intertongues with black shale to the west. These younger volcaniclastic rocks are correlated with the oldest rocks in the Lead area, where a tuff near the contact of quartzite in the Ellison Formation and carbonate–silicate–sulfide facies iron formation of the Homestake Formation has an age of 1.97 Ga. Quartzite of the Ellison Formation can be traced to the south into the Rochford area, where it changes laterally to deeper water turbidites. Similar early turbidites overlie the basalt section of the central Black Hills.

In the east-central Black Hills, an unconformity separates early turbidite deposits from younger conglomerate, debris flows, and alkalic basalts. Similar volcanism occurs in the Rochford area, where an alkalic tuff associated with carbonate–silicate facies iron formation has an age of 1.88 Ga. This approximately 90 Ma age gap with underlying rocks is correlated with the unconformity of the east-central Black Hills and indicates a similar unconformity or tectonic break in the Lead area. Debris flow deposits of the central and eastern Black Hills are overlain by turbidite deposits and pinch out to the west so that the depositional break cannot be followed everywhere. Tuffaceous sediment, converted to micaceous schist, forms a laterally extensive unit intertonguing with greywacke in the south-central Black Hills and is believed to be equivalent to the Rochford volcanism. The younger turbidite deposits fine upward to a thick monotonous shale, which is the youngest metasedimentary unit.

Following sedimentation, major east-trending, north-vergent nappes (F_1) developed. These were overprinted by north-northwest-trending folds (F_2), characterized by steeply dipping axial plane foliation and well-developed mineral lineation. A 1.85 Ga disturbance of the Little Elk Granite may indicate the age of F_2 folding and metamorphism. This regional metamorphism may have culminated in development and emplacement of the S-type Harney Peak Granite and associated pegmatite, at 1.71 Ga, which resulted in a major doming event in the Harney Peak area and superposition of extensive thermal metamorphism on earlier regional metamorphism. Rb–Sr whole-rock equilibration of alkalic metatuff and turbidite units, possibly related to a widespread nonpenetrative northeast-trending foliation, occurred at 1.55 Ga.

Regional interpretations indicate that the Black Hills area is part of reworked Archean basement of the Wyoming craton, which repeatedly failed, with development of intracontinental or possibly back-arc basins, during the Early Proterozoic. Basin closure directions varied, but largely reflect east–west compressional events associated with the Trans-Hudson Orogen.

Résumé

L'évolution géologique des Black Hills, à l'ouest du Sud Dakota est tributaire d'évènements tectoniques et sédimentologiques complexes d'âge Protérozoïque Inférieur (2,5–1,6 Ga). Dans les "Black Hills uplift," les roches de la province Wyoming sont représentées par des granites tardi-archéens le long de la marge occidentale et près de Nemo dans la partie NE. Près de Nemo, des sédiments grossiers dérivés de l'ouest et déposés dans un bassin en extension révèlent une première dégradation du socle d'âge Protérozoïque Inférieur. Ces dépôts sont recoupés par un sill de méta-gabbro (2,17 Ga) épais d'un kilomètre et différencié par gravité, puis plissés et érodés avant le développement du bassin principale de Black Hills. Près de Nemo, les strates du bassin discordant principal sont représentées par des conglomérats de deltaïques, dérivés de l'est et déposés le long de failles listriques; à Bear Mountain, des roches semblables surmontent le socle archéen. A l'est et sud-est de Black Hills, des quartzites de plateformes, équivalentes ou plus jeunes que les conglomérats de deltas, sont surmontées par des méta-basaltes interdigités à l'ouest avec des shales noirs. Ces roches volcano-clastiques jeunes sont correlées avec les plus vieilles roches de la région de Lead; un tuf près du contact avec les quartzites de la formation d'Ellison et de l'horizon ferrifère à carbonate–silicate–sulfure de la formation Homestake à un âge de 1,97 Ga. Les quartzites de la formation d'Ellison se poursuivent au sud dans la région de Rochford où ils passent latéralement à des turbidites d'eaux profondes. Des turbidites primaires semblables surmontent les basaltes des Black Hills central.

Dans les Black Hills est-central, une discordance sépare les turbidites primaires des conglomérats, coulées de débris et basaltes alcalins. Un volcanisme semblable se retrouve dans la région de Rochford où un tuf alcalin associé à un horizon ferrifère faciès carbonate–silicate est daté de 1,88 Ga. Ce hiatus de 90 Ma avec les roches sous-jaçentes est correlé avec la discordance des Black Hills est-central et indique une discordance similaire ou une rupture tectonique dans la région de Lead. Des dépôts de coulées de débris dans la partie centrale et est des Black Hills sont surmontés par des turbidites qui s'amincissent à l'est de sorte que la rupture de déposition n'est par continue. Dans les Black Hills sud-central, des sédiments tufacés, passant à des mica-schistes forment une unité étendue latéralement interdigitée avec des grauwackes et est considérée équivalente du volcanisme de Rochford. Les turbidites primaires passent vers le haut à des shales monotones, la plus jeune unité méta-sédimentaire.

Après la sédimentation, des nappes orientées est et nord (F_1) se développent. Elles sont reprises par des plis NNW (F_2) à foliation de plan axial redressés accompagnés de linéations. L'évènement de 1,85 Ga dû au granite Little Elk indique un âge pour les plis F_2 et leur métamorphisme. Ce métamorphisme régional pourrait avoir culminé pendant le développement et la mise en place du granite type-S Harney Peak et des pegmatites associées à 1,71 Ga résultant en un dôme avec superposition d'un métamorphisme thermique. Un réajustement Rb/Sr roche totale s'est produit vers 1,55 Ga dans des méta-tufs alcalins et des turbidites possiblement relié à une foliation NE très répandue et non-pénétrative.

Une interprétation régionale indique que les Black Hills font partie d'un socle archéen remanié se dégradant de façon répétitive avec le développement de bassins intra-continentaux et peut-être d'arrière-arcs pendant le Protérozoïque Inférieur. Les directions de fermeture des bassins varient mais reflètent largement une compression EW associée à l'orogène Trans-Hudson.

INTRODUCTION

Summaries of Precambrian tectonic evolution of the Black Hills have been of limited extent (Redden and Norton, 1975; DeWitt *et al.*, 1986), largely because of the complex geology and lack of precise age control for metasedimentary rocks that make up most of the core of the Black Hills. Until recently, ages of the layered rocks were only broadly constrained by geological relationships as Early Proterozoic or Late Archean. More than 60 years ago, Davis (1926) reported a chemically determined U–Pb age of 1667 Ma for uraninite from a zoned pegmatite associated with the Harney Peak Granite in the southern Black Hills, but concluded that the date "appears impossibly high since it is greater than any yet found for even the lower Precambrian." A later isotopic U–Pb age of 1620 ± 20 Ma (Wetherill *et al.*, 1956) on uraninite from the same locality and a Rb–Sr whole-rock age of 1707 ± 12 Ma (Riley, 1970) for the associated Harney Peak Granite indicated that the chemical age was remarkably good for the last Precambrian igneous event in the Black Hills.

Late Archean rocks in the Black Hills were recognized over 20 years ago. The Little Elk Creek Granite near Nemo (Fig. 1) was shown to be 2.56 Ga (Zartman *et al.*, 1964; Zartman and Stern,

1967), and granite at Bear Mountain, west of Harney Peak, was shown to be about 2.5 Ga (Ratté and Zartman, 1970). Both granites are on the edge of the exposed Precambrian core and were considered to be part of an extensive Archean terrane that had been deformed and metamorphosed together with the overlying rocks during the Early Proterozoic.

Additional metamorphism of the Archean rocks was indicated by Rb–Sr whole-rock reequilibration in the Little Elk Creek Granite (Zartman and Stern, 1967) and granite of Bear Mountain (Ratté and Zartman, 1970). A similar reequilibration in stratified rocks was suggested in the Lead area (Wolfgram, 1977, p. 20).

Geochronological data from vein material in the northern Black Hills, including some in the Homestake mine, suggested that some of the layered rocks were Early Proterozoic or Late Archean. Kulp *et al.* (1956) suggested, on the basis of an analysis of galena from the Homestake mine, that mineralization was Proterozoic, rather than Tertiary as previously advocated by Noble (1950), and that the host rocks were at least 1.5–1.7 Ga. Rye *et al.* (1974) interpreted lead isotope data from galena associated with quartz veins in the Homestake mine as indicating that the Homestake Formation contained lead derived from a 2512 ± 125 Ma source. Implicit in this interpretation was their assumption that galena in the quartz veins at the Homestake mine and in the Clover Leaf mine, about 12 km to the southeast, was emplaced at 1600 Ma. Their data, if interpreted using the plumbotectonics model of Doe and Zartman (1979) for Precambrian time (Stacey *et al.*, 1976), is consistent with an age of about 2.2 Ga for lead incorporated into the Homestake Formation. None of these data, however, gave precise ages of the metasedimentary rocks.

Until recently, igneous rocks suitable for dating were not recognized. Within the last 7 years, however, zircon-bearing metatuff and metagabbro sills have been identified and dated. The resultant ages establish that most layered rocks in the Black Hills are Early Proterozoic. In this paper, we report new results of U–Th–Pb zircon and monazite analyses for five Early Proterozoic units and present a preliminary tectonic model for the Black Hills during Late Archean and Early Proterozoic time. Although several metasedimentary sequences remain undated, geological relationships make it unlikely that their ages differ greatly from those of the other Early Proterozoic rocks.

ANALYTICAL PROCEDURE

Uranium–thorium–lead

Sized and abraded fractions of zircon for U–Th–Pb analysis were dissolved using a modified Krogh (1973) method of $HF–HNO_3$ bomb digestions, after which lead was purified by bromide-form anion-exchange resin. Lead concentration was determined by isotope dilution on an aliquot of the sample removed after bomb digestion, but otherwise subjected to parallel chemical treatment. A combined ^{235}U and ^{230}Th tracer was added to the entire sample prior to digestion, and U and Th were separated together on a nitrate-form anion-exchange resin from the elutriant off the lead-isotope composition column. The chemical procedure for the single monazite sample analyzed was similar to that of the zircon, except that initial dissolution was carried out on a hot plate with concentrated HCl + one drop of H_2SO_4.

Lead was loaded with H_3PO_4 and silica gel on a rhenium filament and analyzed by a 30 cm, computer-controlled mass spectrometer providing automated data reduction. Uranium and thorium were analyzed using a triple filament configuration by a 15 cm mass spectrometer. Element concentrations are considered accurate to ±0.5% (2σ), and the following uncertainties are assigned to the measured lead-isotope ratios: $^{206}Pb/^{204}Pb$, ±0.4% (100) to ±2% (10,000); $^{207}Pb/^{206}Pb$, ±0.09%; and $^{208}Pb/^{206}Pb$, ±0.13%. Common lead correcton has been made by first subtracting 0.2 ng of nominal analytical blank and then assuming the remaining ^{204}Pb to be that specified by Stacey and Kramers (1975) for lead of the appropriate age. A more detailed description of the analytical procedure, including results of replicate analyses of a zircon standard, is given in Zartman *et al.* (1986). All discordia intercept ages are based on three or more zircon fractions, and error assignment follows the procedure of Ludwig (1987).

Rubidium–strontium

Analytical procedures used in Rb–Sr analyses are summarized previously (Peterman *et al.*, 1986). Rb and Sr were determined on separate splits by isotope dilution with an HF and H_2SO_4 dissolution. Uncertainties in the concentration ratio ($^{87}Rb/^{86}Sr$) and the isotopic composition ($^{87}Sr/^{86}Sr$) are now estimated to be ±1 and ±0.02% respectively. Decay and isotopic constants used in the calculations are those recommended by the International Union of Geological Sciences (IUGS) Subcommission on Geochronology (Steiger and Jager, 1977). Regressions were made using ISOPLOT (Ludwig, 1987).

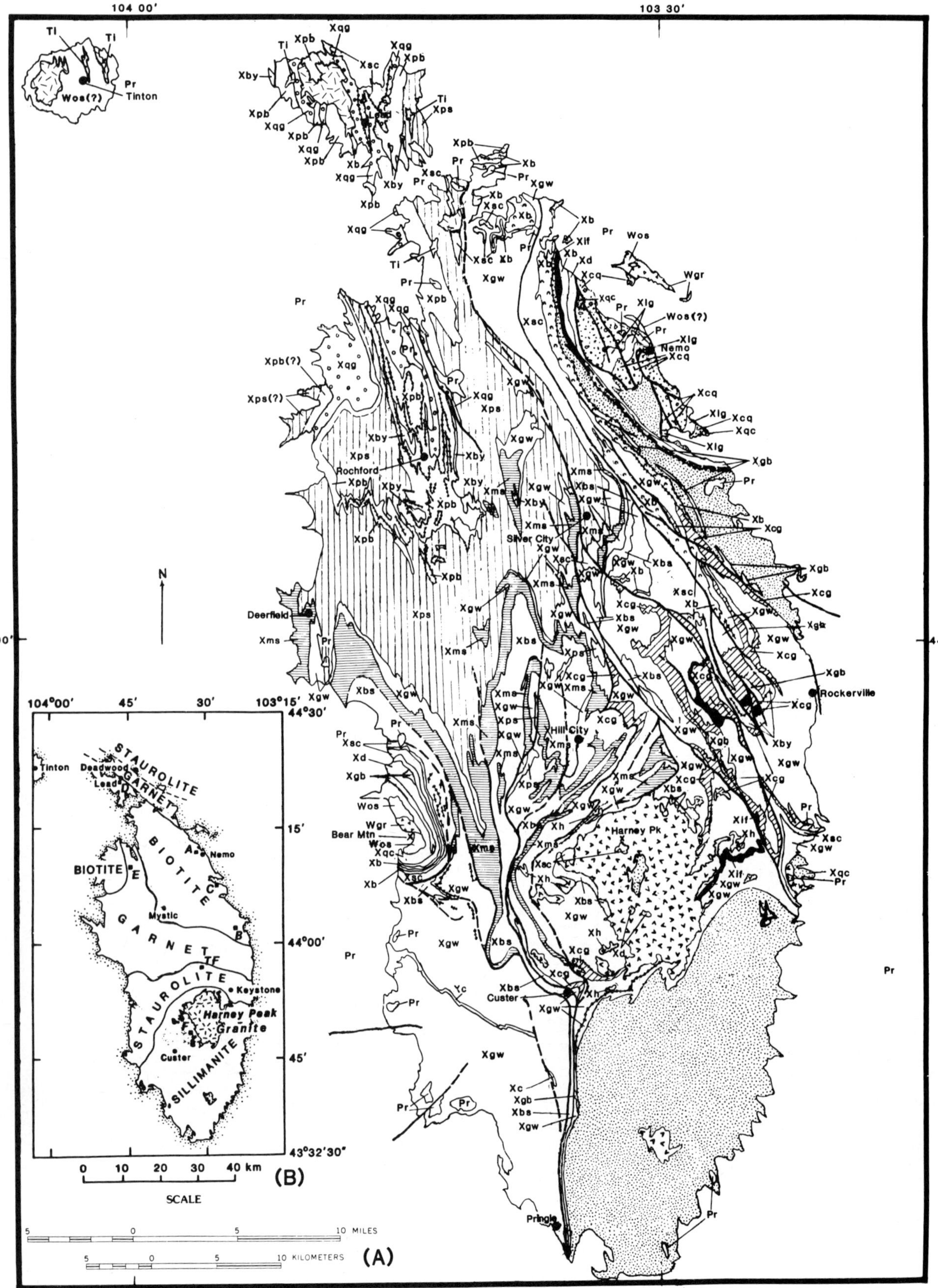
104 00'
103 30'
44 00'
Tinton
Lead
Nemo
Rochford
Silver City
Deerfield
Rockerville
Hill City
Bear Mtn
Harney Pk
Custer
Pringle
N
104°00'
45'
30'
103°15'
44°30'
15'
44°00'
45'
43°32'30"
STAUROLITE
GARNET
BIOTITE
GARNET
STAUROLITE
STAUROLITE
SILLIMANITE
Deadwood
Mystic
Keystone
Harney Peak Granite
0 10 20 30 40 km
(B)
SCALE
5 0 5 10 MILES
5 0 5 10 KILOMETERS
(A)

GEOLOGICAL SETTING

Precambrian rocks of the Black Hills form an approximately 40 km × 80 km oval-shaped domal core rimmed by Paleozoic and Mesozoic rocks (Fig. 1). The region was uplifted during the Laramide orogeny and eroded to about its present level by the close of the Eocene. The Precambrian rocks are predominantly Early Proterozoic metasedimentary units with both northwesterly and northeasterly structural trends and a locally dominant northwest-trending foliation. Locally, the northeasterly structural trend is difficult to recognize, in part because of the northwest-trending foliation. A major granitic dome formed by the Harney Peak Granite disrupts and modifies the structure of the Proterozoic metamorphic rocks in the southern Black Hills.

LATE ARCHEAN (ca. 2.5 Ga) ROCKS

The Archean granite at Bear Mountain (unit Wgr) is exposed in a small dome west of the younger Har-

DESCRIPTION OF MAP UNITS

Ti — Tertiary igneous rocks. Shown only where intruding Precambrian rocks

Pr — Phanerozoic rocks

EARLY PROTEROZOIC

Xh — Harney Peak Granite. Coarse-grained to pegmatitic muscovite granite and pegmatite

Xps — Phyllite and slate. Mica schist in Lead area. Includes Grizzly Formation (Dodge, 1942), Swede Gulch Formation (Bayley, 1972c), part of the Oreville Formation (Ratte' and Wayland, 1969), and a large area of rocks in the core of the Black Hills

Xc — Biotite schist, calcsilicate gneiss, and amphibolite. Crow Formation (Redden, 1963) southwest of Custer.

Xgb — Metagabbro. Sills and dikes of amphibolite or greenstone. At least two separate ages (≈1,883 Ma and 1,964 Ma) known and younger metagabbro intrudes the 1,884 Ma unit Xpb.

Xpb — Phyllite and biotite schist. Contains minor chert and amphibole-bearing rocks. Locally intruded by thin sills of metagabbro. Includes Poverty Gulch Slate, Nahant Schist, and Irish Gulch Slate in the Rochford area (Bayley, 1972c) and Northwestern and Flag Rock Formations in Lead area (Hosted and Wright, 1923)

Xms — Muscovite schist and phyllite. Includes part of the Oreville Formation. Equivalent to unit Xpb in Rochford and Lead areas

Xqq — Quartzite and metagraywacke. Quartzite and siliceous schist. Includes Ellison Formation in the Lead area (Hosted and Wright, 1923), Moonshine Gulch Quartzite north of Rochford (Bayley, 1972c), and siliceous graywacke west of Rochford. Possibly equivalent to part of Xgw and Xcg in central Black Hills. Unconformity or tectonic break at upper contact.

Xbs — Biotite schist or phyllite. Thin bedded and commonly garnet-rich schist. Largely includes parts of the Oreville Formation (Ratte' and Wayland, 1969, Ratte', 1986).

Xcg — Conglomeratic biotite schist and phyllite. Also siliceous biotite phyllite, garnetiferous schist, quartzite, and carbonate-silicate iron-formation. Unconformable lower contact.

Xby — Metabasalt near Rockerville. Includes metabasalt in Rochford, Lead and Rockerville areas

Xgw — Metagraywacke. Siliceous mica schist and impure quartzite at repeated stratigraphic intervals. Includes part of Roubaix Formation (Bayley, 1972b), Mayo and Bugtown Formations (Redden, 1968), and members of the Oreville Formation as well as unamed units in the central Black Hills.

Xb — Metabasalt. Amphibolite, greenstone, and actinolite schist. Included are Hay Creek Greenstone (Bayley, 1972c) and part of the Vanderlehr Formation (Redden, 1968)

Xif — Iron-formation. Iron-formation, ferruginous chert, and minor mica schist. Included are Homestake Formation (Hosted and Wright, 1923), Rochford Formation and Montana Mine Formation (Bayley, 1972c), and iron-formation in the Keystone area and in minor areas throughout the Precambrian core of the Black Hills

Xsc — Siliceous biotite phyllite, calcareous biotite phyllite, and schist. Includes the Poorman Formation (Hosted and Wright, 1923) and the Reausaw Slate (Bayley, 1972b) which intertongues with metabasalt (Xb)

Xd — Dolomitic marble and schist. Marble, phyllite, and calcareous phyllite. Includes the Roberts Draw Formation in the Nemo area (Redden, 1981) and part of the Vanderlehr Formation at Bear Mountain

Xqc — Quartzite and metaconglomerate. Conglomeratic, siliceous schist, feldspathic schist, and minor marble. In the Nemo area, map unit includes taconite conglomerate and is equivalent to the Estes Formation (Redden, 1981). In the eastern and southeastern part of the Black Hills, map unit includes the Buck Mountain Quartzite (Redden, 1981) and Gingrass Draw Slate (Bayley, 1972b) and the extensive quartzite southeast of the main Harney Peak Granite. In the Bear Mountain area, unit includes the lower part of the Vanderlehr Formation.

UNCONFORMITY

Xlg — Layered metagabbro. Gravity-differentiated Blue Draw Metagabbro sill (2,170 Ma, Redden, 1981)

Xcq — Metaconglomerate and quartzite. Conglomerate and feldspathic schist, biotite schist, taconite, and phyllite. Includes the Boxelder Creek Formation, Benchmark Iron-formation (Redden, 1981), and an unnamed carbonate-silicate-oxide-facies iron-formation along the west side of the Nemo area.

UNCONFORMITY

ARCHEAN

Wgr — Granite. Granite and gneissic granite. Includes the Little Elk Granite and the granite of Bear Mountain

Wos — Older metasedimentary rocks. Dominantly quartz-feldspar-biotite schist. Includes Nemo Iron-Formation (Redden, 1981). Schist in Tinton area is Archean (?)

Contact, dashed where approximately located

Fault, dashed where approximately located

Figure 1. *(A) Simplified geological map of Precambrian units in the Black Hills. (B) Map showing metamorphic isograds and dated sample locations in the Black Hills.*

ney Peak Granite (Fig. 1). This unnamed granite is leucocratic, fine grained to pegmatitic, peraluminous, and S-type, and, except for local features indicative of cataclastic deformation, closely resembles the much younger Harney Peak Granite. Its Rb–Sr whole-rock minimum age of 2415 ± 65 Ga (Zartman *in* Ratte, 1986), however, clearly implies an original Archean age. U–Pb ages of zircon from the granite are strongly discordant and have an upper intercept age of 2392 ± 230 Ma (Gosselin *et al.*, 1988). The granite consists of several small sills emplaced in bedded quartz–feldspar–biotite schist (unit Wos) that locally contains garnet, staurolite, and kyanite. Overlying this schist is a thin discontinuous metaconglomerate that contains granite and feldspar pebbles presumed to be derived from the granite. Thus, the metaconglomerate is considered to be Early Proterozoic.

The Little Elk Granite (unit Wgr) in the Nemo area, along the northeastern edge of the Precambrian core (Fig. 1), is dominantly porphyritic, coarse grained, biotitic, and commonly has a well-developed augen structure and pervasive north-northwesterly trending foliation. The granite has a revised U–Pb zircon age of 2549 ± 11 Ma based on the combined data of Zartman and Stern (1967) and more recent analyses by Gosselin *et al.* (1988). Stratified Archean rocks intruded by the Little Elk Granite consist of metamorphosed arkose, quartzite, pelitic schist, and meta-conglomerate (unit Wos). These rocks are intensely feldspathized immediately adjacent to the granite. Bedding in screens of similar rocks within the granite trends northeasterly. Nearby, poorly exposed, oxide-facies banded iron formation (Nemo Iron Formation) is also believed to be Archean, but its age is unproven. The coarse clastic character of most of unit Wos suggests a proximal source and is consistent with deposition in a rift basin. If the Nemo Iron Formation is also Archean, but younger than the clastic rocks, the depositional environment evolved from rift to a more stable basin.

Archean rocks may be present in a small area near the core of the Harney Peak Granite and in the Tinton area (Fig. 1). Mapping near Harney Peak (Redden, 1968, unpublished data, 1986) documents the same sequence of Early Proterozoic rock units that overlie Archean rocks in the Bear Mountain dome, although no basal conglomerate unit unconformably overlying recognizable Archean rocks was seen. However, because lithologic characteristics of the Harney Peak Granite and the granite at Bear Mountain are so similar, small bodies of Archean granite in the core of the former could easily be overlooked. Schistose metagreywacke in the isolated Tinton area (Fig. 1) is believed to be Archean because of its high biotite content and virtual absence of muscovite. High biotite/muscovite ratios are characteristic of Archean schists in the Black Hills, whereas schists derived from Proterozoic metagreywackes commonly have muscovite exceeding biotite.

EARLY PROTEROZOIC (PRE-2.17 Ga) ROCKS

The oldest Early Proterozoic stratigraphic sequence crops out near Nemo (Redden, 1981) and consists of a basal assemblage of chloritic metaconglomerate, metamorphosed fanglomerate, taconite meta-conglomerate, and chloritic quartzite (unit Xcq). Some of these rocks exhibit pronounced lateral facies changes to phyllite containing lenses of dolomitic marble. Overlying and interfingering with this basal assemblage is a tongue of pyritic, Au- and U-bearing fluvial metaconglomerate and quartzite, which grades upward into a thick section (>3300 m) of quartzite. All of these clastic rocks are included in the Boxelder Creek Formation as redefined by Redden (1981). An oxide-facies banded iron formation (Benchmark Iron Formation) at the top of this sequence is the youngest metasedimentary rock exposed below the angular unconformity at the base of unit Xqc (Tables I and II).

Age

The 1000-m-thick, gravity differentiated Blue Draw Metagabbro (Redden, 1981) Nemo sill (unit Xlg) emplaced in the upper part of the quartzite (unit Xcq) indicates that the enclosing rocks were subhorizontal when the sill was intruded. Good exposures of the unconformity at the top of unit Xcq demonstrate that one phase of folding occurred prior to the development of the unconformity (Redden, 1981, unpublished data, 1986), but after intrusion of unit Xlg. Four zircon fractions from coarse-grained granophyre in the upper part of the sill (Table II, Fig. 2A) are discordant, but with $^{207}Pb/^{206}Pb$ ages as old as 2131 Ma. Three fractions give virtually identical U–Pb dates, but the finest fraction, although having an unusually high content of common lead, is notably less discordant. A discordia defined by the four fractions has an upper intercept of 2170 ± 120 Ma. The large uncertainty and mean square of weighted deviates (MSWD) is partly due to excessive scatter of the three data points that are approximately coincident (Fig. 2A). The unconformity at the top of unit Xcq must, therefore, be younger than 2170 ± 120 Ma and is considered to mark the close of earliest Early Proterozoic sedimentation.

Source area and tectonic events

Diverse clasts in metaconglomerate of the lowermost part of the pre-2.17 Ga Early Proterozoic rocks indicate that a continental source area was being eroded during deposition of unit Xcq. Clasts of chert, oxide-facies iron formation, quartzite, and vein quartz are common in the metaconglomerate and there are sparse clasts of ferruginous dolomitic marble (Table II). Detrital chromite in the pyritiferous metaconglomerate indicates that ultramafic rocks were probably exposed in the Archean basement, although none now crop out. No Archean source areas are exposed for some of the detrital quartzite clasts. Feldspathic clasts are similar to some wall rocks of the Little Elk Granite, as are oxide-facies iron formation and chert clasts. Refolded folds in clasts of iron formation in the lower chloritic metaconglomerate of unit Xcq indicate that polyphase metamorphism and deformation took place prior to 2.17 Ga in the source area.

The oldest pre-2.17 Ga tectonic event that can be inferred from the depositional record is normal faulting and associated alluvial fan sedimentation (Table II). The pronounced facies changes in the conglomeratic strata suggest derivation from western source areas, but because these changes are observed only in cross-section and in blocks that have undergone extensive later deformation, a three-dimensional reconstruction of the geometry of the alluvial fans and source areas is not possible. Importantly, however, the metamorphosed fanglomerate and fluvial deposits of unit Xcq indicate an extensional continental environment in the Nemo area and northeastern Black Hills prior to 2.17 Ga. The Nemo area seems to have been on the actively rifting eastern margin of the Wyoming Archean craton. The eastern extent of the basin receiving sediment of unit Xcq is unknown.

A period of folding, which postdated development of the unconformity at the top of unit Xcq, has overturned the unconformity and rotated the plunges of early (preunconformity) folds past 90°. If the unconformity is rotated along an east–west axis to an original subhorizontal attitude, the preunconformity folds have north-northwesterly trends and low plunges. Deformation predating the unconformity at the top of unit Xcq could have been caused by east-northeast compression (Table II).

Pre-2.17 Ga metamorphosed fanglomerate, quartzite, and oxide-facies iron formation as found in the Nemo area are lacking in the Bear Mountain area. Although such deposits could have been eroded at some later date, it is more likely that the western Black Hills area was an exposed part of the Wyoming Archean craton and experienced erosion prior to 2.17 Ga (Table II).

EARLY PROTEROZOIC (1.715–2.17 Ga) ROCKS

Older, shallow-water deposits

Rock units deposited above the unconformity at the top of unit Xcq are all shallow-water deposits. In the Nemo area, the oldest of these units (Xqc) consists of metamorphosed fanglomerate, quartzite, metamorphosed arkose, dolomite-cemented metaconglomerate, clastic dolomitic marble, and slate and phyllite of the Estes Formation (Redden, 1981). The Estes was deposited along north-trending listric growth faults in three separate basins, which are progressively younger to the east. Because of deposition adjacent to actively rotating fault blocks, the Estes Formation ranges in thickness from about 50 m to as much as 3000 m along a 3 km strike length in the most well-exposed middle basin. Metamorphosed arkose in the upper part of unit Xqc is almost certainly derived from the Archean Little Elk Granite, whereas abundant metamorphosed fanglomerate in the lower part is almost totally derived from the Boxelder Creek Formation and Benchmark Iron Formation (unit Xcq).

The Estes grades upward from metamorphosed boulder fanglomerate into finer grained, metamorphosed, shallow-water clastic sediments and carbonate rocks (unit Xd) of the Roberts Draw Formation (Redden, 1981). West of unit Xd in the Nemo area is a generally north-trending belt of vertically dipping metagabbro sills or dykes (unit Xgb) and minor amounts of phyllite and marble. To the north, this belt includes pillowed amygdaloidal basalt (unit Xb).

An inferred fault bounds the western side of this belt of metagabbro intrusives. To the west of the fault, metasedimentary rocks consist of a thin unnamed oxide–carbonate–silicate-facies iron formation (Bayley, 1972b; Fantone, 1983) that is overlain by shelf-facies quartzite, phyllite, and schist of the Buck Mountain Quartzite (Bayley, 1972b). The basal contact of the Buck Mountain Quartzite is disconformable, and the unit onlaps the iron formation and decreases in thickness to the north. Locally, the Buck Mountain contains metaconglomerate and metamorphosed arkose identical to metaconglomerate and metamorphosed arkose in the Estes. Consequently, the Buck Mountain is believed to be stratigraphically equivalent to the Estes, but deposited upon different underlying rocks. Hence, the Buck Mountain is also shown as unit Xqc (Fig. 1, Table I). The unnamed iron formation is tentatively correlated with the oxide-facies iron formation in the pre-2.17 Ga rocks (Benchmark Iron Formation). Unit Xqc, west of the inferred fault, includes rocks of the Gingrass Draw Slate (Bayley, 1972b), a unit of largely mafic tuff presumably deposited in relatively deep water. Quartzite in the

Table I
Tentative correlation of Precambrian rock unit protoliths in various parts of the Black Hills

Bear Mountain (formation names from Ratté (1986) and Redden (1968))	Rochford (formation names from Bayley (1972c))	Lead (formation names from Hosted and Wright (1923))	Central Black Hills (formation names from Redden (1963), Ratté and Wayland (1969), and Bayley (1972b))	Nemo (formation names as redefined by Redden (1981))
			Harney Peak Granite (Xh) 1.715 ± 3 Ma	
Shale (Xps) part of Oreville Formation	Swede Gulch Formation (Xps) shale. Gabbroic sills (Xgb)	Grizzly Formation (Xps) shale	Shale (Xps) Gabbroic sills (Xgb)	
Greywacke (Xgw) Mayo Formation southwest of Custer. Not present east of Bear Mountain				
Crow Formation (Xc) tuffaceous rocks				
Greywacke (Xgw) Bugtown Formation and Zimmer Ridge Member, Oreville Formation			Greywacke (Xgw). Bugtown Formation and part of Oreville Formation	
Altered tuffs (Xms). Part of Oreville Formation	Poverty Gulch Slate (Xpb) shale and chert. Rochford Formation (Xif) carbonate–silicate banded iron formation. Nahant Schist and Rapid Creek Greenstone (Xby) basalt and black shale. Montana Mine Formation (Xif) carbonate–silicate banded iron formation 1884 ± 29 Ma. Irish Gulch Formation (Xpb) shale and tuff	Flag Rock Formation (Xpb, Xby) shale, basalt, chert	Altered tuffs (Xms). Part of Oreville Formation. Gabbroic sills at Prairie Creek (Xgb) 1883 Ma ± 5 Ma	
Black shale and distal greywacke (Xbs). Part of Oreville Formation			Greywacke (Xgw). Part of Oreville Formation. Black shale and distal greywacke (Xbs). Part of Oreville Formation. Greywacke (Xgw). Part of Oreville Formation Debris flow conglomerate to east; sandstone, carbonate–silicate banded	

Unconformity or fault			iron formation and basalt near Rockerville (Xcg, Xby)	
	Unconformity	Unconformity	Unconformity	
	Moonshine Gulch Quartzite (Xqg) sandstone changing to greywacke toward southwest	Northwestern Formation (Xpb) shale. Ellison Formation (Xqg) sandstone and siltstone shelf facies 1974 ± 8 Ma. Homestake Formation (Xif) carbonate–silicate banded iron formation. Poorman Formation (Xsc) black shale and possible basalt	Greywacke (unnamed) (Xgw)	—Fault—
Upper Vanderlehr Formation. Basaltic volcaniclastic rocks (Xb), black shale with carbonate–silicate banded iron formation (Xsc)			Hay Creek Greenstone (Xb) pillowed basalt and lenses of carbonate–silicate banded iron formation. Reausaw Slate (Xsc) black shale. Gabbroic sill at Bogus Jim Creek (Xgb) 1964 ± 15 Ma	Amygdular pillow basalt (Xb)
Lower Vanderlehr Formation. Dolomite and shale (Xd)				Roberts Draw Formation (Xd) dolomite and siltstone
Lower Vanderlehr Formation. Sandstone, arkose, conglomerate (Xqc)			Buck Mountain Quartzite and Gingrass Draw Slate (Xqc) sandstone and siltstone shelf facies	Estes Formation (Xqc) fanglomerate, banded iron formation conglomerate sandstone, arkose, and dolomite
—Unconformity—			—Unconformity—	—Unconformity—
				Blue Draw Metagabbro. Gabbroic sills (Xlg) 2170 ± 120 Ma
			Carbonate–silicate banded iron formation (Xif)	Benchmark Iron Formation (Xcq) oxide-facies banded iron formation
			—Fault—	Boxelder Creek Formation (Xcq) banded iron formation conglomerate, fanglomerate, quartzite, lithic arenite
				—Unconformity—
Granite, pegmatite (Wgr) 2415 ± 65 Ma. Siltstone and greywacke (Wos)				Little Elk Granite (Wgr) 2549 ± 11 Ma. Arkose, conglomerate, quartzite, oxide banded iron formation (Nemo Iron Formation) (Wos)

Note: Lithologic abbreviations from Figure 1.

Table II

Late Archean and Early Proterozoic tectonic events in the Black Hills

Tectonic events	Source area; type of clasts	Depositional setting; type of sediments	Igneous activity	Age constraints (Ma)
Northwest–southeast compression, development of local late foliation		Isotopic reequilibration		1540–1700 (?)
Contact metamorphism doming, attenuation of existing folds			Harney Peak Granite	1715 ± 3
East-northeast–west-southwest compression, isoclinal folding, regional metamorphism, development of F_2				1715–1850 (?), but post-F_1
Northwest–southeast compression, isoclinal folding, development of F_1, thrust faulting?				1715–1850 (?), but pre-F_2
Subsidence and probably local faulting associated with volcanism	Turbidite sources to east and probably southwest; largely sedimentary and metamorphic provenance. Proterozoic source locally; quartzite and basaltic clasts, Rockerville; Archean(?) source to north; cannibalization of early Proterozoic probable	Marine turbidite basin; fining-upward strata include tuffaceous shale; stable marine slope; minor rift; debris flows, and conglomerate in east-central Black Hills; volcanism in Rockerville and Rochford areas	Tuff in Montana Mine Formation. Basalt flows; gabbro sills (Prairie Creek)	1884 ± 29 1883 ± 5
Deformation and uplift; folds poorly constrained, but apparently east–west compression	—Unconformity—		Gabbro sill at Bogus Jim Creek	1964 ± 15
Major subsidence of Black Hills basin	Turbidite sources to east–southeast, possibly Archean craton; sedimentary and metamorphic provenance	Marine slope; turbidite deposits uppermost; equivalent shallow water quartzite in Lead area deepening toward Rochford and central Black Hills	Tuff (Lead area)	1974 ± 8

	Volcanic centres in eastern and northern parts of Black Hills	Deep marine; pillow lavas, black shale	Tholeiitic basalt	No available age
		Marine shelf-slope; tuff and black shale at Bear Mountain		
	Submerged source	Marine shelf; carbonate rocks at Nemo, Bear Mountain, and Harney Peak		
Ensialic rifting; north–south listric faulting in Nemo area, uplift to east	Archean and pre-2.17 Ga; source to east, marine transgression; metamorphic and sedimentary clasts. Little Elk Granite and granite at Bear Mountain exposed	Marine shelf and shelf-slope; fanglomerate at Nemo, sandstone of Buck Mountain Quartzite to west; onlapping marine sandstone and arkose at Bear Mountain		Pre-1964 ± 15
	—Unconformity—			
Northeast–southwest compression, early (pre-F_1) folding				
			Blue Draw metagabbro	2170 ± 110
Ensialic rifting; faulting in Nemo area, uplift to west	Wyoming Archean craton to west with possible ultramafic source; quartzite, chert, dolomite, oxide banded iron formation clasts	Marine alluvial and fluvial fans of Nemo area; fining-upward sequence of sandstone dolomite, and oxide banded iron formation		
	—Unconformity—			
Compression (?), orientation unknown				
			Little Elk Granite	2549 ± 11
			Granite at Bear Mountain	2415 ± 65
Subsidence	Archean craton	Intracratonic basin (?), oxide banded iron formation, arkose, conglomerate (sequence unknown)		

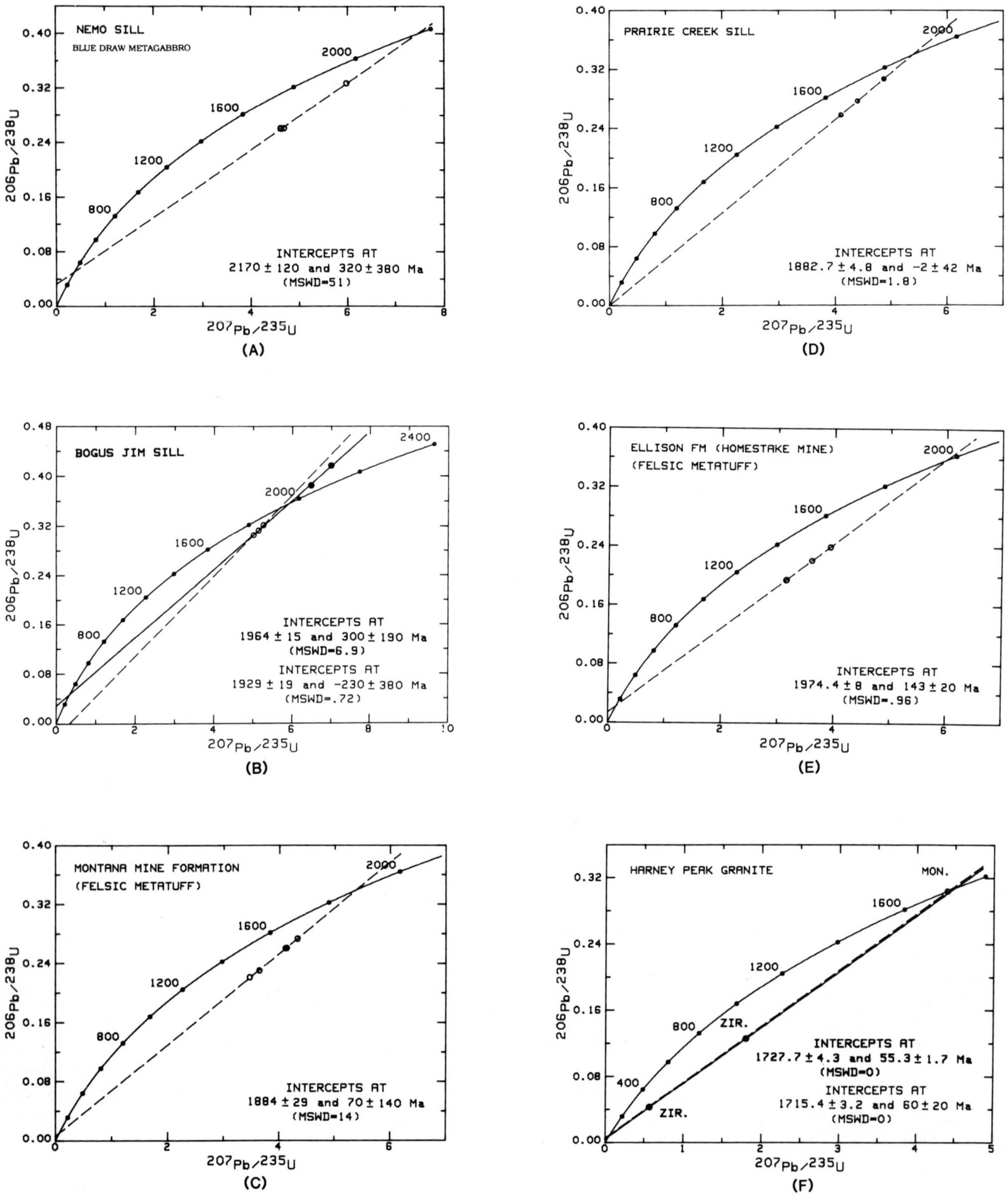

Figure 2. *Concordia diagrams for new zircon and monazite age determinations in Black Hills rocks. (A) Blue Draw Metagabbro sill (unit Xlg) in Nemo area. (B) Metagabbro sill (unit Xgb) near Bogus Jim Creek. (C) Metatuff in Montana Mine Formation (unit Xif) near Rochford. (D) Metagabbro sill (unit Xgb) near Prairie Creek. (E) Metatuff in Ellison Formation (unit Xqg) near Lead. (F) Harney Peak Granite (unit Xh) south of Sylvan Lake.*

geologically poorly known area of Custer State Park, southeast of Harney Peak, is included in unit Xqc even though Paleozoic rocks locally conceal the continuity of the quartzite.

Although no pre-2.17 Ga rocks were deposited in the Bear Mountain area, the post-2.17 Ga section contains a lensoid basal metaconglomerate overlain by metamorphosed arkose and quartzite, also shown as unit Xqc. These clastic rocks represent a transgressive marine section that fines upward into slate, phyllite, and thin dolomitic marble of part of the Vanderlehr Formation (Redden, 1968; Ratté, 1986), shown as unit Xd (Fig. 1). The sequence at Bear Mountain is much thinner than the Estes and Roberts Draw sections at Nemo, but the two are believed to be equivalent because of their lithologic similarity and similar overlying rocks (Table I). Thin screens of units Xqc and Xd are also present in the core of the Harney Peak Granite dome.

Age and tectonic environment. No metasedimentary units in the older, shallow-water sequence have been directly dated. The Buck Mountain Quartzite, however, is intruded by many metagabbro sills, as is the Roberts Draw Formation. These sills are of more than one age, however, as field relations indicate that rock units younger than the Buck Mountain Quartzite and Roberts Draw Formation are also cut by metagabbro dykes and sills and, within the belt of metagabbro, crosscutting relationships suggest two ages of intrusive activity. Sparse blue quartz-bearing phases of some metagabbros typically contain zircon and ages have been determined (Table III) for a sill mapped by Fricke (1982) at Prairie Creek, about 8 km north-northwest of Rockerville (Fig. 1). An age is also determined for another sill along Bogus Jim Creek about 10 km south-southeast of Nemo (Fig. 1). The Prairie Creek sill intrudes the Buck Mountain Quartzite and the Bogus Jim Creek sill is part of the belt of metagabbros that intrudes phyllite and carbonate rocks adjacent to the Nemo area.

Three zircon fractions from the metagabbro sill at Prairie Creek define a discordia with an upper intercept of 1883 ± 5 Ma and a lower intercept of about 0 Ma (Table III, Fig. 2D). As discussed in a later section, this age is close to that of a tuff in the Rochford area (Fig. 1). Because metagabbro bodies cut many of the rocks in the southern and eastern Black Hills, this 1.88 Ga age is interpreted as representing a widespread episode of gabbroic intrusion. Certainly, the host rocks of the Prairie Creek sill are older than 1.88 Ga and younger than 2.17 Ga.

Three zircon fractions from the metagabbro sill at Bogus Jim Creek (Fig. 1) are normally discordant and two are reversely discordant (Table III, Fig. 2B). One of the three normally discordant zircon fractions was abraded, but its U–Pb ages increased only slightly. All five fractions define a chord with an upper intercept at 1964 ± 15 Ma and a lower intercept of 300 ± 190 Ma. If both the normally and the reversely discordant zircons reflect the same event, the upper intercept age is valid. If the reversely discordant points are deleted, the upper intercept is 1930 Ma, but the chord has a lower intercept of -250 Ma. If the lower intercept is forced through zero, the upper intercept is 1940 ± 9 Ma. Thus, the intercept age is relatively insensitive to which regression is used and to whether or not the reversely discordant points are included. The data are interpreted as indicating that the gabbro sill at Bogus Jim Creek was emplaced about 1964 ± 15 Ma. This age indicates that the Roberts Draw Formation, which overlies the Estes Formation in the northeastern Black Hills, is older than 1.96 Ga, but younger than 2.17 Ga. It is unlikely that the Buck Mountain Quartzite is older than the Estes; thus, these zircon results indicate that the older shallow-water deposits (unit Xqc) are between 1.964 ± 15 Ma and 2.17 Ga. Because basalt flows are associated with the belt of metagabbro, the 1.96 Ga age is interpreted as representing a minimum age for the basaltic rocks (unit Xb) overlying the Roberts Draw.

The tectonic environment leading to the development of the Estes Formation is clearly that of an intracontinental rift basin, as indicated by the development of northwest-trending growth faults, uplift to the east, and deposition of marine fanglomerate, which changed to shelf facies sandstone farther to the west and south (Table II). Sediment in the Estes indicates that the northeastern Black Hills was on a rift margin, but a well-developed shelf area existed to the south-southeast and shallow-water conditions extended to the southwest, into the Bear Mountain area. These events in the Nemo area mark the beginning of the main Black Hills depositional basin.

Older volcanic and sedimentary rocks

To the west of, and stratigraphically above, the Buck Mountain Formation and Gingrass Draw Slate (unit Xqc) is a sequence of pillowed, tholeiitic metabasalt (unit Xb); black slate and phyllite (unit Xsc); and associated lenses of carbonate–silicate facies iron formation (unit Xif), which Bayley (1972b) included in the Hay Creek Greenstone and Reausaw Slate. The equivalent metabasaltic unit above the Roberts Draw Formation (unit Xd) in the Nemo area is amygdaloidal, suggesting a shallow water origin. Metabasaltic flows pinch out into unit Xsc north of Silver City, and in a general way the relative amount of basalt increases to the east and north. No pillow lavas are known in the Bear Mountain area, but tuffaceous rocks in a cherty

Table III

U–Th–Pb analytical data and ages for zircon and monazite from stratified and plutonic units in the Black Hills

Mesh size	Concentration (ppm) U	Th	Pb	Isotopic composition of lead (atom per cent) ^{204}Pb	^{206}Pb	^{207}Pb	^{208}Pb	Age (Ma) $\frac{^{206}Pb}{^{238}U}$	$\frac{^{207}Pb}{^{235}U}$	$\frac{^{207}Pb}{^{206}Pb}$	$\frac{^{208}Pb}{^{232}Th}$
Blue Draw Metagabbro sill, Nemo											
-100 +150	846	342	246	0.0445	78.05	10.59	11.32	1496	1752	2072	1538
-200 +250H	1121	339	313	0.0146	80.75	10.71	8.528	1496	1765	2101	1618
-200 +250L	1148	339	317	0.0099	81.22	10.60	8.178	1492	1754	2082	1602
-250 +325	893	nd	450	0.4340	62.60	14.01	22.95	1826	1973	2131	—
Metagabbro sill near Prairie Creek											
-50 +100	279	492	120	0.0359	59.90	7.624	32.44	1668	1795	1945	1650
-100 +200	608	1204	229	0.0210	59.53	7.150	33.30	1481	1656	1886	1356
Metagabbro sill near Bogus Jim Creek											
-50 +100	456	1140	219	0.0060	56.35	6.794	36.85	1754	1842	1943	1533
-50 +100A	403	919	193	0.0077	58.08	6.995	34.92	1798	1863	1936	1584
-100 +150	449	1103	270	0.0080	55.33	6.858	37.80	2100	2043	1986	1980
-150 +250	459	1131	216	0.0108	56.13	6.827	37.03	1717	1821	1942	1527
Metatuff unit of Ellison Formation, Lead											
-100 +150A	281	188	79.1	0.0809	74.32	9.921	15.68	1384	1617	1940	1197
-150 +200	345	238	82.0	0.1249	72.40	10.16	17.31	1147	1444	1914	989
-250 +325	413	265	93.9	0.0717	74.57	9.698	15.66	1144	1442	1913	1032
- 325	346	211	88.3	0.0570	75.26	9.653	15.03	1287	1549	1928	1207
Metatuff unit of Montana Mine Formation near Rochford											
Total	176	161	54.9	0.0425	72.85	8.936	18.17	1496	1661	1877	1254
-100 +150	136	112	42.5	0.0565	72.68	9.074	18.19	1494	1657	1870	1354
-100 +150A	135	114	45.5	0.0917	71.26	9.406	19.24	1558	1698	1875	1398
-200 +250	143	112	39.0	0.0599	73.39	9.225	17.33	1337	1560	1875	1172
- 325	174	145	44.8	0.0232	74.21	8.756	17.01	1288	1521	1861	1110
Harney Peak Granite near Sylvan Lake											
-100 +150S	1234	438	202	0.3397	71.49	12.02	16.15	767	1046	1685	446
-100 +150E	4585	1608	282	0.3651	66.36	11.36	21.92	274	459	1537	361
Monazite	8499	69390	7920	0.0009	28.20	2.975	68.82	1709	1712	1715	1702

Notes: The analyzed mineral is zircon unless otherwise noted. Decay constants: $^{238}U = 1.55125 \times 10^{-10}$ year^{-1}; $^{235}U = 9.8485 \times 10^{-10}$ year^{-1}; $^{232}Th = 4.9375 \times 10^{-11}$ year^{-1}; and $^{238}U/^{235}U = 137.88$. The isotopic composition of common lead is assumed to be $^{204}Pb:^{206}Pb:^{207}Pb:^{208}Pb$ = 1:15.3:15.2:34.9 for all samples except Harney Peak Granite, for which it is assumed to be $^{204}Pb:^{206}Pb:^{207}Pb:^{208}Pb$ = 1:15.7:15.3:35.4. Abbreviations: H, density greater than 4.6; L, density less than 4.6; A, abraded fraction; S, stubby crystals; E, euhedral crystals; and nd, not detected black slate (unit Xbs) that overlies unit Xd are correlated with the older volcanic rocks of the central Black Hills (Table I). Similar cherty black slate and tuffaceous rocks are noted above unit Xd in large inclusions in the centre of the Harney Peak Granite.

A black slate and phyllite unit (Xsc) is also exposed in the Lead area (Table I), where it was named the Poorman Formation (Noble and Harder, 1948). Metabasalt is present within this unit in the Homestake mine area and the Homestake Formation, a carbonate–silicate–sulphide iron formation, host of the gold ore in the mine, overlies the Poorman Formation. Above the iron formation is a quartzite and phyllite/schist unit, named the Ellison Formation by Noble and Harder (1948). This quartzitic unit (unit Xqg) was traced southward from the Lead area by Bayley (1972a) into the Rochford area, where it was correlated with the Moonshine Gulch Quartzite (Bayley, 1972c). Recent mapping by J.A. Redden (unpublished data) and Weissenborn (1987) indicates that the Moonshine Gulch is equivalent to a quartzose metagreywacke mapped by Bayley (1972c) northwest of Rochford (Fig. 1). Apparently, the shallow water Ellison Formation at Lead changes facies to deeper water metagreywacke to the south.

Age and tectonic environment. Although the tholeiitic metabasalt (unit Xb) has not been dated, its age can be approximated if the rock units are correctly correlated in Table I. In the lowest part of the Ellison Formation at Lead, virtually in contact with the Homestake Formation, a bed of metatuff contains zircon. Four zircon fractions from the metatuff, one of which was abraded, are 20–30% discordant, but a discordia defined in the data has an upper intercept of 1974 ± 8 Ma and a lower intercept of 143 ± 20 Ma (Table III, Fig. 2E). The 1974 ± 8 Ma age of the metatuff from the lowermost Ellison is probably essentially the age of the Homestake Formation inasmuch as tuffaceous sediment typically is associated with the carbonate–silicate iron formations elsewhere in the Black Hills. Presumably, the Poorman Formation (unit Xsc) and units Xsc and Xb in the east-central Black Hills are not much older. The apparently equivalent east-central Black Hills rocks are locally cut by metagabbro sills, but none are suitable for dating. Because there are dated events of gabbroic emplacement (1964 and 1883 Ma) and similar gabbroic rocks intrude rocks younger than the basalts, it is likely that the basalt (unit Xb) is at least 1964 Ma old and the 1974 Ma age at Lead is more likely to be its approximate age.

Turbidite and younger rocks

Much of the central and southwestern part of the Black Hills is underlain by metamorphosed deep-water turbidite deposits or metagreywacke (unit Xgw) and lesser amounts of metamorphosed tuffaceous rock and shales (units Xms and Xbs), which overlie units Xb, Xbs, and Xsc (Fig. 1). Excellently preserved primary structures indicate that metagreywacke overlies the older metabasaltic rocks (unit Xb). Metagreywacke in the area north and northeast of the Harney Peak Granite generally thins from east to west. Locally, at least four separate tongues of metagreywacke are documented.

In the Hill City area, a unit (Xbs) of garnetiferous rocks derived from black shale and distal turbidites, which is part of the Oreville Formation of Ratté and Wayland (1969), is present within the metagreywacke section. To the southwest, where intervening metagreywacke lenses out, this unit is in contact with mica schist (unit Xms). This latter unit is widely distributed in the central Black Hills and is believed to be derived from altered tuff. Unit Xms also lenses out to the north and east into metagreywacke. To the northwest, toward Rochford, unit Xms loses its identity as a mappable unit due to intertonguing with metamorphosed coarse volcaniclastic material (unit Xpb). The metagreywacke overlying unit Xms pinches out to the west (central part of Fig. 1). This greywacke may be equivalent to the one southwest of Custer (Bugtown Formation, Table I). Southwest of Custer, the thin Crow Formation (Xc), probably derived in part from alkalic tuffaceous rock, overlies proximal greywacke of the Bugtown Formation and in turn is overlain by a thick, more distal greywacke of the Mayo Formation. Because of a major fault, which passes through Custer (Fig. 1), these units cannot be directly correlated with those in the central Black Hills, but may be equivalent to the upper part of the turbidite section. The Mayo is cut by many metagabbro bodies, but none are suitable for dating.

In the east-central part of the Precambrian core, metaconglomerate, metamorphosed debris flows, and quartzite (unit Xcg) onlap quartzite of the Buck Mountain Quartzite (unit Xqc) and contain clasts of the same quartzite, indicating at least a local unconformity. Unit Xcg pinches out to the west and north in the central Black Hills, where it intertongues with metagreywacke; it reappears south of the Harney Peak Granite dome (Fig. 1) and although locally interrupted by faults, apparently rims the dome around Harney Peak. West of Rockerville, a volcanic centre is indicated by metabasalt (unit Xby) and associated metamorphosed volcaniclastic rocks, debris flows, and carbonate-facies iron formation. These latter lithologies are included with unit

Xcg. To the north, in the area about 9 km east of Silver City (Fig. 1), unit Xcg apparently unconformably overlies inverted greywacke and older metavolcanic rocks, which are on the west limb of a north-trending syncline. This suggests a major depositional/tectonic break. However, the unconformity cannot be mapped in the central Black Hills, where unit Xcg appears to pinch out. In the Rochford area, however, there is an apparent equivalent break in the deposition, and phyllite and garnetiferous schist (unit Xpb), metabasalt and metamorphosed volcaniclastic rocks (unit Xby), and two carbonate–silicate facies iron formations (unit Xif) overlie considerably older quartzites (Moonshine Gulch Quartzite, unit Xqg). These younger rocks are equivalent to the Flagrock, Northwestern, and part of the Grizzly formations of Noble and Harder (1948) in the Lead area (Table I).

Rocks younger than the Moonshine Gulch in the Rochford area are part of a metavolcanic sequence with pronounced lateral facies changes and coarse volcaniclastic debris, which indicates the area near Rochford was a volcanic centre. Metabasaltic pillow lavas are similar in composition to other basalts in the Black Hills, but metamorphosed tuffaceous strata associated with the Montana Mine Formation (Bayley, 1972c) are alkalic, as indicated by high zirconium and niobium concentrations. According to Bayley (1972c), most tuffaceous material is older than the basalt, but the complex structure and lateral facies changes allow other stratigraphic interpretations to be made. These metamorphosed alkalic tuffs and associated volcaniclastic rocks are not shown separately, but are included in units Xpb and Xif (Fig. 1). Unit Xpb and the included iron formation, metatuff, and metabasalt are correlated with unit Xms and the younger greywacke units previously described to the east and south (Table I) and also in a general way with unit Xcg. Sills and dykes of metagabbro intrude rocks as young as unit Xpb in the Rochford area, but are not shown separately.

Above the metavolcanic rocks and iron formation in the Rochford area and above the metagreywacke in the north-central part of the Black Hills (Fig. 1) is a thick monotonous unit of phyllite and slate (unit Xps). This youngest metasedimentary unit has been named the Grizzly and Swede Gulch formations in different parts of the Black Hills (Table I). Except for some of the very thick metagreywacke units of the central Black Hills, unit Xps is the only metasedimentary sequence not known to be cut by metagabbro sills or dykes.

Age and tectonic environment. The oldest metagreywacke that directly overlies the older metavolcanic rocks is younger than the volcanic rocks whose age is considered to be approximately 1.97 Ga. Other turbidite sections in the Black Hills contain no metavolcanic rocks and are undated.

Alkalic metatuff interbedded with the carbonate–silicate facies Montana Mine Formation in the Rochford area (included in unit Xpb, Fig. 1) contains abundant zircon, which yields a rather precise depositional age. Data from five fractions of zircon from two samples, one fraction of which was abraded (Table I, Fig. 2C) define a discordia with an upper intercept of 1984 ± 29 Ma and a lower intercept of 70 ± 140 Ma. This age of 1884 ± 29 Ma is about 90 Ma younger than the nearby underlying Moonshine Gulch Quartzite (unit Xqg), and indicates either a structural break or unconformity between the volcanic rocks and the Moonshine Gulch Quartzite. An unconformity is favoured because of the relationships described relative to the conglomerate and debris flow deposits (unit Xcg) and the younger volcanic rocks in the Rockerville area. Also, Noble and Harder (1948) believed that an unconformity in the Lead area was responsible for the lateral disappearance of the Northwestern Formation (part of unit Xpb), which overlies the Ellison Formation. However, because of incomplete exposures, extreme structural complexity, and abundant facies changes in the Rochford area, an early, unrecognized folded thrust fault cannot be precluded to explain the 90 Ma age discrepancy. The age data and the previously described unconformity below unit Xcg in the east-central Black Hills imply that a tectonic event occurred during this interval. The limited evidence for a preunconformity north-trending fold was noted in the preceding section, but little else is known about the tectonic event. The preceding uncertainties indicate the need for additional dates throughout the turbidite section. At present, however, the upper part of the layered rocks, including the upper turbidites, is interpreted as being approximately 1.87 Ga in age.

HARNEY PEAK GRANITE

The youngest Precambrian rock is the Harney Peak Granite described in detail by Redden *et al.* (1982). The granite is leucocratic, peraluminous, and texturally very heterogeneous, ranging from fine-grained to pegmatoid textured layers, and consists of hundreds of separate intrusive bodies and thousands of granitic and pegmatitic bodies in the surrounding metamorphic rocks. Lithological and chemical characteristics of the Harney Peak fit the classic definition of an S-granite (Chappell and White, 1974).

Emplacement of the granite was accompanied by contact metamorphism and development of a thermal aureole 3-5 km wide on the north side of the main granite mass and possibly much wider on the south side (Fig. 1B). Intrusion of the granite also resulted in a major domal structure near Harney Peak (Fig. 1) and several smaller domes. The country rock generally has a well-developed schistosity characterized by intensely flattened aggregates of sillimanite or andalusite and well-oriented micas within less than a kilometre of the granite contact. The schistosity is believed to have formed by outward and upward growth of the dome during emplacement of many separate granite bodies and contrasts with the more random orientation of many of the micaceous minerals over most of the thermal aureole.

Riley (1970) reported a whole-rock, Rb–Sr isochron age of 1697 ± 33 Ma for the Harney Peak Granite (recalculated using decay constants given in Steiger and Jager (1977). Six whole-rock samples of the Harney Peak collected by Walker *et al.* (1986) had Sm–Nd isotopic ratios that agreed with a 1.7 Ga age for the granite, but two samples had Sm–Nd isotopic ratios of about 2.2 Ga. However, Walker *et al.* (1986) also obtained a Sm–Nd whole-rock and apatite age of 2.0 ± 0.1 Ga for the Tin Mountain pegmatite, a pegmatite body associated with the granite. Because of this apparent age disagreement using Sm–Nd analyses, a sample of the granite was collected about 7 km south of Harney Peak (Fig. 1) that contains both zircon and monazite (Table III). Correlative with their high uranium contents, two zircon fractions are very discordant, with $^{206}Pb/^{238}U$ ages of 274 and 767 Ma (Fig. 2F). By contrast, a monazite fraction has nearly concordant U–Pb and Th–Pb ages and a $^{207}Pb–^{206}Pb$ age of 1715 ± 3 Ma. Despite their discordance, data for the two zircon fractions are aligned on a chord with an upper intercept of 1728 Ma and thus support the monazite age. The monazite age and whole-rock Rb–Sr age (Riley, 1970) are concordant within experimental error. Because our dated sample is slightly foliated, indicating relatively early emplacement, we doubt that valid older emplacement ages will be determined. Thus, the 2.0 ± 0.1 Ga Sm–Nd age of the pegmatite and the older Sm-Nd ages (Walker, 1984) of some of the granite are unexplained anomalies.

The Bear Mountain dome, near the western edge of the Precambrian (Fig. 1), has an Archean granite core, but no Harney Peak Granite exposed at the surface. A Rb-Sr date on muscovite in boudin fillings on the flank of the dome of 1680 ± 25 Ma (Zartman *in* Ratté, 1986) indicates that the dome probably formed at the same time as the Harney Peak dome.

1.87–1.71 Ga TECTONIC EVENTS

Following deposition of the youngest Early Proterozoic rocks (unit Xps) some time after 1884 Ma and before emplacement of the Harney Peak Granite at 1715 Ma, the layered rocks experienced multiple folding, faulting, and metamorphism. The dominant structural trend is now north-northwest (Fig. 1). At several places, however, the stratigraphic sequence indicates major early fold trends at high angles to this north-northwest trend. Where these early folds can be identified, they have, in general, a northeasterly trend, as shown by the major synclinal axis in unit Xps (Figs. 1 and 3). These northeasterly trending structures are referred to as F_1 folds because they represent the earliest folding event that affected all layered rocks of the Black Hills (note that older folding events were recognized in pre-2.17 Ga rocks of the Nemo area and may exist in pre-1.9 Ga rocks of the central and northern Black Hills). These F_1 folds were isoclinal and overturned to the north-northwest as evidenced by dominant southerly dips of bedding and plunges of younger minor folds. Lineation and foliation related to F_1 folding are not visible where distinct F_1 fold closures have been recognized. This suggests folding at low confining pressures, such as in the upper nappe regime. Several faults, which parallel lithologic contacts in the central and south-central part of the Precambrian core (Fig. 1), probably originated as thrusts related to this early F_1 nappe phase of deformation.

The next major tectonic event that affected all layered rocks was the development of north-northwesterly trending F_2 folds, which were commonly accompanied by the formation of steeply dipping axial plane foliation (S_1). This second deformation resulted from major east-northeast–west-southwest compression (Table II) as suggested by the (1) tightness of the folds, (2) warping of some of the F_1 folds into parallelism with F_2, and (3) steepness of the F_1 fold plunges, which were reoriented by F_2. The accompanying regional metamorphism probably resulted in the distribution of the garnet and staurolite isograds more or less as shown in Figure 1B. The sillimanite isograd, however, resulted from the later thermal metamorphism associated with the Harney Peak Granite. Kyanite probably formed in the more deeply buried areas as part of the regional metamorphism because it is still preserved in the central part of the elevated Bear Mountain dome. The effect of F_2 was to produce large-scale, northwest-trending synforms and antiforms across much of the central Black Hills (Figs. 1 and 3). If this terminology is employed, the major folds at the Homestake mine in the Lead area, as described by Noble *et al.* (1949), are F_2 folds rather than F_1 folds. The age of the F_2 folding and metamorphism has not been firmly established. Zartman and Stern (1967) reported a Rb–Sr isochron age of 1840 ± 70 Ma for the Archean Little Elk Granite. Because the granite has a prominent north-northwest foliation, this age is believed to be the age of the F_2 event.

Although of lesser regional extent than the F_2 folding, another folding event (F_3), characterized by folds whose axial planes trend more westerly than the F_2 folds, has affected much of the central Black Hills. These F_3 folds are noted near Deerfield and the Pactola Reservoir area located just east of Silver City (Fig. 1), and have been referred to as cross-folds in the Homestake mine area (Noble *et al.*, 1949). F_3 folds refold earlier F_1 and F_2 structures, and have steep plunges and a generally non-penetrative foliation (S_3). A consistent right lateral slip pattern to the folds suggests they may have formed as the result of a major shear couple involving the entire region. Because of a general similarity in trend, the F_3 folds could also be a late-stage modification of the F_2 deformation.

Local deformation, presumably younger than the F_3 structures, resulted from emplacement of the Harney Peak Granite and creation of the domal

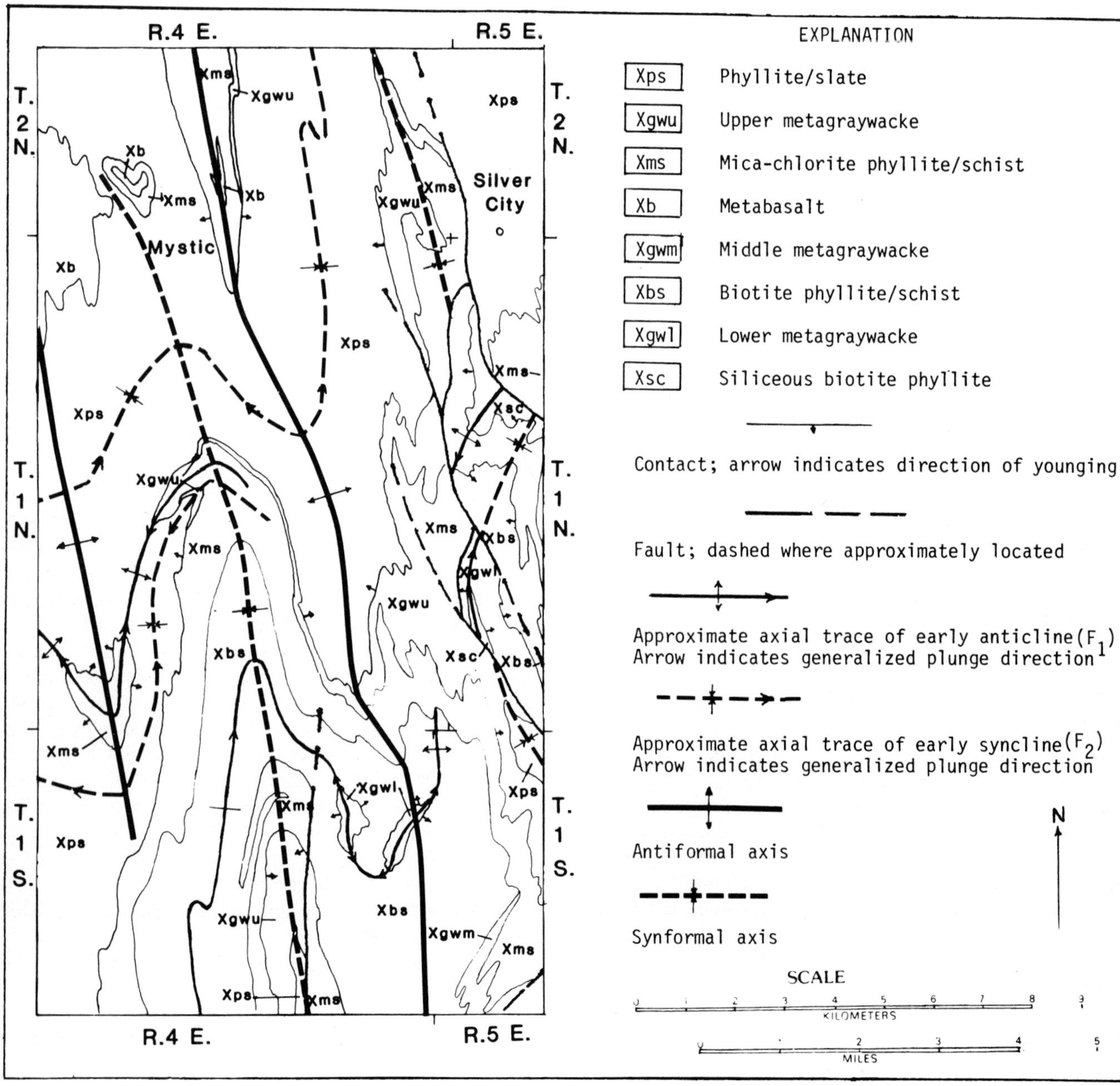

Figure 3. *Simplified geological map of part of the central Black Hills showing early (F_1) and late (F_2) fold axes.*

structure centred on Harney Peak (Fig. 1) and several much smaller domes, such as those in the Berne quadrangle (Redden, 1968). Deformation related to emplacement of the granite is most noticeable in the Hill City area and along the northern perimeter of the Harney Peak Granite where F_1 and F_2 folds are warped into parallelism with the granite contact, and some new folds may have been produced as a result of the intrusion of the granite. If the granite originated by downwarping along major F_2 folds, the F_2 deformation and metamorphism could be much closer in age to the granite than the 1.84 Ga age that, as noted previously, we favour.

A late, northeasterly trending, generally steeply dipping foliation (S_4) deforms sillimanite aggregates southwest of Custer (Redden, 1963) and was originally considered to be related to the granite emplacement. This generally inconspicuous foliation is present over much of the central Black Hills, and has been recognized west of Harney Peak (Redden, 1968), along the eastern edge of the Precambrian core (Marin, 1983), and in the Rochford and Silver City areas (J.A. Redden, unpublished data). The extent and degree of development varies from poorly developed spaced cleavage to well-developed axial plane cleavage in small folds. The northeast-trending foliation has not been observed within the Harney Peak Granite, but it is not clear whether this absence is due to the foliation predating the granite or because the massive granite body was not as easily deformed as the more micaceous metasedimentary rocks. Because the foliation deforms sillimanite that is believed to have formed as a result of the emplacement of the Harney Peak Granite, we favour a post-1715 Ma age for its development.

POST-1.7 Ga ISOTOPIC REEQUILIBRATION

Whole-rock samples of two suites of supracrustal rocks were dated by the Rb–Sr method (Table IV). The isotopic system appears to record a post-Harney Peak Granite disturbance that may correlate with the development of the late northeasterly trending foliation previously described.

One suite comprises small samples obtained from a block of Montana Mine Formation representing approximately 25 cm of section of light- and dark-grey tuffaceous beds (laminations of 1 mm or less to beds of up to 6 cm thick). A strongly developed cleavage at high angles to the bedding cannot be related to the regional foliation because the unoriented sample is from an inaccessible mine shaft. This block is part of the same bulk sample from which zircon was separated and dated (Table III, Fig. 2C). Samples were obtained by drilling holes

Table IV
Rb–Sr whole-rock data for samples of the Montana Mine Formation and for metagreywackes near Three Forks

Sample No.	Rb (ppm)	Sr(ppm)	$^{87}Rb/^{86}Sr$	$^{87}Sr/^{86}Sr$
Montana Mine Formation				
BLH-4 (bulk sample)	191.9	71.65	7.906	0.90334
BLH-4-1	199.8	74.39	7.927	0.90287
BLH-4-2	198.7	74.17	7.910	0.90279
BLH-4-3	169.4	48.70	10.317	0.95365
BLH-4-4	173.3	61.63	8.306	0.90982
BLH-4-5	177.5	85.52	6.102	0.86332
BLH-4-6	167.8	88.92	5.542	0.84916
BLH-4-7	108.3	41.94	7.617	0.89742
Metagreywackes at Three Forks				
516-1-86 (concretion)	21.29	124.7	0.434	0.72582
516-2-86	193.3	128.8	4.395	0.81933
516-3-86	251.9	78.51	9.495	0.92817
516-4-86	210.5	76.48	8.123	0.90336
516-5-86	117.1	107.5	3.179	0.78956

parallel to the beds with a 1.5 cm carbide tipped bit. Fine cuttings were produced that were suitable for analyses without further processing. A split of the bulk sample prepared for zircon separation was also analyzed. The data points are well aligned on an isochron plot (Fig. 4) with minimum dispersion indicated by an MSWD of 0.8. The slope corresponds to an age of 1534 ± 50 Ma, and an intercept of 0.728 ± 0.005. Clearly, the isochron records a post-1.7 Ga event that resulted in isotopic equilibration at the scale represented by the thinly bedded sample. Gross open system behaviour is shown by a model age of 1770 Ma calculated from the bulk sample and an initial Sr-isotope ratio of 0.702, which represents a volume of several cubic decimetres.

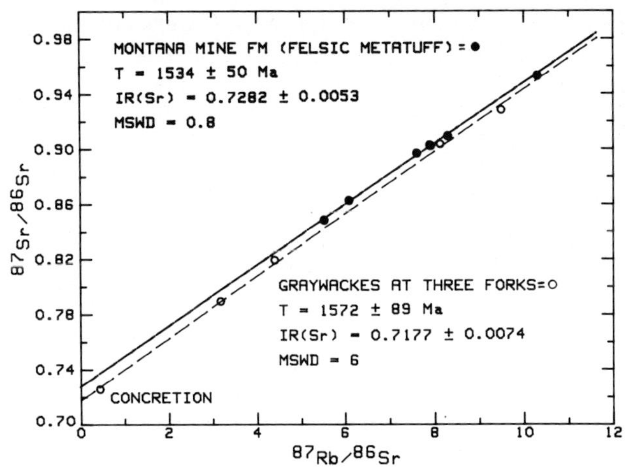

Figure 4. *Rb–Sr isochron plot for samples of the Montana Mine Formation and of metagreywackes near Three Forks.*

The second suite of samples was collected from a roadcut at Three Forks about 8 km northeast of Hill City (TF in Fig. 1B). The cut displays vertically dipping metagreywackes (staurolite grade) with well-preserved primary sedimentary features. Both graded and crossbedded features are still discernable in the coarser beds, which attain thicknesses of up to 6 dm. Finely bedded intervals of Bouma DEF sequences contain laminations a few millimetres thick. Metamorphosed concretions, now triaxial ellipsoids 5–10 dm long, contain poikiloblastic garnet and radiating hornblende in a finer grained matrix of quartz and feldspar. The metagreywackes contain nearly randomly oriented biotite (fine granofelsic texture) characteristic of the outer part of the thermal aureole around the Harney Peak Granite. Five samples, including one concretion, were collected along a stratigraphic interval of about 30 m. The Rb–Sr data define an isochron that is subparallel to the Montana Mine isochron. The date is 1572 ± 89 Ma, with an intercept of 0.718 ± 0.007; minor excessive scatter is indicated by a MSWD of 6. Large-scale open system behaviour is also indicated in this section by a model age of 1760 Ma (initial Sr-isotope ratio of 0.702) for a calculated bulk rock (excluding the concretions, which are volumetrically negligible).

The two isochron ages are the same within error and, as such, can be pooled to yield a single age of 1543 ± 43 Ma. Deducing the tectonic significance of this age is less straightforward. A few other isotopic ages of Precambrian units in the Black Hills exhibit disturbance that could be attributed to Middle Proterozoic events (Zartman and Stern, 1967; Riley, 1970), although minor perturbation of the systems during the Laramide cannot be precluded. The good linearity of the two Rb–Sr whole-rock isochrons and the fact that the ages agree for units that are lithologically different and were sampled at grossly different scales suggest that the 1543 ± 43 Ma age is recording a real event. The only recognized structural event that might fit this date is the development of the northeast-trending S_4 foliation found over much of the Black Hills. We tentatively suggest that the S_4 development occurred at approximately 1.54 Ga.

What this event may have been in a regional sense is problematic. However, Rb–Sr and K–Ar biotite ages of Archean rocks in the southern third of the Wyoming age province record a cryptic event between about 1.4 and 1.6 Ga and the regional pattern in ages has been interpreted to reflect Middle Proterozoic uplift (Peterman and Hildreth, 1978). The late S_4 foliation that strikes northeast in the Black Hills may have been part of a much larger tectonic system, the effects of which are manifested differently across the region.

CONCLUSIONS AND TECTONIC INTERPRETATIONS

Most of the conclusions from the stratigraphic, depositional, tectonic, and chronologic studies on the Black Hills are embodied in Tables I and II. The following general conclusions and tectonic interpretations are possible from the data. Correlation of geological units and events recognized in the Black Hills with those in surrounding areas is hampered by the fact that the Black Hills are more than 300 km away from pertinent outcrops in southern Wyoming and more than 1100 km away from rocks of similar age in Manitoba and Saskatchewan. However, first-order lithostructural domains can be delineated in the subsurface of the northern Great Plains through regional gravity and aeromagnetic data (e.g., Green *et al.*, 1985; Klasner and King, 1986), thus allowing some degree of connectivity with exposed terranes.

(1) The Black Hills are underlain by Archean basement belonging to the Wyoming craton. Where exposed, the basement consists of metamorphosed fine- to coarse-grained epicontinental clastic rocks and granite (2.55 Ga). Coarse clastic rocks in the oldest Proterozoic rocks also indicate that there were areas of ultramafic rocks and dolomite in the nearby Archean basement. The granitic rocks have rare earth element concentrations and patterns similar to those of Late Archean granitoids of the Wyoming province (Gosselin *et al.*, 1988) and have old Nd model ages suggestive of derivation from pre-existing continental crust 3.3 Ga or older (Walker *et al.*, 1986). Middle Archean crust is now known to have been widespread in the Wyoming province, whereas much of the western Superior Province was a juvenile Late Archean crustal addition (Peterman and Futa, 1988). Heat flow measurements in the Black Hills (Drury, 1988) more closely match the heat flow of the Wyoming craton than that of the Superior craton.

The Wyoming craton extends east of the Black Hills, probably to the North American Central Plains Anomaly, which is a deep-crustal conductive zone that extends from southeastern Wyoming through western South Dakota (just east of the Black Hills) to northern Saskatchewan (Camfield and Gough, 1977; Green *et al.*, 1985; Klasner and King, 1986). Within the framework of the exposed Trans-Hudson Orogen in northern Saskatchewan, the terrane represented by the Black Hills is analogous to the Cree Lake Zone, where Archean basement and overlying Early Proterozoic strata were both severely deformed in a ductile manner during the Hudsonian orogeny (Lewry *et al.*, 1985). Like the Black Hills, the Cree Lake Zone is bounded on the west by an Archean block, the Amer Lake zone (Lewry *et al.*, 1985), where Hudsonian tectonic overprinting is much less severe.

(2) Prior to 2.17 Ga, coarse-grained clastic rocks with local anomalous concentrations of uranium and minor chromite were deposited in an extensional setting on continental Archean basement. Growth faults, pronounced facies changes, and sedimentary features suggest a western source and deposition in alluvial fans with a fining-upward sequence ending in oxide facies iron formation that ultimately may have been part of a marine basin to the east. Rocks of similar composition and tectonic setting, and permissibly similar age are exposed in the Sierra Madre and Medicine Bow mountains of southern Wyoming (Graff, 1979; Houston and Karlstrom, 1980). Here, rocks of the Deep Lake Group of the Snowy Pass Supergroup are interpreted to have been deposited on a passive margin of the Wyoming craton between ~2.3 and 2.0 Ga (Karlstrom *et al.*, 1983).

(3) These Early Proterozoic strata and the Archean basement were intruded by 2.17 Ga gabbro sills, folded about north-northwest-trending axes, and then eroded. This folding event, which is bracketed between 1.98 and 2.17 Ga, represents compression along the plate margin, perhaps recording early activity in the development of the Trans-Hudson Orogen. The basement and pre-2.17 Ga Early Proterozoic rocks were then cut by growth faults and quartz-rich, feldspathic sediment derived from the east was deposited in alluvial fans that connected to a marine shelf and tidal flat to the west and south. Clastic rocks were deposited on the Archean basement on the west side of the Black Hills at Bear Mountain in an onlapping sequence. Sedimentation during this interval marked the beginning of the main Black Hills basin. Rocks of approximately similar composition and tectonic setting in southern Wyoming are in the Lower Libby Creek Group of the Snowy Pass Supergroup (Karlstrom *et al.*, 1983).

(4) Deepening of the east-central part of the main Black Hills basin took place at ~2.0 Ga with deposition of black shale accompanied by the submarine eruption of tholeiitic basalt. Locally, deep water turbidite deposition followed. The age of these rocks is based on correlation with similar rocks in the isolated Lead area that are overlain by shallow water shelf quartzite and the gold-bearing Homestake Formation, which was deposited at ~1.97 Ga. Turbidite deposits generally thicken to the south and east, suggesting a northwestern transport direction and possible eastern source. The shallow water quartzite strata of the Lead area are correlated with deeper water turbidite and quartzite deposits in the Rochford area to the southwest. Gabbroic sills (1.96 Ga) intrude these older basin rocks in the eastern Black Hills.

In a plate tectonic scenario, the depositional environment at this time was either an intracontinental or a back-arc basin. The general composition of the turbidites favours the former. The eastern margin of the Wyoming craton was undergoing uplift, volcanism, and possibly deformation while a marine basin was forming to the west in the Black Hills. Water depths increased from east to west and north to south. Volcanism was dominantly in the east. Source areas for earlier shelf quartzites and most of the turbidite section were to the east. Primitive volcanic arcs may have been active at this time in the Trans-Hudson Orogen in Canada (Lewry, 1981; Green *et al.*, 1985; Lewry *et al.*, 1986; Van Schmus *et al.*, 1987).

(5) Following initial development and deepening of the basin, the available age data and stratigraphic interpretations indicate a depositional break, probably preceded by a poorly understood deformational event. At Rochford, quartzose turbidite deposition was interrupted by extrusion of basalt flows, volcanogenic sediments, tuff (1.88 Ga), and carbonate–silicate facies iron formation. The 1.88 Ga alkalic tuff is in strata immediately overlying siliceous strata correlated with the 1.97 Ga Ellison Formation at Lead, thus indicating a 90 Ma time gap in the stratigraphic record. The 1.88 Ga tuff at Rochford is correlated with a widespread mica-rich pelitic unit in the upper part of the turbidite section in the central Black Hills.

A similar break in the east-central Black Hills is indicated by an unconformity separating conglomerates, debris flows, and accompanying minor volcanic rocks and the older turbidite deposits and tholeiitic flows. In one area, the older rocks were apparently folded and overturned prior to deposition of conglomerate and debris flow material. Contact relationships of thin gabbroic intrusions (1.88 Ga) suggest emplacement in previously folded rocks both in the east-central Black Hills (Prairie Creek sill) and in the older rocks at Rochford. The debris flows are overlain by thick younger turbidite deposits, indistinguishable from the older deposits, so that the depositional break cannot be recognized in the absence of the debris flows. Although deformation that preceded this break has not been well documented, the general rock distribution suggests northerly trends of folds unless the rocks were reoriented by later events. The time gap and limited evidence of deformation suggest the Black Hills basin was again subjected to east–west closure, which could be attributed to plate margin effects along the Trans-Hudson axis. However, it is too speculative at this time to attempt correlation with other areas.

(6) After 1.88 Ga, but before 1.71 Ga, a thick, widespread shale unit in the central Black Hills was deposited. The youngest turbidites in the southern Black Hills may be correlative with this shale unit or the upper part of the younger

turbidite section of the central Black Hills. The general lack of shallow-water structures and rocks indicative of stable tectonic conditions strongly suggests that the main Black Hills basin continued to evolve as an intracontinental or back-arc basin from ca. 1.9 to 1.8 Ga.

(7) Sometime between ca. 1.85 and 1.71 Ga, all the stratified rocks and the Archean basement were folded along east-northeast axes. Where this F_1 folding can be readily recognized, there are no recognizable early metamorphic minerals or lineations formed by such minerals. The event responsible for this folding is enigmatic, although it was presumably associated with convergence during the evolution of the Trans-Hudson Orogen.

(8) After F_1 folding, but before 1.71 Ga, the stratified rocks and Archean basement were folded along north-northwest axes (F_2) and regionally metamorphosed. Minor F_3 deformation may have been a continuation of F_2. Rb–Sr data on Archean rocks suggest that metamorphism may have occurred about 1.84 Ga. This compressional event was presumably related to convergence along the Trans-Hudson Orogen.

(9) The 1.71 Ga peraluminous Harney Peak Granite probably formed by melting of Archean crust and minor amounts of Early Proterozoic supracrustal rocks (Redden *et al.*, 1982; Walker *et al.*, 1986). Domes related to emplacement of the granite deform both F_1 and F_2 structures. The main phase of igneous activity associated with the exposed Trans-Hudson Orogen is about 1.91–1.84 Ga (e.g., Van Schmus *et al.*, 1987), and the Harney Peak Granite is probably too young to be related to this activity. More likely, the generation of the Harney Peak was triggered by tectonic activity in the Central Plains Orogen immediately to the south (Bickford *et al.*, 1986; Sims and Peterman, 1986). The Haystack Granite was emplaced at the same time in the southern Hartville uplift in Wyoming, where, in contrast to the Black Hills, the Archean craton was relatively stable during the Early Proterozoic (Peterman, 1982). Thus, older Early Proterozoic and Archean terranes adjacent to the Central Plains Orogen were locally intruded by 1.71–1.72 Ga granites.

(10) Rb–Sr whole-rock systems of metatuff and metagreywacke record a cryptic event at 1.54 Ga, which involved isotopic reequilibration at a scale of centimetres to metres. The regional extent and cause of this isotopic reequilibration are unknown. Late, northeast-trending cleavage (S_4) may be another manifestation of the event that caused the isotopic disturbances.

REFERENCES

Bayley, R.W., 1972a, Geologic field compilation map of the northern Black Hills, South Dakota: United States Geological Survey, Open-file Report 72-29, scale 1:48,000.

Bayley, R.W., 1972b, Preliminary geologic map of the Nemo district, Black Hills, South Dakota: United States Geological Survey, Miscellaneous Geologic Investigations Map I-712, scale 1:24,000.

Bayley, R.W., 1972c, A preliminary report on the geology and gold deposits of the Rochford district, Black Hills, South Dakota: United States Geological Survey, Bulletin 1332-A, 24 p.

Bickford, M.E., Van Schmus, W.R. and Zietz, I., 1986, Proterozoic history of the midcontinent region of North America: Geology, v. 14, p. 492–496.

Camfield, P.A. and Gough, D.I., 1977, A possible Proterozoic plate boundary in North America: Canadian Journal of Earth Sciences, v. 14, p. 1229–1232.

Chappell, B.W. and White, A.J.R., 1974, Two contrasting granite types: Pacific Geology, v. 8, p. 173-174.

Davis, G.L., 1926, The composition and age of uranium minerals from Katanga, South Dakota, and Utah: American Journal of Science, 5th Series, v. 11, p. 201–217.

DeWitt, E., Redden, J.A., Wilson, A.B. and Buscher, D., 1986, Mineral resource potential and geology of the Black Hills National Forest, South Dakota and Wyoming: United States Geological Survey, Bulletin 1580, 135 p. (Includes a section on salable commodities by J.S. Dersch.)

Dodge, T.A., 1942, Amphibolites of the Lead area, northern Black Hills, South Dakota: Geological Society of America Bulletin, v. 53, p. 561–583.

Doe, B.R. and Zartman, R.E., 1979, Plumbotectonics, the Phanerozoic, in Barnes, H.L., ed., Geochemistry of Hydrothermal Ore Deposits, 2nd ed.: John Wiley and Sons, New York, NY, p. 22–70.

Drury, M.J., 1988, Tectonothermics of the North American Great Plains basement: Tectonophysics, v. 148, p. 299–307.

Fantone, K., 1983, Geochemistry and petrology of the carbonate iron formations and ferruginous cherts of the northeastern Black Hills, South Dakota, unpublished M.Sc. thesis, South Dakota School of Mines and Technology, Rapid City, 107 p.

Fricke, J.N., 1982, Geologic investigation of Precambrian mafic intrusive rocks near McGee siding, Pennington County, South Dakota, unpublished M.Sc. thesis, South Dakota School of Mines and Technology, Rapid City, 63 p.

Gosselin, D.C., Papike, J.J., Zartman, R.E., Peterman, Z.E. and Laul, J.C., 1988, Archean rocks of the Black Hills, South Dakota: reworked basement from the southern extension of the Trans-Hudson Orogen: Geological Society of America Bulletin, v. 100, p. 1244–1249.

Graff, P.J., 1979, A review of the stratigraphy and uranium potential of early Proterozoic (Precambrian X) metasediments in the Sierra Madre, Wyoming: Contributions to Geology, v. 17, p. 1047–1069.

Green, A.G., Hajnal, Z. and Weber, W., 1985, An evolutionary model of the western Churchill Province and western margin of the Superior Province in Canada and the north-central United States: Tectonophysics, v. 116, p. 281–322.

Hosted, J.O. and Wright, L.B., 1923, Geology of the Homestake ore bodies and the Lead area of South Dakota: Engineering and Mining Journal, v. 115, p. 793–799, 836–843.

Houston, R.S. and Karlstrom, K.E., 1980, Uranium-bearing quartz-pebble conglomerates: exploration model and United States resource potential: Bendix Field Engineering Corporation, Open-file Report GJBX-1(80), 510 p.

Karlstrom, K.E., Flurkey, A.J. and Houston, R.S., 1983, Stratigraphy and depositional setting of the Proterozoic Snowy Pass Supergroup, southeastern Wyoming: record of an early Proterozoic Atlantic-type cratonic margin: Geological Society of America Bulletin, v. 94, p. 1256–1274.

Klasner, J.S. and King, E.R., 1986, Precambrian basement geology of North and South Dakota: Canadian Journal of Earth Sciences, v. 23, p. 1983-1102.

Krogh, T.E., 1973, A low contamination method for hydrothermal decomposition of zircon and extraction of U and Pb for isotopic age determinations: Geochimica et Cosmochimica Acta, v. 37, p. 485–494.

Kulp, J.L., Eckelmann, W.R., Gast, P.W. and Miller, D.S., 1956, Age of the Black Hills (South Dakota) gold mineralization: Geological Society of America Bulletin, v. 67, p. 1557–1558.

Lewry, J.F., 1981, Lower Proterozoic arc-microcontinent collisional tectonics in the western Churchill Province: Nature, v. 294, p. 69–72.

Lewry, J.F., Sibbald, T.I.I. and Schledewitz, D.C.P., 1985, Variation in character of Archean rocks in the western Churchill Province and its significance, *in* Ayres, L.D., Thurston, P.C., Card, K.D. and Weber, W., eds., Evolution of Archean Supracrustal Sequences: Geological Association of Canada, Special Paper 28, p. 239–261.

Lewry, J.F., Collerson, K.D., Bickford, M.E. and Van Schmus, W.R., 1986, An evolutionary model of the western Churchill Province and western margin of the Superior Province in Canada and the north-central United States – discussion and reply: Tectonophysics, v. 131, p. 183–197.

Ludwig, K.R., 1987, ISOPLOT 200, a plotting and regression program for isotope geochemists for use with HP series 200 computers: United States Geological Survey, Open-file Report 85-513, 47 p.

Marin, J.R., 1983, Structure and depositional environment of the Precambrian rocks in the McGee area, Black Hills, South Dakota, unpublished M.Sc. thesis, South Dakota School of Mines and Technology, Rapid City, 89 p.

Noble, J.A., 1950, Ore mineralization in the Homestake gold mine, Lead, South Dakota: Geological Society of America Bulletin, v. 61, p. 221–225.

Noble, J.A. and Harder, J.O., 1948, Stratigraphy and metamorphism in a part of the northern Black Hills and the Homestake mine, Lead, South Dakota: Geological Society of America Bulletin, v. 59, p. 941–975.

Noble, J.A., Harder, J.O. and Slaughter, A.L., 1949, Structure of a part of the northern Black Hills and the Homestake mine, Lead, South Dakota: Geological Society of America Bulletin, v. 60, p. 321–352.

Peterman, Z.E., 1982, Geochronology of the southern Wyoming age province – summary: 1982 Archean Geochemistry Field Conference, Part 1, Guide to Field Trips, p. 1–23.

Peterman, Z.E. and Futa, K., 1988, Contrasts in Nd crustal residence ages between the Superior and Wyoming cratons: Geological Society of America, Abstracts with Programs, v. 20, p. A137.

Peterman, Z.E. and Hildreth, R.A., 1978, Reconnaissance geology and geochronology of the Precambrian of the Granite Mountain, Wyoming: United States Geological Survey, Professional Paper 1055, 22 p.

Peterman, Z.E., Goldich, S.S., Fischer, L.B. and Futa, K., 1986, Geochronology of basement granites, Stephenson County, Illinois, *in* Peterman, Z.E. and Schnable, D.C., eds., Shorter Contributions to Isotope Research: United States Geological Survey, Bulletin 1622, p. 41–50.

Ratté, J.C., 1986, Geologic map of the Medicine Mountain quadrangle, Pennington County, South Dakota: United States Geological Survey, Miscellaneous Investigation Series Map I-1654, scale 1:24,000.

Ratté, J.C. and Wayland, R.G., 1969, Geology of the Hill City quadrangle, Pennington County, South Dakota – a preliminary report: United States Geological Survey, Bulletin 1271-B, 14 p.

Ratté, J.C. and Zartman, R.E., 1970, Bear Mountain gneiss dome, Black Hills, South Dakota – age and structure: Geological Society of America, Abstracts with Programs, v. 2(5), p. 345.

Redden, J.A., 1963, Geology and pegmatites of the Fourmile quadrangle, Black Hills, South Dakota: United States Geological Survey, Professional Paper 297-D, p. 199–291.

Redden, J.A., 1968, Geology of the Berne quadrangle, Black Hills, South Dakota: United States Geological Survey, Professional Paper 297-F, p. 343-408.

Redden, J.A., 1981, Summary of the geology of the Nemo area, *in* Rich, F.J., ed., Geology of the Black Hills, South Dakota and Wyoming, Geological Society of America Guidebook, Rocky Mountain Section, Annual Meeting, Rapid City, South Dakota, 1981: American Geological Institute, Falls Church, VA, p. 193–210.

Redden, J.A. and Norton, J.J., 1975, Precambrian geology of the Black Hills, *in* United States Congress, Senate Committee on Interior and Insular Affairs, Mineral and Water Resources of South Dakota: United States 94th Congress, 1st Session, p. 21-28.

Redden, J.A., Norton, J.J. and McLaughlin, R.J., 1982, Geology of the Harney Peak Granite, Black Hills, South Dakota: United States Geological Survey, Open-file Report 82-481, 18 p.

Riley, G.H., 1970, Isotopic discrepancies in zoned pegmatites, Black Hills, South Dakota: Geochimica et Cosmochimica Acta, v. 34, p. 713–725.

Rye, D.M., Doe, B.R. and Delevaux, M.H., 1974, Homestake gold mine, South Dakota. Part 2. Lead isotopes, mineralization ages, and source of lead in ores of the northern Black Hills: Economic Geology, v. 69, p. 814–822.

Sims, P.K. and Peterman, Z.E., 1986, Early Proterozoic Central Plains orogen: a major buried structure in the north-central United States: Geology, v. 14, p. 488-491.

Stacey, J.S. and Kramers, J.D., 1975, Approximation of terrestrial lead isotope evolution by a two-stage model: Earth and Planetary Science Letters, v. 26, p. 207–221.

Stacey, J.S., Doe, B.R., Silver, L.T. and Zartman, R.E., 1976, Plumbotectonics IIA, Precambrian massive sulfide deposits: United States Geological Survey, Open-file Report 76-476, 26 p.

Steiger, R.H. and Jager, E., 1977, Subcommission on geochronology; convention and use of decay constants in geo- and cosmochronology: Earth and Planetary Science Letters, v. 36, p. 359-362.

Van Schmus, W.R., Bickford, M.E., Lewry, J.F. and MacDonald, R., 1987, U–Pb geochronology in the Trans-Hudson Orogen, northern Saskatchewan, Canada: Canadian Journal of Earth Sciences, v. 24, p. 407–424.

Walker, R.J., 1984, The origin of the Tin Mountain pegmatite, Black Hills, South Dakota, unpublished Ph.D. dissertation, State University of New York at Stony Brook, Stony Brook, 384 p.

Walker, R.J., Hanson, G.N., Papike, J.J. and O'Neil, J.R., 1986, Nd, O and Sr isotopic constraints on the origin of Precambrian rocks, southern Black Hills, South Dakota: Geochimica et Cosmochimica Acta, v. 50, p. 2833–2846.

Weissenborn, P.R., 1987, The Precambrian geology of the western portion of the Rochford gold mining district, Black Hills, South Dakota, unpublished M.Sc. thesis, South Dakota School of Mines and Technology, Rapid City, 147 p.

Wetherill, G.W., Tilton, G.R., Davis, G.J. and Aldrich, L.T., 1956, New determinations of the age of the Bob Ingersoll pegmatite, Keystone, South Dakota: Geochimica et Cosmochimica Acta, v. 9, p. 292–297.

Wolfgram, D., 1977, Wall rock alteration and the localization of gold in the Homestake mine, Lead, South Dakota, unpublished Ph.D. dissertation, University of California, Berkeley, 139 p.

Zartman, R.E. and Stern, T.W., 1967, Isotopic age and geologic relationships of the Little Elk Granite, northern Black Hills, South Dakota: United States Geological Survey, Professional Paper 575-D, p. D157–D163.

Zartman, R.E., Norton, J.J. and Stern, T.W., 1964, Ancient granite gneiss in the Black Hills, South Dakota: Science, v. 145, p. 479–481.

Zartman, R.E., Kwak, L.M. and Christian, R.D., 1986, U–Pb systematics of a mixed zircon population – the granite at Yale Farm, Berkshire Massif, Connecticut, *in* Peterman, Z.E. and Schnable, D.C., eds., Shorter Contributions to Isotope Research: United States Geological Survey, Bulletin 1622, p. 81–98.

Proterozoic Granite–Pegmatite Magmatism, Black Hills, South Dakota: Structure and Geochemical Zonation

E.F. Duke, C.K. Shearer, J.A. Redden, and J.J. Papike
Institute for the Study of Mineral Deposits, South Dakota School of Mines and Technology, Rapid City, South Dakota 57701, U.S.A.

Duke, E.F., Shearer, C.K., Redden, J.A. and Papike, J.J., Proterozoic granite–pegmatite magmatism, Black Hills, South Dakota: structure and geochemical zonation, in Lewry, J.F. and Stauffer, M.R., eds., The Early Proterozoic Trans-Hudson Orogen of North America: Geological Association of Canada, Special Paper 37, p. 253–269.

Abstract

The peraluminous Harney Peak Granite and associated pegmatites (ca. 1.72 Ga) intrude Lower Proterozoic (2.2-1.88 Ga) metamorphic rocks throughout the southern Black Hills. The granitic rocks are exposed in a large central dome (approximately 100 km^2), in several smaller domelike bodies, and in thousands of small sills and dykes. Many of the latter are transitional to pegmatitic rocks, and a few are rare-element, zoned pegmatites. The main pluton formed by coalescence of dozens of large sill-like intrusions that distended and domed the adjacent country rocks while preserving an internal domal pattern of foliation, textural layering, and country rock inliers. A lack of crosscutting relationships between sill-like bodies and the generally more deformed fabric of intrusive bodies within the central parts of the granite suggest that the dome grew laterally by addition of new magma along the perimeter. This growth model is supported by analogy with structural relationships in the smaller Calamity Peak dome. The absence of a clearly defined gravity anomaly over the main dome and simple models based on available gravity data indicate that the pluton may not possess a deep root. The gravity models and metamorphic zonation suggest, however, that additional concentrations of granitic intrusives may exist at a shallow depth beneath much of the sillimanite-grade terrane. Mineralogical and chemical zonation of the entire granite-pegmatite system is indicated by a nucleus of high K/Rb biotite-bearing granite, which occupies the central and southeastern portions of the main dome. This grades generally outward to lower K/Rb, more fractionated, tourmaline-bearing granite and pegmatite, although structural relations are quite complex in detail. Together with the limited structural evidence that the dome grew outward, this geometry is consistent with derivation of the outer tourmaline granites through fractional crystallization of biotite granites similar to those currently exposed in the central and southeastern portions. Pegmatites in country rocks near the dome are more likely to be simple types, lacking mineralization, and similar to portions of the granite, whereas there are larger numbers of more evolved, zoned pegmatites enriched in Nb, Ta, Be, and Li near the limits of the field.

Résumé

Le granite peralumineux de Harney Peak et les pegmatites associées (1,72 Ga) recoupent des roches métamorphiques d'âge Protérozoïque Inférieur (2,2–1,88 Ga) au sud des Black Hills. Les granites affleurent sours forme d'un large dôme central (vers 100 km²), en petits amas et en milliers de petits sills et dykes. Ces derniers passent graduellement à des pegmatites dont quelques unes sont zonées et contiennent des terres rares. Le pluton principal est formé par agglomération de grands sills ayant distendue les roches encaissantes pour former un dôme en préservant un patron radial des foliations, un litage texturale et des enclaves de l'encaissant. L'absence de recoupement entre les sills et une plus grande déformation des intrusifs dans la partie centrale du granite montre que le dôme a crû latéralement par addition de magma en périphérie. Ce modèle pétrogénétique montre des analogies structurales avec le petit dôme de Calamity Peak. Un modèle de gravité simple ainsi que l'absence d'anomalie bien définie sur le dôme principal indique que le pluton n'est pas enraciné. Les modèles de gravité ainsi que la zonation métamorphique suggère toutefois la présence d'intrusifs à faible profondeur sous des terrains métamorphisés au faciès de la sillimanite. La présence de granites à biotite dans la partie centrale et SE du dôme principal, montrant des rapports élevés de K/Rb indique une zonation chimique et minérale du système granite–pegmatite. Ce dernier montre des rapports K/Rb plus faible et des granites à tourmaline et des pegmatites plus fractionnées en périphérie quoique les relations structurales soient complexes. Bien que les preuves structurales d'une croissance périphérique soient peu convainquantes, ce modèle géométrique est compatible avec une genèse des granites à tourmaline périphériques par fractionnement des granites à biotite similaires à ceux affleurant dans la partie centrale et SE du dôme. Les pegmatites dans l'encaissant près du dôme ne sont pas minéralisées et montrent des similitudes avec certains granites alors que beaucoup d'autres sont plus évoluées et zonées montrant des teneurs élevées en Nb, Ta, Be et Li.

INTRODUCTION

An important manifestation of Early Proterozoic orogenic activity in the Black Hills region was the production and emplacement of large volumes of granitic material consisting of granite, pegmatitic granite, and pegmatite. The Harney Peak Granite and related rocks provide the only exposed record of this Early Proterozoic granitic magmatism in the Dakota segment of the Trans-Hudson Orogen. The granite crops out in an area of about 100 km^2, where it forms the core of a large central dome, as well as in thousands of smaller intrusive bodies exposed throughout the southern Black Hills. Understanding the structure and emplacement of such a major granitic rock series is critical to the interpretation of metamorphism and tectonic processes in this portion of the Trans-Hudson Orogen. This paper reviews the distribution and structural aspects of granitic and pegmatitic intrusives of the Harney Peak Granite system, and also examines current views on the emplacement mechanisms and geochemical zonation of rocks in this series.

Portions of the material presented in this paper have been addressed by Redden and Norton (1975), Redden *et al.* (1985), and DeWitt *et al.* (1986). Early treatments of the structure of the Harney Peak Granite include papers by Darton and Paige (1925), Balk (1931), and Runner (1943). Redden *et al.* (this volume) review age determinations and the tectonic evolution of Precambrian rocks of the Black Hills. For a summary of chemical and petrological data on the granite, the reader is referred to Shearer *et al.* (1987b). Walker *et al.* (1986) discussed Nd, O, and Sr isotopic constraints on the age and origin of the granite and pegmatites. The reader should note, however, that age data on the Harney Peak Granite presented by Walker *et al.* (1986) are contradicted by new data included in Redden *et al.* (this volume). Comprehensive descriptions of the field relations of many of the zoned pegmatites associated with the granite are found in Page *et al.* (1953) and Norton *et al.* (1964).

REGIONAL GEOLOGICAL SETTING

The Black Hills and the Trans-Hudson Orogen

The Black Hills is a north-northwest trending uplift of Laramide age (~60 Ma) located in western South Dakota and eastern Wyoming (Lisenbee, 1978). The Black Hills uplift represents the eastern limit of major Laramide tectonic activity. Erosion of the core of the uplift currently exposes a complex Precambrian terrane dominated by Lower Proterozoic igneous and metasedimentary rocks, but which includes minor areas of Archean rocks.

Current reconstructions of Precambrian age and structural provinces of the midcontinent region place the Black Hills near the western boundary of the ~1.8–1.9 Ga Trans-Hudson orogenic belt, either in the transition into the Archean Wyoming Province (Sims and Peterman, 1986) or in the easternmost Wyoming Province (Bickford *et al.*, 1986). Furthermore, the Black Hills are located near the southern termination of the Trans-Hudson belt, where it is truncated by the ca. 1.6–1.8 Ga Central Plains Orogen. Figure 1 shows the location of the Black Hills in relation to the southern extension of the Trans-Hudson Orogen and adjacent Precambrian age provinces. The dominance of Lower Proterozoic metasedimentary rocks in the Black Hills, with minor amounts of Archean granite and metasedimentary rocks, is an association comparable in its general features to the Cree Lake Zone of northern Saskatchewan (Lewry *et al.*, 1985). An important difference, however, between the Black Hills and exposed Precambrian terranes along strike in northern Saskatchewan and Manitoba is in the apparently younger age of the major plutonic events in the Black Hills, where the Harney Peak Granite and related rocks are dated at about 1.72 Ga. The youngest dated plutonic rock in the Cree Lake Zone is the 1.82 Ga Junction Granite (M.E. Bickford, personal communication, 1988). Elsewhere in the exposed Trans-Hudson belt, late granitic sheets and pegmatites are as young as 1.78 Ga; however, these are undeformed.

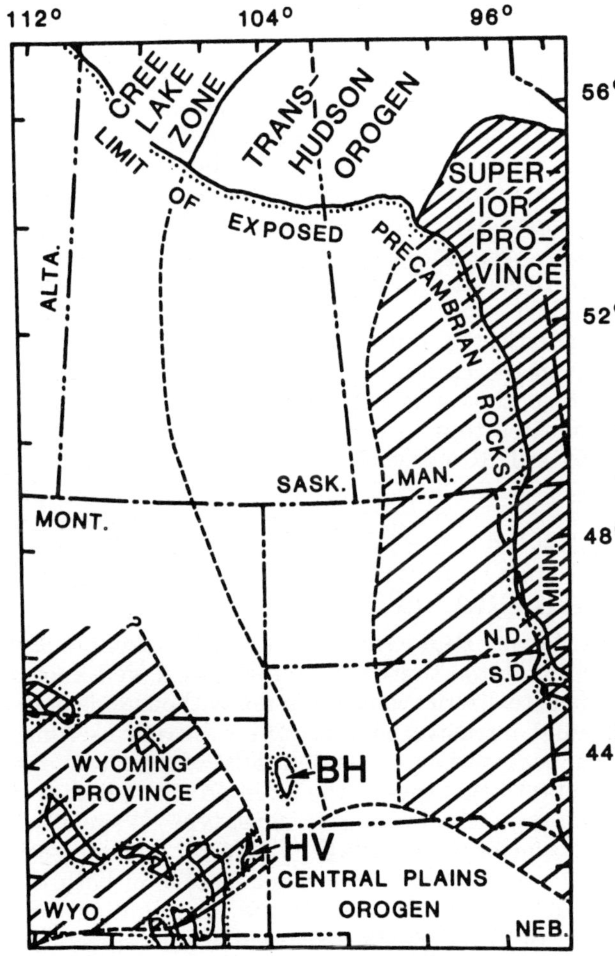

Figure 1. *Map showing the location of the Black Hills (BH) in relation to inferred Precambrian age provinces of the midcontinent region. Close-ruled pattern shows exposed Archean terranes; open-ruled pattern shows inferred Archean in subsurface. Abbreviation: HV, Hartville Uplift.*

Precambrian evolution of the Black Hills region

The Precambrian core of the Black Hills is dominated by metasedimentary rocks deposited in the approximate interval 2.2–1.88 Ga, but includes minor Archean (~2.5 Ga) granite and older metasedimentary rocks (Redden *et al.*, this volume). The Lower Proterozoic metasedimentary strata consist chiefly of metamorphosed greywackes, shales, and quartzites. The lithology and structure of the Archean rocks are reviewed by Gosselin *et al.* (1988).

Early Proterozoic metamorphism commenced at some time after 1.88 Ga and is assumed to have ended with the intrusion of the Harney Peak Granite and related rocks at about 1.72 Ga (Riley, 1970; Redden *et al.*, this volume). The granitic intrusives superposed low-pressure, high-temperature "regional contact" metamorphism on earlier, regional metamorphic assemblages. The available chronology of metamorphic events does not allow detailed resolution of the thermal history during this ca. 160 Ma period; however, resetting of Rb–Sr ages at 1.70–1.75 Ga in other portions of the Dakota segment of the Trans-Hudson Orogen implies a relatively late, widespread thermal event approximately coeval with emplacement of the Harney Peak Granite (Van Schmus *et al.*, 1987). Whether this thermal and plutonic episode records a late tectonic phase of the Trans-Hudson Orogen or, alternatively, magmatic events related to accretion in the Central Plains Orogen remains a fundamental problem in the evolution of this region.

Metamorphic intensity increases from low grade in the central Black Hills southward toward the granite; garnet zone and lower grade assemblages are succeeded by staurolite and a broad area of sillimanite zone metamorphism in the southern Black Hills (Fig. 2). Sillimanite–K-feldspar zone assemblages occur in a restricted area along the southwest flank of the Harney Peak dome and are probably present in local areas to the south, where the regional geology is poorly known. Andalusite and cordierite are common in the staurolite zone, indicating a relatively low pressure regime. A static thermal aureole characterized by late overprinting of metamorphic minerals is evident in many areas a few kilometres from the granite and contrasts with the well-developed schistosity nearer the granite. This difference in the schistosity was attributed by Redden *et al.* (1985) to the late-stage expansion of the dome due to continued emplacement of granitic bodies.

The Proterozoic structural development of the Black Hills has been summarized by Redden and Norton (1975), DeWitt *et al.* (1986), and Redden *et al.* (this volume). Briefly, they believe an early period of north-directed nappelike folding and thrusting along east–west or northeasterly trending axes was followed by further folding along north-northwesterly axes. The main development of metamorphic minerals accompanied the latter event, which also produced the dominant schistosity and north-northwest-trending geological map pattern characteristic of much of the Black Hills. These events were followed by the emplacement of the Harney Peak Granite in the major domal structure, along with the development of several minor intrusive domes and the emplacement of thousands of small pegmatitic and granitic bodies in the adjacent country rocks.

The widespread distribution of granitic rocks in the southern Black Hills south of about latitude 44°N gives way northward to a region of generally lower metamorphic grade and no known Proterozoic granitic intrusions (Fig. 2). Near the northeastern limit of exposure of Proterozoic rocks, near the city of Deadwood, the metamorphic grade increases to above the staurolite isograd (Noble and Harder, 1948). Furthermore, pegmatitic granite resembling Harney Peak Granite is known to occur in this area both in outcrop (Ferguson and Turgeon, 1908) and in the subsurface, where it has been recovered from drill cores. A Rb–Sr model age consistent with the 1.7 Ga age of the Harney Peak Granite was obtained on one such sample by Z.E. Peterman (personal communication, 1987). The geochronologic data and isograd pattern, along with the lithologic similarities, suggest the existence of additional plutons related to the Harney Peak Granite in the extreme northeastern portion of the Proterozoic rocks. These plutons may extend north and east beneath the Phanerozoic cover.

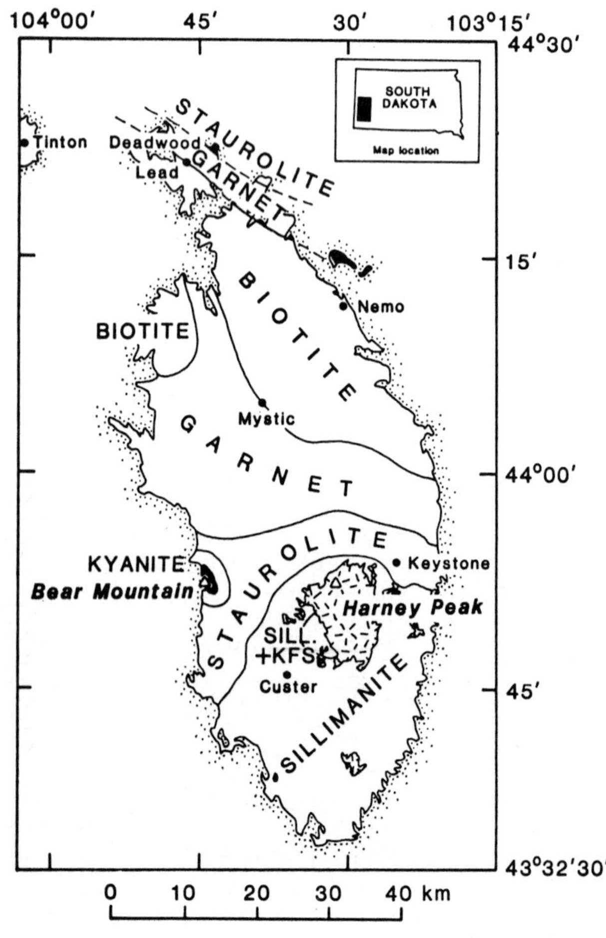

Figure 2. *Map of the Precambrian core of the Black Hills, South Dakota, showing the distribution of known metamorphic zones (identified by key index minerals) and the Harney Peak Granite (random line pattern). Solid black areas are Archean rocks. Stippled pattern indicates Phanerozoic cover. Modified from Redden and Norton (1975).*

Minor exposures of Precambrian pegmatite and granite also occur in the Tinton dome, northwest of the main exposures of Precambrian rocks (Fig. 2). Minor lithologic differences exist, however, between both the granites and the country rocks in this area and those exposed in the rest of the Black Hills region. No age determinations are available on these rocks and it is possible that they are of Archean age.

Pegmatitic granite and zoned pegmatites also occur in the Hartville Uplift approximately 150 km south-southwest of the Black Hills (Fig. 1). Here, pegmatites that intrude Archean rocks have been dated as 1.7 Ga or younger (Snyder, 1980).

GENERAL PETROLOGY AND PETROGRAPHY OF THE HARNEY PEAK GRANITE

The Harney Peak Granite is a classic example of a granitic pluton that is surrounded by a well-known pegmatite district containing many thousands of individual intrusive bodies. These intrusives range from bodies that are essentially identical to the dominant lithologies of the main pluton to bodies of pegmatite including classic zoned types (Page *et al.*, 1953; Norton *et al.*, 1964). Chemical and compositional zonation is a prominent feature of both the central granitic dome and the surrounding pegmatite field and will be discussed in a later section.

The Harney Peak Granite is made up of a variety of textural and chemical phases of overall granitic composition. Some important mineralogical and chemical characteristics of the Harney Peak Granite are summarized in Table I. On the basis of the peraluminous composition of the granite, low Ca content, high values of $^{87}Sr/^{86}Sr_{initial}$ and $\delta^{18}O_{SMOW}$, and low ε_{Nd}^{1715}, it is likely that the granite was derived by anatexis of a metasedimentary source region. However, attempts to model the origin of various portions of the granite system using the compositions of exposed metasedimentary or metaigneous protoliths have failed to account for all the isotopic (Walker *et al.*, 1986) or trace element characteristics (Shearer *et al.*, 1987b) of the granite. In addition, the Nd isotope data of Walker *et al.* (1986) suggest that the protolith was at least in part of Archean age. Redden *et al.* (1985) estimated that final crystallization of the Harney Peak Granite occurred at approximately 3 kbar (300 MPa), corresponding to a crustal depth of 10–11 km.

STRUCTURE OF THE GRANITE–PEGMATITE MASSES

Distribution

The general distribution of granite and pegmatite bodies of the Harney Peak system is shown in Figure 3, along with known metamorphic isograds. The outermost intrusives occur between the staurolite and sillimanite isograds. They extend to the south until concealed by Phanerozoic rocks and their southern limit is not known. The granite contact in Figure 3 is arbitrary in many places in that it is drawn where it is no longer possible to map individual country rock screens on any reasonable scale. It is also evident from Figure 3 that there are additional areas of granite, and that the granitic rocks cannot be considered as a simple system structurally centred around the main Harney Peak dome.

Table I
Chemical and mineralogical characteristics of Harney Peak Granite*

		Source
SiO_2	72.5–77.1 wt.%	Shearer *et al.* (1987b)
Al_2O_3	12.3–17.1 wt.%	Shearer *et al.* (1987b)
Normative corundum	1.7–7.2 wt.%	Shearer *et al.* (1987b), Redden *et al.* (1985)
CaO	0.16–1.20 wt.%	Shearer *et al.* (1987b)
Na_2O	2.5–8.0 wt.%	Shearer *et al.* (1987b)
K_2O	0.8–6.5 wt.%	Shearer *et al.* (1987b)
$K_2O/(K_2O + Na_2O)$	0.06–0.68	Shearer *et al.* (1987b)
$^{87}Sr/^{86}Sr_{initial}$	0.7157	Riley (1970)$^+$
ε_{Nd}^{1715}	-2.0 to -9.9	Walker *et al.* (1986)
$\delta^{18}O_{SMOW}$	10.9–13.8‰	Walker *et al.* (1986)
Varietal minerals	Muscovite ± biotite, tourmaline, garnet, sillimanite	Redden *et al.* (1985)
Accessory minerals	Apatite, zircon, monazite	Redden *et al.* (1985)

* Does not include chemically and mineralogically diverse rare element pegmatites in the Harney Peak system.

$^+$ Recalculated by DeWitt *et al.* (1986) using $\lambda_{Rb} = 1.42 \times 10^{-11}$ year^{-1}

The distribution of granite and pegmatite bodies that are too small to be mapped individually was compiled by Norton (1975) and is shown in Figure 3B by means of contours indicating the number of pegmatites (or small granitic bodies) per square mile (2.5 km^2). Most of the area within the granite–pegmatite field has been mapped at a scale of 1:24,000 or larger; however, the distribution in the area south and southeast of the Harney Peak dome is based on reconnaissance. It has been estimated that in excess of 20,000 mappable bodies make up the exposed granite–pegmatite field, although only about 2% of these are zoned pegmatites with any commercial potential (J.J. Norton, personal communication, 1988). The contours in Figure 3B indicate that much of the granite–pegmatite field is displaced to the south relative to the main Harney Peak pluton, and that the highest concentration of intrusives lies in an area 15 km southwest of the main dome, where there are in excess of 300 intrusions per square mile (2.5 km^2). It should be noted, however, that because the contours are based on numbers of individual bodies they do not necessarily reflect the volumetric distribution of granite and pegmatite.

General structural characteristics of the intrusive bodies

Most of the smaller intrusives in the Harney Peak system are tabular or lensoid bodies concordant with schistosity and bedding in the country rock, although discordant, dykelike bodies and irregular masses with steep to vertical dips are also present. An additional structure, characteristic of some intrusions, consists of fingerlike protrusions of granite, which are similar to those described by Noble (1952) and Pollard *et al.* (1975).

The internal structure of most of the smaller bodies is simple and homogeneous, except for a thin fine-grained border zone. Other intrusives are weakly layered and may show a greater proportion of perthite megacrysts near the hanging-wall contact. Zoned bodies can be exceedingly complex, displaying both concentrically and vertically differentiated zones, replacement bodies, and fracture-filling units (Page *et al.*, 1953).

Detailed examination of the contacts and adjacent country rocks indicates that emplacement of most of the individual granite or pegmatite bodies was accompanied by distension and plastic deformation of the enclosing rocks. Evidence includes induced parallelism of country rock schistosity to intrusive contacts (Kupfer, 1963, p. E16–E17; Redden, 1963, p. 255–256) and reversal of dips of pre-intrusive schistosity above the tops of some vertical dykes and domelike bodies (Redden, 1963, p. 264; Duke *et al.*, 1988).

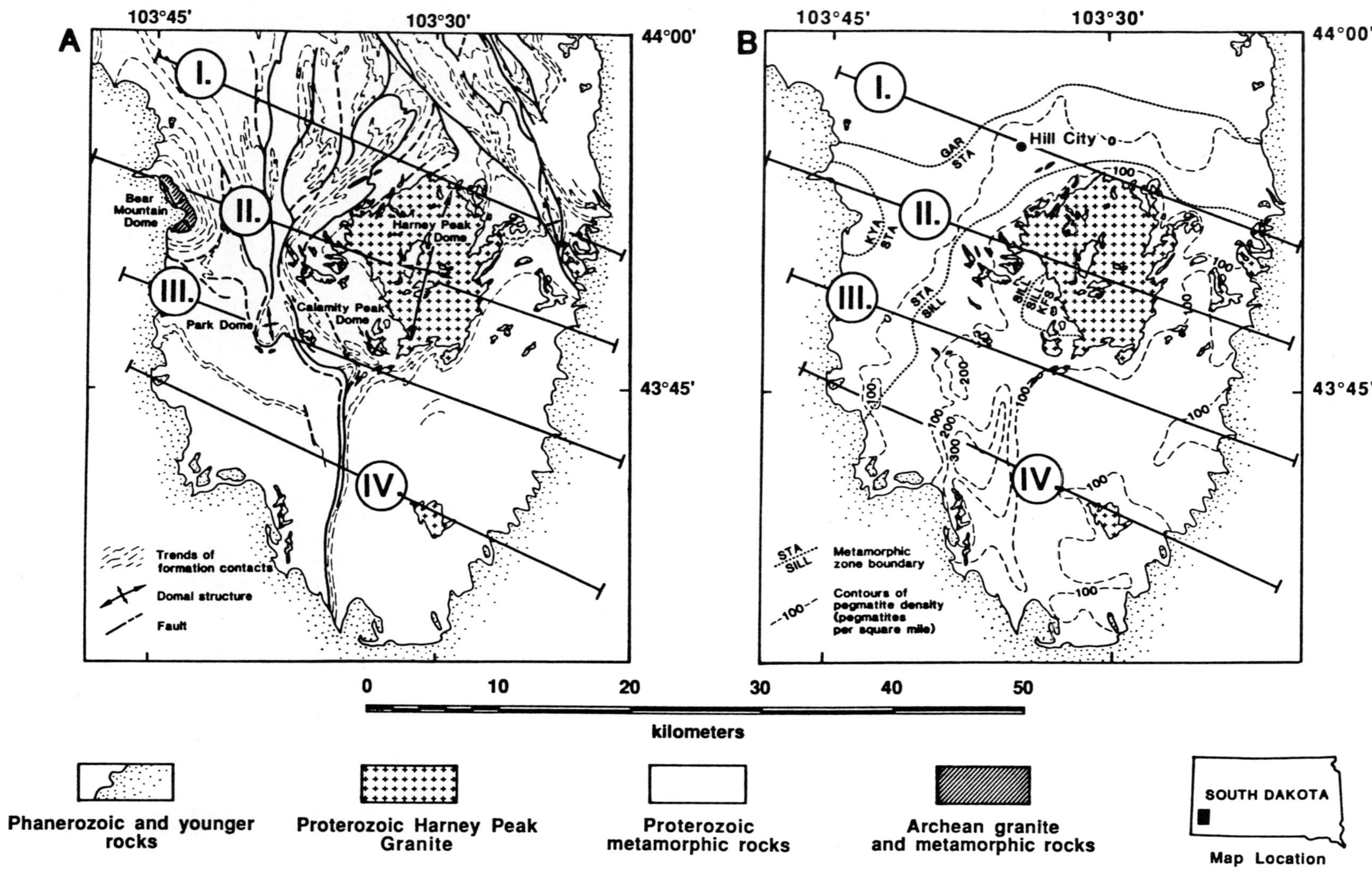

Figure 3. *Maps of the southern Black Hills showing structural and magmatic features of the region. Roman numerals indicate the locations of gravity profiles presented in Figure 7. (A) Generalized structural map showing trends of known lithologic contacts and major domal structures. Most faults are synmetamorphic and predate the doming events. Modified from a compilation by DeWitt* et al. *(1986, Plate 1). (B) Metamorphic and magmatic framework illustrating the distribution of metamorphic zones and estimated density of pegmatites or small granitic bodies. Based on Redden* et al. *(1985, Fig. 1).*

Redden (1963, p. 223) documented a general degree of host rock control on the shape, size, and attitude of the smaller granite–pegmatite bodies in part of the area. Virtually all the bodies in relatively incompetent micaceous schists are tabular sills, whereas discordant irregular bodies are common along fractures in more competent hosts, such as quartzo-feldspathic schists and amphibolites.

Virtually all of the larger granitic masses in the Harney Peak system are multiple intrusions and display complex internal structures. Several of the better documented plutons are described in greater detail in the following sections to illustrate the important structural features of the larger granite bodies.

Harney Peak dome

The main body of the Harney Peak Granite occupies an intrusive dome, roughly 100 km^2 in area, formed by the coalescence of a multitude of large sill-like or sheetlike bodies and subordinate dykes and pluglike masses (Fig. 4). Bedding and schistosity in country rocks surrounding the dome generally dip outward from the granite, although locally the granite does have crosscutting contacts with the metasedimentary rocks (Fig. 3A). Screens and large inliers of metamorphic country rock within the pluton reflect the overall domal structure. This is particularly true of the structurally lowest inliers. These strata form a coherent "xenolith dome" within the complex (Runner, 1943; J.A. Redden, unpublished data) and they are correlated with the

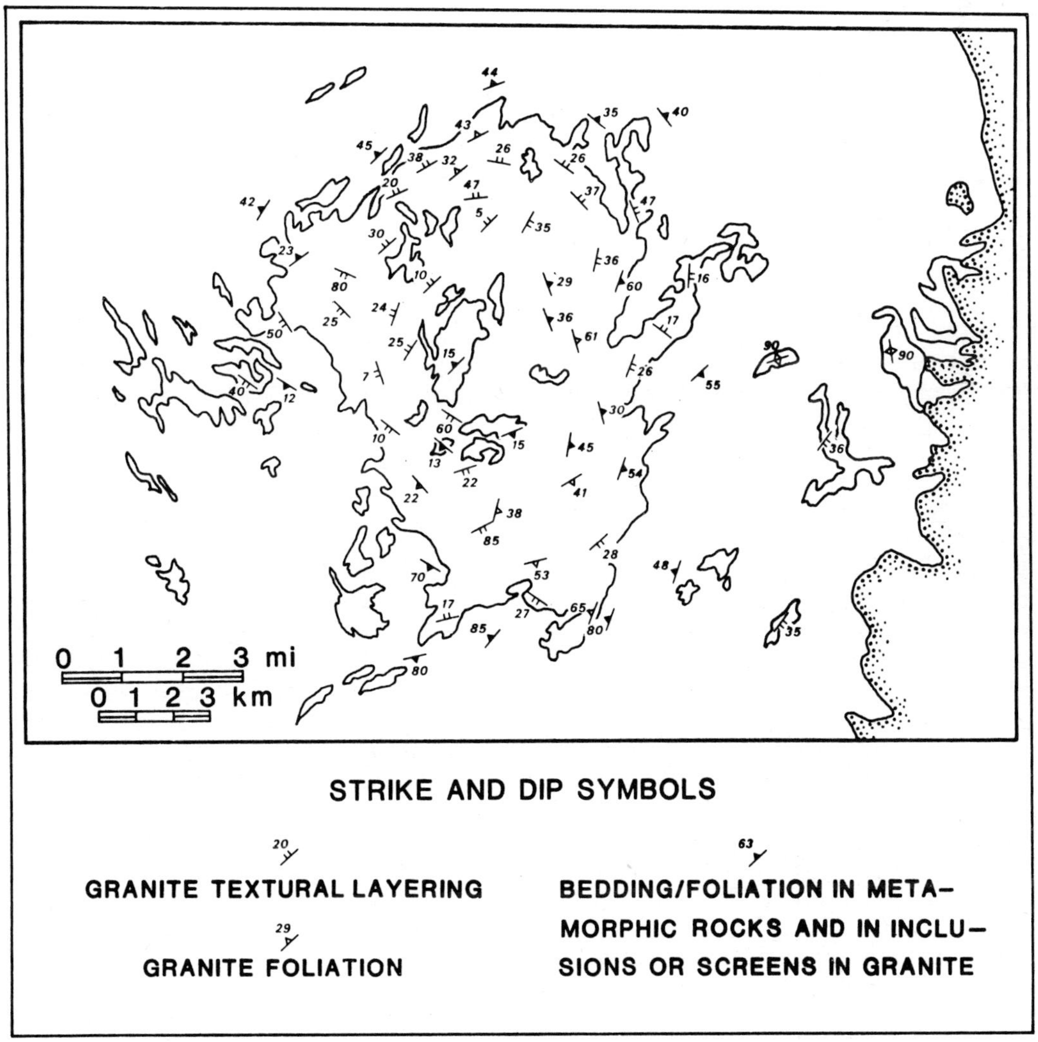

Figure 4. *Generalized structural map of the Harney Peak dome showing the attitudes of planar fabric elements in the granite, host rocks, and inclusions. Based on maps of Ratté and Wayland (1969), Norton (1976), and unpublished maps of the authors.*

oldest parts of the Proterozoic section exposed at Bear Mountain (Figs. 2 and 3A), indicating that Archean rocks may occur at a shallow depth in the Harney Peak dome.

The major sheetlike masses that make up the bulk of the dome are inferred to be 30–100 m thick, although thicker bodies probably occur, and to extend for up to 2000 m along strike. These sheets generally conform with the regional domal structure, although they vary from grossly concordant to discordant with respect to schistosity and bedding in adjacent country rock, and, thus, are not sills in the strict sense. Where numerous sheets coalesce to form the larger plutonic masses, they are commonly separated by a variable thickness of country rock in concordant attitude, although they may occur in direct contact with the adjacent granite sheets. The metasedimentary inliers tend to be poorly exposed. Thus, only the general distribution is shown in Figures 3 and 4. It is evident, however, that on a local scale country rock screens and other inliers make up a significant proportion of the total volume of the intrusive dome.

The major sheetlike bodies are not simple, homogeneous intrusive units as each is generally characterized by numerous internal contacts and textural or compositional variations. Whether the internal contacts are the result of multiple intrusion forming a composite mass or whether they result largely from internal differentiation processes and autoinjection is not always clear. The principal types of internal structures are textural layers and minor fracture-filling units. Textural layering consists of alternating segregations of coarse-grained pegmatitic granite and fine- to medium-grained granite. Although this layering is present to a greater or lesser degree in much of the pluton, it tends to be better developed in bodies of granite near the perimeter of the dome. Layers vary in thickness in different parts of the complex, but a general range between 0.1 and 2.0 m is typical. Individual textural layers virtually always parallel the contacts of the major sheetlike bodies in which they occur, and extend along strike for distances of over 100 m. Fracture-filling bodies include thin aplite dykes and various pegmatitic units 0.1–5.0 m thick, which commonly are oriented roughly normal to the upper and lower contacts of the enclosing sheets with steep to vertical dips but which may occur with other attitudes.

In addition to the planar fabric elements imparted by textural layering, the attitudes of major intrusive sheets, and the orientations of country rock screens in the pluton, many portions of the Harney Peak dome display a weak to moderate foliation due to the alignment of muscovite and biotite. Foliation is observed only in the finer grained varieties of granite, and in these is restricted to muscovite- and/or biotite-rich rocks, being absent in the tourmaline-rich phases. The various planar fabrics are generally parallel and collectively serve to outline the domal structure of the pluton (Fig. 4).

Deformation in country rocks surrounding the main dome is widespread and attests to the intrusive forces accompanying inflation of the dome. On a regional scale, country rock structures generally dip radially away from the dome on all sides and early faults and isoclinal folds have been deformed and rotated into parallelism with the dome. Dynamic recrystallization close to the dome resulted in a well-developed late schistosity subparallel to contacts, which is surrounded by a more typical thermal overprinting as previously described. Stretching on the flanks and locally, where exposed, on the upper part of the dome is indicated by flattening of metamorphic mineral aggregates, abundant boudinage structure, and thinning of stratigraphic units.

Calamity Peak pluton

The Calamity Peak pluton is a small domelike mass located 2 km southwest of the main Harney Peak dome; it has been studied in detail as a possible analogue of the much larger Harney Peak dome (Fig. 5). The pluton covers an area of only about 1 km^2, yet the generally good exposures and approximately 200 m of relief provide three-dimensional control on the internal structure of the intrusive system. Detailed structural studies have been conducted by Kupfer (1960, 1963) and Duke *et al.* (1988). The following description is based largely on the results of Duke *et al.* (1988).

Layered structures similar to those in the main Harney Peak dome are exceptionally well developed at Calamity Peak. Most conspicuous is the metre-scale textural layering, which is prominent throughout most of the pluton (Fig. 6). Coarse, pegmatitic layers rich in perthitic K-feldspar alternate with fine- to medium-grained layers of equigranular granite that are relatively enriched in sodic plagioclase. Kupfer (1963) and Redden (1963) reported correspondingly high K and Ba in coarse layers and Na, Ca, Fe, and Mn in fine layers at Calamity Peak. These results are largely corroborated by the study of Rockhold *et al.* (1987).

Contacts between coarse- and fine-grained layers can be either sharp or very diffuse. Fine-grained layers, however, commonly show a distinctive compositional banding consisting of alternating layers 2–20 mm thick containing concentrations of light and dark minerals (Fig. 6), a feature termed "line rock" by Schaller (1925) in the Pala pegmatite district in California. Similar fine-scale banding is present locally in the main Harney Peak dome and, rarely, in the wall zones of zoned pegmatites. At

Calamity Peak, light layers are enriched in plagioclase and muscovite, whereas dark layers are rich in tourmaline, garnet, and K-feldspar (Rockhold *et al.*, 1987). The line rock layering reflects changes in mineral abundance only, with little or no variation in the chemical composition of the individual phases (Rockhold *et al.*, 1987).

As shown by Kupfer (1960, 1963), the overall pattern of internal layering, as well as the geometry of the pluton contacts, is that of an intrusive dome (Fig. 5). Structural relationships between the pluton and regionally dipping schistose country rocks, however, do not indicate a simple domal structure. Schistosity and bedding conform to the contact only along the southern margin of the pluton. Country rock schistosity and bedding are at a high angle to the contact at the apparent floor of the pluton, which is exposed in the core of the "dome" and on the northern side. Marked discordancy is also indicated by the attitudes of numerous small country rock inclusions in the pluton. These generally lie sandwiched between textural layers and show no evidence of having achieved their present positions through stoping, but instead appear to have been "shouldered" into their present positions due to emplacement of the granite.

A striking structural feature at Calamity Peak is the occurrence at about a dozen sites, of truncated textural layering resembling sedimentary unconformities (Duke *et al.*, 1988). Both metre-scale textural layering and the associated line rock layering, if present, are truncated along knife-edge contacts, above which conformable layering resumes. A significant feature of the unconformities is that all truncations occur upward or outward toward the outer contacts of the pluton. Thus, as with sedimentary unconformities, these structures give a sense of the direction of "topping" or "younging" during the growth of the layered series. This type of truncated layering has not been observed in the larger Harney Peak dome, however, which has not been mapped in nearly as much detail as Calamity Peak.

Park dome

In contrast to the Calamity Peak dome, the Park dome (Fig. 3A), about 8 km to the west, is a small, slightly elongate dome in the metasedimentary rocks with only minor exposures of granitic rocks (Redden, 1968). The dome is defined by reversal of the regional south-southwest dip of the metasedimentary rocks and warps an early low-angle fault into concordancy with the dome. Early folds are refolded by late chevron and more open folds, which trend parallel to the longer axis of the structure and tend to die out down plunge. At the present level of erosion, the structural relief of the dome is on the order of 200–300 m.

Because of the anomalous structure and the presence in the central part of the dome of large K-feldspar porphyroblasts, which postdate the late folding, Redden (1968) believed that the dome is cored at a relatively shallow depth by Harney Peak Granite. Because flattening and local reversals of dip are known above some of the large intrusive bodies in the area and potassic metasomatism is commonly observed near large bodies of granite (Redden *et al.*, 1985), the evidence for a granitic core is strong. The size of the Park dome is comparable to that which must have existed above the Calamity Peak dome, so one might speculate that the structure of the granite below the Park dome is similar to that at Calamity Peak.

Bear Mountain dome

A large and prominent domal structure exists at Bear Mountain 10–15 km west-northwest of the Harney Peak dome, near the limit of exposed Precambrian rock. Small bodies of granite exposed near the core of the dome are similar in appearance to Harney Peak Granite, but are of Archean age (2.5 Ga, Gosselin *et al.*, 1988). Lower Proterozoic metasedimentary and metavolcanic rocks unconformably overlie the Archean section on the flanks of the dome. Structural evidence indicates that formation of the dome was roughly contemporaneous with doming in the Harney Peak area (Ratté, 1986), and muscovite fillings between boudins, which appear to have formed during doming, yielded a Rb–Sr age of 1.68 Ga, consistent with the age of the Harney Peak Granite (R.E. Zartman *in* Ratté, 1986). Thus, it is likely that the Bear Mountain dome is also cored by unexposed Harney Peak Granite.

PRELIMINARY GRAVITY MODELING

Simple two-dimensional gravity models have been constructed for four profiles through the granitic terrane of the southern Black Hills (Figs. 3 and 7). The overall granitic lithology and low content of mafic minerals in the Harney Peak Granite should generally result in a density contrast between the granite and more micaceous wall rocks. Thus, the gravity data might provide information regarding the subsurface distribution of the granitic plutons.

The models presented here are based on the regional Bouguer gravity map of Kleinkopf and Redden (1975) contoured at 1 mGal (note: 1 Gal = 1 cm/s^2) intervals. In the area of interest, Kleinkopf and Redden (1975) corrected their measurements for terrain effects. The gravity stations are rather widely spaced, however, generally with less than

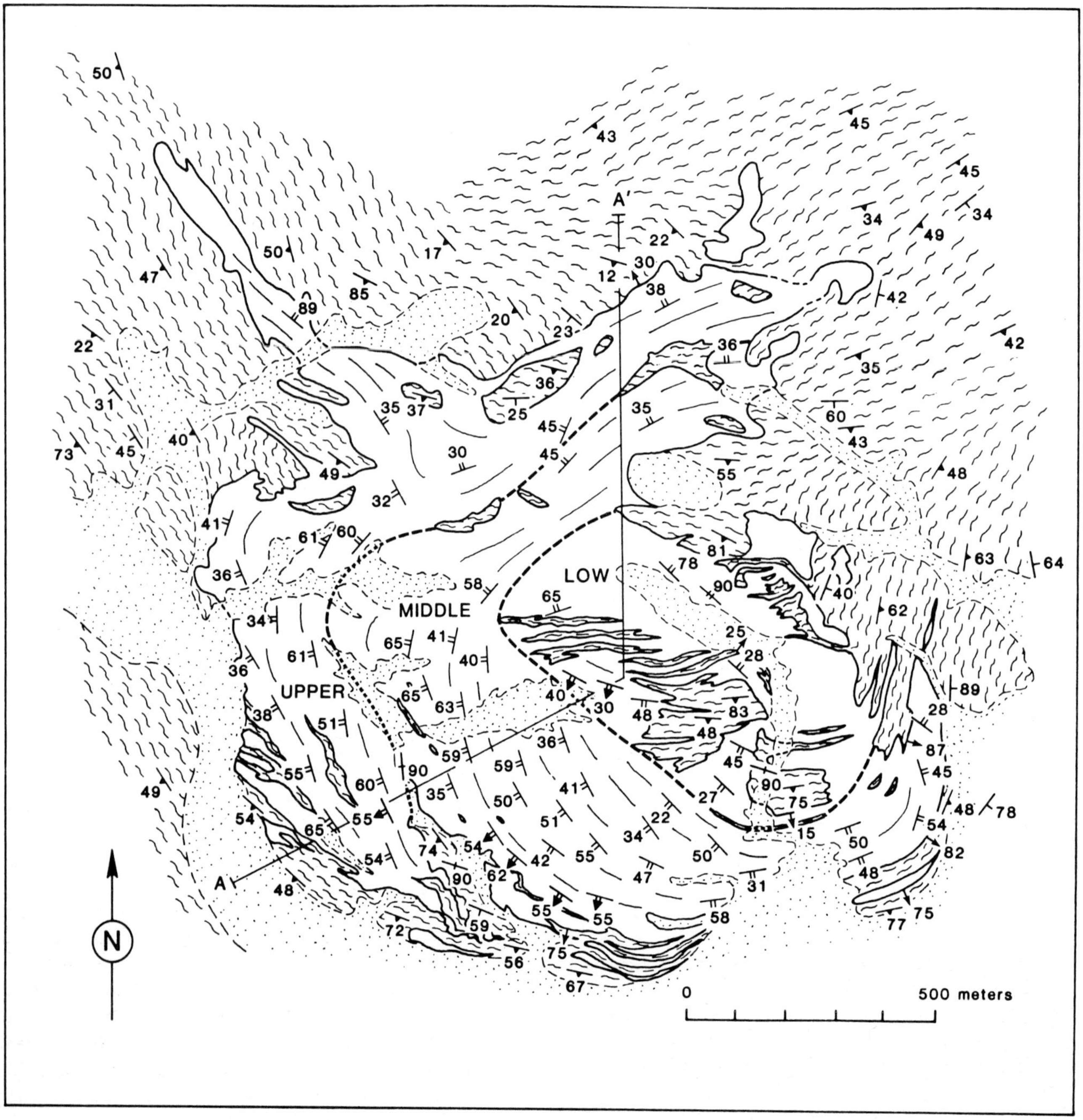

STRUCTURE SECTION ALONG LINE A–A′

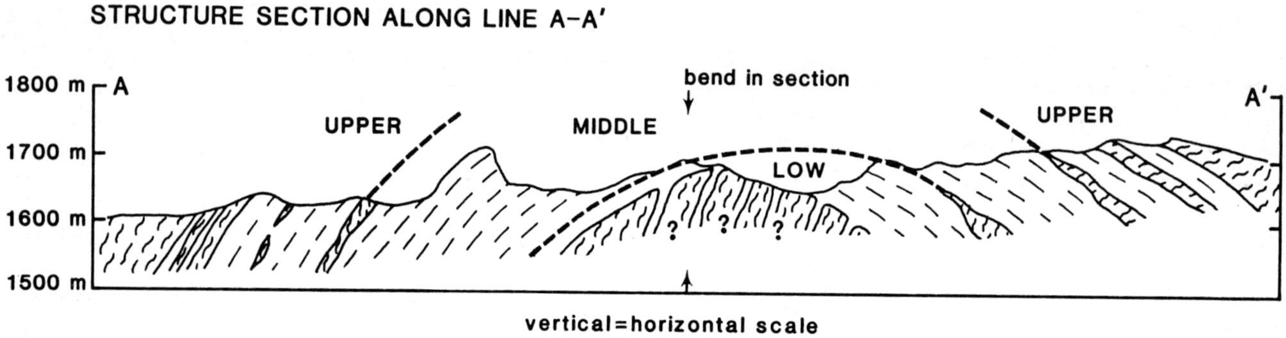

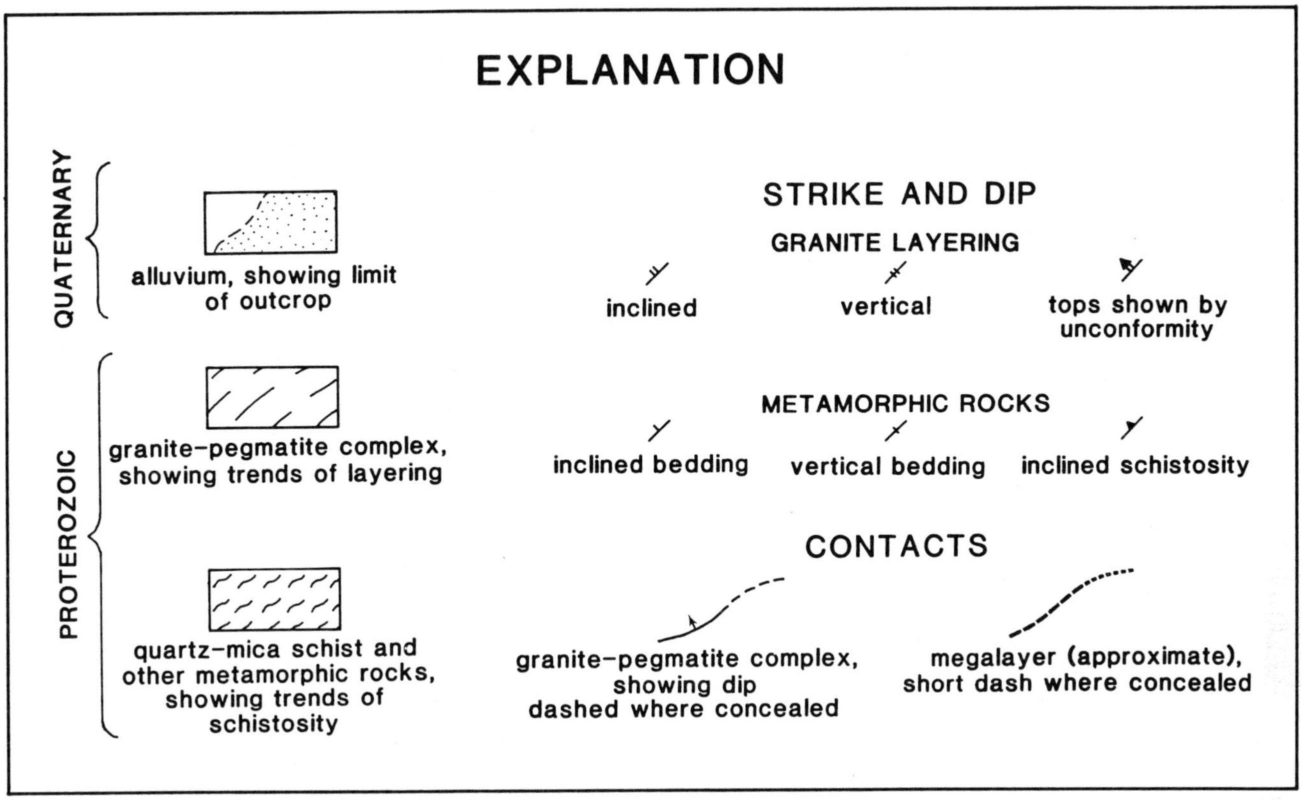

Figure 5. *Geological map and structural section (opposite page) of the Calamity Peak pluton. Three possible major intrusive sheets ("megalayers") are designated low, middle, and upper. From Duke* et al. *(1988).*

one station per 5 km^2 area; this severely restricts the sensitivity of the data and modeling. Residual gravity anomalies were calculated along each profile by subtracting regional gradients of 0.4–0.9 mGal km^{-1}, with gravity values decreasing from west to east across the area. The residual anomalies are small, usually less than 5 mGal, even over the large granitic body in the Harney Peak dome. Hypothetical granitic bodies were modeled in the subsurface assuming an effective density contrast of -0.06 to -0.07 g cm^{-3} with the enclosing country rock. This assumption is considered reasonable for leucocratic granitic rocks, containing minor amounts of country rock inclusions, in contact with the predominant quartz–mica schist lithologies of the metasedimentary rocks. The assumptions clearly are invalid locally where minor amounts of meta-iron formation or amphibolite occur in the country rocks or where the mica schist grades into quartzite.

Figure 6. *Well-developed textural layering in granite in the Calamity Peak dome. Alternating coarse- and fine-grained granite dip away from the viewer at about 65°. Fine-grained layers exhibit millimetre-scale (line rock) banding.*

The model profiles shown in Figure 7 reproduce the general features of the observed gravity data by assuming a series of simple granitic bodies at depth or extensions of surface bodies consistent with available structural data. Because of the sparse distribution of gravity stations and the many simplifying assumptions upon which the models are based, the results shown in Figure 7 are only approximations and cannot be viewed as unique solu-

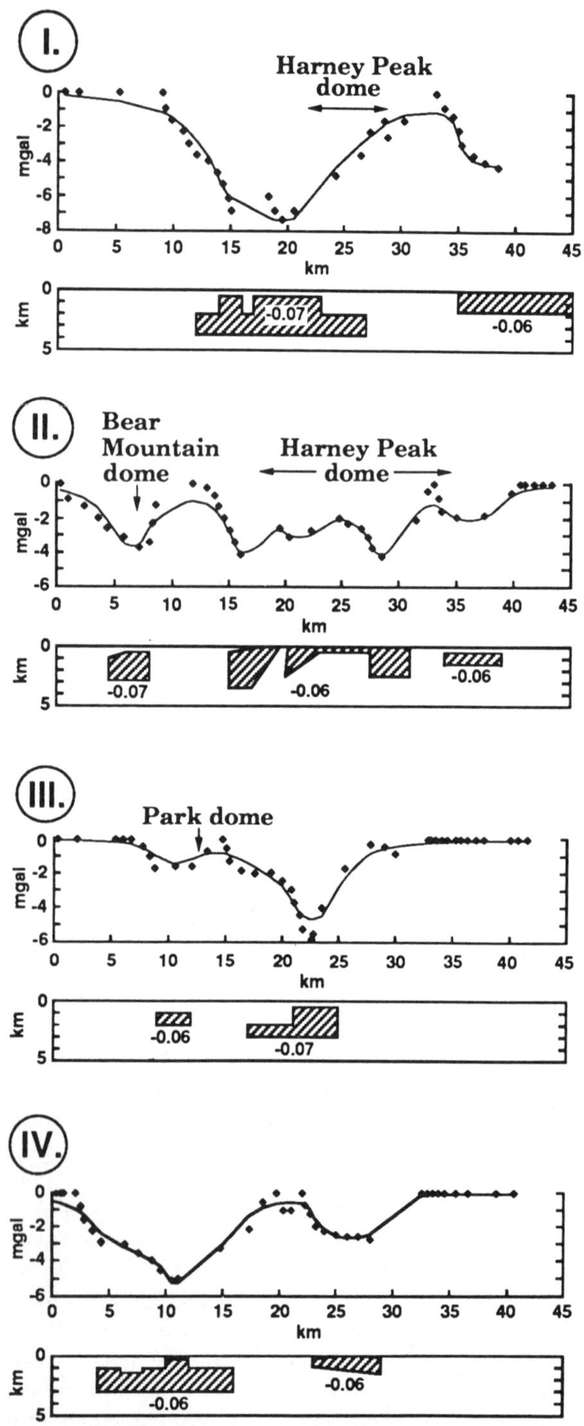

Figure 7. *Observed (dots) and calculated (solid lines) Bouguer gravity along four profiles through the granitic terrane of the southern Black Hills. Hypothetical subsurface granitic bodies are shown by ruled patterns and are assumed to have a density contrast of -0.06 or -0.07 g cm*$^{-3}$ *relative to the metasedimentary country rocks in these models. See Figure 3 for location of gravity profiles. (Note: 1 Gal = 1 cm/s*^{2}.*)*

tions. Important uncertainties exist in the magnitude of the regional gradient and the precise density of the different rock units at depth. Nevertheless, the models do point out some important constraints imposed by the gravity data, which are independent of the specific model parameters. For example, the relatively low magnitude of the residual anomaly associated with the main Harney Peak Granite dome permits no greater than 5 km of granite beneath the dome, assuming any reasonable density contrast. Furthermore, the models predict that the central core of the Harney Peak dome may be underlain by as little as 0.5 km of granitic rock below which lies the "floor" of the intrusion (Fig. 7, line II). A major subsurface extension of the Harney Peak dome is indicated north and northwest of the main body in the vicinity of Hill City (Fig 7, line I), and a lesser extension may occur to the south-southwest (Fig. 7, line III). Because of the low magnitude of the anomalies and the low density contrasts, however, it is impossible to distinguish with any certainty the effects of changes in metasedimentary lithologies, e.g., mica schist versus metagreywacke or quartzite, from the influence of granitic bodies.

Other significant features of the preliminary gravity models include the indication of a sizable granite mass beneath the axis of the Bear Mountain dome (Fig. 7, line II), a smaller body located approximately under the Park dome (Fig. 7, line III), and the probable existence of additional granite bodies near the eastern limit of Proterozoic rocks extending under the Phanerozoic cover (Fig. 7, lines I and II). These features are consistent with the existence of sizable granitic bodies that are either mapped or inferred from independent structural or geochemical evidence. A major, unexpected residual anomaly is centred approximately 15 km southwest of the Harney Peak dome (Fig. 7, line IV) in a general area that has no mapped large granitic bodies, but has greater than 300 pegmatite bodies per square mile (2.5 km^{2}) locally. The modeled causative body, which reaches to a shallow depth below the surface, could be continuous with the main Harney Peak body, but more likely represents a separate intrusive complex. Despite the significant gravity anomaly, there is no structural evidence to support a granitic dome in this area and other causes for the anomaly cannot be ruled out.

In general, the locations of the modeled granitic bodies correlate with the mapped distribution of pegmatites. With the exception of the inferred body beneath the Bear Mountain dome and part of the northwest extension of the Harney Peak body, all modeled bodies are within the limits of occurrences of Proterozoic pegmatites at the surface. The majority of the proposed bodies occur beneath rocks of sillimanite or higher metamorphic grade, although the proposed body in the Hill City area, northwest of the Harney Peak dome may largely lie beneath lower grade rocks of the staurolite zone.

GEOCHEMICAL ZONATION IN THE GRANITE SYSTEM

A distinct geochemical zonation has been recognized for some time in the pegmatite district surrounding the Harney Peak Granite (Norton, 1975; Cerny, 1982; Redden *et al.*, 1985). Although not as well documented, a corresponding chemical and mineralogical zonation is present within the granitic rocks of the main Harney Peak dome and its satellite bodies and has been alluded to by various authors (Runner, 1943; Cerny, 1982; Redden *et al.*, 1985; Shearer *et al.*, 1987b).

As pointed out by Norton (1975) and Redden *et al.* (1985), most of the more highly evolved, zoned pegmatites that have been mined intensively are in the outer parts of the pegmatite district, generally between the 0 and 100 pegmatite bodies per square mile (2.5 km^2) contours of Figure 3B. Virtually all of the most differentiated bodies, including the Be-, Li-, Sn-, and rare element-enriched pegmatites occur below the 100 pegmatite bodies per square mile (2.5 km^2) contour and are, thus, near the limits of the pegmatite district. Clustering of rare element pegmatites of similar geochemical character, however, suggests that zonation in the pegmatite district is not related solely to distance from the main granite mass. Chemically discrete source magmas were apparently present in different parts of the overall Harney Peak system as a result of local variations in melting history or crystallization processes. The tendency for more evolved pegmatites to occur in clusters away from the main Harney Peak pluton or in areas of low pegmatite density is also reflected in the trace element chemistry of K-feldspar (Shearer *et al.*, 1987a). Whereas intrusives in areas of low pegmatite density (<100 per square mile (2.5 km^2)) have more evolved K-feldspar chemistry (Rb>4000 ppm, Cs>500 ppm), intrusives in areas of numerous small intrusions (>200 pegmatites per square mile (2.5 km^2)) tend to have more primitive K-feldspar chemistry (Ba>140 ppm, Rb<1000 ppm, Cs<30 ppm).

An additional significant feature in the overall chemical distribution surrounding the Harney Peak Granite is the localization of Sn–Be pegmatites and W vein mineralization in the Hill City area 6 km north-northwest of the main Harney Peak dome (Page *et al.*, 1953; Redden, 1968; Ratté and Norton, 1975). The pegmatites are dominated by quartz and contain lesser amounts of muscovite with minor albite, cassiterite, and beryl. As such, they contrast markedly with the more feldspathic composition of most Harney Peak pegmatites. It seems likely that both the veins and the pegmatites overlie a cupola of granite or a north-northwest extension of the Harney Peak dome, a possibility suggested by the gravity models presented in this paper and by the flattening of regional dips off the main dome.

Zoning within the main Harney Peak dome is manifested in general changes in texture, mineralogy, mineral chemistry, and whole-rock chemistry. The overall pattern is that more fine-grained, homogeneous, equigranular varieties of granite with less evolved mineral and whole-rock chemical characteristics occupy the central and southeastern portions of the Harney Peak dome (Redden *et al.*, 1985; Shearer *et al.*, 1987b). These varieties locally display a weak to distinct mica foliation. Along the western, northern, and northeastern perimeters of the dome, the granite is typically more heterogeneous and pegmatitic, and displays more differentiated mineral and whole-rock chemical trends. Chemical variations from bodies of granite in the interior of the dome to those near the perimeter include outward decreases in whole-rock K/Rb, An content of plagioclase, and whole-rock Ca, Ba, and Sr compared with increases in Fe/(Fe+Mg) in ferromagnesian silicates and whole-rock Nb, Li, Cs, Be, and B (Shearer *et al.*, 1987b). Airborne surveys indicate Th lows of <7 ppm over much of the more differentiated granite along the northern and western perimeters of the dome compared with values of >12 ppm in the less fractionated granite in the southeastern part of the dome (DeWitt *et al.*, 1986; E. DeWitt, personal communication, 1986). Thorium concentrations in a limited number of surface samples (Shearer *et al.*, 1987b), however, generally do not substantiate the indicated trend.

Shearer *et al.* (1987b) demonstrated that K/Rb was among the most useful parameters for tracking the geochemical evolution in the Harney Peak system. The K/Rb distribution shown in Figure 8 illustrates the major chemical trends outlined above. The highest reported K/Rb values are in excess of 250 and occur near the structural core of the dome. In general, K/Rb values decrease outward to less than 100 in the marginal intrusions along the northern and northwestern flanks. Even lower bulk K/Rb can be calculated for the parent magmas of the more evolved pegmatite bodies near the edge of the pegmatite field by combining modal and mineral–chemical data (C.K. Shearer, unpublished data).

Mineralogical and textural zonation trends generally parallel the chemical data as shown by the K/Rb distribution. Portions of the granite with K/Rb less than about 200 are characterized by increasingly pegmatitic texture, by tourmaline as the most conspicuous ferromagnesian silicate (Fig. 8), and by only sporadic occurrences of biotite (<0.5 wt.% calculated from heavy mineral concentrates; J.J. Norton and R.J. McLaughlin, unpublished data). Granites of the central and southeastern portion of the dome contain >0.5 wt.% biotite and virtually no tourmaline. Sillimanite as an accessory mineral is more common in the biotite-rich, tourmaline-absent assemblages.

On the basis of the variations in mineral assemblages, mineral chemistry, and major and trace element whole-rock chemistry, Shearer *et al.* (1987b) concluded that the various granite phases in the main dome could be related by assuming a Rayleigh fractionation model and a parent composition similar to certain exposed portions of the central biotite granites. The authors suggested further that the products of crystal-melt fractionation may have been subsequently modified through vapour-phase transport of some elements. Alternative models, such as deriving the spectrum of rock types by variable degrees of partial melting of possible protoliths, were shown to be less satisfactory, largely on the basis of trace element modeling. Together with the spatial distribution described above — more fractionated bodies on the outside of the dome — the relative ages of the different granite types predicted by the fractional crystallization model of Shearer *et al.* (1987b) would agree with available textural and structural evidence and would indicate general outward growth of the entire domal complex (cf. Runner, 1943). The addition of younger magmas around an older nucleus would also agree with the model of Duke *et al.* (1988) for the smaller Calamity Peak dome.

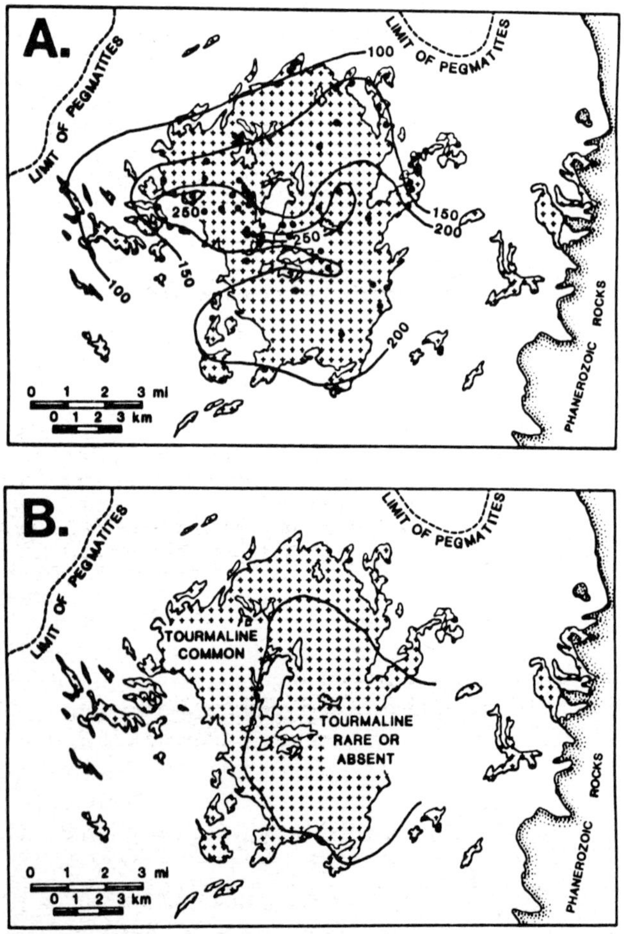

Figure 8. *(A) Map of the main Harney Peak Granite dome contoured with values of whole-rock K/Rb. Based on data of Shearer* et al. *(1987a) and unpublished data of J.J. Norton and R.J. McLaughlin. (B) Map of the main Harney Peak Granite dome showing the distribution of tourmaline as a conspicuous accessory mineral. Biotite is the main ferromagnesian phase in the tourmaline-absent region, but may be present locally in tourmaline-rich portions. Sillimanite is virtually restricted to the tourmaline-absent region.*

SUMMARY AND CONCLUSIONS

The Harney Peak Granite system provides the only detailed record of Early Proterozoic granitic magmatism in the Dakota segment of the Trans-Hudson Orogen. This review of field relations, structural features, and geochemical variations of granitic rocks in the Black Hills serves to highlight intrusive processes that may have been active along more extensive portions of the Trans-Hudson outside the area of basement exposures.

Structural and geochemical features described in the preceding sections for the Harney Peak, Calamity Peak, Park, and Bear Mountain domes are inferred to be representative of different structural levels through Harney Peak Granite plutons and their deformed country rocks. Bear Mountain reflects the highest structural level represented, and is inferred to result from doming over a Harney Peak-type intrusive mass at depth. The present erosional level in the Park dome may be closer to the underlying granite mass, as shown by apparent potassic metasomatism of the country rock, although the development of muscovite between boudin structures in amphibole schist of the Bear Mountain dome may be indicative of the highest level of potassic metasomatism. In both cases, the limited gravity data are consistent with the existence of granitic masses at depth. The Harney Peak and Calamity Peak domes reflect moderately deeply eroded levels exposing the domal granitic plutons along with their deformed country rocks. Collectively, these examples illustrate the formation of domal structures resulting from the emplacement of many of the larger masses of Harney Peak Granite.

These domal masses range from less than 1 km to greater than 10 km in diameter.

In the structures discussed above in which granitic material is exposed (Harney Peak and Calamity Peak), two additional structural characteristics are displayed: major sheetlike granitic bodies and textural layering on a scale of metres.

The structural integrity of some of the major masses of granite — typically 30–100 m thick intrusive sheets separated from adjacent sheets by screens of country rock — suggests that they may not have formed at exactly the same time or from exactly the same sources. Rather, they may have formed separately, either at different evolutionary stages during the growth of the overall intrusive dome or from different magma sources being tapped simultaneously. This contention is supported by the distinct chemical characteristics of each major intrusive sheet (Shearer *et al.*, 1987b). In particular, Shearer *et al.* (1987b) showed that the whole-rock K/Rb effectively discriminates between different sheets in the main Harney Peak dome, although individual samples from a sheet plot at different points along a linear K–Rb array.

Textural layering, typically measuring 0.1–2.0 m in thickness, is the predominant internal structural feature of the Calamity Peak complex and is conspicuous in many portions of the main Harney Peak pluton as well as in numerous other smaller intrusives in the system. On the basis of field relations at two small intrusives, Orville (1960) and Duke *et al.* (1988) concluded that the textural layers result from separate injections of granitic magma that separated into structurally upper, coarse-grained, perthite-rich pegmatite and lower, fine-grained, more sodic granite during crystallization. Superposition of tens, hundreds, or thousands of these layers, owing to repeated injection of magma, could have resulted in the multilayered coarse–fine sequences that make up the larger intrusive sheets and domal complexes. A similar origin for layered pegmatite–aplite intrusions was discussed in detail by Jahns and Wright (1951) and Jahns and Tuttle (1963).

The hundreds or thousands of intrusive pulses recorded by the textural layers in the larger plutons could reflect periodic tapping of a progressively melting source area or a deeper fractionating magma reservoir. Alternatively, the layers may have resulted from a steady-state supply of magma with periodically shifting conduits as earlier conduits became choked off as a result of relatively rapid crystallization. Thus, the major intrusive sheets and domal structures may have grown slowly with the active site of intrusion at any given time restricted to one or more fairly small sheetlike conduits, which are now represented by a texturally layered coarse–fine couplet.

Given the evidence for growth of the major granitic domes by multiple intrusion of magma pulses, which then separated into textural layers, the unconformities between textural layers described in the Calamity Peak complex provide an indication of the sequence of growth of the layered series. The consistent upward truncations observed at Calamity Peak indicate that, in general, new intrusions were added to the outer and upper portions of the dome, outside of previously crystallized layers. Although no such relationships have been documented in the main Harney Peak dome, Runner (1943) speculated that the Harney Peak dome also grew outward on the basis of the generally more deformed fabric of intrusions near the interior of the dome. Outward growth is also consistent with the virtual absence of major crosscutting relationships within the plutons. Whereas systematic chemical changes from the interior to the exterior of the Harney Peak dome, as reported by Shearer *et al.* (1987b), are compatible with progressive outward growth of the dome accompanying fractionation of the granitic magmas, the data do not preclude systematic inward growth.

The major conclusions of this paper are as follows:

(1) Anatectic, S-type granite and associated pegmatite, ~1.72 Ga in age, is widespread in the southern Black Hills, South Dakota, but is also known in the northern Black Hills and in the Hartville Uplift approximately 150 km to the south-southwest. Although basement exposures are restricted, these occurrences support a widespread magmatic episode accompanying crustal anatexis during the late stages of development of the Trans-Hudson Orogen or early stages of the younger Central Plains Orogen.

(2) In the Black Hills, many thousands of small granitic and pegmatitic bodies occur, making their first appearance above the staurolite isograd, with larger bodies restricted to above the sillimanite isograd. This distribution of granitic rocks with respect to metamorphic grade, along with simplified gravity models, suggests that much of the sillimanite-grade terrane is underlain by variable concentrations of large and small granitic bodies.

(3) The smaller granitic bodies dominantly form sills, and to a lesser extent dykes or irregular masses, whereas many of the larger bodies tend to form complex domal intrusions with a layered internal structure composed of hundreds or thousands of individual intrusions. The main Harney Peak dome measures ~100 km^2, but several smaller domes are known or inferred to be cored by pegmatitic granite. The domes show evidence of forcible emplacement late in the metamorphic/tectonic cycle.

(4) Simple two-dimensional models based on available Bouguer gravity data suggest that the main Harney Peak dome does not extend more than

a few kilometres below the surface, although other sizable granitic bodies may occur beneath the sillimanite-grade terrane of the southern Black Hills.

(5) The domelike plutons are made up of a number of large sheetlike granitic bodies that are structurally and chemically distinct (at the trace element level), but show chemical variations in trace elements between different sheets in the pluton. Many of the major sheets are layered on a scale of 0.1–2.0 m, suggesting that they too result from multiple intrusion of tens or hundreds of genetically related magma pulses or from autoinjection within larger bodies. In the small pluton at Calamity Peak, upward and outward truncations in these intrusive layers provide structural evidence that the dome grew by addition of new magma pulses to the outside. A similar model of general outward growth applied to the main Harney Peak dome is consistent with textural relationships in the surrounding metamorphic rocks, with the generally more deformed fabric of interior portions, and with the fractional crystallization model of Shearer *et al.* (1987b).

(6) The main Harney Peak dome is compositionally zoned from chemically unfractionated, high K/Rb biotite granite in the centre and southeast portions to fractionated, low K/Rb tourmaline granite and pegmatite nearer the margins, consistent with outward growth during fractional crystallization of magmas resembling the biotite granites. Further zonation in the Harney Peak system is well documented in the pegmatite field surrounding the main dome with rare element, Li-rich pegmatites concentrated at the edge of the district. Some additional differences in pegmatite characteristics within the field, such as the tendency toward clusters of geochemically similar intrusives, may be attributable to separate parent granitic bodies within the Harney Peak system as a whole.

ACKNOWLEDGEMENTS

We thank the editors of this volume for including Black Hills magmatism in the broader context of the Trans-Hudson Orogen. E. DeWitt, J.J. Norton, and Z.E. Peterman provided unpublished data on the Harney Peak Granite. W.M. Roggenthen assisted with gravity modeling. The comments of M.E. Bickford, J.J. Norton, and an anonymous reviewer helped clarify portions of the manuscript and are greatly appreciated. This research was supported by the United States Department of Energy through Department of Energy Grant DE-FG01-84ER13259 (to J.J. Papike).

REFERENCES

Balk, R., 1931, Inclusions and foliation of the Harney Peak Granite, Black Hills, South Dakota: Journal of Geology, v. 39, p. 736–748.

Bickford, M.E., Van Schmus, W.R. and Zietz, I., 1986, Proterozoic history of the midcontinent region of North America: Geology, v. 14, p. 492–496.

Cerny, P., 1982, Petrogenesis of granitic pegmatites, *in* Cerny, P., ed., Granitic Pegmatites in Science and Industry: Mineralogical Association of Canada, Short Course Handbook, v. 8, p. 405–461.

Darton, N.H. and Paige, S., 1925, Description of the central Black Hills [South Dakota]: United States Geological Survey, Geologic Atlas, Folio 219.

DeWitt, E., Redden, J.A., Wilson, A.B. and Buscher, D., 1986, Mineral resource potential and geology of The Black Hills National Forest, South Dakota and Wyoming: United States Geological Survey, Bulletin 1580, 135 p.

Duke, E.F., Redden, J.A. and Papike, J.J., 1988, Calamity Peak layered granite–pegmatite complex, Black Hills, South Dakota. Part I. Structure and emplacement: Geological Society of America Bulletin, v. 100, p. 825–840.

Ferguson, H.G. and Turgeon, F.A., 1908, An occurrence of Harney granite in the northern Black Hills: Museum of Comparative Zoology Bulletin, v. 49, p. 275–282.

Gosselin, D.C., Papike, J.J., Zartman, R.E. and Peterman, Z.E., 1988, Archean rocks of the Black Hills, South Dakota: reworked basement from the southern extension of the Trans-Hudson Orogen: Geological Society of America Bulletin, v. 100, p. 1244–1259.

Jahns, R.H. and Tuttle, O.F., 1963, Layered pegmatite–aplite intrusives: Mineralogical Society of America, Special Paper 1, p. 78–92.

Jahns, R.H. and Wright, L.A., 1951, Gem- and lithium-bearing pegmatites of the Pala district, San Diego County, California: California Division of Mines, Special Report 7-A, 72 p.

Kleinkopf, M.D. and Redden, J.A., 1975, Bouguer gravity, aeromagnetic, and generalized geologic maps of part of the Black Hills of South Dakota and Wyoming: United States Geological Survey, Map GP-903, 1:250,000, 11 p.

Kupfer, D.H., 1960, Pegmatite–granite relationships in the Calamity Peak area, Black Hills, South Dakota, U.S.A., *in* Neumann, H. and Neuvonen, K., eds., Minerals and Genesis of Pegmatites, Proceedings of Section 17, International Geological Congress, 21st (Norden), Copenhagen, Denmark, p. 77–93.

Kupfer, D.H., 1963, Geology of the Calamity Peak area, Custer County, South Dakota: United States Geological Survey, Bulletin 1142-E, p. E1–E23.

Lewry, J.F., Sibbald, T.I.I. and Schledewitz, D.C.P., 1985, Variation in character of Archean rocks in the western Churchill Province and its significance, *in* Ayres, L.D., Thurston, P.C., Card, K.D. and Weber, W., eds., Evolution of Archean Supracrustal Sequences: Geological Association of Canada, Special Paper 28, p. 239–261.

Lisenbee, A.L., 1978, Laramide structure of the Black Hills Uplift, South Dakota – Wyoming – Montana: Geological Society of America, Memoir 151, p. 165–196.

Noble, J.A., 1952, Evaluation of criteria for the forcible intrusion of magma: Journal of Geology, v. 60, p. 34–57.

Noble, J.A. and Harder, J.O., 1948, Stratigraphy and metamorphism in a part of the northern Black Hills and the Homestake mine, Lead, South Dakota: Geological Society of America Bulletin, v. 59, p. 941–975.

Norton, J.J., 1975, Pegmatite minerals, *in* Mineral and Water Resources of South Dakota. 2nd ed.: South Dakota Geological Survey, Bulletin 16, p. 132–149.

Norton, J.J., 1976, Field compilation map of the geology of the Keystone pegmatite area, Black Hills, South Dakota: United States Geological Survey, Open-file Map 76-297, 1:24,000.

Norton, J.J. *et al.*, 1964, Geology and mineral deposits of some pegmatites in the southern Black Hills, South Dakota: United States Geological Survey, Professional Paper 297-E, p. 293–341.

Orville, P.M., 1960, Petrology of several pegmatites in the Keystone district, Black Hills, South Dakota: Geological Society of America Bulletin, v. 71, p. 1462–1489.

Page, L.R. *et al.*, 1953, Pegmatite investigations, 1942–1945, Black Hills, South Dakota: United States Geological Survey, Professional Paper 247, 228 p.

Pollard, D.D., Muller, O.H. and Dockstader, D.R., 1975, The form and growth of fingered sheet intrusions: Geological Society of America Bulletin, v. 86, p. 351–363.

Ratté, J.C., 1986, Geologic map of the Medicine Mountain quadrangle, Pennington County, South Dakota: United States Geological Survey, Map I-1654, 1:24,000 (includes a section on the age of the granite at Bear Mountain and the Bear Mountain dome by R.E. Zartman).

Ratté, J.C. and Norton, J.J., 1975, Tin, *in* Mineral and Water Resources of South Dakota. 2nd ed.: South Dakota Geological Survey, Bulletin 16, p. 103–104.

Ratté, J.C. and Wayland, R.G., 1969, Geology of the Hill City quadrangle, Pennington County, South Dakota a preliminary report: United States Geological Survey, Bulletin 1271-B, p. B1–B14.

Redden, J.A., 1963, Geology and pegmatites of the Fourmile quadrangle, Black Hills, South Dakota: United States Geological Survey, Professional Paper 297-D, p. 199–291.

Redden, J.A., 1968, Geology of the Berne quadrangle, Black Hills, South Dakota: United States Geological Survey, Professional Paper 297-F, p. 343–408.

Redden, J.A. and Norton, J.J., 1975, Precambrian geology of the Black Hills, *in* Mineral and Water Resources of South Dakota. 2nd ed.: South Dakota Geological Survey, Bulletin 16, p. 21–28.

Redden, J.A., Norton, J.J. and McLaughlin, R.J., 1985, Geology of the Harney Peak Granite, Black Hills, South Dakota, *in* Rich, F.J., ed., Geology of the Black Hills, South Dakota and Wyoming. 2nd ed.: American Geological Institute, Alexandria, VA, p. 225–240.

Redden, J.A., Peterman, Z.E., Zartman, R.E. and DeWitt, E., 1990, U–Th–Pb geochronology and preliminary interpretation of Precambrian tectonic events in the Black Hills, South Dakota, *in* Lewry, J.F. and Stauffer, M.R., eds., The Early Proterozoic Trans-Hudson Orogen of North America: Geological Association of Canada, Special Paper 37, p. 229–251.

Riley, G.H., 1970, Isotopic discrepancies in zoned pegmatites, Black Hills, South Dakota: Geochimica et Cosmochimica Acta, v. 4, p. 713–725.

Rockhold, J.R., Nabelek, P.I. and Glascock, M.D., 1987, Origin of rhythmic layering in the Calamity Peak satellite pluton of the Harney Peak Granite, South Dakota: the role of boron: Geochimica et Cosmochimica Acta, v. 51, p. 487–496.

Runner, J.J., 1943, Structure and origin of Black Hills Precambrian granite domes: Journal of Geology, v. 51, p. 431–457.

Schaller, W.T., 1925, The genesis of lithium pegmatites: American Journal of Science (fifth series), v. 10, p. 269–279.

Shearer, C.K., Papike, J.J., Jolliff, B.L. and Settle, A.L., 1987a, Mineralogical and chemical evolution of a rare-element granite–pegmatite system: evidence from the chemistry of potassium feldspar: Geological Society of America, Abstracts with Programs, v. 19, p. 840.

Shearer, C.K., Papike, J.J. and Laul, J.C., 1987b, Mineralogical and chemical evolution of a rare-element granite–pegmatite system: Harney Peak Granite, Black Hills, South Dakota: Geochimica et Cosmochimica Acta, v. 51, p. 473–486.

Sims, P.K. and Peterman, Z.E., 1986, Early Proterozoic Central Plains Orogen: a major buried structure in the north-central United States: Geology, v. 14, p. 488–491.

Snyder, G.L., 1980, Map of Precambrian and adjacent Phanerozoic rocks of the Hartville Uplift, Goshen, Niobrara, and Platte counties, Wyoming: United States Geological Survey, Open-file Report 80-779.

Van Schmus, W.R., Bickford, M.E., Lewry, J.F. and Macdonald, R., 1987, U–Pb geochronology in the Trans-Hudson Orogen, northern Saskatchewan, Canada: Canadian Journal of Earth Sciences, v. 24, p. 407–424.

Walker, R.J., Hanson, G.N., Papike, J.J. and O'Neil, J.R., 1986, Nd, O and Sr isotope constraints on the origin of Precambrian rocks, southern Black Hills, South Dakota: Geochimica et Cosmochimica Acta, v. 50, p. 2833–2846.

A Model for Tectonic Evolution of the Trans-Hudson Orogen in North and South Dakota

John S. Klasner
Department of Geology, Western Illinois University, and United States Geological Survey, Macomb, Illinois 61455, U.S.A.

Elizabeth R. King
United States Geological Survey, MS 927, National Center, Reston, Virginia 22092, U.S.A.

Klasner, J.S. and King, E.R., A model for tectonic evolution of the Trans-Hudson Orogen in North and South Dakota, in Lewry, J.F. and Stauffer, M.R., eds., The Early Proterozoic Trans-Hudson Orogen of North America: Geological Association of Canada, Special Paper 37, p. 271–285.

Abstract

The Early Proterozoic Trans-Hudson Orogen lies beneath up to 4500 m of Phanerozoic cover in North and South Dakota. Gravity, magnetic, and drillhole data; two-dimensional gravity (density) models of the orogen; and extrapolated data from exposed parts of the Trans-Hudson Orogen in Canada suggest that the buried orogenic belt in the Dakotas is similar to younger orogens. Cross-sectional models of the orogen are developed for both North and South Dakota, incorporating elements typifying such younger orogens. From east to west, such elements comprise a foreland thrust belt (now mostly eroded) emplaced on the Superior craton; an imbricated continental margin, possibly with suture-zone ophiolites; an accreted volcanoplutonic belt and possible accreted microcontinental terrane(s); and a collided continent – the Wyoming craton.

There is evidence of the following sequence of events leading to formation of the Trans-Hudson Orogen in the Dakotas: (1) Rifting occurred along the Superior and Wyoming craton margins prior to 2.0 Ga. (2) Juvenile, magmatic arc crust and arc-related basins developed during oceanic closure between the two Archean continental plates – tonalitic–granodioritic gneisses in the Dakotas are thought to be metamorphic equivalents of such rocks. A prominent gravity low in south-central South Dakota indicates a large granitic pluton in this area. (3) Final oceanic closure, collision, and suturing of the Wyoming and Superior cratons occurred ca. 1.85–1.80 Ga ago. (4) Emplacement of S-type granite plutons in the Black Hills (southeast margin of Wyoming craton) occurred ca. 1.71 Ga ago. The Central Plains Orogen, formed ca. 1.78–1.68 Ga ago, truncates the southern end of the Trans-Hudson Orogen.

Documented Archean gneisses within the Trans-Hudson Orogen and the westward extension of Archean-related geophysical fabrics of the Superior craton suggest underthrusting of Archean crust beneath the fold belt during collisional events.

Résumé

Au Dakota du Nord et du Sud, l'orogène Trans-Hudson d'âge Protérozoïque Inférieur, est recouverte par 4500 m de roches phanérozoïques. Des données de gravité et magnétiques, des forages, des modèles de gravité (densité) bi-dimensionnelle de l'orogène ainsi que des données extrapolées sur des parties de l'orogène Trans-Hudson affleurantes au Canada montrent que la ceinture orogénique enfouie sous les deux Dakotas, est semblable au chaînes réçentes. Des modèles de coupes de l'orogène sont développés pour le Dakota du Nord et du Sud incluant des éléments typiques des chaînes réçentes. D'est en ouest ou trouve une zone externe charriée (maintenant érodée) sur le craton Supérieur; une marge continentale imbriquée, possiblement avec des ophiolites de sutures; une ceinture volcano-plutonique accrétée accompagnée de terranes micro-continentaux et un continent impliqué dans la collision, le craton Wyoming.

Une interprétation des données montrent la séquence suivante conduisant à la formation de l'orogène Trans-Hudson dans les deux Dakotas: (1) ouverture des marges des cratons Supérieur et Wyoming avant 2,0 Ga; (2) formation d'une croûte juvénile d'arcs magmatiques et de bassins pendant la fermeture de l'océan entre les deux plaques continentales archéennes (les gneiss tonalitiques et granodioritiques dans les deux Dakotas sont considérés comme équivalents de ces roches; au centre sud Dakota du Sud une anomalie négative de gravité indique la présence de pluton granitique de grande taille); (3) fin de la fermeture océanique, collision et suture des cratons Wyoming et Supérieur vers 1,85–80 Ga; (4) mise en place des plutons de granites type S dans les Black Hills (marge SE du craton Wyoming) vers 1,71 Ga. L'orogène Central Plains, formée vers 1,78–68 Ga recoupe la partie sud de l'orogène Trans-Hudson. La présence de gneiss archéens dans l'orogène Trans-Hudson et l'extension occidentale des empreintes géophysiques archéennes du craton Supérieur indique un sous-charriage de la croûte archéenne sous la chaîne plissée durant la collison.

INTRODUCTION

North and South Dakota lie within the stable continental platform of North America, where Precambrian basement is covered by up to 4500 m of Phanerozoic strata. Though deeply buried, the basement in the Dakotas (Fig. 1) evidently comprises a belt of Early Proterozoic rocks sandwiched between the Wyoming craton to the west and the Superior craton to the east (for age data, see Peterman and Hedge (1964), Goldich *et al.* (1966), Lidiak (1971), Peterman (1979, 1981), Van Schmus and Bickford (1981), Peterman and Goldich (1982), and Peterman *et al.* (1983)). The nature of basement rocks in South Dakota was characterized by Lidiak (1971); Klasner and King (1986) studied basement rocks in both North and South Dakota. Other studies (Muehlberger *et al.*, 1967; Muehlberger, 1980; Dutch, 1983; Arvidson *et al.*, 1984; Green *et al.*, 1985; Sims and Peterman, 1986; Thomas *et al.*, 1987) covered larger parts of the continent, but all encompass at least part of the Dakotas. These studies show that Early Proterozoic rocks in the Dakotas are part of an ancient orogenic belt that Hoffman (1981) termed the Trans-Hudson Orogen, and that extends southward from exposed Precambrian crust of the western Churchill Province in Canada. They confirm earlier suggestions by Innes (1960) and Wilson and Brisbin (1961) that an orogenic belt lies along the western edge of the Superior Province.

This paper is an interpretive extension of a recent study (Klasner and King, 1986) in which Precambrian basement in the Dakotas was subdivided into several lithotectonic terranes. It focuses on the nature and evolution of elements of the Trans-Hudson Orogen in this region (Fig. 1). We review the character of basement rocks in the

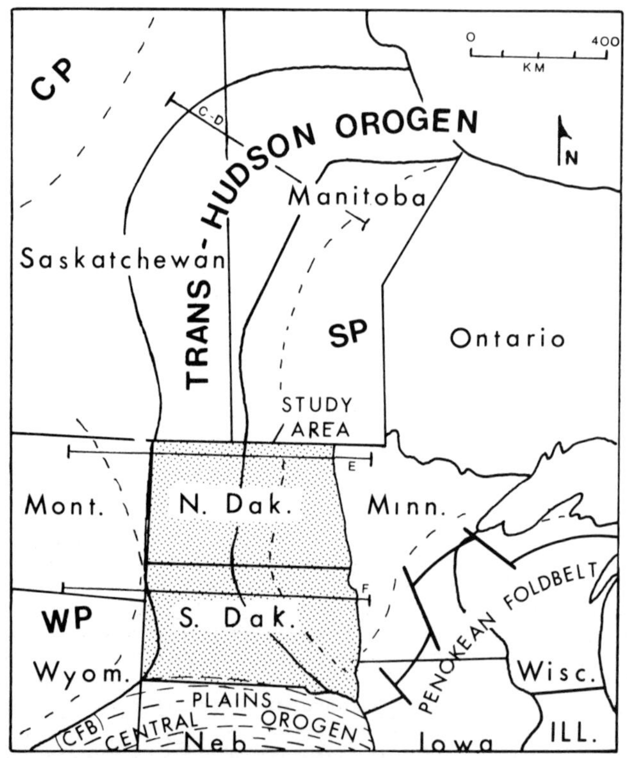

Figure 1. *Location of major tectonic elements as discussed in the text and location of the study area. Dashed lines indicate the general limit of Archean crust reworked or uplifted during formation of the Trans-Hudson Orogen. The location of profiles C–D, E, and F are shown; see text for discussion. Abbreviations: CFB, Cheyenne fold belt; WP, Wyoming craton; SP, Superior craton; and CP, Churchill platform.*

Dakotas and develop tectonic models of the orogen from data compiled in our previous study, from two-dimensional gravity (density) models of the Precambrian crust, from extrapolation of geological data in the exposed part of the orogen to the north, and from analogy with younger orogenic belts.

DISCUSSION OF LITHOTECTONIC TERRANES

Division of Precambrian basement into lithotectonic terranes (Klasner and King, 1986) was based on about 240 drillhole samples (both core and cuttings), widely scattered through an area of over 345,000 km^2, and analysis of gravity and magnetic data. The location of terranes is shown in Figure 2 and their geophysical nature is summarized in Table I. Terrane I comprises largely unreworked Archean continental crust of the Superior craton (cf. King *et al.*, 1984). Terranes II and III both largely comprise Archean crust with a strong Early Proterozoic overprint, marking the westward transition from Superior Province to Early Proterozoic rocks of the Trans-Hudson Orogen. Terrane IV is thought to represent the main accreted part of the orogen. Terrane V encompasses the Black Hills of South Dakota and is thought to consist mainly of Archean crustal remnants overprinted by Early Proterozoic tectonism. The Trans-Hudson Orogen in the Dakotas thus includes terrane IV, the accreted juvenile zone, and terranes II, III, and V, representing variably overprinted Archean crust. Terrane VI contains felsic volcanic rocks and terrane VII comprises a heterolithic rock assemblage that lies on trend with Archean gneisses of northwestern Iowa (W.R. Van Schmus, personal communication, 1985, reported in Klasner and King (1986)).

Selected drillhole data (Fig. 2) highlight key features of the lithotectonic terranes. Terrane I consists mainly of granitoid gneisses and greenstones and reflects the presence of alternating granite and greenstone belts typifying the Superior Province. Terrane II contains granulite facies rocks similar to those in the exposed Pikwitonei subprovince of Manitoba. It also shows two parallel, sinuous magnetic anomalies (A in Fig. 2). Lidiak (1971) obtained a schist from a drillhole that penetrated the easternmost of these, but direct data on the composition of this rock are unavailable. However, the positive magnetic signature suggests that the causative body has high magnetic susceptibility and it may possibly comprise mafic or ultramafic rocks; such rocks, also with a high magnetic signature, occur in several other drillholes in the region. The sinuous shape of the parallel anomalies suggests that the related rock bodies may be folded. Terrane III, in central South Dakota (Fig. 2), includes several positive magnetic anomalies; drillholes in this terrane yield a heterolithic assemblage of metasedimentary, granitic, mafic, and ultramafic rocks. Terrane IV consists mainly of tonalitic, trondhjemitic, and granodioritic gneisses in the north (IVa) and granitic rocks in the south (IVb and IVc). Protoliths of rocks in terrane IVa are uncertain, but they may be largely island arc-related plutons, volcanics, and associated metasediments. Peterman *et al.* (1983) reported high-grade metamorphic Early Proterozoic rocks from this terrane in McKenzie County, North Dakota (C on Fig. 2). However, both Peterman and Goldich (1982) and Peterman and Zartman (1985) also reported Archean rocks within this terrane, in Renville County, North Dakota (B in Fig. 2), that were metamorphosed to granulite grade ca. 1.8 Ga ago. Peterman *et al.* (1983) suggested that "development of this terrane may have been related to tectonic thickening of the crust during convergence of the two marginal Archean plates at about 1.8 Ga following an earlier rift event." The Archean rocks in

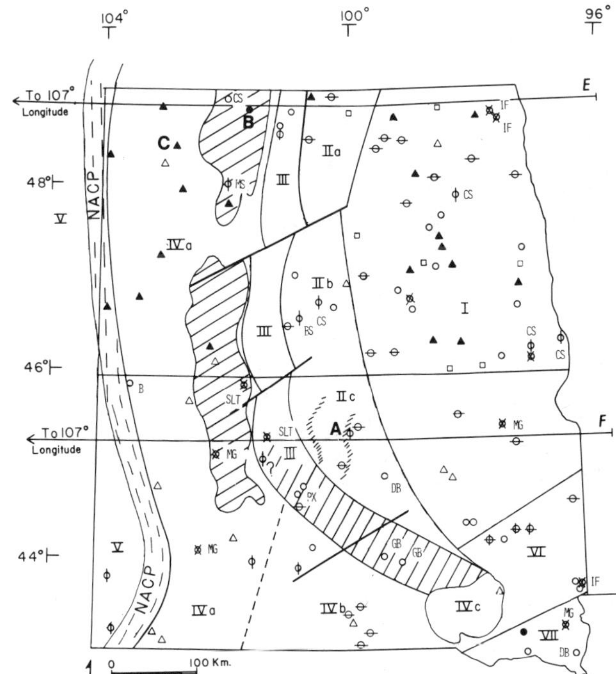

Figure 2. *Lithotectonic terrane divisions of Klasner and King (1986), selected positive magnetic anomalies, and selected drillhole data, as well as the location of gravity profiles E and F. Selected locations of magnetic and drillhole data. Points A, B, and C are discussed in the text.*

Table I
Geophysical characteristics, age, and tectonic significance of the lithotectonic terranes of Klasner and King (1986)

Terrane	Age	Fabric	Tectonic interpretation
I	Archean	Strong northeast trend with weak northwest overprint	Continental foreland of Superior craton
II	Archean	Mixed north-northwest trends with an indication of Archean northeast fabric. Numerous short-wavelength circular and diversely oriented trends	Part of crystalline thrust belt in which Archean crust has been uplifted. May contain supracrustal erosional remnants of overthrust oceanic rock
III	Probably mostly Early Proterozoic and some overprinted Archean	Magnetic low, variable gravity signature. Numerous short-wavelength anomalies	Suture zone. Strongly overprinted Archean crust and heterolithic assemblage of metamorphic and igneous rocks. Strongly sheared, possibly containing hydrated oceanic crust. Same as Thompson–Fox belt to the north
IVa	Early Proterozoic with Archean granulite	Prominent north trend in both gravity and magnetic data	Consists of three parts: East – accreted fore arc oceanic metasediments and metasediments and metavolcanic rocks on trend with the Flin Flon – Snow Lake island arc belt (Green *et al.,* 1985). Central – accreted and uplifted block of Archean granulitic rock. West – probably accreted oceanic island arc volcanic rock or arc-related metasedimentary rocks
IVb and IVc	Early Proterozoic	Prominent west-northwest trending gravity low. IVc elliptical positive magnetic anomaly on east end of gravity low	Accreted magmatic arc that lies on trend with the southern part of Terrane IVa
NACP	Postulated Early Proterozoic	Identified from the work of Camfield and Gough (1977). Marks eastern edge of terrane V	Suture along the eastern edge of Wyoming craton. Possibly sheared and hydrated oceanic crust (ophiolite), as suggested by Green *et al.* (1985)
V	Archean. Mostly Early Proterozoic in the Black Hills area	Overall, the Wyoming craton has a circular magnetic fabric	Part of collided continent
VI	Early Proterozoic	Relatively low gravity and magnetic signatures with short-wavelength north-northwest trending positive anomalies	Epicontinental rhyolite–andesite volcanic province lies mostly north of Penokean fold belt
VII	Probable Archean with Early Proterozoic overprint	Relatively positive gravity and chaotically oriented assemblage of magnetic highs and lows	Similar to terrane III. Upthrust blocks of Archean granulite and supracrustal assemblage of Early Proterozoic rocks that were overthrust onto continental foreland

terrane IV may represent Archean microcontinental remnants or, alternatively, uplifted erosional remnants of more continuous underlying basement, as discussed below. In either case, we interpret the outlined area of positive magnetic signatures on Figure 2, which also has an associated positive gravity signature, as marking the approximate possible extent of Archean crust in terrane IV.

The North American Central Plains (NACP) Conductivity Anomaly (Alabi *et al.*, 1975; Camfield and Gough, 1977; Handa and Camfield, 1984), shown in Figure 2, has been considered to mark the boundary between Archean crust of Wyoming Province to the west and Early Proterozoic crust of the Trans-Hudson Orogen to the east (e.g., Camfield and Gough, 1977; Klasner and King, 1986). However, Jones and Savage (1986) show that just north of the United States – Canadian border the NACP Anomaly lies about 75 km east of the location suggested by Camfield and Gough (1977). Moreover, Sims and Peterman (1986) suggest that the boundary between Wyoming Province and the Trans-Hudson Orogen lies along the Cedar Creek Anticline in eastern Montana, west of the NACP Anomaly as shown in Figure 2 (see dashed line in Fig. 1). The horizontal gradient gravity maps of Thomas *et al.* (1987) also indicate the presence of a significant crustal boundary to the west of the NACP Anomaly; these workers showed that the anomaly departs from their proposed western boundary of the Trans-Hudson Orogen along much of its length and further suggested that the northern extension of the NACP is offset by northeast-trending faults.

The positive magnetic anomalies in terrane IVa (Fig. 2) do not extend into terrane IVb, which is underlain by a prominent negative gravity anomaly interpreted as marking the presence of a large granitic batholith (Klasner and King, 1986). Granitoid rocks occur in drillhole cuttings from this region and granodiorite occurs in one drill core. However, McGinnes *et al.* (1979) suggested that the gravity anomaly, which is a large amplitude, long wavelength feature, may be caused by mass deficiencies in the mantle.

GRAVITY STUDIES

In many respects, the gravity signature across the Trans-Hudson Orogen in the Dakotas is similar to that across the exposed orogen in Saskatchewan and Manitoba, as seen by comparison of three gravity profiles in Saskatchewan–Manitoba, extending from the Superior Province into the orogen (A, B, and C on Fig. 3), with two profiles in the Dakotas (E and F on Fig. 3). Profiles B and C (from Gibb, 1968b) show two parallel, northeast-trending gravity highs, separated by a prominent gravity low, which corresponds to the Thompson belt (Churchill–Superior Boundary Zone, shaded zone in Fig. 3). On the basis of two-dimensional gravity models correlated to surface geology, Gibb (1986b) interpreted the eastern high on profiles B and C as being related to dense granulite gneisses of Pikwitonei subprovince, the central gravity low as being caused by subjacent granitic bodies, and the western gravity high as reflecting relatively dense (Kisseynew) paragneisses in the upper crust. As seen in Figure 3, this relationship does not hold for profile A; in this case, Gibb (1986b) interpreted the southern high as being caused by ultramafic rocks of the Fox River belt and, unlike profiles B and C, there is no well-developed low between this and the Kisseynew gneisses to the north. Gibb (1968b) also constructed a two-dimensional gravity model to explain the long-wavelength regional anomaly (dashed line on profile C, Fig. 3) and suggested that the latter is due to uplift of denser intermediate-level crust; however, he did not preclude uplift of the crust–mantle boundary as an alternative possible explanation for the shape of the regional profile.

Gravity profiles across the Trans-Hudson Orogen in the Dakotas, about 800 km to the south, show similar, though somewhat more subdued, features. Again, two parallel gravity highs and an intervening gravity low occur in the eastern part of the orogen. Figure 3 suggests essential continuity of these features from the exposed Shield, along the length of this part of the orogen. Such an interpretation is supported by previous work (Green *et al.*, 1985), which shows, on the basis of magnetic signatures, that the Thompson belt extends from northern Canada into the United States.

The similarity in gravity signature across widely separated parts of the orogen suggests that the Dakota segment incorporates tectonic elements similar to those seen in exposed parts of the orogen. Although there are some cross-sectional crustal models of parts of the Trans-Hudson Orogen in Canada (e.g., derived from gravity studies by Gibb (1968b), from surface geology by Lewry *et al.* (1985), and from seismic refraction data by Mereu and Hunter (1969), Hall and Hajnal (1969), Green *et al.* (1985), and Morel-à-l'Huissier *et al.* (1987)), none have been constructed across the orogen in the Dakotas. To aid in the development of such sections, we have constructed two, two-dimensional density models (Figs. 4C and 5C; see profiles E and F on Fig. 2 for location) along two profiles in North and South Dakota. We utilized data from previously constructed cross-sections, such as those of Gibb (1968b), for constraints on crustal densities, listed in Table II, and data from Morel-à-l'Huissier *et al.* (1987) for constraints on depths. Geological models (Fig. 3D), such as those of Lewry *et al.* (1985, 1986) and Green *et al.* (1985, 1986), were

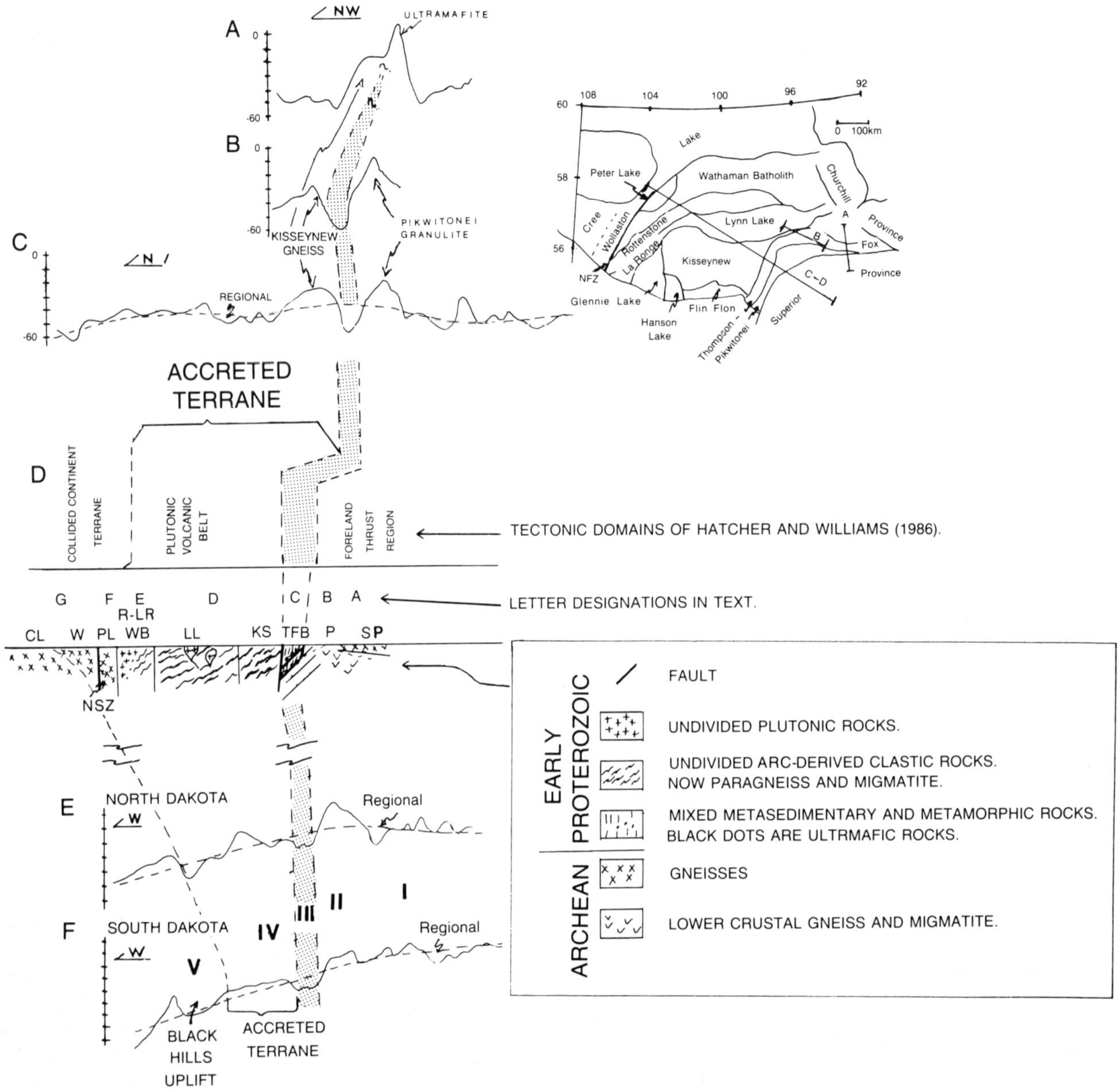

Figure 3. *Gravity profiles A, B, and C in northern Manitoba and Saskatchewan from Gibb (1968b), and profiles E and F in North and South Dakota respectively. Geological cross-section D is from Lewry* et al. *(1985) and Green* et al. *(1985). The location of the cross-section is along profile C–D as shown in the inset. Tectonic domains are from the model of an idealized orogen of Hatcher and Williams (1986). The shaded zone marks the position of a gravity low. Geological domains and features in the crustal section are as follows: CL, Cree Lake; W, Wollaston; Wb, Wathaman Batholith; NSZ, Needle Falls shear zone; PL, Peter Lake; R–LR, Rottenstone – La Ronge; LL, Lynn Lake; KS, Kisseynew; TFB, Thompson–Fox belt; P, Pikwitonei; and SP, Superior Province. Lithotectonic domains I through V of Precambrian basement in the Dakotas are shown between profiles E and F.*

referred to in choosing the type of buried crustal elements that might explain observed gravity anomalies.

Rock densities used in preparing the gravity (density) models (Figs. 4C and 5C) fall well within the ranges shown in Table II, which was prepared from various published sources and from densities obtained from drill cores in the Dakotas. In general, the subsurface crystalline rocks, including intermediate gneisses in terrane IV, are interpreted to increase from a density of 2.75 g/cm^3 near the surface to a lower crustal density of 3.1 g/cm^3. The mantle is assigned a density of 3.35 g/cm^3, well within the range estimated from compressional wave velocities (Table II). There are little data on rock densities in that part of terrane IV (Fig. 2), which has corresponding positive gravity and magnetic anomalies; the only measurement in the area (B in Fig. 2) is of a charnockitic granulite that yields a density of 2.72 g/cm^3 (Z.E. Peterman, personal communication, 1983). This is much lower than the 2.90 g/cm^3 density that we use for granulites in the density model. As mentioned above, because Peterman *et al.* (1983) reported the presence of Archean rocks at location B, within the area of high gravity and magnetic signatures of terrane IVa, we interpret this region as including possible remnants of an Archean microcontinent. The 2.90 g/cm^3 that we use for granulites at this location and elsewhere in the density model explains the presence of positive gravity anomalies above them; the 2.72 g/cm^3 of Peterman would not produce such gravity anomalies. Moreover, the 2.9 g/cm^3 value falls well within the range of densities for granulites given in Gibb (1968a), within the range of seismically derived densities for the lower crust, and within the range of densities of gneisses that occur in the region. In the absence of further data for this area, we suggest that the positive gravity anomaly may be caused by either accreted Archean crust or uplifted lower crustal rocks, either of which could have a relatively high density.

Seismic refraction studies of Morel-à-l'Huissier *et al.* (1987), which crossed the orogen just north of the international border, indicate that depth to the Moho is about 46 km on the west side, 42 km in the centre, and 45 km on the east side of the orogen. These general constraints were adopted for construction of the North and South Dakota models (Figs. 4C and 5C), but crustal thickness was slightly increased in the South Dakota profile to accommodate higher elevations in this area. The seismic refraction models of Morel-à-l'Huissier *et al.* (1987) also suggest that gross velocity and density layering in the crystalline crust are horizontal. Broad wavelength anomalies and a general decrease in the amplitude of the Bouguer anomaly from east to west are due mostly to variations in depth to the Moho, as well as increases in the thickness of Phanerozoic cover from east to west (thicknesses determined from drillhole data). As indicated above, Gibb (1968b) attributed broad regional variation in gravity to density variation in the lower crust, rather than variation in depth to the Moho. We did not model lateral density variations in the lower crust, but cannot rule them out as a cause of regional variation in the Bouguer gravity field; nor can we preclude crustal flexuring as a possible cause of such variation.

In our models, we attribute short wavelength gravity anomalies to lateral variations in upper crustal density. Note again that the models assume an average upper crust density of 2.75 g/cm^3, increasing to a lower crustal density of 3.1 g/cm^3. Gneissic rocks of the central Trans-Hudson Orogen (terrane IVa) also have a density of 2.75 g/cm^3 – the average density of the upper crust; thus, they do not produce gravity anomalies in our models. As shown in Figures 4C and 5C, upper crustal density variations lie above 10 km depth, and are underlain with denser, laterally homogeneous crust.

TECTONIC MODELS

The data used to establish general lithotectonic terranes in basement rocks of the Dakotas (Fig. 2) gave little information on the style and nature of deformation in the rocks concerned. For possible insights into the latter, we referred to work in exposed parts of the Trans-Hudson Orogen and studies of younger orogenic belts.

Proposed correlations of subsurface terranes with exposed Precambrian domains in the north were outlined by Klasner and King (1986). These are summarized in Table I, but are not discussed further here. However, recent U–Pb zircon studies (Van Schmus *et al.*, 1987) and Nd model studies (Chauvel *et al.*, 1987) have indicated that the Glennie Lake domain (Fig. 3, inset) was not Archean, as previously suggested (Lewry, 1981; Lewry *et al.*, 1985) and thus could not correlate with Archean crust in terrane IVa, as proposed in Klasner and King (1986). On the other hand, new isotopic data from the Glennie Lake domain (Chiarenzelli *et al.*, 1987) show that it does include at least some Archean crust, as does the adjacent Hanson Lake block (Van Schmus *et al.*, 1987). Thus, we tentatively retain our proposed general correlation of the Glennie Lake domain with postulated Archean rocks in terrane IVa (see Fig. 6).

Terranes VI and VII, in the southeast part of the region, do not correlate directly with exposed elements of the Trans-Hudson Orogen. We suggest that the heterolithic assemblage of terrane VII may represent reworked Archean foreland and overthrust younger metamorphic rocks related to the

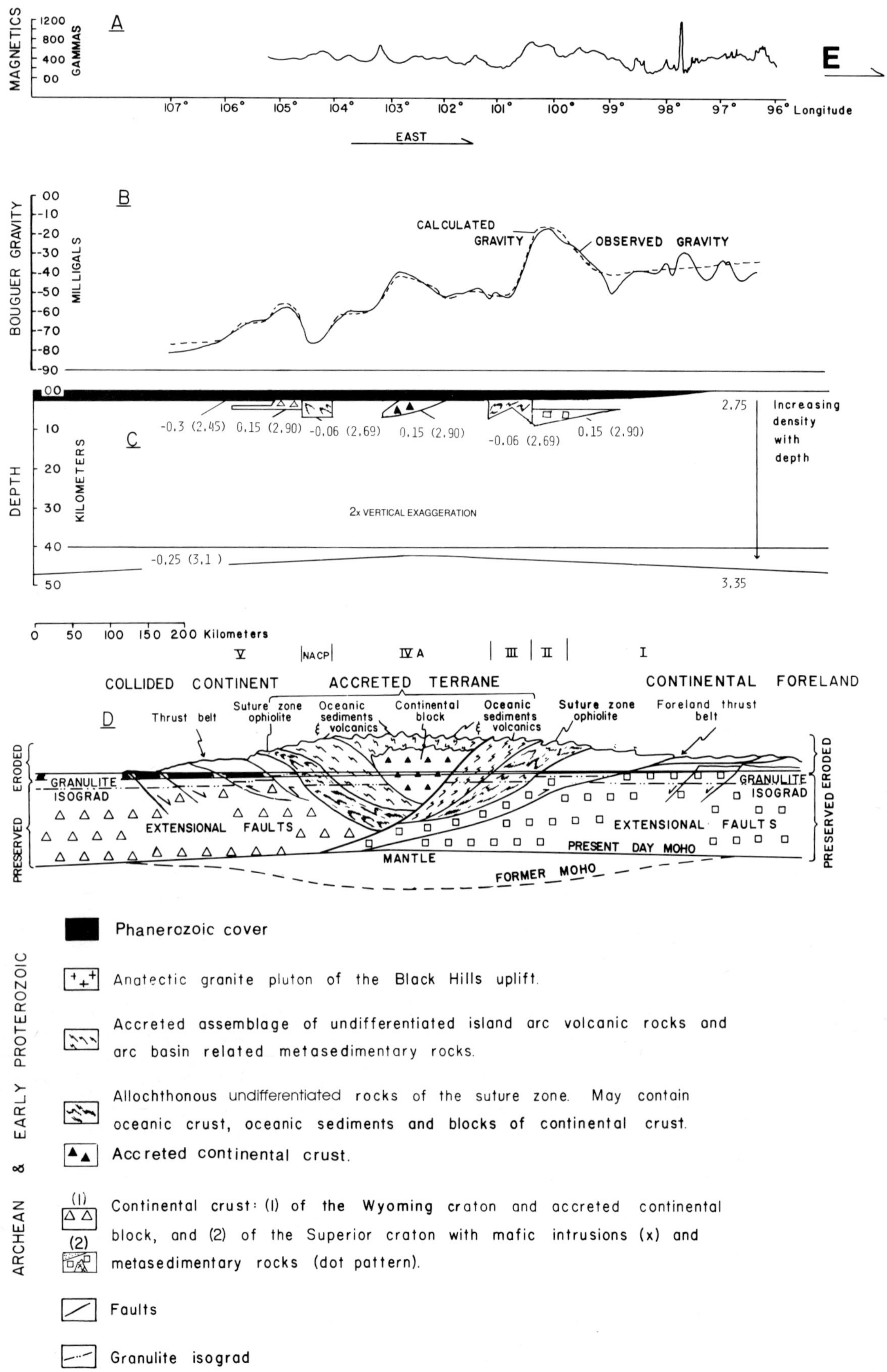

MAGNETICS
GAMMAS
1200
800
400
00
A
E
107° 106° 105° 104° 103° 102° 101° 100° 99° 98° 97° 96° Longitude
EAST
BOUGUER GRAVITY
MILLIGALS
B
CALCULATED GRAVITY
OBSERVED GRAVITY
DEPTH
KILOMETERS
C
-0.3 (2.45) 0.15 (2.90) -0.06 (2.69) 0.15 (2.90) -0.06 (2.69) 0.15 (2.90)
2.75
Increasing density with depth
2x VERTICAL EXAGGERATION
-0.25 (3.1)
3.35
0 50 100 150 200 Kilometers
V NACP IV A III II I
COLLIDED CONTINENT
ACCRETED TERRANE
CONTINENTAL FORELAND
D
Thrust belt
Suture zone ophiolite
Oceanic sediments & volcanics
Continental block
Oceanic sediments & volcanics
Suture zone ophiolite
Foreland thrust belt
ERODED
PRESERVED
GRANULITE ISOGRAD
EXTENSIONAL FAULTS
MANTLE
PRESENT DAY MOHO
FORMER MOHO
Phanerozoic cover
EARLY PROTEROZOIC
ARCHEAN &
Anatectic granite pluton of the Black Hills uplift.
Accreted assemblage of undifferentiated island arc volcanic rocks and arc basin related metasedimentary rocks.
Allochthonous undifferentiated rocks of the suture zone. May contain oceanic crust, oceanic sediments and blocks of continental crust.
Accreted continental crust.
(1)
(2)
Continental crust: (I) of the Wyoming craton and accreted continental block, and (2) of the Superior craton with mafic intrusions (x) and metasedimentary rocks (dot pattern).
Faults
Granulite isograd

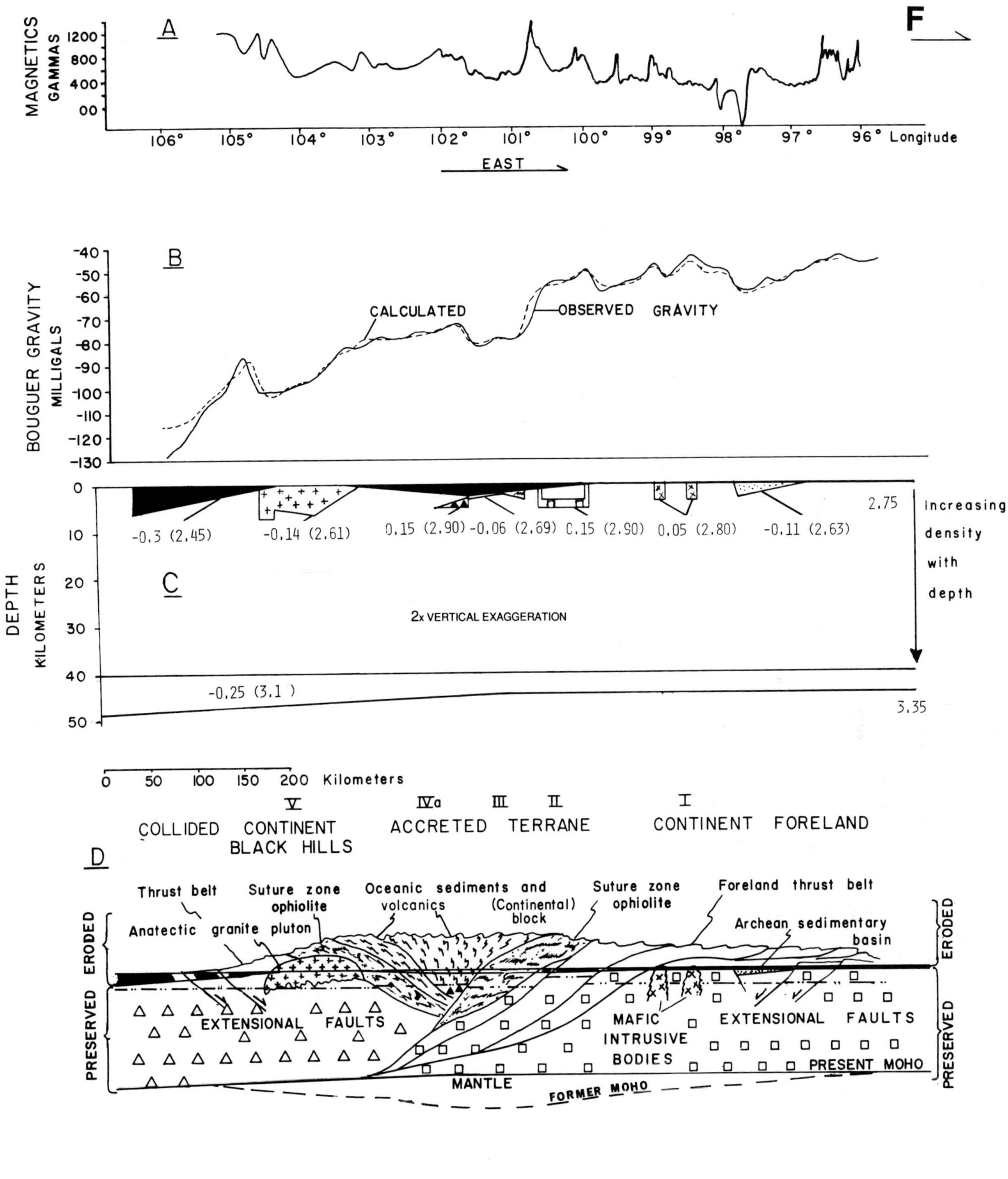

Figure 5. *Profile F. (A) Magnetic profile. (B) Bouguer gravity profile. (C) Two-dimensional gravity (density) model. (D) Tectonic crustal model. See text for discussion and Figure 4 for legend. (Note: 1 δ = 1 μg; 1 Gal = 1 cm / s²*.)

Figure 4. *Profile E. (A) Magnetic profile. (B) Bouguer gravity profile. (C) Two-dimensional gravity (density) model. (D) Tectonic crustal model. See text for discussion. (Note: 1 δ = 1 μg; 1 Gal = 1 cm / s²*.)

Penokean fold belt. Terranes III and VII are similar to each other in that both include heterolithic assemblages lying along the Archean cratonic margin.

In constructing crustal sections in the Dakotas (Figs. 4D and 5D), we used geological and geophysical data from exposed parts of the orogen, as shown on cross-section D in Figure 3 (section lies along line CD on the inset map, Fig. 3). This crosses most major tectonic elements of the orogen in this region and coincides with gravity profile C in Figure 3. The geological cross-section (Fig. 3D) is a composite, generalized mainly from a regional geological map of Lewry *et al.* (1985, Fig. 2) and tectonic models of Green *et al.* (1985) for the eastern margin of the fold belt (see also Lewry *et al.* (1986) and Green *et al.* (1986)). Major tectonic elements indicated in the section, their geotectonic significance, and the postulated correlation with lithotectonic terranes of Klasner and King (1986) are as follows:

(A) Archean crust of the main Superior craton — equivalent to terrane I (Fig. 2).

(B) Uplifted granulite-grade Archean intermediate to lower crust of Pikwitonei subprovince — equivalent to terrane II.

(C) Thompson – Fox River belt (Churchill–Superior Boundary Zone) — equivalent to terrane III. Green *et al.* (1985) interpreted this as a collage of reworked Archean crust, continental margin rocks, and perhaps accreted oceanic sediments and ultramafics, some of which may represent ophiolites. They suggest an eastward sense of vergence in the belt and note that many of the rocks involved are highly sheared and hydrated, giving them a relatively low density and concomitant low gravity anomaly signature. This accords with Table II, which shows low densities for schists and cataclastic rocks relative to the densities of upper crust gneisses.

Table II

Rock densities used in constructing the two-dimensional gravity (density) models (Figs. 4C and 5C) in North and South Dakota

	Gibb (1968a)			Seismic	Drill core data [c]			Densities
Rock type	No.	Range	Mean	correlations[b]	No.	Range	Ave.	used[f]
Phanerozoic sediments				1.8–2.8 (2.4)				2.45
Granite	944	2.50–2.78	2.64		25	2.47–2.77	2.57	2.61
Granitic/ granodioritic/ trondhjemitic/ tonalitic/ gneiss	11	2.69–2.82a	2.75	2.5–3.1 (2.8)	59	2.45–2.93d	2.75	2.75
Granulite: charnockite to ultramafite	214	2.55–3.36	2.73	Same as lower crust	1	2.72e		2.90
Precambrian greywacke and metasediments	162	2.58–3.42	2.76		5	2.45–2.70	2.57	2.63
Shear zone, schists and mylonite					8	2.61–2.71	2.65	2.69
Lower crust				2.8–3.3 (3.0)				2.9–3.1 (estimated)
Mantle				>3.3				3.35

Note: All density values are in g/cm^3.

[a] Quartz diorite.

[b] Correlation of velocity–density curve (Nafe–Drake curve) in Dobrin (1976, p. 53) with seismic velocities in Morel-à-l'-Huissier *et al.* (1987). Range and estimated centre value in parentheses.

[c] Data collected by Klasner from drill cores at state surveys.

[d] Undifferentiated core densities from Peterman (personal communication, 1982).

[e] Pyroxene granulite (charnockite), also from Peterman (personal communication, 1982).

[f] Used a uniform upper crustal density of 2.75 and basal crust density of 3.1.

(D) Rocks of the Kisseynew and Flin Flon – Snow Lake domains include paragneisses and migmatites interpreted as metamorphosed arc volcanics and fore arc or interarc basinal sediments (e.g., Lewry *et al.*, 1981; Stauffer, 1984; Ashton and Wheatley, 1986). They include some mafic pillow lavas and minor ultramafic rocks. Some gneisses in terrane IVa may be equivalent to such rocks.

(E) The Rottenstone – La Ronge – Lynn Lake magmatic belt includes island arc volcanics, migmatites, and abundant tonalitic to granitic plutonic rocks, including the major Wathaman Batholith (Ray, 1980; Ray and Wanless, 1980; Lewry *et al.*, 1985). Tonalitic–granodioritic rocks of terrane IVa

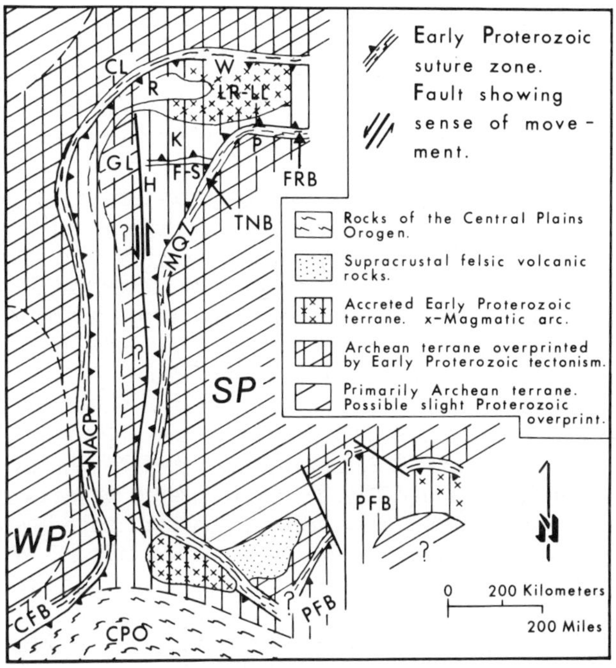

Figure 6. *Interpretive tectonic map of the Trans-Hudson Orogen showing extension of exposed Precambrian rocks into the subsurface of the Dakotas. Modified from Klasner and King (1986). Tectonic zones of Lewry* et al. *(1985) and Green* et al. *(1985) are CL, Cree Lake; K, Kisseynew; FRB, Fox River belt; TNB, Thompson Nickel belt (TNB and FRB are equivalent to the Thompson–Fox belt on Fig. 3); F–S, Flin Flon – Snow Lake belt; W, Wathaman Batholith; R, Rottenstone belt; LR–LL, La Ronge – Lynn Lake belt; GL, Glennie Lake domain; H, Hanson Lake block; and P, Pikwitonei subprovince. Other features include CPO, Central Plains Orogen; CFB, Cheyenne fold belt; PFB, Penokean fold belt; SP, Superior craton; WP, Wyoming craton; NACP, North American Central Plains Anomaly; and MQZ, magnetic quiet zone (Green* et al., *1979).*

may be equivalent to plutonic rocks in the Rottenstone domain; the relation of subterranes IVb and IVc to terrane IVa is presently unknown.

(F) The Peter Lake domain is partly Archean in age (Ray and Wanless, 1980; Bickford *et al.*, 1986) and comprises high-grade felsic/ granitoid gneisses, layered mafic/ultramafic bodies and minor low-grade metasediments, and later granites (Lewry and Thomas, 1987). It is considered (Lewry *et al.*, 1985) to be continuous with the Cree Lake ensialic zone, but may possibly represent an exotic accreted Archean crustal block, perhaps analogous to Archean crust in terrane IVa.

(G) Wollaston domain, forming the eastern part of the Cree Lake zone (Lewry *et al.*, 1985), comprises highly reworked Archean crust and infolded Lower Proterozoic metasediments; it forms the eastern part of the "Churchill platform" of Lewry and Thomas (1987) (Fig. 1). The Cree Lake zone may be equivalent to terrane V.

Collectively, domains in the exposed part of the orogen comprise complexly deformed and metamorphosed rocks representing an accreted crystalline terrane sandwiched between two ancient cratons, the Superior craton to the east and a reworked continental platform to the west (cf. Stauffer, 1984; Lewry *et al.*, 1985). Because this is so similar in many respects to the idealized orogen of Hatcher and Williams (1986), we used the latter model as a guide in constructing our crustal sections for the Dakotas (Figs. 4D and 5D). These sections were developed by correlating some elements of the idealized orogenic model with the density models (Figs. 4C and 5C). We stress that our crustal models are highly diagrammatic and conceptual in nature; aside from quantitative calculations made in constructing the density models, no attempt was made to quantify other parameters, such as the depth to the root of the Early Proterozoic mountain range, the extent of overthrusting or the amount of isostatic rebound.

Essentially in line with the interpretations of Lewry and Collerson (this volume) and Lewry *et al.* (this volume), we recognize, or postulate, the following major elements in the exposed part of the orogen: (a) A presumed, but now completely eroded, foreland fold/thrust belt formerly emplaced on the Superior craton inboard of the Churchill–Superior Boundary Zone (Thompson – Fox River belts) – an imbricated basement/continental margin zone, possibly including ophiolite remnants. (b) The Reindeer Zone (Flin Flon, Kisseynew, La Ronge – Lynn Lake, Glennie Lake, Hanson Lake, and Rottenstone domains), which comprises accreted juvenile rocks, possibly forming a stacked nappe complex, and also the straddling Wathaman Batholith. The Glennie and Hanson Lake domains may in part incorporate, or be underlain by, accreted Archean microcontinental components. (c) The Cree Lake Zone and Peter Lake domain, which include highly reworked continental margin rocks. Lewry *et al.* (1985) suggest that subduction was to the northwest, beneath the Cree Lake Zone, thus causing intense reworking of the latter.

These lithotectonic elements appear to correspond closely to elements of the idealized orogen proposed by Hatcher and Williams (1986) (see letter designations on Fig. 3D). Superior craton (A) and the Pikwitonei subprovince (B) mark the former position of an eroded foreland thrust belt and possible overthrust metamorphic core complex. The imbricated assemblage of the Thompson – Fox River belt marks the approximate position of the suture zone. Domainal components of the Reindeer Zone (Lewry *et al.*, this volume) (D) and Wathaman Batholith (E) represent the accreted juvenile volcanoplutonic belt of Hatcher and Williams (1986), but possibly include accreted microcontinental remnants. The Peter Lake domain (F) and the Cree Lake zone (G) represent the collided continental block of Hatcher and Williams (1986).

In our tectonic models for the Dakotas, the collided continent is represented by Wyoming craton and the continental foreland by Superior craton (see Table I for the tectonic significance of the terranes). In both of our sections, the Archean blocks are portrayed as containing extensional faults (which may have been reactivated as thrusts during later collisional shortening). Klasner and King (1986) noted a weak northwest-trending overprint on northeast-trending fabric of the Superior craton and suggested that this might reflect extensional faulting, as found in the Archean foreland to the Penokean fold belt in northern Michigan. There is some geological evidence for the presence of such extensional structures. The tectonic map of the Black Hills (Lisenbee, 1985; see also Redden *et al.*, this volume) documents pre-2.1 Ga continental rift sedimentary sequences indicating early crustal extension, and Baragar and Scoates (1981) propose Early Proterozoic rifting of the Superior craton in their discussion of the Circum-Superior belt.

In the density models (Figs. 4C and 5C), the high gravity signature of terrane II is interpreted to reflect relatively dense granulite basement gneisses, similar to those of Pikwitonei subprovince (Weber and Scoates, 1978) where formation of the granulites predated Early Proterozoic orogeny. In the tectonic models, we portray these as having been uplifted by basement thrusting during the latter orogeny. Terrane III marks the suture zone, comparable in character to the Thompson – Fox River belt and possibly including overthrust ophiolitic oceanic remnants, as suggested by the sinuous magnetic anomalies discussed above (A on Fig. 2). Outboard of this imbricated continental margin are accreted volcanoplutonic and microcontinental belts of terrane IVa. The postulated batholith of south-central South Dakota, subterranes IVb and IVc, may also be part of this accreted collage. Subterrane IVa is spatially well defined on both gravity and magnetic maps. We interpret the eastern part of this subterrane as being underlain by juvenile arc plutons, volcanics, and associated oceanic/arc metasediments. However, a drillhole in the central part of the subterrane (B on Fig. 2) contains Archean gneiss with an Early Proterozoic granulite facies overprint, and gravity and magnetic data indicate that Archean crust may underlie a large part of the subterrane. This may represent either accreted microcontinents or, as argued below and previously proposed by Green *et al.* (1985), underthrust and uplifted Superior Province basement.

Klasner and King (1986) suggested that the east-northeast trending fabric of terrane I may extend westwards into terrane IV, the central part of the orogen. This might reflect westward underthrusting of Superior Province Archean crust, such that rocks of the central orogen are underlain by terrane I Archean crust, as shown on the geological models (Figs. 4D and 5D) (see also Lewry *et al.*, this volume). Documented Archean crust in terrane IVa may thus represent uplifted blocks of underthrust Superior basement, though this is not specifically represented in the cross-section models.

A zone of intermediate, granodioritic to tonalitic gneisses, indicated by drillhole data, in the western part of terrane IVa, beyond the area of positive magnetic anomaly (Fig. 2), is interpreted as rocks of arc or arc-related basin affinities, which have been accreted to the margin of the Wyoming craton. We follow Green *et al.* (1985) in suggesting that the NACP Conductor may in part reflect the presence of a slice of hydrated, sheared oceanic crust (Figs. 4D and 5D), as in terrane III. The northwest-trending boundary of Sims and Peterman (1986) and Thomas *et al.* (1987), to the west of the NACP in Montana (Fig. 6), may represent the boundary between reworked Archean crust and the less reworked Wyoming craton. The positive gravity anomaly just west of the NACP gives some basis for interpreting the region between the NACP and the boundary of Sims and Peterman (1986) and Thomas *et al.* (1987) as being underlain by uplifted Archean basement rocks.

Terrane V is exposed in the Black Hills uplift. Modelling of this uplift (Fig. 5D) suggests that the attendant gravity low reflects a large felsic intrusive body, represented by the 1.71 Ga Harney Peak Granite. Redden *et al.* (1982) interpret this as a domal intrusion with a base at relatively shallow crustal depth (see also Duke *et al.*, this volume), as indicated on both density and geological models (Figs. 5C and 5D). Huang and Wyllie (1982) suggest that the batholith was emplaced beneath up to 10 km of sediments (see also Lisenbee, 1985); documented age shows that it postdates Early Proterozoic orogeny.

SEQUENCE OF EVENTS

Data presented above, and in Figures 1, 2, and 6, are consistent with the following possible sequence of events during formation of the Trans-Hudson Orogen in the Dakotas:

(1) Both Superior and Wyoming provinces were subjected to rifting, or possibly multiple riftng events, as indicated by 2.1 Ga rift-related metasediments in the Black Hills and evidence of Early Proterozoic rifting in Superior Province (terrane I of Klasner and King (1986) of the eastern Dakotas). Whether the Wyoming and Superior cratons were connected prior to this event is unknown.

(2) Convergence of Wyoming and Superior plates involved subduction of intervening oceanic crust, possibly to the west or northwest, with attendant formation of magmatic arcs and related sedimentary basins. Thomas *et al.* (1987) recognized that subduction polarity is uncertain and requires further analyis.

(3) Oceanic closure occurred ca. 1.85 Ga ago (cf. Van Schmus *et al.*, 1987) or later. The cross-sectional models show the inferred nature of crust formed during attendant collision of Wyoming and Superior cratons. There was probably a regional left-lateral component of movement (Fig. 6) along north or northwest oriented fault zones during or shortly following collision (cf. Lewry, 1981; Green *et al.*, 1985; Thomas *et al.*, 1987).

(4) Posttectonic S-type granitic plutons were emplaced ca. 1.71 Ga in what is now the Black Hills uplift.

(5) The 1.78–1.69 Ga Central Plains Orogen formed south of the Trans-Hudson Orogen (Sims and Peterman, 1986). The Cheyenne fold belt (Houston *et al.*, 1979; Duebendorfer and Houston, 1987) represents part of that orogen accreted to the Wyoming craton.

CONCLUSIONS

We have constructed crustal models for the Trans-Hudson Orogen in the Dakotas and proposed a sequence of events for the evolution of the orogen. The models are based on scattered lithological data from drillholes, the interpretation of geophysical data, extrapolation from the exposed part of the orogen in Canada, and modelling based on analogous younger orogens as proposed in the generic idealized model of Hatcher and Williams (1986). The derived tectonic models are highly speculative and need to be tested. Additional drillhole and seismic reflection data would help, as would further studies of existing data sources; geochemical studies of material from drill cores, similar to those carried out in the Penokean fold belt (Schulz, 1983) would help categorize protoliths of rocks in the orogen. Microstructural studies of drill core material might possibly provide strain data from the different terranes and help determine the nature of deformation within them.

ACKNOWLEDGEMENTS

Reviews of this paper by M. Foose, A.G. Green, G.L. LaBerge, J.F. Lewry, A. Lisenbee, Z. Peterman, K. Schulz, and P.K. Sims resulted in considerable improvements. We thank all for their help, but assume responsibility for any errors in omission or interpretation.

REFERENCES

Alabi, A.O., Camfield, D.I. and Gough, D.I., 1975, The North American Central Plains conductivity anomaly: Royal Astronomical Society Geophysical Journal, v. 43, p. 815–833.

Arvidson, R.E., Bindschadler, D., Bowring, S., Eddy, M., Guinness, E. and Leff, C., 1984, Bouguer image of the North American craton and its structural evolution: Nature (London), v. 311, p. 241–243.

Ashton, K.E. and Wheatley, K.J., 1986, Preliminary report on the Kisseynew gneisses in the Kisseynew–Wildnest lakes area, Saskatchewan: Geological Survey of Canada, Paper 86-1B, p. 305–317.

Baragar, W.R.A. and Scoates, R.F.J., 1981, The Circum-Superior Belt: a Proterozoic plate margin?, *in* Kroner, A., ed., Precambrian Plate Tectonics: Elsevier, Amsterdam, p. 297–330.

Bickford, M.E., Van Schmus, W.R., Macdonald, R., Lewry, J.F. and Pearson, J.G., 1986, U–Pb zircon geochronology project for the Trans-Hudson Orogen: current sampling and recent results: Saskatchewan Geological Survey, Miscellaneous Report 86-4, p. 101–107.

Camfield, P.A. and Gough, D.I., 1977, A possible Proterozoic plate boundary in North America: Canadian Journal of Earth Sciences, v. 14, p. 1229–1238.

Chauvel, C., Arndt, N.T., Kielinzcuk, S. and Thom, A., 1987, Formation of Canadian 1.9 Ga old continental crust. I. Nd isotopic data: Canadian Journal of Earth Sciences, v. 24, p. 396–406.

Chiarenzelli, J.R., Lewry, J.F. and Landon, M., 1987, Bedrock geology, Iskwatikan Lake area: evidence for Hudsonian juxtaposition of Proterozoic and Archean rocks along a ductile detachment surface: Saskatchewan Geological Survey, Miscellaneous Report 87-4, p. 46–51.

Dobrin, M.B., 1976, Introduction to Geophysical Prospecting. 3rd edition: McGraw-Hill Inc., New York, 630 p.

Duebendorfer, E.M. and Houston, R.S., 1987, Proterozoic accretionary tectonics at the southern margin of the Archean Wyoming craton: Geological Society of America Bulletin, v. 98, p. 554–568.

Duke, E.F., Shearer, C.K., Redden, J.A. and Papike, J.J., 1990, Proterozoic granite–pegmatite magmatism, Black Hills, South Dakota: structure and geochemical zonation, *in* Lewry, J.F. and Stauffer, M.R., eds., The Early Proterozoic Trans-Hudson Orogen of North America: Geological Association of Canada, Special Paper 37, p. 253–269.

Dutch, S.I., 1983, Proterozoic structural provinces in the north-central United States: Geology, v. 11, p. 478–481.

Gibb, R.A., 1968a, The densities of Precambrian rocks from northern Manitoba: Canadian Journal of Earth Sciences: v. 5, p. 433–438.

Gibb, R.A., 1968b, A geological interpretation of the Bouguer anomalies adjacent to the Churchill–Superior boundary in northern Manitoba: Canadian Journal of Earth Sciences, v. 5, p. 439–453.

Goldich, S.S., Lidiak, E.G., Hedge, C.E. and Walthall, F.G., 1966, Geochronology of the mid-continent region, United States. 2. Northern area: Journal of Geophysical Research, v. 71, p. 5389–5408.

Green, A.G., Cumming, G.L. and Cederwell, D., 1979, Extension of the Superior–Churchill Boundary Zone into southern Canada: Canadian Journal of Earth Sciences, v. 16, p. 1691–1701.

Green, A.G., Hajnal, A. and Weber, W., 1985, An evolutionary model of the western Churchill Province and western margin of the Superior Province in Canada and the north-central United States: Tectonophysics, v. 116, p. 281–322.

Green, A.G., Hajnal, Z. and Weber, W., 1986, An evolutionary model of the western Churchill Province and western margin of the Superior Province in Canada and the north-central United Sates: reply: Tectonophysics, v. 131, p. 188–197.

Hall, D.H. and Hajnal, Z., 1969, Crustal structure of northwestern Ontario: refraction seismology: Canadian Journal of Earth Sciences, v. 6, p. 81–99.

Handa, S. and Camfield, P.A., 1984, Crustal electrical conductivity in north-central Saskatchewan: the North American Central Plains anomaly and its relation to a Proterozoic plate margin: Canadian Journal of Earth Sciences, v. 21, p. 533–543.

Hatcher, R.D. and Williams, R.T., 1986, Mechanical model for single thrust sheets. Part I. Taxonomy of crystalline thrust sheets and their relationships to mechanical behavior of orogenic belts: Geological Society of America Bulletin, v. 97, p. 975–985.

Hoffman, P.F., 1981, Autopsy of Athapuscow Aulocogen: a failed arm affected by three collisions, in Campbell, F.H.A., Proterozoic Basins of Canada: Geological Survey of Canada, Paper 81-10, p. 97–102.

Houston, R.S., Karlstrom, K.E. and Smithson, S.B., 1979, The Cheyenne belt: the major Precambrian crustal boundary in the western United States: Geological Society of America, Abstracts with Programs, v. 11, p. 446.

Huang, W.L. and Wyllie, P.J., 1982, Phase relationships of S-type granite with H_2O to 35 kbar; muscovite granite from Harney Peak, South Dakota: Journal of Geophysical Research, v. 87, p. 10515–10529.

Innes, M.J.S., 1960, Gravity and isostasy in Manitoba and northern Ontario: Publication of the Dominion Observatory, v. 21, p. 265–338.

Jones, A.G. and Savage, P.J., 1986, North American Central Plains conductivity anomaly goes east: Geophysical Research Letters, v. 13, p. 685–688.

King, E.R., Klasner, J.S., Zietz, I. and Cannon, W.F., 1984, Magnetic data on the Precambrian basement rocks of eastern North Dakota and their regional implications: Geological Society of America, Abstracts with Programs, v. 13, p. 487.

Klasner, J.S. and King, E.R., 1986, Precambrian basement geology of North and South Dakota: Canadian Journal of Earth Sciences, v. 23, p. 1083–1102.

Lewry, J.F., 1981, Lower Proterozoic arc–microcontinent collision tectonics in the western Churchill Province: Nature, v. 294, p. 69–72.

Lewry, J.F. and Collerson, K.D., 1990, The Trans-Hudson Orogen: extent, subdivision, and problems, in Lewry, J.F. and Stauffer, M.R., eds., The Early Proterozoic Trans-Hudson Orogen of North America: Geological Association of Canada, Special Paper 37, p. 1–14.

Lewry, J.F. and Thomas, D.J., 1987, A transect across the Trans-Hudson Orogen, La Ronge – Wollaston Lake: Geological Association of Canada, Field Trip Guidebook, Trip 6, University of Saskatchewan, Saskatoon, 56 p.

Lewry, J.F., Stauffer, M.R. and Fumerton, S., 1981, A Cordilleran-type batholithic belt in the Churchill Province in northern Saskatchewan: Precambrian Research, v. 14, p. 277–313.

Lewry, J.F., Sibbald, T.I.I. and Schledewitz, D.C.P., 1985, Variation in character of Archean rocks in the western Churchill Province and its significance, in Ayres, L.D., Thurston, P.C., Card, K.D. and Weber, W., eds., Evolution of Archean Supracrustal Sequences: Geological Association of Canada, Special Paper 28, p. 239–261.

Lewry, J.F., Collerson, K.D., Bickford, M.E. and Van Schmus, W.R., 1986, An evolutionary model of the western Churchill Province and western margin of the Superior Province in Canada and the north-central United States: discussion: Tectonophysics, v. 131, p. 183–188.

Lewry, J.F., Thomas, D.J., Macdonald, R. and Chiarenzelli, J., 1990, Structural relations in accreted terranes of the Trans-Hudson Orogen, Saskatchewan: telescoping in a collisional regime?, in Lewry, J.F. and Stauffer, M.R., eds., The Early Proterozoic Trans-Hudson Orogen of North America: Geological Association of Canada, Special Paper 37, p. 75–94.

Lidiak, E.G., 1971, Buried Precambrian rocks of South Dakota: Geological Society of America Bulletin, v. 82, p. 1411–1420.

Lisenbee, A.L., 1985, Tectonic map of the Black Hills uplift, Montana, Wyoming, and South Dakota: South Dakota School of Mines and Technology and South Dakota Geological Survey, scale 1:250,000.

McGinnes, L.D., Wolf, M.G., Kohsmann, J.J. and Ervin, C.D., 1979, Regional free air gravity anomalies and tectonic observations in the United States: Journal of Geophysical Research, v. 84, p. 591–601.

Mereu, R.F. and Hunter, J.A., 1969, Crustal and upper mantle structure under the Canadian Shield from Project Early Rise data: Bulletin of the Seismological Society of America, v. 59, p. 147–165.

Morel-à-l'Huissier, P., Green, A.G. and Pike, C.J., 1987, Crustal refraction surveys across the Trans-Hudson Orogen/Williston Basin of south-central Canada: Journal of Geophysical Research, v. 92, p. 6403–6420.

Muehlberger, W.R., 1980, The shape of North America during the Precambrian, in Geophysics Study Committee, eds., Continental Tectonics: National Academy of Sciences, p. 175–183.

Muehlberger, W.R., Denison, R.E. and Lidiak, E.G., 1967, Basement rocks in the continental interior of United States: American Association of Petroleum Geologists Bulletin, v. 51, p. 2351–2380.

Peterman, Z.E., 1979, Geochronology and Archean of the United States: Economic Geology, v. 74, p. 1544–1562.

Peterman, Z.E., 1981, Dating of Archean basement in northeastern Wyoming and southern Montana: Geological Society of America Bulletin, v. 92, p. 139–146.

Peterman, Z.E. and Goldich, S.S., 1982, Archean rocks of the Churchill basement, Williston Basin, North Dakota, in Christopher, J.E. and Kaldi, J., eds., Fourth International Williston Basin Symposium: Saskatchewan Geological Society, Special Publication 6, p. 11–12.

Peterman, Z.E. and Hedge, C.E., 1964, Age of basement rocks from the Williston Basin of North Dakota and adjacent areas: United States Geological Survey, Professional Paper 475-D, p. D100–D104.

Peterman, Z.E. and Zartman, R.E., 1985. The Early Proterozoic Trans-Hudson Orogen in the northern Great Plains of the United States (abstract), in Aldrich, M.J., Jr. and Laughlin, A.W., eds., Proceedings of the Sixth International Conference on Basement Tectonics: International Basement Tectonics Association, Salt Lake City, UT.

Peterman, Z.E., Goldich, S.S. and Zartman, R.E., 1983, High grade reworked Archean rocks in the basement of the Williston Basin, North Dakota: Geological Society of America, Abstracts with Programs, v. 15, p. 660.

Ray, G.E., 1980, Geology of the Parker Lake – Nelson Lake vicinity: Saskatchewan Department of Mineral Resources, Report 190, 40 p.

Ray, G.E. and Wanless, R.K., 1980, The age and geologic history of the Wollaston, Peter Lake, and Rottenstone domains in northern Saskatchewan: Canadian Journal of Earth Sciences, v. 17, p. 333–347.

Redden, J.A., Norton, J.J. and McGlaughlin, R.J., 1982, Geology of the Harney Peak Granite, Black Hills, South Dakota: United States Geological Survey, Open-file Report 82-481, 17 p.

Redden, J.A., Peterman, Z.E., Zartman, R.E. and DeWitt, E., 1990, U–Th–Pb geochronology and preliminary interpretation of Precambrian tectonic events in the Black Hills, South Dakota, *in* Lewry, J.F. and Stauffer, M.R., eds., The Early Proterozoic Trans-Hudson Orogen of North America: Geological Association of Canada, Special Paper 37, p. 229–251.

Schulz, K.J., 1983, Geochemistry of volcanic rocks of northeastern Wisconsin (abstract): 29th Annual meeting of the Institute of Lake Superior Geology, Houghton, MI, p. 39–40.

Sims, P.K. and Peterman, Z.E., 1986, Early Proterozoic Central Plains Orogen: Geology, v. 14, p. 488–491.

Stauffer, M.R., 1984, Manikewan: an Early Proterozoic ocean in central Canada, its igneous history and orogenic closure: Precambrian Research, v. 25, p. 257–281.

Thomas, M.D., Sharpton, V.L. and Grieve, R.A.F., 1987, Gravity patterns and Precambrian structure in the North American Central Plains: Geology, v. 15, p. 489–492.

Van Schmus, W.R. and Bickford, M.E., 1981, Proterozoic chronology and evolution of the midcontinent region of North America, *in* Kroner, A., ed., Precambrian Plate Tectonics: Elsevier, Amsterdam, p. 261–296.

Van Schmus, W.R., Bickford, M.E., Lewry, J.F. and Macdonald, R., 1987, U–Pb geochronology in the Trans-Hudson Orogen, northern Saskatchewan, Canada: Canadian Journal of Earth Sciences, v. 24, p. 407–424.

Weber, W. and Scoates, R.F.J., 1978, Archean and Proterozoic metamorphism in the northwestern Superior Province and along the Churchill–Superior boundary, Manitoba, *in* Fraser, J.A. and Heywood, W.W., Metamorphism in the Canadian Shield: Geological Survey of Canada, Paper 78-10, p. 5–16.

Wilson, H.D.B. and Brisbin, W.C., 1961, Regional structure of the Thompson – Moak Lake nickel belt: Canadian Institute of Mining and Metallurgy Bulletin, v. 54, p. 1–8.

The Archean Wyoming Province and its Relations with Adjacent Proterozoic Provinces

Steven I. Dutch
Natural and Applied Sciences, University of Wisconsin – Green Bay, Green Bay, Wisconsin 54311-7001, U.S.A.

Peter A. Nielsen
Department of Geology, Keene State College, Keene, New Hampshire 03431, U.S.A.

Dutch, S.I. and Nielsen, P.A., The Archean Wyoming Province and its relations with adjacent Proterozoic provinces, in Lewry, J.F. and Stauffer, M.R., eds., The Early Proterozoic Trans-Hudson Orogen of North America: Geological Association of Canada, Special Paper 37, p. 287–300.

Abstract

The Wyoming Province is a block of Archean crust surrounded by Early Proterozoic mobile belts. The boundaries of the province are almost entirely concealed and must be inferred from geophysical and geochronological data. Archean crust is known, or can reasonably be inferred to exist, beneath all but the southeastern corner of Wyoming, beneath most of the southeastern half of Montana, eastward beneath the Black Hills of South Dakota, and westward to the East Humboldt Range in northeastern Nevada. The best exposure of the boundary of the Wyoming Province is the Cheyenne Belt, which has been interpreted as a south-dipping Proterozoic subduction zone. Proceeding counterclockwise around the province boundary from the Cheyenne Belt, there are several localities that constrain the location of the Wyoming Province. Archean rocks are exposed in two inliers in the Black Hills of South Dakota, overlain by a thin Proterozoic shelflike sequence and a thick eugeosynclinal complex. Mixed Archean and Proterozoic ages have been obtained from the Little Rocky Mountains and Little Belt Mountains of north-central Montana. In southwestern Montana, metamorphosed supracrustal rocks once believed to be Proterozoic have been shown to be largely, if not entirely, Archean. Proterozoic rocks have also been postulated in central Idaho, but there is reason to believe they are not early Proterozoic but metamorphosed early Beltian rocks. Several exposures of Proterozoic and Archean crystalline rocks occur in Utah and constrain the location of the southwestern corner of the Wyoming Province. Gravity and magnetic anomaly patterns both indicate northeast structural trends along the northern margin of the Wyoming Province, and north–south trends along the eastern margin. The split between the two trend directions corresponds closely to the North American Central Plains (NACP) Conductivity Anomaly. Archean basement ages have been found in southern Alberta and Saskatchewan, but gravity and magnetic anomaly patterns suggest the existence of a distinct block, here termed the Southern Alberta Block. Paleomagnetic evidence for motion of the Wyoming block relative to the Superior Province is inconclusive. There is no conclusive evidence for very large motion, but motions of over 1000 km are possible and well within the uncertainty of the data.

Résumé

La province de Wyoming est un bloc de croûte archéenne entouré de chaînes plissées d'âge Protérozoïque Inférieur. Les limites de la province ne sont pas visibles et doivent être déduites par géophysique et géochronologie. On a déduit avec suffisamment de certitude la présence de croûte archéenne sous le Wyoming exception faite du coin sud-est, une grande partie de la demie sud-est du Montana, vers l'est sous les Black Hills du Sud Dakota et à l'ouest dans les Humboldt Range dans le nord-est du Nevada. La limite de la province Wyoming la mieux exposée est la chaîne Cheyenne, interprétée comme une zone de subduction protérozoïque à pendage sud. A partir de la chaîne Cheyenne et dans le sens anti-horaire plusieurs localités marquent les limites de la province Wyoming. Deux enclaves archéennes affleurent dans les Black Hills du Sud Dakota, surmontées par une mince séquence d'âge Protérozoïque à caractère de plateforme et par un complexe eugéosynclinal épais. Des âges archéens et protérozoïque proviennent des montagnes Little Rocky et Little Belt du Montana nord central. Au sud-ouest du Montana, des roches considérées d'âge Protérozoïque se sont avérées être presque toutes archéennes. Dans l'Idaho central, les faits portent à croire que des roches d'abord postulées comme étant protérozoïques sont en fait des roches métamorphiques Beltian Inférieur. Des roches cristallines archéennes et protérozoïques affleurent en Utah et marquent le coin sud-ouest de la province Wyoming. Dans la province Wyoming, les patrons d'anomalies magnétiques et de gravité montrent des tendances structurales NE le long de la marge nord et NS le long de la marge est. Le point de séparation des deux directions correspond à l'anomalie de conductivité NACP. Des âges archéens proviennent du sud de l'Alberta et de la Saskatchewan et les patrons d'anomalies magnétiques et de gravité suggèrent l'existence d'un bloc distinct nommé "bloc de l'-Alberta su Sud." Les données paléomagnétiques portant sur un mouvement relatif du bloc Wyoming par rapport à la province Supérieur sont non probantes. On ne peut conclure à de très grands déplacements mais des mouvements de 1000 km sont possibles et en deça des marges d'erreur des données.

INTRODUCTION

The Wyoming Province (Fig. 1) is a block of Archean crust exposed in the cores of Laramide uplifts in Wyoming and adjacent states. The term "Wyoming Province" is a natural appellation, which seems to have independently been invented by several authors. The name has been used at least since publication of a paper by Engel (1963).

Even on the basis of sparse surface data, there is good reason to wonder whether the Archean rocks of Wyoming connect with the Superior Province craton, because directly between the two cratons lie the Black Hills, where part of an obviously substantial Proterozoic mobile belt is exposed. Early workers considered the two cratons part of a single entity, with the Black Hills representing an ensialic or intracontinental mobile belt (e.g., King, 1976). More recent workers, drawing on plate tectonic models, have suggested that the two cratons were once separate, and were joined by collision (Camfield and Gough, 1977; Klasner and King, 1986). This paper surveys the Wyoming Province and its environs and will attempt to synthesize the available information. In particular, we will identify gaps in current knowledge and promising areas for further research. Readers are strongly cautioned to view the result as a compendium of questions rather than answers.

The Wyoming Province (Fig. 1) is a roughly triangular block of Archean crust surrounded by three Early Proterozoic (Precambrian X or Aphebian) mobile belts for which we will use the informal names Thelon, Dakota, and Colorado segments. The Thelon Segment, northwest of the Wyoming Province, and the Dakota Segment, to the east, are branches of the Trans-Hudson Orogen. The Colorado Segment, to the south, is a succession of accretionary terranes somewhat younger than the Thelon and Dakota segments.

On the east, the Archean rocks are bounded by the North American Central Plains (NACP) Conductivity Anomaly and the Dakota Segment of the Churchill Province. Structural trends in the Dakota Segment, based on geophysical interpretation, are northwest in South Dakota, changing to north in North Dakota and southern Manitoba, and then finally to the northeast in northern Manitoba, where they emerge from beneath the Phanerozoic cover. Assemblages identified in northern Manitoba include island-arc rocks, Early Proterozoic oceanic crust, deep marine sediments, and late-stage shallow-water clastic rocks (Lewry *et al.*, 1978; Bailes, 1980; Lewry, 1981; Stauffer, 1984). These assemblages are inferred to continue southward beneath the buried portion of the Dakota Segment (Green *et al.*, 1985).

The northwestern side of the Wyoming Province is bounded by the Thelon Segment, with northeast structural trends. The northwestern portion of this mobile belt is the buried extension of the Archean portion of the Churchill Province, and the Churchill Craton and Cree Lake Zone of Green *et al.* (1985), also termed the Rae and Hearne terranes, respectively, by Hoffman (1987). Between these terranes and the Wyoming Province are a number of smaller blocks of Archean and Proterozoic ages (Ross *et al.*, 1988). Churchill Craton (Rae) rocks, with primary ages greater than 2500 Ma (Baadsgaard and Godfrey, 1972), are characterized by granulite grade assemblages that show an amphibolite facies overprint generated by Early Proterozoic Trans-Hudson orogenic events. They are composed of granodiorite gneisses and subordinate metagabbro, meta-anorthosite, and minor paragneiss.

Leucosomes of Early Proterozoic age are common in areas reworked by Trans-Hudson orogenic events (Lewry *et al.*, 1978). Cree Lake Zone (Hearne) rocks, exposed in northwestern Saskatchewan and the adjacent Northwest Territories, include a number of infolded belts of Early Proterozoic supracrustals and remobilized Archean basement. Both the supracrustal and basement rocks were metamorphosed to upper amphibolite to low-pressure granulite grade during the Trans-Hudson orogenic event (Lewry *et al.*, 1978; Green *et al.*, 1985).

On the south, the Wyoming Province is bounded by the Colorado Segment, which is well exposed in the cores of basement uplifts in the Rocky Mountains. The Colorado Segment consists of a southward-younging succession of volcanic-arc complexes with associated clastic metasedimentary rocks, intruded by Early and Middle Proterozoic granitic rocks (Condie, 1982; Bickford, 1988). The Colorado Segment is younger than the Dakota and Thelon segments; whereas Dakota and Thelon segment metamorphic rocks yield U–Pb metamorphic ages of about 1900 Ma, Colorado Segment rocks average 1800 Ma on the southern margin of the Wyoming Province and 1700 Ma in Arizona and New Mexico. Archean rocks seem to be entirely lacking in the Colorado Segment. Geophysical trends in the Colorado Segment are interrupted in central Colorado by the Front Range uplift, but on either side of the Front Range, gravity and magnetic trends are predominately east–west. In eastern Colorado and western Kansas, the trends appear to curve smoothly into southeast trends of the Central Plains Orogen (Dutch, 1983; Sims and Peterman, 1986). Denison *et al.* (1984) is a good general reference on the Precambrian basement geology of the central plains.

LIMITS OF THE WYOMING PROVINCE

The boundaries of the Wyoming Province are almost entirely concealed and must be inferred from geophysical and geochronological data. Archean crust is widely exposed in Wyoming and in scattered localities in Montana, Utah, Nevada, and Idaho. We can reasonably infer that Archean crust extends beneath all but the southeastern corner of Wyoming, beneath most of the southeastern half of Montana, eastward beneath the Black Hills of South Dakota, and westward to the East Humboldt

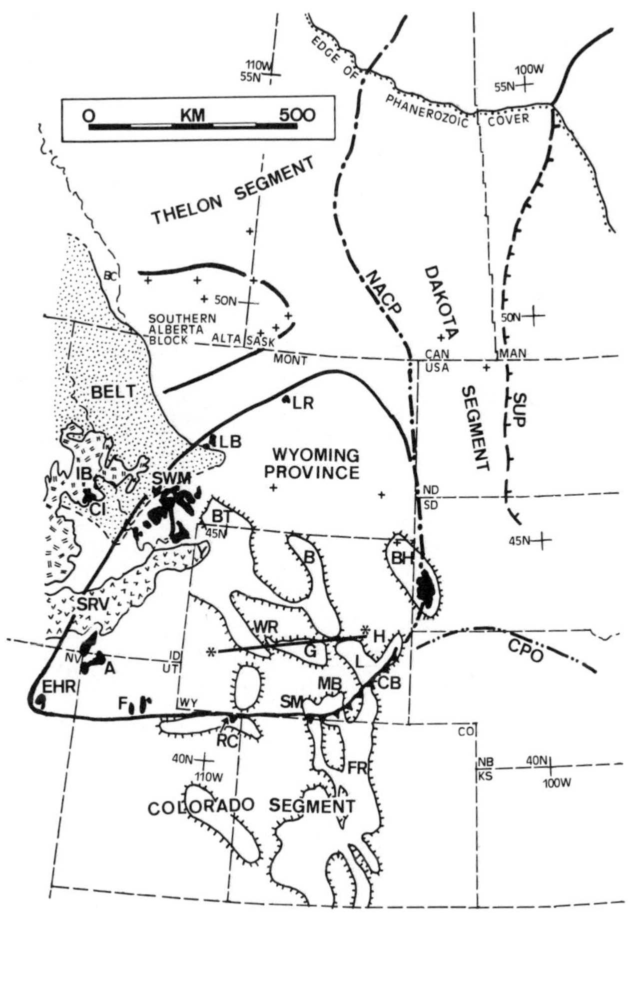

Figure 1. *Tectonic framework and geographic setting of the Wyoming Province. Hachures indicate principal Laramide uplifts: B, Bighorn; BH, Black Hills; BT, Beartooth; FR, Front Range; G, Granite Mountains; H, Hartville Uplift; L, Laramie Range; MB, Medicine Bow; SM, Sierra Madre; and WR, Wind River. The approximate boundary of the Wyoming Province is shown by a solid heavy line with key localities labelled: A, Albion Range; CB, Cheyenne Belt; CI, questionable Early Proterozoic rocks of central Idaho; EHR, East Humboldt Range; F, Farmington Canyon Complex; LB, Little Belt Mountains; LR, Little Rocky Mountains; RC, Red Creek Quartzite; and SWM, Southwestern Montana. Other major geological features: CPO, northern boundary of the Central Plains Orogen; NACP, North American Central Plains Conductivity Anomaly; SUP, western margin of the Superior craton; SRV, Snake River Volcanics; Y, Yellowstone hot spot; IB, Idaho Batholith; and BELT, generalized distribution of Belt Supergroup rocks. Line *–* marks the northern limit of K–Ar age resetting adjacent to the Cheyenne Belt. Crosses mark critical Archean basement age localities (from Peterman (1981a) for Wyoming and Montana, Goldich* et al. *1966) for southeastern Saskatchewan, Frost and Burwash (1986) for Alberta and southwestern Saskatchewan, and Klasner and King (1986) for North Dakota).*

Range in northeastern Nevada (Lush *et al.*, 1988). As far as existing data reveal, the Wyoming Province was a single unit in Proterozoic time, although there seems to be an emerging consensus in favour of Archean accretion of the Wyoming block (Mogk and Henry, 1988; Mogk *et al.*, 1988).

Exposed Archean rocks around the periphery of the Wyoming Province loosely constrain the province's boundary. Generally, outcropping Archean rocks define a minimum outer boundary for the province. Apart from two Archean inliers in the Black Hills, there is no firm information on how far Archean rocks may extend beneath the neighbouring Proterozoic mobile belts.

Definition of the Archean boundaries of blocks like the Wyoming, Slave, Churchill, or Superior provinces is a function of many variables, including the nature of the original boundary and its subsequent thermotectonic history. In the case of the Wyoming craton, relevant factors include the level of the present erosional surface, proximity to the (presumably original) rifted margin, proximity to the convergent margin, and the local degree of crustal shortening during the collisional events of 1900–1600 Ma, which assembled the North American Craton.

Contrasts in metamorphic grade across major shear zones (Karlstrom and Houston, 1984) or overprinting and reworking of pre-existing metamorphic assemblages (Nielsen *et al.*, 1981) may be successfully used to help in defining these provincial boundaries. Case studies of three well-documented areas will serve to illustrate techniques that may aid in locating the Archean age margins of the Wyoming Province.

Karlstrom and Houston (1984), in an investigation of the Cheyenne Belt, noted a strong contrast in metamorphic grade across the belt, with higher grade, island-arc derived rocks thrust northward over lower grade cratonic and miogeoclinal rocks. The present erosional level is shallow enough to deduce the protoliths and their original tectonic setting. Here, the metamorphic data confirm the interpretation available from other geological interpretations. This area will be discussed in greater detail shortly.

A second type of metamorphic and structural pattern at a Precambrian convergent boundary is that described by Thompson *et al.* (1987), James *et al.* (1987), and Thompson (1987). On the Thelon Front at the southern edge of the Slave craton in northwestern Canada, they have documented a complex metamorphic facies distribution accompanied by intense deformation. The deformation records strike-slip motion as well as extensive shortening (thrusting), perhaps indicative of an obliquely directed suturing event with accreted terrain affinities, rather than a Himalayan-type collision and suture with convergence at a high angle to the convergent margin (Hoffman, 1987). None of the exposures near the boundary of the Wyoming Province seems to fit this type of pattern given the present state of understanding.

A third pattern associated with a Precambrian convergent margin was described by Nielsen *et al.* (1981) in the exposed portion of the Athabasca Mobile Belt and adjacent Cree Lake Zone in northern Saskatchewan. Here, the Archean basement complex, originally characterized by high-pressure granulite assemblages, was overprinted by low-pressure granulite to amphibolite assemblages, accompanied by partial melting, to produce the Slave granitoids. They interpret the origin of the Athabasca Mobile Belt as resulting from high heat flow above the northwest-directed subduction zone caused by closure of an Early Proterozoic ocean along what is now called the La Ronge – Lynn Lake arc (Green *et al.*, 1985). This area represents reworked Archean cratonic material that was not rifted and reassembled during the Early Proterozoic assembly of the North American Craton.

Regional metamorphic studies, particularly where variation in calculated *P–T* conditions is documented, are valuable in constraining tectonic models constructed on the basis of other kinds of geological data. Such studies may prove particularly valuable where the rocks of interest are in the subsurface and where basement cores along with geophysical data represent the only available information.

ARCHEAN OF THE WYOMING PROVINCE

The purpose of this summary is to define the Wyoming Province as an entity for contrast and comparison with the surrounding terrains. Most of the Wyoming Province basement is high-grade Archean gneiss and deformed Archean granitic intrusive rocks (Condie, 1976; Wooden *et al.*, 1988). Radiometric dating reveals at least three distinct events. In the Big Horn Mountains, Arth *et al.* (1980) found two events: an early episode of deformation and metamorphism produced banded trondhjemite gneisses that yield U–Pb and Rb–Sr ages of 2950–3000 Ma and a second episode of synkinematic trondhjemitic to granitic intrusion occurred at about 2800 Ma. In the Granite Mountains, an amphibolite-grade metamorphic complex of banded and migmatitic gneisses yields U–Pb ages of 2860 Ma, and a large, undeformed granitic intrusion yields U–Pb ages of 2550 Ma (Peterman and Hildreth, 1978). Peterman (1981b) determined Rb–Sr ages of 2550 Ma in the Little Rocky Mountains, but with substantial scatter in the analytical data.

Exposures in Utah, Nevada, and Idaho define the western margin of the Wyoming Province. The rocks of the Albion Range on the Utah–Nevada border consist of adamellite gneiss, mica schist, amphibolite, and quartzite. Armstrong and Hills (1967) determined a Rb–Sr isochron age for these rocks of 2460 Ma, with K–Ar dates reset by Tertiary metamorphism. The Farmington Canyon Complex in northern Utah, formerly considered to be Proterozoic (Eardley and Hatch, 1940a,b), is now known to be at least partly Archean (Bryant, 1988). Migmatites from this complex yield Rb–Sr and Nd–Sm model ages between 2600 and 3600 Ma. Quartz monzonite gneiss yields U–Pb and Rb–Sr ages of about 1800 Ma. The East Humboldt Range in northeastern Nevada contains biotitic monzogranitic orthogneiss, quartzite, and pelitic schist within a Mesozoic thrust sheet. The orthogneiss yields a U–Pb age of 2520 Ma (Lush *et al.*, 1988). This exposure is the westernmost known Archean rock in the conterminous United States.

It appears that the igneous and metamorphic complex of the northern Wyoming Province experienced three major episodes of deformation and metamorphism centred at about 3000, 2800, and 2500 Ma.

Well-preserved Archean supracrustal rocks are found only in southern Wyoming. These rocks are mostly similar to normal Archean greenstone belt assemblages, but with a lower proportion of volcanic rocks and some quartzite and iron formation. Karlstrom and Houston (1984) suggested that these Archean rocks may be transitional to Proterozoic and later continental-margin assemblages.

The present Wyoming block may have coalesced during multiple Archean accretion events. Karlstrom and Houston (1984) hypothesized that the Archean supracrustal rocks of southeastern Wyoming reflect an Archean suturing. More recently, Mogk *et al.* (1988) and Mogk and Henry (1988) argued for Archean accretion tectonics in the Beartooth Range of northwestern Wyoming.

CHEYENNE BELT

The best exposure of the boundary of the Wyoming Province is the Cheyenne Belt, which has been interpreted as a south-dipping Proterozoic subduction zone. The references cited here (Hills and Houston, 1979; Condie and Shadel, 1984; Johnson *et al.*, 1984; Karlstrom and Houston, 1984; Duebendorfer and Houston, 1987) include basic facts about the Cheyenne Belt, as well as the most essential primary references. It is useful to summarize the geology of this area here because the Cheyenne Belt, which is both well exposed and well studied, can serve as a useful model for interpreting other possible sutures whose geology is less well known.

In the Cheyenne Belt, a shear zone up to 7 km wide separates two markedly different Proterozoic sequences. North of the shear zone, the Snowy Lake Supergroup appears to represent a transgressive continental margin sequence, progressing upward from fluvial facies in the Deep Lake Group to deltaic and finally marine carbonate facies in the overlying Libby Creek Group. Widespread, but poorly dated, dykes and sills appear to represent a rifting episode sometime between 1900 and 2200 Ma. Karlstrom and Houston (1984) interpreted the Deep Lake Group as early rift fill, followed by deposition of the lower Libby Creek Group in a proto-oceanic seaway. They suggest that carbonate rocks in the upper part of the Libby Creek sequence may represent carbonate banks at the edge of a continental shelf.

The assemblage south of the shear zone is dominated by bimodal calc-alkaline volcanic rocks and volcanogenic sediments of the Green Mountain Formation (Condie and Shadel, 1984). These rocks are marked by widespread mafic intrusives, which are unique in the Precambrian of the southwestern United States and may represent dismembered fragments of ophiolites. Syntectonic felsic intrusives are common south of the shear zone, but not north of it, suggesting that the southern assemblage may have been a volcanic arc complex over a south-dipping subduction zone. These intrusives cluster around 1780, 1700, and 1650 Ma in age. Archean basement has not been detected south of the shear zone by radiometric dating or Sm–Nd isotopic studies.

Condie and Shadel (1984) interpreted the Cheyenne Belt as a deeply eroded equivalent of the Southern Appalachians. In this interpretation, the present shear zone is the root of a former thrust complex, which once extended an unknown distance north of the shear zone. Seismic-reflection profiling across the Cheyenne Belt revealed a number of south-dipping reflectors that may represent major shear zones (Allmendinger *et al.*, 1982). Gravity studies by Johnson *et al.* (1984) support this interpretation. These authors examined three north–south gravity profiles across the Cheyenne Belt. By examining profiles along the crests of the Sierra Madre, Medicine Bow, and Laramie Ranges, they were able to isolate the gravity effect of the Wyoming Province boundary from the much larger gravity effect of Laramide block tectonics. In each of the three profiles, the long-wavelength gravity data are adequately modelled by a simple ramp in the Moho about 20–50 km south of the surface expression of the shear zone, with the base of the crust stepped down to the south by 5–8 km. The similarity and simplicity of the models are notable. The location of the step suggests that the suture between the Wyoming Province and the "Colorado

Segment" dips steeply southward, as expected from surface studies and the seismic-reflection profiling.

North of the Cheyenne Belt is a zone about 120 km wide where K–Ar ages have been reset from Archean to Middle Proterozoic (1300–1500 Ma). Peterman and Hildreth (1978) interpreted the reset ages as cooling ages, marking the time when the basement rocks reached a shallow enough level to retain stable K–Ar ages. Assuming a normal geothermal gradient, they estimate 10–12 km of uplift and erosion in this zone. The age-reset boundary coincides with several major Laramide structural features (Fig. 1), suggesting that the boundary is tectonic as well as geochronological.

In the Laramie Range, the Cheyenne Belt is concealed by an intrusive complex consisting of syenite (1435 Ma), the Laramie Anorthosite, and the Sherman Granite (1400 Ma). The extrapolated trend of the Cheyenne Belt passes through the north end of the Laramie Anorthosite and southeast of the Hartville Uplift, where Archean basement is exposed. The significance of the Cheyenne Belt to the 1400 Ma intrusive complex is unknown, but, as Karlstrom and Houston (1984) noted, there are no intrusions of similar age north of the Cheyenne Belt and the only exposed massif anorthosite in the Rocky Mountains sits precisely astride a major structural discontinuity.

RED CREEK QUARTZITE

Although the general progression of this survey is counterclockwise around the Wyoming Province, it is useful to discuss one locality out of sequence. The Red Creek Quartzite, exposed where Utah, Wyoming, and Colorado meet, is unusual among the Middle Proterozoic sequences bordering the Wyoming Province. It consists of great thicknesses of pure quartzite, metamorphosed to amphibolite facies. Schists within the quartzite contain mineral assemblages that Sears *et al.* (1982) interpreted as indicating at least 3.8 kbar (380 MPa) pressure. Amphibolite, the metamorphosed relics of dykes and sills, is volumetrically minor, but widespread. Hansen (1965) interpreted the quartzites as having formed on a stable, slowly subsiding shelf. Folds in the quartzites have generally northeast-trending axes and steeply north-dipping axial planes. Hansen (1965) reported schist in the Red Creek Quartzite with a Rb–Sr age of 2320 Ma. Sears *et al.* (1982) generally concurred with Hansen's interpretation, and in addition reported the discovery of an Archean gneiss complex, the Owiyukuts Complex, at least 2700 Ma in age, over which the Red Creek Quartzite was thrust northward. They concluded that the Red Creek Quartzite is probably equivalent to the Libby Creek Group and that the tectonic setting of the Red Creek – Owiyukuts locality is similar to the Cheyenne Belt: northward-directed overthrusting probably due to collision of the Wyoming block with a south-dipping subduction zone.

CHEYENNE BELT AND PENOKEAN PROVINCE COMPARED

Because the Cheyenne Belt and the Penokean Province south of Lake Superior are the only well-exposed contacts of Archean cratons with younger provinces in the United States, are roughly the same age, and are nearly along strike with each other, many authors have compared them (e.g., Duebendorfer and Houston, 1987). Recent radiometric dating and tectonic interpretations suggest that the two areas are not contemporaneous or correlative, and we stress that this discussion is not meant to imply that they are. Nevertheless, a comparison of tectonic styles in the two belts is illuminating.

In both cases, there is sound reason to believe a Proterozoic volcanic arc complex collided with an older craton. The case for a south-dipping subduction zone in the Cheyenne Belt seems solid. In the Penokean, where exposure is poorer and the state of understanding not as advanced, various authors have postulated subduction zones dipping north (Greenberg and Brown, 1983) and south (Larue, 1983), and two-stage models with both dip directions (Laberge, 1986). Lithologically, both arc complexes are characterized by bimodal volcanic suites and abundant mafic intrusive bodies. As a final provocative note, both areas have large anorogenic intrusive bodies belonging to a widespread 1400 Ma event, consisting of granite, syenite, and anorthosite. It appears that Proterozoic terrane accretion produced a characteristic lithological and structural signature that is represented in both the Cheyenne Belt and the Penokean Province.

BLACK HILLS AND THE DAKOTA SEGMENT

Archean rocks are exposed in inliers in the Black Hills of South Dakota, overlain by a thin Proterozoic shelflike sequence and a thick eugeoclinal complex (Kleinkopf and Redden, 1975; Redden *et al.*, 1987). Volcanic rocks are sparse in the eugeoclinal complex and most abundant in the basal portion, where some mafic intrusive rocks also occur. The Archean inliers have traditionally been interpreted as gneiss domes (Runner, 1943; Zartman *et al.*, 1964; Bayley, 1972; Gosselin *et al.*, 1987) remobilized during the Early Proterozoic. Late- to postorogenic granites and pegmatites about 1720 Ma old intrude the metamorphic rocks of the Black Hills (Duke *et al.*, 1987).

Fold patterns in the Proterozoic rocks are complex. In the northern Black Hills, the principal folds are isoclinal, with axial planes that dip to the east and generally trend north-northwest approximately parallel to the presumed margin of the Wyoming Province (Noble *et al.*, 1949). The isoclinal folds plunge southeastward. In the southern Black Hills, a second set of northeast-trending folds appears to be superimposed on the north-northwest set (Ratte and Wayland, 1969).

The North American Central Plains Conductivity Anomaly (hereafter referred to as NACP) is a critical feature in interpreting the Trans-Hudson Orogen. The NACP was mapped by Alabi *et al.* (1975), Camfield and Gough (1977), and Handa and Camfield (1984) on the basis of magnetometer array studies. The NACP extends from the exposed northeastern end of the Cheyenne Belt, east of the Black Hills, north along the North Dakota – Montana border, and ultimately connects with the Rottenstone – La Ronge Belt in Saskatchewan. Because it connects two major shear zones that are probable sutures, most studies of the NACP have considered it as a probable crustal suture. Thomas *et al.* (1987) located the NACP somewhat east of its location in earlier studies and suggested that it may correlate with a series of en echelon gravity gradients, i.e., it may not be a single, continuous feature.

Two critically placed stations constrain the location of the NACP at its southern end: a station just east of the junction of the NACP with the Cheyenne Belt and another station in the Black Hills. The NACP passes just west of the southernmost station and demonstrates the continuity of the NACP with the Cheyenne Belt. The NACP also passes east of the Black Hills station, implying that the Archean exposures in the Black Hills are Wyoming Province basement rather than allochthonous rocks of the Dakota Segment. Gosselin *et al.* (1987) believed that the petrology of the Archean rocks reflects a late Archean suturing event on the margin of the ancestral Wyoming craton.

The identification of the Dakota Segment as the southern extension of the Churchill Province was suggested by Muehlberger *et al.* (1967) and put on a solid footing by Lidiak (1971). At that time, adequate magnetic data were available for South Dakota, but not for North Dakota. The continuity of the Dakota Segment with the exposed Churchill–Superior boundary, though never seriously in doubt, was confirmed by Green *et al.* (1979) and the magnetic map of Zeitz (1982).

Klasner and King (1986) summarized the current understanding of the Dakota Segment, with emphasis on its relation to the Superior Province. The closest known Archean basement rock east of the NACP is 100 km away; the transition zone between the Churchill and Superior provinces as delineated by Klasner and King (1986) is about 100–150 km east of the NACP. These gaps are not large, and in other geological contexts (such as within the Wyoming Province), one might well be justified in inferring continuous Archean basement across the gaps. Historically, however, the principal reason for postulating continuous Archean basement between the Wyoming and Superior cratons, as in the case of many other mobile belts, was that ensialic mobile belts were the only alternatives to continental collision. With virtually all other "ensialic" mobile belts now known to have formed through plate collision, the theoretical basis for doubting a plate-collision origin for the Dakota Segment has disappeared.

The lithology and structure of the Black Hills can be tentatively interpreted in plate tectonic terms. The lithology of Black Hills rocks suggests initial deposition of a thin cratonic or shelf sedimentary sequence on Archean basement, followed by development of a eugeoclinal (continental rise) sedimentary prism. A westward-dipping subduction zone beneath the Black Hills would be expected to result in abundant volcanic rocks in the eugeoclinal assemblage. Thus, the sparsity of volcanic rocks in the eugeoclinal assemblage and their concentration near the base of the eugeoclinal sequence seems most consistent with vulcanism during continental rifting and the initiation of a new continental margin (Redden *et al.*, 1987).

LITTLE ROCKY MOUNTAINS

Mixed Archean and Proterozoic ages have been obtained from the Little Rocky Mountains of north-central Montana (Peterman, 1981b). The Precambrian rocks consist mainly of biotite gneiss and schist, and amphibolite. The Rb–Sr data set a minimum age of 2550 Ma for the rock, and although the data scatter too widely for a precise isochron, the rocks are definitely established as Archean. K–Ar ages of these rocks fall in the range of 1700–1750 Ma because of Proterozoic resetting. There are no known Proterozoic supracrustal rocks in the Little Rocky Mountains, but the resetting of K–Ar ages implies that the Wyoming Province boundary is not far away.

LITTLE BELT MOUNTAINS

This area was mapped geophysically by Kleinkopf *et al.* (1972). Catanzaro and Kulp (1964) presented the most detailed structural account of this locality. They found four major metamorphic rock units: (1) migmatite with evidence of at least two metamorphisms; (2) grey gneiss, banded but noncontorted; (3) granitic augen gneiss; and (4) phyllonitic schists.

A sheared mafic intrusive, the Pinto Metadiorite, intrudes the gneisses. From concordant and discordant U–Pb ages, Catanzaro and Kulp (1964) determined an age of about 2450 Ma for the migmatite and 1900 Ma for the grey gneiss and Pinto Metadiorite. On the basis of lithology and the evidence for only a single metamorphism, they interpreted the grey gneiss as a metasedimentary unit deposited between 2450 and 1900 Ma.

This sketchy picture suggests that reinvestigation of the Little Belt Mountains may prove fruitful. If the Catanzaro and Kulp (1964) interpretation is correct, the Little Belt Mountains may be an exposure of the northwestern margin of the Wyoming Province. Structural trends as mapped in the transect of Catanzaro and Kulp (1964) vary from easterly to northerly, but have an average strike consistent with the northeast structural trends that would be expected in the Thelon Segment. The extensive cataclasis in the Little Belt rocks also suggests proximity to a major crustal boundary and the Pinto Metadiorite recalls the mafic intrusives that occur south of the Cheyenne Belt. If the Wyoming Province – Thelon Segment boundary is accessible anywhere, the Little Belt Mountains are a leading candidate.

SOUTHWESTERN MONTANA

Crystalline rocks of Archean and perhaps Proterozoic age occur in southwestern Montana. The Beartooth Range consists largely of Archean gneiss and intrusive rocks (Butler, 1966; Mogk *et al.*, 1988). Metamorphosed Precambrian supracrustal rocks occur north and west of the Beartooth Range. Three rock unit names have been in use for the supracrustal rocks: the Pony Group, consisting of mafic gneiss and granulite; the Cherry Creek Group, consisting of pillow lava, mica schist, iron formation, quartzite, and marble; and the Dillon granitic gneiss (Reid, 1955, 1957, 1963; Scholten, 1955; Scholten *et al.*, 1955; Heinrich and Rabbitt, 1960; Burger, 1967; Hadley, 1969a,b; Tysdal, 1976).

Early radiometric dating in this region (summarized by Catanzaro (1966) and King (1976)) suggested an age gradation from Archean in the Beartooth Range to Proterozoic in the Tobacco Root and adjacent ranges, but James and Hedge (1980) found that the supposedly Proterozoic supracrustal rocks fit an Archean Rb–Sr isochron age of 2730 Ma. Vitaliano and Cordua (1979), summarizing the results of extensive field mapping, questioned whether there is any major lithologic distinction between the Cherry Creek and Pony groups, and James and Hedge (1980) regard the existing nomenclature as dubious.

The existing data favour an Archean age for all the crystalline rocks of southwestern Montana. However, the abundance of marble, quartzite, and iron formation in the supracrustal rocks is more typical of Proterozoic than Archean stratigraphy, and unrecognized Proterozoic supracrustal rocks may be present.

Southwestern Montana may be a possible locality for an exposed northwestern margin of the Wyoming Province. To date, no contacts between Archean and Proterozoic rocks have been identified. If the current age data are representative, the boundary of the Wyoming Province passes somewhere northwest of the exposed crystalline rocks. Alternatively, it could be exposed, but as yet unrecognized.

CENTRAL IDAHO

Proterozoic rocks of pre-Beltian age have been reported at localities throughout Idaho (Armstrong, 1975), but the evidence for Thelon Segment rocks in Idaho is equivocal. The Precambrian rocks of Idaho have been overprinted by strong middle Precambrian and Mesozoic metamorphisms so that even Beltian rocks are metamorphosed to gneisses in places. Until recently, some of these rocks were considered gneissic facies of the Idaho Batholith. Ages typical of the Thelon Segment (older than 1800 Ma), if they ever existed, must have been reset in most cases.

Armstrong (1975) argued that pre-Beltian crystalline rocks in central Idaho were exposed in a basement uplift, which he called the Salmon River Arch, during Beltian time and that controlled early Beltian sedimentation. He further argued that widespread quartzites in this region are pre-Beltian. Evans (1981) conducted field and geochronological studies on the presumed rocks of the Salmon River Arch and concluded that there is no evidence for the Salmon River Arch as Armstrong defined it.

The relevance of this region for the present paper is not whether there was a basement high that controlled Beltian sedimentation, but whether the rocks in this region expose Thelon Segment basement northwest of the Wyoming Province. Evans (1981) makes a convincing case that the presumed pre-Beltian quartzites are very early Beltian. The supracrustal rocks are lithologically more similar to Beltian rocks than to the lithology to be expected in Thelon Segment metasediments. The granite that intrudes the quartzites is a rapakivi S-type granite, suggesting an anorogenic environment rather than an active orogenic belt. The zircon ages obtained for the granite cluster around 1350 Ma for the coarsest zircons and grade to about 1500 Ma for the finer fractions. Harrison (1972), Clark (1973), and Evans and Fischer (1986) describe numerous radiometric ages of 1200–1500 Ma in Idaho that may correlate with orogenic events of similar age in the Canadian

Cordillera. Overprinting by Phanerozoic and probably Precambrian metamorphism and deformation makes it unlikely that Thelon Segment rocks in Idaho can easily be recognized or interpreted, if they exist at all. Although it is possible to map the Wyoming Province boundary through central Idaho as some authors, such as Thomas *et al.* (1987), have done, the evidence does not require such an interpretation.

UTAH AND NEVADA

Several exposures of Archean and Proterozoic crystalline rock in Utah and Nevada constrain the location of the southwestern corner of the Wyoming Province. Eardley and Hatch (1940a,b) referred to ancient crystalline rocks in northern Utah as the Farmington Canyon Complex. The Farmington Canyon Complex consists mainly of granitic gneiss, polyphase-deformed banded gneiss, and minor amphibolite (Condie, 1969). Until recently, these rocks were generally regarded as Proterozoic, but Bryant (1988) reported ages greater than 2600 Ma for some of the high-grade metamorphic rocks. However, a quartz monzonite gneiss from the Farmington Canyon Complex yields an age of about 1800 Ma. In addition, lower grade metamorphic assemblages, the Little Willow and Facer formations, were interpreted by Bryant (1988) as a possible Proterozoic ensialic sequence. Thus, there are hints of a Proterozoic continental-margin assemblage in Utah, although tectonic disruption by Phanerozoic thrusting makes it hard to interpret the Precambrian tectonics of this region.

Zartman (1974) mapped a lead-isotope province boundary in the Great Basin that coincides well with the extrapolated southern boundary of the Wyoming Province. Recently, Nelson *et al.* (1983) presented Sm–Nd isotopic evidence for possible Early Proterozoic basement beneath the eastern Great Basin. Nelson and DePaolo (1985) and Bennett and DePaolo (1987) showed a Sm–Nd province with crust-formation age of 2000-2300 Ma covering much of the present Great Basin. This province may represent material accreted to the western side of the Wyoming Province.

CANADIAN CONNECTION

The southern boundary of the Wyoming Province is well located and understood: a south-dipping suture with Proterozoic accreted terranes and no Archean basement to the south. The eastern boundary is less well defined, although the evidence for a Proterozoic rifted continental margin in the Black Hills and a suture related to the NACP is strong.

The least-understood boundary of the Wyoming Province is its northern boundary. The figures in this paper outline the minimum extent of the Wyoming Province, but it is not known how far Archean basement extends to the north. Zartman (1974) outlined an Archean lead-isotope province that extended from the southern boundary of the Wyoming Province to the Canadian border. Frost and Burwash (1986) found Archean Nd–Sm ages from samples across southern Alberta and southwestern Saskatchewan, and Ross *et al.* (1988) reported Archean U–Pb ages from the same region. Thomas *et al.* (1987) suggested that the Wyoming Province boundary follows a prominent gravity gradient across southern Alberta. The Wyoming Province, as outlined in their interpretation, includes most (not all) of the Archean Nd–Sm ages of Frost and Burwash (1986). For reasons discussed below, we suggest that the boundary of Thomas *et al.* (1987) is actually the northern edge of a distinct block, the Southern Alberta Block (Fig. 1).

GEOPHYSICAL SIGNATURE OF THE WYOMING PROVINCE

The Wyoming Province displays a pattern of linear magnetic anomalies roughly concentric with its margins (Figs. 2–4). The magnetic low that outlines the entire eastern half of the Wyoming Province boundary is particularly striking (Fig. 3). Gravity anomalies, where not obscured by Phanerozoic structures, are similar in trend to the magnetic anomaly patterns. This pattern is most conspicuous on the northern and eastern sides of the province, where east-northeast magnetic anomaly trends on the north meet north–south or south-southeast trends on the east. These trends extend into areas known to be underlain by Archean rock. The south-southeast trends on the eastern side of the province are parallel to trends in the Dakota Segment, but occur west of the NACP. The east-northeast trends in Montana are oblique to the provincial boundary as outlined in our figures, and extend into known Archean basement.

The most likely explanations for the anomaly patterns concentric with the Wyoming Province boundary are (1) Archean lithologic zonation, possibly reflecting accretion around a nucleus; (2) Proterozoic deformation and overprinting of Archean basement during a series of accretionary events; or (3) some combination of both.

SOUTHERN ALBERTA BLOCK

North of the zone of northeast-trending anomalies in Montana is a distinct block of north-northwest magnetic and gravity anomalies, which we term the Southern Alberta Block (Fig. 1). The north-northwest anomalies are truncated abruptly on both the north and south by prominent east-northeast magnetic and gravity anomalies. Nd–Sm age data (Frost and Burwash, 1986) strongly favour an Archean age for the basement in this area, but the geophysical data indicate that the Southern Alberta Block is a distinct structural block rather than a simple extension of the Wyoming Province. The Wyoming Province boundary as defined by Thomas *et al.* (1987) coincides with the northern boundary of the Southern Alberta Block and with a major buried rift (Kanasewich *et al.*, 1968). The evidence for a major tectonic boundary in this location is beyond question. Ross *et al.* (1988) mapped this boundary as a north-dipping suture. Because the Southern Alberta Block is entirely concealed, nothing is known of its history relative to adjacent terrains.

REGIONAL GEOPHYSICS AND TECTONIC IMPLICATIONS

Gravity and magnetic anomaly patterns (Figs. 2–4) in the Thelon Segment are generally northeast. Geophysical trends in the Dakota Segment are also northeast in northern Manitoba where the Dakota and Thelon segments are in contact, but change southward to north–south and finally southeast in the Dakotas. The bifurcation between northeast and north–south trends corresponds closely to the NACP Conductivity Anomaly.

The structural pattern appears best explained by accretion of the Wyoming block to the North American Craton during the Proterozoic accretion of the Dakota and Thelon segments. Three different

Figure 2. *Gravity gradients in and near the Wyoming Province. Black denotes gradients greater than 0.6 mGal/km (1 Gal = 1 cm/s*^{2}*). Most of the map is based on the work of Thomas* et al. *(1987); southern Utah, Colorado, and Kansas are derived from the work of Simpson* et al. *(1986). The major tectonic boundaries from Figure 1 are indicated by dotted lines.*

Figure 3. *Magnetic anomalies in and near the Wyoming Province (based on the work of the Committee for the Magnetic Anomaly Map of North America (1987); southern Saskatchewan is based on Green* et al. *(1985)). Black areas are positive anomalies greater than 100 nT; horizontally ruled areas are negative anomalies less than -100 nT. Crosshatched areas lack magnetic coverage or have fitting errors. Major tectonic boundaries from Figure 1 are shown as dotted lines.*

sequences of accretion are possible: (1) Thelon accretes to Superior–Dakota–Wyoming, (2) Thelon–Wyoming accretes to Superior–Dakota,or (3) Thelon–Wyoming–Dakota accretes to Superior. At present, the data are insufficient to show which scenario is correct. The northward-tapering shape of the Wyoming block and its position at the extreme southern end of the Trans-Hudson Orogen also invites speculation about tectonic extrusion in response to collision events elsewhere in the Canadian Shield (Hoffman, 1987).

PALEOMAGNETISM

The analysis of Roy (1983) is enough to inspire humility in any geologist audacious enough to use Precambrian paleomagnetic data. Conflicting analyses of Precambrian pole positions for North America can be used to create a multitude of apparent polar wander (APW) curves, because of sparse data, dating uncertainty, and possible misinterpretation of antipoles.

Figure 4. *Comparison of magnetic and gravity data in and near the Wyoming Province. The heaviest lines indicate the tectonic boundaries from Figure 1. Medium-weight lines indicate geophysical trends. Principal gravity gradients are indicated by solid lines, magnetic highs by long dashes, and magnetic lows by short dashes. T–T–T shows the southern terminus of the Superior craton as suggested by Thomas* et al. *(1987); D-D shows the Superior–Churchill boundary of Dutch (1983).*

Paleomagnetic evidence for motion of the Wyoming block relative to the Superior Province is inconclusive. As noted in the summary of the Wyoming Province, there are only three major events recognized in the Archean of the Wyoming Province. These events are separated by 200 Ma or more, far too sparse to define an APW path. It is possible that careful sampling and dating in the Archean supracrustal rocks north of the Cheyenne Belt might yield paleomagnetic data in sufficient quantity and density to define a pole path for the Wyoming Province, but so far no such studies have been done.

The few pole positions available for the Wyoming Province all fall close to Superior Province poles of comparable age, i.e., within a few tens of degrees. There is no conclusive evidence for very large motions of the Wyoming Province relative to the Superior Province, and the sparse data are consistent with no relative motion. Small motions, on the order of 1000 km, are possible and well within the uncertainty of the data, however.

SUGGESTED FUTURE RESEARCH

(1) The Little Rocky Mountains, the Little Belt Mountains, and southwestern Montana are the most likely areas to find an exposed northwestern boundary of the Wyoming Province. The Little Belt Mountains are the most promising area. Reexamination of the Little Rocky Mountains for previously unrecognized Proterozoic rocks may prove fruitful. The traditional nomenclature in southwestern Montana is largely discredited and reinterpretation of the region is needed.

(2) Much more extensive U–Pb dating of rocks around the periphery of the Wyoming Province is necessary. Much of the presently available dating is of a reconnaissance nature, often based on K–Ar or Rb–Sr methods. U–Pb dating will help establish a rigorous chronology and separate metamorphic events. Much of the U–Pb dating that has been conducted was directed toward verifying the existence of Archean rocks in specific areas. Application of U–Pb dating to the events of the Proterozoic should assume greater importance.

(3) Integration of geophysical data across the United States – Canadian border is essential. The gravity and magnetic maps of North America (Committee for the Gravity Anomaly Map of North America, 1987; Committee for the Magnetic Anomaly Map of North America, 1987) are a step in this

direction, but there are still gaps in coverage, especially in Alberta and Saskatchewan. Completion of the magnetic mapping of southern Saskatchewan and Alberta is vital to a tectonic interpretation of the northern Wyoming Province. Filtering of Canadian gravity data to eliminate the Cordilleran gravity gradient (Thomas *et al.*, 1987) revealed much previously obscured detail. The maps of Thomas *et al.* (1987) are gradient maps; filtered gravity anomaly maps, such as those that exist for the United States (Hildenbrand *et al.*, 1982; Simpson *et al.*, 1986), would also be useful. Ideally, such maps would be integrated across the border and at a generous scale (1:2,500,000 or even 1:1,000,000).

(4) Sm–Nd isotope studies are emerging as a powerful tool in mapping deeply buried basement. Application of these methods to the Wyoming Province could help outline its true extent. The study of Frost and Burwash (1986) is a major step in this direction. Nd–Sm studies in northern Idaho and northwestern Montana could outline the Wyoming Province and clarify the relation between the Wyoming, Thelon, and younger crustal blocks.

(5) Seismic-reflection profiling is another emerging tool of great power. A Consortium for Continental Reflection Profiling (COCORP) transect across the Dakota Segment is already projected, and the results of work in northern Montana have been published (Latham *et al.*, 1988). A southeast transect across the Southern Alberta Block and into central Montana should be revealing, as would transects to locate the western extent of the Wyoming Province. Unfortunately, a COCORP transect across the Great Basin along the 40th parallel is slightly too far south to intersect the Wyoming Province.

ACKNOWLEDGEMENTS

We gratefully acknowledge support from the University of Wisconsin – Green Bay to attend meetings and prepare the final figures.

REFERENCES

Alabi, A.O., Camfield, P.A. and Gough, D.I., 1975, The North American Central Plains Conductivity Anomaly: Geophysical Journal of the Royal Astronomical Society, v. 43, p. 815–833.

Allmendinger, R.W., Brewer, J.A., Brown, L.D., Kaufman, S., Oliver, J.E. and Houston, R.S., 1982, COCORP profiling across the Rocky Mountain Front in southern Wyoming. Part 2. Precambrian basement structure and its influence on Laramide deformation: Geological Society of America Bulletin, v. 99, p. 674–685.

Armstrong, R.L., 1975, Precambrian (1500 m.y. old) rocks of central Idaho – the Salmon River Arch and its role in cordilleran sedimentation and tectonics: American Journal of Science, v. 275–A, p. 437–467.

Armstrong, R.L. and Hills, F.A., 1967, Rb–Sr and K–Ar geochronologic studies of mantled gneiss domes, Albion Range, southern Idaho, USA: Earth and Planetary Science Letters, v. 3, p. 114–124.

Arth, J.G., Barker, F. and Stern, T.W., 1980, Geochronology of Archean gneisses in the Lake Helen area, southwestern Big Horn Mountains, Wyoming: Precambrian Research, v. 11, p. 11–22.

Baadsgaard, H. and Godfrey, J.D., 1972, Geochronology of the Canadian Shield in northeastern Alberta. II. Charles–Andrew–Colin lakes areas: Canadian Journal of Earth Sciences, v. 9, p. 863–881.

Bailes, A.H., 1980, Origin of early Proterozoic volcaniclastic turbidites, southern margin of the Kisseynew sedimentary gneiss belt, File Lakes, Manitoba: Precambrian Research, v. 12, p. 197–225.

Bayley, R.W., 1972, Preliminary geologic map of the Nemo District, Black Hills, South Dakota: United States Geological Survey, Miscellaneous Geological Investigations Map GP-903, scale 1:250,000.

Bennett, V.C. and DePaolo, D.J., 1987, Proterozoic crustal history of the western United States as determined by neodymium isotope mapping: Geological Society of America Bulletin, v. 99, p. 674–685.

Bickford, M.E., 1988, The accretion of Proterozoic crust in Colorado: igneous, sedimentary, deformational and metamorphic history, *in* Ernst, W.G., ed., Metamorphism and Crustal Evolution of the Western United States (Rubey Volume VII): Prentice-Hall, Englewood Cliffs, NJ, p. 411–430.

Bryant, B., 1988, Evolution and early Proterozoic history of the margin of the Archean continent in Utah, *in* Ernst, W.G., ed., Metamorphism and Crustal Evolution of the Western United States (Rubey Volume VII): Prentice-Hall, Englewood Cliffs, NJ, p. 431–445.

Burger, H.R., III, 1967, Bedrock geology of the Sheridan district, Madison County, Montana: Montana Bureau of Mines and Geology, Memoir 41, 22 p.

Butler, J.R., 1966, Geologic evolution of the Beartooth Mountains, Montana and Wyoming. Part 6. Cathedral Peak area, Montana: Geological Society of America Bulletin, v. 77, p. 45–64.

Camfield, P.A. and Gough, D.I., 1977, A possible Proterozoic plate boundary in North America: Canadian Journal of Earth Sciences, v. 14, p. 1229–1238.

Catanzaro, E.J., 1966, Correlation of some Precambrian rocks and metamorphic events in parts of Wyoming and Montana: Mountain, Geologist, v. 4(1), p. 9–21.

Catanzaro, E.J. and Kulp, J.L., 1964, Discordant zircons from the Little Belt (Montana), Beartooth (Montana) and Santa Catalina (Arizona) Mountains: Geochimica et Cosmochimica Acta, v. 28, p. 87–124.

Clark, S.H.B., 1973, Interpretation of a high-grade Precambrian terrane in northern Idaho: Geological Society of America Bulletin, v. 84, p. 1999–2004.

Committee for the Gravity Anomaly Map of North America, 1987, Gravity anomaly map of North America: Geological Society of America, Continent-scale Map-002, scale 1:5,000,000.

Committee for the Magnetic Anomaly Map of North America, 1987, Magnetic anomaly map of North America: Geological Society of America, Continent-scale Map-003, scale 1:5,000,000.

Condie, K.C., 1969, Geologic evolution of the Precambrian rocks of northern Utah and adjacent areas: Utah Geological Survey, Bulletin 82, p. 71–95.

Condie, K.C., 1976, The Wyoming Archean Province in the western United States, *in* Windley, B.F., ed., The Early History of the Earth: Wiley, New York, NY, p. 499–510.

Condie, K.C., 1982, Plate-tectonics model for Proterozoic continental accretion in the southwestern United States: Geology, v. 10, p. 37–42.

Condie, K.C. and Shadel, C.A., 1984, An early Proterozoic volcanic arc succession in southeastern Wyoming: Canadian Journal of Earth Sciences, v. 21, p. 415–427.

Denison, R.E., Lidiak, E.G., Bickford, M.E. and Kisvarsanyi, E.B., 1984, Geology and geochronology of Precambrian rocks in the central interior region of the United States: United States Geological Survey, Professional Paper 1241-C, 20 p.

Duebendorfer, E.M. and Houston, R.S., 1987, Proterozoic accretionary tectonics at the southern margin of the Archean Wyoming craton: Geological Society of America Bulletin, v. 98(5), p. 554–568.

Duke, E.F., Shearer, C.K., Papike, J.J. and Redden, J.A., 1987, Proterozoic granite–pegmatite system of the Black Hills, South Dakota: emplacement mechanisms and geochemical zonation: Geological Association of Canada – Mineralogical Association of Canada, Program with Abstracts, v. 12, p. 38.

Dutch, S.I., 1983, Proterozoic structural provinces in the north-central United States: Geology, v. 11, p. 478–481.

Eardley, A.J. and Hatch, R.A., 1940a, Pre-Cambrian crystalline rocks of north-central Utah: Journal of Geology, v. 48, p. 58–72.

Eardley, A.J. and Hatch, R.A., 1940b, Proterozoic (?) rocks in Utah: Geological Society of America Bulletin, v. 51, p. 795–844.

Engel, A.E.J., 1963, Geologic evolution of North America: Science, v. 140, p. 143–152.

Evans, K.V., 1981, Geology and geochronology of the eastern Salmon River Mountains, Idaho, and implications for regional Precambrian tectonics, unpublished Ph.D. thesis, University of Pennsylvania, 224 p.

Evans, K.V. and Fischer, L.B., 1986, U–Pb geochronology of two augen gneiss terranes, Idaho – new data and tectonic implications: Canadian Journal of Earth Sciences, v. 23, p. 1919–1927.

Frost, C.D. and Burwash, R.A., 1986, Nd evidence for extensive Archean basement in the western Churchill Province, Canada: Canadian Journal of Earth Sciences, v. 23, p. 1433–1437.

Goldich, S.S., Lidiak, E.G., Hedge, C.E. and Walthall, F.G., 1966, Geochronology of the midcontinent region, United States. Part 2. Northern area: Journal of Geophysical Research, v. 71, p. 5389–5408.

Gosselin, D.C., Papike, J.J., Zartman, R.E., Peterman, Z.E. and Laul, J.C., 1987, Archean rocks of the Black Hills, South Dakota: reworked basement from the Dakota Segment of the Trans-Hudson Orogen: Geological Association of Canada – Mineralogical Association of Canada, Program with Abstracts, v. 12, p. 48.

Green, A.G., Cumming, G.L. and Cedarwell, D., 1979, Extension of the Superior–Churchill boundary zone into southern Canada: Canadian Journal of Earth Sciences, v. 16, p. 1691–1701.

Green, A.G., Hajnal, Z. and Weber, W., 1985, An evolutionary model of the western Churchill Province and western margin of the Superior Province in Canada and the north-central United States: Tectonophysics, v. 116, p. 281–322.

Greenberg, J.K. and Brown, B.A., 1983, Lower Proterozoic volcanic rocks and their setting in the southern Lake Superior district, _in_ Medaris, L. G., ed., Early Proterozoic Geology of the Great Lakes Region: Geological Society of America, Memoir 160, p. 67–83.

Hadley, J.B., 1969a, Geologic map of the Cameron Quadrangle, Madison County, Montana: United States Geological Survey, Geologic Quadrangle Map GQ-813, scale 1:62,500.

Hadley, J.B., 1969b, Geologic map of the Varney Quadrangle, Madison County, Montana: United States Geological Survey, Geologic Quadrangle Map GQ-814, scale 1:62,500.

Handa, S. and Camfield, P.A., 1984, Crustal electrical conductivity in north-central Saskatchewan: the North American Central Plains anomaly and its relation to a Proterozoic plate margin: Canadian Journal of Earth Sciences, v. 21, p. 533–543.

Hansen, W.R., 1965, Geology of the Flaming Gorge area, Utah–Colorado–Wyoming: United States Geological Survey, Professional Paper 490, 196 p.

Harrison, J.E., 1972, Precambrian Belt Basin of northwestern United States: its geometry, sedimentation and copper occurrences: Geological Society of America Bulletin, v. 83, p. 1215–1240.

Heinrich, E.W. and Rabbitt, J.C., 1960, Pre-Beltian geology of the Cherry Creek and Ruby Mountains areas, southwestern Montana: Montana Bureau of Mines and Geology, Memoir 38, 39 p.

Hildenbrand, T.G., Simpson, R.W., Godson, R.H. and Kane, M.F., 1982, Digital colored residual and regional Bouguer gravity anomaly maps of the conterminous United States with cut-off wavelengths of 250 and 1000 km: United States Geological Survey, Geophysical Investigations Map GP-953-A, scale 1:7,500,000.

Hills, F.A. and Houston, R.S., 1979, Early Proterozoic tectonics of the central Rocky Mountains, North America: Contributions to Geology, University of Wyoming, v. 17(2), p. 89–109.

Hoffman, P.F., 1987, Continental transform tectonics: Great Slave Lake shear zone (ca. 1.9 Ga), northwest Canada: Geology, v. 15, p. 785–788.

James, D.T., Dixon, J.M., Carmichael, D.M. and Hendersen, J.B., 1987, A transect across the Thelon Front: lithologic, structural and metamorphic relationships between the Slave and Churchill structural provinces: Geological Association of Canada – Mineralogical Association of Canada, Program with Abstracts, v. 12, p. 58.

James, H.L. and Hedge, C.E., 1980, Age of the basement rocks of southwest Montana: Geological Society of America Bulletin, pt. I, v. 91, p. 11–15.

Johnson, R.A., Karlstrom, K.E., Smithson, S.B. and Houston, R.S., 1984, Gravity profiles across the Cheyenne Belt, a Precambrian crustal suture in southeastern Wyoming: Journal of Geodynamics, v. 1, p. 445–472.

Kanasewich, E.R., Clowers, R.M. and McCloughlan, C.H., 1968, A buried rift in western Canada: Tectonophysics, v. 8, p. 513–527.

Karlstrom, K.E. and Houston, R.S., 1984, The Cheyenne Belt: analysis of a Proterozoic suture in southern Wyoming: Precambrian Research, v. 25, p. 415–446.

King, P.B., 1976, Precambrian Geology of the United States: United States Geological Survey, Professional Paper 902, 85 p.

Klasner, J.S. and King, E.R., 1986, Precambrian basement geology of North and South Dakota: Canadian Journal of Earth Sciences, v. 23, p. 1083–1102.

Kleinkopf, M.D. and Redden, J.A., 1975, Bouguer gravity, aeromagnetic, and generalized geologic maps of part of the Black Hills of South Dakota and Wyoming: United States Geological Survey, Geophysical Investigations Map MF-903, scale 1:250,000.

Kleinkopf, M.D., Witkind, I.J. and Keefer, W.R., 1972, Aeromagnetic, Bouguer gravity and generalized geologic maps of the central part of the Little Belt Mountains, Montana: United States Geological Survey, Geophysical Investigations Map GP-837, scale 1:62,500.

Laberge, G.L., 1986, The Proterozoic geology of the Lake Superior region: 50th Annual Tri-State Geological Field Conference Guidebook, v. 1 (papers) p. 1–30.

Larue, D.K., 1983, Early Proterozoic tectonics of the Lake Superior region: tectonostratigraphic terranes near the purported collision zone: Geological Society of America, Memoir 160, p. 33–47.

Latham, T.S., Best, J., Chaimov, T., Oliver, J., Brown, L. and Kaufman, S., 1988, COCORP profiles from the Montana plains: the Archean cratonic crust and a lower crustal anomaly beneath the Williston Basin: Geology, v. 16, p. 1073–1076.

Lewry, J.F., 1981, Lower Proterozoic arc–microcontinent collisional tectonics in the western Churchill Province: Nature, v. 294, p. 69–72.

Lewry, J.F., Sibbald, T.I.I. and Rees, C.R., 1978, Metamorphic patterns and their relationships to tectonics and plutonism in the Churchill Province of northern Saskatchewan, in Fraser, J.A. and Heywood, W.W., eds., Metamorphism of the Canadian Shield: Geological Survey of Canada, Special Paper 78-10, p. 139–154.

Lidiak, E.G., 1971, Buried Precambrian rocks of South Dakota: Geological Society of America Bulletin, v. 82, p. 1411–1420.

Lush, A.P., McGrew, A.J., Snoke, A.W. and Wright, J.E., 1988, Allochthonous Archean basement in the northern East Humboldt Range, Nevada: Geology, v. 16, p. 349–353.

Mogk, D.W. and Henry, D.J., 1988, Metamorphic petrology of the northern Archean Wyoming Province, southwestern Montana: evidence for Archean collisional tectonics, in Ernst, W.G., ed., Metamorphism and Crustal Evolution of the Western United States (Rubey Volume VII), Prentice-Hall, Englewood Cliffs, NJ, p. 362–382.

Mogk, D.W., Mueller, P.A. and Wooden, J.L., 1988, Archean tectonics of the North Snowy Block, Beartooth Mountains, Montana: Journal of Geology, v. 96, p. 125–141.

Muehlberger, W.R., Denison, R.E. and Lidiak, L.G., 1967, Basement rocks in the continental interior of the United States: American Association of Petroleum Geologists Bulletin, v. 51, p. 2351–2380.

Nelson, B.K. and DePaolo, D.J., 1985, Rapid production of continental crust 1.7 to 1.9 b. y. ago: Nd isotopic evidence from the basement of the North American mid-continent: Geological Society of America Bulletin, v. 96, p. 746–754.

Nelson, B.K., Farmer, G.L., Bennett, V.C. and DePaolo, D.J., 1983, Sm–Nd evidence for a possible Penokean-correlative basement terrane in the eastern Great Basin: Eos, Transactions of the American Geophysical Union, v. 64(18), p. 331.

Nielsen, P.A., Langenberg, C.W., Baadsgard, H.B. and Godfrey, J.D., 1981, Precambrian metamorphic conditions and crustal evolution, northeastern Alberta, Canada: Precambrian Research, v. 16, p. 171–193.

Noble, J.A., Harder, J.O. and Slaughter, A.L., 1949, Structure of a part of the northern Black Hills and the Homestake Mine, Lead, South Dakota: Geological Society of America Bulletin, v. 60, p. 321–352.

Peterman, Z.E., 1981a, Dating of Archean basement in northeastern Wyoming and southern Montana: Geological Society of America Bulletin, v. 92, p. 139–146.

Peterman, Z.E., 1981b, Archean gneisses in the Little Rocky Mountains, Montana: United States Geological Survey, Professional Paper 1199-A, p. 1–6.

Peterman, Z.E. and Hildreth, R.A., 1978, Reconnaissance geology and geochronology of the Precambrian of the Granite Mountains, Wyoming: United States Geological Survey, Professional Paper 1055, 22 p.

Ratte, J.C. and Wayland, R.G., 1969, Geology of the Hill City Quadrangle, Pennington County, South Dakota – a preliminary report: United States Geological Survey, Bulletin 1271-B, 14 p.

Redden, J.A., Peterman, Z.E., Zartman, R.E. and DeWitt, E., 1987, U–Th–Pb zircon and monazite ages and preliminary interpretation of a Proterozoic tectonism in the Black Hills, South Dakota: Geological Association of Canada – Mineralogical Association of Canada, Program with Abstracts, v. 12, p. 83.

Reid, R.R., 1955, Pony – Cherry Creek age relations in southwestern Montana: Geological Society of America Bulletin, v. 66, p. 1315–1316.

Reid, R.R., 1957, Bedrock geology of the north end of the Tobacco Root Mountains, Madison County, Montana: Montana Bureau of Mines and Geology, Memoir 36, 25 p.

Reid, R.R., 1963, Metamorphic rocks of the northern Tobacco Root Mountains, Madison County, Montana: Geological Society of America Bulletin, v. 74, p. 293–306.

Ross, G.M., Parrish, R.R., Bowring, S.A. and Tankard, A.J., 1988, Tectonics of the Canadian Shield in the Alberta subsurface: Geological Association of Canada – Mineralogical Association of Canada, Program with Abstracts, v. 13, p. A106.

Roy, J.L., 1983, Paleomagnetism of the North American Precambrian: a look at the data base: Precambrian Research, v. 19, p. 319–348.

Runner, J.J., 1943, Structure and origin of Black Hills Pre-Cambrian granite domes: Journal of Geology, v. 51, p. 431–457.

Scholten, R., 1955, Pony – Cherry Creek age relations in southwestern Montana – a reply: Geological Society of America Bulletin, v. 66, p. 1317–1318.

Scholten, R., Keenmon, K.A. and Kupsch, W.O., 1955, Geology of the Lima region, southwestern Montana and adjacent Idaho: Geological Society of America Bulletin, v. 66, p. 345–404.

Sears, J.W., Graff, P.J. and Holden, G.S., 1982, Tectonic evolution of lower Proterozoic rocks, Uinta Mountains, Utah and Colorado: Geological Society of America Bulletin, v. 93, p. 990–997.

Simpson, R.W., Jachens, R.C., Saltus, R.W. and Blakely, R.J., 1986, Isostatic residual gravity, topographic and first vertical derivative gravity maps of the conterminous United States: United States Geological Survey, Geophysical Investigations Map GP-975, scale 1:7,500,000.

Sims, P.K. and Peterman, Z.E., 1986, Early Proterozoic Central Plains Orogen, a major buried structure in north-central United States: Geology, v. 14, p. 488–491.

Stauffer, M.R., 1984, Manikewan: an early Proterozoic ocean in central Canada, its igneous history and orogenic closure: Precambrian Research, v. 25, p. 257–281.

Thomas, M.D., Sharpton, V.L. and Grieve, R.A.F., 1987, Gravity patterns and Precambrian structure in the North American central plains: Geology, v. 15(6), p. 489–492.

Thompson, D.L., 1987, Structural and metamorphic profile across the Thelon Front in the Ragged Lake – Beaulac area: evidence for major transcurrent motion: Geological Association of Canada - Mineralogical Association of Canada, Program with Abstracts, v. 12, p. 58.

Thompson, P.H., Henderson, J.B. and Frith, R.A., 1987, The Thelon Front – the transition between the Slave Structural Province and the Thelon Tectonic Zone: Geological Association of Canada – Mineralogical Association of Canada, Program with Abstracts, v. 12, p. 58.

Tysdal, R.G., 1976, Geologic map of the northern part of Ruby Range, Madison County, Montana: United States Geological Survey, Miscellaneous Geological Investigations Map MF-951, scale 1:250,000.

Vitaliano, C.J. and Cordua, W.S., 1979, Geologic map of the southern Tobacco Root Mountains, Madison County, Montana: Geological Society of America, Map and Chart Series MC-31, scale 1:62,500.

Wooden, J.L., Mueller, P.A. and Mogk, D.W., 1988, A review of the geochemistry and geochronology of the Archean rocks of the northern part of the Wyoming Province, in Ernst, W. G., ed., Metamorphism and Crustal Evolution of the Western United States (Rubey Volume VII), Prentice-Hall, Englewood Cliffs, NJ, p. 383–410.

Zartman, R.E., 1974, Lead isotopic provinces in the Cordillera of the western United States and their geologic significance: Economic Geology, v. 69, p. 792–805.

Zartman, R.E., Norton, J.J. and Stern, T.W., 1964, Ancient granite gneiss in the Black Hills, South Dakota: Science, v. 145, p. 479–481.

Zeitz, I., *compiler*, 1982, Composite magnetic anomaly map of the United States. Part A. Conterminous United States: United States Geological Survey, Geophysical Investigations Map GP-954-A, scale 1:2,500,000.

Trout Mountain Anomaly, Northern Alberta: its Role in the Northwest Foreland of the Trans-Hudson Orogen

R.A. Burwash and M.A. Power
Department of Geology, University of Alberta,
Edmonton, Alberta T6G 2E3, Canada

Burwash, R.A. and Power, M.A., Trout Mountain anomaly, northern Alberta: its role in the northwest foreland of the Trans-Hudson Orogen, in Lewry, J.F. and Stauffer, M.R., eds., The Early Proterozoic Trans-Hudson Orogen of North America: Geological Association of Canada, Special Paper 37, p. 301–311.

Abstract

Sampling 182 cores from the subsurface Precambrian of northern Alberta resulted in recognition of a widespread occurrence of retrograded granulite and local areas of unretrograded granulite. By analogy with the Mountain Rapids area on the edge of the exposed Shield in northeastern Alberta, the granulite facies metamorphism is inferred to be Archean and the retrogression Hudsonian.

The northwestern foreland of the Trans-Hudson Orogen in northern Alberta was the site of regional K-metasomatism, locally accompanied by the generation of microcline-rich anatectic magma. The Trout Mountain gravity low in the Wabasca River area has a -25 mGal (1 Gal = 1 cm/s^2) residual anomaly after regional gradients are removed. It is best modelled as a 30 km by 150 km microcline granite pluton. Relict blocks of granulite occur on both flanks of the pluton.

Palinspastic reconstruction of the positions of the Bear, Slave, and western Churchill provinces prior to transcurrent movement along the Macdonald fault zone puts a welt of deep crustal material between the Wopmay and Trans-Hudson orogens. The Trout Mountain "pluton" is one of the bodies of anatectic granite in this welt.

Résumé

Des échantillons de forages provenant du Précambrien de l'Alberta du nord ont mis en évidence une rétromorphose très répandue des granulites et localement des roches non rétromorphosées. Le faciès granulite est considéré comme Archéen et la rétromorphose Protérozoïque par analogie avec la région de Mountain Rapids sur la bordure du bouclier au NE de l'Alberta.

Au nord de l'Alberta, la zone externe nord-ouest de l'orogène Trans-Hudson a subi un métasomatisme potassique régional accompagné localement par des magmas anatectiques à microcline. La dépression du champ de gravité de Trout Mountain dans la région de la rivière Wabasca montre une anomalie résiduelle de -25 mGal (1 Gal = 1 cm/s^2) après soustraction du gradient régional. Le modèle correspondant est un pluton de granite à microcline de 30 km × 150 km. Des blocs de granulites se retrouvent sur les deux flancs du pluton.

Une reconstruction palinspastique des positions des provinces Bear, Slave et ouest Churchill avant les décrochements le long de la zone de faille Macdonald implique une bande de matériel crustal profond entre les orogènes Trans-Hudson et Wopmay; le pluton de granite anatectique de Trout Mountain en fait partie.

INTRODUCTION

The northwestern foreland of the Trans-Hudson Orogen, as mapped along the western margin of the exposed Shield, consists of two major tectonic divisions, the Cree Lake zone and the northwest Churchill polymetamorphic terrane (Fig. 1). The lateral extensions of these divisions in the buried Shield are the Cree Lake – Calgary zone and the Athabasca polymetamorphic terrane (Burwash *et al., in press*). The northern and southern boundaries of the Athabasca polymetamorphic terrane were defined by Burwash and Culbert (1976) on the basis of a wide array of geophysical, petrological, and geochemical variables. A pronounced negative Bouguer anomaly on the *Gravity Map of Canada* (Dominion Observatory, 1967), the Fond du Lac Low, was an important factor in locating the southern limit.

In the Wabasca River map area (Fig. 1), an isolated negative Bouguer anomaly was defined and designated the Trout Mountain Low by Walcott and Boyd (1971, Fig. 2). At that time, the principal author of the present paper had 27 basement core samples from this area. Density determinations were undertaken, but there were too few data points to draw any firm conclusions. In 1986, an additional 142 wells from the Wabasca River map area were sampled, providing a suitable suite for a combined geophysical and petrographic study of the Trout Mountain Low.

The Athabasca polymetamorphic terrane is characterized by strong cataclasis and recrystallization, which in many cases obscure the mineralogy and texture of the original igneous or metamorphic rock (Burwash and Krupicka, 1969, 1970). The core samples collected during the 1950s and 1960s contain minor recognizable hypersthene in a few thin sections. Most of these are basic rocks, and an igneous rather than metamorphic overprint was inferred.

The 1986 core collection was examined petrographically in conjunction with studies of Mountain Rapids granulites from exposures along the Slave River in northeastern Alberta (Burwash *et al.*, 1985). At Mountain Rapids, both mafic and felsic granulites occur, with a wide range of second-cycle metamorphic overprint, from unretrograded to fully

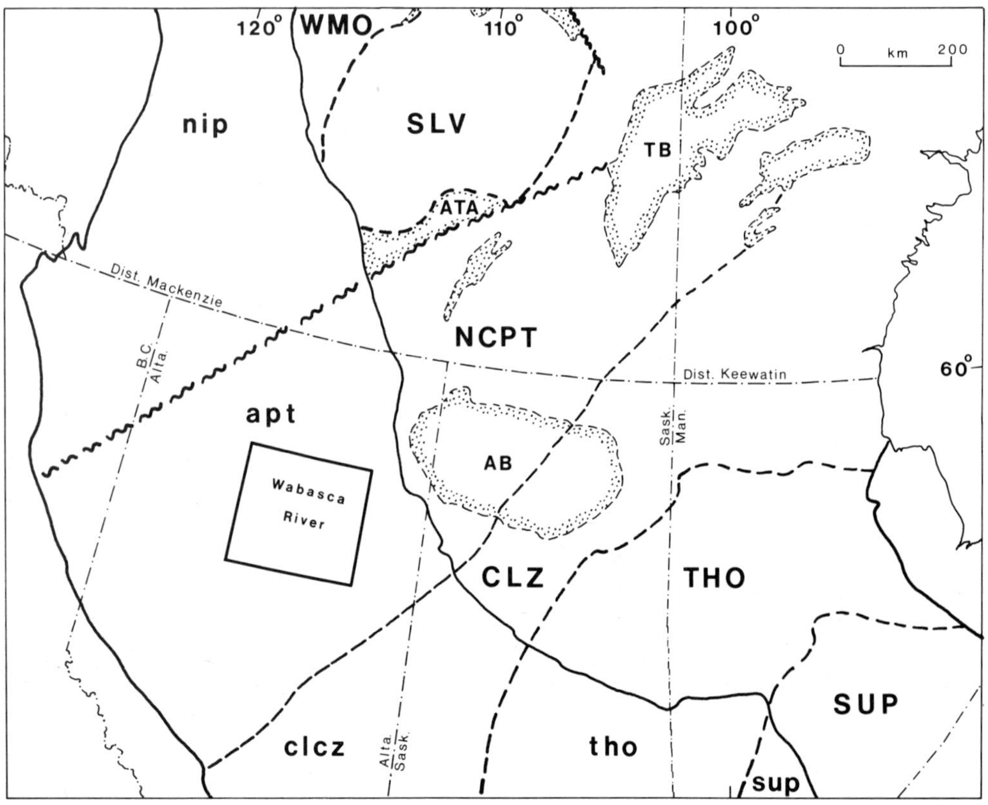

Figure 1. *Lithostructural domains of the western Canadian Shield. Exposed: WMO, Wopmay orogen; SLV, Slave Province; ATA, Athapuskow aulacogen; TB, Thelon basin; NCPT, northwest Churchill polymetamorphic terrane; AB, Athabasca basin; CLZ, Cree Lake zone; THO, Trans-Hudson Orogen; and SUP, Superior Province. Buried: nip, northern interior plains; apt, Athabasca polymetamorphic terrane; clcz, Cree Lake – Calgary zone; tho, Trans-Hudson Orogen; and sup, Superior Province (after Burwash* et al., in press*).*

retrograded. The petrology and chemistry of the new collection of 142 cores are the subject of a paper in preparation by Burwash and Krupicka. Using the texture and mineralogy of the Mountain Rapids rocks as a guide, they recognize nine granulites in these new cores. Pseudomorphs after hypersthene indicate an additional 32 retrogranulites. The abundance of these rock types in the Wabasca River map area gives added tectonic significance to the Trout Mountain gravity anomaly. The present study attempts to derive geophysical/petrological models for the anomaly and then to place it in the tectonic framework of the northwest foreland of the Trans-Hudson Orogen.

TROUT MOUNTAIN GRAVITY ANOMALY

Regional gravity surveys conducted by the Earth Physics Branch (Walcott, 1968; Walcott and Boyd, 1971) covered northern Alberta with a grid of stations with an average spacing of 12 km. Figure 2 is taken from the *Gravity Map of Canada* (Earth Physics Branch, 1980), which incorporated these data. Walcott and Boyd (1971) identified three significant anomalies in this area, the Rocky Mountain, Fond du Lac, and Trout Mountain lows. The Rocky Mountain Low reflects the downward flexure of the crystalline basement toward the Cordillera with the associated stacking of imbricate thrust slices of Late Precambrian and Phanerozoic sedimentary rocks. The Fond du Lac Low or the Edmonton – Kasba Lake lineament (Burwash and Culbert, 1976) apparently reflects a significant domain boundary in the western Churchill Province. The Trout Mountain Low is an isolated feature that straddles the Peace River Arch.

Using a map of gravity stations supplied by the Earth Physics Branch, a grid was plotted for the area 56°00′N to 57°25′N and 114°30′W to 115°45′W, covering the Trout Mountain Low. A Bouguer anomaly map of this study area was prepared with a contour interval of 2 mGal (1 Gal = 1 cm/s^2) (Fig. 3). Regional gradients associated with the Rocky Mountain Low and Peace River Arch were interpolated from the maps of Walcott and Boyd (1971). Removal of the regional effects produced the residual anomaly map shown in Figure 4a.

The size and polarity of the residual anomaly dictate that it is produced by some deficit mass in the basement (Walcott and Boyd, 1971). Densities of 177 cores of Precambrian crystalline rocks from northern Alberta (townships 80 to 126), sampled during the 1986 program, are shown in Figure 5. Core samples from immediately below the Phanerozoic cover, which gave density determinations of less than 2.55, were excluded as being weathered. The total population was divided lithologically into rocks of granulite metamorphic facies, retrograded granulites, and nongranulite gneisses and granites. The granulites and retrogranulites are assigned to the Archean by analogy with the Mountain Rapids granulites of northeastern Alberta (Burwash *et al.*, 1985). This older crust would form the country rock into which the Hudsonian pluton (nongranulite) would be emplaced. Inspection of Figure 5 supports a density contrast of 0.05 g/cm^3 between the granulite/retrogranulite and nongranulite rocks.

Burwash and Krupicka (1970) concluded that a density contrast on the order of 0.05 g/cm^3 exists on a regional basis between the most highly deformed and recrystallized (low density) and the least deformed parts of the western Canadian Shield. The decreased density was matched by increased K_2O and microcline contents. If the K-metasomatism is related to lower crust or upper mantle processes, the vertical persistence of the difference may be considerable. Preservation of recognizable granulite facies mineral assemblages is

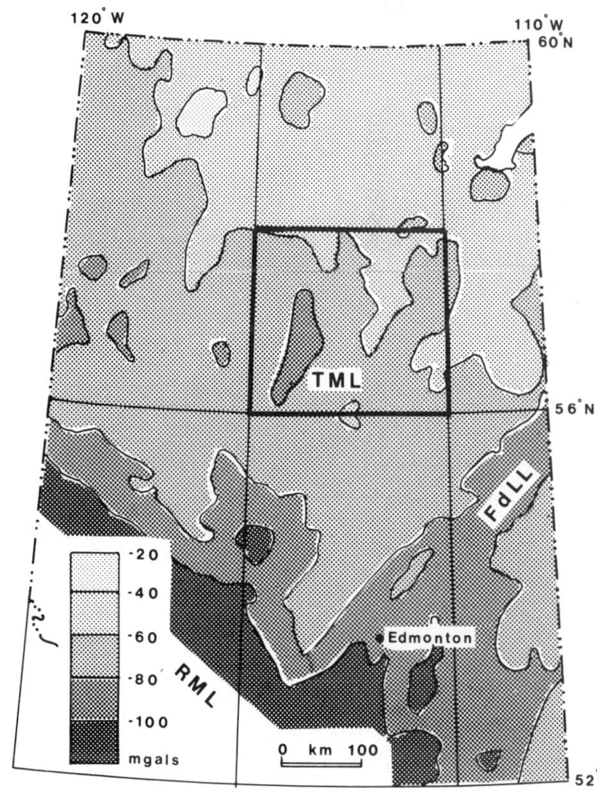

Figure 2. *Bouguer gravity anomaly map of northern Alberta. Simplified from the* Gravity Map of Canada *(Earth Physics Branch, 1980). The rectangle indicates the Wabasca River map area. Abbreviations: TML, Trout Mountain Low; FdLL, Fond du Lac Low; and RML, Rocky Mountain Low. (Note: 1 Gal = 1 cm/s^2.)*

dependent on a low level of postcrystalline deformation.

Three constraints were imposed on the model generated for the Trout Mountain anomaly. A uniform density contrast of 0.05 g/cm^3 between the deficit mass and the surrounding basement is assumed. The top of the model is constrained to the level of the unconformity between the Precambrian and the Phanerozoic cover. The gravity effects of the dipping unconformity are presumed to have been removed in the regional–residual anomaly

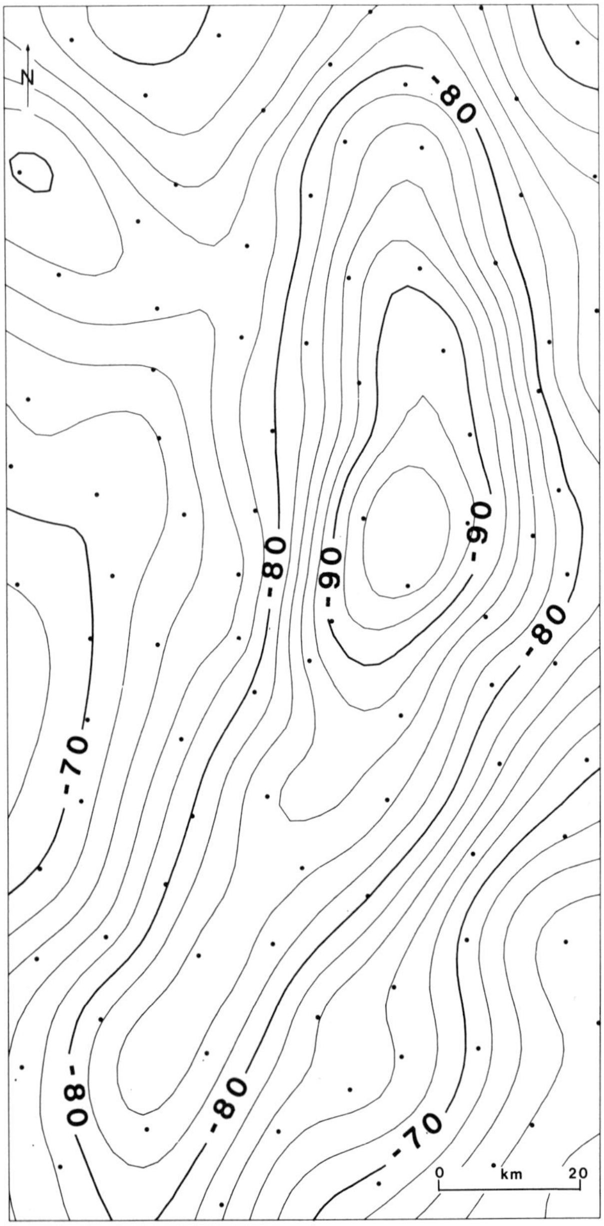

Figure 3. *Bouguer anomaly map of Trout Mountain anomaly. The contour interval is 2 mGal (1 Gal = 1 cm/s^2). The dots indicate gravity stations. Data provided by Earth Physics Branch, Ottawa.*

separation. In this model, this is a horizontal plane set 400 m below sea level. Finally, a flat base is assumed.

There are obvious limitations to these assumptions, but they do not prevent the delineation of the broad outlines of the deficit mass. Figure 5 makes it apparent that the contrast in rock densities may well vary from the stated value. Because information on the subsurface geology is not sufficient to establish the precise location of variations, a density contrast based on an average is all that can be assumed. Given that the assumed average density contrast is small, deviations from this figure will produce models with substantially different deficit masses. Local derivations from the assumed density contrast, particularly in the upper regions of the crust, will also change model geometry. The assumption of a flat base for the body of lower density rock implies that an essentially planar lower crustal substrate exists. If igneous activity is inferred, the upper crust would be the site of a discordant mesozonal pluton with sharply defined, near-vertical contacts; the lower crust would be the catazone, with a more gradational, concordant, sheetlike structure.

The model was derived using the method of Cordell and Henderson (1968). The residual anomaly grid matrix, consisting of 80 × 41 nodes with a nearest neighbour spacing of 2 km, was used as input. The model consists of bundles of equidimensional vertical prisms whose heights are varied until a best-fit model is achieved. Individual prisms are centred beneath the nodes in the gravity matrix and have end-face surface areas equal to the square of the distance between the nodes (i.e., 4 km^2). The centres of prisms are spaced 2 km apart. The height of a prism is initialized using the Bouguer slab thickness for the residual gravity value. The model gravity at that point is then the sum of the contributions from all prisms in the model. The heights are scaled automatically using the ratio of actual residual gravity to that calculated in the model. The root-mean-square (RMS) error and single greatest error were used as measures of the goodness of fit. To circumvent the difficulty of dealing with both positive and negative residual gravity, a 2 km thick Bouguer slab of deficit mass and its gravity effect (-4.188 mGal) were added to the model and residual gravity matrices and subtracted from each when the inversion was complete.

The modelling was performed using an IBM PC and an Amdahl 5870. The University of Alberta mainframe was used in the generation of the residual anomaly grid matrix and in the production of graphical output via the SURFACE II graphics system. The gravity data were interpolated at points spaced 2 km apart within the 160 km × 82 km area shown in Figure 3. Individual

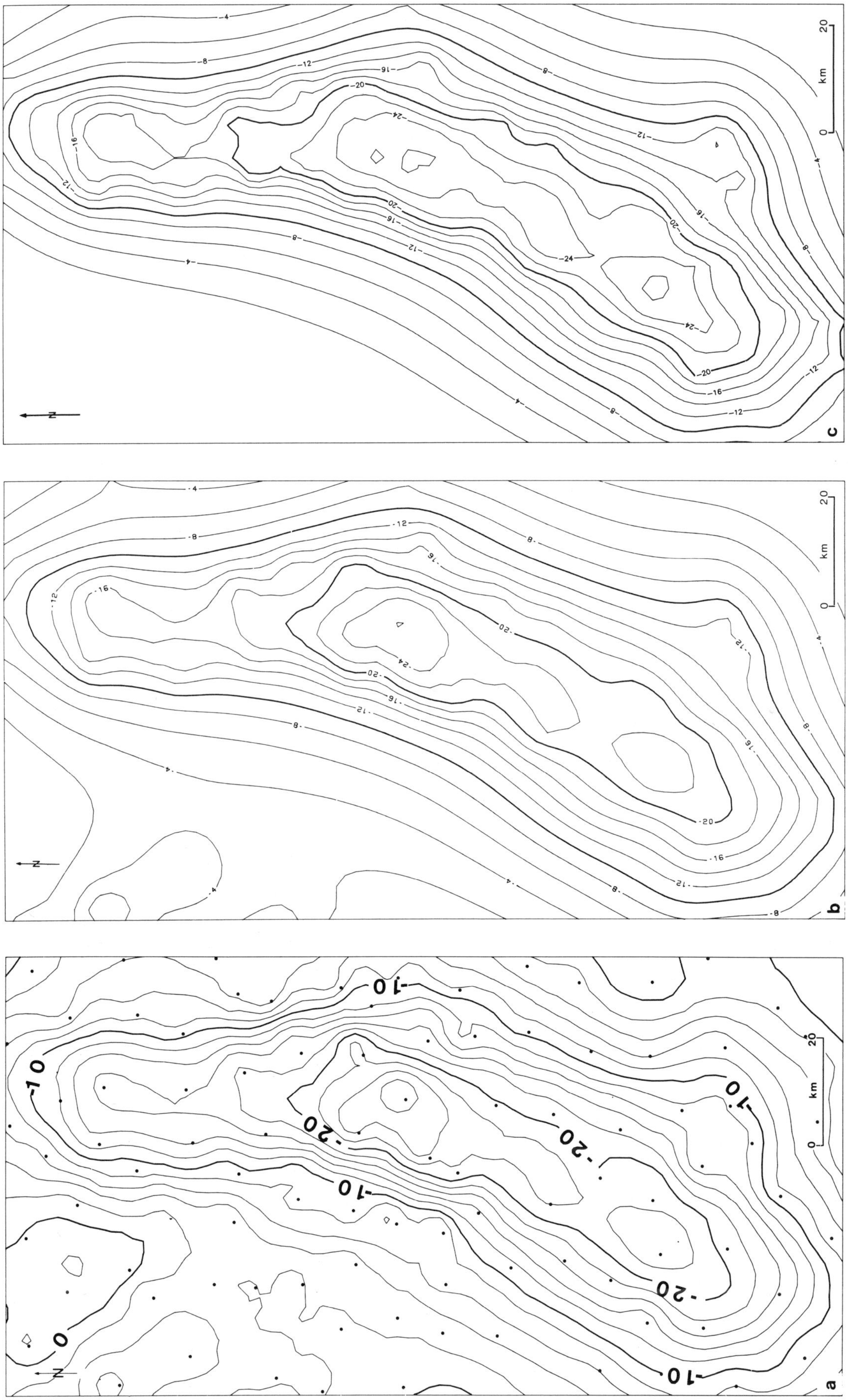

Figure 4. *Residual anomaly maps of the Trout Mountain Low and the model discussed in the text. The contour interval is 2 mGal (1 Gal = 1 cm/s^2). (a) Residual gravity anomaly map of the Trout Mountain Low. (b) Calculated gravity anomaly associated with the model shown in (a). (c) Calculated gravity anomaly associated with the smoothed model shown in Figure 6b.*

runs of 10–14 iterations each were performed on the smaller computer with the base level being adjusted from 18 to 26 km until a best-fit model was achieved at the 25 km level. The calculated model gravity for this solution is shown in Figure 4b and the model itself is shown in Figure 6a. Deficit mass generated to account for residual gravity in the northwest corner of the map has been removed. The associated RMS and single greatest errors were 1.1 and 4.2 mGal, respectively, with the latter occurring off the main anomaly on the east-central edge of the residual map.

A feature of this inversion method is that it may produce more detail than is justified by the quality of the input data or the station spacing. The high-frequency components of the model predictions may alias into the low frequencies that are used to fit the data. Nevertheless, it is one possible solution fitting the interpolated gravity field to the constraints imposed. The development of spires and troughs in the model on the southern and east-central flanks may have resulted from an inability to match the finer variations in the gravity field with the relatively coarse prism dimensions and/or the inability of the model to fit the observed gradient because of a local deviation from the assumed density contrast. This phenomenon does not detract from the usefulness of the model in delineating the overall configuration of the deficit mass. If the actual density contrast shows vertical variation, the model geometry may vary considerably. This effect will be most pronounced at depth.

To attenuate those details in the model that are not justified by the input data, a spatial filter was applied to the model matrix. We first assume that the error in residual gravity at any node is ±-0.02 mGal (i.e., ±-0.01 mGal at any control station) and look at any two average prisms 12.5 km high with a base at 25 km. The minimum spacing necessary to resolve two separate anomalies with a peak to trough amplitude of 0.02 mGal is 21 km. This suggests that any averaging filter should have a search

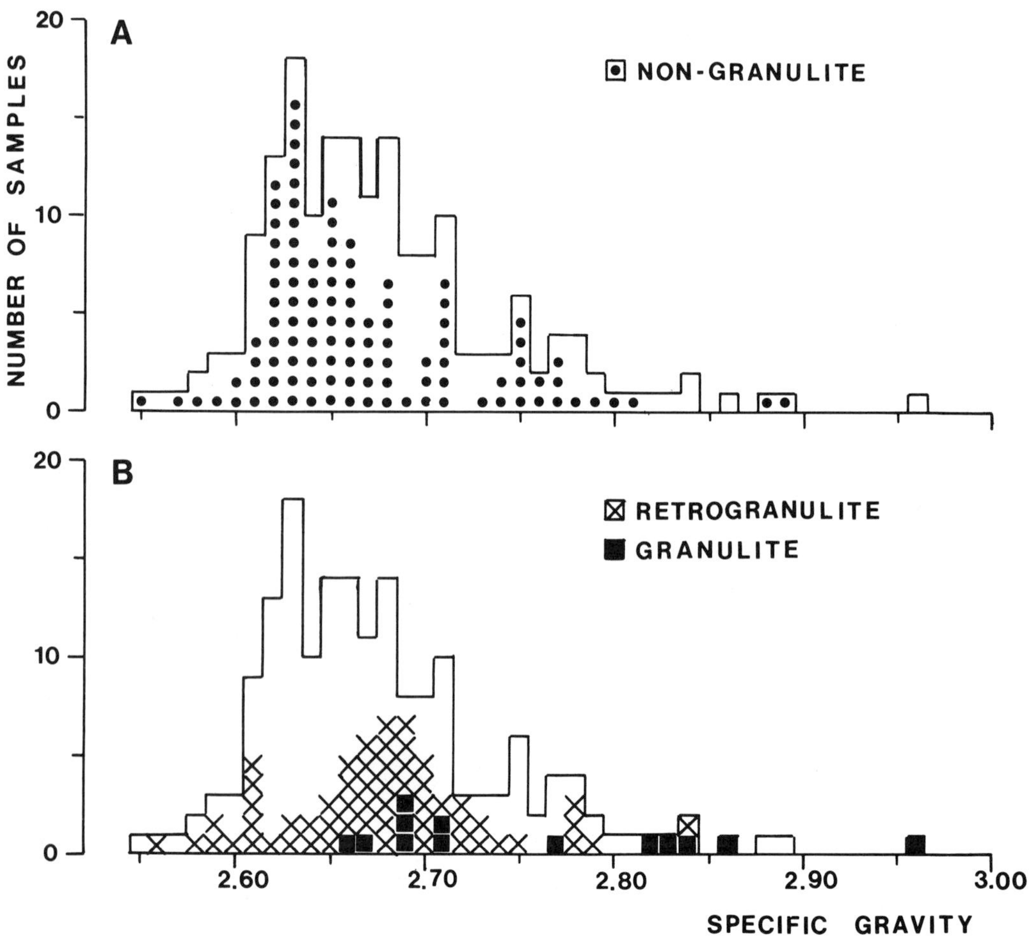

Figure 5. *Histograms of densities of Precambrian basement core samples from northern Alberta (townships 80 to 126; 1986 collection). (A) Density distribution of nongranulite rocks with respect to the total population (n = 177). (B) Densities of granulite facies and retrogranulite samples with respect to the total population.*

radius of 11 km. A weighted moving average filter with a radius of 11 km was applied to the model matrix to produce Figure 6b. A perspective view of the smoothed model is shown in Figure 7 and the residual Bouguer anomaly produced by it is shown in Figure 4c. The smoothed model anomaly matches all of the major details in the Trout Mountain Low, but, as expected, cannot match the finer variations in the latter anomaly. Deficit mass generated to account for subsidiary anomalies in the northwestern corner of the map was removed prior to filtering and accounts for the absence of any residual anomaly in this quadrant.

Only the broad details of the model topography should be discussed in view of the present limitations of the geophysical data. The northern half of

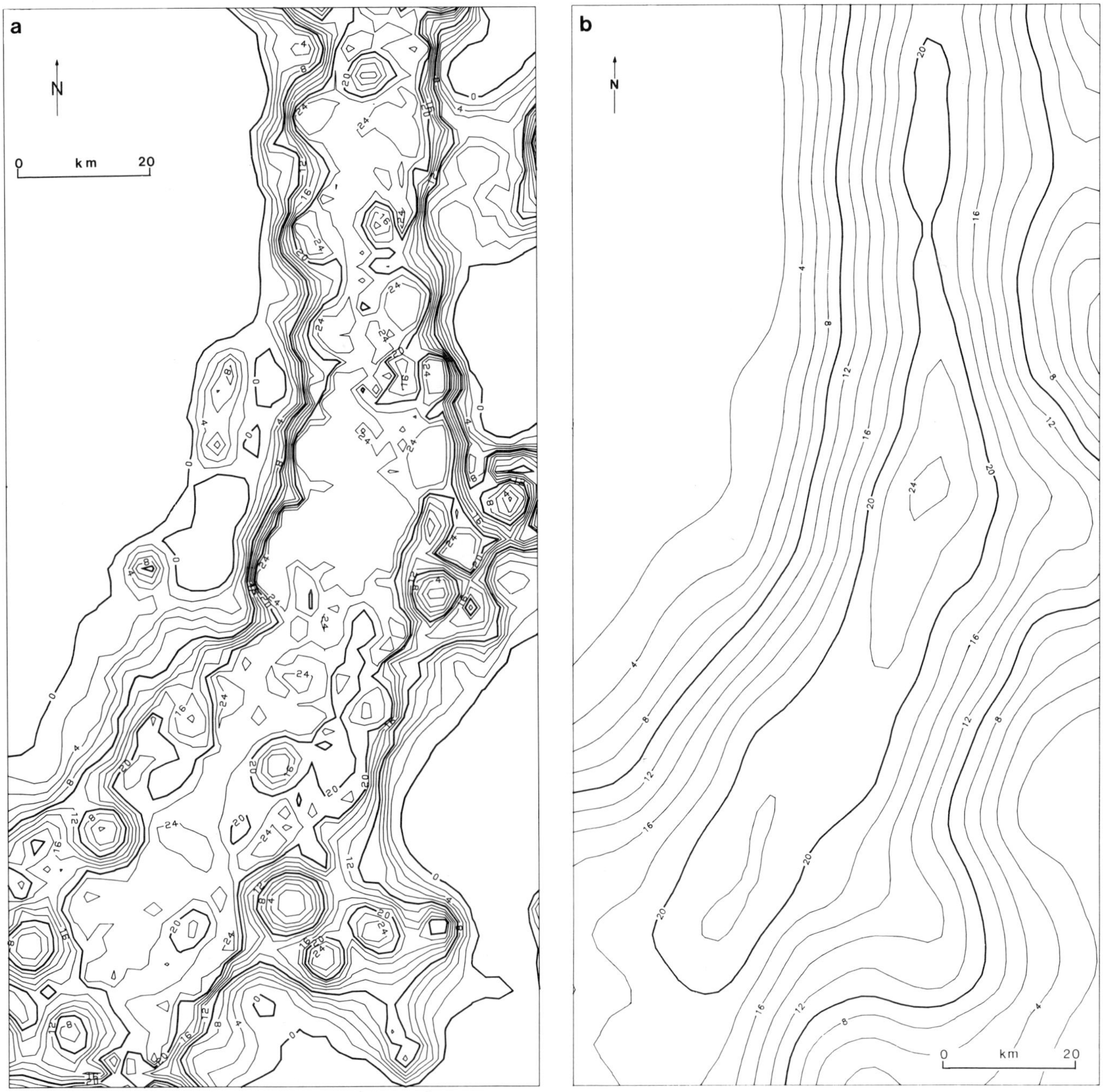

Figure 6. *(a) Model of the low-density rock body associated with the Trout Mountain Low, using vertical prisms 2 km square. The contour interval is 2 km depth. The datum level is at a depth of 25 km below the Paleozoic/Precambrian unconformity. (b) Smoothed contour model equivalent to (a). The configuration of this model closely matches that of Figure 4a,b. Low-amplitude anomalies from the northwestern quadrant of Figure 4 have been removed.*

the model, bounded by steep walls, is approximately 20 km wide and trends due north. The southern half is less sharply defined, more than 30 km wide at the base, and trends N25°E. The orientation and change of trend of the axis of the model closely parallel a number of tectonic lineaments observed in the exposed Shield in northeastern Alberta and the adjacent District of Mackenzie (Bostock, 1981; Langenberg, 1983).

PETROLOGIC MODEL FOR TROUT MOUNTAIN ANOMALY

The distribution of basement rocks with a density of ≤2.65 g/cm³ in the area adjacent to the Trout Mountain Low is shown in Figure 8 and is comparable to that drawn by Sprenke *et al.* (1986) for the exposed Shield of northeastern Alberta; their trend surface map of rock sample densities has contours ranging from 2.62 to 2.71 g/cm³. In Figure 8, the two clusters of points represent the more closely spaced drilling in the Red Earth (56°35′N) and Utikuma (56°05′N) oil fields. The reservoir rock of the Red Earth field is the arkosic basal Paleozoic sand ("granite wash") lying directly on basement. The cluster of cores from this field is close to the geographic centre of the Trout Mountain anomaly and probably provides a representative sample of its lithology.

The characteristic rock type of the Red Earth basement cores is a leucocratic granite gneiss in which microcline is more abundant than plagioclase (Burwash and Krupicka, 1970, Fig. 1). Within the host are bands or schlieren with more abundant (10%) biotite and/or chlorite. Similar rocks are widespread in cores from the Athabasca polymetamorphic terrane. Textural criteria (Burwash and Krupicka, 1969; Burwash, 1978) suggest strong cataclasis accompanied and/or followed by metasomatic recrystallization.

The rock body with the deficit mass can be explained in several ways: metasomatism with limited melting, anatexis with the separation of the leucocratic low-melting point fraction to form a pluton, or a structurally controlled combination of metasomatism and anatexis. The volume of rock represented in the models in Figure 6 seems to argue against a purely metasomatic origin for the low-density rock body. A complete melting would result in rocks with a more normal "igneous aspect" than those found in the cores. With a moderate concentration of volatiles, pegmatites rather than gneisses should have been widespread.

A combination of metasomatism and anatexis could be achieved in a depth-zone melting petrologic model with the base of the low-density rock body being almost entirely magmatic and the upper levels increasingly metasomatic. The upper surface of the gravity models in Figures 6 and 7 would be largely in the depth zone of metasomatism. Basement cores with coarser grained, nonfoliated, microcline-rich rocks would be the upper portion of, or offshoots of, the pluton.

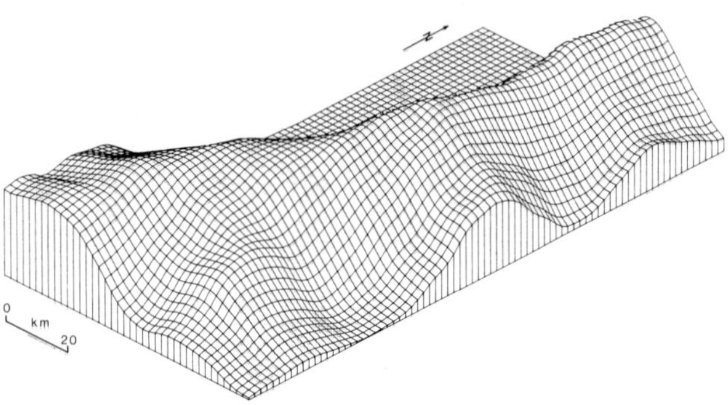

Figure 7. *Three-dimensional presentation of Figure 6b. The base of the block is 79 km × 157 km. Viewed from the southeast.*

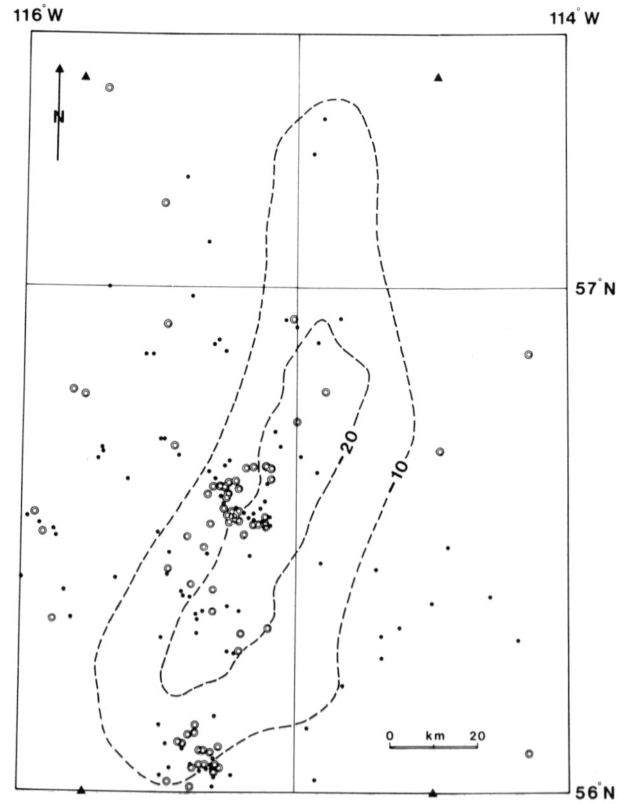

Figure 8. *Location of basement core samples with low densities (≤2.65 g/cm³) in the area adjacent to the Trout Mountain Low. Low density is shown as double rings; other samples are shown as solid dots. Residual gravity anomalies are from Figure 4a. Triangles mark the locations of the corners of Figures 4 and 6.*

A petrologic model in which the degree of partial melting increased with depth would conform to the distribution of granulite facies rocks adjacent to the Trout Mountain gravity low. Figure 9 shows that two areas of unretrograded granulites occur in the Wabasca River map area. One is at the southern end of the anomaly, in the Utikuma oil field. The second is on the western flank of the anomaly, south of the Red Earth field. The geographic position of the latter cluster of granulite core samples is a close analogy to the position of the Mountain Rapids granulites on the western flank of the Slave granitoid body in northeastern Alberta (Burwash *et al.*, 1985). The distribution of retrograded granulites on both sides of the Trout Mountain anomaly is consistent with an Archean crust reactivated and penetrated by a Proterozoic pluton.

RELATIONSHIP OF TROUT MOUNTAIN TO THE TRANS-HUDSON OROGEN

The northwest Churchill and Athabasca polymetamorphic terranes (Fig. 1) are the distal portion

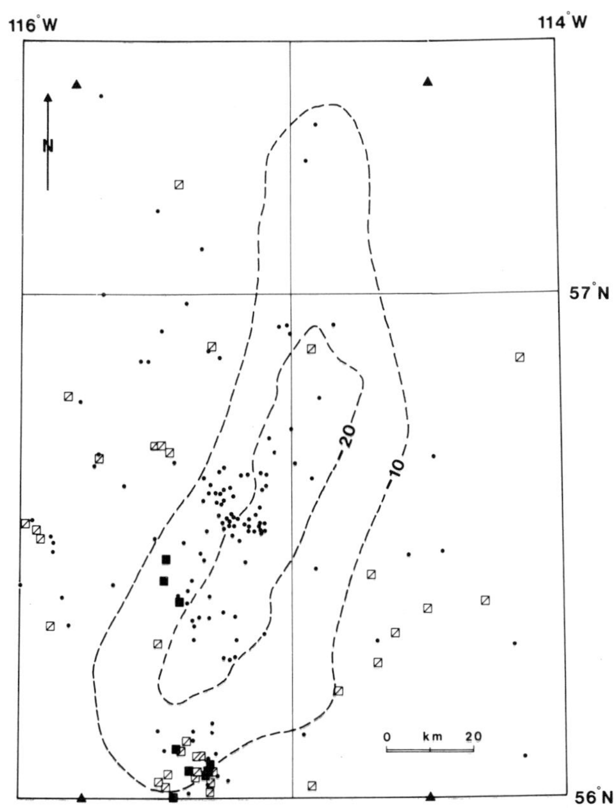

Figure 9. *Location of granulite core sample (solid squares), retrograded granulite samples (open squares with diagonal stroke) adjacent to the Trout Mountain Low, and nongranulite/retrogranulite samples (solid dots).*

of the northwestern foreland of the Trans-Hudson Orogen. The Trout Mountain "pluton" is approximately 500 km west of the Wathaman batholith. From K–Ar dating, it is apparent that the thermal effects of the Hudsonian orogeny extend across the width of Alberta and into northeastern British Columbia (Burwash *et al.*, 1963). Along the western margin of the exposed Shield, magmatism in the Wopmay Orogen (Hoffman and Bowring, 1984) was essentially contemporaneous with the major phase of the Hudsonian (1800–1900 Ma). To test the possibility that the Trout Mountain pluton was related to the Wopmay Orogen, it is necessary to carry out a palinspastic reconstruction of the northwestern Canadian Shield as it may have been prior to transcurrent faulting along the Macdonald fault zone.

Using the large-scale magnetic anomalies over western Canada, Coles *et al.* (1976) suggested that the right-lateral offset along the Macdonald fault zone was approximately 300 km. Hoffman (1987), using more recently published aeromagnetic compilation maps at a scale of 1:1,000,000, places the offset at between 300 and 700 km. A displacement of 500 km is shown in Figure 10. This reconstruction to early Hudsonian time (ca. 1.9 Ga) places the southern extension of the Wopmay Orogen approximately 400 km west of Trout Mountain. The temporal relationship between the Great Bear, Trout Mountain, and Wathaman batholiths awaits geochronologic data from Trout Mountain. This study is under way.

Petrographic study of basement cores from northern Alberta (Fig. 10) indicates that a widespread belt of granulite, now largely retrograded, existed between the Wopmay and Trans-Hudson orogens as magmatic activity was taking place. North-trending shear zones in northeastern Alberta and the adjacent District of Mackenzie may have acted as conduits for metamorphic fluids, resulting in the generation of magmatic/migmatitic bodies, such as the Slave granitoid rocks of the Mountain Rapids area (Bostock, 1981; Goff *et al.*, 1986). The Trout Mountain pluton is inferred to have an analogous origin on the basis of similar lithologic assemblages.

Uplift and erosion of the northwestern foreland of the Trans-Hudson Orogen was rapid prior to the formation of the Helikian basins of the western Shield. Fraser *et al.* (1970) suggested that an isostatically positive area, the "Slave Arch," continued to exist in northern and western Alberta until Cambrian time. The forerunner of the Slave Arch seems to be the welt of granulite facies rocks existing during the Hudsonian, of which the Mountain Rapids and Wabasca terranes are relicts.

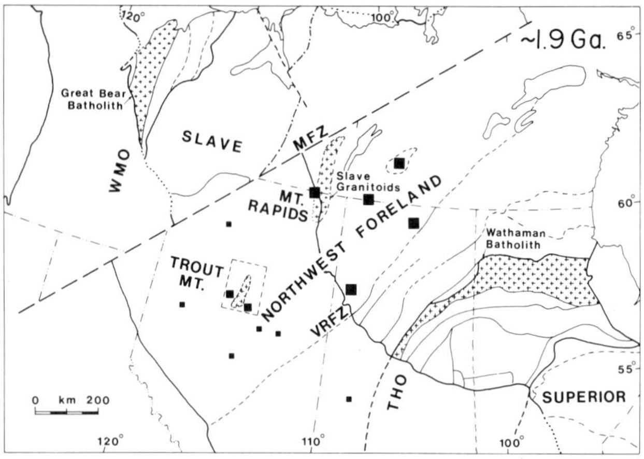

Figure 10. *Palinspastic model of the western Canadian Shield, ca. 1.9 Ga, prior to inferred 500 km right-lateral movement along the Macdonald fault zone. Large squares are recognized granulite / retrograded granulite terranes on exposed Shield; medium squares are clusters of well cores containing granulite; and small squares are individual well samples of granulite. Abbreviations: WMO, Wopmay orogen; THO, Trans-Hudson Orogen; MFZ, Macdonald fault zone; and VRFZ, Virgin River fault zone.*

CONCLUSION

The Trout Mountain gravity anomaly can be closely modelled as a body of low-density rock in the upper 25 km of the crystalline basement. If the density contrast with respect to the country rock is 0.05 g/cm^3, the dimensions of the body are approximately 25 km × 25 km × 160 km. A granitic pluton of these dimensions is inferred. Deep drillholes provide numerous core samples of microcline-rich granitic rocks with densities of ≤2.65 g/cm^3. Other drillholes, on the flank of the anomaly, penetrate granulite facies rocks that are believed to be part of the country rock.

By combining the Trout Mountain geological model with observed field relationships in northeastern Alberta, northwestern Saskatchewan, and the adjacent District of Mackenzie, an extensive relict Archean granulite facies terrane is indicated in the western Churchill Province. This Archean terrane formed a median block between the Superior and Slave provinces. The shape of the Trans-Hudson Orogen in Saskatchewan and Manitoba was controlled in part by the rigid crust in the northwestern foreland. Along the Macdonald fault zone, marginal to the granulite block, the Slave and Bear provinces were translated to the northeast (Hoffman, 1987). Within the block, other early Hudsonian shear zones allowed for the circulation of fluids. These fluids were instrumental in generating anatectic magmas, which were emplaced in bodies such as the Trout Mountain pluton and Slave granitoids.

ACKNOWLEDGEMENTS

The 1986 program of sampling basement cores from Alberta was supported by a grant to R.A. Burwash from the Department of Energy, Mines and Resources. A.H. Shepard and other staff members of the Core Research Centre, Energy Resources Conservation Board, Calgary, were very helpful. Our thanks to H.A.K. Charlesworth for allowing us unlimited use of his personal computer for this project.

Much of the petrological data on the granulites in the 1986 core collection comes from the work of J. Krupicka, who also reviewed an earlier draft of this manuscript. Reviews by E.J. Burwash, F. Langford, D. Gendzwill, and E. Nyland were useful. R.W. Burwash provided technical and editorial assistance.

REFERENCES

Bostock, H.H., 1981, A granitic diapir of batholithic dimensions at the west margin of the Churchill Province: Geological Survey of Canada, Paper 81-1B, p. 73–82.

Burwash, R.A., 1978, Metamorphism of the Athabasca Mobile Belt, a subsurface extension of the Churchill Province, *in* Fraser, J.A. and Heywood, W.W., eds., Metamorphism in the Canadian Shield: Geological Survey of Canada, Paper 78-10, p. 123–127.

Burwash, R.A. and Culbert, R.R., 1976, Multivariate geochemical and mineral patterns in the Precambrian basement of western Canada: Canadian Journal of Earth Sciences, v. 13, p. 1–18.

Burwash, R.A. and Krupicka, J., 1969, Cratonic reactivation in the Precambrian basement of western Canada. 1. Deformation and chemistry: Canadian Journal of Earth Sciences, v. 6, p. 1381–1396.

Burwash, R.A. and Krupicka, J., 1970, Cratonic reactivation in the Precambrian basement of western Canada. 2. Metasomatism and isostasy: Canadian Journal of Earth Sciences, v. 7, p. 1275–1295.

Burwash, R.A., Baadsgaard, H., Campbell, F.A., Cumming, G.L. and Folinsbee, R.E., 1963, Potassium–argon dates of diabase dyke system, District of Mackenzie, N.W.T.: Transactions of the Canadian Institute of Mining and Metallurgy, v. 66, p. 303–307.

Burwash, R.A., Krupicka, J., Basu, A.R. and Wagner, P.A., 1985, Resetting of Nd whole-rock isochrons from polymetamorphic granulites, northeastern Alberta: Canadian Journal of Earth Sciences, v. 22, p. 992–1000.

Burwash, R.A., Green, A.G., Jessop, A.M. and Kanasewich, E.R., *in press*, Geophysical and petrologic characteristics of the basement rocks of the Western Canada Basin, *in* Aitken, J.D. and Scott, D.F., eds., Continental Platforms and Basins of Canada: Decade of North American Geology: Geological Society of America.

Coles, R.L., Haines, G.V. and Hannaford, W., 1976, Large-scale magnetic anomalies over western Canada and the Arctic: a discussion: Canadian Journal of Earth Sciences, v. 13, p. 790–802.

Cordell, L. and Henderson, R.G., 1968, Iterative three-dimensional solution of gravity anomaly data using a digital computer: Geophysics, v. 33, p. 596–601.

Dominion Observatory, 1967, Bouguer gravity anomaly map of Canada: Department of Energy, Mines and Resources, Ottawa, Canada, scale 1:5,000,000.

Earth Physics Branch, 1980, Gravity map of Canada: Gravity Map Series 80-1, Ottawa, Canada.

Fraser, J.A., Donaldson, J.A., Fahrig, W.F. and Tremblay, L.P., 1970, Helikian basins and geosynclines of the northwestern Canadian Shield, in Baer, A.J., ed., Symposium on Basins and Geosynclines of the Canadian Shield: Geological Survey of Canada, Paper 70-40, p. 213–238.

Goff, S.P., Godfrey, J.D. and Holland, J.D., 1986, Petrology and geochemistry of the Canadian Shield of northeastern Alberta: Alberta Research Council, Bulletin 51, 60 p.

Hoffman, P.F., 1987, Continental transform tectonics: Great Slave Lake shear zone (ca. 1.9 Ga), northwest Canada: Geology, v. 15, p. 785–788.

Hoffman, P.F. and Bowring, S.A., 1984, Short-lived 1.9 Ga continental margin and its destruction, Wopmay orogen, northwest Canada: Geology, v. 12, p. 68–72.

Langenberg, C.W., 1983, Polyphase deformation in the Canadian Shield of northeastern Alberta: Alberta Research Council, Bulletin 45, 33 p.

Sprenke, K.F., Wavra, C.S. and Godfrey, J.D., 1986, Geophysical expression of the Canadian Shield of northeastern Alberta: Alberta Research Council, Bulletin 52, 54 p.

Walcott, R.I., 1968, The gravity field of northern Saskatchewan and northeastern Alberta: Dominion Observatory, Ottawa, Canada, Gravity Map Series, Nos. 16–20.

Walcott, R.I. and Boyd, J.B., 1971, The gravity field of northern Alberta, and part of Northwest Territories and Saskatchewan: Department of Energy, Mines and Resources (Canada), Earth Physics Branch, Gravity Map Series, Nos. 103–111.

Evolution of the Cape Smith Belt: Early Proterozoic Continental Underthrusting, Ophiolite Obduction, and Thick-skinned Folding[1]

Marc R. St-Onge and Stephen B. Lucas
Geological Survey of Canada, Continental Geoscience Division, 588 Booth Street, Ottawa, Ontario K1A 0E4, Canada

St-Onge, M.R. and Lucas, S.B., Evolution of the Cape Smith Belt: Early Proterozoic continental underthrusting, ophiolite obduction, and thick-skinned folding, in Lewry, J.F. and Stauffer, M.R., eds., The Early Proterozoic Trans-Hudson Orogen of North America: Geological Association of Canada, Special Paper 37, p. 313–351.

Abstract

The Cape Smith Belt contains evidence for Early Proterozoic plate tectonics in the form of a 2.00 Ga ophiolite and a ca. 1.92–1.84 Ga thrust belt that developed during A-type (continental) subduction. Five distinct tectono-stratigraphic suites have been recognized in the belt: (1) fluviodeltaic sedimentary rocks; (2) proximal to distal fan clastics passing upward and outboard into ca. 1.96 Ga continental rift basalts; (3) transitional-crust basalts intruded by 1.92 Ga layered mafic–ultramafic sills; (4) postrift, deep-water sedimentary rocks; and (5) sheeted dykes, basalts, and mafic–ultramafic cumulates of a 2.00 Ga ophiolite. Suites 1–3 reflect the development of a north-facing rift margin at the edge of the Superior Province. The older age of the ophiolite suggests that it developed independent of the rift margin in a separate basin. The five tectono-stratigraphic suites and the Archean Superior Province basement were deformed during three geometrically distinct deformation episodes (D_1, D_2, and D_3). The D_1 episode involved southward thrusting of suites 2–5 above autochthonous basement gneisses and suite 1 sedimentary rocks. Four stages characterize its tectono-thermal development: (1) piggyback (regular) sequence accretion of Early Proterozoic rocks above a basal décollement; (2) thermal relaxation and prograde metamorphism of the thrust belt, leading to the development of a ductile basal shear zone; (3) out-of-sequence thrusting at thermal peak conditions involving the incorporation of basement slices into the thrust belt; and (4) postthermal peak out-of-sequence thrusting. Thermochemical modelling of garnet zoning profiles in metapelites reveals peak pressures of at least 10 kbar (1 GPa) at the deepest structural levels. Comparison of P–T paths between the out-of-sequence hanging wall and footwall indicates that decompression P–T paths for the hanging wall have steeper dP/dT segments. Syn-D_1 *emplacement of tonalitic to granodioritic plutons in the internal part of the thrust belt constrains the initiation of stage 3 to be younger than 1.84 Ga. Post-D_1 emplacement of syenogranite dykes provides a minimum age of 1.76*

[1]Geological Survey of Canada Contribution No. 42288.

Ga for stage 4. The D_2 and D_3 episodes folded the thrust belt and its footwall basement in a thick-skinned fashion about east–west (D_2) and north–south (D_3) axes. In total, the two folding episodes are responsible for the isolation of the thrust belt in a klippe, and for the regional east–west plunge that yields over 18 km of structural relief at the present erosional surface.

Résumé

La bande du Cap Smith montre les indices d'une tectonique de plaque ayant généré des ophiolites vers 2.00 Ga et une chaîne plissée entre 1.92–1.84 Ga lors d'une subduction continentale de type A. Cinq entités tectono-stratigraphiques ont été reconnues dans la chaîne : (1) des sédiments fluvio-deltaïques; (2) des sédiments deltaïques proximaux et distaux passant vers le haut et l'extérieur à des basaltes de rifts continentaux âgés de 1.96 Ga; (3) des basaltes de croûte transitionnelle recoupés par des sills basiques et ultrabasiques lités de 1.92 Ga; (4) des sédiments profonds post-rift; et (5) des dykes en feuillet et des cumulats basiques et ultrabasiques provenant d'une ophiolite de 2.00 Ga. Les suites 1 et 3 reflètent le développement de la marge d'un rift à vergence nord à la bordure de la province Supérieur. L'âge ancien de l'ophiolite suggère un développement dans un bassin séparé indépendant de la marge du rift. Les cinqs entités tectono-stratigraphiques et le socle archéen du Supérieur ont été déformées durant trois épisodes géométriquement distincts (D_1, D_2, D_3). L'épisode D_1 implique le charriage vers le sud des suites 2 à 5 sur les gneiss de socle autochtones et les sédiments de la suite 1. Le développement tectono-thermal se fait en quatre stades: (1) l'accrétion en séquence des roches Protérozoïques inférieur sur un décollement basale; (2) relaxation thermal et métamorphisme prograde de la chaîne menant à la formation d'un cisaillement ductile basale; (3) charriage hors-séquence durant le paroxisme thermique provoquant un écaillage du socle dans la chaîne plissée; et (4) un charriage hors-séquence synchrone du pic post-thermique. Une modélisation thermo-chimique sur des profiles de grenats zonés provenant de méta-pélites indique des pressions maximum de 10 kbar (1 GPa) aux niveaux structuraux profonds. Une comparaison des trajectoires P–T entre le toit et le mur de l'ensemble des écailles hors-séquence indique que les trajectoires P–T de décompression pour le mur ont des segments dP/dT plus à pic. La mise en place de plutons tonalitiques et granodioritiques syn-D_1 *dans les zones internes de la chaîne marque le début du stade 3 suite à 1.84 Ga. La mise en place de filons de syenogranite post-D_1 place un âge minimum de 1.76 Ga pour le stade 4. Les phases D_2 et D_3 ont plissé la chaîne et son socle selon des axes est–ouest et nord–sud. Les deux phases de plis ont isolé la chaîne plissée sous forme d'une klippe et ont provoqué le plongement régional est–ouest de la chaîne donnant un relief structural d'au moins 18 km au niveau actuel d'érosion.*

INTRODUCTION

Two of the principal arguments against the operation of plate tectonics in the Precambrian have been the lack of evidence for midocean-ridge spreading (e.g., ophiolites) and for subduction (e.g., island arcs, blueschists). However, recent work in the Early Proterozoic Trans-Hudson Orogen has documented both accreted island arc terranes (Stauffer *et al.*, 1975; Chauvel *et al.*, 1987; Thom *et al.*, 1987; Watters *et al.*, 1987) and an obducted ophiolite (St-Onge *et al.*, 1988). These findings strongly argue for the existence of Early Proterozoic lithospheric plates whose evolution was governed by the fundamental processes of seafloor spreading and the subduction of oceanic crust (see Hoffman, 1988). The recognition of A-type subduction (Bally, 1981) and fold-and-thrust belts of Precambrian age (e.g., St-Onge and King, 1987; Hoffman *et al.*, 1988), although not providing direct evidence for plate tectonics, is consistent with the assertion that ancient and modern mountain-building processes are basically the same. One Early Proterozoic belt providing both direct evidence for plate tectonics (an ophiolite) and indirect evidence (A-type subduction) is the Cape Smith Belt (St-Onge *et al.*, 1988). The belt constitutes the northern Quebec segment of the Trans-Hudson Orogen (Fig. 1).

The Cape Smith Belt is a ca. 1.9 Ga thrust belt preserved in a doubly plunging synclinorium, and is

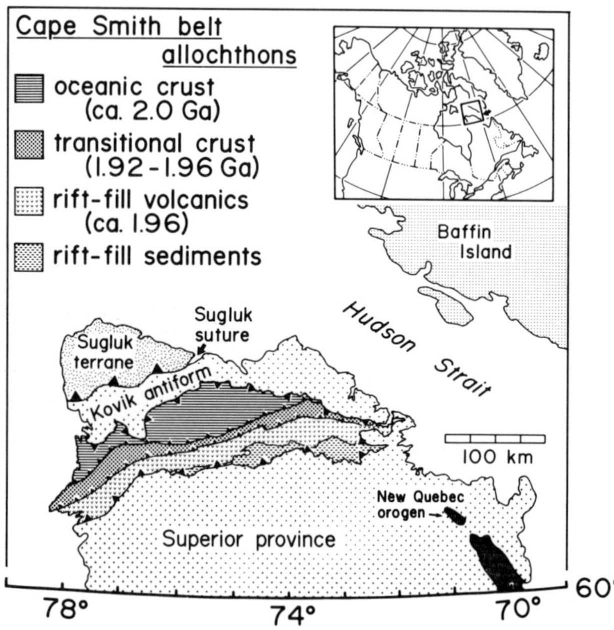

Figure 1. *Location map for the Cape Smith Belt, after Hoffman (1985). The proposed suture between the underthrust Superior Province and the overriding Sugluk continental block is indicated as the "Sugluk suture."*

separated from a proposed root zone to the north by an autochthonous basement antiform (Fig. 1; Hoffman, 1985; St-Onge *et al.*, 1986, 1987, 1988). The southward-verging thrust belt developed during a crustal thickening episode (D_1) above a regional basal décollement separating the underthrust Superior Province basement from the Early Proterozoic allochthons (Fig. 2). Thermal relaxation of the thickened crust led to greenschist–amphibolite facies metamorphism of the thrust belt (Fig. 3) and the development of a ductile shear zone at its base (basal shear zone in Fig. 4) (Lucas and St-Onge, 1988; St-Onge and Lucas, 1988; Lucas, 1989). Late D_1 out-of-sequence thrusts reimbricated the thrust stack and were responsible for further southward translation of the ophiolite thrust sheets relative to the lower plate. D_2 and D_3 represent two episodes of "thick-skinned" folding involving the thrust belt and its footwall basement (D_2: east–west trending axes; D_3: north–south trending axes). The two folding episodes, coupled with uplift and erosion, have resulted in the thrust belt being exposed in an oblique cross-section with over 18 km of cumulative structural relief (Lucas, 1989). Clearly, structural relief of this magnitude provides an opportunity, unparalleled by topographic relief, to study the deep structure of mountain belts (see also King (1986)).

In this paper, the evolution of the Cape Smith Thrust–Fold Belt (Fig. 1) is examined to (1) describe the structure and stratigraphy of a 2.00 Ga ophiolite and a 1.96–1.92 Ga continental rift margin; (2) document the tectono-thermal consequences of northward subduction of the Superior Province margin; and (3) discuss the geometry and mechanisms of "thick-skinned" folding of basement and cover. (Note: All dates presented in this paper are U–Pb zircon on baddeleyite ages derived by Parrish (1989).) In addition, the study constrains part of the history of Early Proterozoic accretion to Laurentia (Hoffman, 1988), and provides an important link between the Trans-Hudson orogenic zone of northern Manitoba and Saskatchewan (Lewry and Sibbald, 1980) and the New Quebec Orogen of Quebec and Labrador (Goulet *et al.*, 1987; van der Leeden *et al.*, this volume). The paper will address the evolution of the Cape Smith Belt in the following format: tectono-stratigraphy, structural history, and tectono-thermal evolution.

PREVIOUS WORK

Geological investigations began in the Cape Smith Belt near the turn of the century (Low, 1902). Early mapping by Bergeron (1957, 1959), Beall (1959, 1960), DeMontigny (1959), Gélinas (1962), Gold (1962), and Stam (1961) established a twofold stratigraphic subdivision of the Early Proterozoic rocks and documented major south-verging thrust faults (Bergeron, 1957; Stam, 1961). Taylor (1982) produced an accurate regional reconnaissance map of the entire northern Ungava Peninsula. Recent work by Lamothe (1986) and co-workers in the west-central portion of the Cape Smith Belt has resulted in the publication of 1:50,000 scale maps covering the belt from 74°15'W to Hudson Bay. The northward and eastward increase in metamorphic grade from subgreenschist on the belt's southern margin to amphibolite grade on the northern margin and at Wakeham Bay (Fig. 3) was first described by Schimann (1978) and Westra (1978).

The stratigraphy and geochemistry of the volumetrically abundant basalts has been studied by numerous workers, including Baragar (1974, 1984), Schwarz and Fujiwara (1977), and recently Hervet (1986) and Picard (1986, 1989a,b). Francis and Hynes (1979) and Hynes and Francis (1982) provide a detailed account of the tectono-stratigraphic relationships and the geochemical characteristics of the mafic to ultramafic flows and sills along a transect in the centre of the belt.

The geotectonic significance of the Cape Smith Belt has long been the subject of controversy (Dimroth *et al.*, 1970; Gibb and Walcott, 1971; Dewey and Burke, 1973; Baer, 1977; Thomas and Gibb, 1977; Baragar and Scoates, 1981; Hoffman, 1985). The principal dispute has centred on whether the belt marks the site of (1) a major collisional geosuture (e.g., Gibb and Walcott, 1971); (2) a geosuture marking the closure of a small, short-lived oceanic basin (e.g., Hynes and Francis, 1982); or (3) an ensialic fold belt (e.g., Baer, 1977). Hoffman (1985) proposed a new tectonic model for the belt based on the following two observations concerning the relationship between the Cape Smith Belt and the underlying basement: (1) basement gneisses along the northern margin record similar Rb–Sr whole-rock ages (2.6–2.9 Ga, Doig (1983)) to the basement along the southern margin (Taylor and Loveridge, 1981); and (2) the basement at the east end of the belt is continuous around the belt from the southern to northern margins. To account for these observations, Hoffman (1985) suggested that the belt is essentially a klippe, isolated from its root zone (geosuture) to the north by a postthrusting basement antiform. In Hoffman's model, the geosuture would be located 50–100 km north of the belt at Sugluk Inlet (Fig. 1), where a major break occurs in the gravity and magnetic fields as well as in metamorphic grade (Westra, 1978). In addition, the granulite grade gneisses north of the proposed geosuture have been dated (Rb–Sr whole rock) at 3.0–3.2 Ga (Doig, 1987), consistent with the hypothesis that the overriding plate is exotic with respect to the underthrust margin of the Superior Province.

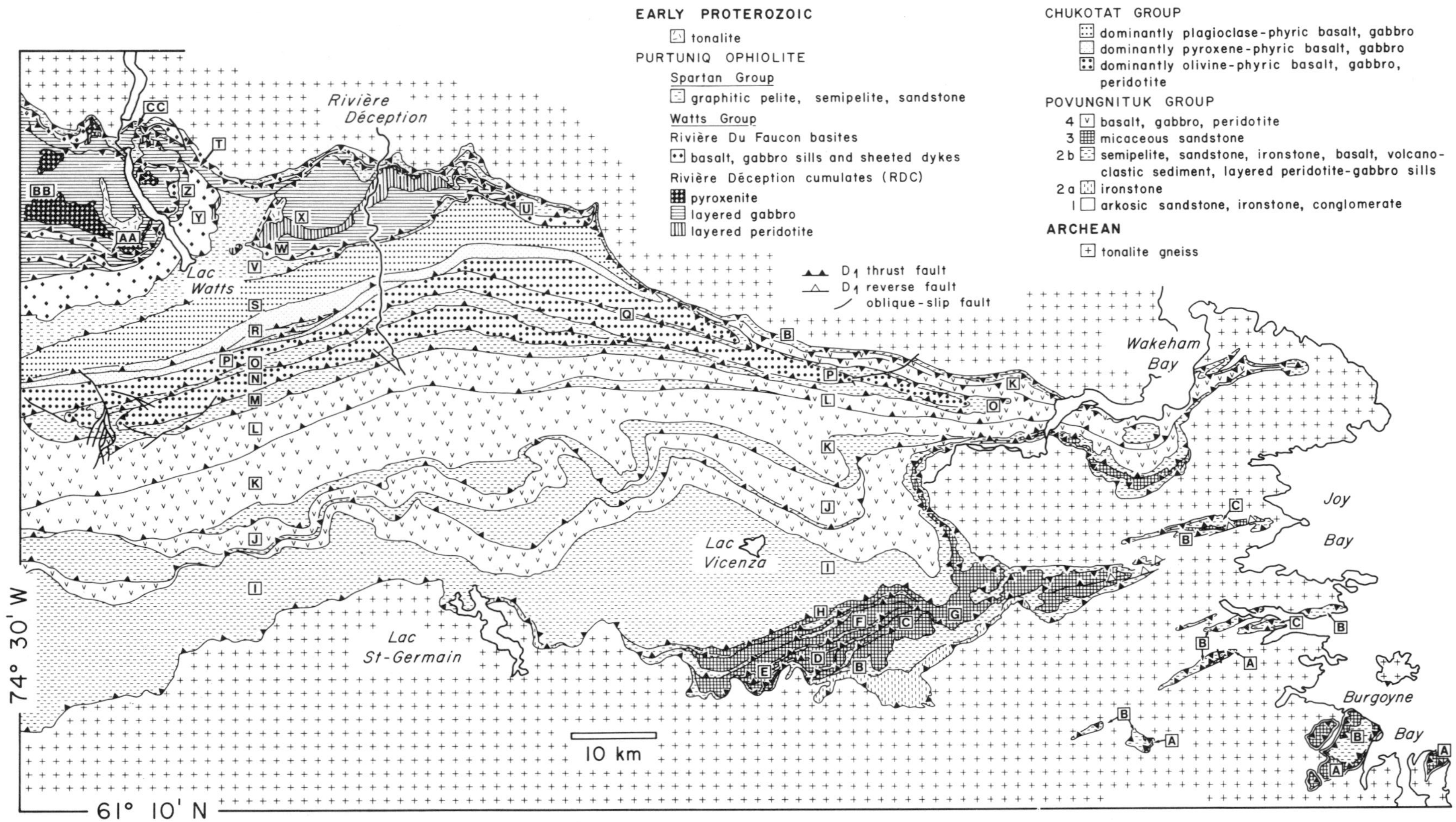

Figure 2. *Geological map of the eastern portion of the Cape Smith Belt. The outliers occur southeast of the conterminous belt in the Burgoyne Bay area. Boxed letters identify thrust sheets referred to in the text. Note that faults are named after the label of the overlying thrust sheet. Numbers (1–4) in the legend for the Povungnituk Group refer to informal formations discussed in the text.*

The regional distribution and timing of Hudsonian deformation recorded at the margins of the Archean Superior craton have been the subject of recent discussion. Following the proposal of Gibb and Walcott (1971) that a Hudsonian-aged (1.8 Ga) suture marked the boundary between the Churchill and Superior provinces, Baragar and Scoates (1981) provided abundant stratigraphic and geochemical evidence of geosutures ringing the margins of the Superior craton. Hoffman (1988) has subsequently clarified the order of suturing of Archean terranes (e.g., Churchill Province) to the Superior craton in the Early Proterozoic using a compilation of precise U–Pb zircon dates. The specific role of the Cape Smith Belt in this protracted period of crustal amalgamation is emerging as a result of U–Pb zircon geochronology by Parrish (1989) and will be discussed below.

TECTONO-STRATIGRAPHIC SUITES

Stratigraphic nomenclature

The sedimentary and volcanic rocks of the Cape Smith Belt have been divided by Bergeron (1959) into a lower Povungnituk Group and an upper Chukotat Group. The contact between the two groups was interpreted as an unconformity by Bergeron (1959), a view supported by Beall (1959, 1960) and Beall *et al.* (1963). However, observations by Taylor (1974), Baragar (1974), and Hynes and Francis (1982) refuted the existence of an unconformity within the Early Proterozoic cover succession. Recent work by St-Onge *et al.* (1987) and St-Onge and Lucas (1989) has demonstrated that the contact between the two groups is clearly tectonic.

Lamothe *et al.* (1984) proposed a revised stratigraphic nomenclature for the Early Proterozoic thrust–fold belt and its underlying Archean basement following geological fieldwork west of lac Watts (Fig. 2). The Povungnituk Group was subdivided into a lower Nituk Subgroup consisting predominantly of clastic metasedimentary rocks (later renamed Lamarche Subgroup (Lamothe, 1986)) and an upper Beauparlant Subgroup comprising mostly continental tholeiitic basalts. Lamothe *et al.* (1984) retained Bergeron's (1959) definition of the Chukotat Group (see section on midbelt thrust sheets). North of thrust V (Fig. 2), two new groups were defined: (1) the Spartan Group, consisting of sedimentary, pyroclastic, volcanic rocks, and (2) the

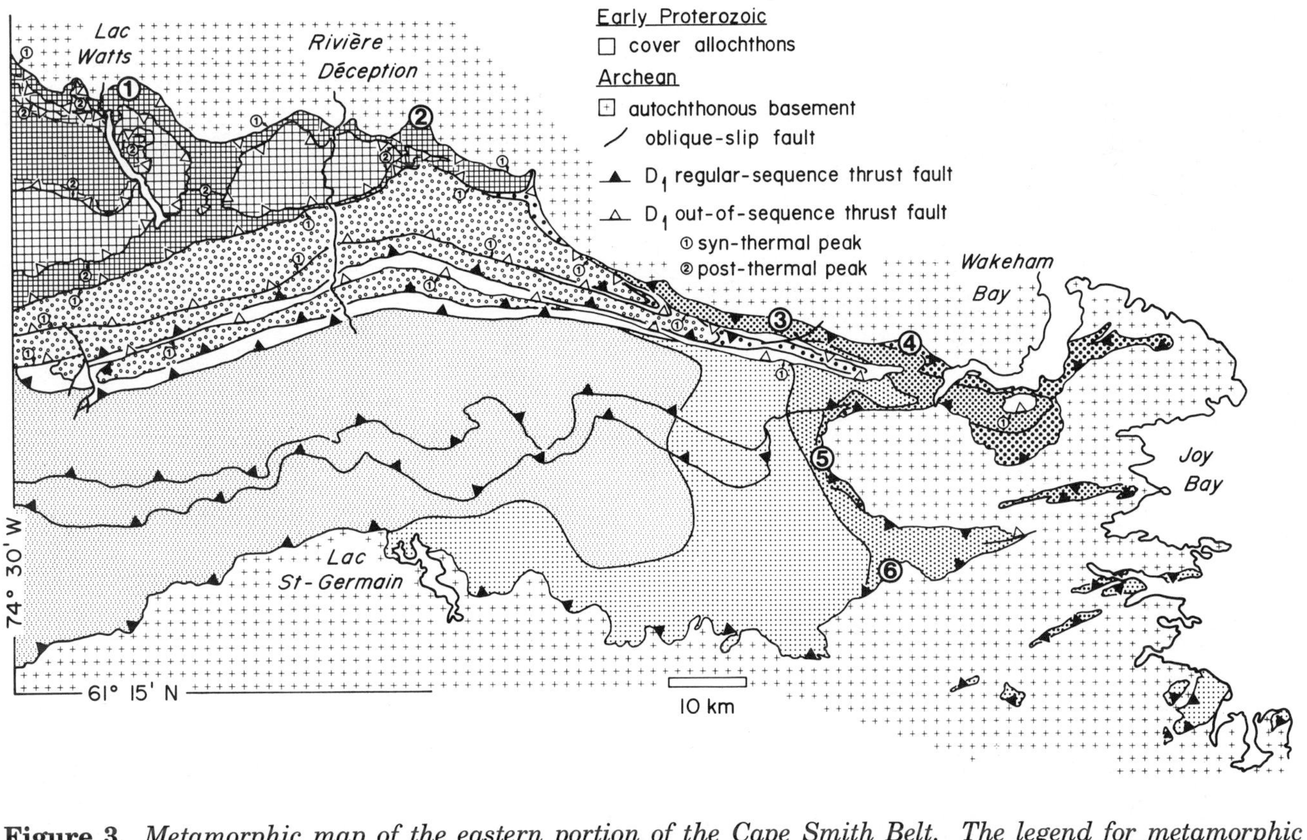

Figure 3. *Metamorphic map of the eastern portion of the Cape Smith Belt. The legend for metamorphic mineral zones (after Bégin* et al., 1988) is given in Figure 16. Large circled numbers (1–6) refer to the locations of samples used in the petrological P–T path study (see Fig. 31).

Watts Group, comprising amphibolite grade metavolcanic rocks and meta-intrusive units (Lamothe *et al.*, 1984).

Geological mapping by St-Onge *et al.* (1986, 1988) and St-Onge and Lucas (1989) has shown that the Archean basement can be mapped continuously around the east end of the belt in the Wakeham Bay area (Fig. 2). The breakdown of the Povungnituk Group into two subgroups reflects a major lithological change from lower quartzo-feldspathic sedimentary rocks (Lamarche Subgroup) to upper tholeiitic basalts (Beauparlant Subgroup). However, for the purposes of this paper, the two subgroups will simply be referred to as the lower and upper Povungnituk groups respectively. Finally, the sedimentary rocks of the lower Povungnituk Group have been further subdivided into three (informal) formations (see below).

The present distribution of Early Proterozoic rocks in the Cape Smith Belt is primarily a function of the age of emplacement, the displacement, and the stacking order of D_1 thrust sheets. A discussion of the tectono-stratigraphic record within the belt must, therefore, take into account the architecture of the thrust–fold belt. Five tectono-stratigraphic units have been recognized in the belt: (1) autochthonous sedimentary rocks of the lower Povungnituk Group; (2) allochthonous sedimentary and volcanic rocks of the Povungnituk Group; (3) allochthonous basalts of the Chukotat Group; (4) allochthonous deep-water sedimentary rocks of the Spartan Group; and (5) allochthonous sheeted dykes, basalts, and mafic–ultramafic cumulates of the Watts Group. The description of these tectono-stratigraphic suites is presented in the order of their occurrence from the external (southern margin, Fig. 2) to internal (northern margin) portions of the thrust belt. All Early Proterozoic rocks within the Cape Smith Belt have been metamorphosed to between greenschist and amphibolite facies. As such, the prefix "meta" applies to all units described in the paper.

Autochthon

Archean basement gneisses. The basement units exposed north, south, and east of the Cape Smith Belt (Fig. 2) consist of migmatitic tonalites locally containing abundant enclaves of amphibolite. Rb–Sr (whole-rock) dating by Doig (1983, 1987) and Taylor and Loveridge (1981) suggests that the age of the gneisses is between 2.6 and 2.9 Ga. The Archean fabrics range from weakly developed foliation

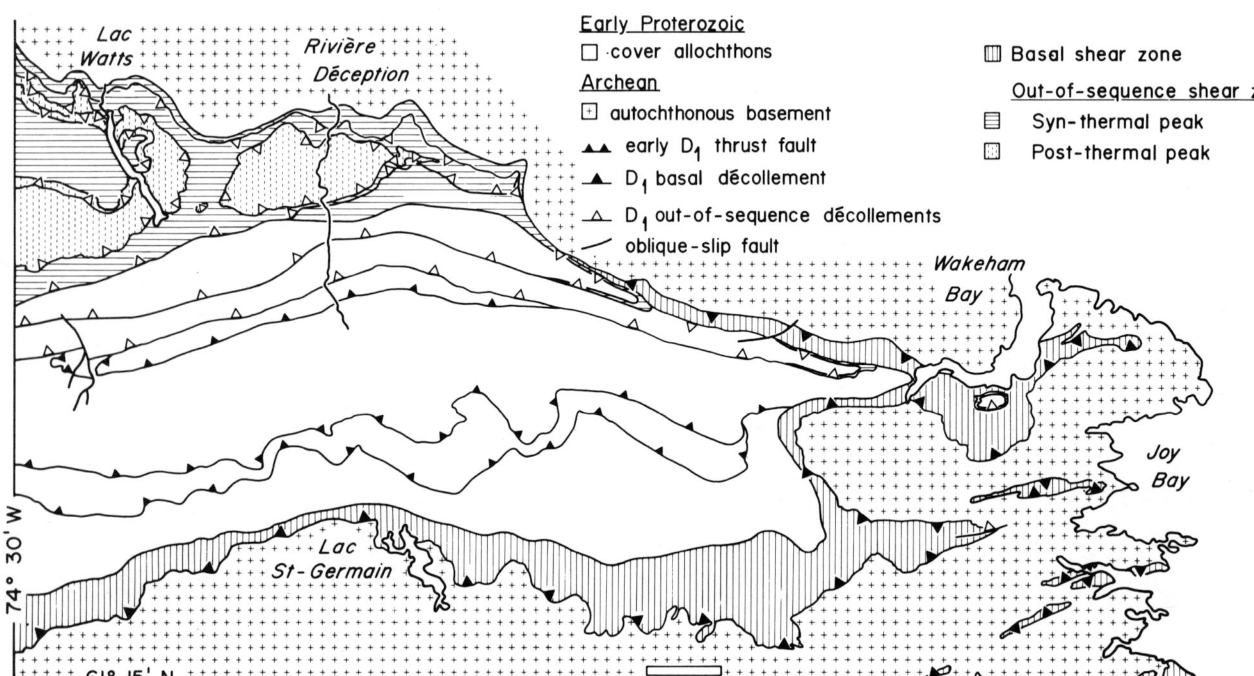

Figure 4. *Map of the eastern portion of the Cape Smith Belt showing the extent of the three shear zone groups: (1) basal shear zone; (2) synthermal peak, out-of-sequence shear zones; and (3) postthermal peak, out-of-sequence shear zones.*

to mylonitic gneissosity. Tonalitic gneisses near the basement/cover contact record the progressive transposition of the Archean fabrics into parallelism with the Proterozoic foliation (Fig. 5).

Lower Povungnituk Group. Autochthonous, Early Proterozoic strata outcrop in two principal modes: (1) as small (<1 m thick) occurrences of clastic sedimentary rocks on the south margin of the main belt; and (2) as more extensive (20–30 m thick) clastic sequences in nine outliers around Burgoyne Bay (Formation 1, Fig. 6). In both the main belt and the outliers, autochthonous units rest unconformably on Archean gneisses that show no evidence of a penetrative Proterozoic fabric. With the exception of one outlier, all autochthonous units are capped by a high strain zone corresponding to the basal décollement of the D_1 thrust belt.

In the main belt, the autochthon comprises pockets of garnetiferous, polymictic conglomerate, quartz pebble-bearing, arkosic grit and/or iron-rich sandstone. In the Burgoyne Bay outliers (Figs. 2 and 6), 2–3 m thick arkosic sandstone beds are an important component of the succession. Rounded to subrounded feldspar and quartz grains vary from less than a millimetre to several millimetres in diameter. Graded bedding is well preserved, with the base of individual beds commonly being a quartz pebble grit. Some of the thinner (less than 1 m thick), arkosic sandstone beds display well-developed trough crossbedding. These are consistent with south to north channel flow in a fluvio-deltaic environment. The arkosic sandstone is interbedded with 1–2 m thick, sandy ironstone beds showing graded bedding and with 2–3 m thick intervals of well-laminated pelite.

Figure 5. *Exposed basement/cover contact in the Wakeham Bay area (Fig. 2). The underlying Archean basement (A) is steeply dipping to the west (right-hand side of photograph). The overridding early Proterozoic allochthonous units (P) are dipping to the south (toward the upper left-hand corner of the photograph). Note that the pre-existing, Archean gneissosity is orthogonal to the basement/cover contact in this locality. Transposition of that fabric in the basal shear zone is restricted to within 3 m of the basement/cover contact. The hammer is 34 cm long.*

External (southern) thrust sheets

Lower Povungnituk Group. Sedimentary rocks preserved in thrust sheets within the outliers southeast of Wakeham Bay (thrust sheets A, B, C, Fig. 2) and along the southern margin of the main belt (thrust sheets B, C, D, Fig. 2) exhibit an internal stratigraphy that is consistent at a regional scale. The lower sedimentary rocks in the southernmost thrust sheets are ironstones, which vary in thickness from 50 to 650 m (Formation 2a, Fig. 2). Different units within the ironstones include (1) ferruginous sandstones; (2) interbedded, massive magnetite/grunerite schist; (3) carbonate-rich grunerite schist; and (4) garnet–biotite–hornblende–grunerite schist. These units are laterally discontinuous; therefore, the ironstones are not subdivided in Figure 2. Overlying the ironstones are 60–960 m of semipelites, arkosic sandstone, and ironstones (Formation 2b, Fig. 2). The semipelites are thinly bedded (centimetre scale), occasionally graded, and characterized by parallel laminations at the top of beds. The arkosic sandstones are composed of subrounded to rounded, millimetre-scale, quartz and feldspar grains. Bed thicknesses range from 15 cm to over 2 m. Graded bedding and tens of metres scale channel structures are commonly preserved in the arkosic sandstones, which are interpreted as proximal fan deposits. The base of the rusty semipelite and arkosic sandstone interval is marked by a narrow, 2–5 m thick, discontinuous interval of dolomite and associated calc-silicate units. Laterally, the semipelites grade into ironstones and volcaniclastic sedimentary rocks interlayered with metabasites. The semipelites are overlain by massive, homogeneous micaceous sandstones (Formation 3, Fig. 2) that vary in thickness from 70 m to 1.7 km. The micaceous sandstone consists of quartz and up to 20% metamorphic biotite and muscovite.

Thrust sheets farther to the north along the southern margin (E to H, Fig. 2) show progressively deeper water (more outboard) facies of Formation 2. This is a reflection of the consistent distal over proximal stacking order for thrust sheets in this part of the belt (see below). In thrust sheets E to H (Fig. 2), the proportion of semipelite, ironstone, and sandstone within Formation 2 changes systematically with each more internal (northerly) thrust sheet. For example, thrust sheet G contains volumetrically less ironstone and arkosic sandstone, but more semipelite, than thrust sheet E.

Thrust sheet H has a larger semipelite to ironstone ratio and laterally less continuous sandstone beds than in structurally lower (more southerly) sheets. The northerly increase in the semipelite to ironstone ratio coupled with the transition to more laterally discontinuous and thinner sandstone beds is interpreted as reflecting a progressive increase in water depth from south to north.

Within the lower thrust sheets (A to H, Fig. 2), the micaceous sandstones (Formation 3) form a laterally extensive blanket overlying Formation 2 deposits. The micaceous sandstones in all thrust sheets show a relatively homogeneous mineral assemblage and uniform grain size both laterally and vertically. Unlike the sedimentary rocks of Formation 2, the micaceous sandstones do not show outboard (south to north) facies changes, and do not interfinger with volcanic rocks. The accumulation of such a thick, homogeneous micaceous sandstone unit lacking channel structures is interpreted to mark the shallowing of a more proximal rift basin, isolated from the main locus of crustal stretching and rift volcanism by syndepositional normal faults.

The Povungnituk Group sedimentary rocks within thrust sheets I, J, K, and L (Fig. 2) consist almost entirely of laminated semipelites interbedded with arkosic sandstones. The micaceous sandstone unit, which caps Formation 2 sediments in more external thrust sheets, is not present. Instead, the semipelites are overlain by pillowed basalts and gabbro sills of the upper Povungnituk Group. The total thickness of lower Povungnituk Group sedimentary

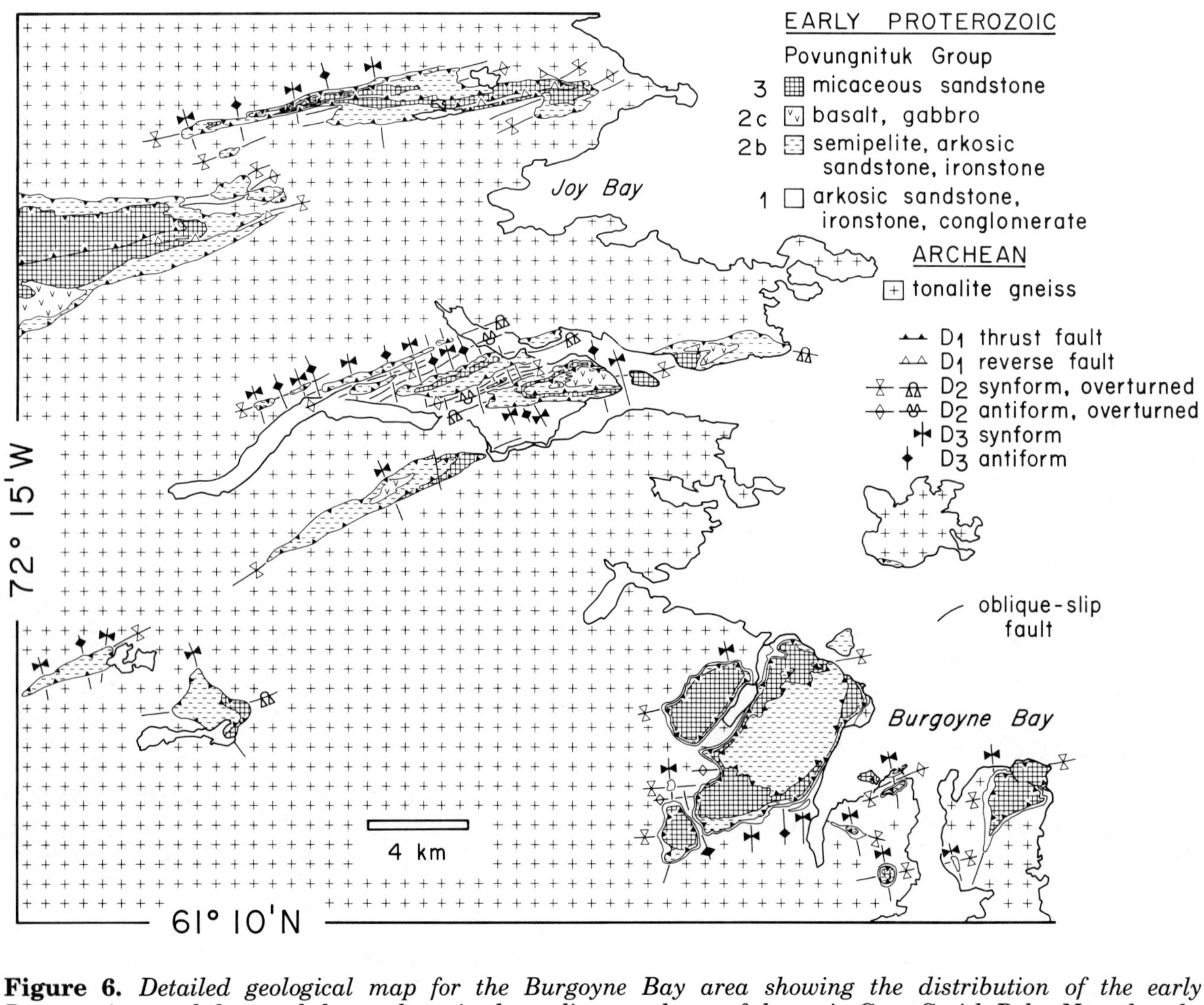

Figure 6. *Detailed geological map for the Burgoyne Bay area showing the distribution of the early Proterozoic autochthon and thrust sheets in the outliers southeast of the main Cape Smith Belt. Note that the basal décollement climbs up from the basement-cover contact into the cover sediments in the nine southern outliers. Numbers (1–3) in the legend for the Povungnituk Group refer to informal formations discussed in the text.*

rocks in thrust sheets I, J, K, and L is greatly inflated by the emplacement of numerous gabbro sills, interpreted to be part of the feeder system to the overlying volcanic rocks. The mafic sills range from 10 to 600 m thick and are essentially conformable with bedding defined by graphitic laminations in the clastic sediments. The gabbro sills have narrow (less than 1 m) chilled margins and rarely show internal layering. This dramatic stratigraphic change suggests that the locus of rift-basin subsidence and active rift volcanism had shifted outboard (northward), which allowed the deposition of micaceous sandstone in an inboard, sediment-dominated basin.

Upper Povungnituk Group. Basalts (Formation 4, Fig. 2) conformably overlie or interfinger with the clastic sedimentary rocks of the lower Povungnituk Group in thrust sheets I, J, K, and L. The volcanic pile has an apparent maximum thickness of 6.7 km, with the top of the Povungnituk Group marked by a tectonic contact. Both thin, pillowed mafic flows and thicker, tabular flows are present in the succession. Dolomite and semipelite rarely occur between flows, and mafic pyroclastic rocks are most common in the lower part of the volcanic section. The base of the mafic pile is also characterized by laterally discontinuous 5–10 m thick volcanic conglomerate units. The latter are typically composed of subrounded to angular blocks of basalt and gabbro that range in diameter from a few centimetres to several metres. The conglomerate is clast-supported and the infilling matrix consists of millimetre-scale basalt rock fragments and carbonate. The immature nature of the clasts, the presence of both basalt and gabbro clasts, and the limited lateral extent of the volcanic conglomerate suggest that this unit is possibly related to ongoing (normal?) faulting during accumulation of the Povungnituk Group.

The mafic volcanic rocks are reported by Hynes and Francis (1982) and Francis *et al.* (1983) to be Fe- and Ti-rich, light rare earth element (LREE) enriched, continental tholeiites. Picard (1989a) reports evidence for alkaline volcanic edifices constructed on top of the tholeiites 60 km northwest of lac St-Germain (Fig. 2). Laterally discontinuous and thin (less than 50 m thick) rhyolite bodies occur within the mafic volcanic pile. A rhyolite sampled 50 km west of lac St-Germain (Fig. 2) has been dated (U/Pb on zircon) at 1.96 Ga.

Midbelt thrust sheets

Lower Povungnituk Group.? Clastic sedimentary rocks of the lower Povungnituk Group crop out within thrust sheets M, O, and Q (Fig. 2). These three thrust sheets occur more internally (i.e., northerly) in the thrust–fold belt and have travelled farther than the southern thrust sheets (Fig. 2). The dominant sedimentary rocks in these thrust sheets are semipelites interbedded with minor distal fan sandstones. The semipelites are thin bedded (centimetre scale) and interfinger with beds of more pelitic composition. The sandstone beds are tens of centimetres thick and are composed chiefly of rounded submillimetre quartz grains. Individual beds are apparently laterally very continuous and show no evidence of channel structures or crossbedding. The finer grained, uniformly bedded sedimentary rocks within the midbelt thrust sheets are compatible with a deeper water (more outboard) facies of the lower Povungnituk Group.

Lower Povungnituk Group sedimentary rocks occur within two fault-bound lozenges ("horses" in the terminology of Boyer and Elliott (1982)) along the thrust fault at the base of thrust sheet R (Fig. 2). The lozenges contain spectacular polymictic conglomerate and local arkosic sandstone beds. The conglomerate is poorly sorted, with subangular to subrounded clasts of tonalite gneiss, basalt, gabbro, and argillite (Fig. 7). The centimetre- to metre-scale clasts form a framework filled with a matrix of quartz, feldspar, biotite, and chlorite. The unsorted nature of the clasts, the presence of both basalt and gabbro clasts, and the similarity of the large granitoid boulders to the basement gneisses suggest that this unit is possibly a fault-scarp deposit related to normal faulting during accumulation of the Povungnituk Group.

In the midbelt area, the lower Povungnituk Group sedimentary rocks are in thrust contact with either (1) footwall upper Povungnituk Group metabasites or (2) hanging wall and footwall Chukotat Group basalts. The apparent thickness of the sedimentary rocks in the midbelt thrust sheets, which is on the order of 1.6 km, is greatly inflated by the emplace-

Figure 7. *Lower Povungnituk Group fault-scarp conglomerate in midbelt thrust sheet R (Fig. 2). Note the presence of large boulders of tonalite gneiss (G) and basalt (B). The height of the person is 1.8 m.*

ment of numerous sills of the Kattiniq intrusive suite (described below).

Chukotat Group. Thrust sheets N, P, and S (Fig. 2) are host to dominantly pillowed basalt flows of the Chukotat Group. The volcanic pile also contains massive flows, and is intruded by thin (5–100 m thick) mafic and layered mafic–ultramafic sills. The lower and upper contacts of the Chukotat Group are tectonic and the thickest preserved stratigraphic section is on the order of 5.6 km. In contrast to the upper Povungnituk Group, the Chukotat Group is essentially devoid of sedimentary rocks. Primary textures and volcanic structures, such as variolitic pillow margins, ropy lava, pillow tubes (Fig. 8), and flow top breccia, are well preserved and suggest submarine accumulation.

Three distinct basalt types can be recognized within the Chukotat Group based on the presence of a dominant pseudomorphed phenocryst type (Francis and Hynes, 1979; Hynes and Francis, 1982; Francis *et al.*, 1983). As a result, dominantly olivine-phyric, pyroxene-phyric, and plagioclase-phyric basalts can be distinguished in the Chukotat Group (Fig. 2). A complete description of field and petrographic characteristics of the Chukotat Group lavas is given in Hynes and Francis (1982). The volcanic rocks range from Mg-rich komatiitic basalts to low-Ti, tholeiitic basalts with affinities to modern midoceanic-ridge basalts (MORBs) (Hynes and Francis, 1982). Based on the chemistry of the volcanic suites, Hynes and Francis (1982) interpreted the Chukotat Group as documenting the change from accumulation of continental rift-type volcanic rocks to the formation of transitional oceanic crust.

Kattiniq intrusive suite. The clastic sedimentary rocks and basalts of the Povungnituk Group in thrust sheets B to Q (Fig. 2) are host to numerous layered peridotite–gabbro sills, which are tentatively interpreted as consanguineous. Individual layered sills range up to 500 m in thickness. The sills are commonly characterized by a basal pyroxenite chilled border phase overlain by columnar-jointed peridotite and capped by gabbro. The gabbroic section in thicker sills is marked by plagioclase cumulate layering, and is roofed by ferrogabbro and quartz ferrogabbro zones. Layered mafic–ultramafic sills emplaced in the Chukotat Group are potentially part of the same suite. However, this correlation requires geochemical and geochronological data. A sample of quartz ferrogabbro from a sill intruding the lower Povungnituk Group sediments of thrust sheet M has yielded a U/Pb age on baddeleyite of 1.92 Ga. The Cu–Ni sulphide deposits and showings, commonly with elevated platinum group element (PGE) values, are predominantly associated with the Kattiniq intrusive suite layered sills in thrust sheet M (Fig. 2).

Internal (northern) thrust sheets

Spartan Group. Clastic sedimentary rocks in the more internal part of the thrust–fold belt (thrust sheets V, U, and the lac Watts – rivière Déception area; Fig. 2) are distinct from the lower Povungnituk Group. The sedimentary rocks, part of Lamothe *et al.*'s (1984) Spartan Group, comprise laminated dark pelites interbedded with semipelite (Fig. 9) and minor sandstone. Beds in the pelites range in thickness from 2 to 3 mm to 1 to 2 cm and are commonly outlined by graphite laminae. Quartz-rich semipelitic beds of similar thicknesses grade upward into graphite-rich pelite. Near the structural top of thrust sheet V, the sedimentary package is characterized by the occurrence of fine-grained sandstone beds generally tens of centimetres thick. The sediments of thrust sheet V are interpreted as deep-water pelagic sediments, with

Figure 8. *Pillowed basalt flows of the Chukotat Group. The basalt is olivine-phyric in this locality. The height of the person is 1.8 m.*

Figure 9. *Laminated graphitic pelites of the Spartan Group. The hammer is 50 cm long.*

the monotony of the sequence interrupted near the top by distal fan deposits.

Watts Group. A distinct assemblage of basalts, sheeted dykes, and layered mafic and ultramafic rocks occurs north of thrust sheet V (Fig. 2). Thrust sheet Y is a 4 km thick sequence of mafic dykes, gabbro sills, and pillow basalts that structurally overlies the Spartan Group sedimentary rocks. The thrust sheet is upright based on pillow facing directions. The base of thrust sheet Y on the east side of lac Watts consists of a sheeted dyke complex (Fig. 10). The largest exposure of the dykes, located 5 km due east of lac Watts, has an exposed width of 60 m, with a strike length exposure of approximately 500 m. Individual half dykes are 20–30 cm wide (Fig. 11) and commonly preserve a 2–3 mm wide one-sided chilled margin. Major element analyses show that the dykes are tholeiitic in composition, whereas rare earth element (REE) patterns indicate that they are flat to LREE enriched [(La_n/Lu_n)

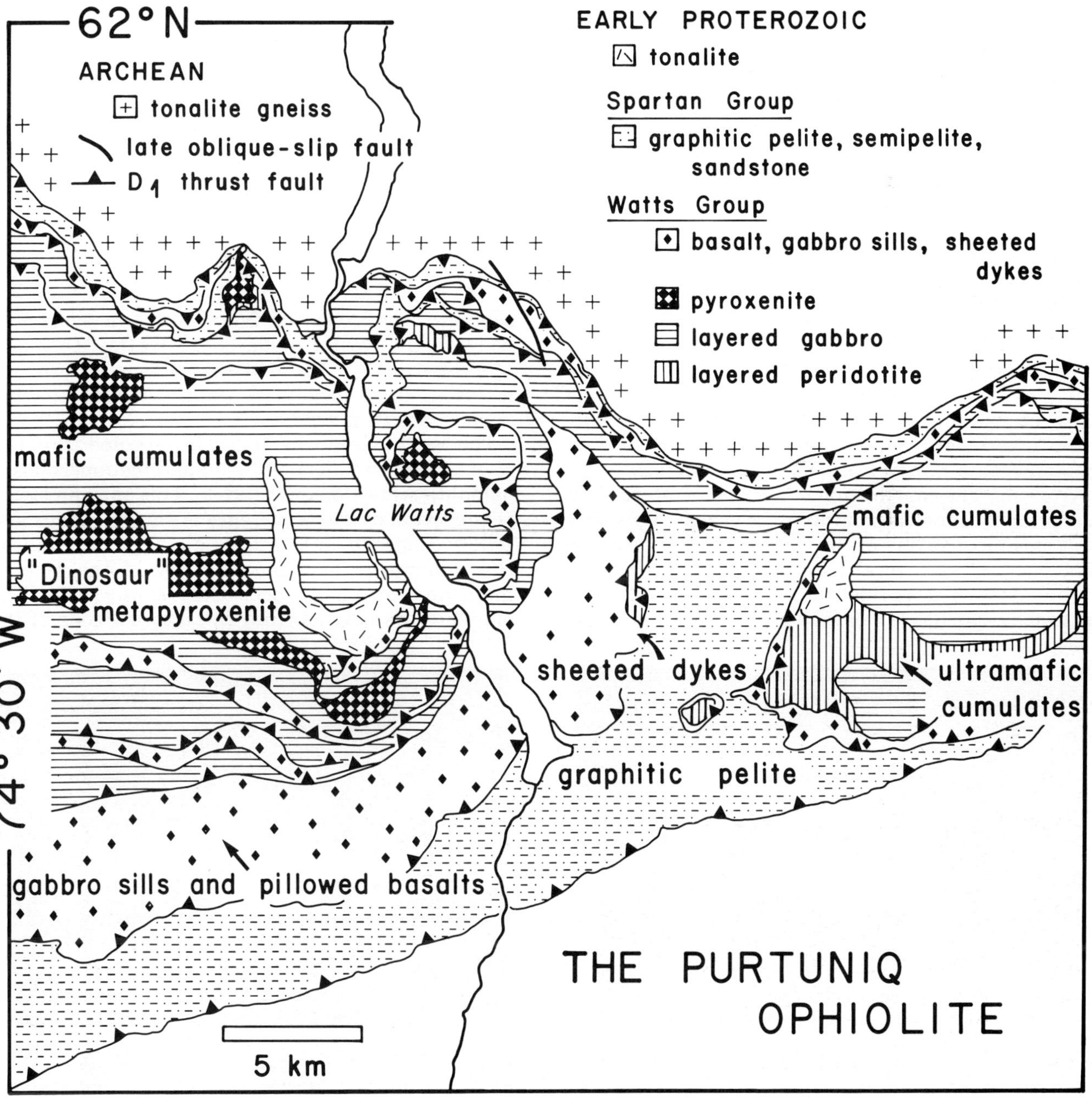

Figure 10. *Detailed geological map of the lac Watts area showing the distribution of the tectonostratigraphic units comprising the Purtuniq ophiolite.*

up to 20] (Scott *et al.*, 1988). It is important to recognize that the sheeted dyke complex is 100% dykes (Fig. 11). This requires multiple dyke emplacement into previously emplaced dykes to produce the sets of half dykes with chilled margins on one side only.

Stratigraphically overlying the sheeted dykes are 2–3 m thick mafic sills and massive basalt flows. This intermediate stratigraphic level between sheeted dykes and basalts is several hundred metres thick. The sills are gabbroic with the grain-size ranging from less than a millimetre to a millimetre. Massive flows and sills are crosscut by discrete single dykes that show the compositional range of the underlying sheeted dykes. Above the mafic sill and massive flow sequence, up to 3.5 km of pillowed basalt and mafic sills are exposed in thrust sheet Y. The basalt flows are tholeiitic with n-MORB type REE patterns (Scott *et al.*, 1988). Pillow tops indicate a stratigraphically upright sequence. A small number of steeply dipping mafic dykes crosscuts both pillows and sills.

A series of extensive, thick (kilometre-scale) imbricates of peridotite, pyroxenite, and layered gabbro occur within thrust sheets T, X, Z, BB, and CC (Fig. 2) in the northern part of the thrust–fold belt. The peridotites show centimetre- to metre-scale layering defined by colour changes (tan versus reddish brown) and differential weathering. The outcrop layering correlates with modal variations in metamorphic mineral assemblages after olivine or pyroxene. The compositional layering is interpreted as primary igneous cumulate layering. Laterally discontinuous layers of chromite, although rare, are documented at several locations in the ultramafic cumulates. The peridotites occur in two different settings (Fig. 10): (1) as fault-bound lozenges (horses) at the base of thrust sheets of layered gabbro or (2) in stratigraphic contact with layered gabbro. In both settings, the peridotites are thought to be part of the crustal cumulate sequence. Tectonized harzburgite (i.e., upper mantle ultramafics) has not been observed in the map area (Scott *et al.*, 1988).

The layered gabbros form the volumetrically most important unit of the Watts Group. They occur in fault-bound sheets ranging in thickness from 100 m to several kilometres (Fig. 10). The centimetre- to metre-scale compositional layering in the gabbros is defined by the modal abundance of two end-member mineral assemblages: (1) hornblende, which forms black metapyroxenite layers, and (2) plagioclase–zoisite, which forms striking white meta-anorthosite layers (Fig. 12). These metamorphic assemblages are interpreted to directly reflect modal variations of primary pyroxene and plagioclase U–Pb analyses of zircons from a layer of gabbroic composition have yielded an age of 1998 ± 2 Ma.

The plutonic rocks of the Watts Group have been interpreted as mafic or ultramafic cumulates (Scott *et al.*, 1988; St-Onge *et al.*, 1988), with only one exception. The exception is an intrusive pyroxenite body on the western side of lac Watts ("Dinosaur Metapyroxenite," Fig. 10) that shows no internal layering and locally contains layered gabbro xenoliths at its margins.

The Watts Group has been interpreted (St-Onge *et al.*, 1988) as a dismembered and metamorphosed ophiolite suite (Moores, 1982). This interpretation is suggested by the occurrence of sheeted dykes and pillowed basalts in stratigraphic sequence, and by the mineralogy and geochemistry of the layered mafic and ultramafic rocks (Scott *et al.*, 1988). The ophiolite preserved within the internal zone of the

Figure 11. *Sheeted dykes of the Watts Group from the base of thrust sheet Y (see Figs. 2 and 10). Note that the outcrop is 100% half-dykes dipping steeply to the east (right-hand side of photograph). The hammer is 34 cm long.*

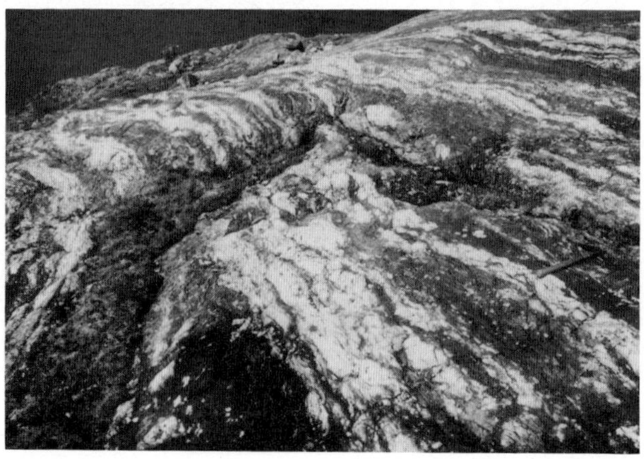

Figure 12. *Layered gabbros of the Watts Group. Meta-anorthositic layers (white) have a nonuniform appearance because of ductile deformation related to out-of-sequence thrusting. The hammer is 34 cm long.*

Cape Smith Belt has been informally named the Purtuniq ophiolite by St-Onge *et al.* (1988). The ophiolite is now, in part, structurally inverted as a result of D_1 thrusting. The ophiolite in the lac Watts area comprises, from structural bottom to top, the following units: (1) sheeted mafic dykes overlain by mafic flows and thin sills, which are intruded by mafic dykes, and (2) layered mafic and ultramafic cumulates. Structurally sandwiched between the ophiolitic thrust sheets (Watts Group) and the underlying transitional crust pillowed basalts (Fig. 10) are the deep-water sedimentary rocks of the Spartan Group. The dominantly graphitic pelites are interpreted to have been deposited on either the transitional crust (Chukotat Group) or the oceanic crust (Purtuniq ophiolite).

CONSTRAINTS ON THE THERMOTECTONIC EVOLUTION OF THE BELT

Introduction

Two important constraints employed to unravel the thermotectonic evolution of the Cape Smith Belt are (1) the relative and absolute ages of various deformational events and (2) the >18 km of structural relief present in the map area. The observational basis and significance of these two constraints are examined in the following two sections.

Relative and absolute timing of deformation episodes

Three geometrically distinct deformation episodes (D_1, D_2, D_3) affected the Superior Province basement, the rift and transitional crust units (1.96–1.92 Ga), and the 2.00 Ga Purtuniq ophiolite. The D_1 episode is bracketed in age between 1.92 Ga (pre-D_1 Kattiniq intrusive suite) and 1.76 Ga (post-D_1 syenogranite dykes). The timing of D_3 deformation is not well constrained, but if the deformation results from the Rae/Superior collision in the northwest–southeast trending New Quebec Orogen (Fig. 1), then it must be older than 1.79 Ga (Hoffman, 1988; Machado *et al.*, 1988).

Employing deformation-metamorphism relationships as a relative chronometer for the D_1 episode (see discussion below), four stages of D_1 deformation have been identified (Lucas, 1989). These are shown in Figure 13 and include (1) a prethermal peak, piggyback (regular-sequence) thrusting stage; (2) a basal shear zone deformation stage, occurring during prograde metamorphism but terminating at thermal peak conditions; (3) a synthermal peak out-of-sequence thrusting stage; and (4) a postthermal peak out-of-sequence thrusting stage. Information about the thrust belt's early evolution is preserved in map-scale structures and fault-zone microstructures found at relatively high structural levels in the belt because these levels escaped basal shear zone deformation (Fig. 4). Thin section- to map-scale structures related to stages 2 to 4 are preserved in the southeastern and northern parts of the map area (Fig. 4). Structures related to each of the deformation stages can be recognized in the metamorphic map (Fig. 3) based on crosscutting relationships between thrust faults and thermal peak mineral isograds. The isograds overprint the piggyback thrust structures and the basal shear zone (Figs. 3 and 4) in the southern (external) part of the belt, but are truncated by syn- and postthermal peak, out-of-sequence thrusts toward the northern (internal) part.

The use of deformation–metamorphism relationships as a relative chronometer for D_1 deformation stages relies on the assumption that thermal peak conditions were reached coevally at comparable structural levels in the thrust belt. Comparable structural levels are found both north (hanging wall) and south (footwall) of the southernmost out-of-sequence thrust (thrust O, Fig. 2). Thrust sheets found on the hanging wall of this thrust are collectively referred to as the out-of-sequence domain, whereas those in the footwall are termed the regular-sequence domain.

Progressive deformation of thermal peak minerals in out-of-sequence shear zones indicates that the out-of-sequence domain reached thermal peak conditions during stage 3 thrusting. Two important observations bear on the timing of the thermal peak in the regular-sequence domain: (1) overturned thermal peak isograds are preserved in metabasalts immediately below thrust O (Fig. 3; Bégin *et al.*, 1988) and (2) basal shear zone deformation terminated at thermal peak conditions after the shear zone was truncated by an out-of-sequence thrust (Fig. 4; St-Onge *et al.*, 1988). These observations can be explained if the footwall to the out-of-sequence domain reached thermal peak conditions as a result of the emplacement, uplift, and erosion of the out-of-sequence thrust sheets. In summary, deformation–metamorphism relationships represent a valid chronometer as comparable structural levels in both the regular-sequence and out-of-sequence domains appear to have reached their thermal peak conditions coevally.

The absolute timing of the different stages of the D_1 episode is constrained by U–Pb zircon dating of syntectonic tonalitic intrusions. A maximum age for stage 1 thrusting is given by the 1.92 Ga age for a layered sill of the Kattiniq intrusive suite. A pluton on the west side of lac Watts (Fig. 10), dated at 1.88 Ga, cuts stage 1 faults. The age of the pluton constitutes a minimum age for stage 1 thrusting at this location. Syn-D_1 tonalite plutons

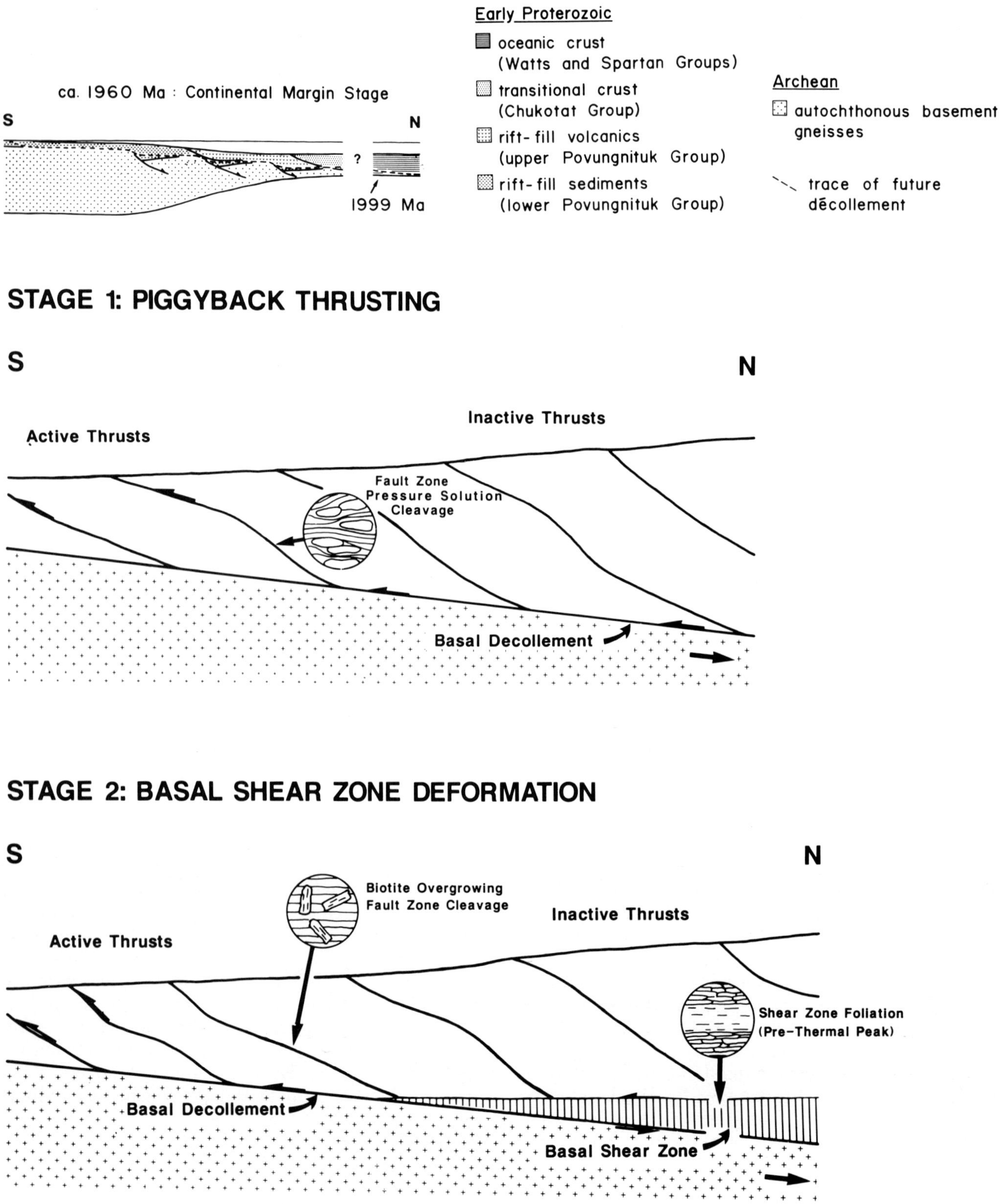

Figure 13. *Schematic representation of the structural evolution of the thrust belt. The inferred paleogeography of the north-facing, Superior Province, continental margin is shown at the top. Postrift sediments are not illustrated. The essential details of the deformation/metamorphism relationships during the four stages of thrust belt development are illustrated in the schematic cross-sections labelled stages 1 to 4.*

in the internal part of the thrust belt constrain the initiation of stage 3 to be younger than 1.84 Ga. Late, crosscutting syenogranite dykes indicate that the termination of stage 4 must predate 1.76 Ga.

Down-plunge constrained cross-sections

Cross-sections of the Cape Smith Belt have been constructed (Lucas, 1989) using the axial or "down-plunge" projection technique (Argand, 1911; Mackin, 1950; Stockwell, 1950; King, 1986). Cross-sections built using down-plunge projections rely on the assumption that the geological information illustrated in the section is predictable at the scale of the study area. The undulating east–west plunge in the Cape Smith Belt can be explained by (1) original variations in the amplitude of D_2 folds and, consequently, D_3 folds (discussed below); and (2) D_3 cross-folding about axes oriented approximately perpendicular to the D_1 and D_2 structures. Lucas (1989) discusses in detail the assumptions and techniques employed in constructing the down-plunge sections.

Strike-parallel sections (east–west) were constructed to establish the variation in structural level exposed along the length of the belt (Fig. 14). The exact geometry of the plunging D_1–D_2 belt was

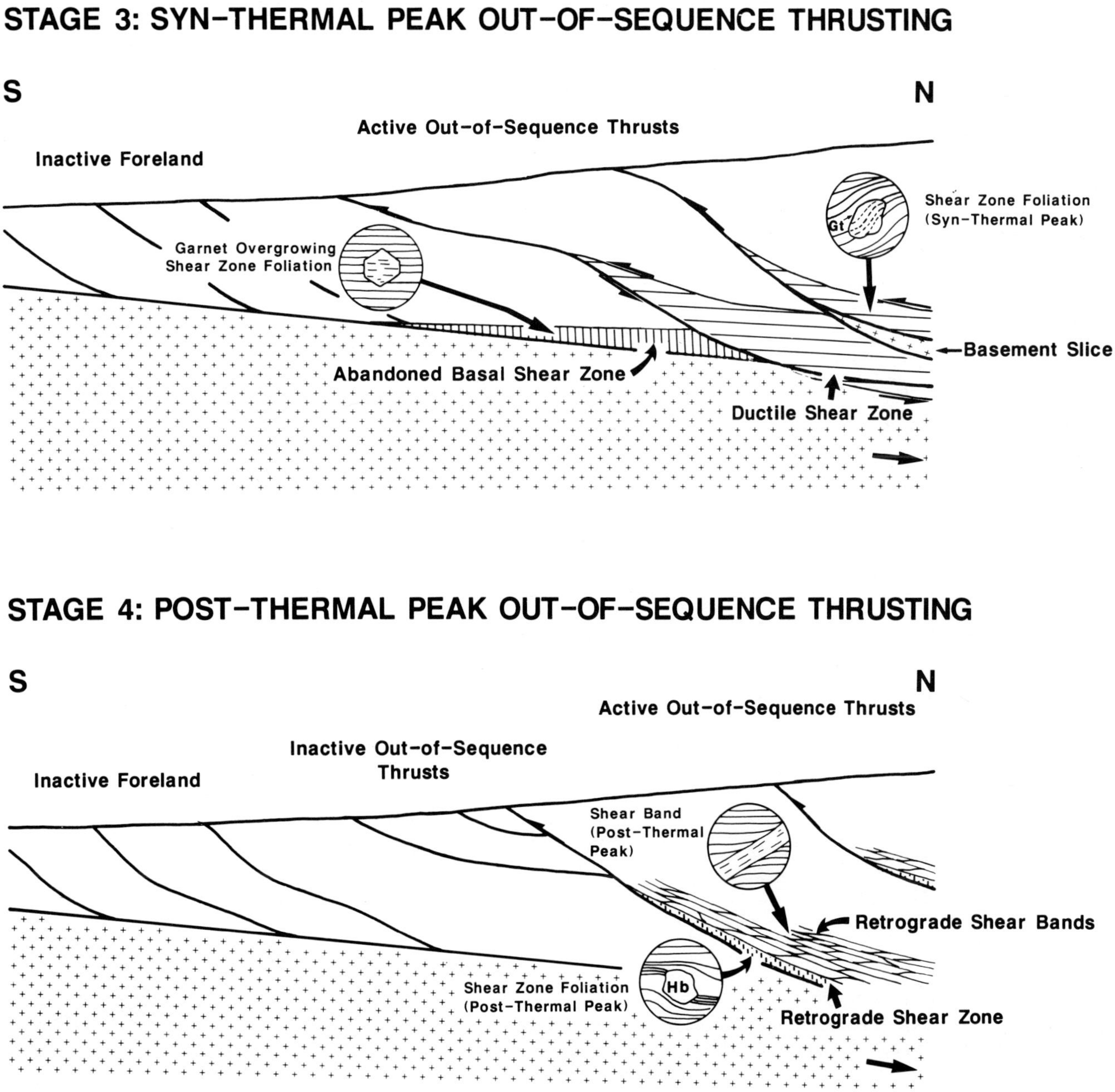

Figure 13. Continued.

established by compiling the orientations of D_1 and D_2 fold axes in domains of subequal plunge. These axes provide the best monitor of east–west structural plunge because they both originated with an east–west trend and a subhorizontal plunge. The position of the exposed basement–cover contact was projected onto vertical, 270°-striking planes using the mean fold axes for the plunge domains encountered along the section lines (Fig. 14).

A strike-perpendicular cross-section (Fig. 15) was built from over 50 independently established down-plunge projections prepared in domains of statistically constant east or west plunge. Individual projections were constructed by projecting the erosion surface geology, exposed in an oblique cross-section, onto the profile plane corresponding to the mean fold axis for that domain. As such, it represents a pre-D_3 view of the D_1 thrust belt, its autochthonous basement, and the D_2 map-scale folds that deform them. Essentially all the information on the section is contained on the geological map of the map area (Fig. 2). It is generalized to the extent that lateral structures and geometries that are not regionally predictable are not included.

Assembly of the composite down-plunge section (Fig. 15) involves stacking the individual projections up from the basement–cover contact in accordance with their relative structural level as predicted by the strike-parallel sections (Fig. 14). Individual projections are stacked up from the basement–cover

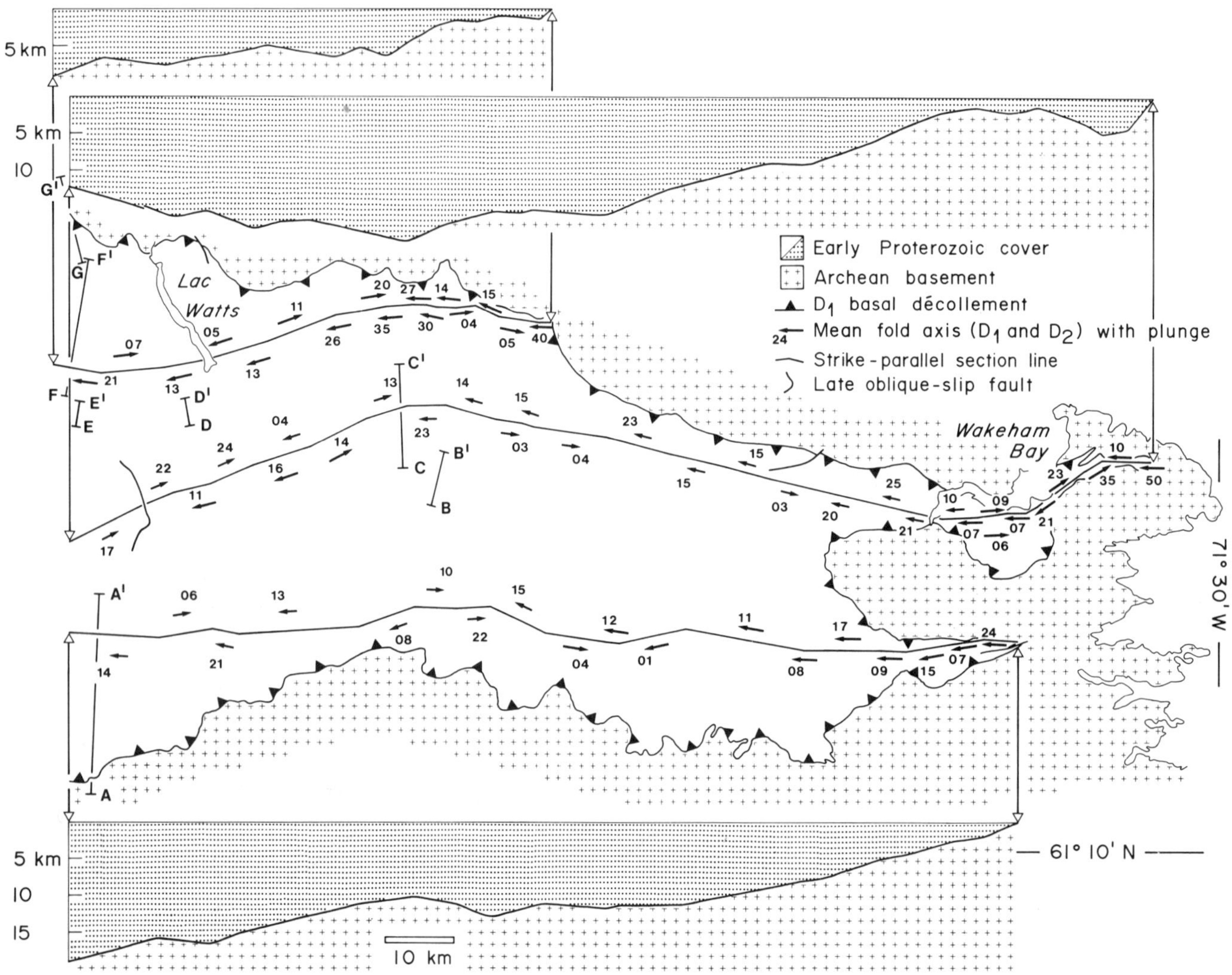

Figure 14. *Strike-parallel cross-sections of the eastern portion of the Cape Smith Belt. The mean D_1 and D_2 fold axis plunges used to construct the sections are shown. Cross-section lines A–A′ to G–G′, representing the structurally highest projections used in the cross-section of Figure 15, are shown in map view. These lines give an erosion surface trace from which to "hang" the cross-section. The remainder of the section lines for the individual projections are not illustrated here or in Figure 15 for reasons of clarity.*

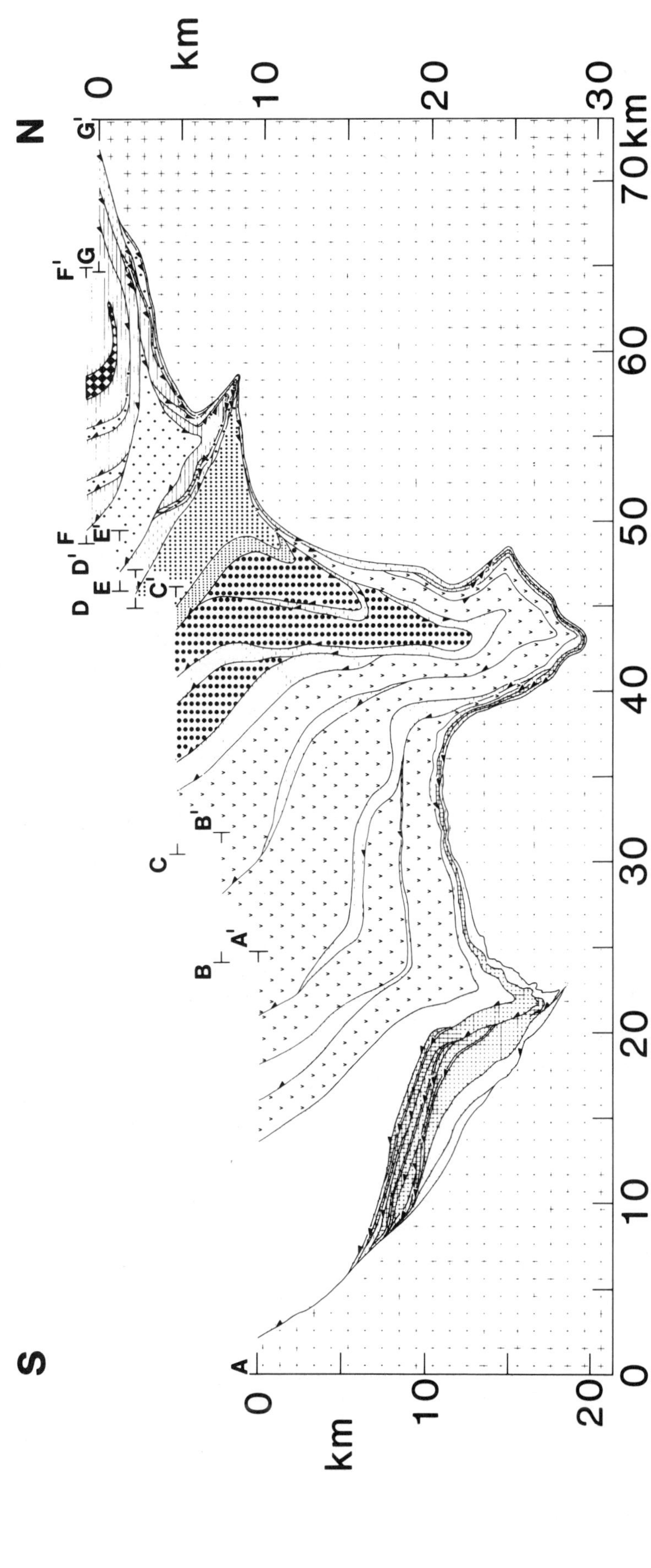

Figure 15. *Down-plunge constrained cross-section of the eastern portion of the Cape Smith Belt. The cross-section is of the D_1 thrust belt following D_2 folding, but prior to D_3 folding. The legend is given in Figure 2. Letters A–A′ to G–G′ identify the section lines for the structurally highest projections used.*

contact because it is the only consistent structural relief datum in the study area. The resulting down-plunge constrained cross-section (Fig. 15) reveals a striking south-to-north increase in the amount of structural relief (from 18 to 30 km). The structural relief gradient is illustrated at the top of the cross-section because the individual projections are stacked up from the basement–cover contact. The gradient is attributed to an increase in the amplitude of the D_2 and D_3 regional-scale folds toward the north, resulting in a continuous south-to-north increase in east or west structural plunge.

The mineral isograds mapped in basalts of the Povungnituk and Chukotat groups (Fig. 3; Bégin *et al.*, 1988; St-Onge and Lucas, 1989) have been projected onto the strike-perpendicular cross-section to form a metamorphic cross-section (Fig. 16). Projection of the isograds is based on the assumption that they are regionally predictable surfaces. The map traces of the Povungnituk Group isograds are approximately parallel to the basal décollement and are folded by D_2 structures (Fig. 3). These observations suggest that the isograds probably plunge parallel to the basal décollement and are thus regionally predictable surfaces. The Chukotat Group isograds are treated similarly as regionally predictable surfaces based on the fact that they are coeval with the Povungnituk Group isograds. An independent test of the pre-D_3 crustal thicknesses portrayed in the cross-section is provided by *P–T* determinations for D_2 assemblages. The D_2 pressure determinations for point "A" (marked on Fig. 16) cluster at 5 kbar (500 MPa) (see below), matching well the 16 km (5 kbar (500 MPa)) of D_2 structural section determined to be above point "A" based on the cross-section.

STRUCTURAL HISTORY

Evolution of the D_1 thrust belt

Stage 1: piggyback (regular-sequence) thrusting. Initial thrust belt growth occurred through accretion of continental rift, transitional crust, and possibly ophiolitic sheets along a system of piggyback-sequence thrusts. These structures are preserved in the footwall (south and east) of the out-of-sequence domain (Figs. 15 and 18). A regional basal décollement developed during stage 1, separating D_1-deformed Early Proterozoic rocks from essentially undeformed Archean basement. The décollement is localized at the basement–cover contact within the conterminous belt and the northern outliers (Fig. 2). In contrast, up to 30 m of undeformed, autochthonous sedimentary rock occur below the décollement in the southern outliers (Figs. 2 and 6). This north-to-south change in the footwall stratigraphic level of the décollement clearly marks the position of a major ramp in the décollement's trajectory (see cross-section on Fig. 17).

The early D_1 geometry of the southern and eastern (more external) portions of the belt is characterized by a thrust system linked to the basal décollement (Fig. 2). The thrusts thicken the rift sequence from at least 7 km to a minimum of 20 km (Fig. 15) by progressively translating more distal rift units onto more proximal ones. Evidence that the fault system originated at the basement–cover contact décollement is found in the Wakeham Bay area (Fig. 18) where a lateral ramp connects thrust K to the basal décollement. A north-to-south movement direction for the thrust system is suggested by two principal observations: (1) most thrust ramps trend east–west and dip northward (Fig. 2); and (2) early D_1 thrusts consistently juxtapose more distal sedimentary and/or volcanic facies onto more proximal facies.

A minimum estimate of the displacement along the regular-sequence décollement at the northern margin (Wakeham Bay area, Fig. 2) can be made by summing displacements on individual faults that root on the décollement, and by constraining its foreland displacement based on stratigraphic overlap. Matching hanging wall and footwall cutoffs for two of the major faults in this system (J and K, Fig. 2) yields minimum displacements for each on the order of 10 km. The décollement in the eastern Wakeham Bay – Burgoyne Bay area has the same tectono-stratigraphic unit in its hanging wall for approximately 70 km parallel to the north–south transport direction. Within this basal rift-fill sedimentary unit (unit 2b of thrust sheet B, Fig. 2), semipelite is consistantly found adjacent to the basal décollement. Employing these constraints on décollement displacement, the minimum displacement at the northern margin is estimated to be on the order of 100 km. This displacement estimate is consistent with the observations that the Povungnituk Group in the thrust belt was deposited on stretched crust, but the footwall basement in the mapped area shows no evidence for crustal stretching and does not contain feeder dykes or sills related to rift volcanism (Hoffman, 1985; St-Onge *et al.*, 1986).

A piggyback-stacking sequence (Elliott and Johnson, 1980; Boyer and Elliott, 1982) is suggested for the early D_1 thrust system based on two important observations. First, thrust faults are deformed by fault-bend folds (Suppe, 1983) associated with ramps in structurally lower thrust faults, indicating that older thrusts are structurally higher in the thrust stack. An example is found southeast of lac Vicenza (Fig. 2), where movement over a ramp in thrust G folded the hanging wall thrust stack, including thrust I. The second observation is that

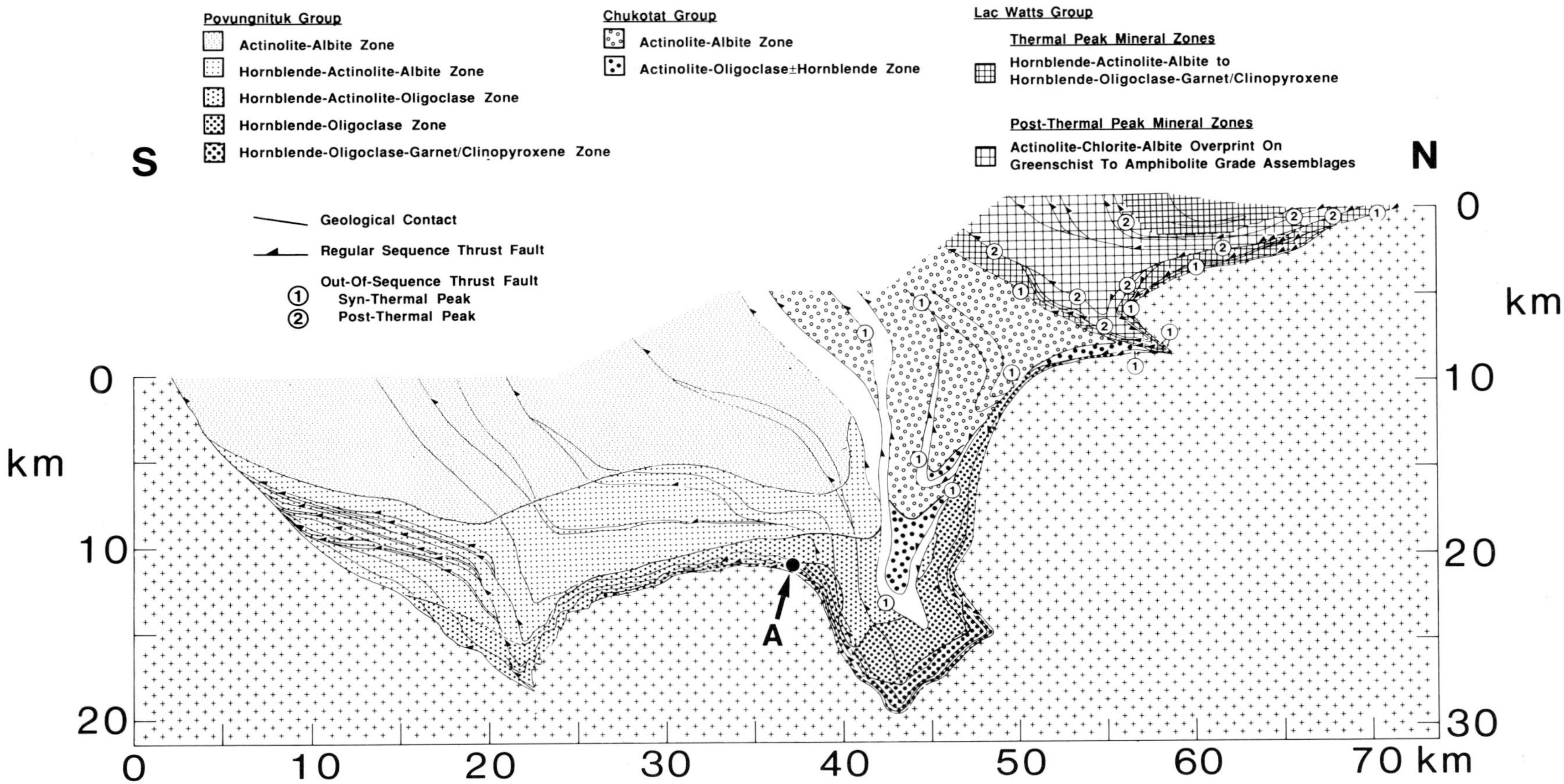

Figure 16. *Down-plunge constrained cross-section showing metamorphic zonation and mineral isograds in the eastern portion of the Cape Smith Belt. Mineral isograds and metamorphic zones are after Bégin* et al. (1988). Point “A” refers to the site of D_2 P–T determinations described in the text.

synmetamorphic movement on the basal décollement (i.e., stage 2 deformation) resulted in deformation of hanging wall thrusts. Figure 19 illustrates how basal shear zone deformation

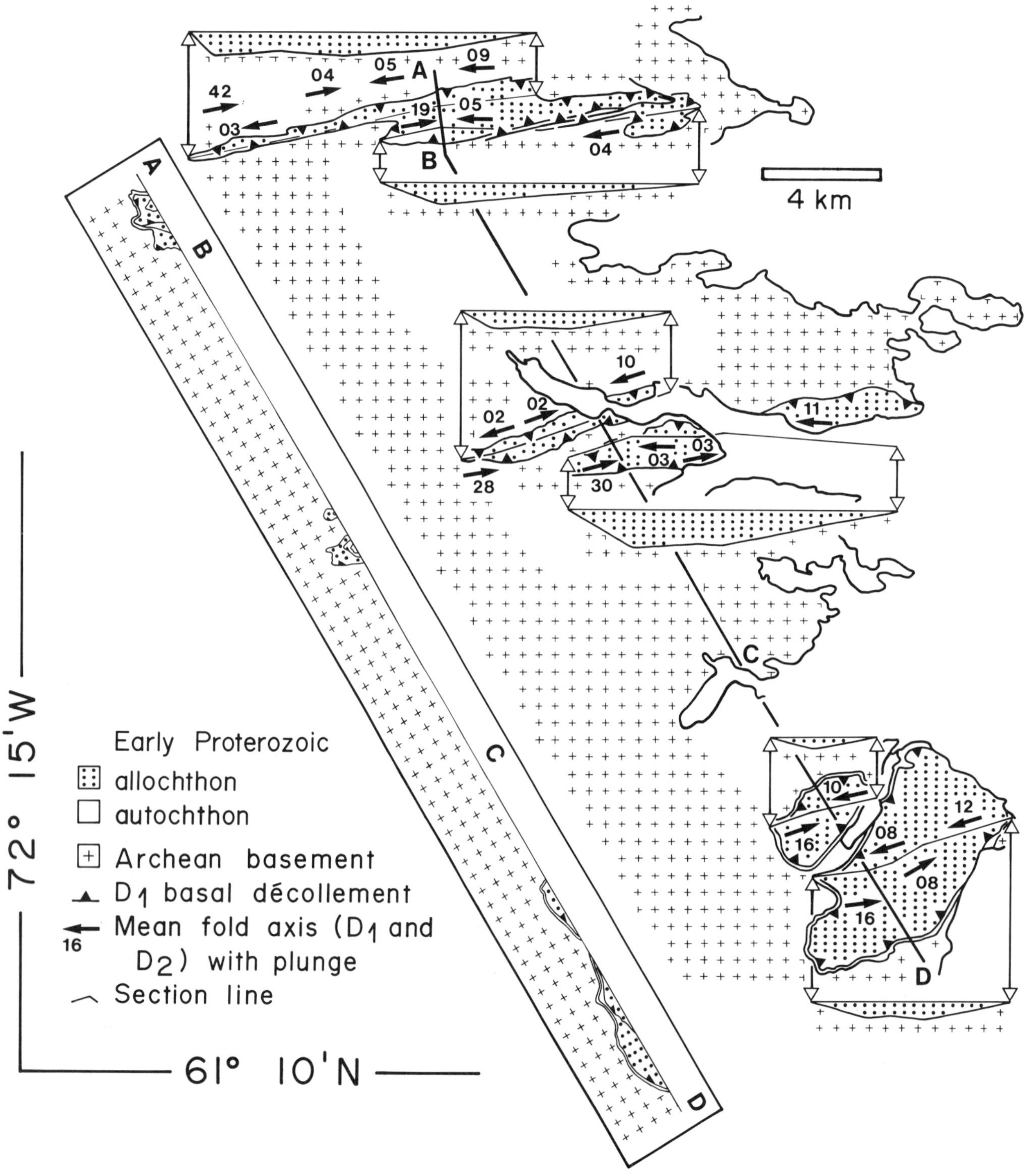

Figure 17. *Strike-parallel cross-section and down-plunge constrained, strike-perpendicular cross-sections of the western outliers shown on Figure 6. The cross-sections document the D_2–D_3 fold interference pattern responsible for the preservation of the early Proterozoic outliers.*

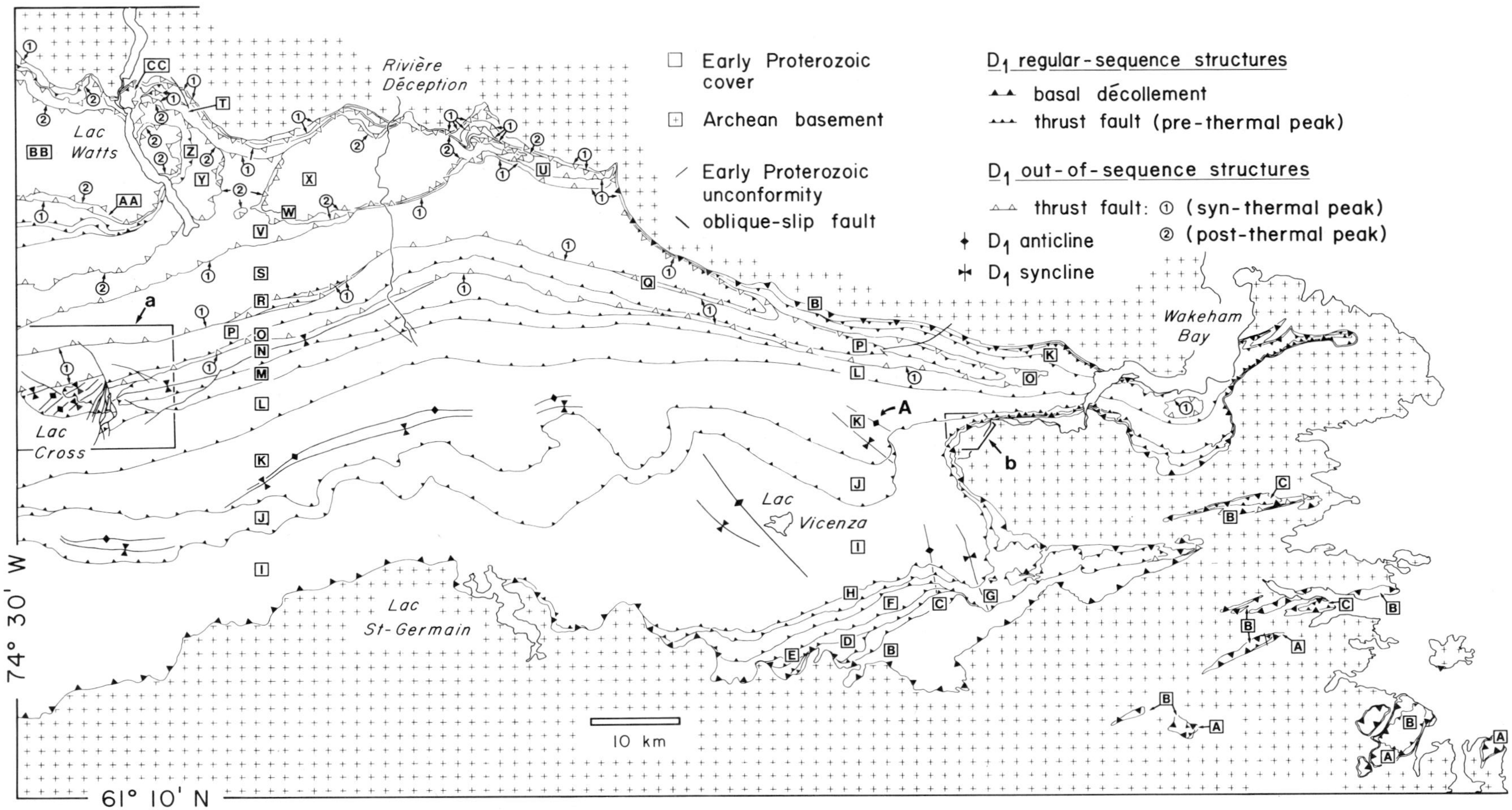

Figure 18. *Compilation of D_1 thrust faults and associated folds for the eastern portion of the Cape Smith Belt. Regular-sequence and out-of-sequence thrust faults are distinguished on this map by closed and open thrust fault teeth respectively. Boxed letters label thrust sheets referred to in the text. Areas "a" and "b" refer to areas shown by Figures 25 and 19 respectively.*

resulted in boudinage of a peridotite sill and consequent flow of the overlying material, including the inactive thrust G (Fig. 18), into the interboudin gaps. These observations indicate that older, inactive thrusts were deformed in the hanging walls of structurally lower faults and, therefore, that the early D_1 thrust belt grew in a piggyback-stacking sequence.

The geometry of the early D_1 thrust belt was largely influenced by the structure and stratigraphy of the underthrust rift margin. Examples include (1) thrusts reutilizing rift margin normal faults (e.g., Jackson, 1980; Gillcrist *et al.*, 1987) and (2) geometry of thrust sheets determined by basalt-to-sedimentary rock ratio in the deforming stratigraphy. The apparent reuse of a north-dipping normal fault by the décollement is suggested by the abrupt truncation of the southern margin rift-sedimentary rock duplexes (thrust sheets C to H, south of lac Vicenza, Fig. 2) against the décollement (Fig. 15). The décollement hanging wall ramp may have developed at the south margin of the rift basin containing the micaceous sandstone unit (unit 3, Fig. 2; see also Fig. 13).

Two distinct thrust-sheet geometries developed in the most external portion of the map area (faults A to L, Fig. 2), depending on whether dominantly sedimentary rocks or basalts were being imbricated. Thrust sheets characterized by rift sedimentary rocks overlain by rift basalts are typically

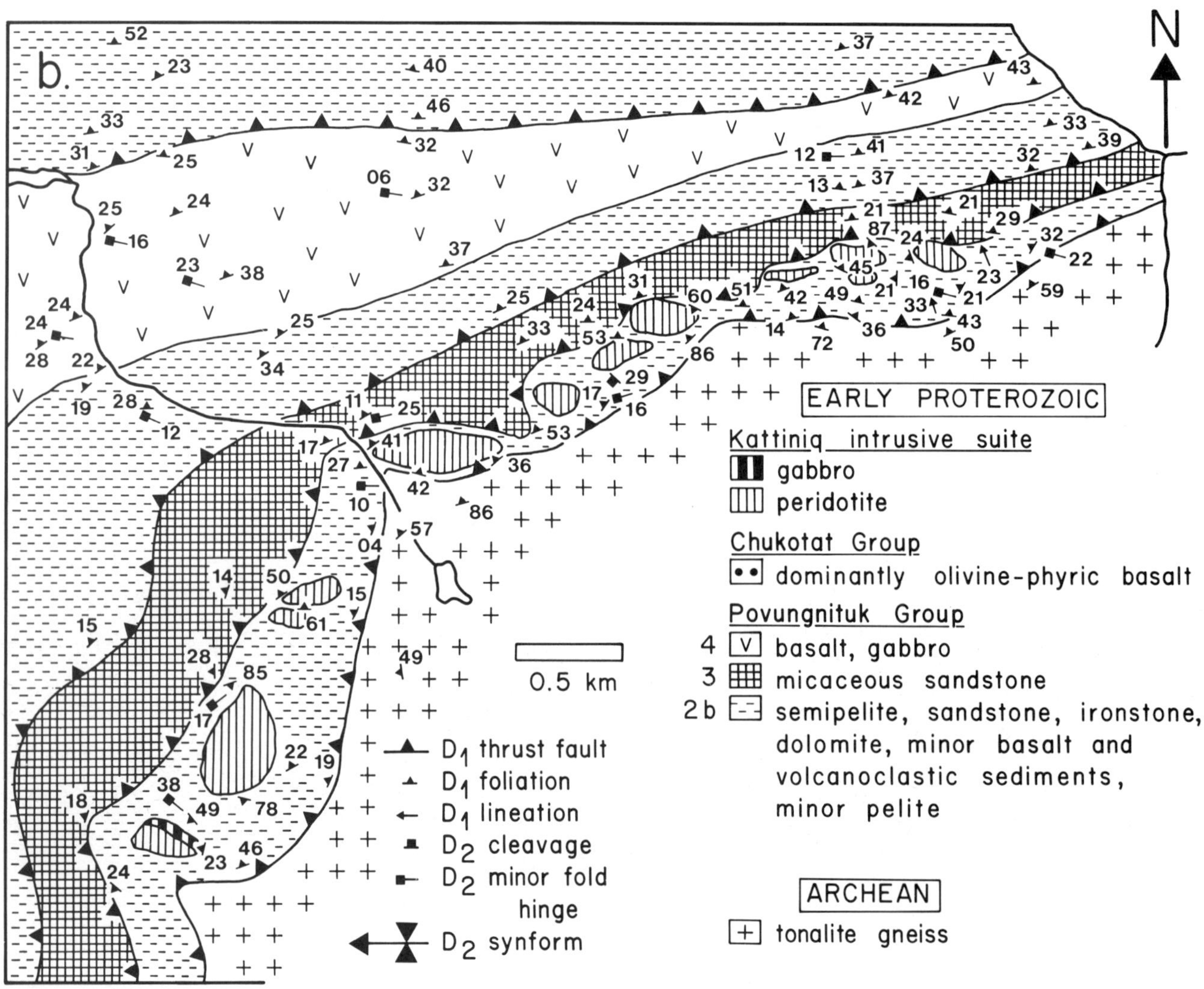

Figure 19. *Detailed geological map of a boudined peridotite sill. Flow of the overlying thrust sheet into the interboudin gaps indicates that this ductile deformation (resulting from synmetamorphic movement along the basal décollement) must have occurred after emplacement of the thrust sheet. The location of the map is indicated in Figure 18 by "b."*

composed of at least 75% basalt. These sheets are 3–7 km thick, laterally extensive, and form an imbricate fan geometry (Fig. 15). In contrast, the imbricates that developed solely in the rift sedimentary rocks (south of lac Vicenza, Fig. 2) are much narrower (0.4–2.5 km), restricted in areal extent by thrust ramps, and form duplex structures (Fig. 15).

Stage 2: basal shear zone deformation. A synmetamorphic shear zone recording top-to-the-south movement has been mapped immediately adjacent to the basal décollement in the southern and eastern parts of the Cape Smith Belt (Lucas and St-Onge, 1986; St-Onge *et al.*, 1986). The extent of the basal shear zone was outlined using several field- and thin section-scale criteria, including (1) the presence of a well-developed tectonic foliation (Fig. 20) and/or stretching lineation; (2) parallelism of all planar features (e.g., bedding, foliation, quartz veins; compare Figs. 20 and 21); (3) demonstrable or inferred strain-induced grain-size reduction; (4) the presence of noncoaxial flow indicators; (5) boudinage of competent bodies (e.g., the boudined peridotite sill illustrated in Fig. 19); and (6) rotation of shear zone fold axes toward the transport direction (sheath fold development, Cobbold and Quinquis (1980)). Applying these criteria, the basal shear zone includes <1 to 5 km of highly strained clastic metasedimentary rocks and metabasalts directly above the basement–cover contact, but only the uppermost 0.5–3 m of footwall basement gneisses (Fig. 4).

Deformation and metamorphic processes associated with shear zone development have been identified through detailed microstructural study (Lucas, 1990). This work has shown that mylonites (Sibson, 1977) were produced in the shear zone from hanging wall sedimentary rocks and basalts and footwall gneisses as a result of grain-size reduction by dislocation creep mechanisms (Tullis and Yund, 1985). The shear zone crosscuts or deforms thrust faults of the early D_1 system (Fig. 4) and, therefore, must postdate fault movement. This is further supported by the observation that while metamorphic minerals overgrow early D_1 fault zone foliations above the basal shear zone, they are also growing and deforming within the shear zone. However, overgrowth of the shear zone foliation by the highest grade minerals indicates that the termination of shear deformation occurred at peak thermal conditions.

The top-to-the-south sense of movement for the basal shear zone has been documented by a suite of kinematic indicators found both above and below the basement–cover contact. A suite of indicators was employed to avoid the potential ambiguities inherent with some individual kinematic indicators (e.g., shear bands; see Harris and Cobbold (1985)). However, all observations consistently gave the same south-directed sense of shear. The kinematic indicators that were used include (1) C/S fabrics (Berthé *et al.*, 1979); (2) shear band foliations (Platt and Vissers, 1980; White *et al.*, 1980); (3) porphyroblasts recording syngrowth rotation in the inclusion trails; and (4) asymmetric folds of the shear zone fabric showing various states of fold axis rotation into the shear direction.

Figure 20. *Sandstone–semipelite sequence with a strong bedding-parallel foliation from the basal shear zone in the Wakeham Bay area. Sedimentary rocks are from the same tectono-stratigraphic unit as those shown in Figure 21. The width of the area shown is 3 m.*

Figure 21. *Siliciclastic sedimentary rocks near the top of the deformation zone developed during movement on the underlying early D_1 thrust fault (approximately 10 m below the bottom of the photograph; thrust J, Fig. 2). Cleavage is absent from bed A (arkosic sandstone), subparallel to bedding in bed B (semipelite), and oblique to bedding in bed C (sandstone), reflecting variations in the grain size and quartz/mica ratio. The thickness of bed A is approximately 0.5 m.*

Stretching lineations are well developed in the basal shear zone and record a mean north-northwest trend where unaffected by subsequent deformation (e.g., along the southern margin and the western boundary of the map area, Fig. 22). The consistent presence and asymmetry of kinematic indicators in the XZ plane of shear zone finite strain ellipsoids demonstrate that the flow direction is parallel to the stretching (i.e., finite extension) direction. The approximate parallelism of the early D_1 thrust transport direction (documented above) and the subsequent shear-flow direction suggest that the south-southeastward tectonic transport direction did not vary significantly throughout the D_1 event. Departures of the stretching lineations from their mean north-northwest trend (Fig. 22) can be attributed to D_2 and D_3 deformation of the thrust belt. D_1 foliations and lineations were passively rotated and transposed into the axial plane of tight D_2 folds during penetrative D_2 deformation. This is well illustrated along the tight northern D_2 synform extending from Wakeham Bay to the west (Fig. 22; see also Fig. 15 for the D_2 fold geometry). In addition, dispersion of lineation trends occurred during D_2 and D_3 folding of the shear zone fabrics even without any significant penetrative strain. A good example of this is found in the lac Watts area (Fig. 22), where the D_2–D_3 dome-and-basin fold interference pattern is well developed near the basement/cover contact.

Basal shear-zone deformation of Archean basement gneisses below the décollement is limited to the uppermost 0.5–3 m (Fig. 5). Within the shear zone, all Archean fabrics are transposed into parallelism with the hanging wall foliation. Basement shear zone metamorphism is characterized by retrogression of feldspars and hornblende to produce the hydrous phases muscovite, biotite, and chlorite. The zone of retrograde basement metamorphism generally extends beyond the zone of ductile deformation (approximately 1 m farther into the basement gneisses). The initial growth of phyllosilicates may have preceded penetrative deformation (Lucas, 1990), suggesting that the weakening effect of phyllosilicates in the quartz–feldspar gneisses (see Tullis and Mardon, 1984; Jordan, 1987) is required to initiate ductile deformation. Consequently, the relatively restricted (<1–3 m) involvement of basement in the basal shear zone may result directly from the limited penetration of the retrogression (hydration) front. Bartley (1982) reached the same conclusion in studying the cause of limited involvement of the un-

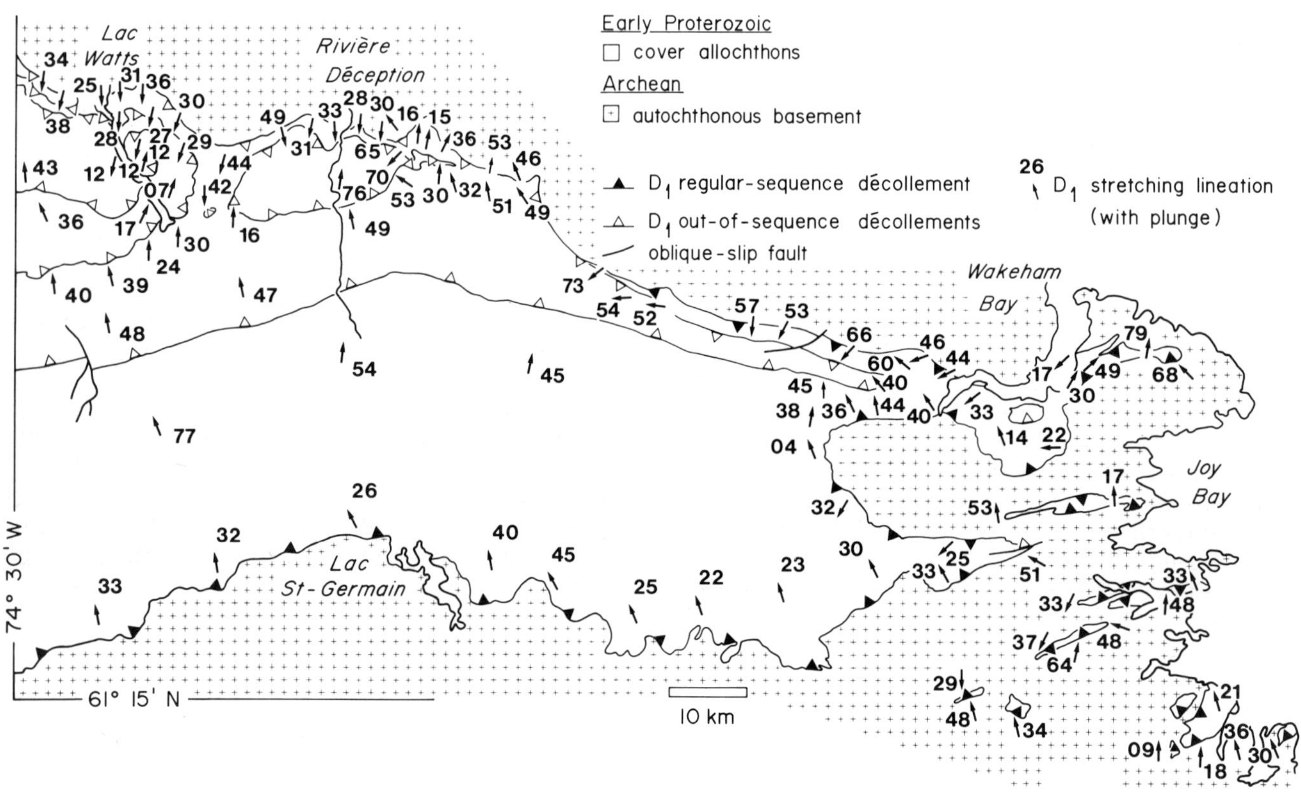

Figure 22. *Map of the mean orientations of D_1 stretching lineations. Each arrow represents the Fisher vector mean of stretching lineation orientations in a domain. Arrows are plotted in the centre of the domain.*

derthrust Baltic Shield basement in the Caledonian orogeny.

Stage 3: synthermal peak out-of-sequence thrusting. Several generations of out-of-sequence thrust faults (Fig. 18) reimbricate the internal part of the thrust belt following development of the basal shear zone. A number of independent criteria have been used to identify out-of-sequence thrust faults (after Searle, 1985): (1) thrusts cutting down section in the transport direction, resulting in elimination of stratigraphic section in the footwall; (2) truncation of previously developed thrust structures in the footwall; (3) syndeformation growth of distinct syn- or postthermal peak metamorphic assemblages in the ductile shear zones associated with the faults; and (4) truncation of metamorphic isograds. Early D_1 thrust faults may have been reused as out-of-sequence faults, although ductile deformation of both the hanging wall and footwall during out-of-sequence fault movement makes this difficult to constrain. In this section, the outcrop- to map-scale geometry of the synthermal peak out-of-sequence faults is examined, whereas that for the postthermal peak faults is discussed in the following section.

Ductile shear zones occur in both the hanging wall and footwall of the out-of-sequence faults (Fig. 4; Lucas, 1989). The out-of-sequence faults are distinguished from their associated shear zones for the following reasons. "Fault" refers to the discrete surface marking the boundary between the hanging wall and footwall (Fig. 18), and is used to describe map-scale structural geometries. In contrast, "shear zone" refers to the zone of penetrative noncoaxial deformation extending above and below the fault that accomplishes the bulk shear displacement of the hanging wall relative to the footwall (Fig. 4).

The synthermal peak shear zones include the entire northern margin domain of composite basement–ophiolite imbricates, and taper toward the south to narrow widths (<5 m) in the transitional crust basalts (Fig. 4). Outcrop- and thin section-scale studies of shear zone lithologies (Lucas, 1990) show that the highest temperature (i.e., thermal peak) minerals are either deformed or growing in the layer-parallel foliation. Ductile deformation of both autochthonous and allochthonous basement gneisses in the synthermal peak shear zones is described below.

The synthermal peak system of out-of-sequence thrust faults extends from the lac Watts – rivière Déception area of the northern margin to the lac Cross – Wakeham Bay area of the midbelt (Fig. 18). A consistent top-to-the-south transport direction is recorded for the out-of-sequence thrusts. The kinematics of the out-of-sequence faults have been documented with an analysis of stretching lineation orientations (Fig. 22) and with observations of shear sense indicators (St-Onge *et al.*, 1988; Lucas, 1989). The synthermal peak system is characterized by two important structural processes: (1) transferral of basement slices into the thrust belt, resulting in reimbrication of the ophiolitic thrust sheets (Figs. 23 and 24); and (2) southward ramping of the lowermost out-of-sequence fault through at least 7.5 km of rift and transitional crust thrust sheets (Figs. 15 and 18). Minimum aggregate displacement estimates for the synthermal peak thrusts are on the order of 50–100 km. These estimates are based on measurements of fault length parallel to the movement direction (i.e., in the cross-section, Fig. 15), given that no hanging wall and footwall cutoffs can be matched. Deformation of early D_1 structures during out-of-sequence thrusting has been documented at both relatively high structural levels (lac Cross area, Fig. 18) and low structural levels (lac Watts – rivière Déception, Fig. 18).

Early D_1 thrust sheets were folded, faulted, and internally strained at relatively high structural levels as a result of out-of-sequence thrusting. In the lower greenschist grade lac Cross area (Figs. 18 and 25), a fault-propagation fold (Suppe, 1983) developed in front of an out-of-sequence fault and folded early D_1 thrust faults. The fault-propagation fold was subsequently sliced by the out-of-sequence fault, stranding the footwall syncline mapped at lac Cross (Fig. 25). Tightening of the footwall syncline during deformation related to further displacement along the out-of-sequence fault resulted in the development of (1) a north-dipping cleavage transecting the fold and (2) a south-verging out-of-syncline thrust in the core of the fold (Fig. 25). In summary, footwall deformation related to the out-of-sequence thrust in the relatively low-grade lac Cross area is primarily limited to the truncated and flattened fault-propagation fold.

The lac Watts – rivière Déception area is characterized by out-of-sequence thrusting of composite ophiolite–basement thrust sheets at amphibolite grade conditions. The synthermal peak thrusts along the northern margin form a relatively complex system of laterally linked imbricates that truncate the old basement–cover décollement in several locations. Five basement imbricates occur east of rivière Déception (Fig. 23), and are interleaved with deep-water sedimentary rocks and ophiolitic basalts and gabbros (Fig. 24). The basement slices are narrow (<200 m thick) and laterally discontinuous, with strike lengths ranging from only several tens of metres to over ten kilometres (Fig. 23). The lowermost out-of-sequence thrusts east of rivière Déception (Fig. 18) are linked to the southward continuation of the synthermal peak fault system (Figs. 16 and 24), which eventually climbs to the

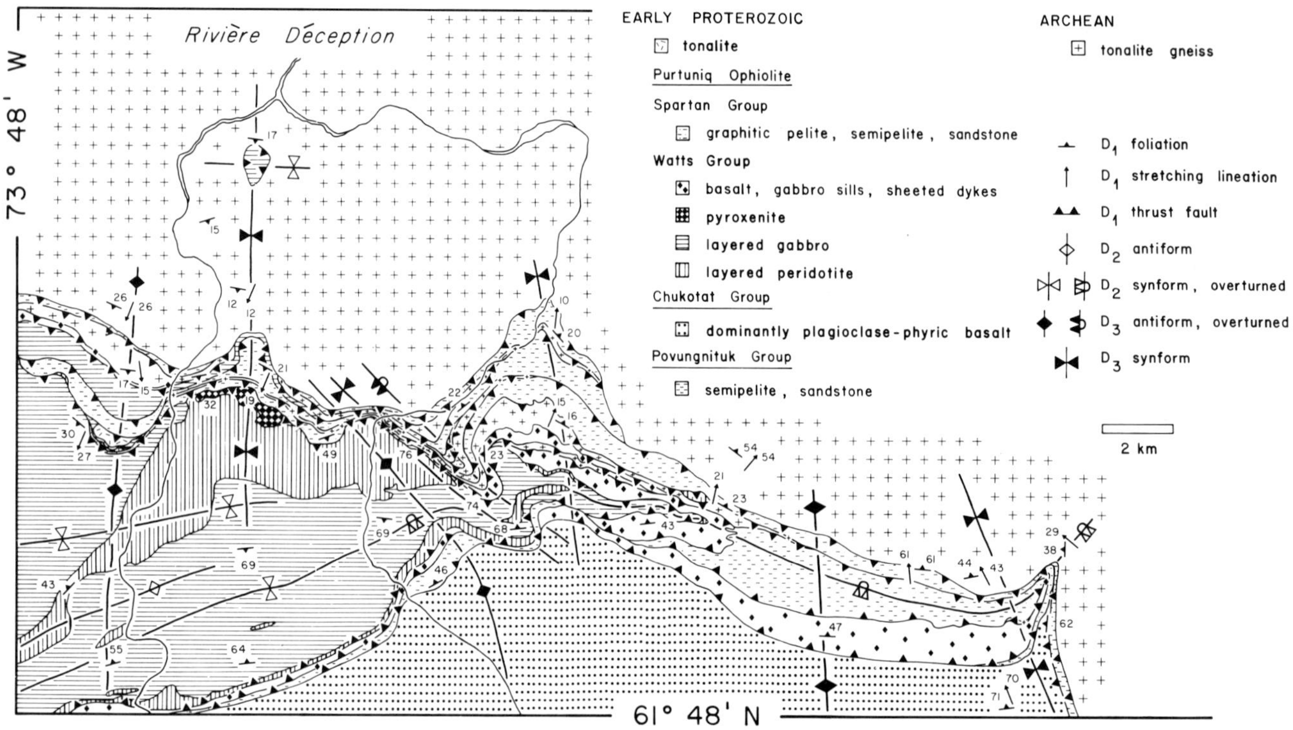

Figure 23. *Detailed geological map of the rivière Déception area (northern margin of the eastern Cape Smith Belt, Fig. 2) showing the basement imbricates associated with the synthermal peak out-of-sequence fault system.*

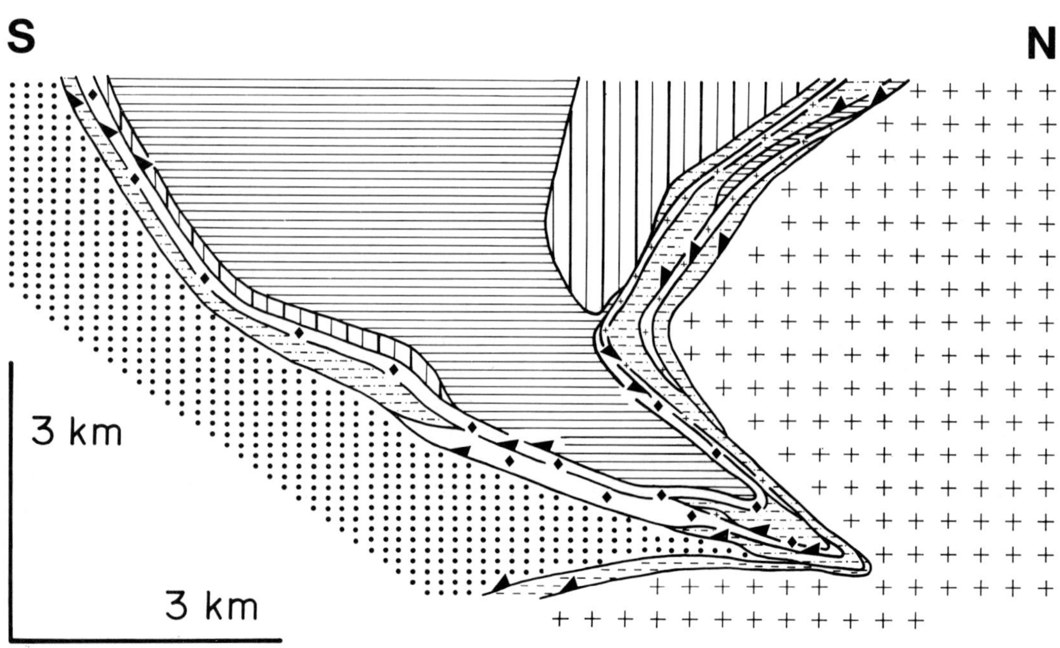

Figure 24. *Detailed down-plunge constrained cross-section of the area illustrated in Figure 23. The legend is given in Figure 23.*

structural level of lac Cross. In contrast, structurally higher synthermal peak thrusts along the northern margin are cut off by an overlying postthermal peak thrust fault (Figs. 23 and 24).

Archean basement is involved in synthermal peak shear zones (Fig. 4) in two settings: (1) allochthonous slices within the thrust belt (Fig. 23) and (2) up to 200 m of parautochthonous basement below the basal out-of-sequence décollement. Ductile strain in both settings probably accumulated during both basal shear zone deformation and out-of-sequence deformation. The allochthonous basement slices retain no Archean fabrics, and are characterized by a protomylonitic foliation in their interior that grades into an ultramylonitic foliation adjacent to their top and bottom contacts. The top contact is interpreted to be the early D_1 basal décollement, whereas the bottom contact of the basement slices is interpreted as an out-of-sequence thrust fault. The basement shear zone below the out-of-sequence décollement (Fig. 4) is marked by a relatively extensive zone (up to 200 m) of protomylonites, contrasting markedly with the relatively narrow shear zone (1–3 m) below the basal décollement in the Wakeham Bay area (Fig. 4).

Stage 4: postthermal peak out-of-sequence thrusting. The final reimbrication and southward displacement of the ophiolitic rocks occurred along two major postthermal peak thrust faults found in the lac Watts – rivière Déception area (Y and BB, Fig. 18). The postthermal peak faults bound klippen of ophiolitic basalts, sheeted dykes, and mafic and ultramafic cumulates, and are the structurally highest faults preserved in the northern out-of-sequence domain (Figs. 15 and 16). The out-of-sequence character of the postthermal peak thrust faults is documented by (1) truncation of synthermal peak thrusts in their footwalls (Figs. 15 and 16); (2) truncation of thermal peak mineral isograds in their footwalls (Bégin *et al.*, 1988); and (3) growth of postthermal peak (retrograde) mineral assemblages in the ductile shear zones adjacent to the faults. The two principal postthermal peak thrusts do not intersect or interact in the study area (Fig. 18). As a result, their relative ages cannot be determined.

Two styles of bulk shear deformation characterize the hanging walls of the postthermal peak thrusts. Ductile shear zones developed directly above the out-of-sequence faults, whereas narrow shear bands deformed the interior of the hanging wall thrust sheets (Fig. 13). The shear zones are marked by 1–10 m of mylonite and an almost complete greenschist overprint on amphibolite grade assemblages. Rotated, internally foliated boudins of anorthosite are found in the gabbroic mylonites of one of these shear zones (Fig. 26). The boudins are interpreted to result from disaggregation of relatively competent anorthositic layers during ductile flow in the enveloping gabbroic layers (Lucas, 1990). The dramatic weakening of the gabbroic layers appears to result from softening due to retrograde reactions producing weaker and finer grained products (actinolite, chlorite) than reactants (hornblende) (Brodie and Rutter, 1985).

The development of retrograde fabrics occurs heterogeneously throughout the hanging wall thrust sheets above the mylonite zones. The synthermal peak foliation in the hanging walls is eworked into a retrograde foliation along small-scale (millimetre to metre) shear zones (Fig. 13; Lucas, 1990). The shear zones anastomose through the ophiolitic rocks and as a result have a shear band geometry (Platt and Vissers, 1980). Retrograde reactions along the shear bands result in partial replacement of amphibolite grade assemblages by greenschist grade minerals (Bégin *et al.*, 1988).

Geometry and mechanisms of D_2 and D_3 basement-involved folding

D_2 east–west trending map-scale folds deform the thrust belt, its footwall basement, and the thermal peak mineral isograds. The map-scale folds are approximately coaxial with the commonly observed outcrop-scale D_2 structures (Fig. 27). All D_2 structures are refolded by north–south trending D_3 folds of basement and cover. The overall effects of the two postthrusting deformation episodes are to (1) isolate the thrust belt from its root zone to the north by an east–west trending basement-cored antiform (Fig. 1) and (2) create a doubly plunging synclinorium containing the thrust belt (Fig. 1).

The geometry of the D_2 map-scale folds is best observed in the down-plunge constrained cross-section (Fig. 15). At the level of the basal décollement in the lac St-Germain – Wakeham Bay area (Fig. 2), the D_2 macroscopic folds produce a striking "W-shaped" geometry. These folds have a relatively high amplitude-to-wavelength ratio (20 km/30 km), and a profile characterized by two cuspate synclinal keels separated by a lobate median antiform. This interface geometry is commonly observed in the deeply eroded cores of ancient mountain belts (e.g., Helvetic Nappes of the Swiss Alps, Ramsay *et al.* (1983); Wopmay Orogen, King, 1986). It has been attributed to finite amplification of buckling instabilities at an interface between two non-Newtonian materials with different viscosities (Ramsay, 1967; Dieterich and Onat, 1969; Smith, 1975). A minimum strike length of 150 km for the "W-shaped" D_2 profile is suggested by gravity data from a detailed survey conducted 10 km west of the map area (Feininger *et al.*, 1985).

The D_2 map-scale geometry at the structural levels exposed in the western and northern parts of

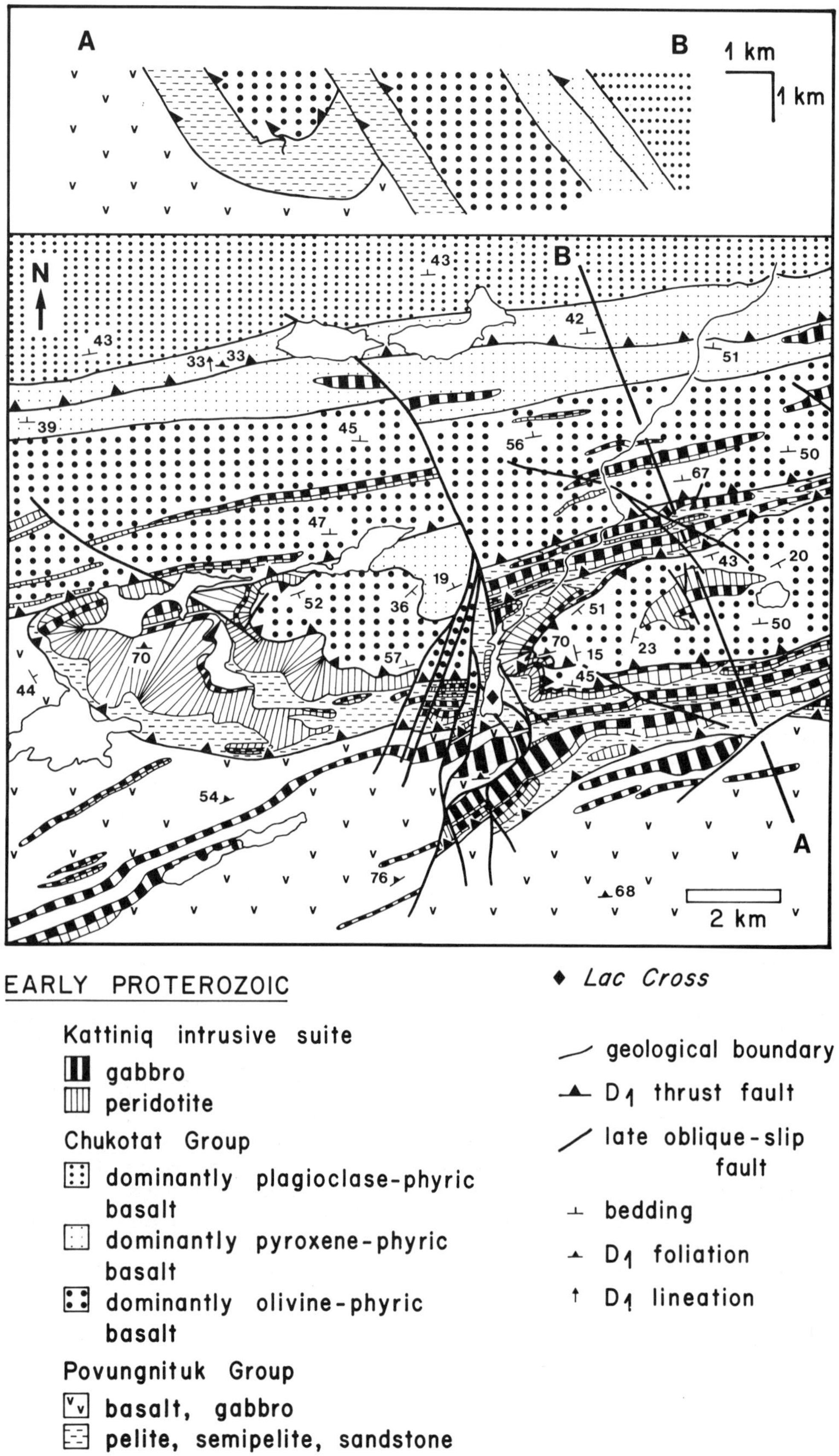
A
B
1 km
1 km
N
B
A
2 km
EARLY PROTEROZOIC
Kattiniq intrusive suite
gabbro
peridotite
Chukotat Group
dominantly plagioclase-phyric basalt
dominantly pyroxene-phyric basalt
dominantly olivine-phyric basalt
Povungnituk Group
basalt, gabbro
pelite, semipelite, sandstone
Lac Cross
geological boundary
D1 thrust fault
late oblique-slip fault
bedding
D1 foliation
D1 lineation

the map area (Figs. 2 and 15) departs significantly from the "W-shaped" profile that underlies the belt. At the relatively high structural level of the lac Cross area (Fig. 18), D_2 map-scale deformation is manifest only by a north-dipping monocline, presumably steepened from its D_1 attitudes. The lac Watts – rivière Déception area (Fig. 2) is characterized by a map-scale D_2 synform that occurs north of the Wakeham Bay area (Fig. 28). The Wakeham Bay synform is a tight to synformal keel isoclinal fold with a steeply dipping axial surface (Fig. 15). The lac Watts – rivière Déception synform changes in character substantially along strike. In the lac Watts area, it is a relatively open structure with an upright axial surface (Fig. 28), whereas east of rivière Déception it becomes overturned with an axial surface dipping at 30° to the north (Figs. 23, 24, and 28). The overturned segments of this synform and the Wakeham Bay synform and the Wakeham Bay synform are laterally discontinuous structures reflecting local oversteepening of the basement–cover contact along the northern limb of the synforms. The variable dip of the northern synformal limbs may result from lateral inhomogeneities in the growth rate of the basement-cored antiform to the north, possibly due to local gravitational enhancement of D_2 buckling.

Figure 26. *Ultramylonite derived from retrograde gabbro with boudins of anorthosite (white, marked "A"). Deformation and retrograde metamorphic reactions occurred in response to movement on the underlying postthermal peak thrust fault BB (<3 m below the photograph; see Fig. 2). Foliation developed during synthermal peak shear is preserved within the large white anorthosite boudin. The chaotic appearance is due to intense microfolding of the retrograde schistosity during the postthrusting D_2 event. The long dimension of the larger plagioclase-rich block is 75 cm.*

An important question concerning D_2 deformation is how the significant strain associated with the map-scale folds (50–100%) is accommodated at the outcrop and grain scales. Foliated lithologies from the D_1 ductile shear zones (Fig. 4) display D_2 minor folds at wavelengths ranging from metres for folded layers (Fig. 29) to less than a millimetre for individual grains. D_2 crenulation cleavages are associated with many of the minor folds (Fig. 29) and can display growth of new phyllosilicates or amphiboles, or evidence of pressure solution strain. In contrast, D_2 minor folds are relatively rare structurally above or below the D_1 shear zones. A plausible mechanical explanation is that lithologies such as basement gneisses or undeformed basalts lack rheological layering of appropriate layer thickness or competency contrast to form buckle folds at the centimetre to metre scale (Johnson, 1980). In these lithologies, D_2 shortening must have been accommodated by some combination of cleavage development, formation of an array of shears, or a more homogeneous grain-scale strain. A detailed study of basement deformation beneath the pinched synforms of the northern outliers (Figs. 6 and 17) has revealed a spaced D_2 cleavage persisting for at least 10 m below the basement–cover contact. In contrast, D_2 fabrics are not observed in the basement underlying the southern outliers, which are characterized by open D_2 map-scale folds (Figs. 6 and 17). This suggests that a minimum amount of D_2 shortening is required to produce penetrative deformation in more competent units, and that below this amount, fold-accommodation strain is accomplished very inhomogeneously.

D_3 deformation produced a suite of north- to northwest-trending folds deforming both basement and cover at wavelengths of 1 cm to 100 km (Fig. 28). As discussed above, erosion following D_3 folding is responsible for the significant structural relief of the D_1–D_2 belt exposed in the eastern Cape Smith map area. Two fold wavelengths dominate the D_3 deformation at map-scale: (1) 50 km wavelength folds, which appear to deform the entire belt tectono-stratigraphy and the basement (e.g., "A," Fig. 28), and (2) folds at 5–10 km wavelengths, which are restricted to the area of the basement–cover contact (e.g., south of lac Vicenza, Fig. 28) or to Povungnituk Group thrust sheets with mafic or ultramafic sills intruding sediments (e.g., "B," Fig. 28). A possible explanation for the limited occurrence of the smaller wavelength folds is that they are restricted to zones in the thrust belt tec-

Figure 25. *Detailed geological map and down-plunge constrained cross-section of the lac Cross area. Note the truncation of the early D_1 (regular-sequence) thrusts by the north-dipping, out-of-sequence thrust fault. The location of the detailed map is indicated on Figure 18 by "a."*

tono-stratigraphy that have appropriate rheological layering to buckle. D_3 strain at structural levels that do not display outcrop- or map-scale folds must have shortened by either grain-scale deformation or small-scale faulting, as in analogous D_2 "undeformed" areas.

Two major hypotheses exist to explain the origin of the D_2 and D_3 map-scale folds of basement and cover: (1) gravitational collapse of density inversions established during D_1 thrusting (i.e., diapirism; Ramberg, 1981) and (2) crustal-scale buckling of rheological interfaces in response to horizontal compression (Biot, 1961; Fletcher, 1974; Zuber *et al.*, 1986). The centrifuge diapir models of Dixon and Summers (1983) predict patterns of finite strain and kinematics that may be found if the gravitational collapse hypothesis is correct. Two predictions of these models with significant implications for the D_2 and D_3 folds are (1) that extensional structures should occur across the crests of the antiforms and (2) that material in a subsiding trough should extend vertically along the basement–cover contact and flow into the hinge.

Field evidence on both mesoscopic and macroscopic D_2 folds appears to refute the two relevant predictions of the gravitational collapse hypothesis, and suggests that the structures may have developed in response to horizontal compression. The occurrence of D_2 outcrop-scale folds and associated crenulation cleavages across the median antiform at the eastern end of the belt (Fig. 27) indicates that the crest of the antiform did not experience extensional strain, as predicted by the diapir models. Furthermore, the conspicuous limb asymmetry of the D_2 outcrop-scale folds bears no direct relation to the map-scale D_2 fold geometry, even though the asymmetry maps out into distinct domains (Fig. 27). A reasonable explanation for the asymmetry changes on the limbs of major D_2 folds (e.g., north of "B" in Fig. 27) is that the minor folds developed prior to final D_2 deformation and were rotated through the hinges of the major D_2 structures. This is supported by a clockwise rotation of initially coplanar D_2 cleavage (axial planar to south-verging folds) from the north to the south limb of the Wakeham Bay syncline (see stereonet A, Fig. 27). Hinge migration of this scale through a synclinal keel is inconsistent with the kinematic patterns predicted by Dixon and Summers' (1983) diapir models. These observations, coupled with

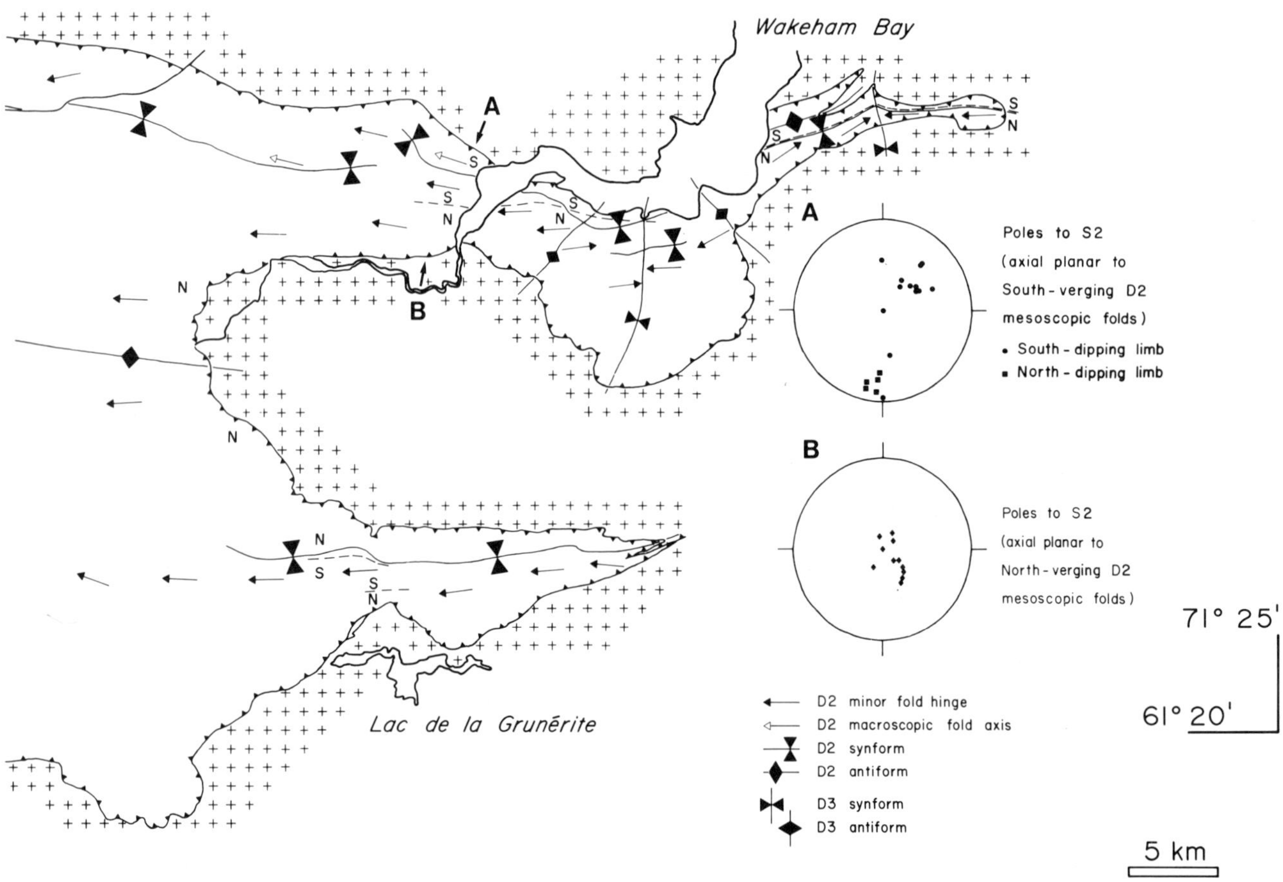

Figure 27. *Compilation of mesoscopic and macroscopic D_2 fold axial traces in the Wakeham Bay area. Hinges are grouped into north (N) or south (S) verging domains based on fold asymmetries.*

the elongate form (150 km strike length) of the "W-shaped" profile, support the conclusion that the D_2 map-scale folds originated from buckling in response to horizontal compression. However, the density inversion established during D_1 (mafic-dominated allochthons overlying tonalitic basement) may have enhanced the amplitude growth rate of buckling instabilities initiated at the basement–cover contact.

THERMAL HISTORY

Early D_1 thermal conditions

The thermal evolution of the Cape Smith Belt started with the initiation of early D_1 regular-sequence thrusting. *P–T* conditions during early D_1 crustal thickening cannot be established at the base of the thrust belt because of overprinting during synmetamorphic basal shear zone deformation. However, at structural levels above the basal shear zone (Fig. 4), the high strain zones associated with individual early D_1 thrust faults are preserved. The mineral assemblages and microstructures found in the early D_1 fault zones north of lac Vicenza (Fig. 18) suggest that rocks were deformed at temperatures of no more than 400°C (Lucas, 1989). Given the 5 km average thickness for thrust sheets in this area (Fig. 16), early D_1 pressure estimates for this structural level range from 3 to 5 kbar (300 to 500 MPa). These constraints on the early D_1 *P–T* conditions suggest that at the onset of D_1 thrusting there was no significant residual heat from lithospheric thinning and/or rift margin magmatism.

Syn-D_1 prograde decompression *P–T* paths

In the regular-sequence thrusting domain, the thermal peak mineral assemblages commonly postdate the establishment of the late D_1 basal shear zone. In the more internal out-of-sequence domain, thermal peak mineral growth is synchronous with deformation in the shear zones associated with the oldest out-of-sequence faults. Quantitative derivation of *P–T* paths of metamorphism from both domains allows the effects of out-of-sequence thrusting on the belt's thermal evolution to be studied.

Metapelitic layers within both thrusting domains contain the following maximum phase assemblage:

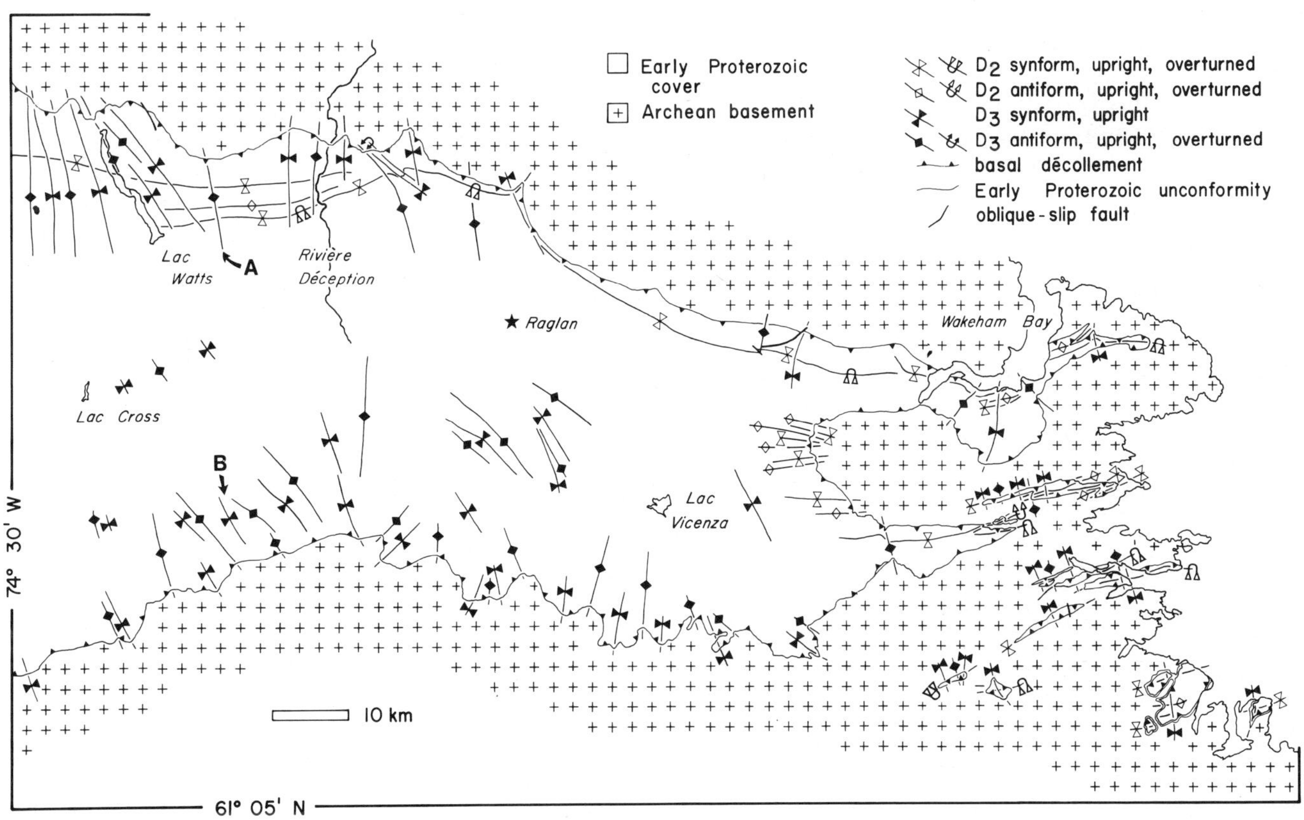

Figure 28. *Compilation of axial traces for D_2 and D_3 macroscopic folds in the eastern Cape Smith Belt map area.*

kyanite – garnet – muscovite – plagioclase – quartz ±staurolite±chlorite. Six-phase subsets of this assemblage (plus H_2O) are univariant in the seven-component system SiO_2–Al_2O_3–FeO–MgO–Na_2DO–K_2O–H_2O. Provided the garnets are chemically zoned, the subset assemblages offer the potential to model changing P and T conditions as recorded during garnet growth (Spear *et al.*, 1984).

Garnets from pelitic samples in the regular-sequence domain overgrow the dominant D_1 foliation in the rock and thus postdate basal shear zone deformation. The garnet interiors are characterized by straight inclusion trails from core to rim. The D_1 foliation in the matrix is defined by the dimensional and/or crystallographic alignment of quartz, tourmaline, opaque minerals, muscovite, and biotite. The fabric in the pelitic matrix is continuous with the inclusion trails within the garnets. Only the late D_2 crenulation of the matrix foliation postdates garnet growth. Other higher temperature minerals, such as kyanite, staurolite, and plagioclase, also overgrow the D_1 matrix foliation. *P–T* estimates derived by analyzing the garnet-bearing assemblages, therefore, constrain physical conditions that followed basal shear-zone deformation.

Garnets from the out-of-sequence domain contain sigmoidal inclusion trails of quartz, tourmaline, and opaque minerals, and are interpreted as syntectonic with respect to out-of-sequence thrusting. The fabric in the matrix is defined by the dimensional and/or crystallographic alignment of quartz, tourmaline, opaque minerals, muscovite, biotite, and kyanite. The matrix fabric is continuous with the inclusion trails in the garnet porphyroblasts, ruling out the possibility that garnet rotation completely postdates growth. *P–T* estimates derived from the garnet-bearing assemblages, therefore, document the evolution of physical conditions during synthermal peak thrusting (stage 3).

Figure 29. *Mesoscopic D_2 folds of in a well-foliated upper Povungnituk Group basalt. The spaced crenulation cleavage is associated with the D_2 folding. The hammer is 34 cm long.*

The samples selected for microprobe analysis were taken from several sites, both in the regular-sequence thrusting domain (samples 3–6, Fig. 3) and the out-of-sequence thrusting domain (samples 1–2, Fig. 3). The set of pelitic samples provides a broad sampling of both external (southern) and internal (northern) portions of the thrust–fold belt. In both thrust domains, the samples were taken from within 200 m of the basement–cover contact. For each of the six sections studied, the garnet with the greatest diameter was selected for analysis to minimize the "sphere effect" (Indares and Martignole, 1985). This approach ensures chemical analyses as close as possible to the true centre of the zoned mineral phase. The garnets were analyzed with an electron microprobe along rim–core–rim traverses at regular (0.1 mm) intervals.

Concentric compositional zoning is well developed in the six garnets analyzed. It is characterized by a continuous decrease in Mn and Ca and by a concomitant increase in Fe, Mg, and Mg/Fe ratio from core to near rim (Fig. 30). The pattern is approximately symmetrical with respect to the garnet core. These compositional trends correspond exactly to those predicted by a partial-equilibrium growth model (Tracy, 1982; Loomis, 1983; and references therein). In this model, the slow rate of volume diffusion in garnet effectively isolates the interior of the growing crystal from the rest of the rock. The absence of breaks or gaps in the zoning trends of the garnet interiors (Fig. 30) suggests that (1) the porphyroblasts are not the result of polymetamorphism (e.g., Rumble and Finnerty, 1974) and (2) that not more than one garnet-producing reaction is required (e.g., Thompson *et al.*, 1977). The univariant, seven-phase assemblage listed above is consistent with a reaction such as

staurolite + muscovite + quartz = kyanite + biotite + garnet + H_2O

being responsible for the garnet growth.

Peak-temperature estimates were obtained from the analyses of the garnet rims and matrix biotites based on the garnet–biotite Fe–Mg exchange equilibrium (Ferry and Spear, 1978; Hodges and Spear, 1982). Thermal peak pressure determinations were calculated with the analyses of garnet rims and matrix plagioclase based on the grossular–kyanite–anorthite–quartz equilibrium. The geobarometric calibration of Newton and Haselton (1981), as corrected by Ganguly and Saxena (1984), was employed, taking into account the modifications presented in Hodges and Spear (1982) and corrected by Hodges and Royden (1984).

With a starting P and T provided by the matrix – garnet rim assemblage for each sample, the thermodynamic modeling technique of Spear and Selverstone (1983) was used to "backtrack" and model the garnet chemical zoning from rim to core in terms of *P* and *T* increments. The resulting *P–T* paths are shown on Figure 31. For each sample, there is a systematic decrease in recorded *P* from high values derived for the garnet core to lower values derived for the garnet rim. In contrast, *T* estimates show a systematic increase from garnet core to a maximum at the rim. The *P–T* paths based on the zoned garnets (Fig. 31) thus record decompression and unroofing of the thrust–fold belt concomitant with prograde metamorphism. The individual *P–T* paths are similar to the initial decompression segment of the general *P–T* trajectory presented in England and Richardson's (1977) model of a tectonically thickened crust undergoing heating, uplift, and erosion.

The four pelites from the regular-sequence thrusting domain record progressively higher thermal peak temperatures (430–590°C) from external (southern) to internal (northern) sample locations (samples 6 to 3, Fig. 31). The maximum pressures recorded for the four samples increase from 6.5 kbar (650 MPa) for the most external sample (6, Fig. 3) to over 9 kbar (900 MPa) for the most internal sample (3, Fig. 3), consistent with a foreland tapering thrust wedge. The four paths are curvilinear and can be described as nested (Fig. 31). The nested *P–T* path morphology suggests that they experienced relatively similar decompression histories, possibly with uplift solely in response to erosion and isostatic unloading.

In contrast, the two samples from the out-of-sequence domain (samples 1–2, Fig. 3) yield significantly distinct *P–T* trajectories. Although paths 1 and 2 are also decompression trajectories (from garnet core to rim), they are characterized by steeper d*P*/d*T* prograde segments than those for the regular-sequence domain (samples 3 to 6, Fig. 31). The steeper d*P*/d*T* segments are probably a reflection of greater uplift and erosion rates (England and Thompson, 1984). Because the two garnets analyzed are syntectonic with respect to out-of-sequence (stage 3) deformation, it can be surmised that the apparently faster uplift rates experienced by these samples are a consequence of stage 3 deformation.

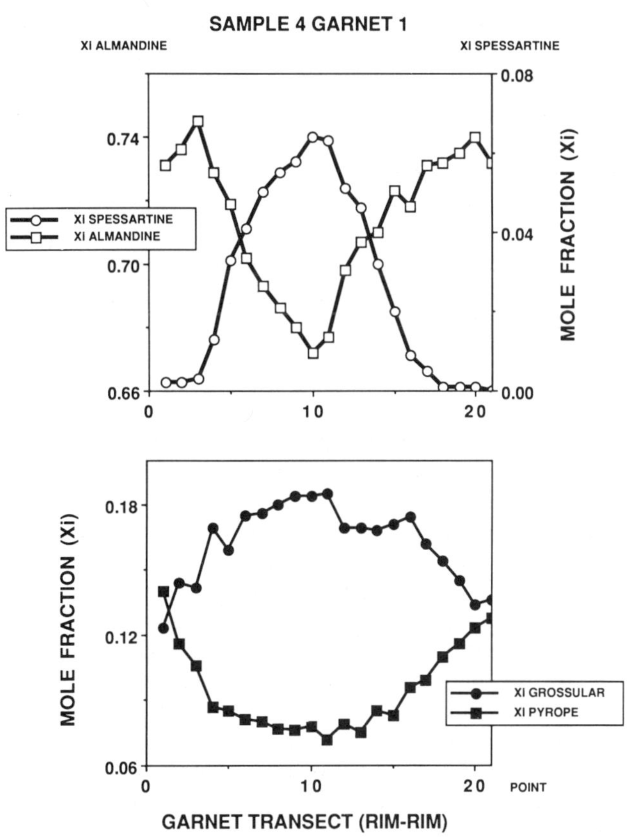

Figure 30. *Garnet zoning profile across sample number 4 (Fig. 3). Numbers on vertical axes represent mole fractions of the four garnet end-member components (almandine, pyrope, spessartine, and grossular). The horizontal axis represents the full diameter of the garnet grain (21 regularly spaced analysis points).*

Syn-D_1 thermal peak *P–T* conditions

The geometry of the thermal peak metamorphic culmination was established at the regional scale through systematic mapping of mineral as-

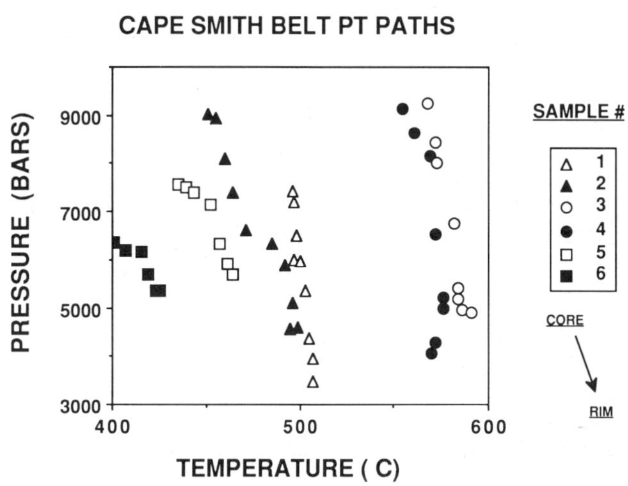

Figure 31. *Site-specific metamorphic* P–T *paths for the six pelite samples described in the text. Samples 3 to 6 are from the regular-sequence domain (Fig. 3). Samples 1 and 2 are from the out-of-sequence domain (Fig. 3). Paths are derived from the garnet compositional zoning profiles and show a pressure decrease and temperature increase from core to rim.*

semblages in the metabasites. Fieldwork (St-Onge *et al.*, 1986, 1987, 1988; St-Onge and Lucas, 1989) was followed by a detailed petrological study of the metamorphic mineral assemblages in mafic units of all tectono-stratigraphic groups (Bégin and Carmichael, 1987; Bégin *et al.*, 1988). The metamorphic culmination is outlined in the regular-sequence thrusting domain (Fig. 3) with the following sequence of mineral zones (Bégin *et al.*, 1988): (1) actinolite–albite, (2) hornblende–actinolite–albite, (3) hornblende–actinolite–oligoclase, (4) hornblende–oligoclase, and (5) hornblende–oligoclase–garnet/clinopyroxene. The five mineral zones are separated by four isograds (hornblende-in, oligoclase-in, actinolite-out, garnet/clinopyroxene-in) shown on Figure 3. The map traces of the hornblende-in, oligoclase-in, and actinolite-out isograds clearly show that they crosscut the early D_1 regular-sequence thrust faults and the basal shear zone (compare Figs. 3 and 4). The mineral isograds, projected on the cross-section (Fig. 16), constrain the possible three-dimensional geometry of the metamorphic culmination. As discussed earlier, the hornblende-in isograd is overturned immediately below a synthermal peak out-of-sequence thrust (O, Fig. 18). This indicates that thermal relaxation of the out-of-sequence thrust stack contributed significantly to the attainment of footwall thermal peak conditions.

In the out-of-sequence domain, the distribution of metamorphic mineral zones is, to a large extent, completely dependent on the location of both the syn- and postthermal peak out-of-sequence faults (Fig. 3). In the Chukotat Group, an oligoclase-in isograd separates an actinolite–albite zone from an actinolite–oligoclase±hornblende zone (Bégin *et al.*, 1988). The isograd is cut in several places by synthermal peak out-of-sequence faults (Fig. 3). In the northern portion of the thrust–fold belt, the Watts Group is characterized by thermal peak mineral zones ranging from hornblende–actinolite–albite to hornblende–oligoclase–garnet/clinopyroxene bearing assemblages (not distinguished at the scale of Fig. 3). In the shear zones associated with postthermal peak out-of-sequence faults, an actinolite–chlorite–albite assemblage overprints all pre-existing greenschist to amphibolite grade assemblages.

Post-D_1 *P–T* conditions

Adjacent to the metapelite employed for *P–T* path number 5 on Figure 31, semipelitic beds contain biotite, muscovite, and chlorite that grew during D_2 microfolding and crenulation cleavage development. In a few samples, garnet is also part of the D_2 mineral assemblage. This overprinting mineral assemblage enables the *P–T* conditions of D_2 deformation to be determined for this location (5 on Fig. 3, "A" on Fig. 16).

Temperature estimates were obtained from microprobe analyses of the (chemically unzoned) garnet and biotite based on the Fe–Mg exchange equilibrium (Ferry and Spear, 1978; Hodges and Spear, 1982). Pressure estimates are based on microprobe analyses of D_2 biotite, muscovite, and chlorite using the calibration of the biotite–muscovite–chlorite–quartz equilibrium by Powell and Evans (1983). Simultaneous solution of the geothermometer and geobarometer equilibria has yielded five estimates of *P* and *T* for five domains of garnet, biotite, chlorite, and muscovite. The *P–T* determinations cluster at 5 kbar (500 MPa) and 450°C, and are interpreted as the ambient conditions during D_2 deformation at this specific locality.

TECTONIC SUMMARY

Tectono-stratigraphic evolution

Paleogeographic relationships both within and between the five tectono-stratigraphic suites of the Cape Smith Belt have been constrained through careful study of stratigraphic facies, geochemistry of the igneous suites, and documentation of the structural geometry of the thrust belt. In particular, defining the stacking order of the D_1 thrust sheets (piggyback and out-of-sequence) has permitted qualitative palinspastic restoration of the pre-D_1 rift-margin (Fig. 13). The following discussion summarizes the evolution of the rift margin and the ophiolite in terms of the five tectono-stratigraphic suites and their paleogeographic relationships.

The tectono-stratigraphic record of (1) the autochthon, (2) external thrust sheets, and (3) midbelt thrust sheets within the Cape Smith Belt documents the evolution of an Early Proterozoic epicontinental rift that ultimately led to the formation of transitional crust (Hynes and Francis, 1982). Deposition of continent-derived sediments (Povungnituk Group) in an approximately east–west-trending basin is indicated by south to north paleocurrent observations in the autochthon. Within the external (southern) thrust sheets of the Povungnituk Group, ironstones, mafic volcaniclastic sedimentary rocks, semipelites, and proximal fan arkosic sandstones comprise a sediment-dominated sequence deposited in a proximal (inboard) rift setting. A more distal (outboard) rift sequence is preserved in more internal Povungnituk Group thrust sheets and includes proximal to distal fan sandstones and semipelites overlain by basalts. The south to north distribution of fluviodeltaic sandstones, proximal to distal fan deposits and pelagic units on the reconstructed paleomargin

(Fig. 13), suggests a northward increase in water depth.

Accumulation of the Povungnituk Group on a continental rift margin is also supported by the similarity of the basalts to modern, within-plate continental tholeiites with respect to overall major element ratios, ranges in TiO_2 content, and trace-element ratios (Hynes and Francis, 1982; Francis *et al.*, 1983). This is corroborated by the observation of the contemporaneous eruption of both basalt and rhyolite, and the occurrence of alkaline basalts within the volcanic sequence (Picard, 1989a). The great thickness (at least 6.7 km) of dominantly submarine Povungnituk Group basalts (Fig. 15) suggests that subsidence kept pace with volcanism, consistent with eruption at an actively rifting margin.

The transition from unstretched to stretched continental crust is preserved in the stratigraphic record of the Povungnituk Group. Deposits on demonstrably unstretched crust are represented by the autochthonous fluviodeltaic sedimentary rocks. Two arguments can be made for the accumulation of the now allochthonous Povungnituk Group sedimentary rocks and basalts on stretched (normal-faulted) crust. First, the presence of fault-bound basins, which evolved independently of each other, is suggested by the limited (proximal) occurrence of the micaceous sandstone unit (Fig. 13). Second, normal faulting during accumulation of the sedimentary and volcanic units is indicated by the presence of potential fault-scarp deposits containing blocks of mafic intrusive rocks and probable Archean basement.

The Chukotat Group, confined to midbelt thrust sheets, records a phase in the magmatic evolution of the Cape Smith Belt that involved the formation of crust transitional between continental rift basalts and MORBs. Chemically, the Chukotat Group ranges from komatiitic basalt compositions to low-Ti tholeiites with MORB affinities (Francis and Hynes, 1979; Hynes and Francis, 1982; Francis *et al.*, 1983). The tholeiitic lavas are low in incompatible elements and have trace element characteristics very similar to those shown by modern ocean-floor basalts (Hynes and Francis, 1982). Chukotat Group magmatism may, therefore, be contemporaneous with the complete rifting apart of continental crust producing transitional crust that evolved to true oceanic crust between 1.96 and 1.92 Ga.

The most internal thrust sheets within the Cape Smith Belt carry the mafic and ultramafic units of the Watts Group. The group has been interpreted by St-Onge *et al.* (1988) as remnants of Early Proterozoic oceanic crust, now obducted onto the northern Superior craton, and preserved in the thrust–fold belt as the Purtuniq ophiolite. Components of the ophiolite include (1) sheeted mafic dykes (2) basaltic flows intruded by mafic dykes and sills, and (3) layered mafic and ultramafic cumulates. The deep-water sedimentary rocks of the Spartan Group are now structurally sandwiched between the obducted ophiolite and underlying Chukotat Group pillowed basalts. Whether the sedimentary rocks were originally deposited on oceanic crust or transitional crust basalts cannot be resolved without further geochronological data.

The 1998 ± 2 Ma age determination on mafic cumulates of the Purtuniq ophiolite eliminates the simple scenario of a Superior Province margin evolving from an epicontinental rift system (Povungnituk Group) to transitional crust (Chukotat Group) and finally to a true oceanic domain (Watts Group). Rather, it is clear that the Watts Group represents older obducted oceanic crust juxtaposed during D_1, against tectono-stratigraphic units belonging to a younger 1.96–1.92 Ga north-facing continental margin. Fieldwork north of the Cape Smith Belt is attempting to identify the crust that must have separated the older oceanic domain (Purtuniq ophiolite) from the younger rift-transitional crust margin.

Tectono-thermal evolution

The study of mountain-building processes and regional tectonics is possible in ancient collisional belts if structural, petrological, and geochronological investigations are integrated to provide a detailed tectono-thermal framework for the belt. In the Cape Smith Belt, the evolution of the thrust belt has been evaluated using a number of techniques, including (1) site-specific *P–T* paths (St-Onge and Lucas, 1988); (2) U–Pb zircon or baddeleyite ages (Parrish, 1989); (3) down-plunge constrained cross-sections (Lucas, 1989); and (4) detailed microstructural studies on the timing of deformation and metamorphism (Lucas, 1990). The following discussion summarizes the tectono-thermal evolution of the D_1 thrust belt, and highlights some important consequences of out-of-sequence thrusting in continental thrust belts.

Development of the thin-skinned thrust belt is interpreted to result from northward underthrusting of the Superior Province continental margin. Underthrusting appears to have been terminated during D_1 by the collision of the upper plate Sugluk terrane (Fig. 1; Hoffman, 1985). However, the plate geometry prior to final collision cannot be clarified without further mapping and geochronology in the hinterland and suture zone north of the belt (Fig. 1). The 2.00 Ga age of the ophiolite indicates that it is exotic with respect to the Superior Province margin, and suggests that it may have been obducted from an upper plate setting.

The basal shear zone developed during prograde metamorphism as a result of an increased importance of thermally activated softening processes (e.g., dislocation creep deformation of quartz and feldspar, Tullis and Yund (1985)). The synmetamorphic change from basal décollement to basal shear zone is interpreted as a transition from discontinuous strain along a narrow fault zone to continuous ductile strain in a broad shear zone (Lucas, 1989). In this model, ductile deformation leading to the development of the basal shear zone initiated at the basement–cover contact (décollement). Widening of the shear zone into the thrust belt and its footwall resulted from strain- and reaction-induced softening at the base of the internal part of the thrust belt (Lucas, 1990). Basal shear zone deformation continued until it was effectively terminated by the propagation of out-of-sequence thrusts from the hinterland décollement into its hanging wall (Fig. 4).

The switch from piggyback-style (regular-sequence) thrusting to out-of-sequence thrusting represents a fundamental change in the style of deformation in the Cape Smith Belt. In general, the out-of-sequence faults were responsible for (1) shifts in the deformation front toward the hinterland; (2) incorporation of footwall basement slices in the thrust belt; and (3) relatively rapid, tectonically driven uplift of metamorphic rocks in the internal zone. Large displacement estimates (50 km) for the syn- and postthermal peak fault systems indicate that they resulted in significant crustal shortening.

Synthermal peak out-of-sequence thrusting had several important consequences on the tectono-thermal evolution of the thrust belt (St-Onge and Lucas, 1988). First, uplift of the out-of-sequence domain during stage 3 deformation resulted in steep dP/dT decompression P–T paths for syntectonic garnets. Second, emplacement of at least 25 km of thrust sheets on the regular-sequence domain resulted in higher temperatures at similar pressures for the regular-sequence domain (footwall) compared with the out-of-sequence domain (hanging wall) (Fig. 31). The overturned isograds in the regular-sequence domain (footwall) below the out-of-sequence thrust stack (Fig. 16) attest to the higher temperature footwall metamorphism being a consequence of emplacement of the hanging wall stack. Finally, the nested decompression P–T paths for the regular-sequence domain (Fig. 31) suggest relatively uniform, isostatically driven uplift, in contrast to the in part tectonically driven uplift of the out-of-sequence domain.

ACKNOWLEDGEMENTS

Dave Scott (Queen's University) and Normand Bégin (University of Calgary) are warmly thanked for their enthusiastic contributions to the Cape Smith Belt project. The project benefited greatly from their careful fieldwork, Ph.D. thesis research, many discussions, and camaraderie. Randy Parrish of the Geological Survey of Canada contributed significantly to the project with precise geochronological data, all of it essential to the understanding of the tectonic evolution of the mountain belt. Randy Parrish, Tim Byrne (Brown University), and Dugald Carmichael and Herb Helmstaedt (both at Queen's University) are thanked for their time spent in the field, where they made fundamental contributions to the mapping program. Herb Helmstaedt is acknowledged for predicting the location of the sheeted dyke complex and for discovering it during his visit to the Cape Smith Belt in 1987. Finally, we would like to acknowledge Bob Baragar, Simon Hanmer, Paul Hoffman, Janet King, John Percival (all at the Geological Survey of Canada), Don Fisher (Penn State), and Jan Tullis (Brown University) for discussions on geological problems and processes of all scales. John Moore and an anonymous reviewer are thanked for careful reviews of a first version of the paper.

REFERENCES

Argand, E., 1911, Les nappes de recouvrement des Alpes Pennines et leurs prolongements structuraux: Matière carte géologique de Suisse, N.S. 31, Berne, 26 p.

Baer, A.J., 1977, Comment: plate tectonics in the Cape Smith Foldbelt: Geology, v. 5, p. 651.

Bally, A.W., 1981, Thoughts on the tectonics of folded belts, in McClay, K. and Price, N.J., eds., Thrust and Nappe Tectonics: Geological Society of London, Special Publication 9, p. 13–32.

Baragar, W.R.A., 1974, Volcanic studies in the Cape Smith – Wakeham Bay belt, New Québec: Geological Survey of Canada, Paper 74-1A, p. 155–157.

Baragar, W.R.A., 1984, Pillow formation and layered flows in the Circum-Superior Belt of eastern Hudson Bay: Canadian Journal of Earth Sciences, v. 21, p. 781–792.

Baragar, W.R.A. and Scoates, R.F.J., 1981, The Circum-Superior Belt: a Proterozoic plate margin?, *in* Kröner, A., ed., Precambrian Plate Tectonics: Elsevier, Amsterdam, p. 297–330.

Bartley, J.M., 1982, Limited basement involvement in Caledonian deformation, Hinnoy, north Norway, and tectonic implications: Tectonophysics, v. 83, p. 185–203.

Beall, G.H., 1959, Cross Lake area, New Québec: Québec Department of Mines, Preliminary Report 396 and Preliminary Map 1267, 9 p.

Beall, G.H., 1960, Laflamme Lake area, New Québec: Québec Department of Mines, Preliminary Report 435 and Preliminary map 1351, 10 p.

Beall, G.H., Hurley, P.M., Fairbairn, H.W. and Pinson, W.H., Jr., 1963, Comparison of K–Ar and whole-rock Rb–Sr dating in New Québec and Labrador: American Journal of Science, v. 261, p. 571–580.

Bégin, N.J. and Carmichael, D.M., 1987, Metabasites in the eastern Cape Smith thrust–fold belt, northern Québec: metamorphic faces, mineral reactions and P–T–CO_2 estimates: Geological Association of Canada – Mineralogical Association of Canada, Program with Abstracts, v. 12, p. 24.

Bégin, N.J., Lucas, S.B. and Carmichael, D.M., 1988, Thermal and tectonic significance of mineral isograds in mafic rocks of the Cape Smith Belt, northern Quebec: Geological Association of Canada – Mineralogical Association of Canada, Program with Abstracts, v. 13, p. 7.

Bergeron, R., 1957, Cape Smith – Wakeham Bay belt, New Québec: Québec Department of Mines, Preliminary Report 355 and Preliminary Maps 1090 and 1196, 8 p.

Bergeron, R., 1959, Povungnituk Range area, New Québec: Québec Department of Mines, Preliminary Report 392 and Preliminary Map 1279, 9 p.

Berthé, D., Choukroune, P. and Jegouzo, P. , 1979, Orthogneiss, mylonite and non-coaxial deformation of granites: the example of the South Armorican Shear Zone: Journal of Structural Geology, v. 1, p. 31–42.

Biot, M.A., 1961, Theory of folding of stratified viscoelastic media and its implications in tectonics and orogenesis: Geological Society of America Bulletin, v. 72, p. 1595–1619.

Boyer, S.E. and Elliott, D., 1982. Thrust systems: Bulletin of the American Association of Petroleum Geologists, v. 66, p. 1196–1230.

Brodie, K.H. and Rutter, E.H., 1985, On the relationship between deformation and metamorphism, with special reference to the behaviour of basic rocks, in Thompson, A.B. and Rubie, D.C., eds., Advances in Physical Geochemistry, v. 4, p. 138–179.

Chauvel, C., Arndt, N.T., Kielinzcuk, S. and Thom, A., 1987, Formation of Canadian 1.9 Ga old continental crust. I. Nd isotopic data: Canadian Journal of Earth Sciences, v. 24, p. 396–406.

Cobbold, P.R. and Quinquis, H., 1980, Development of sheath folds in shear regimes: Journal of Structural Geology, v. 2, p. 119–126.

DeMontigny, P. -A., 1959, Upper Deception River area, New Québec: Québec Department of Mines, Preliminary Paper 398 and Preliminary Map 1285, 8 p.

Dewey, J.F. and Burke, K.C.A., 1973, Tibetan, Variscan, and Precambrian basement reactivation: products of continental collision: Journal of Geology, v. 81, p. 683–692.

Dieterich, J.H. and Onat, N.L., 1969, Slow finite deformation of viscous solids: Journal of Geophysical Research, v. 74, p. 2081–2088.

Dimroth, E., Baragar, W.R.A., Bergeron, R. and Jackson, G.D., 1970, The filling of the Circum-Ungava geosyncline, in Baer, A.J., ed., Symposium of Basins and Geosynclines of the Canadian Shield: Geological Survey of Canada, Paper 70-40, p. 45–144.

Dixon, J.M. and Summers, J.S., 1983, Patterns of total and incremental strain in subsiding troughs: experimental centrifuge models of interdiapir synclines: Canadian Journal of Earth Sciences, v. 20, p. 1843–1861.

Doig, R., 1983, Rb–Sr isotopic study of Archean gneisses north of the Cape Smith Fold Belt, Ungava, Québec: Canadian Journal of Earth Sciences, v. 20, p. 821–829.

Doig, R., 1987, Rb–Sr geochronology and metamorphic history of Proterozoic to early Archean rocks north of the Cape Smith Fold Belt, Quebec: Canadian Journal of Earth Sciences, v. 24, p. 813–825.

Elliott, D. and Johnson, M.R.W., 1980, Structural evolution in the northern part of the Moine thrust belt, NW Scotland: Transactions of the Royal Society of Edinburgh, v. 71, p. 69–96.

England, P.C. and Richardson, S.W., 1977, The influence of erosion upon the mineral facies of rocks from different metamorphic environments: Journal of the Geological Society of London, v. 134, p. 201–213.

England, P.C. and Thompson, A.B., 1984, Pressure–temperature–time paths of regional metamorphism. I. Heat transfer during the evolution of regions of thickened continental crust: Journal of Petrology, v. 25, p. 894–928.

Feininger, T., Lamothe, D., St-Onge, M.R. and Losier, L., 1985, Interprétation gravimétrique de la Fosse de l'Ungava, Ministère de l'Énergie et des Ressources, Québec, DV 85-12.

Ferry, J.M. and Spear, F.S., 1978, Experimental calibration of the partitioning of Fe and Mg between biotite and garnet: Contributions to Mineralogy and Petrology, v. 66, p. 113–117.

Fletcher, R.C., 1974, Wavelength selection in the folding of a single layer with power-law rheology: American Journal of Science, v. 274, p. 1029–1043.

Francis, D.M. and Hynes, A.J., 1979, Komatiite-derived tholeiites in the Proterozoic of New Quebec: Earth and Planetary Science Letters, v. 44, p. 473–481.

Francis, D.M., Ludden, J. and Hynes, A.J., 1983, Magma evolution in a Proterozoic rifting environment: Journal of Petrology, v. 24, p. 556–582.

Ganguly, J. and Saxena, S.K., 1984, Mixing properties of aluminosilicate garnets: constraints from natural and experimental data, and applications to geothermobarometry: American Mineralogist, v. 69, p. 88–97.

Gélinas, L., 1962, Watts Lake area, New Québec: Québec Department of Natural Resources, Preliminary Report 471 and Preliminary Map 1414.

Gibb, R.A. and Walcott, R.I., 1971, A Precambrian suture in the Canadian Shield: Earth and Planetary Science Letters, v. 10, p. 417–422.

Gillcrist, R., Coward, M.P. and Mugnier, J.-L., 1987, Structural inversion and its controls: examples from the Alpine foreland and the French Alps: Geodynamica Acta, v. 1, p. 5–34.

Gold, D.P. , 1962, Brisebois Lake area, New Québec: Québec Department of Natural Resources, Preliminary Report 470 and Preliminary Map 1413.

Goulet, N., Gariépy, C., Mareschal, J.-C. and Machado, N., 1987, Structure, geochronology, gravity, and the evolution of the northern Labrador Trough: Geological Association of Canada – Mineralogical Association of Canada, Program with Abstracts, v. 12, p. 48.

Harris, L.B. and Cobbold, P. R., 1985, Development of conjugate shear bands during bulk simple shearing: Journal of Structural Geology, v. 7, p. 37–44.

Hervet, M., 1986, Géologie de la région de la vallée de Narsajuaq, Fosse de l'Ungava, in Lamothe, D., Gagnon, R. and Clark, T., eds., Exploration en Ungava: données récentes sur la géologie et la gîtologie: Ministère de l'Énergie et des Ressources, Québec, DV 86-16, p. 21–29.

Hodges, K.V. and Royden, L., 1984, Geologic thermobarometry of retrograded metamorphic rocks: an indication of the uplift trajectory of a portion of the northern Scandinavian Caledonides: Journal of Geophysical Research, v. 89, p. 7077–7090.

Hodges, K.V. and Spear, F.S., 1982, Geothermometry, geobarometry and the Al_2SiO_5 triple point at Mt. Moosilauke, New Hampshire: American Mineralogist, v. 67, p. 1118–1134.

Hoffman, P. F., 1985, Is the Cape Smith Belt (northern Quebec) a klippe?: Canadian Journal of Earth Sciences, v. 22, p. 1361–1369.

Hoffman, P. F., 1988, United plates of America, the birth of a craton: Early Proterozoic assembly and growth of Laurentia: Annual Review of Earth and Planetary Sciences, v. 16, p. 543–603.

Hoffman, P. F., Tirrul, R., King, J.E., St-Onge, M.R. and Lucas, S.B., 1988, Axial projections and modes of crustal thickening, eastern Wopmay orogen, northwest Canadian Shield, *in* Clark, S.P. , ed., Processes of Continental Lithospheric Deformation: Geological Society of America, Special Paper 213, p. 1–29.

Hynes, A.J. and Francis, D.M., 1982, A transect of the early Proterozoic Cape Smith foldbelt, New Quebec: Tectonophysics, v. 88, p. 23–59.

Indares, A. and Martignole, J., 1985, Biotite–garnet geothermometry in granulite-facies rocks: evaluation of equilibrium criteria: Canadian Mineralogist, v. 23, p. 187–193.

Jackson, J.A., 1980, Reactivation of basement faults and crustal shortening in orogenic belts: Nature, v. 283, p. 343–346.

Johnson, A.M., 1980, Folding and faulting of strain-hardening sedimentary rocks: Tectonophysics, v. 62, p. 251–278.

Jordan, P. G., 1987, The deformational behaviour of bimineralic limestone–halite aggregates: Tectonophysics, v. 135, p. 185–197.

King, J.E., 1986, The metamorphic internal zone of Wopmay Orogen (early Proterozoic), Canada: 30 km of structural relief in a composite section based on plunge projection: Tectonics, v. 5, p. 973–994.

Lamothe, D., 1986, Développements récents dans la Fosse de l'Ungava, *in* Lamothe, D., Gagnon, R. and Clark, T., eds., Exploration en Ungava: données récentes sur la géologie et la gîtologie: Ministère de l'Énergie et des Ressources, Québec, DV 86-16, p. 1–6.

Lamothe, D., Picard, C. and Moorhead, J., 1984, Région du lac Beauparlant, bande de Cap Smith – Maricourt, Nouveau-Québec: Ministère de l'Énergie et des Ressources, Québec, DP 84-39.

Lewry, J.F. and Sibbald, T.H., 1980, Thermotectonic evolution of the Churchill Province in northern Saskatchewan: Tectonophysics, v. 68, p. 45–82.

Loomis, T.P. , 1983, Compositional zoning of crystals: a record of growth and reaction history, in Saxena, S.K., ed., Kinetics and Equilibrium in Mineral Reactions: Springer-Verlag, New York, p. 1–60.

Low, A.P. , 1902, Report on an exploration of the east coast of Hudson Bay from Cape Wolstenholme to the south end of James Bay: Geological Survey of Canada, Annual Report, v. XIII, pt. D.

Lucas, S.B., 1989, Structural evolution of the Cape Smith Thrust Belt and the role of out-of-sequence faulting in the thickening of mountain belts: Tectonics, v. 8, p. 655–676.

Lucas, S.B., 1990, Relations between thrust belt evolution, grain-scale deformation, and metamorphic processes: Cape Smith Belt, northern Canada: Tectonophysics, v. 178, p. 151–182.

Lucas, S.B. and St-Onge, M.R., 1986, Structural and thermal evolution of the basal shear zone in the early Proterozoic Cape Smith Thrust–Fold Belt, Quebec: Geological Association of Canada – Mineralogical Association of Canada, Program with Abstracts, v. 11, p. 96.

Lucas, S.B. and St-Onge, M.R., 1988, Strain localization and out-of-sequence faulting in thrust belts: examples from the Cape Smith Belt, northern Canada: Geological Society of America, Abstracts with Program, v. 20(7), p. A56.

Machado, N., Goulet, N. and Gariépy, C., 1989, U–Pb ages of reactivated Archean basement and of Hudsonian metamorphism in the northern Labrador Trough: Canadian Journal of Earth Sciences, v. 26, p. 1–15.

Mackin, J.H., 1950, The down-structure method of viewing geologic maps: Journal of Geology, v. 58, p. 55–72.

Moores, E.M., 1982, Origin and emplacement of ophiolites: Reviews of Geophysics and Space Physics, v. 20, p. 735–760.

Newton, R.C. and Haselton, H.T., 1981, Thermodynamics of the garnet–plagioclase–Al_2SiO_5–quartz geobarometry, *in* Newton, R.C., Navrotsky, A. and Wood, B.J., eds., Thermodynamics of Minerals and Melts: Springer-Verlag, Berlin, p. 131–147.

Parrish, R.R., 1989, U–Pb geochronology of the Cape Smith Belt and Sugluk block, northern Québec: Geoscience Canada, v. 16, p. 126–130.

Picard, C., 1986, Lithogéochimie de la partie centrale de la Fosse de l'Ungava, *in* Lamothe, D., Gagnon, R. and Clark, T., eds., Exploration en Ungava: données récentes sur la géologie et la gîtologie: Ministère de l'Énergie et des Ressources du Québec, DV 86-16, p. 57–72.

Picard, C., 1989a, Pétrologie et volcanologie des roches volcaniques de la partie centrale de la Fosse de l'Ungava: Ministère de l'Énergie et des Ressources du Québec, ET 87-07.

Picard, C., 1989b, Lithochimie des roches volcaniques protérozoïques de la partie occidentale de la Fosse de l'Ungava, région au sud du lac Lanyan: Ministère de l'Énergie et des Ressources du Québec, ET 87-14.

Platt, J.P. and Vissers, R.L.M., 1980, Extensional structures in anisotropic rocks: Journal of Structural Geology, v. 2, p. 397–410.

Powell, R. and Evans, J.A., 1983, A new geobarometry for the assemblage biotite–muscovite–chlorite–quartz: Journal of Metamorphic Geology, v. 1, p. 331–336.

Ramberg, H., 1981, Gravity, Deformation and the Earth's Crust: Academic Press, London, 452 p.

Ramsay, J.G., 1967, Folding and Fracturing of Rocks: McGraw-Hill, New York, 555 p.

Ramsay, J.G., Casey, M. and Kligfield, R., 1983, Role of shear in development of the Helvetic fold–thrust belt of Switzerland: Geology, v. 11, p. 439–442.

Rumble, D. and Finnerty, T.A., 1974, Devonian grossularite–spessartine overgrowth on Ordovician almandine from eastern Vermont: American Mineralogist, v. 59, p. 558–562.

Schimann, K., 1978, On regional metamorphism in the Wakeham Bay area, New Québec, in Fraser, J.A. and Heywood, W.W., eds., Metamorphism of the Canadian Shield: Geological Survey of Canada, Paper 78-10, p. 245–248.

Schwarz, E.J. and Fujiwara, Y., 1977, Komatiitic basalts from the Proterozoic Cape Smith Range in northern Québec, Canada, in Baragar, W.R.A., Coleman, L.C. and Hall, J.M., eds., Volcanic Regimes in Canada: Geological Association of Canada, Special Paper 16, p. 193–201.

Scott, D.J., St-Onge, M.R., Lucas, S.B. and Helmstaedt, H., 1988, The 1999 Ma Purtuniq ophiolite: imbricated oceanic crust obliquely exposed in the Cape Smith Thrust–Fold Belt, northern Quebec, Canada: Geological Society of America, Abstracts with Program, v. 20(7), p. A158.

Searle, M.P. , 1985, Sequence of thrusting and origin of culminations in the northern and central Oman Mountains: Journal of Structural Geology, v. 7, p. 129–143.

Sibson, R.H., 1977, Fault rocks and fault mechanisms: Journal of the Geological Society of London, v. 133, p. 191–213.

Smith, R.B., 1975, Unified theory of the onset of folding, boudinage and mullion structure: Geological Society of America Bulletin, v. 86, p. 1601–1609.

Spear, F.S. and Selverstone, J., 1983, Quantitative *P–T* paths from zoned minerals: theory and tectonic applications: Contributions to Mineralogy and Petrology, v. 83, p. 348–357.

Spear, F.S., Selverstone, J., Hiskmott, D., Crowley, P. and Hodges, K.V., 1984, *P–T* paths from garnet zoning: a new technique for deciphering tectonic processes in crystalline terranes: Geology, v. 12, p. 87–90.

Stam, J.C., 1961, On the geology and petrology of the Cape Smith – Wakeham Bay belt, Ungava, Canada: Geologie en Mijnbouw, v. 40, p. 412–421.

Stauffer, M.R., Mukherjee, A.C. and Koo, J., 1975, The Amisk Group: an Aphebian(?) island arc deposit: Canadian Journal of Earth Sciences, v. 12, p. 2021–2035.

Stockwell, C.H., 1950, The use of plunge in the construction of cross-sections of folds: Proceedings of the Geological Association of Canada, v. 3, p. 97–121.

St-Onge, M.R. and King, J.E., 1987, Evolution of regional metamorphism during back-arc stretching and subsequent crustal shortening in the 1.9 Ga Wopmay Orogen, Canada, *in* Oxburgh, E.R., Yardley, B.W.D. and England, P. C., eds., Tectonic Settings of Regional Metamorphism: Philosophical Transactions of the Royal Society of London, Ser. A, v. 321, p. 199–218.

St-Onge, M.R. and Lucas, S.B., 1988, Thermal consequences of tectonically-driven uplift: metamorphic evolution of the Cape Smith Thrust–Fold Belt, Canada: Geological Society of America, Abstracts with Program, v. 20(7), p. A99.

St-Onge, M.R. and Lucas, S.B., 1989, Geology, eastern portion of the Cape Smith Thrust–Fold Belt, parts of the Wakeham Bay, Cratere du Nouveau-Québec and Nuvilik Lakes map areas, northern Québec: Geological Survey of Canada, Maps 1721A to 1735A, scale 1:50,000.

St-Onge, M.R., Lucas, S.B., Scott, D.J. and Bégin, N.J., 1986, Eastern Cape Smith Belt: an early Proterozoic thrust–fold belt and basal shear zone exposed in oblique section, Wakeham Bay and Cratere du Nouveau-Québec map areas, northern Québec: Geological Survey of Canada, Paper 86-1A, p. 1–14.

St-Onge, M.R., Lucas, S.B., Scott, D.J. and Bégin, N.J., 1987, Tectonostratigraphy and structure of the lac Watts – lac Cross – rivière Déception area, central Cape Smith Belt, northern Québec: Geological Survey of Canada, Paper 87-1A, p. 619–632.

St-Onge, M.R., Lucas, S.B., Scott, D.J., Bégin, N.J., Helmsteadt, H. and Carmichael, D.M., 1988, Thin-skinned imbrication and subsequent thick-skinned folding of rift-fill, transitional-crust and ophiolite suites in the 1.9 Ga Cape Smith Belt, northern Quebec: Geological Survey of Canada, Paper 88-1C, p. 1–18.

Suppe, J., 1983, Geometry and kinematics of fault-bend folding: American Journal of Science, v. 283, p. 684–721.

Taylor, F.C., 1974, Reconnaissance geology of a part of the Precambrian Shield, northern Québec and Northwest Territories: Geological Survey of Canada, Paper 74-21, 10 p.

Taylor, F.C., 1982, Reconnaissance geology of a part of the Canadian Shield, northern Québec and Northwest Territories: Geological Survey of Canada, Memoir 399, 32 p.

Taylor, F.C. and Loveridge, W.D., 1981, A Rb–Sr study of a New Québec Archean granodiorite: Geological Survey of Canada, Paper 81-1C, p. 105–106.

Thom, A., Arndt, N.T. and Chauvel, C., 1987, Volcanic arcs and granitoids in the Reindeer zone, Saskatchewan: Geological Association of Canada – Mineralogical Association of Canada, Program with Abstracts, v. 12, p. 95

Thomas, M.D. and Gibb, R.A., 1977, Gravity anomalies and deep structure of the Cape Smith foldbelt, northern Ungava, Québec: Geology, v. 5, p. 169–172.

Thompson, A.B., Tracy, R.J., Lyttle, P. and Thompson, J.B., Jr., 1977, Prograde reaction histories deduced from compositional zonation and mineral inclusions in garnet from the Gassetts schist, Vermont: American Journal of Science, v. 277, p. 1152–1167.

Tracy, R.J., 1982, Compositional zoning and inclusions in metamorphic minerals, *in* Ferry, J.M., ed., Characterization of Metamorphism Through Mineral Equilibria: Mineralogical Society of America, Reviews of Mineralogy, v. 10, p. 355–397.

Tullis, J. and Mardon, D., 1984, Effect of muscovite content on deformation of quartz aggregates: Geological Society of America, Abstracts with Program, v. 16, p. 678.

Tullis, J. and Yund, R.A., 1985, Dynamic recrystallization of feldspar: a mechanism for ductile shear zone formation: Geology, v. 13, p. 238–241.

van der Leeden, J., Bélanger, M., Danis, D., Girard, R. and Martelain, J., 1990, Lithotectonic domains in the high-grade terrain east of the Labrador Trough (Quebec), *in* Lewry, J.F. and Stauffer, M.R., eds., The Early Proterozoic Trans-Hudson Orogen of North America: Geological Association of Canada, Special Paper 37, p. 371–386.

Watters, B.R., Thomas, D.J. and Harper, C.T., 1987, Tectonic setting of early Proterozoic volcanism of the La Ronge domain, northern Saskatchewan: Geological Association of Canada – Mineralogical Association of Canada, Program with Abstracts, v. 12, p. 100.

Westra, L., 1978, Metamorphism in the Cape Smith – Wakeham Bay area north of 61°N, New Québec, *in* Fraser, J.A. and Heywood, W.W., eds., Metamorphism of the Canadian Shield: Geological Survey of Canada, Paper 78-10, p. 237–244.

White, S.H., Burrows, S.E., Carreras, J., Shaw, N.D. and Humphreys, F.J., 1980, On mylonites in ductile shear zones: Journal of Structural Geology, v. 2, p. 175–187.

Zuber, M.T., Parmentier, E.M. and Fletcher, R.C., 1986, Extension of the continental lithosphere: a model for two scales of Basin and Range deformation: Journal of Geophysical Research, v. 91, p. 4826–4838.

Labrador Segment of the Trans-Hudson Orogen: Crustal Development Through Oblique Convergence and Collision

R.J. Wardle, B. Ryan, and G.A.G. Nunn
Newfoundland Department of Mines and Energy, Geological Survey Branch, P.O. Box 8700, St. John's, Newfoundland A1B 4J6, Canada

F.C. Mengel[1]
Department of Earth Sciences, Memorial University of Newfoundland, St. John's, Newfoundland A1B 3X5, Canada

Wardle, R.J., Ryan, B., Nunn, G.A.G. and Mengel, F.C., Labrador segment of the Trans-Hudson Orogen: crustal development through oblique convergence and collision, in Lewry, J.F. and Stauffer, M.R., eds., The Early Proterozoic Trans-Hudson Orogen of North America: Geological Association of Canada, Special Paper 37, p. 353–369.

Abstract

The Labrador segment of the Trans-Hudson Orogen in Labrador is divided into the Western, Central, and Eastern divisions. The Western Division comprises low- to medium-grade supracrustal rocks of the Labrador Trough, which have been transported westward over the Archean Superior Craton in a foreland fold and thrust belt. This division shows an eastward progradation from rocks of miogeoclinal to oceanic aspect.

The Central Division comprises little known Archean to Lower Proterozoic granitoid and supracrustal gneiss intruded by an axial granitoid batholith of Lower Proterozoic age. The division is thrust over the Western Division along a series of major, east-dipping ductile shear zones. Its eastern boundary is a major sinistral ductile shear zone, 5–25 km wide.

The Eastern Division comprises progressively reworked Archean gneiss and lower Proterozoic supracrustal cover of the North Atlantic Craton that have been deformed in both the transcurrent shear zone and in an east-facing fold and thrust belt.

The Labrador segment of the Trans-Hudson Orogen displays marked structural asymmetry and may represent an example of oblique convergence and collision. In a proposed model, this segment developed by early transcurrent motion on the Eastern–Central Division boundary followed by collision with the Superior Craton plate and development of marginal thrust and fold belts during east–west shortening.

[1]Present address: INRS Géoressources, 2700 rue Einstein, C.P. 7500, Sainte-Foy, Québec G1V 4C7, Canada.

Résumé

La partie Labrador de l'orogène Trans-Hudson comporte trois divisions, ouest, centrale et est. La division ouest est composée de roches supracrustales de la fosse du Labrador à métamorphisme faible et moyen poussées vers l'ouest sous forme de chaine plissée sur le craton archéen du Supérieur.

La division centrale comprend des gneiss supracrustaux et granitiques d'âge Archéen et Protérozoïque Inférieur recoupé par un batholite granitique allongé Protérozoïque Inférieur. La division centrale est charriée sur la division ouest le long d'une série de cisaillements ductiles de grandes dimensions pentés vers l'est. La limite orientale est formée par un cisaillement ductile senestre de 5 à 25 km de largeur.

La division est comprend des gneiss Archéen et la couverture protérozoïque du craton Nord Atlantique remaniés progressivement et déformés par un cisaillement décrochant et par une chaîne plissée vers l'est.

Le segment Labrador de l'orogène Trans-Hudson montre une asymétrie structurale marquée et pourrait représenter un exemple de convergence et collision oblique. Le modèle proposé veut que le segment s'est formé par un mouvement primitif décrochant à la limite des divisions est et centrale suivit par la collision du craton Supérieur et le développement d'une chaîne plissée marginale pendant le raccourcissement est–ouest.

INTRODUCTION

One of the most important concepts to have emerged from the study of orogenic belts in recent years is the close causal relationship between deformation, in particular thrusting, and the progressive accretion of crustal elements against continental margins. This understanding has been achieved principally through terrane analysis (Coney *et al.*,1980; Monger *et al.*, 1982) and has led to a reappraisal of Proterozoic orogens, e.g., the Wopmay (Hoffman, 1980; Hoffman and Bowring, 1984) and Trans-Hudson (Lewry, 1981) orogens, and the Precambrian belts of the southwestern United States (Karlstrom and Houston, 1984). It is in this spirit that we attempt an analysis of the Labrador segment of the Trans-Hudson Orogen. We emphasize, however, that as in all attempts to synthesize the geology of the poorly known Arctic and northeastern regions of the Canadian Shield, the task is hindered by a lack of detailed geological mapping and reliable radiometric dates.

The term Trans-Hudson Orogen was originally introduced by Hoffman (1981, p. 97) to refer to a major belt of deformation and metamorphism, with imprecise geographic boundaries, "between the northwest Churchill Province and the Superior Province." Subsequently, Trans-Hudson Orogen was used by J. Lewry and M. Stauffer in the organization of this symposium as a comprehensive term for "much of the Churchill Province" and allied Early Proterozoic terranes in Greenland and central North America. Hoffman (this volume) has now presented a succinct definition of "Trans-Hudson Orogen" arising from concerns that his earlier nomenclature had become synonymous with the whole of Stockwell's (1964) Churchill Province. At the same time, he has introduced new terms for other subzones and events in the Churchill Province (see also Hoffman (1988)), including Rae Province, and Torngat and New Quebec orogens in the eastern and western areas of Labrador and adjacent parts of Quebec, i.e., the "Labrador segment" described in this paper. Although we have no disagreement with Hoffman's terminology for the "Labrador segment," the existence of two separate orogenies has not been unequivocally proven, nor disproven, due to the paucity of geochronological data, and we are hesitant to embrace such a concept at this stage. Consequently, in keeping with the organizational terminology for this symposium, we retain the terms "Labrador segment" (or "the orogen") to refer to the area under discussion, and "Trans-Hudson Orogen" to refer to the whole area of Early Proterozoic deformation and metamorphism of which it is a part (Fig. 1).

From the perspective of tectonic studies, the Labrador segment has the inherent advantages of (1) being exposed in complete cross-section over a comparatively narrow width of 400 km and (2) preserving remnants of continental margin sequen-

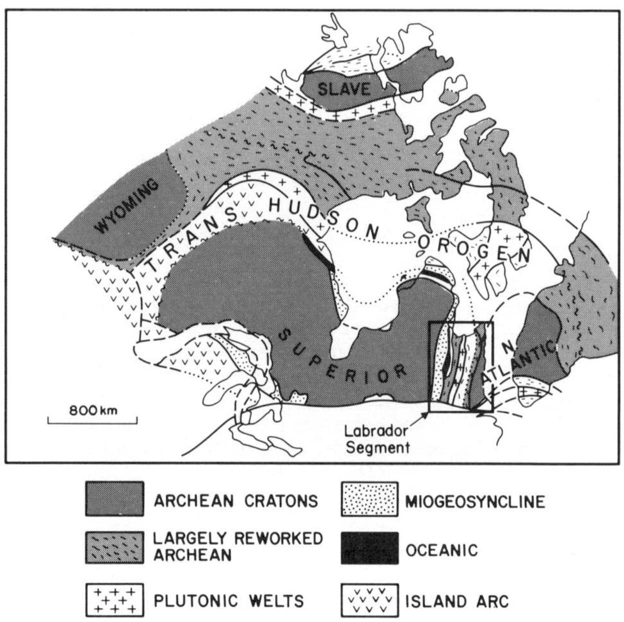

Figure 1. *Tectonic elements of the Trans-Hudson Orogen showing the location of the Labrador segment. Dotted lines signify buried or submarine contacts. After Lewry and Collerson (this volume).*

ces on both bounding cratons. Despite this, however, the orogen is still comparatively poorly known. Detailed mapping has largely been confined to the western margin (e.g., Dimroth *et al.*, 1970; Dimroth, 1978; Dressler, 1979; Wardle, 1982) and parts of the eastern margin (Morgan, 1975; Ryan *et al.*, 1983, 1984). Apart from work by Nunn and Noel (1982) and Hill (1982), little work has been carried out in the interior of the orogen until recently, and much of this region is known only from the reconnaissance mapping of Taylor (1979). Lately, this situation has begun to improve with the initiation of mapping programs in the interior by both the Quebec Ministère de l'Énergie et des Ressources du Québec (van der Leeden, 1983; Bélanger, 1984; Owen, 1984) and the Newfoundland Department of Mines and Energy (Ryan and Lee, 1986; Ryan *et al.*, 1987, 1988). This review, which attempts to provide an integrated picture of the orogen as a whole, is biased by our greater familiarity with the margins of the orogen and in particular those parts that are exposed in Labrador. For a more complete treatment of the interior of the orogen the reader is referred to the accompanying review by van der Leeden *et al.* (this volume).

Geological setting

The Labrador segment of the Trans-Hudson Orogen is bounded to the east and west by the Archean Labrador component of the North Atlantic Craton (NAC) and the Superior Craton respectively (Figs. 1 and 2). In the south, it is truncated by younger (accreted?) crust of ca. 1650 Ma age (Nunn *et al.*, 1985; Wardle *et al.*, 1986) and overprinted by ca. 1000 Ma deformation associated with the development of the Grenville Orogen (Fig. 2).

The Labrador segment is herein divided into three principal divisions, namely the Western, Central, and Eastern divisions (Fig. 2). The Western Division comprises low- to medium-grade metamorphosed sedimentary and volcanic rocks of the Labrador Trough stacked in a west-facing fold and thrust belt and underlain by Archean basement. The Central Division is composed of a variety of high-grade gneissic rocks, likely of mixed Archean and Proterozoic age, intruded by an axial granitoid batholith and characterized by a generally upright structural style. The Eastern Division (Korstgård *et al.*, 1987) represents the deformed western margin of the North Atlantic Craton and its partially preserved supracrustal cover sequence, deformed into both vertical and east-facing structures.

Each division is subdivided into a number of terranes and zones (Fig. 2). Following the usage of Coney *et al.* (1980) and Williams and Hatcher (1982), the term "terrane" is used for crustal elements that are by and large tectonically bounded and that lack any primary stratigraphic affinity with their neighbours. We use the term "zone" for elements that, although tectonically bounded or otherwise, may be paleogeographically linked by virtue of stratigraphy, facies, or other criteria to their neighbours.

WESTERN DIVISION

Basement to the Western Division is formed by the granulite- to upper amphibolite-facies Archean Ashuanipi Metamorphic Complex (Percival, 1987), comprising garnet–biotite (supracrustal?) gneiss, a variety of tonalitic to granitic orthogneiss and plutonic rocks, and very minor nepheline syenite. These rocks are unconformably overlain by the Knob Lake Group of the Labrador Trough, which is divided into two major zones by the Ferrum River thrust (Figs. 3 and 4).

The Schefferville zone (Fig. 4), comprising those rocks west of the Ferrum River thrust, consists of a sequence of continental arkose (Seward Subgroup) at its base passing up through shallow-marine shale into dolostone containing channelized debris-flow breccias (Attikamagen Subgroup). This sequence is terminated by a disconformity–unconformity, above which occurs a deepening-upward succession of littoral to sublittoral quartzite, shallow-marine ironstone locally interbedded with alkalic volcanics, euxinic black shale, and clastic turbidites (Ferriman Subgroup). This is followed conformably by terrestrial arkose. A syenite clast from conglomerate interbedded with the alkalic volcanics has yielded a U–Pb zircon age of $1879^{+5.2}_{-3.5}$ Ma (Birkett *et al.*, *in prep.*) and provides a precise age for both volcanism and coeval ironstone deposition.

The Howse zone (Fig. 3) consists of continental arkose (Seward Subgroup) (Fig. 4) overlain by a thick pile of interstratified turbidite shale and siltstone, basalt flows, and gabbro sills of the Attikamagen Subgroup. The age of these rocks is probably comparable to a U–Pb zircon date of 2142^{+4}_{-2} Ma on a thin rhyolite unit intercalated with abundant basalt flows in the northern part of the Howse zone (Krogh <u>in</u> Clark, 1984). The Attikamagen Subgroup basalts pass up into a sequence of quartzite and ironstone identical to that seen in the Schefferville zone and that provides reliable evidence of paleogeographic linkage between the two zones. The eastern part of the Howse zone consists of turbidite shales and voluminous gabbro sills that were originally considered to stratigraphically overlie ironstone (Wardle, 1982). However, a recent U–Pb zircon date on a gabbro sill (Birkett *et al.* (*in prep.*), also in Hoffman (this volume)) of $1883.8^{+1.6}_{-1.5}$ Ma indicates a pre-ironstone age.

Rocks of the Schefferville and Howse zones have been proposed by Wardle and Bailey (1981) to represent an east-deepening continental margin prism formed during rifting of the Superior Craton (cf. Baragar and Scoates, 1981), a model based predominantly on the presence of voluminous turbidites and basaltic rocks of oceanic tholeiite character (Baragar, 1967; Dimroth, 1971, 1978) within the Howse zone. More recently, Hoffman (1987) has proposed the Ferriman Subgroup to represent foredeep deposition in advance of thrust sheets encroaching from the east. This alternative model is consistent with the presence of an intra-Knob Lake Group unconformity, interpreted to result from passage of a flexural arch. A lower foredeep has also been proposed (Hoffman, 1987) to include the Swampy Bay Subgroup (Fig. 4). This sequence, however, is restricted to the central part of the

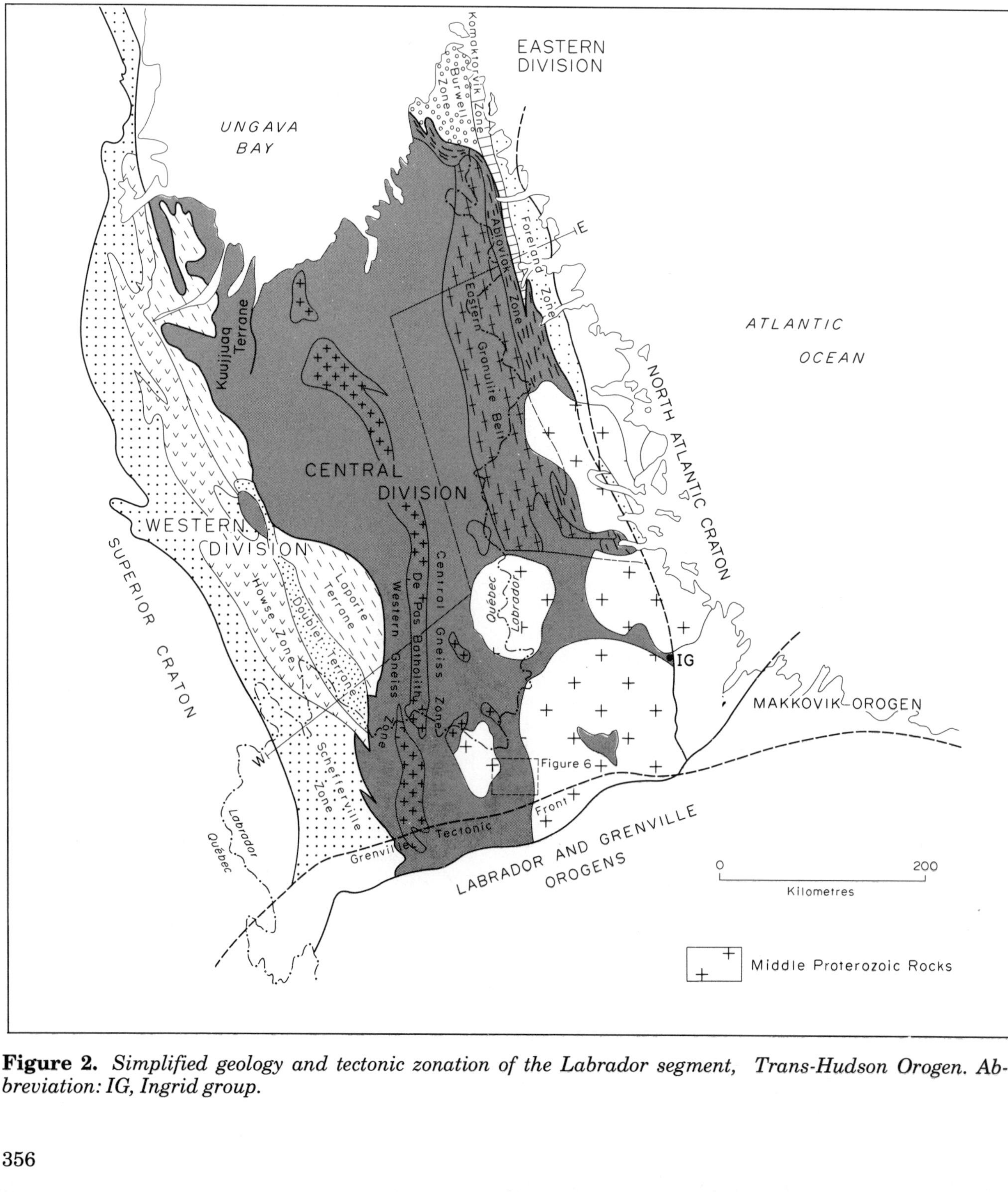

Figure 2. *Simplified geology and tectonic zonation of the Labrador segment, Trans-Hudson Orogen. Abbreviation: IG, Ingrid group.*

Schefferville zone and indicates that foredeep development and causative thrust loading must have been similarly restricted. The reason for this is not clear.

The basaltic flows and sills of the Howse zone are overwhelmingly of oceanic tholeiite character (Baragar, 1967; Dimroth, 1971, 1978) and appear to have formed in two pulses at ca. 2100 and 1880 Ma. The relative proportions of the two pulses have, however, yet to be established. The earlier basaltic rocks rest conformably on redbeds (Seward Subgroup) of the initial rift phase and would likewise appear to be of rift origin (cf. Wardle and Bailey, 1981). The tectonic setting of the younger basaltic rocks is problematic. Wardle and Bailey (1981) suggested an environment of continued rifting associated with alkalic volcanism in the Schefferville zone; Hoffman (1987), however, has proposed an association with foredeep development.

The Doublet terrane (Figs. 3 and 4), separated from the Howse zone by the Walsh Lake thrust, is dominated by mafic pyroclastics and pillow basalt of the Doublet Group and voluminous gabbro and ultramafic sills. Although the Doublet Group has, in the past, been interpreted to be stratigraphically younger than the Knob Lake Group (e.g., Frarey, 1967; Dimroth, 1978), it appears to be entirely fault bounded, has not been reliably dated, and, hence, is given a terrane designation. The primitive oceanic tholeiite character of the sequence and the presence of thick peridotite sills are likely indicative of a mantle source and suggest that the terrane most likely developed during oceanic rifting. The lack of clearly developed ophiolite stratigraphy suggests that it may represent a fragment of the continent–ocean transition rather than true oceanic crust (cf. Wardle and Bailey, 1981).

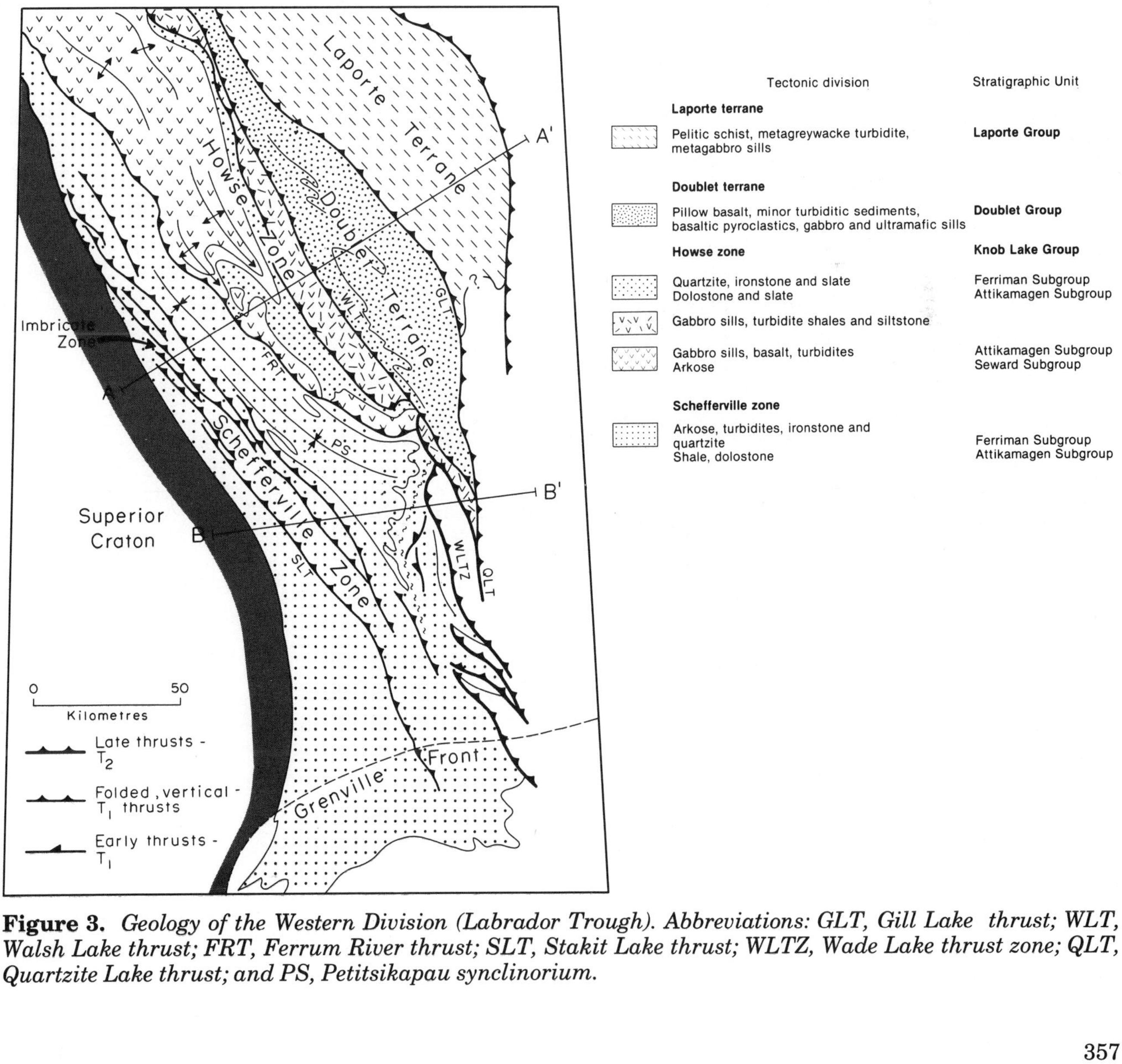

Figure 3. *Geology of the Western Division (Labrador Trough). Abbreviations: GLT, Gill Lake thrust; WLT, Walsh Lake thrust; FRT, Ferrum River thrust; SLT, Stakit Lake thrust; WLTZ, Wade Lake thrust zone; QLT, Quartzite Lake thrust; and PS, Petitsikapau synclinorium.*

The Laporte terrane, which is thrust over the Doublet terrane, has not been mapped in detail, but is known to consist of monotonous biotite schists apparently derived from a greywacke protolith (Baragar, 1967) of presumed turbidite character. Wardle and Bailey (1981) suggested that the sequence formed a distal part of the passive-margin flysch prism around the Superior Craton. However, in the absence of more detailed knowledge, particularly concerning the age and sedimentary source of the terrane, other interpretations, such as fore-arc or back-arc basin, could also be advanced.

Western Division structure

The overall structural profile is that of a west-verging thrust stack (Fig. 5). The Schefferville zone contains a narrow, east-dipping autochthonous sequence interpreted to extend at depth beneath the frontal Stakit Lake thrust. East of this thrust, the zone is dominated by a series of imbricate thrusts rooted in the basal Stakit Lake décollement within shales of the Attikamagen Subgroup (Figs. 3 and 4). The imbricate zone gives way eastward to a structural style dominated by long-wavelength, periclinal folds associated with the widely spaced, Ferrum River and Walsh Lake thrusts. These are interpreted to ramp steeply upward from a master basal décollement lying deep within the stratigraphic sequence, probably at the basement–cover interface (Fig. 4). The westward transition from long-wavelength folding to imbricate thrusting is inferred to be controlled by a basement ramp (cf. Harrison *et al.*, 1972; Fig. 3).

The southeastern part of the Labrador Trough has also been affected by a later period of thrusting involving basement (Fig. 5, section B–B′). The north–south trending Walsh Lake and Quartzite Lake thrusts enclose a basement slice (Wardle, 1982), and are oblique to the overall strike of the Western Division. They have folded the Ferrum River thrust, refolded the associated northwest-trending folds of the Petitsikapau synclinorium from horizontal into vertical to overturned attitudes (Fig. 3), and lifted the Doublet and Laporte terranes above the present erosion surface.

CENTRAL DIVISION

The Central Division comprises the western gneiss zone, the Kuujjuaq terrane, the De Pas batholith, the central gneiss zone, and the Abloviak shear

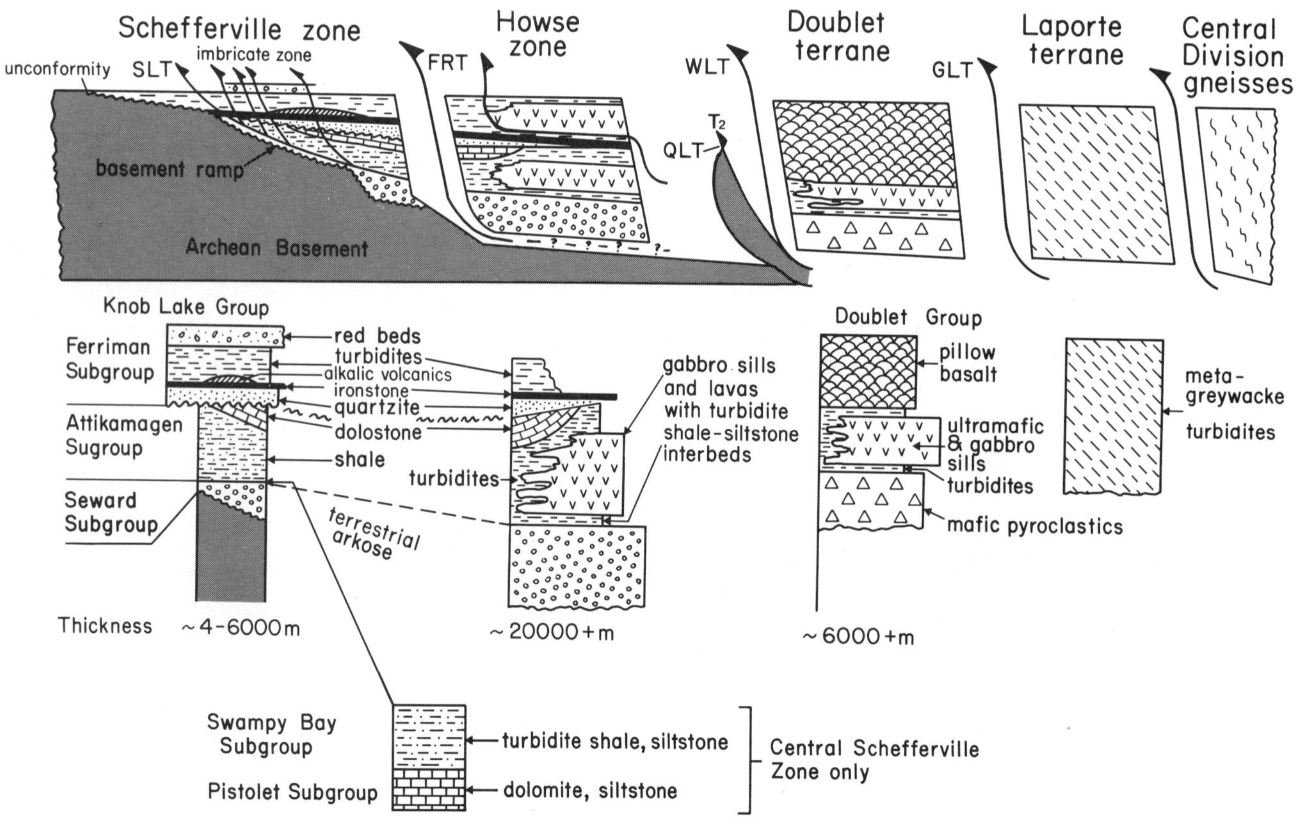

Figure 4. *Schematic stratigraphic sections of tectonic zones and terranes, Western Division. Thrust abbreviations as for Figure 3.*

zone (Fig. 2). On the basis of our present meagre knowledge of this division, there is little evidence to suggest fundamental structural breaks between the subdivisions.

The western gneiss zone (Lac Tudor zone of van der Leeden *et al.* (this volume)), the westernmost unit of the Central Division, is composed typically of granitoid gneiss of tonalitic to granodioritic composition and elongate bands of amphibolite–metagabbro. Metamorphic grade increases eastward from upper amphibolite to granulite facies. Mylonitic fabrics dominate the zone and are the result of ductile shearing in association with west-directed thrusting and dextral transcurrent movement (van der Leeden *et al.*, this volume). Gneiss of the western gneiss zone has not been reliably dated. However, similar gneiss, which appears as domes and thrust slices within the Western Division (Fig. 2), is likely to be of Archean age (Dimroth, 1978; Wardle and Bailey, 1981) and a correlation may be justified. Similarly, Machado *et al.* (1988) have reported an age of ca. 2700 Ma for a gneiss dome within the easternmost Labrador Trough near Ungava Bay.

In the northwestern part of the Central Division, Poirier *et al.* (1987) and Perreault and Hynes (1988) have recognized a basement-cored tectonic block, the Kuujjuaq terrane, composed of metasedimentary rocks, amphibolites (metatuffs), and tonalite. Suturing of the terrane to the Superior Craton occurred in association with tonalite intrusion and granulite-facies metamorphism in the interval 1845–1829 Ma (Machado *et al.*, 1988). Late metamorphism occurred in the interval 1790–1770 Ma (Perreault and Hynes, 1988).

The De Pas batholith (referred to as the Ungava batholith by Wardle *et al.* (1986)) is in intrusive contact with the central gneiss zone (see below) and the southernmost part of the western gneiss zone (Fig. 2). However, the remainder of the contact with the western gneisses is a shear zone. The batholith comprises K-feldspar-megacrystic, orthopyroxene-bearing granite to granodiorite with abundant enclaves of amphibolite and amphibolite–diorite gneiss. Emplacement clearly postdated development of gneissic fabrics in both the western and the central gneiss zones (Wardle 1985; van der Leeden *et al.*, this volume). The batholith has yielded U–Pb

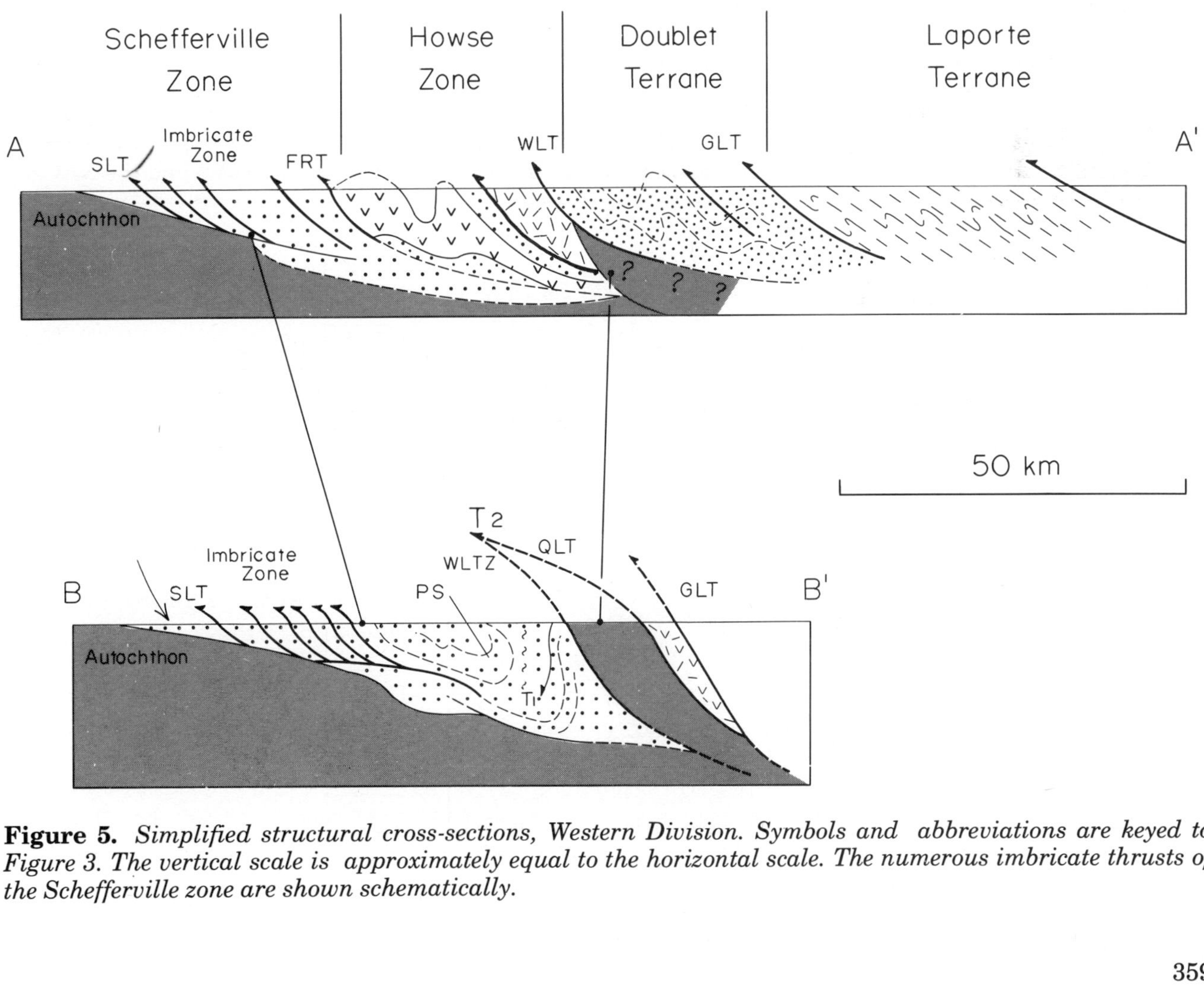

Figure 5. *Simplified structural cross-sections, Western Division. Symbols and abbreviations are keyed to Figure 3. The vertical scale is approximately equal to the horizontal scale. The numerous imbricate thrusts of the Schefferville zone are shown schematically.*

zircon dates of 1811 ± 3 Ma (Krogh, 1986) in its southern part and 1840 Ma (Bowring in van der Leeden *et al.*, this volume) in its central portion.

The central gneiss zone forms a broad expanse of high-grade gneiss lying east of the De Pas batholith and occupying the core of the orogen. In the south, immediately north of the Grenville tectonic front (Figs. 2 and 6), the zone is composed predominantly of tonalite and migmatitic tonalite gneiss intrusive into enclaves of metasedimentary and mafic metavolcanic rocks (Nunn and Noel, 1982). Although previously considered to be Early Proterozoic (e.g., Emslie, 1978), tonalite has recently yielded a U–Pb zircon age of 2663 ± 6 Ma (Krogh, 1986), indicating that much of the central gneiss terrane in this area may be Archean. Furthermore, it is apparent from the greenschist facies character of some supracrustal rocks that this part of the Central Division has suffered only a low-grade Hudsonian overprint.

In the central parts of the zone, van der Leeden (1983), Bélanger (1984), and Owen (1984) have described the occurrence of low- to medium-grade supracrustal rocks, of probable Early Proterozoic age, infolded with high-grade gneissic rocks, which may form an Archean basement. A similar relationship may exist in northeastern Quebec–Labrador where the Lake Harbour Group (Fig. 7) (Jackson and Taylor, 1972), a sequence of amphibolite-facies pelitic schist, dolomitic marble, quartzite, and metabasalt of inferred shallow-water origin, occurs in association with high-grade gneiss of possible basement aspect (Wardle, 1984). The presence of Archean basement has also been inferred by Ashwal *et al.* (1986, 1988) using Nd, Sr, and Pb isotopic data from Middle Proterozoic intrusions in the central gneiss zone.

Structural style in the central gneiss zone is dominated by steep, north-trending structures locally folded into large open structures (van der Leeden *et al.*, this volume).

The eastern granulite belt is a subdivision of the central gneiss zone distinguished by the widespread presence of granulite-facies assemblages; a strong, northwest-trending structural grain; and a pronounced positive aeromagnetic signature (Fig. 7). The zone comprises a variety of orthogneisses, with intercalated belts of supracrustal rocks that include significant amounts of (Early Proterozoic ?) marble and quartzite. There are no reliable ages for the belt. However, Ryan *et al.* (1988) have reported migmatized, gabbroic anorthosite, similar to that of the NAC and other Archean cratons, in western parts of the belt. Gneissic fabrics in these rocks are crosscut by mafic dykes that have suffered amphibolite-facies recrystallization in a subsequent (Early Proterozoic ?) event. If these dykes are correlated with the Early Proterozoic dyke swarm of the Eastern Division (see below), they suggest an Archean protolith age for parts of the eastern granulite belt. Such dykes are not seen in the eastern Superior Craton and suggest a possible original association of this part of the Central Division and the NAC.

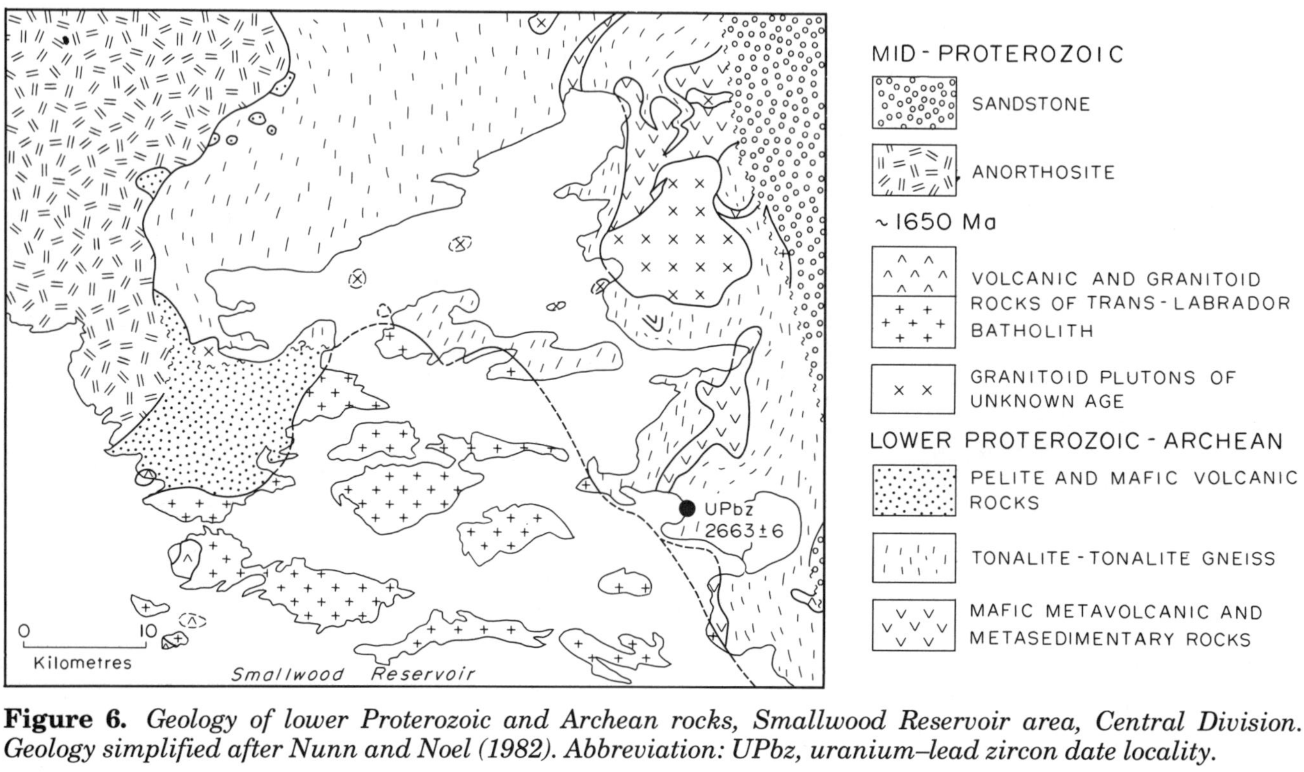

Figure 6. *Geology of lower Proterozoic and Archean rocks, Smallwood Reservoir area, Central Division. Geology simplified after Nunn and Noel (1982). Abbreviation: UPbz, uranium–lead zircon date locality.*

The structure of the granulite belt is characterized by a steeply dipping, north- to northwest-trending gneissosity, with zones of granulite- and amphibolite-facies mylonitic gneiss up to 15 km wide. The majority of the folds and lineations have shallow plunges, and in most cases axial planar fabrics are strongly L-biased.

The Abloviak shear zone forms the easternmost boundary of the Central Division (Fig. 2). It is a major crustal feature strikingly apparent on Landsat imagery and forms a pronounced magnetic low (Fig. 7) that can be traced to Resolution Island (Geological Survey of Canada, 1981) and thence north again to Hall Peninsula on Baffin Island (Hoffman, this volume). The structural style of the Abloviak zone is similar to that of the adjacent Komaktorvik zone (Eastern Division, see below). However, the rocks that comprise the Abloviak zone lack the diagnostic features of NAC crust (see below) and are, hence, assigned to the Central Division.

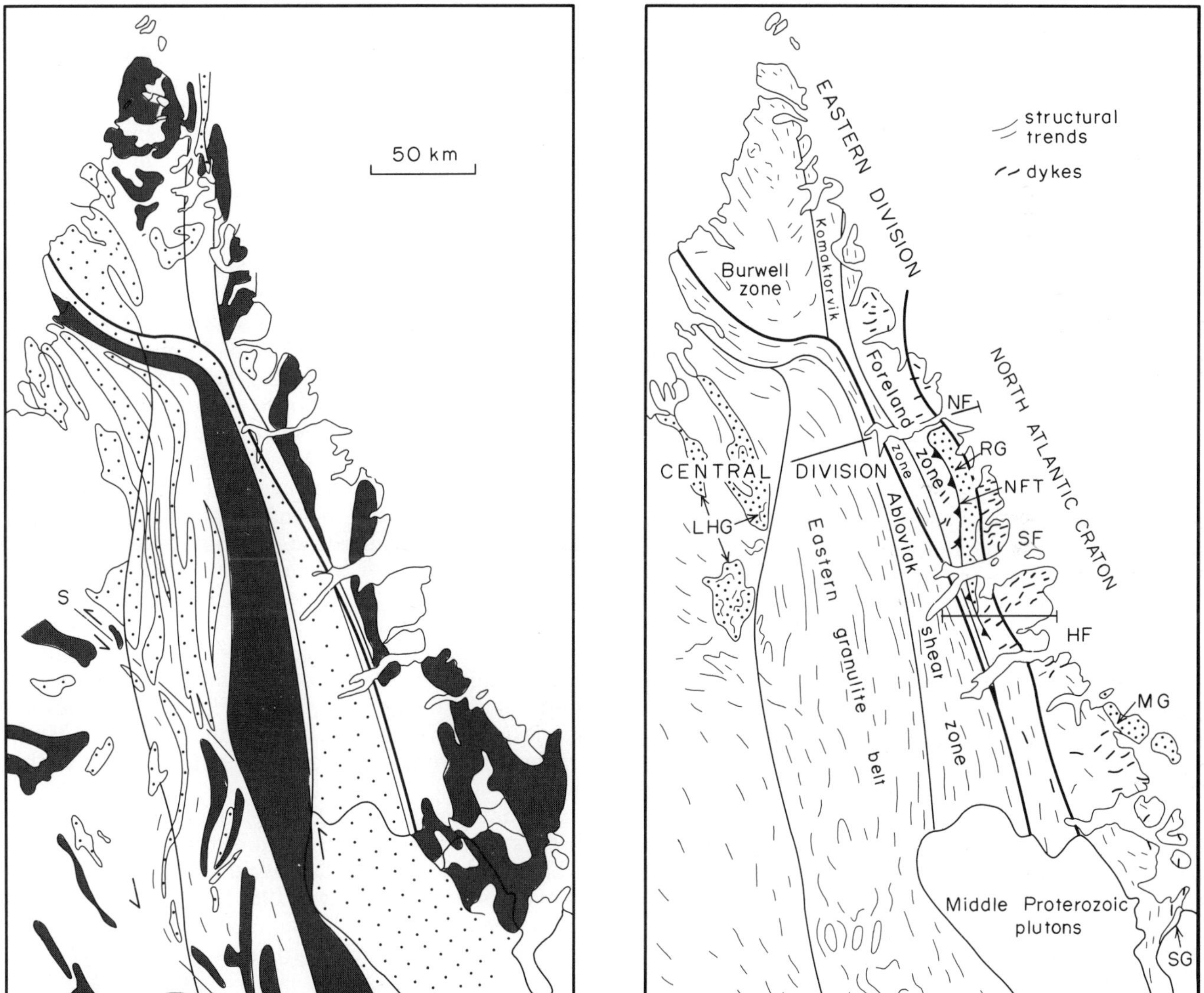

Figure 7. *Geological and aeromagnetic features of the Central Division – Eastern Division – North Atlantic Craton transitions. (a) Aeromagnetic features: highs (above 200 nT) in grey; lows (below -225 nT) in stipple. Adapted from the Geological Survey of Canada (1985). Deflections of linear magnetic anomalies are interpreted to reflect sinistral shear along structures related to transcurrent motion on the Abloviak shear zone. Solid lines are tectonic zone boundaries as on (b). (b) Geological features, including structural trends, distribution of major supracrustal units, and trends of Early Proterozoic mafic dykes. Abbreviations: MG, Mugford Group; RG, Ramah Group; SG, Snyder and Falls Brook groups; LHG, Lake Harbour Group; NF, Nachvak Fiord; SF, Saglek Fiord; HF, Hebron Fiord; and NFT, Nachvak Fiord thrust.*

The Abloviak shear zone coincides with a distinctive garnetiferous gneiss referred to as the Tasiuyak gneiss (Wardle, 1983; Mengel, 1988), but also affects orthopyroxene-bearing granitoid gneisses interlayered with this gneiss. The Tasiuyak gneiss is a banded, quartz–feldspar–garnet–biotite–sillimanite196graphite gneiss interpreted by Ryan *et al.* (1984) and Wardle (1983) to represent a migmatite, and locally a diatexite, derived from an aluminous metasedimentary protolith. The migmatitic banding has been isoclinally folded and transposed into a subvertical, planar, mylonitic fabric containing a strong subhorizontal mineral (quartz dominated) lineation. Narrow bands of mafic granulite indicate that shearing occurred under relatively anhydrous, granulite-facies conditions. The protolith age of the Tasiuyak gneiss is unknown. However, in the southern part of the zone, a lineated orthopyroxene-bearing tonalite gneiss similar to those interlayered with Tasiuyak gneiss has yielded a U–Pb zircon upper intercept discordia age of 1909^{+31}_{-21} Ma (Krogh and Scharer, 1987). This is believed to date the time of granulite-facies metamorphism and lineation development.

EASTERN DIVISION

The Eastern Division comprises tectonically reworked gneisses of the Labrador component of the NAC, divided into the Foreland, Komaktorvik, and Burwell zones. The Foreland and Komaktorvik zones are unified by the common presence of distinctive Archean gneiss associations (see below) and a Late Archean – Early Proterozoic dyke swarm.

The NAC geology in Labrador (e.g., Bridgwater *et al.*, 1975; Collerson *et al.*, 1982; Ermanovics *et al.*, 1982; Ryan *et al.*, 1983, 1984) comprises a polycyclic complex of granitoid and supracrustal gneiss formed between 3800 and 2800 Ma in the north and 3100 and 2700 Ma in the south. Granulite-facies rocks predominate in the north (except in the Saglek area), whereas the area south of the Middle Proterozoic intrusions (Fig. 2) consists of upper amphibolite-facies gneiss with lesser areas of granulite-facies gneiss and belts of lower grade volcanic and granitoid plutonic rocks (Ermanovics *et al.*, 1982). Cratonization was followed by the intrusion of east- to northeast-trending diabase dykes circa 2400 Ma (Taylor, 1979) (Figs. 7 and 8), then by uplift, erosion, and deposition of the Early Proterozoic Mugford, Snyder, Falls Brook, Ramah, and Ingrid groups (Fig. 7b). The Mugford Group (>1800 m thick, Smyth (1976)) consists of a lower siliciclastic sequence that passes upward into a thick pile of subaqueous to subaerial basalt flows and breccias. Interpretations of the tectonic significance of these rocks are hindered by a lack of precise age and chemical data. The Snyder Group (Speer, 1976, 1978) and overlying Falls Brook Group (Berg, 1981) are preserved only as a 300 m thick sliver of contact-metamorphosed siliciclastic rocks, calc-silicate and mafic pyroxene hornfels (metabasalt). The Ramah Group is described below. Hudsonian effects on the NAC are restricted to localized static retrogression, brittle faulting, and biotite cooling ages of ca. 1650–1700 Ma (Collerson, 1983).

The Foreland zone consists of variably retrogressed granulite-facies Archean crust, locally overlain by the Ramah Group. This is a synformal succession of sedimentary rocks that unconformably overlies peneplained Archean gneiss and Proterozoic dykes (Morgan, 1975; Knight and Morgan, 1981; Fig. 7b). The group has been interpreted by Knight and Morgan (1981) to record a shelf to basin transition, the basal shelf component of which has been correlated with the basal Mugford and Snyder groups by Smyth and Knight (1978). Hoffman (1987) has interpreted the basinal upper Ramah Group as the product of foredeep deposition, analogous with the situation in the Western Division.

The eastern margin of the Ramah Group is generally gently west-dipping and little disturbed. However, the western margin of the group is highly deformed and is overthrust by Komaktorvik zone gneiss (Fig. 7) along the Nachvak Fiord thrust. From Saglek Fiord southward, the group is very strongly tectonized, as manifested by numerous intra-Ramah thrusts and thrusts that interleave basement and cover (Ryan *et al.*, 1983, 1984; Mengel, 1985) (Saglek Fiord section, Fig. 8). Hudsonian effects in the foreland zone are best exemplified by the Proterozoic dykes that become progressively recrystallized from greenschist facies in the east to amphibolite facies in the west (Fig. 8). The Archean gneisses have been affected by inhomogeneous Hudsonian shearing, which increases in intensity to the west.

A small remnant of Proterozoic cover rocks also occurs near the southern end of the Foreland zone. These rocks, termed the Ingrid group (Fig. 2) (Ermanovics and Korstgård, 1981), comprise mafic to intermediate volcanic rocks, and conglomerate and clastic rocks, and are overthrust from the west by Archean gneiss.

The Komaktorvik zone is characterized by pervasive, Hudsonian, amphibolite- to granulite-facies reworking of Archean rocks of the NAC to produce finely banded, straightened, and mylonitized gneiss. The diabase dyke swarm dies out within this zone (Figs. 7 and 8) and, where present, is largely transposed into the foliation (Mengel, 1985). The Komaktorvik zone from Nachvak northward is also characterized by a series of elongate, variably deformed gabbro–anorthosite bodies (Taylor, 1979) of probable Archean age (Wardle, 1983). The western margin of the Komaktorvik zone is characterized by subhorizontal lineation of presumed similar age and origin to that in the Abloviak zone.

The Burwell zone (Burwell domain of Korstgård *et al.* (1987)) is a triangular-shaped zone isolated by the divergence of the Abloviak zone and Komaktorvik zone in northernmost Labrador. Taylor (1979) shows this area as being underlain by a mixture of granulite- and amphibolite-facies granitoid gneiss interspersed with bands of supracrustal gneiss, all interpreted to be of Early Proterozoic age. We have not seen this area, but on the basis of its irregular aeromagnetic signature (Geological Survey of Canada, 1985; Fig. 7), we venture to suggest that it is probably an Archean domain, which may have been only moderately affected by Proterozoic tectonism.

STRUCTURAL DEVELOPMENT OF THE CENTRAL–EASTERN DIVISION BOUNDARY

The intense subhorizontal lineation within the Abloviak zone and western part of the Komaktorvik zone implies transcurrent shearing. Isoclinal folds associated with the development of this fabric have a similar subhorizontal plunge and are interpreted to have been rotated into the extension direction by extreme simple shear. Mengel (1988) has described counterclockwise rotation of dykes within the Komaktorvik zone between Nachvak and Saglek fiords, which indicates a sinistral shear sense. A similar sense of shear is indicated by deflections of aeromagnetic anomalies west of the Abloviak zone (Fig. 7), which are interpreted to reflect sinistral motion both on the Abloviak zone and on shear splays that diverge to the northwest. It is also conceivable that these splays are relatively late structures related to movement on the Burwell–Komaktorvik boundary shear, a mechanism proposed by Hoffman (this volume) to explain the large S-sense flexure located at the northern end of the Abloviak zone (Fig. 7). Further criteria for the establishment of shear sense, such as rotated porphyroclasts and C–S fabric relationships, have not yet been found.

In contrast to the Abloviak and western Komaktorvik zones, the bulk of the Eastern Division is dominated by vertical to west-dipping structures, generally with downdip lineation, which are associated with east-directed thrusting. There are few reliable data to constrain the timing of this thrusting versus transcurrent shearing. Morgan (1978) has reported an apparent metamorphic Rb-Sr errorchron age of 1958 Ma (recalculated using a decay constant of $\lambda(^{87}Rb) = 1.42\times10^{-11}$ year^{-1}) for a thin, altered volcanic flow near the base of the Ramah

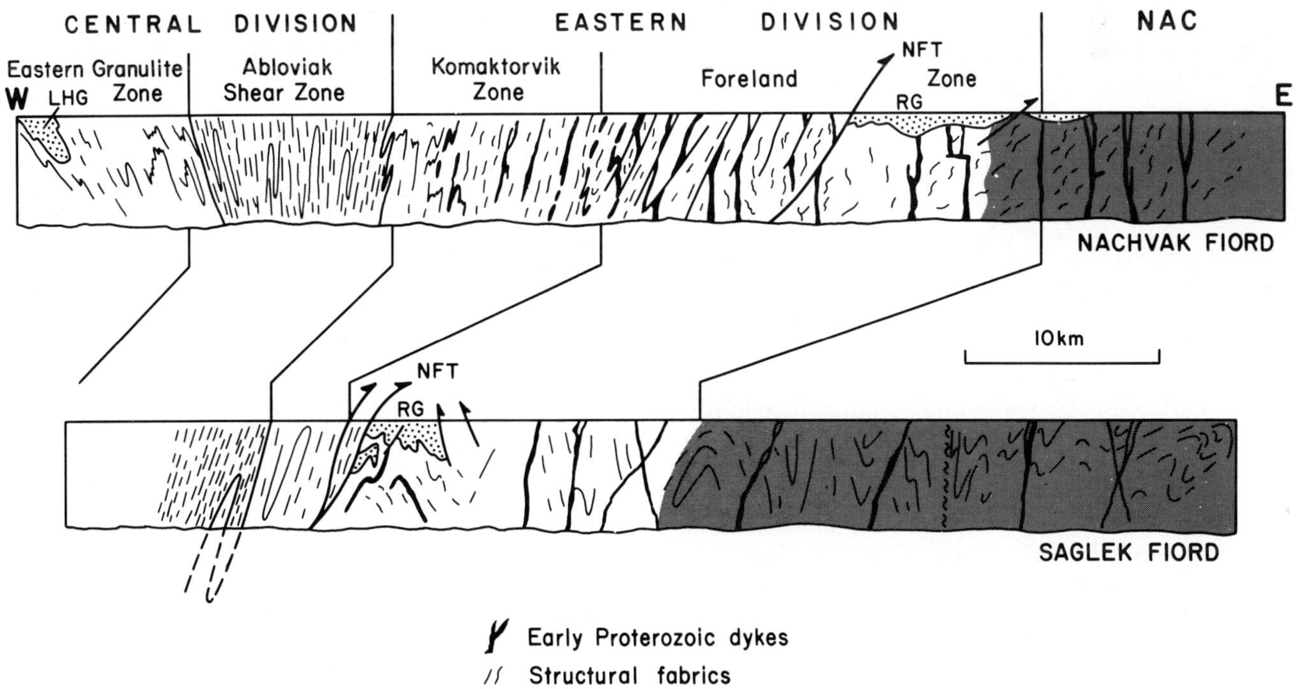

Figure 8. *Schematic structural sections across the eastern margin of the orogen. Section lines and abbreviations are keyed to Figure 7b. Early Proterozoic dykes are shown schematically. The shaded area represents fresh to weakly retrogressed Archean crust of North Atlantic Craton (NAC) affinity. The sections show progressive reworking of this crust, and Proterozoic dykes, westward toward the Abloviak shear zone.*

Group. This may, however, be a mixed age due to partial retention of primary components. Collerson (1983), also using Rb–Sr data, has inferred a thermal disturbance at ca. 1800 Ma in the Labrador part of NAC that is similar to the age of late metamorphism on the western margin of the Central Division (Perreault and Hynes, 1988). On the basis of this meagre evidence, a considerable time interval appears to have elapsed between deformation in the Abloviak zone (1900 Ma) and thermotectonism in the Eastern Division. Mengel (1988) has suggested that the Ramah Group may have been deposited in this interval. Confirmation of this chronology through further high-precision dating is an immediate priority for future work.

DISCUSSION

The structure of the Labrador segment of the Trans-Hudson Orogen is summarized in cross-section in Figure 9, together with Bouguer gravity and aeromagnetic profiles. Although exhibiting some degree of symmetry, such as cratonward-verging, marginal fold and thrust belts, overall the structural profile is asymmetric; in particular, the 150 km wide western fold and thrust belt contrasts with the narrow, 15 km wide, eastern fold and thrust belt, and the vertical Abloviak and Komaktorvik zones. The deep crustal structure of the orogen has been modelled from gravity data by Thomas *et al.* (1978), who proposed that the Central Division, on the basis of its elevated gravity values, comprised thicker, denser crust than the adjacent Western Division and underlying Superior Craton. In the absence of direct constraints on crustal thickness (e.g., seismic refraction), however, the gravity expression could also be modelled in terms of a thin Central Division crust (in comparison with the Superior Craton) and elevated Moho. The gravity profile in Figure 9b shows several regional positive anomalies, which correlate well with the tectonic zones outlined above. The Doublet terrane, not surprisingly, is associated with a pronounced gravity high. Of potentially more significance is the high developed over the western gneiss zone and De Pas batholith. This rises to -30 mGal (1 Gal = 1 cm/s^2) to the north of the section line on Figure 2, and coincides with the western edge of thickened crust as modelled by Thomas *et al.* (1978). Furthermore, it coincides with numerous ductile shear zones and the appearance of granulite-facies rocks (lower crust ?) in the western gneiss zone. Therefore, we suggest that the gravity high implies uplift of dense granulites along major listric thrusts penetrating to lower crustal depths (Fig. 9a). A pronounced gravity high over the Abloviak zone has also been modelled by Thomas *et al.* (1978) to reflect crustal thickening in combination with block faulting. We suggest a refinement (which has not been tested by modelling) of this model that involves a component of west-side-up displacement of the base of the crust along the Abloviak zone (Fig. 9a). A further gravity high, present over the centre of the division (Fig. 9b) is of uncertain origin.

With regard to overall tectonic development of the orogen, it is apparent that the Western Division developed initially as a passive-margin sequence that may have evolved into a foredeep environment with the encroachment of the Central Division from the east. The lack of granites or any significant felsic volcanic component in the Schefferville and Howse zones requires that any subduction at this time must have been directed to the east. The Doublet terrane is tentatively interpreted as a fragment of the continent–ocean transition and the Laporte terrane as either a part of the rise–slope prism or an arc-related basin. Both were thrust onto the Superior margin during collisional accretion. Evidence has been presented suggesting that the Central Division is an Archean block, or plate, with an infolded veneer of Proterozoic supracrustal rocks. Apart from the Doublet terrane and De Pas batholith, there is little preserved evidence for the generation of substantial amounts of new Proterozoic crust in the Central Division such as is seen in the island-arc terranes of the western Trans-Hudson Orogen (Lewry, 1981, 1987). Lithological similarities exist with the NAC, but these may be coincidental. Ashwal *et al.* (1988) suggest, on the basis of Pb, Nd, and Sr data, that there is little, if any, early Archean (i.e., NAC-like) crustal material within the Central Division. The parentage of the Central Division must, therefore, be regarded as uncertain.

The location of the western boundary of the Central Division plate is not clear. On the basis of their gravity data, Thomas and Kearey (1980) suggested a plate suture located along the eastern edge of the Labrador Trough (Western Division in this paper). This suggestion is strengthened by the presence of abundant shear zones within the western gneiss zone. Perreault and Hynes (1988) have also suggested a suture located along the western edge of the Kuujjuaq terrane. Recognition of sutures in surface geology is hindered by the expected similarity between reworked Archean gneiss of the Superior Craton and that of the western gneiss zone. The suture, therefore, may truly be cryptic. An alternative possibility is that it may lie along the site of the De Pas batholith. This has been suggested by Thomas and Kearey (1980) to represent the root zone of a magmatic arc related to eastward subduction under the Central Division. However, the overall composition of the batholith appears to be granitic, as opposed to the quartz diorite – tonalite – granodiorite compositions typical of Cenozoic arc-related batholiths (Pitcher, 1983; Brown *et al.*, 1984). It is also significant that the batholith is a relatively late feature that was intruded after ambient high-grade metamorphism. In view of these characteristics, it is suggested that the De Pas batholith is more likely to have developed as a late-orogenic suite produced by crustal anatexis following collision of the Central Division and the Superior plate. Similar models have been postulated for plutonic welts of the Canadian Cordillera (Monger *et al.*, 1982), and the Chipewyan batholith of the western Trans-Hudson Orogen (Halden *et al.*, 1987). This does not, however, preclude the possibility that the De Pas batholith was intruded along the site of an earlier arc.

The eastern boundary of the Central Division, the Abloviak zone, is also a potential plate boundary; however, there has yet to be recognized any major magmatic belt that would indicate subduction activity across this zone. Whether the Abloviak zone developed as a transcurrent fault separating two fundamentally different plates or split an originally continuous NAC plate is open to speculation.

Final assembly of the orogen is interpreted to have occurred in an environment of overall oblique convergence between the Superior and NAC plates (Fig. 10), resulting in frontal thrusting on the Superior margin and transpressional deformation in the east (Korstgård *et al.*, 1987). These ideas are similar to those put forward by Watterson (1978) who, however, envisaged a largely intraplate setting for Early Proterozoic orogenesis. The consequences of the relative northward motion of the NAC plate are seen as sinistral displacement along the Abloviak zone, dextral shear along the eastern margin of the NAC in the Laxfordian belt of Scotland (Park and Tarney, 1987), and frontal, south-directed thrusting in the Nagssugtoqidian Belt of Greenland (Escher *et al.*, 1976; Myers, 1987; Fig. 10). Abloviak deformation is thought to have commenced ca. 1910 Ma on the basis of the single U–Pb zircon date. Metamorphic ages for the Nagssugtoqidian Mobile Belt, however, cluster around 1850 Ma (Kalsbeek, 1986), indicating that

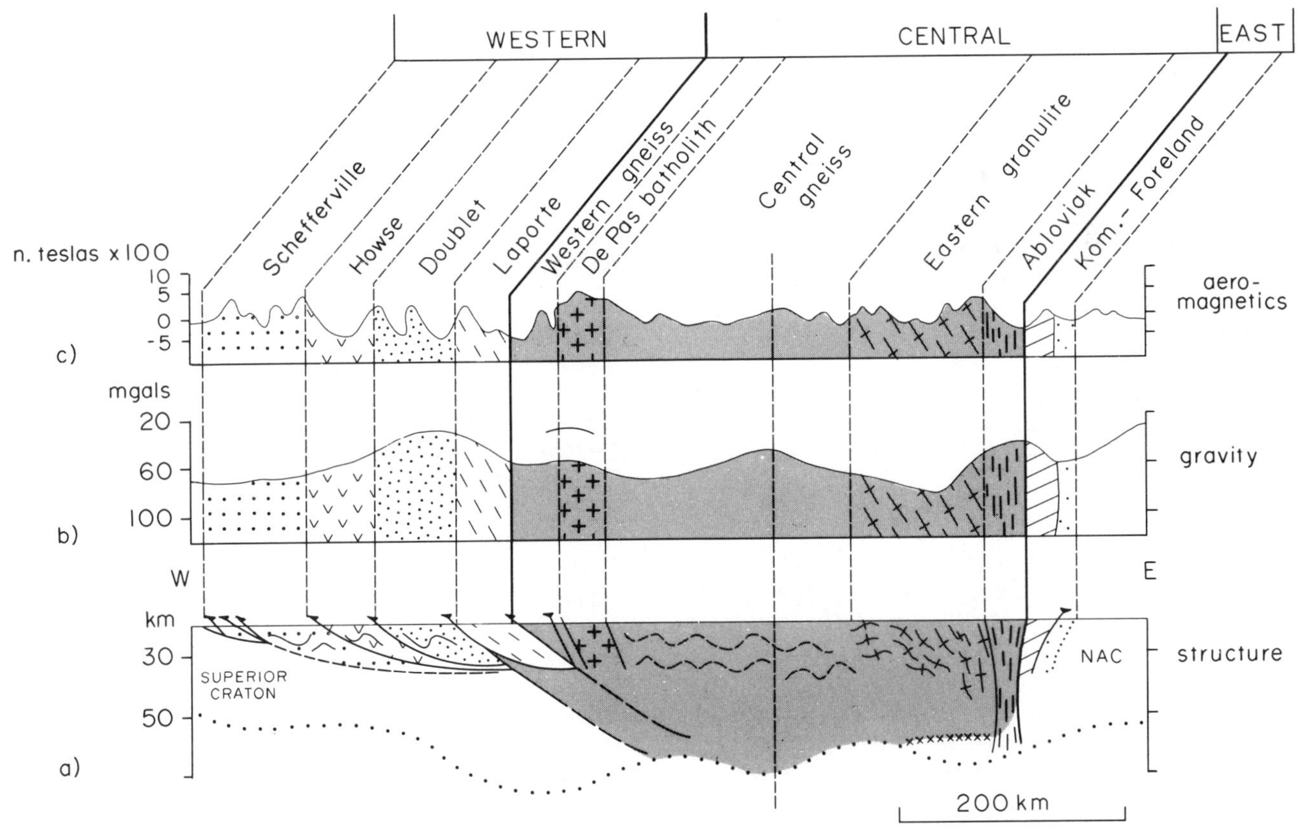

Figure 9. *(a) Structural, (b) Bouguer gravity, and (c) magnetic profiles across the Labrador segment, Trans-Hudson Orogen. The dotted line in (a) is the base of the crust as calculated from gravity data by Thomas and Kearey (1980). The line of crosses indicates a possible alternative position for the base of the crust west of the Abloviak zone. Magnetic data are taken from the Geological Survey of Canada (1985); gravity data are from Thomas* et al. (1978). Abbreviation: NAC, North Atlantic Craton.

movement on the Abloviak zone likely continued to at least this time. The early age for Abloviak deformation contrasts with the 1845–1829 Ma range for high-grade metamorphism and inferred collision on the western margin (Machado *et al.*, 1988), which are more comparable with Nagssugtoqidian ages.

This suggests that the overall transpressional model advocated by Korstgård *et al.* (1987) may have been a two-stage process consisting of early transpressional deformation on the Nain – Central Division boundary followed by collision of the Superior plate and Central Division, and consequent east–west shortening across the orogen. This later collisional event may also have involved a component of oblique convergence that resulted in dextral shear within the western gneiss zone (van der Leeden *et al.*, this volume). The Central Division, therefore, appears to have acted as a promontory (suggested by Hoffman (this volume) to be the southern end of his Archean Rae Craton) that has moved south relative to the bounding Superior and NAC plates (Fig. 10).

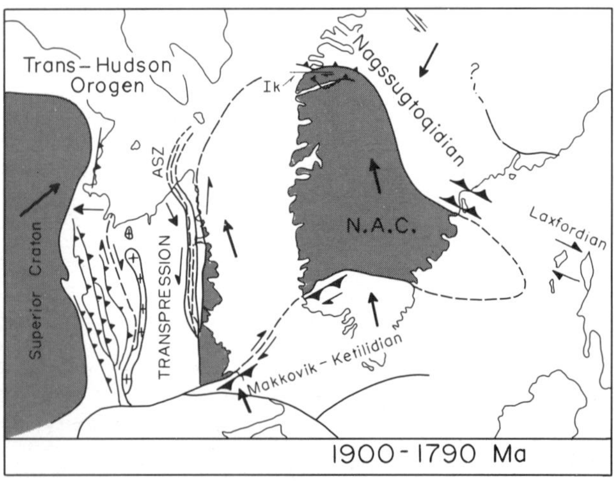

Figure 10. *Model showing the development of the Labrador segment, Trans-Hudson Orogen, by oblique convergence and collision of the Superior and North Atlantic plates. The deformation sequence is proposed to have involved (1) early transcurrent shear along the Abloviak zone in response to Central Division – NAC interaction continuing to ca. 1850 Ma (northwestward motion of NAC was accommodated by dextral shear and thrusting on its eastern and northern margins respectively) and (2) east- and west-directed thrusting on the margins of the orogen in response to Superior – Central Division collision and east–west shortening. This probably also involved oblique convergence and produced dextral shearing in the western gneiss zone.*

CONCLUSIONS

The Labrador segment of the Trans-Hudson Orogen is divided into (1) a Western Division, recording rifting and collisional events on the Superior Craton margin; (2) a Central Division, which appears to be an Archean plate of uncertain affinity; and (3) an Eastern Division forming part of the NAC. Vestiges of the continent–ocean transition may be represented by the Doublet Group. Apart from the De Pas batholith, however, there is little evidence for the preservation of substantial amounts of juvenile Proterozoic crust.

The assembly of the orogen was controlled by (1) oblique convergence of Central Division and NAC plates leading to sinistral transcurrent shear along the Central Division – NAC boundary ca. 1900 Ma and possibly younger, and (2) collision of the Superior plate with the combined Central Division – NAC crust, resulting in east–west shortening, development of marginal, foreland fold and thrust belts (1845–1829 Ma), dextral shearing, and intrusion of the De Pas batholith (1811 Ma).

Many of these ideas are obviously speculative and based on a meagre data base. They underscore, however, the need for additional mapping and associated studies, particularly geochronology, in an orogen that has the potential to yield valuable new insights into the fundamental processes of terrane suturing and transcurrent shearing in the middle and lower crust.

ACKNOWLEDGEMENTS

Fieldwork upon which the first three authors' experience is based was financed, in part, through jointly funded projects of the Department of Energy, Mines and Resources, Canada, and the Province of Newfoundland and Labrador. The work of F. Mengel was, in part, financed by the Department of Energy, Mines and Resources, Canada, and by the Natural Sciences and Engineering Research Council of Canada grant of T. Rivers, Department of Earth Sciences, Memorial University.

The authors wish to thank N. Goulet of Université du Québec a Montréal and P. Hoffman of the Geological Survey of Canada, and also our counterparts in the Québec Ministère des l'Énergie et des Ressources, particularly T. Clark, M. Bélanger, J. van der Leeden, and V. Owen, for valuable field discussions and access to unpublished information. T. Birkett is thanked for his views and for providing unpublished information on the Labrador Trough.

The manuscript has benefitted considerably from comments by P. Hoffman and T. Rivers. Formal review was provided by K. Collerson and an anonymous reviewer. Tony Paltanavage, David Leonard, and Ken Byrne of the Newfoundland Department of Mines are thanked for drafting the figures, and Joanne Matthews for word processing. Publication is with the permission of the Assistant Deputy Minister, Geological Survey Branch, Newfoundland Department of Mines.

REFERENCES

Ashwal, L.D., Wooden, J.L. and Emslie, R.F.,1986, Sr, Nd and Pb isotopes in Proterozoic intrusives astride the Grenville Front in Labrador: implications for crustal contamination and basement mapping: Geochimica et Cosmochimica Acta, v. 50, p. 2571–2585.

Ashwal, L.D., Wooden, J.L., Wiebe, R.A. and Emslie, R.F., 1988, Isotopic signatures of Proterozoic intrusives in the eastern Canadian Shield as probes of older basement types: Geological Association of Canada – Mineralogical Association of Canada – Canadian Society of Petroleum Geologists, Program with Abstracts, v. 12, p. A3.

Baragar, W.R.A., 1967, Wakuach map area: Geological Survey of Canada, Memoir 344, 174 p.

Baragar, W.R.A. and Scoates, R.F.J., 1981, The Circum-Superior belt: a Proterozoic plate margin? in Kröner, A., ed., Precambrian Plate Tectonics. Developments in Precambrian Geology 4: Elsevier, p. 297–330.

Bélanger, M., 1984, Région du Lac Brisson, Territoire-du-Nouveau Québec: Gouvernment du Québec, Ministère de l'Énergie et des Ressources, Direction générale de l'-Exploration géologique et minérale, Service de la Géologie, DV 84-20.

Berg, J.H., 1981, Stratigraphic comparison of the Snyder and Falls Brook groups with the Mugford Group, in Morse, S.A., ed., The Nain Anorthosite Project, Labrador: Field Report 1981: Department of Geology and Geography, University of Massachusetts, Amherst, MA, Contribution No. 40, p. 17–22.

Birkett, T.C., Parrish, R., Watanabe, D. and Findlay, J.M., *in prep.*, Age of the Howse Lake Sill, and related sediments and volcanics of the Labrador Trough.

Bridgwater, D., Collerson, K.D., Hurst, R.W. and Jesseau, C.W., 1975, Field characteristics of early Precambrian rocks from Saglek, coast of Labrador: Geological Survey of Canada, Paper 75-1A, p. 287–296.

Brown, G.C., Thorpe, R.S. and Webb, P.C., 1984, The geochemical characteristics of granitoids in contrasting arcs and comments on magma sources: Geological Society of London, v. 141, p. 413–426.

Clark, T., 1984, Géologie de la région du Lac Cambrien, Territoire-du-Nouveau Québec: Ministère de l'Énergie et des Ressources, Direction générale de l'Exploration géologique et minérale, Service de la Géologie, ET 83-02.

Collerson, K.D., 1983, The Archean gneiss complex of northern Labrador. 2. Mineral ages, secondary isochrons, and diffusion of strontium during polymetamorphism of the Uivak gneisses: Canadian Journal of Earth Sciences, v. 20, p. 707–718.

Collerson, K.D., Kerr, A., Vocke, R.D. and Hanson, G.N., 1982, Reworking of sialic crust as represented in late Archean-age gneisses, northern Labrador: Geology, v. 10, p. 202–208.

Coney, P.J., Jones, D.L. and Monger, J.W.H., 1980, Cordilleran suspect terranes: Nature, v. 288, p. 329–333.

Dimroth, E., 1971, The evolution of the central segment of the Labrador Geosyncline. Part II. The ophiolitic suite: Neues Jahrbuch fur Geologie und Paläontologie, Abhandlungen, v. 137, p. 209–248.

Dimroth, E., 1978, Labrador Trough area (54°30′–56°30′): Ministère des Richesses Naturelles, Québec, Rapport Géologique 193, 396 p.

Dimroth, E., Baragar, W.R.A., Bergeron, R. and Jackson, G.D., 1970, The filling of the Circum-Ungava Geosyncline, in Baer, A.J., ed., Symposium on Basins and Geosynclines of the Canadian Shield: Geological Survey of Canada, Paper 70-40, p. 45–142.

Dressler, B., 1979, Région de la Fosse du Labrador: Gouvernment du Québec, Ministère des Richesses Naturelles du Québec, Direction générale de la Recherche géologique et minérale, Service de la Géologie, Rapport Géologique 195, 117 p.

Emslie, R.F., 1978, Elsonian magmatism in Labrador: age, characteristics and tectonic setting: Canadian Journal of Earth Sciences, v. 15, p. 438–453.

Ermanovics, I. and Korstgård, J., 1981, Geology of the Hopedale block and adjacent areas, Labrador: report 3: Geological Survey of Canada, Paper 81-1A, p. 69–76.

Ermanovics, I.F., Korstgård, J.A. and Bridgwater, D., 1982, Structural and lithological chronology of the Archean Hopedale block and the adjacent Proterozoic Makkovik Subprovince, Labrador: report 4: Geological Survey of Canada, Paper 82-1B, p. 153–165.

Escher, A., Sørensen, K. and Zeck, H.P., 1976, Nagssugtoqidian mobile belt in West Greenland, in Escher, A. and Watt, W.S., eds., Geology of Greenland: Grønlands Geologiske Undersøgelse, Copenhagen, p. 77–95.

Frarey, M.J., 1967, Willbob and Thompson Lake map areas, Quebec and Newfoundland: Geological Survey of Canada, Memoir 348, 73 p.

Geological Survey of Canada, 1981, Magnetic anomaly map of Arctic Canada: Geological Survey of Canada, Map 1512A.

Geological Survey of Canada, 1985, Magnetic anomaly maps, Torngat Mountains, Churchill Falls and Schefferville: Geological Survey of Canada, Maps NO-20-M, NN-20-M, and NN-19-M, scale 1:1,000,000.

Halden, N.M., Clark, G.S., Lenton, P.G., Corkery, M.T. and Schledewitz, D., 1987, Geochemical and isotopic constraints on the tectonic significance of the Chipewyan batholith, and Proterozoic magma genesis: Geological Association of Canada – Mineralogical Association of Canada, Program with Abstracts, v. 12, p. 51.

Harrison, J.M., Howell, J.E. and Fahrig, W.F., 1972, A geological cross-section of the Labrador miogeosyncline near Schefferville, Quebec: Geological Survey of Canada, Paper 70-37, 34 p.

Hill, J.D., 1982, Geology of the Flowers River – Notakwanon River area, Labrador: Newfoundland Department of Mines and Energy, Mineral Development Division, Report 82-6, 139 p.

Hoffman, P.F., 1980, Wopmay Orogen: a Wilson cycle of early Proterozoic age in the northwest of the Canadian Shield, in Strangway, D.W., ed., The Continental Crust and its Mineral Deposits: Geological Association of Canada, Special Paper 20, p. 523–549.

Hoffman, P.F., 1981, Autopsy of Athapuscow Aulacogen: a failed arm affected by three collisions, in Campbell, F.H.A., ed., Proterozoic Basins of Canada: Geological Survey of Canada, Paper 81-10, p. 97–101.

Hoffman, P.F., 1987, Proterozoic foredeeps, foredeep magmatism and giant iron-formation, in Kröner, A., ed., Proterozoic Lithospheric Evolution: American Geophysical Union, Geodynamics Series, No. 17, p. 85–98.

Hoffman, P.F., 1988, United plates of America, the birth of a craton: Early Proterozoic assembly and growth of Laurentia: Annual Review of Earth and Planetary Sciences, No. 16, p. 543–603.

Hoffman, P.F., 1990, Subdivision of the Churchill Province and extent of the Trans-Hudson Orogen, in Lewry, J.F. and Stauffer, M.R., eds., The Early Proterozoic Trans-Hudson Orogen of North America: Geological Association of Canada, Special Paper 37, p. 15–39.

Hoffman, P. and Bowring, S.A., 1984, A short lived 1.9 Ga continental margin and its destruction, Wopmay Orogen, northwest Canada: Geology, v. 12, p. 68–72.

Jackson, G.D. and Taylor, F.C., 1972, Correlation of major Aphebian rock units in the northeastern Canadian Shield: Canadian Journal of Earth Sciences, v. 9, p. 1659–1669.

Kalsbeek, F., 1986, The tectonic framework of the Precambrian shield of Greenland. A review of new isotopic evidence, in Developments in Greenland Geology: Grønlands Geologiske Undersøgelse, Rapport No. 128, p. 55–64.

Karlstrom, K.E. and Houston, R.S., 1984, The Cheyenne belt: analysis of a Proterozoic suture in southern Wyoming: Precambrian Research, v. 25, p. 415–416.

Knight, I. and Morgan, W.C., 1981, The Aphebian Ramah Group, northern Labrador, in Campbell, F.H.A., ed., Proterozoic Basins of Canada: Geological Survey of Canada, Paper 81-10, p. 313–330.

Korstgård, J., Ryan, B. and Wardle, R.J., 1987, The boundary between Proterozoic and Archean crustal blocks in West Greenland and northern Labrador, in Park, R.G. and Tarney, J., eds., Evolution of the Lewisian and Comparable Precambrian High Grade Terrains: Geological Society of London, Special Publication No. 27, p. 247–259.

Krogh, T.E., 1986, Report to Newfoundland Department of Mines and Energy on isotopic dating results from the 1985–1986 geological research agreement: Newfoundland Department of Mines and Energy, Mineral Development Division, Open File Report LAB 707, 106 p.

Krogh, T.E. and Scharer, U., 1987, Labrador geochronology. Report for Newfoundland Department of Mines and Energy: Newfoundland Department of Mines and Energy, Mineral Development Division, Open File, 47 p.

Lewry, J.F., 1981, Lower Proterozoic arc–microcontinent collisional tectonics in the western Churchill Province: Nature, v. 294, p. 69–72.

Lewry, J.F., 1987, The Trans-Hudson Orogen: extent, subdivision and problems: Geological Association of Canada – Mineralogical Association of Canada, Program with Abstracts, v. 12, p. 67.

Lewry, J.F. and Collerson, K.D., 1990, The Trans-Hudson Orogen: extent, subdivision, and problems, in Lewry, J.F. and Stauffer, M.R., eds., The Early Proterozoic Trans-Hudson Orogen of North America: Geological Association of Canada, Special Paper 37, p. 1–14.

Machado, N., Perreault, S. and Hynes, A., 1988, Timing of continental collision in the northern Labrador Trough, Quebec: evidence from U–Pb geochronology: Geological Association of Canada – Mineralogical Association of Canada – Canadian Society of Petroleum Geologists, Program with Abstracts, v. 13, p. A76.

Mengel, F.C., 1985, Nain–Churchill Province boundary: a preliminary report on a cross-section through the Hudsonian front in the Saglek Fiord area, northern Labrador: Newfoundland Department of Mines and Energy, Mineral Development Division, Report 85-1, p. 33–42.

Mengel, F.C., 1988, Thermotectonic evolution of the Proterozoic–Archean boundary in the Saglek area, northern Labrador, unpublished Ph.D. thesis, Memorial University of Newfoundland.

Monger, J.W.H., Price, R.A. and Templeman-Kluit, D.J., 1982, Tectonic accretion and the origin of the two major metamorphic and plutonic welts in the Canadian Cordillera: Geology, v. 10, p. 70–75.

Morgan, W.C., 1975, Geology of the Precambrian Ramah Group and basement rocks in the Nachvak Fiord – lek Fiord area, northern Labrador: Geological Survey of Canada, Paper 74-54.

Morgan, W.C., 1978, Ramah Group volcanics, Labrador, in Wanless, R.K. and Loveridge, W.D., eds., Rubidium–Strontium Isotopic Age Studies. Report 2 (Canadian Shield), p. 56–61.

Myers, J.S., 1987, The East Greenland Nagssugtoqidian mobile belt compared with the Lewisian Complex, in Park, R.G. and Tarney, J., eds., Evolution of the Lewisian and Comparable Precambrian High Grade Terrains: Geological Society of London, Special Publication No. 27, p. 235-246.

Nunn, G.A.G. and Noel, N., 1982, Regional geology east of Michikamau Lake, central Labrador: Newfoundland Department of Mines and Energy, Mineral Development Division, Report 82-1, p. 149–167.

Nunn, G.A.G., Thomas, A. and Krogh, T.E., 1985, The Labradorian Orogeny: geochronological database: Newfoundland Department of Mines and Energy, Mineral Development Division, Report 85-1, p. 43–54.

Owen, V., 1984, Région du Lac Leif: Gouvernment du Québec, Ministère de l'Énergie et des Ressources, Direction générale de l'Exploration géologique et minérale, Service de la Géologie, DP 85-14.

Park, R.G. and Tarney, J., 1987, The Lewisian Complex: a typical Precambrian high-grade terrain?, in Park, R.G. and Tarney, J., eds., Evolution of the Lewisian and Comparable Precambrian High Grade Terrains: Geological Society of London, Special Publication No. 27, p. 13–25.

Percival, J.A., 1987, Geology of the Ashuanipi granulite complex in the Schefferville area, Québec: Geological Survey of Canada, Paper 87-1A, p. 1–10.

Perreault, S. and Hynes, A., 1988, Timing in the tectonic evolution of the northeast segment of the Labrador Trough, Kuujjuaq, northern Quebec: Geological Association of Canada – Mineralogical Association of Canada–Canadian Society of Petroleum Geologists, Program with Abstracts, v. 13, p. A97–A98.

Pitcher, W.S., 1983, Granite type and tectonic environment, in Hsu, K., ed., Mountain Building Processes: Academic Press, p. 19–40.

Poirier, G., Perreault, S. and Hynes, A., 1987, The nature of the eastern boundary of the Labrador Trough near Kuujjuaq, Quebec: Geological Association of Canada – Mineralogical Association of Canada, Program with Abstracts, v. 12, p. 81.

Ryan, A.B., Martineau, Y., Korstgård, J. and Lee, D., 1984, The Archean–Proterozoic boundary in northern Labrador. Report 2: Newfoundland Department of Mines and Energy, Mineral Development Division, Report 84-1, p. 12–20 (also Geological Survey of Canada, Paper 84-1A, p. 545–551).

Ryan, A.B., Lee, D. and Dunphy, D., 1988, The discovery of probable Archean rocks within the Labrador arm of the Trans-Hudson Orogen near the Labrador-Quebec border: Newfoundland Department of Mines, Mineral Development Division, Report 88-1, p. 1–14.

Ryan, B. and Lee, D., 1986, Gneiss–anorthosite relationships in the Anaktalik Brook – Kogaluk River area (NTS 14D/1,8) Labrador: Newfoundland Department of Mines and Energy, Mineral Development Division, Report 86-1, p. 79–88.

Ryan, B., Martineau, Y., Bridgwater, D., Schiotte, L. and Lewry, J., 1983, The Archean–Proterozoic boundary in the Saglek Fiord area. Report 1: Newfoundland Department of Mines and Energy, Mineral Development Division, Report 83-1, p. 91-98 (also Geological Survey of Canada, Paper 83-1A, p. 297-304).

Ryan, B., Lee, D. and Corriveau, L., 1987, Geology of the eastern Churchill Province between Anaktalik Brook and Cabot Lake (NTS 14D/2,6,7), Labrador: Newfoundland Department of Mines, Mineral Development Division, Report 87-1, p. 155-159.

Smyth, W.R., 1976, Geology of the Mugford Group, northern Labrador: Newfoundland Department of Mines and Energy, Mineral Development Division, Report 76-1, p. 72–79.

Smyth, W.R. and Knight, I., 1978, Correlation of the Aphebian supracrustal sequences, Nain Province, northern Labrador: Newfoundland Department of Mines and Energy, Mineral Development Division, Report 78-1, p. 59–64.

Speer, J., 1976, The stratigraphy and metamorphism of the Snyder Group, Labrador, unpublished Ph.D. thesis, Virginia Polytechnic Institute and State University.

Speer, J., 1978, The stratigraphy and depositional environment of the Aphebian Snyder Group, Labrador: Canadian Journal of Earth Sciences, v. 15, p. 52–68.

Stockwell, C.H., 1964, Fourth report on structural provinces, orogenies and time-classification of rocks of the Canadian Precambrian Shield: Geological Survey of Canada, Paper 64-17, pt. 2, p. 1–21.

Taylor, F.C., 1979, Reconnaissance geology of a part of the Precambrian Shield, northeastern Québec, northern Labrador and Northwest Territories: Geological Survey of Canada, Memoir 393, 99 p.

Thomas, M.D. and Kearey, P., 1980, Gravity anomalies, block faulting and Andean-type tectonism in the eastern Churchill Province: Nature, Physical Science, v. 282, p. 61–63.

Thomas, M.D., Halliday, D.W. and Stephenson, R., 1978, Gravity anomalies and geological structure in northern Labrador and northeastern Québec: Energy, Mines and Resources Canada, Earth Physics Branch, Gravity Map Series, Numbers 157–161, 38 p.

van der Leeden, J., 1983: Géologie de la région du lac de la Hutte Sauvage: Gouvernment du Québec, Ministère de l'Énergie et des Ressources, Direction générale de l'Exploration géologique et minérale, Service de la Géologie, DP 84-29.

van der Leeden, J., Bélanger, M. and Davis, D., 1990, Lithotectonic domains in the high-grade terrain east of the Labrador Trough (Québec), in Lewry, J.F. and Stauffer, M.R., eds., The Early Proterozoic Trans-Hudson Orogen of North America: Geological Association of Canada, Special Paper 37, p. 371–386.

Wardle, R.J., 1982, Geology of the south-central Labrador Trough: Newfoundland Department of Mines and Energy, Mineral Development Division, Maps 82-5 and 82-6.

Wardle, R.J., 1983, Nain–Churchill Province cross-section, Nachvak Fiord, northern Labrador: Newfoundland Department of Mines, Mineral Development Division, Report 83-1, p. 68–90.

Wardle, R.J., 1984, Nain–Churchill Province cross-section, Nachvak Fiord, northern Labrador: Newfoundland Department of Mines and Energy, Mineral Development Division, Report 84-1, p. 1–11.

Wardle, R.J., 1985, Geology of the Churchill Falls area: Newfoundland Department of Mines and Energy, Mineral Development Division, Report 85-2, 70 p.

Wardle, R.J. and Bailey, D.G., 1981, Early Proterozoic sequences in Labrador, in Campbell, F.H.A., ed., Proterozoic Basins of Canada: Geological Survey of Canada, Paper 81-10, Supplement, p. 331–358.

Wardle, R.J., Rivers, T., Gower, C.F., Nunn, G.A.G. and Thomas, A., 1986, The northeastern Grenville Province: new insights, in Moore, J.M., Davidson, A. and Baer, A.J., eds., The Grenville Province: Geological Association of Canada, Special Paper 31, p. 13–29.

Watterson, J., 1978, Proterozoic intraplate deformation in the light of southeast Asian neotectonics: Nature, London, v. 273, p. 636–640.

Williams, H. and Hatcher, R.D., Jr., 1982, Suspect terranes and accretionary history of the Appalachian orogen: Geology, v. 10, p. 530–536.

Lithotectonic Domains in the High-grade Terrain East of the Labrador Trough (Quebec)[1]

J. Van der Leeden, M. Bélanger, and D. Danis
Ministère du l'Énergie et des Ressources du Québec, 5700, 4e Avenue, 2e étage, Charlesbourg, Québec G1H 6R1, Canada

R. Girard
Module des Sciences de la Terre,Université du Québec à Chicoutimi, Chicoutimi, Quebec G7H 2B1, Canada

J. Martelain
Département des Sciences de la Terre, Université de Lyon 1, 69622 Villerbanne Cedex, France

Van der Leeden, J., Bélanger, M., Danis, D., Girard, R. and Martelain, J., Lithotectonic domains in the high-grade terrain east of the Labrador Trough (Quebec), in Lewry, J.F. and Stauffer, M.R., eds., The Early Proterozoic Trans-Hudson Orogen of North America: Geological Association of Canada, Special Paper 37, p. 371–386.

Abstract

Recent mapping in the high-grade terrain of Quebec east of the Labrador Trough, northeast of Schefferville, permits recognition of six distinct lithotectonic domains. These are, from west to east, the Laporte, Lac Tudor Shear Zone, De Pas, Rivière George Shear Zone, Mistinibi–Raude, and Mistastin domains. The Laporte (northwest trends) and Lac Tudor Shear Zone (north to north-northwest trends) domains contain immature volcano-sedimentary rocks, marginal basin sediments, and possible basement gneisses; these domains bear some resemblance to more recent fore-arc regions. Metamorphic grade increases eastward, from lower amphibolite facies to granulite facies. The De Pas Domain consists of the north- to north-northwest-trending late- to posttectonic granodioritic De Pas Batholith, which resembles an Andean-type magmatic arc; the boundaries of this domain comprise basement gneiss. To the east lies the Rivière George Shear Zone (north to north-northwest trends) and Mistinibi–Raude (north to northwest and east trends) domains. Three lithologic assemblages are found in these domains: (1) a late metasedimentary assemblage of cratonic origin, (2) a gneissic terrain, and (3) a magmatic assemblage of mafic and intermediate volcanics and intrusions, and associated metasedimentary rocks, intruded by pre- and late- to posttectonic granites. These domains contain features also found in modern back-arc and plateau regions. The easternmost domain, the Mistastin Domain, comprises posttectonic Elsonian plutonic rocks and intervening granulites.

[1]Published with the permission of the Ministère de l'Énergie et des Ressources du Québec. Contribution No. 90-5110-16.

Numerous features of the lithotectonic domains have plate tectonic analogues: the area probably has had a complex tectonic history involving Early Proterozoic rifting, sedimentation and volcanism, subduction, arc formation, and continental collision, implying extensional, contractional, and transverse movements. These events were followed by an important Middle Proterozoic anorogenic magmatic episode.

Résumé

La cartographie récente faite dans les terrains hautement métamorphisés à l'est de la Fosse du Labrador, au nord-est de Schefferville a permis de reconnaître six domaines lithotectoniques. Ces domaines sont d'ouest en est: Laporte, lac Tudor, De Pas, Rivière George, Mistinibi–Raude et Mistastin. Le Laporte est caractérisé par un grain tectonique nord-ouest et le lac Tudor par une zone de cisaillement d'attitude nord à nord-nord-ouest. Ces deux domaines contiennent des séquences volcano-sédimentaires immatures aussi bien que des roches sédimentaires de bassins marginaux ou des gneiss probablement issus d'un socle. Un métamorphisme prograde d'ouest en est, marque une évolution du faciès amphibolite inférieur au faciès granulite. Le domaine de De Pas est formé d'un batholite de roche granodioritique tardi- à post-tectonique d'attitude nord à nord-nord-ouest; ses caractéristiques lui confèrent des ressemblances avec un arc magmatique de type Andin. Ce batholite est limité de part et d'autre par des gneiss de socle. À l'est, la zone de cisaillement de Rivière George marque un domaine d'attitude nord à nord-nord-ouest, elle sépare les roches du batholite de De Pas du domaine de Mistinibi–Raude. Ces deux domaines sont formés de trois assemblages lithologiques: (1) un assemblage tardif de métasédiments d'origine cratonique, (2) un terrain gneissique, (3) un assemblage de roches intrusives et volcaniques de composition intermédiaire à mafique et de roches métasédimentaires associées, lequel est recoupé par des granites pré- et tardi- à post-tectoniques. Ces domaines contiennent des éléments caractéristiques de régions d'arrière-arc et de plateau. Le domaine de Mistastin constitue le domaine le plus à l'est de la région étudiée; il est formé des plutons elsoniens, post-tectoniques et des granulites associées.

Plusieurs éléments des domaines lithotectoniques présentent des analogies avec une tectonique de plaque, et l'histoire tectonique de la région, bien que très complexe peut se résumer par : la formation de rift au début de l'ère Protérozoïque, avec sédimentation et volcanisme, subduction, formation d'un arc magmatique et collision continentale, ces événements se traduisant par des mouvements d'extension, de contraction et de cisaillements obliques. Un magmatisme anorogénique a marqué le dernier événement important de l'histoire géologique durant le Protérozoïque moyen.

INTRODUCTION

In the first part of this paper, we present a synthesis of the results of a regional field-mapping project conducted by the Quebec government; in the second part, we present an interpretative model for a major orogenic event in the area studied.

In 1983, the Quebec Ministry of Energy and Resources undertook a 5 year regional mapping project east of the central Labrador Trough (the "Rivière George Project"). The mapping was done at a scale of 1:50,000, and over 25,000 km^2 were covered from the continental divide in the east and south, northward to latitude 56°30′, and westward to longitude 66° (Fig. 1). At the time of writing, final compilation work for the project was incomplete; consequently, the present study is of a preliminary nature. Reference to past and ongoing work of the Rivière George Project can be found at the end of this paper.

In the east, the Mistastin Batholith was previously mapped by Emslie *et al.* (1980) at a scale of 1:100,000, and the Michikamau Anorthosite was mapped in detail by Emslie (1970); the entire area also formed part of an earlier, larger reconnaissance mapping project, at a scale of 1:250,000, by Taylor (1979).

The area covered by the Rivière George Project lies totally within the Churchill Province (as interpreted by Taylor (1971)), and forms part of what has recently been called the "Labrador Segment of the Trans-Hudson Orogen" (Van Schmus *et al.*, 1987) and the "George River Terrane" (Hoffman, 1987), a southeastern branch of the "Rae Province" (Hoffman, 1988). As a consensus has not yet been reached, we will informally use the name "Labrador segment" for this branch (Fig. 2).

This mapping has led to the recognition of several distinctive lithotectonic domains. Much of the area shows evidence of an Hudsonian orogenic event, for which we here propose a plate tectonic model.

LITHOTECTONIC DOMAINS

The geology of the area is characterized by an asymmetric distribution of lithologies, which can be grouped into several distinctive north–south trending lithotectonic domains; from west to east, these are the Laporte Domain, the Lac Tudor Shear Zone Domain, the De Pas Domain, the Rivière George Shear Zone Domain, the Mistinibi–Raude Domain, and the Mistastin Domain (Figs. 1 and 3).

Laporte Domain

The Laporte Domain is bounded to the southwest by the Labrador Trough, which has previously been described in detail (e.g., Dimroth *et al.*, 1970; Dimroth, 1981). Relevant features of the Labrador Trough are discussed, in relation to a general tectonic model for the part of the Churchill Province under discussion, after the lithotectonic domain descriptions.

The Laporte Domain consists mainly of rocks of the informal unit, the Laporte group (including pelitic and semipelitic schists, gneisses, meta-arkoses, mafic metavolcanics and metagabbros, minor quartzite, metaconglomerate, marble, and meta-ultramafite), along with several regional lenses of weakly foliated, augen-bearing, and strongly migmatized ortho- and paragneisses of granodioritic compositions. There are also some mafic and ultramafic lenses. Contact relations with the encompassing Laporte group have not been observed; however, the granodiorite gneisses may be older than the Laporte group.

The dominant formations comprise immature metasedimentary rocks typical of successions deposited in distal deep-water basins. A period of relative uplift is attested to by other, less abundant, metasedimentary rocks; for example, there is an informal formational unit (indicated in Fig. 3 by a circle pattern) that may have had a fluvial origin. This is composed mainly of meta-arkoses with muscovite-bearing, hematite-stained, and specularite-laminated facies, minor quartzite, quartz wacke, calcarenite, and conglomerate, and very minor pelitic schist and amphibolite. Part of this unit resembles the Hutte Sauvage Group, east of the Rivière George.

The metabasalts are low-K, tholeiitic (Fig. 4a,b), ocean-floor metabasalts (Fig. 4c). Though some samples plot as within-plate basalts in a (Ti/100)–Zr–(Y × 3) diagram (Fig. 4d), all have a strong oceanic affinity (Fig. 4e) (also noted by Wardle and Bailey (1981) for similarly situated metabasalts south of the present area), showing geochemical traits similar to those of midoceanic-ridge basalt (MORB) and back-arc basin basalts (Fig. 4f,g).

Second-phase, tight to open regional folds have imparted a northwest-trending tectonic grain to the Laporte Domain. First-phase schistosity and minor folds are refolded by mesoscopic second-phase folds. The latter range from moderately to steeply inclined and upright to overturned to the southwest.

Lithological units can be traced eastward into the Lac Tudor Shear Zone; metamorphic grade increases from lower amphibolite facies to upper amphibolite and granulite facies.

Lac Tudor Shear Zone Domain

This domain represents a regional shear zone, the Lac Tudor Shear Zone (LTSZ), that affects rocks of

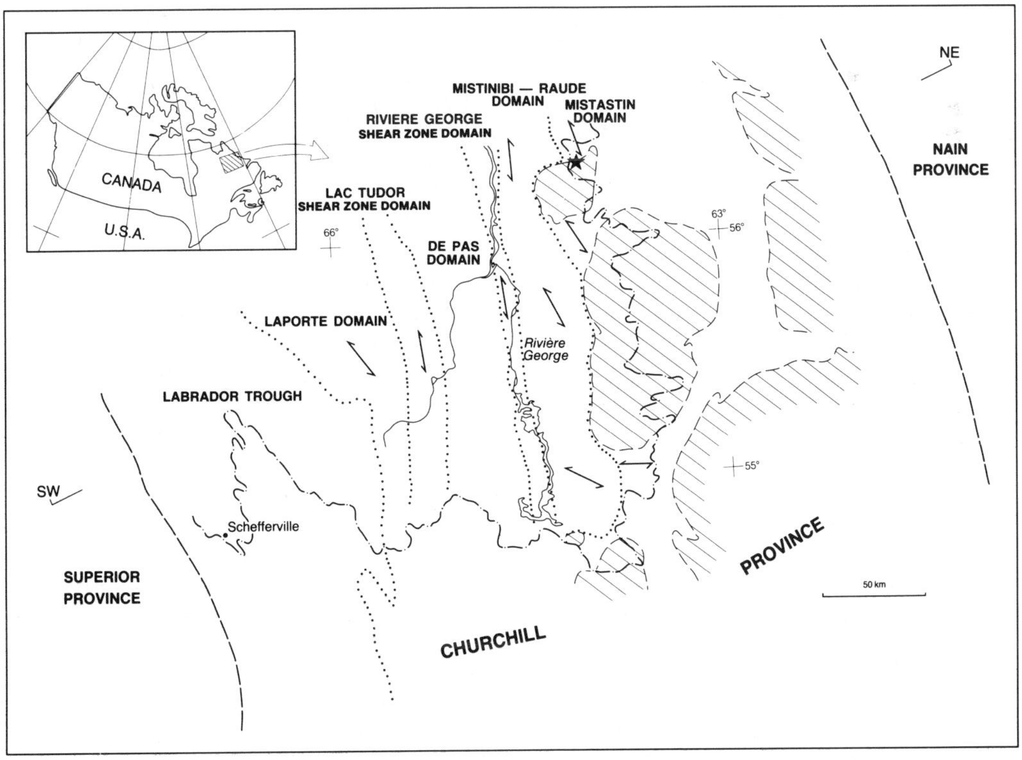

Figure 1. *Location map and regional context of lithotectonic domains of the Rivière George Project area. The double-barbed arrows represent generalized orientations of the dominant tectonic grain. Elsonian intrusions have a slanted bar pattern. The star represents the location of the Lac Brisson mineral deposit. The southwest–northeast (SW–NE) line of cross-section shown in Figure 6 is indicated. The continental divide is indicated by a dash–dot pattern.*

the Laporte Domain to the west (with minor amounts of pre- to syntectonic granitoids in Laporte group rocks) and some lithologies of the De Pas Domain to the east (i.e., granulite facies quartzofeldspathic gneisses and some charnockitic suite rocks).

The LTSZ is up to 20 km wide and is traceable for over 150 km along strike. It is marked by reorientation, steepening, and thinning of lithological units of adjoining domains as they enter the zone. The reorientation reflects a component of regional dextral shear. Deformation is heterogeneous and a mylonitic fabric, commonly with a low-plunging elongation lineation, is restricted to several bands. Mesoscopic kinematic indicators (e.g. C–S fabrics, etc.) are complex: there is evidence for an important dextral shear component (Fig. 5a), as well as contraction across the zone, bringing rocks in the east over rocks to the west.

De Pas Domain

The De Pas Domain is composed essentially of a granitic core (the De Pas Batholith) flanked by enveloping gneissic rocks to the east and west.

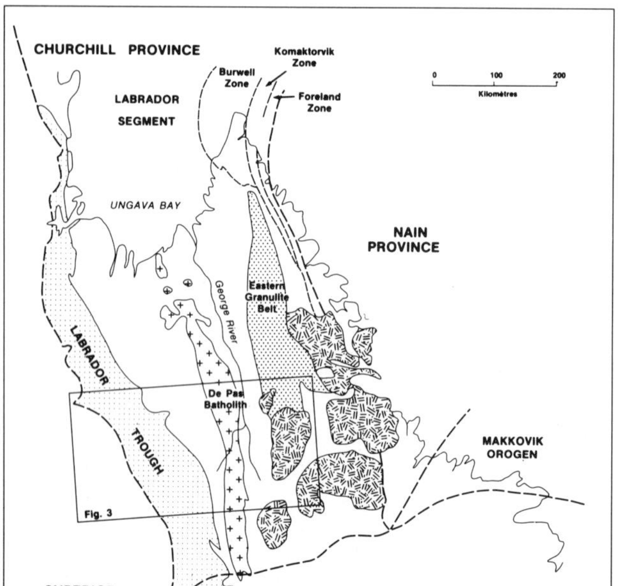

Figure 2. *The Labrador segment is bounded by the Superior, Grenville, and Nain provinces. Elsonian intrusions are represented by a random, short, parallel double line pattern. The generalized unit outlines are after Avramtchev (1985), Hoffman (1988), Taylor (1979), Wardle* et al. *(this volume), and published and unpublished interpretations in reports of the Rivière George Project.*

The De Pas Batholith extends throughout the study area, ranging in apparent width from about 20 to 50 km. The presence of probably correlative granitoids to the south (Wardle *et al.*, 1982, 1987) and to the north (Taylor, 1979) extends the total length of the batholith to over 500 km (Fig. 2). Several satellite plutons occur east of the main batholithic mass. The batholith is a composite body, consisting of late- to posttectonic granitoids that include abundant K-feldspar porphyritic varieties. In the west, a charnockitic–opdalitic suite has been identified, and in the east, a granitic–granodioritic suite. Field relations and geochemistry indicate that the latter is, at least in part, younger and more evolved than the former. Harker diagrams for combined data from the two suites show good coefficients of correlation, and the suites are considered to be comagmatic. Mineralogically and chemically, the De Pas granitoids are "I"-type granites with a well-marked calc-alkaline trend (Fig. 6a) and volcanic-arc affinity (Fig. 6b).

The rocks that envelop the batholith, and occur as inclusions within it, consist of three main quartzofeldspathic gneissic groups: (1) granodioritic to tonalitic rocks of ambiguous parentage, (2) granitic or arkosic rocks, and (3) migmatitic semipelitic rocks. Subordinate metamafites and meta-ultramafites are also present. West of the granitoids, the enveloping rocks are granulites; in the east, they have been metamorphosed to amphibolite facies grade. The Atshakash Complex, occurring at the eastern boundary of (but still within) the De Pas Domain, can be distinguished from other informally grouped eastern gneisses on the basis of lithology and metamorphic grade: it mainly comprises upper greenschist to amphibolite facies interbedded graphitic schists, notable cherty quartzarenites, schistose metatuffs, and metabasalts, possibly representing a starved volcano-sedimentary assemblage formed in a closed basin. This narrow unit, up to 2 km wide, has been followed for over 100 km along strike.

Rivière George Shear Zone Domain

This is a heterogeneous regional shear zone, the Rivière George Shear Zone (RGSZ), composed principally of the George Complex, a well-mixed assortment of metabasalts, metagabbros, metadiorites, calcic paragneisses, both early and posttectonic leucogranites, and some metavolcaniclastics and meta-ultramafites. It also incorporates parts of the Mistinibi Complex, the Pallatin Intrusive Suite, the Ntshuku Complex, and several granite–granodiorite plutons, all of which have been defined in the Mistinibi–Raude Domain to the east, as has the Zeni Complex (a probable correlative with the

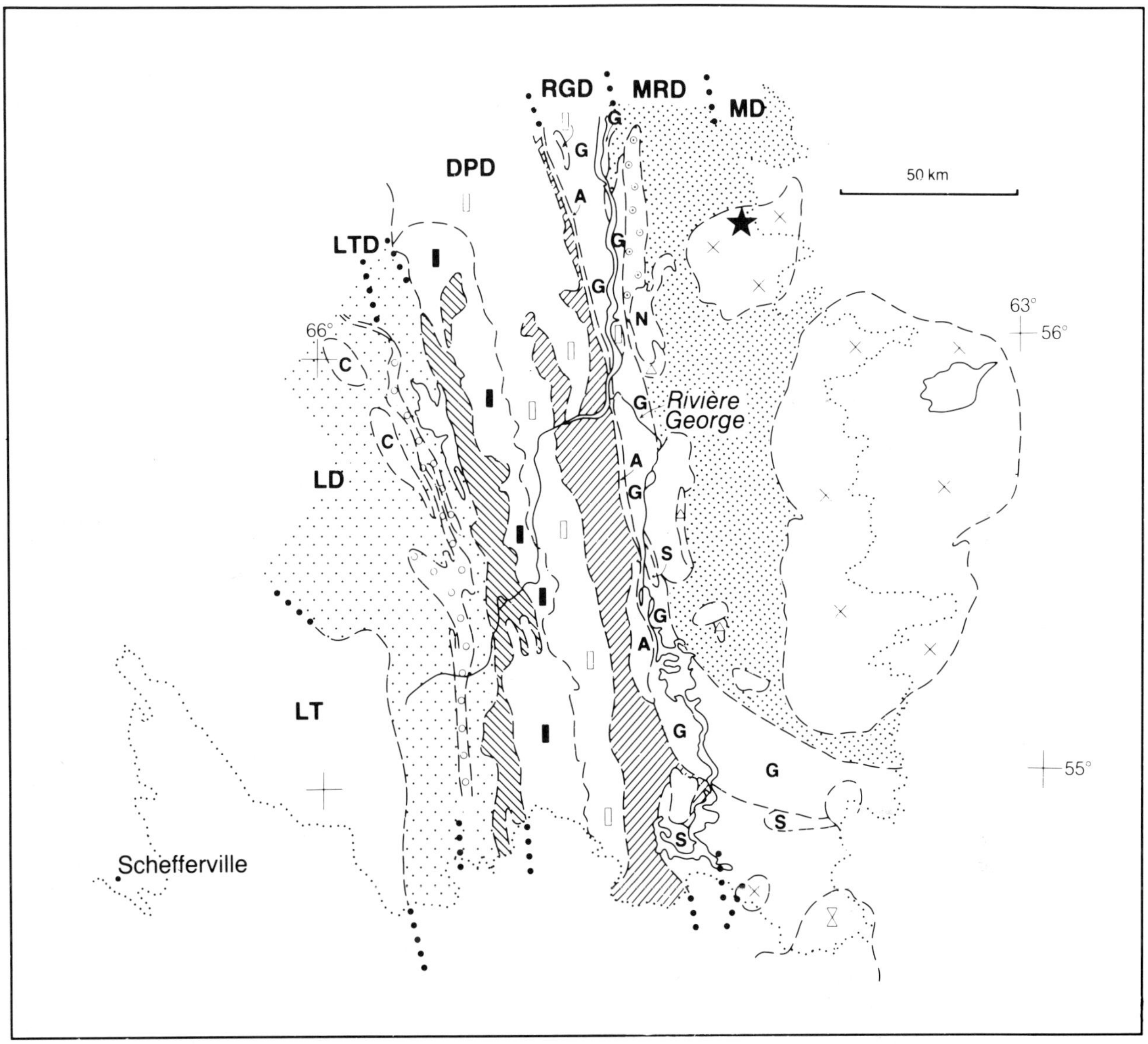

LITHOTECTONIC DOMAINS

LT = Labrador Trough
LD = Laporte Domain
LTD = Lac Tudor Shear Zone Domain
DPD = De Pas Domain
RGD = Rivière George Shear Zone Domain
MRD = Mistinibi – Raude Domain
MD = Mistastin Domain

LITHOLOGIES

★ = Lac Brisson mineral deposit
⧗ = Michikamau Anorthosite
× = Elsonian intrusions
⊙⊙ = Hutte Sauvage Group
○○ = An informal arkosic unit
A = Atshakash Complex
▯ = Granitic suite of the De Pas Batholith
▮ = Charnockitic suite of the De Pas Batholith
∴ = Laporte group
++ = Granitic and granodioritic plutons
S = supracrustal assemblages
△ = Pallatin Intrusive Suite
N = Ntshuku Complex
G = George and Zeni Complexes
C = Granodioritic gneiss
⁞⁞ = Mistinibi Complex
▧ = Gneiss of the western envelope of the De Pas Batholith
▨ = Gneiss of the eastern envelope of the De Pas Batholith

Figure 3. *Generalized lithologies of the lithotectonic domains of the Rivière George Project area. The lithologies are listed in an approximate chronological order.*

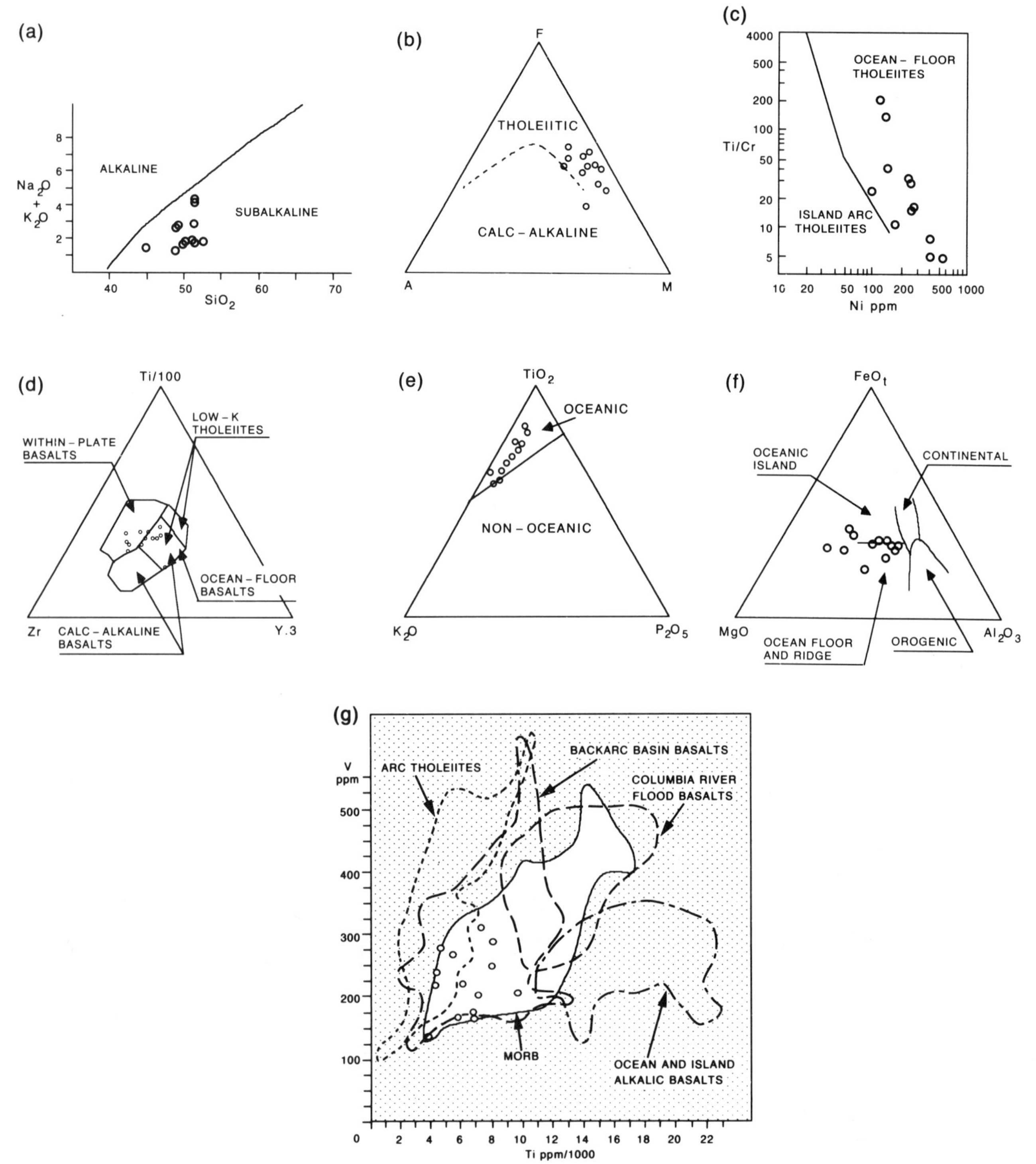

Figure 4. *Geochemical diagrams displaying data from metabasalts of the Laporte group (data from Danis, 1988). (a) ($Na_2O + K_2O$) versus SiO_2 diagram (Irvine and Baragar, 1971), all in wt.%. (b)AFM diagram (Irvine and Baragar, 1971):A = ($Na_2O + K_2O$), F = total iron as FeO, M = MgO, all in wt.%. (c) (Ti/Cr) versus Ni diagram (Beccaluva* et al.*, 1979), all in ppm. (d) (Ti/100)–Zr–(3 × Y) diagram (Pearce and Cann, 1973). (e) TiO_2–K_2O–P_2O_5 diagram (Pearce* et al.*, 1975), all in wt.%. (f) FeO_t (total iron as FeO)–MgO–Al_2O_3 diagram (Pearce* et al.*, 1977); all in wt.%. (g) V versus Ti diagram (after Shervais, 1982), all in ppm; fields drawn by eye using the data of Shervais.*

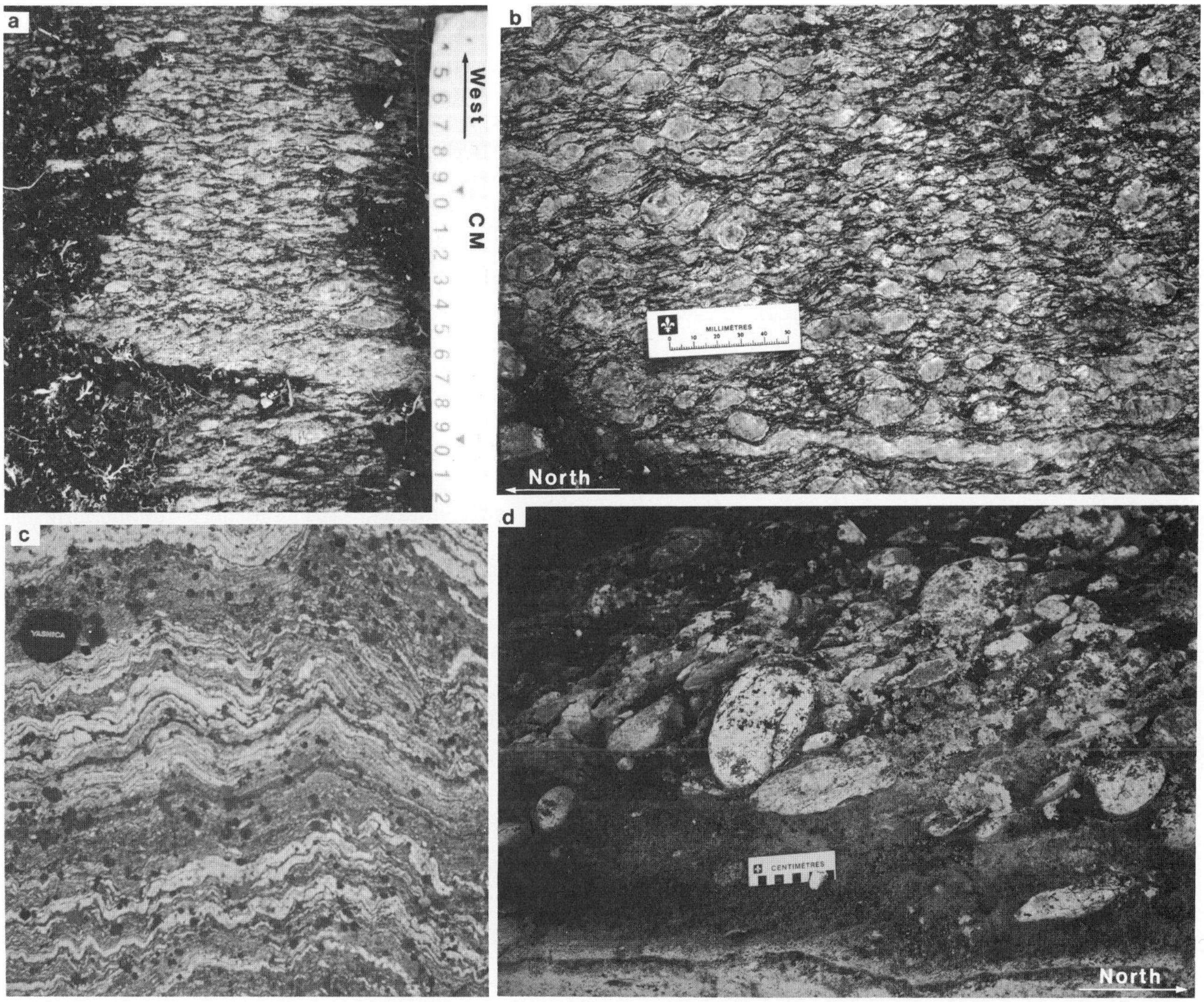

Figure 5. *(a) C–S–C′ mylonite fabric indicative of a component of dextral shear in a porphyritic unit (cut by a leucocratic vein, also deformed) of the charnockitic suite of the De Pas Batholith within the Lac Tudor Shear Zone. The photo is of a horizontal surface. (b) Similar mylonitic fabric to that shown in (a), but developed in a porphyritic unit of the Pallatin Intrusive Suite within the Rivière George Shear Zone. The photo is of a horizontal surface. (c) Mistinibi Complex migmatitic gneiss folded about northwest to north-northwest trending axial surfaces of at least second-phase folds. The photo is of a gently inclined surface. (d) Conglomeratic calcarenite, with imbricated cobbles and pebbles, of the Hutte Sauvage Group. The photo is of a gently inclined surface.*

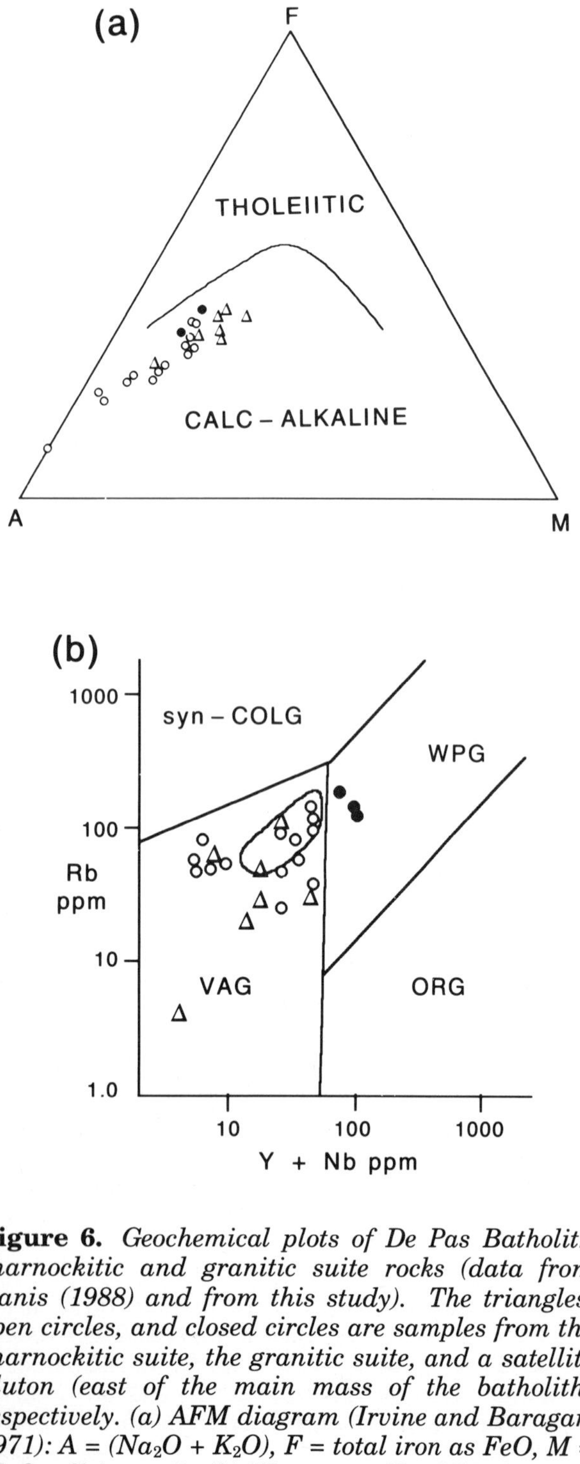

Figure 6. *Geochemical plots of De Pas Batholith charnockitic and granitic suite rocks (data from Danis (1988) and from this study). The triangles, open circles, and closed circles are samples from the charnockitic suite, the granitic suite, and a satellite pluton (east of the main mass of the batholith) respectively. (a) AFM diagram (Irvine and Baragar, 1971): A = (Na_2O + K_2O), F = total iron as FeO, M = MgO, all in wt.%. (b) Rb versus (Y + Nb) diagram (Pearce* et al., *1984): VAG = volcanic arc granites, syn-COLG = syntectonic collision granites, WPG = within-plate granites, and ORG = ocean-ridge granites. The enclosed area represents a field of granites from central Chile (Baldwin and Pearce, 1982). The triangles are samples from a satellite pluton east of the main mass of the De Pas Batholith.*

George Complex). Part of the George Complex may also correlate with some of the eastern gneisses of the De Pas Domain.

The RGSZ can be recognized on aeromagnetic maps, and by the reorientation, thinning, and breakup of map units traced into the zone. It extends over 200 km north–south, with an apparent width varying from 6 km to over 20 km. Mylonites, cataclasites, and less deformed rocks are present. A complex kinematic history revealed by mesoscopic indicators (S–C fabrics, etc.) includes the following: early lateral ductile movements, right-handed (Fig. 5b) and some left-handed; extensional movements (perhaps oblique to allow for rapid deposition of polymictic conglomerates of the Hutte Sauvage Group found to the east); and late contractional brittle–ductile movements. There is evidence, in the east, of thrusting eastward over the Mistinibi–Raude Domain.

Mistinibi–Raude Domain

The Mistinibi–Raude Domain can be divided into three subdomains. A southern subdomain is marked by a westerly to northwesterly trending tectonic grain. Numerous mylonites occur in the northern part. In the far south, outcrop is sparse; some granitoids and paragneiss have been found. The rest of the subdomain comprises the Zeni Complex, which is tentatively correlated with the George Complex in the Rivière George Shear Zone Domain because of the similarity in lithologies.

The central, and largest, subdomain of the Mistinibi–Raude Domain has a dominantly northwest-trending tectonic grain. It consists of the Mistinibi Complex, which comprises abundant migmatitic semipelitic paragneiss (Fig. 5c) with varying amounts of pre- and late- to posttectonic granite, plus diorite, metamafite, and some meta-ultramafites. Near the contact between the Mistinibi and George–Zeni complexes, other intrusive rocks have been identified, including the Pallatin Intrusive Suite, which comprises gabbronorites and "I"-type granitoids and is associated with the Ntshuku Complex (metavolcanics including metavolcaniclastics and metabasalts, with some metasediments). A preliminary U–Pb zircon age (S. Bowring, personal communication, 1987) of about 2.32 Ga on a porphyry from the Pallatin Suite (about 10 km south of the Hutte Sauvage Group) has been obtained.

A small northwestern subdomain displays a northwesterly to northerly trending tectonic grain. A cratonic rift sedimentary assemblage, the Hutte Sauvage Group, occurs in the west, at the contact with the RGSZ. This group is composed principally of meta-arkose (including crossbedded, muscovite-bearing, specularite-laminated, and hematite-stained meta-arkose), quartzitic wacke (locally conglomeratic), and quartzite, with minor metasiltstone and calcarenite (locally conglomeratic, Fig. 5d). A regionally important lens of polymictic metaconglomerate and meta-arkose occurs in the west, where locally it is sheared within the Rivière George Shear Zone. The Hutte Sauvage Group has been interpreted to lie unconformably over the Mistinibi Complex; it has been folded with the underlying rocks, giving it a catamaran-like synclinal keel shape with inward overturned sides. To the east, the Mistinibi Complex increases in metamorphic grade through a low-pressure intermediate metamorphic facies series, from middle amphibolite facies up to granulites of the Mistastin Domain.

Mistastin Domain

The Mistastin Domain includes several units defined in the Mistinibi–Raude Domain to the west and intervening major Elsonian intrusions. The former include probable equivalents of diorites and the Mistinibi Complex, all at granulite facies metamorphic grade and, in part, bounded to the west by a narrow strike-parallel shear zone. These rocks may possibly form part of the "eastern granulite belt" of Wardle *et al.* (this volume), which contains Archean as well as Proterozoic rocks and extends some 80 km to the east (Fig. 2). The tectonic grain of foliated units in the Mistastin Domain is crosscut by Elsonian plutons. These posttectonic intrusions include parts of the Mistastin Batholith (Emslie *et al.*, 1980), which has yielded a preliminary U–Pb zircon age of ca. 1440 Ma (R.F. Emslie, personal communication, 1984), the Michikamau Anorthosite (Emslie (1970); dated by U–Pb zircon methods at 1460 Ma by Krogh and Davis (1973)), and some rapakivi granites. A further posttectonic event is represented by the Lac Brisson peralkaline granite (Currie, 1985; Pillet, 1985; Pillet and Chenevoy, 1985), which hosts the Lac Brisson rare-element deposit (Zajac *et al.*, 1984). This granite, located northwest of the Mistastin Batholith within what is probably another Elsonian pluton, has yielded a Rb–Sr whole-rock age of 1189 Ma (Duthou *et al.*, 1986).

Late diabases

The Mistastin, Mistinibi–Raude, Rivière George, and De Pas domains are crossed by undeformed diabase dykes of variable composition and orientation.

PROPOSED TECTONIC MODEL

The Rivière George area has been the site of several rock-forming events, including those of Archean, Early Proterozoic, and Middle Proterozoic (terms of Harrison and Peterman (1980)) age. The following discussion focuses on the Hudsonian orogeny, which has left an important imprint on the area.

On the basis of similarities in lithotectonic, structural, geophysical, and geographical characteristics of the Rivière George area and those of present-day plate tectonic settings, we propose an Early Proterozoic plate tectonic scenario for a major part of the "Labrador Segment of the Trans-Hudson Orogen" (Van Schmus *et al.*, 1987) of the Churchill Province.

Plate tectonics were invoked nearly 20 years ago to explain some of the rocks of the Churchill Province of Québec; Wilson (1968a), Gibb and Walcott (1971), Dewey and Burke (1973), and Burke and Dewey (1973) all drew sutures within, or at the eastern boundary of, the Labrador Trough. The latter two authors envisaged the collision of two continental masses, represented by the presence of three zones: a western zone formed of Labrador Trough sediments deposited on an "Atlantic-type" continental shelf of the nonreactivated Superior Province basement; a central zone formed of mafic and ultramafic rocks of the Labrador Trough, obducted oceanic crust, and mantle; and an eastern zone formed of the granites and gneisses east of the Labrador Trough, a reactivated basement. Negative and positive gravity anomalies in transects at high angles to the contact between the Superior and eastern Churchill provinces were interpreted as further evidence for continental collision after the closing of intervening ocean basins (Keary, 1976). Thus, in accordance with a model that was used to explain the Himalayas and the Tibetan Plateau (model of basement reactivation due to continental collision of Dewey and Burke (1973)), Keary proposed that the major negative anomaly could represent "an initial downwarp after continental collision," and "the gravity highs would represent undulations on an elevated Conrad discontinuity" (Keary, 1976, p. 376) – elevated by isostatic response of a crust thickened by ductile creep resulting from continental collision; the denser material causing the gravity anomalies would have formed by a process of crustal differentiation resulting from raised geoisotherms and the rise of partial melts from the lower part of the thickened crust.

Several authors (Thomas *et al.*, 1978; Thomas and Keary, 1980; Thomas and Gibb, 1983, 1985) have proposed that the Churchill Province east of the Superior Province has an apparent analogue in the central Andes of Peru, as modeled by Dewey and Bird (1970), Dewey and Burke (1973), Myers (1975), and Barazangi and Isacks (1979). They have cited several features found in both areas: a batholith, major faults and fault-bounded blocks, thrusting in the western and eastern border zones away from the centre of the belt, and ensialic sedimentary basins.

The model presented here represents the area as a zone in which there first occurred continental rifting and ocean formation, followed by ocean closure and collision with broadly concomitant formation of an Andean-type magmatic arc. Figure 7(b) is a schematic cross-section illustrating features of the model. From west to east, there is the Superior Craton, a passive margin wedge, a suture, a fore-arc region, a magmatic arc, a back-arc region, and a plateau region; farther east is a Churchill Inner Zone, an eastern Churchill Border Zone, and, finally, the North Atlantic Craton.

Superior Craton

The Superior Craton represents essentially unreactivated Archean continental crust. Dates of ca. 2.65 Ga have been obtained from preliminary U–Pb geochronology by Mortensen and Percival (1987) on monazites in the migmatized terrain; the terrain also yields K–Ar ages between 2.4 and 2.5 Ga, as noted by Dimroth (1978).

Passive margin wedge

The central Labrador Trough (west of the present map area) has previously been interpreted as an Aphebian continental passive margin wedge (Wardle and Bailey (1981); "Atlantic-type continental margin" of Dewey and Burke (1973)). This wedge comprises two distinctive lithotectonic assemblages. The Labrador Trough is formed mainly of the Kaniapiskau Supergroup, which in generalized fashion can be considered to include (1) a western, dominantly sedimentary succession, the Knob Lake Group, consisting of an early cratonic rift assemblage of continental sediments and some alkali basalts, followed by a shelf–slope–rise sedimentary sequence that includes slivers of base-

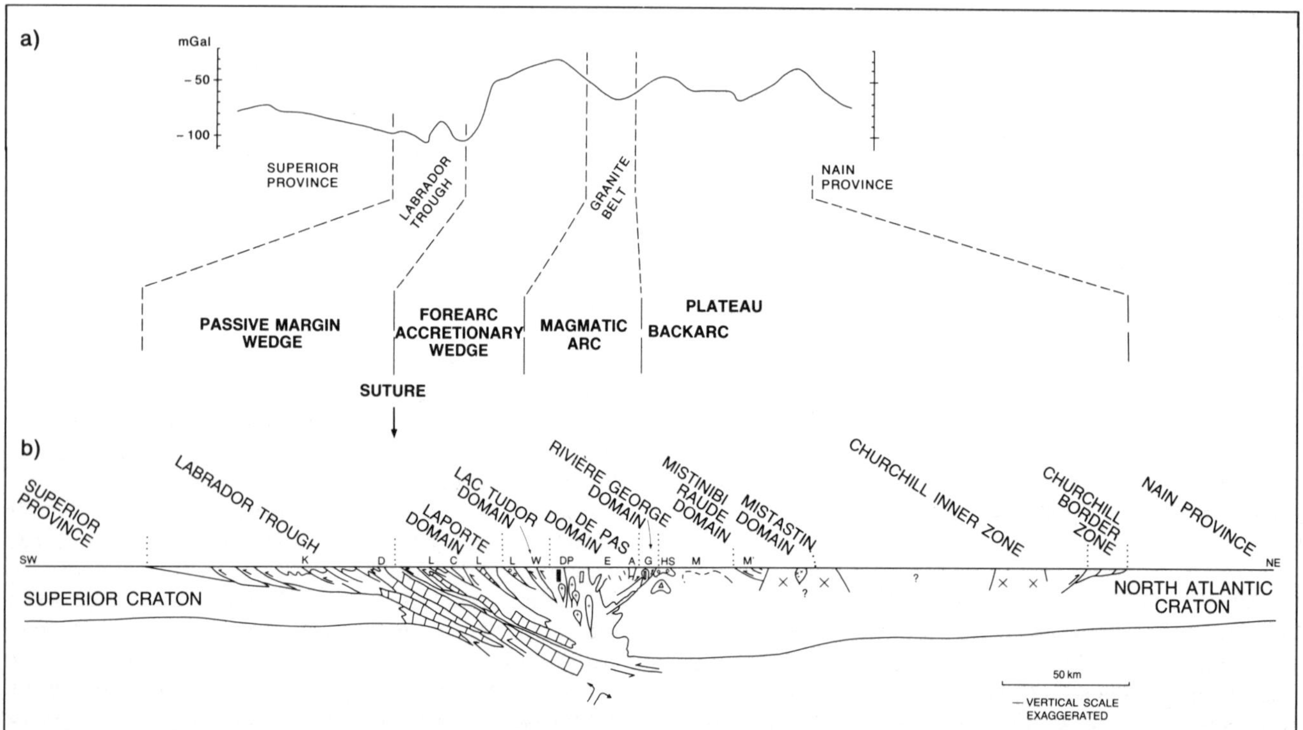

Figure 7. *(a) Bouguer gravity profile and selected generalized lithotectonic entities; after Thomas* et al. *(1978) and Thomas and Keary (1980). The profile lies 150 km northwest of the cross-section shown in (b). The sections and geology are correlated in a preliminary fashion along the dashed lines between (a) and (b). (b) Schematic cross-section across part of the eastern Churchill Province (the line of section, "SW–NE," is indicated in Fig. 1). Legend: K = Knob Lake Group, D = Doublet Group, L = Laporte Group, C = granodiorite gneiss, W and E = gneisses of the western and eastern envelopes of the De Pas Batholith, DP = De Pas Batholith (dark and light rectangles represent charnockitic and granitic suites), A = Atshakash Complex, G = George Complex, HS = Hutte Sauvage Group, M = Mistinibi Complex, M′ = granulite facies Mistinibi Complex, and crosses (x) = Elsonian intrusions.*

ment horsts (Dimroth *et al.*, 1970; Dimroth, 1978; Wardle and Bailey, 1981), and (2) an eastern, generally younger, dominantly mafic to ultramafic igneous succession, the Doublet Group, consisting of low-K, low-Ti tholeiitic basalts, gabbros of similar chemical composition, spilites, and ultramafics (Baragar, 1967; Dimroth, 1970, 1971, 1978; Wardle and Bailey, 1981). This has been compared to an embryonic rift assemblage, related to possible formation of an ocean basin to the east (Wardle and Bailey, 1981).

Structures typical of numerous younger orogenic belts are also found here, and can be explained in terms of plate tectonics (e.g., Coney, 1973). The most important of these is a foreland fold and thrust belt overturned to the southwest toward the largely unreactivated Superior Craton (Dimroth, 1970, 1972, 1981; Le Gallais and Lavoie, 1982; Wardle *et al.*, this volume).

The suture

The line of suture (Fig. 7) is placed at the juxtaposition of the easternmost section of oceanic-type mafics of the Labrador Trough passive margin wedge, and the westernmost section of abundant schists of the Laporte group (interpreted as the main part of a fore-arc accretionary wedge). It is represented by the Keato Lake fault (Baragar, 1967). Dimroth and Dressler (1978) proposed that rocks in the east have been uplifted 10–15 km with respect to those in the west of the trough. Fahrig (1957, p. 116) made the following comment 30 years ago: "For many years it has been assumed that the eastern 'boundary' of the Labrador Trough consists at least in part of a great thrust fault (see Tectonic Map of Canada, 1950)."

In the present model, presumed oceanic crust and the spreading centre, formerly located east of the trough, have been subducted eastward.

Fore-arc accretionary wedge

The fore-arc accretionary wedge is represented principally by the Laporte group. Tholeiitic low-K metabasalts geochemically similar to ocean-ridge basalts with minor tendencies toward within-plate basalt compositions (Fig. 4) are interstratified with schists and metagreywackes, and minor metaultramafites – an association often found in fore-arc regions (Dickinson and Seely, 1979) where the igneous rocks may be emplaced as a result of having been scraped off, or sliced and thrust up from, a subducting slab. The unit of continental arkoses, which are similar to rocks of the Hutte Sauvage Group (here interpreted as a back-arc sedimentary assemblage), east of the magmatic arc, would represent a fore-arc equivalent sedimentation. We suggest that the few regional lenses of granodioritic gneiss might be exotic accreted continental fragments; recent analogues exist (e.g., Nur and Ben-Avraham, 1983). A similar possible explanation for some Archean basement domes near Ungava Bay, on strike with the fore-arc area proposed here, has also been suggested (Goulet *et al.*, 1987; Poirier *et al.*, 1987).

There are also structural similarities to modern fore-arc regions. The wedge contains westerly (suture) directed thrusts and overturned folds. Eastward, toward the magmatic arc, dips steepen and there is increased overturning toward the west. Repetition of local structural sequences indicates the presence of imbricate thrusts. Transcurrent motion has also occurred, as well as contraction across a shear zone, the Lac Tudor Shear Zone, on the arc side of the fore-arc region. Metamorphism increases toward the east, up the structural succession.

Magmatic arc

Farther east, "I"-type calc-alkaline granitoids in batholithic proportions (the De Pas Batholith) are considered to represent a magmatic arc. Plots of data from these granitoids on discriminant diagrams of tectonic setting (Fig. 6b) lie in the field of volcanic-arc granites, comparing favourably with central Chilean active continental margin granitoids and late posttectonic collision granites.

Preliminary U–Pb zircon geochronology by S. Bowring (personal communication, 1987) gives a 1.84 Ga age for a sample from the batholith, about 30 km southwest of the confluence of the George and De Pas rivers. This is close to the age of presumed Hudsonian deformation in the Labrador Trough. However, Krogh (1986), also using the U–Pb zircon method, has dated rocks of the southernmost extension of the De Pas Batholith (Wardle *et al.*, this volume) at 1811 ± 3 Ma. The batholith is evidently a complex, probably compound body and may include a significant proportion of postcontinental collision intrusions. "Basement" gneisses, many of them of granodioritic to tonalitic composition but with uncertain protoliths, as well as migmatites with structural and metamorphic histories older than the granitoids, envelop, and occur as inclusions in, the batholith, as might occur in a continental margin magmatic arc.

In the east, the Atshakash Complex is interpreted here as an intra-arc restricted volcano-sedimentary basin. Modern-day analogues exist (e.g., Dickinson and Suczek, 1979).

Structurally, local foliation trends and the general elongation of granitic units parallel the overall batholithic mass, i.e., north-northwesterly. Late conjugate shears compatible with compression along southwest or west-southwest axes are best developed in the De Pas Domain, although they are also present in the Laporte, Lac Tudor Shear Zone, Rivière George Shear Zone, and Mistinibi–Raude domains.

Rocks of the De Pas Domain are bounded and locally affected by two major shear zones, the Lac Tudor Shear Zone and the Rivière George Shear Zone. Magmatic arc formation with bounding regional shear zones has also been described in the Peruvian Andes (e.g., Myers, 1975).

The back arc

Farther east, the cratonic rift assemblage of the Hutte Sauvage Group is interpreted to represent back-arc sedimentation; modern analogues exist (e.g., Condie, 1982). This sedimentation possibly occurred in a tectonic regime of oblique extension, as the group is bounded by the RGSZ, which contains evidence of lateral displacement.

Southwest of the Hutte Sauvage Group lies a satellite pluton of the De Pas Batholith. Preliminary geochemical data show "I"-type and "A"-type characteristics, as well as certain within-plate granite affinities (Fig. 6), such as might be found in a back-arc intrusion. The pluton gives a preliminary U–Pb zircon date of 1.84 Ga (S. Bowring, personal communication, 1987).

Though it is tempting to place some of the George Complex mafics and ultramafics in a Hudsonian back-arc setting, caution is prescribed because (1) correlatives of these rocks in the south are outside the back-arc geographical position, and (2) the complex occurs in the mixed zone of the Rivière George Shear Zone, which includes part of the Pallatin Intrusive Suite, for which a preliminary date of 2.32 Ga has been obtained (S. Bowring, personal communication, 1987). Thus, it is possible that the complex contains rocks formed prior to the Hudsonian orogeny.

The presence of eastward-directed thrusts is consistent with the back-arc interpretation and, as in some modern analogues (e.g., Fitch, 1972), there is important lateral shearing parallel to the Rivière George Shear Zone.

The plateau

The back arc impinges on a "plateau-type" domain, the "George Plateau," which is still an elevated topographic feature today. It is characterized by a migmatitic terrain with granulitic components and, in much of the northern area, is marked by a low-pressure intermediate metamorphic facies series. More recent analogues exist (e.g., Dewey and Burke, 1973; Toksoz and Bird, 1977).

Hudsonian reworking is indicated by (1) reorientation of these rocks in the Rivière George Shear Zone, and (2) the folded unconformity with the overlying Hutte Sauvage Group. To the east, in Labrador, Newfoundland geologists (Wardle, 1983) have documented a "Churchill Inner Zone" of possible reworked Archean rocks; thus, unrecognized Archean rocks might also occur in the area.

Eastern Churchill Border Zone

To the east, lies a Churchill Border Zone (Wardle, 1983) of Hudsonian reworking, including eastward thrusting, which affects Archean basement and Proterozoic cover ("Eastern Division" of the "Labrador segment of the Trans-Hudson Orogen" defined by Wardle *et al.* (this volume)).

North Atlantic Craton

This area represents an essentially unreactivated Archean granitoid and supracrustal gneiss terrain, the Nain Province (Taylor, 1971; Wardle, 1983), forming part of the North Atlantic Craton (Wardle *et al.*, this volume).

GEOPHYSICAL EVIDENCE

Further evidence favouring a plate tectonic interpretation has come from geophysicists, notably from Keary (1976), Thomas *et al.* (1978), Thomas and Keary (1980), and Thomas and Gibb (1983, 1985). They noted the presence of generalized elongate gravity anomalies parallel to the length of the Churchill Province. Figure 7a represents a northeasterly trending Bouguer gravity profile, after Thomas *et al.* (1978) and Thomas and Keary (1980), taken about 150 km northwest of the schematic geological cross-section of the area shown in Figure 7b. The profile serves as a preliminary approximation of gravity data across the eastern Churchill Province.

With respect to the gravity expression of the Superior Province to the west, the Labrador Trough is an area with a negative gravity anomaly related to downwarped Superior Province continental crust (e.g., Thomas and Keary, 1980); in the present context, possibly also a result of the "lighter" overlying supracrustal thrust stack of the Labrador Trough passive margin wedge. The local gravity high (Fig. 7a) has been related to mafic igneous rocks of the Labrador Trough by Thomas *et al.* (1978).

There is a marked positive anomaly, with some second-order positive and negative zones, across the remainder of the Churchill Province to the east. Some earlier workers interpreted this anomaly as being caused by the presence of dense lower crust formed in a thickened and reactivated Churchill continental crust (in the sense of Dewey and Burke's (1973) model) east of a former Benioff zone (e.g., Thomas and Gibb, 1985); thus, a proposed Hudsonian trench was thought to have been present near the contact between the regional negative and positive gravity anomalies. In the present context, the regional positive gravity anomaly roughly corresponds with the Laporte group fore-arc wedge, and its presumed denser underplated oceanic crust.

Farther east, a second-order negative gravity anomaly is present, which Thomas *et al.* (1978) showed to be roughly coincident with a relatively less dense granite belt; in the present context, this is the De Pas Batholith magmatic arc.

Still farther east, there is a second-order positive gravity anomaly (Thomas *et al.*, 1978), which in the present context corresponds approximately to the back-arc region, where we speculate that there was crustal extension and thinning of continental crust.

PAST OBJECTIONS TO PLATE TECTONICS

Some authors have argued against B-type subduction in the area (Dimroth, 1972, 1981), citing the absence of (1) ophiolites, (2) calc-alkaline magmatism, and (3) blueschist facies metamorphism.

In rebuttal, the following comments are pertinent. First, we concur with the tentative proposal of Wardle and Bailey (1981) that the eastern Labrador Trough may represent the western edge of a modified ophiolitic succession, and that oceanic crust was subducted to the east. Secondly, extensive calc-alkaline magmatism can now be documented east of the Labrador Trough, in the form of the De Pas Batholith. Finally, although blueschists have not been found, the literature on metamorphism in the Labrador Trough documents the presence of (a) inverted metamorphic isograds (Dimroth, 1970), also noted in the fore-arc region defined here and a characteristic feature of younger collision terranes (e.g., Le Fort, 1975), and (b) barroisite (noted by Dimroth (1970)), an intermediate- to high-pressure amphibole (Miyashiro, 1968). Tentatively, one might postulate the presence of a "modified" Hudsonian paired metamorphic belt – with a low pressure – intermediate pressure facies series in the east, and an intermediate pressure facies series (possibly mixed with some high pressure – intermediate pressure facies series rocks) in the west (higher pressure metamorphism than in the east and including high-grade kyanite-bearing rocks; Dimroth and Dressler (1978)).

CONCLUSIONS

Mapping of terrain east of the central Labrador Trough has yielded new geological information that permits the definition of distinct lithotectonic domains. Numerous similarities between features of these domains and those of younger orogens permit a plate tectonic model to be proposed for a major portion of the Labrador segment. This apparent compatibility, in our view, suggests that one need not appeal to nonplate tectonic (e.g., Dimroth *et al.*, 1970) or modified plate tectonic (Barager and Scoates, 1981; Dimroth, 1981) interpretations for the area.

The western part of the Labrador segment shows evidence for a Proterozoic Wilson cycle (e.g., Dewey and Burke, 1974; after Wilson, 1966, 1968b). Evolution of the area involved the following sequence: (1) Aphebian rifting of Archean continental crust, (2) subduction eastward and Andean-type magmatic-arc formation, and (3) Hudsonian oblique continental collision.

The eastern part of the Labrador segment is more complex, and its evolution involved (1) possible Archean rock-forming events, (2) Early Aphebian subduction and magmatic arc-formation – possibly a eugeoclinal island-arc complex (the Pallatin Intrusive Suite), (3) possible plate collisions, (4) Hudsonian reworking, (5) Elsonian crustal extension and intrusion, and (6) later, posttectonic diabases and Paleohelikian peralkaline magmatism.

Extension of the proposed model to the south and north is hampered by the lack of detailed geological mapping and radiometric dates at present (though work is in preparation). In favour of a larger scale application of the model, however, is the observation of several authors (Thomas *et al.*, 1978; Thomas and Keary, 1980; Thomas and Gibb, 1983, 1985) that there is apparent general continuity of Bouguer gravity anomalies and of some lithologies along what is termed here the Labrador segment. The lithologies noted were the Labrador Trough rocks, a central belt of granitic rocks mapped by Taylor (1979), and a belt of granulites now designated the "Eastern Granulite Belt" by Wardle *et al.*, this volume). These lithological features, and accompanying structural characteristics, correspond to parts of our present plate tectonic model as significant petrotectonic assemblages: (1) passive margin wedge, (2) the magmatic arc, and (3) part of the plateau-type domain.

In detail, however, there are lateral changes and discontinuities of units, as one might expect along the strike of any orogen. For instance, some calc-alkaline intrusives and voluminous metatuffs occur near Ungava Bay, west of a series of discontinuously outcropping granitoids on strike with the De Pas Batholith. Poirier *et al.* (1987) suggested that these may represent part of an arc complex (this is on strike with our fore-arc accretionary wedge). The relatively simple fore-arc region of our model probably has evolved differently in the north, where it may contain arc complexes, further fore-arc, and possibly even back-arc relict zones.

Detailed discussion of the possible application (and its consequences) of the present model to the rest of the Labrador segment of the "Trans-Hudson Orogen" (Van Schmus *et al.*, 1987), and around the "Circum-Ungava geosyncline" (Dimroth *et al.*, 1970) or "Circum-Superior Belt" (Gibb and Walcott, 1971) and adjacent Churchill Province, or to the rest of the "Rae Province" (as defined by Hoffman (1988)) is beyond the scope of this report.

Regardless of the speculative nature of our conclusions, we hope that the description of defined lithotectonic elements and their interpretation in plate tectonic terms will serve to guide and stimulate further work leading to a further understanding of the tectonic evolution of the Labrador segment.

ACKNOWLEDGEMENTS

This study draws on the work of numerous individuals (published and unpublished) who have contributed to the "Rivière George Project," undertaken by the Quebec government during the period 1983–1987. Systematic mapping was conducted by Bélanger (1983, 1984a, final report in preparation), Bourque (1985, 1986, final report submitted to editor), Danis (1985, 1986a,b, 1988, preliminary and final reports submitted to editor), Owen (1984, 1985, final report submitted to editor), Taner (1986, 1987, final report submitted to editor), and van der Leeden (1983, 1984a,b, 1986a, final reports submitted to editor). Specialized studies include those by Bélanger (1984b), Bowring (1986), Giroux (1985a,b), Pouliot (1986, final report in preparation), and van der Leeden (1985, 1986b,c, 1987, final reports in preparation). Several university papers and theses written on subjects in the area include those by Asselin (1985), Chabot (1986), Duthou *et al.* (1986), Girard (thesis in preparation, 1985, 1986a,b,c), Martelain (thesis in preparation, 1985, 1986a,b), Martelain *et al.* (1987), Pillet (thesis in preparation, 1983, 1984, 1985), and Pillet and Chenevoy (1985).

The authors are grateful to T. Clark for useful comments on this article. Reviews by C. Gower, J. Lewry, and an anonymous reviewer also improved the content and presentation of the final manuscript; naturally, we accept full responsibility for any remaining incongruities.

REFERENCES

Asselin, R., 1985, Étude du métamorphisme de la région du lac Mistinibi (rivière George), unpublished B.Sc. thesis, Université Laval, 46 p.

Avramtchev, L., 1985, Carte géologique du Québec: Ministère de l'Énergie et des Ressources, Québec, DV 84-02 (map no. 2000).

Baldwin, A.J. and Pearce, J.A., 1982, Discrimination of productive and non-productive porphyritic intrusions in the Chilean Andes: Economic Geology, v. 77, p. 664–674.

Baragar, W.R.A., 1967, Wakuach lake map-area, Québec–Labrador: Geological Survey of Canada, Memoir 344, 174 p.

Baragar, W.R.A. and Scoates, R.F.J., 1981, The Circum-Superior Belt: a Proterozoic plate margin?, in Kroner, A., ed., Precambrian Plate Tectonics: Elsevier, Amsterdam, p. 297–330.

Barazangi, M. and Isacks, B.L., 1979, Subduction of the Nazca Plate beneath Peru: evidence from spatial distribution of earthquakes: Geophysical Journal of the Royal Astronomical Society, v. 57, p. 537–555.

Beccaluva, L., Ohnenstetter, D. and Ohnenstetter, M., 1979, Geochemical discrimination between ocean floor and island arc tholeiites: application to some ophiolites: Canadian Journal of Earth Sciences, v. 16, p. 1874–1882.

Bélanger, M., 1983, Région du lac Brisson: Ministère de l'Énergie et des Ressources, Québec, DV 83-13, p. 50–52.

Bélanger, M., 1984a, Région du lac Brisson: Ministère de l'Énergie et des Ressoruces, Québec, DP 84-20 (annotated map).

Bélanger, M., 1984b, Secteur rivière Déat-sud, indice Poitras: Ministère de l'Énergie et des Ressources, Québec, internal report and map.

Bourque, Y., 1985, Région du lac Mina: Ministère de l'Énergie et des Ressources, Québec, DV 85-12, p. 46.

Bourque, Y., 1986, Géologie de la région du lac Mina: Ministère de l'Énergie et des Ressources, Québec, DP 85-30 (annotated map).

Bowring, S., 1986, Géochronologie dans l'arrière-pays et l'avant-pays de la Fosse du Labrador: Ministère de l'Énergie et des Ressources, Québec, DV 86-14, p. 49.

Burke, K.C.A. and Dewey, J.F., 1973, An outline of Precambrian plate development, in Tarling, D.H. and Runcorn, S.K., eds., Implications of Continental Drift to the Earth Sciences. Volume 2: Academic Press, London, p. 1035–1045.

Chabot, F., 1986, Géologie et métallogénie de l'indice van der Leeden: Université Laval, Québec, Québec, Mémoire de fin d'études, 76 p.

Condie, K.C., 1982, Plate Tectonics and Crustal Evolution. 2nd edition: Pergamon Press, Toronto, 310 p.

Coney, P.J., 1973, Plate tectonics of marginal foreland thrust–fold belts: Geology, v. 1, p. 131–134.

Currie, K.L., 1985, An unusual peralkaline granite near Lac Brisson, Québec–Labrador: Geological Survey of Canada, Paper 85-1A, p. 73–80.

Danis, D., 1985, Région du lac Recouet: Ministère de l'Énergie et des Ressources, Québec, DV 85-12, p. 46–47.

Danis, D., 1986a, Géologie de la région du lac Recouet: Ministère de l'Énergie et des Ressources, DP 85-31 (annotated map).

Danis, D., 1986b, Région du lac Raude: Ministère de l'Énergie et des Ressources, Québec, DV 86-14, p. 46.

Danis, D., 1988, Géologie de la région du lac Recouet: Ministère de l'Énergie et des Ressources, Québec, ET 86-11. (In press.)

Dewey, J.F. and Bird, J.M., 1970, Mountain belts and the new global tectonics: Journal of Geophysical Research, v. 75, p. 2625–2647.

Dewey, J.F. and Burke, K.C.A., 1973, Tibetan, Variscan, and Precambrian basement reactivation: products of continental collision: Journal of Geology, v. 81, p. 683–692.

Dewey, J.F. and Burke, K.C.A., 1974, Hot spots and continental breakup: some implications for collisional orogeny: Geology, v. 2, p. 57–60.

Dickinson, W.R. and Seely, D.R., 1979, Structure and stratigraphy of forearc regions: American Association of Petroleum Geologists Bulletin, v. 63, p. 2–31.

Dickinson, W.R. and Suczek, C.A., 1979, Plate tectonics and sandstone compositions: American Association of Petroleum Geologists Bulletin, v. 63, p. 2164–2182.

Dimroth, E., 1970, Evolution of the Labrador geosyncline: Geological Society of America Bulletin, v. 81, p. 2717–2742.

Dimroth, E., 1971, The evolution of the central segment of the Labrador geosyncline. Part II. The ophiolite suite: Neues Jahrbuch fur Geologie und Palaeontologie Abhandlungen, v. 137, p. 209–248.

Dimroth, E., 1972, The Labrador geosyncline revisited: American Journal of Science, v. 272, p. 487–506.

Dimroth, E., 1978, Labrador Trough area: Ministère des Richesses naturelles, RG-193, 396 p.

Dimroth, E., 1981, Labrador geosyncline: type example of early Proterozoic cratonic reactivation, in Kroner, A., ed., Precambrian Plate Tectonics: Elsevier, Amsterdam, p. 331–352.

Dimroth, E. and Dressler, B., 1978, Metamorphism of the Labrador Trough, in Fraser, J.A. and Heywood, W.W., eds., Metamorphism in the Canadian Shield: Geological Survey of Canada, Paper 78-10, p. 215–236.

Dimroth, E., Baragar, W.R.A., Bergeron, R. and Jackson, G.D., 1970, The filling of the Circum-Ungava Geosyncline, in Baer, A.J., ed., Symposium on Basins and Geosynclines of the Canadian Shield: Geological Survey of Canada, Paper 70-40, p. 45–142.

Duthou, J.L., Pillet, D. and Chenevoy, M., 1986, Rb–Sr geochronology of the lac Brisson peralkaline granite (Labrador, N.E. Canada): Terra Cognita, v. 6, p. 142.

Emslie, R.F., 1970, The geology of the Michikamau intrusion, Labrador: Geological Survey of Canada, Paper 68-57, 85 p.

Emslie, R.F., Cousens, B., Hamblin, C. and Bielecki, J., 1980, The Mistastin Batholith, Labrador–Québec: an Elsonian composite rapakivi suite: Geological Survey of Canada, Paper 80-1A, p. 95–100.

Fahrig, W.F., 1957, Geology of certain Proterozoic rocks in Québec and Labrador: Royal Society of Canada, Special Publication, v. 2, p. 112–123.

Fitch, T.J., 1972, Plate convergence, transcurrent faults, and internal deformation adjacent to southeast Asia and the western Pacific: Journal of Geophysical Research, v. 77, p. 4432–4460.

Gibb, R.A. and Walcott, R.I., 1971, A Precambrian suture in the Canadian Shield: Earth and Planetary Science Letters, v. 10, p. 417–422.

Girard, R., 1985, Région de la rivière Déat: Ministère de l'Énergie et des Ressources, Québec, DV 85-12, p. 48.

Girard, R., 1986a, Géologie de la région de la rivière Déat: Ministère de l'Énergie et des Ressources, Québec, DP 85-27 (annotated map).

Girard, R., 1986b, Région de la rivière Déat, Nouveau Québec: Ministère de l'Énergie et des Ressources, Québec, DV 86-14, p. 48.

Girard, R., 1986c, Géologie, tectonique, et métallogénie de la région de la rivière Déat: Ministère de l'Énergie et des Ressources, Québec, internal report, 35 p.

Giroux, J., 1985a, Traitement numérique d'images Landsat du secteur de la rivière George: Ministère de l'Énergie et des Ressources, Québec, DV 85-12, p. 47–48.

Giroux, J., 1985b, Contribution du traitment numérique des images Landsat du secteur de rivière George dans le cadre de la cartographie géologique régionale: Ministère de l'Énergie et des Ressources, internal report, 77 p.

Goulet, N., Gariépy, C., Mareschal, J.-C. and Machado, N., 1987, Structure, geochronology, gravity, and the tectonic evolution of the northern Labrador Trough: Geological Association of Canada – Mineralogical Association of Canada, Program with Abstracts, v. 12, p. 48.

Harrison, J.E. and Peterman, Z.E., 1980, North American Commission on Stratigraphic Nomenclature. Note 52 – A preliminary proposal for a chronometric time scale for the Precambrian of the United States and Mexico: Geological Society of America Bulletin, v. 91, pt. I, p. 377–380.

Hoffman, P.F., 1987, Birth of a craton: Geological Association of Canada – Mineralogical Association of Canada, Program with Abstracts, v. 12, p. 56.

Hoffman, P.F., 1988, United Plates of America, the birth of a craton: Early Proterozoic Assembly and Growth of Laurentia: Annual Review of Earth and Planetary Sciences, v. 16, p. 543–603.

Irvine, T.N. and Baragar, W.R.A., 1971, A guide to the chemical classification of the common volcanic rocks: Canadian Journal of Earth Sciences, v. 8, p. 523–548.

Keary, P., 1976, A regional structural model of the Labrador Trough, northern Québec, from gravity studies, and its relevance to continent collision in the Precambrian: Earth and Planetary Science Letters, v. 28, p. 371–378.

Krogh, T.E., 1986, Report to Newfoundland Department of Mines and Energy on isotopic dating results from the 1985–1986 geological research agreement: Newfoundland Department of Mines and Energy, Mineral Development Division, Open File LAB 707, 106 p.

Krogh, T.E. and Davis, G.L., 1973, The significance of inherited zircons on the age and origin of igneous rocks – an investigation of the Labrador adamellites: Carnegie Institute of Washington, Yearbook 72, p. 610–613.

Le Fort, P., 1975, Himalayas: the collided range. Present knowledge of the continental arc: American Journal of Science, v. 275A, p. 1–44.

Le Gallais, C.J. and Lavoie, S., 1982, Basin evolution of the Lower Proterozoic Kaniapiskau Supergroup, central Labrador Miogeocline (Trough), Québec: Canadian Petroleum Geology Bulletin, v. 30, p. 150–166.

Martelain, J., 1985, Batholite d'Ungava: Ministère de l'Énergie et des Ressources, Québec, DV 85-12, p. 48.

Martelain, J., 1986a, Le Batholite d'Ungava et son environnement – Labrador central – données préliminaires: Ministère de l'Énergie et des Resources, Québec, MB 86-71, 53 p.

Martelain, J., 1986b, Batholite de De Pas: Ministère de l'Énergie et des Ressources, Québec, DV 86-14, p. 48–49.

Martelain, J., Bélanger, M. and Chenevoy, M., 1987, Petrology and geochemistry of the De Pas Batholith, central Labrador, Québec: Terra Cognita, v. 7, p. 354.

Miyashiro, A., 1968, Metamorphism of mafic rocks, in Hess, H.H. and Poldervaart, A., eds., Basalts, the Poldervaart Treatise on Rocks of Basaltic Compositions: John Wiley and Sons, New York, p. 799–834.

Mortensen, J.K. and Percival, J.A., 1987, Reconnaissance U–Pb zircon and monazite geochronology of the Lac Clairambault area, Ashuanipi complex, Québec: Geological Survey of Canada, Paper 87-2, p. 135–142.

Myers, J.S., 1975, Vertical crustal movements of the Andes in Peru: Nature, v. 254, p. 672–674.

Nur, A. and Ben-Avraham, Z., 1983, Break-up and accretion tectonics, in Hashimoto, M. and Uyeda, S., eds., Accretion Tectonics in the Circum-Pacific Regions: Terra Scientific Publishing Company, Tokyo, p. 3–18.

Owen, V., 1984, Région du lac Leif: Ministère de l'Énergie et des Ressources, Québec, DV 84-17, p. 50–51.

Owen, V., 1985, Région du lac Leif: Ministère de l'Énergie et des Ressources, Québec, DP 85-14 (annotated map).

Pearce, J.A. and Cann, J.R., 1973, Tectonic setting of basic volcanic rocks determined using trace element analyses: Earth and Planetary Science Letters, v. 19, p. 290–300.

Pearce, J.A., Harris, N.B.W. and Tindle, A.G., 1984, Trace element discrimination diagrams for the tectonic interpretation of granitic rocks: Journal of Petrology, v. 25, p. 956–983.

Pearce, T.H., Gorman, B.E. and Birkett, T.C., 1975, The TiO_2–K_2O–P_2O_5 diagram: a method of discriminating between oceanic and non-oceanic basalts: Earth and Planetary Science Letters, v. 24, p. 419–426.

Pearce, T.H., Gorman, B.E. and Birkett, T.C., 1977, The relationship between major element chemistry and tectonic environment of basic and intermediate volcanic rocks: Earth and Planetary Science Letters, v. 36, p. 121–132.

Pillet, D., 1983, Granite hyperalcalin du lac Brisson: Ministère de l'Énergie et des Ressources, Québec, DV 83-13, p. 53.

Pillet, D., 1984, Granite hyperalcalin du lac Brisson: Ministère de l'Énergie et des Ressources, Québec, DV 84-17, p. 51–52.

Pillet, D., 1985, Le granite peralcalin du lac Brisson, Territoire-du-Nouveau-Québec, résultats préliminaires: Ministère de l'Énergie et des Ressources, Québec, MB 85-37, 51 p.

Pillet, D. and Chenevoy, M., 1985, Le granite peralcalin du lac Brisson (Labrador): Terra Cognita, v. 5, p. 319.

Poirier, G., Perrault, S. and Hynes, A., 1987, The nature of the eastern boundary of the Labrador-Trough near Kuujjuaq, Québec: Geological Association of Canada – Mineralogical Association of Canada, Program with Abstracts, v. 12, p. 81.

Pouliot, G., 1986, Télédétection dans les régions adjacentes à la Fosse du Labrador, partie sud: Ministère de l'Énergie et des Ressources, Québec, DV 86-14, p. 49–50.

Shervais, J.W., 1982, Ti–V plots and the petrogenesis of modern and ophiolitic lavas: Earth and Planetary Science Letters, v. 59, p. 101–118.

Taner, M.F., 1986, Reconnaissance géologique dans la région du lac Juillet: Ministère de l'Énergie et des Ressources, Québec, DV 86-14, p. 46–47.

Taner, M.F., 1987, Reconnaissance géologique de la région du lac Juillet, Nouveau-Québec: Ministère de l'Énergie et des Ressources, DP 87-11, 23 p.

Taylor, F.C., 1971, A revision of Precambrian structural provinces in northeastern Québec and northern Labrador: Canadian Journal of Earth Sciences, v. 8, p. 579–584.

Taylor, F.C., 1979, Reconnaissance geology of a part of the Precambrian shield, northeastern Québec, northern Labrador and Northwest Territories: Geological Survey of Canada, Memoir 393, 99 p.

Thomas, M.D. and Gibb, R.A., 1983, Convergent plate tectonics and related faults in the Canadian Shield, in Gabrielsen, R.H., Ramberg, I.B., Roberts, D. and Steinlein, O.A., eds., Proceedings of the Fourth International Conference on Basement Tectonics, Oslo, Norway: International Basement Tectonics Association, Salt Lake City, Utah, p. 115–134.

Thomas, M.D. and Gibb, R.A., 1985, Proterozoic plate subduction and collision: processes for reactivation of Archean crust in the Churchill Province, in Ayres, L.D., Thurston, P.C., Card, K.D. and Weber, W., eds., Evolution of Archean Supracrustal Sequences: Geological Association of Canada, Special Paper 28, p. 263–279.

Thomas, M.D. and Keary, P., 1980, Gravity anomalies, block-faulting and Andean-type tectonism in the eastern Churchill Province: Nature, v. 283, p. 61–63.

Thomas, M.D., Halliday, D.W. and Stephenson, R., 1978, Gravity anomalies and geological structure in northern Labrador and northeastern Québec: Energy, Mines, and Resources, Canada, Gravity Map Series, Nos. 157–161, 38 p.

Toksoz, M.N. and Bird, P., 1977, Formation and evolution of marginal basins and continental plateaus, in Talwani, M. and Pitman, W.C., III, eds., Island Arcs, Deep Sea Trenches, and Back-arc Basins: American Geophysical Union, Ann Arbor, Michigan, p. 379–393.

van der Leeden, J., 1983, Région du lac de la Hutte Sauvage: Ministère de l'Énergie et des Ressources, Québec, DV 83-13, p. 52–53.

van der Leeden, J., 1984a, Région du lac de la Hutte Sauvage. Ministère de l'Énergie et des Ressources, Québec, DP 84-29 (annotated map).

van der Leeden, J., 1984b, Région du lac Mistinibi: Ministère de l'Énergie et des Ressources, Québec, DV 84-17, p. 49–50.

van der Leeden, J., 1985, Synthèse tectonique de la région de la rivière George: Ministère de l'Énergie et des Ressources, Québec, DV 85-12, p. 48–49.

van der Leeden, J., 1986a, Géologie de la région du lac Mistinibi, Nouveau-Québec: Ministère de l'Énergie et des Ressources, Québec, DP 86-12 (annotated map).

van der Leeden, J., 1986b, Compte rendu des travaux sur le terrain – 1985, région de la rivière George: Ministère de l'Énergie et des Ressources, Québec, internal report, 78 p.

van der Leeden, J., 1986c, Synthèse tectonique de la région de la rivière George: Ministère de l'Énergie et des Ressources, Québec, DV 86-14, p. 47–48.

van der Leeden, J., 1987, Synthèse géologique de la région de la rivière George: Ministère de l'Énergie et des Ressources, Québec, DV 87-24, p. 31.

Van Schmus, W.R., Bickford, M.E., Lewry, J.F. and Macdonald, R., 1987, U–Pb geochronology in the Trans-Hudson Orogen, northern Saskatchewan, Canada: Canadian Journal of Earth Sciences, v. 24, p. 407–424.

Wardle, R.J., 1983, Nain–Churchill Province cross-section, Nachvak Fiord, northern Labrador: Newfoundland Department of Mines and Energy, Mineral Development Division, Report 83-1, p. 68–90.

Wardle, R.J. and Bailey, 1981, Early Proterozoic sequences in Labrador: Geological Survey of Canada, Paper 81-10, p. 331–358.

Wardle, R.J. and Colleagues of the Mineral Development Division of the Department of Mines and Energy of Newfoundland, 1982, The Trans-Labrador Batholith; a major pre-Grenvillian feature of the eastern Grenville Province: Friends of the Grenville Workshop, Program with Abstracts, p. 11.

Wardle, R.J., Ryan, B. and Nunn, G.A.G., 1987, Labrador segment of the Trans-Hudson Orogen: crustal development through oblique convergence and collision: Geological Association of Canada – Mineralogical Association of Canada, Program with Abstracts, v. 12, p. 99.

Wardle, R.J., Ryan, B., Nunn, G.A.G. and Mengel, F.C., 1990, Labrador segment of the Trans-Hudson Orogen: crustal development through oblique convergence and collision, in Lewry, J.F. and Stauffer, M.R., eds., The Early Proterozoic Trans-Hudson Orogen of North America: Geological Association of Canada, Special Paper 37, p. 353–369.

Wilson, J.T., 1966, Did the Atlantic close and then re-open?: Nature, v. 211, p. 676–681.

Wilson, J.T., 1968a, Comparison of the Hudson Bay Arc with some other features, in Beals, C.S., ed., Science, History, and Hudson Bay. Volume 2: Canadian Department of Energy, Mines, and Resources, Ottawa, p. 1015–1033.

Wilson, J.T., 1968b, Static or mobile earth: the current scientific revolution: Proceedings of the American Philosophical Society, v. 112, p. 309–320.

Zajac, I.S., Miller, R., Birkett, T. and Nantel, S., 1984, Le gîte de Zr, Y, Nb, et Be du complexe alcalin de Strange Lake, Québec–Labrador: Ministère de l'Énergie et des Ressources, Québec, DV 84-18, p. 127–142.

A Structural Cross-section of the Northern Labrador Trough, New Quebec

E. Boone and A. Hynes
Department of Geological Sciences,
McGill University, 3450 University St., Montreal, Quebec H3A 2A7, Canada

Boone, E. and Hynes, A., A structural cross-section of the Northern Labrador Trough, New Quebec, in Lewry, J.F. and Stauffer, M.R., eds., The Early Proterozoic Trans-Hudson Orogen of North America: Geological Association of Canada, Special Paper 37, p. 387–396.

Abstract

The Northern Labrador Trough contains a succession of Aphebian metasediments, gabbroic intrusives, and metavolcanic rocks deformed during the Hudsonian orogeny. Sediments unconformably overlie the Superior craton in the west and are overthrust from the east by a correlative, eastward-thickening sedimentary sequence with abundant gabbroic intrusives and a thick mafic volcanic unit (the Hellancourt Formation) near its top. The eastern overthrust zone is part of a classic foreland fold-and-thrust belt.

The Hellancourt volcanics are geochemically uniform low-K tholeiitic basalts. They are more evolved and enriched in incompatible elements than N-type midoceanic-ridge basalt (MORB). They have incompatible-element enrichments similar to those of P-type MORB and are in many ways transitional in character between P-type MORB and continental tholeiites.

A structural cross-section of the Northern Labrador Trough, based on data from existing geological maps, indicates at least 65% shortening of the supracrustal packet during the Hudsonian orogeny. Palinspastic restoration of the section places the westernmost mafic volcanic rocks east of the easternmost exposed Archean basement. They may have been erupted in a region where continental crust was thin or absent. All of these features are consistent with development of the Labrador Trough through collapse of an Atlantic-type passive margin.

Résumé

Le nord de la fosse du Labrador contient une succession de meta-sédiments Protérozoïque Moyen, d'intrusifs gabbroïques et de roches méta-volcaniques déformés pendant l'orogénèse Hudsonienne. A l'ouest, des sédiments discordants sur le craton Supérieur sont surmontés de l'est par une séquence sédimentaire corrélative qui s'épaissit vers l'est et contient de nombreux intrusifs gabbroïques et une unité épaisse de volcanites basiques (la formation d'Hellancourt) près du sommet. La zone de charriage est est caractéristique d'une chaîne plissée.

Les volcanites d'Hellancourt sont des basaltes tholéiitiques pauvre en K géochimiquement homogènes et sont plus èvolués et enrichis en éléments incompatibles que les MORB de type N. En outre, ils montrent un enrichissement en éléments incompatibles comparables au MORB type-P et sont à maints égards transitionnels entre les MORB type-P et les tholéiites continentales.

Une coupe structurale du nord de la fosse du Labrador, basée sur les cartes géologiques existantes, montrent un raccourcissement de l'édifice supracrustal d'au moins 65% pendant l'orogénèse Hudsonienne. Une reconstitution palinspastique de la coupe situe les roches volcaniques basiques de l'ouest à l'est du socle archéen affleurant du côté est. Elles ont peut-être été mises en place dans une région à croûte continentale mince ou absente. Ces caractères concordent avec le modèle de développement de la fosse du Labrador dans une marge passive effondrée de type Atlantique.

INTRODUCTION

For many years, the Labrador Trough served as the type example of a Proterozoic "ensialic" orogenic belt (e.g., Dimroth, 1972a,b). Recently, the basis for this interpretation has been called into question, and models for its evolution as a collapsed Atlantic-type continental margin have been proposed (e.g., Wardle and Bailey, 1981; LeGallais and Lavoie, 1982). Low-K tholeiites are conspicuous in the supracrustal succession of the Labrador Trough (Dimroth *et al.*, 1970). Their resemblance to modern abyssal tholeiites and their presence in a supracrustal succession apparently developed on continental crust have remained two of the more puzzling features of the Labrador Trough. In this paper, we address the question of the true geochemical character of the low-K tholeiites and present a structural cross-section of the Northern Labrador Trough, which may reconcile some of the apparent contradictions in models for its evolution.

GEOLOGY OF THE LABRADOR TROUGH

The Labrador Trough contains a succession of sedimentary and volcanic rocks of Early Proterozoic age. It extends from the Grenville province, in the south (53°N, 67°W), north-northwest toward Lac Berthier (Nagvaraaluk), west of Ungava Bay (71°00'W, 60°45'N) (Fig. 1). It is commonly divided into a southern segment, south of the Grenville Front (Gross, 1968); a central segment; and a northern segment, north of 57°N (Dimroth *et al.*, 1970). Most tectonic interpretations of the Labrador Trough have been based on observations in the central segment. Relationships in the southern segment have been complicated by the Grenville event. Although the northern segment is well exposed and well mapped, no stratigraphic or tectonic synthesis of it has been published. The data presented here are from the northern segment, but the general features of the Labrador Trough succession are summarized most simply from the much better studied central segment.

The rocks of the trough belong to the Kaniapiskau Supergroup. The supergroup is characterized by two (or one, in the northern segment) large-scale sedimentary sequences separated by a regional unconformity. In the central segment, the two sequences are similar, each reflecting a deepening-upward environment of sedimentation. The sequences begin with basal stable-shelf deposits, such as orthoquartzites, dolomites, and iron formation, and conclude with deep-water deposits, such as shales, greywackes, and mafic volcanics (Dimroth *et al.*, 1970; Dimroth, 1981; LeGallais and Lavoie, 1982).

The sedimentary and volcanic pile, when reconstructed from the map data, is considerably thickened toward the east, from 100 to 800 m at the western edge to 4000 m in the east (LeGallais and Lavoie, 1982). The abundance of pillowed volcanics and the proportion of shales and deep-water sediments increases in the eastern portion of the trough, indicating that there must have been a deep, magmatically active basin to the east of the autochthonous supracrustals.

The rocks of the Kaniapiskau Supergroup were deformed and metamorphosed during the Hudsonian orogeny (Dimroth *et al.*, 1970). The trough sediments were folded, overturned, and thrust toward the western craton. The metamorphic grade increases from subgreenschist facies in the west to

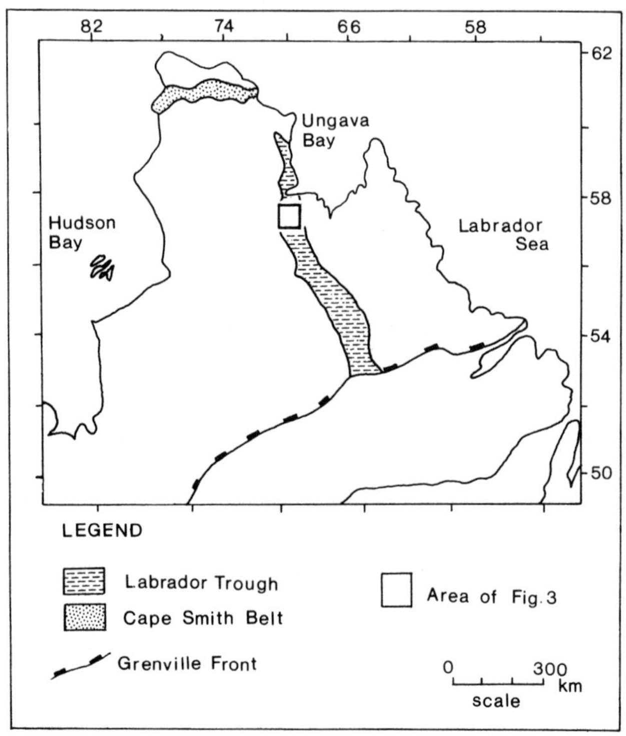

Figure 1. *Location map of the Labrador Trough. The region that is the subject of this paper is outlined, and depicted in greater detail on Figure 3.*

amphibolite and, locally, granulite facies in the eastern portion of the trough (e.g., Dimroth *et al.*, 1970; Baragar and Scoates, 1981; Dimroth, 1981).

NORTHERN LABRADOR TROUGH

The Northern Labrador Trough has all been mapped, generally at a scale of 1:63,360 (Sauvé, 1956, 1957, 1958, 1959; Bergeron, 1957; Gelinas, 1958, 1959, 1960; Gold, 1962; Bérard, 1965; Sauvé and Bergeron, 1965; Hardy, 1976; Clark, 1977, 1978, 1979, 1980; Belanger, 1982). It is similar in many respects to the Central Trough. The supracrustal succession comprises one deepening-upward sedimentary sequence with a thick volcanic unit, the Hellancourt Formation, at or near its top.

As in the Central Trough (LeGallais and Lavoie, 1982), the Northern Trough may be divided into a western autochthonous zone and an eastern allochthonous zone. In the western zone, there are relatively undeformed autochthonous supracrustals, which rest unconformably on the Archean Ungava craton. Locally, these units have been thickened, by thrusting from the east and by folding, but no major décollement is apparent between the basement and the lowermost sediments at the western edge of the trough.

The allochthonous zone is a fold-and-thrust belt, involving sediments of various kinds, gabbroic intrusives, and mafic volcanics. This zone includes the westernmost appearance of the volcanics (Hellancourt Formation) at this latitude, and probable distal correlatives of the sedimentary units identified farther west. It is limited by a major frontal thrust in the west and a steeply eastward-dipping fault, the Lac Rachel Fault, in the east.

Near the Lac Rachel Fault, the metamorphic grade jumps markedly and the terrain east of it is in amphibolite and granulite facies. Traditionally, this terrain has been regarded as "hinterland" to the trough sensu stricto, although it has been suggested that correlatives of the Kaniapiskau Supergroup are identifiable there (Moorhead and Hynes, 1986; Poirier *et al.*, this volume). In this paper, we are concerned only with the Labrador Trough west of the Lac Rachel Fault.

Stratigraphy

Although the entire Northern Labrador Trough has been mapped in detail, there is no comprehensive correlation of stratigraphy between map sheets. In the west, Bérard (1965) has established a stratigraphy for the sediments that lie directly on the Archean basement (Fig. 2). In the middle portion of the trough, Sauvé and Bergeron (1965) and Clark (1980) describe a stratigraphy with similar lithologies, but assign different names (Fig. 2). Based on the lithological descriptions and their locations within the stratigraphic pile, several units may be correlated between map sheets. Thus, the "Lower Dolomite" of the western section established by Bérard (1965) is correlated with the Abner Formation farther east, the Fenimore iron formation is correlated with "iron-bearing rocks" of the middle member of the Baby Formation of Sauvé and Bergeron (1965), and the fine-grained clastics of the Larch River Formation are correlated with similar rocks of the Baby Formation (Fig. 2). These correlations are based solely on the data available from the geological reports.

Two units, the Baby and the Thévenet formations, pose problems. The Baby Formation has been extensively intruded by gabbros. The Larch River Formation, with which it is correlated, has not. Sediments of the two formations are similar and the presence of the intrusives may just reflect localization of the gabbros laterally within the Baby/Larch River Formation. The Thévenet Formation cannot be correlated with any unit to the west of its exposure. It has been suggested that the boundary between the Thévenet and the underlying Hellancourt Formation is a thrust fault (Goulet, 1986). East of Lac Rachel, however, the volcanics are apparently conformably overlain by a unit similar to the Thévenet Formation (J. Moorhead, personal communication, 1987). In our interpretation, no thrust is put between these units. Consequently, the major volcanic formation, the Hellancourt Formation, does not occupy the highest position in the stratigraphic sequence.

Character and geochemistry of the Hellancourt Formation

The Hellancourt Formation is considered to be correlative to the uppermost tholeiitic basalts in the Central Labrador Trough (Doublet Group, Wilbob Formation; Dimroth *et al.* (1970)). In the course of this study, we mapped four sections through the Hellancourt Formation in the Northern Trough in detail. In each case, the base of the formation is sheared, but the underlying Baby Formation is extensively intruded by gabbroic sills that may represent feeders to the flows. We do not interpret the lower contacts to be major thrusts. Stratigraphic thicknesses of the sections vary from 200 to 1200 m. In all except the easternmost section, the top contact of the formation is a thrust. For the easternmost section, the top boundary is a probable conformable contact with the Thévenet Formation. The sections are essentially indistinguishable in field characteristics.

The Hellancourt Formation consists of massive and pillowed flows. As noted by Sauvé and Bergeron (1965), massive flows predominate toward the base of the formation and pillowed flows toward the top. This characteristic is repeated on the formation scale as well as on a scale of tens of metres. The formation consists of a suite of low-K tholeiitic basalts, which preliminary geochemical studies suggested were comparable with modern midoceanic-ridge basalt (MORB) and continental tholeiitic basalts (Sauvé and Bergeron, 1965; Dimroth *et al.*, 1970; Wardle and Bailey, 1981). Based on geochemical analysis of 38 samples, the Hellancourt basalts are geochemically remarkably homogeneous (geochemical data are available from the authors upon request). They appear to be more evolved and enriched than N-type MORB (Schilling, 1975) and have incompatible-element ratios similar to P-type (enriched; Schilling, 1975) MORB. In many ways, the Hellancourt basalts are transitional in character between P-type MORB and continental tholeiites, being more evolved than the former and less enriched than the latter.

Given their apparent eruption within a sedimentary succession, it is not surprising that the Hellancourt basalts are not typical MORB. Furthermore, gabbroic sills in the underlying Baby Formation, which were probably part of the feeder system for the volcanics, are in places markedly glomeroporphyritic (Sauvé and Bergeron, 1965), a feature consistent with contamination of the magmas by underlying sialic crust. However, the degree of enrichment in incompatible elements exhibited by the Hellancourt basalts is relatively low for continental tholeiites, indicating relatively little crustal contamination. We argue on the basis of our cross-section that this may reflect the presence of only very thin or discontinuous continental crust under the sites at which the Hellancourt Formation volcanics were erupted.

STRUCTURAL GEOLOGY

The structure of the fold-and-thrust belt is dominated by shallow-dipping thrusts and north-northwest-trending, shallow-plunging folds. There are both tight, subrecumbent and more open, almost upright folds, and in some places it is clear that folding affects both bedding and an early cleavage. Furthermore, map patterns indicate that some of the major thrusts, and a slaty cleavage, have been folded in places. It is possible, therefore, that the belt has experienced multiple deformation events. However, bedding and cleavage data on the published geological maps indicate that there is no significant departure from cylindrical geometries over broad domains in the fold belt, so that if there

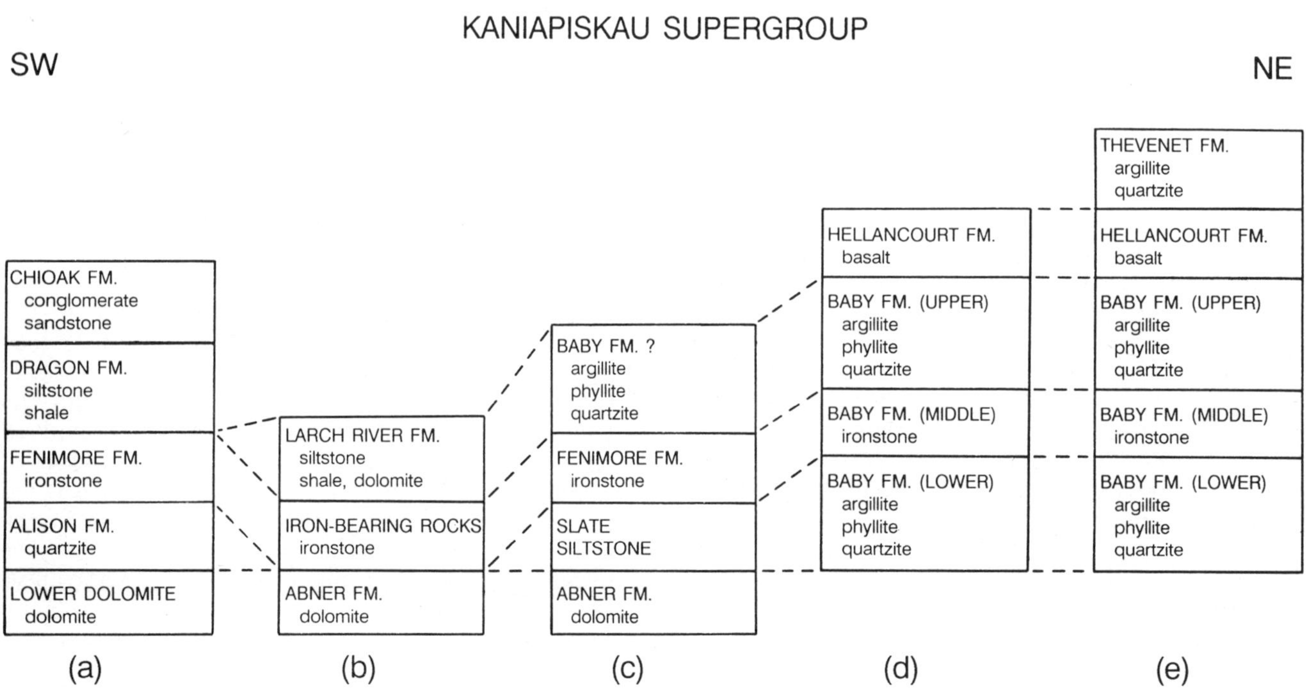

Figure 2. *Stratigraphic sections for the Proterozoic Kaniapiskau Supergroup in various parts of the Northern Labrador Trough, and the correlations between them used for this paper. (a) West of Lac Merchère (Bérard (1965), Clark (1980); see Fig. 3 for geographic localities); (b) between Lac Merchère and the thrust at Lac Couteau (Bérard, 1965; Clark, 1980); (c) Lac Couteau to Lac Rasle (Bérard, 1965; Clark, 1980); (d) Lac Rasle to Lac Rougement (Sauvé and Bergeron, 1965); and (e) east of Lac Rougement (Sauvé and Bergeron, 1965).*

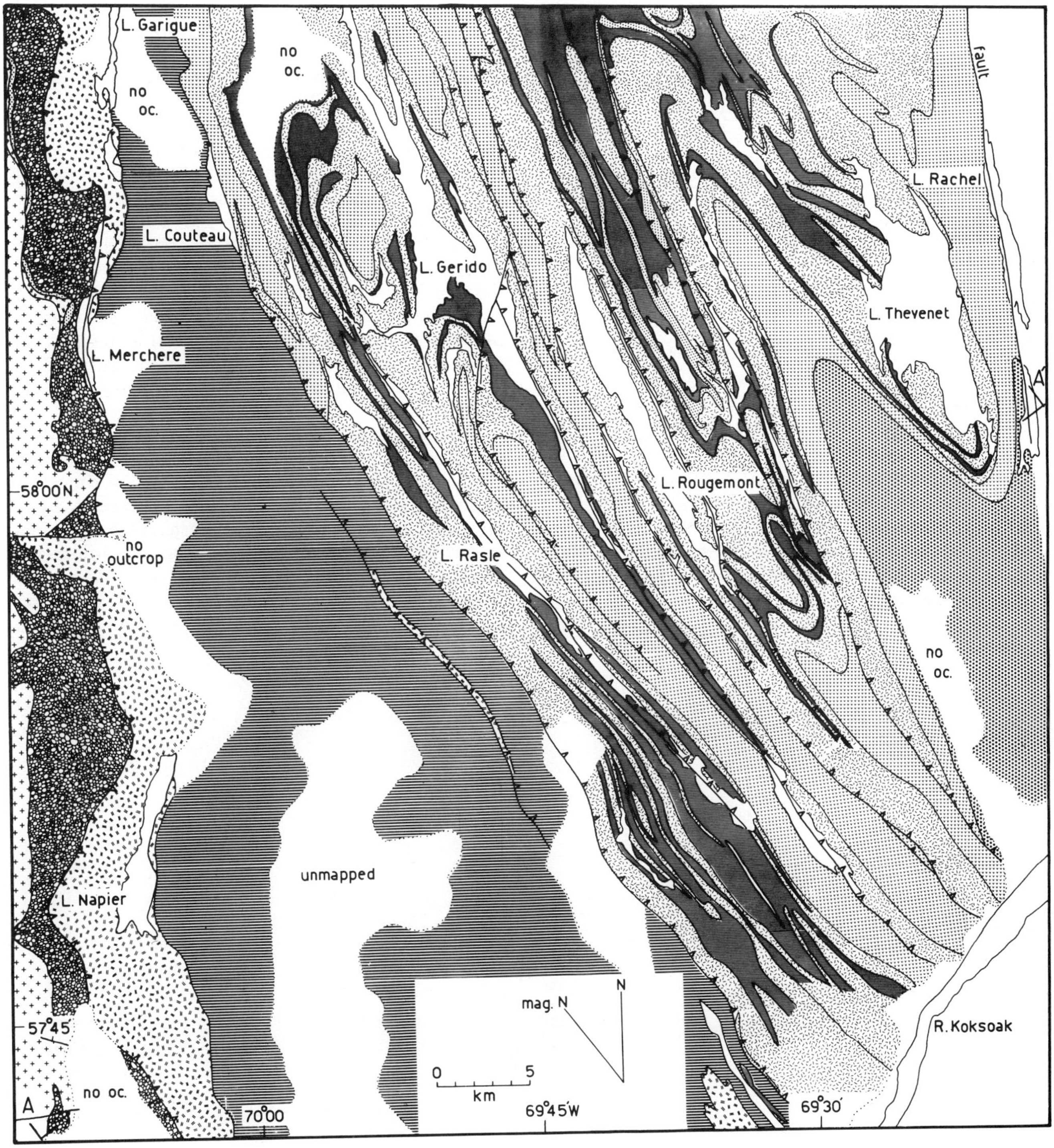

Figure 3. *Geological map of the Northern Labrador Trough, simplified after maps by Sauvé (1956, 1957, 1959), Gelinas (1958), Bérard (1965), Sauvé and Bergeron (1965), Clark (1979, 1980), and Belanger (1982). Legend as on Figure 4. A and A' are the end points for the section shown on Figure 4. The Lac Rachel Fault, in the northeast corner, is steeply dipping and juxtaposes high-grade rocks to the east with rocks of the Kaniapiskau Supergroup to the west. Its slip direction is uncertain. The Fenimore Formation and its correlatives are too thin in most places to be shown as separate units.*

are two or more sets of folds represented, they are essentially coaxial. It is, therefore, equally possible that the structures all result from one deformation event, involving the progressive forelandward collapse of the supracrustal package, with refolding of thrusts and early folds resulting from such processes as passage of the thrust packages over ramps (cf. Suppe, 1983). Without detailed mapping, we cannot choose between these alternatives, but the generally cylindrical character of the folds allows projection parallel to the fold plunge regardless of which scenario is correct, and this, in turn, permits the construction of a structural cross-section for the region.

We prepared a cross-section based on structural data available on the published geological maps and reports for the Northern Labrador Trough cited above, passing through the four transects of the Hellancourt Formation we had studied. The line of section is perpendicular to the north-northwest trend of the major folds and parallel to the probable direction of tectonic transport. It includes the basement–cover contact at the western limit of the Labrador Trough and extends to the Lac Rachel Fault in the east (Fig. 3).

The systematics for building a balanced cross-section from map data are well described (e.g., Gwinn, 1970; Price and Mountjoy, 1970; Woodward *et al.*, 1985). Some of the requirements for a completely restorable balanced cross-section, such as a well-constrained stratigraphy, stratigraphic thicknesses, well-log data, and seismic, gravity, and aeromagnetic data for the area (Elliott, 1983), are not fully met here. A major problem is caused by the abundant gabbroic intrusives in the Baby Formation. These are not always stratiform and create uncertainties in thickness estimates. As a result, the data for the Labrador Trough are not amenable to the construction of a truly balanced cross-section, but a down- (and up-) plunge projection of the structural and stratigraphic data is possible, resulting in a "restorable and admissible cross-section" (Elliott, 1983).

A further uncertainty in the stratigraphy is the thickness of the Larch River Formation. Bérard (1965) acknowledged that this is difficult to measure and used 10–80 m as an estimate, based on measurements at Lac Merchère. This is difficult to reconcile with a unit that has several kilometres of continuous exposure at the surface. The Larch River Formation is correlated with the Baby For-

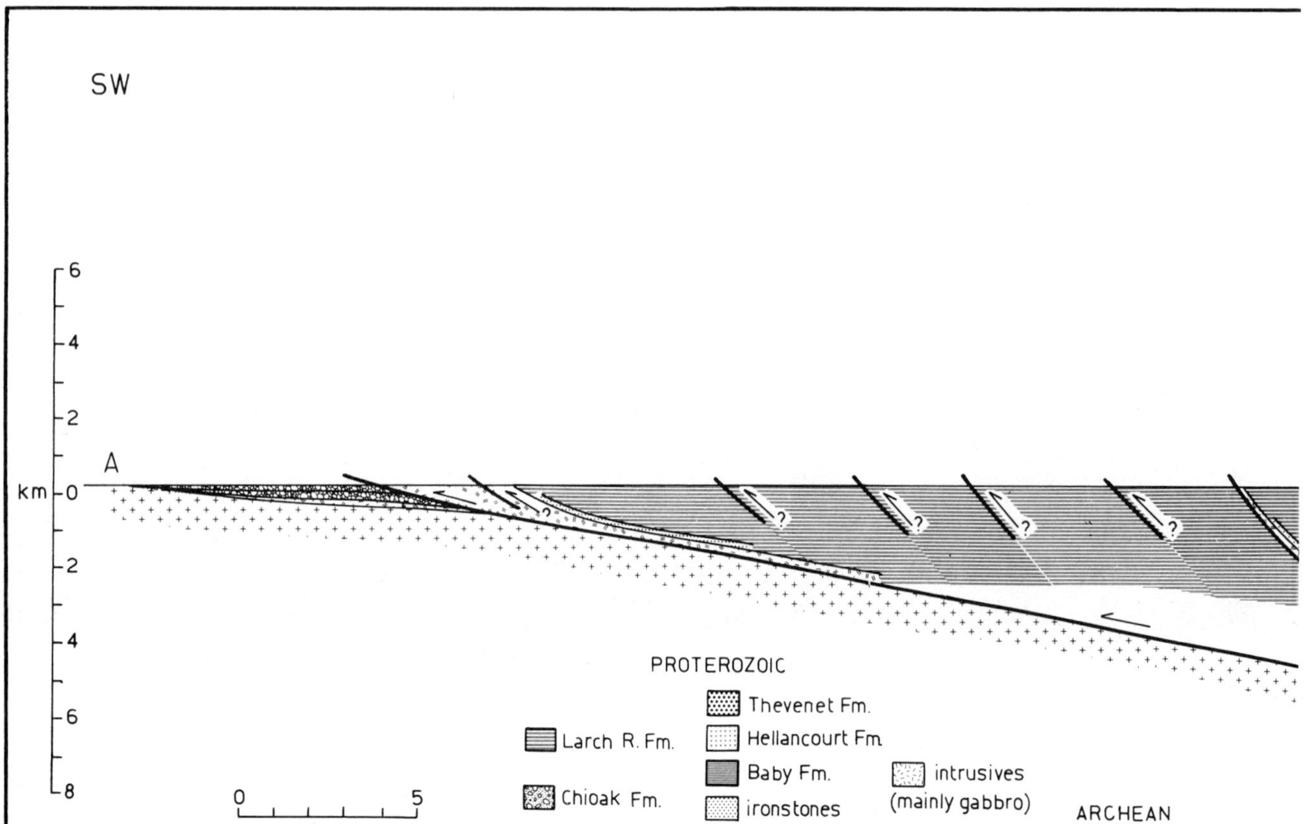

Figure 4. *Interpreted cross-section A-A′ from Figure 3.*

mation, which is 1100 m thick in the east. This is probably a maximum thickness for the Larch River Formation, and we have assumed 600 m for the thickness of the Larch River Formation in the west.

In preparing the cross-section, we projected data from as far away as 25 km on either side of the line of section. The distance chosen was dictated by a change in the direction of fold plunge roughly 25 km north of the line, which renders projection from farther out very unreliable; the same distance was selected as the limit to the south. Fold plunges are generally shallow throughout the region (less than 20°; typically less than 10°), but they vary in magnitude and direction even within 25 km of the line of section. There is, therefore, no universal fold axis for the projection in the line of section. In consequence, the region was divided by inspection into domains for which the plunge was essentially uniform, and a plunge was calculated for each domain by determining the pole to the best-fit great circle to poles of bedding or, in some cases, bedding and cleavage, and comparing it with measured fold axes where given. A total of 1500 measurements of bedding and cleavage were available, and the region was divided into 76 domains using these data. Surface features, such as folds, faults, and formational contacts, were then projected into the line of section sequentially through all the domains between them and the line of section, using the plunges appropriate for each domain. Thus, the procedure was projection through a series of straight-line segments, providing an approximation to a smooth curve that was assumed to be the trajectory for the feature in the subsurface or above ground. The assumption of a smooth trajectory would be incorrect if there were transverse features, such as side-ramps, of which we have no evidence in the region, but which we cannot entirely preclude.

The line-segment approximation may give rise to significant errors in the depths of some features on the line of the section. However, the cross-section is severely constrained by the projection of the fold southwest of Lac Thévenet, at the eastern end of the section (Figs. 3 and 4), which is large, essentially cylindrical, and overlaps the line of section. The projection of this fold constrains the minimum depth of the basement at the eastern end of the section. The section was constructed using this minimum depth to basement because it provides the most conservative estimate of the total amount of shortening achieved across the belt. As there is no

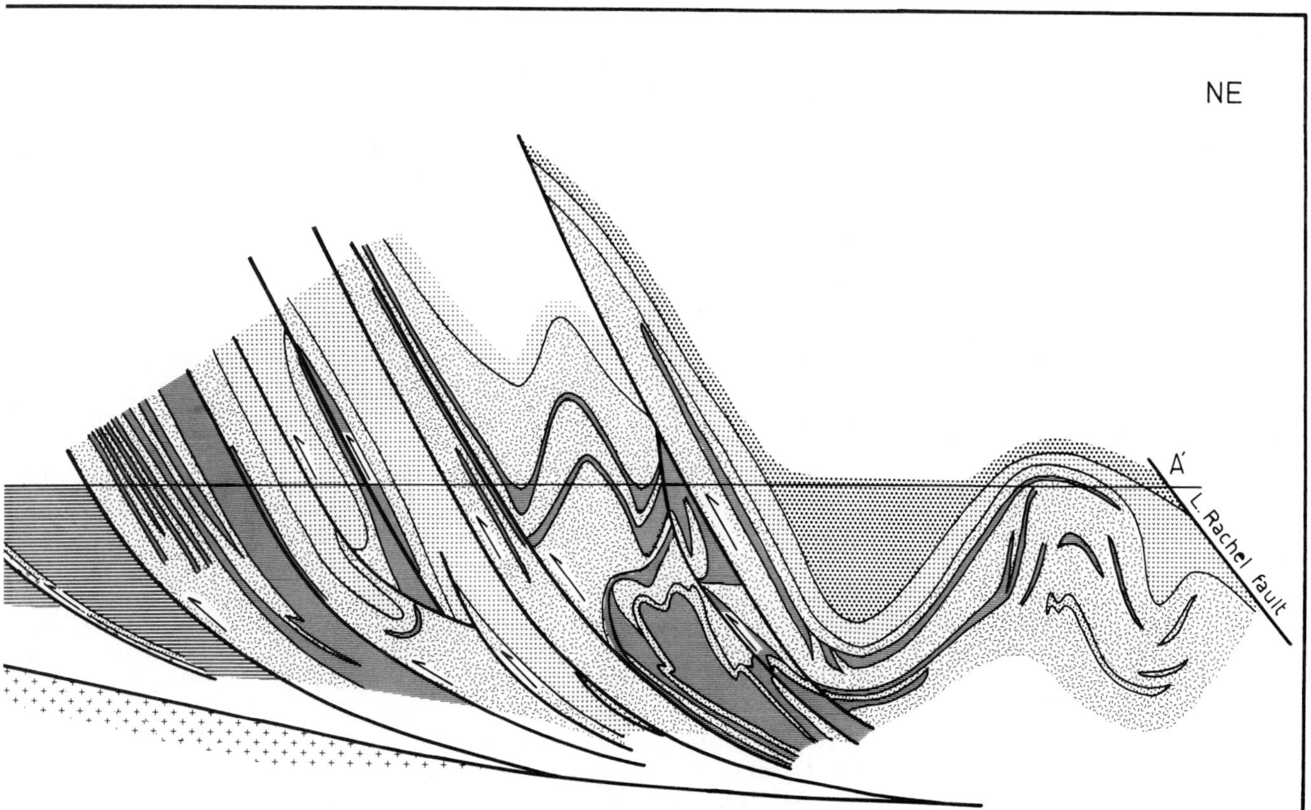

Figure 4. Continued.

evidence of basement involvement west of the Lac Rachel Fault anywhere in the Northern Labrador Trough, we assume that the belt is "thin skinned" and, by analogy with the Canadian Rockies (Price and Mountjoy, 1970), that the basement surface beneath the belt is smooth. On this assumption, a straight basement/cover boundary may be drawn between the projected depth of the Thévenet fold and the position of the exposed basement/cover contact at the western end of the section. Once this lower bound was supplied to the section, projection of the surface features from 25 km either side of the line of section by the method outlined above gave rise to a reasonably tightly constrained subsurface geometry, with the positions of features on Figure 4 correct to within 500 m. It is important to emphasize, however, that the depth to basement in the footwall to the Lac Rachel Fault may be considerably greater than that depicted on Figure 4, in which case the precision of the section at depth becomes far poorer.

The cross-section is similar in many respects to those drawn for other fold-and-thrust belts. The major difference between this section and those published for the Canadian Cordillera is the dip of the basement. In the Canadian Cordillera, the dip of the basement is 1–2° (Price and Mountjoy, 1970), whereas in the Labrador Trough the dip is 5°. The minimum angle of the basement dip is well constrained by the depth to basement for the deepest portion of the cross-section underneath the Thévenet fold. Work in the Appalachians indicates that the dip of the basement there is comparable with that interpreted here (Roeder *et al.*, 1978; Kulander and Dean, 1986).

In view of the uncertainties in the deeper part of the section of Figure 4, and the absence of equivalent footwall and hanging wall truncations for many of the faults, no retrodeformation of the section is possible. However, in the eastern part of the section the bed length of the Hellancourt Formation can be used to provide a minimum distance necessary for palinspastic restoration. This minimum distance is 80 km for the region between the Lac Rachel Fault and the westernmost thrust of Hellancourt Formation rocks, compared with a present distance of 30 km. The minimum shortening required by the section, where Hellancourt Formation is present, is, therefore, 50 km and, because many of the faults do no truncate the Hellancourt top and bottom, the total shortening may be considerably greater. In the west, where the Hellancourt Formation is not present, shortening was estimated by assigning an a priori thickness to the Larch River Formation and calculating the quantity of undeformed Larch River Formation required to fill the thrust sheet. Three thicknesses of Larch were used in the calculations: 600, 900, and 1100 m; resulting in restored lengths for the Larch River part of the cross-section of 118, 79, and 65 km, respectively, compared with a present length of 25 km. It is evident that the assumed thickness of the Larch River Formation influences the amount of shortening for the section considerably. The stratigraphic pile is thought to thin to the west (Dimroth *et al.*, 1970) and, thus, the lowest estimated thickness (600 m) is considered the most reasonable. Palinspastic restoration of a 600 m Larch River Formation and the observed Hellancourt would, therefore, give a total restored section length of 198 km, to which a further minimum 5 km must be added for the frontal thrusts of the Abner Formation. The section has, therefore, been shortened from an initial length of more than 200 km to its present 58 km.

DISCUSSION AND CONCLUSIONS

The section of Figure 4 illustrates very clearly a problem with the Thévenet Formation. This unit of massive and interbedded immature quartzites and argillites occurs stratigraphically above the Hellancourt Formation only in the easternmost region. Clearly, one way out of this problem is to place a thrust at the base of the Thévenet Formation, in which case the Thévenet Formation would be a distal correlative of the lower Baby Formation (Goulet, 1986). Alternatively, local highs within the rift package, such as upfaulted basement, which may be common in conjunction with extreme continental extension (Lister *et al.*, 1986), may have contributed the sediment of the Thévenet Formation, or the Thévenet Formation may be a local flysch formed during the deformation of the trough sediments. There are no data at present that allow us to choose between these alternatives.

Palinspastic restoration would place the Hellancourt Formation well east of the present position of the Lac Rachel Fault. There is no clear evidence of sialic basement in this position. Although there is known Archean basement east of Lac Rachel (Machado *et al.*, 1987), it is not clear that this has not been subsequently transported along strike (N. Goulet, personal communication, 1987). Furthermore, even if the gneissic basement now present there did originate east of Lac Rachel, the supracrustals may have been transported onto it early in the tectonic history (J. Moorhead, personal communication, 1987). Thus, it is likely that the volcanics of the Hellancourt Formation were erupted on very much thinned continental crust, or even at an ocean–continent boundary. This is in every way compatible with their geochemistry.

There are several uncertainties in the cross-section. That with the greatest potential to affect the restoration is the thickness of the Larch River Formation. Even if we adopt the largest estimate for its thickness, restoration of the section places the westernmost Hellancourt Formation rocks only 15 km west of the Lac Rachel Fault; they were still developed far from the craton. Furthermore, we have throughout adopted the most conservative depth-to-basement estimate at the eastern edge of the section. Were we to insert further supracrustals between the Thévenet fold and its underlying basement, restoration of the section would place the Hellancourt Formation even farther east.

Thus, notwithstanding uncertainties in the reconstruction presented here, may of which might be resolved by more detailed mapping, we believe the structure of the region requires considerable tectonic shortening of the supracrustal succession. The overall characteristics of the region are compatible with the preservation of a collapsed Atlantic-style passive margin. The geochemistry of the volcanic rocks is not easily characterized, but is consistent in a general way with what might be expected on very-thinned continental crust rather than in the more stable environment in which typical continental tholeiites are developed. The structural section bears strong resemblance to that of other proposed collapsed continental margins exhibiting "thin-skinned" tectonics, and its restoration places the volcanic rocks well outboard of the exposed and demonstrably unrifted continental basement of the Ungava craton.

ACKNOWLEDGEMENTS

This research was supported by the Quebec Ministry of Education, the Natural Sciences and Engineering Research Council of Canada, and Energy, Mines and Resources Canada. We are grateful to Tom Clark, Normand Goulet, James Moorhead, and Bob Wares for their helpful criticism of the section, and Charles Boone for his assistance in the field. Constructive criticism by the reviewers for this volume resulted in significant improvements to the manuscript.

REFERENCES

Baragar, W.R.A. and Scoates, R.F.J., 1981, The Circum-Superior belt, a Proterozoic plate margin?, *in* Kroner, A., ed., Precambrian Plate Tectonics: Elsevier, Amsterdam, p. 297–330.

Belanger, M., 1982, Région de Lac Faujas: Ministère des Richesses Naturelles Québec, Rapport Préliminaire DP 82-06.

Bérard, J., 1965, Bérard Lake area, New Québec: Quebec Department of Natural Resources, Geological Report 111, 139 p.

Bergeron, R., 1957, Région de Brochant de Bonnard, Nouveau Québec: Ministère des Mines Québec, Rapport Préliminaire 348, 6 p.

Clark, T., 1977, Forbes Lake area, New Québec: Ministère des Mines Québec, Rapport Préliminaire DPV-452, 16 p.

Clark, T., 1978, Région du Lac Hérodier, Nouveau Québec: Ministère des Mines Québec, Rapport Préliminaire DPV-568, 42 p.

Clark, T., 1979, Région du Lac Napier, Territoire du Nouveau Québec: Ministère des Richesses Naturelles Québec, Rapport Préliminaire DPV-663, 22 p.

Clark, T., 1980, Région de la Rivière Koksoak, Territoire du Nouveau Québec: Ministère des Richesses Naturelles Québec, Rapport Préliminaire DPV-781, 18 p.

Dimroth, E., 1972a, Stratigraphy of part of the Central Labrador Trough (between latitude 56°30′ and the height of land): Ministère des Richesses Naturelles Québec, Open File GM-28691.

Dimroth, E., 1972b, The Labrador geosyncline revisited: American Journal of Science, v. 272, p. 487–506.

Dimroth, E., 1981, Labrador geosyncline: type example of early Proterozoic cratonic reactivation, *in* Kroner, A., ed., Precambrian Plate Tectonics: Elsevier, Amsterdam, p. 331–352.

Dimroth, E., Baragar, W.R.A., Bergeron, R. and Jackson, G.D., 1970, The filling of the circum-Ungava geosyncline, *in* Baer, A.J., ed., Precambrian Basins and Geosynclines of the Canadian Shield: Geological Survey of Canada, Paper 70-40, p. 45-142.

Elliott, D., 1983, The construction of balanced cross-sections: Journal of Structural Geology, v. 5, p. 101.

Gelinas, L., 1958, Thévenet Lake area (east half), New Québec: Quebec Department of Mines, Preliminary Report 363, 8 p.

Gelinas, L., 1959, Région du Lac Gabriel (partie est) et la Région du Fort Chimo (partie ouest), Nouveau Québec: Ministère des Mines Québec, Rapport Préliminaire 407, 9 p.

Gelinas, L., 1960, Région du Fort Chimo (partie est), Nouveau Québec: Ministère des Mines Québec, Rapport Préliminaire 418, 6 p.

Gold, D.P., 1962, Hopes Advance Bay area, New Quebec: Quebec Department of Natural Resources, Preliminary Report 442, 9 p.

Goulet, N., 1986, Étude tectonique et stratigraphique de la partie nord de la Fosse du Labrador — Région de la baie aux Feuilles et du lac Bérard: Ministère de l'énergie et des Ressources Naturelles, Québec, Manuscrit Brut 86-27, 22 p.

Gross, G.A., 1968, Geology of iron deposits in Canada. Volume III. Iron ranges of the Labrador geosyncline: Geological Survey of Canada, Economic Geology Report 22, 179 p.

Gwinn, V.E., 1970, Kinematic patterns and estimates of lateral shortening, Valley and Ridge and Great Valley provinces, central Appalachians, south-central Pennsylvania, *in* Fisher, G.W., Pettijohn, F.J., Reed, J.C. and Weaver, K.N., eds., Studies in Appalachian Geology: Central and Southern: Interscience, p. 127–146.

Hardy, R., 1976, Région Lac Roberts des Chefs: Ministère des Richesses Naturelles Québec, Rapport Géologique 171, 99 p.

Kulander, B.R. and Dean, S.L., 1986, Structure and tectonics of central and southern Appalachian Valley and Ridge and Plateau provinces, West Virginia and Virginia: American Association of Petroleum Geologists Bulletin, v. 70, p. 1674–1684.

LeGallais, C.J. and Lavoie, S., 1982, Basin evolution of the Lower Proterozoic Kaniapiskau Supergroup, central Labrador miogeocline (trough) Québec: Bulletin of Canadian Petroleum Geology, v. 30, p. 150–166.

Lister, G.S., Etheridge, M.A. and Symonds, P.A., 1986, Detachment faulting and the evolution of passive continental margins: Geology, v. 14, p. 246–250.

Machado, N., Goulet, N. and Gariepy, C., 1987, Evolution of the Northern Labrador Trough basement: evidence from U–Pb geochronology: Geological Association of Canada – Mineralogical Association of Canada, Program with Abstracts, v. 12, p. 69.

Moorhead, J. and Hynes, A.J., 1986, Structure and metamorphism of the eastern flank of the Renia gneiss dome: Geological Association of Canada – Mineralogical Association of Canada, Program with Abstracts, v. 11, p. 103.

Poirier, G., Perreault, S. and Hynes, A., 1990, Nature of the eastern boundary of the Labrador Trough near Kuujjuaq, Quebec, *in* Lewry, J.F. and Stauffer, M.R., eds., The Early Proterozoic Trans-Hudson Orogen of North America: Geological Association of Canada, Special Paper 37, p. 397–411.

Price, R.A. and Mountjoy, E.W., 1970, Geologic structure of the Canadian Rocky Mountains between Bow and Athabasca rivers – a progress report, *in* Wheeler, J.O., ed., Structure of the Southern Canadian Cordillera: Geological Association of Canada, Special Paper 6, p. 7–25.

Roeder, D., Gilbert, D.E. and Witherspoon, W.D., 1978, Evolution and macroscopic structure of valley and ridge thrust belt, Tennessee and Virginia: University of Tennessee, Studies in Geology, v. 2, p. 14–17.

Sauvé, P., 1956, De Freneuse Lake area (west half), New Québec: Quebec Department of Mines, Preliminary Report 332, 6 p.

Sauvé, P., 1957, De Freneuse Lake area (east half), New Québec: Quebec Department of Mines, Preliminary Report 358, 7 p.

Sauvé, P., 1958, The geology of the eastern border of the Labrador Trough near Fort Chimo, northern Québec: Canadian Mining Journal, v. 79, p. 123–124.

Sauvé, P., 1959, Leaf Bay area, New Québec: Quebec Department of Mines, Preliminary Report 399, 10 p.

Sauvé, P. and Bergeron, R., 1965, Gerido Lake – Thévenet Lake area, New Québec: Quebec Department of Natural Resources, Geological Report 104, 110 p.

Schilling, J.G., 1975, Rare earth variations across "normal sections" of the Reykjanes Ridge, 60°–53°N, Mid-Atlantic Ridge, 29°S, and East Pacific Rise, 2°–19°S, and evidence on the composition of the underlying low-velocity layer: Journal of Geophysical Research, v. 80, p. 1449–1473.

Suppe, J., 1983, Geometry of fault-bend folding: American Journal of Science, v. 283, p. 684–721.

Wardle, R.J. and Bailey, P.G., 1981, Early Proterozoic sequences in Labrador, *in* Campbell, F.H.A., ed., Proterozoic Basins of Canada: Geological Survey of Canada, Paper 81-10, p. 331–359.

Woodward, N.B., Boyer, S.E. and Suppe, J., 1985, An outline of balanced cross-sections, 2nd edition: University of Tennessee, Studies in Geology, v. 11, 170 p.

Nature of the Eastern Boundary of the Labrador Trough Near Kuujjuaq, Quebec

Glenn Poirier, Serge Perreault, and Andrew Hynes
Department of Geological Sciences,
McGill University, 3450 University St., Montreal, Quebec, H3A 2A7, Canada

Poirier, G., Perreault, S. and Hynes, A., Nature of the eastern boundary of the Labrador Trough near Kuujjuaq, Quebec, in Lewry, J.F. and Stauffer, M.R., eds., The Early Proterozoic Trans-Hudson Orogen of North America: Geological Association of Canada, Special Paper 37, p. 397–411.

Abstract

Mapping in the eastern hinterland of the Labrador Trough near Kuujjuaq has identified three distinct lithotectonic blocks, each bounded by major faults. The westernmost block consists of a succession of pelitic and semipelitic schists and basalt-derived amphibolites, which lie on the flanks of a series of doubly plunging antiforms cored by Archean gneisses. The supracrustals of this block are interpreted as stratigraphically correlative to the Kaniapiskau Supergroup of the Labrador Trough. Rocks of the two eastern blocks, however, differ significantly from those of the Labrador Trough. These blocks are dominated by thick successions of gneissic metasandstone and para-amphibolite, with minor paragneiss, calc-silicate, and semipelitic schist and gneiss. One of the eastern blocks also contains a large (40 km × 4 km) body of pre- to syntectonic tonalitic and granodioritic gneisses that shows many similarities to the granitoids of modern arc terrains. Juxtaposition of these two blocks with the western block occurred early in the tectonic history.

Characteristics of the metasediments of the two eastern blocks indicate deposition in a higher energy environment than most sediments of the Kaniapiskau Supergroup. The presence of the granitoid intrusives may indicate that these blocks are part of an arc complex. Whether the blocks represent an uplifted part of the eastern margin of the trough or a true exotic terrain is uncertain.

Résumé

La cartographie de l'arrière-pays à l'est de la fosse du Labrador près de Kuujjuaq a permis d'identifier trois blocs lithotectoniques distincts, tous bordés par des failles majeures. Le bloc ouest consiste en une succession de schistes pélitiques et semi-pélitiques et d'amphibolites dérivées de basaltes, reposant sur les flancs d'une série d'antiformes à double plongée noyautés par des gneiss archéens. Les roches supracrustales appartenant à ce bloc sont considérées comme corrélatives avec le supergroupe de Kaniapiscau de la fosse du Labrador. Les roches des deux blocs est diffèrent considérablement de celles de la fosse du Labrador; des successions épaisses de grès métamorphisés et de para-amphibolites dominent avec des paragneiss mineurs, des roches calco-silicatées et des schistes et gneiss semi-pélitiques. Un des blocs est contient une grande masse (40 km x 4 km) de gneiss tonalitiques et granodioritiques pré- à syn-tectoniques montrant des similitudes avec les granitoïdes d'arcs modernes. La juxtaposition des deux blocs est avec le bloc ouest s'est produite au début de l'histoire tectonique.

Les caractères des méta-sédiments des deux blocs est indiquent un milieu de sédimentation à plus haute énergie que celui de la plupart des sédiments du supergroupe de Kaniapiscau. La présence d'intrusifs de granitoïdes pourrait indiquer que ces blocs sont partie d'un complexe d'arc. Il n'est pas possible de déterminer si ces blocs représentent la partie remontée de la marge est ou un terrain exotique.

INTRODUCTION

Tectonic processes in the Proterozoic have been the subject of debate for many years, particularly since the advent of plate tectonics (cf. papers in Kroner (1981)). Many workers have championed an ensialic origin for Proterozoic orogens, as distinct from an origin through destruction of ocean basins. The principal data in favour of ensialic models have been paleomagnetic signatures and the apparent absence of the subduction-related rocks that might be expected in conjunction with the closure of ocean basins. The Labrador Trough was presented as the type example of an ensialic orogen (Dimroth, 1981) based primarily on observations in the relatively weakly metamorphosed fold–thrust belt that forms most of the trough. Recent mapping in broadly coeval Proterozoic belts in North America has shown, however, that they, at least, are not characterized by a dearth of potential subduction-related material (Hoffman, 1980; Lewry *et al.*, 1981). In this paper, we show that the Labrador Trough, or at least the hinterland beyond its eastern margin, contains some lithotectonic assemblages compatible with an origin above a subduction zone.

GEOLOGICAL SETTING

The Labrador Trough is part of the Trans-Hudson Orogen (Hoffman, 1981), which consists of a series of structural and lithological domains that show the imprint of thermotectonic events dated at 2000–1700 Ma (Lewry *et al.*, 1985; Van Schmus *et al.*, 1987; see Fig. 1). The Labrador Trough segment of the Trans-Hudson Orogen is a southward extension of the Churchill Province (Stockwell, 1963), which occurs between the Archean Superior and Nain provinces in northern Quebec and Labrador respectively. It consists of a north-northwest-trending belt of Aphebian sedimentary and volcanic rocks, which extends 800 km from southern Quebec and Labrador to the western shores of Ungava Bay (Fig. 2). Most authors believe that the sedimentary and volcanic rocks of the trough, the Kaniapiskau Supergroup (Frarey and Duffell, 1964), were deposited in an Aphebian rifting environment (Baragar and Scoates, 1981; Wardle and Bailey, 1981; LeGallais and Lavoie, 1982), although Hoffman (1987) has suggested that the sediments and volcanics were deposited in a foredeep. These rocks were subsequently deformed and metamorphosed during the Hudsonian orogeny (Dimroth, 1978). The eastern boundary of the trough has traditionally been placed where Archean basement rocks first appeared at the surface and the metamorphic grade increased sharply, but recent studies have indicated that sediments of equivalent age may occur far to the east of this boundary (Taylor, 1979; LeGallais and Lavoie, 1982).

The Labrador Trough may be divided into three distinct lithotectonic domains (Dimroth, 1972, 1978; Fig. 2). The westernmost domain is a miogeoclinal zone consisting mainly of sediments with minor amounts of volcanic rock, resting uncomformably on Archean basement. The western parts of this zone are undeformed, but minor deformation is observed in the east. To the east is a typical foreland fold-and-thrust belt consisting of thrust slices of sedimentary and basaltic volcanic rocks, but with no evidence of basement involvement (Boone and Hynes, 1987).

These two zones make up the Labrador Trough sensu stricto. They are separated from the third, or "hinterland," zone by high-angle faults (LeGallais and Lavoie, 1982). The hinterland zone contains sequences of sedimentary and volcanic rocks that have been subjected to intense deformation and medium- to high-grade metamorphism. Relationships between the rocks of this zone and those of the Labrador Trough sensu stricto are obscured by deformation and metamorphism, although some correlations have been made with the Kaniapiskau Supergroup (Taylor, 1979; Wardle and Bailey, 1981).

This paper is concerned with the geology of the hinterland zone near Kuujjuaq, Quebec (58°N), and in particular with several fault-bounded blocks in this region whose relationship to the Kaniapiskau Supergroup is at least suspect.

GENERAL GEOLOGY

The area considered extends from the eastern boundary of the Labrador Trough sensu stricto, which at this latitude is defined by the high-angle reverse Lac Rachel Fault (LRF), to the Koksoak estuary near Kuujjuaq. It consists of three distinct lithostructural zones or blocks, all of which show the effects of medium- to high-grade metamorphism and several episodes of deformation. From west to east, these are the Lac Rachel – Lac Murray block (LRLMB), the Lac Gabriel block (LGB), and the Lac Berthet block (LBB) (Figs. 3 and 4).

The Lac Rachel – Lac Murray block, east of the Lac Rachel Fault, is composed of pelitic to semi-pelitic schists and amphibolites with minor marble, metaconglomerate, and reworked Archean tonalitic gneiss (Gélinas, 1965; Moorhead and Hynes, 1986; Perreault *et al.,* 1988). It is juxtaposed with the Lac Gabriel block along the Lac Turcotte Fault (LTF). The Lac Gabriel block consists of a wide variety of gneissic metasandstones, amphibolites, and pelitic gneisses. It also contains a complex suite of syntectonic granitoids and associated gabbroic rocks. The Lac Berthet block, which is separated from the Lac Gabriel block by the Lac Pingiajjulik Fault (LPF), consists of a thick series of para-amphibolites and arkosic paragneisses mixed with minor amounts of calc-silicate gneiss, quartzite, marble, and semi-pelitic gneiss resting on an Archean migmatitic basement (Perreault *et al.,* 1988).

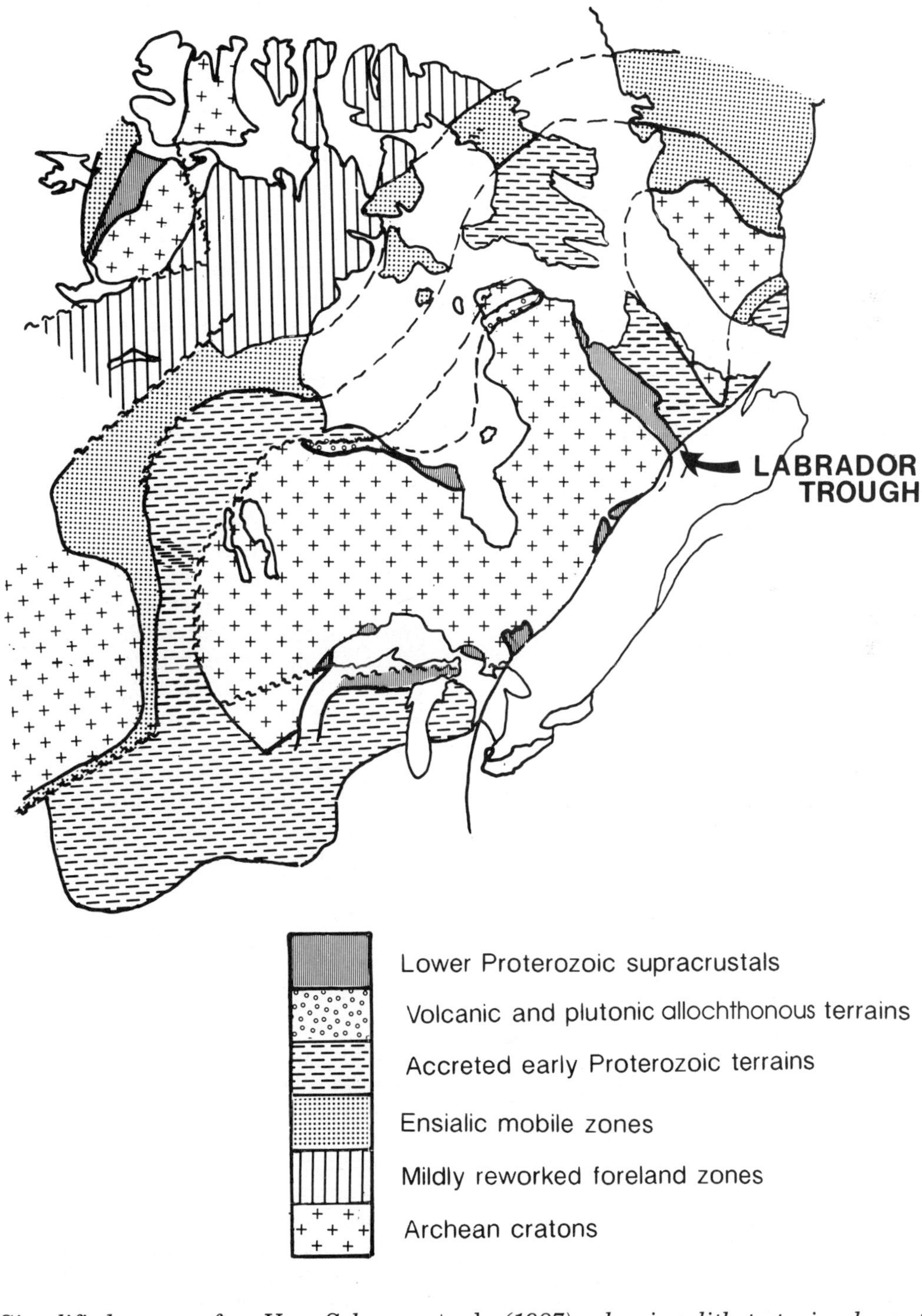

Figure 1. *Simplified map, after Van Schmus* et al. *(1987), showing lithotectonic elements and early Proterozoic terrains of North America. Wavy broken lines indicate major shear zones.*

STRUCTURE AND METAMORPHISM

The area experienced polyphase deformation during the Hudsonian orogeny. The first deformation event, D_1, was the most intense, and produced the main bedding-parallel foliation in all the metasediments of the area. No regional-scale F_1 folds are observed, but minor intrafolial, isoclinal F_1 folds are occasionally noted. The only large-scale structure associated with D_1 is the bedding-parallel Lac Turcotte Fault. The D_2 event produced regional- and outcrop-scale, tight, upright to recumbent folds with axial planes generally trending east to northeast. S_2 foliation is not common and is observed only in pelitic schists in the westernmost parts of the area. The D_3 deformation event produced open to tight folds with highly variable axial plane and fold axis orientations. In the northern parts of the field area, F_3 folds have upright axial planes and plunge to the northeast, whereas in the southern parts of the study area they are commonly overturned with axial planes trending southeast. The intensity of D_3 deformation varies. In some parts of the study area

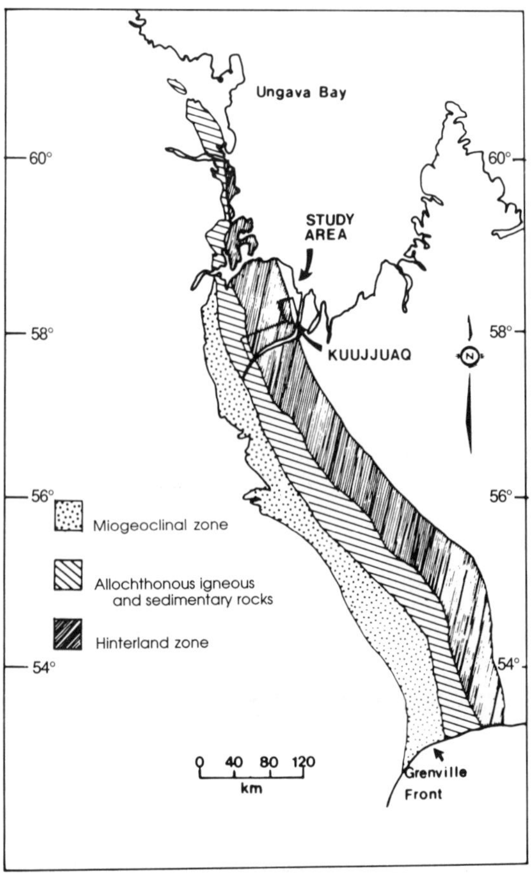

Figure 2. *Major subdivisions of the Labrador Trough. The location of the study area is outlined. The diagram is modified after Dimroth and Dressler (1978).*

D_3 is recognized only by the presence of a gentle, outcrop-scale warping, whereas in other parts it is responsible for the regional tectonic grain (e.g., the doubly plunging antiforms of the LRLMB). The youngest structural features in the area are a number of north-northwest–south-southeast and northeast–southwest faults, which are late to postmetamorphic and have variable slip.

Hudsonian metamorphic conditions in the area range from lower amphibolite facies in the west to granulite facies in the north and along the Koksoak River. Two metamorphic events are recognized in the Lac Gabriel and Lac Berthet blocks, but only one is observed in the LRLMB (Perreault *et al.*, 1988).

The first metamorphic event (M_1) is recorded in the LGB and LBB where it produced upper amphibolite and granulite facies assemblages. This event occurred during D_2 and has been dated at 1830 Ma using U–Pb zircon methods (Machado *et al.*, 1988; Perreault *et al.*, 1988). Typical assemblages are two pyroxenes plus quartz, with or without garnet, in mafic rocks and cordierite, alkali feldspar, garnet, and sillimanite in more aluminous rocks. M_2 metamorphism, which occurred syn- to post-D_3, locally overprinted these assemblages, reequilibrating these rocks to upper amphibolite facies. In the LGB, where M_1 and M_2 were both of upper amphibolite facies, the two events may be distinguished by relationships between porphyroblast growth and deformation (Perreault et al., 1988).

The single metamorphic event recorded in the LRLMB is syn- to post-D_3; contemporaneous with M_2 in the LGB and LBB. It is characterized by medium to high pressure and temperature conditions and increases in grade from west to east (lower to upper amphibolite facies). Isogradic surfaces in the block have moderate westerly dips and trend north-northwest. Peak metamorphic conditions increase from 500 MPa and 500°C near the LRF to 700 MPa and 750°C near Kuujjuaq and 800 MPa at 850°C in the Lac Berthet area (Perreault et al., 1987).

Several lines of evidence (e.g., isograds that crosscut D_3 structural features and undeformed mineral grains in the noses of F_3 folds) indicate that peak metamorphic conditions, which have been dated at 1790 Ma (Machado *et al.*, 1987), postdate major ductile deformation in all blocks. Pressure–temperature paths determined from mineral assemblages in the LGB and in the eastern part of the LRLMB are characterized by periods of isobaric heating after peak pressures were attained. These pressure–temperature paths and the postdeformation peak of metamorphic conditions indicate that D_3 may reflect a major crustal thickening event that was then followed by thermal reequilibration (cf.

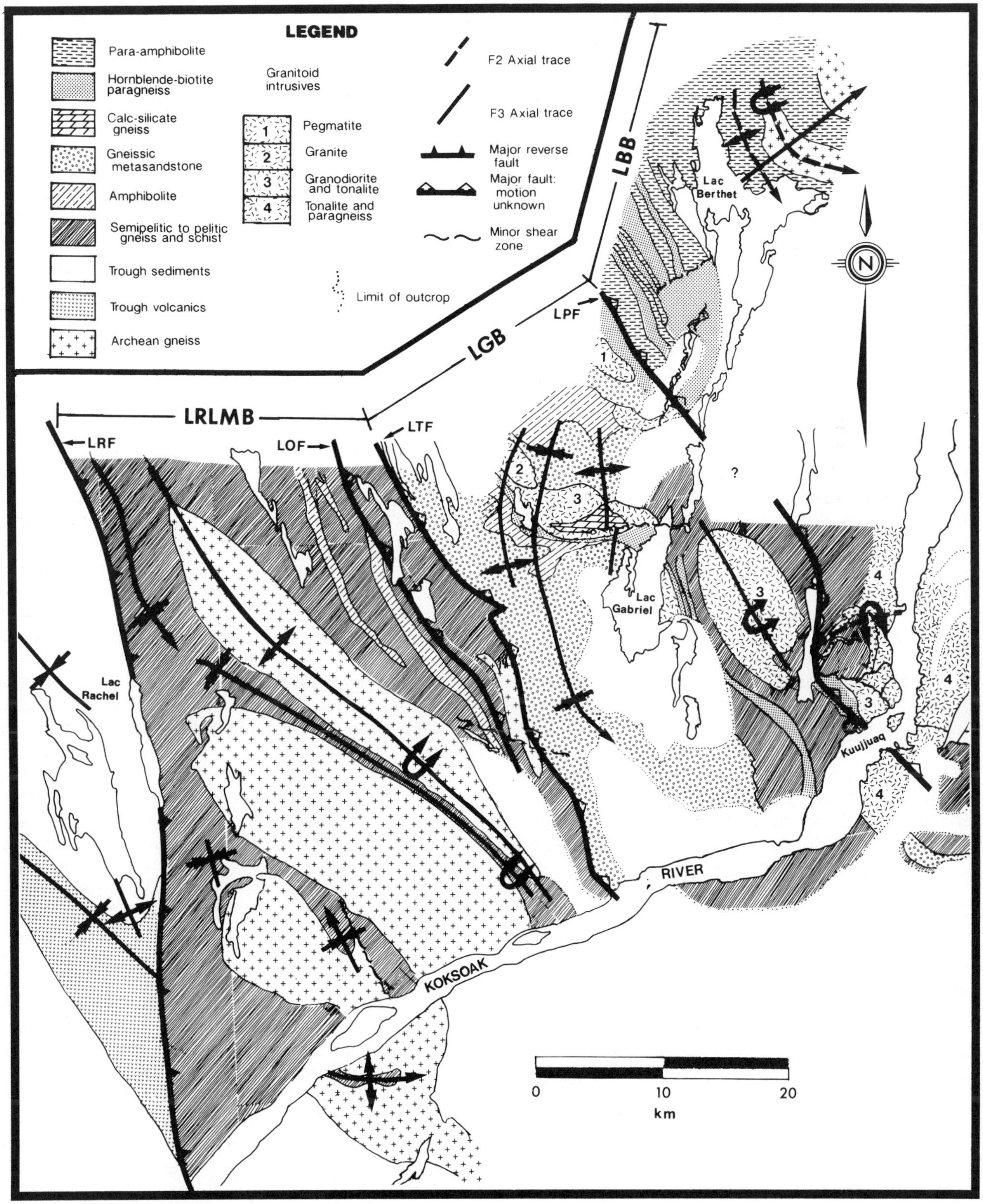

Figure 3. *Map of study area showing major lithological units. Abbreviations: LRLMB, Lac Rachel – Lac Murray block; LGB, Lac Gabriel block; LBB, Lac Berthet block; LRF, Lac Rachel Fault; LOF, Lac Olmstead Fault; LTF, Lac Turcotte Fault; and LPF, Lac Pingiajjulik Fault. The map includes data from Gélinas (1958, 1959, 1960) and Taylor and Skinner (1963).*

Thompson and England, 1984; Perreault *et al.*, 1987).

Major faults in the area have probably had complex movement histories. The LRF juxtaposes two lithotectonic blocks that record very different metamorphic conditions. Peak metamorphic conditions increase from 500°C and 500 MPa on the west side of the fault to 600°C and 800 MPa on the east side of the fault. This change is consistent with a postmetamorphic high-angle reverse motion on the fault, although a strike-slip component cannot be precluded. The nature of pre- to synmetamorphic motion on the LRF is unknown. The LRLMB is cut by a major fault, which probably represents the southern extension of the Lac Olmstead Fault (LOF) mapped by Goulet *et al.* (1987). Changes in peak metamorphic conditions are also observed across this fault. Peak conditions increase from 600°C on the west side of the fault to 700°C on the east, with no increase in pressure. Local shear zones associated with this fault contain C–S fabrics, indicating dextral motion. Fibrous sillimanite in these zones is deformed in contrast to the undeformed sillimanite observed elsewhere in the area, suggesting post-D_3 motion. These features suggest that movement was mainly strike slip and occurred during or after peak metamorphism, but earlier motion cannot be discounted.

The surface trace of the LTF is folded by an F_2 fold, indicating that it may have moved during D_1. Little is known about the type of motion on this fault; prograde metamorphism later in the structural history of the area has overprinted any possible metamorphic break associated with it. The northernmost Lac Pingiajjulik Fault is characterized by a band of rocks displaying a highly recrystallized mylonitic texture. A D_3 fold, which deflects the trace of this fault, indicates that motion occurred prior to D_3.

STRATIGRAPHY

The three blocks in the area contain many mappable units with highly varied lithologies. In this section, we deal with the stratigraphy of each block in turn. Although reliable younging indicators are rare in the area, all evidence indicates that younging is to the northeast. In the absence of more reliable younging indicators, it will be assumed for the purposes of this discussion that younging is to the northeast throughout the region.

Lac Rachel – Lac Murray block

The LRLMB is characterized by a sequence of metasedimentary rocks lying on the flanks of a series of doubly plunging, gneiss-cored antiforms (Figs. 3 and 4). These gneisses have been dated at 2850 Ma using U–Pb zircon methods (Machado *et al.*, 1987) and are interpreted as Archean basement rocks that were folded late in the Hudsonian orogeny. Some of these gneisses, and thin amphibolite units within them, retain granulite facies assemblages, which were partially retrograded to amphibolite facies during Hudsonian metamorphism.

Bedding-parallel foliation in the overlying metasedimentary rocks wraps around the antiforms. Although there is shearing at the contacts with the underlying gneisses, it is not clear whether the sedimentary rocks were originally unconformable on the gneisses or have been transported into their present position on a basement–cover décollement. The present thickness of the cover succession on

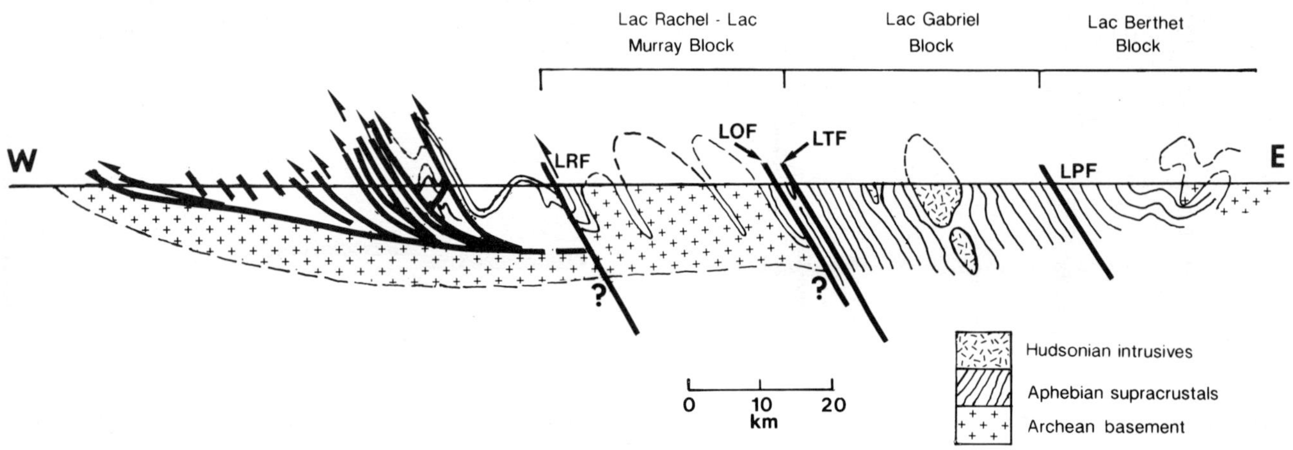

Figure 4. *Schematic cross-section of the Labrador Trough and its hinterland at the latitude of Kuujjuaq, Quebec. Abbreviations as in Figure 3. The cross-section of the foreland fold-and-thrust belt is after Boone and Hynes (1987).*

these domes is approximately 4 km (Fig. 5), but the possibility of tectonic thickening cannot be excluded.

The lowest stratigraphic unit in the cover succession is a highly deformed and locally brecciated marble. Thin dolomitic marble units also occur at higher levels in the stratigraphic sequence and are commonly characterized by distinctive quartz-rich bands rimmed by calc-silicate assemblages (Gélinas, 1965).

The most common rocks in the block are pelitic and semipelitic schists. The contacts between these two rock types are gradational; as the schists become more pelitic, they develop prominent sillimanite or kyanite and staurolite. In the upper parts of the stratigraphic sequence, the pelitic schists have a very distinctive texture consisting of garnet-cored sillimanite nodules 1–2 cm in diameter in a schistose matrix.

Amphibolites in the LRLMB are of two kinds. The more common are derived from basaltic flows and are conformable with surrounding metasediments. These amphibolites are either massive or finely laminated and rarely display deformed pillow structures. Minor pelitic schists and iron-rich quartzites are commonly found in association with them. The remainder of the amphibolites are intrusive in origin and probably represent metamorphosed gabbroic sills. These rocks have a coarser grain size than the metavolcanics and some exhibit glomeroporphyritic textures.

Metaconglomerates occur at several levels in the block. The lowermost is in a sequence of pelites near the base of the succession. It consists mainly of clasts of granitic gneiss in a matrix of actinolite and biotite, although clasts of dolomitic marble, pelitic schist, and amphibolite have also been identified. Another conglomerate unit mapped at the top of the succession contains a similar clast population, but in a matrix of quartz and plagioclase.

On the eastern side of the block, an elongate body of homogeneous, leucocratic, tonalitic gneiss is observed. This gneiss is composed mainly of plagioclase and quartz, with up to 15% biotite and minor sillimanite. "Xenoliths" are common, and are mostly metapelitic and amphibolitic. The body has

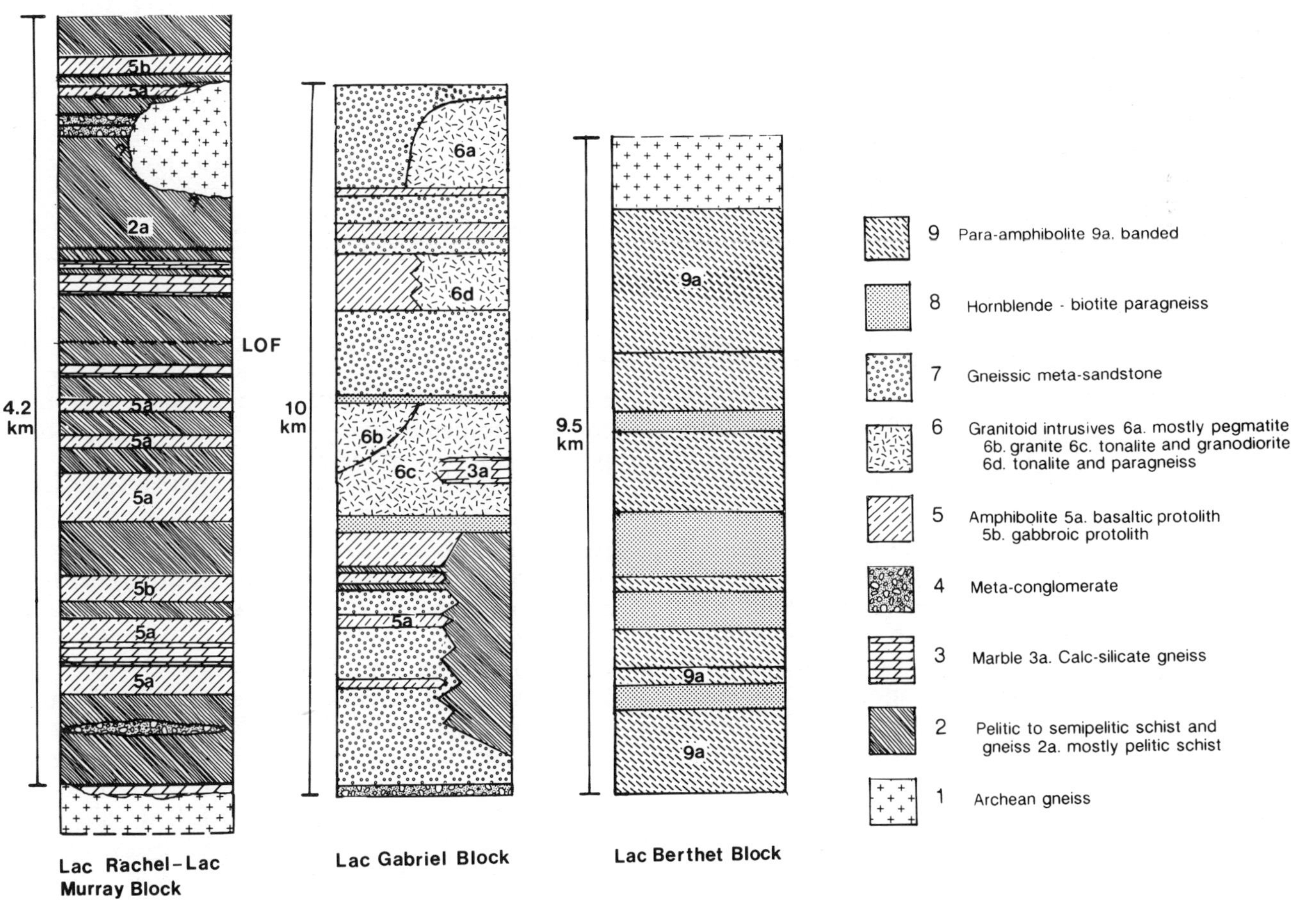

Figure 5. *Simplified "stratigraphic" columns for the three tectonic blocks. Abbreviation: LOF, Lac Olmstead Fault.*

igneous contacts and occasional apophyses. It was emplaced during D_3 in a zone bounded by major ductile faulting. U–Pb zircon ages of 2717 Ma (Machado *et al.*, 1988) indicate that this body probably represents reworked Archean basement.

Lac Gabriel block

The LGB is composed mainly of coarse gneissic metasandstones, semipelitic and pelitic gneisses, and granitoid intrusives with subordinate amounts of biotite–hornblende paragneiss, amphibolite, and calc-silicate gneiss. The overall thickness of the block is approximately 10 km (Fig. 5).

The stratigraphic sequence begins with a thick gneissic metasandstone unit (Fig. 5) containing a basal metaconglomerate composed mainly of clasts of granitic gneiss and lesser amounts of amphibolitic and pelitic clasts. The metaconglomerate fines upward into a medium-grained plagioclase–biotite–quartz–alkali-feldspar gneiss. The gneiss contains numerous minor layers of amphibolite, calc-silicate gneiss, and silicate iron formation.

The gneissic metasandstones grade into a series of semipelitic gneisses containing minor lenses of pelitic gneiss. The transition zone between the gneissic metasandstones and pelitic gneisses commonly contains minor calc-silicate units composed mainly of diopside, tremolite, and plagioclase.

Amphibolites within the gneissic metasandstone and semipelitic units are generally thin (2–3 m), but higher in the sequence they become thicker and are easily traced. Most amphibolites are massive or finely banded and are believed to be metamorphosed basaltic flows, as indicated by rare deformed pillow structures. Other amphibolites show coarse banding and are believed to have epiclastic or pyroclastic origins. Amphibolites with a gabbroic protolith also occur, particularly in the vicinity of Kuujjuaq.

A distinctive amphibolite unit, which is interpreted as a metatuff, is also found in this block. It is a well-foliated rock with alternating hornblende-rich and quartzo-feldspathic layers, commonly with large augen of alkali feldspar. These metatuffaceous units are commonly found on the contacts between biotite–hornblende paragneisses and semipelitic gneisses.

The LGB contains a suite of intrusive rocks ranging in composition from granitic to gabbroic, occurring in three separate complexes: the Lac Gabriel complex, the Elbow Island complex, and the Lac Stewart complex. These complexes are restricted to the same general stratigraphic level and all contacts are roughly parallel to the local foliation. They form a curved belt approximately 30 km long, whose shape suggests that the complexes are megaboudins that were once part of a single batholith (Fig. 3).

The Lac Gabriel complex (located to the west of the north end of Lac Gabriel) consists predominantly of granodioritic and tonalitic gneisses, with lesser amounts of granitic gneiss, hornblende diorite, and gabbro. The gabbroic rocks are composed almost entirely of hornblende, with minor amounts of biotite. Interstitial plagioclase, apatite, and magnetite are found as common accessories. These rocks (gabbros and diorites) are of restricted size and generally do not form mappable units. The granodioritic and tonalitic rocks consist of plagioclase, hornblende, and quartz, with interstitial biotite and alkali feldspar. These rocks show a large variation in colour index, which ranges from 20 to 50. Magnetite is a common accessory mineral in all rock types, but also occurs as cognate xenoliths up to 20 cm in diameter. Xenoliths in the complex are mainly gabbro or hornblende diorite, but a few are probably metasedimentary in origin. The complex shows no evidence of zoning, but a general increase in the mafic content of the granitoids occurs near the south side (lower level) of the complex. Crosscutting relationships generally indicate that the gabbroic and dioritic intrusives predate the granodiorites and tonalites.

Deformation intensity is a function of position within the complex. At its edges, the intrusives are well foliated and interleaved with the surrounding metasediments. Toward the centre of the complex, very little foliation is evident; however, the rare xenoliths of country rock are well foliated. The presence of well-foliated inclusions of country rock in the main parts of the complex, along with the commonly observed gabbroic dykes that cut well-foliated country rock, indicates emplacement of the complex after D_1. F_2 and F_3 folds noted within the complex indicate that emplacement was probably post-D_1, but pre- to syn-D_2.

The Elbow Island complex (located on the eastern and western sides of the Koksoak River near Kuujjuaq), like the Lac Gabriel complex, is syntectonic, but it exhibits some compositional differences. The main rock type of the complex is a hornblende–biotite tonalite. The complex may be divided into two main parts. The first is a large dyke (at least 200 m wide) of tonalite gneiss containing country rock xenoliths up to 25 m^2. The second part is a pluton of plagioclase porphyroblastic tonalite gneiss, with a mylonite zone at its northern contact. Spatially associated with the Elbow Island intrusions are large areas along the Koksoak River underlain by tonalitic gneiss mixed with up to 50% amphibolite and semipelitic gneiss. Igneous ages of 1845 and 1840 Ma (U–Pb, Machado *et al.*, 1988) and metamorphic ages of 1830 Ma have been obtained from these rocks.

The Lac Stewart complex (located approximately 3 km east of Lac Gabriel) is more felsic than the Lac Gabriel and Elbow Island complexes. The main phase of this pluton is a biotite–hornblende granodiorite. In its central part, it commonly contains alkali-feldspar porphyroblasts. The granodiorite is accompanied by lesser amounts of hornblende tonalite gneiss and by amphibolites and ultramafic units near its contacts. It is well foliated along its contacts and does not display any kind of crosscutting relationships with the surrounding metasediments. Although the Lac Stewart pluton has been previously interpreted as a basement dome, there are several features usually associated with known basement that are not observed in the Lac Stewart complex. Most notable is the complete absence of migmatites, which are common features of the basement antiforms to the west. Basement antiforms are also associated with a prominent positive aeromagnetic anomaly, which is also lacking in the Lac Stewart complex. U–Pb zircon dating of a pegmatitic phase of the hornblende granodiorite gives an igneous age of 1833 Ma and a metamorphic age of 1833–1830 Ma (Machado *et al.*, 1988).

Immediately above (east of) the syntectonic intrusions, there is a thin hornblende–biotite paragneiss and another gneissic metasandstone sequence that grades up into semipelites. This gneissic metasandstone sequence is coarser grained and richer in quartz than that below the intrusions. At the top of the LGB, the gneissic metasandstone sequence is interlayered with several amphibolite units of uncertain origin. The uppermost unit of the LGB is a hornblende–biotite paragneiss containing minor marbles, calc-silicate gneisses, and quartzites.

Along the Koksoak River north of Kuujjuaq, posttectonic and postmetamorphic pegmatite veins account for up to 80% of the outcrop. The age of these veins has been determined to be 1775 Ma (Machado *et al.*, 1988). The pegmatites cut late-tectonic and late-metamorphic monzogranite dykes, and are in turn cut by minor dykes and stocks of pink granite. The block was also intruded by a series of postorogenic diabase dykes that cut all structures and exhibit chilled margins. These dykes resemble, and are probably related to, a series of Neohelikian diabase dykes mapped by Taylor (1979) on the east side of Ungava Bay.

Lac Berthet block

The LBB is composed largely of a highly variable sequence of para-amphibolites and hornblende–biotite paragneisses. The great thickness of this sequence (Fig. 5) may reflect some repetition of strata by faulting or isoclinal folding, although no clear evidence for this was found during fieldwork.

The para-amphibolites exhibit broad variation in modal mineralogy. Generally, they are composed mainly of hornblende, plagioclase, and quartz, with minor biotite and common garnet porphyroblasts. They are well foliated, but most lack compositional layering. Most of the block is made up of this type of amphibolite, but there are areas underlain by a very distinctive amphibolite with pronounced compositional layering consisting of thin (1–3 cm) bands of hornblende and/or biotite, plagioclase, and minor cummingtonite, alternating with quartzo-feldspathic bands of the same size. The hornblende–biotite layers commonly contain garnet porphyroblasts, whereas the quartzo-feldspathic layers commonly have 1–2 cm augen of alkali feldspar.

Minor units of quartzite, calc-silicate gneiss, pelitic gneiss, and forsterite-marble also occur. On the eastern side of the Koksoak River, minor tonalitic and granitic gneisses have been mapped. These gneisses are cut by the southeast extension of the LPF and are believed to represent the remnants of a pre- to syn-D_2 pluton. As in the LGB, there was widespread intrusion of syn- to posttectonic pegmatite and later intrusion of diabase dykes.

The block is capped by a large nappe, cored by highly deformed tonalitic to granitic gneisses (Fig. 3). The gneisses are extensively migmatized and are surrounded by a halo of migmatized metasediment. They are, in general, much coarser grained than the metasediments of the block and are believed to represent Archean basement rocks that were emplaced onto the cover rocks during D_1. There are insufficient facing data in the metasediments beneath the gneisses to determine whether the gneisses were thrust onto the metasediments or form the core of a large overturned fold.

GEOCHEMISTRY OF THE GRANITOID INTRUSIVES

Samples from the Lac Gabriel complex were analyzed for major, minor, and trace element compositions using X-ray fluorescence techniques (Table I). The samples exhibit a wide range in SiO_2 content (40–73 wt.%), with the majority of the granitoids containing less than 60 wt.% silica. Harker plots (Fig. 6) show smooth trends for total iron, CaO, and TiO_2, with more scatter in plots of MgO and Al_2O_3 and considerable overlap in the silica content of the tonalites and granodiorites. These features are compatible with derivation of the granitoids from a common source, but trace elements provide no conclusive evidence for this. On Y/Zr plots (Fig. 7), there is a poorly defined trend connecting the hornblende gabbros to the

Table I
Whole-rock analyses of rocks from the Lac Gabriel complex

	Hornblende gabbros									Diorites		Tonalites							
	A24H2	A27G	A2M	A25E2	J118K1	A25E1	J20F1	A24H1	A3K2	A27F	A27C	J20F3	A26LL	A24D2	A24A1	A3K1	J20EC3	A26A	SP8690
SiO_2 (wt.%)	38.92	39.50	40.23	40.86	40.76	41.14	41.56	44.22	44.98	46.87	48.52	50.86	51.61	52.37	53.61	56.25	56.34	57.28	61.47
TiO_2	2.60	2.74	2.83	1.88	2.33	1.76	2.18	2.05	1.47	1.81	1.19	1.52	1.33	1.51	0.99	1.10	1.02	0.85	1.20
Al_2O_3	14.76	13.65	14.95	14.39	14.33	12.37	12.57	17.64	13.78	16.15	14.46	12.24	17.71	15.93	19.07	18.96	14.86	16.67	16.24
Fe_2O_3	20.16	19.87	19.76	16.76	18.05	17.28	18.03	15.18	15.51	13.85	12.48	12.07	8.88	10.89	7.71	7.99	11.12	7.80	8.53
MnO	0.24	0.24	0.29	0.24	0.22	0.29	0.19	0.18	0.21	0.19	0.16	0.09	0.12	0.14	0.10	0.09	0.08	0.13	0.12
MgO	8.44	8.28	6.22	12.44	9.14	11.49	9.90	6.02	9.36	5.73	8.57	8.69	4.96	5.09	3.65	2.32	5.00	4.30	2.24
CaO	11.67	10.60	12.85	7.53	9.12	11.22	11.91	9.83	9.88	10.27	10.52	10.17	9.28	8.38	8.27	6.34	5.43	6.46	4.11
Na_2O	1.58	1.02	1.20	0.71	0.86	1.10	1.24	2.71	1.50	2.58	2.59	2.63	3.73	3.32	4.32	4.80	2.80	3.42	3.09
K_2O	1.14	2.80	1.19	4.39	3.62	1.96	1.49	1.16	2.36	1.33	0.66	0.63	0.98	1.26	1.05	1.12	2.60	1.88	2.25
P_2O_5	0.72	0.75	1.00	0.60	0.81	0.59	0.82	0.56	0.85	0.56	0.35	0.19	0.49	0.36	0.48	0.43	0.34	0.40	0.42
LOI	0.53	0.76	0.44	0.93	0.95	0.67	0.62	0.43	0.63	0.62	0.53	0.71	0.65	0.86	1.06	0.44	0.51	0.83	0.52
Total	100.88	100.38	101.01	101.00	100.39	100.07	100.65	100.10	100.79	99.96	100.03	99.80	99.74	100.11	100.31	99.84	100.10	100.02	100.49
Ni (ppm)	74	54	<10	181	111	166	104	49	158	41	56	92	56	33	28	10	77	53	<10
Cr_2O_3	279	267	41	730	401	802	416	201	570	198	450	788	229	141	125	60	391	224	29
V	388	392	382	354	331	315	375	304	253	288	266	235	173	224	138	133	177	142	137
Y	67	96	55	41	71	46	62	44	37	58	34	29	35	46	26	16	12	23	33
Zr	751	771	201	274	519	257	502	486	168	485	108	148	305	301	124	99	188	154	210
Nb	17	43	16	15	26	15	18	13	17	29	13	9	18	19	11	9	14	13	23
Rb	<5	72	<5	159	125	46	11	10	56	12	<5	<5	<5	6	<5	19	74	40	32
Sr	225	170	131	119	131	95	176	848	264	506	475	415	888	774	1043	1153	495	899	570
BaO	460	1060	300	1190	1280	440	480	680	650	420	200	160	340	580	520	550	1300	790	1530
Pb	<5	10	<5	<5	13	5	<5	<5	<5	<5	<5	<5	<5	<5	<5	7	<5	<5	10
Th	33	66	<5	23	41	16	<5	7	<5	27	<5	<5	<5	<5	<5	<5	<5	<5	<5
U	<5	<5	9	<5	<5	6	7	9	6	12	14	13	14	12	15	15	7	7	11

	Granodiorites								Granites			
	JL18L	J28C1	A25I	A26K1	A24D	A25A	A24L	A27A	J222D2	J21C	J28G2	SP86334
SiO_2 (wt.%)	54.34	55.76	56.33	58.36	59.68	61.11	61.29	66.66	67.80	69.56	72.48	76.15
TiO_2	1.16	0.97	1.06	0.99	1.08	0.88	0.88	0.69	0.56	0.66	0.44	0.04
Al_2O_3	17.45	14.86	19.84	15.79	17.00	15.39	15.73	14.47	15.75	14.67	13.41	14.38
Fe_2O_3	9.52	9.03	6.93	8.07	6.93	8.28	7.45	5.12	4.09	4.58	3.38	0.64
MnO	0.12	0.13	0.09	0.11	0.10	0.14	0.12	0.10	0.05	0.04	0.05	0.08
MgO	4.05	5.57	2.17	3.18	2.14	2.70	2.10	2.05	1.21	0.77	0.55	0.08
CaO	5.98	7.12	6.12	7.00	5.51	5.12	4.92	2.69	2.91	2.06	1.90	1.30
Na_2O	3.72	2.76	4.57	2.93	3.76	2.98	2.94	3.08	3.23	2.21	2.31	3.18
K_2O	2.12	3.28	1.72	2.12	2.34	2.82	3.71	4.24	3.54	5.01	4.86	4.02
P_2O_5	0.45	0.42	0.41	0.31	0.47	0.26	0.26	0.29	0.24	0.23	0.10	0.01
LOI	0.67	0.59	0.65	0.88	0.59	0.44	0.32	0.36	0.51	0.25	0.27	0.15
Total	99.58	99.62	99.89	99.74	99.60	100.12	99.70	99.77	99.89	100.04	99.75	100.03
Ni (ppm)	38	84	11	31	<10	13	<10	12	10	10	10	10
Cr_2O_3	154	353	39	124	58	66	100	86	53	54	15	15
V	184	161	114	157	133	139	134	378	68	46	19	10
Y	35	26	28	32	30	22	30	22	17	10	26	19
Zr	233	189	324	189	169	256	250	323	162	181	250	57
Nb	18	13	13	17	17	16	19	18	10	15	16	11
Rb	65	63	50	27	34	50	70	119	67	157	77	83
Sr	655	710	848	650	582	533	434	562	703	254	302	205
BaO	830	1160	520	1220	910	1630	1330	1440	1420	1350	980	950
Pb	8	38	<5	<5	23	<5	12	8	8	19	7	26
Th	7	10	<5	<5	<5	<5	17	10	<5	8	13	<5
U	12	11	10	7	7	6	6	6	10	<5	6	<5

Notes: All analyses were carried out on a Phillips PW 1400 XRF unit. Major elements plus BaO, Cr_2O_3, Ni, and V were analyzed using fused beads prepared from ignited samples. Raw data were corrected using an alpha technique (Ahmedali, 1983). Detection limits are 0.01 wt.% for major elements, 15 ppm for Cr_2O_3, and 10 ppm for V, Ni, and BaO. Analytical precision for these elements (one standard deviation in wt.%) was calculated from 20 replicate analyses and is as follows: Si, 0.05; Al, 0.03; Cr, 0.001; Fe, 0.01; Mn, 0.001; Mg, 0.058; Ca, 0.01; Na, 0.06; K, 0.001; and P, 0.004. Analytical precision for Ti, Ba, and V is estimated to be 5%. Nb, Zr, Y, Sr, Rb, Pb, Th, and U were analyzed using pressed powder pellets. Raw data were corrected using a Rh Kb Compton scatter procedure. The detection limit for these elements is 5 ppm. Analytical precision is estimated to be 5%.

granodiorites and granites, but other trace element plots (Rb/Sr, Rb/Zr, see Fig. 7) show a large degree of scatter and no observable trends. Most of the rocks of this complex are slightly peraluminous. The samples are also calc-alkaline, having a Peacock index of 59 and showing a good calc-alkaline trend on AFM and Na_2O–K_2O–CaO diagrams (Fig. 8). The Harker plots also show calc-alkaline trends, with MgO, CaO, TiO_2 and total Fe decreasing and Na_2O and K_2O increasing with increasing silica content (cf. Atherton and Sanderson, 1987).

DISCUSSION AND INTERPRETATION

The great predominance of semipelitic and pelitic rocks in the LRLMB suggests that it reflects deposition in a deep-water basinal environment subjected to periods of basaltic volcanism and intrusion. The presence of Archean gneisses underlying these Aphebian metasediments and metavolcanics and the common occurrence of fragments of Archean gneisses in conglomerate indicate that the sequence was probably deposited on granitoid crust. The block is separated from the rocks of the Labrador Trough sensu stricto by the LRF, a postmetamorphic, high-angle reverse fault. The metasediments and metavolcanics of this block are very similar to those of the Labrador Trough sensu stricto and are probably equivalent to parts of the Kaniapiskau Supergroup (Moorhead and Hynes, 1986). These rocks are also similar to and probably correlative with a belt of schistose metasediments and metavolcanics known as the Laporte schists or the "Laporte Group" (Dimroth *et al.*, 1971; Taylor, 1979; Wardle and Bailey, 1981), which are found in the hinterland of the Central Labrador Trough (i.e., the part of the Labrador Trough south of 57°N and north of the Grenville front). These rocks are interpreted as the metamorphosed equivalents of a deep basinal shale–greywacke sequence and have been correlated with the upper portions of the Knob Lake Group of the Kaniapiskau Supergroup (Dimroth, 1978; Wardle and Bailey, 1981).

There is some evidence that the sediments of this block and of the eastern margin of the Labrador Trough in general were deposited on thin continental crust. The mafic volcanics of the Labrador Trough, when palinspastically restored, are located at or near the present eastern boundary of the Labrador Trough (Boone and Hynes, 1987). They are highly uniform in volcanic characteristics and chemical composition and are petrochemically intermediate in character between continental tholeiites and midoceanic-ridge basalt (Boone, 1987). Thus, the supracrustal sequence of the LRLMB is compatible in many ways with a con-

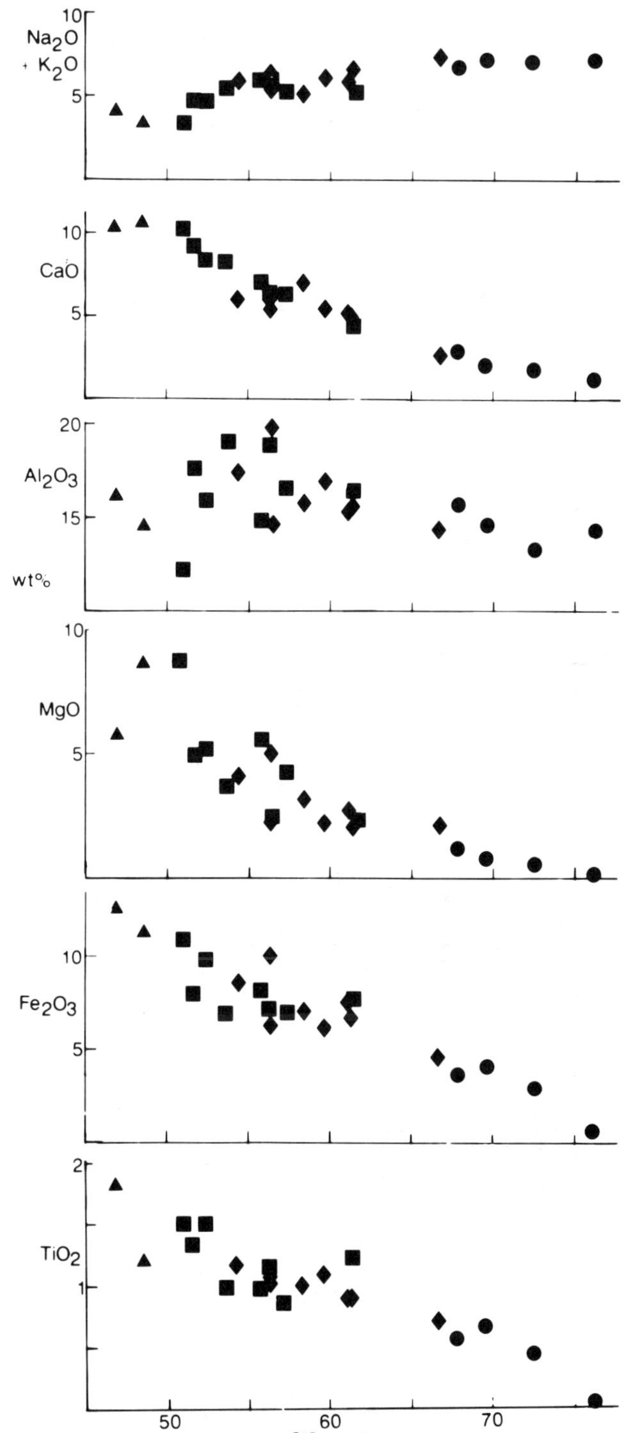

Figure 6. *Major element Harker plots for granitoids of the Lac Gabriel complex. Legend: triangles, diorite; squares, tonalite; diamonds, granodiorite; and circles, granite.*

tinental rise origin, representing distal equivalents of the supracrustal succession preserved in the Labrador Trough, farther west.

In contrast, the eastern two blocks consist of thick sequences of immature clastic metasediments and possible pyroclastics. Although the Lac Gabriel and Lac Berthet blocks each have distinctive lithological features, their depositional environments appear to have been fairly similar, and they are tentatively considered correlative.

The lithological features of these two blocks, such as quartz-rich and arkosic sediments, para-amphibolites, and possible pyroclastics, are compatible with deposition in a slope–shelf environment on a continental margin. The predominance of gneissic metasandstones and arkosic rocks favours a continental sedimentary source (Dickinson, 1980), but whether these metasediments were deposited on continental crust is uncertain. If the gneissic nappe in the northern section of the LBB is Archean, it may represent their overthrust basement.

The relationship of these blocks to the Labrador Trough is not well constrained, as there is no obvious way to correlate their stratigraphy to that of the Kaniapiskau Supergroup. There are two main possibilities for their origin. They may represent exotic blocks, which were juxtaposed with the rocks of the Labrador Trough along the D_1 LTF early in the Hudsonian orogeny, or uplifted portions of the eastern margin of the Labrador Trough. The abrupt changes in lithology across the Lac Turcotte Fault suggest major offset along it, but at present there is no evidence that clearly favours either hypothesis.

Using relative dating of metamorphic and structural events and available isotopic dates, it is possible to develop a fairly detailed geological history of

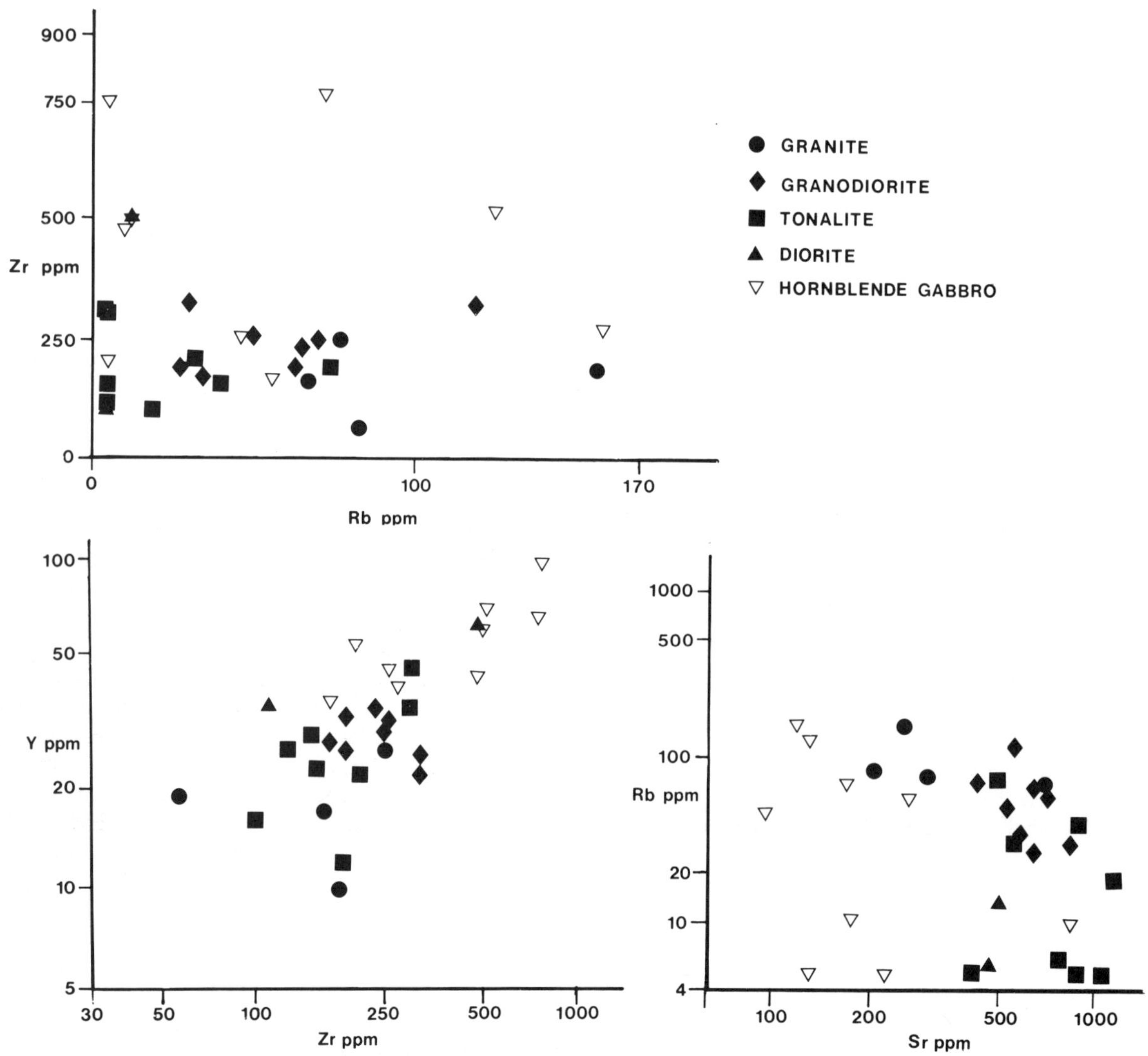

Figure 7. *Rb–Sr, Rb–Zr, and Y–Zr variation for phases in the Lac Gabriel complex.*

the area. This history can then be used to provide constraints on a tectonic interpretation.

The first event recognized after sediment deposition is D_1. This event took the form of a major shortening event, which resulted in the formation of the main foliation (S_1) and the juxtaposition of the LRLMB and LGB along the LTF (pre-1845 Ma).

The temporal and spatial relationships between the LTF and the intrusive complexes of the LGB (i.e., emplacement after D_1 and within 15 km of the LTF) indicate that these complexes probably result from D_1 processes, possibly being the result of melting in the lower plate of the LTF (cf. England and Thompson, 1986). The timing and regional distribution (i.e., restricted to the east side of the LTF) of M_1 metamorphism also suggests a link to D_1 deformation (specifically the LTF). The exact nature of this event is, however, difficult to determine due to extensive overprinting by M_2 metamorphism.

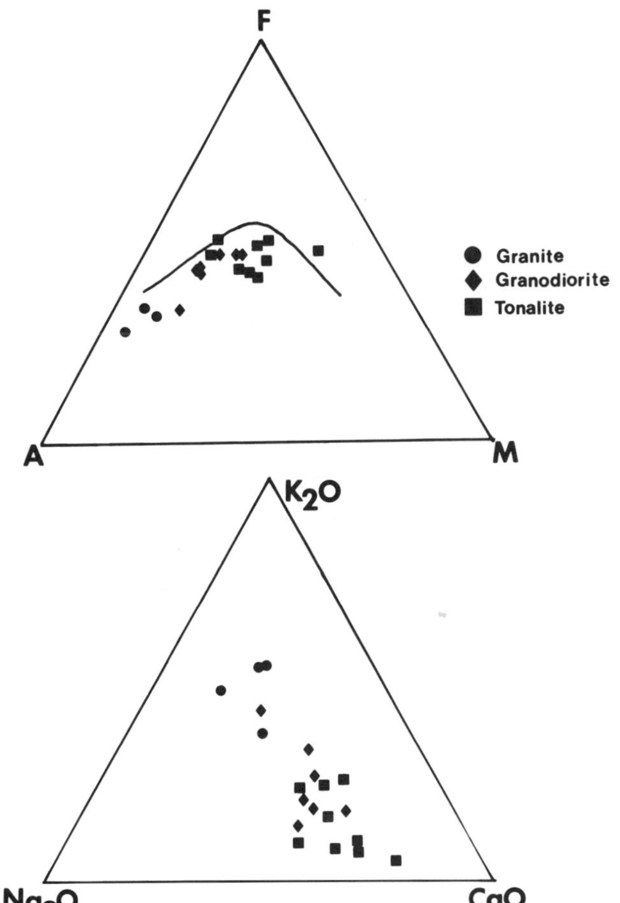

Figure 8. *Granitoids from the Lac Gabriel complex plotted on AFM and Na_2O–K_2O–CaO diagrams. The line separating tholeiitic and calc-alkaline compositions is from Irvine and Baragar (1971).*

D_2 occurred during or after the emplacement of the igneous complexes. West-directed thrusting on the LPF may be responsible for the presence of M_1 granulite facies assemblages in the LBB. D_3 was the last regional deformation event in the area. Work by Perreault *et al.* (1987, 1988) has shown that the M_2 event occurred in response to major D_3 thrusting and subsequent thermal reequilibration. In the LRLMB, M_2 resulted in the medium temperature, high to medium pressure metamorphism of the rocks in the block. In the LBB and LGB, M_2 overprinted M_1 assemblages and produced uniform upper amphibolite assemblages.

The above geological history shows that the structural events of the area are mainly the result of west-directed transport, with metamorphic and igneous features occurring in response to thrusting related to this transport. The long history of shortening and west-directed transport could possibly be the result of a continental collision.

Further information about the tectonic framework of the study area comes from the synkinematic granitic intrusives of the LGB. Various studies have shown that the characteristics of granitoid intrusives are closely related to the tectonic environment of their emplacement. Most of these studies have used both physical and chemical characteristics, (e.g., White and Chappell, 1977; Pitcher, 1983), but purely geochemical parameters have also been used (Pearce *et al.*, 1984; Harris *et al.*, 1986). The intrusives of the Lac Gabriel block are calc-alkaline and, although they are slightly peraluminous, they are essentially I-type granitoids (Chappell, 1978; Hine *et al.*, 1978). On the tectonic discrimination diagram of Pearce *et al.* (1984), the granitoids all fall within the field of volcanic arc granites (Fig. 9). A problem with this mode of discrimination is its incapacity to distinguish volcanic arc granitoids from postcollision granitoids, which may share many of the same characteristics (Pitcher, 1983). However, other features of the granitoids, such as a wide range in composition and SiO_2 content, the dominance of hornblende over biotite as a mafic component, the common association with magnetite, and the presence of mafic xenoliths and mafic intrusives, are more typical of "Cordilleran" I-type granitoids, characteristic of continental arc terrains, than of "Caledonian" I-types, which are characteristic of postorogenic uplift (Pitcher, 1983). One feature of "Cordilleran" I-type granitoids absent here is their common association with large amounts of andesite and dacite. A likely explanation for this is that deep erosion, as evidenced by the high metamorphic grades, has removed formerly associated volcanics.

If the Lac Gabriel and Lac Berthet blocks do represent an arc terrain, this raises questions about what was actually subducted and where this sub-

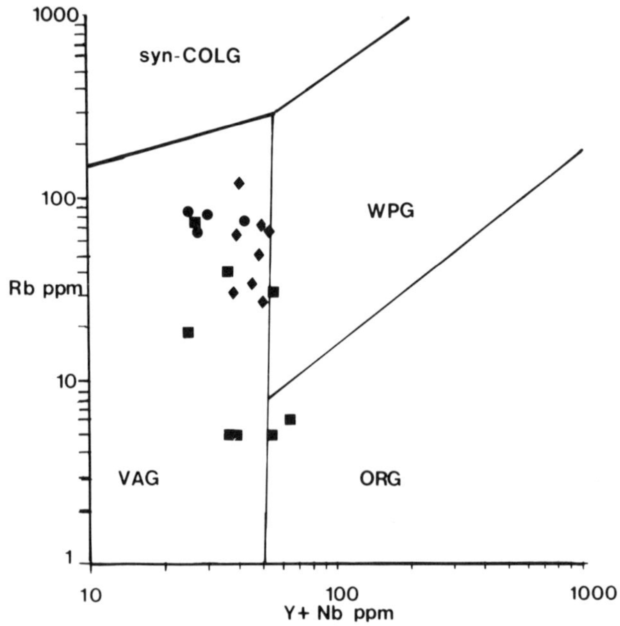

Figure 9. *Granitoids from the Lac Gabriel complex plotted on a Rb / (Y + Nb) discrimination plot (Pearce* et al., *1984). Symbols as for Figure 6. Abbreviations: VAG, volcanic arc granitoids; syn-COLG, syncollisional granitoids; WPG, within-plate granitoids; and ORG, ocean-ridge granitoids.*

duction occurred. Because the timing and type of motion associated with the LTF are so poorly understood, we can only speculate about this. Evidence that the continental crust was substantially thinned or absent on the eastern margin of the Labrador Trough raises the possibility that juxtaposition of the LRLMB with the LGB and LBB may have been accompanied by a large amount of subduction, making the Lac Turcotte Fault a major suture. An alternative possibility is that the Lac Gabriel and Lac Berthet blocks were formed near the eastern boundary of the Labrador Trough and reflect subduction of oceanic crust from farther east, with the granitoids of the LBB simply representing melts generated on the lower plate of the LTF. In either case, it is difficult at this stage to make the argument that the geology of the Labrador Trough is compatible only with ensialic processes. It has many of the hallmarks of a collisional orogen.

ACKNOWLEDGEMENTS

Funding for this work was provided primarily through research grants from the Natural Sciences and Engineering Research Council of Canada and the Quebec Ministry of Education, and through a research agreement with Energy, Mines and Resources Canada. Northern Studies and Research Grants to Poirier (1986) and Perreault (1987) are also gratefully acknowledged. The paper benefited greatly from discussions with James Moorhead, James Gebert, and Roland Deschenes. The paper was much improved through critical reviews by E. Nisbet and P. Simony. Richard Yates and Pat Cerri are thanked for their help with photography and drafting respectively.

REFERENCES

Ahmedali, T., 1983, XRF procedures circular # 1: Department of Geological Sciences, McGill University, Montreal, Quebec.

Atherton, M.P. and Sanderson, L.M., 1987, The Cordillera Blanca Batholith: a study of granite intrusion and the relation of crustal thickening to peraluminosity: Geologische Rundschau, v. 76, p. 213–232.

Baragar, W.R.A. and Scoates, R.F.J., 1981, The Circum-Superior Belt: a Proterozoic plate margin?, in Kroner, A., ed., Precambrian Plate Tectonics: Elsevier, Amsterdam, p. 297–330.

Boone, E., 1987, Petrology and tectonic implications of the Hellancourt volcanics, Northern Labrador Trough, Quebec, unpublished M.Sc. thesis, McGill University, Montreal, Quebec.

Boone, E. and Hynes, A., 1987, A structural cross section of the Northern Labrador Trough, New Quebec: Geological Association of Canada – Mineralogical Association of Canada, Program with Abstracts, v. 12, p. 26.

Chappell, B.W., 1978, Granitoids from the Moonbi District, New England Batholith, Eastern Australia: Journal of the Geological Society of Australia, v. 25, p. 267–283.

Dickinson, W.R., 1980, Plate tectonics and key petrological associations, in Strangway, D.W., ed., The Continental Crust and its Mineral Deposits: Geological Association of Canada, Special Paper 20, p. 341–360.

Dimroth, E., 1972, The Labrador Geosyncline revisited: American Journal of Science, v. 272, p. 487–506.

Dimroth, E., 1978, Labrador Trough area (54°30′-56°30′): Ministère des Richesses Naturelles, Québec, Rapport Géologique 193, 346 p.

Dimroth, E., 1981, Labrador Geosyncline: a type example of early Proterozoic cratonic reactivation, in Kroner, A., ed., Precambrian Plate Tectonics: Elsevier, Amsterdam, p. 331–352.

Dimroth, E. and Dressler, B., 1978, Metamorphism of the Labrador Trough, in Fraser, J.A. and Heywood, W.W., eds., Metamorphism in the Canadian Shield: Geological Survey of Canada, Paper 78-10, p. 215–236.

Dimroth, E., Baragar, W.R.A., Bergeron, R. and Jackson, G.D., 1971, The filling of the Labrador Trough, in Baer, A.J., ed., Basins and Geosynclines of the Canadian Shield: Geological Survey of Canada, Paper 70-40, p. 45–142.

England, P.C. and Thompson, A.B., 1986, Some thermal and tectonic models for crustal melting in continental collision zones, in Coward, M.P. and Ries, A.C., eds., Collision Tectonics: Geological Society Special Publication No. 19, p. 83–94.

Frarey, M.J. and Duffell, S., 1964, Revised stratigraphic nomenclature for the central part of the Labrador Trough: Geological Survey of Canada, Paper 64-25, 13 p.

Gélinas, L., 1958, Thevenet Lake area (east half), New Quebec: Quebec Department of Mines, Preliminary Report 363, 8 p.

Gélinas, L., 1959, Région du Lac Gabriel (partie est), et la région de Fort Chimo (partie ouest), Nouveau Québec: Ministère des Mines, Québec, Rapport préliminaire 407.

Gélinas, L., 1960, Région de Fort Chimo (partie est), Nouveau Québec: Ministère des Mines, Québec, Rapport préliminaire 418, 5 p.

Gélinas, L., 1965, Géologie de la région de Fort Chimo et des Lacs Gabriel et Thévenet, Nouveau Québec, unpublished D.Sc. thesis, Université Laval, Québec, 212 p.

Goulet, N., Gariépy, L., Mareschal, J.-C. and Machado, N., 1987, Structure, geochronology, gravity and tectonics of the Northern Labrador Trough: Geological Association of Canada – Mineralogical Association of Canada, Program with Abstracts, v. 12, p. 48.

Harris, N.B.W., Pearce, J.A. and Tindle, A.G., 1986, Geochemical characteristics of collision-zone magmatism, in Coward, M.P. and Ries, J.A., eds., Collision Tectonics: Geological Society Special Publication No. 19, p. 67–81.

Hine, R., Williams, I.S., Chappell, B.W. and White, A.J.R., 1978, Contrasts between I-type and S-type granitoids of the Kosciusko Batholith: Journal of the Geological Society of Australia, v. 25, p. 235–247.

Hoffman, P.F., 1980, Wopmay Orogen: a Wilson cycle of early Proterozoic age in the northwest of the Canadian Shield, in Strangway, D.W., ed., The Continental Crust and its Mineral Deposits: Geological Association of Canada, Special Paper 20, p. 523–549.

Hoffman, P.F., 1981, Autopsy of Athapuscau Aulacogen: a failed arm affected by three collisions, in Campbell, F.H.A., ed., Proterozoic Basins of Canada: Geological Survey of Canada, Paper 81-10, p. 97–102.

Hoffman, P.F., 1987, Early Proterozoic foredeeps, foredeep magmatism and Superior-type iron-formations of the Canadian Shield, in Kroner, A., ed., Proterozoic Lithospheric Evolution: American Geophysical Union, Geodynamics Series, No. 17, p. 85–98.

Irvine, T.N. and Baragar, W.R.A., 1971, A guide to the chemical classification of common volcanic rocks: Canadian Journal of Earth Sciences, v. 8, p. 523–548.

Kroner, A., *editor,* 1981, Precambrian Plate Tectonics: Elsevier, Amsterdam, 781 p.

LeGallais, C.J. and Lavoie, S., 1982, Basinal evolution of the lower Proterozoic Kaniapiskau Supergroup, Central Labrador Miogeocline (Trough), Quebec: Bulletin of Canadian Petroleum Geology, v. 30, p. 150–166.

Lewry, J.F., Sibbald, T.I.I. and Schledwitz, D.C.P., 1985, Variation in the character of Archean rocks in the western Churchill Province and its significance, in Ayres, L.D., Thurston, P.C., Card, K.D. and Weber, W., eds., Evolution of Archean Supracrustal Sequences: Geological Association of Canada, Special Paper 28, p. 239–261.

Machado, N., Goulet, N. and Gariepy, C., 1987, Evolution of the Northern Labrador Trough basement: evidence from U–Pb geochronology: Geological Association of Canada – Mineralogical Association of Canada, Program with Abstracts, v. 12, p. 69.

Machado, N., Perreault, S. and Hynes, A., 1988, Timing of continental collision in the Northern Labrador Trough, Quebec: evidence from U–Pb geochronology: Geological Association of Canada – Mineralogical Association of Canada, Program with Abstracts, v. 13, p. 76.

Moorhead, J. and Hynes, A., 1986, Structure and metamorphism of the eastern flank of the Renia gneiss dome: Geological Association of Canada – Mineralogical Association of Canada, Program with Abstracts, v. 11, p. 103.

Pearce, J.A., Harris, N.B.W. and Tindle, A.G., 1984, Trace element discrimination diagrams for the tectonic interpretation of granitic rocks: Journal of Petrology, v. 25, p. 956–983.

Perreault, S., Hynes, A. and Moorhead, J., 1987, Metamorphism of the eastern flank of the Labrador Trough, Ungava, Northern Quebec: Geological Association of Canada – Mineralogical Association of Canada, Program with Abstracts, v. 12, p. 80.

Perreault, S., Hynes, A., Machado, N. and Hynes, A., 1988, Timing in the NE segment of the Labrador Trough, Kuujjuaq, Northern Quebec: Geological Association of Canada – Mineralogical Association of Canada, Program with Abstracts, v. 13, p. 97.

Pitcher, W.S., 1983, Granite type and tectonic environment, in Hsu, K.J., ed., Mountain Building Processes: Academic Press, London, p. 19–40.

Stockwell, C.H., 1963, Third report on structural provinces, orogenies and time classification of rocks of the Canadian Precambrian Shield: Geological Survey of Canada, Paper 63-17, p. 125–131.

Taylor, F.C., 1979, Reconnaissance geology of a part of the Precambrian Shield, northeastern Quebec, northern Labrador and Northwest Territories: Geological Survey of Canada, Memoir 393, 99 p.

Taylor, F.C. and Skinner, R., 1963, Fort Chimo, New Quebec: Geological Survey of Canada, Paper 63-47, 4 p.

Thompson, A.B. and England, P.C., 1984, Pressure–temperature–time paths of regional metamorphism. II. Their inference and interpretation using mineral assemblages in metamorphic rocks: Journal of Petrology, v. 25, p. 929–955.

Van Schmus, W.R., Bickford, J.F., Lewry, J.F. and MacDonald, R., 1987, U–Pb geochronology in the Trans-Hudsonian orogeny, northern Saskatchewan, Canada: Canadian Journal of Earth Sciences, v. 24 , p. 407–424.

Wardle, R.J. and Bailey, D.G., 1981, Early Proterozoic sequences in Labrador, in Campbell, F.H.A., ed., Proterozoic Basins of Canada: Geological Survey of Canada, Paper 81-10, p. 331–359.

White, A.J.R. and Chappell, B.W., 1977, Ultrametamorphism and granitoid gneiss: Tectonophysics, v. 43, p. 7–22.

Model Lead Ages from the Labrador Trough and Their Stratigraphic Implications[1,2]

T. Clark
Ministère de l'Énergie et des Ressources,
5700, 4e avenue ouest, A-209, Charlesbourg, Québec G1H 6R1, Canada

R.I. Thorpe
Geological Survey of Canada,
601 Booth Street, Ottawa, Ontario K1A 0E8, Canada

Clark, T. and Thorpe, R.I., Model lead ages from the Labrador Trough and their stratigraphic implications, in Lewry, J.F. and Stauffer, M.R., eds., The Early Proterozoic Trans-Hudson Orogen of North America: Geological Association of Canada, Special Paper 37, p. 413–431.

Abstract

Ten samples, mostly galena, from the Labrador Trough were analyzed for their lead isotope compositions. Six samples are hosted by sedimentary rocks and four by gabbro sills. Model ages were calculated using two models: one derived on the basis of lead isotope compositions and ages for Archean volcanogenic massive sulphide deposits in western Superior Province (Western Superior model) and the other obtained from the isotopic and age data for volcanogenic massive sulphide deposits of Proterozoic age in Wisconsin and adjusted to take into account certain other Proterozoic deposits in Australia and Canada ("Proterozoic" model). The two models give results differing, depending on the sample, by 126–178 Ma. The model ages obtained range from 1793 to 2309 Ma with the Western Superior model and from 1935 to 2446 Ma with the "Proterozoic" model. In addition, anomalous model ages of 1022–1148 Ma were obtained for one sample that had been subjected to the addition of uranogenic lead. Based on the calculated Th/U ratio of the lead source, the samples from five sites (six samples) may have had single-stage evolution histories; the others may have had more complex origins. Compared with western Superior Province massive sulphide deposits and some other Canadian Proterozoic deposits, the Labrador Trough specimens have high calculated μ values ($^{238}U/^{204}Pb$), which are, however, comparable with those of certain Scandinavian deposits of Early Proterozoic age.

---

[1]Publication authorized by the Ministère de l'Énergie et des Ressources, Québec. Contribution No. 90-5110-15.

[2]The contribution of R.I. Thorpe to this manuscript is recorded as Geological Survey of Canada Contribution No. 49188.

The Western Superior model appears to give model ages more consistent with available U–Pb zircon ages and known stratigraphic relations than the "Proterozoic" model. Using a published zircon age for the Sokoman Formation as a calibration point, galenas from stratigraphically lower levels give model ages greater than the age of the Sokoman, even though some of these galenas are epigenetic. The Bertin Lake galenas suggest that rocks at least as old as 2300 Ma may exist in the Labrador Trough. Model ages (1992 or 2170 Ma) determined from galena in the Abner Formation of the northern trough are consistent with recent field studies suggesting that the so-called third depositional cycle in the northern trough does not exist, and is an artifact of regional thrusting and stratigraphic repetition. The Abner Dolomite, previously thought to be at the base of the third cycle, is, in fact, older than the Sokoman Formation and is probably equivalent to the Denault Dolomite of the south-central trough. A time gap of at least 110 Ma appears to exist between the deposition of the Abner–Denault and the Sokoman formations. Although regional stratigraphic work suggests that the Sokoman and Baby iron formations are correlative, the model lead ages obtained result in contradictory interpretations. Based on the spread of the model ages, deposition in the Labrador Trough spanned a period perhaps as long as 450 Ma.

Résumé

Le plomb isotopique de dix échantillons provenant de la fosse du Labrador et composés principalement de galène a été analysé. Six échantillons proviennent de sédiments et quatre de sills de gabbro. Les âges modèles ont été calculés selon deux techniques: l'une tenant compte des taux de plomb isotopique et tenant compte des âges des dépôts de sulfures massifs volcanogéniques archéens dans l'ouest du Supérieur (modèle Supérieur ouest) et l'autre basée sur les données isotopiques et les âges des dépôts de sulfures massifs volcanogéniques protérozoïques du Wisconsin et ajustés pour d'autres dépôts protérozoïques en Australie et au Canada (modèle "Protérozoïque"). Les deux modèles donnent des résultats différents; la différence varie de 126 à 178 Ma. Les âges modèles varient de 1793 à 2309 Ma avec le modèle Supérieur ouest et de 1935 à 2466 Ma avec le modèle "Protérozoïque". Des âges modèles anomaux de 1022–1148 Ma ont été obtenus sur un échantillon ayant subi un enrichissement en plomb uranogénique. En se basant sur les rapports Th/U de la source du plomb radiogénique on déduit que les échantillons de cinq sites (six échantillons) ont été formés en une étape alors que les autres ont des origines plus complexes. Comparés avec les dépôts de sulfures massifs de l'ouest du Supérieur et certains autres dépôts protérozoïques canadiens, les échantillons de la fosse du Labrador ont des valeurs calculées élevées de mu ($^{238}U/^{204}Pb$), qui sont cependant comparables avec des dépôts scandinaves d'âge protérozoïque inférieur.

Le modèle Supérieur ouest semble donner des âges plus comparables avec les âges U/Pb zircon et les relations stratigraphiques connues que le modèle "Protérozoïque". En utilisant un âge zircon de la formation de Sokoman comme calibration, les galènes des niveaux stratigraphiques inférieurs donnent des âges modèles plus vieux que la formation de Sokoman, même si certaines de ces galènes sont épigénétiques. Les galènes du lac Bertin indiquent que des roches aussi vieilles que 2300 Ma pourraient exister dans la fosse du Labrador. Des âges modèles (1992 ou 2170 Ma) sur des galènes de la formation d'Abner concordent avec les données de terrain voulant que le soi-disant troisième cycle de dépôt dans le nord de la fosse est en fait un artefact dû à des charriages à grande échelle provoquant des répétitions stratigraphiques. Les dolomies d'Abner qu'on croyait être à la base du troisième cycle sont en fait plus vieilles que la formation de Sokoman et probablement équivalentes aux dolomies de Denault de la fosse sud centrale. Il semble qu'un écart de 110 Ma existe entre le dépôt de la formation de Denault–Abner et celui de la formation de Sokoman. Bien que la stratigraphie régionale suggère que les formations de fer de Baby et de Sokoman soient corrélatives, les âges modèles de plomb mènent à des interprétations contraires. Les âges modèles indiquent que la sédimentation dans la fosse du Labrador s'est poursuivie sur 450 Ma.

INTRODUCTION

The Labrador Trough is the eroded remnant of an Early Proterozoic (Aphebian) thrust and fold belt. It consists of two parallel belts: a western belt formed mainly of platform and shallow basin sedimentary rocks, and an eastern belt formed of a mixture of basalt, gabbro, and associated sedimentary rocks (Fig. 1). To the east, the immediate hinterland consists of amphibolite-grade schists and gneisses, in part equivalent to rocks of the trough, as well as reworked Archean basement rocks. Most of the trough sequence experienced westerly directed thrusting, folding, and metamorphism during the Hudsonian Orogeny at the close of the Aphebian, about 1700–1800 Ma ago.

Few isotopic age data are available for the trough. Most previous radiometric dating was done by the K–Ar method, giving results that reflect the effects of greenschist- to amphibolite-grade metamorphism. A few U–Pb zircon ages, as noted below, have recently become available.

In this paper, we present model lead ages for 10 samples, mostly galena, six of which are hosted by sedimentary rocks of the trough. The remainder of the samples come from various gabbro sills injected into the sedimentary rocks. The purpose of this paper is to examine the stratigraphy of the trough, as established mainly by field studies, in the light of these data. It will be shown that certain regional correlations indicated by recent field studies are supported by the lead isotope data, even though most of the data were derived from epigenetic occurrences. The results are in agreement with a simplified stratigraphy for the northern part of the trough, and are in accord with the general stratigraphic picture as it is known in the southern trough.

STRATIGRAPHY

The general stratigraphy of the Labrador Trough has been summarized by Dimroth (1970a), Dimroth *et al.* (1970), Wardle and Bailey (1981), and Le-Gallais and Lavoie (1982). Dimroth (1970a) and Dimroth *et al.* (1970) attempted to correlate formations in the northern part of the trough with those in the southern part.

In the central trough, Dimroth (1970a) postulated the existence of two major depositional cycles (Fig. 2). He identified a possible third cycle in the northern trough. A typical cycle consists of siliciclastic and chemical sediments at the base, overlain by deeper water clastic sediments and, at the top, voluminous basalts and associated sediments. The volcano-sedimentary pile of the trough is intruded at several levels by mafic and, locally, ultramafic–mafic sills, which are lumped together under the term "Montagnais Group."

Hoffman (1987) interpreted parts of the trough sequence as foredeep deposits, laid down in one or more westwardly migrating basins forming in advance of a developing pile of thrust sheets. He suggested that the Sokoman Iron Formation, the Menihek turbidites, and the coeval mafic volcanic rocks of Dimroth's second cycle are representative of such a basin. He also raised the possibility that the first-cycle detrital sediments and volcanic rocks of the Swampy Bay and Attikamagen subgroups are

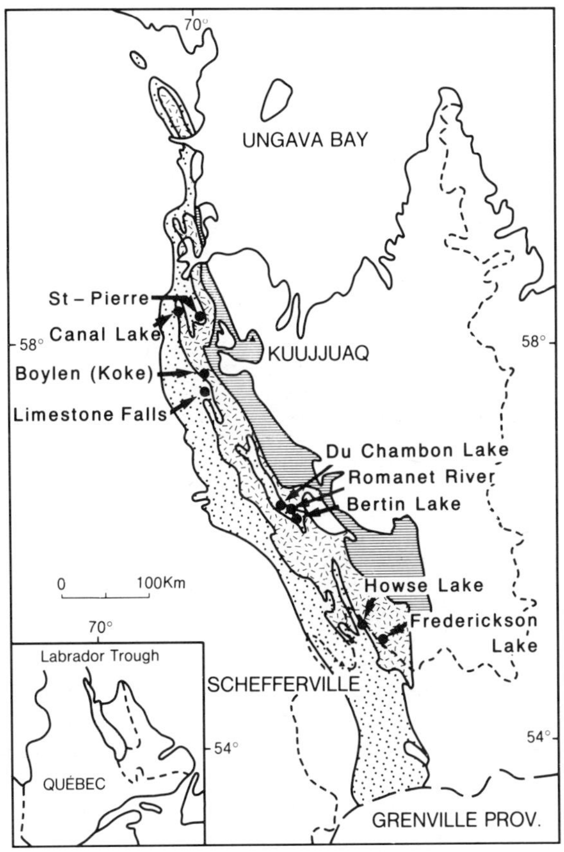

Figure 1. *General geology and location of sample sites.*

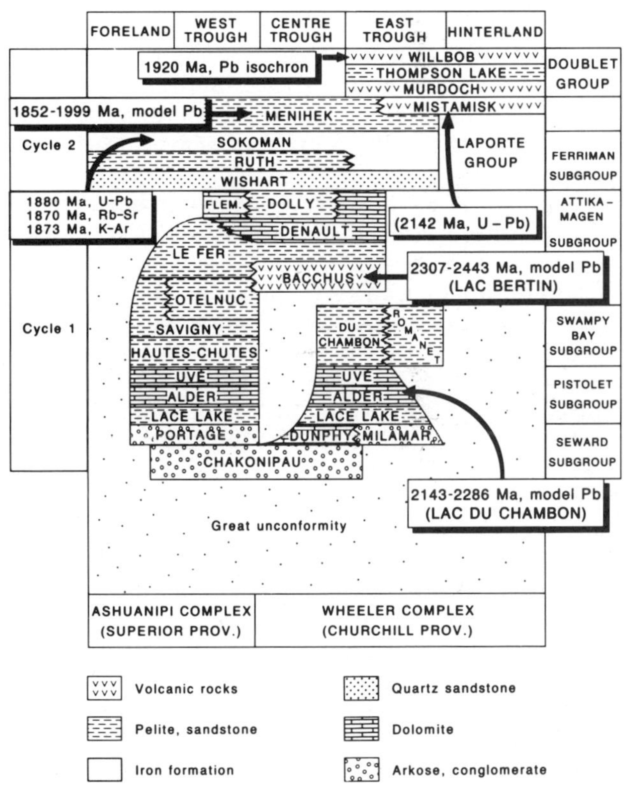

Figure 2. *Schematic stratigraphic column for the south-central Labrador Trough as proposed by Dimroth (1970a). Nimish Formation volcanic rocks (not shown) are locally intercalated with rocks of the Wishart, Ruth, and Sokoman formations. The Irene Lake Iron Formation (not shown) is of very limited distribution at the base of the Thompson Lake Formation. Shown are pertinent age determinations from this study and others (Fryer, 1972; Dressler, 1975; Dressler and Krogh in Clark, 1984; Rohon, 1987; Chevé and Machado, 1988).*

foredeep deposits, although he was unable to cite evidence for significant pre-Wishart compressional deformation. Chevé (1987), however, showed that compressional deformation affected rocks of the central trough prior to deposition of the Sokoman Formation.

LeGallais and Lavoie (1982) described the structural pattern in terms of lithotectonic zones oriented parallel to the trough. These zones are separated from each other by thrust faults, now shallow to moderately dipping in the western trough, and moderately to steeply dipping in the eastern trough. In certain sectors, the thrust faults are closely spaced and imbricate.

Cycle 1 has been recognized mainly in the southern half of the trough, whereas cycle 2 extends throughout its length. Rocks attributed to cycle 3 are a feature of the northern trough. The difficulty in recognizing the major depositional cycles is a result of the uncertainties, due to repetition by thrusting, in the lateral stratigraphic correlations. These uncertainties are at the root of the problem of whether or not a third cycle of sedimentation occurred.

The postulated third cycle of the northern trough (Fig. 3) was described by Dimroth (1970a) as consisting, at the base, of chemical sediments (Abner Dolomite, upper iron formation, i.e., the Baby Iron Formation) and, at the top, of shale and greywacke (Larch River Formation) overlain to the east by basalt and minor associated sediments (Hellancourt Formation). These formations are of major areal importance, and underlie most of the Labrador Trough north of latitude 57°N. The existence of a similar third cycle in the southern trough is suggested by the succession Irene Lake Iron Formation, Thompson Lake Slate, and Willbob Basalt. It is clear, then, that establishing whether an upper iron formation, stratigraphically distinct from the Sokoman Iron Formation, exists is crucial to the recognition of a third depositional cycle and to the establishment of sound regional stratigraphic correlations.

As discussed in more detail later, recent work (Clark, 1988) suggests that the upper iron formation (i.e., the ferriferous middle member of the Baby Formation) is stratigraphically equivalent to the Sokoman Iron Formation of cycle 2, and the Abner Dolomite to the Denault Dolomite near the top of cycle 1. The Larch River shales and greywackes, whose distribution and stratigraphic position have been poorly defined (see, for example, Fahrig (1965), Bérard (1965), and Sauvé and Bergeron (1965)), are thought to be equivalent to the detrital rocks of the lower Baby Formation (cf. Dimroth, 1970a). It appears, therefore, that the postulated third cycle in the northern trough is in large part equivalent to cycle 2.

PREVIOUS GEOCHRONOLOGICAL STUDIES

The early geochronological data for the Labrador Trough and vicinity were acquired using the K–Ar method. Biotite schists from the trough northwest of Kuujjuaq (Fort Chimo) have given K–Ar whole-rock ages ranging from 1400 to 2060 Ma, and averaging about 1600 Ma (Beall *et al.*, 1963). The 2060 Ma age suggested to them that some sedimentation occurred before this date. Biotite schists

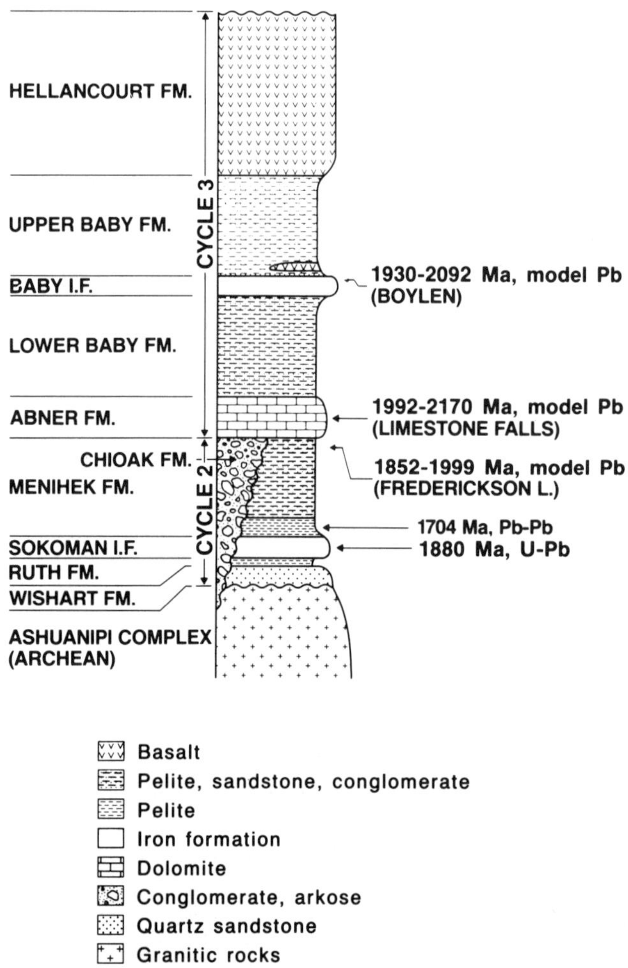

Figure 3. *Stratigraphy of the northern Labrador Trough as established by early workers (e.g., Bérard, 1965; Sauvé and Bergeron, 1965). The Larch River Formation (not shown) had previously been placed either above or below the Baby Iron Formation. Shown are pertinent Pb model ages (this study), one Pb isochron age (Bell and Thorpe, 1986), and one U–Pb zircon age (Chevé and Machado, 1988). In addition, U–Pb zircon ages obtained by Goulet* et al. *(1987) establish maximum and minimum ages for sedimentation in the northern trough of* 2721^{+5}_{-3} *Ma and* 1769 ± 15 *Ma respectively.*

north of Leaf Bay gave K–Ar mica ages of about 1560 Ma (Lowden *et al.*, 1963). The sectors from which these samples were collected were metamorphosed to greenschist or lower amphibolite grade during the Hudsonian Orogeny. Thus, most of these rocks likely experienced isotopic modification during metamorphism, so that dates are probably more representative of the cooling ages of the rocks than of their depositional ages (Stockwell, 1982).

Whole-rock Rb–Sr isochron ages have been determined from some sedimentary rocks of the Labrador Trough (Fryer, 1972). Samples of slate from two different formations differ only slightly in age (Attikamagen Slate: 1900 Ma; Menihek Slate: 1855 ± 74 Ma). The age of the Sokoman Iron Formation is thus bracketed at about 1870 ± 50 Ma (Fig. 3) according to Fryer. Although Fryer felt that these are depositional ages, Stockwell (1982) regarded them as having been modified by metamorphism.

Bell and Thorpe (1986) obtained a Pb–Pb isochron age of 1704 ± 7 Ma for uraniferous phosphorites at the base of the Menihek Formation near Schefferville. They interpreted this as a depositional age and considered that deposition immediately preceded the onset of orogeny. They noted the conflict between their results and those of Fryer (1972) for the Menihek slates. However, they recalculated Fryer's Rb–Sr data (using four of his seven points) and obtained an age of 1713 ± 74 Ma, in better agreement with their own results.

Dressler (1975) obtained a whole-rock K–Ar age of 1873 ± 53 Ma from a lamprophyre dyke in the north-central trough (Fig. 2). The dyke is related to the carbonatites, meimechites, and melilite tuffs of the Castignon Lake alkalic complex (Dimroth, 1970b), interpreted by Dressler to be of pre-Menihek (i.e., Sokoman) age. This age agrees well with that obtained by Fryer for the Sokoman Formation.

Chevé and Machado (1988) reported a U–Pb zircon age of 1880 ± 2 Ma for the Castignon Lake complex. The dated rock is a carbonatite feeder dyke related to diatremes and pyroclastic rocks, the latter being interbedded with the Sokoman Formation. These authors consider that the alkalic complex marks a volcanic event contemporaneous with the deposition of the Sokoman Formation. The zircon age is thus interpreted as the age of the Sokoman Formation, and provides a very important calibration point for the chronostratigraphy of the trough. It is worth noting that the zircon age is very close to ages obtained earlier by Rb–Sr and K–Ar methods (Fryer, 1972; Dressler, 1975). Chevé and Machado's (1988) age determination is virtually identical, within uncertainty limits, to the approximately 1880 Ma U–Pb age obtained from zircons of the Nimish Formation (T. Birkett, personal communication, 1989; also cited in Hoffman (1988)). Volcanic rocks of the Nimish Formation are intercalated with iron formation of the Sokoman Formation in the Schefferville region.

A spread of 95–125 Ma exists between the ages of the Sokoman Formation and the peak of Hudsonian regional metamorphism determined by Goulet *et al.* (1987; see below). During this time interval, the Menihek Formation turbidites were deposited.

Dressler and Krogh (as reported in Clark (1984)) obtained a U–Pb zircon age of 2142^{+4}_{-2} Ma from rhyolite of the Mistamisk Formation at Colombet Lake in the central trough (Fig. 2). Regional mapping by Dimroth (1978) showed that the Mistamisk Formation could correlate with either the Menihek Formation (the interpretation he favoured) or the upper part of the Bacchus Formation. The age of the rhyolite, when compared with the zircon age of the Sokoman-equivalent Castignon Lake complex, suggests that the Mistamisk correlates better with the upper part of the Bacchus. The rhyolite date is of great importance because it is the oldest reliable age thus far published for rocks of the Kaniapiskau Supergroup. An interval of about 262 Ma exists between the ages of the Sokoman and Mistamisk formations, indicating the importance of the sub-Wishart Formation unconformity at the base of the second depositional cycle in the trough. The zircon data also suggest that the minimum time required to deposit all of the rocks in the trough is about 370 Ma, corresponding to the time between the Mistamisk deposition and the peak of regional metamorphism. This time period will require expansion depending on the age of the pre-Mistamisk rocks, which include a great thickness of Bacchus Formation volcanics and Seward subgroup rift-type sediments.

Rohon (1987) obtained a whole-rock lead isochron age (using the Stacey–Kramers model) of 1920 ± 100 Ma for comagmatic basalt, gabbro, and ultramafic rock of the southern trough (Fig. 2). The basalts belong to the Willbob Formation, situated structurally at the top of the succession in this part of the trough and probably correlative with the Hellancourt volcanic rocks in the northern trough. The gabbro and ultramafic rocks are assigned to the Montagnais Group.

A U–Pb age of about 1880 Ma (T. Birkett, personal communication, 1989; also cited in Hoffman (1988)) was obtained from zircons in a gabbro sill at Howse Lake, 40 km northeast of Schefferville. The sill intrudes Menihek Formation pelitic rocks according to Frarey (1967) and Dimroth (1978), but on the basis of the zircon age and its proximity to the age of the Sokoman Formation, this correlation is doubtful (T. Birkett, personal communication, 1989).

Kish and Cuney (1981) obtained a U–Pb concordia-intercept age of about 1800 Ma for uranium mineralization in the Mistamisk Lake area of the central trough. The mineralization is found in albite-rich veins penetrating Lace Lake Formation mudstone and dolomite in the lower part of the first depositional cycle. From their fluid inclusion study, these authors concluded that the mineralization was emplaced at a temperature of about 300–350°C toward the end of the Hudsonian metamorphic event.

Several geochronological studies have been carried out on rocks from the gneissic terrane immediately east of the Labrador Trough (e.g., Beall *et al.*, 1963; Wanless *et al.*, 1965; Goulet *et al.*, 1987; Ciesielski and Loveridge, personal communication, 1983). In general, they confirm that Archean rocks are present in the hinterland to the trough, and that Hudsonian metamorphism partially updated them. Goulet *et al.* (1987) reported U–Pb zircon ages of 2721 ± 4 Ma for a gneissic dome northwest of Kuujjuaq and 2721^{+5}_{-3} Ma for a granodiorite pluton immediately west of the trough. Their results also place the time of peak Hudsonian metamorphism at 1769 ± 15 Ma, some 170 Ma earlier than the 1600 Ma K–Ar cooling age determined by Beall *et al.* (1963). These dates establish maximum and minimum ages for sedimentation in the northern trough.

NEW LEAD ISOTOPE DATA

Methods of analysis

Most of the specimens were analyzed by Geospec Consultants Ltd., Edmonton, under contract to the Geological Survey of Canada. Two analyses were carried out, courtesy of Dr. B.R. Doe, United States Geological Survey, at the mass spectrometry laboratory in Denver, Colorado, and one analysis was carried out at the Geochronology Laboratory of the Geological Survey of Canada in Ottawa. All measured isotope ratios have been normalized to approximately absolute values on the basis of numerous measurements of the United States National Bureau of Standards reference lead SRM-981.

Analyses by Geospec Consultants Ltd. were performed on a Micromass MM-30 mass spectrometer. Specimens were prepared for analysis by dissolving a small fragment of the galena or clausthalite in 10 mL of high purity 2*N* HCl, and evaporating the solution slowly until $PbCl_2$ crystallized. The crystals were extracted from the solution, washed in 4*N* HCl, dried, and dissolved in water. An aliquot of 1–2 μg was then loaded on a rhenium filament in a silica gel – phosphoric acid mixture. The analytical procedures followed by the Geological Survey of Canada were as outlined by Thorpe *et al.* (1984).

The quoted 2σ analytical uncertainties, based on reproducibility of results on reference lead SRM-981, for the analyses conducted by Geospec Consultants Ltd. from 1983 to date are 0.042, 0.052, and 0.058% for the ratios $^{206}Pb/^{204}Pb$, $^{207}Pb/^{204}Pb$, and $^{208}Pb/^{204}Pb$ respectively. However, on the basis of duplicate analyses and the results for specimens that can reasonably be presumed to be isotopically identical, although not strictly duplicates, the calculated uncertainties are 0.064, 0.068, and 0.075% for these ratios. The analyses provided by the United States Geological Survey, courtesy of B.R. Doe, have analytical uncertainties that are considered to be less than 0.1%. Analytical uncertainties have been calculated at about 0.18, 0.24, and 0.30%, respectively, for the three ratios for the analysis conducted by the Geological Survey of Canada.

One set of model ages presented here (Table I, Fig. 4) for Proterozoic leads from the Labrador Trough has been calculated using the Western Superior model derived by Thorpe (*in prep.*) on the basis of lead isotope compositions and ages for Archean volcanogenic massive sulphide deposits of western Superior Province. An Abitibi model, which has been found necessary to account for the isotopic compositions of massive sulphide deposits in the Abitibi greenstone belt, yields model ages that are about 40 Ma less than those given by the Western Superior model, but the Abitibi model has not been shown to be applicable to any Proterozoic deposits.

In comparison with syngenetic Archean leads, those from Proterozoic deposits show considerably greater scatter when their compositions are recalculated to a constant age. In many cases, the sources of Proterozoic ore leads appear to have had an accelerated evolution (Richards, 1986; Thorpe, *in prep.*) to yield ratios that, at a given age, are greater than accounted for by models for Archean leads. As noted elsewhere (Deb *et al.*, 1989), a general "Proterozoic" model can be derived that is based largely on the isotopic and age data for volcanogenic massive sulphide deposits of Proterozoic age in Wisconsin (Van Schmus, 1980; Afifi *et al.*, 1984). ("Proterozoic" is in quotation marks to indicate that the model, although based on data for Proterozoic deposits, is probably most applicable to some suites of deposits in sedimentary sequences, but is not necessarily adequate for Proterozoic deposits of other types.) This model, however, yields model ages for the Mt. Isa, McArthur River, Broken Hill, and Sullivan sediment-hosted massive sulphide deposits that are a little younger than their known ages (LeCouteur, 1973, 1979; Page, 1981; Zartman *et al.*, 1982; Gulson, 1984, 1985; Gulson *et al.*, 1985). Consequently, model parameters have been adjusted slightly to yield a general

"Proterozoic" model that results in model ages in good agreement with the known ages of the deposits.

It is unfortunate that, at the present state of knowledge regarding the isotopic evolution of Proterozoic ore leads, the ranges of calculated model ages are so broad. This is particularly true for those leads with relatively high calculated source μ (i.e., $^{238}U/^{204}Pb$) values. Nevertheless, it is anticipated that the relative ages of the leads should be indicated correctly by their model ages.

The model parameters for the Western Superior and "Proterozoic" models are as follows:

	Western Superior model	"Proterozoic" model
t_0	4315 Ma	3810 Ma
A_0	9.6088	10.844
B_0	10.9722	12.632
C_0	29.819	31

Results of isotopic analyses

The 12 lead isotope analyses obtained in this study from nine different sites are listed in Table I and are illustrated in Figures 2,3,4, and 5. Summaries of the characteristics of the mineral deposits and occurrences represented by these analyses are presented below in the order of the oldest to youngest model age. Two of these deposits, Boylen (Koke) and Frederickson Lake, are of sediment-hosted massive sulphide type. Samples of disseminated galena in dolomite were collected at Du Chambon Lake and Limestone Falls. Other samples came from various epigenetic vein and breccia deposits, such as St. Pierre, Bertin Lake, Romanet River, Howse Lake, and Canal Lake.

Two model ages are quoted for each sample, one for the Western Superior model and one for the general "Proterozoic" model. The two models give model ages differing by 126–178 Ma. Excluding the anomalous model age for the Romanet River sample, the model ages obtained range from 1793 to

Table I
Analytical results

Sample no.	Locality or deposit	$^{206}Pb/^{204}Pb$	$^{207}Pb/^{204}Pb$	$^{208}Pb/^{204}Pb$	West Superior model	
					Calculated μ	Calculated Th/U
TQ 84-76	St. Pierre	15.001	15.203	35.296	9.329	4.447
RF 66-837	Canal Lake	15.004	15.197	35.256	9.308	4.410
RF 66-837	Canal Lake	14.996	15.189	35.249	9.288	4.410
TQ 78-27	Boylen (Koke)	15.222	15.245	35.143	9.294	4.127
TQ 84-75	Limestone Falls	15.195	15.286	35.276	9.443	4.263
TQ 84-75	Limestone Falls	15.185	15.284	35.273	9.444	4.270
TQ 84-165	Romanet River	16.533	15.301	34.621	8.863	2.890
TQ 84-74	Du Chambon Lake	14.751	15.100	34.669	9.206	4.147
TQ 84-73	Bertin Lake	14.411	14.998	34.273	9.197	4.112
TQ 84-72	Bertin Lake	14.413	14.994	34.263	9.180	4.100
TQ 85-135	Howse Lake	15.370	15.217	35.354	9.112	4.153
RF 66-836	Frederickson L.	15.298	15.222	35.222	9.172	4.117

Sample no.	Model ages (Ma)		Laboratory (date)
	West Superior model	"Proterozoic" model	
TQ 84-76	2053	2214	Geospec Consultants Ltd. (1985)
RF 66-837	2045	2204	United States Geological Survey (1980)
RF 66-837	2043	2200	Geospec Consultants Ltd. (1983)
TQ 78-27	1930	2092	Geological Survey of Canada (1979)
TQ 84-75	1989	2167	Geospec Consultants Ltd. (1986)
TQ 84-75	1994	2172	Geospec Consultants Ltd. (1986)
TQ 84-165	1022	1148	Geospec Consultants Ltd. (1985)
TQ 84-74	2143	2286	Geospec Consultants Ltd. (1986)
TQ 84-73	2309	2446	Geospec Consultants Ltd. (1986)
TQ 84-72	2304	2439	Geospec Consultants Ltd. (1986)
TQ 85-135	1793	1935	Geospec Consultants Ltd. (1986)
RF 66-836	1852	1999	United States Geological Survey (1980)

2309 Ma with the Western Superior model and from 1935 to 2446 Ma with the "Proterozoic" model. These ages are in general accord with the established age constraints for sedimentation within the Labrador Trough.

The calculated Th/U ratios for the sources of the lead constitute a parameter that, although not highly diagnostic, is still of considerable interest. Thorpe (*in prep.*) has observed that volcanogenic massive sulphide deposits and other leads that have had, at most, short crustal histories all yield calculated Th/U ratios ranging approximately from 4.04 to 4.16. Leads yielding ratios in this range have not necessarily evolved in a single-stage manner, but significantly higher and lower ratios almost certainly reflect more complex lead evolution. According to this criterion, the Frederickson Lake, Boylen, Du Chambon Lake, Bertin Lake, and Howse Lake leads may have had basically single-stage evolution histories.

Bertin Lake. At Bertin Lake, west of Romanet Lake in the central trough, the Bertin Lake Fault juxtaposes Dunphy Formation dolomite (to the north) and Montagnais Group gabbro sills intercalated with Bacchus Formation basalts and associated sedimentary rocks (to the south). The Bertin Lake Fault is a major normal fault, which forms the southern limit of the Mistamisk–Romanet horst, a late-tectonic feature that tran-

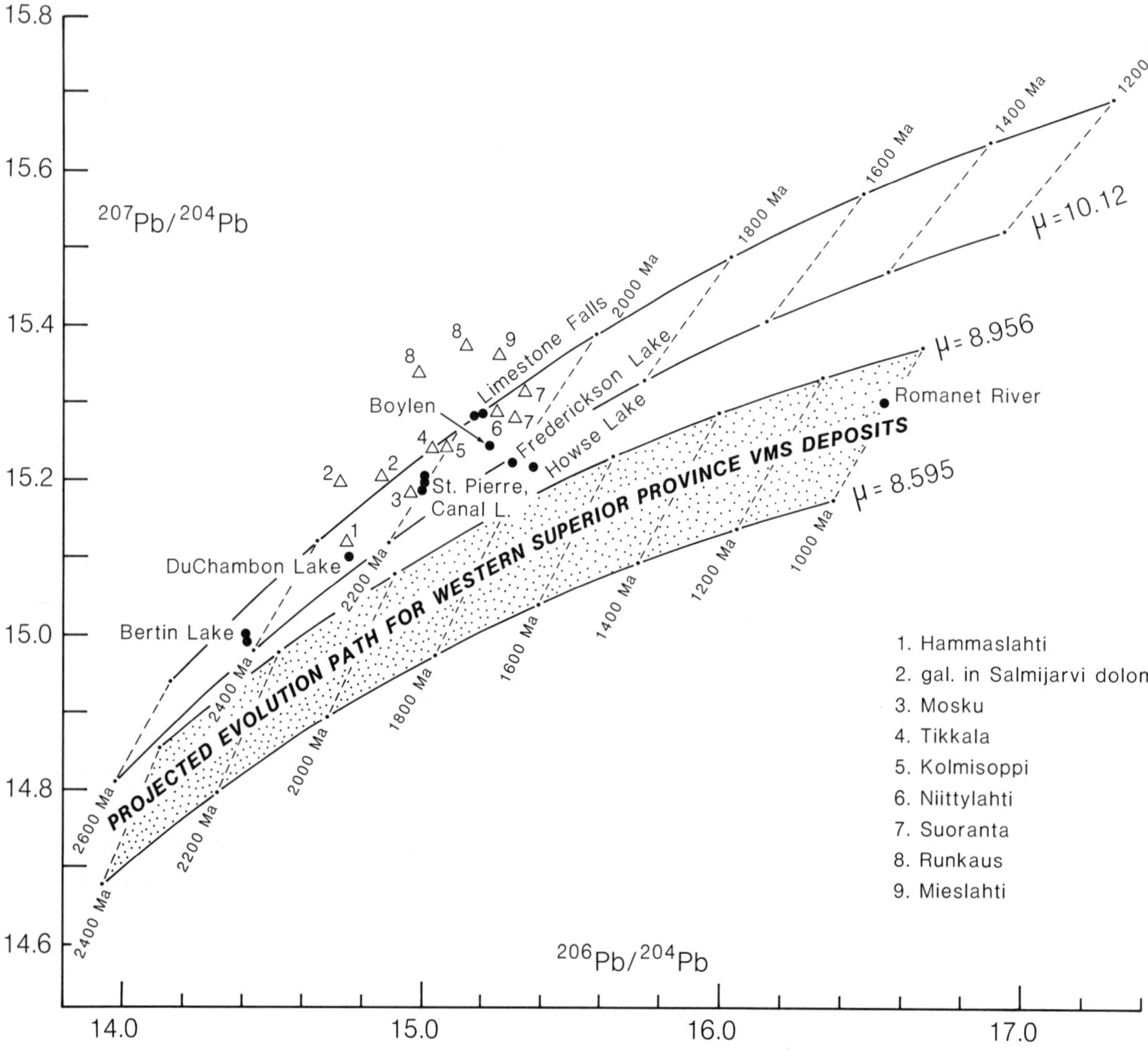

Figure 4. *Results of isotopic analyses shown on a plot of $^{207}Pb/^{204}Pb$ versus $^{206}Pb/^{204}Pb$. Some analyses of galenas from Finland by Vaasjoki (1981) and Sakko and Laajoki (1975) are also shown (numbered triangles). The occurrences in Finland are briefly characterized in the references cited. Abbreviation: VMS, volcanogenic massive sulphide.*

sects, in a northwest–southeast direction, lithological and earlier structural trends (Chevé, 1985; Clark, 1986). The horst forms a tectonic window exposing platform sedimentary rocks of the Seward and Pistolet subgroups (Fig. 2) as well as deeper water, basinal sedimentary rocks of the Swampy Bay subgroup.

Near the Bertin Lake site, the fault is marked by a tectonic breccia about 10 m wide (Chevé, 1985) consisting of angular fragments of silicified gabbro cemented by calcite. Calcite veins also penetrate the fragments. Millimetre-scale masses of pyrrhotite, pyrite, sphalerite, and chalcopyrite occur in the calcite cement and, to a lesser extent, in the gabbro fragments. South of the breccia, the gabbro is intensely fractured and penetrated by abundant calcite veins containing sphalerite, chalcopyrite, and pyrite. About 200 m south of the breccia, the gabbro is cut by a calcite-filled vertical tension gash 50 m long, oriented subparallel to the Bertin Lake Fault. Galena, sphalerite, pyrite, and chalcopyrite occur in the immediate wall rocks of the tension gash and in calcite cementing wall rock fragments. The maximum abundance of sulphides is 3%.

Two samples of galena yielded model ages of 2304–2439 and 2309–2446 Ma respectively. The calculated Th/U ratio (Table I) suggests that this lead may have had a single-stage evolution history.

Du Chambon Lake. Galena, accompanied locally by a little chalcopyrite and sphalerite, forms small, lenticular concentrations and fine disseminations in sandy dolomite of the Alder Formation at Du Chambon Lake in the central trough (Clark, 1986). The sample site is located in the Mistamisk–Romanet horst discussed in the preceding section.

The galena gives a model age of 2143–2286 Ma. The calculated Th/U ratio (Table I) suggests that this may be a single-stage lead.

St. Pierre deposit. This small deposit, located 75 km west of Kuujjuaq, consists of a 3 m long lens of massive sulphides within phyllites of the lower member of the Baby Formation, beneath the Baby Iron Formation (Bélanger, 1982; Fournier, 1983; Wares and Berger, 1987; Wares *et al.*, 1988). A grab sample of massive sulphides analyzed 17.4% Zn, 4.5% Cu, 22.4% Pb, 532 g/t Ag, and 91 g/t Au (Clark *et al.*, 1986). Wares *et al.* (1988) noted that sphalerite, pyrite, galena, and chalcopyrite form a discordant body cutting and brecciating barren, fine-grained pyrite. They concluded that the base and precious metals are epigenetic, and were emplaced during metamorphism. They also speculated that the source of the base and precious metals was a nearby, hidden, volcanogenic sulphide body, similar in origin to the Boylen deposit. The

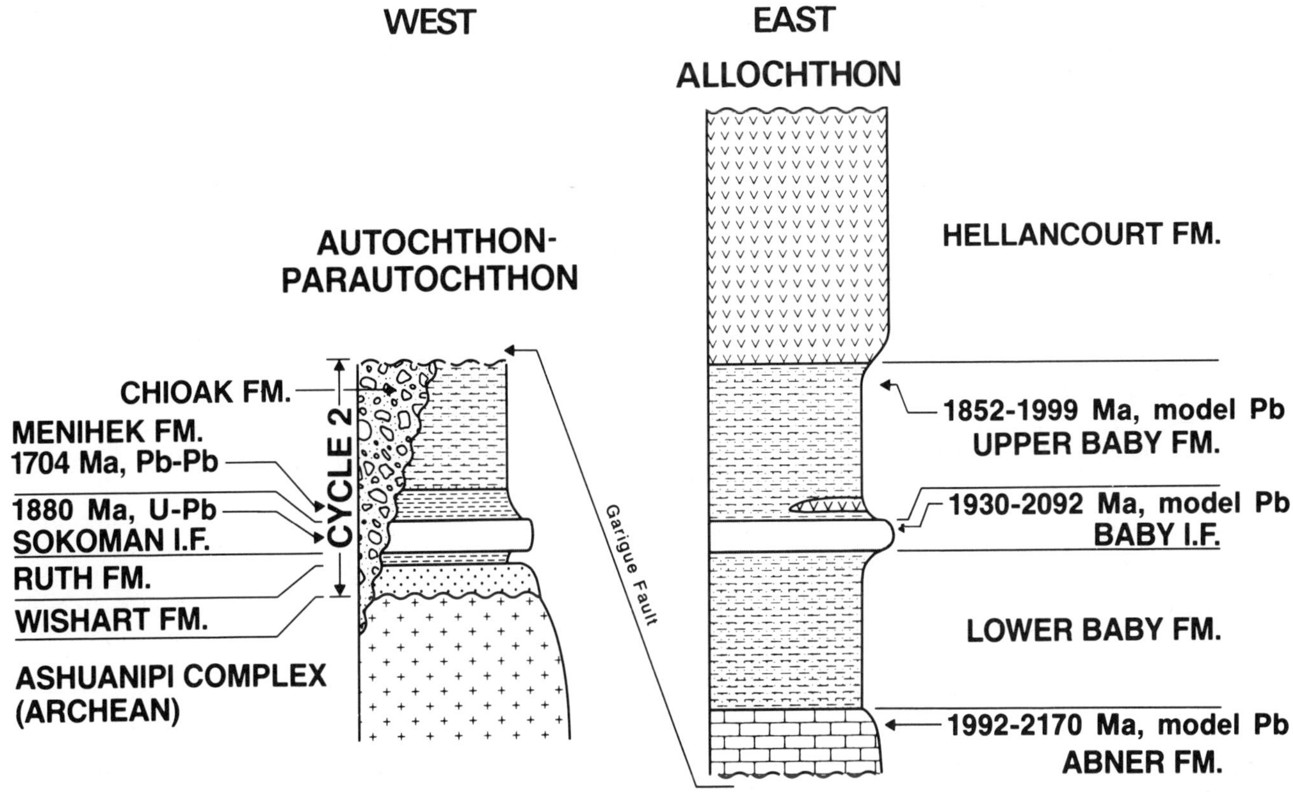

Figure 5. *Tectono-stratigraphic relations in the northern Labrador Trough. Some pertinent ages are also shown. See Figure 3 for age references and legend.*

remobilization would have caused metal enrichment and changes in metal ratios.

The lead isotope model age for this deposit is 2053–2214 Ma. The calculated Th/U ratio for the lead source (Table I) is very high, suggesting an evolution history of more than one stage for this lead.

Canal Lake. A specimen of galena associated with pyrrhotite and chalcopyrite from a gabbro sill in the Canal Lake area gives duplicate model ages of 2043–2200 and 2045–2204 Ma. The precise location of the sample site, unfortunately, is uncertain. The isotopic composition of the sample is identical, within analytical uncertainty, to that of the St. Pierre deposit. Nevertheless, the geographic information provided by Franklin *et al.* (1983) for the Canal Lake sample suggests that it was collected at a site some 27 km northwest of the St. Pierre deposit.

The fact that the galena comes from a gabbro sill points to a complex history for the mineralization. This possibility is further suggested by the high calculated Th/U ratios for the lead source (Table I).

Limestone Falls. In the north-central trough, southwest of Kuujjuaq, the Abner Formation underlies a wide area west of the Caniapiscau River. It consists of tightly folded interbeds of massive to thin-bedded, grey dolomite, grey stromatolitic dolomite, sandy dolomite, dolomitic sandstone, and intraformational conglomerate. At a site 4.5 km northwest of Limestone Falls, galena occurs in a 3 m thick bed of siliceous dolomite breccia. The galena forms disseminated grains and patches, and accounts for less than 1% of the rock. It is associated with disseminated pyrite, which appears to have experienced minor remobilization during deformation.

Duplicate analyses of the galena have yielded model ages of 1989–2167 and 1994–2172 Ma. The calculated Th/U ratio (Table I) is somewhat high, suggesting that this lead evolved during more than one stage.

Boylen (Koke) deposit. This important base metal massive sulphide deposit, situated near the Koksoak River about 80 km southwest of Kuujjuaq, has been evaluated at 1.2 Mt, grading 6.9% Zn, 0.7% Cu, 1.1% Pb, 54.5 g/t Ag, and 1.0 g/t Au (Armstrong, 1976). It has been described by Fournier (1982, 1983), Wares and Berger (1987), and Wares *et al.* (1988). The massive sulphides, which occupy the core of a refolded, isoclinal F_1 fold, consist of pyrite, sphalerite, chalcopyrite, pyrrhotite, and a little galena in a quartz-rich (chert) gangue. The stratiform body is enveloped in black slate at the top of carbonate facies iron formation of the Baby Formation. Wares *et al.* (1988) interpreted the deposit to be syngenetic and of exhalative origin. They note, however, that the deposit is highly deformed and entirely recrystallized.

A sample of massive sulphides gives lead isotope model ages of 1930–2092 Ma. The calculated Th/U ratio (Table I) is indicative of a possible one-stage evolution history.

Frederickson Lake deposit. The Frederickson Lake base metal deposit (Fournier, 1982, 1983; Gebert, 1986), located about 50 km northeast of Schefferville, consists of a stratiform sulphide lens within pyritic pelites in what was mapped as the top of the Menihek Formation (Frarey, 1967; Dimroth, 1978), in close proximity to a glomeroporphyritic gabbro sill. An uncertainty has recently been introduced as to the age of the host sedimentary rocks (Fig. 3) because of an approximately 1880 Ma U–Pb age determined on zircons from a gabbro sill cutting the formation at Howse Lake (T. Birkett, personal communication, 1989; also cited in Hoffman (1988)). Hence, the host rocks may be older than the Menihek, the pelites being part of an older unit that has been overthrust.

The deposit has been evaluated at 279,000 t grading 4.4% Zn, 0.8% Cu, 42.2 g/t Ag, and 0.7 g/t Au. The sulphides consist principally of pyrrhotite, pyrite, sphalerite, and chalcopyrite, accompanied by minor galena. They exhibit a parallel lamination due to an alternation of bands rich in sphalerite and pyrrhotite. Fournier (1983) postulated that initially the sulphides formed in the gabbro sill and were subsequently ejected from the sill during its emplacement. Gebert (1986) and Gebert *et al.* (1987), on the other hand, concluded that the sulphides were formed as a result of hydrothermal, exhalative activity prior to the intrusion of the mafic magma, and that the origin of the deposit is similar to that of the Boylen deposit in the northern trough.

The sulphide sample analyzed in the present study gives model ages of 1852–1999 Ma and a calculated Th/U ratio (Table I) that suggests a possible single-stage evolution.

Howse Lake. Epigenetic galena occurs in a vein cutting a glomeroporphyritic gabbro sill at Howse Lake, 40 km northeast of Schefferville. The gabbro sill intrudes pelitic sedimentary rocks considered by Frarey (1967) and Dimroth (1978) to belong to the Menihek Formation, but which, as noted above, may belong to an older formation.

The galena gives model ages of 1793–1935 Ma and a calculated Th/U ratio that points to a single-stage evolution for the lead. The model ages span the age of the host gabbro sill, determined to be about 1880 Ma, as noted above.

Romanet River. Clausthalite (PbSe) from the G.M. showing on the Romanet River gives model ages of 1022–1143 Ma and a very low calculated Th/U ratio for the lead source (Table I). This age is anomalous, the sample clearly having been subjected to the addition of a uranogenic lead component. The clausthalite is associated with pitchblende and a little altaite in a pod of dolomitic carbonate emplaced during an episode of sodic alteration (Bell and Ruzicka, 1985; Clark, 1986). This locality is near the uraniferous deposits for which Kish and Cuney (1981) reported a U–Pb concordia-intercept age of about 1800 Ma.

If it is assumed that the initial isotopic composition of the clausthalite was in the range between those for Bertin Lake and Du Chambon Lake galenas, the calculated range of slopes is 0.144–0.113, corresponding to continuous addition ages of 2276–1849 Ma. Because the $^{208}Pb/^{204}Pb$ of the Romanet River specimen, by itself, suggests a very approximate mineralization age of 1900–2150 Ma, it would appear that a continuous addition interpretation for the uranogenic component is justified. This is in accord with the association of pitchblende with the clausthalite analyzed.

DISCUSSION

The lead isotope compositions reported here for galena specimens from the Labrador Trough are relatively distinct in comparison with other Proterozoic leads in Canada. They yield Early to Middle Proterozoic ages and plot in a high-μ position on the standard lead evolution plot (Fig. 4). In the latter respect, they are very distinct from a suite of leads from the Central Mineral Belt, Labrador, Newfoundland (Geological Survey of Canada, unpublished data), and from the Cu–Au deposits of the Chibougamau District (Thorpe *et al.*, 1984), which plot along the general projected evolution path for Superior Province massive sulphide deposits (Fig. 4). Only a few deposits elsewhere in Canada are isotopically similar to those of the Labrador Trough. These include Thubun Lake, south of the East Arm of Great Slave Lake, which yields analyses plotting between those for the St. Pierre and Frederickson Lake occurrences (Thorpe, 1982), and leads from the silver-arsenide vein deposits of the Cobalt district, Ontario (Thorpe *et al.*, 1986), which are comparable to the Du Chambon and Hammaslahti analyses (see discussion below).

The lead isotope data are discussed further with regard to their significance to the interpretation of stratigraphy in the Labrador Trough and with regard to their similarity to data for deposits and occurrences in Karelian rocks in Finland.

Stratigraphic relations in the Labrador Trough in view of the Pb isotope data

Our isotopic data are consistent with the hypothesis that only two depositional cycles are present in the Labrador Trough. The third depositional cycle is an artifact of the effects of regional thrusting and resultant stratigraphic repetition (Clark, 1988). This conclusion is not affected by the great uncertainty in the lead model ages. The relative ages should likely be correct, but, as indicated earlier, model ages are not closely defined due principally to the lack, to date, of rigid modeling of such leads. It is possible that the uncertainty arises because of a dispersion of isotopic compositions resulting from the complexity of underlying geological processes and that future modeling will fail, unless the leads being studied can be subdivided on a tectonic environment or similar basis, to define more precise models.

Our oldest model age was obtained from galena in a fracture in a gabbro sill near Bertin Lake (2307–2443 Ma, average of two samples). The galena appears to be related in origin to a regional normal fault active late in the orogenic episode (Chevé, 1985). Possible explanations that can be considered regarding the fact that the model age of the Bertin Lake lead appears to be older than the fault are as follows:

(1) The lead was remobilized from a sulphide deposit, most likely of syngenetic massive sulphide type, in stratigraphically lower strata. Militating against this interpretation, however, is the fact that the calculated Th/U ratio for the source provides no evidence for a complex history for this lead, and also that the remobilization would need to have occurred with little or no addition of radiogenic lead, an unlikely scenario.

(2) The galena is not related to the Bertin Lake Fault, but was, in fact, emplaced soon after intrusion of the host gabbro. Model ages of 2307–2443 Ma are, by this scheme, assumed valid and provide minimum ages for the Bacchus Formation volcanic and sedimentary rocks with which the gabbros are intercalated.

The assumption that the approximate model lead ages for the Bertin Lake galena are valid has other stratigraphic implications. The Bacchus Formation, for example, would be at least 160 Ma older than the lead hosted by the Alder Formation at Du Chambon Lake. This would suggest (Fig. 6) that the Bacchus is correlative with the sediments of the lower part of the first cycle, as proposed by LeGallais and Lavoie (1982) (see also Dimroth's stratigraphy, Fig. 2). The whole first cycle sequence would then seem to be older than previously considered, deposition having begun very early in the Aphebian. Knowledge of the true age of the Bacchus is also important for understanding the stratigraphic context of the Mistamisk Formation. The Mistamisk has yielded a 2142^{+4}_{-2} Ma zircon age from rhyolite (Dressler and Krogh cited in Clark (1984)). As noted earlier, Dimroth (1978) did not determine the precise stratigraphic position of the Mistamisk in his regional work. Chevé and Machado's (1988) zircon age of 1880 ± 2 Ma for the Sokoman Formation suggests the Mistamisk, at 2142^{+4}_{-2} Ma, is correlative with the upper part of the Bacchus in the lower depositional cycle. Our lead isotope data for the Bertin Lake galena, if assumed to give a valid minimum age for the Bacchus, are in broad agreement with the chronostratigraphy as suggested by the zircon data, and would point to an extended period of volcanism during Bacchus–Mistamisk time.

The Alder Formation occupies a low stratigraphic position in the lower cycle. The 2143–2286 Ma model age for galena in the Alder Formation at Du Chambon Lake is older than all previous K–Ar and Rb–Sr ages for trough rocks. The model age provides a minimum age for the host Alder Dolomite. The 2143 Ma model age (Western Superior model) is close to the 2142^{+4}_{-2} Ma age obtained by Dressler and Krogh for zircons from Mistamisk Formation rhyolite, suggesting that the Alder is approximately correlative with the upper part of the Bacchus Formation (Fig. 6).

The 1992–2170 Ma model age (average of two analyses) obtained for galena in the Abner Formation near Limestone Falls is important for the stratigraphic position of the Abner. The age is significantly older than the 1880 ± 2 Ma zircon age for the Sokoman Formation. The model age represents a minimum age for the Abner Dolomite because the galena may be epigenetic, as suggested by textures and the somewhat high calculated Th/U ratio for the lead source of the galena, which indicates that the lead may have experienced more than one stage in its evolution. Until recently, the Abner has been regarded as forming the base of the supposed third depositional cycle in the northern trough, stratigraphically higher than the Menihek (Fig. 3) (Bérard, 1965; Dimroth, 1970a). This cycle consists of the Abner Dolomite; overlain by Baby Formation slates, schists, and iron formation; and Hellancourt Formation basalt. Larch River Formation rocks, identified only in the northern trough, were interpreted by Clark (1988) as being equivalent to the lower member of the Baby Formation.

Recent field studies in the Napier Lake and Bérard Lake areas in the northern trough suggest a different stratigraphic arrangement. A regionally important thrust fault, the Garigue Fault, was identified below the Abner in the Napier Lake region (Clark, 1979). A northward extension of this fault to Bérard Lake was mapped by Goulet (1986) and Budkewitsch (1986). The fault marks the western limit of an allochthonous terrane, as defined by Goulet (1986). The Abner, Baby, and Hellancourt formations were thrust over the parautochthonous rocks of the Ferriman subgroup and the Menihek and Chioak formations (Fig. 5; see also Clark (1988)). This interpretation suggests that the Abner Dolomite of the northern trough is correlative with the Denault Dolomite of the southern trough, as has been postulated by Hoffman and Grotzinger (1988). The Denault, which is stratigraphically lower than the Sokoman Formation, had previously not been recognized north of

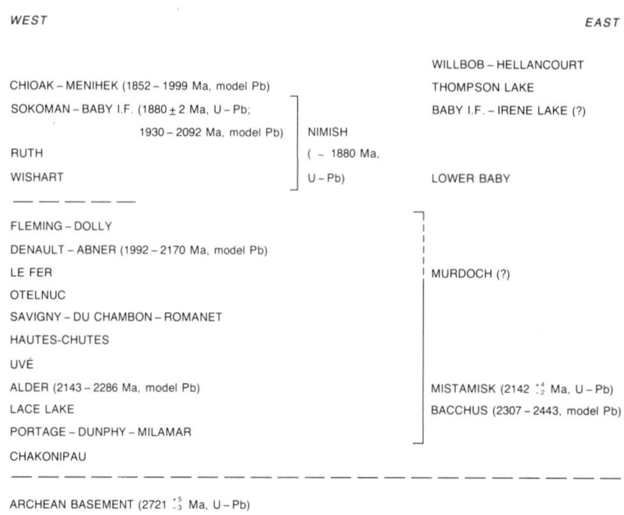

Figure 6. *Schematic representation of the interpreted stratigraphy of the Labrador Trough, combining the main stratigraphic elements of the northern and southern sectors. Shown are pertinent lead model ages (this study) and U–Pb zircon ages (see text for references). The horizontal dashed lines represent unconformities.*

latitude 56°03′N in regional studies (e.g., Dimroth, 1978; Clark, 1979; Dressler, 1979). The model age of galena from the Abner (1992–2170 Ma) supports the proposition that the Abner is older than the Sokoman (1880 ± 2 Ma). This galena appears to be younger, however, than that from the Alder Formation, in the lower part of the first cycle, which gives a model age of 2143–2286 Ma. Our results, therefore, agree with the hypothesis that the Abner and Denault formations are equivalent (Fig. 6).

In a regional study of the Baby Iron Formation in the Hérodier Lake area, southwest of Kuujjuaq, Clark (1988) concluded that the Sokoman and Baby iron formations are correlative. There, the Baby Iron Formation overlies the Abner Dolomite and is separated from the latter by a thin unit of dark grey mudstone in the west and by a thick sequence of pelites and sandstones farther east. Systematic lateral facies variations in the Baby Iron Formation suggest a gradual change in depositional environment from shallow water and energetic conditions in the west to deeper water and tranquil conditions in the east. The westernmost outcrops of the Baby Iron Formation are lithologically very similar to the Sokoman Iron Formation (oxide and carbonate facies; shallow-water depositional environment), which occurs in the parautochthonous zone farther west. Thus, in passing from the Sokoman to the Baby Iron Formation (west to east), the oxide facies, abundant in the Sokoman, gives way to the silicate–carbonate facies of the Baby. The carbonate facies is present in both units. The sulphide facies appears to be unique to the Baby Iron Formation. The facies transition occurs near where the Garigue Fault causes structural repetition (Fig. 7).

The 1930–2092 Ma model age determined for galena in the Boylen (Koke) massive sulphide deposit in the northern trough broadly supports this correlation when compared with other model ages. This deposit, thought to be syngenetic (Wares *et al.*, 1988), occurs within the upper part of the Baby Iron Formation. The model age is younger than that of galena from the Abner Dolomite, but older than that of galena from the Frederickson Lake deposit in what may be the upper Menihek Formation.

The 1930 Ma model age given by the Western Superior model for the Boylen lead is 50 Ma greater than the U–Pb zircon age for the Sokoman determined by Chevé and Machado (1988). The 2092 Ma age obtained using the "Proterozoic" model is much older. The large discrepancy involving the "Proterozoic" model age may mean that the latter is very inaccurate, or if it is accurate, that the Baby

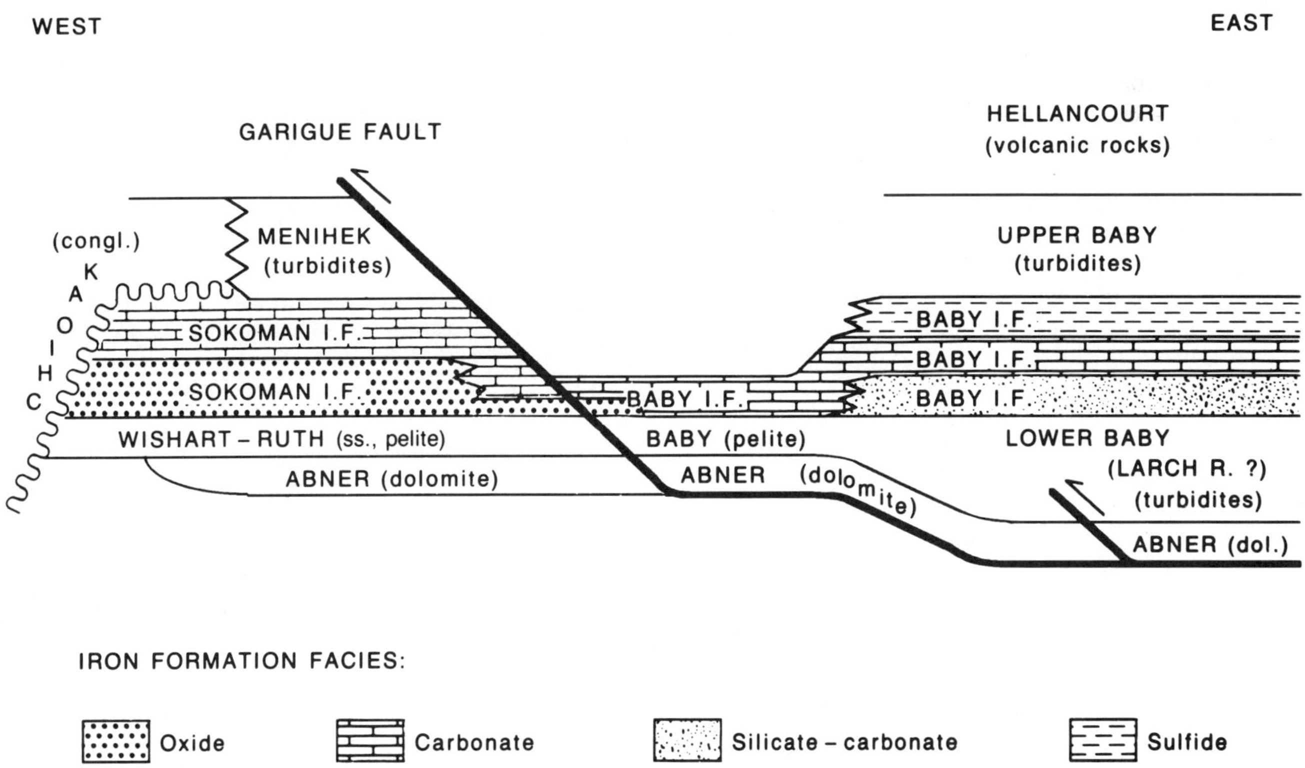

Figure 7. *Facies variations in the Sokoman and Baby iron formations and stratigraphy, Hérodier Lake area (southwest of Kuujjuaq), Labrador Trough. The Chioak conglomerates are not represented in the figure. Formation thicknesses are schematic.*

and Sokoman iron formations are not correlative in spite of the field evidence. Although Wares *et al.* (1988) interpreted the Boylen deposit as syngenetic, they noted that it has been highly deformed and totally recrystallized, and that metal ratios have probably been changed. It seems unlikely, however, that a primitive lead component was added during this deformation because the calculated source Th/U ratio is compatible with single-stage lead evolution. If the Western Superior model age is reliable, and in addition indicates the approximate age of the Baby Iron Formation, then it would appear that the Sokoman and Baby are not precisely correlative or that the iron formation is diachronous. Alternatively, the 50 Ma difference may simply mean that the Western Superior model is not totally satisfactory for the Boylen lead.

Hellancourt Formation basalts conformably overlie the Baby Formation in the northern trough (Sauvé and Bergeron, 1965; Wares and Goutier, 1989). Assuming the equivalence of the Sokoman and Baby iron formations, the Hellancourt basalts are homotaxial with basalts of the upper part of the Menihek in the southern trough. The Hellancourt basalts would thus belong to the second depositional cycle.

The 2053–2214 Ma model age range obtained for galena of the St. Pierre deposit is problematic. The deposit is situated in lower Baby Formation rocks, below the iron formation, according to Bélanger (1982) and Wares *et al.* (1988). The model age is, however, older than that of galena from the Abner Dolomite at Limestone Falls. A possible explanation is that the disseminated galena at Limestone Falls is epigenetic, as the somewhat high calculated Th/U ratio of the lead source suggests. As an alternative explanation, it may be possible that the St. Pierre host rocks are, in fact, older than the Abner, and that this has not been recognized because of the severe faulting in the area. On the other hand, the very high calculated Th/U ratio for the St. Pierre lead source could indicate that this source was depleted in U relative to Pb (and Th), and thus the model ages could be too great. If this is the case and if the lead component that had a retarded evolution was derived from sedimentary materials within the host sequence, these sediments must have been derived from granulite facies terranes.

The difference between the minimum age of the Abner Formation, as indicated by the model lead age for galena from Limestone Falls, and the age of the Sokoman Formation (1880 ± 2 Ma) is about 110 Ma (Western Superior model) or about 290 Ma ("Proterozoic" model). The time span between the two formations would be even greater if the Abner galena is epigenetic. This difference would, in part, be accounted for in the stratigraphic record by the unconformity at the top of the Denault (which we believe is equivalent to the Abner) in the western trough (Fig. 6). There appears to be no evidence for an unconformity above the Abner in the eastern trough (Clark, 1978, 1979), but the overlying lower Baby pelites and sandstones are very thick (at least 600 m according to Sauvé and Bergeron (1965)), suggesting deposition over a long period of time.

Galena from the Frederickson Lake base metal sulphide deposit gives model ages of 1852–1999 Ma. The Western Superior model age is consistent with a syngenetic interpretation for the galena and with a Menihek age for the host sediments. However, should the host rocks be older than the Menihek Formation, as suggested by a zircon age from a gabbro sill intruding these rocks at Howse Lake (see below), then the model age determined using the "Proterozoic" model might be permissible (if the host rocks were considerably older than the Menihek), or else the age using the Western Superior model represents a minimum age for the host rocks.

Galena from the Howse Lake area gives a model age range of 1793–1935 Ma. This galena is epigenetic and occurs in a vein cutting glomeroporphyritic gabbro. A recently obtained U–Pb zircon age for the host gabbro sill is about 1880 Ma (T. Birkett, personal communication, 1989; also cited in Hoffman (1988)). In view of the zircon age of the sill, the model age for the galena using the Western Superior model is permissible, but the model age using the "Proterozoic" model is not.

The 1704 Ma lead isochron age determined by Bell and Thorpe (1986) for uraniferous phosphorite from the base of the Menihek Formation does not agree with the model ages presented here or with other ages. It is very much younger, for example, than the subjacent Sokoman Formation and is also younger than the age of peak regional metamorphism, i.e., 1769 ± 15 Ma (Goulet *et al.*, 1987). Therefore, the 1704 Ma isochron age almost certainly does not represent the depositional age of the Menihek Formation.

According to Hoffman (1987), the Chioak Formation conglomerates and sandstones are molasse deposits, laid down in a foredeep during crustal readjustments in the foreland of the trough in response to the westward advance of thrust sheets during the Hudsonian Orogeny. The reconstruction illustrated in Figure 5 shows the Chioak as the youngest formation in the western trough. In view of its post-Sokoman stratigraphic position, the age of the Sokoman, and the age of peak regional metamorphism, the Chioak must have been deposited just before regional deformation and metamorphism began to affect the western parts of the trough.

A whole-rock lead isochron age of 1920 ± 100 Ma has been determined by Rohon (1987) for a suite of comagmatic basic igneous rocks from the Willbob Formation and associated gabbros of the southern trough (Fig. 2). The Willbob volcanics have been considered to be stratigraphically higher than the Menihek Formation (e.g., Baragar, 1967; Wardle and Bailey, 1981). This lead isochron age is, within the given uncertainty limits, in agreement with both the 1852 Ma (Western Superior) and 1999 Ma ("Proterozoic") model ages for the Frederickson Lake galena situated near the top of a sequence generally considered to be the Menihek.

The few U–Pb zircon dates available from the Labrador Trough provide some indication of the accuracy of the model lead ages. The 1880 ± 2 Ma age (Chevé and Machado, 1988) for the Sokoman Formation shows, using either the Western Superior or the “Proterozoic” model, that galenas from stratigraphically lower levels (Bertin Lake, Du Chambon Lake, Limestone Falls) give model ages greater than the age of the Sokoman. This is true even though the Bertin Lake and Limestone Falls galenas are epigenetic. Conversely, syngenetic galena from Frederickson Lake, using the Western Superior model, gives a model age younger than the age of the Sokoman Formation; using the “Proterozoic” model, the same galena gives a model age greater than that of the Sokoman. If the host rocks belong to the Menihek Formation, as generally thought, then only the Western Superior model age is permissible. However, if it is considered possible that the host rocks are older than the Sokoman, then either model age would be permissible.

The Western Superior model age (1992 Ma) for galena in the Abner is consistent with known stratigraphic and U–Pb zircon data. The 2170 Ma model age given by the “Proterozoic” model, on the other hand, is probably too old, in view of the 2142^{+4}_{-2} Ma zircon age for the Mistamisk Formation. The latter is likely correlative with the upper part of the Bacchus Formation; thus, it should be older than the Abner Formation.

The “Proterozoic” model gives an age of 2092 Ma for the Boylen (Koke) massive sulphide deposit at the top of the Baby Iron Formation. The latter has, from regional stratigraphic considerations (Clark, 1988), been correlated with the Sokoman Iron Formation. The “Proterozoic” model age is thus about 210 Ma too old, judging by the U–Pb date for the Sokoman; alternatively, the Sokoman and Baby iron formations are not equivalent. The Western Superior model age (1930 Ma) is much closer to the U–Pb age, but is also too great (by 50 Ma).

Finally, the “Proterozoic” model gives an age (1935 Ma) for epigenetic galena that is greater than that of the host gabbro (about 1880 Ma) at Howse Lake. The 1793 Ma model age by the Western Superior model is consistent with a metamorphic origin for this galena.

Overall, the Western Superior model gives ages more consistent with known stratigraphic relations and available U–Pb zircon ages than does the “Proterozoic” model.

The Scandinavian connection

The Labrador Trough leads are very similar to one set of data for Proterozoic deposits and occurrences, that provided by Vaasjoki (1981) for leads from the Karelian schists of Finland. For the most part, these leads are comparable, in terms of source μ values, to those from the Labrador Trough, except that the Runkaus and Mieslahti leads were derived from sources with even higher μ values (Fig. 4). The Hammaslahti composition is similar to that reported in this study for a specimen from Du Chambon Lake; Mosku, Tikkala, and Kolmisoppi are similar to St. Pierre and Canal Lake; and Niittylahti and Suoranta are similar to Boylen.

It is of interest to pursue some aspects of the isotopic similarities noted above between the Labrador Trough and Karelian sequences of northern Europe a little further for two reasons: (1) because there are significant parallels in the lithologies, in some aspects of the stratigraphic relations, and in the tectonic environments in which sequences were deposited and subsequently deformed in the two regions, and (2) because more geochronological data are available for the Karelian rocks.

The Karelian rocks of the Svecokarelide belt extend for about 850 km along the southwestern margin of the Archean Baltic Shield from the U.S.S.R., through Finland, to northern Sweden. The nomenclature used for Karelian sequences is not consistent, especially from region to region in Finland and Karelia, U.S.S.R. According to one general scheme, these are subdivided into the Sariola, Jatuli, and Kaleva groups. The craton is overlain by the Sariola Group, a sequence consisting of immature to moderately mature continental arkose and conglomerate that both overlie and underlie a unit of mafic volcanic rock. The succeeding Jatuli Group consists of more mature quartzite and conglomerate, and has shallow-water marine sediments, including dolomite, pelite, black schist, and iron formation, in its upper part. Metavolcanic rocks and diabase are widely intercalated with the sedimentary rocks (Simonen, 1980). Jatuli Group rocks are overlain by turbidites (greywacke and pelite) and ophiolite-like ultramafic–mafic complexes of the Kaleva Group. Most of the Karelian succession is thought to be intracratonic and of anorogenic origin. The Jatulian rocks were likely deposited during rifting of the Archean craton

(Gaál, 1982). To the southwest, the Karelian rocks are succeeded by those of the Svecofennian province, consisting of orogenic-type sedimentary and volcanic rocks deposited in an island-arc setting (Gaál and Gorbatschev, 1987).

The age range for Karelian sedimentation has been moderately well defined. The Sariola rocks unconformably overlie mafic layered complexes that have ages of about 2430–2450 Ma (Simonen, 1980), which establishes an older limit. A Pb–Pb age of 2455 ± 45 Ma for felsic volcanic rocks of the Sumi Group, considered to underlie the Sariola Group, in Karelia, U.S.S.R. (Negrutza and Negrutza, 1981), is in accord with this maximum age restriction. As noted by Laajoki (1986), at Kuusamo a zircon U–Pb age for a quartz porphyry clast indicates that Karelian deposition began there after 2404 Ma. Lava from the lower part of the Kurkikylä Group, equivalent to the rocks of the Sumi Group and/or Sariola Group, has yielded a Pb–Pb age of 2271 ± 93 Ma (Sakko in Strand, 1988). If the glaciogenic sedimentary rocks of the Urkkavaara Formation, which form the lower part of the upper Sariola Group (Marmo and Ojakangas, 1984), are time equivalent to any of the glacially generated units (Ramsay Lake, Bruce, and Gowganda) within the Huronian Supergroup in Canada, as suggested by Marmo and Ojakangas, then a minimum age of 2219 Ma is indicated (Corfu and Andrews, 1986) for this unit. The age of the base of the Jatuli Group must be at least as great as about 2250 Ma because (1) a Pb–Pb isochron age of 2331 ± 98 Ma has been reported (Laajoki, 1980) for metavolcanic rocks from its basal part in the Puolanka area of Finland; (2) stromatolitic dolomite within its upper part has yielded a Pb–Pb isochron age of 2300 ± 140 Ma (Salop, 1983); (3) U–Pb zircon ages of Jatulian metavolcanic rocks and metadiabases are in the age range 2000–2200 Ma (Simonen, 1980); and (4) diabase dykes in eastern Finland and the U.S.S.R. that cut Jatulian rocks yield U–Pb zircon ages of 2160–2180 Ma (Sakko, 1971; Salop, 1983). The Pääkkö iron formation in the upper part of the Jatuli Group has yielded a Pb–Pb isochron age of 2080 ± 45 Ma (Sakko and Laajoki, 1975). Huhma (1986) obtained a Sm–Nd isochron age of 2090 ± 70 Ma for Jatulian basalts from the Jouttiaapa Formation in northern Finland. The sedimentary rocks of the Kaleva Group are younger than about 2100 Ma (Huhma, 1986), and they have a minimum age of about 1925 Ma, the age of the oldest plutons emplaced during the Svecokarelian Orogeny (Welin, 1987). The upper part of the Kaleva sequence contains distal turbidites and slices of ophiolites about 1970 Ma old (Huhma, 1986; Kontinen, 1987). In summary, the Sariola Group may be about 2400-2300 Ma, the Jatuli Group about 2300–2100 Ma, and the Kaleva Group about 2100–1925 Ma in age.

Elsewhere in Scandinavia and the U.S.S.R., a sequence of immature sedimentary and volcanic rocks (Lower Lapponian Group) overlying the Baltic Shield in central Lapland has been assigned an age of 2500–2600 Ma (Gaál and Gorbatschev, 1987), thus extending to an even earlier time the onset of clastic sedimentation. The overlying Upper Lapponian Group rocks are considered to be 2300–2500 Ma old (Perttunen in Sjostrand *et al.*, 1986). Farther south in Finland and the U.S.S.R., arkoses, conglomerates, and associated volcanic rocks of the Sumi Group are also thought to be 2300–2500 Ma old (Gaál and Gorbatschev, 1987).

The minimum age for Karelian sedimentation is perhaps best established by ages of 1870–1930 Ma for the Svecokarelian orogeny, with the oldest of these ages being for plutons from near the Archean craton in eastern Finland (Welin, 1987). According to Gaál and Gorbatschev (1987), however, the "Karelian" granites are outlying manifestations of Svecofennian plutonism associated with the development of new continental Svecofennian crust at 1930–1870 Ma outboard from the Archean craton and its Karelian cover sequences.

Karelian lead isotope analyses by Vaasjoki (1981) and Sakko and Laajoki (1975), as shown in Figure 4, are apparently all for epigenetic sulphide occurrences. The most primitive Runkaus analysis is for a vein in albite diabase beneath Karelian quartzite. The other Runkaus analysis and that for Mieslahti represent occurrences in Sariola Group rocks. The Salmijärvi, Mosku, and Tikkala leads are hosted by Jatuli Group rocks, and the Hammaslahti and Suoranta leads are apparently hosted by Kaleva Group rocks. Because of the epigenetic character of these leads, the model ages can be viewed as providing only minimum ages for the hosting sequences.

Model ages for these Karelian leads range from 2439 Ma (carbonate–quartz–sulphide vein cutting Salmijärvi dolomite) to 2070 (Suoranta quartz vein) by the "Proterozoic" model, and from 2252 to 1900 Ma by the Western Superior model. The model ages for the Hammaslahti breccia/shear zone copper deposit are too great by both models, suggesting that its stratigraphic assignment may be questionable. Other model ages by the Western Superior model are permissible, but that by the "Proterozoic" model for Salmijärvi is too great, and that for Suoranta may also be too great. This suggests that the Western Superior model is more suitable for Karelian leads, as was also concluded for the Labrador Trough leads, than a model based primarily on sediment-hosted massive sulphide deposits of Proterozoic age in Australia.

CONCLUSIONS

The relative chronology suggested by our model lead ages is probably correct, but the absolute ages may require modification as improvements are made to lead evolution models. Most of our results give at least minimum ages for the lead mineralizations and for their host rocks.

The calculated Th/U ratios for the lead sources suggest that several of our leads may have experienced only a single-stage evolution history (Bertin Lake, Du Chambon Lake, Boylen (Koke), Frederickson Lake, and Howse Lake leads). More complex histories are possible for the St. Pierre, Canal Lake, and Limestone Falls leads. Addition of uranogenic lead has resulted in an anomalous model age for the Romanet River specimen.

The Western Superior model appears to give results more consistent with known stratigraphic relations and available U–Pb zircon ages than the "Proterozoic" model. Also, it should be noted that early estimates of the age of the Sokoman Formation by K–Ar and Rb–Sr methods gave results that are surprisingly close to the recently determined U–Pb zircon age.

With respect to Labrador Trough stratigraphy, our model lead ages suggest or are consistent with the following major points: (1) rocks at least as old as 2300 Ma probably exist in the Labrador Trough, based on the model age of galena from the vein at Bertin Lake; (2) the Mistamisk Formation is equivalent to the upper part of the Bacchus Formation; (3) the third depositional cycle in the northern trough does not exist, and is an artifact of regional thrusting and stratigraphic repetition; (4) the Abner Dolomite is older than the Sokoman Iron Formation and is probably equivalent to the Denault Dolomite of the southern trough; (5) a difference of at least 110 Ma is indicated between the age of the Abner (Denault) Dolomite and the Sokoman Iron Formation; this difference may correspond to the time span represented by the unconformity at the top of the Denault in the western trough; (6) regional stratigraphic work suggests that the Sokoman and Baby iron formations are correlative; however, the lead models used in the present study give contradictory results concerning their time equivalence; and (7) based on the spread of our model ages, two major episodes of deposition in the Labrador Trough spanned a period of perhaps as much as 450 Ma, in comparison with a minimum span of 260 Ma based on two previously determined U–Pb zircon ages.

The lead isotope data for deposits and occurrences in the Labrador Trough are comparable to those for leads from the Sariolan, Jatulian, and Kalevan sequences of Early Proterozoic age in Finland (Sakko and Laajoki, 1975; Vaasjoki, 1981). Basement–cover and sedimentary–volcanic sequences in the two regions show similarities in lithologic characteristics, tectonic settings, deformational history and style, and the probable derivation of lead predominantly from recycled material from underlying and subjacent Archean terranes.

ACKNOWLEDGEMENTS

We are grateful to M. Bélanger, R.T. Bell, T. Birkett, and S. Chevé for kindly providing samples. Clark acknowledges numerous fruitful discussions with M. Bélanger on the stratigraphy of the Labrador Trough, and with N. Goulet and R. Wares on the tectonics and stratigraphy of the northern trough. Critical reading of the manuscript by M. Bélanger, R.T. Bell, T. Birkett, and D.F. Sangster led to significant improvements, for which we extend our gratitude.

REFERENCES

Afifi, A., Doe, B.R., Sims, P.K. and Delevaux, M.H., 1984, U–Th–Pb isotope chronology of sulfide ores and rocks in the Early Proterozoic metavolcanic belt of northern Wisconsin: Economic Geology, v. 79, p. 338–353.

Armstrong, R.C., 1976, Koke project, Cominco: Ministère de l'Énergie et des Ressources, Québec, Report GM-32180.

Baragar, W.R.A., 1967, Wakuach Lake map-area, Quebec–Labrador: Geological Survey of Canada, Memoir 344.

Beall, G.H., Hurley, P.M., Fairbairn, H.W. and Pinson, W.H., 1963, Comparison of K–Ar and whole-rock Rb–Sr dating in New Quebec and Labrador: American Journal of Science, v. 261, p. 571–580.

Bélanger, M., 1982, Région du lac Faujas, Nouveau-Québec: Ministère de l'Énergie et des Ressources, Québec, Report DP 82-06.

Bell, R.T. and Ruzicka, V., 1985, Uranium in the Circum-Ungava belt, northern Quebec and Labrador: new information from the central Labrador Trough: Geological Survey of Canada, Paper 85-1A, p. 145–149.

Bell, R.T. and Thorpe, R.I., 1986, Pb–Pb isochron age of uraniferous phosphorite at the base of the Menihek Formation, Labrador Trough: Geological Survey of Canada, Paper 86-1B, p. 585–589.

Bérard, J., 1965, Région du lac Bérard, Nouveau-Québec: Ministère des Richesses naturelles, Québec, Report RG-111.

Budkewitsch, P., 1986, A structural study of the Chioak–Abner formation contact, northern part of the Labrador Trough, New Quebec, unpublished B.Sc. thesis, Concordia University, Montreal, 61 p.

Chevé, S., 1985, Les indices minéralisés du lac Romanet, Fosse du Labrador: Ministère de l'Énergie et des Ressources, Québec, Report ET 83-13.

Chevé, S., 1987, Le complexe carbonatitique du lac Castignon, Fosse du Labrador: Ministère de l'Énergie et des Ressources, Québec, Report DP 87-10.

Chevé, S. and Machado, N., 1988, Reinvestigation of the Castignon Lake Carbonatite Complex, Labrador Trough, New Quebec: Geological Association of Canada – Mineralogical Association of Canada, Program with Abstracts, v. 13, p. A20.

Clark, T., 1978, Région du lac Hérodier, Nouveau-Québec: Ministère de l'Énergie et des Ressources, Québec, Report DPV-568.

Clark, T., 1979, Région du lac Napier, Territoire du Nouveau-Québec: Ministère des Richesses naturelles, Québec, Report DPV-663.
Clark, T., 1984, Géologie de la région du lac Cambrien, Territoire du Nouveau-Québec: Ministère de l'Énergie et des Ressources, Québec, Report ET 83-02.
Clark, T., 1986, Géologie et minéralisations de la région du lac Mistamisk et de la rivière Romanet: Ministère de l'Énergie et des Ressources, Québec, Report ET 83-22.
Clark, T., 1988, Stratigraphie, pétrographie et pétrochimie de la formation de fer de Baby dans la région du lac Hérodier, Fosse du Labrador: Ministère de l'Énergie et des Ressources, Québec, Report ET 87-13.
Clark, T., Bélanger, M. and Giovenazzo, D., 1986, Métaux précieux au Nouveau-Québec/Precious metals in New-Québec: Ministère de l'Énergie et des Ressources, Québec, Direction générale de l'Exploration géologique et minérale, PRO-6.
Corfu, F. and Andrews, A.J., 1986, A U–Pb age for mineralized Nipissing diabase, Gowganda, Ontario: Canadian Journal of Earth Sciences, v. 23, p. 107–109.
Deb, M., Thorpe, R.I., Cumming, G.L. and Wagner, P.A., 1989, Age, source and stratigraphic implications of lead isotope data for conformable sediment-hosted base metal deposits in the Proterozoic Aravalli–Delhi orogenic belt, northwestern India: Precambrian Research, v. 43, p. 1–22.
Dimroth, E., 1970a, Evolution of the Labrador Geosyncline: Geological Society of America Bulletin, v. 81, p. 2717–2742.
Dimroth, E., 1970b, Meimechites and carbonatites of the Castignon Lake Complex, New Quebec: Neues Jahrbuch für Mineralogie, Abhandlungen, v. 112, p. 239–278.
Dimroth, E., 1978, Région de la Fosse du Labrador (54°30′-56°30′): Ministère de l'Énergie et des Ressources, Québec, Report RG-193.
Dimroth, E., Baragar, W.R.A., Bergeron, R. and Jackson, G.D., 1970, The filling of the Circum-Ungava geosyncline, in Baer, A.J., ed., Symposium on basins and Geosynclines of the Canadian Shield: Geological Survey of Canada, Paper 70-40, p. 45–142.
Dressler, B., 1975, Lamprophyres of the north-central Labrador Trough, Quebec, Canada: Neues Jahrbuch für Mineralogie, Monatshefte, v. 6, p. 245–290.
Dressler, B., 1979, Région de la Fosse du Labrador (56°30′-57°15′): Ministère de l'Énergie et des Ressources, Québec, Report RG-195.
Fahrig, W.F., 1965, Lac Hérodier, Québec: Geological Survey of Canada, Map 1146A.
Fournier, D., 1982, Gîtes de Cu–Zn et Cu–Ni dans la partie centrale de la Fosse du Labrador: Ministère de l'Énergie et des Ressources, Québec, Report DPV-929.
Fournier, D., 1983, Gîtes de Cu–Zn et de Cu–Ni de la Fosse du Labrador, Québec, unpublished Doctoral thesis, Université Pierre et Marie Curie, Paris VI, Paris, France.
Franklin, J.M., Roscoe, S.M., Loveridge, W.D. and Sangster, D.F., 1983, Lead isotope studies in Superior and Southern provinces: Geological Survey of Canada, Bulletin 351.
Frarey, M.J., 1967, Willbob Lake and Thompson Lake map-areas, Québec and Newfoundland (23 O/1 and 23 O/8): Geological Survey of Canada, Memoir 348.
Fryer, B.J., 1972, Age determinations in the Circum-Ungava geosyncline and the evolution of Precambrian banded iron-formations: Canadian Journal of Earth Sciences, v. 9, p. 652–663.
Gaál, G., 1982, Proterozoic tectonic evolution and late Svecokarelian plate deformation of the Central Baltic Shield: Geologische Rundschau, v. 71, p. 158–170.
Gaál, G. and Gorbatschev, R., 1987, An outline of the Precambrian evolution of the Baltic Shield: Precambrian Research, v. 35, p. 15–52.
Gebert, J., 1986, Métallogénie des indices de Cu–Zn–Pb–Ag–Au dans la région du lac Frédérickson, Fosse du Labrador: Ministère de l'Énergie et des Ressources, Québec, Report MB 86-20.
Gebert, J., Fox, J.S. and Barrett, T., 1987, A re-interpretation of the Zn–Cu and Ni–Cu deposits of the Frederickson Lake area, central Labrador Trough: Geological Association of Canada – Mineralogical Association of Canada, Program with Abstracts, v. 12, p. 45.
Goulet, N., 1986, Étude tectonique et stratigraphique de la partie nord de la Fosse du Labrador, région de la baie aux Feuilles et du lac Bérard: Ministère de l'Énergie et des Ressources, Québec, Report MB 86-27.
Goulet, N., Gariépy, C., Mareschal, J.-C. and Machado, N., 1987, Structure, geochronology, gravity and the tectonic evolution of the northern Labrador Trough: Geological Association of Canada – Mineralogical Association of Canada, Program with Abstracts, v. 12, p. 48.
Gulson, B.L., 1984, Uranium–lead and lead–lead investigations of minerals from the Broken Hill lodes and mine sequence rocks: Economic Geology, v. 79, p. 476–490.
Gulson, B.L., 1985, Shale-hosted lead–zinc deposits in northern Australia: lead isotope variations: Economic Geology, v. 80, p. 2001–2012.
Gulson, B.L., Porritt, P.M., Mizon, K.J. and Barnes, R.G., 1985, Lead isotope signatures of stratiform and strata-bound mineralization in the Broken Hill Block, New South Wales, Australia: Economic Geology, v. 80, p. 488–496.
Hoffman, P.F., 1987, Early Proterozoic foredeeps, foredeep magmatism and Superior-type iron-formations of the Canadian Shield, in Kroner, A., ed., Proterozoic Lithospheric Evolution: American Geophysical Union Geodynamics Series, v. 17, p. 85–98.
Hoffman, P.F., 1988, United plates of America, the birth of a craton: Early Proterozoic assembly and growth of Laurentia: Annual Review of Earth and Planetary Sciences, v. 16, p. 543–603.
Hoffman. P.F. and Grotzinger, J.P., 1988, Abner/Denault reef complex (2.1 Ga), Labrador Trough, N.E. Quebec, in Geldsetzer, H.H.J., James, N.P. and Tebbutt, G.E., eds., Reefs – Canada and Adjacent Areas: Canadian Society of Petroleum Geologists, Memoir 13.
Huhma, H., 1986, Sm–Nd, U–Pb and Pb–Pb isotopic evidence for the origin of the Early Proterozoic Svecokarelian crust in Finland: Geological Survey of Finland, Bulletin 337, 48 p.
Kish, L. and Cuney, M., 1981, Uraninite–albite veins from the Mistamisk valley of the Labrador Trough, Quebec: Mineralogical Magazine, v. 44, p. 471–483.
Kontinen, A., 1987, An early Proterozoic ophiolite – the Jormua mafic–ultramafic complex, northeastern Finland: Precambrian Research, v. 35, p. 313–341.
Laajoki, K., 1980, Karjalaisten Muodostumien Alaosat Puolangan Tutkimusten Valossa: Geologi, v. 32 (4–5), p. 45–49.
Laajoki, K., 1986, The Precambrian supracrustal rocks of Finland and their tectono-exogenic evolution: Precambrian Research, v. 33, p. 67–85.
LeCouteur, P.C., 1973, A study of lead isotopes from mineral deposits in southeastern British Columbia and in the Anvil Range, Yukon Territory, unpublished Ph.D. thesis, University of British Columbia, 142 p.
LeCouteur, P.C., 1979, Age of the Sullivan lead–zinc deposit: Geological Association of Canada, Cordilleran Section, Abstracts, p. 19.
LeGallais, C.J. and Lavoie, S., 1982, Basin evolution of the Lower Proterozoic Kaniapiskau Supergroup, central Labrador Miogeosyncline (Trough), Québec: Bulletin of Canadian Petroleum Geology, v. 30, p. 150–166.
Lowden, J.A., Stockwell, C.H., Tipper, H.W. and Wanless, R.K., 1963, Age determinations and geological studies: Geological Survey of Canada, Paper 62-17, p. 100–103.

Marmo, J.S. and Ojakangas, R.W., 1984, Lower Proterozoic glaciogenic deposits, eastern Finland: Geological Society of America, Bulletin 95, p. 1055–1062.

Negrutza, T.F. and Negrutza, V.Z., 1981, Early Proterozoic Sarioli tilloids in the eastern part of the Baltic Shield, U.S.S.R., in Hambrey, M.J. and Harland, W.B., eds., Earth's Pre-Pleistocene Glacial Record: Cambridge University Press, p. 683–686.

Page, R.W., 1981, Depositional ages of the stratiform base metal deposits at Mount Isa and McArthur River, Australia, based on U–Pb zircon dating of concordant tuff horizons: Economic Geology, v. 76, p. 648–658.

Richards, J.R., 1986, Lead isotope signatures: further examination of comparisons between South Africa and Western Australia: Transactions of the Geological Society of South Africa, v. 89, p. 285–304.

Rohon, M.-L., 1987, Indices de Cu–Ni et de pyrite–pyrrhotite entre les lacs Retty et Low, Fosse du Labrador: Ministère de l'Énergie et des Ressources, Québec, Report MB 87-44.

Sakko, M., 1971, Varhais-Karjalaisten metadiabaasien radiometrisiä zirconi-ikiä (Radiometric zircon ages on the Early Karelian metadiabases): Geologi, v. 23, p. 117–118.

Sakko, M. and Laajoki, K., 1975, Whole rock Pb–Pb isochron age for the Paakko iron formation in Vayrylankyla, south Puolanka area, Finland: Bulletin of the Geological Society of Finland, v. 47, p. 113–116.

Salop, L.J., 1983, Geological evolution of the Earth during the Precambrian: Springer-Verlag, 459 p.

Sauvé, P. and Bergeron, R., 1965, Région des lacs Gerido et Thévenet, Nouveau-Québec: Ministère des Richesses naturelles, Québec, Report RG-104.

Simonen, A., 1980, The Precambrian in Finland: Geological Survey of Finland, Bulletin 304, 58 p.

Sjostrand, T., Often, M., Perttunen, V. and Krill, A., 1986, Archaean and Proterozoic geology in northern Finland, Norway and Sweden: 7th. IAGOD Symposium and Nordkalott Project Meeting, Sveriges Geologiska Undersokning, Ca. 66, p. 1–84, preliminary edition.

Stockwell, C.H., 1982, Proposals for time-classification and correlation of Precambrian rocks and events in Canada and adjacent areas of the Canadian Shield. Part 1. A time classification of Precambrian rocks and events: Geological Survey of Canada, Paper 80-19.

Strand, K., 1988, Alluvial sedimentation and tectonic setting of the Early Proterozoic Kurkikylä and Kainuu groups in northern Finland, in Laajoki, K. and Paakkola, J., eds., Sedimentology of the Precambrian Formations in Eastern and Northern Finland: Geological Survey of Finland, Special Paper 5, p. 75–90.

Thorpe, R.I., 1982, Lead isotope evidence regarding Archean and Proterozoic metallogeny in Canada: Revista Brasileira de Geociencias, v. 12 (1–3), p. 510–521.

Thorpe, R.I., Guha, J., Franklin, J.M. and Loveridge, W.D., 1984, Use of the Superior Province lead isotope framework in interpreting mineralization stages in the Chibougamau District, in Guha, J. and Chown, E.H., eds., Chibougamau – Stratigraphy and Mineralization: Canadian Institute of Mining and Metallurgy, Special Volume 34, p. 496–516.

Thorpe, R.I., Goodz, M.D., Jonasson, I.R. and Blenkinsop, J., 1986, Lead-isotope study of mineralization in the Cobalt district, Ontario: Canadian Journal of Earth Sciences, v. 23, p. 1568–1575.

Vaasjoki, M., 1981, The lead isotopic composition of some Finnish galenas: Geological Survey of Finland, Bulletin 316, 30 p.

Van Schmus, W.R., 1980, Chronology of igneous rocks associated with the Penokean orogeny in Wisconsin, in Morey, G.B. and Hanson, G.N., eds., Selected Studies of Archean Gneisses and Lower Proterozoic Rocks, Southern Canadian Shield: Geological Society of America, Special Paper 182, p. 159–168.

Wanless, R.K., Stevens, R.D., Lachance, G.R. and Rimsaite, R.Y.H., 1965, Age determinations and geological studies. Part 1. Isotopic ages, report 5: Geological Survey of Canada, Paper 64-17.

Wardle, R.J. and Bailey, D.G., 1981, Early Proterozoic sequences in Labrador, in Campbell, F.H.A., ed., Proterozoic Basins of Canada: Geological Survey of Canada, Paper 81-10 (supplement), p. 331–358.

Wares, R. and Berger, J., 1987, Synthèse métallogénique des indices de sulfures au nord du 57e parallèle, Fosse du Labrador: Ministère de l'Énergie et des Ressources, Québec, Report MB 87-19.

Wares, R. and Goutier, J., 1989, Synthèse métallogénique des indices de sulfures au nord du 57e parallèle, Fosse du Labrador (Étape II): Ministère de l'Énergie et des Ressources, Québec, Report MB 89-38.

Wares, R., Berger, J. and St-Seymour, K., 1988, Synthèse métallogénique des indices de sulfure au nord du 57e parallèle, Fosse du Labrador (Étape I): Ministère de l'Énergie et des Ressources, Québec, Report MB 88-05.

Welin, E., 1987, The depositional evolution of the Svecofennian supracrustal sequence in Finland and Sweden: Precambrian Research, v. 35, p. 95–113.

Zartman, R.E., Peterman, Z.E., Obradovich, J.D., Gallego, M.D. and Bishop, D.T., 1982, Age of the Crossport C sill near Eastport, Idaho, in Reid, R.R. and Williams, G.A., eds., Society of Economic Geologists' Coeur d'Alene Field Conference, Idaho, 1977: Idaho Bureau of Mines and Geology, Bulletin 24, p. 61–69.

Timing of Collisional Events in the Trans-Hudson Orogen: Evidence from U–Pb Geochronology for the New Quebec Orogen, the Thompson Belt, and the Reindeer Zone (Manitoba and Saskatchewan)

N. Machado[1]

Department of Geology, Royal Ontario Museum, 100 Queen's Park, Toronto, Ontario M5S 2C6, Canada

Machado, N., Timing of collisional events in the Trans-Hudson Orogen: evidence from U–Pb geochronology for the New Quebec Orogen, the Thompson Belt, and the Reindeer Zone (Manitoba and Saskatchewan), in Lewry, J.F. and Stauffer, M.R., eds., The Early Proterozoic Trans-Hudson Orogen of North America: Geological Association of Canada, Special Paper 37, p. 433–441.

Abstract

Comparison of available U–Pb ages for the New Quebec Orogen, the Thompson Belt in Manitoba, and the Reindeer Zone in Manitoba and Saskatchewan leads to the following conclusions: (1) Magmatic activity was well established in the internal zone of the Trans-Hudson Orogen and along the western and eastern margins of Superior Province by 1890–1870 Ma. (2) Rae–Superior collision, dated at 1840–1829 Ma in the northern part of the New Quebec Orogen, is slightly younger (5 Ma) than accretion of the Reindeer Zone to the Archean Hearne Province. (3) The youngest magmatism in the Flin Flon Belt (1836–1830 Ma) is contemporaneous with granitic plutonism in the Kisseynew Gneiss Belt and coeval with Rae–Superior collision. (4) In the Thompson Belt, granitic magmatism at 1822 Ma and high-grade metamorphism at 1809 Ma are approximately coeval with peak metamorphism (1815 Ma) and granitoid emplacement in the Kisseynew Belt and represent an early phase of collision between the Reindeer Zone and the Superior Province. (5) Amphibolite facies metamorphism is coeval in the New Quebec Orogen and the Thompson Belt, at 1793–1769 Ma. In addition, both belts experienced similar thermal regimes for 50–60 Ma. (6) Although in the New Quebec Orogen the last Hudsonian manifestations are low temperature hydrothermal activity (1746–1740 Ma), in the Thompson Belt they are marked by granitoid intrusion and renewed amphibolite facies metamorphism at 1726–1720 Ma, possibly representing a second phase of the Reindeer–Superior collision. (7) The absence of U–Pb ages around 1720 Ma in the Reindeer Zone suggests that the latest episodes of collision with the Superior craton are recorded in the Thompson Belt. Together with the relatively young age of the Flin Flon Belt, this might indicate an eastward or southeastward migration of thermotectonic activity in the western Trans-Hudson Orogen.

[1]Present address: GEOTOP, Université du Québec à Montréal, C.P. 8888, Succ. A, Montréal, Québec H3C 3P8, Canada.

Résumé

Les âges U–Pb disponibles pour l'orogène du Nouveau Québec, la ceinture de Thompson et la Zone de Reindeer appartenant à la chaîne Trans-Hudsonienne permettent d'arriver aux conclusions suivantes: (1) L'activité magmatique dans la Zone de Reindeer et dans les marges continentales est et ouest de la Province du Supérieur s'étend de 1890 à 1870 Ma. (2) La collision entre les provinces du Supérieur et celle de Rae a été datée à 1840–1829 Ma dans la partie nord de l'orogène du Nouveau Québec et a donc eu lieu peu après (5 Ma) la collision entre la Zone de Reindeer et la Province de Hearne. (3) Le magmatisme le plus jeune de la ceinture de Flin Flon (1836–1830 Ma) est contemporain du plutonisme granitique de la ceinture de Kisseynew et de la collision Supérieur–Rae. (4) Dans la ceinture de Thompson, le magmatisme granitique le plus ancien (1822 Ma) et le métamorphisme granulitique (1809 Ma) sont contemporains du métamorphisme et de l'emplacement de magmas anatectiques dans la ceinture de Kisseynew; ils pourraient représenter la première phase de collision entre la Zone de Reindeer et la Province du Supérieur. (5) L'orogène du Nouveau Québec et la ceinture de Thompson ont subi en même temps un métamorphisme au facies amphibolite. De plus, ces régions ont subi des régimes thermiques identiques pendant 50–60 Ma. (6) Alors que la dernière phase de l'orogenèse hudsonienne au Nouveau Québec est représentée par un événement hydrothermal daté à 1746–1740 Ma, le dernier événement reconnu dans la ceinture de Thompson est caractérisé par l'intrusion de granitoides associée à un métamorphisme au faciès amphibolite datés à 1726–1720 Ma. Ces phénomènes sont probablement l'expression d'une deuxième phase de collision entre la Zone de Reindeer et la Province du Supérieur et représenteraient des ajustements le long de la zone de suture. (7) L'absence d'âges U–Pb autour de 1720 Ma dans la Zone de Reindeer laisse supposer que les derniers épisodes de convergence entre la Zone de Reindeer et la Province du Supérieur ont été enregistrées dans la ceinture de Thompson. Ce dernier fait ainsi que les âges relativement plus jeunes de la ceinture de Flin Flon suggèrent une migration vers le sud ou le sud-est de l'activité thermotectonique du secteur occidental de l'orogène Trans-Hudsonien.

INTRODUCTION

The North American Shield comprises Archean cratons sutured along a system of Early Proterozoic orogenic belts of which the Trans-Hudson Orogen is the main component (Hoffman, 1988; Lewry and Collerson, this volume). This orogen extends from South Dakota to Saskatchewan, Manitoba, and across Hudson Bay into Quebec and Labrador, and includes areas traditionally incorporated in the Circum-Superior Belt (Baragar and Scoates (1981) and references therein), such as the Thompson Nickel Belt, Fox River Belt, Sutton Inlier, Belcher and Ottawa islands, Cape Smith Belt, and Labrador Trough.

Recent increases in geological, geochemical, and geochronological data bases indicate that the Trans-Hudson Orogen evolved in a collisional plate tectonic regime (Lewry and Collerson, this volume), similar to that prevailing in the Phanerozoic, and suggest that different segments underwent specific tectonic histories within the same time frame. The aim of this study is to compare the evolution of three sectors of the Trans-Hudson Orogen to correlate major tectonic events across the orogen. U–Pb ages for the Thompson Belt (Manitoba), the adjacent Reindeer Zone (Manitoba and Saskatchewan), and the Labrador Trough (Quebec) are presented and similar Hudsonian events are equated.

NEW QUEBEC OROGEN

This orogen is an Early Proterozoic foreland fold and thrust belt bordering the eastern margin of the Archean Superior Province. In the west, the belt comprises autochthonous platformal quartzites, carbonates, iron formation, and shales lying unconformably on the Archean foreland (Fig. 1a; Dimroth (1981) and references therein). To the east, a series of thrust sheets include thick sequences of tholeiitic flows and sills intercalated with sediments in part correlative with the autochthonous sequences. This is followed farther to the east by extensive metapelites (Laporte schists). In the central sector of the orogen, the whole sequence has been interpreted as two successive foredeeps developed on initial rift and passive margin sediments (Hoffman, 1987).

In the northern sector of the orogen, north of Koksoaq River (Fig. 1b), the Laporte schists occupy tight synclines between Archean gneiss domes (Moyer, Renia, and Boulder). This area is bounded by faults parallel to the general orientation of the belt: the Lake Rachel fault to the west and the Lake Olmstead – Lake Turcotte fault zone in the east (Goulet *et al.*, 1987; Perrault *et al.*, 1987). The southward extension of the latter (here informally called the Laporte fault) marks the boundary between the Laporte schists and remobilized Archean gneisses. The hinterland farther east is a complex zone of Archean gneiss domes in the west (Turcotte and Olmstead) and metasedimentary rocks of uncertain origin, cut by pegmatites and metatonalitic intrusions (Stewart and Elbow Island) near Kuujjuaq (Perrault *et al.*, 1987).

Metamorphic grade increases eastward (Dimroth, 1972; Dimroth and Dressler, 1978), reaching granulite facies at Kuujjuaq (Perrault *et al.*, 1987). In the area of the Archean domes, the Laporte schists reached amphibolite facies and the domal inliers were retrogressed to the same facies. Machado *et al.* (1989a) show that the granulite

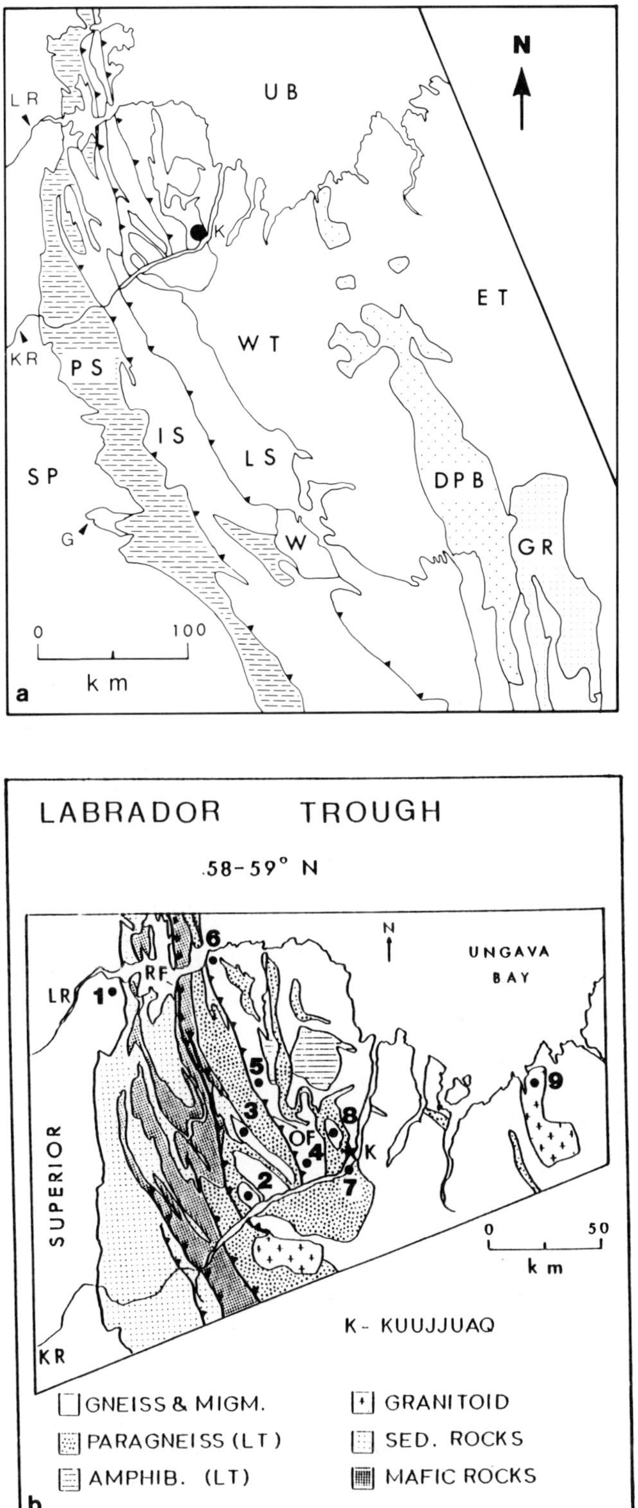

facies metamorphism is 2883–2868 Ma old in the domes located between the Lake Rachel and the Lake Olmstead faults, but ca. 2720 Ma in the domes immediately to the east of the latter. These authors suggest that fault-bounded basement blocks of differing Archean histories were uplifted and retrogressed during the Hudsonian orogeny. The granulite facies metamorphism in the Kuujjuaq area has been dated at 1833–1829 Ma (Machado *et al.*, 1988).

Three phases of deformation have been recognized in the area between the Lake Rachel fault and Kuujjuaq (Perrault *et al.*, 1987; Poirier *et al.*, this volume), followed by a postmetamorphic dextral wrench fault system (Goulet *et al.*, 1987). Peak metamorphic pressures of 0.8 GPa were followed by a thermal peak of about 600°C (Perrault *et al.*, 1987). The Hudsonian tectonic–metamorphic activity lasted from 1840 to 1740 Ma (Machado *et al.*, 1988, 1989a) as discussed below.

In some areas of the Trans-Hudson Orogen, such as the eastern sector of the orogen, the nomenclature is in a state of flux (Lewry and Collerson, this volume). The area between the Laporte fault and the Nain Province has been named the Tunulic Subprovince of the Churchill Province by Stockwell (1982), is included in the Archean Rae Province by Hoffman (1988), and corresponds to the Labrador Segment (subdivided into Eastern, Central, and Western divisions) of Wardle *et al.* (this volume). Because this poorly mapped area comprises reworked Archean gneisses and Early Proterozoic rocks of diverse origin and uncertain significance, and to avoid confusion with "Labrador Orogen" (Thomas *et al.*, 1985) and "Trans-Labrador

Figure 1. *(a) Schematic geological map of part of the Ungava Segment of the Trans-Hudson Orogen. Legend: DPB, De Pas Batholith; ET, George River Zone; G, Castignon Lake; GR, George River Domain; IS, igneous sequence; K, Kuujjuaq; KR, Koksoaq River; LR, Leaf River; LS, Laporte schists; PS, platformal sequence; SP, Superior Province; UB, Ungava Bay; W, Wheeler Dome; and WT, Kuujjuaq Zone. PS, IS, and LS compose the Kaniapiscau Zone. The New Quebec Orogen comprises the Kaniapiscau Zone and at least part of the Kuujjuaq Zone. Adapted from Avramtchev (1985). (b) Schematic geological map of the northern New Quebec Orogen between Koksoaq River (KR) and Leaf River (LR). Legend: 1, Leaf River intrusion; 2 and 3, Moyer and Boulder domes, respectively, with the Renia Dome between them (to the west is the Lake Rachel fault (RF) and to the east the Turcotte–Olmstead fault (OF) system); 4–6, Turcotte, Olmstead, and Leaf Strait domes respectively; 7 and 8, Elbow Island and Lake Stuart intrusions; and 9, northern De Pas Batholith. Adapted from Avramtchev (1985).*

Batholith" (Gower and Owen, 1984), the region is here called the Ungava Segment. It can be subdivided into several major lithotectonic domains or zones (Lewry and Collerson, this volume). From west to east, the Ungava Segment includes the Kaniapiscau Zone (between the Superior Province and the Laporte fault), traditionally known as "Labrador Trough" and part of the New Quebec Orogen as defined by Hoffman (1988), and the Kuujjuaq Zone (between the Laporte fault and De Pas Batholith), which at least in part represents the internides of the New Quebec Orogen. The De Pas Batholith, in the central Ungava Segment, is followed by the George River Zone, bounded to the east by the Abloviak shear zone, and by the Torngat Orogen (Hoffman, 1988).

East of the Kaniapiscau Zone, the Ungava Segment is underlain by paragneisses and remobilized Archean gneisses and is intruded by the north-trending De Pas Batholith, of calc-alkaline affinity (Van der Leeden *et al.*, this volume). In the south, the batholith is associated with metasediments and calc-alkaline metavolcanics and it has been interpreted as being related to a Proterozoic continental collision zone (Van der Leeden *et al.*, this volume). The eastward increase in metamorphic grade and ages of ca. 1810 Ma for granitoid intrusion in the south (T.E. Krogh, personal communication) and for reactivated basement in the north (Machado *et al.*, 1989a) led these authors to suggest that the New Quebec Orogen may represent the external section of an older orogen represented in part by the De Pas Batholith. Further evidence is provided by ages of ca. 2.3 Ga for igneous rocks from the George River Zone (Fig. 1a; Girard *et al.*, 1988). The presence of Archean rocks in the Kuujjuaq Zone indicates that the New Quebec Orogen resulted from convergence between Superior Province and an Archean block of uncertain extent, which, for descriptive convenience, is here referred to as the Rae "craton" (or "platform" as defined by Lewry and Collerson, this volume).

THOMPSON BELT

This belt is a north-northeast trending zone of retrogressed Archean gneisses at the northwestern margin of the Superior Province (Fig. 2a; Bell, 1971; Peredery *et al.*, 1982; Green *et al.*, 1985; Weber, this volume). The Thompson Belt is the western part of the Churchill–Superior Boundary Zone (Weber, this volume). If the "Churchill Province" terminology is to be discontinued, Nelson Zone is a more logical name for the latter, as proposed by Lewry and Collerson (this volume).

Metasedimentary and mafic–ultramafic rocks were tightly infolded with the gneisses during the Hudsonian orogeny. The Setting Lake fault zone marks the western limit of the Thompson Belt (Fig. 2b) where it is in contact with the Kisseynew Domain of the Reindeer Zone. The eastern limit of the Thompson Belt is a poorly defined transitional zone to the Archean Pikwitonei Granulite Domain and the Sachigo Subprovince of the Superior Province. The transition zone is marked by retrogression and transposition of the east-trending Archean granulites to northeast–southwest trending amphibolites. In the central Thompson Belt, the Setting Lake fault splays to the northeast, and farther north this fault system turns eastward (Fig. 2b). The faults and the foliation in both the gneisses and supracrustal rocks dip steeply to the southeast (Bell, 1971; Fueten *et al.*, 1986; Bleeker, this volume).

Two Archean age groups have been recognized in the Thompson Belt gneisses; one at 2.65–2.77 Ga and the other older than 3.0 Ga (Machado *et al.*, 1990). The gneisses were remobilized and intruded by granitic stocks and pegmatites between 1822 and 1720 Ma (Machado *et al.*, 1987) as a result of the collision between the Superior Province and the Reindeer Zone.

TIMING OF HUDSONIAN EVENTS IN THE NEW QUEBEC OROGEN AND THE THOMPSON BELT

The areas of study in the Thompson Belt and in the New Quebec Orogen represent different tectonic settings. Whereas the Thompson Belt is a reworked continental margin of the Trans-Hudson Orogen, the area studied in the New Quebec Orogen is in the hinterland of this orogen. Although this difference introduces some uncertainty in the relatively chronology of precollisional events in the two belts, it becomes less relevant for later events.

Mafic and alkalic magmatism

In the northern New Quebec Orogen, a glomeroporphyritic sill near the top of the Hellancourt Formation was dated at 1874 ± 3 Ma (N. Machado and N. Goulet, unpublished data). In the central sector of the orogen, a carbonatite from the Castignon Lake area has an age of 1880 ± 2 Ma (Chevé and Machado, 1988). These data show the presence of foredeep mafic magmatism and continental alkaline magmatism in the New Quebec Orogen by 1874–1880 Ma.

The age of the mafic–ultramafic rocks of the Thompson Belt is unknown. However, mafic magmatism related to the evolution of this belt is represented by the Fox River sill and the Molson dykes, dated at 1883 ± 2 Ma (Heaman *et al.*, 1986). Therefore, the available data suggest that magmatism is approximately coeval in both belts.

Granitoid intrusion and granulite facies metamorphism

The tonalitic gneisses of Elbow Island and Lake Stewart intruded granulite facies gneisses near Kuujjuaq, in the hinterland of the New Quebec Orogen, at 1840 ± 2 and 1833 ± 2 Ma respectively (Machado *et al.*, 1988). The first intrusion also contains zircon dated at 1829 ± 2 Ma, interpreted as the age of granulite facies metamorphism; in the latter intrusion, the presence of three generations of zircon, all dated at 1833 ± 2 Ma, led those authors to conclude that emplacement coincided with granulite facies metamorphism. These data show that peak metamorphic conditions are coeval with, or slightly younger than, tonalite intrusion. The Elbow Island intrusion contains gneissic xenoliths and abundant Archean zircons, indicating that it represents remobilized Archean basement. The structural characteristics of the Kuujjuaq area, together with the presence of remobilized basement, tonalitic intrusions, and granulite facies metamorphism, indicate that this region is part of a collision zone established at 1840–1829 Ma (Machado *et al.*, 1988).

The Wintering Lake granodiorite, located in the north-central Thompson Belt, close to the boundary with the Pikwitonei Granulite Domain, was dated at 1822 ± 3 Ma (Machado *et al.*, 1987). The abundance of inherited Archean zircon indicates a significant crustal component in the granodioritic magma.

In the Sasagiu Rapids area in the south-central Thompson Belt, granitic sheets conformably to disconformably intrude banded grey gneisses. A U–Pb study revealed that in addition to zircons older than 3.0 Ga, the gneiss also contained 1809 ± 14 Ma zircons, possibly dating high-grade metamorphism (Machado *et al.*, 1990). The granite intruded at 1726 ± 12 Ma and contains abundant inherited zircon 2.65 Ga old.

Granitoid intrusion and high-grade metamorphism are interpreted as resulting from a first collisional stage between the Reindeer Zone and the Superior Province at 1822–1809 Ma. The 1726 Ma granitoid intrusion and 1720 Ma metamorphism (see below) may represent a second phase of this collision.

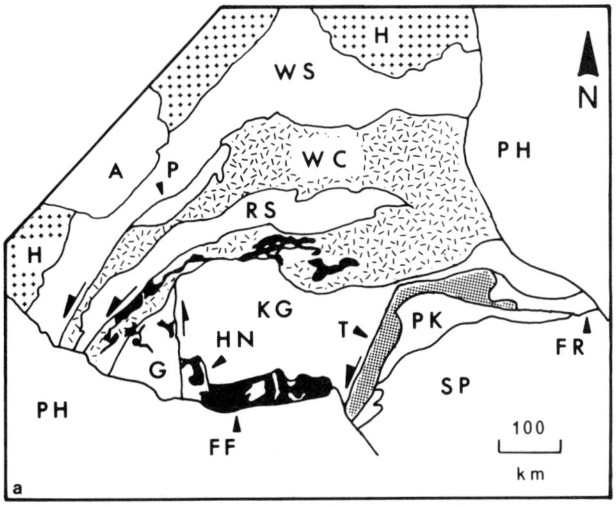

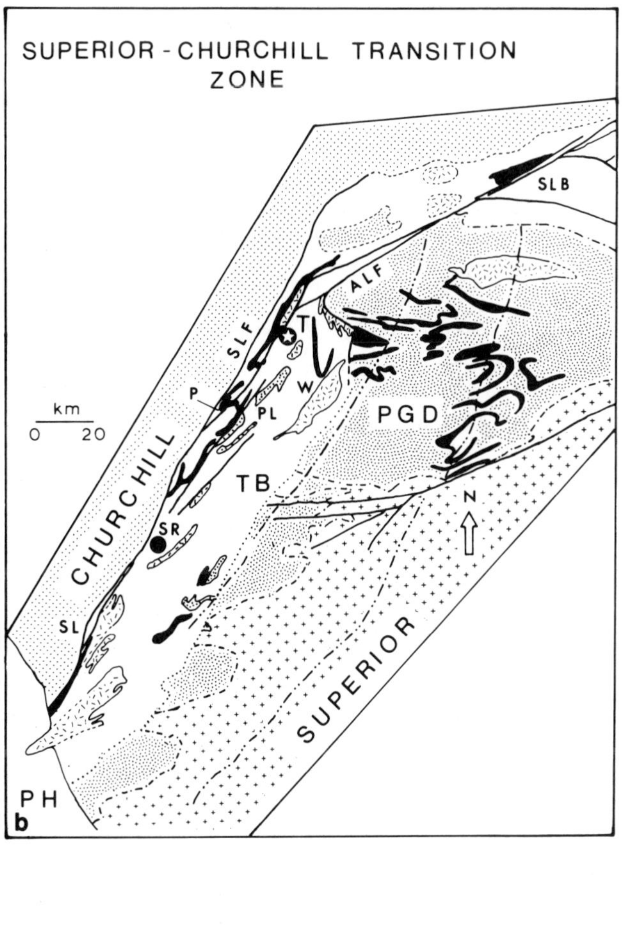

Figure 2. *(a) Schematic geological map of the Reindeer Zone (Manitoba–Saskatchewan). Metavolcanic rocks are shown in black. Legend: A, Athabasca Basin; FF, Flin Flon Belt; FR, Fox River Belt; G, Glennie Domain; H, Hearne Province; HN, Hanson Block; KG, Kisseynew Gneiss Belt; P, Peter Lake Domain; PH, Phanerozoic cover; PK, Pikwitonei Granulite Domain (part of the Superior Province); RS, Rottenstone Domain; SP, Superior Province; T, Thompson Belt; WC, Wathaman–Chipewyan Batholith; and WS, Wollaston Domain. Adapted from Hoffman (1988). (b) Schematic geological map of the Thompson Belt (TB). Metavolcanic and metasedimentary rocks are shown in black. Legend: ALF, Apussigamassi Lake fault; P, Pipe pit; PGD, Pikwitonei Granulite Domain; PH, Phanerozoic cover; PL, Paint Lake; SL, Setting Lake; SLB, Split Lake Block; SLF, Setting Lake fault zone; SR, Sasagiu Rapids; T, Thompson; and W, Wintering Lake intrusion.*

Amphibolite facies metamorphism

Monazite and titanite from gneisses are coeval in both belts: they range from 1793 ± 2 to 1769 ± 5 Ma in the northern New Quebec Orogen and from 1786 ± 2 to 1770 ± 2 Ma in the Thompson Belt. The ages of pegmatite emplacement are included in these intervals: at 1774 ± 4 Ma in the New Quebec Orogen and at 1786 ± 2 and 1771 ± 2 Ma in the Thompson Belt (Machado *et al.*, 1987, 1988).

It is interesting to note that both belts reached amphibolite facies simultaneously and that they stayed in it for comparable periods of time: 24 Ma (New Quebec Orogen) and 16 Ma (Thompson Belt). Moreover, the time span between the latest intrusion and the onset of amphibolite facies metamorphism lasted 40 Ma in the New Quebec Orogen and 36 Ma in the Thompson Belt. These observations suggest that both areas underwent similar and coeval thermal regimes during about 50–60 Ma.

Late events in the New Quebec Orogen and the Thompson Belt

The final stage of Hudsonian metamorphism in the New Quebec Orogen is represented by rutile ages of ca. 1746 and 1740 ± 5 Ma, tentatively interpreted as those of hydrothermal activity (Machado *et al.*, 1989a). No comparable ages have been found in the Thompson Belt, probably because of the absence of rutile in the gneisses of this belt.

The 1726 ± 12 Ma age for the granite in the Sasagiu Rapids area is identical to a 1720 ± 5 Ma titanite age from the Thompson Pit gneiss in the northern sector of the Thompson Belt (Machado *et al.*, 1987). These ages are ca. 45 Ma younger than monazite and titanite ages accompanying amphibolite facies metamorphism, and indicate renewed granitoid intrusion in the south-central Thompson Belt and amphibolite facies metamorphism in the northern sector of the belt.

Data for the Thompson Belt indicate two periods of metamorphism and granitoid intrusion approximately 100 Ma apart (1822–1809 Ma and 1726–1720 Ma). These events are tentatively interpreted as representing two phases of the Reindeer–Superior collision, the youngest probably related to late adjustments along the collisional suture.

TEMPORAL RELATIONSHIPS BETWEEN THE REINDEER ZONE, THOMPSON BELT, AND NEW QUEBEC OROGEN

The region of the Trans-Hudson Orogen between the Thompson Belt and the reworked Archean Hearne Province (Hoffman, 1988) to the northwest, termed the Reindeer Zone (Stauffer, 1984; Lewry and Collerson, this volume), is a collage of lithotectonic domains (Fig. 2a) consisting mainly of arc-related metavolcanic, plutonic, and metasedimentary rocks of Early Proterozoic age (Lewry *et al.*, this volume).

The formation of volcanic arcs in the Reindeer Zone at 1890–1870 Ma (Baldwin *et al.*, 1987; Van Schmus *et al.*, 1987; Gordon *et al.*, this volume) is coeval with emplacement of the Molson dykes and the Fox River sill (1883 Ma) and with foredeep and alkalic magmatism in the New Quebec Orogen at 1880–1874 Ma (Fig. 3). These data indicate extensive magmatic activity both offshore and on the eastern and northwestern margins of the Superior craton in that period. Older magmatism at 1915 and 1920–2000 Ma has been reported for the Lynn Lake belt (Baldwin *et al.*, 1987) and for the Cape Smith Belt (Parrish, 1989), respectively, suggesting protracted activity in widely separated regions. Ages of 1995–2000 Ma for the Purtuniq ophiolite in the Cape Smith Belt, together with ongoing deformation at 1880 Ma (Parrish, 1989), suggest that the early evolutionary stages of this belt are older than

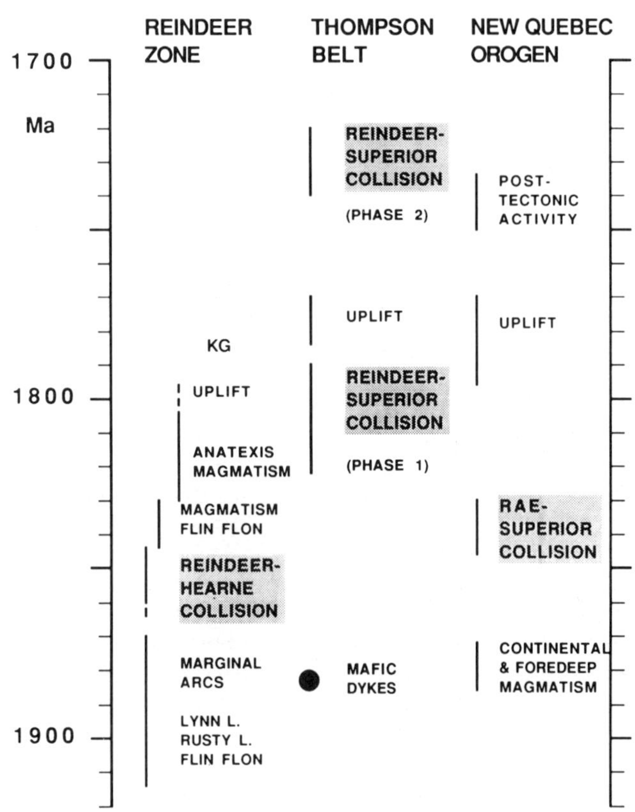

Figure 3. *Chronogram derived from U–Pb ages for the Reindeer Zone, the Thompson Belt, and the New Quebec Orogen. Data sources are indicated in the text.*

their equivalents in other segments of the Trans-Hudson Orogen.

Accretion of the Reindeer Zone to the Hearne Province, possibly represented by the 1860–1845 Ma old Wathaman–Chipewyan Batholith (Van Schmus *et al.*, 1987), is older than the Rae–Superior collision at 1840–1829 Ma (Machado *et al.*, 1988) and also older than the Reindeer–Superior collision at 1822–1809 Ma (Fig. 3).

The youngest volcano-plutonic event in the Flin Flon Belt in Manitoba, dated at 1836–1830 Ma (Gordon *et al.*, this volume), is younger than the Reindeer–Hearne collision (Fig. 3) and was in part coeval with granitic plutonism in the Kisseynew Gneiss Belt, which started at 1831 Ma and lasted until 1809 Ma (Gordon *et al.*, this volume). Peak metamorphism and generation of anatectic granitoids at 1815 Ma and a monazite age of 1805 Ma in this belt (Gordon *et al.*, this volume) are close to the 1822 Ma age of the Wintering Lake granodiorite and to the 1809 Ma age of high-grade metamorphism in the Thompson Belt. These ages of plutonic–metamorphic events in the Kisseynew Belt and the Thompson Belt may date the initial stage of amalgamation of the eastern Reindeer Zone to the Superior Province. Granitoid intrusion and metamorphism at 1726–1720 Ma in the Thompson Belt are interpreted as representing continuing convergence after the main Reindeer–Superior collision. Although no equivalent ages have been found in the Kisseynew Belt, the small granitoid bodies located in the eastern sector of this belt may be of similar age.

The fact that the youngest magmatism in the Flin Flon Belt postdates Reindeer–Hearne collision, together with still younger ages for granitoid intrusion and metamorphism in the Kisseynew terrain and the Thompson Belt, point to an eastward or southeastward migration of the thermotectonic activity. Therefore, only the latest phases of stabilization of the western Trans-Hudson Orogen and its convergence with the Superior Province are recorded in the Thompson Belt.

It is interesting to note that U–Pb ages in the range of 1726–1720 Ma have not been reported from the Reindeer Zone. However, identical ages have been found for metamorphism and granitoid intrusion in the Black Hills, South Dakota (Redden *et al.*, this volume), in the southern extension of the Trans-Hudson Orogen (Green *et al.*, 1985; Lewry and Collerson, this volume), in agreement with a southeastern migration of thermotectonic activity in the western Trans-Hudson Orogen.

The 1746–1740 Ma age for the latest Hudsonian activity in the New Quebec Orogen indicates that stabilization of this belt occurred about 20 Ma earlier than the final stabilization of the western Trans-Hudson Orogen.

CONCLUSIONS

Comparison of available U–Pb mineral ages for the New Quebec Orogen, the Thompson Belt, and the Reindeer Zone of the Trans-Hudson Orogen shows the following:

(1) Formation of island arcs in the Reindeer Zone, intrusion of the Molson dykes and the Fox River sill in the Reindeer–Superior Boundary Zone, and foredeep and continental magmatism in the New Quebec Orogen occurred between 1890 and 1870 Ma.

(2) The Rae–Superior collision at 1840–1829 Ma is at least 10 Ma younger than the Reindeer–Hearne collision, dated at 1860–1850 Ma.

(3) The youngest volcanic–plutonic activity in the Flin Flon Belt (1836–1830 Ma) overlaps granitic plutonism in the Kisseynew Gneiss Belt (1831–1809 Ma) and is coeval with the Rae–Superior collision.

(4) In the Thompson Belt, the first phase of granitic magmatism at 1822 Ma and granulite facies metamorphism at 1809 Ma are approximately coeval with peak metamorphism (1815 Ma) and granitoid intrusion in the Kisseynew Belt in Manitoba and mark the initial phase of the Reindeer–Superior collision.

(5) Amphibolite facies metamorphism and pegmatite emplacement are coeval in the New Quebec Orogen and the Thompson Belt at 1793–1769 Ma and both were under similar thermal regimes for 50–60 Ma.

(6) The youngest Hudsonian activity in the Thompson Belt is represented by amphibolite facies metamorphism and granitoid intrusion at 1726–1720 Ma and may be an expression of continuing Reindeer–Superior convergence after initial collision.

(7) The absence of comparable U–Pb ages in the Reindeer Zone suggests that the Thompson Belt recorded the latest phases of collision between the Trans-Hudson Orogen and the Superior Province.

(8) Although the main collisional events took place in a relatively short interval, 1860–1809 Ma, they are diachronous in different segments of the Trans-Hudson Orogen, probably as a result of plate tectonic regimes specific to each segment.

ACKNOWLEDGEMENTS

Partial funding for this study was provided by Natural Sciences and Engineering Research Council of Canada grants and a Canada–Manitoba Mineral Development contract to T. Krogh. Comments by F. Corfu, G. Dunning, W. Bleeker and S. Philippe and reviews by R. Parrish, G. Clark, and J. Lewry are greatly appreciated.

REFERENCES

Avramtchev, L., 1985, Carte geologique du Quebec, 1:1,500,000: Ministere de l'Energie et Ressources, Quebec, carte no. 2000 du DV-84-02.

Baldwin, D.A., Syme, E.C., Zwanzig, H.V., Gordon, T.M., Hunt, P.A. and Stevens, R.D., 1987, U–Pb ages from the Lynn Lake and Rusty Lake metavolcanic belts, Manitoba: two ages of Proterozoic magmatism: Canadian Journal of Earth Sciences, v. 24, p. 1053–1063.

Baragar, W.R.A. and Scoates, R.F.J., 1981, The Circum-Superior Belt: a Proterozoic plate margin?, *in* Kroner, A., ed., Precambrain Plate Tectonics: Elsevier, Amsterdam, p. 297–330.

Bell, C.K., 1971, History of the Superior–Churchill boundary in Manitoba, *in* Turnock, A.C., ed., Geoscience Studies in Manitoba, Geological Association of Canada, Special Paper 9, p. 5–10.

Bleeker, W., 1990, New structural–metamorphic constraints on Early Proterozoic oblique collision along the Thompson Nickel Belt, Manitoba, Canada, *in* Lewry, J.F. and Stauffer, M.R., eds., The Early Proterozoic Trans-Hudson Orogen of North America: Geological Association of Canada, Special Paper 37, p. 57–73.

Chevé, S. and Machado, N., 1988, Reinvestigation of the Castignon Lake carbonatite complex, Labrador Trough, New Quebec: Geological Association of Canada – Mineralogical Association of Canada, Program with Abstracts, v. 13, p. A20.

Dimroth, E., 1972, The Labrador Geosyncline revisited: American Journal of Science, v. 272, p. 487–506.

Dimroth, E., 1981, Labrador Geosyncline: type example of Early Proterozoic cratonic reactivation, *in* Kroner, A., ed., Precambrian Plate Tectonics: Elsevier, Amsterdam, p. 331–352.

Dimroth, E. and Dressler, B., 1978, Metamorphism of the Labrador Trough: Geological Survey of Canada, Paper 78-10, p. 215–236.

Fueten, F., Robin, P.-Y. and Pickering, M.E., 1986, Deformation in the Thompson Belt, central Manitoba: a progress report: Geological Survey of Canada, Paper 86-1B, p. 797–809.

Girard, R., Woussen, G., Belanger, M. and Bowring, S., 1988, The Ntshukian magmatic event and its implications on the Labrador Trough tectonic evolution: Geological Association of Canada – Mineralogical Association of Canada, Program with Abstracts, v. 13, p. A45.

Gordon, T.M., Hunt, P.A., Bailes, A.H. and Syme, E.C., 1990, U–Pb ages from the Flin Flon and Kisseynew belts, Manitoba: chronology of crust formation at an Early Proterozoic accretionary margin, *in* Lewry, J.F. and Stauffer, M.R., eds., The Early Proterozoic Trans-Hudson Orogen of North America: Geological Association of Canada, Special Paper 37, p. 177–199.

Goulet, N., Gariepy, C., Mareschal, J.-C. and Machado, N., 1987, Structure, geochronology, gravity and the tectonic evolution of the northern Labrador Trough: Geological Association of Canada – Mineralogical Association of Canada, Program with Abstracts, v. 12, p. 48.

Gower, C.F. and Owen, V., 1984, Pre-Grenvillian and Grenvillian lithotectonic regions in eastern Labrador – correlations with the Sveconorwegian Orogenic Belt in Sweden: Canadian Journal of Earth Sciences, v. 21, p. 678–693.

Green, A.G., Hajnal, Z. and Weber, W., 1985, An evolutionary model of the western Churchill Province and western margin of the Superior Province in Canada and the north-central United States: Tectonophysics, v. 116, p. 281–322.

Heaman, L., Machado, N., Krogh, T.E. and Weber, W., 1986, Precise U–Pb zircon ages for the Molson dyke swarm and the Fox River sill: constraints for Early Proterozoic crustal evolution in northeastern Manitoba, Canada: Contributions to Mineralogy and Petrology, v. 94, p. 82–89.

Hoffman, P.F., 1987, Early Proterozoic foredeeps, foredeep magmatism and Superior-type iron formations of the Canadian Shield, *in* Kroner, A., ed., Proterozoic Lithospheric Evolution: American Geophysical Union, Geodynamics Series Volume 17, p. 85–98.

Hoffman, P.F., 1988, United plates of America, the birth of a craton: Early Proterozoic assembly and growth of Laurentia: Annual Reviews of Earth and Planetary Sciences, v. 16, p. 543–603.

Lewry, J.F. and Collerson, K.D., 1990, The Trans-Hudson Orogen: extent, subdivision, and problems, *in* Lewry, J.F. and Stauffer, M.R., eds., The Early Proterozoic Trans-Hudson Orogen of North America: Geological Association of Canada, Special Paper 37, p. 1–14.

Lewry, J.F., Thomas, D.J., Macdonald, R. and Chiarenzelli, J., 1990, Structural relations in accreted terranes of the Trans-Hudson Orogen, Saskatchewan: telescoping in a collisional regime?, *in* Lewry, J.F. and Stauffer, M.R., eds., The Early Proterozoic Trans-Hudson Orogen of North America: Geological Association of Canada, Special Paper 37, p. 75–94.

Machado, N., Heaman, L., Krogh, T.E. and Weber, W., 1987, U–Pb geochronology progam: Thompson Belt and northern Superior Province. Manitoba Energy and Mines, Report of Field Activities, p. 145–147.

Machado, N., Perrault, S. and Hynes, A., 1988, Timing of continental collision in the northern Labrador Trough: evidence from U–Pb geochronology: Geological Association of Canada – Mineralogical Association of Canada, Program with Abstracts, v. 13, p. A76.

Machado, N., Goulet, N. and Gariepy, C., 1989a, U–Pb geochronology of reactivated Archean basement and of Hudsonian metamorphism in the northern Labrador Trough: Canadian Journal of Earth Sciences, v. 26, p. 1–15.

Machado, N., Krogh, T.E. and Weber, W., 1990, U–Pb geochronology of the Thompson Belt (Manitoba): evidence for pre-Kenoran and Pikwitonei-type crust and Early Proterozoic basement reactivation in the western margin of the Archean Superior Province: Canadian Journal of Earth Sciences, v. 27, p. 794–802.

Parrish, R.R., 1989, U–Pb geochronology of the Cape Smith Belt and Sugluk block, northern Quebec: Geoscience Canada, v. 16, p. 126–130.

Peredery, W.V. and Geological Staff, 1982, Geology and nickel sulphide deposits of the Thompson Belt, Manitoba, *in* Hutchinson, R.W., Spence, C.D. and Franklin, J.M., eds., Precambrian Sulphide Deposits (H.S. Robinson Memorial Volume): Geological Association of Canada, Special Paper 25, p. 165–209.

Perreault, S., Hynes, A. and Moorhead, J., 1987, Metamorphism of the eastern flank of the Labrador Trough, Kuujjuaq, Ungava, northern Quebec: Geological Association of Canada – Mineralogical Association of Canada, Program with Abstracts, v. 12, p. 80.

Poirier, G., Perreault, S. and Hynes, A., 1990, Nature of the eastern boundary of the Labrador Trough near Kuujjuaq, Quebec, *in* Lewry, J.F. and Stauffer, M.R., eds., The Early Proterozoic Trans-Hudson Orogen of North America: Geological Association of Canada, Special Paper 37, p. 397–411.

Redden, J.A., Peterman, Z.E., Zartman, R.E. and DeWitt, E., 1990, U–Th–Pb geochronology and preliminary interpretation of Precambrian tectonic events in the Black Hills, South Dakota, *in* Lewry, J.F. and Stauffer, M.R., eds., The Early Proterozoic Trans-Hudson Orogen of North America: Geological Association of Canada, Special Paper 37, p. 229–251.

Stauffer, M.R., 1984, Manikewan: an early Proterozoic ocean in central Canada, its igneous history and orogenic closure: Precambrian Research, v. 25, p. 257–281.

Stockwell, C.H., 1982, Proposals for time classification and correlation of Precambrian rocks and events in Canada and adjacent areas of the Canadian Shield: Geological Survey of Canada, Paper 80-19.

Thomas, A., Nunn, G.A.G. and Wardle, R.J., 1985, A 1650 Ma orogenic belt within the Grenville Province of northwestern Canada, in Tobi, A.C. and Touret, J.L.R., eds., The Deep Proterozoic Crust in the North Atlantic Provinces, Reidel, Dordrecht, Netherlands, 603 p.

Van der Leeden, J., Bélanger, M., Danis, D., Girard, R. and Martelain, J., 1990, Lithotectonic domains in the high-grade terrain east of the Labrador Trough (Quebec), in Lewry, J.F. and Stauffer, M.R., eds., The Early Proterozoic Trans-Hudson Orogen of North America: Geological Association of Canada, Special Paper 37, p. 371–386.

Van Schmus, W.R., Bickford, M., Lewry, J. and MacDonald, R., 1987, U–Pb geochronology in the Trans-Hudson Orogen, northern Saskatchewan, Canada: Canadian Journal of Earth Sciences, v. 24, p. 407–424.

Wardle, R.J., Ryan, B., Nunn, G.A.G. and Mengel, F.C., 1990, Labrador segment of the Trans-Hudson Orogen: crustal development through oblique convergence and collision, in Lewry, J.F. and Stauffer, M.R., eds., The Early Proterozoic Trans-Hudson Orogen of North America: Geological Association of Canada, Special Paper 37, p. 353–369.

Weber, W., 1990, The Churchill–Superior Boundary Zone, southeast margin of the Trans-Hudson Orogen: a review, in Lewry, J.F. and Stauffer, M.R., eds., The Early Proterozoic Trans-Hudson Orogen of North America: Geological Association of Canada, Special Paper 37, p. 41–55.

The Early Proterozoic Rinkian Belt of Central West Greenland

J. Grocott
School of Geological Sciences, Kingston Polytechnic,
Kingston upon Thames, United Kingdom

T.C.R. Pulvertaft
Geologisk Centralinstitut, Københavns Universitet,
DK-1350 København, Denmark

Grocott, J. and Pulvertaft, T.C.R., The Early Proterozoic Rinkian belt of central West Greenland, in Lewry, J.F. and Stauffer, M.R., eds., The Early Proterozoic Trans-Hudson Orogen of North America: Geological Association of Canada, Special Paper 37, p. 443–463.

Abstract

The Rinkian belt developed on the site of an Early Proterozoic epicontinental marginal basin, which formed in response to northward-dipping subduction at a convergent margin in the Nagssugtoqidian belt to the south. The tectonic setting was unstable, and seven regionally important phases are recognized in the structural and metamorphic evolution of the belt. Three of these phases involved extension and four involved shortening of the lithosphere.

Initial rifting in response to a small stretching factor led to deposition of widespread sequences of shelf-type sediments. Thicknesses exceeding 1.5 km only occur in three isolated fault-controlled basins. The succeeding thermal subsidence phase is represented by a starved basin facies. Overlying this are flysch turbidites that show remarkable lateral and vertical uniformity, suggesting deposition during renewed extension. There was no metamorphism of sediments at the base of the sequence during initial rifting, and the following phase of thrusting interleaved the sequence under nonmetamorphic conditions.

The Rinkian belt is characterized by metamorphism under high temperature – low pressure facies conditions. This metamorphism was accompanied by west-northwest–east-southeast extension, during which low-angle ductile detachments were initiated. Growth of diapiric mantled gneiss antiforms accompanied extension. Reversal of displacement sense on these low-angle detachments occurred before thermal relaxation was achieved and ductile thrusting caused interleaving of basement and cover during continued high temperature – low pressure metamorphism.

Following this phase of overthrusting to the west-northwest, a third phase of high temperature ductile deformation occurred, with renewed displacements on low-angle detachment zones. Displacements were to the west and may have been responsible for the formation of the westerly verging sheath-like fold nappes and asymmetric mantled gneiss domes that characterize particularly the northern Rinkian. Final displacements on the low-angle detachments were due to brittle thrusting to the north, during which thrusts exploited anisotropic mylonitic rocks in the zones of intense ductile strain.

Résumé

La ceinture Rinkian s'est développée dans un bassin épicontinental Protérozoïque formé par une zone de subduction à vergence nord située près des marges convergentes de la ceinture Nagssugtoqidian au sud. La tectonique est instable et l'évolution métamorphique et structurale comportent sept phases régionales importantes. Trois de ces phases sont en extension alors que quatre impliquent un raccourcissement de la croûte.

Des séquences de plateformes se sont déposées dans un contexte de rift avec des épaisseurs excédant 1,5 km dans trois bassins isolés bordés par des failles. Des faciès euxéniques sont le résultat d'une phase de subsidence thermique subséquente. Les flysh turbiditiques qui succèdent montrent une grande uniformité latérale et verticale et sont déposés dans un environnement en extension. Initialement, le métamorphisme est absent de la base de la séquence qui est charriée dans un environnement non métamorphique par une phase subséquente.

La ceinture Rinkian se caractérise par un métamorphisme haute température basse pression. Le métamorphisme est produit pendant une phase d'extension ESE provoquant des décollements ductiles accompagnés par la croissance d'antiformes de gneiss diapiriques. Une inversion du sens du mouvement le long de ces décollements s'est produite avant la fin de la relaxation thermique et des chevauchements ductiles ont imbriqué socle et couverture pendant une phase continue de métamorphisme haute température basse pression.

Suivant cette phase de charriage ONO, une troisième phase de déformation ductile à haute température provoque de nouveaux déplacements le long des zones de décollements. Des déplacements vers l'ouest pourraient être à l'origine de nappes "en fourreaux" à vergence ouest et de dômes de gneiss asymétriques qui caractérisent le nord de la ceinture Rinkian. Les derniers charriages cassants à vergence nord provoquent des mouvements le long des zones de décollements et se retrouvent préférentiellement sur les mylonites anisotropiques des zones à déformation ductile.

EARLY PROTEROZOIC BELTS OF WEST AND SOUTH GREENLAND

The Archean craton of West Greenland is flanked by Early Proterozoic belts of deformation and metamorphism formed 1.9–1.7 Ga ago (Fig. 1). The Ketilidian mobile belt lies to the south of the craton and contains crust generated 1.9–1.7 Ga ago (Patchett and Bridgwater, 1984; Kalsbeek and Taylor, 1985). This belt is characterized by intense calc-alkaline igneous activity and formed at an Early Proterozoic convergent plate boundary (Bridgwater *et al.*, 1973; Watterson, 1978).

The Nagssugtoqidian belt lies to the north of the craton. Its boundary was defined as the place where Kangâmiut dolerite dykes, which are undeformed in the craton, become intensely deformed (Ramberg, 1948; Escher *et al.*, 1975). Deformed Kangâmiut dykes only occur in the southern part of the belt, but the main elements of Nagssugtoqidian structural style – northeast trending linear belts separated by less deformed areas – can be recognized as far north as Jakobshavn (69°13′N). The presence of low deformation areas in the belt, with structural styles similar to those in the craton, was taken to indicate that the Nagssugtoqidian included large amounts of reworked Archean rocks. Accordingly, a model was developed in which the linear belts represented intraplate ductile shear zones, which, by analogy with southeast Asian neotectonics, were slip-line faults formed in response to continental collison at or near the site of the Ketilidian mobile belt (Bak *et al.*, 1975; Watterson, 1978).

Recently, it has been shown that not all of the rocks in the Nagssugtoqidian belt are Archean. Significant proportions of the supracrustal rocks in the belt are likely to be Early Proterozoic (Kalsbeek *et*

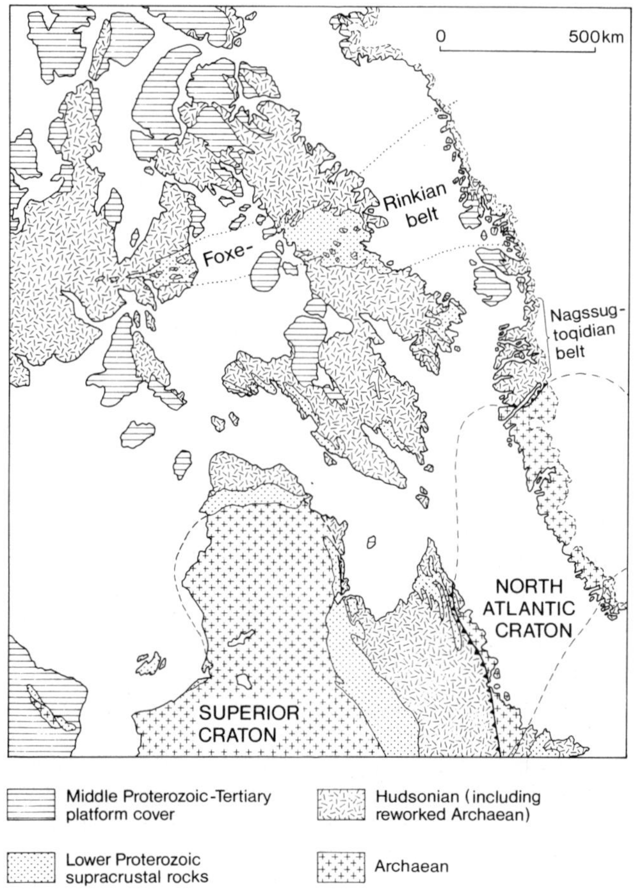

Figure 1. *Map showing the main divisions in the Precambrian of northeast Canada and West Greenland. The pre-drift fit is a compromise among many different proposals, designed to reduce the lateral displacement and gap in Nares Strait to a minimum.*

al., 1984) and some of the gneisses have also been proven to be Proterozoic (Kalsbeek *et al.*, 1987). In the Nordre Strømfjord area (68°N), the latter can be traced into an undeformed quartz diorite, the emplacement of which took place 1920 Ma ago. Isotopic evidence (Sr_i about 0.703, model μ_1 values about 8.0, initial ε_{Nd} about 0) shows that these rocks contain no significant Archean crustal contribution. Furthermore, their major element chemistry is calc-alkaline and in trace element discrimination diagrams they plot in the "volcanic arc granite" field. Despite the modest volume of the rocks in question, Kalsbeek *et al.* (1987) conclude that a northward-dipping subduction zone existed below Nordre Strømfjord around 1900 Ma ago, and that a suture must exist somewhere between Nordre Strømfjord and the southern margin of the Nagssugtoqidian belt. These conclusions are entirely consistent with the tectonic setting for the Rinkian belt proposed in this paper.

The Rinkian belt lies to the north of the Nagssugtoqidian belt and is markedly different in structural style (Escher and Pulvertaft, 1976). The purpose of this paper is to describe the structural and metamorphic development of the Rinkian belt, to discuss the kinematics that led to the development of the structural and metamorphic patterns, and, finally, to consider the likely tectonic setting of the Rinkian belt within the Early Proterozoic tectonic framework of West and South Greenland.

DEFINITION OF THE RINKIAN BELT

The Rinkian belt is described in the literature as being characterized by fold nappes with sheath-like geometry and flat-lying axial surfaces that affect both Archean basement gneisses and Proterozoic cover sediments. These structures are refolded by upright to slightly overturned mantled gneiss domes. Although the contrast between the Rinkian and Nagssugtoqidian belts seems clear in terms of structural style and was the original basis for defining the Rinkian (Escher and Pulvertaft, 1976), neither the area between Jakobshavn and Nûgssuaq nor that north of Red Head (Fig. 2a) is well enough known to allow precise boundaries to be drawn on the basis of structural style – assuming that they exist at all. Indeed, in the Atâ Sund area (68°50′N), which lies just south of Nûgssuaq and north of the Rinkian boundary as originally defined by Escher and Pulvertaft (1976), there is a domain of Archean granite that is virtually unaffected by Proterozoic deformation and that borders very low grade, possibly Lower Proterozoic, metasediments of the Anap nunâ Group (Kalsbeek *et al.*, 1988).

For the purpose of this paper, the unifying factor that is used to arbitrarily define the Rinkian belt is the proven extent of a distinctive Lower Proterozoic cover sequence known as the Karrat Group (Henderson and Pulvertaft, 1967). This outcrops extensively in a 430 km wide belt extending from Mârmorilik to Holm Ø and has been identified as far south as Nûgssuaq and as far north as Red Head (Fig. 2a). Similar successions have been mapped on Baffin Island and in the Melville peninsula (Fig. 1; Henderson *et al.*, 1979; Henderson, 1983). Although Early Proterozoic (Hudsonian) deformation and metamorphism are known to have profoundly affected the area north of Red Head (Dawes *et al.*, 1988), the difficulty in distinguishing Lower Proterozoic from Archean supracrustal sequences in this area has led to uncertainty as to which structures are Early Proterozoic. For this reason, nothing more will be said about the area north of Red Head in this paper.

TECTONIC SIGNIFICANCE OF RINKIAN METAMORPHISM

The Rinkian belt is characterized almost throughout by high temperature – low pressure metamorphism. It is now a well-established belief that extension of the lithosphere can lead to high temperature – low pressure metamorphism in the crust (Thompson, 1981; Wickham and Oxburgh, 1985; Sandiford and Powell, 1986). Extension of the lithosphere occurs during rifting in the early stages of continental breakup, at convergent plate margins in areas of back-arc extension, or in transtensional strike-slip fault systems. In all cases, it is likely that extension will ultimately be followed by shortening. The structure of the lithosphere in ancient back-arc regions is likely to be complex because several extensional and shortening phases can be superimposed in response to changes in subduction geometry and kinematics. The pressure–temperature–time paths of crust underlying marginal basins can also be expected to yield complex patterns and to depend on whether or not thermal equilibrium was reached following each phase of contraction or extension.

A number of elegant analyses of pressure–temperature–time paths of rocks that have undergone regional metamorphism have now been conducted (England and Thompson, 1984; Thompson and England, 1984). This research tells us exactly what to expect only in situations constrained by assumptions made to make the problem tractable. To improve the models, constant reference must be made to information contained in the rocks themselves. In this paper, we discuss the kinematics of development of an Early Proterozoic high temperature – low pressure regime and the relationship between deformation and metamorphism within this regime. It is suggested that the regime developed on the site of an epicontinental back-arc basin.

Recognition of extensional tectonics in the middle crust

Where stratigraphic control is good in the upper crust, recognition of extensional faults is commonly straightforward, even where partial or complete reversal of the displacement has occurred during later inversion tectonics (see, for example, Baird *et al.* (1987)). Extensional fault systems in the upper crust are commonly linked to low-angle lithosphere-scale shear zones, which are the major structures accommodating lithospheric extension (Wernicke, 1985; Lister *et al.*, 1986). Much of the data on which models of the geometry of such extensional fault systems are based come from the Basin and Range

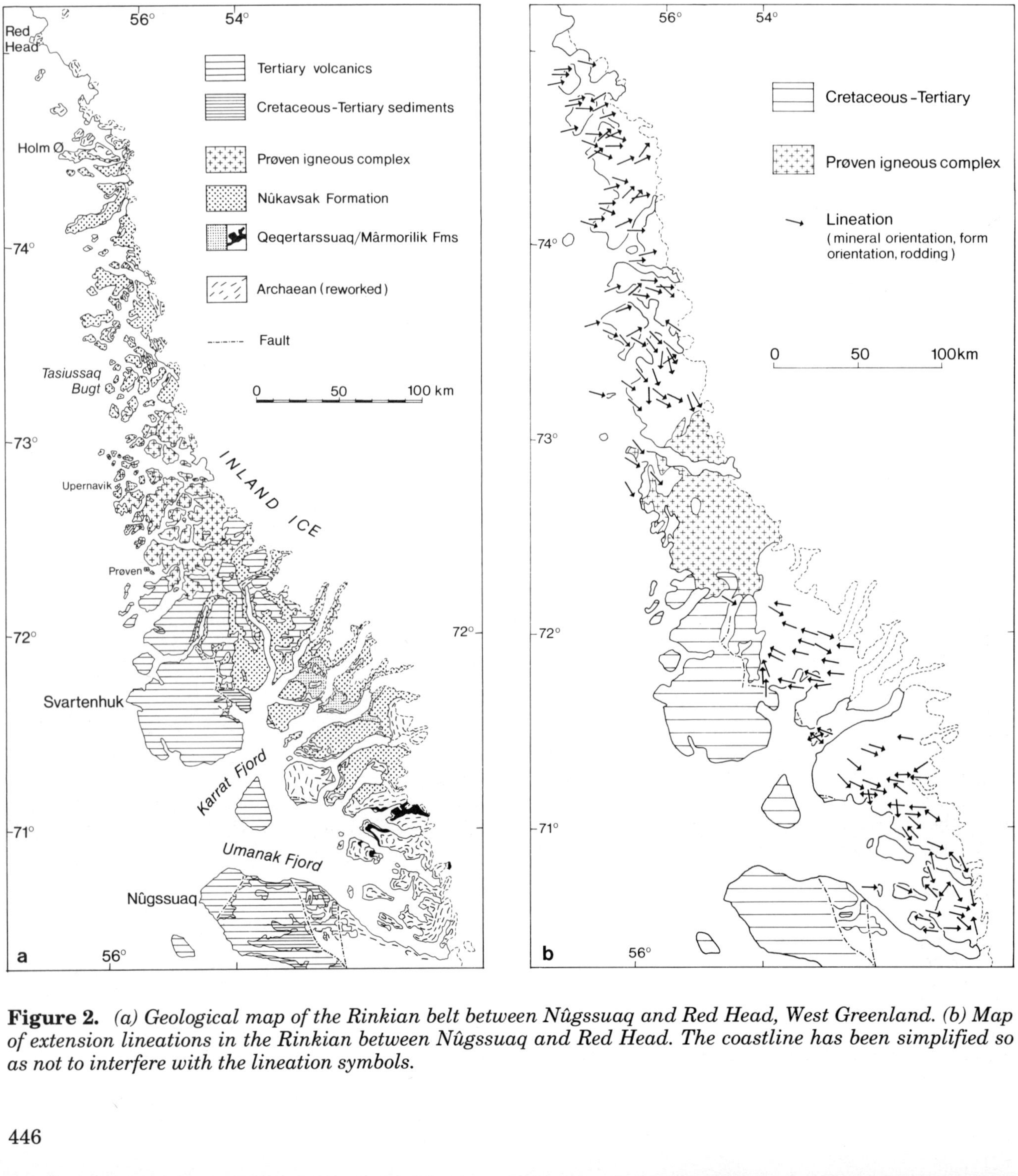

Figure 2. *(a) Geological map of the Rinkian belt between Nûgssuaq and Red Head, West Greenland. (b) Map of extension lineations in the Rinkian between Nûgssuaq and Red Head. The coastline has been simplified so as not to interfere with the lineation symbols.*

province, in which the extensional structures have not (yet) been inverted during shortening of the lithosphere.

Wernicke (1985) has shown that extensional fault systems are frequently asymmetric. An extensional imbricate fan forms in the proximal part of the upper plate and is separated from large block-faulted ranges of the distal part of the upper plate by a metamorphic core complex. High temperature – low pressure metamorphic rocks occur in the upper part of the metamorphic core complex below the detachment, but middle crustal metamorphic rocks are generally not exposed to view in the Basin and Range (Coney, 1980). This leaves us with the paradox that middle crustal metamorphic rocks formed in extensional regimes are only likely to be exposed due to compressional tectonics during crustal shortening, which may have a profound influence on the finite structural and metamorphic pattern.

When kinematics have changed from extension to shortening in back-arc basinal settings, the evidence quoted for extension is frequently indirect. In the Early Proterozoic Wopmay Orogen, St-Onge and King (1987) recognize three stages, each linked to a distinct tectono-metamorphic phase in the development of a continental magmatic arc. The first of these regimes, epicontinental back-arc rifting, is followed by a phase of shortening and intrusion of plutons and, finally, by uplift and erosion. Evidence for the rifting stage is not structural; rather, it concerns high temperature – low pressure metamorphism of the syn-rift sediments, which occurred before major tectonic thickening, and was associated with steep geothermal gradients. Early rifting was accompanied by volcanic activity and emplacement of alkaline complexes. The argument is convincing for the Wopmay Orogen, but the lack of structural evidence showing that high temperature – low pressure metamorphism is, indeed, related to extension during initial rifting can lead to an oversimplified view of the relationship between deformation and metamorphism in high temperature – low pressure belts. For example, it is not enough to argue, as Wickam and Oxburgh (1985, 1987) have done in the Hercynian Pyrenees (an orogen that has many parallels with the Rinkian), that high temperature – low pressure metamorphism is evidence per se of the extensional or transtensional nature of an orogen. This is especially so when there is evidence that a regional episode of shortening preceded widespread high temperature – low pressure metamorphism, as has been documented both in the Pyrenees (van den Eeckhout and Zwart, 1988) and in the Rinkian (Grocott and Vissers, 1984). In this paper, we relate the high temperature – low pressure metamorphism in the Rinkian belt to lithospheric extension in an orogenic setting in which several regionally important phases of extension and shortening alternated. Nevertheless, structures attributable to middle crustal extension are not easily identified even in this area of alpine relief and perfect exposure.

THE RINKIAN BELT OF WEST GREENLAND: AN INVERTED EPICONTINENTAL MARGINAL BASIN?

The Rinkian belt forms part of the Foxe–Rinkian belt, in which a remarkably uniform sequence of Early Proterozoic supracrustal rocks was deposited and is now preserved in a west-southwest–east-northeast trending belt across the Melville peninsula, Baffin Island, and West Greenland (Fig. 1). The belt is interpreted as an epicontinental marginal (back-arc) basin developed in the upper plate above a subduction zone that lay in the Nagssugtoqidian belt to the south (Grocott and Pulvertaft, 1987; Kalsbeek *et al.*, 1987).

Stratigraphy and depositional history of the Lower Proterozoic sequence

In West Greenland, the Lower Proterozoic sediments of the Foxe–Rinkian belt are designated the Karrat Group (Henderson and Pulvertaft, 1967). The Karrat Group was deposited unconformably on an Archean basement that apparently forms a continuous floor to the sedimentary basin throughout the Rinkian belt. Basement granitoids yield Rb–Sr isochron and zircon U–Pb ages between 2710 and 2570 Ma, even in areas where Rinkian recumbent sheath folds have developed (Kalsbeek, 1981, 1986; Andersen and Pulvertaft, 1985). In the Upernavik area, the Karrat Group sediments are cut by the syntectonic Prøven igneous complex, which has a Rb–Sr isochron age of 1860 ± 25 Ma (Kalsbeek, 1981). This establishes the minimum age of the Karrat Group.

The stratigraphy of the Karrat Group is simple in that the number of constituent formations is small and the formations are easily distinguished (Fig. 3). The two lowest formations, the Qeqertarssuaq and Mârmorilik formations, consist of shelf-type sequences that are believed to be correlatives originally separated by a basement high (Garde and Pulvertaft, 1976; Henderson and Pulvertaft, 1987). The uppermost formation of the Karrat Group, the Nûkavsak Formation, is a thick, uniform turbidite flysch sequence that once blanketed the entire area from Mârmorilik to Red Head (Fig. 2a).

The Qeqertarssuaq Formation shows considerable variation in stratigraphic thickness. It is thickest in the Ingia Fjord – Kangigdleq area and near Upernavik (Figs. 2a and 3). The formation consists mostly of siliciclastic rocks – white quartzites, graphitic quartzites, calc-silicates (formerly marls), pelites and semi-pelites, and rare marbles. The stratigraphically highest unit in the formation is a basic volcanic member that originally consisted of both lava flows and pyroclastic rocks. Lenses of ultrabasic rocks within the Qeqertarssuaq Formation may have been tectonically emplaced; hence, strictly speaking, they are not part of the stratigraphy.

Where rocks of the Qeqertarssuaq Formation occur in their normal stratigraphic position, immediately above and close to the gneiss basement, they are high-grade metamorphic rocks belonging to a high temperature – low pressure facies series, except where late retrogression and mylonitization have taken place. They are generally highly deformed and primary structures are rare, although on Svartenhuk Halvø basic volcanics of the Qeqertarssuaq Formation, occurring as early thrust sheets emplaced in the Nûkavsak Formation, exhibit well-preserved pillows and hyaloclastites (Grocott and Vissers, 1984). Areas where the Qeqertarssuaq Formation is thickest are believed to represent sites of former rift basins.

The Mârmorilik Formation is dominated by carbonate rocks. It reaches a structural thickness of 1.6 km at Mârmorilik. Surprisingly, in view of the intense deformation of the formation at Mârmorilik, clastic rocks at the base locally preserve sedimentary structures and an unconformable relationship with the basement gneisses (Garde and Pulvertaft, 1976). The basal clastic unit is overlain by dolomitic marbles that give way to calcite marbles in the upper part of the formation (Garde, 1978). The Black Angel Zn–Pb orebody occurs in the upper calcite marble. The carbonates are overlain by the Nûkavsak Formation; deformation at the boundary is strong, but there is no suggestion of there having been an unconformity here. Metamorphic grade at Mârmorilik is in the greenschist facies (see Garde (1978) for details) and the formation here constitutes the largest area of low-grade rocks exposed in the Rinkian belt.

The upper formation of the Karrat Group is the Nûkavsak Formation. Away from Mârmorilik and Kangigdleq (Figs. 3 and 4), the Nûkavsak Formation overlaps both the Mârmorilik and the Qeqertarssuaq formations and lies almost directly on basement gneisses. However, a thin veneer of Qeqertarssuaq Formation intervenes between the gneisses and Nûkavsak Formation in most places.

The Nûkavsak Formation is a turbidite flysch sequence with a minimum structural thickness of 5 km. The rocks are intensely folded and no estimate of stratigraphic thickness has been made due to the lack of mappable marker horizons in this very monotonous formation. Graphitic pelites occur at or near the base of the formation. These are overlain by a thick stack of metagreywacke–pelite couplets interpreted as turbidites of facies C and D using the terminology of Walker and Mutti (1973) (Henderson and Pulvertaft, 1987). Individual metagreywacke layers commonly show grading, with Bouma A or B

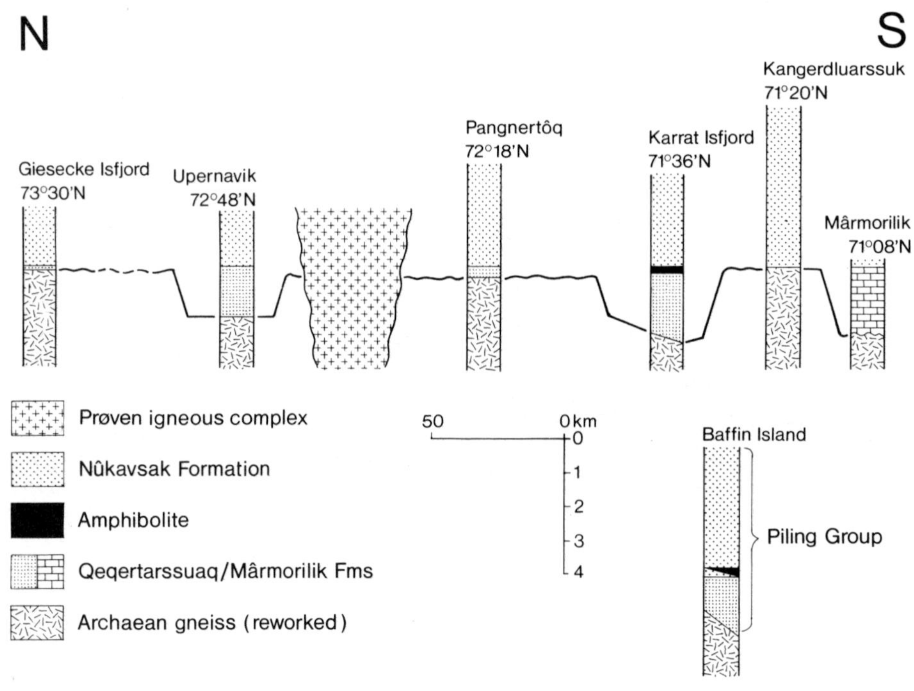

Figure 3. *Stratigraphic columns for the Lower Proterozoic Karrat Group, and a column for the Piling Group on Baffin Island for comparison.*

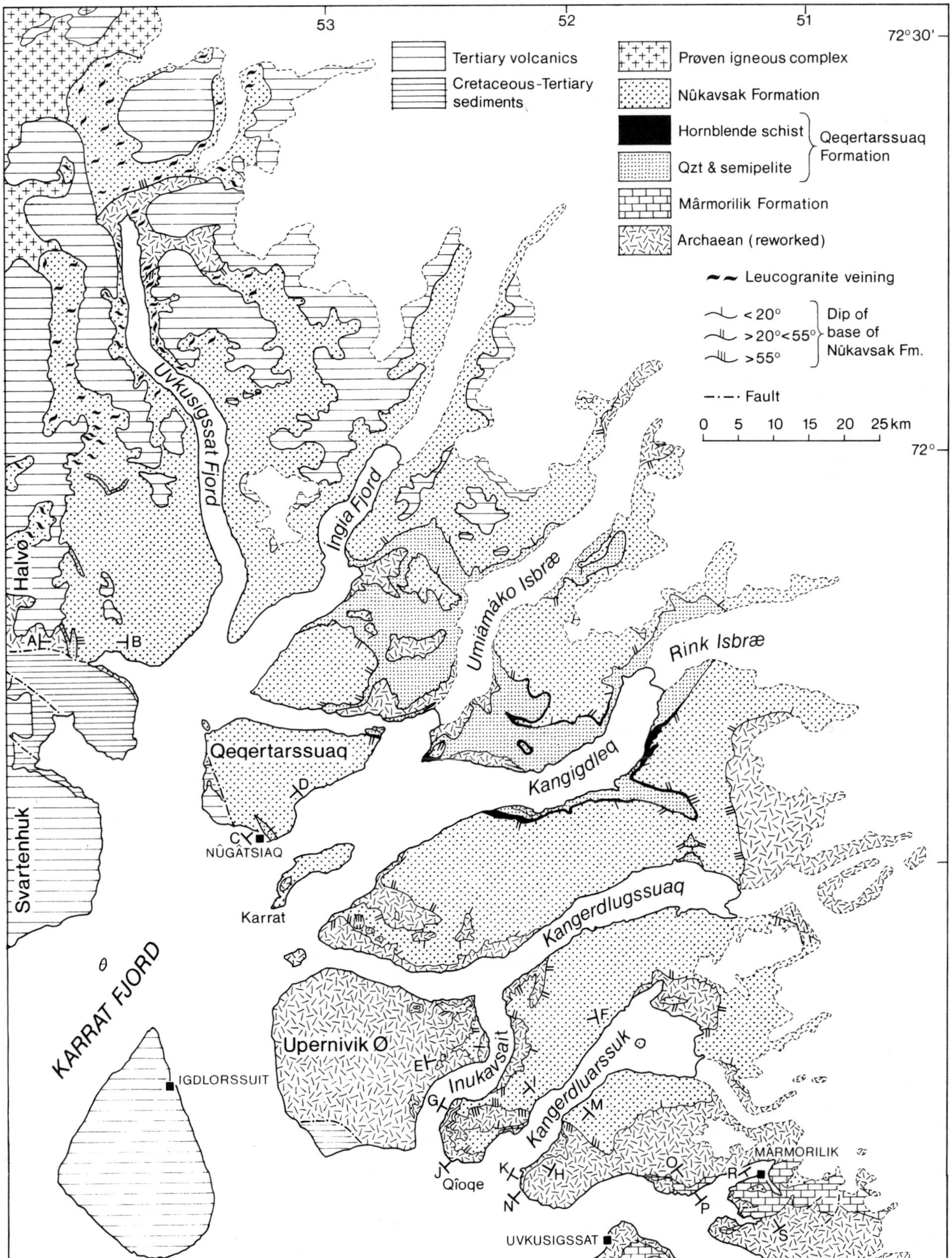

Figure 4. *Geological map of the Rinkian belt between Mârmorilik and Pangnertôq, indicating also the lines of sections shown in Figure 7.*

through E divisions present, but sometimes Bouma divisions C and D are absent so that grading can hardly be seen. No systematic vertical or lateral variation in grain size, thickness of metagreywacke–pelite couplets, or metagreywacke:pelite ratio has been noted in the formation. A trough rather than a fan setting has been suggested by Henderson and Pulvertaft (1987). Unfortunately, with few exceptions, sole structures have been planed off by bedding-plane slip, so nothing is known about the turbidity current transport direction. There is no evidence of synsedimentary slumping, or the presence of seismoturbidites.

The Nûkavsak Formation is metamorphosed in the lower amphibolite facies over wide areas in the south of the belt, with metamorphic grade increasing towards the Archean basement through a high temperature – low pressure facies series until anatexis occurs close to the basement–cover contact in places where the Qeqertarssuaq Formation is thin. Metamorphic grade also increases towards the Prøven igneous complex from the south and particularly from the north (Fig. 5).

Depositional history. Marked variations in the thickness of the Qeqertarssuaq Formation suggest that deposition was concentrated in basins that may have been fault bounded, although syndepositional faults have not yet been recognized (Henderson and Pulvertaft, 1987). Both areas where the Qeqertarssuaq Formation is thickest are now domes, so the basins in which the sediments accumulated have been inverted in the true sense. Thrust interleaving of the Qeqertarssuaq Formation with the gneiss basement can be seen in the Umiámáko Isbrae – Kangigdleq area (Fig. 4). In areas of such extensive later tectonic activity, any faults that were active during deposition will be difficult to recognize today.

The complete contrast in lithology between the siliciclastic Qeqertarssuaq Formation and the contemporary thick carbonates of the Mârmorilik Formation, combined with rapid thinning of the Qeqertarssuaq Formation towards Mârmorilik, suggests that a barrier of some kind, probably a basement high, separated the basins in which the two formations were deposited. It is shown later that the Mârmorilik Formation was probably deposited in a half graben in the area where the formation reaches its greatest structural thickness at Mârmorilik. In other words, deposition of the Qeqertarssuaq and Mârmorilik formations is believed to have taken place during a phase of rifting. The areas where the formations are thickest represent the positions of former basins that were most likely fault bounded. Both formations overlap these basins, the Qeqertarssuaq Formation to the north, as far as Holm Ø (Escher and Stecher, 1980), and the Mârmorilik Formation to the south, as far as the centre of Nûgssuaq peninsula.

Following deposition of sediments of the Qeqertarssuaq Formation, there was an episode of basic volcanism concentrated in the Ingia Fjord – Kangigdleq area and on Svartenhuk Halvø. After this, the depositional area widened and deepened rapidly, apparently without further faulting, and a starved basin developed in which black shales accumulated. This starved basin facies probably reflects a phase of thermal subsidence following stretching. After the interlude of reduced sedimentation, represented by the black shales, turbidity currents poured arenaceous material across the entire basin. Subsidence seems to have kept pace with deposition, as no systematic vertical changes have been noted in the Nûkavsak Formation. The situation invites comparison with foreland basins. However, slumped horizons, seismoturbidites, and lateral facies changes have not been recorded in the Nûkavsak Formation, and there is no evidence of the intricate relationship between sedimentation and tectonics that typifies many foreland basins. We suggest that deposition of the Nûkavsak Formation represents a phase of renewed extension. Figure 6a illustrates the situation as it might have been at the beginning of deposition of the Nûkavsak Formation.

STRUCTURAL DOMAINS

Four structural domains are recognized in the Rinkian belt between Nûgssuaq and Holm Ø. From south to north, these are the (1) Nûgssuaq–Mârmorilik domain, (2) Mârmorilik–Pangnertôq domain, (3) Prøven igneous complex, and (4) Upernavik – Red Head domain. The Mârmorilik–Pangnertôq domain is described first because ideas on the structural evolution were mainly developed here.

Mârmorilik–Pangnertôq structural domain (Fig. 4)

The Ingia Fjord – Kangigdleq dome complex occurs in the centre of this domain. A structural thickness of almost 3 km of Qeqertarssuaq Formation occurs in the dome complex and is succeeded with apparent conformity by the Nûkavsak Formation, which overlaps it to the south away from the dome complex. Other mantled gneiss domes occur in the domain; their distribution and shape are best illustrated in structural contour maps drawn for the base of the Nûkavsak Formation (Henderson, 1969; Henderson and Pulvertaft, 1987). The rather simple pattern revealed by the structural contours was taken by Henderson and Pulvertaft (1967) to preclude significant crustal shortening in this

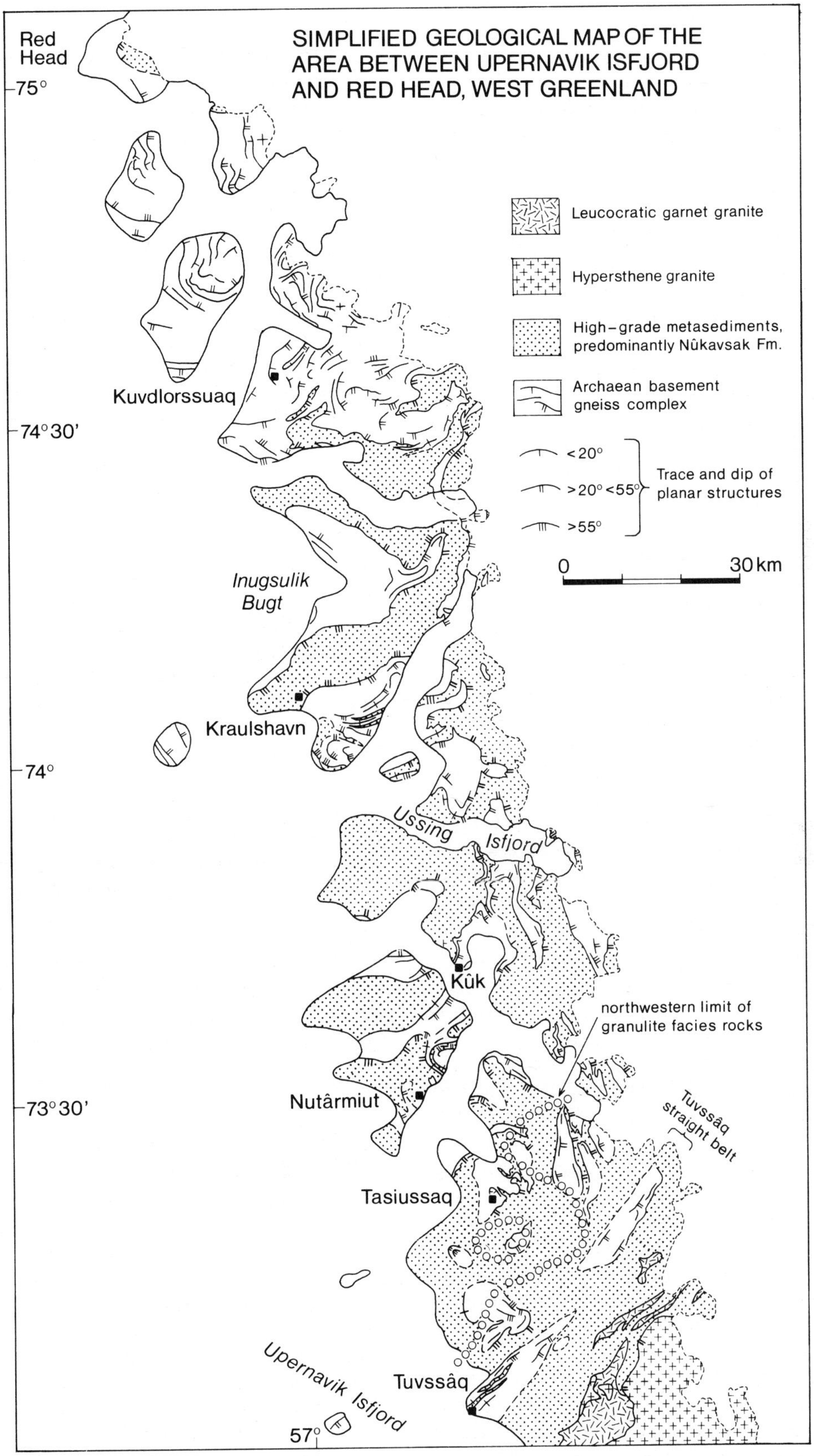

Figure 5. *Geological map of the Rinkian belt between Upernavik Isfjord and Red Head. To show the structural pattern clearly, the coastline has been generalized, and skerries and small islands have been grouped and shown as single larger areas of land.*

domain, and they favoured diapirism driven by Raleigh–Taylor instability as the cause of the development of the mantled gneiss domes. In the 1:100,000 map sheet descriptions (Henderson and Pulvertaft, 1987), this idea is developed further to embrace the formation of the Kigarsima nappe, a gneiss-cored recumbent sheath structure (see below). However, thrust intercalation of basement and Qeqertarssuaq Formation in the Umiámáko Isbrae – Kangigdleq area was also recognized, and in the present work we identify a number of major low-angle shear zones, thereby establishing the importance of lateral displacements in the evolution of this part of the Rinkian belt. Henderson and Pulvertaft (1987) contemplated the possibility that the whole of the Nûkavsak Formation in this domain is allochthonous, but discarded the idea because it seemed incompatible with the apparent transition of the Mârmorilik Formation into the Nûkavsak Formation in the Mârmorilik area; they had not noticed that a shear zone at the base of the Nûkavsak Formation could descend into the basement at the southern end of the Qîoqe peninsula.

A preliminary estimate of peak metamorphic conditions in the Mârmorilik–Pangnertôq structural domain (Henderson and Pulvertaft, 1987) suggested that temperatures reached 600–650°C, but P_{H_2O} did not exceed 4.5 kbar (450 MPa) in the lowest parts of the Karrat Group. These conditions were attained within a high temperature – low pressure facies series. However, sillimanite has been found replacing kyanite in metasediments of the Qeqertarssuaq Formation (Henderson and Pulvertaft, 1987), suggesting that at least some rocks in the belt were metamorphosed at medium pressures during early metamorphism.

On Svartenhuk Halvø (Fig. 4), most of the Nûkavsak Formation is in the biotite–andalusite zone, but approaching the basement, both here and elsewhere, a series of thin metamorphic zones can be mapped, culminating in a zone of anatexis that includes the Qeqertarssuaq Formation adjacent to the gneiss and extends into the basement. Growth of staurolite, andalusite, cordierite, and garnet porphyroblasts postdates the main foliation at distances greater than 2 km from the gneisses, but sillimanite growth and anatexis is syn-main phase near the basement–cover contact (Grocott *et al.*, 1987). This high temperature – low pressure facies series continues into the gneisses, which are granitic to tonalitic in composition. Migmatites occur close to the contact with the metasediments, but gneisses more than 1 km below this contact are not migmatitic and contain only weak Proterozoic tectonic fabrics. Granulite facies mineralogy, which may be Archean, is preserved in low-deformation enclaves within the tonalites. Where these occur in otherwise migmatitic rocks, hypersthene has recrystallized during Proterozoic deformation. This suggests strongly that migmatization occurred due to ingression of water into gneisses at temperatures of granulite facies metamorphism.

Despite the simple map pattern of the Mârmorilik–Pangnertôq structural domain, in which Proterozoic sediments overlie a system of basement domes, the structural history is complex. Six important stages in the structural and metamorphic evolution of the domain can be recognized and they provide important clues to understanding Rinkian tectonics.

Stage 1: deposition of the Karrat Group. This began with simultaneous deposition of the Qeqertarssuaq and Mârmorilik Formations. Garde (1978)

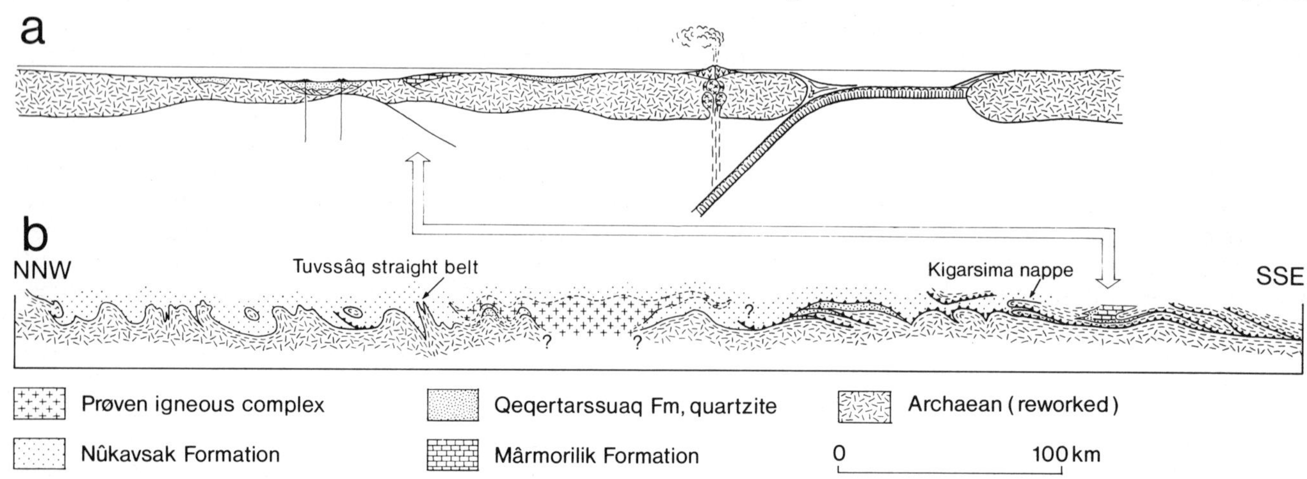

Figure 6. *(a) Cartoon showing the tectonic setting for central and northern West Greenland immediately prior to deposition of the Nûkavsak Formation. (b) Generalized cross-section through the Rinkian belt from Nûgssuaq to Holm Ø.*

has demonstrated that the maximum stratigraphic thickness of the Mârmorilik Formation occurs at Mârmorilik and coincides with the maximum structural thickness, so that the thickness variations shown on Figure 7h may reflect primary asymmetry of the basin in which the Mârmorilik Formation was deposited. Such asymmetry suggests a half-graben geometry and the likely presence of a southerly dipping listric normal growth fault at the northern margin of the formation. In the Magdlak area (Fig. 7h), the basement–cover contact was reworked during ductile thrusting, and this, coupled with substantial ductile deformation within the formation, makes such a fault difficult to recognize.

No evidence for the existence of faults active during deposition has yet been recognized in the Qeqertarssuaq Formation. In the Ingia–Kangigdleq area, the Qeqertarssuaq Formation is overlapped by the Nûkavsak Formation, and to the east this overlap is abrupt (Fig. 4). Cross-sections published by Henderson and Pulvertaft (1987) are consistent with omission of the Qeqertarssuaq Formation at the inner end of Kangerdlugssuaq on a north–south trending normal fault with downthrow to the west. However, subsequent reworking at the contact between the Nûkavsak Formation and basement has obliterated direct evidence for such a fault. The east side of the Ingia Fjord – Kangigdleq dome complex and the Kangerdlugssuaq dome might be rollover anticlines related to this fault.

The small number of widely separated basins in which the Qeqertarssuaq Formation and the Mârmorilik Formation were deposited suggests that the crustal stretching factor (β) was small.

Stage 2: pre-main phase deformation. The Qeqertarssuaq and Nûkavsak Formations were interleaved by thrusting under nonmetamorphic conditions, generally without development of cleavages. Thrust slices of Qeqertarssuaq Formation occur in the Nûkavsak Formation east of Nûgâtsiaq and on Svartenhuk Halvø (Fig. 4). Folds related to this deformation occur widely in these formations, where they are overprinted by the main phase cleavage and refolded by main phase folds (Grocott and Vissers, 1984). They are class 1B to 1C, tight, chevron-like folds identical in style to folds formed during main phase deformation (stage 4), so that no assumption about the age of folds can be made without detailed surface examination, which handicaps rapid interpretation of the spectacular cliff outcrops typical of the area. On Qeqertarssuaq, pre-main phase thrusting was to the east-northeast (inversion on north–south trending extensional faults?), and on Svartenhuk Halvø to the northeast. Rocks of the Qeqertarssuaq Formation were not metamorphosed during or prior to this shortening event (Grocott and Vissers, 1984). This is important because it means that the base of the sedimentary pile could not have undergone metamorphism during deposition of the Nûkavsak Formation: the high temperature – low pressure metamorphism that characterizes the belt is later and unrelated to the extension(s) that led to deposition of the Karrat Group. There is no evidence that basement gneiss was involved in pre-main phase thrusting in this structural domain.

Stage 3: extensional deformation during high temperature – low pressure metamorphism (main phase deformation). Structural analysis of Svartenhuk Halvø, in the northern part of the Mârmorilik–Pangnertôq structural domain, has concentrated on the patterns of strain variation across the basement–cover contact of a mantled gneiss antiform (Van den Eeckhout and Grocott, 1982; Grocott *et al.*, 1987). This work indicates that diapirism has played a role in Rinkian structural development. A key structure in the metasediments of Svartenhuk Halvø is a cleavage fan in which the main phase cleavage (a term used by Grocott *et al.* (1987) and retained here) steepens progressively as the gneiss–metasediment contact is approached (Fig. 7a). The geometry of the structures and the strain variation along the line of section shown in Figure 7a are consistent with the view that the gneisses rose diapirically into the sediments during the deformation that produced the main cleavage of Svartenhuk Halvø. Important in reaching this conclusion was the argument of Grocott *et al.* (1987) that the cleavage fan is a feature of the main phase strain pattern rather than the result of refolding because consistent patterns of crenulation and crenulation cleavage are absent in pelitic rocks within the fan.

Although a case has been made for diapirism having played an important role in the structural development of Svartenhuk Halvø, not all the data can be accommodated in such a model. Downdip extension lineations occur only within the high strain zone at the basement–cover contact; to the east, metasediments contain a north-northwest to northwest trending extension lineation, which lies parallel with the basement–cover contact (Fig. 2b). Peak metamorphism was syn-main phase in a zone of anatexis developed in both gneisses and metasediments close to the basement–cover contact. In lower grade metasediments, more than 2 km from the contact, peak metamorphism was post-main phase (Grocott *et al.*, 1987). The distribution of metamorphic minerals and the timing of their growth imply narrow metamorphic zones and a steep geothermal gradient during main phase deformation. These conditions, coupled with the low viscosity of the migmatitic gneisses, favour diapirism (cf. Talbot, 1979). Nevertheless, diapirism, if it occurred, is likely to be a consequence of a steep

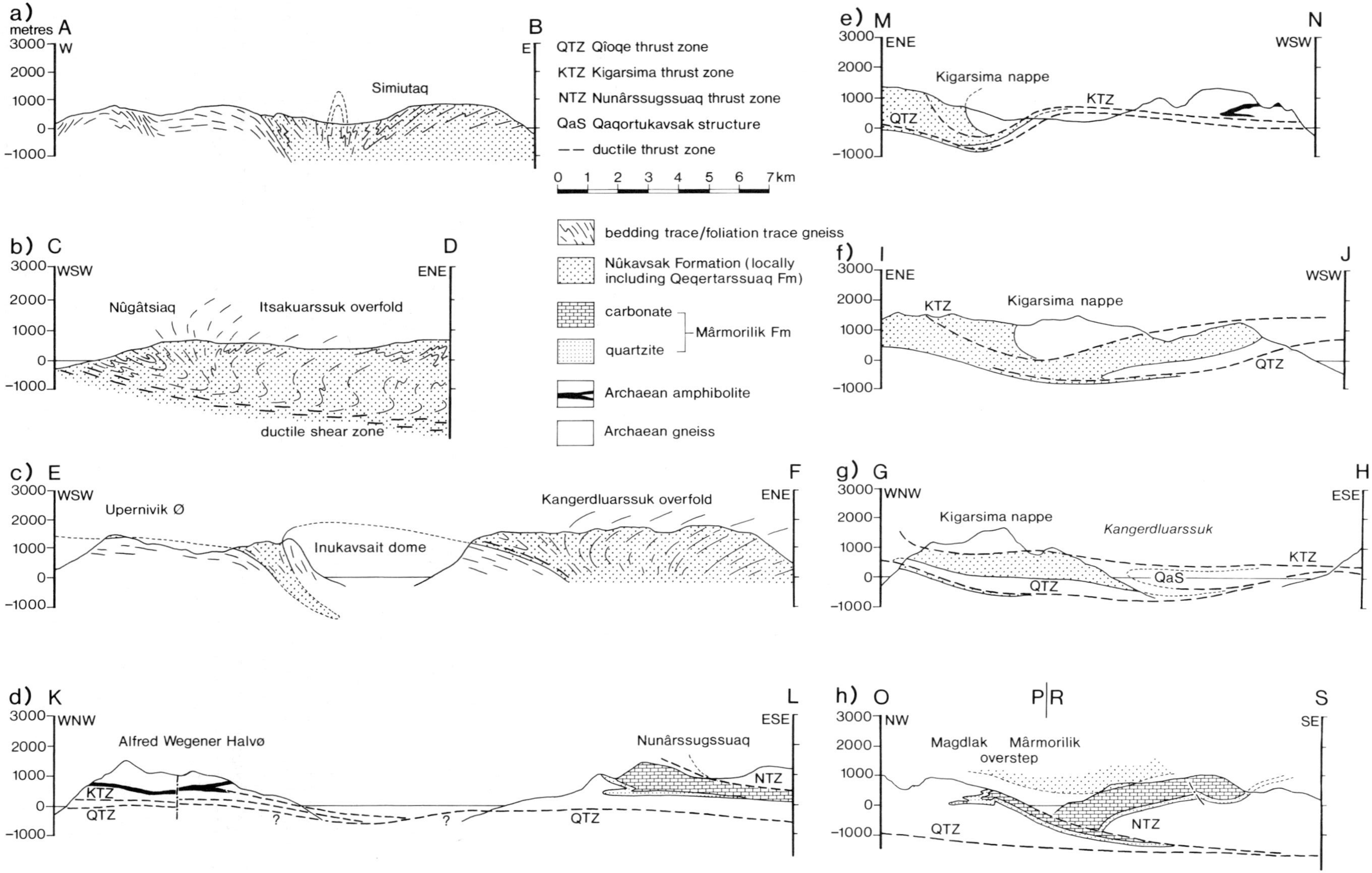

Figure 7. *Cross-sections illustrating important features of the structure in the area between Nunârssugssuaq and Svartenhuk Halvø. The lines of the sections are shown in Figure 4. Horizontal and vertical scales are the same.*

geothermal gradient rather than its cause. Therefore, it is the identification of the tectonics responsible for the steep geothermal gradient and the high temperature – low pressure metamorphism on Svartenhuk Halvø, and elsewhere in the Mârmorilik–Pangnertôq structural domain, that is of paramount importance.

The basement–metasediment contact on Svartenhuk Halvø (Fig. 7a) is an extensional shear zone (Grocott *et al.*, 1987). These authors argued that the shear zone formed on one side of a mantled gneiss antiform that grew by diapirism. However, the downdip stretching lineation that characterizes the contact gives way to a gently plunging, northwest–southeast trending, extension lineation that defines the usual orientation of the x-axis of the strain ellipsoid in the Nûkavsak Formation in Svartenhuk Halvø and to the east (Fig. 2b). The cleavage associated with this extension lineation has a low dip. Moreover, it intensifies downwards into a major low-angle shear zone exposed in the central reaches of Uvkusigssat Fjord, remote from mantled gneiss domes (Figs. 4 and 6). None of the major ductile shear zones that affect basement and lie at the basement–cover contact, or those within the Nûkavsak Formation on Svartenhuk Halvø and immediately to the east, introduce basement slices into the cover. This contrasts with the obvious duplication of the basement–cover contact on low-angle shear zones in the Nûgssuaq–Mârmorilik domain to the south (Figs. 7d–h and 8), and may indicate that the main phase deformation is extensional in the central and northwestern parts of the Mârmorilik–Pangnertôq domain.

The main phase cleavage fan on Svartenhuk Halvø is accompanied by a fan in the axial surfaces of main phase folds (Fig. 7a). These folds, like the cleavage, may reflect 40–50% horizontal shortening of the cover on the flank of a diapiric mantled gneiss antiform (Grocott *et al.*, 1987). Twenty kilometres to the northeast, in Uvkusigssat Fjord, chevron-like folds remain responsible for 40% horizontal shortening of the Nûkavsak Formation. Their remoteness from the mantled gneiss antiform makes these folds difficult to relate to diapirism. Hinge lines are parallel to the west-northwest–east-southeast oriented extension lineation, and the folds become tighter towards deeper structural levels and towards the major low-angle shear zone in the central reaches of Uvkusiggsat Fjord. If these folds are main phase folds, then it is necessary to explain how they formed in an extensional regime.

Unfortunately, the relationship between these folds and the main phase cleavage has not yet been examined in Uvkusigssat Fjord, so the possibility that they are pre-main phase folds cannot be ruled out. This possibility is underlined by observations made on the south coast of Qeqertarssuaq (Fig. 4) in an area similarly remote from mantled gneiss domes or antiforms. The main phase cleavage is strong and overprints many chevron-like pre-main phase folds, whereas main phase folds are rare. These pre-main phase folds are refolded by the Itsakuarssuk overfold (Fig. 7b), and have an identical style to chevron-like folds in Uvkusigssat Fjord and elsewhere in the domain.

At present, there is a problem with assigning such folds to phases in the Mârmorilik–Pangnertôq domain. More information is required on the relationship between chevron-like folds and cleavage, and on fold attitude and distribution throughout the domain, before we can discuss the relationship between folds and low-angle shear zones in detail. The "zigzag folds" illustrated by Henderson and Pulvertaft (1987, Figs. 30, 32, and 36) may well be pre-main phase structures.

The idea that main phase deformation is extensional receives support from the timing of metamorphism. The main phase deformation responsible for the formation of the northwest–southeast trending extension lineations predates peak metamorphism in the Nûkavsak Formation above the low-angle shear zone exposed in Uvkusigssat Fjord. This situation is similar to that described for the Hospitalet massif in the Pyrenees by van den Eeckhout and Zwart (1988), who show that such a metamorphic pattern can occur in rocks above an extensional shear zone. Following these authors, we suggest that development of low-angle stretching lineations, which predate peak metamorphism in a high temperature – low pressure facies series and are not associated with duplication by overthrusting or with important igneous activity, is best interpreted as having taken place in an extensional regime. Thus, it is concluded that structural development on Svartenhuk Halvø during the main phase involved diapirism during crustal extension on low-angle ductile shear zones. It may be possible to separate the component of the finite strain due to regional extension from that due to diapirism on Svartenhuk Halvø, but this has not yet been attempted.

Stage 4: post-main phase low-angle ductile shear zones. The fold defined by the cleavage fan and steepening of bedding towards the gneiss–metasediment contact on Svartenhuk Halvø can be traced southeastwards into Qeqertarssuaq. From east to west across this island, the same steepening of cleavages, accompanied by tightening of main phase and pre-main phase folds, can be observed (Fig. 7b). This structure was called the Itsakuarssuk overfold by Henderson and Pulvertaft (1967, 1987). Despite appearances, the situation is different from that on Svartenhuk Halvø. West of Nûgâtsiaq, the steep limb of the cleavage fan overlies a low-angle shear zone represented by a zone of

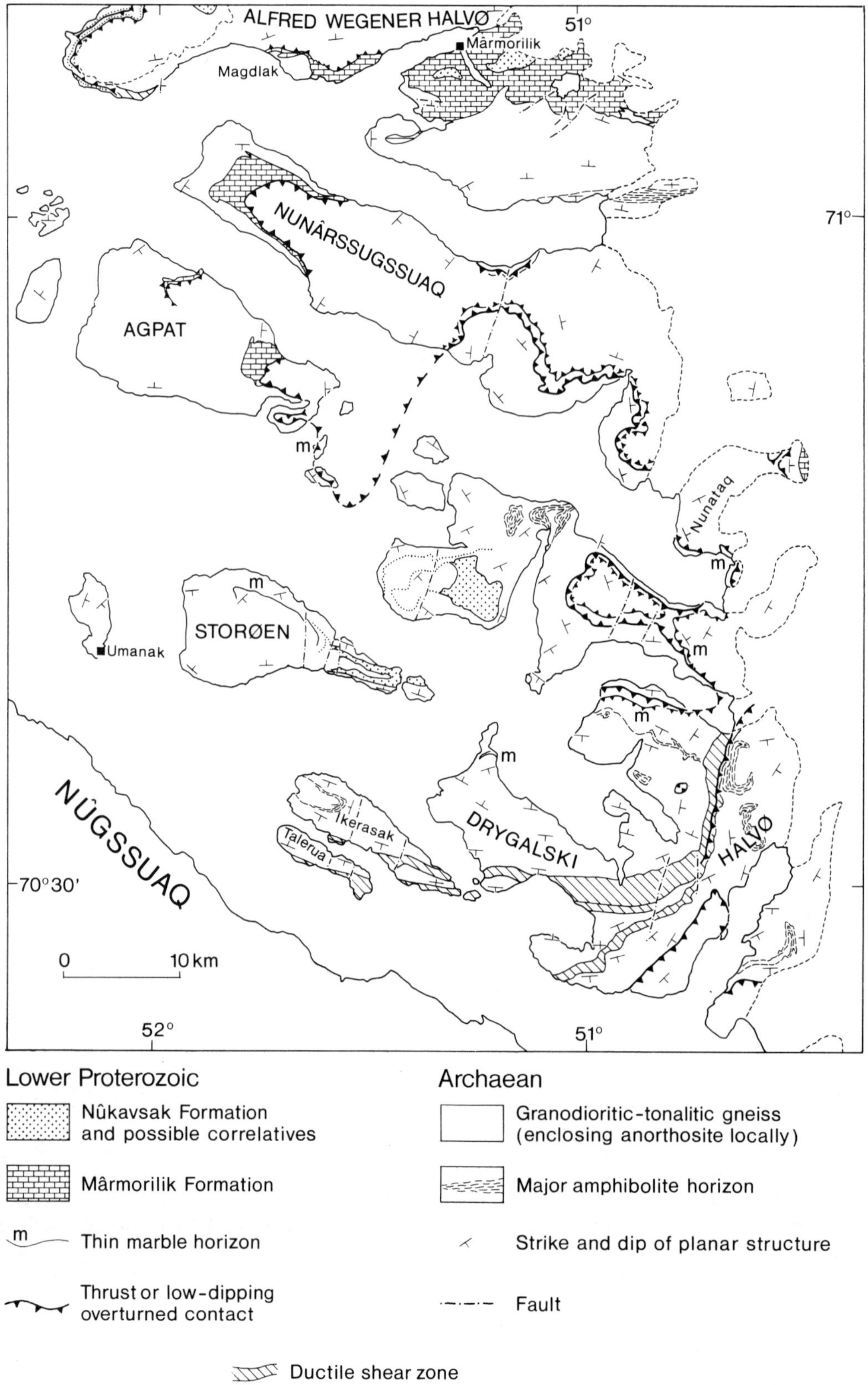

Figure 8. *Map of the main geological features in the area between Nûgssuaq and Alfred Wegener Halvø.*

strongly deformed migmatitic pelites resembling those that elsewhere occur near the base of the Nûkavsak Formation. Above the shear zone, bedding and main phase cleavage in the steep limb of the overfold are folded on shallowly east-dipping axial planes and both fabrics are transposed in the shear zone (Fig. 7b). The shear plane dips gently east and the extension lineation in the most strongly deformed rocks plunges 1–15° towards 130°. This direction is taken to be close to the movement direction of the shear zone. Rotation of the axial planes of the crenulations of the main phase cleavage into the zone of high strain shows that the shear zone has a sinistral horizontal component. Coupled with the southeast plunging lineation, this indicates that the hanging wall moved towards the northwest during post-main phase displacements.

The shear zone is also exposed beneath the Itsakuarssuk overfold in eastern Karrat (Fig. 4). In western Karrat, deeper levels of the shear zone are exposed in a stack of thrust slices formed during phase 6, a phase of brittle thrusting and folding. A structural thickness of 1.2 km of intensely ductilely deformed, flat-lying Nûkavsak Formation is exposed in western Karrat. The protolith includes pelitic rocks, but psammites within the Nûkavsak Formation have become mylonites with an unusual porphyroclast assemblage of diopside and quartz derived from diopside–quartz veins formed during high-temperature metamorphism. The extension lineation in the mylonites trends 310–130° within a subhorizontal shear plane. The intense ductile strain postdates migmatization and growth of diopside in the psammites, but retrogression of the high-temperature parageneses has not been observed.

The mylonites of west Karrat are separated from the shear zone that underlies the Itsakuarssuk overfold in east Karrat by a brittle thrust, which creates difficulties in structural correlation. In less deformed zones in west Karrat, migmatization in pelitic rocks predated or was coeval with the development of a strong foliation, which does not appear to be a crenulation cleavage. This foliation overprints folds in exactly the same manner as the main phase foliation overprints pre-main phase folds elsewhere in the domain. The foliation intensifies into zones with mylonitic foliation in western Karrat and has not been observed to be crenulated. It appears that the structural history of the highly deformed rocks in west Karrat did not involve first crenulation and then transposition of the main phase fabric as was the case in east Karrat and Qeqertarssuaq. Therefore, it is suggested that formation of the Itsakuarssuk overfold involved renewed displacements only in the upper part of a main-phase low-angle detachment, the lower parts of which, unaffected by post-main phase reworking, were later thrust to the north (phase 6) to form the stack of thrust slices of mylonite now exposed in western Karrat.

An argument has been made, based on evidence from Svartenhuk Halvø, that main phase deformation in the Mârmorilik–Pangnertôq structural domain was extensional. Post-main phase deformation involved displacements of rocks to the northwest in the hanging wall of a major shear zone. Whether this occurred during a phase of extensional tectonics or during shortening is discussed below, after the geology of the Nûgssuaq–Mârmorilik structural domain has been reviewed. It is clear, however, that a zone of intense ductile deformation exists at or near the basement–cover contact in the Qeqertarssuaq–Karrat area. A zone of high strain also underlies the Kangerdluarssuk overfold, an overfold of similar style and scale as the Itsakuarssuk overfold, between Kangerdluarssuk and Kangerdlugssuaq and probably also that at Johannes Brae (Fig. 7c; see also Henderson and Pulvertaft (1987, Fig. 30)). Recognition of these zones has established that a major décollement does occur at or near the base of the Nûkavsak Formation throughout most of the Mârmorilik–Pangnertôq structural domain, a conclusion that Henderson and Pulvertaft (1987) were reluctant to draw.

Stage 5: mantled gneiss domes. In the Inukavsait area (Fig. 7c), the basement–cover contact and strongly deformed rocks in the ductile shear zone beneath the Kangerdluarssuk–Kangerdlugssuaq overfold are refolded by a system of west-northwest–east-southeast gneiss domes first illustrated in the stratum contour maps of Henderson (1969). Henderson and Pulvertaft (1967) found no evidence that these structures were formed by fold interference, and favoured a diapiric origin. The strain patterns of these mantled gneiss domes have not been studied, and little is known about the kinematics of their formation.

The mantled gneiss domes and antiforms of the Mârmorilik–Pangnertôq domain occur in the footwall of major west-northwest directed thrusts that characterize the Nûgssuaq–Mârmorilik structural domain and are described below. The Inukavsait dome (Fig. 7c) could be interpreted as a hanging-wall fold overlying yet another easterly dipping thrust zone. However, the dome is circular in plan and a complex shape would be required for the underlying thrust zone involving a frontal and two lateral ramps. This hypothesis is tenable, and might also apply to the other mantled gneiss domes and antiforms in the domain, but it has not been tested by detailed observations on the ground. The suggestion does not rule out diapirism as having had a role in the origin of the domes.

Stage 6: brittle fold and thrust system. Brittle thrusts occur in a well-developed thrust system in Karrat, but have not been reported elsewhere in the Mârmorilik–Pangnertôq structural domain. The thrust system is obvious in southwest Karrat and duplex structure is common. North-overturned buckle and bending folds occur within the thrust sheets. The direction of transport is to the north. In Karrat, a 5 km scale fold related to the thrust system refolds the Itsakuarssuk overfold. The amount of transport on this thrust system is unknown. The system is so well developed in Karrat that it seems inconceivable that structures of this type do not occur elsewhere. One candidate is the duplex-containing north-directed thrusts that occur on the eastern coast of Kangigdleq north of Johannes Brae. Several Middle Proterozoic postmetamorphic basic dykes cross the area (Kalsbeek and Taylor, 1986), none of which has been shown to have been displaced by thrusts. This suggests that the brittle thrust system may be a late Rinkian event.

Nûgssuaq–Mârmorilik structural domain

The conclusion that the domain north of Mârmorilik is characterized by large low-angle detachments in the cover immediately raises the question of where these shear zones pass into basement. The area south of Mârmorilik is dominated by recumbent folds and flat-lying structure (Pulvertaft, 1973), including thin but very extensive thrust sheets (Fig. 8; Grocott, 1984; Pulvertaft, 1986). Pulvertaft (1986) has traced a single ductile thrust zone – the Nunârssugssuaq thrust zone – over much of this domain. Cross-sections through the Nunârssugssuaq thrust sheet in the Mârmorilik district show that it is responsible for emplacing a basement sheet over the Mârmorilik Formation (Fig. 7d). Extension lineations in the Nunârssugssuaq thrust zone trend northwest–southeast and extension lineations with similar orientation occur throughout this domain (Grocott, 1984). The thrust zone cuts upsection in a northwesterly direction, and it is clear that the movement direction was to the northwest.

In Nunârssugssuaq, Archean gneisses with a cover of Mârmorilik Formation overlie another ductile thrust zone – the Qîoqe thrust zone – which passes just below the western tip of the Nunârssugssuaq peninsula. This zone is exposed to the northwest as a major low-angle ductile shear zone on Alfred Wegener Halvø (Fig. 7d). The Nunârssugssuaq thrust zone and the Qîoqe thrust zone may share a branch line east of Nunârssugssuaq, but the position of this structure has not yet been identified. Marble outcrops in the footwall of the Nunârssugssuaq thrust zone at Nunataq and in the hanging wall just to the east (Fig. 8). However, little is known at present about the geometry after deposition of the Mârmorilik Formation away from Mârmorilik, and no attempt can be made to balance or restore the cross-sections. In addition, no cutoff in the hanging wall of the thrust zone can be matched with a footwall cutoff, so that displacement cannot be estimated. The thin marble in the footwall of the major ductile thrust zone at Nunataq seems to indicate that at least 70 km of displacement occurred on this zone because there is no corresponding hanging wall cutoff exposed to the northwest. Finally, a third major thrust zone – the Ikerasak thrust zone – has been mapped in the southern part of this structural domain above the Nunârssugssuaq thrust sheet (Fig. 8).

Large-scale recumbent sheath folds, defined by Proterozoic and Archean supracrustal horizons, are characteristic of the gneisses between the major thrust zones. The similar orientation of extension lineations in ductile thrust zones and the x-axis of sheath folds suggests that the structures formed during a progressive Proterozoic deformation with a fixed kinematic framework. However, intense deformation, characterized by mylonitic foliations, occurs within supracrustal units, especially at supracrustal–gneiss contacts. These mylonitic fabrics are folded by the sheath-like recumbent fold nappes, the axial planes of which are rotated into the ductile thrust zones (Grocott, 1984). Thus, although an interpretation involving progressive deformation is simplest, we stress that this is not unequivocally established. This uncertainty becomes important when correlation between structural domains is considered.

Metamorphic grade is at upper amphibolite facies throughout most of the Nûgssuaq–Mârmorilik structural domain. The high temperature – low pressure parageneses are syn- to postkinematic (Grocott, 1984). Near Mârmorilik, the Mârmorilik Formation is metamorphosed in the greenschist facies (Garde, 1978), with the grade increasing downwards and reaching amphibolite facies in the Archean basement. This arrangement of metamorphic facies suggests that the Mârmorilik Formation at Mârmorilik has remained at a shallow level in the crust and was not overridden by higher thrust sheets before being carried northwards in the hanging wall of the Qîoqe thrust zone.

The final phase of deformation in this domain involves steepening of all the structures southwards towards the Talerua shear zone (Fig. 8). This vertical shear zone has a subhorizontal extension lineation, but the movement pattern is uncertain. Metamorphic grade within the shear zone is lower in amphibolite facies than elsewhere in this domain (Grocott, 1984).

Relation between low-angle detachments in the Nûgssuaq–Mârmorilik and Mârmorilik–Pangnertôq domains

The essential contrast between these structural domains is that, in the north, Proterozoic cover overlies Archean basement with low-angle detachments near the base of the cover. In the south, basement and cover are interleaved by thrusting to the northwest and relatively little cover is exposed. Passage from one structural domain to the other occurs across the peninsulas of Alfred Wegener Halvø and Qîoqe (Figs. 4 and 8). On Alfred Wegener Halvø, the Qîoqe thrust zone, which to the southeast underlies the Mârmorilik Formation and its basement, is a zone of ductile deformation that skirts the base of the prominent cliffs at the western end of this peninsula and climbs laterally into the Nûkavsak Formation to the northeast (Fig. 7d,e). The cross-sections in Figure 7e and 7f also show an overlying thrust sheet, the southern part of the Kigarsima nappe of Henderson and Pulvertaft (1967, 1987). The Kigarsima thrust zone (at the base of the Kigarsima nappe) probably originates at a branch line with the Qîoqe thrust zone south of Alfred Wegener Halvø, and is responsible for the cutoff of a prominent Archean amphibolite in the hanging wall at the west end of this peninsula (Fig. 7e). Cross-sections across and parallel to the movement direction (Fig. 7e–g) show that the basement–cover contact defines large-scale sheath folds. The thrust zones climb laterally and frontally from basement to cover in the Alfred Wegener Halvø – Qîoqe area. Several illustrations of the structures in this area are provided by Henderson and Pulvertaft (1987, Figs. 19–25); note that the "augen biotite schist" layers in Henderson and Pulvertaft's Figure 19 are zones of mylonitized Nûkavsak Formation and are the sites of the Kigarsima (upper) and Qîoqe (lower) thrust zones. The Qaqortukavsak overfold of Henderson and Pulvertaft (1987, Fig. 29) is a small gneiss-cored sheath fold trapped between the Kigarsima and Qîoqe thrust zones in the area between section lines I–J and M–N (Figs. 4 and 7g).

The cross-sections show that finite displacements of the ductile shear zones at the base of the Qîoqe and Kigarsima thrust sheets are contractional, with the direction of thrusting to the west-northwest. However, the propagation sequence is less certain. The Kigarsima and Qîoqe thrust zones and the shear zone at the base of the Kangerdluarssuk overfold all predate the mantled gneiss domes in the Inukavsait area because they are folded above them (Fig. 7c, f). The Inukavsait dome may be a hanging-wall fold above a southeasterly diping thrust zone in the footwall of the major nappes. This would imply a normal piggyback sequence of propagation, with thrusts becoming younger towards the west-northwest in the transport direction.

As already metioned, the Kangerdluarssuk overfold is a major structure in the Nûkavsak Formation, exposed in the peninsulas east of Upernivik , overlying a shear zone at the basement–cover contact (Fig. 7c). It is identical in style, and may be continuous with, the Itsakuarssuk overfold, which overlies a pre-main phase shear zone with a west-northwest transport direction on Qeqertarssuaq (Fig. 7b). Although it seems likely that both overfolds are associated with transport to the west-northwest, the relationship between the Kangerdluarssuk overfold and the Kigarsima thrust zone and nappe does not support a simple piggyback sequence of thrust propagation. Between Inukavsait and Kangerdluarssuk (Fig. 4), structures associated with the Kigarsima nappe appear to be superimposed on the overfold. This indicates that this overfold existed in the footwall before the nappe was emplaced, and implies that the Kigarsima thrust zone formed out of sequence.

Klippen of a thrust zone and overlying basement-cored nappe occur on high ground in the northeast of Upernivik and as far north as Karrat. They were attributed to the Kigarsima nappe by Henderson and Pulvertaft (1967). However, the sheath-like geometry now recognized in the nappe on Qîoqe precludes this interpretation. It is likely that these klippen represent the remains of a higher nappe that has been largely eroded away above the mantled gneiss domes of the Inukavsait–Karrat area (Fig. 4b).

In spite of the complications described above, the general conclusion is that the west-northwest directed movement of the hanging wall above the post-main phase shear zone on eastern Karrat and Qeqertarssuaq correlates with the west-northwest thrusting in the Nûgssuaq–Mârmorilik structural domain. On western Karrat, a large thickness of mylonite was formed during earlier displacements, and it appears that only the upper part of a main phase low-angle shear zone was reworked during overthrusting to the west-northwest. There is no evidence for a major post-main phase ductile shear zone on Svartenhuk Halvø, and its seems that overthrusting to the west-northwest decreases in importance northwards across the Mârmorilik–Pangnertôq structural domain. The main phase deformation is virtually unaffected by post-main phase strain on Svartenhuk Halvø, substantially overprinted by post-main phase shear zones on Qeqertarssuaq and Karrat, and so overprinted farther south that it has not yet been recognized here at all.

If main phase deformation is linked with crustal extension, this extension may also be responsible for the syn- to post-main phase high temperature –

low pressure metamorphism on Svartenhuk Halvø and as far south as Karrat. Such extension may also be responsible for syn- to post-main phase high temperature – low pressure metamorphism farther south if reversal of displacements on low-angle main phase detachments in this area occurred before thermal relaxation of the attenuated continental crust.

Prøven igneous complex domain

Two main rock types occur in the Prøven igneous complex (Fig. 5): hypersthene granite with potash feldspar megacrysts, and a garnet leucogranite (Escher and Pulvertaft, 1968). The igneous rocks are foliated, but detailed structural analysis of the complex has not been carried out.

Both the northern and southern boundaries of the intrusion dip inwards (Fig. 4b). A northeast–southwest trending zone of garnet leucocratic sheets in northeast Svartenhuk Halvø is related to the Prøven igneous complex. This zone dips gently to the north and the sheets were emplaced at a late stage of the main phase deformation in conjugate sets of extension shear fractures (Grocott and Vissers, 1984). Hypersthene granite from the complex has a Rb–Sr whole rock isochron age of 1860 ± 25 Ma and an initial $^{87}Sr/^{86}Sr$ ratio of 0.7067 (Kalsbeek, 1981). The age is interpreted as an intrusion age by Kalsbeek (1981), whereas the high initial $^{87}Sr/^{86}Sr$ ratio is consistent with involvement of older continental crustal material in petrogenesis. The data are scanty, but if the garnet leucogranite sheets and hypersthene granite are indeed coeval, the 1860 ± 25 Ma age apparently dates the main phase deformation.

North of Upernavik, the northeast–southwest trending, 6–8 km wide, Tuvssâq straight belt includes both metasediments and basement gneisses. Foliation in this belt dips steeply to the southeast, below the Prøven igneous complex. Lineations in a wider zone northwest of the igneous complex plunge southeast (Fig. 2b), but it is not known whether these reflect extension or contractional displacements.

South of the Tuvssâq straight belt, the Prøven igneous complex forms a thick folded sheet; erosion has exposed windows through the sheet into underlying sediments of the Karrat Group. Thus, the igneous rocks overlap sediments of both the Nûkavsak Formation and the Qeqertarssuaq Formation by at least 45 km on the north side of the intrusion (Figs. 4b and 5). There is no evidence of such a broad overlap on the south side of the Prøven igneous complex, where the boundary with the Karrat Group dips steeply north. The Prøven igneous complex is surrounded by a zone of hypersthene-bearing rocks in the Karrat Group. On the north side of the intrusion, hypersthene-bearing rocks occur throughout the Tuvssâq straight belt and as far as 73°30′N (Fig. 5), 40 km from the main outcrop of hypersthene granite. To the south of the intrusion, the zone of hypersthene-bearing supracrustal rocks is much narrower.

Upernavik – Red Head structural domain

The Geological Survey of Greenland has recently published a series of maps of this area (Geological Survey of Greenland, 1984, 1985a,b), and a few notes on the geology have been published in survey reports (Escher and Pulvertaft, 1968; Escher and Stecher, 1978, 1980). The domain is dominated by a series of overturned to nearly recumbent sheath-like folds involving gneiss and metasediments (Figs. 4b and 5). All of these folds show westerly or northwesterly vergence. From Nutârmiut northwards, fold axes, mineral lineations, and rodding plunge gently to moderately east (Fig. 2b). These data strongly suggest that the sheath folds developed in a subhorizontal shear regime with the upper plate moving west. Southeast of Nutârmiut, a southeast-plunging lineation becomes dominant and persits into the Tuvssâq straight belt. These two directions have been recognized in the major detachment zone in the Mârmorilik–Pangnertôq structural domain on Karrat and are assumed to have kinematic significance. On Karrat, a 250–070° lineation overprints the 310–130° lineation. Consequently, the northwest–trending mantled gneiss domes occurring in the Inukavsait area could be related to displacement in a westerly direction. The Inukavsait area may thus represent an immature version of the structure occurring north of the Tuvssâq straight belt and as far north as Holm Ø.

There appears to have been early interleaving along the basement–cover boundary prior to generation of the sheath-like folds. This is particularly clear northeast of Nûtarmiut and in the inner part of Inugsulik Bugt. Between Holm Ø and Red Head, the Nûkavsak Formation is virtually absent. The lithologies present in the gneisses resemble those of the Umanak area, but little is known about the structure.

CONCLUSION

The initial rift phase in the Rinkian belt was responsible for deposition of the Qeqertarssuaq and Mârmorilik Formations over most of the area between Nûgssuaq and Red Head, with thick accumulations in three basins at Upernavik, Ingia–Kangigdleq, and Mârmorilik. A starved basin facies at the base of the succeeding Nûkavsak Formation represents thermal subsidence following this active extension. Deposition of the thick sequence of overlying flysch turbidites, characterized by lateral and vertical uniformity coupled with a lack of slumped units and seismoturbidites, favours deposition in a basin created by renewed extension.

The succeeding phase of shortening occurred prior to metamorphism. Therefore, the high temperature – low pressure metamorphism that characterizes the belt is not a syn-rift metamorphism. Important deformations were superimposed on this early phase of thrusting. The main phase deformation is synchronous with high temperature – low pressure metamorphism in low-angle high-strain zones, and predates peak metamorphism in domains lying structurally above these zones. Main phase deformation is believed to have involved extension of the belt because in Svartenhuk Halvø, where subsequent overprinting is light, there is no evidence of duplication of the stratigraphy during the main phase and no suggestion that the metamorphism can be attributed to the addition of large volumes of igneous rock to the crust. On Svartenhuk Halvø, main phase extension was accompanied by diapirism to form a mantled gneiss antiform.

Throughout most of the southern part of the belt, main phase structures are overprinted by post-main phase ductile overthrusting to the west-northwest. The movement direction involved is very close to that associated with the main phase deformation. Ductile thrust zones in the southern part of the belt are responsible for extensive interleaving of basement and cover, and carry large-scale recumbent sheath folds in their hanging walls. These thrusts climb from basement into cover northwards, where they can be shown to partially rework main phase low-angle ductile shear zones. Post-main phase deformation dies out to the north and is weak on Svartenhuk Halvø.

Low-angle ductile shear zones are refolded by a system of northwest trending mantled gneiss domes and antiforms in the southern part of the belt. North of Upernavik, asymmetric gneiss-cored domes and antiforms showing a consitent westerly vergence dominate the structural pattern. These structures are interpreted as incipient sheath folds formed in a subhorizontal shear regime.

Diapirism is no longer believed to have been the fundamental driving force in the formation of structures in the Rinkian, even though it almost certainly has played a role in the formation of some of the mantled gneiss domes and antiforms. The discovery of a zone of intense ductile shearing at or near the base of the flysch turbidites of the Nûkavsak Formation and of extensive ductile thrust zones that have led to interleaving of basement and cover in the southern part of the belt has led to a new view of the Rinkian as primarily a belt of major lateral movements. This view is consistent with some hitherto unexplained features of the Rinkian belt; in particular, the northwest–southeast to east–west trend of extension lineations that characterizes large tracts of the belt.

The alternating phases of extension and shortening that have led to the complex history of the Rinkian belt are consistent with the proposal that the belt represents an epicontinental marginal basin. Figure 4a shows the situation in the early Proterozoic of West Greenland just after deposition of the Qeqertarssuaq and Mârmorilik Formations. Subsequent structural and metamorphic evolution of the belt depended on the kinematics and dynamics of the convergent plate margin in the Nagssugtoqidian, and the alternation of extensional and compressive phases described here is to be expected in this type of setting.

ACKNOWLEDGEMENTS

Most of the authors' fieldwork in the Rinkian has been carried out as part of the systematic mapping of West Greenland by the Geological Survey of Greenland. Cooperation and discussions with F. Kalsbeek and J.C. Escher are gratefully acknowledged. The paper is published with the permission of the Director of the Geological Survey of Greenland.

Fieldwork conducted in 1986 by JG was supported by a Royal Society of London Overseas Field Research Grant and made possible through equipment grants 1983/01 and 1984/05 provided by the Stichting Dr. Schurmannfonds (Leiden, Netherlands). Reinoud Vissers and Bas van den Eeckhout are thanked for many hours of discussion on Rinkian geology and for their cheerful companionship in the field.

REFERENCES

Andersen, M.C. and Pulvertaft, T.C.R., 1985, Rb–Sr whole rock "ages" from reworked basement gneisses in the Umanak area, central West Greenland: Bulletin of the Geological Society of Denmark, v. 34, p. 205–212.

Baird, A.W., Grocott, J. and Moody, R.T.J., 1987, Tertiary tectonics in central Tunisia and its Mediterranean plate tectonic significance: Terra Cognita, v. 7, p. 107.

Bak, J., Sørensen, K., Grocott, J., Korstgaard, J.A., Nash, D. and Watterson, J., 1975, Tectonic implications of Precambrian shear belts in western Greenland: Nature, v. 254, p. 566–569.

Bridgwater, D., Escher, A. and Watterson, J., 1973, Tectonic displacements and thermal activity in two contrasting Proterozoic mobile belts from Greenland: Philosophical Transactions of the Royal Society of London, Series A, v. 273, p. 513–533.

Coney, P.J., 1980, Cordilleran metamorphic core complexes: an overview: Geological Society of America, Memoir 153, p. 7–31.

Dawes, P.R., Larsen, O. and Kalsbeek, F., 1988, Archaean and Proterozoic crust in north West Greenland: evidence from Rb–Sr whole-rock age determinations: Canadian Journal of Earth Sciences, v. 25, p. 1365–1373.

England, P.C. and Thompson, A.B., 1984, Pressure–temperature–time paths of regional metamorphism. 1. Heat transfer during the evolution of regions of thickened continental crust: Journal of Petrology, v. 25, p. 894–928.

Escher, A. and Pulvertaft, T.C.R., 1968, The Precambrian rocks of the Upernavik–Kraulshavn area (72°00′-74°15′N), West Greenland: Geological Survey of Greenland, Report 15, p. 11–14.

Escher, A. and Pulvertaft, T.C.R., 1976, Rinkian mobile belt of West Greenland, in Escher, A. and Watt, W.S., eds., Geology of Greenland: Geological Survey of Greenland, Copenhagen, p. 105–119.

Escher, J. and Stecher, O., 1978, Precambrian geology of the Upernavik – Red Head region (72°15′-75°15′N), northern West Greenland: Geological Survey of Greenland, Report 90, p. 23–26.

Escher, J. and Stecher, O., 1980, Field work on Precambrian granites and metasediments in the Søndre Upernavik – Kuvdlorssuaq region (72°00′-74°40′), northern West Greenland: Geological Survey of Greenland, Report 100, p. 38–41.

Escher, A., Escher, J.C. and Watterson, J., 1975, The reorientation of the Kangâmiut dyke swarm, West Greenland: Canadian Journal of Earth Sciences, v. 12, p. 158–173.

Garde, A., 1978, The Lower Proterozoic Mârmorilik Formation, east of Mârmorilik, West Greenland: Meddelelser om Grønland, v. 200(3), p. 1–71.

Garde, A. and Pulvertaft, T.C.R., 1976, Age relations of the Precambrian Mârmorilik Marble Formation, central West Greenland: Geological Survey of Greenland, Report 80, p. 49–53.

Geological Survey of Greenland, 1984, Geological map of Greenland, scale 1:100,000, sheet Tasiussaq (73 V. 1 Syd).

Geological Survey of Greenland, 1985a, Geological map of Greenland, scale 1:100,000, sheets Ussing Isfjord (73 V. 1 Nord) and Kuvdlorssuaq (74 V. 1 Nord/Syd).

Geological Survey of Greenland, 1985b, Geological map of Greenland, scale 1:500,000, Upernavik Isfjord.

Grocott, J., 1984, Geometry of superimposed colinear deformations in the Ikerasak area, Umanak district, central West Greenland: Precambrian Research, v. 26, p. 235–263.

Grocott, J. and Pulvertaft, T.C.R., 1987, The early Proterozoic Rinkian belt of central West Greenland: Geological Association of Canada – Mineralogical Association of Canada, Program with Abstracts, v. 12, p. 49.

Grocott, J. and Vissers, R.L.M., 1984, Field mapping of the early Proterozoic Karrat Group on Svartenhuk Halvø, central West Greenland: Geological Survey of Greenland, Report 120, p. 25–31.

Grocott, J., van den Eeckhout, B. and Vissers, R.L.M., 1987, Mantled gneiss antiforms and fold nappes in the Rinkian belt, West Greenland: diapiric structures or structures formed in a thrust system?: Journal of the Geological Society of London, v. 144, p. 723–734.

Henderson, G., 1969, The use of structural contour maps in the study of gneiss–metasediment relations in the Umanak area, West Greenland: Geological Association of Canada, Special Paper 5, p. 129–142.

Henderson, G. and Pulvertaft, T.C.R., 1967, The stratigraphy and structure of the Precambrian rocks of the Umanak area, West Greenland: Meddelelser fra Dansk Geologisk Forening, v. 17, p. 1–20.

Henderson, G. and Pulvertaft, T.C.R., 1987, The lithostratigraphy and structure of a Lower Proterozoic dome and nappe complex: descriptive text to 1:100,000 sheets Mârmorilik, Nûgâtsiaq and Pangnertôq, West Greenland: Geological Survey of Greenland, Copenhagen, p. 1–72.

Henderson, J.R., 1983, Structure and metamorphism of the Aphebian Penrhyn Group and its Archaean basement complex in the Lyon Inlet area, Melville Peninsula, District of Franklin: Geological Survey of Canada, Bulletin 324, p. 1–50.

Henderson, J.R., Shaw, D., Mazurki, M., Henderson, M., Green, R. and Brisbin, D., 1979, Geology of part of the Foxe Fold Belt, central Baffin Island, District of Franklin: Geological Survey of Canada, Paper 79-1A, p. 95–99.

Kalsbeek, F., 1981, The northward extent of the Archaean basement of Greenland – a review of Rb–Sr whole rock ages: Precambrian Research, v. 14, p. 203–219.

Kalsbeek, F., 1986, The tectonic framework of the Precambrian shield in Greenland: a review of new isotopic evidence: Geological Survey of Greenland, Report 128, p. 55–64.

Kalsbeek, F. and Taylor, P.N., 1985, Isotopic and chemical variation in granites across a Proterozoic continental margin – the Ketilidian mobile belt of South Greenland: Earth and Planetary Science Letters, v. 73, p. 65–80.

Kalsbeek, F. and Taylor, P.N., 1986, Chemical and isotopic homogeneity of a 400 km long basic dyke in central West Greenland: Contributions to Mineralogy and Petrology, v. 93, p. 439–448.

Kalsbeek, F., Taylor, P.N. and Henriksen, N., 1984, Age of rocks, structures, and metamorphism in the Nagssugtoqidian mobile belt, West Greenland – field and Pb-isotope evidence: Canadian Journal of Earth Sciences, v. 21, p. 1126–1131.

Kalsbeek, F., Pidgeon, R.T. and Taylor, P.N., 1987, Nagssugtoqidian mobile belt of West Greenland: an 1850 Ma suture between two Archaean continents – chemical and isotopic evidence: Earth and Planetary Science Letters, v. 85, p. 365–385.

Kalsbeek, F., Taylor, P.N. and Pidgeon, R.T., 1988, Well-preserved Archaean and Proterozoic rocks in the north-eastern part of Disko Bugt, West Greenland, with implications for the nature of Proterozoic mobile belts in Greenland: Canadian Journal of Earth Sciences, v. 25, p. 773–782.

Lister, G.S., Etheridge, M.A. and Symonds, P.A., 1986, Detachment faulting and the evolution of passive continental margins: Geology, v. 14, p. 246–250.

Patchett, P.J. and Bridgwater, D., 1984, Origin of continental crust of 1.9–1.7 Ga defined by Nd isotopes in the Ketilidian terrain of South Greenland: Contributions to Mineralogy and Petrology, v. 87, p. 311–318.

Pulvertaft, T.C.R., 1973, Recumbent folding and flat-lying structure in the Precambrian of northern West Greenland: Philosophical Transactions of the Royal Society of London, Series A, v. 273, p. 535–545.

Pulvertaft, T.C.R., 1986, The development of thin thrust sheets and basement–cover sandwiches in the southern part of the Rinkian belt, Umanak district, West Greenland: Geological Survey of Greenland, Report 128, p. 75–87.

Ramberg, H., 1948, On the petrogenesis of the gneiss complexes between Sukkertoppen and Christianshaab, West Greenland: Meddelelser fra Dansk Geologisk Forening, v. 11, p. 312–327.

Sandiford, M. and Powell, R., 1986, Deep crustal metamorphism during continental extension: modern and ancient examples: Earth and Planetary Science Letters, v. 79, p. 151–158.

St-Onge, M.R. and King, J.E., 1987, Evolution of regional metamorphism during back-arc stretching and crustal shortening in the 1.9 Ga Wopmay orogen, Canada: Philosophical Transactions of the Royal Society, Series A, v. 321, p. 199–218.

Talbot, C.J., 1979, Infrastructural migmatitic upwelling in East Greenland interpreted as thermal convective structures: Precambrian Research, v. 8, p. 77–93.

Thompson, A.B., 1981, The pressure–temperature (*P–T*) plane viewed by geophysicists and petrologists: Terra Cognita, v. 1, p. 12–20.

Thompson, A.B. and England, P.C., 1984, Pressure–temperature–time paths of regional metamorphism. 2. Their inference and interpretation using mineral assemblages in metamorphic rocks: Journal of Petrology, v. 25, p. 929–955.

Van den Eeckhout, B. and Grocott, J., 1982, Investigations on the Precambrian rocks of Svartenhuk Halvø, central West Greenland: Geological Survey of Greenland, Report 110, p. 22–26.

Van den Eeckhout, B. and Zwart, H.J., 1988, A Hercynian crustal-scale extensional shear zone in the Pyrenees: Geology, v. 16, p. 135–138.

Walker, R.G. and Mutti, E., 1973, Turbidite facies and facies associations, *in* Middleton, G.V. and Bouma, A.H., eds., Turbidites and Deep Water Sedimentations: Pacific Section, Society of Economic Paleontologists and Mineralogists, Short Course Notes, p. 119–157.

Watterson, J., 1978, Proterozoic intraplate deformation in the light of south-east Asian neotectonics: Nature, v. 273, p. 636–640.

Wernicke, B., 1985, Uniform-sense normal simple shear of the continental lithosphere: Canadian Journal of Earth Sciences, v. 22, p. 108–125.

Wickham, S.M. and Oxburgh, E.R., 1985, Continental rifts as a setting for regional metamorphism: Nature, v. 318, p. 330–333.

Wickham, S.M. and Oxburgh, E.R., 1987, Low-pressure regional metamorphism in the Pyrenees and its implications for the thermal evolution of rifted continental crust: Philosophical Transactions of the Royal Society of London, Series A, v. 321, p. 219–242.

Structural Styles in the Deep Levels of an Early Proterozoic Foreland Thrust Belt

Adrian F. Park[1] and J.J. Doody
Department of Geology, University of Glasgow,
Glasgow G12 8QQ, Scotland, United Kingdom

Park, A.F. and Doody, J.J., Structural styles in the deep levels of an Early Proterozoic foreland thrust belt, in Lewry, J.F. and Stauffer, M.R., eds., The Early Proterozoic Trans-Hudson Orogen of North America: Geological Association of Canada, Special Paper 37, p. 465–482

Abstract

Concepts derived from thin-skinned tectonic models for thrust belts can be applied to deformation and structural styles in all levels of a crustal stack. In the Kaavi area of the Svecokarelides of eastern Finland, the deep levels of a thrust belt are exposed: frontal and sidewall ramps, duplex zones, and nappe sheets are recognized. High metamorphic grade (600–750 °C, 3–5 kbar (300–500 MPa)) coeval with thrust propagation means that ductile processes dominated; synkinematic magmatism was pervasive. Thus, the thrust belt includes such features as ductile shear zones, sheath folds, and mantled gneiss domes, which are characteristic of high-grade terrains throughout Proterozoic and Archean tectonic provinces. All of these can be related kinematically to thrusting, rather than some uniquely early Precambrian crustal process.

Résumé

Le modèle tectonique par écaillage pour la formation des chaînes plissées peut être appliqué aux styles structuraux propres à tous les niveaux de la croûte. Dans les Svecokarelides de l'est de la Finlande, la région de Kaavi montre les niveaux profonds d'une chaîne plissée: des rampes frontales et latérales, des zones de duplex et des nappes ont été reconnues. Les charriages se produisent en régime ductile avec un métamorphisme entre 600–750 °C et 3–5 kbar (300–500 MPa) et un magmatisme syn-cinématique pénétratif. La chaîne plissée comportent des zones de cisaillements ductiles, des plis en fourreaux et des dômes de gneiss caractéristiques des terrains à haut métamorphisme des provinces tectoniques archéennes et protérozoïques. Ces structures peuvent être reliées cinématiquement aux charriages plutôt qu'uniquement à des processus crustaux Précambrien Inférieur.

---

[1] Present address: Department of Geology, University of New Brunswick, P.O. Box 4400, Fredericton, New Brunswick E3B 5A3, Canada.

INTRODUCTION

Thin-skinned tectonic models for foreland thrust belts have been developed largely from studies of Phanerozoic orogens. Such thrust belts involve shortening, by predominantly brittle processes, of sedimentary sequences that never underwent prolonged thermal reequilibration in high-grade regimes (e.g., Dahlstrom, 1970), or are areas where high-grade allochthons were partially retrogressed during exhumation along continuations of high-level thrust belts propagated to depth (e.g., Hossack, 1979; Coward, 1983). Detailed geometric analysis has largely been restricted to such high-level segments (e.g., Elliot and Johnson, 1980). Although theoretical models, reconnaissance studies, and geophysical studies have considered the nature of thrust-belt related shortening beneath the "elastic lid" (approximately 10 km depth), detailed analysis of the geometric form and style of deformation in the ductile zone has been limited in scope (cf. Coward, 1983, 1984), or has been concerned with large-scale considerations (cf. Daly, 1986).

Regional studies of deformation in high-grade Precambrian terrains have used two approaches. One method tacitly accepts the apparent difference in style seen in such belts as evidence that thin-skinned tectonic models do not apply and proceeds to propose fundamentally different mechanisms (e.g., diapiric reactivation). Alternatively, thin-skinned models are applied without considering the relevance of a polyphase history or the high metamorphic grades developed during deformation.

In part of the Svecokarelides (Early Proterozoic) of eastern Finland, one phase of deformation in a polyphase history involved shortening largely taken up in discrete shear zones, many of which are demonstrably thrusts or geometrically related elements, such as ramps and sidewalls. Peak metamorphism, synchronous with this deformation phase, was uniformly high and, superficially, thermal equilibration between overthrust units appears to have been almost complete. Thus, the deep levels of this foreland thrust belt include not only most of the geometric elements of a truly thin-skinned shortening zone but also many of the classic features of high-grade Precambrian terrains, namely synkinematic granitoids and migmatites, ductile shear zones, variably reworked basement allochthons, and mantled gneiss domes. In this paper, part of the deformation history of a typical high-grade terrain is reviewed, with the demonstration that many of these characteristic features are consequences of crustal shortening by thrusting under a ductile regime.

GEOLOGICAL BACKGROUND

General description

The Karelian and Svecofennian supergroups (Meriläinen, 1980) form an Early Proterozoic supracrustal cover sequence overlying the Presvecokarelian Archean basement gneisses in eastern and central Finland. To the southwest, these basement and cover rocks have an ambiguous relationship with the Early Proterozoic Svecofennian complex, which is highly disturbed by Early to Middle Proterozoic strike-slip movements along shear zones of the Raahe-Ladoga lineament (Figs. 1 and 2). To the north and east, the Karelian supracrustal formations lap onto the basement with a contact that is essentially a complex, variably tectonized unconformity (see Gaál *et al.*, 1975; Pekkarinen, 1979; Simonen, 1980; Koistinen, 1981). The Kalevian mica schists and gneisses of the Svecofennian Supergroup form the youngest part of this cover sequence, but are separated from the older Karelian Supergroup by an unconformity, locally overstepping onto the Archean basement (Luukkonen and Lukkarinen, 1985). All the supracrustal rocks display the effects of a complex, polyphase deformation history, involving metamorphic reconstitution in the amphibolite facies in most areas, though locally greenschist and granulite facies are preserved (Koistinen, 1981; Treloar *et al.*, 1981; Halden and Bowes, 1984; Korsman *et al.*, 1984). Deposition belongs to the period 2.20–1.97 Ga; late- and posttectonic granites yield ages in the range 1.89–1.85 Ga (see review and references in Gaál (1986) and Park (1988b)). The tectono-thermal climax is generally regarded as occurring ca. 1.90 Ga.

Tectono-stratigraphy

The area covered by Figure 2 is divisible into distinct tectonic provinces with differing lithofacies assemblages and successions. The whole area is bounded on the west by the Savo suture and the Kotalahti nickel belt, and is subdivided fundamentally by the Suhmura thrust zone.

East of the Suhmura thrust zone, Archean basement gneisses are overlain locally by the Sariolan lithofacies, an assemblage of poorly sorted arenites and rudites, with paleoregoliths and paleosols (satrolite), mafic lavas, calc-silicate skarns, and carbonates. This is overlain by the autochthonous, epicontinental Jatulian lithofacies, which oversteps directly onto the basement. The lower Jatulian consists largely of metaquartzites, but includes a major development of mafic lavas, related pyroclastic rocks, and hypabyssal intrusions. The upper Jatulian is more "marine," containing metaquartzites, metagreywackes, phyllites, banded ironstones, and carbonates. The Höytiäinen succession, a thick sequence dominated by metagreywackes and chemogenic deposits, is considered to be the distal equivalent of the lower, and possibly part of the upper, Jatulian lithofacies (Pekkarinen, 1979; Meriläinen, 1980; Ward, 1987, 1988).

West of the Suhmura thrust belt, the lithofacies succession differs. No definite Sariolan lithofacies is seen, but locally the basement gneisses are overlain by diverse lower Jatulian lithologies. These consist predominantly of massive metaquartzites with mafic lavas, but locally a thinned succession is dominated by arkosites, calc-silicate rocks, black schists, carbonates, and metabasites (e.g., west of the Maarianvaara basement high, and around the Juojärvi, Sotkuma, Liperi and Oravisalo domes (cf. Huhma, 1971, 1975; Ward, 1987, 1988)). Upper Jatulian lithofacies, the greywacke–carbonate assemblage, is of restricted occurrence, found only northeast of Kajoo and near Kaavi (Fig. 2). The Jatulian lithofacies are succeeded by the autochthonous Kalevian psammites and pelites of the Svecofennian Supergroup (Gaál and Gorbatschev, 1987). This group locally oversteps directly onto the Archean basement, and in these localities a basal arkosite–calc-silicate skarn lithofacies is recognized (Huhma, 1971, 1975; Park, 1988a).

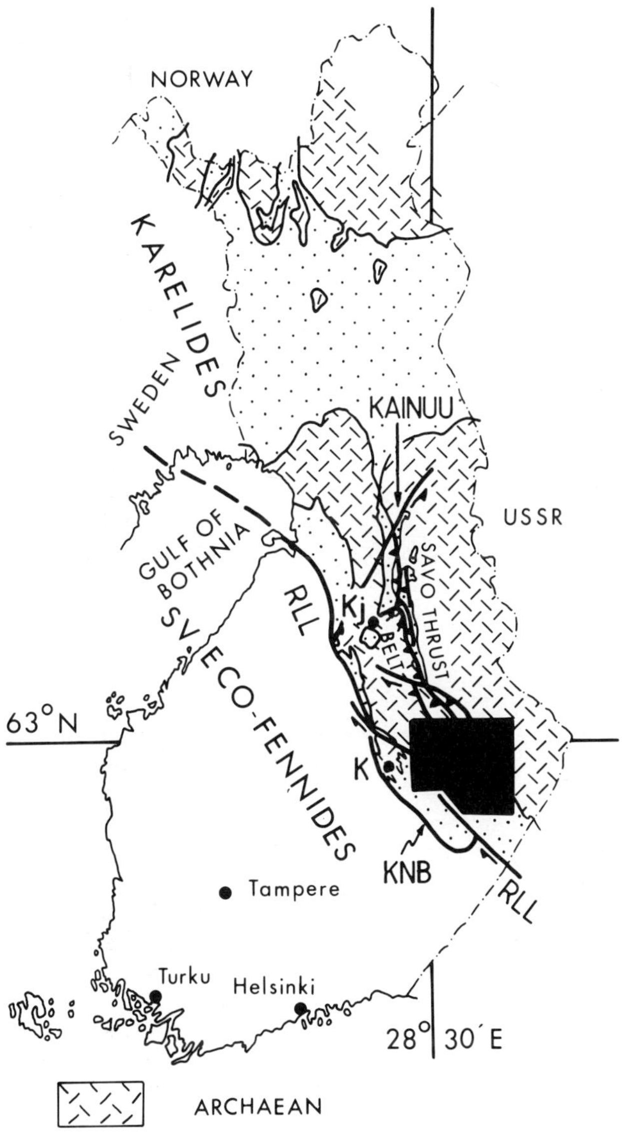

Figure 1. *Geological map of Finland showing the location of some principal features discussed in the text. The location of Figure 2 is blacked out. Abbreviations: K, Kuopio; Kj, Kajaani; KNB, Kotalahti nickel belt; and RLL, Raahe–Ladoga lineament.*

The area west of the Suhmura thrust, termed the Savo province, contains a number of allochthons within the supracrustal cover, the largest being the Outokumpu nappe. The stratigraphy in this nappe differs considerably from that in the autochthon, consisting entirely of supracrustal rocks of the Svecofennian Supergroup, which unconformably overlie Karelian Supergroup rocks in the autochthon. At the stratigraphic and structural base of the Outokumpu nappe, the Kalevian sediments overlie, without any recognized break, a distinctive group of black schists, chemogenic sediments, mafic volcanic rocks, and altered ultramafic plutonic rocks known as the Outokumpu assemblage. This group of rocks hosts the Cu–Ni–Co–Zn sulphide ores of the Outokumpu mining district, and no equivalent is recognized in the autochthon (Park, 1988b).

According to Kontinen (1987), the Kalevian lithofacies can be further subdivided into an upper and lower division. The lower Kalevian contains quartzarenites, psammites, arkosites, and conglomerates, whereas the upper Kalevian consists of a uniform sequence of psammites and pelites of flyschoid aspect. No lower Kalevian is recognized in the Outokumpu nappe, but both subdivisions are recognized in the autochthon.

Deformation and metamorphism

A full descriptive account of the polyphase deformation history can be found in Park and Bowes (1983) and Bowes *et al.* (1984); a summary is included here as Table I. Briefly, deformation commenced with the emplacement of the Outokumpu nappe; thickening of the supracrustal pile initiated metamorphic reconstitution, producing a ubiquitous metamorphic banding in the cover rocks. This was originally a pressure solution segregation in psammites, recrystallized as a mineral segregation banding (S_1) as temperatures rose. Locally, e.g., around Outokumpu, S_1 is axial planar to now isoclinal folds (F_1) that also influence the outcrop of the Outokumpu nappe (hence, the nappe is regarded as a pre-D_1 structure). S_1 is itself folded by a set of more open folds

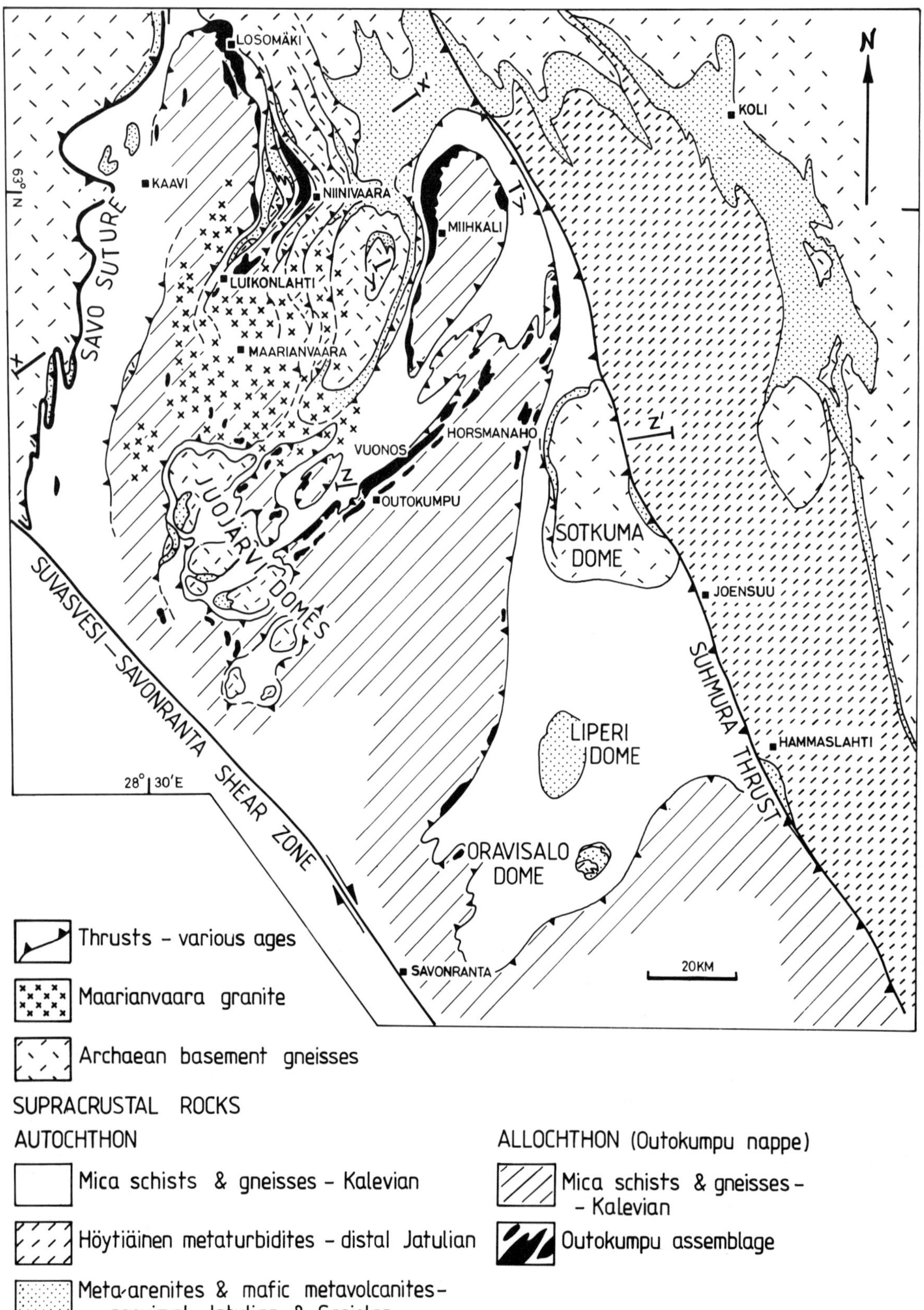

Figure 2. *Geological map of part of the Savo thrust belt in eastern Finland showing the location of regional features mentioned in the text. The lines x–x′, y–y′, and z–z′ are the approximate lines of sections used to construct the composite shown in Figure 12. (After Huhma, 1971; Koistinen, 1981, 1987; Park and Bowes, 1983; Bowes* et al., *1984; and Ward, 1987, 1988.)*

(F_2), with a pronounced axial planar fabric (S_2). In the Savo thrust belt, these folds are usually asymmetrical structures showing strong transposition of S_1 and S_0 (bedding) in one, usually the overturned, limb. This limb is usually subparallel to the axial planar fabric and is often the site of a D_2 slide or even a thrust (Park and Bowes, 1983).

Peak metamorphic conditions occurred either during or after development of mica alignment in S_2 because locally the M_2 mineral assemblages are either aligned and define S_2 (e.g., at Kaavi) or form a static overgrowth (e.g., around Outokumpu (Park, 1983; Treloar *et al.*, 1981)). Likewise, where partial melting is evident (e.g., west of Outokumpu and southwest of Kaavi), neosome veins show a variety of relationships to D_2 features, but are usually consistent with development during or immediately after D_2 deformation (Park and Bowes, 1983). Geothermometric and barometric data are scarce,

Table I

Svecokarelian events in the Losomäki–Maarianvaara segment of the Savo thrust belt (see Fig. 2) (after Huhma, 1981, 1986; Koistinen, 1981; Park and Bowes, 1983; Park, 1988c)

	Emplacement of the Outokumpu nappe (post-1.97 Ga). Development of pressure solution fabric-segregation
D_1	Rare isoclinal folds (F_1), recrystallization of segregation (S_1) with biotite growth in mica schists and gneisses, transposition of S_0 produces composite fabric S_{0-1}. Intense shearing of basement–cover interface, but no pervasive deformation of basement gneisses
D_2	Open to tight folds (F_2) overturned to the east; axial planar schistosity (S_2); thrusts and slides in overturned limbs. Peak of metamorphism (600–700°C, 2–4 kbar (200–400 MPa)), partial melting produces neosomes in mica schists and gneisses. Earliest Maarianvaara granitoids (quartz diorite, granodiorite, granite) intruded. Thrusts penetrate basement to depth, deformation intense, but in restricted zones. Basement–cover intercalated stack produced
D_3	Open, upright folds trending north–south (F_3); axial planar cleavage (S_3), locally crenulation. Onset of metamorphic retrogression. Probable maximum intrusion of Maarianvaara granitoids
D_4	Open folds plunging southwest (F_4); axial planar cleavage (S_4), locally crenula-tion. $F_3 \times F_4$ interference pattern produced. Continued metamorphic retrogression. Last Maarianvaara granitoids intruded (ca. 1.87 Ga)
D_{late}	Several sets of late fractures Posttectonic microtonalite and lamprophyre dykes intruded (1.86–1.83 Ga)

but a peak of 600–650°C at 3.5–4.5 kbar (350–450 MPa) is determined from Outokumpu (Treloar *et al.*, 1981), and similar temperatures with lower pressures (2–4 kbar (200–400 MPa)) prevailed around Niinivaara (Park, 1983). Temperatures may have been lower still at Losomäki, whereas the coexistence of kyanite with granitic neosomes at Luikonlahti (Väyrynen, 1939) suggests both pressures and temperatures were higher in the southwest of the Kaavi district and, by inference, west of Outokumpu.

Post-D_2 events and their associated fabrics are largely related to regional retrogression: D_3 and D_4 are the most important of these in the Savo thrust belt. F_3 folds deform S_2 into open, upright structures, with subhorizontal axes, trending generally north–south. This trend usually swings to north-northeast–south-southwest or even northeast–southwest as the Suvasvesi–Savonranta shear zone is approached (Fig. 2). The axial planar fabric (S_3) is either a fracture cleavage or crenulation depending on the degree of anisotropy produced by the S_2 schistosity. M_3 mineral assemblages in the Kaavi and Miihkali areas indicate greenschist facies, but near Outokumpu, *P–T* conditions of the amphibolite or epidote–amphibolite facies may still have prevailed.

F_4 folds are also upright, open structures of consistent trend, plunging southwest. The axial planar fabric (S_4) is generally a fracture cleavage and, in suitable lithologies, a well-developed crenulation (L_4). F_3 and F_4 interfere, producing a dome and basin pattern that dominates much of the Outokumpu region, forming features such as the basin-like closure of Outokumpu assemblage around Miihkali, and contributing a component to the Maarianvaara dome (Figs. 2 and 3) (Gaál *et al.*, 1975; Koistinen, 1981; Bowes *et al.*, 1984).

Reworking of the basement gneisses during Svecokarelian deformation and metamorphism varies enormously in its style and extent. In the east, between Koli and the Soviet border, the emplacement of a ca. 2.5 Ga old suite of granites effectively marks the end of major reconstitution of the Presvecokarelian rocks. In the Savo thrust belt, reworking is more extensive. Firstly, the exposed basement west of the Suhmura thrust zone is contained within thrust-bounded units, so that strictly it is all allochthonous. Secondly, Svecokarelian granites like the Maarianvaara pluton and a late porphyritic granite around Savonranta have Sr and Nd ratios consistent with partial derivation by melting of an evolved Archean source (Halden, 1982; Huhma, 1986). Thirdly, U–Pb isotopic zircon dates from basement gneisses west of the Savo suture, primarily around Kuopio (Fig. 1), show evidence of thermal resetting by Pb leakage at around 1.86 Ga (Kouvo and Tilton, 1966).

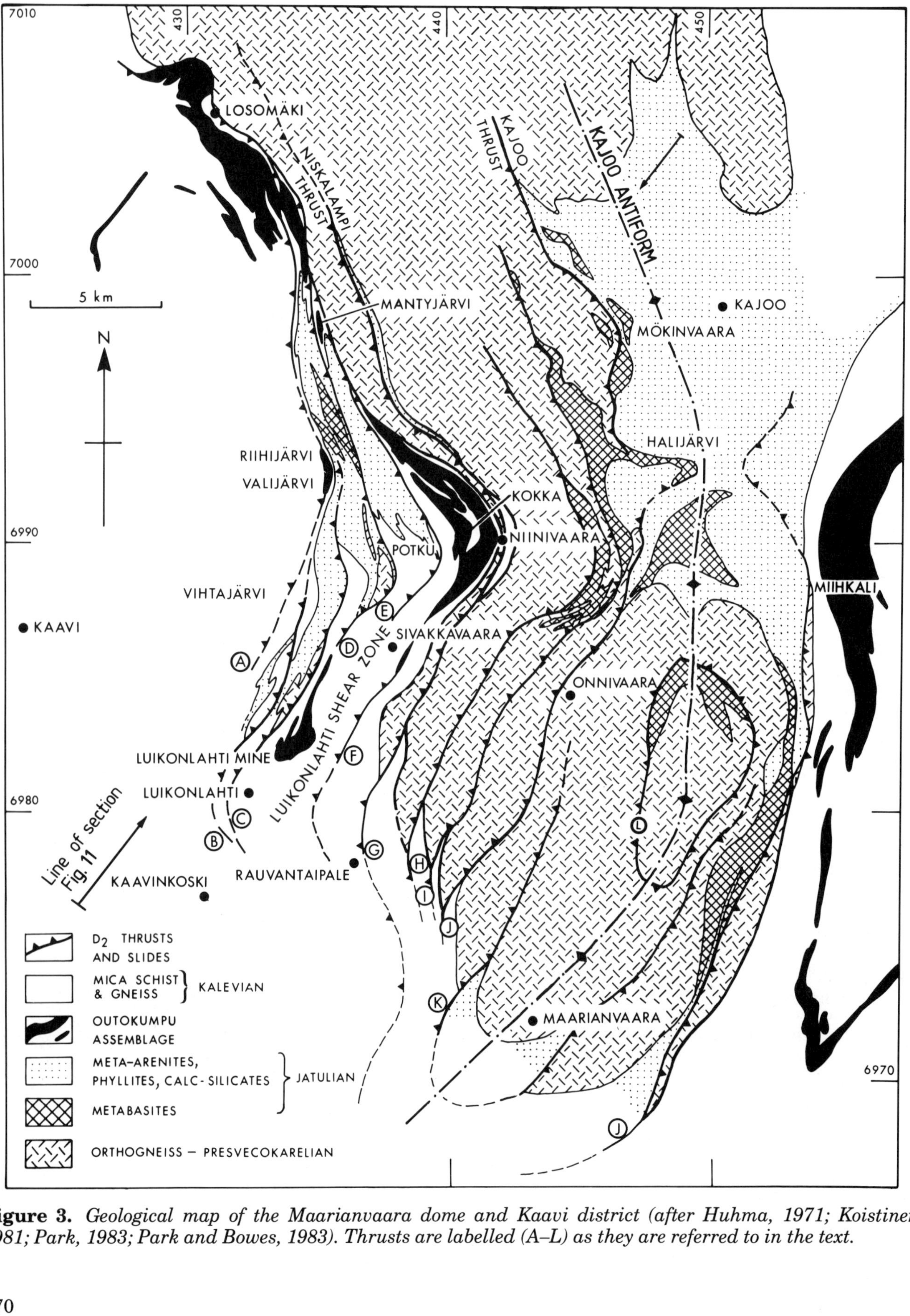

Figure 3. *Geological map of the Maarianvaara dome and Kaavi district (after Huhma, 1971; Koistinen, 1981; Park, 1983; Park and Bowes, 1983). Thrusts are labelled (A–L) as they are referred to in the text.*

SAVO THRUST BELT

The Savo thrust belt can be traced north from Outokumpu for some 200 km into the Kajaani area, and possibly as far as Kainuu (Fig. 1). To date, only the first 50 km north from Outokumpu have been subjected to detailed analysis. This part of the belt has the most complete coverage on published 1:100,000 scale geological maps (Huhma, 1971; Paavola, 1980).

Losomäki–Luikonlahti area

This segment of the Savo thrust belt (Figs. 1, 2, and 3) is the best exposed and contains the least development of Maarianvaara granite, which obscures much detail farther south. It is delimited by the shear zones at Losomäki (Poskijärvi shear zone) and Luikonlahti (Luikonlahti shear zone). Between these two northeast–southwest trending structures, the outcrop is dominated by a lithological pattern largely controlled by D_2 structures.

D_1 features throughout this zone are dominated by the S_1 metamorphic segregation banding, which usually lies close to S_0, producing a composite, transposed fabric (S_{0-1}). This S_1 fabric is discordant only in the rare closures of F_1 folds, to which it is axial planar. No large-scale F_1 folds are evident (cf. Park and Bowes, 1983). Wherever sedimentary younging can be demonstrated in this area, bedding faces upward on F_2 folds, suggesting that the absence of large F_1 folds is a real feature of the bedrock, and not a consequence of erratically distributed exposure.

D_1 fabrics penetrate the basement gneisses only to a limited extent, producing a sheared zone immediately beneath the intact sub-Karelian unconformity, e.g., around Niinivaara. Here, the sheared zone is some 10 m thick, possibly reflecting the prior development of paleoregolith (Park, 1988a). No deep penetration of the basement is evident, and S_1 or S_{0-1} usually lie parallel to the basement–cover interface.

In the Outokumpu area, F_1 folds deform the outcrop pattern of the Outokumpu nappe, but in the Kaavi area the base of this nappe lies parallel to S_1 and S_0; no large-scale F_1 folds affect its outcrop pattern.

D_2 structures and fabric elements have variable relationships to the F_2 folds. The latter occur on all scales, and are invariably open to tight asymmetrical structures, overturned toward the east. S_2 is an axial planar fabric, usually formed by reorientation of micas in the S_1 banding, but is locally a transposing penetrative schistosity. Axial planes are usually asymmetrically disposed within the F_2 folds, being parallel or subparallel to the overturned limbs, where transposition and attenuation (slides) are evident (Fig. 4a). Locally, the removal of layers or the presence of phyllonite or mylonite marks the presence of a thrust. Even where thrust contacts are clearly exposed, as at Mökinvaara (west of Kajoo), where basement gneiss lies above Jatulian metaquartzite with sharp contact, high-strain zones several metres wide are evident (Park and Bowes, 1983). Slides and thrusts in the cover rocks can be traced as discrete entities for considerable distances: locally, these anastomose around intervening, rootless F_2 folds, defining ductile duplexes (Fig. 4b). Such features are developed on all scales, from a single outcrop up to several kilometres in extent, and are particularly distinctive where there is lithological diversity. Examples from the Potku and Kokka duplex zones are illustrated in Figure 5.

The orientation of S_2, and the F_2 axial plane, changes systematically from west to east, defining a broad antiform whose crest lies in the Kajoo area

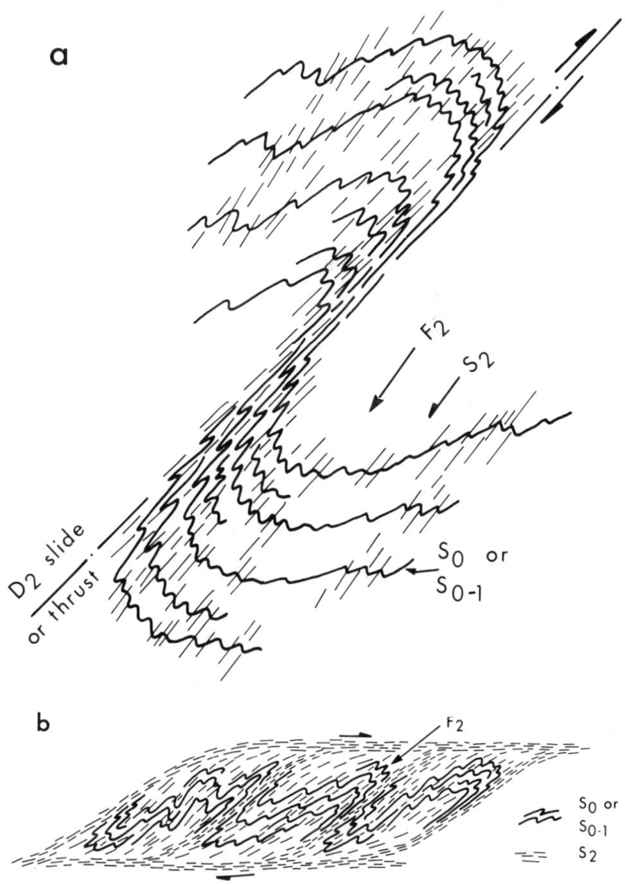

Figure 4. *(a) Idealized relationship of F_2 folds and axial planar fabric (S_2) to zones of D_2 transposition and D_2 slides. (b) Fabric relationships and slide–fold/thrust–fold geometry in an idealized D_2 ductile duplex. These features are developed on several scales.*

(Kajoo antiform). Generally, the thrusts and slides of D_2 age are concordant to this arched surface.

F_2 axes, either directly observed or inferred from the L_2 intersection lineations, have a generally consistent plunge of 20–40° toward the south of south-southwest from Losomäki southward to the Riihijärvi–Vihtajärvi area. Farther south, the plunge steepens and the trend swings around to the southwest in the vicinity of the Luikonlahti shear zone, through a belt up to 3 km wide. This swing parallels the outcrop trend of the basement–cover contact, and of the D_2 thrusts. Locally, some curvilinearity is evident in the F_2 hinges, e.g., around Potku (Figs. 3 and 5b). This becomes more pronounced as the Luikonlahti shear zone is approached, and true sheath folds, with paraboloidal hinge lines, are present at the Luikonlahti mine (Park, 1988c).

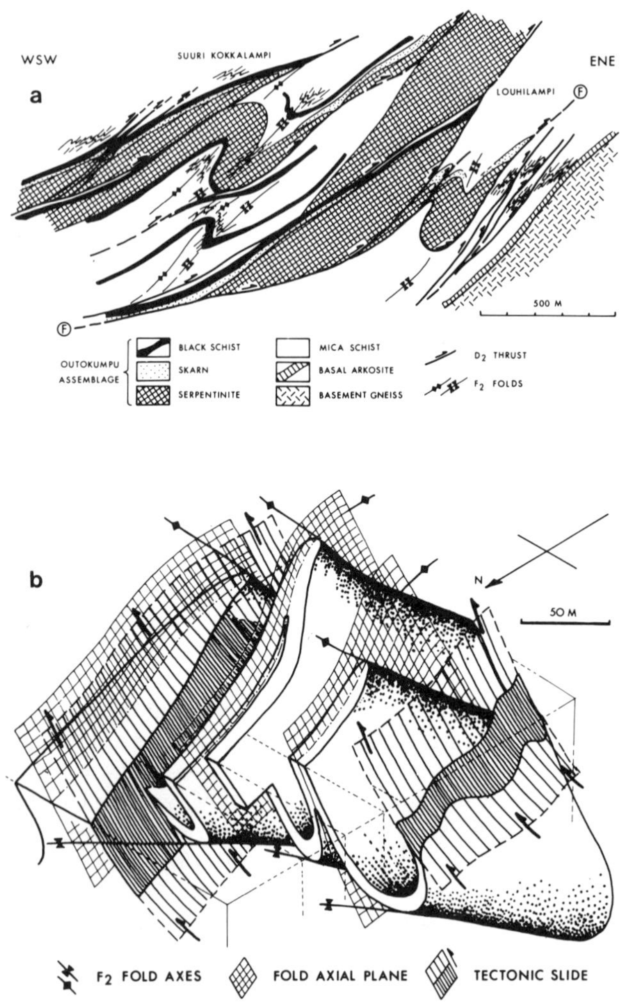

Figure 5. *Details of fabrics, folds, slides, and thrusts in D_2 ductile duplex zones. (a) Cross-section in the Kokka duplex between Suuri Kokkalampi and Louhilampi, 2 km northwest of Niinivaara. F is the thrust so marked on Figures 3, 10, and 11 (after Park (1983) and unpublished material). (b) Block diagram (isometric projection from the northwest) of fold–slide relationships picked out by the upper surface of a Jatulian arkosite in the Potku duplex, Potkulampi, 5 km northwest of Niinivaara (from Park and Bowes, 1983).*

D_2 structures, unlike those of D_1, penetrate the basement gneisses to considerable depths. This is particularly the case around Niinivaara and Kajoo, where D_2 thrusts bring basement gneisses over cover rocks. Other examples occur at Karpalolampi, where an F_2 antiformal core carries basement gneiss, bounded on the east by a westward-dipping thrust. In the absence of intercalated cover rocks, thrusts can be detected in the basement gneisses by the presence of linear high-strain zones, within which the Presvecokarelian fabric elements and structures are transposed into a new fabric, often with mylonitic or phyllonitic characteristics, depending on whether or not hydration has accompanied grain-size reduction. Pre-D_2 pegmatoids are developed into coarse rods, and a hornblende stretching lineation develops in the granodioritic gneisses. A locally developed "big feldspar gneiss," a foliated granitoid with rapakivi characteristics, undergoes spectacular changes. The feldspar megacrysts suffered varying forms of deformation, both oblate and strongly prolate forms being observed at Kajoo and Niinivaara respectively.

A number of stretching lineations are developed, no single form being widespread. The rodlike pre-D_2 pegmatoids in basement gneiss are one form; prolate feldspar porphyroclasts are another. In the arkosic Jatulian metaquartzites, feldspar clasts are reduced to linear aggregates of muscovite; these are conspicuously developed around Halijärvi. There is widespread linear growth of hornblende in some highly strained gneisses around Niinivaara. Such markers are rare in the cover rocks west of the main basement outcrop, but examples include microboudinage of linear cummingtonites in metaserpentinite at Suuri Kokkalampi, and prolate deformed pillows at Losomäki. General shape fabric markers include the sheath folds at the Luikonlahti mine.

The general form of S_2 through this zone changes most strikingly in two well-defined shear zones. The Luikonlahti shear zone (Fig. 3) is a 2–3 km wide D_2 structure in which S_2 trends northeast–southwest and dips about 80° to the northwest. In the smaller Poskijärvi shear zone, at Losomäki, which is only 300–500 m wide (Fig. 6), S_2 trends northeast–southwest, dipping southeast at 45–50°. The basement–cover contact is sinistrally displaced by some 10 km at Luikonlahti, whereas a dextral shift of a little over 1 km is evident at Losomäki.

The most conspicuous features of the Luikonlahti shear zone are the big F_2 sheath folds at the Luikonlahti mine (Park, 1988c). The sheaths have an X-axis length in excess of 3 km and truly paraboloidal hinge lines. The X-axes here plunge 20–30° to the southwest, all in the S_2 plane. No such features can be seen at the surface in the Poskijärvi shear zone, but pillows develop a prolate form around the northeast end of the shear zone here, and the stretching direction so defined plunges southwest at about 30° in the plane of S_2.

The Poskijärvi shear zone cannot be traced very far. To the southwest, it is lost in poorly exposed ground only 1.5 km to the southwest of Ala-Poskijärvi (Fig. 6). To the northeast, despite good exposure, it is not evident in the basement gneisses beyond Ylä-Poskijärvi; it dies out across the strike of S_2. The Luikonlahti shear zone is larger, its effects being visible for at least 10 km southwest of the Luikonlahti mine, though it is locally obscured by Maarianvaara granite. To the northeast, its effects are present in the basement northeast of Niinivaara, but between Kajoo and Halijärvi, where it should continue, there is no sign of it (Fig. 3).

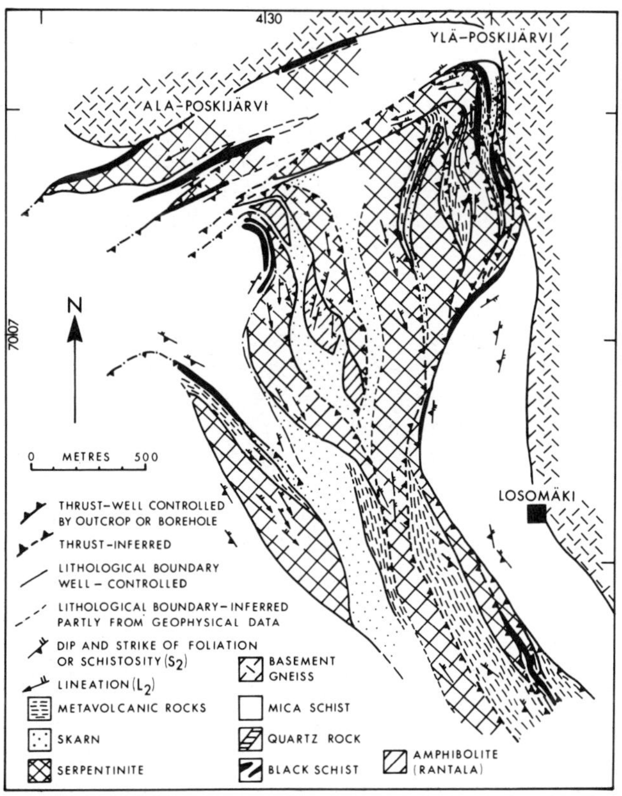

Figure 6. *Geology and structure of the Losomäki area (for location see Fig. 3) showing the termination of D_2 thrusts and imbricate slices (horses) against the Poskijärvi shear zone, interpreted as an oblique sidewall ramp (after Park, 1984).*

Both shear zones die out downward in the thrust stack.

Luikonlahti–Maarianvaara area (the Maarianvaara dome)

The overall geometry and geological relationships of the Maarianvaara dome are consistent with the concept of a "mantled gneiss dome" as defined by Eskola (1949). An ovoid core of basement gneiss is partially surrounded by a "mantle" of early Proterozoic supracrustal rocks whose primary layering dips radially outward from the core, and whose strike is crudely tangential. Within the gneiss core, one of the more prominent fabrics also displays a crude concordance to the basement–cover contact, especially in close proximity to that contact. Furthermore, the Maarianvaara granite, which invaded both basement and cover rocks, has many of the features of Eskola's (1949) palingenetic granites, whose genesis was crucial to the "inflation" of the gneiss diapir his model demanded. The D_2 high-strain zones seen in the basement gneisses here superficially conform to this pattern, though they become hard to follow precisely in the southwest corner of the dome, where the Maarianvaara granite obscures much detail.

The Maarianvaara granite is a diffuse, complex pluton consisting of a plexus of sheets, veins, and larger bosses that are both mutually discordant and, in the area south of Luikonlahti, generally discordant to both the layering and most D_1 and D_2 fabric elements and structures. The sheeted mode of emplacement ensured that the plentiful enclaves and pendants of country rock have undergone relatively little rotation, and "ghost stratigraphy" and structures can be traced right through the granite outcrop. Poor exposure of nongranitoid cover lithologies precludes detailed definition of such features. Both granite and basement gneiss are generally well exposed.

The crude north–south trend of S_2 and S_{0-1} seen between Losomäki and Luikonlahti is restored south of the Luikonlahti shear zone, and the general pattern of asymmetrical F_2 folds with thrusts or slides in their overturned limbs is still apparent.

To the east of the dome, exposure is generally poor, especially in the cover lithologies in the Miihkali area. Sufficient exposure exists, when augmented by borehole and geophysical information, to permit definition of major lithological units and the attitude of the principal fabrics (S_{0-1} and S_2). Both fabrics dip steeply eastward. The most striking feature of the eastern margin of the dome is the presence of slices of basement gneiss that appear to die out or close with depth.

As in the traverse from Kaavi to Kajoo, between the area west of Maarianvaara and east of Miihkali, S_2 is concordant to a single surface arched in a broad antiformal warp whose crest runs through the core of the Maarianvaara domes east of Onnivaara.

Post-D_2 structures and fabrics

Apart from the broad antiformal warping over the Maarianvaara dome and the basement ridge between Halijärvi and Kajoo (Kajoo antiform), a series of later structures and related fabrics affect the S_2 surface and, on a smaller scale, the S_2 fabric. Two of these affect the regional outcrop pattern in fundamental ways.

D_3 structures include open, upright folds (F_3) whose axes trend roughly north–south and plunge about the horizontal to both north and south (Fig. 7b). An axial planar fabric (S_3) is expressed as a fracture cleavage, locally quartz filled in psammites, or as a crenulation in more anisotropic and schistose lithologies, particularly the chlorite and talc schists of the Outokumpu assemblage. Fabric intersections and F_3 minor folds permit the vergence of major F_3 folds to be determined. Around Kaavi, north of the Kaavi–Niinivaara line, L_3 intersection lineations plunge south and F_3 minor folds indicate a major antiform to the east. D_2 fabric elements (S_2 poles, S_{0-1} poles, and L_2 lineations) lie on a single girdle whose pole coincides with the L_3 plunge and trend. This is consistent with interpretation of the broad flexure (Kajoo antiform) as a major F_3 antiform (Figs. 3 and 7b).

D_4 structures include upright, open folds (F_4) whose axes plunge consistently to the southwest across most of the area. A hinge-parallel lineation (L_4) is frequently a crenulation, and the axial planar fabric (S_4), like S_3, varies from a quartz-filled fracture cleavage in psammite to a crenulate cleavage in schistose units. Limited vergence data suggest a major antiformal F_4 culmination running through the core of the Maarianvaara dome (Fig. 7c).

Unlike the northern area, D_2 and D_1 fabric elements from the Maarianvaara dome show evidence of more complex rotation. These data are consistent with rotation about a north–south axis (D_3), and then about a southwest-plunging axis (D_4). This implies that one component of the geometry of the Maarianvaara dome is produced by interfering F_3 and F_4 folds.

GEOMETRIC INTERPRETATION OF THE SAVO THRUST BELT

In "thin-skinned" terminology, the Poskijärvi and Luikonlahti shear zones are sidewall ramps, with the Poskijärvi structure having the form of an oblique ramp. Between Losomäki and Niinivaara, and

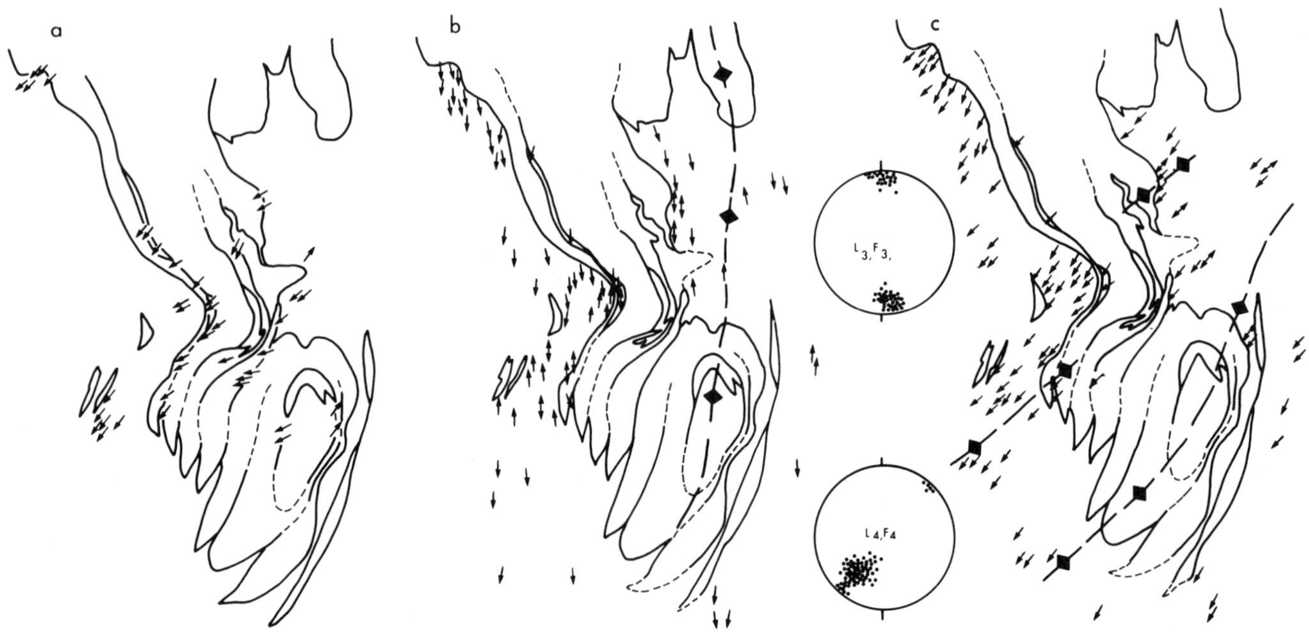

Figure 7. *Lineation maps of the Maarianvaara dome and Kaavi districts. (a) D_2 stretching lineations (L_2) and derived stretching directions (see text). (b) L_3 ($S_3 : S_2$, $S_3 : S_{0-1}$) intersection lineations and F_3 fold axes plunge and trend. (c) L_4($S_4 : S_3$, $S_4 : S_2$, $S_4 : S_{0-1}$) intersection lineations and F_4 fold axes plunge and trend. Includes data from Huhma (1971), Koistinen (1981), Park (1983, 1988c), and Park and Bowes (1983).*

from Kaavi to Kajoo, the outcrop defines horses contained between a series of frontal ramps, the whole zone overlying an unexposed flat. A pre-D_3 restoration, removing the Kajoo antiform, leaves a nearly uniform attitude to the axial planes of F_2 folds. No truly autochthonous basement to this thrust stack is seen in the area under consideration. The Poskijärvi and Luikonlahti sidewall shear zones die out with depth in the visible sections (i.e., within the allochthon). Thus, neither can be regarded as being related to fundamental features of the present autochthon. One candidate for a regional floor thrust or basal detachment now manifest at the surface is the Suhmura thrust, east of the Sotkuma dome (cf. Väyrynen, 1939; Ward, 1987, 1988). If this structure is extrapolated westward around the underside of the Viinijärvi synform and into the Kajoo antiform, it should underlie the Maarianvaara area, between 5 and 10 km beneath the present surface (see below).

STRAIN CONSIDERATIONS

Strain markers are not abundant in any lithology in the area. Only the pillow lavas in the Outokumpu assemblage at Losomäki and the feldspar porphyroclasts (megacrysts) in the Presvecokarelian "big feldspar gneiss" at Mökinvaara and Niinivaara have proved quantifiable to date. Elsewhere, erratically distributed stretching markers can be used to indicate stretching directions, and features like the Luikonlahti shear zone are amenable to larger scale considerations of regional bulk strain.

Both the Losomäki pillows and the feldspar megacrysts produce shape fabrics related to D_2 thrusts. Figure 8 illustrates this point; in all cases, the strain evident in the ellipsoid shape increases toward D_2 shears (measured as k values versus proximity to D_2 shears). Progressively more prolate ellipsoids develop in the proximity of the sidewall ramps. At Niinivaara, this prolate form is comparable to the sheath folds developed at the Luikonlahti mine. The sheath fold X-axes and the prolate long axes share the same orientation once the rotation due to D_3 is removed. This reinforces the conclusion (Park, 1988c) that simple shear mechanisms largely account for features observed in this shear zone. On the basis of this limited data set, a gross partitioning of shear strain seems evident; simple shear dominates the sidewall shear zones, whilst pure shear dominates the frontal ramp.

Passive rotation of the S_{0-1} layering into the simple shear dominated sidewall shear zones may also be used as a first approximation to simple shear strain partitioning on the scale of kilometres. Figure 9 is constructed using this as a crude measure of γ, utilizing angular distance in the horizontal plane of the azimuth to S_{0-1} and the azimuth of S_2 in the Luikonlahti sidewall ramp. Given that the sidewall is a nearly vertical feature (dip about 80°), and over much of this area the dip of S_{0-1} is >20°, the map is a semiquantitative representation of the nucleation of simple shear related to this structure. Only in the Kajoo–Halijärvi area, where S_{0-1} dips <10°, do errors swamp the resolution.

Figure 9 illustrates the splaying of D_2 simple shear zones above and below the sub-Karelian unconformity at Niinivaara. Through the area from the Luikonlahti mine to just east of Niinivaara, simple shear as a deformation mechanism nucleates on a series of en echelon zones, each identified on the ground as a zone where D_2 thrusts run into and merge with the Luikonlahti sidewall shear zone. This pattern of splays dies out upward into the metasedimentary cover southwest of Luikonlahti, where eventually the Maarianvaara granite

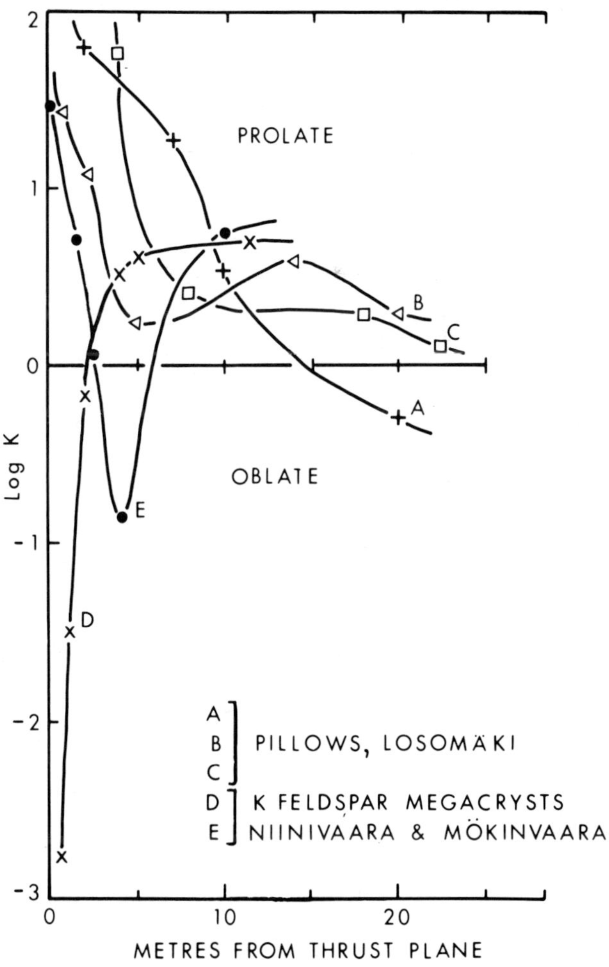

Figure 8. *Plot of Flinn's* k *values versus proximity to D_2 thrust planes showing progressive development of prolate or oblate ellipsoids (see text for discussion).*

obscures the detail. To the northeast and east of Niinivaara, the splays die out abruptly against the Kajoo thrust, which seems to be the deepest level in the thrust stack to which the Luikonlahti sidewall shear zone can be traced.

STRETCHING LINEATIONS AND DIRECTION

Like strain markers, stretching lineations are neither abundant nor well developed. A number of diverse examples are recognized locally, but no one form is ubiquitous. Examples were listed earlier under strain markers, and their orientations are represented on Figure 7a. In addition, stretching directions can be determined from D_2 boudins, making the not unreasonable assumption that D_2 boudin long axes lie in the plane of S_2 and the extension direction lies perpendicular to this also in the plane of S_2. A similar stretching direction is also indicated by X-axis orientation in sheath folds and the two demonstrably coincide in the area of the Luikonlahti mine (Park, 1988c).

Allowing for some scatter by rotation in D_3 and D_4 deformation, a very uniform stretching direction emerges over the area of Figure 7a, with a southwest–northeast or west-southwest–east-northeast orientation. When considered alongside the geometry of F_2 folds and related D_2 thrusts, this identifies tectonic transport from the southwest–west-southwest toward northeast–east-northeast, almost parallel to the Luikonlahti sidewall shear zone.

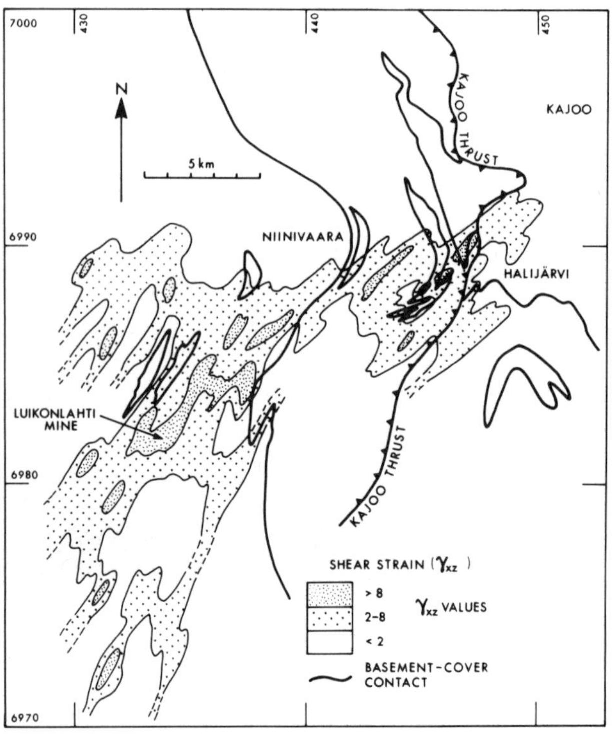

Figure 9. *Approximate distribution of shear strain (γ_{xz}) associated with the Luikonlahti shear zone (sidewall) (see text for method and discussion).*

ALONG-STRIKE AND DOWNDIP SECTIONS

Figures 10 and 11 represent first-order constructions taken downdip to the west from a line along the crest of the Kajoo F_3 antiform, and looking north/northeast from the line of the Luikonlahti sidewall shear zone respectively.

The along-strike section (Fig. 10) reveals a number of important features. All the thrust planes in this section eventually emerge from a basement–cover slice stack into cover alone toward the south (Outokumpu). The Luikonlahti sidewall dominates the central part of this section. Its boundary on the south is apparently very abrupt, but northward it relates to thrusts in the cover in a more gradational fashion. A number of thrust planes enter it at variable levels, while the vertical structures splay out upward. This sidewall dies out upward as the simple shear dissipates through multiple splays. Downward, in contrast, though some splays are evident (cf. Figs. 9 and 10), the whole structure dies out rather abruptly on the hanging wall of the Kajoo thrust. This position also corresponds with the lowest recognized presence of Kalevian overstep on the sub-Karelian unconformity.

By contrast, the Poskijärvi sidewall shear zone is relatively insignificant; it can only be traced through a few thrust slices. Definition becomes poor quite rapidly upward, but the structure dies out abruptly downward on the hanging wall of the Niskalampi thrust. Its position coincides with the convergence of a number of thrusts in the Riihijärvi area.

The downdip section taken by projection onto the Luikonlahti sidewall from the northwest (Fig. 11) is essentially a modified form of section A–B from Park and Bowes (1983, Fig. 3). As the Luikonlahti sidewall shear zone represents a zone in which simple shear dominates D_2 deformation mechanisms, this parallel profile represents the one section where assumptions about plane strain are most likely to be valid. As such, it is the only section in which even partial restoration to a pre-D_2 geometry can be attempted. With the tectonic transport direction identified earlier, it forms the basis for the restorations made below. The area delimited by the box (in Fig. 11) is that part of the section produced by direct projection down the nearly cylindrical F_2 folds. Other parts of the section, such as the Losomäki duplex, the Miihkali mass, and the continuation of thrust A, are projected with greater uncertainty to show their approximate positions and relationships. The larger scale heterogeneity of D_2 strain is particularly evident.

Duplex zones can be recognized on all scales, including those too small to be represented in Figure 11. The geometrical analogy between this ductile thrust stack, and those seen in brittle regimes is particularly emphasized. This has implications for the interpretation of the geometrical relationships and mechanisms of D_2 deformation.

In the finer resolution of this section, the two duplex zones at Potku and Kokka stand out. Each has recognizable floor and roof thrusts, and only involves cover rocks. A third duplex exists beneath the Kokka floor thrust (F), but the separation of the roof and floor thrusts is less than 20 m (cf. Fig. 5a). Other duplexes probably exist in the cover rocks of the Karpalolampi and lower Luikonlahti horses (between C and D, and B and C thrusts respectively),

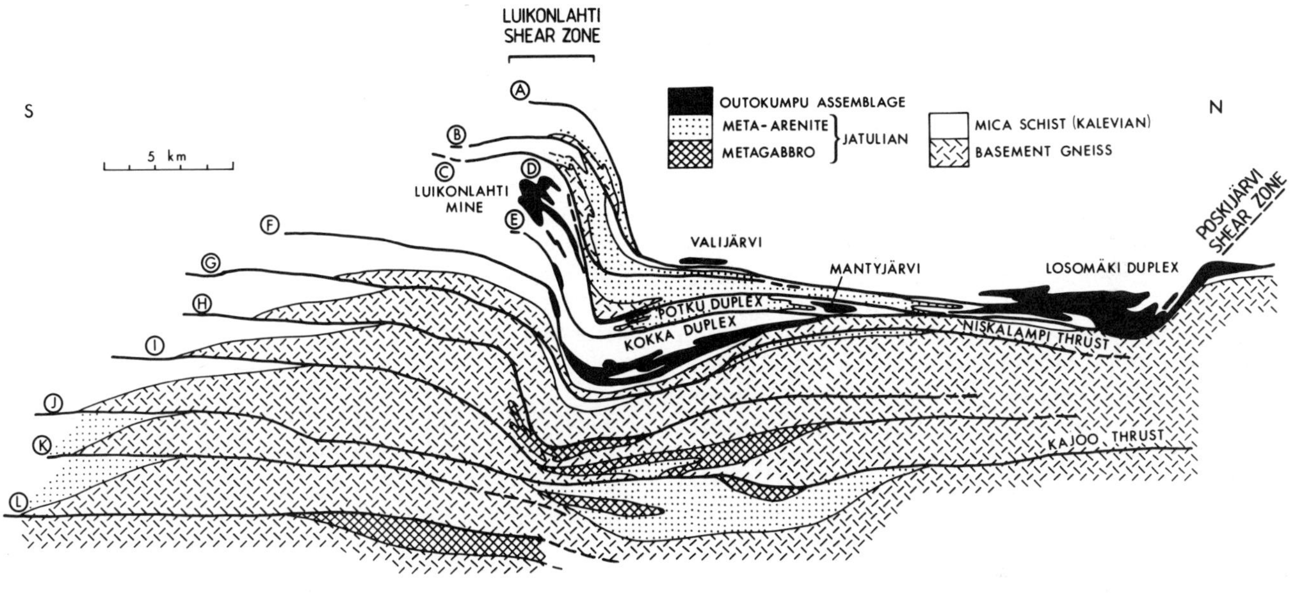

Figure 10. *Along-strike section constructed looking west from the crest of the F_3 Kajoo antiform, along the strike of the D_2 thrusts, essentially orthogonal to the transport direction (the line of section projection is approximately parallel to the crest of the Kajoo antiform, see Fig. 3). Thrusts are identified by letters, as on Figures 3 and 11.*

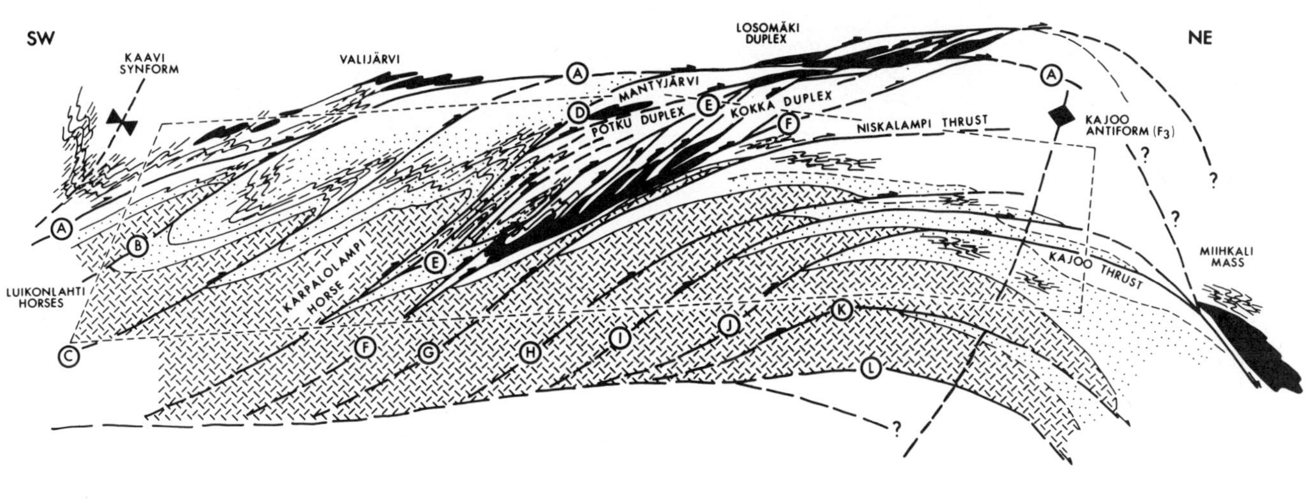

Figure 11. *Downdip section constructed parallel to, and along, the Luikonlahti shear zone (see Fig. 3). The area delimited by the box represents relationships projected directly down plunge on F_2 folds into the line of section using the least curvilinear hinges. Outside this box the relationships are truly represented, but the geometry is more approximate. Letters denoting thrusts are the same as on Figures 3 and 10.*

but a lack of outcrop and abundant sheets of Maarianvaara granite obscure the detail. Significantly, duplexes on this scale are absent in the slices of basement gneiss, and there appears to be a minimum size for horses of basement gneiss. This is reflected in the restricted nature of D_2 deformation within the main basement gneiss outcrop area.

On a larger scale, thrusts A to F converge upward and eastward, defining a definite roof thrust, whereas the projected convergence of thrusts D to L at depth defines a floor thrust. These two thrusts would, in turn, circumscribe a larger scale duplex, whose eastern extremity dips over the F_3 Kajoo antiform, and beneath the Miihkali serpentinite mass on the western edge of the Viinijärvi synform (Fig. 12).

In Figure 12, thrust A is speculatively projected over the Kajoo antiform and beneath the Miihkali mass. Both the Losomäki duplex and the Miihkali mass (also probably a duplex; though very poorly exposed, low-angle shear zones, probably thrusts, are found in borehole intersections; see Koistinen (1981) and Park (1983)) belong to a higher level than the duplex between thrusts A and L. By implication, the Outokumpu–Polvijärvi serpentinite belt and the rocks of the Viinijärvi synform (most of the outcrop of the Outokumpu nappe) lie above roof thrust A. This is consistent with the views expressed in Bowes *et al.* (1984) describing the Outokumpu area (Viinijärvi synform) as an area dominated by open F_3 folds affecting F_1 major folds, with F_2 having only limited effects on outcrop pattern. On this projection, the A roof thrust would reemerge between Polvijärvi and the western edge of the Sotkuma basement dome, where thrusts are described (Gaál *et al.*, 1975; Koistinen, 1987).

The depth at which thrusts D–L (and probably A–C) merge is unconstrained, as thrust L only crops out in the window at the core of the Maarianvaara dome. Thrusts A to L may actually rise off the continuation of the Suhmura thrust, consistent with its position and distance from the projected reappearance of thrust A near Polvijärvi. In which case, the Suhmura structure is probably the floor thrust to this entire segment of the Savo thrust belt.

The pre-D_2 nature of the Outokumpu nappe is also emphasized, in that the outcrops of Outokumpu assemblage, the defined base of the allochthon, in the Kokka duplex and at Mantyjärvi lie beneath the roof thrust A.

RECONSTRUCTION OF THE SUB-KARELIAN UNCONFORMITY

In the absence of a more widespread and comprehensive data set for D_2 strain, it is premature to

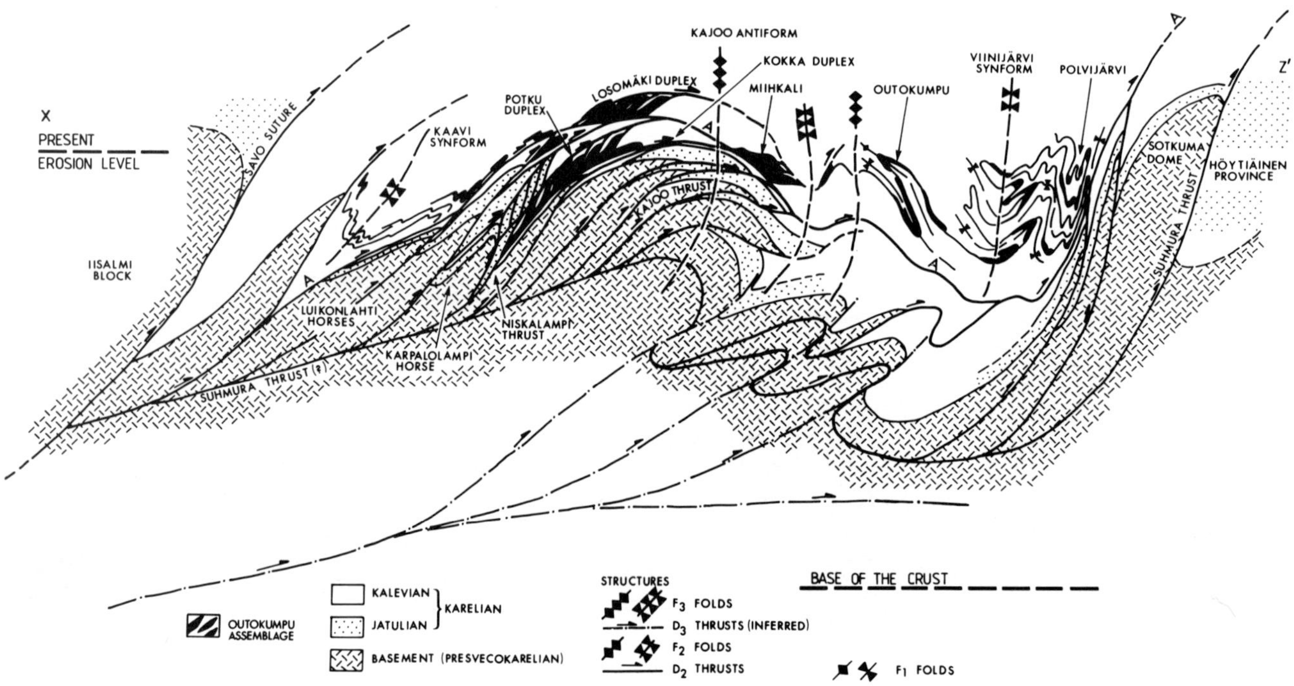

Figure 12. *Composite cross-section illustrating the structure of the Losomäki–Maarianvaara segment of the Savo thrust belt, showing its probable relationship to the rocks of the Vinnijärvi synform and Sotkuma dome to the east. The major F_3 antiforms are interpreted as hanging wall or tip-line antiforms above conjectural D_3 thrusts and a deep crust detachment. The Suhmura thrust is interpreted as a floor thrust to the largest scale D_2 duplex in Figure 11. Lines of the component sections in this composite are marked on Figure 2.*

attempt even a minimal estimate of D_2 shortening or to attempt a balanced reconstruction across this segment of the Savo thrust belt. Nevertheless, on a local scale, some partial reconstructions are valid.

Though tectonized to varying degrees within the area under consideration, the depositional unconformity of the Karelian Supergroup and Kalevian metasediments on Archean Presvecokarelian gneisses can be identified on most of the exposed contacts (Figs. 3 and 13). It possesses three important characteristics.

(1) For most of its extent, a paleoregolith horizon can be identified even when substantially deformed and reconstituted, rendering such para-allochthonous relationships quite distinct to truly allochthonous thrust contacts.

(2) A major overstep is evident. Around Niinivaara and south of Luikonlahti, Kalevian mica schists and psammites rest directly on basement gneiss, with locally developed paleosols and arkosites. Elsewhere, Jatulian meta-arenites are present.

(3) The sub-Jatulian unconformity was the preferred site for the intrusion of a suite of gabbro bodies, now preserved as metagabbro–amphibolite slivers of varying size throughout the area.

Huhma (1975) presented some evidence for another unconformity within the Jatulian formations around Riihijärvi, where conglomerates, breccias or meta-arenite, amphibolite, and gneiss clasts occur in upper Jatulian phyllites. This is consistent with the less deformed successions seen farther east (Pekkarinen, 1979; Ward, 1988), but it cannot be traced as a feature here.

A Kalevian overstep onto basement around Niinivaara is evident (Fig. 13). If this area is subjected to a first-order palinspastic reconstruction, drawing back the D_2 thrusts parallel to the Luikonlahti sidewall and the D_2 tectonic transport direction, the area of overstep emerges more clearly. Most striking is the form of the northwestern edge of the area of overstep; a nearly straight edge lying parallel to, and almost coincident with, the Luikonlahti sidewall shear zone (Figs. 3 and 13). The Jatulian lithofacies north and south of this overstepped feature are quite distinct, especially the lower Jatulian meta-arenites. To the south and east of the overstep lies a thick succession of pure metaquartzites with minor arkosites and calc-silicate layers. To the north and west, the Jatulian succession is possibly an order of magnitude thinner; is dominated by arkosites, calc-arkosites, and calc-silicate skarns; and contains only minor pure metaquartzites. The upper Jatulian, north of the overstep, also contains Huhma's (1975) intraformational breccias.

The interpretation offered here is that this edge represents an upper Jatulian fault scarp above which any pre-existing Jatulian formations had been stripped prior to Kalevian deposition. The fault possibly exercised some control on Jatulian deposition and represents a fundamental basement structure exploited during D_2 thrusting as the feature on which the Luikonlahti sidewall shear zone nucleated.

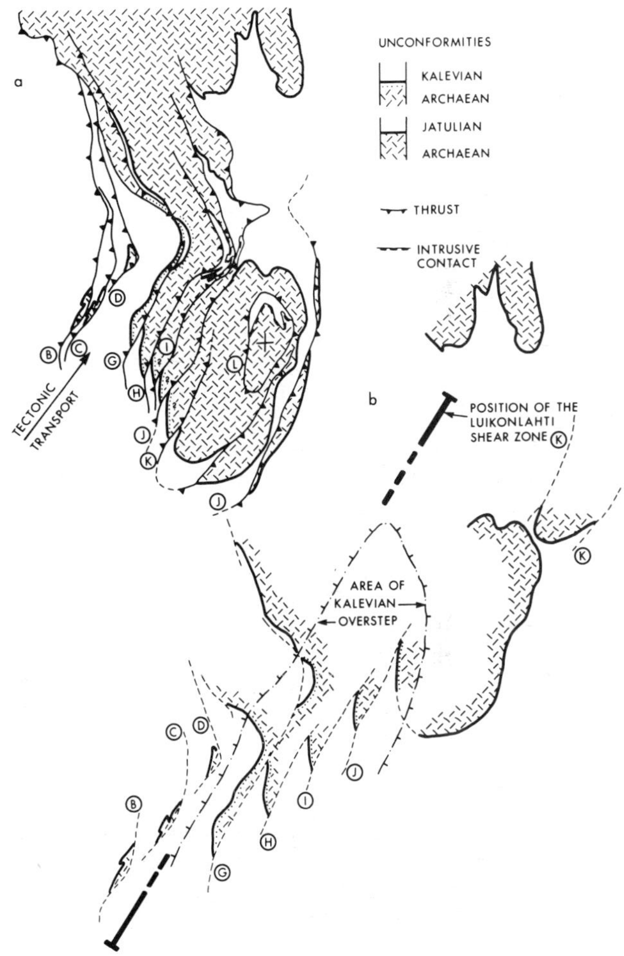

Figure 13. *Partially balanced pre-D_2 strain reconstruction of the overstep of Kalevian metasediments on the sub-Karelian unconformity. Reconstruction is controlled by estimates of simple shear strain on the Luikonlahti shear zone (sidewall) and the L_2 stretching lineation, interpreted as the tectonic transport direction. (a) Present outcrop configuration showing the distribution of the various unconformities. Letters identifying thrusts are the same as on Figures 3, 10, and 11. (b) Pre-D_2 strain reconstruction showing the relationship of the edge of the Kalevian overstep to the position of the Luikonlahti shear zone (sidewall).*

CONCLUSIONS

(1) In the entirely ductile thrust regime in this area of the Savo thrust belt, geometries analogous to those in "brittle" thin-skinned regimes can be recognized. These geometries are similar on several scales.

(2) The expression of thrust geometries is a function of strain partitioning in a heterogeneous crustal stack; particularly the partitioning of predominantly simple and predominantly pure shear strain into the sidewalls and ramps respectively. Such geometrical complexity is a function of lithological heterogeneity, well displayed by the differing scale of ductile duplex development in cover and basement rocks. This complexity should be expected wherever the crust (or lithosphere) retains shear strength and is probably a common feature on crustal ramps.

(3) Pure shear and simple shear dominate in differing zones: pure shear on the frontal ramps and simple shear along the sidewalls. Identification of this partitioning reflects the orientation of the various planar surfaces to the principal compression direction.

(4) Major sidewalls, such as the Luikonlahti shear zone, nucleate on points where major asperities (the scarp edge of the Kalevian overstep) coincide with a major boundary discontinuity (the basement–cover interface), and where the orientation of linear asperities lies close to parallelism with the principal compression direction.

(5) The overall geometry of thrusts in this segment of the Savo thrust belt implies progressive deformation related to thrusts sequentially developed from east to west, with the Suhmura thrust the first to move, and the Savo suture the last. All of this shortening was in response to compression in a northeast–southwest direction, from the southwest against the main cratonic foreland block to the east (Fig. 1).

(6) Recognition of the geometrical analogue provides a framework for D_2 deformation, but it is the crucial recognition of a nearly vertical sidewall ramp parallel to the tectonic transport direction that will facilitate the use of the techniques of section balancing to attempt realistic minimal restorations of discrete strain increments in a polyphase deformation history.

DISCUSSION

Though D_3–D_4 fold interference has a marked effect on outcrop pattern through this part of the Savo thrust belt, it is the D_2 structures, dominated by thrust–fold relationships, that determine the three-dimensional framework. The pattern and style of D_2 are largely an interaction between a geometrical analogue of a thin-skinned thrust regime and earlier features, the most fundamental of which are boundary discontinuities and other major asperities in the pre-D_2 layering. Once established, the D_2 structural framework exerted a fundamental control on subsequent activity.

Recognition of the D_2 geometry as an analogue of a thin-skinned thrust regime should permit limited reconstructions, allowing the removal of the effects of a major strain and shortening increment in a polyphase sequence. Three requirements of section balancing (Boyer and Elliot, 1982; Elliot, 1983) assume (1) plane strain in the line of section, (2) that change in layer length has been achieved largely by buckle folding, and (3) the line of section is either parallel to the tectonic transport direction or has a known relationship to it. Neither of the first two assumptions are valid in a ductile regime. This is why the identification of zones like the Luikonlahti sidewall shear zone, dominated by simple shear strain and parallel to the tectonic transport direction, is so important. These are the only zones where a parallel section can be constructed with a plane strain assumption that is at all realistic. This remains so, even when complicating factors like the sheath folds at the Luikonlahti mine are taken into account, for such sections will still represent an improvement in precision over approximations based on area balancing in pure shear dominated regimes.

As an example of the implications of this approach to Precambrian high-grade terrains, we can cite the interpretation of the history of the Maarianvaara dome. This is one of Eskola's (1949) type "mantled gneiss domes" or reactivated gneiss diapirs, though Frosterus and Wilkman (1920) and Wegmann (1928) recognized the importance of thrusts in lithological relationships within the feature (a relationship now recognized as commonplace; cf. van Staal and Williams, 1983). Any analysis of strain around such features (cf. Brun, 1980; Brun *et al.*, 1981; Gapais and Brun, 1981) must take into account the polyphase nature of their deformational history and, therefore, of the strain increments within and around them. We recognize in the history of the Maarianvaara dome (1) preserved Presvecokarelian fabrics and structures; (2) a syndepositional asperity, probably a middle or upper Jatulian fault scarp; (3) imposed increments of Svecokarelian deformation related to shear along the basement–cover interface and high-strain zones related to basement excavating thrusts; (4) an F_3–D_4 fold interference; and (5) generation and emplacement of the Maarianvaara granite.

All of these features contribute to present outcrop geometry, but others can be inferred. For instance, if the Suhmura thrust represents the surface expression of a D_2 basal detachment, it must continue at depth beneath the Maarianvaara dome. Either the dome is the surface expression of activity beneath this basal detachment, such as a deep crustal diapir, or it represents a tip-line antiform developed over a D_3 thrust propagating upward from the same basal detachment. Whichever of these models is correct, it is no longer valid to make simplistic analyses of strain that ignore the polyphase nature of its increments.

The establishment of a geometrical framework for each strain increment and the development of a temporal framework permit a more realistic approach to the structure of high-grade terrains. This dispenses with the need to invoke uniquely Precambrian mechanisms, e.g., necessarily diapiric mantled gneiss domes. Such high-grade terrains provide an analogue for the deep crust. If so, the degree of complexity evident in this segment of a deep crustal foreland thrust belt requires the reassessment of inclined, superficially simple, deep crust reflectors seen in seismic reflection profiles across such terrains.

ACKNOWLEDGEMENTS

This paper formed the basis for a presentation by one of us (AFP) to the Orogenic Fronts special session at the Geological Association of Canada – Mineralogical Association of Canada joint annual meeting in Saskatoon in May 1987. We would like to thank Don Rousell for the invitation to submit a contribution, and John Lewry and Mel Stauffer for considering its inclusion in this volume. AFP also acknowledges financial support from the University of Glasgow Research Support Fund and Clydesdale Bank plc.

REFERENCES

Bowes, D.R., Halden, N.M., Koistinen, T.J. and Park, A.F., 1984, Structural features of basement and cover rocks in the eastern Svecokarelides, *in* Kröner, A. and Greiling, R., eds., Precambrian Tectonics Illustrated: Schweizerbart'sche Verlagsbuchhandlung, Stuttgart, p. 147–171.

Boyer, S.M. and Elliot, D., 1982, Thrust systems: American Association of Petroleum Geologists Bulletin, v. 66, p. 1196–1230.

Brun, J.-P., 1980, The cluster-ridge pattern of mantled gneiss domes in eastern Finland: evidence for large scale gravitational instability in the Proterozoic crust: Earth and Planetary Science Letters, v. 47, p. 441–449.

Brun, J.-P., Gapais, D. and Le Theoff, B., 1981, The mantled gneiss domes of Kuopio (Finland): interfering diapirs: Tectonophysics, v. 74, p. 283–304.

Coward, M.J., 1983, Thrust tectonics, thin-skinned or thick-skinned, and the continuation of thrusts to deep in the crust: Journal of Structural Geology, v. 5, p. 113–124.

Coward, M.J., 1984, Major shear zones in the Precambrian crust; examples from NW Scotland and southern Africa and their significance, *in* Kröner, A. and Greiling, R., eds. Precambrian Tectonics Illustrated: Schweizerbart'sche Verlagsbuchhandlung, Stuttgart, p. 207–236.

Dahlstrom, C.D.A., 1970, Structural geology in the eastern margin of the Canadian Rocky Mountains: Canadian Petroleum Geologists Bulletin, v. 18, p. 332–406.

Daly, M.C., 1986, The intracratonic Irumide Belt of Zambia and its bearing on collision orogeny during the Proterozoic of Africa, *in* Coward, M.P. and Ries, A.C. eds., Collision Tectonics: Geological Society of London Special Publication No. 19, p. 321–328.

Elliot, D., 1983, The construction of balanced cross-sections: Journal of Structural Geology, v. 5, p. 101.

Elliot, D. and Johnson, M.R.W., 1980, The structural evolution of the northern part of the Moine thrust zone: Transactions of the Royal Society of Edinburgh: Earth Sciences, v. 71, p. 69–96.

Eskola, P.E., 1949, On the problem of mantled gneiss domes: Quarterly Journal of the Geological Society of London, v. 104, p. 461–476.

Frosterus, B. and Wilkman, W., 1920, Suomen Geologinen Yleiskartta, Lehti D3; Joensuu: Suomen Geologinen Tutkimuslaitos, Helsinki.

Gaál, G., 1986, 2200 million years of crustal evolution: the Baltic Shield: Geological Society of Finland Bulletin, v. 58, p. 149–168.

Gaál, G. and Gorbatschev, R., 1987, An outline of the Precambrian evolution of the Baltic Shield: Precambrian Research, v. 35, p. 15–52.

Gaál, G., Koistinen, T.J. and Mattila, E., 1975, Tectonics and stratigraphy of the vicinity of Outokumpu, N. Karelia, Finland: Geological Survey of Finland Bulletin, No. 271.

Gapais, D. and Brun, J.-P., 1981, A comparison of mineral grain fabrics and finite strain in amphibolites from eastern Finland: Canadian Journal of Earth Sciences, v. 18, p. 995–1003.

Halden, N.M., 1982, Structural, metamorphic and igneous history of migmatites in the deep levels of a wrench fault regime, Savonranta, eastern Finland: Transactions of the Royal Society of Edinburgh: Earth Sciences, v. 73, p. 17–30.

Halden, N.M. and Bowes, D.R., 1984, Metamorphic development of cordierite-bearing layered schist and mica schist in the vicinity of Savonranta, eastern Finland: Geological Society of Finland Bulletin, v. 56, p. 3–23.

Hossack, J.R., 1979, The use of balanced cross-sections in the calculation of orogenic contraction: Journal of the Geological Society of London, v. 136, p. 705–711.

Huhma, A., 1971, Suomen Geologinen Ylieskartta 1:100,000 Lehti 4222, 4224, 4311, Outokumpu, Polvijärvi, Sivakkavaara: Suomen Geologinen Tutkimuslaitos, Espoo.

Huhma, A., 1975, Precambrian rocks of the Outokumpu, Polvijärvi and Sivakkavaara map sheets. 1:100,000 sheets 4222, 4224, 4311: Suomen Geologinen Tutkimuslaitos, Espoo. (Summary in English.)

Huhma, A., 1981, Youngest Precambrian dyke rocks in northern Karelia, eastern Finland: Geological Society of Finland Bulletin, v. 53, p. 67–82.

Huhma, H., 1986, Sm–Nd, U–Pb and Pb–Pb isotopic evidence for the origin of the Early Proterozoic Svecokarelian crust in Finland: Geological Survey of Finland Bulletin, No. 337.

Koistinen, T.J., 1981, Structural evolution of an early Proterozoic stratabound Cu-Co-Zn deposit, Outokumpu, Finland: Transactions of the Royal Society of Edinburgh: Earth Sciences, v. 72, p. 115–158.

Koistinen, T.J., 1987, The Outokumpu ore district, east Finland, a closed basin, *in* Finnish–Soviet Symposium, The Early Proterozoic of the Baltic Shield: Commission for Scientific Co-operation Between the Soviet Union and Finland, Petroskoi. (In press.)

Kontinen, A., 1987, An early Proterozoic ophiolite – the Jormua mafic–ultramafic complex, northeastern Finland: Precambrian Research, v. 35, p. 313–341.

Korsman, K., Hölttä, P., Hautala, T. and Wasenius, P., 1984, Metamorphism as an indicator or evolution and structure of the crust in eastern Finland: Geological Survey of Finland Bulletin, No. 328.

Kouvo, O. and Tilton, G.R., 1966, Mineral ages from the Finnish Precambrian: Journal of Geology, v. 71, p. 421–442.

Luukkonen, E. and Lukkarinen, H., 1986, Explanation to the stratigraphic map of Middle Finland: Suomen Geologian Tutkimuskeskus, Tutkimusraportti No. 74, 47 p.

Meriläinen, K., 1980, On the stratigraphy of the Karelian formations, in Silvennoinen, A., ed., Jatulian Geology in the Eastern Part of the Baltic Shield: Committee for Scientific and Technical Cooperation Between Finland and the Soviet Union, Rovaniemi, p. 97–112.

Paavola, J., 1980, Suomen Geologinen Ylieskartta, 1:100,000. Lehti 3334; Nilsiä: Suomen Geologinen Tutkimuslaitos, Espoo.

Park, A.F., 1983, Sequential development of metamorphic fabrics and structural elements in polyphase deformed serpentinites in the Svecokarelides, eastern Finland: Transactions of the Royal Society of Edinburgh: Earth Sciences, v. 74, p. 33–60.

Park, A.F., 1984, Nature, affinities and significance of metavolcanic rocks in the Outokumpu assemblage, eastern Finland: Geological Society of Finland Bulletin, v. 56, p. 25–52.

Park, A.F., 1988a, A new section through the sub-Karelian unconformity at Niinivaara, Savo, eastern Finland: Geological Society of Finland Bulletin, v. 60, p. 67–73.

Park, A.F., 1988b, Nature of the early Proterozoic Outokumpu assemblage, eastern Finland: Precambrian Research, v. 38, p. 131–146.

Park, A.F., 1988c, Structure of the Luikonlahti mine: sheath folds in the Svecokarelides of eastern Finland: Journal of Structural Geology, v. 10, p. 487–498.

Park, A.F. and Bowes, D.R., 1983, Basement–cover relationships during polyphase deformation in the Svecokarelides of the Kaavi district, eastern Finland: Transactions of the Royal Society of Edinburgh: Earth Sciences, v. 74, p. 95–118.

Pekkarinen, L.J., 1979, The Karelian formations and their depositional basement in the Kiihtelysvaara–Vartsila area, eastern Finland: Geological Survey of Finland Bulletin, No. 301.

Simonen, A., 1980, The Precambrian in Finland: Geological Survey of Finland Bulletin, No. 304.

Treloar, P.J., Koistinen, T.J. and Bowes, D.R., 1981, Metamorphic development of cordierite–amphibole rocks and mica schists in the vicinity of the Outokumpu ore deposit, Finland: Transactions of the Royal Society of Edinburgh: Earth Sciences, v. 72, p. 201–215.

van Staal, C.R. and Williams, P.F., 1983, Evolution of a Svecofennian mantled gneiss dome in SW Finland, with evidence for thrusting: Precambrian Research, v. 21, p. 101–128.

Väyrynen, H., 1939, On the geology and tectonics of the Outokumpu ore field and region: Commission Géologique du Finlande Bulletin, No. 124.

Ward, P.G., 1987, Early Proterozoic deposition and deformation at the Karelian craton margin in southeastern Finland: Precambrian Research, v. 35, p. 71–93.

Ward, P.G., 1988, Early Proterozoic Kalevian lithofacies and their interpretation in the Hammaslahti–Rääkkylä area, eastern Finland, in Laajoki, K. and Paakkola, J., eds., Sedimentology of the Precambrian Formations in Eastern and Northern Finland: Geological Survey of Finland Special Paper 5, p. 29–48.

Wegmann, C.E., 1928, Über die Tektonik der jüngeren Faltung in Ostfinnland: Fennia, No. 50(16), p. 1–22.

Cameroon: a Tectonic Keystone in the Pan-African Network[1]

S.F. Toteu
Centre de Recherches Géologiques et Minières, BP 333, Garoua, Cameroon

J.M. Bertrand
Centre de Recherches Pétrographiques et Géochimiques, BP 20, 54501 Vandoeuvre lès Nancy Cedex, France

J. Penaye
Centre de Recherches Géologiques et Minières, BP 333, Garoua, Cameroon, and Ecole Nationale Superieure de Géologie, BP 452, 54001, Nancy Cedex, France

J. Macaudiere
Centre de Recherches Pétrographiques et Géochimiques, BP 20, 54501 Vandoeuvre lès Nancy Cedex, France, and Ecole Nationale Superieure de Géologie, BP 452, 54001, Nancy Cedex, France

S. Angoua
Centre de Recherches Pétrographiques et Géochimiques, BP 20, 54501 Vandoeuvre lès Nancy Cedex, France, and Faculté des Sciences, BP 239, 54506, Vandoeuvre lès Nancy Cedex, France

P. Barbey
Centre de Recherches Pétrographiques et Géochimiques, BP 20, 54501 Vandoeuvre lès Nancy Cedex, France, and Faculté des Sciences, BP 239, 54506, Vandoeuvre lès Nancy Cedex, France

[1]Centre de Recherches Pétrographiques et Géochimiques Contribution No. 791.

Toteu, S.F., Bertrand, J.M., Penaye, J., Macaudiere, J., Angoua, S. and Barbey, P., Cameroon: a tectonic keystone in the Pan-African network, in Lewry, J.F. and Stauffer, M.R., eds., The Early Proterozoic Trans-Hudson Orogen of North America: Geological Association of Canada, Special Paper 37, p. 483–496.

Abstract

The place of Cameroon in the Pan-African network of orogenic belts is examined in terms of a detailed survey of the Poli and Yaoundé regions located in the northern and southern parts of the country respectively. In the Poli region, although the first stage (D1) of the evolution is still poorly understood, the younger D2 and later events involved northeast–southwest movement, resulting in wrench faulting and thrusting. The migmatized gneissic formations of the Poli region, previously interpreted as an old basement, now appear more likely to represent partly high-grade equivalents of the Poli schists (800 Ma). The abundance of younger Pan-African calc-alkaline orthogneisses interleaved both in the schists and in the gneisses (630 Ma) is a salient characteristic of the Poli region. Comparison with the Yaoundé region, where strongly metamorphosed platform sediments are thrust to the south-southwest onto the Congo craton, suggests that the Sanaga–Adamaoua region, situated between the two areas studied, represents a boundary zone between a stable craton-controlled zone to the south and a partly accreted crust to the north. The ubiquity of northeast–southwest directions of tectonic movement in southern and western Africa suggests that the model of continental transform faults controlling Pan-African tectonics in southern Africa can be applied to central and western Africa.

Résumé

La situation du Cameroun au sein du réseau des chaînes Pan-Africaines est discutée d'après les résultats d'études détaillées menées dans les régions de Poli et de Yaoundé, au nord et au sud du pays. Dans la région de Poli, la première étape (D1) de l'évolution est encore mal comprise; l'évènement D2 et les évènements tardifs correspondent à une direction de mouvement nord-est–sud-ouest responsable de charriages et de décrochements. Les formations gneissiques migmatisées de la région de Poli, interprétées jusqu'à présent comme représentant un vieux socle, sont maintenant considérées en partie, comme des équivalents métamorphiques de haut degré des schistes de Poli datés à 800 Ma. Une autre caractéristique importante de la région de Poli est un important volume d'orthogneiss de composition calco-alcaline, plus jeunes (630 Ma), interstratifiés aussi bien dans les gneiss que dans les schistes. La comparaison avec la région de Yaoundé, au sud, où des sédiments de type plateforme, très métamorphiques ont été charriés vers le sud-sud-ouest, sur le craton du Congo, suggère qu'il existe, dans la région intermédiaire de la Sanaga–Adamaoua située entre les deux régions étudiées, une frontière majeure: au sud, il s'agit d'un domaine influencé par la proximité du craton stable, tandis qu'au nord, une partie importante de la croûte résulte d'une accrétion d'âge Protérozoïque supérieur. La similitude des directions de mouvement d'âge Pan-Africain entre les parties méridionales et occidentales de l'Afrique suggère que le modèle impliquant que la tectonique Pan-Africaine de la partie sud de l'Afrique est commandée par une grande zone de cisaillement à caractère transformant, peut aussi s'appliquer aux parties centrales et occidentales du continent.

INTRODUCTION

Proterozoic orogenic belts occur as sinuous zones forming complex nets around older cratons in all continents. In North America, there are several belts that evolved during the 2000–1600 Ma time range; whereas in Africa, a similar belt pattern is the result of Pan-African tectonic activity, which occurred about 600 Ma ago. Both Pan-American and Pan-African activity occurred through continental accretion, island-arc and continental-margin magmatism, and possibly accretion of displaced terranes (Black *et al.*, 1979; Lewry, 1981; Condie, 1982; Stoeser and Camp, 1985).

In Africa, well-defined linear belts were formed by ensialic processes, Wilson-cycle-type evolution, or thermal reactivation of old terranes; the latter interpretation is generally applied to less well-known regions where geochronological data are scarce and structural information is lacking. In Cameroon, although evidence for a Pan-African imprint was well documented in terms of Rb/Sr age determinations, most of the metamorphic units were considered, until recently, to represent a Lower Proterozoic cover with respect to the Archean granulites and granites of the Congo craton (Bessoles and Lasserre, 1977).

The location of Cameroon in the Pan-African network is of prime importance as it forms the link between the Transaharan belt (Cahen *et al.*, 1984), the Mayombe – West Congo belt, and northeastern Brazil (Fig. 1). Intervening cratons are the West African craton, the eastern limit of which is well defined in Togo, Benin, Mali, and Algeria (Black *et al.*, 1979; Trompette, 1979; Affaton *et al.*, 1980); the Congo–Chaillu craton, cropping out in southern Cameroon, Centrafrican Republic, Gabon, Congo, and Zaïre; and a still ill-defined Nile craton to the northeast (Rocci, 1965).

Two areas are currently being studied in detail: northern Cameroon (Ngako, 1986; Njel, 1986; Toteu *et al.*, 1986) and the Yaoundé region in southern Cameroon (Nzenti *et al.*, 1984, 1988; Nedelec *et al.*,

1986; Nzenti, 1987). They are separated by the Sanaga–Adamaoua region, where assumed Pan-African mylonites are largely hidden by Mesozoic to recent uplift and associated volcanism and faulting (Fig. 1).

In northern Cameroon, several areas scattered along the southwestern end of the Poli schist belt have been investigated. This schist belt belongs to the Bessoles' "Séries intermédiaires," classically attributed to the Lower or Middle Proterozoic. However, new geochronological data (Toteu *et al.*, 1987) indicate that it represents a Late Proterozoic volcaniclastic assemblage and that some gneisses and a large part of the orthogneisses, previously attributed to the basement complex (Bessoles and Trompette, 1980), are also Late Proterozoic in age.

In southern Cameroon, a study of the Yaoundé migmatites and granulites has confirmed Bessoles and Lasserre's (1977) hypothesis of large-scale Pan-African nappes emplaced onto the Congo craton, but has also shown that the granulite facies metamorphism of the Yaoundé migmatites is Pan-African in age (Ball *et al.*, 1984; Jegouzo, 1984; Nzenti *et al.*, 1984, 1988).

The aim of this paper is to compare the tectonic evolutions of northern and southern Cameroon and to discuss their places in the Pan-African orogenic network.

TECTONIC AND METAMORPHIC EVOLUTION OF NORTHERN CAMEROON

Geological setting of the Poli schist belt

The most detailed maps available are those of Koch (1959), Schwoerer (1965), and Le Fur (1971). Le Fur (1971) proposed a lithostratigraphic sequence for the southwestern part of the schist belt (Fig. 2). From bottom to top, the sequence is as follows: (1) gneisses and migmatites of the "basement complex"; (2) the Sakje Formation, composed of leptynites, mica schists, quartzites, amphibolites, calc-schists, with some conglomerates; (3) a lower volcanic unit composed of tuffs of intermediate composition,

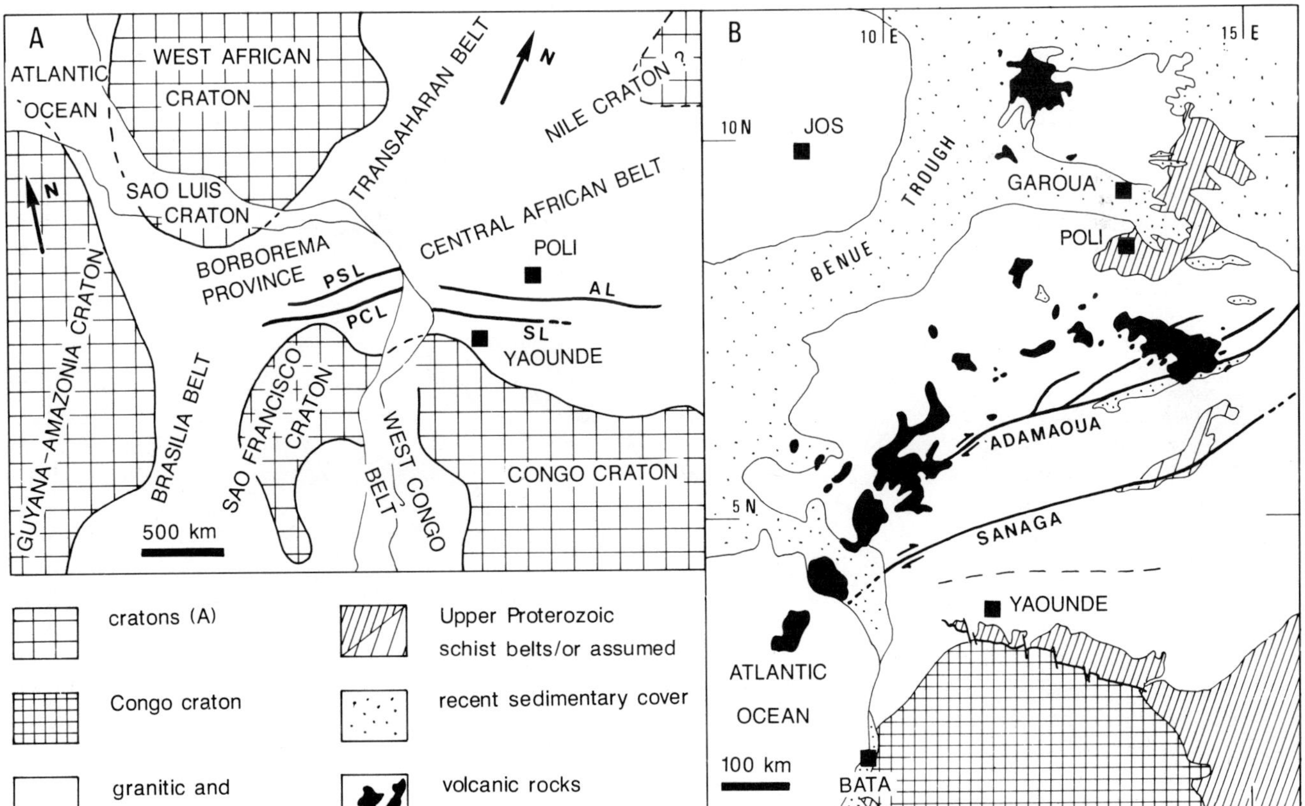

Figure 1. *(A) Location of the main cratonic areas and fold belts in Africa and South America (fit of continents from Bullard* et al. *(1965)). Abbreviations: PSL, Patos lineament; PCL, Pernambucco lineament; AL, Adamaoua lineament; and SL, Sanaga lineament. (B) Sketch map of major tectonic features of Cameroon. Adamaoua and Sanaga lineaments are indicated; the broken line corresponds to the northern limit of the Congo craton from gravimetric data (*in *Dumont, 1986).*

keratophyric tuffs and lavas, spilitic tuffs and lavas with pillow lavas; and (4) an upper clastic formation with metapelites.

The stratigraphic and tectonic relationships of these formations, however, are not clear because, according to the map, the older formation crops out more frequently towards the centre of the belt, whereas the younger pelitic formation shows, in places, transitional metamorphic relationships with the "basement complex." It now appears that this stratigraphic succession was based mainly upon differences in metamorphic grade and the presence of recognizable volcanics. Nowadays, the gneissic and migmatitic units are no longer assumed to represent only an old basement with respect to the schists (Toteu *et al.*, 1987; Toteu, 1987). New critical observations are listed below.

(1) The occurrence in both schists and gneisses of calc-alkaline orthogneisses (the "basic to intermediate plutonic" (BIP) suite), which are broadly synchronous with the older recognized deformation event (D1). These orthogneisses locally form more than half the volume of the previous "basement complex"; they are well dated in the 660–630 Ma range.

(2) The tectonic evolutions of both groups are comparable in spite of contrasting metamorphic grades, i.e., tangential deformation involving the development of a gently dipping schistosity (D1), followed by a compressional folding with subvertical axial planes (D2). Metamorphic grade corresponding to the second phase is of MP/LT type in the Poli schists and of MP/HT type in the gneisses. The age of the older deformation is defined only by the D1 pre- to syntectonic emplacement of 630 Ma old basic to intermediate plutons. The D2 event corresponds to a major foliation transposition, commonly accompanied by widespread migmatization and culminating in the late-tectonic emplacement of large granitic batholiths with an age of about 580 Ma.

(3) There is evidence for a volcanic event characterized by tholeiitic (Njel, 1986) and alkaline felsic volcanics in the schist belt. These volcanic rocks are 830 Ma old (Toteu *et al.*, 1987), thus confirming

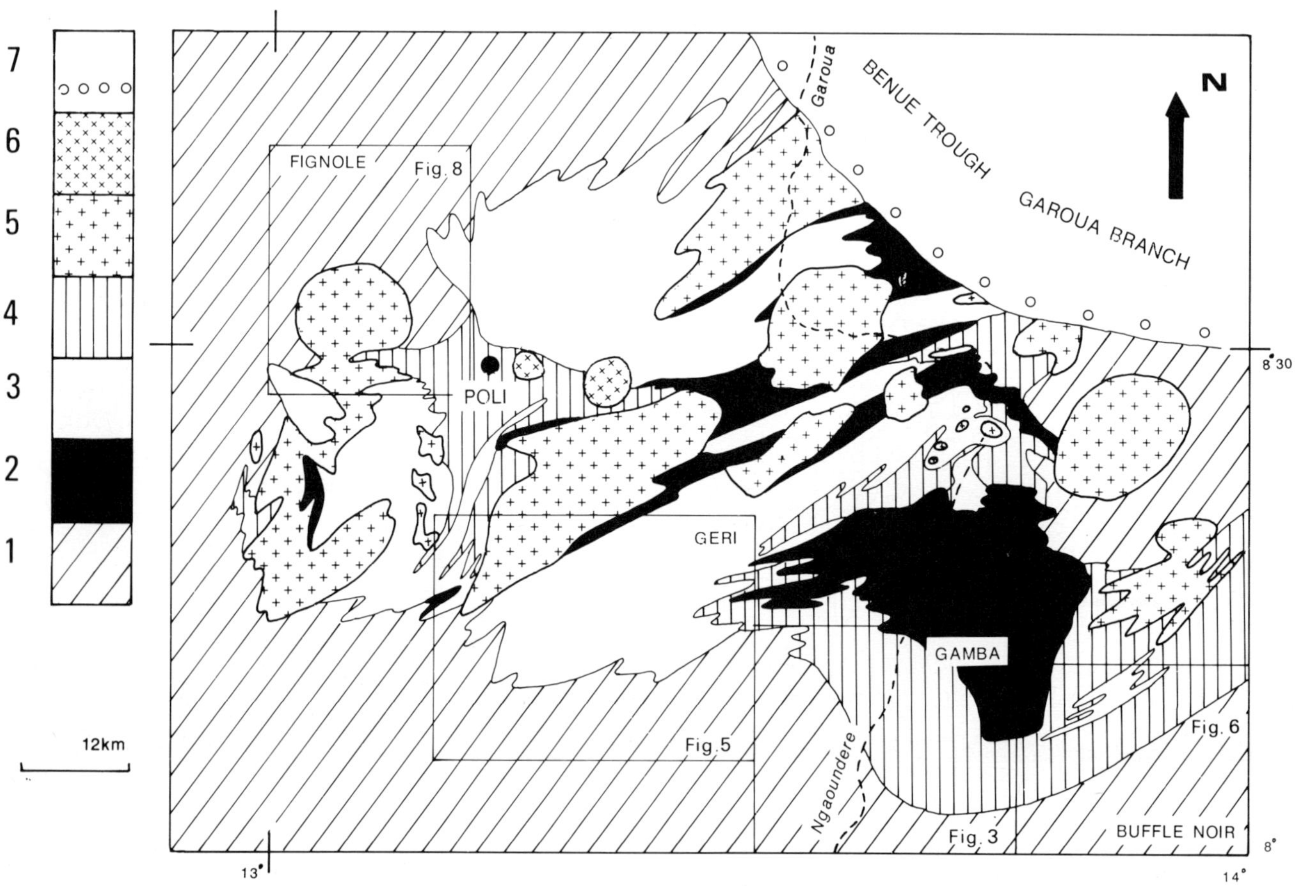

Figure 2. *Geological sketch map of the Poli region (modified from Le Fur (1971)). Study areas are indicated with the corresponding figure number. Legend: (1) gneisses and migmatites, (2) Sakje Formation, (3) lower volcanic unit, (4) upper clastic unit, (5) undifferentiated syn- and post-D2 granites, (6) Tertiary younger plutons, and (7) sedimentary cover of the Benue trough.*

the Upper Proterozoic age of the Poli schists (see also Ekwueme (1987) for the adjacent Nigerian basement). Some amphibolites belonging to the gneissic units have chemical compositions similar to the metavolcanics associated with the Poli schists.

(4) No clear petrographic or structural criteria have been found to support a basement–cover relationship, but both metamorphic and tectonic transitional contacts have been recognized. Nevertheless, some 2000 Ma old inherited zircons in the gneisses (Toteu *et al.*, 1987; Penaye, 1988) indicate either that part of the banded gneisses and migmatites represents genuine basement or, in some cases, that a basement source was present in the vicinity to provide material for clastic sediments interleaved with the volcanics.

Structural evolution of the southwestern end of the Poli schist belt, Gamba and Geri areas

The Gamba area (Figs. 3 and 4) is currently being studied by one of us (SA). Map and structural observations indicate a sharp tectonic and metamorphic contact between schists and gneisses. This contact is marked by retrogressive mylonitic gneisses. Along the northwestern contact (Fig. 3), vertical mylonites show a horizontal lineation interpreted to result from dextral wrench faulting.

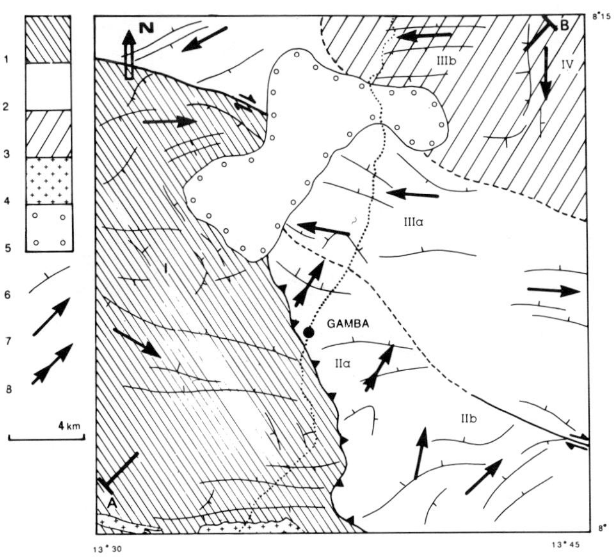

Figure 3. *Structural sketch map of the Gamba area. Legend: (1) gneisses and migmatites, (2) schists, (3) mica schists, (4) late-D2 granites, (5) Lower Paleozoic (Mangbei Formation), (6) foliation trajectories, (7) plunge direction of D2 fold axes, and (8) plunge of L2 stretching lineations. The bold line with black triangles represents a thrust plane. Line A–B is the cross-section shown in Figure 4.*

Toward the southeast, northeast-dipping (50°) mylonites and schists, together with northeast-plunging fold axes, suggest that schists overlie the gneissic domain (domain I on Fig. 3) along a thrust contact.

The gneisses contain an important proportion of orthogneisses and exhibit what was originally a flat-lying foliation (S1), parallel to a compositional banding, associated with conformable migmatic mobilizates. Except for its commonly well-defined foliation and the presence of relict minerals such as garnet, the geometry and thermal regime of the D1 event cannot be reconstructed. During the D2 event, D1 textures have been almost completely recrystallized and S1 heterogeneously refolded by trains of tight, upright to overturned, east–west trending folds with subhorizontal axes. Anatexis became more important with increasing intensity of D2 deformation and culminated approximately within the most tightly folded zones where the S1/S2 transposition is most intense. Anatexis partly erased previous planar and linear structures, and grades to a subhomogeneous anatectic granite that crops out in the southern part of the area studied.

In the schists, three domains (numbered II–IV on Fig. 3) have been defined with similar lithologies but slight differences in metamorphic grade. In domain II, which follows the schist–gneiss contact, D2 folds refold S1 foliation and have north-northeast trends and varying plunges (IIb); whereas in the IIa domain, a downdip stretching lineation is associated with a steeper S2 cleavage (50° to the northeast). In mylonites close to the contact, shear-sense criteria indicate a northward, down to northeast, movement direction along the thrust plane. Domain III is characterized by east–west trending, overturned (IIIa) to upright (IIIb) folds, with subhorizontal axis and associated cleavage; these F2

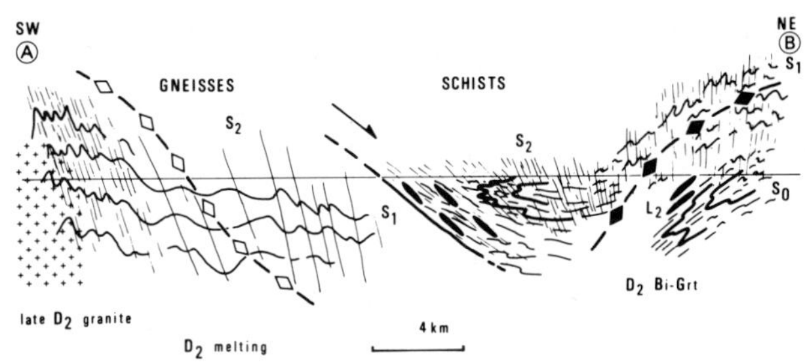

Figure 4. *Gamba area cross-section. S0, S1, and S2 are indicated. Solid diamonds correspond to the D2 biotite isograd; open diamonds correspond to D2 melt-rocks. The arrow indicates the latest retrogressive movement along the thrust plane; the apparent normal movement is probably related to late-D2 doming and granite emplacement.*

folds refold S1. Domain IIIb shows a northward increase in metamorphic grade. S0/S1 relationships suggest the existence of an F1 large-scale recumbent fold. Domain IV, where fold axes trend north–south, is less well understood; together with the IIIb domain, it shows a higher metamorphic grade.

In the schematic section shown in Figure 4, domain II is interpreted as representing the hanging-wall block of a nappe, deformed under simple shear conditions, whereas compressive east–west folding and north–south flattening characterizes both the underlying gneisses and the schists situated above the sheared sole. The thrust itself is likely to correspond to a late increment in the D2 progressive deformation as indicated by the metamorphic break between schists and gneisses and by the retrogression shown by the mylonites.

The Geri area (Ngako, 1986) is situated west of the Gamba area, where the schist–gneiss contact runs east–west (Fig. 5). Ngako's interpretation of the structure of the area is based on the contention that the gneisses may represent a basement because of the presence of granulite-facies remnants. The most important features in this area are as follows:

(1) A strip of retrogressed gneisses follows the contact between schists and gneisses, as in the Gamba area where the eastern extent of this contact shows a horizontal stretching lineation. A strike-slip fault may explain part of the variation of L2 trajectories shown on the map.

(2) The Bantadje unit, considered by Ngako as belonging to the basement, corresponds to foliated granites similar to those of the Bakonwa type (660 Ma old, Toteu *et al.* (1987)).

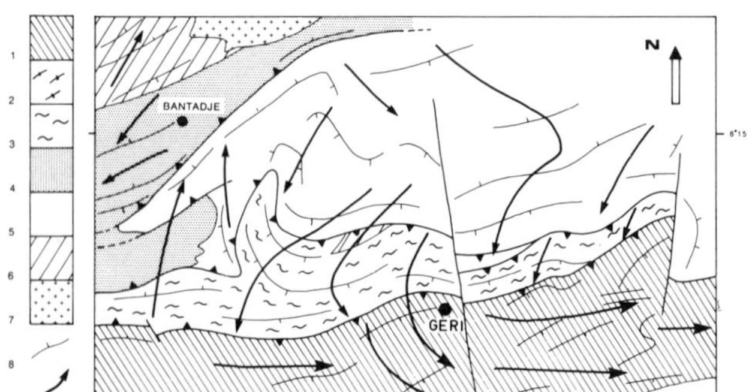

Figure 5. *Structural sketch map of the Geri area (modified and reinterpreted from Ngako (1986)). Legend: (1) gneisses and migmatites; (2) metadiorites; (3) retrogressed gneisses; (4) Bantadje Formation, foliated granites; (5) schists; (6) mica schists; (7) late-D2 granite (Kogue granite); (8) foliation trajectories; and (9) D2 fold axes and L2 lineation plunges. Bold lines with black triangles represent thrust planes.*

Structural evolution of the eastern edge of the Poli schist belt, Buffle Noir game reserve area

The Buffle Noir area (Figs. 6 and 7) is currently being studied by one of us (JP). Metamorphic terranes here exhibit a regional northeast–southwest trending steep foliation. Two main formations have been defined.

(1) In the north, a low- to medium-grade schist formation (Poli schists, 3 and 4 on Fig. 6) is made of volcaniclastic rocks (greywackes) with minor limestones and felsic to mafic metavolcanics.

(2) In the south, a heterogeneous high-grade gneissic formation contains two separated units of metagreywackes (1 on Fig. 6) with mafic volcanics and interleaved metadiorites. Rare layers of metapelite are also present. The southernmost unit contains many relicts of granulitic association. The two units are separated by a laccolith of syntectonic

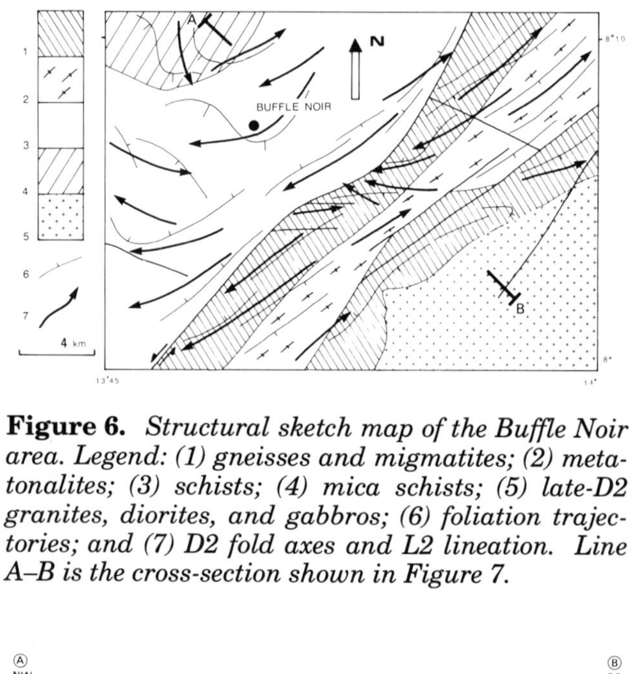

Figure 6. *Structural sketch map of the Buffle Noir area. Legend: (1) gneisses and migmatites; (2) metatonalites; (3) schists; (4) mica schists; (5) late-D2 granites, diorites, and gabbros; (6) foliation trajectories; and (7) D2 fold axes and L2 lineation. Line A–B is the cross-section shown in Figure 7.*

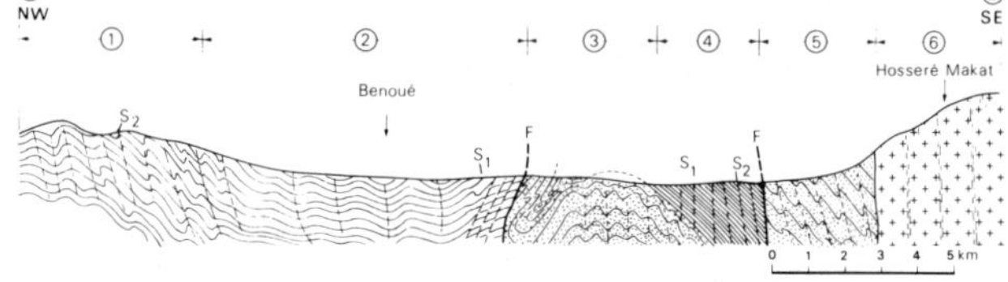

Figure 7. *Buffle Noir cross-section. Legend: (1) mica schists, (2) schists, (3) northern gneissic unit, (4) metatonalites, (5) southern gneissic unit, and (6) late-D2 granites, diorites, and gabbros.*

metatonalites (2 on Fig. 6) containing large enclaves (1 m to 1 km in size) of gneisses and tentatively interpreted as deep-seated equivalents of the Poli tholeiitic volcanics (Penaye, 1988). This formation is affected by southward-increasing migmatization.

The contact zone between the Poli schists and the gneisses is a N40°E to N55°E trending sinistral strike-slip fault, which is outlined in the field by low-grade mylonites. Faults with the same trend are scattered across the whole region and late dextral faults are also found with the same direction. A syn- to late-tectonic subalkaline plutonic complex (gabbros, diorites, and porphyritic granites; 5 on Fig. 6) intrudes the gneissic formation in the southeastern corner of the area.

The structural evolution of this area can be summarized as follows:

(1) The older S1 surface was probably originally flat lying and locally still has this attitude. In the gneisses, the foliation is defined as a compositional banding with granulite-facies relicts, and, in the schists, as the main schistosity transposing both a faint earlier planar structure and the stratigraphic markers. Rare F1 isoclinal folds show various axial orientations.

(2) The younger S2 surfaces are parallel to the axial planes of a new set of decimetre to decametre F2 folds. In the gneisses, S2 is a biotite-bearing schistosity or a migmatitic foliation. F2 folds are characterized by nearly horizontal axes and steep axial planes. Gently plunging intersection lineation and local mineral lineation are both parallel to fold axes. Migmatization is widespread and small-scale conjugate shear zones (N0°E and N110°E) are linked to the end of this event. In the schists, most F2 folds also have steep axial planes, but some are nearly flat-lying. F2 fold axes vary both in direction and in axial plunge. The S2 axial plane direction is marked by a strain-slip cleavage in the low-grade rocks, grading into a biotite transposition schistosity in the medium-grade micaschists.

The main sinistral strike-slip movements occurred during the D2 phase. Dextral movements appear to have taken place later, under lower metamorphic conditions. The most simple interpretation to explain the D2 patterns (Fig. 6) is to consider the F2 folds as en echelon folds with respect to the sinistral shear zone. However, the axial direction of F2 folds shows extension amounts (indicated by rare stretched minerals, pancake shapes of feldspars, symmetrical boudinage, and conjugate shear zones) too moderate, considering the angle between F2 axes and the shear zone (about 15–20°), to be explained adequately in this manner.

Structural evolution of the western edge of the Poli schist belt, Fignole area

The main feature of this area (Fig. 8), currently being studied by SFT, is a continuous increase in metamorphic grade from the Poli schists to the gneisses. The area is characterized by vertical north-northwest–south-southeast trending foliations. However, northeast–southwest to east–west trends occur locally in the schists (Fig. 8). Isoclinal D1 folds associated with an originally flat-lying foliation and rotational structures observed in garnets suggest a simple-shear regime. In both schists and gneisses, D1 lineations are scarce; their relict character is due to strong transposition and recrystallization during D2. Where preserved, D1 trends are N110°E–N130°E, with varying plunges.

The D2 event is characterized by tight, upright folds with a vertical axial plane foliation. D2 axes and lineations are parallel and show variable plunge (0 to 50°S or 50°N). D2 migmatization increases toward the west, where late-tectonic granites associated with agmatites are emplaced. On the contrary, in the easternmost part of the area, the S1 foliation is commonly outlined by early

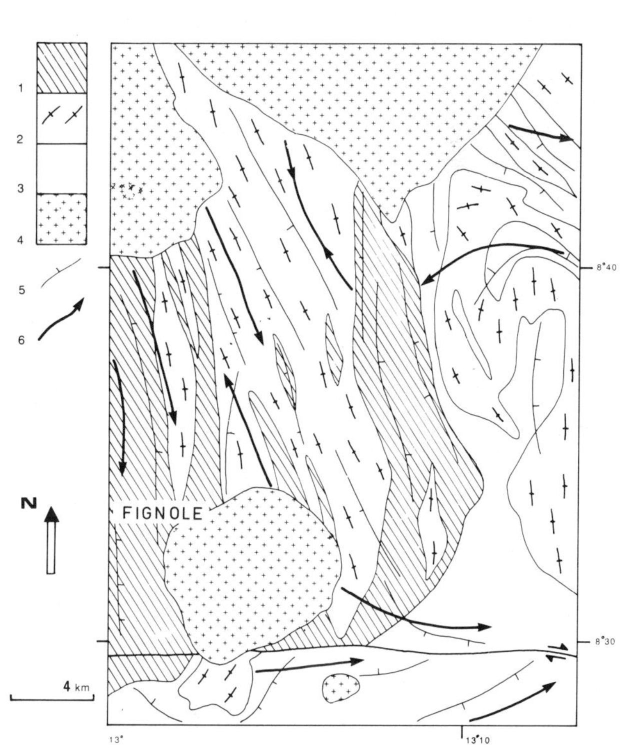

Figure 8. *Structural sketch map of the Fignole area. Legend: (1) gneisses and migmatites, (2) metadiorites (BIP), (3) schists, (4) late- and post-D2 granites, (5) foliation trajectories, and (6) D2 fold axes plunge direction.*

melts, but D2 metamorphic associations are always retrograde: in gneisses and mica schists, garnet and kyanite are retrogressed into biotite and muscovite; in the schists, epidote is fractured. This zonality suggests that D2 isograds dip eastward.

Late D2 shear zones display N80°E–N110°E (dextral) and 0–N160°E (sinistral) conjugate directions.

The Fignole area is thus dominated by the existence of a vertical S2 foliation showing an east–west shortening direction, but there is no evidence of strong north–south stretching.

Tectonic evolution of north Cameroon

The Upper Proterozoic age of the Poli schists is supported by the 800 Ma age of a major volcanic event showing alkaline to tholeiitic affinities; this event could be interpreted as an extensional stage (Toteu *et al.*, 1987). The calc-alkaline basic to intermediate plutonic suite is characterized by low initial ratios (0.7043–0.7045; Toteu *et al.*, 1987) and was interpreted as resulting from a 630 Ma old collision; it is clearly related to the first deformational phase.

No basement can clearly be delineated on a map, although it is assumed from the preservation of granulitic relicts. The similarity of the structural evolution of both schists and gneisses must be stressed as it does not allow a distinction between old and new material to be made. However, petrographical criteria (granulitic remnants in the gneisses) and geochronological evidence favour the presence of ca. 2000 Ma old material, possibly mixed with younger detrital material.

D1 seems to be part of a major nappe-forming event. However, no clear stretching lineation has been observed and axes of early isoclinal folds are highly variable in direction, even though a N100°E to N130°E maximum has been observed in some places. Consequently, it is premature to propose a vergence or movement direction for D1.

D2 tectonic evolution can be related to the development of a system of major wrench faults parallel or subparallel to a northeast–southwest movement direction (Fig. 9). A similar interpretation, involving tear faults or intracontinental transform faults, has already been proposed to explain the tectonic evolution of the Irumides belt in Zambia (Daly, 1986b). Most of the F2 fold axial directions may be explained either as en echelon folds associated with a sinistral wrench fault (Buffle Noir) or as compressional folds, at right angles to the movement direction (Fignole). In restricted areas (nappe contact of Gamba), F2 folds may have developed parallel to the movement direction. In the same way, the north-south trending F2 folds of the Fignole–Kogue area could also be en echelon folds associated with a dextral wrench fault; however, this cannot explain the F2 pattern displayed south of Kogue, in the Geri area. Such a dextral shear zone was proposed recently to explain the emplacement of the Kogue granite pluton (Bassahak, 1988). In the Buffle Noir area, the low angle between fold axes and the shear zone implies a

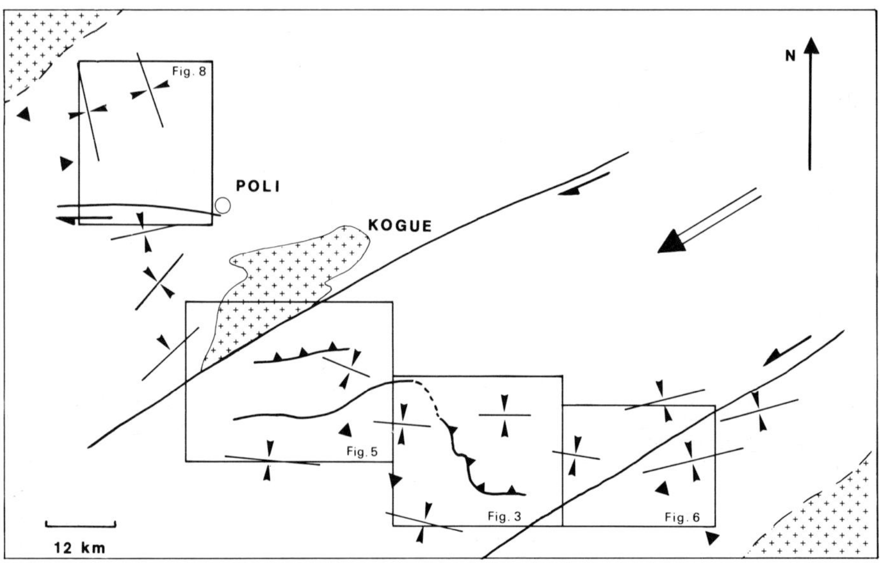

Figure 9. *D2 tectonic evolution of the Poli area. Opposite arrows represent local shortening directions and corresponding D2 schistosities. Opposite triangles represent late-D2, northwest–southeast shortening directions expressed by discrete conjugate shear planes. The distribution of D2 structures is controlled by a southwest movement direction that induced major tear faults and associated thrusts; emplacement of syn- to late-D2 granitic plutons may be controlled by the late-D2 shortening direction.*

large amount of strain, which would be accompanied by a shear component in the folds, a feature that has not been recognized. However, if deformation was initiated under conditions close to melting, the total finite strain may not have been entirely recorded. Alternatively, if the F2 axial direction is considered as representing the perpendicular to the shortening direction in each point, the observed regional variation may be the result of the impingement of an unmelted block, situated to the northeast below the Poli schist domain, against a melting zone. Such a model postulates a movement direction similar to that of the tear fault model, and is consistent with the location of D2 migmatites.

The late D2 compressional event, shown by the ubiquitous presence of conjugate shear zones with the same direction throughout the region, cannot be interpreted in terms of the above models. It corresponds to a northwest–southeast shortening direction, at right angles to the major wrench faults. As this direction is perpendicular to the alignment of late-tectonic granitic batholiths, it may reflect the emplacement of these commonly large plutons. This assumption is supported by the increased melt material associated with late D2 shear zones close to batholiths (e.g., west of Fignole).

TECTONIC AND METAMORPHIC EVOLUTION OF SOUTHERN CAMEROON

South of the Adamaoua mylonitic zone (Fig. 1), the Pan-African fold belt is composed of reworked Archean formations in the west (Nyong and Lokundje series) and of low- to high-grade metasedimentary formations in the east (Mbalmayo and Yaoundé series) (Fig. 10).

From field, structural, and chemical data, both the Yaoundé and Mbalmayo series (Nedelec *et al.*, 1986; Nzenti *et al.*, 1988) are interpreted as shallow-water sedimentary sequences composed mainly of shales and volcanogenic greywackes with dolomitic marls, dolomites, evaporitic sediments, quartzites, and iron-rich beds. These sequences were probably deposited in intracontinental extensional basins or on an epicontinental platform at the northern passive margin of the Congo craton; there is no evidence of subduction-related volcano-sedimentary formations.

The tectono-metamorphic evolution of the Yaoundé series is monocyclic (Ball *et al.*, 1984) and results from the succession of two deformation stages (Nzenti *et al.*, 1988). The first stage (D1) is marked by the development of an S1 surface representing the tectonic transposition of a compositional layering. D1-related prograde metamorphism reached high-pressure granulite facies conditions (800°C, 12 kbar (1.2 GPa)) and triggered widespread migmatization. The subsequent deformation (D2) is characterized by strong shear movements, first enhanced by melting and then evolving under retrograde conditions. The direction of movement indicated by the stretching lineation parallel to fold axes is N20°E–N40°E, with a southward transport direction. The retrograde metamorphic assemblages indicate an inverted metamorphic gradient: metamorphic grade increases from the foreland (Congo craton) to the core of the nappe (Yaoundé region), with a north-dipping foliation (Fig. 11). The age of this D2 event is estimated at 565 Ma (Lasserre and Soba, 1979).

The Nyong series has been interpreted recently as an early Proterozoic nappe, thrust to the south onto the Congo craton (Feybesse *et al.*, 1986). However, it seems more likely that it is a basement nappe of Pan-African age as the direction of movement and the foliation trajectories are in continuity with those of the Yaoundé series. Gravimetric data (Dumont, 1986) suggest that the Congo craton extents to the north beneath the Yaoundé nappe (Figs. 10 and 11). Thus, the nature and structure of the Pan-African formations south of the Adamaoua line are governed by the presence of an old continental crust and correspond to southward-moving nappe sheets. Such an environment precludes the presence of a suture in this region.

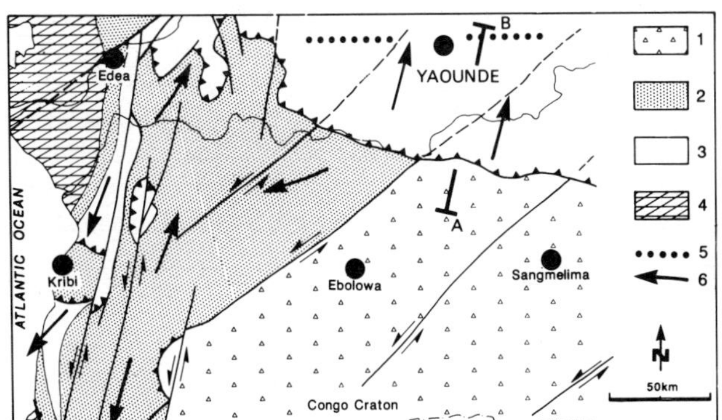

Figure 10. *Structural sketch map of southern Cameroon (data from Nzenti* et al. *(1988) and Feybesse* et al. *(1986)). Legend: (1) Ntem granulites (Archean Congo craton), (2) Nyong series (reworked Archean), (3) Yaoundé and Mbalmayo series (Upper Proterozoic), (4) Cretaceous to Recent formations, (5) assumed northern limit of underlying Archean craton (from Dumont, 1986), and (6) plunge direction of the stretching lineation. Line A–B is the cross-section shown in Figure 11.*

DISCUSSION AND CONCLUSION

Tectonic evolution

Our work has focused on the northern (Poli region) and southern (Yaoundé region) areas of Cameroon. There is a critical area between that is still poorly known, the Sanaga–Adamaoua region. After comparing the northern and southern regions, we will discuss possible working hypotheses concerning the Sanaga–Adamaoua region.

The average movement directions are similar in both northern and southern regions (N20°E in the south, N50°E in the north). In the Yaoundé region, this corresponds to a true transport direction, verging toward the Archean craton, with a large amount of stretching; in the north, the northeasterly direction corresponds to a wrench-fault system with associated folds, interpreted as representing tear faults. The corresponding S2 foliations are flat lying in the south and steeply dipping in the north, although an interesting exception is the wrench and thrust system developed along the western edge of the Congo craton (Fig. 10; Jegouzo, 1984). During the D2 event, high-grade metamorphic conditions seem to be higher pressure in the south (Nzenti *et al.*, 1988). From available geochronological data, the end of metamorphism is identical in both regions, around 570 Ma (Lasserre and Soba, 1979; Toteu *et al.*, 1987).

All of these features confirm that most of Cameroon belongs to the same Pan-African belt (Bessoles and Lasserre, 1977). However, the geodynamic environments are different: platform sediments thrust onto the stable craton in the south, and Upper Proterozoic crustal thinning linked to emplacement of a large volume of juvenile magmas in the north, followed by collision-related calc-alkaline magmas (BIP). Thus, the intermediate region is truly critical. The following observations may be useful for a better understanding of this region.

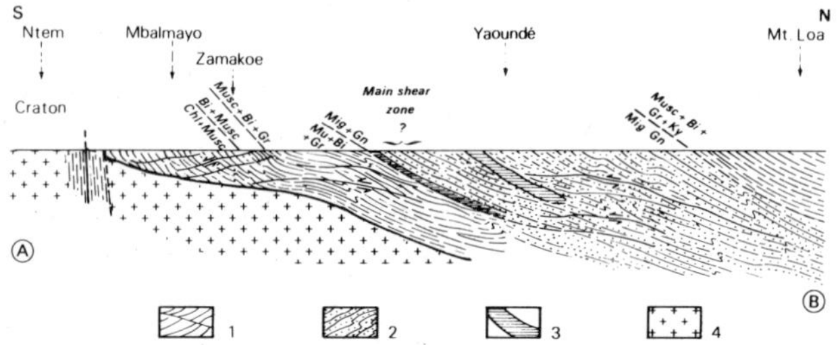

Figure 11. *Ntem–Yaoundé cross-section. Legend: (1) low-grade metamorphic formations, (2) high-grade granulitic and migmatitic formations, (3) pyriclasites, and (4) Ntem Archean granulites.*

(1) The limit between the areas dominated by N50°E and N20°E directions is situated between the Adamaoua line and the Sanaga wrench fault (Fig. 1). It is locally outlined by a narrow, low-grade, volcano-sedimentary belt (Lom series of unknown age). It is also, based on gravity data, the northernmost possible extent of an underlying stable craton (Dumont, 1986).

(2) Along the northern edge of the Adamaoua line, gneisses similar to those of the Poli region contain premigmatitic granulitic remnants in gently dipping mylonites (E. Ngangom, personal communication, 1987).

(3) A 10-km step in crustal thickness has been determined from seismic data (Stuart *et al.*, 1985), with a thin crust (23 km) north of the Adamaoua; it was considered as a young feature associated either with the Cameroon volcanic line or with the Benue trough, but, based on the presence of an undeformed Lower Paleozoic formation (Mangbei series), this northern Cameroon thin crust could be a Pan-African inheritance.

The suture problem

In West and Central Africa (Fig. 12), well-defined cratons comprise both Archean nuclei and Lower Proterozoic assemblages whose internal pattern in terms of orogenic belts and/or domains is still poorly understood. This is also the case in South America, where Archean and Lower Proterozoic formations are both present in the Guyana and Sao Francisco cratons.

The only well-identified Pan-African suture zone is known along the eastern edge of the West African craton (Black *et al.*, 1979; Lesquer and Louis, 1982). The westernmost tectonic domain of the Transaharan belt, east of the craton, is considered to represent a complete Wilson cycle from oceanic opening to major collision, probably involving several microcontinents. Other possible sutures shown on Figure 12 are highly speculative, and are discussed below.

(1) Along the eastern edge of the Air mountains (Niger), an ophiolitic assemblage has been studied recently (Cosson *et al.*, 1987) in a thrust belt first mapped by Black *et al.* (1967). This thrust belt forms a northeast-verging tectonic contact between overthrust high-grade gneisses and low-grade metavolcanics and ultramafic bodies. This entire package has been thrust onto a basement (and its undetached Late Proterozoic cover (Proche Tenere series)). The rocks here may be comparable with gneissic assemblages of Lower Proterozoic polycyclic rocks in central and eastern Hoggar (Bertrand *et al.*, 1986). This likely suture has been extended on Figure 12 to the Tiririne belt in Algeria, and toward the south to the Damagaran

region, where highly deformed rocks have been noted (Liegeois *et al.*, 1985). The Tiririne belt has been previously interpreted as an intracontinental linear fold belt (Bertrand *et al.*, 1978).

(2) East of Lake Chad, a major gravity anomaly, unfortunately occurring in a region entirely covered by recent formations, has been delineated by Louis (1970), who, owing to the similarity in pattern with anomalies in the circum-West African craton, tentatively proposed that it could represent another suture zone. This hypothesis was recently discussed by Freeth (1984). Such a suture may well explain the geodynamic evolution of northern Cameroon and also the calc-alkaline granitoids occurring in Tibesti region (Pegram *et al.*, 1976).

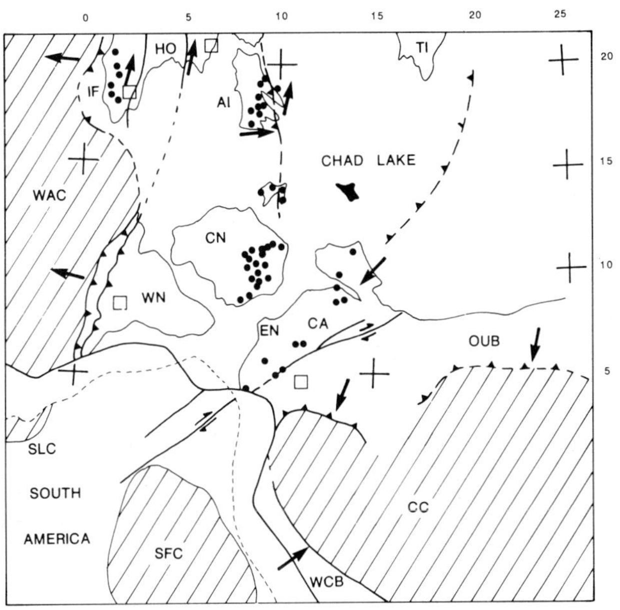

Figure 12. *Cameroon geodynamic environment. The hatched areas correspond to Archean and/or Lower Proterozoic cratons (contacts are modified from Louis (1970)). Lines with triangles correspond to well-defined or assumed suture zones or to major thrust zones. Arrows indicate the mean movement directions: in the West African craton (WAC), arrows are located for clarity on the stable foreland, but actually represent the stretching lineation occurring in the westernmost thrust sheet; elsewhere, arrows represent the average direction of the stretching lineations. Dots indicate younger alkaline granites. Open squares correspond to areas where Eburnean or older ages have been determined. Cratonic areas: WAC, West African craton; CC, Congo craton; SFC, Sao Francisco craton; and SLC, Sao Luis craton. Pan-African belts: WCB, West Congo belt; OUB, Oubanguides belt; IF, Adrar des Iforas; HO, Hoggar; AI, Air; TI, Tibesti; WN, CN, and EN, western, central, and eastern Nigeria; and CA: Cameroon.*

(3) In the Centrafrican Republic, Poidevin (1983) recently proposed the existence of another Pan-African belt, the Oubanguides, which, albeit an area of uncertainty, extends eastward the major thrust contact of the Yaoundé gneisses onto the Congo craton. The Oubanguides belt is still poorly documented and, as for southern Cameroon, there is no direct evidence for a suture.

(4) There is no major gap with northeastern Brazil (Borborema province), especially if most of the structures of this area are Brazilian in age (i.e., Pan-African) according to Caby and Arthaud (1986). However, there is no consensus about the tectonic evolution of northeastern Brazil and many Brazilian geologists favour interpretations where pre-Upper Proterozoic features play an important role (Jardim de Sa *et al.*, 1987).

A general remark must be made concerning the location in western and central Africa of alkaline "anorogenic" complexes of any age, from Late Pan-African to Recent, with respect to possible Pan-African sutures. In most cases, the alignments of younger granites follow identified or suspected sutures on their active margin side. In southern Cameroon, however, Late Pan-African alkaline complexes (A. Michard, unpublished data; E.N. N'Sifa, personal communication; R.W. Van Schmus, personal communication) are far from any suture and occur in the southwestern corner of the thrust and wrench-fault belt. The assumption of a partly asthenospheric origin for some alkaline ring complexes (Liegeois and Black, 1984; Liegeois, 1987) and the control of their emplacement by the latest wrench faults (accompanied with a reversal of movement), which ended the collision process in the Iforas region (Boullier *et al.*, 1986), suggest a severe disturbance of the lithosphere–asthenosphere boundary ("harpoon effect" of Liegeois and Black, 1984). Such a disturbance may have lasted for hundreds of millions of years and controlled the "anorogenic" intrusives.

The Pan-African network

On Figure 13, major Pan-African tectonic features have been indicated at the continental scale. As synchronism is not certain, we have considered all the proven or assumed movement directions in the 700–500 Ma time span. Data are from Bertrand *et al.* (1986), Black *et al.* (1979), Boullier (1982), Cosson *et al.* (1987), Coward (1981), Daly (1986a), Leblanc (1975), Lecorche (1985), Shackleton (1986), and Villeneuve (1987). The similarity of northeast–southwest lineament directions in South Africa and West Africa indicates that Daly's model (1986a), suggesting continental transform fault control, may also be applied to Central and West Africa. However, the Mozambique belt and the Arabo–Nubian shield in northeast Africa and Arabia seem to have evolved differently. Thus, the Adamaoua fault zone and its Northeast Brazil counterparts (Patos and Pernambuco lineaments) may have, on a large scale, the same significance as the Schliesen–Mwembeshi–Chimaliro zone in South Africa (Daly, 1986a). Areas of assumed reworked crust have been delineated within Pan-African domains (Fig. 13). In the Transaharan belt, such areas (Central Hoggar – Nigeria and Western Hoggar granulites) suggest that old, pre-Pan-African microcontinents, incorporated into the belt, separate linear zones of younger Upper Proterozoic crust. The main aim of future research will be to determine the extent of crustal accretion during the Pan-African orogenesis. In this respect, much work remains to be done in the central part of Cameroon. Furthermore, as we cannot be sure of the existence of a major suture along the Chad gravity anomaly, Pan-African interaction between the Congo craton and the Nigeria Eburnean reactivated basement is still a major problem on which depends any interpretation of the largely unknown regions of southern Sudan and Libya (toward the hypothetical Nile craton).

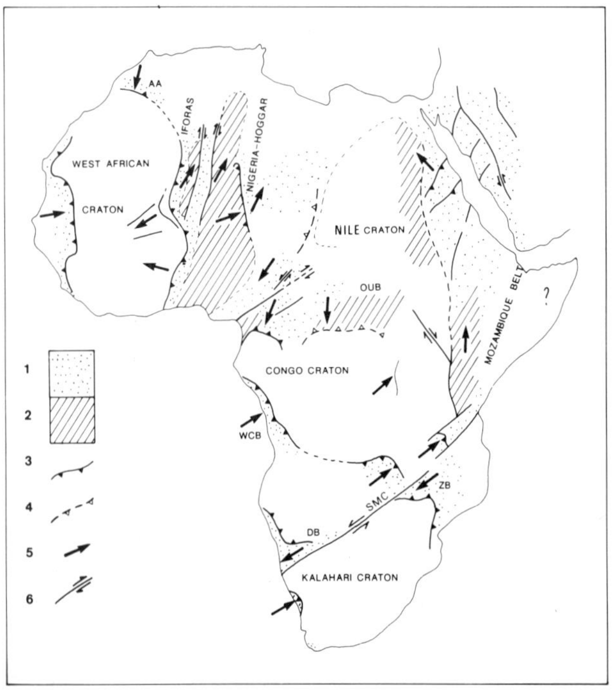

Figure 13. *The Pan-African network. Legend: (1) Pan-African domains, (2) reworked Lower Proterozoic and Archean basement within Pan-African domains; (3) major sutures and thrust zones, (4) assumed sutures and thrust zones, (5) mean relative movement directions between blocks (from literature interpretation) and/or average stretching lineation with recognized or apparent movement (see text for references), and (6) main wrench-fault systems. Abbreviations: AA, Anti-Atlas (Morocco); WCB, West Congolian belt; DB, Damara belt; ZB, Zambezi belt; OUB, Oubanguides belt; and SMC, Schliesen–Mwembeshi–Chimaliro shear zone.*

ACKNOWLEDGEMENTS

This work has been supported in Cameroon by the Institut de Recherches Géologiques et Minières; the authors gratefully thank its director D. Soba. The agreement between the University of Yaoundé and the Institut National Polytechnique de Lorraine, Nancy, allowed some of us to travel to Cameroon. C. Devey is thanked for checking the English text. Thanks also to W.R. Church and an anonymous reviewer for constructive comments.

REFERENCES

Affaton, P., Sougy, J. and Trompette, R., 1980, The tectonostratigraphic relationships between the Upper Precambrian and Lower Paleozoic Volta basin and the Pan-African, Dahomeyide orogenic belt (West Africa): American Journal of Science, v. 280, p. 224–248.

Ball, E., Bard, J.P. and Soba, D., 1984, Tectonique tangentielle dans la catazone pan-africaine du Cameroun: les gneiss de Yaoundé: Journal of African Earth Sciences, v. 2, p. 91–95.

Bassahak, J., 1988, Le complexe plutonique du massif de Kogue (Poli, Nord Cameroun), unpublished thesis, Université de Nancy.

Bertrand, J.M., Caby, R., Ducrot, J., Lancelot, J.R., Moussine-Pouchkine, A. and Saadallah, A., 1978, The late Pan-African ensialic linear fold belt of eastern Hoggar (Algeria); geology, structural development, U/Pb geochronology, tectonic implications for the Hoggar shield: Precambrian Research, v. 7, p. 349–376.

Bertrand, J.M., Michard, A., Boullier, A.M. and Dautel, D., 1986, Structure and U/Pb geochronology of the Central Hoggar (Algeria). A reappraisal of its Pan-African evolution: Tectonics, v. 5, p. 955–972.

Bessoles, B. and Lasserre, M., 1977, Le complexe de base du Cameroun: Bulletin de la Société géologique de France, v. 19, p. 1085–1092.

Bessoles, B. and Trompette, R., 1980, Géologie de l'Afrique. La chaîne panafricaine "zone mobile d'Afrique centrale (partie sud) et zone mobile soudanaise": Mémoire du Bureau de recherches géologiques et minières no. 92, Orléans, 394 p.

Black, R., Jaujou, M. and Pellaton, C., 1967, Carte géologique du massif de l'Aïr (1/500.000). Notice explicative: Bureau de recherches géologiques et minières, Orléans, 57 p.

Black, R., Caby, R., Moussine-Pouchkine, A., Bayer, R., Bertrand, J.M., Boullier, A.M. and Fabre, J., 1979, Evidence for late Precambrian plate tectonics in West Africa: Nature, London, v. 278, p. 223–227.

Boullier, A.M., 1982, Étude structurale du centre de l'Adrar des Iforas (Mali). Mylonites et tectogénèse: Thèse d'État, Institut National Polytechnique de Lorraine, Nancy, inédit, 327 p.

Boullier, A.M., Liegeois, J.P., Black, R., Fabre, J., Sauvage, M. and Bertrand, J.M., 1986, Late Pan-African tectonics marking the transition from subduction-related calc-alkaline magmatism to within-plate alkaline granitoids (Adrar des Iforas): Tectonophysics, v. 132, p. 233–246.

Bullard, E., Everett, J.E. and Smith, A.G., 1965, The fit of continents around the Atlantic: Philosophical Transactions of the Royal Society, London, v. 258, p. 41–51.

Caby, R. and Arthaud, M., 1986, Major Precambrian nappes of the Brazilian belt, Ceara, northeast Brazil: Geology, v. 14, p. 871–874.

Cahen, L., Snelling, N.J., Delhal, J. and Vail, J.R., 1984, The Geochronology and Evolution of Africa: Clarendon Press, Oxford, 512 p.

Condie, K.C., 1982, Plate tectonic models for Proterozoic continental accretion in the SW United States: Geology, v. 10, p. 37–42.

Cosson, Y., Rocci, G. and Boullier, A.M., 1987, Rock deformation at a Pan-African plate boundary outlined by ophiolites in Air (Niger) (abstract): EUG IV, Terra cognita, p. 101.

Coward, M.P., 1981, The Moabi syncline: the West Congolian of Gabon and its tectonic significance: Annales de la Société géologique de Belgique, v. 104, p. 225–259.

Daly, M.C., 1986a, Crustal shear zones and thrust belts: their geometry and continuity in Central Africa: Philosophical Transactions of the Royal Society, London, v. A317, p. 111–128.

Daly, M.C., 1986b, The intracratonic Irumides belt of Zambia and its bearing on collision orogeny during the Proterozoic of Africa, in Coward, M.P. and Ries, A.C., eds., Geological Society Special Publication No. 19, London, p. 321–328.

Dumont, J.F., 1986, Identification par télédétection de l'accident de la Sanaga (Cameroun). Sa position dans le contexte des grands accidents d'Afrique Centrale et de la limite nord du craton congolais: Géodynamique, v. 1, p. 13–19.

Ekwueme, B.N., 1987, Structural orientations and Precambrian deformation episodes of Uwet area, Oban massif, SE Nigeria: Precambrian Research, v. 34, p. 269–289.

Feybesse, J.L., Johan, V., Maurizot, P. and Abessolo, A., 1986, Mise en évidence d'une nappe synmétamorphe d'âge éburnéen dans la partie NW du craton zaïrois (SW Cameroun), in Les formations birrimiennes en Afrique de l'-Ouest: Centre International de Formation et d'Études Géologiques, Paris, Publication occasionnelle no. 10, p. 105–111.

Freeth, S.J., 1984, How many rifts are there in West Africa?: Earth and Planetary Science Letters, v. 67, p. 219–227.

Jardim de Sa, E.F., Macedo, M.H.F., Legrand, J.M., McReath, I. Galindo, A.C. and Sa, J.M., 1987, Proterozoic granitoids in a polycyclic setting: the Serido region, NE Brazil, in International Symposium on Granites and Associated Mineralizations: Secretary of Mines and Energy, Bahia State, Salvador, Extended Abstracts, p. 103–110.

Jegouzo, P., 1984, Évolution structurale du Sud-Ouest du Cameroun durant l'orogénèse pan-africaine (abstract): Association de tectonique cisaillante et chevauchante, Colloque chevauchement et déformation, Toulouse, p. 23.

Koch, P., 1959, Carte géologique de reconnaissance du Cameroun à l'échelle du 1/500.000, coupure Garoua-Ouest avec notice explicative: Direction des Mines et de la Géologie, Cameroun, 48 p.

Lasserre, M. and Soba, D., 1979, Migmatisation d'âge Panafricain au sein des formations camerounaises appartenant à la zone mobile d'Afrique Centrale: Comptes rendus sommaires de la Société géologique de France, fascicule 2, p. 64–68.

Leblanc, M., 1975, Ophiolites précambriennes et gites arseniés de cobalt (Bou Azzer, Maroc): Centre Géologique et Géophysique de Montpellier, France, Mémoire hors série.

Lecorche, J.P., 1985, Les Mauritanides face au craton Ouest Africain, structure d'un secteur clef: la région d'Ijibiten (Est d'Akjoujt, République Islamique de Mauritanie): Travaux des Laboratoires de Sciences de la Terre de l'Université de Marseille-Saint Jérome, série B, no. 26, 446 p.

Le Fur, Y., 1971, Les indices de cuivre du groupe volcano-sédimentaire de Poli (Cameroun): Bulletin du Bureau de recherches géologiques et minières, Orléans, v. 6, p. 79–91.

Lesquer, A. and Louis, P., 1982, Anomalies gravimétriques et collision continentale au Précambrien: Géoexploration, v. 20, p. 275–293.

Lewry, J.F., 1981, Lower Proterozoic arc–microcontinent collisional tectonics in the western Churchill Province: Nature, v. 294, p. 69–72.

Liegeois, J.P., 1987, Le batholithe composite de l'Adrar des Iforas (Mali): Géochimie et géochronologie d'une succession magmatique du calco-alcalin à l'alcalin dans le cadre de l'orogénèse panafricaine, unpublished thesis, Université Libre de Bruxelles, 335 p.

Liegeois, J.P. and Black, R., 1984, Petrology and Rb/Sr geochronology of the late Pan-African calc-alkaline–alkaline transition in the Adrar des Iforas (Mali): Upper Precambrian crustal accretion, in Klerkx, J. and Michot, J., eds., African Geology: Musée royal de l'Afrique centrale, Tervuren, p. 115–145.

Liegeois, J.P., Black, R., Franconi, A. and Zeinelabadin, M., 1985, Pan-African magmatism in the Damagaran region (southern Niger): preliminary results (abstract): 13th Colloquium on African Geology, St. Andrews, Scotland, p. 123.

Louis, P., 1970, Contribution géophysique à la connaissance du bassin du lac Tchad: Mémoire de l'Office de recherches scientifiques et techniques d'Outre-Mer no. 42, 311 p.

Nedelec, A., Macaudiere, J., Nzenti, J.P. and Barbey, P., 1986, Évolution structurale et métamorphique des schistes de Mbalmayo (Cameroun). Implications pour la structure de la zone mobile pan-africaine d'Afrique centrale, au contact du Craton du Congo: Comptes rendus de l'Académie des Sciences, Paris, Série II, v. 303, p. 75–80.

Ngako, V., 1986, Évolution métamorphique et structurale de la bordure Sud-Ouest de la "Série de Poli" (Segment camerounais de la chaîne Panafricaine): Mémoires et documents du Centre armoricain d'étude structurale des socles no. 5, Rennes, 185 p.

Njel, U.O., 1986, Paléogéographie d'un segment de l'orogène pan-africain: la ceinture volcano-sédimentaire de Poli (Nord-Cameroun): Comptes rendus de l'Académie des Sciences, Paris, Série II, v. 303, p. 1737–1742.

Nzenti, J.P., 1987, Pétrogénèse des migmatites de Yaoundé (Cameroun). Eléments pour un modèle géodynamique de la chaîne pan-africaine nord-équatoriale, unpublished thesis, Université de Nancy, 154 p.

Nzenti, J.P., Barbey, P., Jegouzo, P. and Moreau, C., 1984, Un nouvel exemple de ceinture granulitique dans une chaîne protérozoïque de collision: les migmatites de Yaoundé au Cameroun: Comptes rendus de l'Académie des Sciences, Paris, Série II, v. 299, p. 1197–1199.

Nzenti, J.P., Barbey, P., Macaudiere, J. and Soba, D., 1988, Origin and evolution of the high-grade Yaoundé gneisses (Cameroun): Precambrian Research, v. 38, p. 91–109.

Pegram, W.J., Register, J.K., Fullagar, P.D., Ghuma, M.A. and Rogers, J.J.W., 1976, Pan-African ages from a Tibesti massif batholith, southern Libya: Earth and Planetary Science Letters, v. 30, p. 128–132.

Penaye, J., 1988, Pétrologie et structure des ensembles métamorphiques au Sud-Est de Poli (Nord Cameroun), unpublished Ph.D. thesis, Institut National Polytechnique de Lorraine, Nancy, 196 p.

Poidevin, J.L., 1983, La tectonique pan-africaine à la bordure nord du Craton congolais: l'orogénèse des Oubanguides (abstract): 12th Colloque on African Geology, Musée Royal de l'Afrique Centrale, Tervuren, Belgium, and Université Libre de Bruxelles, p. 75.

Rocci, G., 1965, Essai d'interprétation de mesures géochronologiques: la structure de l'Ouest Africain: Sciences de la terre, Nancy, v. 3–4, p. 461–478.

Schwoerer, P., 1965, Carte géologique de reconnaissance du Cameroun à l'échelle du 1/500.000, coupure Garoua-Est avec notice explicative: Direction des Mines et de la Géologie du Cameroun, 48 p.

Shackleton, R.M., 1986, Precambrian collision tectonics in Africa, _in_ Coward, M.P. and Ries, A.C., eds., Collision Tectonics: Geological Society Special Publication no. 19, London, p. 329–349.

Stoeser, D.B. and Camp, V.E., 1985, Pan-African microplate accretion of the Arabian shield: Geological Society of America Bulletin, v. 97, p. 817–826.

Stuart, G.W., Fairhead, J.D., Dorbath, L. and Dorbath, C., 1985, A seismic refraction study of the crustal structure associated with the Adamaoua Plateau and Garoua Rift, Cameroon, West Africa: Geophysical Journal of the Royal Astronomical Society, v. 81, p. 1–12.

Toteu, S.F., 1987, Chronologie des grands ensembles structuraux de la région de Poli. Accrétion crustale dans la chaîne panafricaine du Nord-Cameroun: Thèse de Doctorat d'État, Université de Nancy, inédite, 197 p.

Toteu, S.F., Michard, A., Macaudiere, J., Bertrand, J.M. and Penaye, J., 1986, Données géochronologiques nouvelles (U/Pb et Rb/Sr) sur la zone mobile pan-africaine du nord Cameroun: Comptes rendus de l'Académie des Sciences, Paris, Série II, v. 303, p. 375–378.

Toteu, S.F., Michard, A., Bertrand, J.M. and Rocci, G., 1987, U/Pb dating of Precambrian rocks from northern Cameroon, orogenic evolution and chronology of the Pan-African belt of Central Africa: Precambrian Research, v. 37, p. 71–87.

Trompette, R., 1979, The Pan-African Dahomeyide fold belt: a collisionnal orogen (abstract): 10ème Colloque International de Géologie africaine, Université de Montpellier, p. 72–73.

Villeneuve, M., 1987, Le sillon plissé de l'Itomwe-Kadubu (Zaïre oriental): un témoin probable de l'orogénèse panafricaine au coeur du craton zaïrois: Comptes rendus de l'Académie des Sciences, Paris, Série II, v. 304, p. 835–840.

Index